Engineering Physics

D.K. Bhattacharya

Associate Director
Solid State Physics Laboratory Delhi, DRDO

Poonam Tandon

Associate Professor
Maharaja Agrasen Institute of Technology, New Delhi

OXFORD
UNIVERSITY PRESS

OXFORD
UNIVERSITY PRESS

Oxford University Press is a department of the University of Oxford.
It furthers the University's objective of excellence in research, scholarship,
and education by publishing worldwide. Oxford is a registered trade mark of
Oxford University Press in the UK and in certain other countries.

Published in India by
Oxford University Press
Ground Floor, 2/11, Ansari Road, Daryaganj, New Delhi, 110 002, India

First published in 2015
Fourth impression 2017

ISBN-13: 978-0-19-945281-1
ISBN-10: 0-19-945281-4

Typeset in Times New RomanMTStd
by Welkyn Software Solutions Pvt. Ltd
Printed in India by Chennai Micro Print (P) Ltd., Chennai – 600 029.

To my wife
For continuing to inspire

–D.K. Bhattacharya

To my parents and family

–Poonam Tandon

Preface

Science and engineering form the backbone of any technological innovation. Engineering focuses on the conversion of scientific ideas into viable products and technologies. Physics is a fundamental aspect of science. Therefore, knowledge of physics relevant to engineering is critical for converting ideas to products. An understanding of physics also helps engineers understand the working and limitations of existing devices and techniques, which eventually leads to new innovations and improvements.

It is interesting to note that in spite of the complexities of modern technology, the underlying principle behind these still remain simple. In fact, it would not be wrong to say that unless the basic physics behind a technology is fully understood, it would be impossible to implement the full potential of the technology. The fundamental concepts of physics have laid the foundation for advances in engineering technology.

ABOUT THE BOOK

Engineering Physics is primarily designed to serve as a textbook for undergraduate students of engineering. However, the scope of the book has been expanded to go beyond the needs of an undergraduate engineering student. It will also serve as a reference book for undergraduate science (B Sc) students, scientists, technologists, and practitioners of various branches of engineering.

The book thoroughly explains all relevant and important topics in a student-friendly manner. The language and approach towards understanding the fundamental topics of physics is clear. The mathematics has been kept simple and understandable, enabling readers to easily understand the principle and idea behind a concept. The book lays emphasis on explaining the principles as well as the applications of a given topic using numerous solved examples and self-explanatory figures and diagrams.

The book begins with a detailed coverage of optics, which includes topics such as interference, diffraction, polarization, lasers, holography, and optical fibre systems. It then delves into discussions on waves and oscillations, architectural acoustics, and ultrasonic waves and their applications. The basic principles of classical mechanics are elucidated for easy understanding of the complex concepts discussed later. Relativistic mechanics has been explained by comparing it with the principles of classical mechanics. The fundamentals of quantum mechanics and statistical mechanics are then presented. In the subsequent chapters, electromagnetism, including dielectric properties, magnetic properties, and electromagnetic field theory are explained. In-depth treatment of topics such as X-rays, crystal physics, band theory of solids, and semiconductor physics is provided in the book. This is followed by a discussion on conducting and superconducting materials.

A chapter on nuclear physics and radioactivity has been included as there is renewed interest in energy generation through nuclear reactors. After this, the basic ideas in the field of new engineering materials and nanotechnology are discussed.

KEY FEATURES

- Dedicated chapters on band theory of solids and statistical mechanics
- Detailed discussions on optics, acoustics, and electromagnetism
- Advanced topics such as new engineering materials, nanotechnology, and non-destructive testing
- List of symbols at the beginning of each chapter
- Close to 450 solved examples and over 400 self-explanatory illustrations which aid in easy understanding of concepts
- Summary of concepts, applications, and key formulae at the end of each chapter for quick recapitulation
- Around 1400 chapter-end exercises including multiple-choice questions, review questions, and numerical problems listed under the self-assessment section for practice
- Answers to multiple-choice questions and numerical problems at the end of the book

ORGANIZATION OF THE BOOK

Engineering Physics consists of 23 chapters and is divided into five parts. Part I discusses optics, waves, and acoustics. Part II includes topics related to mechanics. Part III covers electromagnetism. Parts IV and V cover solid state physics, and nuclear physics and new engineering materials respectively. A chapter-wise scheme of the book is presented here.

Part I: Optics, Waves, and Acoustics

Chapter 1 explains superposition and coherence followed by a detailed discussion on the methodologies for the observation of interference. The different types of interferences are also covered.

Chapter 2 presents the ab-initio Huygen's principle and Fresnel's and Fraunhoffer's diffraction in detail. The resolving power of important optical instruments such as plane diffraction grating, telescope, and microscope has been explained.

Chapter 3 describes the phenomenon of polarization, types of polarization, and methods of producing polarization. Topics such as Malus law, Nicol prism as a polarizer and analyzer, quarter- and half-wave plates, Fresnel's theory of optical rotation, and polarimeter are discussed in detail. A basic idea of photo elasticity is also presented.

Chapter 4 discusses the ordered excited state—lasers. The various components, types, and industrial applications of lasers are discussed in detail. The technique of holography has also been covered in this chapter.

Chapter 5 explains the propagation of light waves in an optical fibre system. The chapter discusses the numerical aperture of optical fibre systems, the various types of optical fibres and their classification, fibre drawings, splicing, LEDs, detectors, and endoscopes.

Chapter 6 on waves and oscillations elucidates the concept of potential energy, the linear restoring force resulting in linear harmonic oscillations, damped harmonic oscillations, quality factor, and forced vibrations and its phase characteristics. The contemporary research topics of chaos and coupled oscillations have been introduced in this chapter. The conventional definition of waves, and waves in closed and open pipes are discussed for completeness.

Chapter 7 covers architectural acoustics which has become an important feature of building design. This chapter details the classification of sound, characteristics of musical sounds, intensity of sound, reverberation, Sabine's formula, room acoustics, absorption coefficient, and acoustic quieting along with methods of quieting.

Ultrasonic waves are used in non-destructive testing techniques and are produced by the magnetostriction method and piezoelectric effect. *Chapter 8* deals with the properties and detection of ultrasonic waves, cavitation, acoustic grating, SONAR, and the industrial and medical applications.

Part II: Mechanics

Chapter 9 explains the fundamentals of classical mechanics such as vector analysis, rotational motion, central forces, elastic properties, and fluid dynamics.

Chapter 10 throws light on relativistic mechanics. Topics such as inertial and non-inertial frames, Galilean transformations, and Michelson–Morley experiment have been covered in detail. The chapter also covers the special theory of relativity, Lorentz transformation, variation of mass leading to the mass–energy equivalence, energy–momentum relations, and twin paradox.

Chapter 11 lays emphasis on quantum mechanics. The chapter covers photoelectric effect, Einstein's photoelectric equation, Milikan's experiment, De Broglie's experiment, Thomson experiment, and the two-slit interference experiment. Phase and group velocity, matter waves, Heisenberg's uncertainty principle, Schrodinger's wave equation, quantum tunnelling, and quantum or discrete behaviour of material particles are explained in this chapter.

Chapter 12 expounds on the behaviour of an assembly of particles for an ideal gas. The quantum concept related to discrete behaviour of particles introduces energy levels, energy states, and degeneracy. The various ways of stacking particles in an assembly are categorized using the Bose–Einstein, Fermi–Dirac, and Maxwell–Boltzmann statistics. The Planck's quantum theory of radiation encompasses the two independent classical radiation laws—Rayleigh and Wein's laws—under special limits.

Part III: Electromagnetism

Chapter 13 on dielectric properties of materials includes important topics such as electric dipole, dipole moment, dielectric constant, and polarizability. The different types of polarizations in dielectrics, their frequency and temperature dependence, and Claussius–Mossotti equation are presented in detail. Dielectric losses, their breakdown, and the applications of dielectric materials are also covered.

Chapter 14 discusses the magnetic properties of materials such as dia, para, and ferromagnetism in detail. The phenomenon of hysteresis, ferrites, and important applications of magnetic materials are also included in this chapter.

Chapter 15 presents the mathematical concepts of gradient, divergence, and curl along with their physical significance. The Maxwell's contribution of combining both electrostatics and magnetostatics laws has been detailed both mathematically and physically. The application of Maxwell's equations in understanding the guided waves for a free space, dielectric medium, and conducting medium with the idea of skin penetration depth has been explained exhaustively.

Part IV: Solid State Physics

X-rays, to date, have made useful contributions towards material analysis and medical applications. *Chapter 16* presents a discussion on diffraction of X-rays, X-ray spectrum, the different methods of production of X-rays, and its important applications. The Compton effect has been explained in this chapter to further strengthen the basis of quantum mechanics.

Chapter 17 on crystal physics introduces lattices, miller indices, atomic radius, coordination number, and packing factor. Crystal structures, polymorphism, and allotropy have been explained. The different types of crystal defects are also presented.

Chapter 18 introduces the band theory of solids using quantum mechanics. The Kronig–Penney model for solids is explained in detail. The Fermi–Dirac statistics has been used to explain the density of states and work function.

Chapter 19 on semiconductor physics begins with an introduction to intrinsic and extrinsic types of semiconductors. Fermi level has been physically explained for understanding the electron or hole assembly. The p–n junction has been detailed in the form of the diode equation. The Hall effect has been discussed. Some of the special reverse bias devices such as LED, LCD, photodiodes, solar cells, and Zener diodes are explained. The topic of photoconductivity has also been included.

Chapter 20 on conducting materials includes Drude's theory, thermal conductivity and Wiedemann–Franz law, Lorentz number, free electron gas, Fermi energy, and carrier concentration, in addition to a discussion on conducting polymers.

Superconductors fall into a special class of materials. *Chapter 21* describes the special characteristics of superconductors such as Meissner effect, coherence length, isotope effect, and energy gap, among others. Josephson tunnelling, BCS theory, and London's equation have also been explained. The types and applications of superconductors have been covered in this chapter.

Part V: Nuclear Physics and New Engineering Materials

Chapter 22 on nuclear physics and radioactivity has been included in the book, in view of its importance in energy generation, in addition to the use of fossil fuels. The chapter covers nuclear forces, conservation laws, and radioactive laws. The theory of nuclear fusion and fission has been explained with a mention of nuclear reactors.

Chapter 23 discusses new engineering materials and nanotechnology. Important topics such as metallic glasses, shape memory alloys, and nanotechnology are presented. Carbon nanotubes, fullerenes, and graphene are also discussed.

The book contains six appendices which include SI units, the periodic table, physical constants, lattice constants, properties and band gaps of semiconductors, and properties of silicon, germanium, and gallium arsenide. The answers to multiple-choice questions and numerical problems are also provided at the end of the book.

ONLINE RESOURCES

For the benefit of faculty and students reading this book, additional resources are available online at http://oupinheonline.com/book/bhattacharya-tandon-engineering-physics/9780199452811.

For Faculty

- Solutions manual
- Chapter-wise PPTs

For Students

- Test generator
- Model question papers
- Links to interactive animations (indicated with 🐭 in text)

ACKNOWLEDGEMENTS

Dr Bhattacharya is grateful to his family for being extremely cooperative and tolerant during the entire period of developing this book.

Dr Poonam is grateful to Dr Nand Kishore Garg, Chairman, Maharaja Agrasen Society of Education, for his ongoing support and encouragement for creating an academic environment in the pursuit of higher education. She would like to express her gratitude to Prof. B.N. Mishra (Director Emeritus), Sh. Gyanendra Srivastava (CEO), Prof. M.L. Goyal (Director), and Prof. Ram Kishore (HOD, Applied Science) for always motivating and guiding her during the development of this book. She is also thankful to her colleagues and students for their healthy interaction and critical suggestions. Dr Poonam is indebted to her family, without whom this endeavour would not have been possible.

Last but not the least, Dr Bhattacharya and Dr Poonam would also like to acknowledge with gratitude the support and guidance provided by the editorial team at Oxford University Press, India, without whom the book would not have been a reality.

The authors would be grateful to receive suggestions and feedback for the book at dk_bhattacharya@yahoo.co.uk and spuat@yahoo.com.

D.K. Bhattacharya
Poonam Tandon

FEATURES OF

Statistical Mechanics **12** CHAPTER

Band Theory of Solids **18** CHAPTER

New Engineering Materials and Nanotechnology **23** CHAPTER

Comprehensive Coverage

The book includes dedicated chapters on Statistical Mechanics and Band Theory of Solids among others. Advanced topics like non-destructive testing are explained in detail.

List of Symbols

A list of symbols is provided to facilitate easier understanding of the figures and equations used in the text.

List of Symbols

f = Frequency of alternating current
v = Natural frequency
E = Young's modulus
ρ = Density
λ = Wavelength
V = Velocity

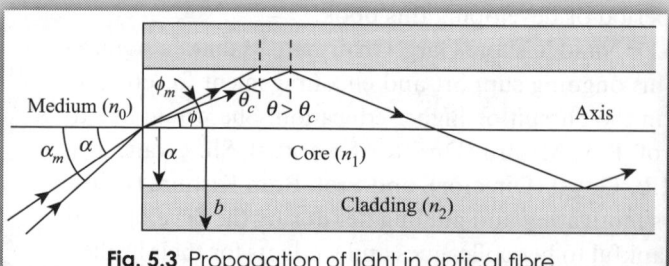

Fig. 5.3 Propagation of light in optical fibre

Colour	Wavelength range (nm)	Typical efficacy (lm/W)
Red	$620 < \lambda < 645$	152
Red-orange	$610 < \lambda < 620$	98
Green	$520 < \lambda < 550$	93
Cyan	$490 < \lambda < 520$	155
Blue	$460 < \lambda < 490$	315

Table 19.3 Lumiled catalogue

Figures and Tables

More than 400 well-labelled figures and tables help users visualize the concepts and principles of engineering physics.

THE BOOK

Solved Examples

Close to 450 solved examples interspersed across chapters help readers understand the application of concepts in engineering problems.

Example 11.13 Compare the uncertainties in velocity of a proton contained in a 10 Å box.

Solution The following is the dimension of the box: $\Delta x = 10 \times 10^{-10}$

$$\Delta p \Delta x = \frac{h}{4\pi}$$

IMPORTANT CONCEPTS

1. X-ray diffraction is actually strong X-ray reflections at certain specific angles is referred to as the Bragg's law:

$$2d \sin\theta = n\lambda$$

IMPORTANT FORMULAE

1. Magnetization: $M = \dfrac{evr}{2}$ 2. Bohr magneton: $\mu_B = \dfrac{eh}{4\pi m}$

APPLICATIONS

1. An optical fibre consists of a cylindrical glass core surrounded by a layer of a slightly lower refractive index.
2. Optical fibres are generally made from glass or plastic.

Takeaways

Enumeration of important concepts helps in quick revision. A summary of important formulae helps readers easily solve numerical problems. The practical applications of the concepts studied aid in correlating these in real-world scenarios.

Multiple-choice Questions

17.1 Grain boundaries are present in
    (a) polycrystalline materials
    (b) crystalline materials
    (c) amorphous material

Chapter-end Self-assessment Section

This section comprises around 1400 questions including Multiple-choice Questions, an assortment of critical-thinking questions and concept-based questions under Review Exercises, and Numerical Problems. Answers to multiple-choice questions and numerical problems are given at the end of the book.

Review Questions

9.1 What are vectors?
9.2 What is a unit vector?
9.3 When are two vectors considered equal?

Numerical Problems

6.1 The differential equation of an oscillator is

$$\frac{d^2x}{dt^2} + 2\gamma\frac{dx}{dt} + \omega_0^2 x = 0$$

Brief Contents

Detailed Contents

PART III: ELECTROMAGNETISM   507

PART V: NUCLEAR PHYSICS AND NEW ENGINEERING MATERIALS — 839

PART I
OPTICS, WAVES, AND ACOUSTICS

Interference

Learning objectives

After studying this chapter, students will be able to

- understand the concept of superposition of waves
- comprehend the meaning of coherence
- understand the physics of interference from two-point sources
- elucidate Young's double-slit interference experiment in detail
- explain different types of interference
- understand interference obtained using a Fresnel's biprism
- realize interference patterns obtained using a Llyod's mirror
- explain the formation of interference pattern in thin films
- comprehend the formation of interference pattern in a wedge-shaped film
- understand the formation of Newton's rings
- solve numerical problems based on interference from two-point sources, Young's double-slit experiment, Fresnel's biprism, interference in thin films, wedge-shaped film, and Newton's rings

List of Symbols

A = Amplitude
δ = Phase
ω = Angular frequency
I = Intensity

c = Velocity of light
ΔL = Coherence length
λ = Wavelength
ψ = Wave function

β = Fringe width
μ = Refractive index
t = Thickness
R = Radius

1.1 INTRODUCTION

When two or more waves travel simultaneously through a medium, the resultant displacement at any point of the medium is given by the vector sum of the displacements of the individual waves. This is called the principle of superposition of waves. In sound waves, this results in two interesting consequences: stationary waves and beats. In the case of light, one such interesting consequence of the principle of superposition is *interference*.

Application of the principle of interference can easily be observed in nature. Waves in water get superimposed and result in an interference pattern. Such interference patterns can be observed when small pebbles are thrown into lakes or ponds. A single stone will create waves in water, but if a second stone is dropped at a small distance from the first one, the waves generated by the two stones will interact and create interference patterns due to superposition. At regions of constructive interference, water waves can reach great heights, and create hazardous conditions and extreme damage. Superposition has been discussed in detail in this chapter. Concepts of superimposition have then been used for understanding the phenomenon of interference. Different types of interference patterns observed in thin films have also been discussed in detail. Some applications of interference have been presented towards the end of this chapter.

1.2 SUPERPOSITION OF WAVES

A wave represents a travelling disturbance. It is represented mathematically in the following form:

$$\psi = A\sin(\omega t + \delta) \tag{1.1}$$

where A represents the amplitude, w the angular frequency, and δ the phase of a wave with respect to some reference. As the wave travels through a medium, the particles of the medium get acted upon by the wave.

According to the principle of superposition, the resultant displacement of a particle of the medium acted upon by two or more waves simultaneously is given by the algebraic sum of the displacements produced by the individual waves.

Thus, if two waves are represented by

$$\psi_1 = A_1 \sin \omega t \tag{1.2}$$

and $$\psi_2 = A_2 \sin(\omega t + \delta) \tag{1.3}$$

where δ represents the phase difference between the two waves, then the resultant displacement, y, is given by the following equation:

$$\psi = \psi_1 + \psi_2 = A_1 \sin \omega t + A_2 \sin(\omega t + \delta) \tag{1.4}$$

When more than two waves are involved, the general expression for resultant displacement becomes as follows:

$$\psi = \psi_1 + \psi_2 + \psi_3 + \ldots + \psi_n \tag{1.5}$$

$$= A_1 \sin(\omega t + \delta_1) + A_2 \sin(\omega t + \delta_2) + \ldots + A_n \sin(\omega t + \delta_n) \tag{1.6}$$

where $A_1, A_2, \ldots, A_n$ represent the amplitudes and $\delta_1, \delta_2, \ldots, \delta_n$ represent the phases of the waves. Equation (1.6) is not easy to solve for general situations. For some special situations, however, a solution can be visualized. The easiest situation to visualize is the one in which all the individual waves have the same amplitude, say A, and also all the waves are in phase. The amplitude of the resultant of n waves in this case will be nA, and the corresponding intensity

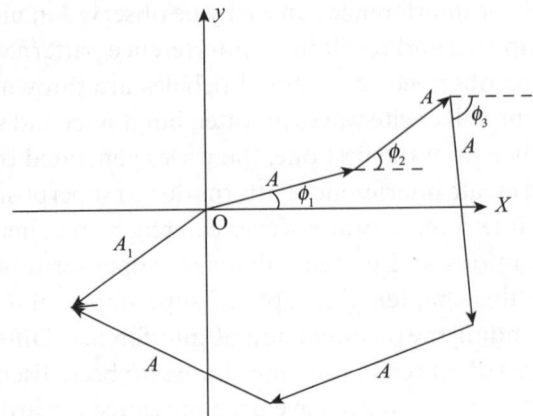

Fig. 1.1 Evaluation of resultant amplitude

will be $n^2 A^2$. For a random distribution of phases, a graphical method has to be used for evaluating the resultant amplitude. A typical figure is shown in Fig. 1.1, where A_1 represents the amplitude of the resultant.

Summation of projections of amplitudes along the x-direction results in the following equation:

$$A_{1x} = A(\cos\delta_1 + \cos\delta_2 + \ldots + \cos\delta_n) \tag{1.7}$$

To calculate the intensity, square of Eq. (1.7) should be estimated. This would result in terms such as $\cos^2\delta_1$, $2\cos(\delta_1)\cos(\delta_2)$, and $\cos^2(\delta_2)$. For the superimposition of a large number of waves, the terms having products such as $\cos\delta_1\cos\delta_2$ will average out to zero. Thus, only terms such as $\cos^2\delta_1$ and $\cos^2\delta_2$ can be assumed to survive. Similarly, the y-components would lead to the following relation:

$$A_{1y} = A(\sin\delta_1 + \sin\delta_2 + \ldots + \sin\delta_n) \tag{1.8}$$

For evaluating intensity, square of Eq. (1.8) would also be needed. Once again, this would involve terms such as $\sin^2\delta_1$, $\sin^2\delta_2$, and $2\sin\delta_1\sin\delta_2$. For a large number of waves, the cross-terms like $\sin\delta_1\sin\delta_2$ can be assumed to result in zero average value, leaving out terms such as $\sin^2\delta_1$ and $\sin^2\delta_2$. The total resultant intensity can be obtained by adding contribution from Eqs (1.7) and (1.8). Thus, we have the following relation:

$$I = A_1^2 = A^2\left(\cos^2\delta_1 + \cos^2\delta_2 + \cos^2\delta_3 + \ldots + \cos^2\delta_n\right)$$
$$+ A^2\left(\sin^2\delta_1 + \sin^2\delta_2 + \sin^2\delta_3 + \ldots + \sin^2\delta_n\right) \tag{1.9}$$

Since $(\cos^2\delta + \sin^2\delta) = 1$, we get the following expression:

$$I = A_1^2 = nA^2 \tag{1.10}$$

The average resultant intensity is, therefore, n times the average intensity of a single wave. From Eq. (1.10) it can also be concluded that the resultant average amplitude is proportional to $\sqrt{n}$, where n represents the number of waves.

Example 1.1   Two coherent sources whose intensity ratio is 64:1 produce interference fringes. Deduce the ratio of maximum intensity to minimum intensity.

Solution Suppose A_1 and A_2 represent the amplitudes of waves emitted by the two sources. Then we have

$$\frac{I_1}{I_2} = \frac{A_1^2}{A_2^2} \tag{1.11}$$

which yields the following equation:

$$\frac{A_1}{A_2} = \sqrt{\frac{I_1}{I_2}} = \frac{8}{1} \tag{1.12}$$

giving $\quad A_1 = 8A_2 \tag{1.13}$

We also know that

$$\frac{I_{max}}{I_{min}} = \frac{\left(A_1 + A_2\right)^2}{\left(A_1 - A_2\right)^2} \tag{1.14}$$

Using Eq. (1.13) in Eq. (1.14), we get the following relation:

$$\frac{I_{max}}{I_{min}} = \frac{\left(8A_2 + A_2\right)^2}{\left(8A_2 - A_2\right)^2} = \frac{\left(9A_2\right)^2}{\left(7A_2\right)^2} = \frac{81}{49}$$

Thus, $\quad I_{max} : I_{min} = 81:49$

Example 1.2 Two coherent sources whose intensity ratio is 4:1 produce interference fringes. Deduce the ratio of maximum intensity to minimum intensity.

Solution Let us consider the following equation: $\dfrac{I_1}{I_2} = \dfrac{A_1^2}{A_2^2}$

which yields $\dfrac{A_1}{A_2} = \sqrt{\dfrac{I_1}{I_2}} = \dfrac{2}{1}$

giving, $\quad A_1 = 2A_2$

We also know that $\dfrac{I_{max}}{I_{min}} = \dfrac{\left(A_1 + A_2\right)^2}{\left(A_1 - A_2\right)^2} = \dfrac{\left(2A_2 + A_2\right)^2}{\left(2A_2 - A_2\right)^2} = \dfrac{3^2}{1^2} = 9$

Therefore, the required ratio is 9:1.

1.3 COHERENCE

Two waves are coherent if they have a constant phase between them; coherent waves also have the same frequency.

1.3.1 Temporal Coherence

A wave travels along its direction of propagation. Different points along the direction of propagation have a phase associated with them. If the phase difference between any two points along the direction of propagation is independent of time, then the wave is said to be temporaly coherent. *Temporal coherence* is also called *longitudinal coherence*. A time-independent phase difference also implies that the wave is monochromatic or of one wavelength. Suppose δ_1 and δ_2 represent the phases at points 1 and 2 for a wave at a particular instant t_1. In addition, $\delta_1{}'$ and $\delta_2{}'$ represent the phases at the same two points 1 and 2 at a different time t_2. For a wave having temporal coherence, the following relation must hold:

$$\delta_2 - \delta_1 = \delta_2' - \delta_1' \tag{1.15}$$

1.3.2 Spatial Coherence

Spatial coherence is related to the phase of a wave at different points that are transverse to the direction of propagation. If the phase difference between any points located transverse to the direction of propagation is independent of time, then the wave is said to be spatially coherent.

1.3.3 Coherence Time and Coherence Length

The time interval over which the phase of a wave remains constant is called the *coherence time*. For a perfectly monochromatic sinusoidal wave, the coherence time is infinity. In reality, no wave is perfectly monochromatic, and therefore a finite coherence time exists. The coherence time is generally represented by a symbol Δt. The distance travelled by light during one coherence time is called the coherence length and is represented by the symbol ΔL for light waves:

$$\Delta L = C\Delta t \tag{1.16}$$

where c represents the velocity of light.

1.4 INTERFERENCE OF LIGHT FROM TWO POINT SOURCES

Sustainable interference patterns will occur only when overlapping waves satisfy the following conditions:
1. The waves must be of similar types (e.g., both of them are either light or sound waves).
2. Wave sources must be coherent.
3. The waves must have comparable (but not necessarily equal) amplitudes.

If two coherent, monochromatic point sources are set up, then an interference pattern is formed in the region where their waves overlap. Assume that the two sources are in phase with one another. Maxima will be formed where the waves from both sources arrive exactly half a cycle (π) out of phase. Phase differences are caused by the different distances travelled by the waves from the source to the point concerned. An extra path of one half-wavelength from one source will introduce a phase difference of π radians, resulting in cancellation (a minimum), whereas a path difference of any whole-number multiple of wavelength results in the waves arriving in phase and adding (a maximum). The situation is depicted schematically in Fig. 1.2. Figure 1.2(a) depicts the situation where the path difference between the waves originating from S_1 and S_2 is $\lambda/2$, resulting in a phase difference of $\pi/2$. The situation in which the path difference and the phase difference are zero is depicted in Fig. 1.2(b)

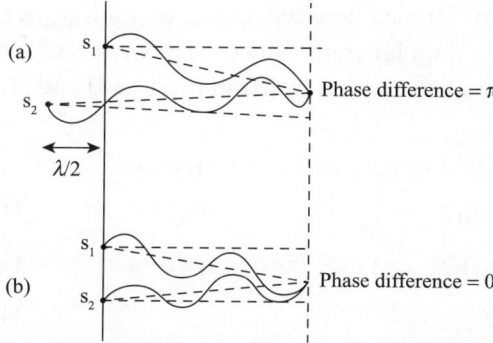

(a)

S_1

S_2

Phase difference = π

λ/2

(b)

S_1

S_2

Phase difference = 0

Fig. 1.2 Interference from two point sources (a) Path difference is λ/2 (b) Path difference and the phase difference are zero

Path difference x is the difference in distance from each source to a particular point and δ represents the difference in phase of the waves at a point. In general,

$$\delta = \frac{2\pi x}{\lambda} \qquad (1.17)$$

Equation (1.17) implies that $x = \lambda$; a path difference of a whole wavelength leads to a phase difference of $\delta = 2\pi$.

Maxima are not regions with a permanent large disturbance—they oscillate like any other part of the wave, passing through zero to negative values every cycle. They represent the positions where this oscillation has the maximum amplitude.

Constructive interference occurs due to the superposition of two waves at a point such that the crest of one wave falls on the crest of the other, that is, *the path difference between two waves is an integral multiple of the wavelength ($n\lambda$). Intensity is maximum at these points (n is an integer or zero).*

Destructive interference occurs due to the superposition of two waves at a point such that the crest of one wave falls on the trough of the other, that is, the path difference between the two waves is $\left(n + \dfrac{1}{2}\right)\lambda$, where λ is the wavelength and n is an integer or zero. *Intensity is minimum at these points.*

Monochromatic sources are sources of light waves having the same wavelength or frequency. In Section 1.3, we have learnt that coherent sources are sources of light waves having the same wavelength or frequency and a constant phase difference. *Coherent sources* are obtained by splitting a light source into parts using various phenomenon of light like reflection, and refraction. Some of the common methods used for the generation of coherent waves are as follows:

1. Young's double-slit experiment
2. Fresnel's biprism
3. Llyod's mirror

1.5 MATHEMATICAL TREATMENT OF INTERFERENCE

Let the waves from two coherence sources be represented as follows:

$$\psi_1 = \psi_{10} \sin(\omega t - kx) \qquad (1.18)$$

$$\psi_2 = \psi_{20} \sin(\omega t - kx + \delta) \qquad (1.19)$$

where δ is the constant phase difference between them, ψ_{10} and ψ_{20} are the amplitudes, and w represents the angular frequency of the two waves.

The resultant of their superposition is given by [from Eqs (1.18) and (1.19)] the following relation:

$$\psi = \psi_1 + \psi_2 = \psi_{10} \sin(wt - kx) + \psi_{20} \sin(wt - kx) \cos \delta$$
$$+ \psi_{20} \cos(wt - kx) \sin \delta \tag{1.20}$$

or $\quad \psi = (\psi_{10} + \psi_{20} \cos \delta) \sin(wt - kx) + \psi_{20} \sin \delta \cos(wt - kx) \tag{1.21}$

Let $\quad A\cos\theta = \psi_{10} + \psi_{20} \cos \delta \tag{1.22}$

and $\quad A\sin\theta = \psi_{20} \sin \delta \tag{1.23}$

Therefore, $A^2 = \psi_{10}{}^2 + \psi_{20}{}^2 + 2\psi_{10}\psi_{20} \cos \delta \tag{1.24}$

and $\quad \tan\theta = \dfrac{\psi_{20} \sin \delta}{\psi_{10} + \psi_{20} \cos \delta} \tag{1.25}$

Using Eqs (1.21)–(1.23) in Eq. (1.24), the following relation can be obtained:

$$\psi = A\sin(wt - kx)\cos\theta + A\cos(wt - kx)\sin\theta \tag{1.26}$$

which gives the following expression:

$$\psi = A\sin(wt - kx + \theta) \tag{1.27}$$

We see that the resultant wave has an amplitude A, given by Eq. (1.24), and a phase angle of θ, given by Eq. (1.25), with respect to the wave of source [Eq. (1.18)].

1.5.1 Constructive Interference

From Eq. (1.24), A^2 is maximum when

$$\cos\delta = 1 \text{ or } \delta = 0, \dots, 2n\pi \tag{1.28}$$

$$A_{max}^2 = \psi_{10}^2 + \psi_{20}^2 + 2\psi_{10}\psi_{20}$$

or $\quad A_{max}^2 = (\psi_{10} + \psi_{20})^2$

or $\quad A_{max} = \psi_{10} + \psi_{20} \tag{1.29}$

Since intensity $I \sim A^2$, the maximum intensity is expressed as follows:

$$I_{max} = k_1 \left(\psi_{10}^2 + \psi_{20}^2 + 2\psi_{10}\psi_{20} \right), \tag{1.30}$$

where k_1 is a constant of proportionality

1.5.2 Destructive Interference

Again from Eq. (1.24), A^2 is minimum when $\cos\delta = -1$ or

$$\delta = (2n+1)\pi \tag{1.31}$$

$$\therefore A_{min}^2 = \psi_{10}^2 + \psi_{20}^2 - 2\psi_{10}\psi_{20} = (\psi_{10} - \psi_{20})^2$$

or $\quad A_{min} = \psi_{10} - \psi_{20} \tag{1.32}$

$$\therefore \text{ Minimum intensity } I_{min} = k \left(\psi_{10}^2 + \psi_{20}^2 - 2\psi_{10}\psi_{20} \right) \tag{1.33}$$

where k is a constant of proportionality.

Thus, for constructive interference, the following relations hold:

Phase difference $= \delta = 0, ..., 2n\pi$

Path difference $= \dfrac{\lambda}{2\pi}\delta = 0, ..., n\lambda$

Whereas for destructive interference, the following relations hold:

Phase difference $= \delta = (2n+1)\pi$

Path difference $= \dfrac{\lambda}{2\pi}\delta = \left(n+\dfrac{1}{2}\right)\lambda$

From Eqs (1.29) and (1.32) we can conclude that A_{max} will be the greatest and A_{min} the least when $\psi_{10} = \psi_{20} = \psi_0$, that is, the two superposing waves have equal amplitude, because in that case the following relations are true:

$A_{max} = 2\psi_0$

$A_{min} = 0$

$I = k_1\left(\psi_0^2 + \psi_0^2 + 2\psi_0^2\cos\delta\right) = 2\psi_0^2\left(1+\cos\delta\right)k_1$

Example 1.3 Determine the ratio of intensity at the centre of a bright fringe to the intensity found at a point one-quarter of the distance between two fringes from the centre.

Solution From Eq. (1.24), wet get the following relation:

$$I = A^2 = \psi_{10}^2 + \psi_{20}^2 + 2\psi_{10}\psi_{20}\cos\delta \tag{1.34}$$

When $\psi_{10} = \psi_{20}$, Eq. (1.34) can be rewritten in the following form:

$$I = \psi_{10}^2 + \psi_{10}^2 + 2\psi_{10}^2\cos\delta \tag{1.35}$$

$$= 2\psi_{10}^2\left(1+\cos\delta\right) \tag{1.36}$$

At the centre, $\delta = 0$. Using Eq. (1.36), we get the following expression:

$$I_0 = 2\psi_{10}^2(1+1) = 4\psi_{10}^2 \tag{1.37}$$

The phase difference between two consecutive fringes is 2π. Thus, the phase difference at a distance that is one-quarter of the distance between two fringes will be $\dfrac{2\pi}{4} = \dfrac{\pi}{2}$.

Suppose I_1 represents the intensity at a distance that is one-quarter of the distance between two fringes, then using Eq. (1.36) we get the following relation:

$$I_1 = 2\psi_{10}^2\left(1+\cos\dfrac{\pi}{2}\right) = 2\psi_{10}^2 \tag{1.38}$$

Using Eqs (1.37) and (1.38), we obtain the following relation: $\dfrac{I_0}{I_1} = \dfrac{4\psi_{10}^2}{2\psi_{10}^2} = 2$

Example 1.4 Determine the ratio of intensity at the centre of a bright fringe to the intensity found at a point one-third of the distance between two fringes from the centre.

Solution Let us consider the following relation: Phase difference $= \dfrac{2\pi}{3}$

Thus, $I_1 = 2\psi_{10}^2\left(1+\cos\frac{2\pi}{3}\right) = 2\psi_{10}^2(1+\cos120) = 2\psi_{10}^2(1-0.5) = 2\psi_{10}^2 \times (0.5) = \psi_{10}^2$

$$\therefore \frac{I_0}{I_1} = \frac{4\psi_{10}^2}{\psi_{10}^2} = 4$$

1.6 YOUNG'S DOUBLE-SLIT INTERFERENCE

The phenomenon of interference was first demonstrated experimentally by Thomas Young in the year 1801. A schematic of the phenomenon is shown in Fig. 1.3.

Sunlight is made to pass through the pin hole S. Two closely spaced pin holes S_1 and S_2 are placed on the way, and the interference pattern was observed on screen XY. Young observed a few coloured bright and dark bands on the screen. Some modern modifications in the original set-up use narrow slits in place of pin holes S_1 and S_2, and sunlight is replaced by monochromatic light. As a result of these modifications, the interference pattern consists of equally spaced bright and dark bands.

As sunlight passes through the pin hole S, spherical waves originating from the pin holes start spreading out, as shown in Fig. 1.3. These spherical waves are incident on pin holes S_1 and S_2. According to the Huygen's principle, each point on the wavefront is a centre of secondary wavelets. Thus, secondary waves start spreading out from pin holes S_1 and S_2, as shown in Fig. 1.3. As the secondary waves spread out, their radii increase and they superimpose on each other. In Fig. 1.3, crests and troughs are represented, respectively, by continuous and dotted circular arcs. There are points at which the crest

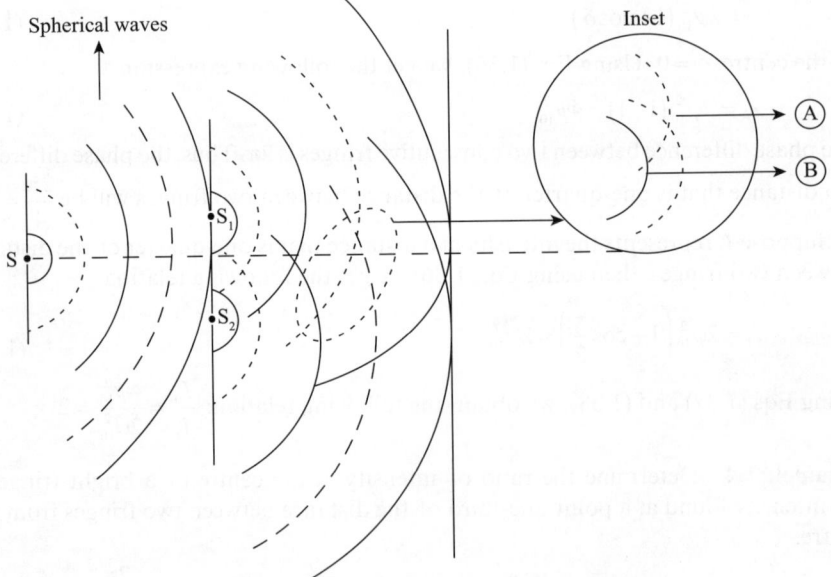

Fig. 1.3 Schematic of Young's double-slit experiment

of one of the secondary waves falls on that of another wave or the trough of one of the waves coincides with that of another wave. One such point is point A (shown in the inset of Fig. 1.3). The resultant amplitude is the sum of the amplitudes of the two individual waves, and therefore at such points, the resultant amplitude increases. This phenomenon, as mentioned earlier, is known as *constructive interference*. Since intensity is proportional to the square of the amplitude, the resultant intensity at points undergoing constructive interference also increases. On the contrary, there are points where the crest of one wave falls on the trough of another wave or vice versa. One such point is point B (shown in the inset of Fig. 1.3). The resultant amplitude at these points is the difference between the two individual amplitudes and is therefore minimum. Such points are called regions of destructive interference. Since intensity is proportional to the square of amplitude, intensity gets minimized at points having destructive interference.

1.6.1 Theory of Fringe Formation

Figure 1.4 shows the source slit S and the two slits S_1 and S_2 that are equidistant from S.

The distance between slits S_1 and S_2 is $2d$, and the screen is assumed to be at distance D from the plane containing slits S_1 and S_2. Point O on the screen is equidistant from S_1 and S_2. Therefore, the path difference between the waves reaching point O from S_1 and S_2 is zero. This also means that there is no phase difference between the waves reaching point O from S_1 and S_2. The intensity at point O is, therefore, maximum. Let us now consider a point R at a distance

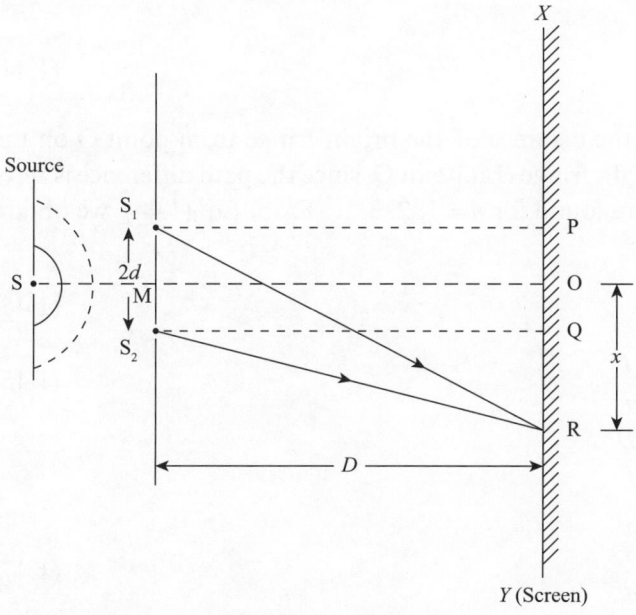

Fig. 1.4 Condition for fringe formation on screen

x from O, as shown in Fig. 1.4. From the right-angle triangle S_1PR, we get the following relation:

$$(S_1R)^2 = (S_1P)^2 + (PR)^2$$

yielding $(S_1R)^2 = D^2 + (x + d)^2$ (1.39)

In addition, from the right-angle triangle S_2QR we get the following relation:

$$(S_2R)^2 = (S_2Q)^2 + (QR)^2$$

yielding $(S_2R)^2 = D^2 + (x - d)^2$ (1.40)

Using Eqs (1.39) and (1.40), we get the following relation:

$$(S_1R)^2 - (S_2R)^2 = (x + d)^2 - (x - d)^2 = 4xd$$

leading to $(S_1R - S_2R)(S_1R + S_2R) = 4xd$ (1.41)

In Young's set-up, the distance between the screen and the plane containing slits S_1 and S_2, D, is much greater than the distance between the slits, $2d$ or x; therefore, $(S_1R + S_2R)$ can be replaced by $2D$ without introducing an appreciable error. Thus, Eq. (1.41) can be rewritten as follows:

$$(S_1R - S_2R) 2D = 4xd$$

Or $(S_1R - S_2R) = 4xd/2D = 2xd/D$ (1.42)

Let us now determine the location of bright and dark fringes.

Bright fringes   Point P is bright if the path difference is a whole-number multiple of wavelength λ. Thus,

$$S_2P - S_1P = n\lambda \quad \text{where} \quad n = 0, 1, 2, \dots$$ (1.43)

Substitution of Eq. (1.43) into Eq. (1.42) leads to the following expression:

$$\frac{2xd}{D} = n\lambda$$

or $$x = \frac{n\lambda D}{2d}$$ (1.44)

Equation (1.44) gives the distance of the bright fringe from point O on the screen. The central bright fringe is at point O, since the path difference is zero. Other bright fringes are found for $n = 1, 2, 3, \dots$ From Eq. (1.44), we obtain the following relations:

$$n = 1, \ x_1 = \frac{\lambda D}{2d}$$ (1.45)

$$n = 2, \ x_2 = \frac{2\lambda D}{2d}$$ (1.46)

$$n = 3, \ x_3 = \frac{3\lambda D}{2d}$$

$$\vdots$$

$$n = n, \ x_n = \frac{n\lambda D}{2d}$$ (1.47)

The linear distance between any two consecutive fringes is given as follows:

$$x_2 - x_1 = \frac{2\lambda D}{2d} = \frac{\lambda D}{2d} = \frac{\lambda D}{2d} \tag{1.48}$$

Dark fringes Point P is dark if the path difference is an odd number multiple of a half-wavelength. In this case,

$$(S_2 P - S_1 P) = (2n+1)\frac{\lambda}{2} \tag{1.49}$$

where $n = 0, 1, 2, 3, \ldots$

Using Eqs (1.42) and (1.49), we get the following relation: $\dfrac{2xd}{D} = \dfrac{(2n+1)}{2}\lambda$ which implies that

$$x = \frac{(2n+1)\lambda D}{4d} \tag{1.50}$$

From Eq. (1.50), we get the following relations for the dark fringes:

$$n = 0 \text{ gives } x_0 = \frac{\lambda D}{4d} \tag{1.51}$$

$$n = 1 \text{ gives } x_1 = \frac{3\lambda D}{4d} \tag{1.52}$$

$$n = 2 \text{ gives } x_2 = \frac{5\lambda D}{4d} \tag{1.53}$$

$$\vdots$$

$$n = n \text{ gives } x_n = \frac{(2n+1)\lambda D}{4d} \tag{1.54}$$

The distance between two consecutive dark fringes is given as follows:

$$x_2 - x_1 = \frac{5\lambda D}{4d} - \frac{3\lambda D}{4d} = \frac{2\lambda D}{4d} = \frac{\lambda D}{2d} \tag{1.55}$$

From Eqs (1.48) and (1.55), it is clear that the spacing between two consecutive bright fringes (maxima) is the same as the distance between two consecutive dark fringes (minima). This expression is also called the fringe width, and a symbol β is often used to represent it. From Eqs (1.48) and (1.55), one can conclude that the fringe width is directly proportional to D and λ, and inversely proportional to the distance between the two slits, $2d$.

Example 1.5 Two straight and narrow parallel slits 0.9 mm apart are illuminated using a monochromatic light source. A screen placed at a distance of 90 cm is used to obtain fringes. It is found that the distance between consecutive fringes is 0.4 mm. Determine the wavelength of light.

Solution Using Eq. (1.48), we can write the following expression for fringe width β:

$$\beta = \frac{\lambda D}{2d} \tag{1.56}$$

We can rewrite Eq. (1.56) in the following form:

$$\lambda = \frac{\beta \times 2d}{D} \tag{1.57}$$

Using the given values in Eq. (1.57), we get the following relation:

$$\lambda = \frac{0.04 \times 0.09}{90} = 4 \times 10^{-5} \text{ cm} \text{ or } \lambda = 4000\text{Å}.$$

Example 1.6 Two coherent sources are placed 1 mm apart and generate interference fringes on a screen 0.9 m away. The second dark fringe is formed at a distance of 0.9 mm from the central fringe. Determine the wavelength of the monochromatic light used.

Solution From Eq. (1.50), we get the following expression:

$$x_n = \frac{(2n+1)\lambda D}{4d} \tag{1.58}$$

which implies that

$$\lambda = 4dx_n/[D(2n + 1)] \tag{1.59}$$

Substitution of the given values in Eq. (1.59) yields the following relation:

$$\lambda = \frac{0.1 \times 2 \times 0.09}{90 \times 5} = 4 \times 10^{-5} \text{ cm}$$

Example 1.7 In Young's experiment, let a light of wavelengths 5.4×10^{-7} m and 6.85×10^{-8} m be used in turn, keeping the geometry same. Compare the fringe widths in the two cases.

Solution From Eq. (1.55), we have the following relation:

$$\beta = \frac{\lambda D}{2d} \tag{1.60}$$

For the two wavelengths λ_1 and λ_2, we can write the following equations:

$$\beta_1 = \frac{\lambda_1 D}{2d} \quad \text{and} \quad \beta_2 = \frac{\lambda_2 D}{2d} \tag{1.61}$$

Using Eq. (1.61), we can write the following expression: $\dfrac{\beta_1}{\beta_2} = \dfrac{\dfrac{D}{2d} \times 5.4 \times 10^{-7}}{\dfrac{D}{2d} \times 6.85 \times 10^{-8}} = 8$

Thus, $\beta_1 = 8\beta_2$

Example 1.8 In a Young's double-slit experiment, the slits are separated by 0.28 mm and the screen is placed 1.4 m away. The distance between the central bright fringe and the fourth bright fringe has been measured to be 1.2 cm. Determine the wavelength of light.

Solution From Eq. (1.55), we get the following relation: $\beta = \dfrac{\lambda D}{2d}$ which yields the following equation:

$$\lambda = \frac{\beta(2d)}{D} \tag{1.62}$$

Use of these values in Eq. (1.62) leads to the following expression:

$$\lambda = \left(\frac{1 \cdot 2}{4}\right) \times \frac{0.28 \times 10^{-5}}{1.4} = 600 \times 10^{-9} \text{ m}$$

Example 1.9 Two straight and narrow parallel slits located 1 mm apart are illuminated using a monochromatic light source. A screen placed at a distance of 100 cm is used to obtain fringes. It is found that the distance between consecutive fringes is 0.5 mm. Determine the wavelength of light.

Solution The wavelength of light is $\lambda = \dfrac{\beta \times 2d}{D} = \dfrac{0.05 \times 0.1}{100} = 5 \times 10^{-5}$ cm

Example 1.10 Two coherent sources are placed 0.9 mm apart and generate interference fringes on a screen 1 m away. The second dark fringe is formed at a distance of 0.08 cm from the central fringe. Determine the wavelength of the monochromatic light used.

Solution Wavelength of light, $\lambda = \dfrac{4dx_n}{D(2n+1)} = \dfrac{0.09 \times 2 \times 0.08}{100(5)} = 2.9 \times 10^{-5}$ cm

Example 1.11 In the Young's experiment, let light of wavelengths 6.2×10^{-7} m and 7.1×10^{-8} m be used in turn, keeping the geometry same. Compare the fringe widths in these two cases.

Solution On comparing fringe widths, $\dfrac{\beta_1}{\beta_2} = \dfrac{\lambda_1}{\lambda_2} = \dfrac{6.2 \times 10^{-7}}{7.1 \times 10^{-8}} = \dfrac{62}{7.1} = 8.73$

Example 1.12 In a Young's double-slit experiment, the slits are separated by 0.32 mm and the screen is placed 1.5 m away. The distance between the central bright fringe and the fourth bright fringe is measured to be 1.3 cm. Determine the wavelength of light.

Solution The wavelength is calculated as follows:

$$\lambda = \frac{\beta(2d)}{D} = \left(\frac{1.3}{4}\right) \times 10^{-2} \times \frac{0.32 \times 10^{-3}}{1.5} = \left(\frac{1.3}{4}\right) \times \frac{6.32 \times 10^{-5}}{1.5} = 693.3 \times 10^{-9} \text{ m.}$$

1.7 TYPES OF INTERFERENCE

The phenomenon of interference requires two wavefronts to interact. These wavefronts can be obtained in two different ways, resulting in two different classes of interference: (a) division of wavefront and (b) division of amplitude.

1.7.1 Division of Wavefront

In this type of interference, the incident wavefront is divided into two parts using the phenomenon of reflection or refraction, or simple splitting of a wavefront at an obstruction (diffraction). The two parts of the wavefronts are then made to travel unequal distances before reuniting at some angle. This process leads to the production of an interference pattern. Lloyd's mirror, Fresnel's biprism, YDSE are examples of this type of interference respectively.

1.7.1.1 Fresnel's Biprism

A biprism is obtained by combining two acute prisms base to base. The obtuse angle of the prism is about 179° and the other angles are about 30° each. Fresnel used the biprism to demonstrate the phenomenon of interference. The biprism was used to produce two coherent images of a given slit, as shown in Fig. 1.5.

These two coherent images serve as coherent sources and therefore fulfil a condition necessary for observing interference. The main slit S is illuminated

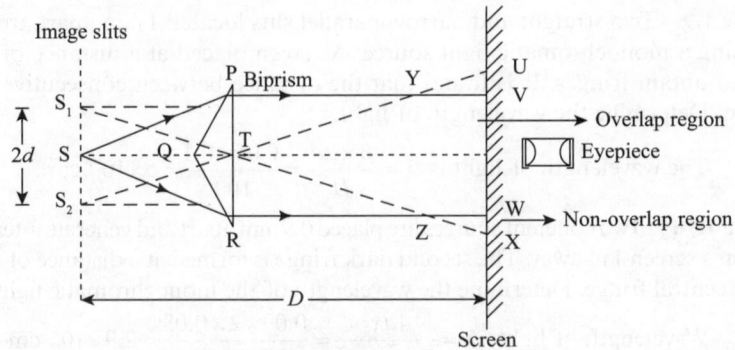

Fig. 1.5 Schematic of biprism experiment

by a monochromatic light source like sodium vapour lamp. The light emerging out of the slit S falls on the biprism. The wavefront gets divided into two parts at the edge PR. One part of the wavefront passes through the upper half PTQ of the biprism. This wavefront gets deviated through a small angle towards edge R of the biprism and appears to diverge from the virtual image S_1. Similarly, the other part of the wavefront passes through the lower half RTQ of the biprism and appears to diverge from the virtual image S_2. Interference fringes are observed in the overlapping region TZWVYT and can be observed with the help of an eyepiece shown in Fig. 1.5.

Fresnel's biprism set-up can be used in interesting applications. Some of these are as follows:

1. Determination of the wavelength of light
2. Fringes using white light
3. Determination of the thickness of a thin sheet of transparent material

Determination of wavelength of light   From Eq. (1.55), we get the following relation: $\beta = \dfrac{\lambda D}{2d}$

which can be rewritten as follows:

$$\lambda = \frac{\beta(2d)}{D} \tag{1.63}$$

Thus, the wavelength of incident light can be determined if the fringe width (β), distance between the two virtual sources ($2d$), and the horizontal distance between the slit S and the eyepiece are known.

Fringes using white light   White light consists of wavelengths in the range 4400 Å (violet) to 7500 Å (red). If white light is made to fall on the main slit S of the biprism set-up, then all the constituent wavelengths produce their own interference pattern. The resultant interference pattern has the following characteristics:

1. A zero-order fringe at the centre of the interference pattern is white.
2. On either side of the zero-order fringe, a few coloured fringes are obtained. The inner edges of the coloured fringes are reddish, whereas the outer edges are violet in colour.

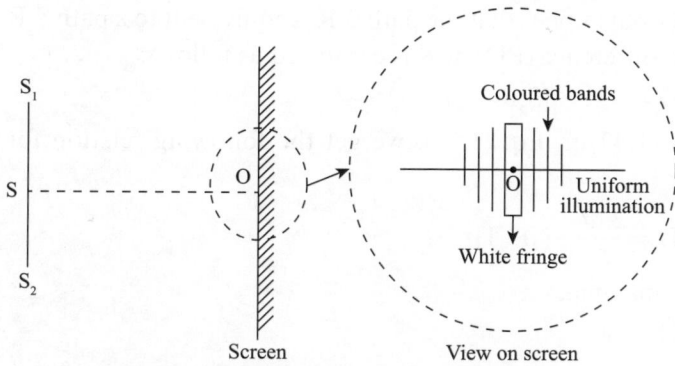

Fig. 1.6 Fringe pattern using white light

3. The regions beyond the coloured fringes have uniform illumination on either side.

Figure 1.6 shows a schematic representation of the fringe pattern obtained using a white light source. The inset shows a small portion of the screen after rotating the diagram through an angle of 90°.

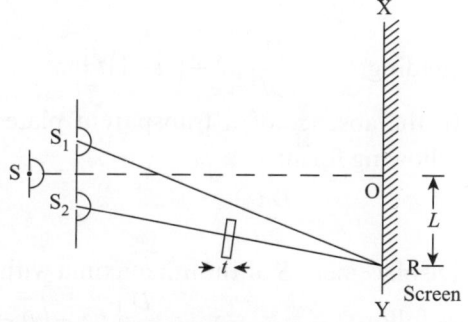

Fig. 1.7 Alteration of interference pattern with introduction of thin sheet

Determination of thickness of thin sheet of transparent material Fresnel's biprism set-up can also be used to determine the thickness of a thin sheet of transparent material. For this, a thin sheet of transparent material is introduced in the path of the interfering beams, as shown in Fig. 1.7.

Let us assume that the thickness of the sheet is t and the refractive index of the material of the sheet is μ. Introduction of the transparent sheet displaces the entire fringe system without changing the fringe width. For the path S_2R, a distance $(S_2R - t)$ is traversed in air and a distance t is traversed in a medium of refractive index μ. The time taken by light to cover the distance S_2R is thus given by the following expression:

$$T = (S_2R - t)/c + t/c_p \tag{1.64}$$

where c_p represents the velocity of light in the material of the thin sheet. We know that

$$\frac{c}{c_p} = \mu \tag{1.65}$$

Substitution of Eq. (1.65) in Eq. (1.64) results in the following relation:

$$T = (S_2R - t)/c + t\mu/c$$

giving the following relation:

$$T = [S_2R + (\mu - 1)t]/c \tag{1.66}$$

Equation (1.66) implies that the path S_2R is equivalent to a path $S_2R + (\mu - 1)t$ in air. Path difference (PD) at R is expressed as follows:

$$PD = S_1R - [S_2R + (\mu - 1)t] \tag{1.67}$$

Using Eq. (1.42) in Eq. (1.67), we get the following relation for the path difference:

$$PD = \frac{2xd}{D} - (\mu - 1)t \tag{1.68}$$

for the nth maxima,

$$PD = n\lambda \tag{1.69}$$

Combining Eqs. (1.68) and (1.69), we get the following relation:

$$\frac{2x_n d}{D} - (\mu - 1)t = n\lambda$$

yielding, $x_n = \dfrac{D}{2d}\left[n\lambda + (\mu - 1)t\right]$ \hfill (1.70)

In the absence of a transparent plate, $t = 0$ and Eq. (1.70) reduces to the following form:

$$x_{n0} = \frac{Dn\lambda}{2d} \tag{1.71}$$

Displacement S of the nth maxima with the introduction of the plate becomes as follows: $S = x_n - x_{n0} = \dfrac{D}{2d}\left[n\lambda + (\mu - 1)t\right] - \dfrac{Dn\lambda}{2d}$

resulting in the following relation:

$$S = \frac{D}{2d}(\mu - 1)t \tag{1.72}$$

The expression for S is independent of the order n of the fringe.
Equation (1.72) can be rewritten in the following form:

$$t = \frac{S(2d)}{D(\mu - 1)} \tag{1.73}$$

Equation (1.73) can be used to determine the value of thickness of the transparent sheet.

1.7.1.2 Lloyd's Mirror

In the year 1834, Lloyd used a set-up to demonstrate an interference pattern that can be observed due to the division of a wavefront. Figure 1.8 shows the formation of an interference pattern schematically.

PQ is a mirror that is polished on the front surface and blackened at the back, to avoid multiple reflections. The narrow slit S_1 is illuminated using monochromatic light. Light from slit S_1 is incident on the mirror at grazing angle. At the grazing incidence, almost the entire incident light is reflected by the mirror. The reflected beams appear to diverge from S_2; therefore, S_2 and S_1 act as two coherent sources. BS_1G represents the direct cone of light and BS_2E

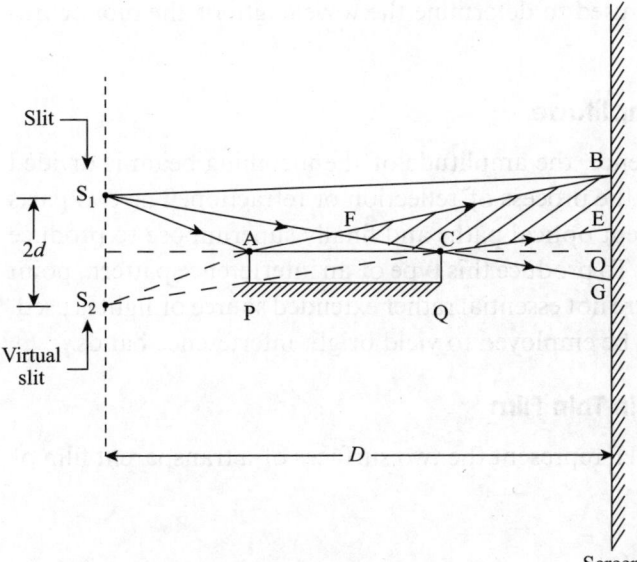

Fig. 1.8 Interferences pattern formation using Lloyd's mirror

the reflected cone of light. From Fig. 1.8, one can observe that the two cones superimpose in the region BFCE, and interference fringes are observed in this overlapping region.

The central zero-order fringe is expected to lie at O along the perpendicular bisector of $S_1 S_2$. In Fig. 1.8, point O lies outside the region of overlap. However, the central fringe can be observed by moving mirror PQ towards the screen, so that point O on the screen coincides with the end Q of the mirror. The central fringe observed at O under these circumstances is, however, found to be dark. This observation can only be explained if one assumes that the *light suffers a phase change of π when it gets reflected from a denser medium*. In fact, this conclusion is one major contribution of the Lloyd's mirror experiment.

There is another interesting method of making the central fringe visible. A thin film of a transparent material like mica is introduced in the path of the direct ray, and slit S_1 is illuminated by a white light source. The entire fringe pattern is then displaced in the upward direction. The central fringe also shifts upwards and is found to be dark. Coloured fringes are found on either side of the central dark fringe.

The fringe width for the interference pattern obtained in Lloyd's single mirror set-up is given by the following relation:

$$\beta = \frac{\lambda D}{2d} \tag{1.74}$$

Equation (1.74) can be rewritten in the following form:

$$\lambda = \frac{\beta(2d)}{D} \tag{1.75}$$

Equation (1.74) can be used to determine the wavelength of the monochromatic source.

1.7.2 Division of Amplitude

In this type of interference, the amplitude of the incoming beam is divided into two parts through the process of reflection or refraction. The two parts then travel along different optical paths and finally superimpose to produce an interference pattern. To produce this type of an interference pattern, point or narrow line sources are not essential rather extended source of light is used. Broad *light sources* can be employed to yield bright interference bands.

1.7.2.1 Interference in Thin Film

In Fig. 1.9, GH and G_1H_1 represent the two surfaces of a transparent film of uniform thickness:

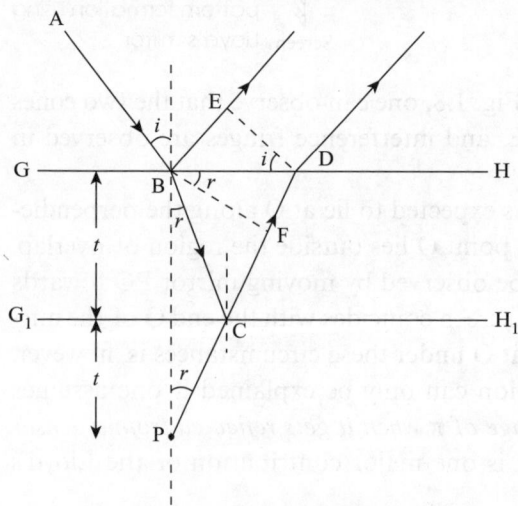

Fig. 1.9 Interference pattern in reflected light

$$\text{Path difference, PD} = \mu(\text{BC} + \text{CD}) - \text{BE} \tag{1.76}$$

Using Snell's law at the interface, we get

$$\mu = \frac{\sin i}{\sin r} = \frac{\text{BE}/\text{BD}}{\text{FD}/\text{BD}} = \frac{\text{BE}}{\text{FD}}$$

giving the following relation:

$$\text{BE} = \mu(\text{FD}) \tag{1.77}$$

Using Eq. (1.77) in Eq. (1.76), we obtain the following expression:

$$\text{PD} = \mu(\text{BC}+\text{CD}) - \mu(\text{FD})$$

which results in the following relation: $\text{PD} = \mu(\text{BC}+\text{CF}+\text{FD}) - \mu(\text{FD})$

yielding $\text{PD} = \mu(\text{BC}+\text{CF})$ \hfill (1.78)

Since BC = PC, Eq. (1.78) leads to the following form:

$$PD = \mu(PF) \tag{1.79}$$

From triangle BPF, we get the following relation: $\cos r = \dfrac{PF}{BP}$
resulting in

$$PF = BP\cos r = 2t\cos r \tag{1.80}$$

Substitution of the expression for PF [Eq. (1.80)] into Eq. (1.79) yields the following expression:

$$PD = \mu \times 2t\cos r = 2\mu t\cos r \tag{1.81}$$

We have learnt in Section 1.7.1.2 that a ray reflected from a denser medium suffers a phase change of π, which corresponds to a path difference of $\dfrac{\lambda}{2}$.

The effective path differences $(PD)_{\text{eff}}$ thus becomes as follows:

$$(PD)_{\text{eff}} = (2\mu t\cos r \pm \lambda/2) \tag{1.82}$$

For maxima, the following relation must hold: $2\mu t\cos r \pm \dfrac{\lambda}{2} = n\lambda$
which gives the following expression:

$$2\mu t\cos r = (2n \pm 1)\lambda/2 \tag{1.83}$$

When Eq. (1.83) is fulfilled, the thin film would appear bright in the reflected pattern.

For minima, we must satisfy the following relation: $2\mu t\cos r \pm \dfrac{\lambda}{2} = (2n \pm 1)\dfrac{\lambda}{2}$

yielding $2\mu t\cos r = (2n \pm 1)\dfrac{\lambda}{2} \pm \dfrac{\lambda}{2}$ $\hfill(1.84)$

Since $(n+1)$ or $(n-1)$ can also be taken as an integer, we can rewrite Eq. (1.84) in the following form:

$$2\mu t\cos r = n\lambda \tag{1.85}$$

where $n = 0, 1, 2, 3, \ldots$

When Eq. (1.85) is fulfilled, the films appear dark in the reflected pattern.

An interference pattern can also be observed in the transmitted light. Figure 1.10 is a schematic representation of this situation.

Figure 1.10 shows two transmitted rays, BT and DS, which are obtained after reflection and refraction of the corresponding incident rays. BF and DE are normals drawn on DC and BT, respectively. When DC is extended in the backward direction, it meets the extended BH at I.

The effective path difference $(PD)_{\text{eff}}$ is given as follows:

$$(PD)_{\text{eff}} = \mu(BC + CD) - BE \tag{1.86}$$

We know that $\mu = \sin i/\sin r = [BE/BD]/[FD/BD]$

Thus, $\quad BE = \mu FD \tag{1.87}$

Using Eq. (1.87) in Eq. (1.86), we get the following relation:

$$(PD)_{\text{eff}} = \mu(BC + CF + FD) - \mu FD$$

leading to $(PD)_{\text{eff}} = \mu(BC + CF) = \mu(BI)$

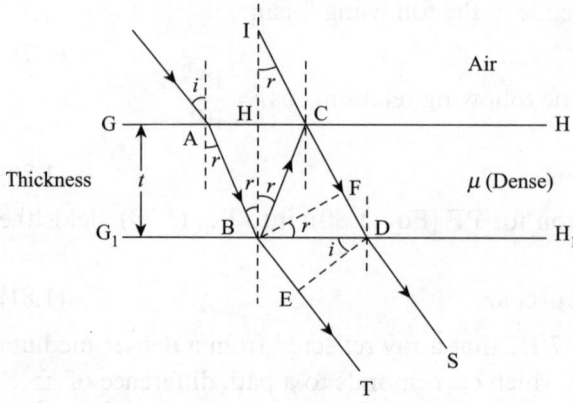

Fig. 1.10 Interference pattern in transmitted light

implying that

$$(PD)_{eff} = 2\mu t \cos r \qquad (1.88)$$

In this case, reflections have taken place inside the film, and therefore the ray travels from a denser medium to a rarer medium, that is, air. Thus, no additional phase change of π is involved.

For maxima,

$$2\mu t \cos r = n\lambda \qquad (1.89)$$

The condition indicated by Eq. (1.89) results in the film appearing bright in the transmitted light.

For minima,

$$2\mu t \cos r = (2n \pm 1)\frac{\lambda}{2} \qquad (1.90)$$

where n can take values 1, 2, 3, ...

The condition indicated by Eq. (1.90) results in the film appearing dark in the transmitted light.

A comparison of Eqs (1.83), (1.85), (1.89), and (1.90) reveals that the conditions of maxima and minima get reversed as we change from reflected light to transmitted light.

Let us now see what happens if we replace monochromatic light with white light. The effective path difference is dependent upon μ, which in turn depends upon the wavelength of the incident light. For any particular region on the film and for a particular viewing position, the condition for maxima would be satisfied only for some wavelengths. Bright-coloured fringes would appear in this position. Neighbouring wavelengths would result in reduced intensity. Wavelengths for which the condition for minima is satisfied would be absent in the observed pattern. The basic pattern would remain the same as we change the position of the eye or the region of the film that we are looking at. We also see that conditions for maxima and minima get reversed as we go from the reflected to the transmitted light. Therefore, the colours absent in the reflected light are visible in the transmitted light and vice versa. Thus, the colours of the reflected and transmitted light are complementary to each other.

1.7.2.2 Wedge-shaped Film

Suppose we have two plane surfaces AB and A_1B_1 that are inclined at an angle θ with respect to each other. The plane surfaces then enclose a wedge-shaped film within their inner surfaces. A schematic representation is provided in Fig. 1.11.

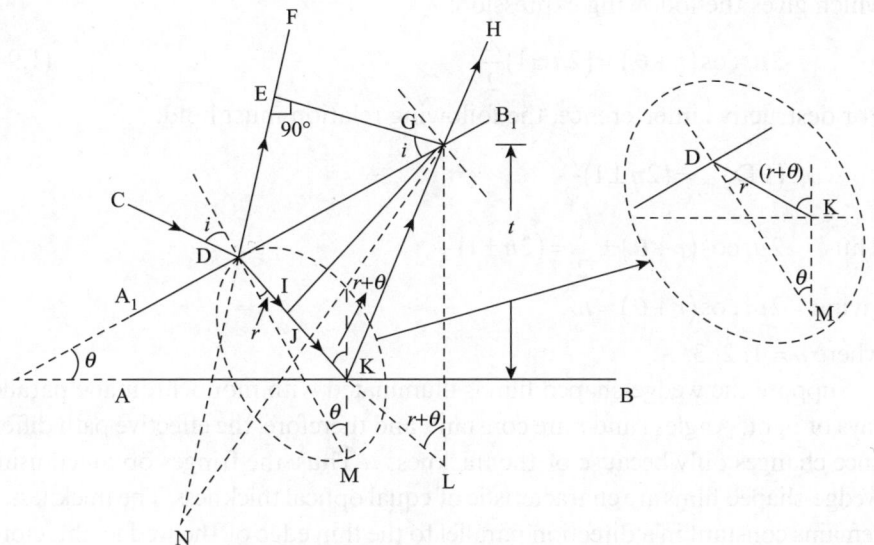

Fig. 1.11 Interference in wedge-shaped film

It is clear from Fig. 1.11 that the film obtained within the two surfaces AB and A_1B_1 is of variable thickness, with the thickness increasing as one goes from end A towards end B. Let μ represent the refractive index of the film. Ray DF is obtained due to reflection from the top surface A_1B_1, whereas ray GH is obtained by a reflection at the bottom surface AB and a refraction at the top surface A_1B_1. Interference is obtained between rays DF and GH, which appear to diverge from point N. Thus, interference occurs at the virtual point N. The optical path difference PD between the two rays DF and GH is given by the following expression:

$$(PD) = \mu(DK + KG) - DE \tag{1.91}$$

Since $DE = \mu(DI)$, Eq. (1.91) can be rewritten in the following form:

$$(PD) = \mu(DI + IJ + KG) - \mu\,(DI)$$

which leads to the following expression: $(PD) = \mu(IJ + KG) = \mu(IL)$

yielding $(PD) = 2\mu t \cos(r + \theta)$ $\tag{1.92}$

Ray DF is obtained by reflection from the surface of a denser medium and incorporates an additional path difference of $\dfrac{\lambda}{2}$ due to this reflection. Thus, the effective path difference $(PD)_{eff}$ is expressed as follows:

$$(PD)_{eff} = 2\mu t \cos(r+\theta) \pm \frac{\lambda}{2} \tag{1.93}$$

For constructive interference, the following relation must hold:

$$(PD)_{eff} = 2\mu t \cos(r+\theta) \pm \frac{\lambda}{2} = n\lambda$$

which gives the following expression:

$$2\mu t \cos(r+\theta) = (2n \pm 1)\frac{\lambda}{2} \tag{1.94}$$

For destructive interference, the following relation must hold:

$$(PD)_{eff} = (2n \pm 1)\frac{\lambda}{2}$$

Thus, $2\mu t \cos(r+\theta) \pm \dfrac{\lambda}{2} = (2n \pm 1)\dfrac{\lambda}{2}$

giving $2\mu t \cos(r+\theta) = n\lambda$

where $n = 1, 2, 3, \ldots$

Suppose the wedge-shaped film is illuminated with monochromatic parallel rays of light. Angles i and r are constant, and therefore, the effective path difference changes only because of the thickness t. Thus, the fringes obtained using wedge-shaped films are characteristic of equal optical thickness. The thickness, t, remains constant in a direction parallel to the thin edge of the wedge; therefore, we obtain straight fringes that are parallel to the edge of the wedge.

For the nth maxima, we can write the following equation:

$$2\mu t \cos(r+\theta) = (2n+1)\frac{\lambda}{2} \tag{1.95}$$

For normal incidence and air film, we have $r = 0$ and $\mu = 1$

Using these values in Eq. (1.95), we get the following expression:

$$2t \cos\theta = (2n+1)\frac{\lambda}{2} \tag{1.96}$$

Suppose this fringe is obtained at a distance x_n from the thin edge, as shown in Fig. 1.12.

From Fig. 1.12, we obtain the following relation:

$$t = x_n \tan\theta \tag{1.97}$$

Using Eqs (1.96) and (1.97), we get the following expression:

$$2x_n \tan\theta \cos\theta = (2n+1)\frac{\lambda}{2}$$

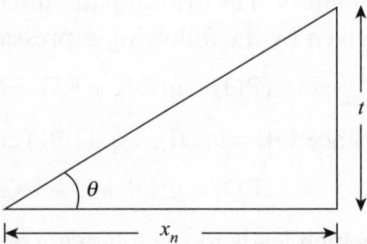

Fig. 1.12 Maxima at distance x_n from edge

giving $\quad 2x_n \sin\theta = (2n+1)\dfrac{\lambda}{2}$ $\hfill (1.98)$

If the $(n+1)$th maxima is obtained at a distance x_{n+1} from the thin edge, then for this fringe, we can similarly write the following expression:

$$2x_{n+1}\sin\theta = \left[2(n+1)+1\right]\frac{\lambda}{2} = (2n+3)\frac{\lambda}{2} \hfill (1.99)$$

Subtracting Eq. (1.98) from Eq. (1.99), we get the following relation:

$$2(x_{n+1}-x_n)\sin\theta = \lambda$$

Therefore, fringe spacing can be written as follows:

$$\beta = x_{n+1} - x_n = \frac{\lambda}{2\sin\theta} \approx \frac{\lambda}{2\theta} \hfill (1.100)$$

Equation (1.100) is valid for small values of θ, measured in radians.

1.7.2.3 Newton's Rings

Figure 1.13 shows a schematic representation of a plano-convex lens L kept on a plane glass plate GP. An air film of variable thickness is then formed between the bottom surface of the lens and the top surface of the glass plate. From Fig. 1.13, it is clear that the thickness of the film increases as we move away from the point of contact. Thickness of the air film is zero at the point of contact and is constant along the circles drawn using the point of contact as the centre. The resultant interference pattern thus consists of alternate dark and bright rings that are concentric around the point of contact. These rings are also known as Newton's rings, as they were first analysed by Newton.

A monochromatic ray QR is incident on the plane surface of the plano-convex lens L. RS represents the refracted ray. At the glass–air interface, a portion gets reflected and comes out of the lens in the form of ray 1. The portion transmitted at point S gets reflected at point T on the top surface of the glass plate and finally comes out of the lens as ray 2. Since ray 2 results due to reflection at an air (rarer)–glass (denser) interface, it undergoes a phase change of π. The rays are coherent and produce an interference pattern.

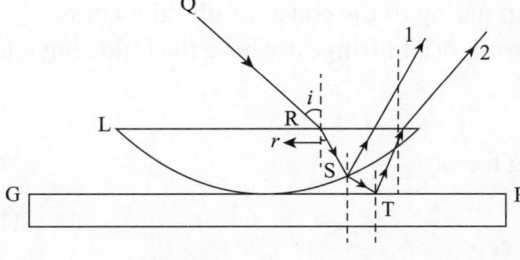

Fig. 1.13 Formation of Newton's rings

The effective path difference between rays 1 and 2 is given by expression (1.93), which is valid for the interference pattern obtained with films of variable thickness, namely

$$(PD)_{eff} = 2\mu t \cos(r + \theta) + \frac{\lambda}{2} \tag{1.101}$$

If the plano-convex lens has a large radius of curvature, the angle θ is extremely small and can be neglected. Equation (1.101) can then be written as follows:

$$(PD)_{eff} = 2\mu t \cos r + \frac{\lambda}{2} \tag{1.102}$$

For an air film, $\mu = 1$ and, for normal incidence, $r = 0$. Under these conditions, Eq. (1.102) reduces to the following form:

$$(PD)_{eff} = 2t + \frac{\lambda}{2} \tag{1.103}$$

Figure 1.14 shows the curved surface of the lens as a part of a circle with centre C_1.

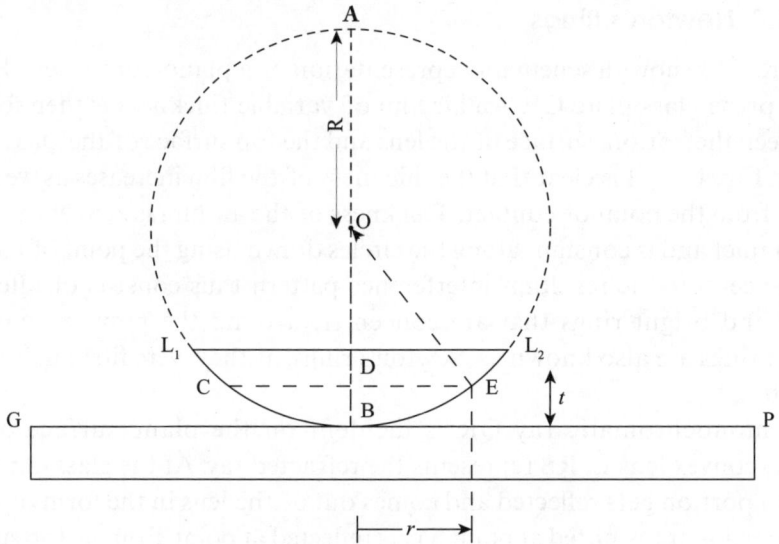

Fig. 1.14 Schematic of curved surface of lens in contact with plane glass plate

L_1BL_2 represents the lens placed on the glass plate GP. The curved surface L_1BL_2 of the lens is part of the spherical surface shown as a dotted circle with centre O in Fig. 1.14. Let R represent the radius of curvature and r the radius of the Newton's ring corresponding to the constant film thickness t.

Using Eq. (1.103), for the nth bright fringe, we have the following relation:

$$2t + \frac{\lambda}{2} = n\lambda$$

which leads to the following form:

$$2t = (2n - 1)\frac{\lambda}{2} \tag{1.104}$$

where, $n = 1, 2, 3, \ldots$

For the nth dark ring, we have

$$2t = n\lambda \tag{1.105}$$

where, $n = 0, 1, 2, 3, \ldots$

From the property of the circle, for the circle shown in Fig. 1.13, we can write the following expression: $NP \times NQ = NO \times ND$

Substituting values, we get the following equation:

$$r \times r = t(2R - t) = 2Rt - t^2 \approx 2Rt$$

which gives the following relation:

$$r^2 = 2Rt \quad \text{or} \quad t = \frac{r^2}{2R} \tag{1.106}$$

Using Eqs (1.104) and (1.106), we get the following expression for a bright ring:

$$2 \cdot \frac{r^2}{2R} = (2n - 1)\frac{\lambda}{2}$$

yielding $r^2 = \dfrac{(2n - 1)\lambda R}{2}$

Substituting $r = \dfrac{D}{2}$, we get the following equation: $\dfrac{D^2}{4} = \dfrac{(2n-1)\lambda R}{2}$

which yields the following expression:

$$D = \sqrt{(2\lambda R)(2n - 1)} \tag{1.107}$$

or $\qquad D \propto \sqrt{2n - 1} \tag{1.108}$

Thus, diameters of bright rings are proportional to the square roots of the odd numbers $(2n - 1)$.

Using Eqs (1.105) and (1.106), we get the following relation for the nth dark ring: $2 \cdot \dfrac{r^2}{2R} = n\lambda$

which leads to the following expression: $r^2 = n\lambda R$

or $\qquad D^2 = 4n\lambda R \tag{1.109}$

Thus, $\quad D = 2\sqrt{n\lambda R} \propto \sqrt{n} \tag{1.110}$

Diameters of dark rings are proportional to the square roots of natural numbers. Figure 1.15 shows a schematic representation of Newton's rings as seen in reflected light.

Newton's rings can also be observed in the transmitted light. In this case, for bright rings we have the following condition:

$$2t = n\lambda \tag{1.111}$$

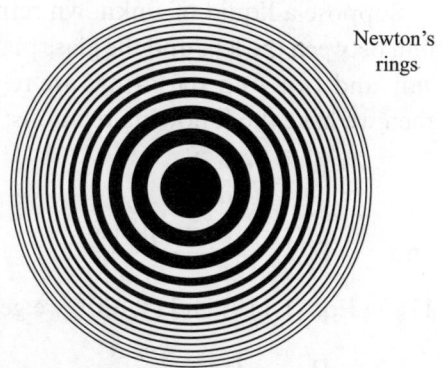

Newton's rings

Fig. 1.15 Newton's rings in reflected light

and for dark rings we have the following condition:

$$2t = (2n-1)\frac{\lambda}{2} \tag{1.112}$$

Combining Eqs (1.106) and (1.111), we obtain the following relation for bright rings: $2 \times \dfrac{r^2}{2R} = n\lambda$

or $r^2 = n\lambda R$

which results in the following equation:

$$D = 2\sqrt{n\lambda R} \propto \sqrt{n} \tag{1.113}$$

Combining Eqs (1.112) and (1.106), we get the following expression for dark rings: $2 \times \dfrac{r^2}{2R} = (2n-1)\dfrac{\lambda}{2}$

which gives the following relation: $r^2 = \dfrac{(2n-1)\lambda R}{2}$

Thus, $D = \sqrt{2\lambda R} \times \sqrt{2n-1} \propto \sqrt{(2n-1)} \tag{1.114}$

The central ring is bright in the transmitted pattern, whereas it is dark in the reflected pattern.

Let D_n and D_{n+p} represent, respectively, the diameters of the nth and $(n+p)$th dark rings obtained in the reflected pattern. Using Eq. (1.109), we get the following relation:

$$D_n^{\,2} = 4n\lambda R \tag{1.115}$$

and $D_{n+p}^2 = 4(n+p)R\lambda \tag{1.116}$

From this, we get the following equation:

$$D_{n+p}^2 - D_n^2 = 4pR\lambda \tag{1.117}$$

which results in the following expression:

$$\lambda = \frac{D_{n+p}^2 - D_n^2}{4pR} \tag{1.118}$$

Wavelength λ can be determined using Eq. (1.118).

Suppose a liquid of unknown refractive index μ is used to replace the air film between the lens and the glass plate. If the corresponding diameters of the nth and $(n+p)$th dark rings are represented by D_n' and $D_{(n+p)}'$, respectively, then we have the following relations:

$$D_n'^{\,2} = \frac{4n\lambda R}{\mu} \tag{1.119}$$

and $D_{(n+p)}'^{\,2} = \dfrac{4(n+p)\lambda R}{\mu} \tag{1.120}$

Using Eqs (1.119) and (1.120), we get the following expression:

$$D_{n+p}'^{\,2} - D_n'^{\,2} = \frac{4p\lambda R}{\mu}$$

or $\qquad \mu = \dfrac{4p\lambda R}{D_{n+p}^{\prime 2} - D_{n'}^2}$ $\hfill$ (1.121)

Using Eq. (1.118) in Eq. (1.121), we get the following expression:

$$\mu = \frac{D_{(n+p)}^2 - D_n^{\,2}}{D_{n+p}^{\prime 2} - D_n^{\prime 2}}$$ $\hfill$ (1.122)

Refractive index μ of the liquid can be determined using Eq. (1.122).

Example 1.13   A biprism is kept at a distance of 5 cm from a slit illuminated by a sodium light ($\lambda = 5890$ Å). Fringes are obtained on a screen placed at a distance of 80 cm from the biprism. Fringe width for the pattern obtained on the screen is 9.514×10^{-2} cm. Calculate the distance between the two coherent sources.

Solution   From Eq. (1.55), we have the following relation:

$$\beta = \frac{\lambda D}{2d}$$ $\hfill$ (1.123)

Rearrangement of Eq. (1.123) gives the following expression:

$$2d = \frac{\lambda D}{\beta}$$ $\hfill$ (1.124)

where $D = (80 + 5) = 85$ cm.

Substitution of these values into Eq. (1.124) yields the following value:

$$2d = \frac{5890 \times 10^{-8} \times 85}{9.514 \times 10^{-2}} = 0.053 \text{ cm}$$

Example 1.14   Sodium light ($\lambda = 5893$ Å) is used first in a Fresnel's biprism set-up. A total of 60 fringes are observed in the field of view of the eyepiece. Calculate the number of fringes that would be observed in the same field of view if the sodium light is replaced by a mercury vapour lamp with $\lambda = 5461$ Å.

Solution   From Eq. (1.55), we have the following expression:

$$\beta = \frac{\lambda D}{2d}$$ $\hfill$ (1.125)

If n fringes are observed in a field of view of length λ, then we can write the following formula:

$$\beta = \frac{\lambda D}{2d} = \frac{l}{n}$$ $\hfill$ (1.126)

Equation (1.126) can be rewritten as follows:

$$n\lambda = \left(\frac{2d}{D}\right) l = \text{constant}$$ $\hfill$ (1.127)

Therefore, $n_1 \lambda_1 = n_2 \lambda_2$ $\hfill$ (1.128)

For the given situation, we have the following relation: $60 \times 5893 = n_2 \times 5461$

which yields the following value: $n_2 = \dfrac{60 \times 5893}{5461} \cong 65$

Example 1.15   In a thin film, between points A and B, six fringes are seen with a light of wavelength 6000 Å. If the light used is of wavelength 4500 Å, what are the number of fringes observed between A and B?

Solution If t represents the thickness of the film between points A and B, then from Eq. (1.85), we get the following relation:

$$2\mu t \cos r = n\lambda \tag{1.129}$$

Since two wavelengths are involved, we can write the following formula:

$$2\mu t \cos r = n_1\lambda_1 = n_2\lambda_2 \tag{1.130}$$

or $6 \times 6000 = n_2 \times 4500$, that is, $n_2 = 8$

Example 1.16 An interference pattern is first obtained using a biprism set-up. When a thin sheet of glass ($\mu = 1.5$) of 5 μm thickness is introduced in the path of one of the interfering rays, the central fringe is shifted to a position normally occupied by the fifth fringe. Calculate the wavelength of light used.

Solution Using Eq. (1.72), we can write the following expression:

$$S = \frac{D}{2d}(\mu - 1)t \tag{1.131}$$

For the present problem, $S = 5\beta$; therefore, Eq. (1.131) can be rewritten as follows:

$$5\beta = \frac{D}{2d}(\mu - 1)t \tag{1.132}$$

From Eq. (1.55), we know that

$$\beta = \frac{\lambda D}{2d} \tag{1.133}$$

Using Eq. (1.133) in Eq. (1.132), we get the following expression: $5\frac{\lambda D}{2d} = \frac{D}{2d}(\mu - 1)t$ which yields the following relation:

$$\lambda = \frac{(\mu - 1)t}{5} \tag{1.134}$$

Substituting these values into Eq. (1.134), we get the following relation:

$$\lambda = \frac{(1.5 - 1)5 \times 10^{-4}}{5} = 5 \times 10^{-5} \text{ cm}$$

Example 1.17 In a Newton's ring set-up, the diameter of the fourth ring was found to be 0.4 cm and that of the 24th ring was 0.8 cm. The radius of curvature of the plano-convex lens is 100 cm. Calculate the wavelength of light used.

Solution Using Eq. (1.118), we get the following relation:

$$\lambda = \frac{D_{n+p}^2 - D_n^2}{4pR} \tag{1.135}$$

In the given problem, $n + p = 24$ and $n = 4$. Thus $p = 20$.
Substituting these values into Eq. (1.135), we get the following value:

$$\lambda = \frac{(0.8)^2 - (0.4)^2}{(4 \times 20 \times 100)} = 6 \times 10^{-5} \text{ cm}$$

Example 1.18 In a Newton's ring experimental set-up, the diameter of the ninth ring changes from 1.42 to 1.28 cm when a liquid of refractive index μ replaces air in the space between the lens and the plate. Determine the refractive index of the liquid.

Solution From Eq. (1.119), we get the following relation:

$$D_n'^2 = \frac{4n\lambda R}{\mu}$$

(1.136)

With the liquid occupying the space, we get the following equation:

$$D_n'^2 = \frac{4 \times 9 \times \lambda R}{\mu}$$

(1.137)

For air as a medium, we have the following expression:

$$D_9^2 = 4 \times 9 \times \lambda R$$

(1.138)

Using Eqs (1.137) and (1.138), we can obtain the following value: $\mu = \dfrac{D_9^2}{D_9'^2} = \dfrac{(1.42)^2}{(1.28)^2} = 1.231$

Example 1.19 A biprism is kept at a distance of 6 cm from a slit illuminated by a light of wavelength $\lambda = 6000$ Å. Fringes are obtained on a screen placed at a distance of 80 cm from the biprism. Fringe width for the pattern obtained on the screen is 8.214×10^{-2} cm. Calculate the distance between the two coherent sources.

Solution The distance is calculated as follows:

$$2d = \frac{\lambda D}{\beta} = \frac{6000 \times 10^{-8} \times 86}{8.214 \times 10^{-2}} = 0.063 \text{ cm}$$

Example 1.20 A light of wavelength $\lambda = 5800$ Å is first used in a Fresnel's biprism set-up. A total of 50 fringes are observed in the field of view of the eyepiece. Calculate the number of fringes that would be observed in the same field of view if the light is replaced by a light of wavelength $\lambda = 5100$ Å.

Solution Let us consider the following expression: $50 \times 5800 = n_2 \times 5200$

which yields the following value: $n_2 = \dfrac{5 \times 5800}{5200} \cong 56.$

Example 1.21 In a thin film, between points A and B, five fringes are seen with a light of wavelength 5000 Å. What are the number of fringes observed between A and B?

Solution Let us consider the following expression: $50 \times 5800 = n_2 \times 5000$

This yields the following value: $n_2 = \dfrac{5 \times 5800}{5000} \cong 6$

Example 1.22 An interference pattern is first obtained using a biprism set-up. When a thin sheet of glass ($\mu = 1.5$) of 6 µm thickness is introduced in the path of one of the interfering rays, the central fringe is shifted to a position normally occupied by the fourth fringe. Calculate the wavelength of the light used.

Solution Let us consider the following expression:

$$\lambda = \frac{(\mu - 1)t}{4} = \frac{(1.5 - 1) \times 6 \times 10^{-4}}{4} = 7.5 \times 10^{-5} \text{ cm}$$

Example 1.23 In a Newton's ring set-up, the diameter of the third ring has been found to be 0.2 cm and that of the 20th ring 0.7 cm. The radius of curvature of the plano-convex lens is 90 cm. Calculate the wavelength of the light used.

Solution   Let us consider the following expression: $\lambda = \dfrac{D_{n+p}^2 - D_n^2}{4pR}$

Where $n + p = 20;\ n = 3$ giving $p = 17$

Substituting these values, we get the following result:

$$\lambda = \frac{(0.7)^2 - (0.2)^2}{4 \times 17 \times 90} = \frac{(0.49 \times 0.04)}{4 \times 17 \times 90} = 7.35 \times 10^{-5}\,\text{cm}$$

Example 1.24   In a Newton's ring experimental set-up, the diameter of the eighth ring changes from 1.25 to 1.14 cm when a liquid of refractive index μ replaces air in the space between the lens and the plate. Determine the refractive index of the liquid.

Solution   The refractive index is $\mu = \dfrac{D_8^2}{D_8'^2} = \dfrac{(1.25)^2}{(1.14)^2} = 1.20$

IMPORTANT CONCEPTS

1. When two or more waves travel simultaneously through a medium, the resultant displacement at any point of the medium is given by the vector sum of the displacements of the individual waves. This is called the principle of superposition of waves.
2. Two waves are coherent if they have a constant phase between them and also have the same frequency.
3. If the phase difference between any two points along the direction of propagation is independent of time, then the wave is said to be temporarily coherent.
4. If the phase difference between any points located transverse to the direction of propagation is independent of time, then the wave is said to be spatially coherent.
5. Constructive interference occurs due to the superposition of two waves at a point such that the crest of one wave falls on the crest of the other, that is, the path difference between two waves is an integral multiple of the wavelength ($n\lambda$). The intensity is maximum at these points (n is an integer or zero).
6. Destructive interference occurs due to the superposition of two waves at a point such that the crest of one wave falls on the trough of the other, that is, the path difference between the two waves is $\left(n + \dfrac{1}{2}\right)\lambda$, where λ is the wavelength and n is an integer or zero. The intensity is minimum at these points.
7. The phenomenon of interference requires two wavefronts to interact. These wavefronts can be obtained in two different ways, resulting in two different types of interference: (a) division of wavefront and (b) division of amplitude.

APPLICATIONS

1. Many optical coatings use optical interference to deliver specific properties. One important example is the use of antireflection coatings. Destructive interference of reflected rays ensures the absence of chosen wavelengths in the reflected light. A similar principle is also used to fabricate narrow-bandpass or band-reject filters. These filters are extensively used in optical systems.

2. Another interesting application of destructive interference can be observed in noise cancelling headphones. These headphones have an inbuilt mechanism and circuitry, which produce their own sound waves that imitate the incoming noise in every respect, except that the sound waves produced by the headphone circuitry is 180° out of phase with the intruding waves.

IMPORTANT FORMULAE

1. A wave is represented as follows:
$$\psi = A\sin\omega t$$

2. $I = A_1^2 = nA^2$

3. $\Delta L = c\,\Delta t$

4. $\delta = \dfrac{2\pi x}{L}$

5. $A^2 = \psi_{10}^2 + \psi_{20}^2 + 2\psi_{10}\psi_{20}\cos\delta$
$$\tan\theta = \frac{\psi_{20}\sin\delta}{\psi_{10} + \psi_{20}\cos\delta}$$

6. $I_{\max} = k\left(\psi_{10}^2 + \psi_{20}^2 + 2\psi_{10}\psi_{20}\right)$

7. $I_{\min} = k\left(\psi_{10}^2 + \psi_{20}^2 - 2\psi_{10}\psi_{20}\right)$

8. For constructive interference:
$$PD = 0, \ldots, n\lambda$$

9. For destructive interference:
$$PD = \left(n + \frac{1}{2}\right)\lambda$$

10. $\dfrac{I_{\max}}{I_{\min}} = \dfrac{(\psi_{10} + \psi_{20})^2}{(\psi_{10} - \psi_{20})^2}$

11. $\beta = $ fringe width $= \dfrac{\lambda D}{2d}$

12. Displacement S of the nth maxima due to a plate of thickness t and refractive index μ is as follows: $S = \dfrac{D}{2d}(\mu - 1)t$

13. For interference from thin films:
$$2\mu t\cos r = n\lambda$$

14. For wedge-shaped films:
$$F = \frac{\lambda}{2\sin\theta} \approx \frac{\lambda}{2\theta}$$

15. For Newton's rings, for the nth dark ring in reflected light:
$$D^2 = 4n\lambda R$$
For nth bright ring in reflected light:
$$D = \sqrt{(2\lambda R)(2n - 1)}$$

16. For Newton's rings:
$$\mu = \frac{D_{(n+p)}^2 - D_n^2}{D_{(n+p)}'^2 - D_n'^2}$$

SELF-ASSESSMENT

Multiple-choice Questions

1.1 A travelling disturbance is represented by
   (a) $\psi = A\sin(\omega t + \delta)$
   (b) $\psi = A\sin^2(\omega t + \delta)$
   (c) $\psi = A\cos(\omega t + \delta)$
   (d) $\psi = A\cos^2(\omega t + \delta)$

1.2 For superposition of n waves of equal amplitude, we have
   (a) $I = n^2 A^2$
   (b) $I = n^3 A^2$
   (c) $I = nA^2$
   (d) $I = nA$

1.3 Coherence length and coherence time are related through the expression
   (a) $\Delta L = \dfrac{c}{\Delta t}$
   (b) $\Delta L = c^2 \Delta t$
   (c) $\Delta L = (c\Delta t)^2$
   (d) $\Delta L = c\Delta t$

1.4 For constructive interference, we have

(a) $I_{max} \propto \left(\psi_{10}^2 + \psi_{20}^2 + 2\psi_{10}\psi_{20} \right)$

(c) $I_{max} \propto \left(\psi_{10}^2 + \psi_{20}^2 \right)$

(b) $I_{max} \propto \left(\psi_{10}^2 + \psi_{20}^2 - 2\psi_{10}\psi_{20} \right)$

(d) $I_{max} \propto \left(\psi_{10}^2 - \psi_{20}^2 \right)$

1.5 Linear distance between two consecutive fringes is given by

(a) $\dfrac{\lambda(2d)}{D}$

(b) $\dfrac{\lambda D}{d}$

(c) $\dfrac{\lambda D}{2d}$

(d) $\lambda D d$

1.6 For a biprism, the obtuse angle is around

(a) 190°

(b) 119°

(c) 100°

(d) 120°

1.7 Displacement of the nth maxima when a transparent plate is kept in the path of one of two interfering rays is

(a) $S = \dfrac{2d}{D}(\mu - 1)t$

(c) $S = \dfrac{D}{2d}(\mu - 1)t$

(b) $S = \dfrac{d}{D}(\mu - 1)$

(d) $S = \dfrac{D}{2d}\mu t$

1.8 For interferences in thin films, the condition for bright fringe is

(a) $2\mu t \cos r = n\lambda$

(c) $2\mu t \cos r \pm \lambda = n\lambda$

(b) $2\mu t \cos r \pm \dfrac{\lambda}{2} = n\lambda$

(d) $2\mu t \cos r - \lambda = n\lambda$

1.9 Fringe width β for the interference pattern obtained with a wedge-shaped film is given by

(a) $\dfrac{\lambda}{\sin\theta}$

(b) $\dfrac{\lambda}{\cos\theta}$

(c) $\dfrac{\lambda}{2\sin\theta}$

(d) $\dfrac{\lambda}{2\cos\theta}$

1.10 Dark rings in a Newton's ring set-up obey the relation

(a) $D \propto \sqrt{n}$

(c) $D \propto \sqrt{(2n-2)}$

(b) $D \propto \sqrt{2n}$

(d) $D \propto \sqrt{(2n-1)}$

1.11 Intensity of a travelling disturbance with amplitude B is proportional to

(a) B

(b) $\dfrac{1}{B}$

(c) $\dfrac{1}{B^2}$

(d) B^2

1.12 Two overlapping waves produce a stable interference pattern; their amplitudes must be

(a) vastly different

(c) comparable

(b) equal

(d) unrated

1.13 Condition for temporal coherence is

(a) $\delta_2 - \delta_1 = \delta_2' + \delta_1'$

(c) $\delta_2 - \delta_1 = 2(\delta_2' - \delta_1')$

(b) $\delta_2 - \delta_1 = \delta_2' - \delta_1'$

(d) $\delta_2 - \delta_2 = 2(\delta_1' - \delta_2')$

1.14 Coherence time for a perfectly monochromatic sinusoidal wave is

(a) infinity

(b) 0

(c) 1

(d) 2

1.15 A path difference of one half-wavelength introduces a phase difference of

(a) 2π

(b) π

(c) $\dfrac{3\pi}{2}$

(d) $\pi/2$

1.16 Constructive interference will not take place for a phase value equal to
    (a) 0   (b) 2π (c) π (d) 4π

1.17 In a Young's double-slit experiment, the position of bright fringes is given by
    (a) $x = \dfrac{ndD}{2\lambda}$ (b) $x = \dfrac{ndD}{\lambda}$ (c) $x = \dfrac{n\lambda d}{D}$ (d) $x = \dfrac{n\lambda D}{2d}$

1.18 Fresnel's biprism experiment can be used to determine the wavelength of light using the expression
    (a) $\lambda = \dfrac{\beta(2d)}{D}$ (b) $\lambda = \dfrac{\beta d}{D}$ (c) $\lambda = \dfrac{\beta D}{d}$ (d) $\lambda = \dfrac{\beta D}{d}$

1.19 Fresnel's biprism experiment is being conducted using a white light source. The zero-order fringe obtained is
    (a) black   (c) dependent on λ
    (b) white   (d) bright

1.20 When light gets reflected from a denser medium, it suffers a phase change of
    (a) 2π (b) $\pi/2$ (c) π (d) 3π

1.21 In an interference pattern produced by identical coherent sources of monochromatic light, the intensity at the site of central maximum is I. If intensity at the same spot when either of the two slits is closed is I_0, we must have the condition that
    (a) $I = I_0$ (c) $I = 4I_0$
    (b) $I = 2I_0$ (d) I and I_0 are not related

1.22 What happens when monochromatic light used in Young's slit experiment is replaced by white light?
    (a) Bright fringes become white.   (c) All fringes are coloured.
    (b) The central fringe is white and   (d) No fringes are observed.
        all other are coloured.

1.23 A path difference of $3\pi/2$ between two waves corresponds to a phase difference of
    (a) $3\pi/2$ (b) $\pi/3$ (c) 3π (d) $2\pi/3$

1.24 In a biprism experiment, 5 mm wide fringes are obtained on a screen placed 1.0 m away from coherent sources using a light of wavelength 5000 Å. The separation between the two coherent resources is
    (a) 1.0 mm   (b) 0.1 mm   (c) 0.01 mm   (d) 0.05 mm

1.25 Newton's ring experiment is based on
    (a) division of amplitude   (c) none of these
    (b) division of wavefront   (d) combination of (a) and (b)

Review Questions

1.1 What is the difference between temporal coherence and spatial coherence?

1.2 If the amplitudes of two coherent light waves are in the ratio 1:4, find the ratio of maximum to minimum intensity in the interference pattern.

1.3 Find an expression for the intensity distribution when two sinusoidal coherent waves with amplitudes A_1 and A_2 and a phase difference of ϕ superpose to produce interference.

1.4 Find an expression for the fringe width in the interference pattern of Young's double-slit experiment.

1.5 Two independent sources of light of the same wavelength cannot produce interference. Justify.

1.6 Explain why an extended source of light is required for fringes in a Newton's ring experiment. When white light is used in place of a monochromatic light, what change is expected?

1.7 Can you measure the refractive index of a liquid by Newton's ring experiment? Explain.

1.8 Explain interference of light due to thin films.

1.9 Explain superposition of waves.

1.10 Derive an expression for interference in thin films due to reflection.

1.11 Explain why a convex lens is placed between a monochromatic light source and a microscope while performing experiments on Newton's rings.

1.12 Describe in detail, with the necessary theory, an experiment to determine the refractive index of a transparent liquid using Newton's rings.

1.13 Why are very narrow slits used in Young's double-slit interference experiment?

1.14 Describe, with the necessary equation, how you will determine the refractive index of water using Newton's ring apparatus.

1.15 With the help of a suitable ray diagram, describe the production of Newton's rings.

1.16 What is a coherent source? Explain the different methods used to obtain coherent sources.

1.17 Find out the similarities and dissimilarities between a Newton's ring and a wedge-shaped film.

1.18 Prove that the diameter of the nth dark ring in a Newton's ring set-up is directly proportional to the square root of the ring number.

1.19 Explain why the thin-film interference pattern for a wedge-shaped film are parallel lines, whereas for Newton's ring it is circular.

1.20 Describe the origin of colour on a thin film, along with the derivation of constructive and destructive conditions.

1.21 What is interference?

1.22 Explain the principle of superposition.

1.23 Show that for n interfering waves, $I = A_1^2 = nA^2$.

1.24 What is coherence?

1.25 How many types of coherence are generally observed?

1.26 Define coherence time and coherence length.

1.27 Show that $I_{max} = k\left(\psi_{10}^2 + \psi_{20}^2 + 2\psi_{10}\psi_{20}\right)$

1.28 Derive the following expression: $\dfrac{I_{max}}{I_{min}} = \dfrac{\left(\psi_{10} + \psi_{20}\right)^2}{\left(\psi_{10} - \psi_{20}\right)^2}$.

1.29 Derive an expression for fringe width for a Young's double-slit experiment.

1.30 How can one use a Fresnel's biprism set-up to determine the wavelength of light?

1.31 Describe the fringe pattern obtained using a Fresnel's biprism set-up when white light is used.

1.32 How can one determine the thickness of a thin sheet of transparent material using a Fresnel's biprism set-up?

1.33 How does the Lloyd's mirror set-up confirm a phase change of π as light waves get reflected from a denser medium?

1.34 Derive the expression $2\mu t \cos r = n\lambda$ for interference patterns observed in thin films.

1.35 What type of interference pattern does one observe using wedge-shaped films?

1.36 Show that for interference from wedge-shaped thin films, the following relation

holds: $\beta = \dfrac{\lambda}{2\sin\theta} \approx \dfrac{\lambda}{2\theta}$

1.37 Describe a set-up that can be used to observe Newton's rings.

1.38 Show that for the nth bright ring in a Newton's ring set-up in reflected light the
diameter is given by the following expression: $D = \sqrt{(2\lambda R)(2n-1)}$

1.39 Show that for Newton's rings, $\mu = \dfrac{D_{(n+p)}^2 - D_n^2}{D_{(n+p)}'^2 - D_n'^2}$

Numerical Problems

1.1 Two coherent sources whose intensity ratio is 49:1 produce interference fringes.
Deduce the ratio of maximum intensity to minimum intensity.

$$\left[Hint: \frac{I_{max}}{I_{min}} = \frac{(A_1 + A_2)^2}{(A_1 - A_2)^2} \right]$$

1.2 Determine the ratio of intensity of the centre of a bright fringe to the intensity
found at a point $\dfrac{1}{6}$ of the distance between two fringes from the centre.

$$\left[Hint: I = 2\psi_{10}^2 (1 + \cos\delta) \right]$$

1.3 Two straight and narrow parallel slits 0.8 mm apart are illuminated using a
monochromatic light source. A screen placed at a distance of 100 cm is used to
obtain fringes. It is found that the distance between consecutive fringes is 0.5 mm.
Determine the wavelength of light. $\left[Hint: \beta = \dfrac{\lambda D}{2d} \right]$

1.4 Two coherent sources are placed 1.2 mm apart, which generate interference
fringes on a screen 0.9 m away. The second dark fringe is formed at a distance
of 1 mm from the central fringe. Calculate the wavelength of the monochromatic
light used. $\left[Hint: x_n = \dfrac{(2n+1)\lambda D}{4d} \right]$

1.5 A biprism is kept at a distance of 5 cm from a slit illuminated by a sodium light (λ
= 5890 Å). Fringes are obtained on a screen placed at a distance of 90 cm from the
biprism. Fringe width for the pattern obtained on the screen is 9.213×10^{-2} cm.
Determine the distance between the two coherent sources. $\left[Hint: \beta = \dfrac{\lambda D}{2d} \right]$

1.6 Sodium vapour light (λ = 5893 Å) is used first in a Fresnel's biprism set-up. A
total of 60 fringes are observed in the field of view of the eyepiece. Determine the
number of fringes that would be observed in the same field of view if the source
is replaced with another source of wavelength λ = 5300 Å. [$Hint: n_1\lambda_1 = n_2\lambda_2$]

1.7 In Young's experiment, let light of wavelengths 5×10^{-7} m and 8×10^{-8} m be used
in turn, keeping the geometry same. Compare the fringe width in the two cases.

$$\left[Hint: \beta = \dfrac{\lambda D}{2d} \right]$$

1.8 In a thin film, between points A and B, six fringes are seen with a light of wavelength 5400 Å. If the light used is of wavelength 4100 Å, what are the number
of fringes obtained between A and B? [$Hint: 2\mu t \cos r = n\lambda$]

1.9 In a Young's double-slit experiment, the slits are separated by a distance of 0.3 mm and the screen is placed 1.42 m away. The distance between the central bright fringe and the fourth bright fringe is measured to be 1.1 cm. Calculate the wavelength of light. $\left[Hint:\ \beta = \dfrac{\lambda D}{2d} \right]$

1.10 An interference pattern is first obtained using a biprism set-up. When a thin sheet of glass ($\mu = 1.5$) of 4 μm thickness is introduced in the path of one of the interfering rays, the central fringe is shifted to a position normally occupied by the fourth fringe. Determine the wavelength of the light used. $\left[Hint:\ S = \dfrac{D}{2d}(\mu - 1)t \right]$

1.11 In a Newton's ring set-up, the diameter of the eighth ring has been found to be 0.42 cm and that of the 25th ring 0.84 cm. The radius of curvature of the plano-convex lens is 95 cm. Determine the wavelength of the light used.
$$\left[Hint:\ \lambda = \frac{D_{n+p}^{2} - D_{n}^{2}}{4pR} \right]$$

1.12 In a Newton's ring experimental set-up, the diameter of the eighth ring changes from 1.35 to 1.17 cm when a liquid of refractive index μ replaces air in the space between the lens and the plate. Calculate the refractive index of the liquid.
$$\left[Hint:\ D_{8}^{'2} = \frac{4 \times 8 \times \lambda R}{\mu} \right]$$

1.13 Two coherent sources have their intensities in the ratio 81:9. An interference pattern is obtained using these two sources. Calculate the ratio of maximum intensity to minimum intensity. $\left[Hint:\ \dfrac{I_{max}}{I_{min}} = \dfrac{(A_1 + A_2)^2}{(A_1 - A_2)^2} \right]$

1.14 In Problem 1.5, the wavelength of light source is changed to 5400 Å. Calculate the distance between consecutive fringes. $\left[Hint:\ \beta = \dfrac{\lambda D}{2d} \right]$

1.15 The wavelength of the monochromatic light source in Problem 1.4 is changed to 6000 Å. Calculate the new distance of the second dark fringe from the central fringe. $\left[Hint:\ x_n = \dfrac{(2n+1)\lambda D}{4d} \right]$

Diffraction

Learning Objectives

After studying this chapter, students will be able to

- understand Huygen's principle
- comprehend the differences between interference and diffraction fringes
- realize the concept of zone plates
- explain the diffraction pattern due to a straight edge
- understand how to obtain the resultant of multiple simple harmonic motions
- elucidate Fraunhofer diffraction at a single slit and a circular aperture
- explain the working of a diffraction grating
- comprehend the meaning of resolving power and derive expressions for the resolving power of some optical instruments

List of Symbols

λ = Wavelength	I = Intensity	d_1 = Slit width
T = Time period	r = Radius	d_2 = Width of opaque
d = Distance	δ = Phase difference	part
R = Amplitude	f = Focal length	

2.1 INTRODUCTION

In Chapter 1, we learnt about the interference of light from two coherent sources. What happens when light from a narrow source (point source) falls on a narrow slit? According to the corpuscular theory of light, if objects are placed in the path of light, sharp shadows are to be expected. However, if one places a thin wire or a razor blade under a bright, point source of light, one may observe the edges of the shadow to be somewhat fuzzy (not very sharp). Similarly, for the light passing through a narrow slit and being received on a screen, the edges of the slit may not appear to be very sharp on the screen. Is it possible that light bends around the edges of the object placed in its path?

In fact, it does, and this bending of light around the edge of an obstacle is called *diffraction*. The phenomenon of diffraction was first discovered by Grimaldi in 1665. In fact, this bending is not restricted to light waves only and can happen even for sound waves, as illustrated in the adjacent image. The reason we can still hear a person, even when we cannot see him/her, is that for diffraction to be effective the wavelength of light/sound needs to be of the same order as the size of the obstacle. Sound has a higher wavelength and, therefore, can get diffracted across the relatively larger-size wall. We will study diffraction in detail in this chapter.

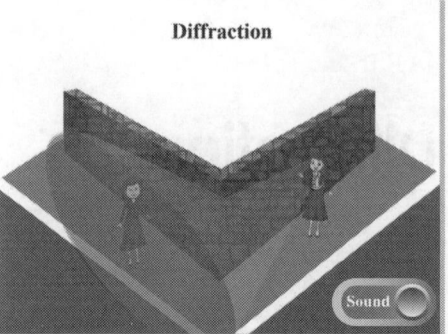

Sound waves bend round objects of a similar size to their wavelength. The wall below has a similar size to the sound's wavelength. The effect is called diffraction.

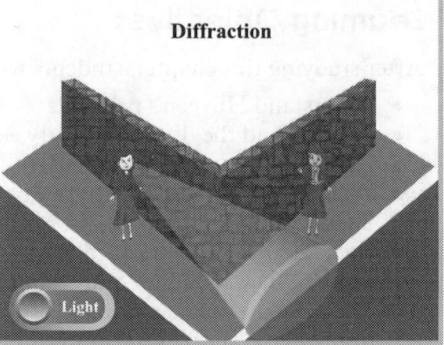

Light has a very small wavelength. so only very small objects or gaps can effect its direction. The wall blocks the light and the person can't see round the corner.

2.2 HUYGEN'S PRINCIPLE

Huygen postulated a model for light waves, similar to water waves. On the basis of this model, he proposed a hypothesis for geometrically constructing the position of a wavefront.

2.2.1 Postulates of Huygen's Principle

Huygen offered two postulates for constructing the position of a wavefront:

Postulate 1   Each point on a wavefront (primary wavefront) acts as a source of new disturbances, called *secondary wavelets*, which travel in all directions with the velocity of light in that medium.

Postulate 2   The surface touching these secondary wavelets tangentially in the forward direction at any instant gives a new wavefront, called a *secondary wavefront at that instant*.

A *ray of light* is a straight line normal to the wavefront, representing the direction of propagation of the wavefront.

2.2.2 Construction of Secondary Wavefronts

Let AB be a section of the primary wavefront at any instant of time (Fig. 2.1). To find the position of the secondary wavefront at time *t*, consider points 1, 2, 3, 4, ... on the wavefront AB. Taking each point as the centre, draw spheres of radius ct, where c is the speed of light in the medium. These spherical surfaces represent the secondary wavelets at time t. Draw surface $A_F B_F$, touching all the secondary wavelets tangentially in the forward direction; $A_F B_F$ is the secondary wavefront.

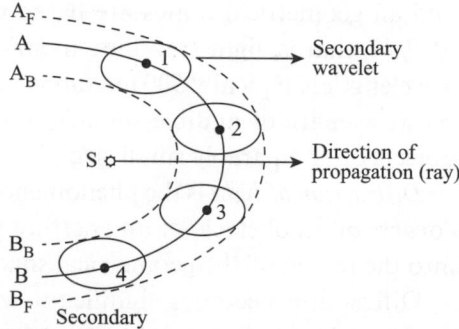

Fig. 2.1 Construction of secondary wavefront

The surface $A_B B_B$, touching all the secondary wavelets in the backward direction would give the backward secondary wavefront. However, Huygen postulated that the action of the secondary wavelets was confined only to the points at which they touched the forward envelope, and thus no backward wavefront exists.

According to Stoke's law, the intensity of the spherical wavelets varies as $\left(\dfrac{1+\cos\theta}{2}\right)$, where θ is the angle between the direction of propagation of the wavelet and the normal at that point. For backward direction, $\theta = 180°$; therefore, intensity = 0.

2.3 PHENOMENON OF DIFFRACTION

To listen to someone, you do not need to point your ears at them, and if you speak to a friend in front of you, your words are also heard by someone standing beside you. This occurs because of *diffraction*, the spreading of waves as they pass through an aperture (i.e., a hole, like your mouth). Sound waves will also diffract, but the extent of diffraction will depend on the wavelength λ, and the size a of the diffracting object.

If $\lambda/a \geq 1$, very significant diffraction occurs. When $\dfrac{\lambda}{a} < 1$, less significant diffraction occurs, where the centre part of the wave remains more or less unaffected. If $\lambda/a \ll 1$, no significant diffraction occurs.

Diffraction is a characteristic exhibited by waves of all types. TV and VHF radio use short waves (a few metres), which do not diffract significantly around natural barriers such as hills, buildings, and cliffs. This means that receiving aerials must be aligned to the transmitters—this is also true for much shorter wavelengths used in microwave links. On the other hand, long-wave radio signals (wavelengths around 1 km) diffract around most objects and so can be received even at places that lie within a short-wave 'shadow'.

To see what someone is doing, you have to look at them. We usually do not experience diffraction of light in our everyday life. In fact, ray diagrams and all geometrical optics are based on the idea of *rectilinear propagation of light*, that is, light travelling in straight lines. Light does diffract, but its wavelength is typically 500 nm (about a million times smaller than the shortest wavelength of audible sound), so light is only significantly diffracted by objects of comparably small size.

Diffraction of light is the phenomenon of bending of light waves around the corners of an obstacle or an aperture placed in its path, and their spreading into the region of the geometrical shadow.

Diffraction becomes significant when the dimensions of the aperture or obstacle are comparable to the wavelength of the light. Diffraction occurs due to mutual interference of the secondary wavelets starting from portions of the primary wavefront, which are allowed to pass through the aperture.

Diffraction is of two types:

Fresnel diffraction This type of diffraction occurs when the source and the screen are at a finite distance from the diffracting aperture, such that the incident and diffracted wavefronts are spherical or cylindrical.

Fraunhofer diffraction This type of diffraction occurs when the source and the screen are at an infinitely large distance from the diffracting aperture, such that incident and diffracted wavefronts are plane.

2.4 DIFFERENCE BETWEEN INTERFERENCE AND DIFFRACTION FRINGES

Table 2.1 lists the differences between interference and diffraction fringes.

Table 2.1 Interference and diffraction fringes

Interference fringes	Diffraction fringes
All fringes are of equal width.	The width of the central maximum is twice that of the other fringes.
All bright fringes are of equal intensity.	The central bright fringe is of maximum intensity, and the secondary maxima are of decreasing intensity.
The maxima occur at path differences between the two waves of $n\lambda$ and the minima at path difference of $\left(n+\dfrac{1}{2}\right)\lambda$.	The minima occur at path differences of $n\lambda$, and the maxima at the path difference of $\left(n+\dfrac{1}{2}\right)\lambda$ between the waves from the ends of the slit.
It is the result of interference of two different wavefronts of two coherent sources.	It is the result of interference of waves from different parts of the same wavefront.
Minima have zero intensity.	Minima are never perfectly dark.

2.5 FRESNEL'S HALF-PERIOD ZONES

In Fig. 2.2, $A_1A_2A_3A_4$ represents a monochromatic plane wavefront of wavelength λ. The wavefront is travelling towards point Q. Calculation of the resultant intensity at point Q was simplified by Fresnel by dividing the wavefront into a number of half-period zones.

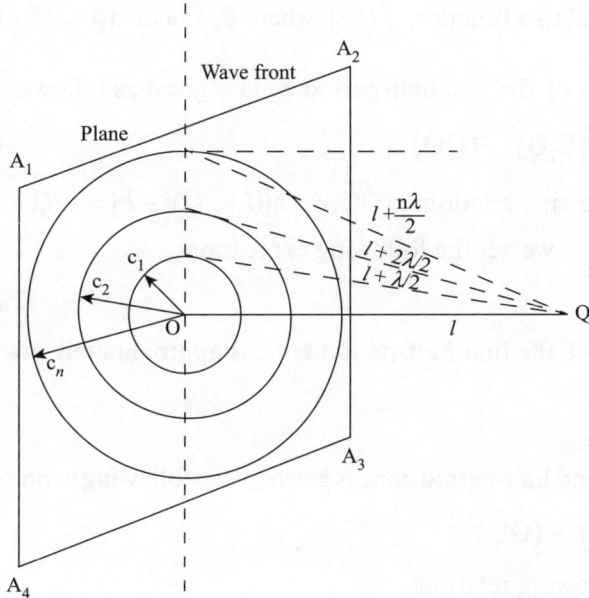

Fig. 2.2 Concept of Fresnel's half-period zones

QO is a normal drawn on the wavefront from point Q, with QO =1. Taking Q as the centre, spheres are drawn with radii equal to $[1 + \lambda/2]$, $[1 + 2\lambda/2]$, ..., $[1 + n\lambda/2]$. The plane $A_1A_2A_3A_4$ cuts these spheres in concentric circles centred at O with radii $OC_1, OC_2, ..., OC_n$. Area of the first innermost circle is called the first half-period zone, the area enclosed between the first and the second circle is called the second half-period zone, and so on, with the area between the $(n-1)$th and nth circle being the nth half-period zone.

Since $C_1Q = OC + \dfrac{\lambda}{2}$, the path difference between the ray reaching Q from O and the ray reaching Q from the circumference of the circle with radius OC_1 is $\dfrac{\lambda}{2}$. The path difference means a phase difference of π. The phase difference due to points lying with the circular region covered by the circle of radius OC_1 will vary from O to π. Thus, the mean phase difference for the secondary wavelets originating from the first zone is given as $(O + \pi)/2 = \pi/2$. The phase difference between zones 1 and 2 is π. Thus, the mean phase for the secondary wavelets originating from the second zone is given as $(\pi + 2\pi)/2 = 3\pi/2$. On the same lines, it can be shown that the mean phase of secondary wavelets

originating from the third, fourth, ... zones are $\dfrac{5\pi}{2}, \dfrac{7\pi}{2}, ...$ *Thus, we can see that successive zones differ in phase π that implies a half-period,* that is, $T/2$. This is the reason that these zones are called *half-period zones. Amplitude of disturbance at point Q due to a particular zone is*
1. directly proportional to the area of the zone;
2. inversely proportional to the average distance of the zone from Q; and
3. directly proportional to a function $f(\theta_n)$, where θ_n is a measure of obliquity of the nth zone

From Fig. 2.2, the area of the first half-period zone is given as follows:

$$\pi(OC_1)^2 = \pi\left[(C_1Q)^2 - (OQ)^2\right] \tag{2.1}$$

which leads to the following relation: $\pi(OC_1)^2 = \pi[(l + \lambda/2)^2 - l^2] = \pi\,[l\lambda + \lambda^2/4]$
Neglecting the term $\dfrac{\lambda^2}{4}$, we get the following expression:

$$\pi(OC_1)^2 \approx nl\lambda \tag{2.2}$$

Thus, the radius OC_1 of the first half-period zone is approximately given as follows:

$$OC_1 \approx \sqrt{(l\lambda)} \tag{2.3}$$

Radius OC_2 of the second half-period zone is given by the following expression:

$$OC_2 = \left[(C_2Q)^2 - (OQ)^2\right]$$

which leads to the following relation:

$$OC_2 = \left[(l + \lambda)^2 - l^2\right]^{1/2} \cong \sqrt{2l\lambda} \tag{2.4}$$

Using Eqs (2.3) and (2.4), area of the second half-period zone can be given as follows:

$$\pi[(OC_2)^2 - (OC_1)^2] = \pi[2l\lambda - l\lambda] = \pi l\lambda \tag{2.5}$$

From Eqs (2.2) and (2.5), it is clear that the approximate expression for area of the first and second half-period zones are same.

The exact expression for area of the nth half-period zone is given by the following expression: $\pi\left[(OC_n)^2 - (OC_{n-1})^2\right]$

which results in the following relation: $\pi[\{(l + n\lambda/2)^2 - l^2\} - \{[l + (n-1)\lambda/2]^2 - l^2\}]$

yielding $\pi[\{l^2 + nl\lambda + n^2\lambda^2/4 - l^2\} - \{l^2 + (n-1)\,\lambda l + [(n-1)\lambda/2]^2 - l^2\}]$

giving $\pi\,[\lambda l + n^2\lambda^2/4 - (n^2\lambda^2/4 + \lambda^2/4 - 2n\lambda^2/4)]$

leading to $\pi\,[\lambda l + 2n\lambda^2/4 - \lambda^2/4] = \pi[\lambda l + \lambda.\lambda/4.(2n - 1)]$

or $\qquad \pi\lambda[l + (2n - 1)\lambda/4] \tag{2.6}$

The average distance of the nth zone from Q is given by the following expression:

$$d_{AV} = \{[l + n\lambda/2] + [l + (n-1)\,\lambda/2]\}/2 = l + (2n - 1)\lambda/4 \tag{2.7}$$

The angle between the normal to a particular zone and the line joining the zone to Q is called the obliquity angle, and the obliquity factor is represented by $f(\theta_n)$.

Amplitude due to the nth zone using Eqs (2.6) and (2.7) can be seen to be of the following form:

$$\alpha = \pi[\{[l + (2n - 1)\ \lambda/4]\lambda\}/[l + (2n - 1)\lambda/4]]\ f(\theta_n)$$
$$\alpha = \pi\lambda f(\theta_n) \tag{2.8}$$

From Fig. 2.2, we can observe that $f(\theta_n)$ decreases as the order of the zone increases. From Eq. (2.8), we can conclude that the amplitude of the wave reaching point Q decreases as the order n of the zone increases.

Suppose R_1, R_2, R_3, ..., R_n represent the amplitudes of the secondary waves reaching point Q from the first, second, third, ..., nth zones, respectively. From Eq. (2.8), we have seen that the amplitude of waves from successive zones decreases as the order of zones becomes higher. The amplitude of the secondary wave reaching from a particular zone can be approximated to be the mean of the amplitudes due to the zones preceding and succeeding it. Thus, for example, $R_2 \cong \dfrac{(R_1 + R_3)}{2}$; $R_4 \cong \dfrac{(R_3 + R_5)}{2}$, etc.

We must also remember that a phase difference of π exists between the waves reaching Q from consecutive zones.

Figure 2.3 is a schematic representation of the amplitudes of waves reaching point Q from successive zones.

Lengths of arrows in Fig. 2.3 are measures of amplitudes of the waves, and a change in direction of an arrow represents a phase change of π between consecutive zones.

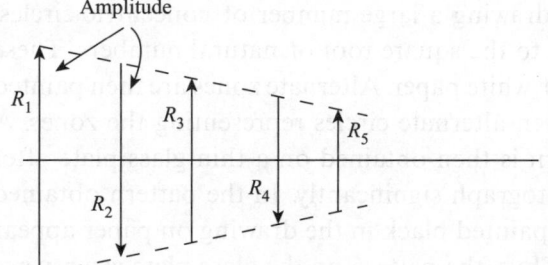

Fig. 2.3 Schematic representation of secondary waves from different zones

The resultant amplitude at Q at a given instant is given as follows:

$$R_T = R_1 - R_2 + R_3 - R_4 + \dots + R_n \tag{2.9}$$
$$\text{if } n \text{ is odd}$$

or
$$R_T = R_1 - R_2 + R_3 - R_4 \dots - R_n \tag{2.10}$$
$$\text{if } n \text{ is even}$$

For an odd n, we can write using Eq. (2.9) that

$$R_T = \frac{R_1}{2} + \left[\frac{R_1}{2} - R_2 + \frac{R_3}{2}\right] + \left[\frac{R_3}{2} - R_4 + \frac{R_5}{2}\right] + \dots \tag{2.11}$$

However, we know that

$$R_2 = \frac{R_1 + R_3}{2} \text{ and } R_4 = \frac{R_3 + R_5}{2} \tag{2.12}$$

Using Eq. (2.12) in Eq. (2.11), we get the following relation:

$$R_T = \frac{R_1}{2} + \frac{R_n}{2} \tag{2.13}$$

For an even n, we can write the following relation:

$$R_T = \frac{R_1}{2} + \frac{R_{n-1}}{2} - R_n \tag{2.14}$$

We have already learnt that the amplitude due to secondary wavelets decreases with increasing n. Therefore, as $n \to \infty$, R_n and R_{n-1} tend to zero. Using Eqs (2.13) and (2.14), the resultant amplitude at point Q is given as follows:

$$R_T = \frac{R_1}{2} \tag{2.15}$$

The resultant intensity, I_T, can be written using Eq. (2.15) as follows:

$$I_T = R_T{}^2 = \frac{R_1{}^2}{4} \tag{2.16}$$

2.6 ZONE PLATE

Fresnel's method of dividing a plane wavefront into different half-period zones is used in the fabrication of an optical device called a *zone plate*. A zone plate can also be used to prove rectilinear propagation of light. A zone plate is constructed by first drawing a large number of concentric circles, with their radii proportional to the square root of natural numbers. These circles are drawn on a sheet of white paper. Alternate zones are then painted black, with the region between alternate circles representing the zones. A photograph of such a pattern is then obtained on a thin glass plate after reducing the size of the photograph significantly. In the pattern obtained on the glass plate, the zones painted black in the drawing on paper appear transparent and vice versa. Thus, the pattern on the glass plate is a reverse of the pattern on the paper. This pattern obtained on the glass plate is called a *zone plate*. There can be positive or negative zone plates. In a positive zone plate, the central zone is transparent, whereas in the negative zone plate, the central zone is opaque. A schematic of the two types of zone plates is shown in Fig. 2.4.

A section of a zone plate perpendicular to the plane of paper is shown in Fig. 2.5. The section is represented by $X_1 X_2$, and S represents the monochromatic point source.

Point Q represents a screen. OC_1, OC_2, ... OC_n provide the radii of the first, second, ..., nth half-period zones, and are represented by r_1, r_2, ..., r_n.

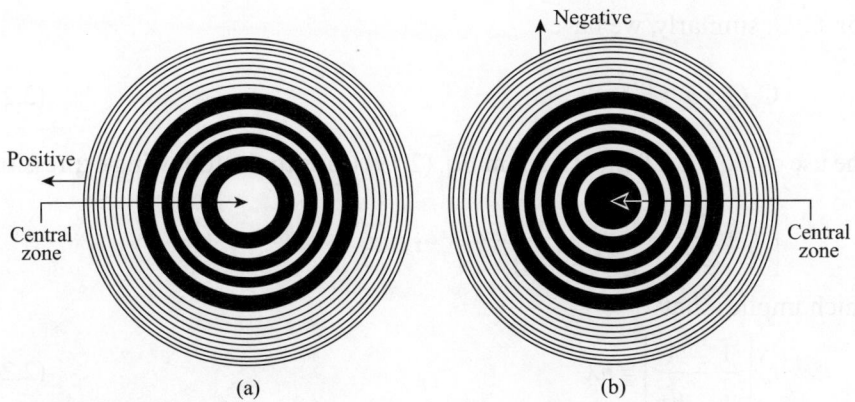

Fig. 2.4 Schematic representation of two types of zone plates (a) Positive (b) Negative

The source is at a distance of a_1 from the zone plate, and the screen is at a distance of a_2 from the zone plate. The screen is located in such a way that a path difference of $\lambda/2$ exists between consecutive zones. We therefore have the following relations:

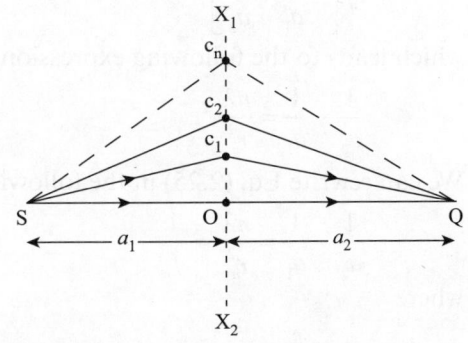

Fig. 2.5 Schematic of section of zone plate

$$SC_1 + C_1Q = SO + OQ + \frac{\lambda}{2}$$

$$= a_1 + a_2 + \frac{\lambda}{2}$$

(2.17)

and

$$SC_2 + C_2Q = SO + OQ + \frac{2\lambda}{2} = a_1 + a_2 + \frac{2\lambda}{2}$$

(2.18)

and so on, until

$$SC_n + C_nQ = SO + OQ + \frac{n\lambda}{2} = a_1 + a_2 + \frac{n\lambda}{2}$$

(2.19)

From Fig. 2.5, we can see that

$$\left(SC_n\right)^2 = a_1^2 + r_n^2$$

(2.20)

leading to

$$SC_n = \left[a_1^2 + r_n^2\right]^{1/2} = a_1\left[1 + \frac{r_n^2}{a_1^2}\right]^{1/2} \cong a_1\left[1 + \frac{r_n^2}{2a_1^2}\right]$$

(2.21)

implying that

$$SC_n = a_1 + \frac{r_n^2}{2a_1}$$

(2.22)

For $C_n Q$, similarly, we have

$$C_n Q = a_2 + \frac{r_n^2}{2a_2} \tag{2.23}$$

The use of Eqs (2.22) and (2.23) in Eq. (2.19) leads to the following expression:

$$\left[a_1 + \frac{r_n^2}{2a_1} \right] + \left[a_2 + \frac{r_n^2}{2a_2} \right] = a_1 + a_2 + \frac{n\lambda}{2}$$

which implies that

$$r_n^2 \left[\frac{1}{a_1} + \frac{1}{a_2} \right] = n\lambda \tag{2.24}$$

Using sign conventions, we can rewrite Eq. (2.24) in the following form:

$$r_n^2 \left[-\frac{1}{a_1} + \frac{1}{a_2} \right] = n\lambda$$

which leads to the following expression:

$$\frac{1}{a_2} - \frac{1}{a_1} = \frac{n\lambda}{r_n^2} \tag{2.25}$$

We can rewrite Eq. (2.25) in the following form:

$$\frac{1}{a_2} - \frac{1}{a_1} = \frac{n\lambda}{r_n^2} = \frac{1}{f_n} \tag{2.26}$$

where

$$f_n = \frac{r_n^2}{n\lambda} \tag{2.27}$$

Equation (2.27) is similar to the lens formula:

$$\frac{1}{v} - \frac{1}{u} = \frac{1}{f} \tag{2.28}$$

Thus, the zone plate serves as a convergent lens, with a_2 and a_1 representing the image and object distance, respectively, and f_n representing the focal length.

Let $R_1, R_2, R_3, \ldots, R_n$ represent the amplitudes of the first, second, third, …, nth zones, respectively. Waves reaching point Q from alternate zones (successive transparent zones) have a path difference of λ, that is, the waves are in phase. The resultant amplitude R_T at Q is given as follows:

$$R_T = R_1 + R_3 + R_5 + \ldots \tag{2.29}$$

Thus, the resultant amplitude is many times greater than the resultant amplitude due to all zones, namely $R_{1/2}$. Point Q is thus extremely bright and can be said to be the image of the source S. The zone plate thus demonstrates a convergent focusing action.

Example 2.1   The first ring of a zone plate has a diameter of 1.2 mm. Plane waves of wavelength 5000 Å fall on the plate. Determine the location of the screen so that light is focused on the brightest spot.

Solution From Eq. (2.27), we get the following relation:

$$fn = \frac{r_n^2}{n\lambda} \tag{2.30}$$

which implies that

$$f_1 = \frac{r_1^2}{\lambda} \tag{2.31}$$

Substituting these values in Eq. (2.31), we have the following result:

$$f_1 = \frac{(0.12)^2}{5000 \times 10^{-8}} = 288 \text{ cm}$$

Example 2.2 A zone plate is being used with a point source of light. Along the axis, the strongest and the second strongest images are formed at distances of 30 and 5 cm from the zone plate, respectively. Both the images are formed on the same side of the source. Determine the distance of the source from the zone plate. Assume $\lambda = 5.6 \times 10^{-5}$ cm.

Solution Equation (2.27) implies that

$$f_n = \frac{r_n^2}{n\lambda} \tag{2.32}$$

Equation (2.28) implies that

$$\frac{1}{v} - \frac{1}{u} = \frac{1}{f} \tag{2.33}$$

Using Eqs (2.32) and (2.33), we can write the following expression:

$$\frac{1}{v} - \frac{1}{4} = \frac{n\lambda}{r^2}$$

$$1/v - 1/u = n\lambda/r^2 \tag{2.34}$$

where v and u are the distances of image and object from the zone plate, respectively. For the first image, we have

$$\frac{1}{30} - \frac{1}{u} = \frac{\lambda}{r^2} \tag{2.35}$$

and for the second image, we have

$$\frac{1}{5} - \frac{1}{u} = \frac{3\lambda}{r^2} \tag{2.36}$$

Using Eqs (2.35) and (2.36), we get the following relation:

$$3\left(\frac{1}{30} - \frac{1}{u}\right) = \frac{1}{5} - \frac{1}{u}$$

which gives the following expression:

$$\frac{1}{10} - \frac{3}{u} = \frac{1}{5} - \frac{1}{u}$$

implying that

$$\frac{1}{10} - \frac{1}{5} = \frac{3}{u} - \frac{1}{u} \quad \text{or} \quad -\frac{1}{10} = \frac{2}{u}$$

that is, $u = -20$ cm

The negative sign indicates that the point source lies towards the left of the zone plate.

Example 2.3 Plane waves of wavelength 5000 Å fall on a zone plate. Light is focused on the brightest spot when a screen is placed at a distance of 350 cm. Calculate the diameter of the first ring of the zone plate.

Solution Equation (2.27) gives the following relation:

$$f_n = \frac{r_n^2}{n\lambda} \tag{2.37}$$

resulting in

$$f_1 = \frac{r_1^2}{\lambda} \tag{2.38}$$

Equation (2.38) can be rewritten as follows:

$$r_1 = \sqrt{f_1\lambda} \tag{2.39}$$

Substituting these values, we get $r_1 = \sqrt{350 \times 5000 \times 10^{-8}} = \sqrt{0.0175} = 0.132$ cm

Example 2.4 The wavelength in Example 2.3 is changed to 6000 Å, keeping the other parameters same. Calculate the new diameter of the first ring of the zone plate.

Solution From Eq. (2.39), we get the following relation: $r_1 = \sqrt{f_1\lambda}$
Using the given values, we get $r_1 = \sqrt{350 \times 6000 \times 10^{-8}} = \sqrt{0.021} = 0.145$ cm

Example 2.5 The first ring of a zone plate has a diameter of 1.1 mm. Plane waves fall on the plate and produce the brightest focused spot on a screen placed at a distance of 300 cm. Calculate the wavelength of the source used.

Solution From Eq. (2.38), one obtains the following expression: $f_1 = \frac{r_1^2}{\lambda}$
resulting in

$$\lambda = \frac{r_1^2}{f_1} \tag{2.40}$$

Using the values in Eq. (2.40), we get the following value: $\lambda = \frac{(0.11)^2}{300} = 4033$ Å

Example 2.6 A zone plate is being used with a point source of light. Along the axis, the strongest image is formed at a distance of 40 cm from the zone plate. The strongest and the second strongest images are formed on the same side of the source. The source is at a distance of 20 cm to the left of the zone plate. Assuming that $\lambda = 6 \times 10^{-5}$ cm, calculate the location of the second strongest image.

Solution Let x represent the location of the second strongest image w.r.t. the zone plate. Using Eq. (2.34), we can write the following expression for the strongest image:

$$\frac{1}{40} + \frac{1}{20} = \frac{\lambda}{r^2} \tag{2.41}$$

Similarly, for the second strongest image, we have the following relation:

$$\frac{1}{x} + \frac{1}{20} = \frac{3\lambda}{r^2} \tag{2.42}$$

Using Eqs (2.41) and (2.42), we can write the following expression:

$$3\left(\frac{1}{40} + \frac{1}{20}\right) = \frac{1}{x} + \frac{1}{20}$$

leading to $\dfrac{9}{40} = \dfrac{1}{x} + \dfrac{1}{20}$

yielding $\dfrac{1}{x} = \dfrac{9}{40} - \dfrac{1}{20} = \dfrac{9-2}{40} = \dfrac{7}{40}$

resulting in $x = \dfrac{40}{7} = 5.71\,\text{cm}$

Example 2.7   The location of the source in Example 2.6 is changed so that it is at a distance of 10 cm, to the left of the zone plate. All other parameters remain the same. Calculate the new location of the second strongest image.

Solution   Equation (2.41) gets modified to the following form:

$$\frac{1}{40} + \frac{1}{10} = \frac{\lambda}{r^2} \tag{2.43}$$

In addition, Eq. (2.42) gets modified to the following form:

$$\frac{1}{x} + \frac{1}{10} = \frac{3\lambda}{r^2} \tag{2.44}$$

Using Eqs (2.43) and (2.44), we can write the following relation: $3\left(\dfrac{1}{40} + \dfrac{1}{10}\right) = \dfrac{1}{x} + \dfrac{1}{10}$

resulting in $\dfrac{15}{40} = \dfrac{1}{x} + \dfrac{1}{10}$

leading to $\dfrac{1}{x} = \dfrac{15}{40} - \dfrac{1}{10} = \dfrac{11}{40}$

that is, $x = \dfrac{40}{11} = 3.64\,\text{cm}$

Example 2.8   A zone plate is being used with a point source of light of wavelength $\lambda = 5 \times 10^{-5}$ cm. The strongest and the second strongest images are formed on the same side of the source. The second strongest image is formed at a distance of 4 cm from the zone plate. The source is kept at a distance of 20 cm towards the left of the zone plate. Calculate the location of the strongest image.

Solution   Equation (2.28) implies that

$$\frac{1}{v} - \frac{1}{u} = \frac{1}{f} \tag{2.45}$$

where $f_n = \dfrac{r_n^2}{n\lambda}$ \hspace{2cm} (2.46)

Using Eqs (2.45) and (2.46), we get for the strongest image

$$\frac{1}{x} + \frac{1}{20} = \frac{\lambda}{r^2} \tag{2.47}$$

For the second strongest image, we have the following relation:

$$\frac{1}{4} + \frac{1}{20} = \frac{3\lambda}{r^2} \tag{2.48}$$

Using Eqs (2.47) and (2.48), we can write the following expression:

$$3\left(\frac{1}{x} + \frac{1}{20}\right) = \frac{1}{4} + \frac{1}{20} = \frac{6}{20} = \frac{3}{10}$$

resulting in $\dfrac{1}{x} + \dfrac{1}{20} = \dfrac{1}{10}$

yielding $\dfrac{1}{x} = \dfrac{1}{10} - \dfrac{1}{20} = \dfrac{2-1}{20} = \dfrac{1}{20}$

that is, $x = 20$ cm

2.7 DIFFRACTION AT STRAIGHT EDGE

In Fig. 2.6, S represents a narrow slit placed perpendicular to the plane of the paper.

A monochromatic source is incident on slit S. The cylindrical wavefront W_1W_2 originating from the slit is incident upon the straight edge A_1A_2. A screen X_1X_2 is placed at a distance of a_2 from the straight edge. A general point on the screen is represented by P such that SA_1P is perpendicular to the screen. Pure geometrical optics would predict uniform illumination above point P and no illumination below P. The experiment, however, reveals something completely different. A series of maxima and minima regions are observed in the space above P [see Fig. 2.6(a)]. These maxima and minima regions are called diffraction fringes. The intensity of the maxima regions is found to decrease as we travel away from the mid-point on the screen. What is most interesting is that light also penetrates into the region on the screen below point P. In fact, the intensity of light falls off gradually as we go away from point P, as shown in Fig. 2.6(b).

To evaluate the detailed intensity pattern observed on the screen, we will consider some special locations on the screen.

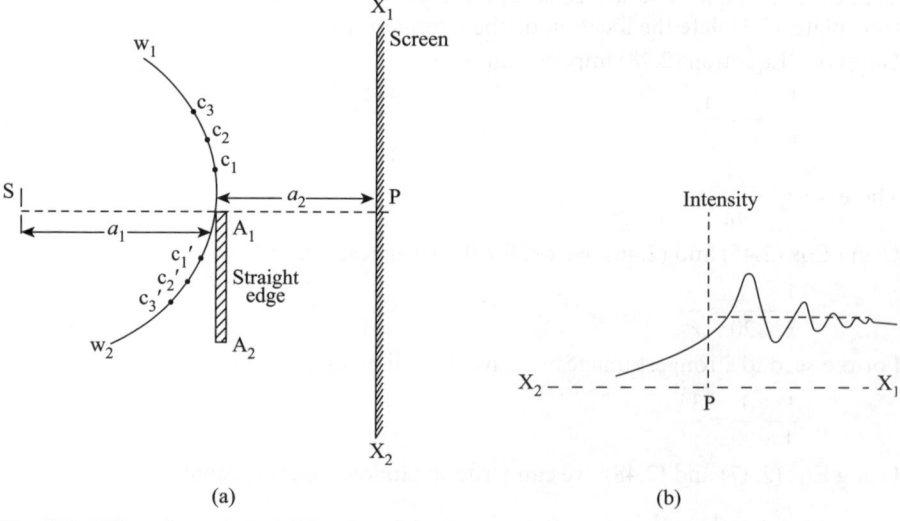

(a)   (b)

Fig. 2.6 Diffraction at straight edge (a) schematic representation (b) intensity distribution

Case 1 Point P is located at the mid-point of the screen. This point is fully exposed to the upper half of the cylindrical wavefront AW_1. Let R_1, R_2, R_3, etc., represent the amplitudes of the waves received from half-period strips C_0C_1, C_1C_2, C_2C_3, etc., at point P. Waves from alternate strips are out of phase by π. The resultant amplitude R_T at P is given as follows:

$$R_T = R_1 - R_2 + R_3 - R_4 + \ldots = R_{1/2} \qquad (2.49)$$

The intensity, I_T, at P is thus

$$I_T \propto \left(\frac{R_1}{2}\right)^2 \propto \frac{R_1^2}{4} \qquad (2.50)$$

Case 2 Let us now consider a point P_1 that is above point P, as shown in Fig. 2.7. A straight line joining S and P_1 meets the cylindrical wavefront W_1W_2 at C_0. The point C_0 thus represents the pole for point P_1.

Suppose the location of P_1 is such that it is exposed to the complete upper half wavefront and one half period C_0C_1' from the lower half of the cylindrical wavefront. The resultant amplitude R_T' is then given as follows:

$$R_T' = \frac{R_1}{2} + R_1 = \frac{3R_1}{2} \qquad (2.51)$$

Intensity I_T' is given as follows:

$$I_T' \propto R_{T'}^2 \propto \frac{9R_1^2}{4} \qquad (2.52)$$

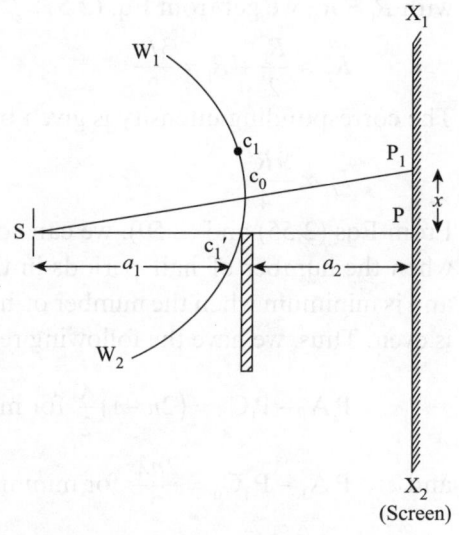

Fig. 2.7 Diffraction pattern due to straight edge

A comparison of Eqs (2.50) and (2.52) reveals that the intensity at P_1 is nine times that at P.

If the location of P_1 is such that it is exposed to the complete upper half wavefront and two half periods C_0C_1' and $C_1'C_2'$, then the resultant amplitude R_T'' becomes

$$R_T'' = \frac{R_1}{2} + R_1 - R_2 \qquad (2.53)$$

Since $R_1 \approx R_2$, Eq. (2.53) leads to the following form:

$$R_T'' \approx \frac{R_1}{2} \qquad (2.54)$$

The corresponding intensity is given by the following equation:

$$I_T'' \approx \frac{R_1^2}{4} \qquad (2.55)$$

Let us now consider the situation where point P is exposed to three strips C_0C_1', $C_1'C_2'$, and $C_2'C_3'$ in the lower half wavefront along with the complete upper half wavefront. The resultant amplitude R_T''' is given by

$$R''' = \left(\frac{R_1}{2}\right) + (R_1 - R_2 + R_3) \tag{2.56}$$

Since, $R_2 = \dfrac{R_1 + R_3}{2}$, we can rewrite Eq. (2.56) in the following form:

$$R_T''' = \left(\frac{R_2}{2}\right) + \left(\frac{R_1}{2} + \frac{R_3}{2}\right) \tag{2.57}$$

with $R_1 \approx R_3$, we get from Eq. (2.57)

$$R_T'' \approx \frac{R_1}{2} + R_1 = \frac{3R_1}{2} \tag{2.58}$$

The corresponding intensity is given by the following expression:

$$I_T \propto \frac{9R_1^2}{4} \tag{2.59}$$

From Eqs (2.55) and (2.59), we can conclude that the intensity is maximum when the number of half periods in the lower half of the wavefront is odd and is minimum when the number of half periods in the lower half wavefront is even. Thus, we have the following relation:

$$P_1A_1 - P_1C_0 = (2n-1)\frac{\lambda}{2} \quad \text{for maxima} \tag{2.60}$$

and $\quad P_1A_1 - P_1C_0 = \dfrac{2n\lambda}{2} \quad \text{for minima} \tag{2.61}$

where $n = 1, 2, 3, \ldots$
From Fig. 2.7, we can see that

$$(P_1A_1)^2 = (PA_1)^2 + (PP_1)^2 = a_2^2 + x^2 \tag{2.62}$$

or $\quad (P_1A_1) = (a_2^2 + x^2)^{1/2} = a_2\left(1 + \dfrac{x^2}{a_2^2}\right)^{1/2}$

which can be approximated to the following form: $(P_1A_1) = a_2\left(1 + \dfrac{x^2}{2a_2^2}\right)$

or $\quad (P_1A_1) = a_2 + \dfrac{x^2}{2a_2} \tag{2.63}$

Similarly,

$$P_1S = (a_1 + a_2) + \frac{x^2}{2(a_1 + a_2)} \tag{2.64}$$

In addition, $P_1C_0 = P_1S - C_0S$

giving $\quad P_1C_0 = \left[(a_1 + a_2) + \dfrac{x^2}{2(a_1 + a_2)} - a_1\right]$

leading to

$$P_1 C_0 = a_2 + \frac{x^2}{2(a_1 + a_2)} \tag{2.65}$$

Using Eqs (2.63) and (2.65), we have the following relation:

$$P_1 A_1 - P_1 C_0 = \left[a_2 + \frac{x^2}{2a_2} \right] - \left[a_2 + \frac{x^2}{2(a_1 + a_2)} \right]$$

which leads to the following expression:

$$P_1 A_1 - P_1 C_0 = \frac{x^2}{2a_2} - \frac{x^2}{2(a_1 + a_2)} = \frac{a_1 x^2}{2a_2 (a_1 + a_2)} \tag{2.66}$$

Substitution of Eq. (2.66) into Eq. (2.60) leads to the following relation:

$$\frac{a_1 x_n^2}{2a_2 (a_1 + a_2)} = (2n - 1)\frac{\lambda}{2}$$

yielding $x_n^2 = \dfrac{(2n-1)a_2 (a_1 + a_2)\lambda}{a_1}$

resulting in

$$x_n = \left[\frac{(2n-1)(a_1 + a_2)a_2 \lambda}{a_1} \right]^{1/2} \tag{2.67}$$

Equation (2.67) represents the condition required to achieve the maximum intensity at point P_1.

Similarly, using Eq. (2.61) leads to the following form: $\dfrac{a_1 x_n^2}{2a_2 (a_1 + a_2)} = \dfrac{2n\lambda}{2}$

which results in the following form: $x_n^2 = \dfrac{2n(a_1 + a_2)a_2 \lambda}{a_1}$

yielding

$$x_n = \left[\frac{2n(a_1 + a_2)a_2 \lambda}{a_1} \right]^{1/2} \tag{2.68}$$

Equation (2.68) gives the condition required for obtaining a minimum at point P_1.

Let us now consider an arbitrary point P_2 in the geometrical shadow region. The situation is depicted schematically in Fig. 2.8.

Suppose the position of P_2 is such that the entire lower half of the wavefront and the first strip $C_0 C_1$ from the upper half are obstructed by the obstacle $A_1 A_2$. The resultant R'_{T_1} is given by the following equation:

$$R'_{T_1} = -R_2 + R_3 - R_4 + R_5 - \ldots = -\frac{R_2}{2} \tag{2.69}$$

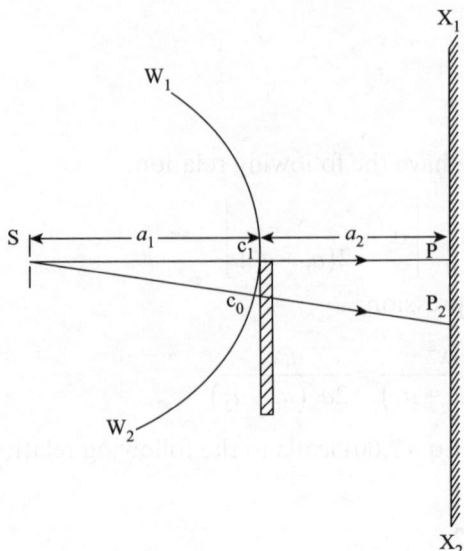

Fig. 2.8 Schematic of pattern due to a point in geometrical shadow

The corresponding intensity is expressed as follows:

$$I'_{T_1} \propto \frac{R_2^2}{4} \tag{2.70}$$

Thus, I'_{T_1} is very close to the intensity at P, since $R_1 \approx R_2$.

Suppose the location P_2 is such that the complete lower half of the wave-front and two strips C_0C_1 and C_0C_2 of the upper half of the wavefront are obstructed. Then the resultant amplitude R'_{T_1} is given as follows:

$$R''_{T_1} = R_3 - R_4 + R_5 - R_6 + \ldots = \frac{R_3}{2} \tag{2.71}$$

The corresponding intensity is expressed as follows:

$$I''_{T_1} \propto \frac{R_3^2}{4} \tag{2.72}$$

A comparison of Eqs (2.70) and (2.72) helps us to conclude that the intensity I''_{T_1} is much less than the intensity at P.

In general, the intensity of light gradually reduces to zero in the geometrical shadow region.

For Eq. (2.67), the position of first maximum is given by the following relation:

$$x_1 = \left[\frac{(a_1 + a_2)a_2\lambda}{a_1} \right]^{1/2} \tag{2.73}$$

The position of fourth maximum is given by the following relation:

$$x_4 = \left[\frac{7(a_1 + a_2)a_1\lambda}{a_1} \right]^{1/2} \tag{2.74}$$

Subtraction of Eq. (2.73) from Eq. (2.74) leads to the following expression:

$$x_4 - x_1 = \left[\frac{7(a_1 + a_2)a_2\lambda}{a_1}\right]^{\frac{1}{2}} - \left[\frac{(a_1 + a_2)a_2\lambda}{a_1}\right]^{\frac{1}{2}}$$

which implies that

$$x_4 - x_1 = \left[\frac{(a_1 + a_2)a_2\lambda}{a_1}\right]^{\frac{1}{2}}\left[\sqrt{7} - 1\right] \tag{2.75}$$

Equation (2.75) can be used to evaluate the wavelength λ of the monochromatic source.

Example 2.9 A narrow slit gets illuminated by a light of wavelength 6000 Å. The slit is placed at a distance of 20 cm from a straight edge. Measurements are carried out at a distance of 160 cm from the straight edge. Determine the distance between the first and second dark bands.

Solution Distance x_n of the nth dark band (minima) outside the geometrical shadow is given by Eq. (2.68), that is,

$$x_n = \left[\frac{2n(a_1 + a_2)a_2\lambda}{a_1}\right]^{\frac{1}{2}} \tag{2.76}$$

For the first dark band, we have $n = 1$ and Eq. (2.76) yields the following relation:

$$x_1 = \left[\frac{2 \times 1 \times (20 + 160) \times 160 \times 6000 \times 10^{-8}}{20}\right]^{\frac{1}{2}}$$

giving $x_1 = \left[\dfrac{2 \times 180 \times 160 \times 6000 \times 10^{-8}}{20}\right]^{\frac{1}{2}} = 0.4157 \text{ cm} \tag{2.77}$

For the second dark band, $n = 2$ and Eq. (2.76) yields the following expression:

$$x_2 = \left[\frac{2 \times 2 \times (20 + 160) \times 160 \times 6000 \times 10^{-8}}{20}\right]^{\frac{1}{2}}$$

giving $x_2 = 0.5879 \text{ cm} \tag{2.78}$

Using Eqs (2.77) and (2.78), we get $x_2 - x_1 = 0.5879 - 0.4157 = 0.1722 \text{ cm}$
Hence, the distance between the first and second dark bands is 0.1722 cm.

Example 2.10 A narrow slit gets illuminated by a light of wavelength 5000 Å. The slit is placed at a distance of 15 cm from a straight edge. Measurements are carried out at a distance of 150 cm from the straight edge. Calculate the distance of the first dark band outside the geometrical shadow.

Solution Distance x_n of the nth dark band is given by the following expression:

$$x_n = \left[\frac{2n(a_1 + a_2)a_2\lambda}{a_1}\right]^{\frac{1}{2}} \tag{2.79}$$

Substituting $n = 1$ and the given values, we get the following relation:

$$x_1 = \left[\frac{2(15 + 150) \times 150 \times 5000 \times 10^{-8}}{15}\right]^{\frac{1}{2}}$$

resulting in $x_1 = \left[\dfrac{2 \times 165 \times 150 \times 5000 \times 10^{-8}}{15}\right]^{1/2}$

giving $\quad x_1 = (0.165)^{1/2} = 0.4062$

Example 2.11 A narrow slit is illuminated by a monochromatic light. The slit is placed at a distance of 15 cm from a straight edge. Measurements are carried out at a distance of 150 cm from the straight edge. The distance of the second dark band is found to be 0.5472 cm. Calculate the wavelength of light used.

Solution Distance x_n of the nth dark band outside the geometrical shadow is given as follows:

$$x_n = \left[\frac{2n(a_1 + a_2)a_2\lambda}{a_1}\right]^{1/2} \tag{2.80}$$

Using $n = 2$ and the given values, we get the following relation:

$$0.5472 = \left[\frac{2 \times 2\,(15 + 150) \times 150 \times \lambda}{15}\right]^{1/2} \Rightarrow 0.5472 = \left[\frac{2 \times 2 \times 165 \times 150}{15}\right]^{1/2} \lambda^{1/2}$$

yielding $0.5472 = (6600)^{1/2}\,\lambda^{1/2}$, that is, $0.5472 = 81.24\,\lambda^{1/2}$

giving $\quad \lambda = \left(\dfrac{0.5472}{81.24}\right)^2 = 4537 \times 10^{-8}\,\text{cm}$

Example 2.12 For the set-up given in Example 2.11, where would the first dark band be formed?

Solution For the first dark band: $x_1 = \left[\dfrac{2(a_1 + a_2)a_2\lambda}{a_1}\right]^{1/2}$

Using the given values, we get the following relation:

$$x_1 = \left[\frac{2 \times (15 + 150) \times 150 \times 4537 \times 10^{-8}}{15}\right]^{1/2}$$

yielding $x_1 = \left[\dfrac{2 \times 165 \times 150 \times 4537 \times 10^{-8}}{15}\right]^{1/2} = (0.1497)^{1/2} = 0.3869\,\text{cm}$

Example 2.13 What would be the width of a slit if the wavelength of light used is 5×10^{-5} cm, keeping all other parameters same as in Example 2.20?

Solution From Eq. (2.122), we have the following relation: $d \cong \dfrac{\lambda}{\theta_1}$

Using the given values, we get the following result: $d \cong \dfrac{5 \times 10^{-5} \times 1000}{8} \cong 0.0625\,\text{cm}$

Example 2.14 A light of wavelength 6×10^{-5} cm falls on a narrow slit. The pattern formed is observed on a screen placed at a distance of 100 cm from the slit. The width of the slit is 0.07 cm. Calculate the location of the first minima w.r.t. the central maxima.

Solution Suppose the first minimum lies at x cm on either side of the central maximum for a small angle, θ_1. Using Eq. (2.119), we can write the following relation:

$$\theta_1 = \frac{x}{100} \tag{2.81}$$

Equation (2.108) yields the following expression:

$$d \sin\theta_1 = 1 \times \lambda \tag{2.82}$$

which, for small angles, gets simplified to the following form:

$$d \cong \frac{\lambda}{\theta_1} \tag{2.83}$$

Using Eq. (2.81) and the given value of λ in Eq. (2.83), we get the following relation:

$$0.07 = \frac{6 \times 10^{-5} \times 100}{x}$$

leading to $x = \dfrac{6 \times 10^{-5} \times 100}{0.07} = 0.086$ cm

Example 2.15 Light of wavelength 5×10^{-5} cm falls on a narrow slit. The pattern formed is observed on a screen. The width of the slit is 0.065 cm and the first minimum lies 0.07 cm on either side of the central maximum. Calculate the location of the screen with reference to the slit.

Solution Using Eq. (2.119), for a small angle θ_1, we can write the following relation:

$$\theta_1 = \frac{0.07}{x} \tag{2.84}$$

where x represents the location of the screen.

From Eq. (2.122), we can write the following expression: $0.065 \cong \dfrac{5 \times 10^{-5} \times x}{0.07}$

yielding $x \cong \dfrac{0.065 \times 0.07}{5 \times 10^{-5}} \cong 91$ cm

Example 2.16 The source wavelength in Example 2.21 is changed to 6500 Å. Calculate the new angular separation between the first-order minima on either side of the central maximum.

Solution Using Eq. (2.124), we get the following relation:

$$\sin\theta = \frac{\lambda}{d} \tag{2.85}$$

Using the given value in Eq. (2.85), we get the following expression:

$$\sin\theta = \frac{6500 \times 10^{-8}}{6.2 \times 10^{-4}} = 0.105$$

yielding $\theta = 6° 2'$

Example 2.17 A slit is illuminated with a light of 5500 Å. The angular separation between the first-order minima on either side of the central maximum is 6°. Calculate the slit width.

Solution Using Eq. (2.85) we can write the following relation: $\sin\theta = \dfrac{\lambda}{d}$

giving $\quad \sin 6 = \dfrac{5500 \times 10^{-8}}{d}$

resulting in $d = \dfrac{5500 \times 10^{-8}}{\sin 6} \cong \dfrac{5500 \times 10^{-8}}{0.105} \cong 5.24 \times 10^{-4}\,\text{cm}$

Example 2.18 A slit of width 5.8×10^{-4} cm is illuminated with a monochromatic light. The angular separation between the first-order minima on either side of the central maximum is found to be $5°$. Calculate the wavelength of the light used.

Solution Equation (2.85) gives the following relation: $\sin\theta = \dfrac{\lambda}{d}$

yielding $\lambda = \sin\theta \times d$

Using the given values, we get the following expression: $\lambda = \sin 5 \times 5.8 \times 10^{-4}$

giving $\quad \lambda = 0.087 \times 5.8 \times 10^{-4} = 5046 \times 10^{-8}$ cm

Example 2.19 The wavelength of light is changed to 5500 Å in the set-up indicated in Example 2.22. Calculate the new angle at which the first dark band is formed.

Solution From Eq. (2.126), we have

$$d \sin\theta = \lambda \tag{2.86}$$

resulting in

$$\sin\theta = \frac{\lambda}{d} \tag{2.87}$$

Substituting the given values in Eq. (2.87), we get the following relation:

$$\sin\theta = \frac{\lambda}{d} = \frac{5500 \times 10^{-8}}{0.028} = 0.00196$$

leading to $\theta = \sin^{-1} 0.00196 = 0.112° = 6.7'$

2.8 RESULTANT OF MULTIPLE SIMPLE HARMONIC MOTIONS

Let us assume that a particle is simultaneously acted upon by n SHM vibrations. All the vibrations have the same amplitude A, and δ represents the phase difference between successive vibrations. Figure 2.9 is a schematic representation of these vibrations.

A phase difference δ exists between bc and ab, bc and cd, etc. The phase difference between cd and ab is 2δ, between de and ab is 3δ, etc. To determine R (resultant) and θ, let us resolve the individual amplitudes along ab and those perpendicular to ab. The resultant component along ab is given as follows:

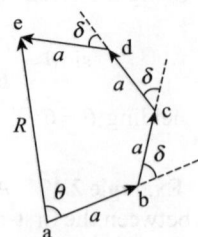

$$R\cos\theta = A\begin{bmatrix} 1+\cos\delta+\cos 2\delta \\ +\ldots+\cos(n-1)\delta \end{bmatrix} \tag{2.88}$$

and $\quad R\sin\theta = A\big[\sin\delta+\sin 2\delta+\ldots+\sin(n-1)\delta\big] \tag{2.89}$

Fig. 2.9 Resultant of multiple SHMs

Multiplying Eq. (2.88) by $2\sin\dfrac{\delta}{2}$, we get the following relation:

$$2R\cos\theta\sin\frac{\delta}{2} = A\left[2\sin\frac{\delta}{2} + 2\cos\delta\sin\frac{\delta}{2} + \ldots + 2\cos(n-1)\delta\sin\frac{\delta}{2}\right]$$

leading to

$$2R\cos\theta\sin\frac{\delta}{2} = A\left[\begin{array}{l}\dfrac{2\sin\delta}{2} + \left\{\sin\dfrac{3\delta}{2} - \sin\dfrac{\delta}{2}\right\} + \ldots \\[2mm] + \left\{\sin\left(n-\dfrac{1}{2}\right)\delta - \sin\left(n-\dfrac{3}{2}\right)\delta\right\}\end{array}\right]$$

resulting in $2R\cos\theta\sin\dfrac{\delta}{2} = A\left[\sin\dfrac{\delta}{2} + \sin\left(n-\dfrac{1}{2}\right)\delta\right]$

yielding $2R\cos\theta\sin\dfrac{\delta}{2} = 2A\sin\dfrac{n\delta}{2}\cos\left(\dfrac{n-1\delta}{2}\right)$

that is,

$$R\cos\theta = A\frac{\sin n\frac{\delta}{2}}{\sin \frac{\delta}{2}}\cos\frac{(n-1)\delta}{2} \tag{2.90}$$

Multiplying Eq. (2.89) by $2\sin\dfrac{\delta}{2}$ and going through the same simplifying process, we get the following relation:

$$R\sin\theta = A\frac{\sin n\frac{\delta}{2}}{\sin \frac{\delta}{2}}\sin\frac{(n-1)\delta}{2} \tag{2.91}$$

Squaring Eqs (2.90) and (2.91) and adding, we obtain the following relation:

$$R^2 = A^2\frac{\sin^2 n\frac{\delta}{2}}{\sin^2 \frac{\delta}{2}} \tag{2.92}$$

which results in the following expression:

$$R = A\frac{\sin n\frac{\delta}{2}}{\sin \frac{\delta}{2}} \tag{2.93}$$

Dividing Eq. (2.91) by Eq. (2.90), the following relation can be obtained:

$$\tan\theta = \tan\frac{(n-1)\delta}{2}$$

Thus,

$$\theta = \frac{(n-1)\delta}{2} \tag{2.94}$$

If n is infinitely large and its amplitude A and phase difference δ are extremely small quantities, we can rewrite Eq. (2.93) in the following form:

$$R = A\frac{\sin\delta_1}{\sin\delta\frac{1}{n}} = A\frac{\sin\delta_1}{\delta\frac{1}{n}} \tag{2.95}$$

where $n\delta = 2\delta_1$ and $\dfrac{\delta_1}{n}$ is an extremely small quantity.

Assuming that $nA = A_1$, we can rewrite Eq. (2.95) as follows:

$$R = A_1 \frac{\sin \delta_1}{\delta_1} \tag{2.96}$$

and $\quad \theta = \dfrac{(n-1)\delta}{2} \approx \dfrac{n\delta}{2} = \delta_1 \tag{2.97}$

2.9 FRAUNHOFER DIFFRACTION AT SINGLE SLIT

If the source and the screen are at infinitely large distances from the diffracting aperture, both the incident and the diffracted wavefronts can then be assumed to be plane wavefronts. The resulting diffraction is called Fraunhofer diffraction. We will describe the Fraunhofer diffraction in detail in this section.

Figure 2.10 is a schematic representation of single slit diffraction set-up.

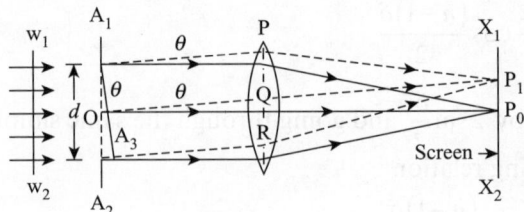

Fig. 2.10 Schematic of single-slit diffraction

A_1A_2 represents a section of a narrow slit of width d perpendicular to the plane of the paper. A plane wavefront W_1W_2 of monochromatic light of wavelength λ is incident on the slit. The diffracted light is focused using a convex lens onto a screen X_1X_2 placed at the focal plane of the lens. According to the Huygen's principle, each point of the principal wavefront results in secondary spherical wavelets that spread out in all directions. The secondary wavelets travelling perpendicular to the slit are brought to focus at point P_0 by the lens. The secondary wavelets travelling at an angle θ with respect to the normal get focused at point P_1 on the screen. A_1A_3 is perpendicular on A_2R.

The path difference, PD, between the secondary wavelets originating from A_1 and A_2 in the direction θ is given as follows:

$$A_2A_3 = A_1A_2 \sin\theta = d\sin\theta \tag{2.98}$$

The corresponding phase difference, δ_1, is given as follows:

$$\delta_1 = \frac{2\pi}{\lambda} d\sin\theta \tag{2.99}$$

Let us assume that the width d of the slit is divided into n equal parts. Let the amplitude of the wave originating from each part be represented by A.

The phase difference δ between waves originating from any two consecutive parts is then expressed as follows:

$$\delta = \frac{1}{n}\left(\frac{2\pi}{\lambda}\right) d \sin\theta \tag{2.100}$$

Using Eqs (2.93) and (2.100), the resultant amplitude R can be expressed in the following form:

$$R = \frac{A\sin\dfrac{n\delta}{2}}{\sin\dfrac{\delta}{2}} = A\frac{\sin\left(\pi d \sin\dfrac{\theta}{\lambda}\right)}{\sin\left(\pi d \sin\dfrac{\theta}{n\lambda}\right)} \tag{2.101}$$

Substitution of $\phi = \pi d \sin\dfrac{\theta}{\lambda}$ in Eq. (2.101) leads to the following relation:

$$R = A\frac{\sin\phi}{\sin\dfrac{\phi}{n}} \tag{2.102}$$

Since $\dfrac{\phi}{n}$ is an extremely small quantity, Eq. (2.102) can be rewritten in the following form:

$$R = A\frac{\sin\phi}{\dfrac{\phi}{n}} = nA\frac{\sin\phi}{\phi} = \frac{A_T \sin\phi}{\phi} \tag{2.103}$$

One must remember that in Eq. (2.103), $n \to \infty$, $A \to 0$ and the product $nA = A_T$ always remains finite.

Using Eq. (2.102), we can write the following expression for intensity:

$$I = R^2 = A_T^2\left[\frac{\sin\phi}{\phi}\right]^2 \tag{2.104}$$

2.9.1 Principal Maxima

Using the expanded form of $\sin\phi$, we can rewrite Eq. (2.103) in the following form: $R = \dfrac{A_T}{\phi}\left[\phi - \dfrac{\phi^3}{3!} + \dfrac{\phi^5}{5!} - \dfrac{\phi^7}{7!} + \ldots\right]$

which leads to the following expression:

$$R = A_T\left[1 - \frac{\phi^2}{3!} + \frac{\phi^4}{5!} - \frac{\phi^6}{7!} + \ldots\right] \tag{2.105}$$

R will maximize to A_T for $\phi = 0$, that is,

$$\phi = \frac{\pi d \sin\theta}{\lambda} = 0 \quad \text{or} \quad \sin\theta = 0 \quad \text{or} \quad \theta = 0 \tag{2.106}$$

Thus, $\theta = 0$ or the secondary wavelets that travel normal to the slit result in maxima on the screen. These are known as the *principal maxima*. The intensity would be proportional to the square of the amplitude.

2.9.2 Position for Minimum Intensity

From Eq. (2.104), we can see that the intensity on the screen would be minimum if $\sin\phi = 0$. This leads to the following formula:

$$\phi = \pm\pi, \pm 2\pi, \pm 3\pi, \pm 4\pi, \ldots = \pm m\pi \tag{2.107}$$

Equation (2.107) implies that

$$\frac{\pi d \sin\theta}{\lambda} = \pm m\pi \text{ or } d\sin\theta = \pm m\lambda \tag{2.108}$$

where $m = 1, 2, 3, \ldots$ The value of $m = 0$ leads to $\theta = 0$, that is, the principal maximum.

2.9.3 Secondary Maxima

In addition to the principal maxima for $\theta = 0$, weak secondary maxima are also observed between equally spaced minima given by Eq. (2.107).

Differentiation of Eq. (2.104) leads to the following equation:

$$\frac{dI}{d\phi} = \frac{d}{d\phi}\left[A_T^2 \left(\frac{\sin\phi}{\phi} \right)^2 \right] \tag{2.109}$$

For minima, we have

$$\frac{dI}{d\phi} = \frac{d}{d\phi}\left[A_T^2 \left(\frac{\sin\phi}{\phi} \right)^2 \right] = 0 \tag{2.110}$$

yielding $A_T^2 \cdot \dfrac{2\sin\phi}{\phi} \cdot \dfrac{(\phi\cos\phi - \sin\phi)}{\phi^2} = 0$ \qquad (2.111)

Equation (2.111) implies that

$$\text{either } \sin\phi = 0 \text{ or } (\phi\cos\phi - \sin\phi) = 0 \tag{2.112}$$

Here, $\sin\phi = 0$ gives the value of ϕ for which minima with zero intensity are obtained on the screen. Secondary maxima are, therefore, given by solutions of the following equation:

$$\phi\cos\phi - \sin\phi = 0 \text{ or } \phi = \tan\phi \tag{2.113}$$

Equation (2.113) can be solved for ϕ graphically by simultaneously plotting $y = \phi$ and $y = \tan\phi$ on the same graph. Figure 2.11 shows these plots and also the points of intersection of the two plots.

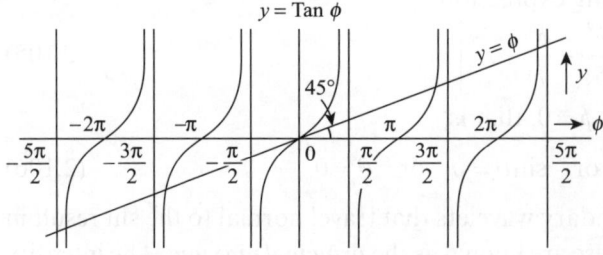

Fig. 2.11 Plots of $y = \phi$ and $y = \tan\phi$

The points of intersections are given by the following expression:

$$\phi - 0, \pm\frac{3\pi}{2}, \pm\frac{5\pi}{2}, \text{ etc.} \tag{2.114}$$

More exact values can be shown to be of the following form:

$$\phi = 0, \pm 1.430\pi, \pm 2.462\pi, \pm 3.471\pi, \text{ etc.} \tag{2.115}$$

Using Eqs (2.104) and (2.114), we obtain the following expression:

$$I_1 = A_T^2 \left[\frac{\sin\left(3\frac{\pi}{2}\right)}{3\frac{\pi}{2}} \right]^2 \cong \frac{A_T^2}{22} \tag{2.116}$$

where I_1 represents the intensity of the first secondary maximum.
Similarly, the intensity of the second secondary maxima I_2 is given by the following equation:

$$I_2 = A_T^2 \left[\frac{\sin\left(5\frac{\pi}{2}\right)}{\left(5\frac{\pi}{2}\right)} \right] \cong \frac{A_T^2}{62} \tag{2.117}$$

Similarly, intensity of the other secondary maxima can be calculated. Intensity of the principal maxima using Eq. (2.105) is given as follows:

$$I = R^2 = A_T^2 \tag{2.118}$$

From Eqs (2.116)–(2.118), it is clear that the intensity of secondary maxima decreases progressively and most of the incident light energy is concentrated in the principal maxima.

2.9.4 Intensity Distribution Graph

Figure 2.12 is a schematic representation of the variation of intensity as a function of ϕ.

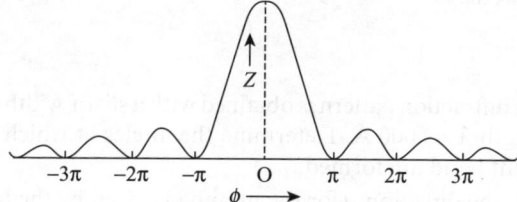

Fig. 2.12 Schematic representation of variation of intensity versus ϕ

The principal maximum is at $\phi = 0$ and the secondary maxima occur at $\phi = \pm\frac{3\pi}{2}, \pm\frac{5\pi}{2}$, and so on. Minima positions that lie between secondary maxima are given as $\phi = \pm\pi, \pm 2\pi, \pm 3\pi$, and so on. Thus, the secondary maxima are not exactly midway between two minima but are displaced towards the centre of the diffraction pattern.

Example 2.20 Light of wavelength 5.8×10^{-5} cm falls on a narrow slit. The pattern formed is observed on a screen placed at a distance of 100 cm from the slit. Determine the width of the slit if the first minimum lies 0.08 cm on either side of the central maximum.

Solution From Fig. 2.10, we can see that

$$\sin\theta_n = \frac{\text{Distance of first minima from the central maxima}}{\text{Distance of the screen from the slit}} \tag{2.119}$$

For small θ:

$$\theta_1 = \frac{0.08}{100} = \frac{0.8}{1000} \qquad (2.120)$$

From Eq. (2.108), we get the following relation:

$$d \sin\theta_1 = 1 \times \lambda \qquad (2.121)$$

which leads to the following expression:

$$d \cong \frac{\lambda}{\theta_1} \qquad (2.122)$$

Using the given value of λ and Eq. (2.120) in Eq. (2.122), we get the following result:

$$d \cong \frac{5.8 \times 10^{-5} \times 1000}{0.8} = 0.0725 \text{ cm}$$

Example 2.21   A slit of width 6.2×10^{-4} cm is illuminated with light of wavelength 5800 Å. Determine the angular separation between the first-order minima on either side of central maximum.

Solution   From Eq. (2.108), we have the following expression:

$$d \sin\theta = m\lambda \qquad (2.123)$$

where θ represents the angular separation of the nth-order minimum from the central maximum, d represents the slit width, and λ represents the wavelength of light.
For $m = 1$ Eq. (2.123) leads to the following expression:

$$\sin\theta = \frac{\lambda}{d} \qquad (2.124)$$

Substitution of the given values in Eq. (2.124) yields the following relation:

$$\sin\theta = \frac{5800 \times 10^{-8}}{6.2 \times 10^{-4}} = 0.09$$

resulting in $\theta = 5°10'$

Example 2.22   Fraunhofer diffraction pattern is obtained with a slit of width 0.28 mm and light source of wavelength $\lambda = 6000$ Å. Determine the angles at which the first dark band and the next bright band are formed.

Solution   From Eq. (2.108), the direction θ for the minima is given by the following relation:

$$d \sin\theta = m\lambda, \quad m = 1, 2, 3, \ldots \qquad (2.125)$$

For the first dark band, $m = 1$. Hence,

$$d \sin\theta = \lambda \qquad (2.126)$$

or $\sin\theta = \dfrac{\lambda}{d}$

Substituting these values in Eq. (2.126), we get the following expression:

$$\sin\theta = \frac{6000 \times 10^{-8}}{0.028} = 0.00214$$

which leads to the following result:

$$\theta = \sin^{-1}(0.00214) = 7.2' \qquad (2.127)$$

Angle of diffraction θ_1 corresponding to the first bright band on either side of the central maximum is given by Eq. (2.114), that is,

$$d \sin \theta_1 = \frac{3\lambda}{2} \tag{2.128}$$

yielding $\sin \theta_1 = \dfrac{3\lambda}{2d}$ (2.129)

Substitution of these values in Eq. (2.129) results in the following value:

$$\sin \theta_1 = \frac{3}{2} \times 0.00214 = 0.0032$$

Thus, $\quad \theta_1 = \sin^{-1}(0.00321) = 11$

Example 2.23 Fraunhofer diffraction pattern is obtained with a slit of width 0.25 mm and a monochromatic light. The angle at which the first dark band is formed is 6′. Calculate the wavelength of light used.

Solution The calculations are as follows:

$$6' = \frac{6}{60} = 0.1^\circ \tag{2.130}$$

From Eq. (2.87), we get the following relation: $\sin \theta = \dfrac{\lambda}{d}$
leading to

$$\lambda = \sin \theta \times d \tag{2.131}$$

Using Eq. (2.130) and the given value of d, we get the following result:

$$\lambda = \sin 0.1^\circ \times 0.025 = 0.00175 \times 0.025$$

leading to $\lambda = 4375 \text{Å}$

Example 2.24 Fraunhofer diffraction pattern is observed using light of wavelength 5500 Å and a single slit. The bright band next to the first dark band is formed at an angle of 15′. Calculate the width of the slit.

Solution

$$15' = \frac{15}{60} = \frac{1}{4} = 0.25^\circ \tag{2.132}$$

From Eq. (2.129), we have the following relation: $\sin \theta_1 = \dfrac{3}{2} \dfrac{\lambda}{d}$

leading to $d = \dfrac{3\lambda}{2 \sin \theta_1}$

Using the given values, we get the following expression: $d = \dfrac{3 \times 5500 \times 10^{-8}}{2 \times \sin 0.25^\circ}$

yielding $d = \dfrac{3 \times 5500 \times 10^{-8}}{2 \times 0.00436} = 0.019$ cm

2.10 FRAUNHOFER DIFFRACTION FOR CIRCULAR APERTURE

Let us now replace the single slit of Fig. 2.10 with a circular aperture. Figure 2.13 is a schematic representation of such a situation.

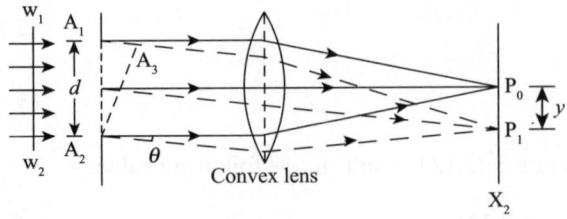

Fig. 2.13 Fraunhofer diffraction for circular aperture

AB represents a circular aperture of diameter d and $W_1 W_2$ represents a plane wavefront of monochromatic light of wavelength λ. The wavefront is assumed to be normally incident on the circular aperture. The diffracted beam is focused on the screen using a convex lens of focal length f. In the plane of the circular aperture, every point becomes a source of secondary wavelets that spread out in all directions towards the right. The secondary circular aperture comes to focus at point P_0. Since all the secondary wavelets coming to point P_0 travel the same distance, they all reinforce each other and result in the central maximum.

Let us now consider the waves travelling at an angle θ with respect to the normal to the aperture. The secondary wavelets travelling at angle θ come to focus at point P_1 on the screen. Let us also assume that $P_0 P_1 = y$.

The path difference PD between waves originating from points A_1 and A_2 is given as follows:

$$PD = A_1 A_3 = A_1 A_2 \sin\theta = d\sin\theta \qquad (2.133)$$

The intensity at point P_1 will be minimum if PD is an integral multiple of λ and maximum if PD is an odd multiple of $\frac{\lambda}{2}$, Thus,

$$d\sin\theta = n\lambda \text{ (minima)} \qquad (2.134)$$

and

$$d\sin\theta = (2n+1)\frac{\lambda}{2} \text{ (maxima)} \qquad (2.135)$$

where n can take values $1, 2, 3, \dots$ The term corresponding to $n = 0$ results in the *central maxima*.

If point P_1 is a point of minimum intensity, then all the points lying on a circle of radius y will have the minimum intensity. In general, then, the diffraction pattern consists of a central bright disc surrounded by alternate dark and bright concentric rings. The central disc is called the *Airy's disc* and the concentric rings are called *Airy's rings*. The dark rings have zero intensity, and the intensity of bright rings gradually reduces to zero as we go away from point P_0 on the screen.

Let us assume that either the lens is kept very close to the circular aperture or the screen is held at a very large distance from the lens. Under both these conditions, we have the following relation:

$$\sin 0 = \theta = \frac{y}{f} \qquad (2.136)$$

For the first minimum, from Eq. (2.134), we obtain the following expression:

$$d\sin\theta = 1\cdot\lambda, \text{ that is, } \sin\theta = \frac{\lambda}{d} \tag{2.137}$$

Using Eqs (2.136) and (2.137), we get the following relation:

$$\frac{y}{f} = \frac{\lambda}{d} \tag{2.138}$$

which implies that

$$y = \frac{f\lambda}{d} \tag{2.139}$$

where y represents the radius of the Airy's disc. It was shown by Airy that the exact expression for y is the following:

$$y = 1.22\frac{f\lambda}{d} \tag{2.140}$$

Thus, the radius of the Airy's disc is inversely proportional to the diameter of the circular aperture.

Example 2.25   Fraunhofer diffraction pattern due to a circular aperture is obtained on a screen placed at a distance of 90 cm from the lens. The light used has a wavelength of 6000 Å and the diameter of the aperture is 0.1 mm. Determine the separation between the central disc and the first secondary minimum.

Solution   From Eq. (2.139), we get the following expression:

$$y = \frac{f\lambda}{d} \tag{2.141}$$

Substituting these values, we get $y = \dfrac{90 \times 6000 \times 10^{-8}}{0.01} = 0.54$ cm

2.11 PLANE DIFFRACTION GRATING

A diffraction grating consists of a large number of parallel slits of the same width that are separated by equal opaque spaces. One of the ways in which gratings are constructed involves ruling of equidistant parallel lines on a transparent material like glass using a fine diamond point. The ruled lines are opaque to light, whereas the space between any two lines is transparent to light and act as slits. Such a grating is called a *plane transmission grating*. Another method of producing gratings is by drawing lines on a plane or a concave silvered surface. Light then gets reflected from a point situated between two lines. Such gratings are called *plane or concave reflection gratings*. Figure 2.14 is a schematic representation of a section of plane transmission grating that is placed perpendicular to the plane of the paper.

Let d_1 represent the width of each slit and d_2 the width of each opaque part. The sum $(d_1 + d_2)$ is called the grating element. X_1X_2 is a screen placed perpendicular to the plane of paper. A parallel beam of monochromatic

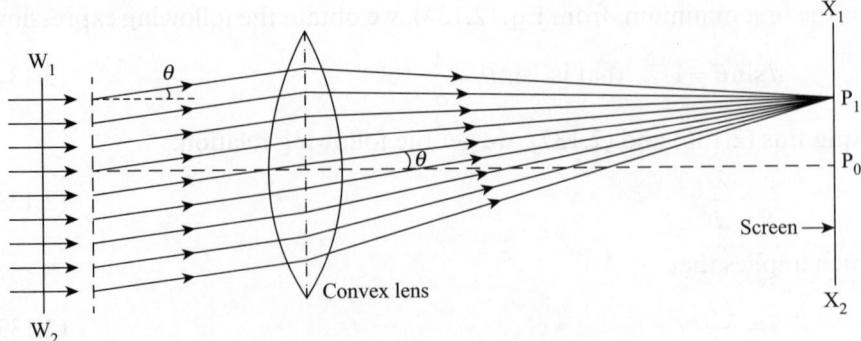

Fig. 2.14 Diffraction due to plane grating

light of wavelength λ is incident normally on the grating. Each of the infinite points in the slit becomes sources of secondary wavelets. Secondary wavelets travelling in the direction of incident light come to focus at point P_0 on the screen due to the focusing action of the convex lens. Point P_0 then becomes the central maximum. Secondary waves travelling at an angle θ with respect to the direction of the incident light reach point P_1 after passing through the lens, in different phases. This results in dark and bright bands on both sides of the central maximum.

We can apply the theory of Fraunhofer diffraction due to a single slit in order to determine the intensity at point P_1. For any one slit, wavelets originating from all the points of the slit can be considered equivalent to a simple wave of amplitude $\left(\dfrac{A_T \sin \phi}{\phi} \right)$ starting from the middle point of the slit with $\phi = \left(\dfrac{nd_1 \sin \theta}{\lambda} \right)$.

If there are N slits, we get diffracted waves from the middle point of all the slits. The path difference between waves from consecutive slits is $(d_1 + d_2) \sin \theta$. The corresponding phase difference is given by $(2\pi/\lambda)(d_1 + d_2) \sin \theta$. Let us represent this phase difference by 2α.

Thus, the amplitude in a direction θ is the resultant amplitude of N vibrations, each of amplitude $\left(\dfrac{A_T \sin \phi}{\phi} \right)$, with a common phase difference given by the following equation:

$$\frac{2\pi}{\lambda}(d_1 + d_2) \sin \theta = 2\alpha \tag{2.142}$$

Using Eq. (2.102), the resultant amplitude R_T can be written as follows:

$$R_T = \left(\frac{A_T \sin \phi}{\phi} \right) \frac{\sin N\alpha}{\sin \alpha} \tag{2.143}$$

The corresponding intensity I_T is written as follows:

$$I_T = R_T^{\ 2} = \left(\frac{A_T \sin \phi}{\phi} \right)^2 \left(\frac{\sin N\alpha}{\sin \alpha} \right)^2 \tag{2.144}$$

In Eq. (2.144), the term $\left(\dfrac{A_T \sin\phi}{\phi}\right)^2$ gives the distribution of intensity due

to a single slit, and the term $\left(\dfrac{\sin^2 N\alpha}{\sin^2 \alpha}\right)$ represents the interaction between

the different slits.

The intensity given by Eq. (2.144) would be maximum for the following condition:

$$\sin\alpha = 0 \text{ or } \alpha = \pm n\pi \text{ where } n = 0, 1, 2, 3, \ldots \tag{2.145}$$

For these values of α, $(\sin N\alpha)$ also becomes zero; therefore, the ratio $\left(\dfrac{\sin N\alpha}{\sin \alpha}\right)$ becomes indeterminable. The magnitude of $\left(\dfrac{\sin N\alpha}{\sin \alpha}\right)$ is, there-

fore, to be evaluated using the L'Hospital rule, that is, by differentiating the numerator and the denominator. Thus, we have the following equation:

$$\underset{\alpha \to \pm n\pi}{\text{Lim}} \left(\frac{\sin N\alpha}{\sin \alpha}\right) = \underset{\alpha \to \pm n\pi}{\text{Lim}} \frac{\dfrac{d}{d\alpha}(\sin N\alpha)}{\dfrac{d}{d\alpha}\sin \alpha} \tag{2.146}$$

which results in the following expression:

$$\underset{\alpha \to \pm n\pi}{\text{Lim}} \frac{\sin N\alpha}{\sin \alpha} = \underset{\alpha \to \pm n\pi}{\text{Lim}} \frac{N\cos N\alpha}{\cos \alpha} = \pm N \tag{2.147}$$

Therefore,

$$\underset{\alpha \to \pm n\pi}{\text{Lim}} \left(\frac{\sin N\alpha}{\sin \alpha}\right)^2 = N^2 \tag{2.148}$$

From Eq. (2.144), we get the following relation:

$$I_T = \left(\frac{A_T \sin\phi}{\phi}\right)^2 N^2 \tag{2.149}$$

These maxima are called the *principal maxima* and occur when the following

relation holds: $\dfrac{n}{\lambda}(d_1 + d_2)\sin\theta = \pm n\pi$

This implies that

$$(d_1 + d_2)\sin\theta = \pm n\lambda \tag{2.150}$$

where $n = 0, 1, 2, 3, \ldots$

The maxima corresponding to $n = 0$ are called the zero-order maxima and $n = 1, 2, 3$, etc., result in the first order, second order, third order, etc., of the

Minima are obtained under the following condition:

$$\sin N\alpha = 0 \text{ but } \sin\alpha \neq 0 \tag{2.151}$$

Thus, for minima

$$N\beta = \pm m\pi \tag{2.152}$$

yielding $N \cdot \dfrac{\pi}{\lambda}(d_1 + d_2)\sin\theta = \pm m\pi$ (2.153)

or $\quad N(d_1 + d_2)\sin\theta = \pm m\lambda$ (2.154)

where m can take all integral values except $0, N, 2N, \ldots, nN$, because for these values of m, $\sin\alpha$ becomes zero and one gets the principal maxima. Thus, m can take values $1, 2, 3, \ldots, (N-1)$.

Hence, minima form at locations that are adjacent to the principal maxima. In addition to minima existing between two principal maxima, there also exist secondary maxima. To determine the location of these secondary maxima, let us differentiate Eq. (2.144) with respect to α and equate the result to zero. Thus, we have the following equation:

$$\frac{dI_T}{d\alpha} = \left(\frac{A_T \sin\phi}{\phi}\right)^2 2\left(\frac{\sin N\alpha}{\sin\alpha}\right) \times \left[\frac{N\cos N\alpha \sin\alpha - \sin N\alpha \cos\alpha}{\sin^2\alpha}\right] = 0$$

which yields the following relation:

$$N\cos N\alpha \sin\alpha - \sin N\alpha \cos\alpha = 0$$

that is, $N\tan\alpha = \tan N\alpha$ (2.155)

Roots of Eq. (2.155) other than those for which $\alpha = \pm n\pi$ give the location of secondary maxima. Figure 2.15 is a schematic representation of Eq. (2.155).

From Fig. 2.15, we can write the following equation:

Fig. 2.15 Schematic representation of Eq. (2.131)

$$\sin N\alpha = \frac{N}{\sqrt{N^2 + \cot^2\alpha}} \tag{2.156}$$

Thus, $\dfrac{\sin^2 N\alpha}{\sin^2\alpha} = \dfrac{N^2}{(N^2 + \cot^2\alpha) \times \sin^2\alpha}$

On simplification, we get the following relation:

$$\frac{\sin^2 N\alpha}{\sin^2\alpha} = \frac{N^2}{N^2\sin^2\alpha + \cos^2\alpha} = \frac{N^2}{1 + (N^2 - 1)\sin^2\alpha} \tag{2.157}$$

Using Eqs (2.148) and (2.157), we get the following expression:

$$\frac{\text{Intensity of secondary maxima}}{\text{Intensity of principal maxima}} = \frac{1}{1 + (N^2 - 1)\sin^2\alpha} \tag{2.158}$$

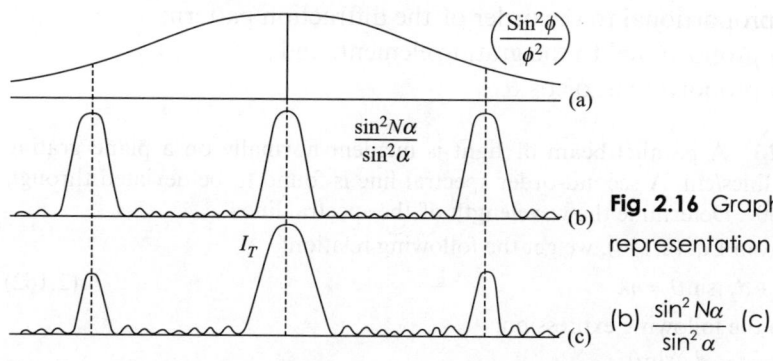

Fig. 2.16 Graphical representation (a) $\left(\dfrac{\sin^2 \phi}{\phi^2}\right)$ (b) $\dfrac{\sin^2 N\alpha}{\sin^2 \alpha}$ (c) I_T

From Eq. (2.158), we can see that as N increases, the intensity of secondary maxima relative to the principal maxima decreases and finally becomes negligible. Figure 2.16 is a graphical representation of variation of the resultant intensity as seen on the screen; it also shows individual contributions due to the terms $\left(\dfrac{\sin^2 \phi}{\phi^2}\right)$ and $\dfrac{\sin^2 N\alpha}{\sin^2 \alpha}$, respectively.

From Eq. (2.150), the condition for obtaining the principal maxima is as follows:

$$(d_1 + d_2)\sin\theta = n\lambda \tag{2.159}$$

The quantity $(d_1 + d_2)$ is called the grating element; n is the order of the maximum and θ is the angle of diffraction corresponding to a particular wavelength. Manufacturers of gratings usually supply the number (n_1) of lines that are to be ruled per inch on a grating. Thus, we have

$$n_1 \times (d_1 + d_2) = 1'' = 2.54 \text{ cm}$$

yielding $(d_1 + d_2) = \dfrac{2.54}{n_1} \text{ cm}$ $\tag{2.160}$

Using Eqs (2.159) and (2.160), λ can be determined using the known values of θ and n.

From Eq. (2.159), it is clear that the angle of diffraction is dependent on the wavelength λ. Equation (2.159) can be rewritten in the following form:

$$(d_1 + d_2)\sin\theta = n\lambda$$

Differentiating both sides, we get the following relation:

$$(d_1 + d_2)\cos\theta \, d\theta = n d\lambda$$

which results in the following expression:

$$\frac{d\theta}{d\lambda} = \frac{n}{(d_1 + d_2)\cos\theta} \tag{2.161}$$

Equation (2.161) expresses the dispersive power $\dfrac{d\theta}{d\lambda}$ for a grating. It is clear that the dispersive power is

1. directly proportional to the order of the diffraction pattern;
2. inversely proportional to the grating element; and
3. inversely proportional to cos θ.

Example 2.26 A parallel beam of light is incident normally on a plane grating having 4300 lines/cm. A second-order spectral line is found to be deviated through an angle of 30°. Determine the wavelength of the spectral line.

Solution From Eq. (2.150), we get the following relation:

$$(d_1 + d_2)\sin\theta = n\lambda \tag{2.162}$$

This leads to the following expression:

$$\lambda = \frac{(d_1 + d_2)\sin\theta}{n} \tag{2.163}$$

Substitution of the given values in Eq. (2.163) yields the following expression:

$$\lambda = \frac{1}{4300} \times \frac{\sin 30}{2} = \frac{1}{4300 \times 2 \times 2}$$

giving $\lambda = 5814 \times 10^{-8}$ cm

Example 2.27 A monochromatic light of wavelength 5860×10^{-8} cm is incident normally on a 2 cm wide grating. The first-order spectrum is produced at an angle of 20° with respect to the normal. Determine the total number of lines on the grating.

Solution From Eq. (2.150), we get the following expression:

$$(d_1 + d_2)\sin\theta = n\lambda$$

which implies that $(d_1 + d_2) = \dfrac{n\lambda}{\sin\theta}$

For $n = 1$, we have the following expression:

$$(d_1 + d_2) = \frac{\lambda}{\sin\theta} \tag{2.164}$$

The number of lines per cm $= \dfrac{1}{(d_1 + d_2)} = \dfrac{\sin\theta}{\lambda}$ $\tag{2.165}$

The total number of lines: $N = \dfrac{2}{(d_1 + d_2)} = \dfrac{2 \times \sin\theta}{\lambda}$ $\tag{2.166}$

Substituting these values in Eq. (2.166), we get the following result:

$$N = \frac{2 \times \sin(20)}{5860 \times 10^{-8}} = \frac{2 \times 0.3420}{5860 \times 10^{-8}} = 11{,}672$$

2.12 RESOLVING POWER

Our eyes are able to distinguish between two objects if they are either not too close to each other or not too far from our eye. When our eyes cannot differentiate between two objects, we can use optical instruments such as a telescope or a microscope for this purpose.

For distinguishing between closely spaced spectral lines, a prism or grating is employed. The image of a point object formed by an optical instrument is actually a diffraction pattern with a bright central maximum, secondary maxima of decreasing intensity, and minima in between. Any optical instrument is said to be able to resolve two point objects if the corresponding diffraction patterns can be distinguished from each other. *The ability of an optical instrument to form two separate diffraction patterns of two objects is called its resolving power.*

A criterion for resolution of two point sources by an optical instrument was first proposed by Rayleigh.

According to the Rayleigh's criterion, two point sources are resolvable by an optical instrument if the central maximum of the diffraction pattern of one of them falls over the first minimum of the diffraction pattern of the other.

Similarly, two spectral lines are resolvable if the central maximum of the diffraction pattern of one of the wavelengths falls over the first minimum due to the other wavelength, or vice versa.

Figure 2.17 is a schematic representation of the Rayleigh's criterion. In Fig. 2.17(a), the principal maxima in the diffraction patterns obtained using a

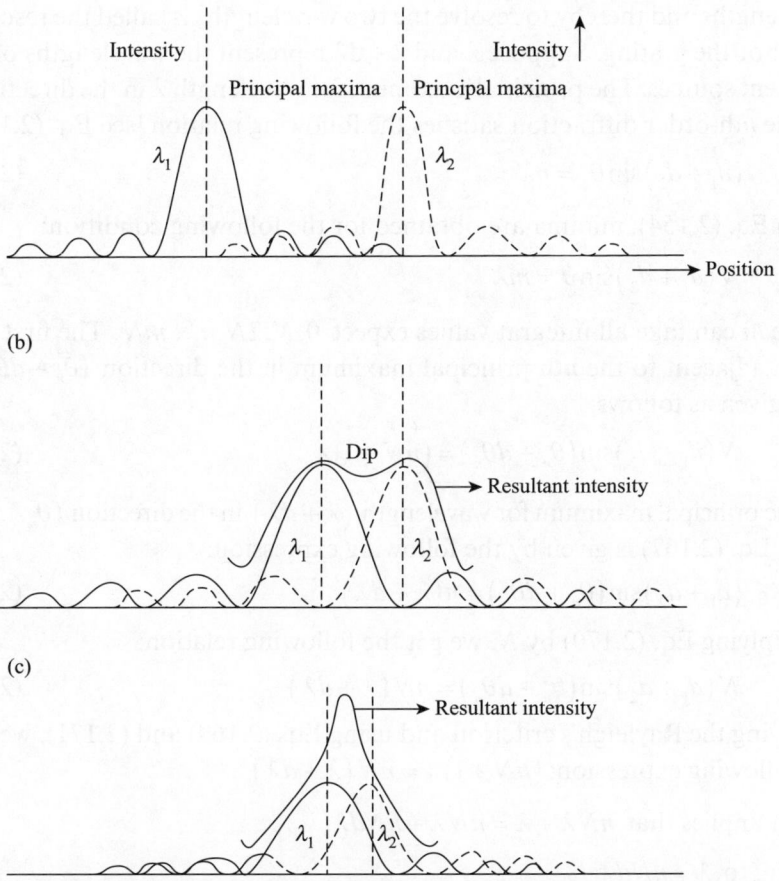

Fig. 2.17 Schematic representation of Rayleigh's criterion (a) Widely separated wavelengths (b) Resultant showing dip (c) Very close wavelengths

grating for two different wavelengths λ_1 and λ_2 have been sketched. The difference in wavelengths is such that the two maxima are separated in space by a considerable distance. A sufficiently large region of zero intensity exists between the two maxima, and therefore the two wavelengths can be resolved well by the grating.

In Fig. 2.17(b), the difference between the two wavelengths λ_1 and λ_2 is such that the principal maximum due to any one wavelength coincides with the first minimum due to the other wavelength. The resultant intensity shows a dip, and therefore the two wavelengths are still resolved. Figure 2.17(b) is in fact the schematic representation of the Rayleigh's criterion. Fig. 2.17(c) shows the situation when the two wavelengths come even closer. The resultant intensity in this case does not show any dip and gives the impression of diffraction pattern due to a single source. Thus, the two wavelengths in Fig. 2.17(c) are not resolved.

2.12.1 Resolving Power of Grating

The ability of a grating to form separate diffraction maxima for two very close wavelengths and thereby to resolve the two wavelengths is called the resolving power of the grating. Suppose λ and $\lambda + d\lambda$ represent the wavelengths of two coherent sources. The principal maximum for wavelength λ in the direction θ_n for the nth-order diffraction satisfies the following relation [see Eq. (2.150)]:

$$(d_1 + d_2)\sin\theta_n = n\lambda \tag{2.167}$$

From Eq. (2.154), minima are obtained for the following condition:

$$N(d_1 + d_2)\sin\theta = m\lambda \tag{2.168}$$

where m can take all integral values expect $0, N, 2N, \ldots, mN$. The first minimum adjacent to the nth principal maximum in the direction $(\theta_n + d\theta_n)$ is then given as follows:

$$N(d_1 + d_2)\sin(\theta_n + d\theta_n) = (nN + 1)\lambda \tag{2.169}$$

The principal maximum for wavelength $(\lambda + d\lambda)$ in the direction $(\theta_n + d\theta_n)$ using Eq. (2.167) is given by the following expression:

$$(d_1 + d_2)\sin(\theta_n + d\theta_n) = n(\lambda + d\lambda) \tag{2.170}$$

Multiplying Eq. (2.170) by N, we get the following relation:

$$N(d_1 + d_2)\sin(\theta_n + d\theta_n) = nN(\lambda + d\lambda) \tag{2.171}$$

Applying the Rayleigh's criterion and using Eqs (2.169) and (2.171), we have the following expression: $(nN + 1)\lambda = nN(\lambda + d\lambda)$

which implies that $nN\lambda + \lambda = nN\lambda + nNd\lambda$

leading to $\lambda = nNd\lambda$

yielding $\dfrac{\lambda}{d\lambda} = nN \tag{2.172}$

Substitution of the expression for n from Eq. (2.167) into Eq. (2.172) results in the following expression:

$$\frac{\lambda}{d\lambda} = \frac{N(d_1 + d_2)\sin\theta_n}{\lambda}$$ (2.173)

2.12.2 Resolving Power of Telescope

A telescope is an optical device that is used to view distant objects. *The resolving power of a telescope is defined as the reciprocal of the smallest angle subtended at the objective lens by two distant objects that can just be seen as separate objects through the telescope.*

Figure 2.18 is a schematic representation of formation of diffraction pattern through a telescope. The diameter of the objective lens is assumed to be d. Let us assume that two neighbouring distant objects subtend at an angle $d\theta$ at the objective of the telescope. The lens of the objective and the ring supporting it act like a circular aperture and produce a Fraunhofer diffraction pattern in the focal plane of the objective. In Fig. 2.18, a screen has been assumed to be placed at the focal plane of the objective lens.

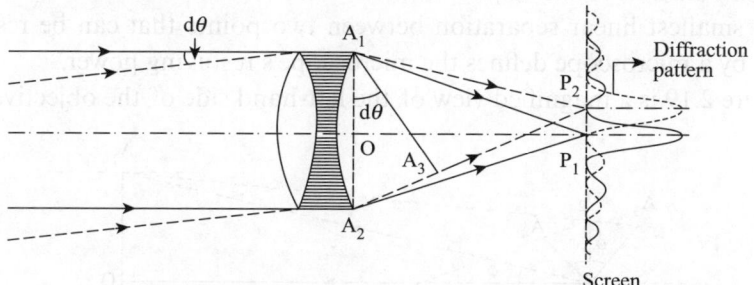

Fig. 2.18 Resolving power of telescope

P_1 and P_2 represent the position of central maxima in the images of the two objects. The secondary waves travelling in the directions A_1P_2 and A_2P_2 meet at P_2 and have a path difference PD, expressed as follows:

$$PD = (A_2P_2 - A_1P_2) = A_2A_3$$

From Fig. 2.18, we get the following relation:

$$A_2A_3 = A_1A_2 \sin(d\theta) = A_1A_2\, d\theta \text{ (for small angles)}$$ (2.174)

A_1A_2 is the diameter of the objective lens; therefore, Eq. (2.174) can be rewritten as follows:

$$A_2A_3 = d(d\theta)$$ (2.175)

From Eqs (2.134) and (2.175), we know that $d(d\theta) = \lambda$ implies that position P_2 corresponds to the first minimum of the image formed at the first source. In Fig. 2.18, P_2 is also the position of the central maxima for the image formed at the second source. This is Rayleigh's condition for resolution of two neighbouring distant sources; the condition is fulfilled if

$$d(\mathrm{d}\theta) = \lambda \quad \text{or} \quad \mathrm{d}\theta = \frac{\lambda}{d} \tag{2.176}$$

Equation (2.176) is valid for a rectangular aperture.

According to Airy, for the case of a circular aperture, Eq. (2.176) gets modified to the following form:

$$\mathrm{d}\theta = \frac{1.22\lambda}{d} \tag{2.177}$$

Resolving powers of a telescope is then given by the following expression:

$$\frac{1}{\mathrm{d}\theta} = \frac{d}{1.22\lambda} \tag{2.178}$$

Thus, a large-diameter objective results in high resolving powers.

2.12.3 Resolving Power of Microscope

A microscope is used to magnify an object to see its finer details. *The ability of a microscope to form distinctly separate images of two neighbouring small objects is called its resolving power.*

The smallest linear separation between two points that can be resolved clearly by a microscope defines the microscope's resolving power.

Figure 2.19 is a magnified view of the left-hand side of the objective lens.

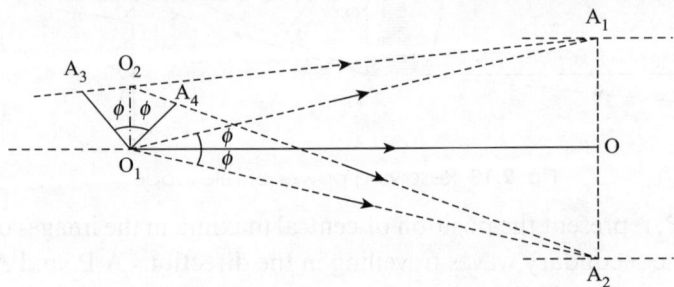

Fig. 2.19 Magnified view of image formation by objective lens

From Fig. 2.19, we have the following relation:

$$PD = (O_2 A_2 + A_2 O_1) - (O_2 A_1 + A_1 O_1) \tag{2.179}$$

Since $A_2 O_1 = A_1 O_1$, Eq. (2.179) can be rewritten in the following form:

$$PD = O_2 A_2 - O_2 A_1 \tag{2.180}$$

In Fig. 2.19, $O_1 A_3$ is a perpendicular drawn on $A_3 A_1$ and $O_1 A_4$ is a perpendicular drawn on $O_2 A_2$; therefore,

$$O_2 A_2 - O_2 A_1 = (O_2 A_4 + A_4 A_2) - (A_3 A_1 - A_3 O_2) \tag{2.181}$$

Since $A_4 A_2 = O_1 A_2 = O_1 A_1 = A_3 A_1,$

Equation (2.181) can be rewritten in the following form:
$$O_2A_2 = O_2A_1 = O_2A_4 + A_3O_2 \qquad (2.182)$$
Using Eq. (2.180) in Eq. (2.182), we obtain the following relation:
$$PD = O_2A_4 + A_3O_2 = S_1\sin\phi + S_1\sin\phi = 2S_1\sin\phi \qquad (2.183)$$
According to Rayleigh's criterion, the path difference needs to be equal to λ for the central maxima of the image of the first object to coincide with the first minima of the second object for the two objects to be resolved by the objective of the microscope. The smaller this linear separation, the higher the resolving power of the microscope. Figure 2.20 is a schematic representation of the image formation of two near objects by a microscope.

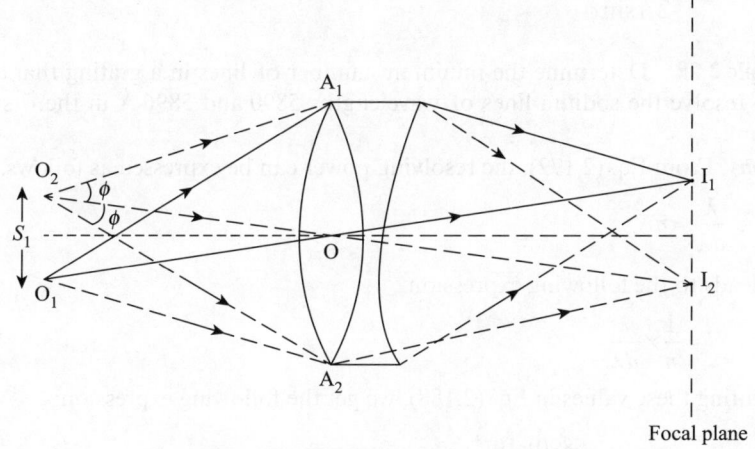

Fig. 2.20 Image formation by microscope

A_1A_2 represents the aperture of the objective of the microscope. O_1 and O_2 represent two closely placed self-luminous point objects that are separated by a linear distance S_1. The periphery of the objective acts as a circular aperture, and the sources O_1 and O_2 produce diffraction patterns at the focal length of the objective lens. The diffraction pattern consists of a central bright disc surrounded by alternate dark and bright rings. The images I_1 and I_2 shown in Fig. 2.20 represent the central maxima of the respective diffraction patterns.

According to the Rayleigh's criterion, the two objects would be fully resolved if the central maxima of one of the diffraction patterns coincide with the first minima of the other. This will happen if the path difference between the $O_2A_2I_1$ extreme rays and $O_2A_1I_2$ equals λ. Thus, using Eq. (2.183), we get the following expression:
$$2S_1\sin\phi = \lambda \qquad (2.184)$$
which yields the following relation:
$$S_1 = \frac{\lambda}{2\sin\phi} \qquad (2.185)$$

first minima of the other. This will happen if the path difference between the $O_2A_2I_1$ extreme rays and $O_2A_1I_2$ equals λ. Thus, using Eq. (2.183), we get the following expression:

$$2S_1 \sin\phi = \lambda \tag{2.184}$$

which yields the following relation:

$$S_1 = \frac{\lambda}{2\sin\phi} \tag{2.185}$$

A microscopes with a high resolving power uses objectives immersed in an oil of refractive index μ. For this situation, Eq. (2.185) gets modified to the following form:

$$S_1 = \frac{\lambda}{2\mu\sin\phi} \tag{2.186}$$

Example 2.28 Determine the minimum number of lines in a grating that are just able to resolve the sodium lines of wavelengths 5890 and 5896 Å in the first-order spectrum.

Solution From Eq. (2.172), the resolving power can be expressed as follows:

$$\frac{\lambda}{d\lambda} = nN \tag{2.187}$$

which leads to the following expression:

$$N = \frac{1}{n} \times \frac{\lambda}{d\lambda} \tag{2.188}$$

Substituting these values in Eq. (2.188), we get the following expression:

$$N = \frac{5890 \times 10^{-8}}{1 \times (5896 \times 10^{-8} - 5890 \times 10^{-8})}$$

which leads to the following result: $N = \dfrac{5890 \times 10^{-8}}{1 \times 6 \times 10^{-8}} \cong 982$

IMPORTANT CONCEPTS

1. According to Huygen's first postulate, each point on a wavefront (primary wave-front) acts as a source of new disturbances, called *secondary wavelets*, which travel in all directions with the velocity of light in that medium.
2. According to Huygen's second postulate, the surface touching these secondary wavelets tangentially in the forward direction at any instant gives a new wavefront, called the *secondary wavefront at that instant*.
3. *Diffraction of light* is the phenomenon of bending of light waves around the corners of an obstacle or an aperture placed in its path, and its spreading into the region of the geometrical shadow.
4. Diffraction is the result of interference of waves from different parts of the same wavefront.
5. Amplitude of disturbance at a point Q due to a particular zone is
   (a) directly proportional to the area of the zone, and

point sources are resolvable by an optical instrument if the central maximum of the diffraction pattern of one of them falls over the first minimum of the diffraction pattern of the other.

10. The resolving power of a telescope is defined as the reciprocal of the smallest angle subtended at the objective lens by the two distant objects that can just be seen as separate objects through the telescope.

11. The ability of a microscope to form distinctly separate images of two neighbouring small objects is called its resolving power.

APPLICATIONS

1. Diffraction gratings are capable of breaking white light into its constituent colours. Prisms also carry out the same task. Unlike in prisms, however, in a diffraction grating, deflection of any specific colour is proportional to its wavelength. The spectra produced by gratings are, therefore, easier to calibrate compared to that produced by prisms.

2. Diffractive optics is also used in holography for reconstructing three-dimensional images of objects using laser light.

IMPORTANT FORMULAE

1. For a zone plate,

$$\frac{1}{a_2} - \frac{1}{a_1} = \frac{n\lambda}{r_n^2} = \frac{1}{f_n}$$

For diffraction at a straight edge, the condition for a point to be of maximum intensity is as follows:

$$x_n = \left[\frac{(2n-1)(a_1 + a_2)a_2\lambda}{a_1} \right]^{1/2}$$

and the condition for achieving minima is as follows:

$$x_n = \left[2n\frac{(a_1 + a_2)a_2\lambda}{a_1} \right]^{1/2}$$

2. The resultant of multiple SHMs is given by the following relation:

$$R = \frac{A\sin\dfrac{n\delta}{2}}{\sin\dfrac{\delta}{2}}$$

3. For Fraunhofer diffraction at a single slit:

$$d \sin \theta = \pm m\lambda \ (m = 1, 2, 3 \ldots)$$

which leads to minima.

For Fraunhofer diffraction at a circular aperture:

$$d \sin \theta = m\lambda \text{ (minima)}$$

$$d \sin \theta = (2n+1)\frac{\lambda}{2}\text{(maxima)}$$

4. The radius, y, of the Airy's disc is given by the following relation:

$$y = 1.22\frac{f\lambda}{d}$$

5. For a plane diffraction grating:

$$I_T = RT^2 = \left(\frac{A_T \sin\phi}{\phi} \right)^2 \left(\frac{\sin N x}{\sin x} \right)^2$$

6. The dispersive power $\dfrac{d\theta}{d\lambda}$ of a plane diffraction grating is given by the following expression:

$$\frac{d\theta}{d\lambda} = \frac{n}{(d_1 + d_2)\cos\theta}$$

7. The resolving power of a grating is written as follows:

$$\frac{\lambda}{d\lambda} = nN = N\frac{(d_1 + d_2)\sin\theta_n}{\lambda}$$

8. The resolving power of a telescope is expressed as follows:

$$\frac{1}{d\theta} = \frac{d}{1.22\lambda}$$

9. For a microscope, the minimum resolvable linear distance between two self-luminous point objects is given by the following relation:

$$S_1 = \lambda/2\sin\phi$$

SELF-ASSESSMENT

Multiple-choice Questions

2.1 A ray of light is a straight line that is
  (a) parallel to the wavefront
  (b) aligned at 45° to the wavefront
  (c) normal to the wavefront
  (d) aligned at 120° to the wavefront

2.2 According to Stoke's law, intensity of spherical wavelets varies as
  (a) $\cos\theta$
  (b) $\sin\theta$
  (c) $\left(\dfrac{1+\sin\theta}{2}\right)$
  (d) $\left(\dfrac{1+\cos\theta}{2}\right)$

2.3 The condition for achieving very significant diffraction is
  (a) $\dfrac{\lambda}{a} = 1$
  (b) $\dfrac{\lambda}{a} < 1$
  (c) $\dfrac{\lambda}{a} \ll 1$
  (d) $\dfrac{\lambda}{a} = \infty$

2.4 Fresnel diffraction takes place when the
  (a) source is at a finite distance, but the screen is at an infinite distance from the aperture
  (b) source is at an infinite distance, but the screen is at a finite distance from the aperture
  (c) source and the screen are at finite distances from the aperture
  (d) source and the screen are at infinite distances from the aperture

2.5 Interference fringes have
  (a) equal width
  (b) unequal width
  (c) irregular width
  (d) infinite width

2.6 The mean phase for the secondary wavelets originating from the second half-period zone is
  (a) $\dfrac{\pi}{2}$
  (b) π
  (c) 2π
  (d) $\dfrac{3\pi}{2}$

2.7 For a zone plate
  (a) $fn = n\lambda$
  (b) $fn = \dfrac{1}{n\lambda}$
  (c) $fn = \dfrac{r_n}{n_\lambda}$
  (d) $fn = \dfrac{r_n^2}{n_\lambda}$

2.8 The condition for obtaining minima in Fraunhofer diffraction patterns due to a single slit is
  (a) $d\cos\theta = \dfrac{1}{\lambda}$
  (b) $d\sin\theta = \dfrac{1}{\lambda}$
  (c) $d\sin\theta = \pm m\lambda$
  (d) $d\sin\theta = \dfrac{1}{\pm m\lambda}$

2.9 The condition for obtaining maxima in Fraunhofer diffraction patterns due to a circular aperture is

(a) $d \sin \theta = n\lambda$

(c) $d \sin \theta = (n+1)\lambda$

(b) $d \sin \theta = (2n+1)\dfrac{\lambda}{2}$

(d) $d \sin \theta = (n+1)\dfrac{\lambda}{2}$

2.10 The principal maxima for a diffraction grating are given by the condition

(a) $(d_1 + d_2)\sin\theta = \pm\left(n + \dfrac{1}{2}\right)\lambda$

(c) $(d_1 + d_2)\sin\theta = \pm n\lambda$

(b) $(d_1 + d_2)\sin\theta = \pm\left(2n + \dfrac{1}{2}\right)\lambda$

(d) $(d_1 + d_2)\sin\theta = \pm\left(n + \dfrac{3}{2}\right)\lambda$

2.11 Bending of light around the edge of an obstacle is called

(a) diffraction   (b) dispersion   (c) diffusion   (d) drift

2.12 Secondary wavelets travel with a speed that is

(a) $<c$ (b) $= c$ (c) $>c$ (d) $\geq c$

2.13 Action of the secondary wavelets is confined to

(a) backward direction

(b) both backward and forward directions

(c) forward direction

(d) in all directions

2.14 Huygen's conception of secondary waves

(a) helps us to find the focal length of a thick lens

(b) is a geometrical method to find a wavefront

(c) is used to determine the velocity of light

(d) is used to explain polarization of light

2.15 A diffraction pattern is obtained using a beam of red light. If the red light is replaced by a blue light, then

(a) the diffraction pattern remains unchanged

(b) diffraction bands become narrow and crowded together

(c) bands become broader and farther apart

(d) bands disappear

2.16 The aperture of the human eye is 2 mm. Assuming the mean wavelength of light to be 5000 Å, the angular resolution limit of the eye is nearly

(a) 2 min   (b) 0.5 min   (c) 1 min   (d) 1.5 min

2.17 Diffraction pattern of a single slit consists of a central bright band, which is

(a) wide, and is flanked by alternate dark and bright bands of decreasing intensity

(b) narrow, and is flanked by alternate dark and bright bonds of equal intensity

(c) wide, and is flanked by alternate dark and bright bands of equal intensity

(d) narrow, and is flanked by alternate dark and bright bands of decreasing intensity

2.18 In Huygen's wave theory, the locus of all the points in the same state of vibration is called a

(a) half-period zone

(c) wavefront

(b) vibrator

(d) ray

2.19 A high-resolving-power telescope should use an

(a) objective of large focal length

(c) eyepiece of small focal length

(b) eyepiece of large focal length

(d) objective of large aperture

2.20 For a microscope, the minimum resolvable linear distance between two self-luminous point objects is given by

(a) $\dfrac{\lambda}{\sin\phi}$ (b) $\dfrac{\lambda}{3\sin\phi}$ (c) $\dfrac{\lambda}{2\sin\phi}$ (d) $\dfrac{2\lambda}{\sin\phi}$

2.21 If the objective of a microscope is immersed in oil, its resolving power
    (a) reduces   (c) increases
    (b) remains the same   (d) can increase or decrease

2.22 The resolving power of a microscope is
    (a) directly proportional to λ (c) directly proportional to λ^2
    (b) inversely proportional to λ (d) inversely proportional to λ^2

2.23 For a rectangular aperture, Rayleigh's condition leads to

(a) $d\theta = \dfrac{\lambda}{d}$ (b) $d\theta = \dfrac{d}{\lambda}$ (c) $d\theta = \dfrac{2\lambda}{d}$ (d) $d\theta = \dfrac{\lambda}{2d}$

2.24 For a circular aperture, Rayleigh's condition leads to

(a) $d\theta = \dfrac{1.5\lambda}{d}$ (b) $d\theta = \dfrac{1.3\lambda}{d}$ (c) $d\theta = \dfrac{1.22\lambda}{d}$ (d) $d\theta = \dfrac{1.32\lambda}{d}$

2.25 The resolving power of a telescope is given by

(a) $\dfrac{1.22\lambda}{d}$ (b) $\dfrac{1.22d}{\lambda}$ (c) $\dfrac{d}{1.22\lambda}$ (d) $\dfrac{1.22\lambda}{d}$

2.26 The dispersive power of a grating is
    (a) inversely proportional to the order of the diffraction pattern
    (b) directly proportional to the order of the diffraction pattern
    (c) inversely proportional to the square of the order of the diffraction pattern
    (d) directly proportional to the square of the order of the diffraction pattern

2.27 Dispersive power of a grating is
    (a) directly proportional to the grating element
    (b) not dependent upon the grating element
    (c) inversely proportional to the grating element
    (d) directly proportional to the square of the grating element

2.28 For the diffraction pattern obtained using a circular aperture, the central disc is
    (a) dark   (c) may be dark or bright
    (b) bright   (d) coloured

2.29 For Fraunhofer diffraction at a single slit, the intensity is proportional to

(a) $\left(\dfrac{\sin\phi}{\phi}\right)^2$ (b) $\dfrac{\sin\phi}{\phi}$ (c) $\dfrac{\phi}{\sin\phi}$ (d) $\left(\dfrac{\phi}{\sin\phi}\right)^2$

2.30 The equivalent focal length of a zone plate is
    (a) directly proportional to λ
    (b) inversely proportional to λ
    (c) directly proportional to square of λ
    (d) inversely proportional to square of λ

2.31 According to the corpuscular theory of light, if an object is placed in the path of light, then
    (a) no shadow is expected   (c) a sharp shadow is expected
    (b) light will bend   (d) multiple shadows are expected

2.32 Short waves used in a VHF radio
    (a) diffract significantly around natural barriers
    (b) do not diffract significantly around natural barriers

(c) sometimes diffract significantly around natural barriers

(d) produce shadow of natural barriers

2.33 In Fresnel diffraction, the incident and diffracted wavefronts can be considered

    (a) spherical   (b) plane   (c) elliptical   (d) circular

2.34 In Fraunhofer diffraction, the incident and diffracted wavefronts can be considered

    (a) spherical   (b) plane   (c) elliptical   (d) circular

2.35 In diffraction fringes, the central maximum is

    (a) equal in width to that of other fringes

    (b) half the width of other fringes

    (c) twice the width of other fringes

    (d) thrice the width of other fringes

Review Questions

2.1 Give the two postulates of Huygens's principle.

2.2 Differentiate between Fresnel diffraction and Fraunhofer diffraction.

2.3 List five differences between interference and diffraction fringes.

2.4 What are half-period zones?

2.5 Show that the amplitude due to the nth zone is proportional to $\pi\lambda f(\theta_n)$.

2.6 What is a zone plate?

2.7 Show that for a zone plate

$$\frac{1}{a_2} - \frac{1}{a_1} = \frac{m\lambda}{r_n 2}$$

where a_1 and a_2 represent, respectively, the distances of the source and the screen from the zone plate.

2.8 Describe the diffraction pattern obtained when light encounters a straight edge.

2.9 Derive an expression for the condition for obtaining minima for diffraction at a straight edge.

2.10 Show that the resultant of multiple SHM vibrations occurring simultaneously is given by the following expression:

$$R = A\frac{\sin\dfrac{n\delta}{2}}{\sin\dfrac{\delta}{2}}$$

2.11 Derive the condition for obtaining minima for the diffraction pattern due to a single slit.

2.12 Derive an expression for the radius of Airy's disc in the diffraction pattern due to a circular aperture.

2.13 Show that for a plane diffraction grating:

$$I_T = \left(\frac{A_T \sin\theta}{\phi}\right)\left(\frac{\sin Nx}{\sin x}\right)$$

2.14 Derive the condition for obtaining the principal maxima in a plane diffraction grating.

2.15 Derive an expression for the dispersive power of a plane diffraction grating.

2.16 State and explain Rayleigh's criterion for resolution.

2.17 Derive an expression for the resolving power of a plane diffraction grating.

2.18 Derive an expression for the resolving power of a telescope.

2.19   Derive an expression for the resolving power of a microscope.

2.20   Distinguish between X-ray diffraction (XRD) and light-ray diffraction.

2.21   What is meant by diffraction grating? How is it useful for the determination of wavelength of a monochromatic resource?

2.22   What are the types of diffraction? Differentiate between them.

2.23   What is a grating element?

2.24   What do you mean by diffraction? State the different types of diffraction.

2.25   What is diffraction grating? What is the advantage of increasing the number of lines in a grating?

2.26   Explain the experimental method of determination of the wavelength of a spectral line using a diffraction grating.

2.27   Define diffraction of light. Why is it not evident in daily experience?

2.28   Describe the construction of a diffraction grating.

2.29   What is Rayleigh's criterion of resolution for diffraction? Write an expression for the resolving power of a grating.

2.30   State the two types of diffraction and differentiate between them.

2.31   Explain how the number of lines ruled per centimetre on a plane transmission grating decides the maximum number of orders of diffraction.

2.32   Explain Fraunhofer diffraction at a single slit.

2.33   Obtain the condition for obtaining the principal maxima and minima for Fraunhofer diffraction at a single slit.

2.34   Draw the intensity distribution curve for Fraunhofer diffraction at a single slit.

2.35   What are the dissimilarities between a zone plate and a convex lens?

2.36   In a grating, the opaque space is three times the slit width. Which order will be absent in the grating spectrum?

2.37   Write the similarities between a zone plate and a converging lens.

2.38   What is the resolving power of a grating having N number of rulings, in the nth order?

2.39   What happens to the width of the central maximum in a single-slit Fraunhofer diffraction pattern when the slit width is increased?

2.40   Write down the expression for intensity distribution of a single-slit diffraction pattern.

2.41   How does the intensity of the single-slit diffraction pattern vary with the angle of diffraction?

2.42   Find out the condition that θ (angle of diffraction) needs to satisfy in order to be the position of a secondary maximum.

2.43   What is the condition for obtaining the first minimum in the diffraction pattern due to a single slit?

2.44   How does the resolving power of a grating depend upon the total number of rulings on the grating?

2.45   Under what situations are certain orders in optical gratings missing?

2.46   Describe Rayleigh's criterion of resolution.

2.47   Derive an expression for the intensity distribution of Fraunhofer-type diffraction pattern in case of a single slit.

2.48   What do you understand by a missing order spectrum?

2.49   Prove that the intensity of the secondary maxima formed for Fraunhofer diffraction at a single slit is of decreasing order.

2.50   Write the basic difference between Fresnel- and Fraunhofer-type diffraction.

2.51   A diffraction pattern is observed using a beam of red light. What happens if the red light is replaced by blue light?

2.52 What is the difference between interference and diffraction?

2.53 What are the factors on which the intensity of a point due to a Fresnel half-period zone depends?

Numerical Problems

2.1 The first ring of a zone plate has a diameter of 1.1 mm. Plane waves of wavelength 5100 Å fall on the plate. Find the location of the screen such that light is focused on the brightest spot. $\left[Hint:\ fn = \dfrac{r_n^2}{nl} \right]$

2.2 The diameter of the first ring of the zone plate in Problem 2.1 is changed to 0.9 mm. Calculate the shift in the location of the screen to obtain a focused bright spot on the screen. $\left[Hint:\ fn = \dfrac{r_n^2}{n\lambda} \right]$

2.3 A zone plate is being used with a point source of light. Along the axis, the strongest and the second strongest images are formed at distances of 21 and 3 cm from the zone plate, respectively. Both the images are formed on the same side of the source. Calculate the distance of the source from the zone plate. Assume that $\lambda = 6.0 \times 10^{-5}$ cm $\left[Hint:\ \dfrac{1}{v} - \dfrac{1}{u} = \dfrac{n\lambda}{r^2} \right]$

2.4 How much does the distance of the source change if the wavelength of the source is reduced to $\lambda = 5.8 \times 10^{-5}$ cm

2.5 A narrow slit gets illuminated by a light of wavelength 5800 Å. The slit is placed at a distance of 20 cm from the straight edge. Measurements are carried out at a distance of 180 cm from the straight edge. Calculate the distance between the first and second dark bands. $\left[Hint:\ x_n = \left[\dfrac{2n(a_1 + a_2)a_2\lambda}{a_1} \right]^{\frac{1}{2}} \right]$

2.6 Light of wavelength 6×10^{-5} cm falls on a narrow slit. The pattern formed is observed on a screen placed at a distance of 90 cm from the slit. Calculate the width of the slit if the first minima lies 0.09 mm on the other side of the central maximum.

$\left[Hint:\ \sin\theta n = \dfrac{\text{Distance of first minima from the central maxima}}{\text{Distance of the screen from the slit}} \right]$

2.7 A slit of width 6×10^{-4} cm is illuminated with a light of wavelength 6000 Å. Calculate the angular separation between the first-order minima on either side of the central maximum. $\left[Hint:\ d\sin\theta = \dfrac{\lambda}{d} \right]$

2.8 Fraunhofer diffraction pattern is obtained with a slit of width 0.3 mm and a light of wavelength $\lambda = 5860$ Å. Calculate the angles at which the first dark band and the next bright band are formed. [*Hint: $d \sin\theta = m\lambda$*]

2.9 Fraunhofer diffraction pattern due to a circular aperture is obtained on a screen placed at a distance of 100 cm from the lens. The light used has a wavelength of 5830 Å, and the diameter of the aperture is 0.09 mm. Calculate the separation between the central disc and the first secondary minimum. $\left[Hint:\ y = \dfrac{f\lambda}{d} \right]$

2.10   A parallel beam of light is incident normally on a plane grating having 3900 lines/cm. The second-order spectral line is found to be deviated through 30°. Calculate the wavelength of the spectral line.   [*Hint*: $(d_1 + d_2)\sin\theta = n\lambda$]

2.11   A monochromatic light of wavelength 6100×10^{-8} cm is incident normally on a 1.8 cm wide grating. The first-order spectrum is produced at an angle of 18° with respect to the normal. Calculate the total number of lines on the grating.
[*Hint*: $(d_1 + d_2)\sin\theta = n\lambda$]

2.12   Calculate the minimum number of lines in a grating that is just able to resolve two sources of light of wavelengths 6000 and 6010 Å. $\left[Hint:\ \dfrac{\lambda}{d\lambda} = nN \right]$

2.13   Fraunhofer diffraction pattern is obtained using a light of wavelength 6000 Å and a single slit. When the source of light is changed to another wavelength, the angular width of the central maxima decreases by 20%. Calculate the new wavelength. $\left[Hint:\ \text{Angular width of the central maximum } \omega = \dfrac{2\lambda}{d} \right]$

2.14   Fraunhofer diffraction pattern is obtained using light of wavelength 6000 Å and a single slit. The source wavelength is changed to 5000 Å. Calculate the percentage reduction in the angular width of the central maxima. $\left[Hint:\ \dfrac{\omega'}{\omega} = \dfrac{\lambda'}{\lambda} \right]$

2.15   Fraunhofer diffraction pattern is obtained using a single slit and light of wavelength 6000 Å. When the whole apparatus is immersed in a liquid, the angular width of the central maxima reduces by 20%. Calculate the refractive index of the liquid. $\left[Hint:\ n = \dfrac{\lambda}{\lambda_{medium}} \right]$

2.16   Fraunhofer diffraction pattern is obtained using a single slit and light of wavelength 6000 Å. The whole apparatus is then immersed in a liquid of refractive index 1.3. Calculate the reduction in the width of the central maxima. $\left[Hint:\ n = \dfrac{\lambda}{\lambda_{medium}} \right]$

2.17   The aperture of the human eye is 2 mm. Calculate the angular resolution of the eye for an incident wavelength of 6000 Å. $\left[Hint:\ \text{Angular resolution} = \dfrac{1.22\lambda}{d} \right]$

Polarization

Learning Objectives

After studying this chapter, students will be able to

- understand the concept of polarization
- explain the process of polarization by reflection
- state Malus law
- demonstrate the process of polarization by scattering
- explain the concept of birefringence
- describe the construction, principle, and working of Nicol prism
- state the working of quarter- and half-wave plates
- realize the concept of optical activity
- comprehend the meaning of specific rotation
- describe the Fresnel's theory of optical rotation
- demonstrate the working of different types of polarimeters
- explain the concept of photoelasticity
- elucidate the working of photoelastic bench
- solve numerical problems based on the angle of polarization, Malus law, Nicol prism, o-ray, e-ray, specific rotation, and different types of polarized light

List of Symbols

A = Amplitude	I = Intensity	v = Velocity
ω = Angular frequency	C_1 = Critical angle	σ = Stress
μ = Refractive index	δ = Phase difference	c = Velocity of light
p = Angle of polarization	t_1 = Thickness	Δ = Phase shift
	S = Specific rotation	E = Electric field
r = Angle of refraction	C = Concentration	

3.1 INTRODUCTION

The phenomena of interference and diffraction establish the wave nature of light. A wave motion can be either longitudinal or transverse. In a longitudinal wave motion, the disturbance produced is along the direction of propagation, whereas in a transverse wave motion, the disturbance is perpendicular to the direction of propagation. Ordinary light is transverse in nature. Many physical phenomena depend on the transverse nature of light. In fact, the existence of this phenomenon is proof of the transverse nature of light. One such phenomenon is polarization. In this chapter, we would study the phenomenon of polarization in detail, including an understanding of the phenomenon, methods of producing polarized light, and some uses and applications of this interesting phenomenon.

3.2 PHENOMENON OF POLARIZATION

Light undergoes some changes in characteristics as it passes through some materials like crystalline tourmaline. Suppose light is made to pass through a pair of tourmaline crystals C_1 and C_2 that are kept perpendicular to the direction of propagation of light. If we fix the position of C_1 but rotate crystal C_2, the intensity of light coming out of the second crystal exhibits a systematic pattern. When the planes of both the crystals are held at right angles to the direction of propagation, the output intensity is maximum. As crystal C_2 is rotated, a situation comes when the plane of crystal C_2 becomes perpendicular to that of crystal C_1. In this situation, the intensity of the outgoing light reaches its minimum value. Ordinary light consists of disturbance or vibration in a plane perpendicular to the direction of propagation. Such light is also called *unpolarized light*. The tourmaline crystal only allows vibrations along its plane to pass through. The output from C_1 thus has vibrations only along the plane of the crystal. Such light is called polarized light. Figure 3.1 is a schematic representation of the process of polarization for two different orientations of crystal C_2.

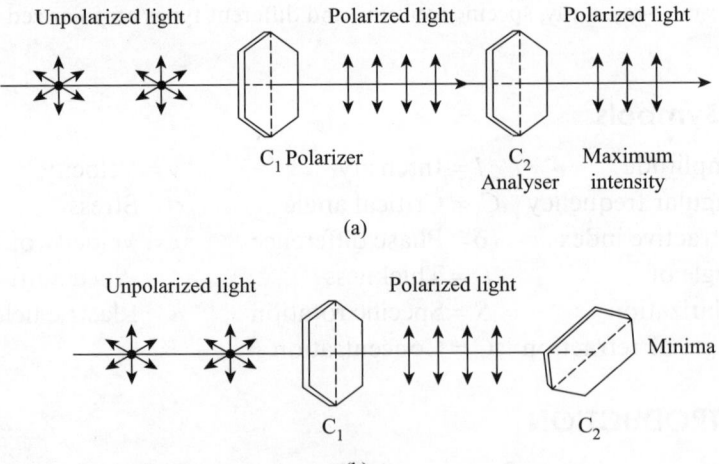

Fig. 3.1 Two important polarizer–analyser positions (a) Maximum intensity condition (b) Minimum intensity condition

Figure 3.1(a) represents the situation when the planes of crystals C_2 and C_1 are parallel to each other. The light coming out of C_1 passes through C_2 without restriction, and the output intensity is maximum.

The phenomenon of polarization also reveals that light shows a transverse wave motion. Due to the phenomenon of polarization, light is symmetric with respect to the direction of propagation.

3.2.1 Types of Polarization

As discussed in Section 3.2, polarization takes place due to the transverse nature of light waves. Light is a form of electromagnetic wave, which consists of sinusoidally oscillating electric and magnetic fields, perpendicular to each other and also perpendicular to the direction of propagation. Although electric and magnetic fields can be used to indicate the state of polarization, traditionally, the orientation of the electric field is accepted as the direction of polarization.

3.2.1.1 Unpolarized light

Ordinary light consists of an electric field oscillating in all planes, as shown in Fig. 3.2. Such light is called unpolarized light.

3.2.1.2 Plane Polarized Light

Light that consists of an electric field oscillating in only one direction is called plane polarized light. Such polarized light is shown schematically in Fig. 3.3.

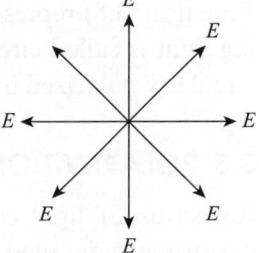

Fig. 3.2 Electric field oscillating in all planes for unpolarized light

3.2.1.3 Elliptically Polarized Light

Suppose two waves are perpendicular to each other and have a phase difference of $\pi/2$ simultaneously. These waves (electric fields in the case of light) can be represented in the following form:

$$x = A_1 \sin\left(wt + \pi/2\right) \qquad (3.1)$$

and

$$y = A_2 \sin\left(wt\right) \qquad (3.2)$$

Equations (3.1) and (3.2) can be rewritten in the following forms:

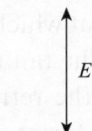

Fig. 3.3 Plane polarized light

$$\frac{x}{A_1} = \cos wt \qquad (3.3)$$

and

$$\frac{y}{A_2} = \sin wt \qquad (3.4)$$

Equations (3.3) and (3.4) can be combined to yield the following relation:

$$\frac{x^2}{A_1^2} + \frac{y^2}{A_2^2} = 1 \qquad (3.5)$$

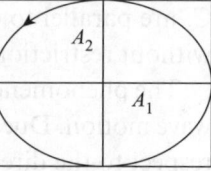

Equation (3.5) represents a symmetrical ellipse. Thus, the electric field traces an ellipse and the light is said to be elliptically polarized. A schematic of elliptically polarized light is shown in Fig. 3.4.

Fig. 3.4 Elliptically polarized light

3.2.1.4 Circularly Polarized Light

If two electric fields, assumed to be having a phase difference of $\pi/2$ and perpendicular to each other, have the same amplitude, then we can write the following equations:

$$x = A\sin\left(wt + \pi/2\right) \qquad (3.6)$$

and

$$y = A\sin\left(wt\right) \qquad (3.7)$$

Combining Eqs (3.6) and (3.7), we can write the following relation:

$$x^2 + y^2 = A^2 \qquad (3.8)$$

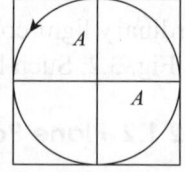

Equation (3.8) represents a circle, and the corresponding light is called circularly polarized. A schematic of circularly polarized light is shown in Fig. 3.5.

Fig. 3.5 Circularly polarized light

3.3 POLARIZATION BY REFLECTION

Reflection of light can produce polarization. In the year 1812, Brewster performed a detailed study on polarization due to reflection using different interfaces. His study revealed that at a particular value of angle of incidence, the reflected light gets completely polarized in a plane that is perpendicular to the plane of incidence. This angle of incidence is dependent upon the interface at which the reflection takes place and is called the *angle of polarization*. The tan of the angle of polarization was found to be numerically equal to the refractive index μ; the relationship can be expressed in the following form:

$$\mu = \tan p \qquad (3.9)$$

where p represents the angle of polarization. Equation (3.9) is also known as Brewster's law.

For the air–glass interface, the polarizing angle is 57°. The process of polarization at the air–glass interface is shown in Fig. 3.6.

The unpolarized incident light is represented by dots and arrows, indicating electric field components that are parallel and perpendicular to the plane

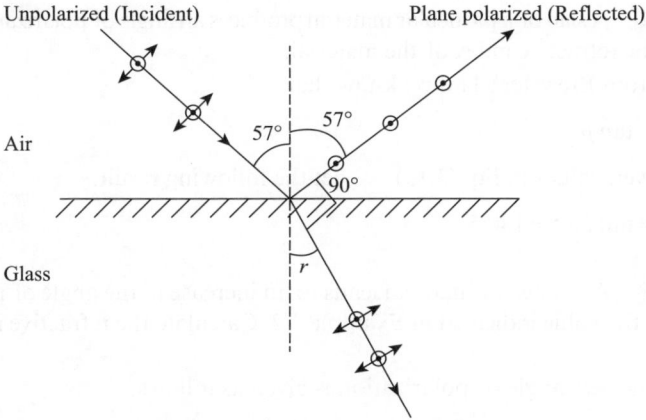

Fig. 3.6 Polarization at air–glass interface

of paper. A portion of the incident light gets reflected and another portion gets refracted. From Brewster's law, we know that

$$\mu = \tan p \qquad (3.10)$$

If r is the angle of refraction, using Snell's law, we can write the following equation:

$$\mu = \frac{\sin p}{\sin r} \qquad (3.11)$$

Equating Eqs (3.10) and (3.11), we get the following relation:

$$\tan p = \frac{\sin p}{\sin r}$$

which can be rewritten in the following form:

$$\frac{\sin p}{\cos p} = \frac{\sin p}{\sin r}$$

leading to $\sin r = \cos p = \sin (\pi/2 - p)$

yielding $r = \pi/2 - p$, that is, $r + p = \pi/2$

Thus, the reflected and refracted rays are perpendicular to each other, as shown in Fig. 3.6. It must be kept in mind that the refractive index of a material is wavelength dependent. Thus, any interface will have different polarizing angles for different wavelengths.

Example 3.1 A slab of a special type of glass produces an angle of polarization of 60°20′. Determine the refractive index of the glass.

Solution From Brewster's law, we know that

$$\mu = \tan p \qquad (3.12)$$

Use of the value of p in Eq. (3.12) leads to the following result:

$$\mu = \tan 60°20′ = 1.7556$$

Example 3.2   A slab of a particular material produces an angle of polarization of 55.5°. Determine the refractive index of the material.

Solution   From Brewster's law, we know that

$$\mu = \tan p \qquad (3.13)$$

Using the given values in Eq. (3.13), we get the following result:

$$\mu = \tan 55.5 = 1.455$$

Example 3.3   A change of material leads to an increase in the angle of polarization by 20% over the value indicated in Example 3.2. Calculate the refractive index of the material.

Solution   The new angle of polarization is given as follows:

$$p = 55.5 + \left(\frac{55.5 \times 20}{100}\right) = 66.6$$

Using this value in Brewster's law, we get the following result:

$$\mu = \tan 66.6 = 2.3109$$

Example 3.4   For a particular material, the refractive index is found to be 1.6826. Find the angle of polarization for a slab of this material.

Solution   From Brewster's law, we have $1.6826 = \tan p$

leading to $p = \tan^{-1} 1.6826 = 59.28°$

3.4 MALUS LAW

In the set-up shown in Fig. 3.1, crystal C_1 is designated as a polarizer and C_2 as an analyser. Suppose a plane polarized light coming out of the polarizer is incident on an analyser. Figure 3.7 is a schematic representation of that situation.

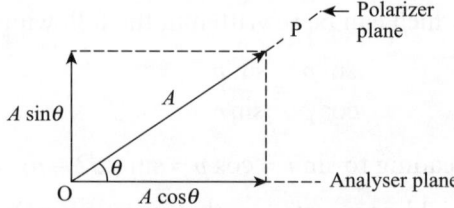

Let us also assume that the light ray can be represented by amplitude A. An angle θ is assumed to exist between the planes of the polarizer

Fig. 3.7 Schematic representation of Malus law

and the analyser. The component of amplitude along the plane of the analyser is $A \cos\theta$, and the component perpendicular to the plane of the analyser is $A \sin\theta$. Due to the property of the analyser, the component $A \cos\theta$ is transmitted through the analyser and the component $A \sin\theta$ is blocked. Intensity I_A transmitted through the analyser is given by the following equation:

$$I_A = (A\cos\theta)^2 = A^2 \cos^2\theta \qquad (3.14)$$

The intensity I_p of the incident polarized light can be written as follows:

$$I_p = (A\cos\theta)^2 + (A\sin\theta)^2 = A^2 \qquad (3.15)$$

Combining Eqs (3.14) and (3.15), one gets the following expression:

$$I_A = I_p \cos^2 \theta \qquad (3.16)$$

From Eq. (3.16), we can conclude that

$$I_A \propto \cos^2 \theta \qquad (3.17)$$

Thus, *the intensity of polarized light transmitted through an analyser is proportional to the square of the cosine of the angle between the planes of transmission of the analyser and the polarizer.* This is Malus law.

Let us look at two special cases for Eq. (3.17).

Case 1 $\theta = 0$

Planes of the polarizer and analyser are parallel in this case. From Eq. (3.16), we can write that

$$I_A = I_p \qquad (3.18)$$

Case 2 $\theta = \pi/2$

Planes of the polarizer and analyser are perpendicular in this case. Once again, using Eq. (3.16), we now have the following relation:

$$I_A = 0 \qquad (3.19)$$

Example 3.5 Two polarizing sheets are initially adjusted so that their polarizing directions are parallel. This orientation results in maximization of the transmitted light intensity. Indicate the angle through which either sheet should be turned so that the transmitted intensity becomes 1/4 of the original value.

Solution From Malus law, we know that

$$I_A = I_p \cos^2 \theta \qquad (3.20)$$

In the given situation, Eq. (3.20) can be rewritten in the following form:

$$\frac{I_p}{4} = I_p \cos^2 \theta$$

leading to $\cos^2 \theta = \dfrac{1}{4}$, yielding $\cos \theta = \pm\dfrac{1}{2}$

Thus, $\theta = \pm\, 60°,\ \pm\, 120°$

Example 3.6 Two polarizing sheets are initially adjusted so that their polarizing directions are parallel. This orientation results in the maximization of the transmitted light intensity. Indicate the angle through which either sheet should be turned so that the transmitted intensity becomes 1/3.5 of the original value.

Solution From Malus law, we know that

$$I_A = I_p \cos^2 \theta \qquad (3.21)$$

For the given situation, Eq. (3.21) can be rewritten in the following form:

$$\frac{I_p}{3.5} = I_p \cos^2 \theta$$

resulting in $\cos^2\theta = \dfrac{1}{3.5}$ that is, $\theta = \cos^{-1} 0.535$ giving $\theta = 57.7°, 122.3°$

Example 3.7 Two polarizing sheets are initially adjusted so that their polarizing directions are parallel. This orientation results in the maximization of the transmitted light intensity. One of the sheets is then turned through an angle of 42°. Find the ratio I_A/I_p.

Solution From Malus law, we know that

$$\frac{I_A}{I_p} = \cos^2\theta \tag{3.22}$$

Using the given values in Eq. (3.22), we get the following result:

$$\frac{I_A}{I_p} = \cos^2 42 = 0.55$$

Example 3.8 Two polarizing sheets are initially adjusted so that their polarizing directions are parallel. One of the sheets is then turned through an angle of 45.2°. Calculate the ratio I_p/I_A.

Solution From Malus law, we have

$$\frac{I_A}{I_p} = \cos^2 45.2$$

leading to $\dfrac{I_A}{I_p} = 0.49 \Rightarrow \dfrac{I_p}{I_A} = 2.04$

3.5 POLARIZATION BY SCATTERING

Let P be a particle of the medium whose size is of the same order as the wavelength of the light passing through the medium. In this case, particle P causes scattering of light. Let us assume that unpolarized light falls on P, along the z-axis, as shown in Fig. 3.8.

If the scattered light is viewed along the x-axis, perpendicular to the direction of incidence, only the electric field vector parallel to the y-axis will be seen (as light is a transverse wave). Therefore, light will be seen to be polarized parallel to the y-axis.

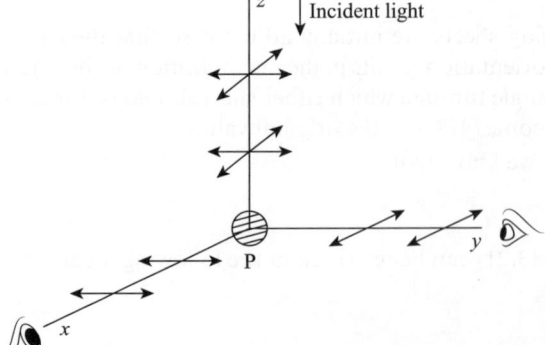

Fig. 3.8 Polarization by scattering

If the scattered light is viewed along the *y*-axis, perpendicular to the direction of incidence, only the electric field vector parallel to the *x*-axis will be seen. Therefore, light is seen to be polarized parallel to the *x*-axis.

3.6 BIREFRINGENCE

The refractive index of a medium is dependent upon the velocity of light in the medium. In many materials, the velocity of light is independent of the direction of propagation or the state of polarization. Such materials are said to be *optically isotropic*. Materials that do not display this property are said to be *optically anisotropic*. Some examples of anisotropic materials are calcite, quartz, tourmaline, etc. If a beam of unpolarized light is incident on a slab of an anisotropic material, splitting of the beam takes place at the crystal surface. This *double bending* of the beam transmitted through the anisotropic material is called double refraction.

Detailed experimentation involving anisotropic materials was carried out by *Erasmus Bartholinus* around 1669. His experiments primarily involved calcite crystals. Out of the two refracted rays, one is found to obey the *Snell's* law and is called the *ordinary ray* or *o-ray*. The other ray that does not obey ordinary laws of refraction is termed *extraordinary ray* or *e-ray*. The phenomenon is shown in Fig. 3.9.

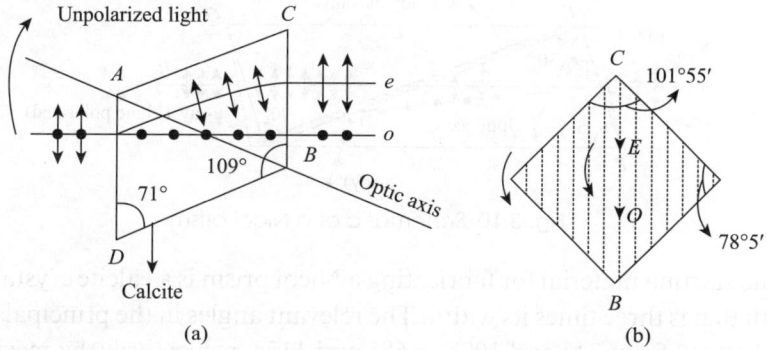

Fig. 3.9 Creation of o-ray and e-ray in calcite (a) Side view (b) Top view

Furthermore, the o-ray is found to have an electric field perpendicular to the principal section of the calcite crystal shown in Fig. 3.9. The e-ray, on the other hand, has an electric field vector along the principal section of the calcite crystal. Thus, the following are the characteristics of the two rays:

1. The o-ray travels through a crystal, with the same speed v_o in all directions. You will recollect that the index of refraction, n, is the ratio of the velocity of light in vacuum to its velocity in the medium. The o-ray, thus, has a single index of refraction, n_o.

2. The e-ray travels through the medium with different speeds in different directions. For calcite, the range of speeds is between v_o and a larger value v_e. The corresponding index of refraction varies from n_o to a smaller value n_e.

Crystals displaying double refraction can be classified into two types, namely (a) uniaxial and (b) biaxial. For uniaxial crystals, there exists only one direction (optic axis) along which the o- and e-ray travel with the same velocity. Some examples of uniaxial crystals are calcite, tourmaline, and quartz. For biaxial crystals, two such directions exist along which the velocities of o- and e-ray are the same. Some examples of biaxial crystals are topaz, aragonite, etc.

3.7 NICOL PRISM

We have seen that unpolarized light incident upon a calcite crystal results in the production of o- and e-ray. Furthermore, both the rays are plane polarized, with electric vectors mutually perpendicular to each other. If some technique is used to eliminate one of the rays, the ray emerging from the system would then be plane polarized. A Nicol prism can eliminate the o-ray by utilizing the phenomenon of total reflection at the surface of a thin film of Canada balsam that separates two pieces of calcite.

3.7.1 Construction

Figure 3.10 shows a schematic of a Nicol prism.

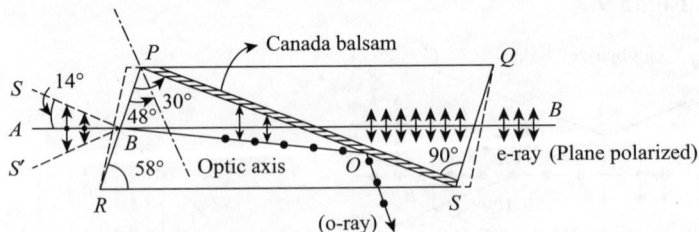

Fig. 3.10 Schematic of a Nicol prism

The starting material for fabricating a Nicol prism is a calcite crystal with a length that is three times its width. The relevant angles in the principal section are changed from 71° and 109° to 68° and 112°, respectively, by mechanical grinding of the end faces. This is shown schematically in Fig. 3.10. The ground calcite crystal is then cut into two pieces through a plane that is perpendicular to the principal section and the end faces PR and QS. After grinding and polishing the cut faces to obtain optically flat surfaces, they are joined together with Canada balsam. Canada balsam is chosen because its refractive index falls between the refractive indices for the o- and e-ray for calcite. For example, for the sodium D line, the refractive index values are as follows:

μ_o = Refractive index for the o-ray = 1.658

μ_c = Refractive index for Canada balsam = 1.55

and

μ_e = Refractive index for the e-ray = 1.486

Thus, the refractive index for o-ray with respect to that of Canada balsam is given as follows:

$$\mu_{oc} = \frac{1.658}{1.550}$$

yielding $\sin C_1 = \dfrac{1}{\mu_{oc}} = \dfrac{1.550}{1.658}$

resulting in

$$C_1 = \text{critical angle} = 69° \tag{3.23}$$

3.7.2 Principle

Ray AB in Fig. 3.10 is incident on face PR in a direction that is parallel to side PQ. A double refraction takes place at the surface of the calcite crystal, leading to the production of o- and e-ray. Both o- and e-ray are plane polarized, but in mutually perpendicular directions. The angle of incidence of the o-ray at the interface between the calcite and Canada balsam is greater than the critical angle of 69° due to the dimensions of the crystal. The o-ray, therefore, gets totally reflected and absorbed within the tube containing the Nicol prism. The e-ray, on the other hand, travels from a rarer (calcite) to a denser (Canada balsam) medium and does not suffer total internal reflection. The emerging e-ray is laterally displaced with respect to the original direction. The e-ray transmitted out of the Nicol prism is plane polarized. Thus, the Nicol prism is able to convert unpolarized light to plane polarized light. The polarizing property of a Nicol prism is dependent upon the incident angle of the o-ray on the calcite–Canada balsam interface. If this angle of incidence on the calcite–Canada balsam interface is less than 69°, no total reflection of the o-ray takes place, and this in turn leads to an unpolarized emergent ray. It can be shown that this results in the range of the angle of incidence ABS′ being limited to 14°, as indicated in Fig. 3.10. With the increase in the angle formed by the incident ray AB on the face PR of the calcite crystal, the e-ray gets more refracted towards the optic axis. It is to be remembered that along the optic axis, the o- and e-ray travel with the same velocity and, therefore, have the same refractive index. Thus, with an increase in refraction, the e-ray tends to become more and more parallel to the optic axis, and this in turn results in its refractive index becoming closer to the o-ray.

As μ_e approaches μ_o, the e-ray also gets internally reflected at the calcite–Canada balsam interface. In this situation, no light emerges from the Nicol prism.

3.7.3 Applications

A Nicol prism can be used to polarize an incident unpolarized light, as described in Section 3.7.1. A set of two Nicol prisms can be used to set up a polarizer–analyser combination. Such a set-up is shown in Fig. 3.11.

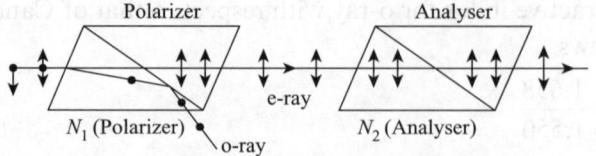

Fig. 3.11 Polarizer–Analyser combination

The first Nicol prism N_1 is used to produce polarized light, whereas the second Nicol prism N_2 is used to analyse the light emerging from N_1. If the principal sections of the two Nicol prisms are parallel to each other, the e-ray emerging out of N_1 passes un-attenuated through the Nicol prism. This is the orientation sketched in Fig. 3.11. As N_2 is rotated, the intensity of light emerging out of N_2 reduces and ultimately reaches zero. The zero-intensity position corresponds to the situation where the principal section of N_1 and N_2 are perpendicular to each other. This orientation of N_1 and N_2 is also called the crossed position.

3.7.4 Production of Elliptically and Circularly Polarized Light

Figure 3.12 shows a set-up comprising a Nicol prism and a calcite crystal.

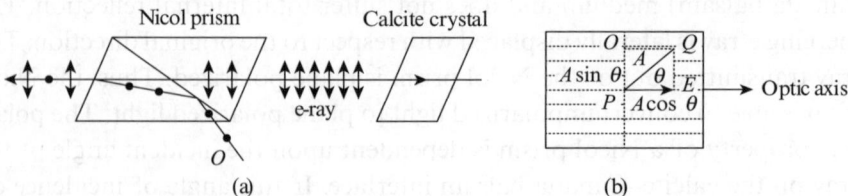

Fig. 3.12 Set-up using Nicol prism and calcite crystal (a) Side view (b) Components of amplitude

The Nicol prism converts unpolarized light into plane polarized light. This plane polarized light is incident normally on a calcite crystal that is cut in such a manner that its optic axis is parallel to the faces shown in Fig. 3.12. The amplitude vector A makes an angle θ with the optic axis. The incident plane polarized light gets split inside the calcite crystal into two components, namely ordinary and extraordinary. The component $A \sin \theta$, perpendicular to the optic axis, is the amplitude of the o-ray. The amplitude of e-ray is $A \cos \theta$ along the optic axis. The o- and e-ray travel with different velocities and therefore develop a phase difference, δ, as they traverse the thickness of the calcite crystal. The equation for e-ray can be written in the following form:

$$x = A \cos \theta \sin (\omega t + \delta) \tag{3.24}$$

and the o-ray can be expressed as follows:

$$y = A \sin \theta \sin \omega t \tag{3.25}$$

Let us assume that

$$A \cos \theta = A_1 \tag{3.26}$$

and

$$A \sin \theta = A_2 \tag{3.27}$$

Substitution of Eqs (3.26) and (3.27) in Eqs (3.24) and (3.25) leads to the following relations:

$$x = A_1 \sin (\omega t + \delta) \tag{3.28}$$

and

$$y = A_2 \sin (\omega t) \tag{3.29}$$

Equation (3.29) results in the following expression:

$$\sin \omega t = \frac{y}{A_2} \tag{3.30}$$

which leads to the following expression:

$$\cos \omega t = \sqrt{1 - \left(\frac{y}{A_2}\right)^2} \tag{3.31}$$

Equation (3.28) can be rewritten as follows:

$$\frac{x}{A_1} = \sin \omega t \cos \delta + \cos \omega t \sin \delta \tag{3.32}$$

Using Eqs (3.30) and (3.31) in Eq. (3.32), one gets the following relation:

$$\frac{x}{A_1} = \frac{y}{A_2} \cos \delta + \sqrt{\left(1 - \frac{y^2}{A_2^2}\right)} \sin \delta$$

Squaring both the sides, we get the following expression:

$$\frac{x^2}{A_1^2} + \frac{y^2}{A_2^2} \cos^2 \delta - \frac{2x}{A_1} \cdot \frac{y}{A_2} \cos \delta = \left(1 - \frac{y^2}{A_2^2}\right) \sin^2 \delta$$

which can be rewritten in the following form:

$$\frac{x^2}{A_1^2} + \frac{y^2}{A_2^2} \cos^2 \delta + \frac{y^2}{A_2^2} \sin^2 \delta - \frac{2x}{A_1} \frac{y}{A_2} \cos \delta = \sin^2 \delta \tag{3.33}$$

leading to

$$\frac{x^2}{A_1^2} + \frac{y^2}{A_2^2} - \frac{2xy}{A_1 A_2} \cos \delta = \sin^2 \delta \tag{3.34}$$

Equation (3.34) is a general equation of an ellipse.

Let us consider a few special cases:

Case 1

$$\delta = 0 \tag{3.35}$$

This implies that

$$\sin\delta = 0 \text{ and } \cos\delta = 1 \tag{3.36}$$

Use of Eq. (3.36) in Eq. (3.34) leads to the following expression:

$$\frac{x^2}{A_1^2} + \frac{y^2}{A_2^2} - \frac{2xy}{A_1 A_2} = 0$$

which can be rewritten in the following form:

$$\left(\frac{x}{A_1} - \frac{y}{A_2}\right)^2 = 0$$

yielding $\dfrac{x}{A_1} - \dfrac{y}{A_2} = 0$

which results in the following relation:

$$y = \frac{A_2}{A_1}x \tag{3.37}$$

Equation (3.37) is the equation of a straight line ($Y = mx + c$). The light emerging from the calcite crystal is, therefore, plane polarized, with the electric field in the same plane as that of the incident light. This plane polarized light is shown schematically in Fig. 3.13.

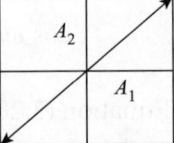

Fig. 3.13 Plane polarized light for $\delta = 0$

Case 2

$$\delta = (2n+1)\frac{\pi}{2}, \ \left(\text{where } n = 0, 1, 2, 3, \ldots\right) \tag{3.38}$$

Under this condition,

$$\cos\delta = 0, \ \sin\delta = 1 \tag{3.39}$$

Using Eq. (3.39) in Eq. (3.34), we get the following relation:

$$\frac{x^2}{A_1^2} + \frac{y^2}{A_2^2} = 1 \tag{3.40}$$

Equation (3.40) represents an ellipse. The light emerging from the calcite crystal is, thus, elliptically polarized, as shown in Fig. 3.14. The electric field vector rotates along an ellipse, as shown in Fig. 3.14.

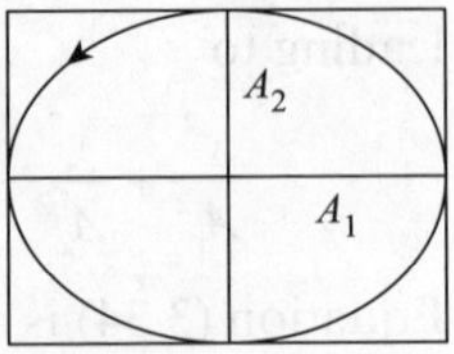

Case 3

$$\delta = \frac{\pi}{2} \text{ with } A_1 = A_2 \tag{3.41}$$

Under these conditions, Eq. (3.34) reduces to the following form:

$$x^2 + y^2 = A_1^2 \tag{3.42}$$

Fig. 3.14 Elliptically polarized light for $\delta = (2n+1)\dfrac{\pi}{2}$

Equation (3.42) is the equation of a circle. The light emerging from the calcite crystal is thus circularly polarized. This situation prevails when the plane polarized light incident on the calcite crystal makes an angle of 45° with its optic axis. This circularly polarized light is shown schematically in Fig. 3.15. The electric field vector rotates along a circle, as shown in Fig. 3.15.

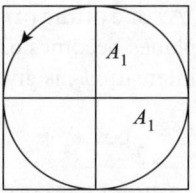

Fig. 3.15 Circularly polarized light for $\delta = \pi/2$ and $A_1 = A_2$

Example 3.9 Two Nicol prisms are adjusted to be in the crossed position. One of the Nicol prisms is then rotated through 30°. Calculate the percentage of incident unpolarized light that will pass through the system of two Nicol prisms.

Solution In the crossed position, the angle between their principal planes is 90°. As one of the Nicol prisms is rotated through 30°, the final angle between their principal planes becomes 60°. The first Nicol prism splits the incident light into ordinary and extraordinary components. The ordinary component is not transmitted. With no absorption loss, the intensity of transmitted light is $\frac{1}{2}$.

The transmitted e-component is once again split into extraordinary and ordinary components by the second Nicol prism. The o-components have an intensity $(\frac{1}{2})\sin^2\theta$ and the e-components have an intensity $(\frac{1}{2})\cos^2\theta$, where θ is the angle between the two Nicol prisms.

The extraordinary component emerges from the second Nicol prism. Thus, the final intensity, I_f, is given as follows:

$$I_f = \frac{I}{2}\cos^2\theta \tag{3.43}$$

For $\theta = 60°$, Eq. (3.43) leads to the following form:

$$\frac{I_f}{I} = \frac{\cos^2 60}{2} = \frac{\left(\frac{1}{2}\right)^2}{2} = \frac{1}{8} \tag{3.44}$$

The percentage transmission is then $\dfrac{I_f}{I} \times 100 = \dfrac{100}{8} = 12.5\%$

Example 3.10 A plane polarized light is incident on a quartz plate that is cut parallel to the axis. Calculate the least thickness of the plate for which the o- and e-ray recombine to form a plane polarized light. Assume that $\mu_e = 1.5533$, $\mu_o = 1.5442$, and $\lambda = 5.4 \times 10^{-5}$ cm.

Solution The calculations are as follows:

$$t_1 = \frac{\lambda}{2(\mu_e - \mu_o)} \tag{3.45}$$

Substitution of the given values in Eq. (3.45) leads to the following result:

$$t_1 = \frac{5.4 \times 10^{-5}}{2(1.5533 - 1.5442)} = 2.97 \times 10^{-3} \text{ cm}$$

Example 3.11 Two Nicol prisms are adjusted to be in the crossed position. One of the Nicol prisms is then rotated through 26.5°. Calculate the percentage of incident unpolarized light that will pass through the system of two Nicol prisms.

Solution As one of the prisms is rotated through 26.5°, the final angle between their principal planes becomes 63.5°.
The final intensity, I_f, is given as follows:

$$I_f = \frac{I}{2}\cos^2\theta \qquad (3.46)$$

For, $\theta = 63.5$, Eq. (3.46) yields the following result:

$$\frac{I_f}{I} = \frac{\cos^2 63.5}{2} = \frac{(0.45)^2}{2} = 0.10$$

resulting in a percentage of 10.

Example 3.12 If the Nicol prism indicated in Example 3.11 was rotated through 36.5°, what would have been the percentage of incident unpolarized light that would have passed through the system of two Nicol prisms?

Solution As one of the prisms is rotated through 36.5°, the final angle between their principal planes becomes 53.5°. The final intensity, I_f, is given as follows:

$$I_f = \frac{I}{2}\cos^2\theta \qquad (3.47)$$

For $\theta = 53.5$, Eq. (3.47) results in the following value:

$$\frac{I_f}{I} = \frac{\cos^2 53.5}{2} = \frac{(0.59)^2}{2} = 0.17$$

resulting in a percentage of 17.

Example 3.13 If the Nicol prism indicated in Example 3.11 was rotated by 16.5°, what would have been the percentage of incident unpolarized light that would have passed through the system of two Nicol prisms?

Solution As one of the prisms is rotated through 16.5°, the final angle between their principal planes becomes 73.5°.
The final intensity, I_f, is given by the following relation:

$$I_f = \frac{I}{2}\cos^2\theta \qquad (3.48)$$

For $\theta = 73.5°$, Eq. (3.48) leads to the following result:

$$\frac{I_f}{I} = \frac{\cos^2 73.5}{2} = \frac{(0.28)^2}{2} = 0.04$$

giving a percentage value of 4.

Example 3.14 Two Nicol prisms are adjusted to be in the crossed position. One of the Nicol prisms is rotated through some angle. It is found that 22.5% of the incident unpolarized light passes through the system of two Nicol prisms. What was the angle through which the Nicol prism was rotated?

Solution For the given situation, $\dfrac{I_f}{I} = \dfrac{\cos^2\theta}{2} = 0.225$

resulting in $\cos^2\theta = 0.450$

yielding $\theta = \cos^{-1} 0.67 = 47.9°$

The angle through which the Nicol prism was turned was $90 - 47.9 = 42.1°$.

Example 3.15 If the percentage of the incident unpolarized light in Example 3.14 was found to be 42.6%, what was the angle through which the Nicol prism was rotated?

Solution For the given situation, $\dfrac{I_f}{I} = \dfrac{\cos^2\theta}{2} = 0.426$

resulting in $\cos^2\theta = 0.852$, yielding $\theta = \cos^{-1} 0.92 = 23.1°$

The angle through which the Nicol prism was turned was $90 - 23.1 = 66.9°$.

Example 3.16 Two Nicol prisms are adjusted to be in the crossed position. On rotating one of the prisms the ratio $2I_f/I$ is found to be $1/6.6$. What was the angle through which the Nicol prism was rotated?

Solution The ratio I_f/I is given by the following relation: $\dfrac{2I_f}{I} = \cos^2\theta = \dfrac{1}{6.6}$

resulting in $\cos\theta = 0.55$

or $\theta = \cos^{-1} 0.55 = 56.6°$

The angle through which the Nicol prism was rotated was $90 - 56.6 = 33.4°$.

Example 3.17 A plane polarized light is incident on a quartz plate that is cut parallel to the axis. The minimum thickness of the plate for which the o- and e-ray recombine to form a plane polarized light is 2.65×10^{-3} cm. If $\mu_e = 1.5500$, calculate the value of μ_o, assuming that $\lambda = 5.2 \times 10^{-5}$ cm.

Solution μ_o is calculated as follows:

$$t_1 = \frac{\lambda}{2(\mu_e - \mu_o)} \tag{3.49}$$

Rearranging Eq. (3.49), we get the following relation:

$$\mu_o = \mu_e - \frac{\lambda}{2t_1} \tag{3.50}$$

Use of the given values in Eq. (3.50) results in the following expression:

$$\mu_o = 1.5500 - \left(\frac{5.2 \times 10^{-5}}{2 \times 2.65 \times 10^{-3}} \right) = 1.5402$$

Example 3.18 If the wavelength in Example 3.17 was $\lambda = 4.8 \times 10^{-5}$ cm keeping all other parameters the same, calculate the value of μ_o.

Solution We have

$$t_1 = \frac{\lambda}{2(\mu_e - \mu_o)} \tag{3.51}$$

leading to

$$\mu_o = \mu_e - \frac{\lambda}{2t_1} \tag{3.52}$$

Using the given values in Eq. (3.52), we get the following relation:

$$\mu_o = 1.5500 - \left(\frac{4.8 \times 10^{-5}}{2 \times 2.65 \times 10^{-3}} \right) = 1.5409$$

Example 3.19   A plane polarized light is incident on a quartz plate that is cut parallel to the axis. The least thickness of the plate for which the o- and e-ray recombine to form a plane polarized light is 2.01×10^{-3} cm. If $\mu_e = 1.5432$ and $\mu_o = 1.5380$ for the material, calculate the wavelength of light.

Solution   We have $t_1 = \dfrac{\lambda}{2(\mu_e - \mu_o)}$

yielding

$$\lambda = 2t_1(\mu_e - \mu_o) \tag{3.53}$$

Using the given values in Eq. (3.53), we get the following relation:

$$\lambda = 2 \times 2.01 \times 10^{-3}(1.5432 - 1.5380)$$

resulting in $\lambda = 2.09 \times 10^{-5}$ cm

Example 3.20   If the least thickness of the plate for which the o- and e-ray recombine to form a plane polarized light was found to be 2.16×10^{-3} cm in Example 3.19, calculate the wavelength of the light used.

Solution   The wavelength is calculated as follows:

$$\lambda = 2t_1(\mu_e - \mu_o) \tag{3.54}$$

Using the given values in Eq. (3.54), we get the following expression:

$$\lambda = 2 \times 2.16 \times 10^{-3}(1.5432 - 1.5380) = 2.25 \times 10^{-5}\text{ cm}$$

Example 3.21   A plane polarized light is incident on a quartz plate that is cut parallel to the axis. The minimum thickness of the plate for which the o- and e-ray recombine to form a plane polarized light is found to be 2.55×10^{-3} cm. If $\mu_o = 1.5400$, calculate the value of μ_e, assuming that $\lambda = 5 \times 10^{-5}$ cm.

Solution   We have $t_1 = \dfrac{\lambda}{2(\mu_e - \mu_o)}$

yielding

$$\mu_e = \mu_o + \frac{\lambda}{2t_1} \tag{3.55}$$

Using the values in Eq. (3.55), we get the following relation:

$$\mu_e = 1.5400 + \left(\frac{5 \times 10^{-5}}{2 \times 2.55 \times 10^{-3}} \right) = 1.5498$$

Example 3.22   If the minimum thickness of the plate for which the o- and e-ray recombine to form a plane polarized light was found to be 2.41×10^{-3} cm in Example 3.21, calculate μ_e.

Solution   We have

$$\mu_e = \mu_o + \frac{\lambda}{2t_1} \tag{3.56}$$

Using the given values in Eq. (3.56), we get the following relation:

$$\mu_e = 1.5400 + \left(\frac{5 \times 10^{-5}}{2 \times 2.41 \times 10^{-3}} \right)$$

leading to $\mu_e = 1.5504$

3.8 QUARTER- AND HALF-WAVE PLATES

We have learnt that a calcite crystal splits an incident unpolarized light into o- and e-ray. The o- and e-ray travel through the calcite crystal with different velocities, with the velocity of the e-ray being greater than that of the o-ray. A calcite crystal of suitable thickness can, therefore, be used to introduce controlled amount of phase difference between the two rays.

Figure 3.16 shows a schematic representation of unpolarized light falling normally on a calcite crystal that is cut parallel to its optic axis.

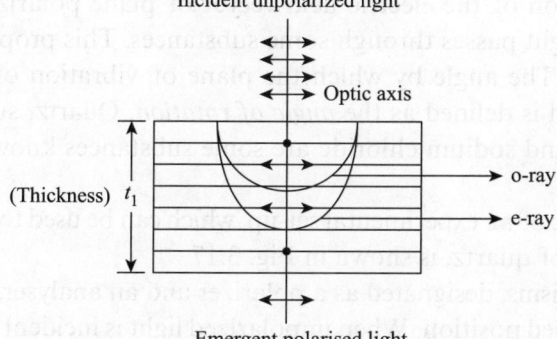

Fig. 3.16 Unpolarized light incident normally on calcite crystal

The incident light gets split into o- and e-ray, and these rays travel along the same path with different velocities. If t_1 represents the thickness of the calcite plate and μ_e and μ_o represent the refractive indices of the e- and o-ray, respectively, then the path difference, P–D, between the two rays can be expressed in the following form:

$$P - D = \mu_o t_1 - \mu_e t_1 = (\mu_o - \mu_e) t \tag{3.57}$$

Two special cases deserve specific attention.

Case 1

$$P - D = \frac{\lambda}{4} = (\mu_o - \mu_e) t_1 \tag{3.58}$$

Such a plate is called a quarter-wave plate, since it introduces a path difference of $\lambda/4$ between the o- and e-ray. The thickness, as evaluated from Eq. (3.58), is given by the following relation:

$$t_1 = \frac{\lambda}{4(\mu_o - \mu_e)} \tag{3.59}$$

A quarter-wave plate introduces a phase difference of $\frac{\pi}{2}$ radians between the o- and e-ray.

Case 2

$$P - D = \frac{\lambda}{2} = (\mu_o - \mu_e)t_1 \qquad (3.60)$$

Such a plate is called a half-wave plate since it introduces a path difference of $\lambda/2$ between the o- and e-ray. The corresponding thickness is given by the following relation:

$$t_1 = \frac{\lambda}{2(\mu_o - \mu_e)} \qquad (3.61)$$

A half-wave plate introduces a phase difference of π between the o- and e-ray.

3.9 OPTICAL ACTIVITY

Plane of vibration of the electric field vector of plane polarized light gets rotated as the light passes through some substances. This property is called *optical activity*. The angle by which the plane of vibration of the electric field gets rotated is defined as the *angle of rotation*. Quartz, sugar crystals, turpentine oil, and sodium chloride are some substances known to display this property.

The schematic of an experimental set-up, which can be used to demonstrate optical activity of quartz, is shown in Fig. 3.17.

Two Nicol prisms, designated as a polarizer and an analyser, are adjusted to be in the crossed position. When unpolarized light is incident on the polarizer face, no light comes out of the analyser. This results in darkness or no illumination condition at the output, as indicated in Fig. 3.17(a). A quartz plate is then introduced between the polarizer and the analyser. The plane

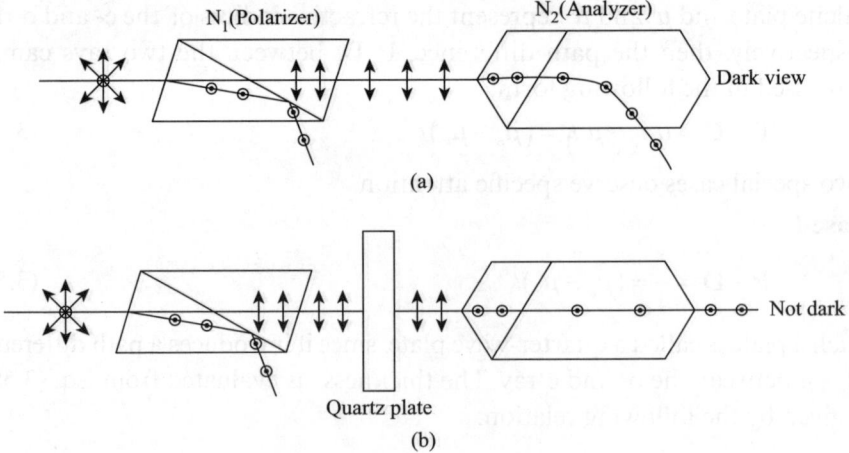

Fig. 3.17 Experimental set-up for demonstrating optical activity of quartz (a) No illumination condition (b) Not complete dark condition

of polarization gets rotated as light comes out of the quartz plate due to the optical activity of quartz. The output from the analyser now has some illumination and is not dark, as shown in Fig. 3.17(b). The analyser needs to be rotated to get the crossed position of no illumination at the output again. This demonstrates that the quartz plate is able to rotate the plane of polarization and is therefore an optically active substance. Optical activity has the following characteristics:

1. The plane of polarization can get rotated in either the clockwise direction (as observed in a direction opposite to the direction of propagation) or the anticlockwise direction. If the plane of polarization is rotated in the clockwise direction as plane polarized light passes through it, such a substance is called *dextrorotatory* or *right handed*. If the substance results in a rotation of the plane of polarization in the anticlockwise direction, the substance is called *laevorotatory* or *left handed*.
2. The amount of rotation produced, θ, is proportional to the thickness t_1 of the optically active substance.
3. The amount of rotation produced is proportional to the concentration of the optically active substance in the solution or the vapour.
4. The amount of rotation produced is inversely proportional to the square of wavelength of the incident light.
5. Suppose plane polarized light passes through a number of optically active substances sequentially. Then the total rotation is found to be the algebraic sum of individual rotations. If the individual rotations are θ_1, θ_2, θ_3, etc., then the rotation θ_T is given by the following relation:

$$\theta_T = \theta_1 + \theta_2 + \theta_3 + \dots \tag{3.62}$$

In Eq. (3.62), the individual angles are given signs depending upon the direction of rotation of the plane of polarization. As a convention, a rotation in the anticlockwise direction is assigned a positive sign and that in the clockwise direction a negative sign.

3.10 SPECIFIC ROTATION

The magnitude of optical activity of a substance is expressed in terms of its *specific rotation*. *The specific rotation of a substance for a particular temperature and wavelength is defined as the angle of rotation produced by 1-decimetre thickness of the substance having a concentration of 1 g/cm³.* Thus, specific rotation, S, is given by the following relation

$$S = \frac{\theta}{t_1 \times C} \tag{3.63}$$

where θ is the angle of rotation in degrees, t_1 is the thickness in decimetres, and C is the concentration in g/cm³.

3.11 FRESNEL'S THEORY OF OPTICAL ROTATION

A theory explaining the phenomenon of optical rotation was developed by Fresnel. Fresnel's theory uses the fact that *a linearly polarized light can be visualized as a resultant of two oppositely directed circularly polarized vibrations (actually electric fields) that have the same frequency but half the amplitude.* Fresnel made the following assumptions:

1. A plane polarized light entering a crystal along its optic axis gets split into two circularly polarized components. One of these components is right-handed and the other is left-handed.

2. The two circularly polarized components travel with the same angular velocity in an optically inactive substance like calcite.

3. The two circularly polarized components travel with different velocities in an optically active substance like quartz. If the substance happens to be dextrorotatory, the right-handed circularly polarized component travels faster than the left-handed circularly polarized component. For a laevorotatory substance, on the other hand, the left-handed circularly polarized component travels faster than the right-handed circularly polarized component.

4. The two circularly polarized components recombine on emerging from the optically active substance. Due to the different velocities of the two components, a phase difference develops between the two circularly polarized components. This phase difference results in a rotation of the plane of polarization.

Figure 3.18 shows a schematic representation of Fresnel's theory for an optically inactive substance.

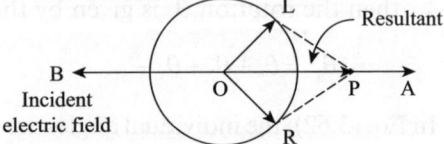

Fig. 3.18 Fresnel's theory for optically inactive substance

The incident plane polarized light gets split into two circularly polarized components. OR represents the electric field vector for the circularly polarized light, rotating in the clockwise direction. Similarly, OL represents the electric field vector for the circularly polarized light rotating in the anticlockwise direction. OP is the resultant of OL and OR. From Fig. 3.18, one can see that OP is in the same plane as the electric field vector of the incident light. The two circularly polarized components recombine on emerging from the optically inactive substance and result in a plane polarized light with no rotation of the plane of polarization.

Figure 3.19 is a schematic representation of the Fresnel's theory for a dextrorotatory optically active substance.

On entering the optically active substance, the incident plane polarized light is split into two circularly *polarized* components. Component OR travels faster than component OL. As light emerges from the optically active substance, the clockwise component traverses a greater angle δ than the anticlockwise component. The resultant of these two components, OA', is a plane polarized light along $A'B'$. From Fig. 3.19, it is clear that the plane of vibration gets rotated through an angle $\delta/2$. The angle of rotation of the plane of polarization is due to the difference in velocities of the two circularly polarized components. Thus, the exact magnitude of the angle of rotation is dependent upon the thickness of the optically active substance.

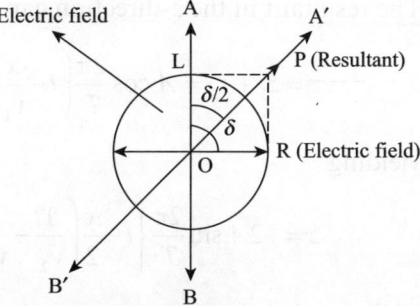

Fig. 3.19 Fresnel's theory for dextrorotatory substance

We will now offer a mathematical treatment for the rotation of the plane of polarization. Suppose a plane polarized wave travels through an optically active substance in the positive x-direction. This plane polarized wave can be represented by two circularly polarized components in the YZ plane. The right-handed circularly polarized wave can be represented by the following expressions:

$$y_1 = A \sin \frac{2\pi}{T}\left(t - \frac{x}{v_1}\right) \tag{3.64}$$

and

$$z_1 = A \cos \frac{2\pi}{T}\left(t - \frac{x}{v_1}\right) \tag{3.65}$$

where v_1 is the velocity of the right-handed circularly polarized wave.

Similarly, for the left-handed circularly polarized wave, we can write the following equations:

$$y_2 = A \sin \frac{2\pi}{T}\left(t - \frac{x}{v_2}\right) \tag{3.66}$$

and

$$z_2 = -A \cos \frac{2\pi}{T}\left(t - \frac{x}{v_2}\right) \tag{3.67}$$

The resultant in the y-direction can be expressed as follows:

$$y = y_1 + y_2 = A \sin \frac{2\pi}{T}\left(t - \frac{x}{v_1}\right) + A \sin \frac{2\pi}{T}\left(t - \frac{x}{v_2}\right)$$

resulting in

$$y = 2A \sin \frac{2\pi}{T}\left\{t - \frac{x}{2}\left(\frac{1}{v_1} + \frac{1}{v_2}\right)\right\} \times \cos \frac{\pi x}{T}\left(\frac{1}{v_2} - \frac{1}{v_1}\right) \tag{3.68}$$

The resultant in the z-direction can be expressed in the following form:

$$z = z_1 + z_2 = A \cos\frac{2\pi}{T}\left(t - \frac{x}{v_1}\right) - A \cos\frac{2\pi}{T}\left(t - \frac{x}{v_2}\right)$$

yielding

$$z = -2A \sin\frac{2\pi}{T}\left\{t - \frac{x}{2}\left(\frac{1}{v_2} - \frac{1}{v_1}\right)\right\} \times \sin\frac{\pi x}{T}\left(\frac{1}{v_2} - \frac{1}{v_1}\right) \tag{3.69}$$

Dividing Eq. (3.69) by Eq. (3.68), one gets the following expression:

$$\frac{z}{y} = -\tan\frac{\pi x}{T}\left(\frac{1}{v_2} - \frac{1}{v_1}\right) = \tan\theta \tag{3.70}$$

where

$$\theta = -\frac{\pi x}{T}\left(\frac{1}{v_2} - \frac{1}{v_1}\right) \tag{3.71}$$

The light emerging from the optically active substance is a plane polarized light, with the plane of polarization making an angel θ with the y-axis, as shown in Fig. 3.20.

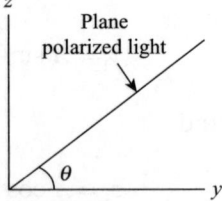

Fig. 3.20 Light emerging from optically active substance

The following conclusions can be drawn from Eq. (3.70):

1. The relation $v_1 > v_2$ implies a negative θ, which represents a clockwise rotation or a dextrorotatory substance.
2. The relation $v_1 < v_2$ implies a positive θ, which represents an anticlockwise rotation or a laevorotatory substance.
3. The relation $v_1 = v_2$ implies that $\theta = 0$. This situation indicates an optically inactive substance.

Example 3.23 The plane of polarization gets rotated through 23.8° as light travels through an 18 cm-long column of 20% sugar solution. Determine the specific rotation of the solution.

Solution Specific rotation, S, is given as follows:

$$S = \frac{\theta°/\text{dm g/cm}^3}{t_1 C} \tag{3.72}$$

where t_1 is the length of column in decimetres. Using the given values in Eq. (3.72), we get the following result: $S = \dfrac{23.8}{1.8 \times 0.2} = 66.1°/\text{dm g/cm}^3$.

Example 3.24 The plane of polarization gets rotated through 23.4° as light travels through a 22% solution of a substance. If the specific rotation of the substance is 64°, calculate the length of the column.

Solution Specific rotation, S, is given by the following relation: $S = \dfrac{\theta}{t_1 c}$

yielding

$$t_1 = \frac{\theta}{Sc} \tag{3.73}$$

Using the given values in Eq. (3.73), we get the following result:

$$t_1 = \frac{23.4}{64 \times 0.22} = 1.66\,\text{dm} = 16.6\,\text{cm}$$

Example 3.25 The plane of polarization gets rotated as light travels through an 18% solution of a substance. If the specific rotation of the substance is 61° and the length of the column is 17 cm, calculate the angle through which the plane of polarization gets rotated.

Solution Specific rotation, S, is given as follows: $S = \dfrac{\theta}{t_1 c}$

leading to

$$\theta = St_1 c \tag{3.74}$$

Using the given values in Eq. (3.74), we get the following result

$$\theta = 61 \times 1.7 \times 0.18 = 18.7°$$

3.12 POLARIMETER

The angle of rotation produced by an optically active substance is measured by a *polarimeter*. We will discuss some important types of polarimeters in this section.

3.12.1 Laurent's Half-shade Polarimeter

Figure 3.21 shows a schematic of a Laurent's half-shade polarimeter.

Light from a monochrome source is incident on a convex lens, which results in a parallel beam. The first Nicol prism N_1 acts as a polarizer for the incident beam of light. The plane polarized light that emerges from N_1 passes through a half-shade device before travelling through a tube containing an optically active liquid. Light coming out of the tube is incident on a second Nicol prism N_2, which acts as an analyser. Any rotation of the analyser can be quantified using a circular scale, which also has a Vernier scale for reading the fractions of a degree. The light emerging from analyser N_2 is seen using a telescope, with the telescope being focused on the half-shade device.

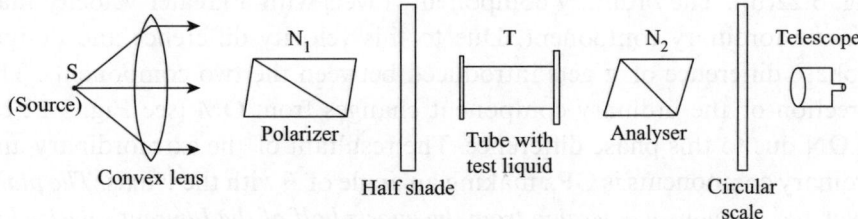

Fig. 3.21 Schematic of Laurent's half-shade polarimeter

3.12.1.1　Role of Half-shade Device

The two Nicol prisms N_1 and N_2 form a polarizer–analyser pair. In the crossed position, the field of view is dark. When an optically active substance is kept in between the two Nicol prisms, the field of view is not dark any more. A considerable rotation of the analyser is required to restore the dark field of view. There is some ambiguity in relocating the dark field of view, and this, in turn, leads to inaccuracy in the exact extent of rotation produced by the optically active substance. This difficulty is overcome using a Laurent's half-shade device.

A Laurent's half-shade device is shown schematically in Fig. 3.22(a).

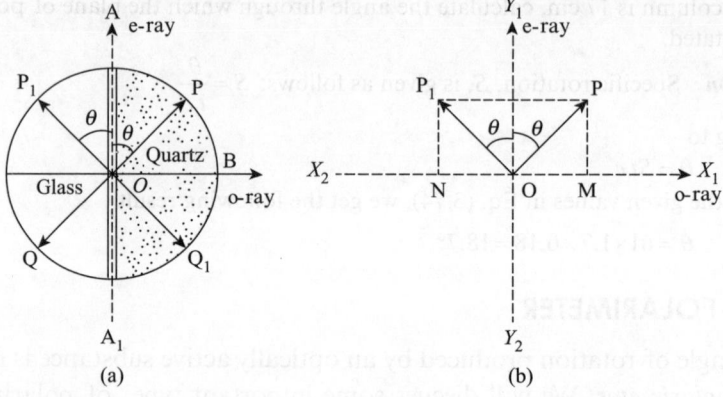

Fig. 3.22 Role of half-shade device (a) Schematic representation (b) Ordinary and extraordinary components

The device consists of two semicircular plates that are cemented together along the diameter AA_1. The semicircular half-plate ABA_1 is made of quartz and is cut parallel to the optic axis. The quartz plate introduces a phase difference of π between the e- and o-ray. The thickness of the glass plate ABA_1 is chosen such that it absorbs the same amount of light as the quartz plate.

Suppose a plane polarized light is incident upon the half-shade device, with the plane of polarization being along PQ. PQ makes an angel of θ with AA_1. The plane of polarization of the light emerging from the glass half of the circular plate continues to be along PQ. The light gets split into two components inside the quartz plate. The ordinary component is along X_1X_2 and the extraordinary component is along Y_1Y_2 or the optic axis [see Fig. 3.22(b)]. The ordinary component travels with a greater velocity than the extraordinary component. Due to this velocity difference and design, a phase difference of π gets introduced between the two components. The direction of the ordinary component changes from OM [see Fig. 3.22(b)] to ON due to this phase difference. The resultant of the extraordinary and ordinary components is OP_1, making an angle of θ with the y-axis. *The plane polarized component emerging from the quartz half of the Laurent's device has a plane of polarization along P_1Q_1.*

If the principal plane of N_2 is parallel to PQ, then the light emerging from the glass portion of the Laurent's plate would pass unobstructed through N_2. The light emerging from the quartz portion would, however, be partially obstructed. The field of view would then show a brighter glass half than the quartz half. On the other hand, when the principal plane of N_2 is parallel to P_1Q_1, the light from the quartz portion passes unobstructed, making it appear brighter. If the principal plane of the analyser N_2 is parallel to y-axis (AA$_1$), then the plane polarized light emerging from the glass and quartz portions are equally inclined with respect to the principal plane. This ensures that the field of view has equal brightness over the entire circular plate. *Any deviation from this position of equal brightness is detected very accurately by the eye, as it is very sensitive to brightness differences in a simultaneously projected visual stimuli.* Thus, the Laurent's half-shade device helps in improving the sensitivity of a polarimeter.

3.12.1.2 Determining Specific Rotation

The tube of a polarimeter is initially filled with water. The telescope is then focused on the half-shade plate, and the analyser N_2 is rotated till the position of equal brightness is obtained. The tube is then filled with an optically active substance. The position of equal brightness gets disturbed. Analyser N_2 is then rotated till the position of equal brightness is restored. If θ_1 represents the angular position of the analyser for water and θ_2 that for the optically active substance, the specific rotation is determined using the following relation:

$$S = \frac{(\theta_2 - \theta_1)}{t_1 \times C} \tag{3.75}$$

where t_1 is the length of the optically active solution in decimetres and C is the concentration of the solution in g/cc.

3.12.1.3 Specific Rotation of Sugar Solution

From Eq. (3.75), it is clear that a plot of $(\theta_2 - \theta_1)$ versus C would be a straight line. The tube is first filled with water, and the initial reading θ_1 of the analyser is noted for the equal brightness position using the circular scale. Water in the tube is then replaced with a sugar solution of a known concentration. The position of equal brightness is restored by rotating the analyser. The new position θ_2 of the analyser is then noted. A similar experiment is then carried out using sugar solutions of different concentrations. A plot similar to that shown in Fig. 3.23 is then obtained.

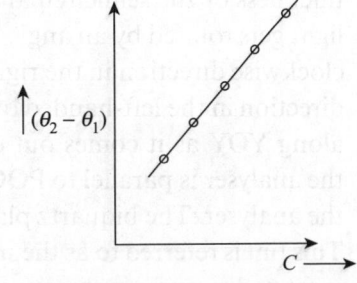

Fig. 3.23 Plot of $(\theta_2 - \theta_1)$ vs C

The slope of the curve gives the magnitude of $(\theta_2 - \theta_1)/C$. This information is then used to evaluate specific rotation of sugar using the following expression:

$$S = \frac{(\theta_2 - \theta_1) \times 10}{t_1 \times C} \tag{3.76}$$

where t_1 is the length of the tube in centimetres.

3.12.2 Biquartz Polarimeter

The half-shade device used in the Laurent's polarimeter is replaced with a biquartz plate in a biquartz polarimeter. The monochromatic source is also replaced with a white light source. Two semicircular quartz plates are joined together to form a biquartz plate. One semicircular plate is made out of left-handed quartz and the other out of right-handed quartz, as shown schematically in Fig. 3.24.

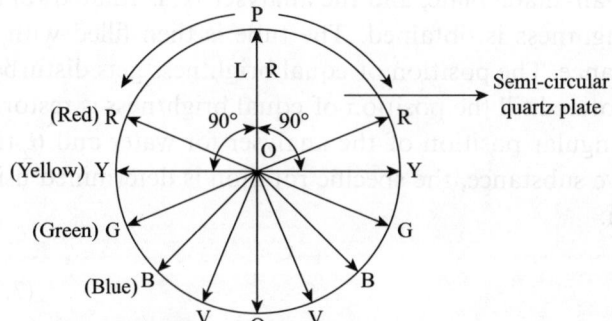

Fig. 3.24 Schematic of biquartz plate

The two semicircular quartz plates are cut in a direction that is perpendicular to the optic axis. They are then joined together along the diameter PQ. The incident white light gets converted to a plane polarized light by the polarizer in the polarimeter (refer to Fig. 3.21). After entering the biquartz plate, the planes of polarization of different wavelengths of the white light undergo varying magnitudes of rotation, as indicated in Fig. 3.24. For example, if the thickness of the semicircular quartz plate is 3.75 mm, the yellow-wavelength light gets rotated by an angle of 90°. Moreover, the rotation takes place in the clockwise direction in the right-handed quartz plate and in the anticlockwise direction in the left-handed quartz plate. Thus, the yellow light gets polarized along YOY as it comes out of the biquartz plate. If the principal plane of the analyser is parallel to POQ, the yellow light is then unable to come out of the analyser. The biquartz plate then has a greyish-violet tint at this position. This tint is referred to as the *sensitive tint* or the *tint of passage*. As the analyser is rotated away from the position of *sensitive tint*, one half of the field of view appears blue and the other half appears red. Rotation of the analyser in the opposite direction results in an interchange of the appearance of the field of view due to the two semicircular half quartz plates.

The position of the sensitive tint is used as a reference in a biquartz polarimeter to determine the rotation produced by any optically active substance.

3.13 PHOTOELASTICITY

Some substances such as glass, celluloid, and bakelite are normally doubly refracting. They, however, display the phenomenon of double refraction under the influence of stress. This phenomenon of double refraction under stress can be used to analyse stress distribution in mechanical components and material structures. *The development of double refraction due to applied stress is called photoelasticity.* Suppose we take two Nicol prisms to form a pair of polarizer and analyser. Let us also assume that a substance displaying photoelasticity is placed in between them after adjusting the analyser and the polarizer to obtain the crossed position. In the absence of any applied stress, the crossed position is not disturbed due to the presence of the substance displaying photoelasticity. When stress is applied on the substance, birefringence leads to the emergence of an elliptically polarized light from the substance. This elliptically polarized light is partially transmitted through the analyser, and the crossed position is accordingly disturbed. *The phenomenon of appearance of optical anisotropy in a normally isotropic substance under the influence of external forces is called induced birefringence or artificial double refraction.*

3.13.1 Theory of Photoelasticity

A material showing birefringence displays a stress-dependent change in the refractive index. Mathematically, this can be expressed in the following form:

$$\mu_1 - \mu_0 = c_1\sigma_1 + c_2(\sigma_2 + \sigma_3) \tag{3.77}$$

$$\mu_2 - \mu_0 = c_1\sigma_2 + c_2(\sigma_3 + \sigma_1) \tag{3.78}$$

$$\mu_3 - \mu_0 = c_1\sigma_3 + c_2(\sigma_1 + \sigma_2) \tag{3.79}$$

where μ_1, μ_2, and μ_3 represent the principal refractive indices of the photoelastic material under the respective principal stresses σ_1, σ_2, and σ_3, and μ_0 represents the refractive index under no stress condition. The coefficients c_1 and c_2 are referred to as the stress-optic coefficients. These coefficients are material dependent.

Equations (3.77)–(3.79) are also referred to as the stress-optic law. For a two-dimensional situation, and considering the x–y plane, we can rewrite Eqs (3.77)–(3.79) in the following forms:

$$\mu_1 - \mu_0 = c_1\sigma_1 + c_2\sigma_2 \tag{3.80}$$

and

$$\mu_2 - \mu_0 = c_1\sigma_2 + c_2\sigma_1 \tag{3.81}$$

Using Eqs (3.77)–(3.79), the relative refractive index can be expressed as follows:

$$\mu_2 - \mu_1 = (c_2 - c_1)(\sigma_1 - \sigma_2) = c_{21}(\sigma_1 - \sigma_2) \tag{3.82}$$

$$\mu_3 - \mu_2 = (c_2 - c_1)(\sigma_2 - \sigma_3) = c_{21}(\sigma_2 - \sigma_3) \tag{3.83}$$

and

$$\mu_1 - \mu_3 = (c_2 - c_1)(\sigma_3 - \sigma_1) = c_{21}(\sigma_3 - \sigma_1) \tag{3.84}$$

An electric field is associated with a light ray. For a light ray travelling in the z-direction, the electric field can be written in the following form:

$$E = A\cos\frac{2\pi}{\lambda}(z - ct) \tag{3.85}$$

where c represents the velocity of light. Equation (3.85) can be rewritten as follows:

$$E = A\cos\varphi \tag{3.86}$$

According to the stress-optic law, introduction of a photoelastic substance of thickness t_1 in the path of a light ray introduces phases given by the following relations:

$$\phi_1 = \frac{2\pi}{\lambda} \times t_1 \times (\mu_1 - \mu_0) \tag{3.87}$$

and

$$\phi_2 = \frac{2\pi}{\lambda} \times t_1 \times (\mu_2 - \mu_0) \tag{3.88}$$

Equations (3.87) and (3.88) can be combined to result in the following expression:

$$\phi_2 - \phi_1 = \frac{2\pi}{\lambda} \times t_1 \times (\mu_2 - \mu_1) = \Delta_{12} \tag{3.89}$$

where Δ_{12} is the angular phase shift. Using Eq. (3.82) in Eq. (3.89), we can write the following relation:

$$\Delta_{12} = \frac{2\pi}{\lambda} \times t_1 \times C_{21}(\sigma_1 - \sigma_2) \tag{3.90}$$

Equations similar to Eq. (3.90) can also be written for other directions. These equations would be as follows:

$$\Delta_{23} = \frac{2\pi}{\lambda} \times t_1 \times C_{21}(\sigma_2 - \sigma_3) \tag{3.91}$$

and

$$\Delta_{31} = \frac{2\pi}{\lambda} \times t_1 \times C_{21}(\sigma_3 - \sigma_1) \tag{3.92}$$

If we restrict ourselves to a two-dimensional situation, using $\sigma_3 = 0$ in Eqs (3.90)–(3.92), we get the following expression:

$$\Delta_{12} = \frac{2\pi}{\lambda} \times t_1 \times C_{21} \left(\sigma_1 - \sigma_2 \right) \tag{3.93}$$

yielding

$$\left(\sigma_1 - \sigma_2 \right) = \frac{\Delta_{12}}{2\pi} \times \frac{1}{t_1} \times \frac{\lambda}{C_{21}} \tag{3.94}$$

which can be rewritten in the following form:

$$\left(\sigma_1 - \sigma_2 \right) = N \times \frac{\lambda}{t_1 C_{21}} \tag{3.95}$$

where N is known as the fringe order.

From Eq. (3.93), we can conclude that the angular phase difference is directly proportional to the difference between the principal stresses. At points having $\sigma_1 = \sigma_2$, the angular phase difference becomes zero and black dots appear at these points. Such points are called *isotropic points*. If $\sigma_1 = \sigma_2 = 0$, then also the fringe order becomes zero, and such points are called *singular points*.

3.13.2 Fringe Pattern

Let us now use the theory of photoelasticity to look at the fringe pattern obtained using a *polariscope*. A polariscope is an instrument that consists of a polarizer and an analyser, and has the provision to keep a photoelastic substance in between. The polarizer and analyser are initially adjusted to be in the crossed position. The photoelastic material is then kept in between them. The light coming out of the polarizer is plane polarized along the optic axis, as shown in Fig. 3.25.

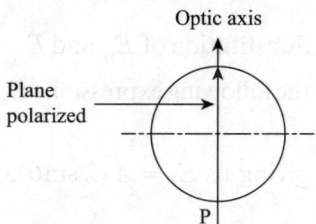

Fig. 3.25 Plane polarized light from polarizer

The electric field of the plane polarized light can be expressed in the following form:

$$E = A \cos wt \tag{3.96}$$

This plane polarized light enters a photoelastic material. Suppose σ_1 and σ_2 represent the principal stress directions. Components of the electric field in these two directions can be written as follows:

$$E_{1i} = A \cos wt \cos \theta \tag{3.97}$$

$$E_{2i} = A \cos wt \sin \theta \tag{3.98}$$

Components E_{1i} and E_{2i} are shown schematically in Fig. 3.26(a).

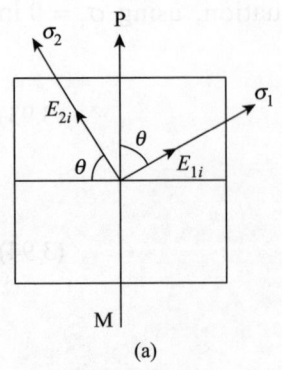

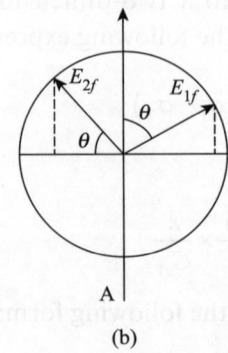

Fig. 3.26 Plane polarized light incident on photoelastic material (a) Component of incident electric field (b) Components of outgoing electric field

If $\sigma_1 \neq \sigma_2$, then a phase difference, δ, gets introduced between the two components of the electric field. The light coming out of the photoelastic material has electric field components given by the following equations:

$$E_{1f} = A\cos(wt+\delta)\,\cos\theta \qquad (3.99)$$

and

$$E_{2f} = A\cos wt \sin\theta \qquad (3.100)$$

In the crossed position, the optic axis of the analyser is perpendicular to the polarizer. The electric field transmitted through the analyser has the following form:

$$E_T = (E_{1f})\sin\theta - (E_{2f})\cos\theta \qquad (3.101)$$

Substitution of E_{1f} and E_{2f} from Eqs (3.99) and (3.100) into Eq. (3.101) results in the following expression: $E_T = [A\cos(wt+\delta)\cos\theta]\sin\theta - [A\cos wt \sin\theta]\cos\theta$

giving us $E_T = A\,(2\sin\theta\cos\theta)\left[\dfrac{\cos(wt+\delta)-\cos wt}{2}\right]$

yielding $E_T = A\sin 2\theta\left[-\sin\left(wt+\dfrac{\delta}{2}\right)\sin\dfrac{\delta}{2}\right]$

implying that

$$E_T = -A\sin 2\theta \sin\left[wt+\dfrac{\delta}{2}\right]\sin\dfrac{\delta}{2} \qquad (3.102)$$

Intensity, I, is proportional to the square of E_T, leading to the following relation: $I = A^2\sin^2 2\theta \sin^2\left[wt+\dfrac{\delta}{2}\right]\sin^2\dfrac{\delta}{2}$

yielding

$$I = I_0\,\sin^2 2\theta \sin^2\left[wt+\dfrac{\delta}{2}\right]\sin^2\dfrac{\delta}{2} \qquad (3.103)$$

where $I_0 = A^2$ is the maximum transmitted light intensity.

Two special cases deserve mention.

Case 1 $2\theta = n\pi, n = 0, 1, 2, \ldots$

This condition implies that $\sin^2 2\theta = 0$

which, when used in Eq. (3.103), leads to the following form:

$$I = 0 \qquad\qquad (3.104)$$

Thus, if one of the principal stress directions is along the optic axis of the polarizer, extinction of intensity occurs. Such a fringe pattern is called an *isoclinic fringe pattern* and can be used to determine the principal stress directions.

Case 2 $\delta/2 = n\pi, n = 0, 1, 2, 3, \ldots$

This condition implies that $\sin^2 \delta/2 = 0$

which, when used in Eq. (3.103), leads to the following expression:

$$I = 0 \qquad\qquad (3.105)$$

Thus, if the principal phase difference is either zero ($n = 0$) or an integral multiple of wavelength, extinction of intensity takes place. Such a fringe pattern is called an *isochromatic fringe pattern*, since the extinction is specific to wavelength.

3.13.3 Photoelastic Bench

A photoelastic bench is used to study the photoelastic nature of materials. Figure 3.27 shows a schematic representation of a photoelastic bench.

Main components of a photoelastic bench are as follows:

Monochromatic source   The incident unpolarized light is supplied by the monochromatic source.

Lens (L_1)   A parallel beam of light is produced by the lens L_1.

Polarizer   The unpolarized light is converted into plane polarized light by the polarizer.

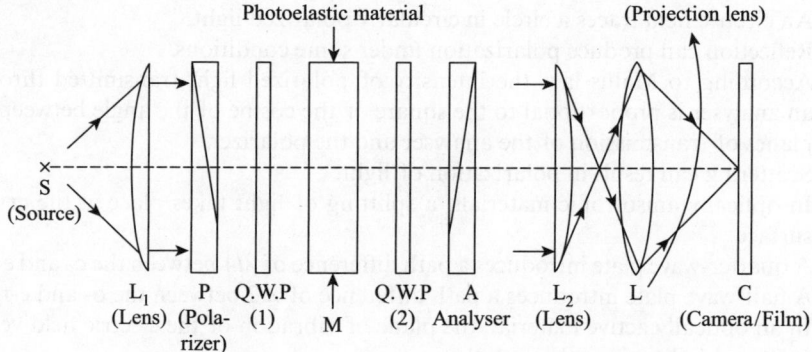

Fig. 3.27 Schematic layout of photoelastic bench

First quarter-wave plate　The plane polarized light is converted into circularly polarized light by the first quarter-wave plate.

Test material　The material to be tested is kept after the first quarter-wave plate. Provision exists to apply stress on the test material.

Second quarter-wave plate　The circularly polarized light is converted back into plane polarized light by the second quarter-wave plate.

Analyser　The light exiting the test material is studied using the analyser.

Lens (L_2)　The parallel beam of light is projected on the projection lens by lens L_2.

Projection lens　The fringe pattern obtained is projected on a screen or a camera using this arrangement.

Notes:
1. Polarizing glasses can be used to cut out the reflected glare. Many of the polarizing glasses sold in the market are actually fake. One easy way to verify whether they are genuine or not is to look through them at the light that is being reflected from the windows of a building. If the glasses are rotated, the brightness must vary, going from bright to dark twice during the course of a complete rotation.
2. Light from clouds is unpolarized, whereas that from the blue sky is partially plane polarized. This feature can be used to enhance contrast by reducing the intensity of light received from the sky. This is achieved by keeping a polarizing film in front of the camera lens.

IMPORTANT CONCEPTS

1. The phenomenon of polarization proves the transverse nature of light.
2. The ordinary light is unpolarized.
3. Light that consists of an electric field oscillating in only one direction is called plane polarized light.
4. An electric field traces an ellipse in elliptically polarized light.
5. An electric field traces a circle in circularly polarized light.
6. Reflection can produce polarization under some conditions.
7. According to Malus law, the intensity of polarized light transmitted through an analyser is proportional to the square of the cosine of the angle between the planes of transmission of the analyser and the polarizer.
8. Scattering can result in polarization of light.
9. In optically anisotropic materials, a splitting of light takes place at the crystal surface.
10. A quarter-wave plate introduces a path difference of $\lambda/4$ between the o- and e-ray.
11. A half-wave plate introduces a path difference of $\lambda/2$ between the o- and e-ray.
12. In an optically active material, the plane of vibration of the electric field vector rotates as light passes through the material.
13. The development of double refraction due to applied stress is called photoelasticity.

APPLICATIONS

1. Polarizing films are used in photography for a variety of reasons. Some of the reasons include enhancement of blue light, cutting unnecessary light reflected from the surface of a river or a shop window, etc. The photograph of a blue sky appears whitish without the use of a polarizing filter.
2. Sunglasses having a polarizing film can reduce the glare and dazzle due to reflection. Such sunglasses are, therefore, employed by skiers, athletes, and pilots for clarity of vision.
3. Polarizing films control the light that is transmitted through liquid crystal molecules. The liquid crystal molecules act as shutters that change direction with the application of a voltage. The liquid crystals are sandwiched between two polarizing films whose polarization directions are arranged to be orthogonal with respect to each other.

IMPORTANT FORMULAE

1. Brewster's law implies that

$$\mu = \tan p$$

2. According to Malus law,

$$I_A \propto \cos^2\theta$$

3. Thickness of a quarter-wave plate:

$$t = \frac{\lambda}{4(\mu_e - \mu_0)}$$

4. Thickness of a half-wave plate:

$$t = \frac{\lambda}{2(\mu_e - \mu_0)}$$

5. Specific rotation, S:

$$S = \frac{\theta}{t_1 C}$$

6. The stress-optic law in two dimension has the following form:

$$\mu_1 - \mu_0 = c_1\sigma_1 + c_2\sigma_2$$
and $\mu_2 - \mu_0 = c_1\sigma_2 + c_2\sigma_1$

7. For a photoelastic material,

$$(\sigma_1 - \sigma_2) = \frac{\Delta_{12}}{2\pi} \times \frac{1}{t_1} \times \frac{\lambda}{c_{21}}$$

SELF-ASSESSMENT

Multiple-choice Questions

3.1 The phenomenon of polarization proves
   (a) particle nature of light
   (b) wave nature of light
   (c) that light has energy
   (d) that light is characterized by momentum
3.2 In a plane polarized light, the electric field oscillates
   (a) in two mutually perpendicular directions
   (b) in the direction of travel
   (c) in only one direction
   (d) perpendicular to the direction of travel

3.3 A symmetrical ellipse is represented by

(a) $\dfrac{x^2}{A_1^2} + \dfrac{y^2}{A_2^2} = 1$

(c) $\dfrac{x^2}{A_1^2} + \dfrac{y^2}{A_2^2} = 0$

(b) $\dfrac{x^2}{A_1^2} + \dfrac{y^2}{A_2^2} = x^2$

(d) $\dfrac{x^2}{A_1^2} + \dfrac{y^2}{A_2^2} = A_1^2$

3.4 A circularly polarized light is represented by

(a) $x^2 + y^2 = z^2$

(c) $x^2 + y^2 = y^2$

(b) $x^2 + y^2 = A^2$

(d) $x^2 + y^2 = Ax^2 + B$

3.5 Brewster's law is expressed as

(a) $p = \tan \mu$

(c) $\mu = \tan p$

(b) $p \tan \mu = 1$

(d) $\mu_1 = \mu_2 \tan c$

3.6 Malus law implies that

(a) $I_A \infty \cos \theta$ (b) $I_A \infty \sin \theta$ (c) $I_A \infty \sin^2 \theta$ (d) $I_A \infty \cos^2 \theta$

3.7 Snell's law is obeyed by

(a) e-ray

(c) both e- and o-ray

(b) o-ray

(d) neither e- nor o-ray

3.8 The critical angle for the o-ray for calcite–Canada balsam interface for sodium D-line is

(a) 60°   (b) 67°   (c) 68°   (d) 69°

3.9 For a half-wave plate

(a) $t_1 = \dfrac{\lambda}{(\mu_0 - \mu_e)}$

(c) $t_1 = \dfrac{\lambda}{4(\mu_0 - \mu_e)}$

(b) $t_1 = \dfrac{\lambda}{2(\mu_0 - \mu_e)}$

(d) $t_1 = \dfrac{3\lambda}{4(\mu_0 - \mu_e)}$

3.10 In the expression, $S = \dfrac{\theta}{t_1 \times c}$, the thickness t_1 is measured in

(a) m   (b) mm   (c) cm   (d) dm

3.11 A photoelastic bench uses a

(a) monochromatic source

(b) white light source

(c) both monochromatic and white light sources

(d) halogen lamp source

3.12 A quarter-wave plate converts plane polarized light into

(a) linearly polarized light

(b) circularly polarized light

(c) elliptically polarized light

(d) plane polarized light

3.13 A photoelastic bench must have the provision to apply

(a) an electric field

(c) stress

(b) a magnetic field

(d) thermal stress

3.14 The Projection lens in a photoelastic bench projects the fringe pattern on the

(a) screen

(c) eye

(b) camera

(d) (a) or (b)

3.15 For the isoclinic fringe pattern,

(a) $2\theta = n\pi$

(c) $\theta = n\pi$

(b) $2\theta = (2n + 1)\pi$

(d) $2\theta = (n + 1)\pi$

3.16 For the isochromatic fringe pattern,
    (a) $\delta = n\pi$
    (c) $\delta = (n+1)\pi$

    (b) $\dfrac{\delta}{2} = n\pi$
    (d) $\dfrac{\delta}{2} = (2n+1)\dfrac{\pi}{2}$

3.17 For isotropic points,
    (a) $\sigma_1 > \sigma_2$
    (c) $\sigma_1 = \sigma_2$
    (b) $\sigma_1 < \sigma_2$
    (d) $\sigma_1 \geq \sigma_2$

3.18 For singular points,
    (a) $\sigma_1 = \sigma_2$
    (c) $\sigma_2 > \sigma_1$
    (b) $\sigma_1 > \sigma_2$
    (d) $\sigma_1 = \sigma_2 = 0$

3.19 The angular phase shift Δ_{12} is proportional to
    (a) $\dfrac{1}{\lambda}$
    (b) λ
    (c) $\dfrac{1}{\lambda^2}$
    (d) λ^2

3.20 Choose the correct form of two-dimensional stress-optic law:
    (a) $\mu_1 - \mu_0 = \sigma_1 + \sigma_2$
    (c) $\mu_1 - \mu_0 = \dfrac{\sigma_1}{\sigma_2}$

    (b) $\mu_1 - \mu_0 = c_1\sigma_1 + c_2\sigma_2$
    (d) $\mu_1 - \mu_0 = \sigma_1\sigma_2$

3.21 The stress-optic coefficients are
    (a) same for all materials
    (c) material dependent
    (b) zero for all materials
    (d) variable quantities

3.22 Optical anisotropy under the influence of external forces is called
    (a) induced birefringence
    (c) neither (a) nor (b)
    (b) artificial double refraction
    (d) both (a) and (b)

3.23 The sensitive tint in a biquartz plate has a
    (a) greyish-violet tint
    (c) bluish-grey tint
    (b) blue tint
    (d) red tint

3.24 Laurent's polarimeter uses a
    (a) biquartz plate
    (c) quartz plate
    (b) half-shade device
    (d) quarter plate

3.25 Specific rotation has units of
    (a) cm
    (b) radian
    (c) degree
    (d) arc sec

Review Questions

3.1 What is polarization of light waves?
3.2 Represent plane polarized light schematically.
3.3 Explain Malus law.
3.4 Explain the working principle of a Nicol prism.
3.5 How is a Nicol prism used to produce circularly polarized light?
3.6 Explain the working of a half-wave plate.
3.7 What are dextrorotatory materials?
3.8 Use Fresnel's theory to explain the dextrorotatory behaviour of some materials.
3.9 Explain the working of a biquartz polarimeter.
3.10 Show that $(\sigma_1 - \sigma_2)$ is inversely proportional to the thickness of a photoelastic material.
3.11 Explain the important parts of a photoelastic bench.
3.12 What are optically anisotropic media?
3.13 Differentiate between uniaxial and biaxial crystals.
3.14 Describe the construction of a Nicol prism with the help of a suitable schematic.

3.15 Give some applications of a Nicol prism.
3.16 Differentiate between plane polarized, circularly polarized, and elliptically polarized lights.
3.17 Explain the working of a quarter-wave plate.
3.18 What are laevorotatory materials?
3.19 Write the salient features of Fresnel's half-shade polarization.
3.20 Describe the working of the Laurent's half-shade polarimeter.
3.21 Define specific rotation of an optically active material.
3.22 What are photoelastic materials?

Numerical Problems

3.1 A slab of a specific material produces an angle of polarization of $59°10'$. Calculate the refraction index of the material. [*Hint: $\mu = \tan p$*]

3.2 Determine the angle of polarization for a material with refractive index 1.5. [*Hint: $\mu = \tan p$*]

3.3 Two polarizing sheets are first adjusted to make their polarizing direction parallel. This adjustment leads to the maximization of the transmitted light intensity. Calculate the angle through which either sheet should be turned so that the transmitted intensity becomes ½ of the original value.

[*Hint: $I_A = I_P \cos^2 \theta$*]

3.4 Two polarizers are initially adjusted to make their polarizing directions parallel and thus maximize the transmitted light. One of the polarizers is then rotated through an angle of $30°$. Calculate the intensity of transmitted light in the new orientation. [*Hint: $I_A = I_P \cos^2\theta$*]

3.5 Two Nicol prisms are initially adjusted to be in the crossed position. One of the Nicol prisms is then rotated through $60°$. Determine the percentage of incident unpolarized light that will pass through the system of two Nicol prisms.

$$\left[Hint: I_F = \frac{I}{2}\cos^2\theta \right]$$

3.6 A plane polarized light is incident on a quartz plate that is cut parallel to the axis. Determine the least thickness of the plate that ensures that the o- and e-ray recombine to form a plane polarized light. Assume that $\mu_e = 1.5533, \mu_o = 1.5442$,

and $\lambda = 5 \times 10^{-5}$ cm. $$\left[Hint: t_1 = \frac{\lambda}{2(\mu_e - \mu_0)} \right]$$

3.7 Calculate the thickness of a quarter-wave plate for the situation presented in

Problem 3.6. $$\left[Hint: t_1 = \frac{\lambda}{4(\mu_e - \mu_0)} \right]$$

3.8 The plane of polarization is rotated through an angle of $25°$ as light passes through a 20 cm-long column of 20% solution of an unknown constituent.

Calculate the specific rotation of the solution. $$\left[Hint: S = \frac{\theta}{t_1 C} \right]$$

3.9 A slab of a particular material produces an angle of polarization of $53.6°$. Determine the refractive index of the material. [*Hint: $\mu = \tan p$*]

3.10 A change of material leads to an increase in the angle of polarization by 25% over the value indicated in Problem 3.9. Calculate the refractive index of the material. [*Hint:* $\mu = \tan p$]

3.11 For a particular material, the refractive index is found to be 1.6287. Find the angle of polarization for a slab of this material. [*Hint:* $\mu = \tan p$]

3.12 Two polarizing sheets are initially adjusted so that their polarizing directions are parallel. This orientation results in the maximization of the transmitted light intensity. Indicate the angle through which either sheet should be turned so that the transmitted intensity becomes 1/4.3 of the original value.

$$[Hint: I_A = I_p \cos^2 \theta]$$

3.13 Two polarizing sheets are initially adjusted so that their polarizing directions are parallel. This orientation results in maximization of the transmitted light intensity. One of the sheets is then turned through an angle of 38.7°. Find the

ratio I_A/I_p. $$\left[Hint: \frac{I_A}{I_p} = \cos^2 \theta \right]$$

3.14 Two polarizing sheets are initially adjusted so that their polarizing directions are parallel. One of the sheets is then turned through an angle 41.8°. Calculate

the ratio I_p/I_A. $$\left[Hint: \frac{I_A}{I_p} = \cos^2 \theta \right]$$

3.15 Two Nicol prisms are adjusted to be in the crossed position. One of the Nicol prisms is then rotated through 28.1°. Calculate the percentage of incident unpolarized light that will pass through the system of two Nicol prisms.

$$\left[Hint: I_f = \frac{I}{2} \cos^2 \theta \right]$$

3.16 If the Nicol prism indicated in Problem 3.15 was rotated through 31.8°, what would have been the percentage of incident unpolarized light that would have

passed through the system of two Nicol prisms? $$\left[Hint: I_f = \frac{I}{2} \cos^2 \theta \right]$$

3.17 If the Nicol prism indicated in Problem 3.15 was rotated by 17.2°, what would have been the percentage of incident unpolarized light that would have passed

through the system of two Nicol prisms? $$\left[Hint: I_f = \frac{I}{2} \cos^2 \theta \right]$$

3.18 Two Nicol prisms are adjusted to be in the crossed position. One of the Nicol prisms is rotated through some angle. It is found that 23.8% of the incident unpolarized light passes through the system of two Nicol prisms. What was the

angle through which the Nicol prism was rotated? $$\left[Hint: \frac{I_f}{I} = \frac{\cos^2 \theta}{2} \right]$$

3.19 If the percentage of incident unpolarized light in Problem 3.18 was found to be 41.3%, what was the angle through which the Nicol prism was rotated?

$$\left[Hint: \frac{I_f}{I} = \frac{\cos^2 \theta}{2} \right]$$

3.20 Two Nicol prisms are adjusted to be in the crossed position. On rotating one of the prisms, the ratio $2I_f/I$ is found to be 1/5.3. What was the angle through which the Nicol prism was rotated?

$$\left[Hint: \frac{I_f}{I} = \frac{\cos^2 \theta}{2} \right]$$

3.21 A plane polarized light is incident on a quartz plate that is cut parallel to the axis. The minimum thickness of the plate for which the o- and e-ray recombine to form a plane polarized light is 2.42×10^{-3} cm. If $\mu_e = 1.5406$, calculate the value of μ_o, assuming that $\lambda = 5.2 \times 10^{-5}$ cm.

$$\left[Hint: t_1 = \frac{\lambda}{2(\mu_e - \mu_o)} \right]$$

3.22 If the wavelength in Problem 3.21 was $\lambda = 4.3 \times 10^{-5}$ cm keeping all other parameters same, calculate the value of μ_o.

$$\left[Hint: t_1 = \frac{\lambda}{2(\mu_e - \mu_o)} \right]$$

3.23 A plane polarized light is incident on a quartz plate that is cut parallel to the axis. The least thickness of the plate for which the o- and e-ray recombine to form a plane polarized light is 2.21×10^{-3} cm. If $\mu_e = 1.5512$ and $\mu_o = 1.5462$ for the material, calculate the wavelength of light.

$$\left[Hint: t_1 = \frac{\lambda}{2(\mu_e - \mu_o)} \right]$$

3.24 If the least thickness of the plate for which the o- and e-ray recombine to form a plane polarized light was found to be 1.98×10^{-3} cm in Problem 3.23, calculate the wavelength of the light used. $[Hint: \lambda = 2t_1(\mu_e - \mu_o)]$

3.25 A plane polarized light is incident on a quartz plate that is cut parallel to the axis. The minimum thickness of the plate for which the o- and e-ray recombine to form a plane polarized light is found to be 2.41×10^{-3} cm. If $\mu_o = 1.5460$, calculate the value of μ_e, assuming that $\lambda = 5.8 \times 10^{-5}$ cm.

$$\left[Hint: t_1 = \frac{\lambda}{2(\mu_e - \mu_o)} \right]$$

3.26 If the minimum thickness of the plate for which the o- and e-ray recombine to form a plane polarized light was found to be 2.82×10^{-3} cm in Problem 3.25, then calculate μ_e.

$$\left[Hint: \mu_e = \mu_o + \frac{\lambda}{2t_1} \right]$$

3.27 The plane of polarization gets rotated through 21.8° as light travels through a 24.5% solution of a substance. If the specific rotation of the substance is 61.5°, calculate the length of column.

$$\left[Hint: S = \frac{\theta}{t_1 C} \right]$$

3.28 The plane of polarization gets rotated as light travels through a 16.2% solution of a substance. If the specific rotation of the substance is 58.3° and the length of column is 18.2 cm, calculate the angle through which the plane of polarization gets rotated.

$$\left[Hint: S = \frac{\theta}{t_1 C} \right]$$

Lasers

Learning Objectives

After studying this chapter, students will be able to

- understand the concepts of spontaneous emission, stimulated emission, and population inversion
- comprehend the principle of operation and construction of He–Ne laser, CO_2 laser, and Nd–YAG laser
- realize the principle of operation and structure of a semiconductor laser
- elucidate important industrial and medical applications of lasers
- explain important aspects of holography and its applications

List of Symbols

$\Delta\theta$	= Angular divergence	A_{21}	= Probability of spontaneous transition
λ	= Wavelength		
D	= Diameter of aperture	B_{12}	= Probability of induced transition
f	= Focal length of lens		
τ_C	= Coherence time	B_{21}	= Probability of stimulated transition

4.1 INTRODUCTION

The fact that light travels in a straight line and transition of energy takes place led to the thought of creating order in disorder. Particles or molecules present in ordinary light have random motion, although the exhibition of light is primarily due to an atom being de-excited. To get a focused intense beam, the process of de-excitation has to be scrutinized and designed as per our needs. The laser invention happened because of the order created in the excitation of atoms, thereby intensifying the radiation due to emission. Thus, the introduction of lasers brought about an era of advanced, highly programmed beam of optical rays. Ordinary energy transitions have been

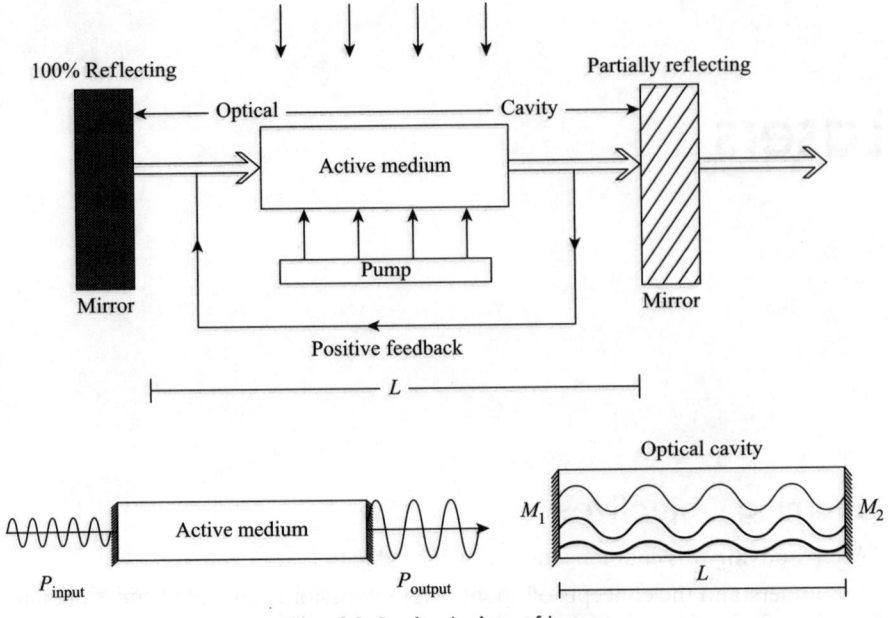

Fig. 4.1 Basic design of laser

transformed into a designed, tailor-made beam of radiation. This beam is intense, focused, and monochromatic in nature. The journey of the discovery of light amplification through stimulated emission of radiation began when Einstein made a theoretical contribution in the year 1916 (discussed in detail in Section 4.5). This was followed by a practical shaping of the device by C.H. Townes and his co-workers in the year 1954. A. Scahwlow, T.H. Maiman, and A. Javan are some of the scientists whose contributions to this field of physics are indeed par excellence. These contributions led to the manifold applications of lasers ranging from biological surgeries, entertainment, and communications, to lethal weaponry. The active medium has atoms in the state of population inversion which amplifies the input light beam by stimulated emission. This process is referred to as optical amplification as depicted in the basic diagram (refer to Fig. 4.1) of a laser.

4.2 COMPONENTS OF LASER

Before we go into the details of how a laser works, let us first have a look at its major components. The basic components of a laser have been shown in Fig. 4.2.

4.2.1 Active Medium

The term 'active medium' in a laser device refers to a conglomeration of atoms, molecules, or ions in any state of matter (i.e., a solid, liquid, or gas) that creates a situation wherein the number of atoms in the higher energy state

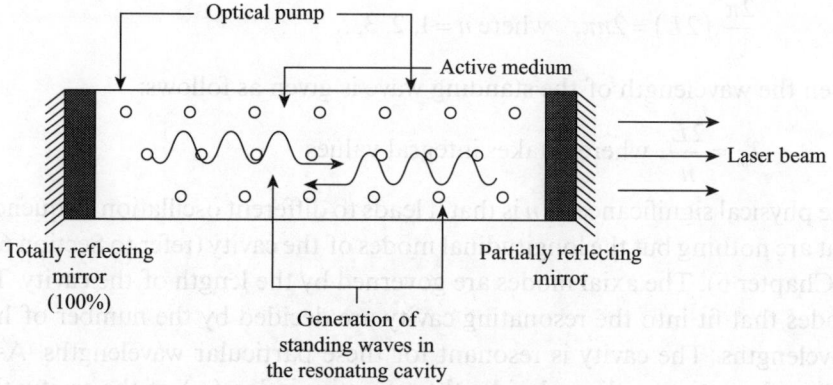

Fig. 4.2 Components of laser

is more than that in the lower energy state. This happens by the choice of the constituents of the active medium. The active medium is 'photon sensitive'. As an electromagnetic radiation is incident on the active medium, amplification of the incident radiation takes place, contrary to the ordinary medium. In the ordinary medium, de-excitation of atoms takes place due to a spontaneous emission, whereas in the active medium, de-excitation takes place through a stimulated emission. Optical amplification of the incident radiation is a phenomenon characteristic to the design of a laser, which occurs owing to population inversion in an active medium. The nature of the active medium classifies the lasers as gas laser, solid-state laser, and semiconductor laser. As we proceed to the detailed design of various laser devices, this aspect will become clearer.

4.2.2 Pumping Source

The pumping mechanism in a laser device helps to achieve the state of population inversion. The pumping source keeps a sustained supply of atoms, molecules, or ions (in any particular state of matter) to the active medium.

4.2.3 Optical Resonator

An optical resonator helps to further intensify and amplify the input as well as the output radiation. The incident electromagnetic radiation needs to be amplified to a certain suitable level to get a desired intense laser output. In addition, for a sustained amplified output radiation from the laser, a part of the output energy is fed back into the system. This is achieved by keeping the active medium in a resonating cavity, which can be as simple as a pair of mirrors. One of the mirrors is totally (100%) reflecting, whereas the other is partially transparent such that the laser output can pass through. Simple parallel plane mirrors form a cavity, with the nodes at the ends. Thus, if L is the length of the cavity, then the going and coming of the wave will result in a phase change of '2π'; hence,

$$\frac{2\pi}{\lambda_m}.(2L) = 2n\pi, \quad \text{where } n = 1, 2, 3, \ldots$$

Then the wavelength of the standing wave is given as follows:

$$\lambda_m = \frac{2L}{n}, \text{ where } n \text{ takes integral values}$$

The physical significance of n is that it leads to different oscillation frequencies that are nothing but the longitudinal modes of the cavity (refer to Section 6.14 in Chapter 6). The axial modes are governed by the length of the cavity. The modes that fit into the resonating cavity are decided by the number of half wavelengths. The cavity is resonant for these particular wavelengths. As in optics, the active medium decides the refractive index (μ) of the cavity; this, in turn, modifies the preceding equation into the following form:

$$\lambda_m = \frac{2L\mu}{n}, \text{ where } n = 1, 2, 3, \ldots$$

Correspondingly, $\vartheta = \dfrac{c}{\lambda} = \dfrac{nc}{2L\mu}$

The frequency difference between the consecutive modes is as follows:

$$\Delta\vartheta = \vartheta - \vartheta_{n-1} = \frac{nc}{2L\mu} - \frac{(n-1)c}{2L\mu} = \frac{c}{2L\mu}$$

In addition to the axial modes, there are transverse modes, which are governed by the cross-sectional dimension of the cavity. Transverse modes have the same phase across the aperture and are few in number. A transverse mode is usually focused on the spot of least size, giving the highest intensity. As we see the design of the cavities, the undesired transitions are suppressed by absorption at the mirrors.

Now that we are aware of the components of a laser, let us see on what principle it works.

4.3 PRINCIPLE OF LASER ACTION

We know that a laser beam is an intense, concentrated, monochromatic beam of light. The principle of a laser is that, through artificial means (at a given time), the number of atoms in an excited state (usually referred to as the metastable state) of an element is made to be much higher than that in the ground state, resulting in the emission of a coherent, monochromatic laser as these atoms de-excite together. The term laser has been expanded as *light amplification by stimulated emission of radiation*. The process of light amplification is to be designed using various media and optical pumping devices such that an amplified simulated beam of light is obtained. The main components of a laser are (a) an active medium, (b) a pumping source, and (c) an optical resonator. In this chapter, we will attempt to explain the basic concepts behind the operation of a laser.

4.4 PROPERTIES OF LASERS

The light emitted from a laser is different from that obtained from a normal source of light where de-excitation takes place. Some of the salient features of a designed lasing device are as follows:

Angular divergence   The lasing beam is observed to have very low angular divergence of the order of 10^{-5} radians. Diffraction of a laser beam gives its divergence. Such a beam whose divergence is given by its diffraction is said to be a diffraction limited beam. The angular divergence $(\Delta\theta)$ of a beam due to an aperture of diameter D, illuminated by a wavelength λ, is given as follows:

$$\Delta\theta = \frac{\lambda}{D} \tag{4.1}$$

Focusing   As has been stated in the definition for angular divergence, the laser beam is diffraction limited; thus, the radius of the focused spot is given as follows:

$$r = \frac{\lambda_l f}{a} \tag{4.2}$$

$$\Rightarrow A = \pi r^2 = \pi \left(\frac{\lambda_l f}{a}\right)^2 \tag{4.3}$$

where l_1 is the wavelength of the laser light, f is the focal length of the lens, a is the radius of the aperture, and A gives the area of the focused spot.

Numerically, from Eq. (4.3), the area of the focused spot is calculated; it is nearly of the order of $10^{-10}\,\text{m}^2$ (refer to the following examples for exact numerical input). Thus, laser beams can be focused to extremely small regions with a very high intensity.

Intensity   Owing to these two properties of angular divergence and focusing, a lasing beam becomes highly intense. The power of a laser beam is of the order of $10^5\,\text{W}$ for a continuous laser, hence the manifold industrial and biomedical applications of laser.

Monochromaticity   'Chrome' means colour and 'monochrome' means only one colour, that is, a laser beam is characterized by a single wavelength. Lasing beams have a pure spectrum, having a small spectral width $\Delta\lambda \sim 10^{-6}\,\text{Å}$. This property leads to the vast applications of lasers in holography, optical communications, spectroscopy, and other fields. The spectral width is related to the coherence length of a given radiation (i.e., $L = c\tau_c$, where τ_c is the coherence time, the time for which the wave remains oscillatory):

$$\Delta\lambda \sim \frac{\lambda^2}{L} = \frac{\lambda^2}{c\tau_c} = \frac{\Delta\vartheta}{\vartheta} \tag{4.4}$$

Hence, it is evident from this mathematical relation that, for that specific spectral width, purity of the spectrum and coherence of time and space are interrelated. It is worth mentioning here that the spectral width is an indication of the monochromaticity of a given radiation.

Example 4.1 Show that the angular spread of a filament is more than that of a laser. Assume that the length of the filament is 1 mm, focal length 10 cm, and diameter of the aperture 5 cm.

Solution   The angular spread due to a filament is calculated as follows:

$$\Delta\theta = \frac{l}{f} = \frac{1\,\text{mm}}{100\,\text{mm}} = 0.01\,\text{radians}$$

The angular spread of the aperture for a laser is calculated to be the following:

$$\Delta\theta = \frac{\lambda}{D} = \frac{5x \times 10^{-5}}{5} = 10^{-5}\,\text{radians}$$

Hence, it is evident that the angular divergence of a laser beam is much smaller than that of a filament (of an ordinary bulb).

Example 4.2   Find the intensity of a 1 mW laser beam $(\lambda_l = 6 \times 10^{-5}\,\text{cm})$ that is incident on an aperture whose focal length is nearly 2.0 cm and whose diameter is 2.0 mm.

Solution *Area of the focused spot* $A = \pi \left(\dfrac{\lambda_l f}{a}\right)^2 = \left(\dfrac{22}{7}\right) \times \left(\dfrac{6 \times 10^{-5} \times 2.0}{0.1}\right)^2 = 4.52 \times 10^{-6}\,\text{cm}^2$

The intensity is given as follows: $I = \dfrac{P}{A} = \dfrac{1 \times 10^{-3}\,\text{W}}{4.52 \times 10^{-10}\,\text{m}^2} = 2.2 \times 10^{6}\,\text{W/m}^2$

Thus, one can observe that even a low-power laser can produce intensity of the order of 10^6 SI units.

Example 4.3   Calculate the electric field corresponding to a 1.0 MW laser beam incident on a lens of focal length 2.0 cm $(\lambda_l = 6 \times 10^{-5}\,\text{cm}$, with a beam width of 2 mm).

Solution   Area of the focused spot is calculated as follows:

$$A = \pi\left(\frac{\lambda_l f}{a}\right)^2 = \frac{22}{7} \times \left(\frac{6 \times 10^{-5} \times 2.0}{0.1}\right)^2 \approx 10^{-6}\,\text{cm}^2 = 10^{-10}\,\text{m}^2$$

The intensity is given as follows:

$$I = \frac{P}{A} = \frac{1 \times 10^6\,\text{W}}{4.52 \times 10^{-10}\,\text{m}^2} = 2.2 \times 10^{15}\,\text{W/m}^2$$

Now, the intensity of the electric field is given as follows:

$$I = \frac{1}{2}\epsilon_0 c\,E_0^2 \Rightarrow 2.2 \times 10^{15} = \frac{1}{2} \times 8.854 \times 10^{-12} \times 3 \times 10^8\,E_0^2 \text{ (SI units)}$$

$$E_0 = 4.06 \times 10^9\,\text{V/m}$$

Thus, the laser beam has high-power electric fields. These high electric fields are used for welding, drilling, cutting materials, and other such applications.

Example 4.4   Find the spectral width of a He–Ne laser having a coherent time of the order of nanoseconds.

Solution   The spectral width of the laser is calculated as follows:

$$\Delta\lambda \sim \frac{\lambda^2}{c\tau_c} = \frac{(6.438 \times 10^{-5})^2}{3 \times 10^{10} \times 10^{-9}} \sim 0.01\,\text{Å}$$

Example 4.5   Compare the monochromaticity of an ordinary light $(\Delta\vartheta \sim 10^{10}\,\text{Hz})$ and a laser light $(\Delta\vartheta \sim 100\,\text{Hz})$, for a light of frequency $10^{14}\,\text{Hz}$.

Solution For laser light, $\dfrac{\Delta \vartheta}{\vartheta} = \dfrac{100}{10^{14}} = 10^{-12}$

For ordinary light, $\dfrac{\Delta \vartheta}{\vartheta} = \dfrac{10^{10}}{10^{14}} = 10^{-4}$

It is evident that the monochromaticity of a laser light is much more than that of an ordinary light.

4.5 PRINCIPLES OF SPONTANEOUS EMISSION AND STIMULATED EMISSION

After having studied the principles of laser and its properties in the previous sections, it now becomes increasingly imperative for us to understand how transitions take place in elements. Under normal conditions, an atom always tries to be in the state of minimum energy, which is referred to as the lowest energy state or the ground state. From the preliminary ideas of Bohr's quantum theory, transitions are referred to as 'quantum jumps' taken by a photon. Or simply stated, an atom jumps from its ground state to the excited states by absorbing a photon. There are selection rules that govern this process; therefore, we can only talk about the 'allowed transitions'. Contrary to this phenomenon of allowed transitions is excitation, which occurs due to the collision of electrons in their random motion, leading to non-radiative transitions. We will not dwell much on this, as this has been the principle for all electrical (solid state) discharges. As an illustrative example, one can see Fig. 4.3, where both the phenomena of absorption and emission are explained in a ruby crystal.

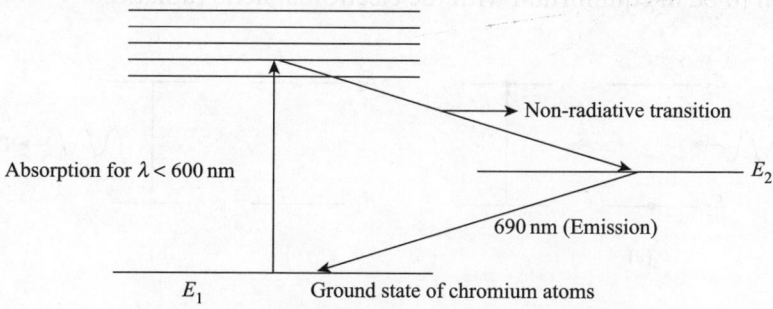

Fig. 4.3 Absorption and emission in ruby crystal

Our primary interest in this chapter is to understand the emission spectrum, which is a fall back of the phenomenon of absorption of light. Obeying the conservation rules of transitions, the number of atoms that emit photons during time Δt is directly proportional to the number of excited atoms, as illustrated in Fig. 4.4. As we study the theory of photon jumps, we can observe that the lifetime of an atom in the excited state plays a key role in transitions.

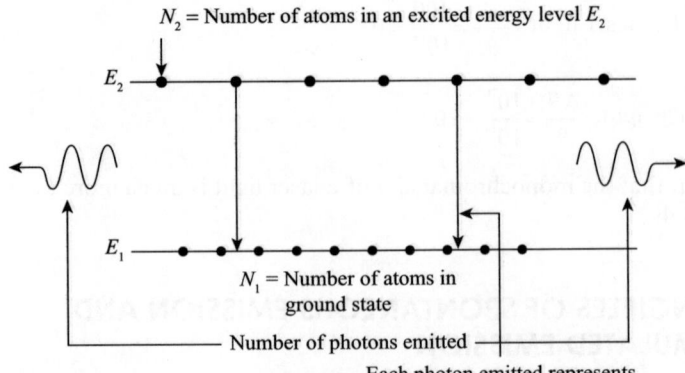

Fig. 4.4 Conservation rules of transitions, which state that the number of atoms that emit photons during time Δt is directly proportional to the number of excited atoms

In 1917, Einstein developed the theory of transitions among different atomic energy states under the influence of an electromagnetic radiation. Three types of radiative transitions can take place: (a) absorption, (b) spontaneous emission, and (c) stimulated emission (alternatively referred to as reverse absorption); these processes can be seen in Fig. 4.5. The first two radiative transitions are well established. Einstein introduced the new concept of stimulated emission, which is actually reverse absorption, such that during the emission process, the characteristics of the second photon become identical to those of the first absorbed photon. This theory helps us calculate the probability of a radiative transition occurring between two states, assuming the atomic system to be in equilibrium with the electromagnetic radiation.

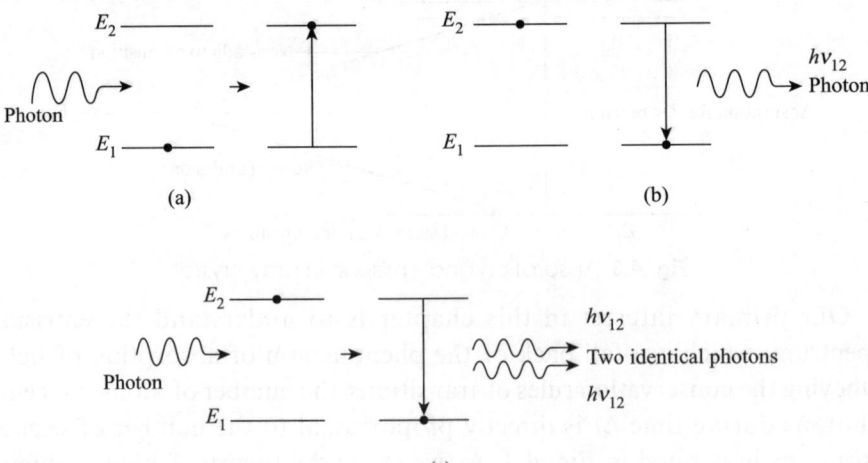

Fig. 4.5 Radiative transitions (a) Absorption (b) Spontaneous emission (c) Stimulated emission

Suppose a system consists of an assembly of a large number of atoms in different energy states. Assume that the system is at an absolute temperature T. The number of atoms, n, per unit volume in an energy state E is given by the Boltzmann distribution law:

$$n = n_0 \exp\left(-\frac{E}{k_B T}\right) \tag{4.5}$$

where k_B is the Boltzmann constant ($k_B = 1.38 \times 10^{-23}$ J/K) and n_0 represents the number of atoms per unit volume in the ground state. Let n_1 and n_2 represent the number of atoms per unit volume in the two energy states E_1 and E_2, respectively, with $E_2 > E_1$. Using Eq. (4.5), we can write the following expression:

$$\frac{n_2}{n_1} = \exp\left(-\frac{E_2 - E_1}{k_B T}\right) \tag{4.6}$$

From Eq. (4.6) it can be concluded that $n_2 < n_1$ for $E_2 > E_1$.

When an atom is excited to a higher energy level, it returns to the ground-state energy level in approximately 10^{-8} seconds by releasing a photon of energy hv, where $hv = E_2 - E_1$. Transition of atoms from a higher energy level to a lower energy level without any external assistance is called *spontaneous transition*.

The following are the characteristics of a spontaneous emission:
1. It is probabilistic in nature. The probabilistic behaviour indicates a disorder in the de-excitation process. Spontaneous transitions cannot be controlled.
2. The frequency of the emitted radiation is identical to that of the photon/atom that was originally absorbed.
3. Directions and phases of the emitted photons/atoms are random in nature.
4. The light emitted is incoherent and is not necessarily monochromatic in nature, as it is a superposition of random phases.
5. The net intensity is proportional to the number of radiating atoms. If I is the intensity of the light emitted by one atom and N is the number of atoms, then the total intensity I_{total} is given as follows: $I_{total} = NI$
6. The rate equation that governs the spontaneous emission process can be represented as follows: $R_{sp} = -\left[\dfrac{dn_2}{dt}\right]_{sp} = A_{21} n_2$

Thus, atoms in energy state E_2 undergo spontaneous transition to lower energy state E_1. The number of such spontaneous transitions per unit volume per unit time, n_{21}, is given by the following equation:

$$n_{21} = A_{21} n_2 \tag{4.7}$$

where A_{21} is the probability of spontaneous transition per unit time from level E_2 to level E_1. These transitions from higher to lower energy states result in emission of an electromagnetic radiation of energy $h\vartheta$, given as follows:

$$h\vartheta = E_2 - E_1 \tag{4.8}$$

It is clear that $1/A_{21}$ is a measure of the lifetime of the upper state before spontaneous decay to the lower state. Assume the atoms of the system to be in thermal equilibrium with the electromagnetic radiation of frequency = $(E_2 - E_1)/h$. Some of the atoms in lower energy state E_1 will undergo transitions to higher energy state E_2 due to absorption of this radiation. The number of such transitions (n_{12}) per unit volume per unit time is given by the following equation:

$$n_{12} = B_{12}\, n_1 u_9 \qquad (4.9)$$

Here, B_{12} is the probability of radiation-induced transitions per unit time from lower energy state E_1 to higher energy state E_2, and u_9 represents the energy density of the electromagnetic radiation.

As we study the emission spectra in detail, we observe that the interaction of an electromagnetic radiation with a system can induce transition of atoms from a higher energy state E_2 to a lower energy state E_1. These result in stimulated or induced emission of radiation of frequency n, given by the relation $(E_2 - E_1)/h$. The number of stimulated transitions per unit volume per unit time, n'_{21}, is given by the following expression:

$$n'_{21} = B_{21}\, n_1 u_9 \qquad (4.10)$$

where B_{21} represents the probability per unit time for a stimulated emission from energy level E_2 to energy level E_1.

The characteristics of a stimulated emission are as follows:

1. A stimulated emission is induced by the interaction of a system with an electromagnetic radiation.
2. The electromagnetic radiation must have a resonance effect with the system, which means that both must possess the same frequency, direction, phase, and polarization.
3. The emitted photon is identical to the absorbed photon of the electromagnetic radiation, resulting in 'light amplification' in the emission spectra. The phenomenon of light amplification means that when a single photon interacts with an excited atom, two photons will emerge, which in turn will create two more photons, and so on, thus creating a chain reaction of photons (as illustrated in Fig. 4.6).
4. All atoms in the emission spectra have the same phase as they emerge together as a 'coherent radiation'.
5. The intensity of light will be proportional to the square of the number of atoms radiating the light. Thus, if N is the number of atoms and I is the intensity, then $I_{\text{total}} = N^2 I$
6. The rate equation that governs the stimulated emission process can be written as follows: $R_{\text{st}} = -\left[\dfrac{dn_2}{dt}\right]_{\text{st}} = B_{21} u_9\, n_2$

Coefficients A_{21}, B_{12}, and B_{21} are known as Einstein's coefficients or Einstein's A and B coefficients. In equilibrium, the time rate of transition from energy level E_2 to energy level E_1 must equal the time rate of transition from energy level E_1 to energy level E_2. Thus, using Eqs (4.5–4.10), we can write the following expression:

$$n_{21} + n'_{21} = n_{12} \tag{4.11}$$

which can be expressed in the following form:

$$A_{21}n_2 + B_{21}n_1u_9 = B_{12}n_1u_9 \tag{4.12}$$

This yields the following equation:

$$u_9 = \frac{A_{21}n_2}{(B_{12}n_1 - B_{21}n_1)} \tag{4.13}$$

which can be rewritten as follows:

$$u_9 = \frac{A_{21}}{B_{21}}\left[\frac{1}{\left(\left(\frac{B_{12}}{B_{21}}\right)\left(\frac{n_1}{n_2}\right)\right) - 1}\right] \tag{4.14}$$

Using Eq. (4.6) in Eq. (4.14), we get the following expression:

$$u_9 = \frac{A_{21}}{B_{21}}\left[\frac{1}{\left(\left(\frac{B_{12}}{B_{21}}\right)\exp\left[\frac{(E_2 - E_1)}{kT}\right]\right) - 1}\right] \tag{4.15}$$

Further, using Eq. (4.6) in Eq. (4.11), one gets the following relation:

$$u_9 = \frac{A_{21}}{B_{21}}\left[\frac{1}{\left(\left(\frac{B_{12}}{B_{21}}\right)\exp\left[\frac{(h9)}{kT}\right]\right) - 1}\right] \tag{4.16}$$

Now, according to *Planck's* radiation formula:

$$u_9 = \frac{8\pi h 9^3}{c^3}\left[\frac{1}{\exp\left(\frac{h9}{kT}\right) - 1}\right] \tag{4.17}$$

where c is the velocity of light. Equations (4.16) and (4.17) must be in agreement; therefore, comparing the two,

$$B_{12} = B_{21} \tag{4.18}$$

$$\frac{A_{21}}{B_{21}} = \frac{8\pi h 9^3}{c^3} \tag{4.19}$$

Equations (4.18) and (4.19) are also known as Einstein relations. Equation (4.16) implies that the probability of radiation-induced transitions per unit time equals the probability of stimulated emissions per unit time.

Using simple quantum-mechanical radiation formula and equations (4.7), (4.9), (4.19) and (4.20), it can be shown that

$$\frac{n_{21}}{n'_{21}} = \frac{A_{21}}{B_{21}u_9} = \exp\left(\frac{h9}{kT}\right) - 1 \qquad (4.20)$$

Thus, if $h9 \gg kT$, spontaneous emission is more probable than stimulated emission. However, if $h9 \ll kT$, stimulated emission starts playing an important role. *Such a condition exists in the microwave region of a spectrum.*

Thus, the attainment of stimulated or induced emission is possible in elements, as a system may have more than one way to attain a given energy level; each different way is a different state.

Example 4.6 Show that the fraction of atoms in the excited state is much smaller than that in the ground state at a temperature of $3000\,\text{K}$ (nearly that of a filament bulb) and an energy gap of $2\,\text{eV}$.

Solution We know from our discussion in Section 4.5 that the ratio of atoms in the higher energy level to that in the lower energy level is given as follows:

$$\frac{n_2}{n_1} = \exp\left(-\frac{E_2 - E_1}{k_\text{B}T}\right)$$

$$\frac{n_2}{n_1} = \exp\left(-\frac{2.0 \times 1.6 \times 10^{-19}\,J}{(1.38062 \times 10^{-23}\,\text{J/K}) \times (3000\,\text{K})}\right)$$

$$\frac{n_2}{n_1} = \exp(-7.73) = 0.00044$$

Thus, even at such a high temperature, the probability of finding electrons in the excited state is very small. Hence, it is extremely difficult to attain the state wherein a stimulated emission can take place between an electromagnetic radiation and the system.

4.6 POPULATION INVERSION

A laser uses the phenomenon of stimulated emission of radiation. Suppose E_2 and E_1 represent two allowed energy levels, where $E_2 > E_1$. As discussed in the previous section, normally an atom in an excited energy state E_2 relaxes by the process of spontaneous emission to the state having energy level E_1. Normally, the number of atoms in the excited energy state E_2 is much lower than that in the lower energy state E_1. The mean life of excited atoms before they undergo a spontaneous emission is around 10^{-8} s. Lasers, however, utilize the phenomenon of stimulated emission. For some excited states, the mean life of excited atoms can increase up to 10^5 times the normal value. In some mechanisms, the rate of excitation of atoms from the ground state to a higher energy state is greater than the rate of decay of atoms from a higher energy state to a lower energy state. Under these conditions, the number of atoms (n_2) in the higher energy state becomes greater than that in the lower energy states. This condition is essential for laser action and is known as *population inversion*.

A photon of energy $h9_{12} = E_2 - E_1$ is incident on an atom in an excited state with energy level E_2. This photon stimulates the atom to go down to energy level E_1 and in the process releases a photon of energy $h9_{12}$. The electromagnetic

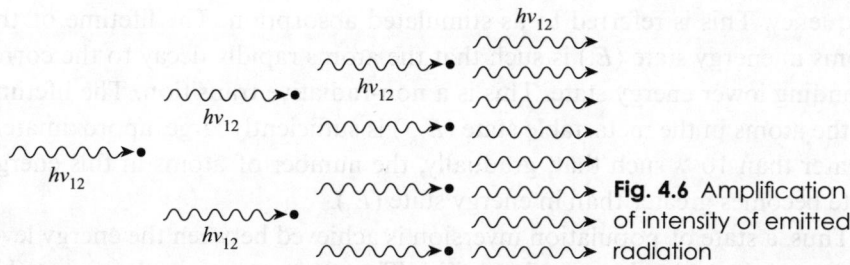

Fig. 4.6 Amplification of intensity of emitted radiation

wave corresponding to the emitted photon has the same energy, phase, polarization, and direction as that corresponding to the incident photon. Two photons result from this interaction. These two photons can in turn be incident on other atoms and release four photons, and the process can continue indefinitely. This results in amplification of the intensity of the emitted radiation, as shown schematically in Fig. 4.6.

4.6.1 Population Inversion by Pumping

The condition of population inversion is created by the process of pumping, which can be achieved in a variety of ways. A number of methods are capable of creating population inversion, which are as follows:

Optical pumping Photons of a chosen frequency are used for the process of optical pumping.

Three-level laser system We can assume that the transitions between all the three energy levels are permissible (Fig. 4.7). For a transition from energy level E_G to E_2,

$$h\vartheta_{13} = E_3 - E_1 \tag{4.21}$$

The atoms in the ground state (E_G) absorb the photons to reach the higher energy state. This is achieved by pumping photons of nearly compatible

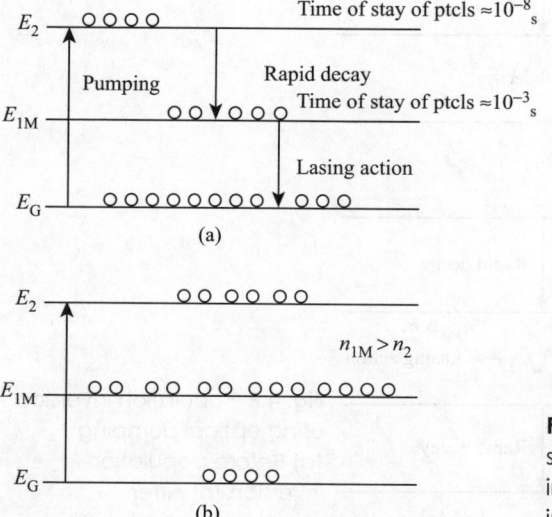

Fig. 4.7 Three-energy-level system (a) Before population inversion (b) After population inversion

frequency. This is referred to as stimulated absorption. The lifetime of the atoms in energy state (E_2) is such that the atoms rapidly decay to the corresponding lower energy state. This is a non-radiative transition. The lifetime of the atoms in the metastable state (E_{1M}) is sufficiently large, approximately greater than $10^{-8}s$ such that, gradually, the number of atoms in this energy state becomes greater than in energy state (E_2).

Thus, a state of population inversion is achieved between the energy level (E_{1M}) and the ground energy level (E_G). The photons stay in the metastable energy state for a longer duration of time. As the photons possess nearly the same frequency, stimulated emission take place almost simultaneously thus creating conditions for obtaining a lasing beam.

Four-level laser system   In this technique of four-level system (Fig. 4.8), the main characteristic of the energy levels is that the lifetime of the energy bands labelled 'M' (metastable) have a lifetime greater than that of an ordinary excited energy level. Incident photons with compatible frequency to the energy levels of the system are used to excite the atoms from the ground state to the higher energy state. Most of the excited atoms are already existing in the higher energy states, in this case E_2 and E_4. Minimum energy pumping leads to stimulated absorption such that photon transition takes place from the lower energy state to higher energy state. As the lifetime of atoms in the metastable state is relatively high, over a period of time the number of atoms in this state becomes greater than that in the ordinary higher state, that is, the condition of

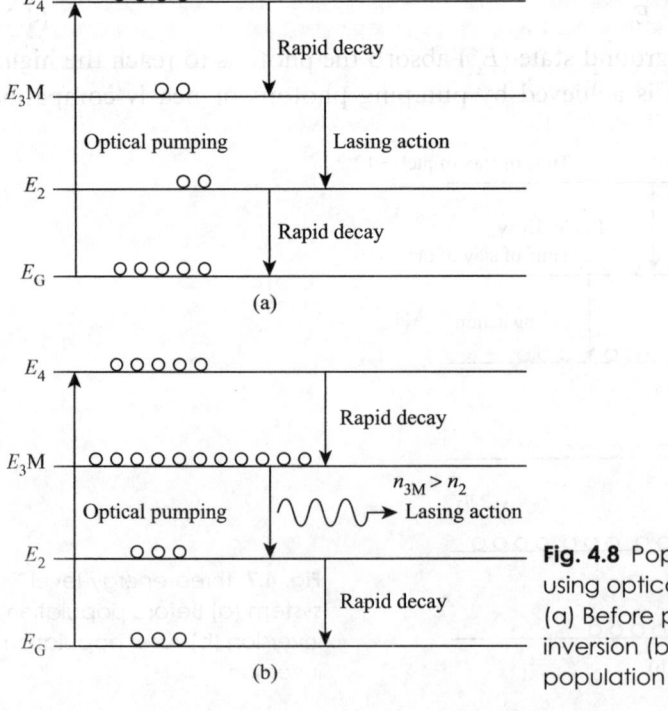

Fig. 4.8 Population inversion using optical pumping (a) Before population inversion (b) After population inversion

population inversion is achieved between the metastable states and the ordinary ground states. The advantage of the four-level laser system over the three-level laser system is that by giving minimum energy to the incident photons, one can create population inversion. The stimulated emission takes place from the metastable energy level to the ground state energy level E_G, or energy level E_2.

Electrical pumping An intense electrical discharge in the active medium can also be used to raise atoms to their higher-energy-level states. This technique is more commonly used in gases, where the discharge creates a plasma state in which active centres collide inelastically with the available free electrons, leading to a predominant population of higher-energy-level atoms. Sometimes, other atoms and molecules are deliberately introduced into the gas to facilitate the process.

Chemical pumping In this technique, exothermic chemical reactions are used to raise the energy of active centres to higher levels. This process leads to population inversion.

Thermal pumping In this technique, the active material is first raised to a high temperature. This leads to an increase in the energy levels. The temperature of the active material is then rapidly brought down, giving rise to population inversion.

4.7 TYPES OF LASERS

Lasers are of different types. Some of the more common types are solid-state, gas, and semiconductor lasers. The active material used in a laser decides the type of a particular laser. We will study some important laser systems in this section.

4.7.1 He–Ne Laser

He–Ne laser is the first functional gas laser. Gas lasers are supposed to emit light that is monochromatic, directional, and continuous. It is discussed in detail in the following sections.

4.7.1.1 Principle of Operation

As mentioned in Section 4.2.1, an active medium refers to a combination of gases (in case of gas laser) that are capable of triggering population inversion as light is incident on it. In a He–Ne laser, a mixture of helium (He) and neon (Ne) in the ratio of 10:1 is taken for the active medium. Energy levels of He and Ne are shown in Fig. 4.9. As is evident from the ratio of the mixtures, the amount of He gas is much more than the Ne gas, which is required for the He atoms to be excited. These excited He atoms collide with the Ne atoms, thus raising them to higher energy levels. If He gas is not present, then Ne atoms would not be excited to higher energy levels, and no coherent lasing beam would be formed. Only Ne laser would have a very low pumping efficiency and hence there would be no lase.

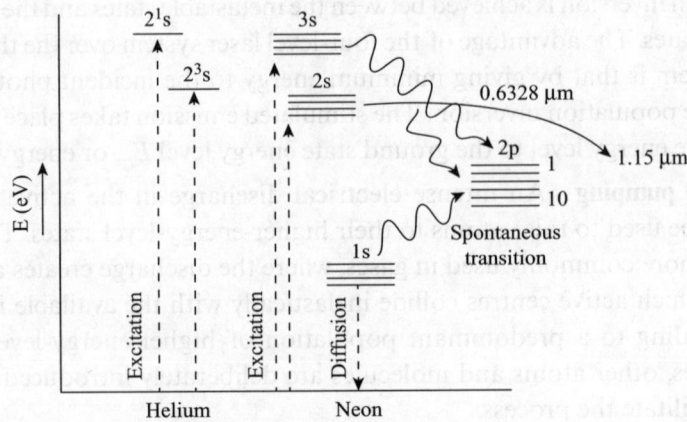

Fig. 4.9 Energy levels of He and Ne

A discharge passing through the gas mixture raises He atoms to higher energy levels 2^3s and 2^1s through collisions with accelerated particles.

Levels 2^3s and 2^1s are metastable states, and no allowed transitions are possible from such states. The He atoms in these excited states collide inelastically with the Ne atoms in the ground state. The He atoms transfer their energy to the Ne atoms and return to the ground state. The Ne atoms are promoted to higher metastable states 3s and 2s due to the gained energy. Since the metastable states 3s and 2s have higher lifetimes than the 2p state, population inversion occurs.

$$He^*\left(2^3S_1\right)+Ne^1S_0 \rightarrow He(^1S_0)+Ne^*2S_2 + \Delta E$$

$$He^*\left(2^1S\ \right)+Ne^1S_0 + \Delta E \rightarrow He(^1S_0)+Ne^*3S_2$$

The states marked with a '*' symbol represent the excited states. The energy difference ΔE between the excited states is nearly of the order of 0.05 eV, which is supplied by the kinetic energy.

4.7.1.2 Pumping

The energy of the pumping source in a laser is provided by a high voltage electrical discharge. This discharge is passed through the gas mixture between the electrodes, that is, anode and cathode, within the tube. A radio frequency generator is used to excite the active material used in the He–Ne laser.

Optical resonator　In an optical resonator, the resonating cavity consists of two concave mirrors at each end (one mirror can be plane also). One of the mirrors is 100% reflecting at the lasing frequency, whereas the other is partially reflecting. The reflecting mirrors amplify the longitudinal mode radiation by reflecting the mode back upon itself, gaining more power in each turn than is lost. The mirrors may be sealed inside the tube; however, in the recent fabrication there is an external mirror arrangement, which is preferred. The typical arrangement is shown in Fig. 4.10.

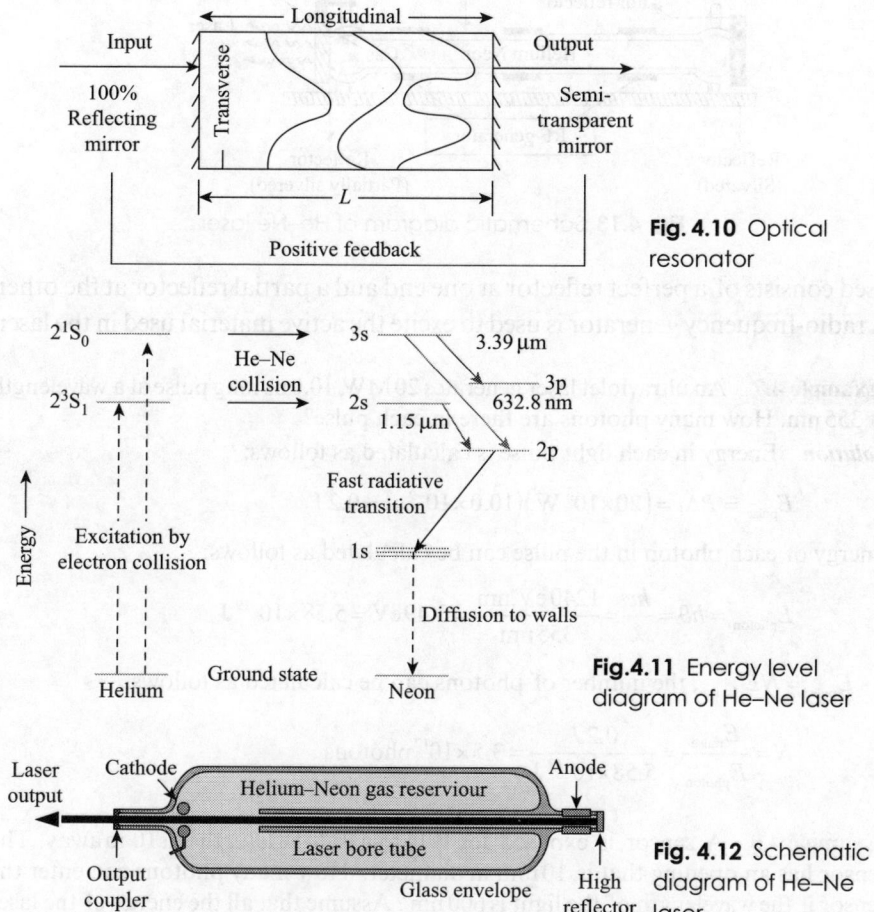

Fig. 4.10 Optical resonator

Fig.4.11 Energy level diagram of He–Ne laser

Fig. 4.12 Schematic diagram of He–Ne laser

Lasing beam As the excited Ne atoms relax from the metastable state 3s to state 2p, as shown in the energy level diagram in Fig. 4.11, photons of wavelength 0.6328 μm are emitted. Similarly, when excited Ne atoms relax from the metastable state 2s to state 2p, photons of wavelength 1.15 μm are emitted, as shown in Fig. 4.12. These photons can, in turn, be used to create a stimulated emission of photons of the same wavelength using suitably designed optical resonators. This process has a cascading effect and ultimately produces an intense beam that exits through the partially silvered end of the optical resonator.

4.7.1.3 Construction of He–Ne Laser

Figure 4.13 shows a schematic diagram of a He–Ne laser.

An He–Ne laser consists of a fused quartz tube that is filled with a mixture of Ne and He gases. The quartz tube is about 80 cm long, with a diameter of about 1.5 cm. The quartz tube is filled with a mixture of He (at a pressure of 1 mmHg) and Ne (at a pressure of 0.1 mmHg) gases. The optical resonator

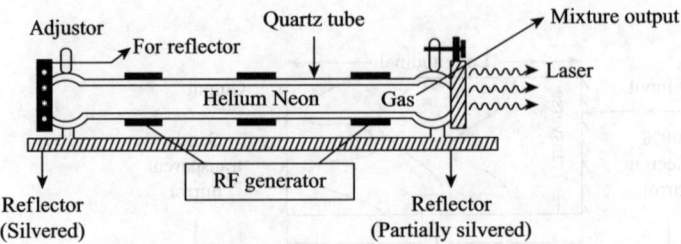

Fig. 4.13 Schematic diagram of He–Ne laser

used consists of a perfect reflector at one end and a partial reflector at the other. A radio-frequency generator is used to excite the active material used in the laser.

Example 4.7 An ultraviolet laser generates 20 MW, 10.0 ns long pulse at a wavelength of 355 nm. How many photons are there in each pulse?

Solution Energy in each light pulse is calculated as follows:

$$E_{pulse} = P\Delta t = \left(20 \times 10^6 \text{ W}\right)\left(10.0 \times 10^{-9} \text{ s}\right) = 0.2 \text{ J}$$

Energy of each photon in the pulse can be calculated as follows:

$$E_{photon} = h\vartheta = \frac{hc}{\lambda} = \frac{1240 \text{ eV.nm}}{355 \text{ nm}} = 3.49 \text{ eV} = 5.58 \times 10^{-19} \text{ J}$$

As $E_{pulse} = NE_{photon}$, the number of photons can be calculated as follows:

$$N = \frac{E_{pulse}}{E_{photon}} = \frac{0.2 \text{ J}}{5.58 \times 10^{-19} \text{ J}} = 3.5 \times 10^{17} \text{ photons}$$

Example 4.8 A sensor is exposed for 0.1 s to a 10 MW laser that is 10 m away. The sensor has an opening that is 10 mm in diameter. How many photons can enter the sensor if the wavelength of the light is 600 nm? Assume that all the energy of the laser is given off as light.

Solution The energy of a photon of the light is calculated as follows:

$$E = \frac{hc}{\lambda} = \frac{(6.634 \times 10^{-34} \text{ J.s} \times 3 \times 10^8 \text{ m/s})}{600 \times 10^{-9}} = 3.3 \times 10^{-19} \text{ J}$$

The laser has 10 MW power. The number of photons emitted per second is, thus,

$$N = \frac{10 \times 10^6 \text{ J/s}}{3.3 \times 10^{-19} \text{ J}} = 3.03 \times 10^{25} \text{ photons/s}$$

As the radiation is spherically symmetrical, the number of photons entering the sensor per second is N multiplied by the ratio of the aperture area to the area of a sphere of radius 10 mm:

$$(3.03 \times 10^{25}) \frac{\pi(0.010)^2}{4\pi(10)^2} = 0.7575 \times 10^{19} \text{ photons/s}$$

The number of photons that enter the sensor in 0.1 s is $(0.1) \times (0.7575 \times 10^{19}) = 7.575 \times 10^{17}$ photons.

We observe the operation of sensors in our daily life. These sensors will be categorized as optical sensors or light sensors.

4.7.2 Ruby Laser

Ruby laser is a solid-state laser that produces pulses of visible light at a wavelength of 694.3 nm, which corresponds to the red colour. The pulses of lasing beam are of the order of a millisecond.

4.7.2.1 Principle of Operation

The amplification medium (or the laser gain medium) is a cylindrical crystal of ruby. Ruby possesses the property of fluorescence, which causes stimulated emission. The ends of the crystal are flat; one of the ends is completely silvered, whereas the other is partially silvered. Ruby, which basically is an aluminium oxide (Al_2O_3) crystal, is suspended (doped) with particles of chromium. The energy levels of the activator chromium atoms are broadened due to characteristics of the aluminium oxide crystal. Chromium atoms possess two energy levels that correspond to a lifetime of 10^{-8} s. The metastable state is also available in the energy band structure of chromium atoms with a lifetime of 3×10^{-3} s. As is the characteristic of a metastable state, the particle stays in this state for a relatively longer period than in an ordinary excited energy state.

4.7.2.2 Pumping Source

Chromium ions account for about 0.05% of the weight of a ruby crystal. Al^{3+} ions are replaced by Cr^{3+} ions. A flash lamp, filled usually by xenon gas, acts as a source of photons for chromium atoms. Chromium ions present in the ground state absorb these photons to make a transition from the ground state to band $E_1 (\lambda \sim 6600\,\text{Å})$ or band E_2 $(\lambda \sim 4000\,\text{Å})$. As mentioned earlier, chromium atoms have a metastable state; thus, there is non-radiative transition from state E_1 and E_2 to the metastable state (M), the extra energy being absorbed by the lattice (in designing a laser, liquid nitrogen is used as a cooling agent to remove extra heat from the ruby rod). As the metastable state has a longer lifetime, these atoms continue to be collected at this state. At a given point of time, the number of atoms in the metastable state of chromium is more than that in its ground state. This state is referred to as population inversion, in a lasing system.

4.7.2.3 Optical Resonator

In optical amplification, a medium with population inversion is required. Chromium atoms in the state of population inversion are now capable of amplifying the input light beam by their stimulated emission. Thus, this whole active medium is placed in a resonating (optical) cavity, created by the use of mirrors that are facing each other. The two reflecting ends of the ruby rod form a cavity.

Ruby has broad and powerful absorption bands in the visual spectrum at around 400–550 nm, with a very long fluorescence lifetime of 3 ms. The lasing action is triggered by spontaneously emitted photons with a deep red colour at a wavelength of 694.3 nm, with a very narrow line width of 0.53 nm.

As the population inversion reaches its threshold value due to increased pumping energy, both spontaneous and stimulated emissions start with a sharp peak at 694.3 nm. The lased light that is produced is coherent in nature. As the energy level depopulates through stimulated emission, the lasing action stops. As the pumping is going on, before the output goes to zero, the atoms again settle in the metastable energy level to generate a LASE. Thus, the output obtained is in the form of a discontinuous lasing beam having spikes at a wavelength of 694.3 nm. As the flash lamp operation is pulsed the lasing output is also pulsed.

Example 4.9 The metastable state of a ruby laser is at 1.786 eV. Calculate the wavelength of light emitted.

Solution Energy level of the metastable state for a ruby laser is indicative of the difference in the energy levels:

$$\Rightarrow E_2 - E_1 = 1.786 \, \text{eV}$$

$$E_2 - E_1 = \frac{hc}{\lambda} = \frac{1240 \, \text{eV.m}}{\lambda} \Rightarrow \lambda = \frac{1240 \, \text{eV.nm}}{1.786 \, \text{eV}} = 694.3 \, \text{nm}$$

Thus, the emitted light has a wavelength of 694.3 nm.

Example 4.10 Calculate the pulse energy in eV, if 2 moles of Cr^{+3} ions are involved in the population inversion process in a ruby laser that emits a light of wavelength 694.3 nm.

Solution As calculated from the preceding example, 1.786 eV is the energy of the single photon in the metastable level. As indicated, 2 moles of Cr^{+3} ions are involved in the population inversion process that produces a laser pulse. Hence, the pulse energy can be evaluated as follows:

$$2 \times 6.023 \times 10^{23} \times 1.786 \, \text{eV} = 21.514 \times 10^{23} \, \text{eV} = 2.1514 \times 10^{24} \, \text{eV}$$

Example 4.11 The light output of a typical laser is 694.3 nm. What is the difference between the energy levels of the excited and the metastable states? What will be the energy of the photon emitted? What is the frequency associated with the photon? If two moles of photons are emitted per second, what is the power of the laser output?

Solution The calculations are as follows:

$$\lambda = 694.3 \, \text{nm}$$

$$h\vartheta = E_2 - E_1 = \frac{hc}{\lambda} = \frac{1240 \, \text{eV.nm}}{694.3 \, \text{nm}} = 1.785 \, \text{eV}$$

Thus, the photon emitted has an energy of 1.785 eV, which is also the difference between the metastable state and the excited levels.

$$\text{Frequency of the photon} = \vartheta = \frac{c}{\lambda} = \frac{\dfrac{3 \times 10^8 \, \text{m}}{\text{s}}}{694.3 \, \text{nm}} = \frac{\dfrac{3 \times 10^8 \, \text{m}}{\text{s}}}{694.3 \times 10^{-9} \, \text{m}}$$

$$= 0.00432 \times 10^{17} \, \text{Hz} = 4.32 \times 10^{14} \, \text{Hz}$$

Energy in joules is $= h\vartheta = 6.634 \times 10^{-34} \, \text{J.s} \times 4.32 \times 10^{14} \, \text{Hz} = 28.658 \times 10^{-20} \, \text{J}$

Two mole of photons $= 2 \times 6.023 \times 10^{23} \times 28.658 \times 10^{-20} \, \mathrm{J} = 345.60 \times 10^3 \, \mathrm{J} = 345 \, \mathrm{kJ}$

Thus, the power output is about 0.35 MW or 0.35×10^6 W. This is nearly equal to the electrical power used by a small city. Therefore, is it possible to use a laser to finish our city power woes? No, this is a pulse laser and lasts for just about a few nanoseconds. The laser has to be recharged using a capacitor and the laser rod has to be cooled, so it is not a continuous process.

4.7.3 Nd–YAG Laser

Nd–YAG laser uses neodymium (rare earth element) and yttrium aluminium garnet (YAG) for the lasing action. It is a four-level solid-state laser. Its details are discussed in the following sections.

4.7.3.1 Principle of Operation

Energy levels of Nd in an Nd–YAG system are shown in Fig. 4.14. Photons from a krypton flash lamp are used to impart energy to Nd atoms. This raises the Nd atoms from the ground level to higher energy levels E_3 and E_4.

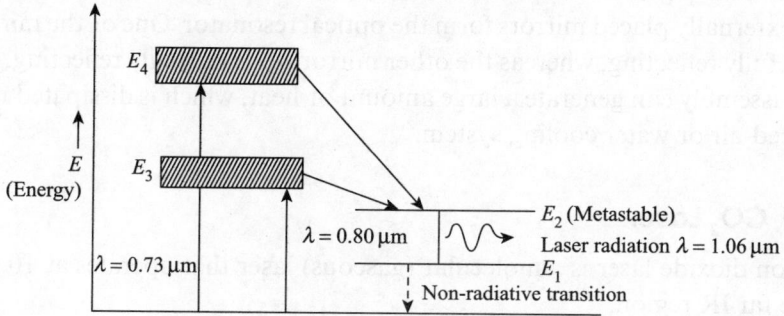

Fig. 4.14 Energy levels in Nd–YAG system

The corresponding wavelengths for these transitions are $0.73\,\mu\mathrm{m}$ and $0.80\,\mu\mathrm{m}$. Nd atoms undergo transitions from energy levels E_3 and E_4 to the metastable state E_2. These transitions are non-radiative in nature. Accumulation of atoms at energy level E_2 leads to population inversion between levels E_2 and E_1, resulting in spontaneous transition from level E_2 and E_1. The emitted photon triggers a chain of stimulated photons with energy corresponding to the difference between energy levels E_2 and E_1. The photon number multiplies rapidly due to reflection from the oppositely placed mirrors. The complete process leads to an intense laser radiation of wavelength 1.06 μm. Transition of Nd atoms from energy level E_1 to the ground level is rapid and non-radiative.

4.7.3.2 Construction

Figure 4.15 is a schematic diagram showing the construction of an Nd–YAG laser.

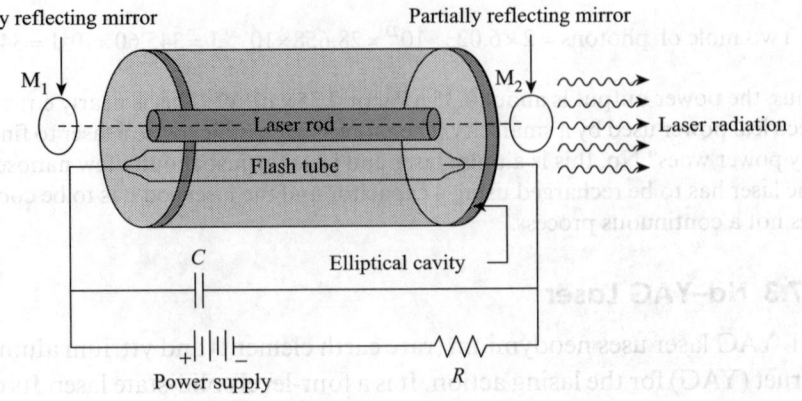

Fig. 4.15 Construction of Nd–YAG laser

A small amount of yttrium ions (Y^{3+}) is replaced by neodymium ions (Nd^{3+}) in the active material used in this laser. This active material is then shaped as a cylindrical rod to form the active element or the laser rod. The laser rod and a flash tube are then placed inside an elliptical reflector cavity. The entire operation of the flash tube is controlled using a capacitor. The ends of the laser rod are polished to optical flatness level and held parallel to each other. Two externally placed mirrors form the optical resonator. One of the mirrors, M_1, is fully reflecting, whereas the other mirror M_2 is partially reflecting. This laser assembly can generate a large amount of heat, which is dissipated using a forced-air or water cooling system.

4.7.4 CO₂ Laser

Carbon dioxide laser is a molecular (gaseous) laser that operates at $10.6\,\mu m$ in the far IR region.

4.7.4.1 Principle of Operation

A CO_2 laser utilizes a gaseous mixture of CO_2, He, and N_2. Vibration levels of the ground state of CO_2 and N_2 molecules are shown in Fig. 4.16. An electric discharge through the gas results in a collision of electrons with N_2 molecules. This excites the N_2 molecules to higher energy states. The excited N_2 molecules in turn collide with CO_2 molecules and raise them from ground state to higher energy levels. Population of the E_5 level of CO_2, therefore, starts increasing. This process ultimately leads to population inversion, resulting in two spontaneous emission transitions. These are as follows:

1. Transition from E_5 level to E_4 level, resulting in a laser beam of wavelength $10.6\,\mu m$.
2. Transition from E_5 level to E_3 level, resulting in a laser beam of wavelength $9.6\,\mu m$.

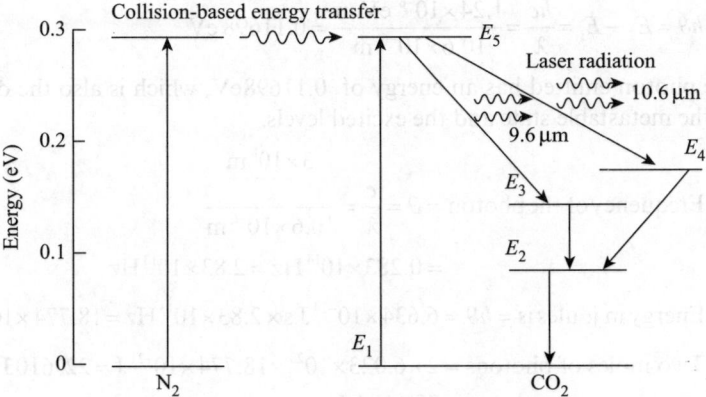

Fig. 4.16 Vibration levels of CO_2 and H_2

4.7.4.2 Construction

Figure 4.17 illustrates a schematic diagram showing the basic construction of a CO_2 laser.

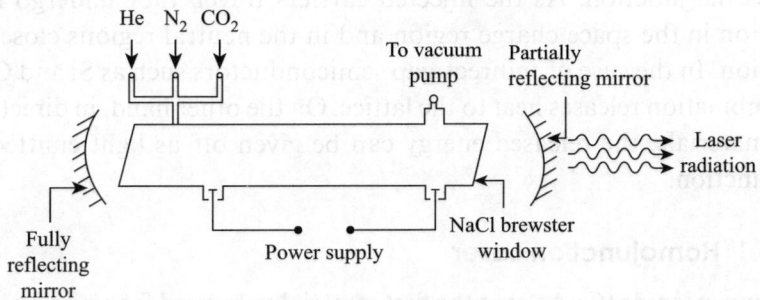

Fig. 4.17 Schematic diagram showing basic construction of CO_2 laser

The laser set-up consists of a quartz discharge tube that is around 5 m long and 2.5 cm in diameter. The discharge tube has inlets for CO_2, He, and N_2, with provision for adjusting the partial pressures of these gases individually. Terminals of the discharge tube are connected to a power supply that provides the required DC voltage. NaCl brewster windows are provided at the ends of the discharge tube, to ensure polarization of the generated laser light. A fully and a partially reflecting concave mirror constitute the optical resonator. The gas fed into the system helps conduct the heat generated within the central region of the discharge tube to its walls.

Example 4.12 The light output of a typical laser is 10.6 μm. What is the difference between the energy levels of the excited state and the metastable state? What will be the energy of the photon emitted? What is the frequency associated with the photon? If two moles of photons are emitted per second, what is the power of the laser output?

Solution The calculations are as follows:

$$\lambda = 10.6 \, \mu m$$

$$h\vartheta = E_2 - E_1 = \frac{hc}{\lambda} = \frac{1.24 \times 10^{-6} \text{ eV.m}}{10.6 \times 10^{-6} \text{ m}} = 0.11698 \text{ eV}$$

Thus, the photon emitted has an energy of 0.11698eV, which is also the difference between the metastable state and the excited levels.

$$\text{Frequency of the photon} = \vartheta = \frac{c}{\lambda} = \frac{3 \times 10^8 \frac{m}{s}}{10.6 \times 10^{-6} \text{ m}}$$

$$= 0.283 \times 10^{14} \text{ Hz} = 2.83 \times 10^{13} \text{ Hz}$$

$$\text{Energy in joules is} = h\vartheta = 6.634 \times 10^{-34} \text{ J.s} \times 2.83 \times 10^{13} \text{ Hz} = 18.774 \times 10^{-21} \text{ J}$$

$$\text{Two moles of photons} = 2 \times 6.023 \times 10^{23} \times 18.774 \times 10^{-21} \text{ J} = 22,610 \text{ J}$$

$$= 22.610 \text{ kJ}$$

Thus, the power output is about 22.6kW. At this infrared wavelength, it has to be a CO_2 laser.

4.7.5 Semiconductor Laser

When a p–n junction is forward biased, injection of carriers takes place across the junction. As the injected carriers travel, they undergo recombination in the space charge region and in the neutral regions close to the junction. In the case of indirect-gap semiconductors such as Si and Ge, this recombination releases heat to the lattice. On the other hand, in direct-band-gap materials, the released energy can be given off as light emitted from the junction.

4.7.5.1 Homojunction Laser

Gallium arsenide (GaAs) was the first material to be used for emission of laser radiations and continues to be used along with other related III–V compound alloys. A homojunction laser is shown schematically in Fig. 4.18.

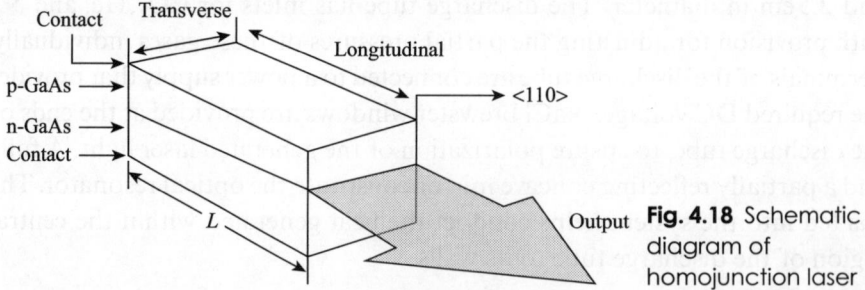

Fig. 4.18 Schematic diagram of homojunction laser

A pair of parallel faces are cleaved or polished perpendicular to the <110> axis. Suitable biasing conditions lead to the emission of laser light from these planes. The two remaining faces are roughened. This leads to inefficient reflection; as a result, lasing action is eliminated in these directions. As you know, population inversion is an essential requirement for the lasing action

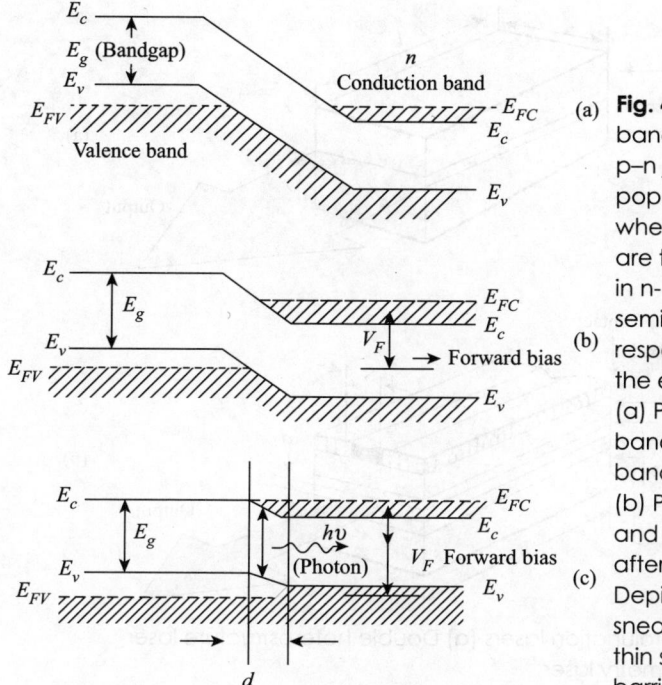

Fig. 4.19 Energy band diagram for p–n junction, showing population inversion, where E_{Fc} and E_{FV} are the Fermi levels in n- and p-type semiconductors, respectively, and E_g is the energy bandgap (a) Position of valence band and conduction band before biasing (b) Position of valence and conduction band after forward bias (c) Depiction of photon sneaking through the thin semiconductor barrier

to take place. Figure 4.19 shows the energy band diagram for a p–n junction formed between degenerate semiconductors.

As the semiconductors have degenerate energy levels, the Fermi levels are below the valence band edge and above the conduction band edge, for p and n sides, respectively. When a forward bias V_F is applied, electrons are injected from the n-side and holes are injected from the p-side. At a high enough forward bias (Fig. 4.19c), the device reaches a high-injection condition, with a large concentration of holes and electrons in the transition region. The region d shown in Fig. 4.19(c) contains a large concentration of electrons in the conduction band and a large concentration of holes in the valence band. Thus, population inversion is achieved. From Fig. 4.19, it is clear that the following is the essential condition for population inversion:

$$(E_{FC} - E_{FV}) > E_g \tag{4.22}$$

4.7.5.2 Heterojunction Laser

A junction formed between two semiconductors having different band gaps is called a heterojunction. Heterojunction structures can also be used to fabricate semiconductor lasers. Such heterostructures must have negligible interface traps and, therefore, must use semiconductors with very low lattice mismatches. For GaAs ($a = 5.6533$ Å) substrates, the ternary compound Al_xGa_{1-x} As with lattice mismatch <0.1% is a good choice. Similarly, for InP

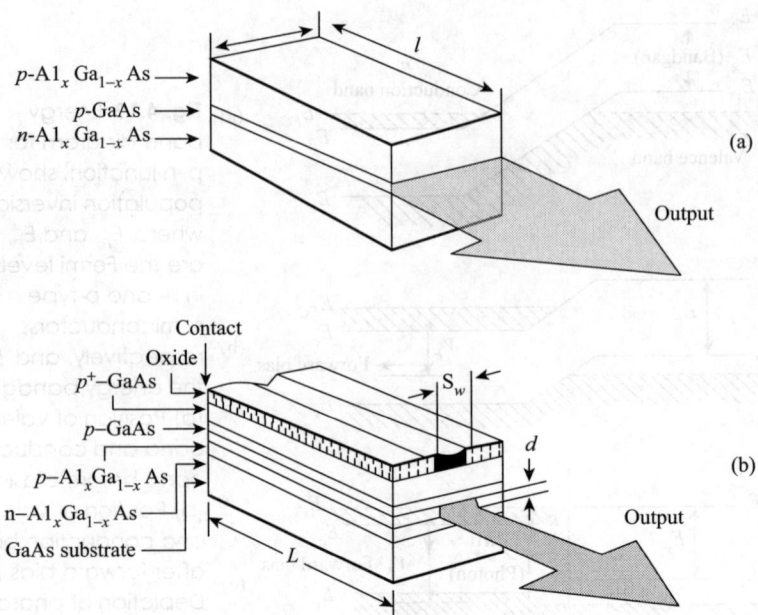

Fig. 4.20 Heterojunction lasers (a) Double heterostructure laser (b) Stripe geometry laser

($a = 5.8686$ Å) substrates, the quaternary compound ($Al_xGa_{1-x}As_yP_{1-y}$) with a nearly identical lattice constant is a perfect match. Figure 4.20(a) shows a double-heterostructure laser. The structure consists of a thin layer of a semi-conductor (e.g., GaAs) sandwiched between layers of another semiconductor (e.g., $Al_xGa_{1-x}As$). This type of laser requires much less current to operate than a homojunction laser with a similar geometry.

Homojunction and double-heterostructure laser structures are large-area lasers, since the entire area along the plane of the junction can participate in the emission of radiation. Figure 4.20(b) shows a stripe geometry laser. An oxide layer is used in this structure to isolate all but the stripe contact. The lasing area gets restricted to a narrow region under the contact. The use of this structure leads to reduced operating current, elimination of multiple-emission areas, and better reliability.

4.8 INDUSTRIAL APPLICATIONS OF LASER

Some important industrial applications of lasers are discussed in this section.

4.8.1 Lasers in Welding

Laser beams are highly coherent. This coherency is used to focus a laser beam on a very small area with the help of a proper lens system. This results in extremely high energy density within the focused area. Such focused laser beams are extremely useful in spot-welding applications.

Laser-based spot welding has the following main advantages:
1. No physical contact of the source with the material being welded is required.
2. Welding can be performed in air.
3. A negligible rise in temperature occurs in the surrounding environment.
4. A variety of materials can be welded.
5. Welding of a joint is possible even after glass sealing.

Laser-based spot welding is used to weld electronic components, thermocouple junctions, and electronic micromodules.

4.8.2 Lasers in Heat Treatment

Laser beams possess a large amount of energy. This energy can be used to give heat treatment to materials. Heat treatment is generally used to increase the strength of materials. Since laser beams can be focused to a very small area and can also be scanned over the surface of the material, specific weak locations can be targeted for strengthening. This type of heat treatment is used in automobile and aerospace industries. Critical components undergoing stress during their operation are strengthened to reduce the failure rates. A good structural analysis both at the surface and at the bulk level is usually useful to provide an input regarding the zones requiring strengthening.

4.8.3 Lasers in Cutting

The energy carried by a laser beam can be utilized for cutting materials. Even materials as hard as diamond can easily be cut using the energy of a laser. Metals (as thick as 0.005 inch) can also be cut using lasers. The process of cutting involves melting, burning, and vaporization. The edges obtained by cutting are quite sharp and well defined. Industrial laser cutters are quite often employed to cut flat-sheet of materials as well as structural materials. Laser cutting has several advantages over conventional mechanical cutting. Some of these advantages are as follows:
1. No physical contact
2. Very high precision
3. Negligible warping due to small heat-affected zones

4.9 MEDICAL APPLICATIONS OF LASER

Laser-based systems are used for tooth drilling, eye surgery, cancer treatment, and precision cardiovascular surgery. Laser-based surgery is accurate, painless, and faster than conventional surgery. Lasers are particularly well suited for microsurgery due to their high power density. This leads to rapid tissue vaporization without affecting the surrounding healthy tissues. Microsurgery is usually performed with pulsed excimer lasers. One example of microsurgery is laser ablation of the cornea. Ultraviolet light from a laser selectively breaks covalent bonds and leads to explosive evaporation from a small region of the surface of the eye. The depth of the cut due to each pulse is less than half a micrometre. A laser beam is scanned across the cornea for adjusting its curvature and thus its focal length. This procedure can correct defects of vision. Laser is thus used essentially to form an additional lens on the surface of the cornea. This additional lens takes over the role of a contact lens or glasses.

4.10 HOLOGRAPHY

The word 'holography' has its origin in Greek; the word *holos* means 'the whole' and *graphy* means 'the writing'. Thus, holography refers to 'complete recording', whereas conventional photography gives us a two-dimensional image of an object. This two-dimensional image is actually a distribution of the square of the amplitude of the light projected on a photographic film or any other sensing medium. Three-dimensional effect is missing in a conventional photograph. Thus, viewing of the photograph in different directions does not produce any difference, and no characteristic of depth is included. Holography records the light waves reflected from an object. This photographic record is called a *hologram*. A hologram contains all the optical information about the object. In contrast to a conventional photograph, a hologram contains information about the amplitude as well as the phase of the light reflected from the object. The holograph needs to be illuminated with a laser source to reveal a three-dimensional image of the object. This method of image formation is called the *reconstruction process*. We will discuss some important aspects of holography in this section.

4.10.1 Principle of Holography

Suppose a beam of coherent light is incident on a point object from the left to the right direction, as shown in Fig. 4.21. This incident wavefront encounters point object A as it travels. The point object scatters the incident wavefront, resulting in spherical scattered wavefronts, as shown in Fig. 4.21. These spherical wavefronts have A as their centre. The scattered or diffracted wavefronts and the original incident wavefront interact with each other, producing an interference pattern on the photographic plate $P_1 P_2$. Constructive and destructive interferences lead to a pattern of bright and dark concentric circles.

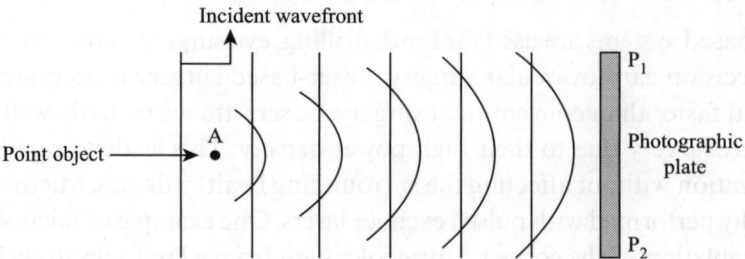

Fig. 4.21 Wavefront modification due to point object

Fig. 4.22 Wavefront modification due to multiple point objects

Let us now generalize the situation to a three-dimensional object placed in the way of the incident wavefront, as shown in Fig. 4.22. The three-dimensional object can be assumed to be made up of a large number of point objects. Each point generates its own characteristic scattered or diffracted spherical wavefronts, as depicted in Fig. 4.22. The diffracted wavefronts and the incident wavefront interact with each other to form an interference pattern on the photographic plate P_1P_2.

This photographic impression is a hologram. The object can be reconstructed by illuminating the hologram with a suitable laser beam.

4.10.2 Recording of Hologram

Figure 4.23 shows a schematic diagram of an arrangement for recording a hologram. The incident laser beam is first split into two components. One of the components serves as a reference beam. The other component interacts with the three-dimensional object. The beam diffracted or scattered from the object and the reference beam are made to fall on the photographic plate. The two beams produce an interference pattern on the photographic plate, producing the hologram.

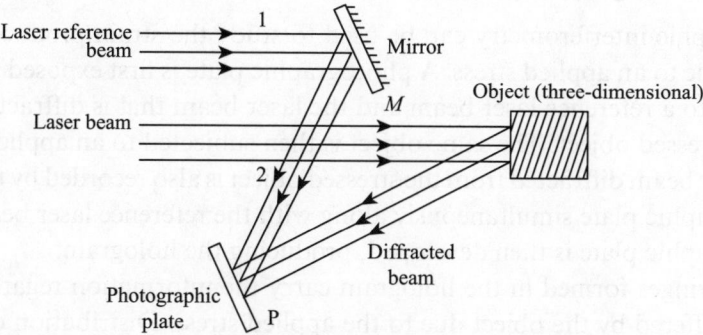

Fig. 4.23 Schematic diagram showing the recording of hologram

4.10.3 Reconstruction of Hologram

An arrangement for the reconstruction of an object from the information contained in the hologram is shown in Fig. 4.24.

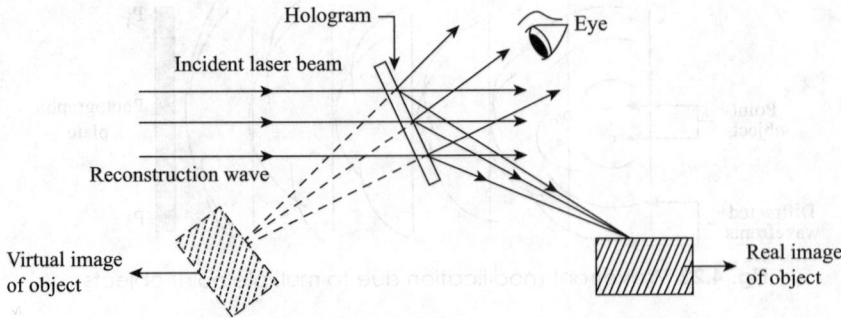

Fig. 4.24 Reconstruction of hologram

The process of reconstruction of the image of an object is actually a reversal of the process followed for generating the hologram. The hologram is illuminated with a laser beam that is identical to the beam used to generate the hologram. The hologram acts like a complex grating and produces a diffraction pattern. The diffracted waves carry all the information that was contained in the waves diffracted by the object when the hologram was being generated. As shown in Fig. 4.24, two images are created by the diffracted waves. One of these images is a virtual image, whereas the other is the real image. Different perspectives of the object can be visualized by changing the angle of viewing. A light-sensitive medium kept at the location of the real image can be used to photograph the object. Although the virtual image cannot be photographed, it has all the characteristics of the object, similar to a parallax.

4.10.4 Applications of Holography

We will discuss some important applications of holography in this section.

4.10.4.1 Holographic Interferometry

Holographic interferometry can be used to study the strain produced in an object due to an applied stress. A photographic plate is first exposed simultaneously to a reference laser beam and the laser beam that is diffracted from the unstressed object. The same object is then subjected to an applied stress. The laser beam diffracted from the stressed object is also recorded by the same photographic plate simultaneously along with the reference laser beam. The photographic plate is then developed, producing the hologram.

The fringes formed in the hologram carry the information regarding the strain suffered by the object due to the applied stress. Distribution of strain within the object can be studied using the holographic interferometry technique.

4.10.4.2 Acoustic Holography

Acoustic holography utilizes sound waves to generate a hologram. It has many interesting applications. One of them is for observing the image of an object placed in water. An ultrasonic transducer generates sound waves inside water. A part of these waves is used as a reference beam. The other part is made to fall on the object. The sound waves reflected by the object are made to interfere with the reference beam, producing ripples on the surface of water. This ripple pattern is photographed and the hologram is formed.

Acoustic holography is also used to image our internal organs to assist physicians. A coherent ultrasonic beam is split into two parts. One part serves as a reference beam and the other gets scattered by the internal organ being studied. The reference and scattered beams interfere with each other to form a hologram. This hologram can then be used to reconstruct a three-dimensional image of the internal organ.

4.10.4.3 Holographic Microscopy

In holographic microscopy, a laser beam passes through the specimen being studied and then through a microscope. This beam interferes with the reference beam to produce a hologram. Any cross section of the specimen can be viewed after reconstruction of the image. Thus, the depth of field of a holographic microscope is much higher than that of the conventional high-power microscopes. Time-varying phenomenon can also be studied using holographic microscopy.

4.10.4.4 Other Applications

Holography has other applications such as holographic cinematography, holographic television, holographic character recognition, and holographic optical engineering.

Example 4.13 Calculate the relative population of Na atoms in a sodium lamp in the first excited state and in the ground state at a temperature of 250°C. $[\lambda = 590 \, \text{nm}]$

Solution From Eq. (4.6), we get the following relation:

$$\frac{n_2}{n_1} = \exp\left[-\frac{(E_2 - E_1)}{k_B T}\right]$$

where n_1 and n_2 are the numbers of atoms in the ground and the excited states, respectively, and E_1 and E_2 are the corresponding energies in these states; k_B is the Boltzmann constant and T the absolute temperature. In addition,

$$E_2 - E_1 = h\vartheta$$

where h is the Planck's constant and ϑ the frequency. Using the preceding two equations, the following equation can be written:

$$\frac{n_2}{n_1} = \exp\left[-\frac{h\vartheta}{kT}\right] = \exp\left[-\frac{hc}{\lambda kT}\right]$$

where c is the velocity of light.

Substitution of the given values of λ and T, and the known values of c and k in the preceding equation results in the following value:

$$\frac{n_2}{n_1} = \exp\left[-\frac{6.625 \times 10^{-34} \times 3 \times 10^8}{5.9 \times 10^{-7} \times 1.38 \times 10^{-23} \times 523}\right] = \exp(-46.67) = 5.39 \times 10^{-21}$$

Example 4.14 Calculate the ratio of a stimulated emission to a spontaneous emission for sodium D lines at a temperature of 250°C.

Solution From Eq. (4.13), we get the following expression:

$$\frac{n_{21}'}{n_{21}} = \frac{1}{\exp\left(\dfrac{h\vartheta}{kT}\right) - 1} = \frac{1}{\exp\left(\dfrac{hc}{\lambda k_B T}\right) - 1}$$

Substituting appropriate values in this equation we get the required ratio:

$$\frac{n_{21}'}{n_{21}} = \frac{1}{\left(\exp\left[\dfrac{6.625 \times 10^{-34} \times 3 \times 10^8}{1.38 \times 10^{-23} \times 523 \times 5.9 \times 10^{-7}}\right] - 1\right)} = \frac{1}{(\exp\left[0.00467 \times 10^4\right] - 1)}$$

$$= 5.24 \times 10^{-21}$$

Example 4.15 The wavelength of a He–Ne laser is 632.8 nm. Its output power is 3.147 mW. How many photons are emitted per minute when it is in operation?

Solution Frequency of the emitted photons is given as follows:

$$\vartheta = \frac{c}{\lambda} = \frac{3 \times 10^8}{6328 \times 10^{-10}} = 4.79 \times 10^{14}\, Hz$$

The energy E of each photon is expressed as follows:

$$E = h\vartheta = 6.625 \times 10^{-34} \times 4.74 \times 10^{14} = 3.14 \times 10^{-19}\, J$$

Energy E_m emitted per minute is calculated as follows:

$$E_m = 3.147 \times 10^{-3} \times 60\, J/min$$

Therefore, the number of photons emitted per second, N, is calculated to be as follows:

$$N = \frac{E_m}{E} = \frac{3.147 \times 10^{-3} \times 60}{3.14 \times 10^{-19}} \cong 6 \times 10^{17}$$

Example 4.16 Calculate the mode separation of the longitudinal cavity in terms of frequency if the distance between the plane mirrors is 30 cm. (Assume that the cavity is filled with air.)

Solution The distance between the plane mirrors is = 30 cm. Hence,

$$\vartheta = \frac{c}{\lambda} = \frac{nc}{2L\mu} = \frac{n \times 3 \times 10^8\, m/s}{2 \times 0.30\, m \times 1} = 0.5 \times 10^9\, Hz$$

Example 4.17 Evaluate the value of the integral multiple n, for an optical resonator of length $L = 50$ cm operating at an optical frequency of $\vartheta = 4.0 \times 10^{14}$ Hz, corresonding to $\lambda = 5000$.

Solution We know that $n = \dfrac{2\mu L}{c}$

$$n = \frac{2 \times 1 \times 50\, cm \times 4.0 \times 10^{14}}{3 \times 10^{10}\, cm\, s^{-1}} = 133.3 \times 10^4 = 1.33 \times 10^6$$

Thus, one observes that the integral multiple n, which represents the different oscillation frequencies, is a very large number for ordinary practical resonating cavities in optical resonators.

IMPORTANT CONCEPTS

1. A LASER beam is an intense, concentrated, monochromatic beam of light.
2. The main components of a laser are (a) an active medium, (b) a pumping source and (c) an optical resonator.
3. The salient properties of a laser are as follows:
   (a) Angular divergence: $\Delta\theta = \dfrac{\lambda}{D}$, where λ is the wavelength of light and D is the diameter of the aperture.
   (b) Focussing: $r = \dfrac{\lambda_1 f}{a}$, where f is the focal length of the lens.
   (c) Monochromaticity: $L = c\tau_c$, where L is the coherence length and τ_c is the relaxation time.

   $$\Delta\lambda \sim \frac{\lambda^2}{L} = \frac{\lambda^2}{c\tau_c} = \frac{\Delta\vartheta}{\vartheta}; \quad \Delta\lambda \text{ is the spectral width.}$$

4. The number of atoms per unit volume, n, in an energy state E is written as follows:

   $$n = n_0 \exp\left(-\frac{E}{kT}\right)$$

5. The principles of spontaneous and stimulated emissions state the following formulae for transitions, where A_{21} is the coefficient of a spontaneous emission and B_{21} is the coefficient of a stimulated emission:

   $$\frac{n_{21}}{n'_{21}} = \frac{A_{21}}{B_{21}u_\vartheta} = \exp\left(\frac{h\vartheta}{kT}\right) - 1$$

   $$B_{12} = B_{21} \quad \frac{A_{21}}{B_{21}} = \frac{8\pi h\vartheta^3}{c^3}$$

   Thus, if $h\vartheta \gg kT$, the spontaneous emission is more probable than stimulated emission. However, if $h\vartheta \ll kT$, the stimulated emission starts playing an important role.
6. The number n_2 of atoms in a higher energy state becomes greater than that in lower energy states. This condition is essential for laser action and is known as *population inversion*.

APPLICATIONS

1. Neodymium and carbon dioxide laser is used as range finders in artillery tanks used by defence forces all across the globe.
2. Lasers can be used for communication between submarines underwater, as it ensures that the signals are not intercepted easily.
3. Lasers are used in laser-guided anti-tank missiles and other weaponry.
4. Laser beams are extensively used in optical fibre communication.
5. High-intensity laser beams are used for precise cutting and hardening of metals. It is also used to check alignment of metals, like in case of railway tracks.

6. Laser technology is also used in high-quality printing machines.
7. Holographic scanners are often used as bar code scanners.
8. Holographic windshields are used in military and defence aircraft, where these not only work as windshields, but also form a display system on which information can be relayed, thus enabling the pilot to have a look at necessary information without taking sight off the screen. This technology is often called the head's up display (HUD).
9. Holographic optical elements are used as replacements of minor gratings, lenses, etc.

IMPORTANT FORMULAE

1. Angular divergence: $\Delta\theta = \dfrac{\lambda}{D}$

2. $B_{12} = B_{21} \quad \dfrac{A_{21}}{B_{21}} = \dfrac{8\pi h\vartheta^3}{c^3}$

3. $\dfrac{n_{21}}{n'_{21}} = \dfrac{A_{21}}{B_{21}u_\vartheta} = \left(\exp\left(\dfrac{h\vartheta}{kT}\right) - 1 \right)$

SELF-ASSESSMENT

Multiple-choice Questions

4.1　The term laser stands for
  (a) light amplification by stimulated emission of radiation
  (b) light amplitude of spontaneous emitted radiation
  (c) light ampere of sporadic emission.
  (d) light ampere of spot emission

4.2　In stimulated emission, a photon of
  (a) same frequency, phase, and polarization and direction of propagation is generated
  (b) differing frequency, phase, and polarization and direction of propagation is generated
  (c) same frequency and differing phase, polarization, and direction of propagation is generated
  (d) same frequency and phase, and differing polarization and direction of propagation is generated

4.3　Population inversion is defined as that state in which
  (a) the number of atoms in a higher energy level is more than that in a lower energy level
  (b) the number of atoms in the ground state is more than that in an excited state
  (c) the number of atoms in the ground and excited states is the same
  (d) The number of atoms in any state is of no consequence

4.4　The Einstein's coefficients are
  (a) *A* for emission and *B* for absorption
  (b) *A* for spontaneous emission and B_{12} and B_{21} for absorption and stimulated emission

(c) A for absorption and B_{12} and B_{21} for spontaneous and stimulated emission

(d) A for absorption and B_{12} and B_{21} for stimulated absorption and emission

4.5 Which of the following represents a three-level laser?

(a) He–Ne laser   (c) Semiconductor laser

(b) Ruby laser   (d) All of these

4.6 Which of the following represent a four-level laser?

(a) He–Ne laser   (c) Semiconductor laser

(b) Ruby laser   (d) All of these

4.7 The resonating cavity in a laser design helps

(a) create population inversion

(b) create an amplified, coherent lasing beam

(c) create a three-level laser beam

(d) none of these

4.8 Optical pumping in a laser is done

(a) to create population inversion

(b) to create an amplified, coherent lasing beam

(c) to create a three-level laser beam

(d) none of these

4.9 Gas laser is a modification of the solid-state laser, as

(a) the lasing beam is continuous, coherent, and monochromatic

(b) the lasing beam is discontinuous, incoherent, and directional

(c) the lasing beam is continuous, coherent, and non-directional

(d) none of these

4.10 Holography is the phenomenon of

(a) light waves being reflected from an object

(b) creating a two-dimensional image of an object

(c) generating a two-dimensional regular photograph

(d) none of these

4.11 Activator atoms in a ruby laser is

(a) aluminium   (c) iron

(b) oxygen   (d) chromium

4.12 In a ruby laser, population inversion is achieved by

(a) chemical reactions

(b) inelastic collision between atoms

(c) optical pumping

(d) application of a strong electric field

4.13 In a He–Ne laser, the laser light emits due to

(a) transition from $3s \rightarrow 2p$ (c) transition from $2s \rightarrow 2p$

(b) transition from $3s \rightarrow 3p$ (d) none of these

4.14 A laser requires mirrors because

(a) they provide optical feedback

(b) they invert the population inversion

(c) they determine the wavelength at which lasing occurs

(d) they perform none of these

4.15 In a spontaneous emission, the emitted photon can move

(a) in the direction of the field

(b) in a straight line

(c) in any random direction

(d) opposite to the direction of the field

Review Questions

4.1   Spatial and temporal coherence are major attributes of a lasing beam. Comment.

4.2   Stimulated emission design is the key to good lasing action. Enunciate.

4.3   Differentiate between an LED, a laser, and an ordinary light source.

4.4   Distinguish between photography and holography.

4.5   What are coherent sources?

4.6   Distinguish between spontaneous and stimulated emissions.

4.7   What are the characteristics of a laser?

4.8   What does population inversion mean?

4.9   State some of the applications of lasers in engineering and industry.

4.10   Differentiate between a ruby and an Nd–YAG laser.

4.11   Classify the different types of lasers based on their active medium.

4.12   State the important features of a hologram.

4.13   Differentiate between an LED and a laser.

4.14   What does pumping mean?

4.15   Differentiate between a homojunction and a heterojunction laser.

4.16   List some applications of lasers in communication and military fields.

4.17   What is the use of He in a CO_2 laser?

4.18   Mention a few applications of lasers in the medical field.

4.19   What is holography?

4.20   What is meant by the active material in a laser?

4.21   How is a hologram recorded?

4.22   Describe the construction and working of an Nd–YAG laser.

4.23   Describe the construction and working of a CO_2 laser.

4.24   Describe the construction and working of a semiconductor laser.

4.25   Describe the various modes of vibrations of a CO_2 molecule. Describe the construction and working of a CO_2 laser with necessary diagram.

4.26   Classify lasers based on their active medium, citing one example for each.

4.27   Explain the construction and working of a He–Ne laser. What are the merits of this laser?

Numerical Problems

4.1   Calculate the relative population of Na atoms in a sodium lamp in the first excited state and in the ground state at a temperature of 300°C. Assume that

$l = 590\,\text{nm}.$ $\left[Hint: \dfrac{n_2}{n_1} = \exp\left[-\dfrac{(E_2 - E_1)}{kT} \right] \right]$

4.2   The relative population of atoms in two energy levels E_1 and E_2 is 10^{-20}. Calculate the energy level difference in eV. Assume that $T = 300\,\text{K}$.

$\left[Hint: \dfrac{n_2}{n_1} = \exp\left[-\dfrac{(E_2 - E_1)}{kT} \right] \right]$

4.3   Determine the ratio of a stimulated emission to a spontaneous emission at a temperature of 300°C for sodium D lines. $\left[Hint: \dfrac{n'_{21}}{n_{21}} = \dfrac{1}{\left[\exp\left(\dfrac{h\vartheta}{kT} \right) - 1 \right]} \right]$

4.4 Determine the ratio of a stimulated emission to a spontaneous emission at a temperature of 250°C for a material with an emitted radiation of wavelength 600 nm.

$$\left[Hint: \frac{n'_{21}}{n_{21}} = \frac{1}{\exp\left(\dfrac{h\vartheta}{kT}\right) - 1} \right]$$

4.5 The power output of a particular He–Ne laser is 4 mW. Assuming the wavelength of the He–Ne laser to be 632.8 nm, calculate the number of photons emitted per minute during the laser's operation. $\left[Hint: N = \dfrac{E_m}{E} \right]$

4.6 Show that the angular spread due to a filament is more than that due to a laser. Assume the length of the filament to be 0.5 mm, focal length to be 5 cm, and the diameter of the aperture to be 2.5 cm. $\left[Hint: \Delta\theta = \dfrac{\lambda}{D} \right]$

4.7 Find the intensity of a 2 mW laser beam ($\lambda_1 = 6 \times 10^{-5}$ cm) that is incident on an aperture whose focal length is nearly 1.5 m and whose diameter is 3.0 mm.

$$\left[Hint: I = \frac{P}{A} \right]$$

4.8 Calculate the mode separation of the longitudinal cavity in terms of frequency if the distance between the plane mirrors is 25 cm. (Assume that the cavity is filled with air.) $\left[Hint: \vartheta = \dfrac{c}{\lambda} = \dfrac{nc}{2L\mu} \right]$

4.9 Calculate the electric field corresponding to a 2.0 mW laser beam incident on a lens of focal length 3.0 cm, $\lambda_1 = 6 \times 10^{-5}$ cm, and with a beam width of 2 mm.

4.10 Find the spectral width for a He–Ne laser having a coherent time of the order of 2 nanoseconds.

4.11 Compare the mono-chromaticity of an ordinary light ($\Delta\vartheta \sim 10^{10}$ Hz) and a laser light ($\Delta\vartheta \sim 50$ Hz), for light of frequency 10^{14} Hz.

4.12 Show that the fraction of atoms in the excited state is much smaller than that in the ground state for a temperature of 4000 K (nearly that of a filament bulb) and energy gap of 1 eV.

4.13 Evaluate the value of the integral multiple 'n', for an optical resonator of length $L = 40$ cm operating at an optical frequency of $\vartheta = 2.0 \times 10^{14}$ Hz, corresponding to $\lambda = 5000$Å.

4.14 An ultraviolet laser generates a 10 MW, 10.0 ns-long pulse at a wavelength of 355 nm. How many photons are there in each pulse?

4.15 A sensor is exposed for 0.1 s to a 5 MW laser that is 5 m away. The sensor has an opening that is 5 mm in diameter. How many photons enter the sensor if the wavelength of the light is 600 nm? Assume that all the energy of the laser is given off as light.

4.16 The metastable state of a ruby laser is at 1.8 eV. Calculate the wavelength of light emitted.

4.17 Calculate the pulse energy in eV, if 2 moles of Cr^{+3} ions are involved in the population inversion process in a ruby laser that emits light of wavelength 694.3 nm.

4.18 The light output of a typical laser is 694.3 nm. What is the difference in the energy levels of the excited state and the metastable state? What will be the energy of the photon emitted? What is the frequency associated with the photon? If four moles of photons are emitted per second, what is the power of the laser output?

4.19 The light output of a typical laser is 10.6μm. What is the difference in the energy levels of the excited state and the metastable state? What will be the energy of the photon emitted? What is the frequency associated with the photon? If four moles of photons are emitted per second, what is the power of the laser output?

4.20 Calculate the relative population of Na atoms in a sodium lamp in the first excited state and in the ground state at a temperature of 275°C. ($\lambda = 590$ nm)

4.21 Calculate the ratio of stimulated emission to spontaneous emission for sodium D lines at a temperature of 300°C.

4.22 The wavelength of He–Ne laser is 632.8 nm. Its output power is 7 mW. How many photons are emitted per minute when it is in operation?

Optical Fibre Systems

Learning Objectives

After studying this chapter, students will be able to

- understand the propagation of light in optical fibres
- comprehend the basic principle of operation of optical fibres
- realize step-index optical fibres
- explain the concept of numerical aperture
- elucidate the meaning of multipath time dispersion
- explain graded-index optical fibres
- describe fabrication of optical fibres (double-crucible technique of fibre drawing)
- demonstrate some important applications of optical fibres

List of Symbols

i = Angle of incidence

r = Angle of refraction

n = Refractive index

i_c = Critical angle of incidence

θ_c = Critical angle

α_m = Angle of acceptance

NA = Numerical aperture

Δn_r = Relative refractive index difference

v = Velocity

t = Time

L = Distance

c = Velocity of light in vacuum

P_{in} = Input power

P_{out} = Output power

A = Attenuation loss

λ = Wavelength

h = Planck's constant

E_g = Energy bandgap

I_L = Photocurrent

μ_P = Mobility of holes

μ_n = Mobility of electrons

G_L = Photogeneration rate

τ_p = Excess minority carrier hole lifetime

e = Electronic charge

L_n = Diffusion length for electrons

L_p = Diffusion length for holes

Φ_i = Incident flux

α = Absorption coefficient

5.1 INTRODUCTION

Communication is an important driver of technology. Transfer of information in any communication system uses a carrier wave that is modulated by the information signal. This modulated carrier wave is then demodulated at the receiver end to extract the necessary information. The volume and speed of information transfer are dependent on the frequency of the carrier wave. It is advantageous to use higher frequencies for the carrier wave in order to increase the volume and speed of information that is transferred. Fibre-optic communication utilizes carrier frequency spectra corresponding to frequencies of the order of 10^{14} Hz, which is 10,000 times higher than microwave frequencies. Optical fibres constitute the medium that is used for transmitting such high frequencies. The first low-attenuation optical fibre was developed by Robert Maurer, Donald Keck, and Peter Schultz of Corning Glass Corporation of USA in the year 1970. We will study some important aspects of fibre optics in this chapter, such as characteristics of optical fibres, their fabrication process, and, finally, their applications.

5.2 PROPAGATION OF LIGHT IN OPTICAL FIBRES

We will present the basic operating principle along with the underlying physics of optical fibres in this section.

5.2.1 Total Internal Reflection

A ray of light travelling from a medium of refractive index n_1 to another medium of refractive index n_2 undergoes refraction at the interface. This refraction is governed by Snell's law, given as follows:

$$\frac{\sin i}{\sin r} = \frac{n_2}{n_1} \tag{5.1}$$

where i and r represent the angle of incidence and the angle of refraction, respectively. If a ray of light travels from a denser to a rarer medium, it bends away from the normal. In such a scenario, three different possibilities can result, depending on the incidence angle, as shown in Fig. 5.1.

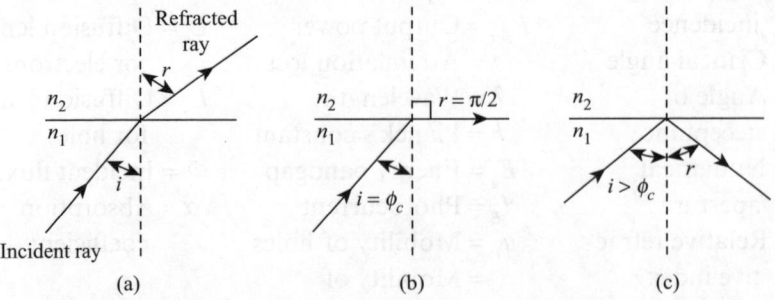

Fig. 5.1 Refraction of light travelling from denser to rarer medium
(a) $i < \phi_c$, (b) $i = \phi_c$, (c) $i > \phi_c$

Figure 5.1(a) shows the refracted ray bending away from the normal in a general case. As the incidence angle increases, the refracted ray bends more and more towards the interface between the two media. When i reaches a certain critical value (the critical angle), the angle of refraction becomes equal to 90° and the refracted ray travels horizontally, along the interface. Any further increase in the angle of incidence results in the refracted ray being reflected back into the medium of refractive index n_1. This phenomenon of re-entering of the refracted ray into the medium from which the ray incident on the interface originated is called *total internal reflection*. From Eq. (5.1) we can write the following expression: $\dfrac{\sin i_c}{\sin 90°} = \dfrac{n_2}{n_1}$ resulting in the following relation:

$$\sin i_c = \frac{n_2}{n_1} \tag{5.2}$$

5.2.2 Principle of Optical Fibres

An optical fibre consists of a cylindrical glass core surrounded by a cladding layer of a slightly lower refractive index. Figure 5.2 shows a schematic diagram of an optical fibre along with the path traversed by a ray of light incident on one end of the optical fibre.

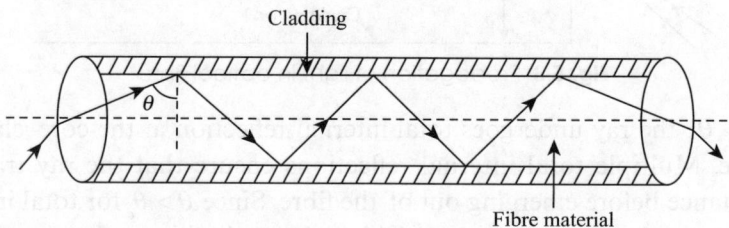

Fig. 5.2 Schematic of ray travelling in optical fibre

The ray of light is incident on the interface between the fibre material and the cladding at an angle of θ. If $\theta \geq \theta_c$, then the incident ray is totally internally reflected. From Snell's law, this implies that

$$\sin \theta_c = \frac{n_2}{n_1} \tag{5.3}$$

The incident ray suffers multiple total internal reflections and can travel long distances before emerging out of the optical fibre. Thus, optical fibres have been proved to be extremely useful for long-distance transmission. It is, however, necessary to first convert the original electrical signals to optical signals using suitable devices. In addition, for transmission to be efficient, signal attenuation due to absorption should be minimized. In practice, this can be achieved by special preparations and the use of purification techniques.

5.2.3 Fractional Refractive Index

Let us assume that the refractive index of the core material is n_1 and that of the cladding material is n_2. A quantity Δn_r is then defined by the following term:

$$\Delta n_r = (n_1 - n_2)/n_1$$

The quantity Δn_r is called the relative refractive index difference or the fractional refractive index.

5.3 NUMERICAL APERTURE AND ACCEPTANCE ANGLE

A ray of light entering the flat end of an optical fibre at an angle α with respect to the axis gets bent towards the normal, as shown in Fig. 5.3. The ray entering the solid core makes an angle ϕ with the axis. The ray travels through the core and is incident at an angle θ on the core–cladding interface. Note that

$$\phi = \frac{\pi}{2} - \theta$$

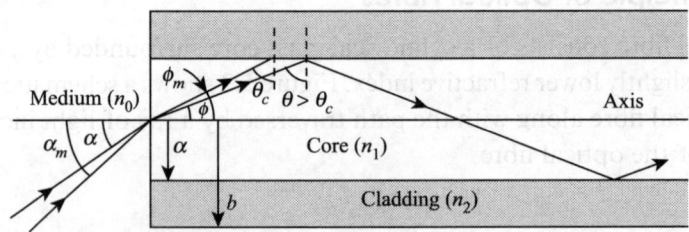

Fig. 5.3 Propagation of light in optical fibre

If $\theta > \theta_c$, the ray undergoes total internal reflection at the core–cladding interface. Multiple total internal reflections ensure that the ray travels a long distance before emerging out of the fibre. Since $\theta > \theta_c$ for total internal reflection to take place, ϕ needs to be less than a limiting value ϕ_m, given as follows: $\phi_m = \frac{\pi}{2} - \theta_c$

In other words, the angle α that the incident ray makes with the axis of the optical fibre should be less than a maximum value α_m for total internal reflection to take place. The angle α_m is called the *angle of acceptance* of the fibre. All the incoming rays that are incident within a cone of half-angle α_m are collected and propagated by the optical fibre.

Application of Snell's law at the core–air interface gives the following equation:

$$n_0 \sin \alpha = n_1 \sin \phi \tag{5.4}$$

Using the limiting values in Eq. (5.4), the following relation is obtained:

$$n_0 \sin \alpha_m = n_1 \sin \phi_m \tag{5.5}$$

We also have the following relation:

$$\phi_m = \frac{\pi}{2} - \theta_c \tag{5.6}$$

The use of Eq. (5.6) in Eq. (5.5) yields the following expression:

$$n_0 \sin \alpha_m = n_1 \cos \theta_c \tag{5.7}$$

We can write that

$$\cos \theta_c = [1 - \sin^2 \theta_c]^{1/2} \tag{5.8}$$

The use of Eq. (5.3) in Eq. (5.8) leads to the following expression:

$$\cos \theta_c = \left[1 - \frac{n_2^2}{n_1^2} \right]^{1/2}$$

which gives the following relation:

$$\cos \theta_c = \frac{\left(n_1^2 - n_2^2 \right)^{1/2}}{n_1} \tag{5.9}$$

Substitution of Eq. (5.9) into Eq. (5.7) results in the following expression:

$$n_0 \sin \alpha_m = n_1 \frac{\left(n_1^2 - n_2^2 \right)^{1/2}}{n_1}$$

which implies that

$$n_0 \sin \alpha_m = (n_1^2 - n_2^2)^{1/2} \tag{5.10}$$

The term $n_0 \sin \alpha_m$ is called the *numerical aperture* (NA) for an optical fibre. The numerical aperture decides the light-gathering capacity of a fibre. Thus, we can write the following expression:

$$NA = (n_1^2 - n_2^2)^{1/2} \tag{5.11}$$

Let us define a quantity called relative refractive index difference Δ by the following expression:

$$\Delta = \frac{n_1^2 - n_2^2}{2n_1^2} \tag{5.12}$$

The use of Eq. (5.12) in Eq. (5.11) leads to the following relation:

$$NA = n_1 \sqrt{2\Delta} \tag{5.13}$$

Numerical aperture is a measure of the quantity of light that can be collected by an optical fibre. The light-gathering capability of an optical fibre increases with an increase in its numerical aperture. From Eqs (5.10) and (5.11), it is clear that $NA = \sin \alpha_m$ for $n_0 = 1$. Since the maximum value of $\sin \alpha_m$ can be 1, NA cannot exceed 1. For $\alpha_m = 90$, NA $= 0$; that is, the fibre completely reflects the incident light. Numerical apertures of practical optical fibres generally fall in the range of 0.2–1.

5.4 TYPES OF OPTICAL FIBRES

Optical fibres are classified into three major categories based on the following parameters:
1. Raw material of the fibre
2. Number of modes of propagation
3. Refractive index profile

We will discuss the various types of optical fibres in this section.

5.4.1 Classification Based on Raw Material of Fibre

Optical fibres are generally made from two basic materials:
1. Glass
2. Plastic

5.4.1.1 Glass Optical Fibres

If an optical fibre is made by fusing mixtures of metal oxides and silica glasses, it is known as a glass fibre. Examples include GeO_2–SiO_2 core and SiO_2 clad; SiO_2 core and P_2O_3–SiO_2 clad.

5.4.1.2 Plastic Optical Fibres

Polystyrene and acrylate compounds are used to fabricate plastic optical fibres. In general, plastic optical fibres are inexpensive and highly flexible. They are extremely tough and have a long life. However, on the negative side, they exhibit greater attenuation compared to glass fibres. Examples are polystyrene core and methylmethacrylate clad; polymethyl-methacrylate core and copolymer clad.

5.4.1.3 Types of Rays

As light propagates through an optical fibre, two possibilities exist (Fig. 5.4):
1. The light ray can be launched in such a manner that it is in a plane that contains the axis of the optical fibre. Even after total internal reflection, the light ray continues to remain in the same plane. The ray thus always crosses the axis of the optical fibre. Such rays are called *meridional rays*.
2. The light ray can also be launched in a manner such that it belongs to a plane that does not contain the fibre axis. This can happen, for example, in a situation where the ray is launched at an angle such that it does not intersect the axis of the fibre. Even after total internal reflection, the light ray would never intersect the axis of the optical fibre. Such rays are called *skew rays*.

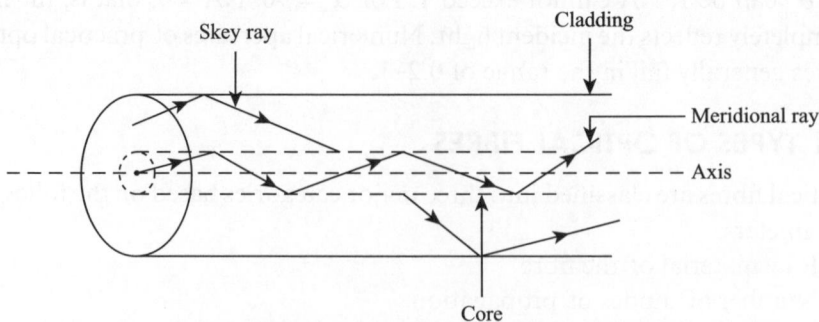

Fig. 5.4 Schematic representation of meridional and skew rays

5.4.2 Classification Based on Number of Modes of Propagation

Depending on the number of modes of propagation of the signal, optical fibres can be divided into two types:
1. Single-mode optical fibres
2. Multimode optical fibres

5.4.2.1 Single-mode Optical Fibres

A single-mode optical fibre is designed to carry only one signal of light (mode). This signal or ray of light often contains a range of different wavelengths. A typical single-mode optical fibre has a core diameter of 8–10 μm and a cladding diameter of 125 μm. It is designed such that only a small refractive index difference exists between the core and the cladding. Some special types of single-mode optical fibres exist that have been chemically or physically altered to give special properties, such as dispersion-shifted fibre and non-zero dispersion-shifted fibre. Only laser light can be used for the signals propagating through single-mode fibres.

5.4.2.2 Multimode Optical Fibres

In a multimode optical fibre, more than one signal can be propagated. Multimode fibres have several advantages over single-mode fibres (Table 5.1). Larger core diameter of a multimode fibre is an added advantage for end-to-end connection of similar fibres. Multimode optical fibres are mostly used for communication over shorter distances, such as within a building or inside a campus. Typical multimode links have data rates of 10 Mbit/s to 10 Gbit/s over link lengths of up to 600 m—more than sufficient for the majority of applications within the premises.

Table 5.1 Differences between single-mode and multimode fibres

Single-mode fibres	Multimode fibres
Only one mode can be propagated.	A large number of modes can be propagated.
Core diameter is small.	Core diameter is large.

5.4.3 Classification Based on Refractive Index Profile

Optical fibres can be divided into two more types based on the refractive index difference between the core and the cladding, which are as follows:
1. Step-index optical fibres
2. Graded-index optical fibres

5.4.3.1 Step-index Optical Fibres

A step-index optical fibre consists of a solid cylindrical core of diameter $2a$ and refractive index n_1, surrounded by a coaxial cylindrical cladding of outer

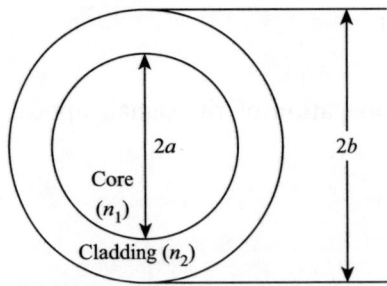

Fig. 5.5 Schematic diagram of step-index fibre

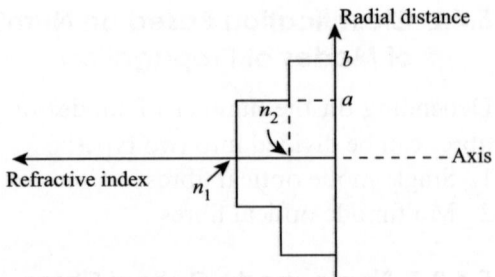

Fig. 5.6 Step function dependence of refractive index

diameter $2b$ and refractive index n_2. A schematic diagram of a step-index fibre is shown in Fig. 5.5.

Typical solid cylindrical core diameters are in the range of 50–100 μm, and cylindrical cladding diameters are in the range of 118–140 μm. Such an optical fibre is called a *step-index fibre* due to the step function dependence of the refractive index on radial distance, as shown in Fig. 5.6.

As mentioned earlier, for light to propagate through a fibre, the incoming rays must be incident within a cone of half-angle α_m around the axis of the optical fibre limit of $\alpha = 0$ to $\alpha = \alpha_m$. Figure 5.7 is a schematic diagram showing the path travelled by (a) a ray incident at one end of the fibre at an angle $\alpha = 0°$ and (b) a ray incident at an angle $\alpha = \alpha_m$.

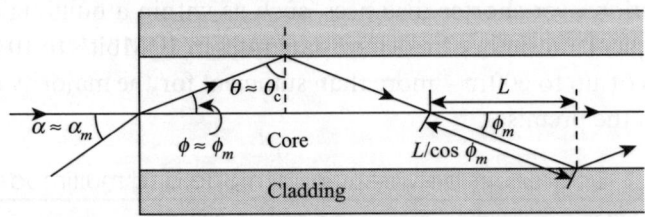

Fig. 5.7 Paths of rays travelling through step-index fibre

The ray incident along the axis of the core traverses a distance L with a velocity v in time t_1, which can be expressed as follows:

$$t_1 = \frac{L}{v} \tag{5.14}$$

However, by definition,

$$n_1 = \frac{c}{v} \tag{5.15}$$

where c is the velocity of light.

Using Eq. (5.15) in Eq. (5.14), we get the following relation:

$$t_1 = \frac{Ln_1}{c} \tag{5.16}$$

Since the ray with $\alpha = \alpha_m$ covers the same distance L, the actual distance traversed is given by $L/\cos \phi_m$, as shown in Fig. 5.7. The time t_2 taken for this travel is given as follows:

$$t_2 = \frac{Ln_1}{c(n_2 / n_1)} = \frac{Ln_1}{c \cos \phi_m} \tag{5.17}$$

Using Eq. (5.6) in Eq. (5.17), we get the following relation:

$$t_2 = \frac{Ln_1}{c \sin \theta_c} \tag{5.18}$$

Using Eq. (5.3), Eq. (5.18) can be rewritten in the following form:

$$t_2 = \frac{Ln_1}{c(n_2/n_1)} = \frac{Ln_1^2}{cn_2} \tag{5.19}$$

The rays entering the optical fibre at $\alpha = 0°$ and $\alpha = \alpha_m$ are launched at the same time but are separated by a time interval Δt after traversing a length L within the optical fibre. This time interval Δt is given by the following relation:

$$\Delta t = t_2 - t_1$$

which, on using Eqs (5.19) and (5.16), yields the following equation:

$$\Delta t = \frac{Ln_1^2}{cn_2} - \frac{Ln_1}{c} = \frac{Ln_1}{c}\left(\frac{n_1 - n_2}{n_2}\right) \tag{5.20}$$

Pulse broadening per unit length can be expressed in the following form:

$$\frac{\Delta t}{L} = \frac{n_1}{n_2}\left(\frac{n_1 - n_2}{c}\right) \tag{5.21}$$

The term $\Delta t/L$ is also referred to as *multipath time dispersion* of an optical fibre.

5.4.3.2 Graded-index Fibres

Multipath time dispersion is one serious disadvantage of step-index fibres, which limits their use in long-distance applications. A fibre in which the refractive index of the core varies with radius helps minimize or even eliminate this disadvantage. Such an optical fibre is called a *graded-index fibre*. The variation of the refractive index with radial distance r from the axis for a graded-index fibre is described by the following equation:

$$n(r) = \begin{cases} n_1 = n_0\left[1 - 2\Delta\left(\dfrac{r}{a}\right)^{\alpha}\right]^{\frac{1}{2}}, \text{ for } r \leq a \\[2mm] n_2 = n_0\left[1 - 2\Delta\right]^{\frac{1}{2}}, \text{ for } b \geq r \geq a \end{cases} \tag{5.22}$$

where $n(r)$ represents the refractive index at a radial distance r and n_0 its maximum value along the axis of the optical fibre (i.e., at $r = 0$), a is the core radius, b is the radius of the cladding, and α is the profile parameter. The value of the profile parameter decides the exact variation of the refractive index within the

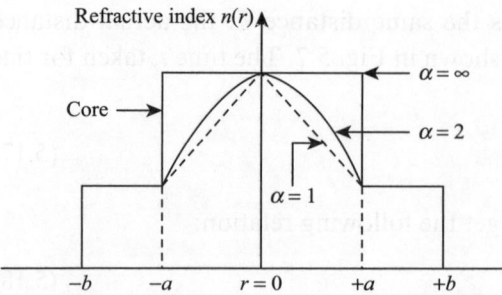

Refractive index $n(r)$

Core

$\alpha = \infty$

$\alpha = 2$

$\alpha = 1$

$-b$ $-a$ $r = 0$ $+a$ $+b$

Fig. 5.8 Refractive index variation for graded-index fibre

core of the fibre. Figure 5.8 shows the variation of the refractive index within the core for some specific values of the profile parameter.

The refractive index variation with radial distance is triangular for $\alpha = 1$, parabolic for $\alpha = 2$, and equivalent to that of a step-index fibre for $\alpha = \infty$.

The numerical aperture NA for a parabolic profile of refractive index is expressed as follows:

$$\text{NA} = \left(n_1^2 - n_2^2 \right)^{\frac{1}{2}} \tag{5.23}$$

Using Eq. (5.22) in Eq. (5.23), we get the following expression:

$$\text{NA} = \left[n_0^2 \left\{ 1 - 2\Delta \left(\frac{r}{a} \right)^2 \right\} - n_0^2 \left(1 - 2\Delta \right) \right]^{\frac{1}{2}}$$

or

$$\text{NA} = n_0 \left[2\Delta \left(\left(1 - \frac{r^2}{a^2} \right) 1 - \frac{r^2}{a^2} \right) \right]^{\frac{1}{2}} \tag{5.24}$$

At $r = 0$, that is, along the axis of the optical fibre,

$$\left(\text{NA} \right)_{\text{axis}} = n_0 \sqrt{2\Delta} \tag{5.25}$$

At $r = a$, that is, at the core–cladding interface,

$$\left(\text{NA} \right)_{r = a} = 0 \tag{5.26}$$

Propagation of the ray through a graded-index optical fibre can be visualized by assuming the fibre to be made up of several coaxial cylindrical layers with progressively decreasing refractive index as one goes away from the axis of the fibre. A schematic representation of this visualization is given in Fig. 5.9.

The changing refractive index of successive coaxial cylindrical layers continues to bend the incident ray till the condition for total internal reflection is met, after which the ray travels back towards the axis of the core. The ray travelling along the axis is not affected during its journey through the optical fibre. The rays near the axis of the core travel shorter paths in comparison with those near the core–cladding interface. Velocity of the rays near the axis of the core is, however, lower than that of the rays near the core–cladding interface. These two opposing conditions help in reducing the multipath time dispersion. The reduction of multipath dispersion to

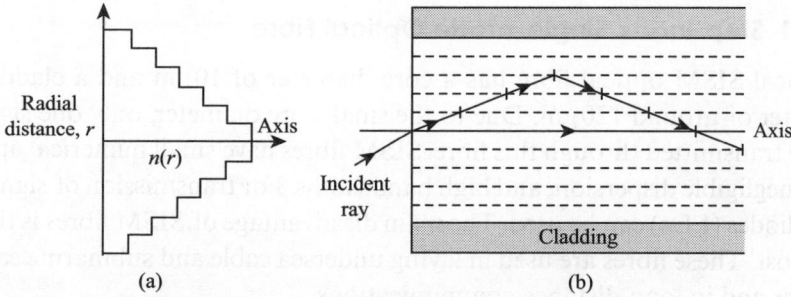

Fig. 5.9 Ray propagation through graded-index optical fibre
(a) Refractive index dependence (b) Ray propagation

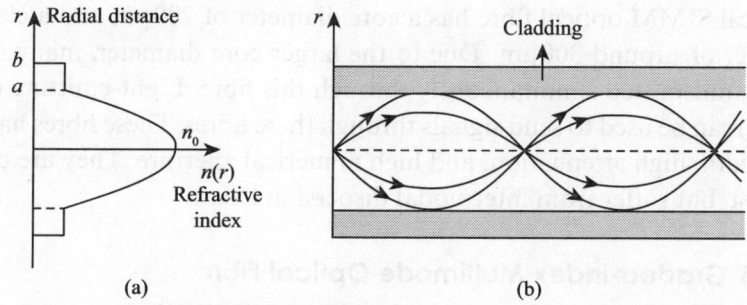

Fig. 5.10 Reduction of multipath dispersion (a) Refractive index
profile (b) Ray propagation

Table 5.2 Differences between step-index and graded-index fibres

Step-index fibre	Graded-index fibre
The refractive index at the core–cladding interface changes suddenly.	The refractive index at the core–cladding interface changes gradually.
Light rays pass through the fibre axis.	Light rays propagate in the form of skew rays.
Light rays follow a zigzag path.	Light rays follow a helical path.
This fibre has a low bandwidth.	This fibre has a high bandwidth.
Distortion is high in multimode type.	Distortion is low.

nearly zero for a parabolic refractive index profile is shown schematically
in Fig. 5.10 (Table 5.2).

5.4.4 Classification Based on Refractive Index Profile and Number of Modes

Optical fibres can be divided into three more categories based on the refractive index profile and modes of propagation of signals, which are as follows:
1. Step-index single-mode (SISM) optical fibre
2. Step-index multimode (SIMM) optical fibre
3. Graded-index multimode (GIMM) optical fibre

5.4.4.1 Step-index Single-mode Optical Fibre

A typical SISM optical fibre has a core diameter of 10 μm and a cladding diameter of around 120 μm. Due to the small core diameter, only one signal can be transmitted through this fibre. SISM fibres have small numerical apertures, negligible dispersion, and high bandwidths. For transmission of signals, laser diodes (LDs) can be used. The main disadvantage of SISM fibres is their high cost. These fibres are used in laying undersea cable and submarine cable systems and in long-distance communications.

5.4.4.2 Step-index Multimode Optical Fibre

A typical SIMM optical fibre has a core diameter of 200 μm and a cladding diameter of around 300 μm. Due to the larger core diameter, many signals can be transmitted simultaneously through this fibre. Light-emitting diodes (LEDs) can be used to send signals through these fibres. These fibres have low bandwidth, high attenuation, and high numerical aperture. They are of very low cost, but suffer from intermodal dispersion loss.

5.4.4.3 Graded-index Multimode Optical Fibre

A typical GIMM optical fibre has a core diameter of 200 μm and a cladding diameter of around 250 μm. As the refractive index of the core decreases gradually from the core axis to the cladding, the intermodal loss is reduced to a great extent. This is the main advantage of a GIMM fibre, compared to an SIMM fibre. These fibres have low signal attenuation, intermediate bandwidth, and small numerical aperture. The cost of GIMM fibres is, however, high. They are used in intercity trunks between telephone offices.

5.5 DOUBLE-CRUCIBLE TECHNIQUE OF FIBRE DRAWING

The conventional glass-refining techniques are used for producing high-purity powders of raw materials such as SiO_2, GeO_2, B_2O_3, and Al_2O_3. An appropriate mixture of these materials is then melted in platinum or silica crucibles at temperatures in the range 1000–1300°C. After suitable processing, the melt is cooled and drawn into rods or tubes. The double-crucible technique is suitable for continuous manufacture of optical fibres. Figure 5.11 shows a schematic diagram of the set-up for the double-crucible method.

The set-up consists of two concentric platinum crucibles mounted within a vertical cylindrical muffle furnace. The temperature within the furnace can be varied and set extremely accurately within the range 800–1200°C. Raw materials for the core and the cladding are fed into the two crucibles in powder form or as preformed rods. The crucibles have nozzles at the base that allow the clad fibre to be drawn from the melt. Refractive index grading is achieved by diffusion of dopant ions across the core–cladding interface. The technique can be used to

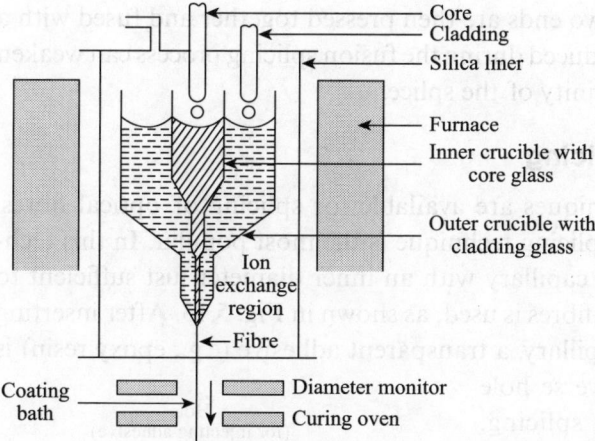

Fig. 5.11 Schematic diagram of the set-up for double-crucible method

produce relatively inexpensive fibres with large core diameters. These in turn result in large numerical apertures. An attenuation level of 3 dB/km has been achieved for sodium borosilicate glass fibres fabricated using this technique.

5.6 SPLICING

The process of forming a permanent joint between two optical fibres is called *splicing*. Splicing is resorted to under two circumstances: (a) to increase the length of the optical fibre and (b) to repair a broken fibre cable. There are two ways in which splicing can be achieved: (a) fusion splicing and (b) mechanical splicing.

5.6.1 Fusion Splicing

Figure 5.12 shows a schematic diagram of a fusion splicing set-up. The set-up uses an electric arc as a heating source. The ends of the prepared fibres requiring splicing are placed in precise alignment.

This is performed using an inspection microscope. A short arc discharge is then used to polish the fibre ends. This fire polishing removes defects due to

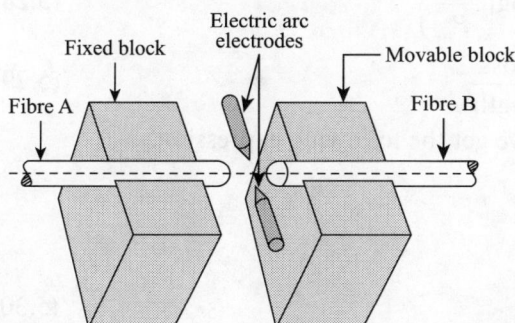

Fig. 5.12 Schematic diagram of fusion splicing set-up

imperfect cleaving. The two ends are then pressed together and fused with a stronger arc. The heat produced during the fusion splicing process can weaken the optical fibre in the vicinity of the splice.

5.6.2 Mechanical Splicing

Several mechanical techniques are available for splicing of optical fibres. Of these, the snug tube splicing technique is the most popular. In this technique, a glass or ceramic capillary with an inner diameter just sufficient to accommodate the optical fibres is used, as shown in Fig. 5.13. After inserting the fibre ends into the capillary, a transparent adhesive (e.g., epoxy resin) is injected through a transverse hole onto the ends requiring splicing. The adhesive produces mechanical bonding as well as index matching. This stable low-loss splicing method places stringent limits on the usable capillary diameters.

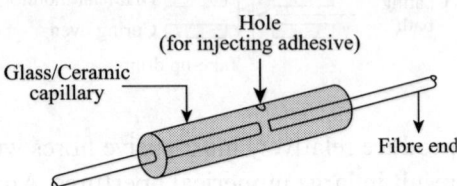

Fig. 5.13 Mechanical splicing technique

5.7 POWER LOSSES IN OPTICAL FIBRES

As an optical signal propagates through an optical fibre, it undergoes a loss in signal strength due to different causes. We will discuss the concept of signal loss in this section.

5.7.1 Losses Due to Attenuation

If P_{in} represents the input power and P_{out} the output power at the receiving end of the optical fibre, the attenuation loss in dB/km is given by the following expression:

$$A = \frac{10}{L} \log_{10} \frac{P_{in}}{P_{out}} \tag{5.27}$$

where L represents the fibre length in km.

Power loss is given by the following expression:

$$\text{Power loss (dB)} = -10 \log \left(\frac{P_{out}}{P_{in}} \right) \tag{5.28}$$

Thus, $$\text{Attenuation} = \frac{\text{Power loss}}{\text{Fibre length}} \tag{5.29}$$

Using Eq. (5.28) in Eq. (5.29), we get the following expression:

$$\log \left(\frac{P_{out}}{P_{in}} \right) = \frac{[-A \times L]}{10}$$

or $$P_{out} = P_{in} \times 10^{-\left(\frac{A \times L}{10} \right)} \tag{5.30}$$

where P_{in} and P_{out} are expressed in W and L is expressed in km.

Conversely, using Eq. (5.30), the length of the optical fibre in km is given by the following expression:

$$L = \frac{10}{A} \log_{10}\left(\frac{P_{in}}{P_{out}}\right) \qquad (5.31)$$

5.7.2 Losses Due to Dispersion

For a step-index optical fibre, pulse broadening per unit length given in Eq. (5.21) is as follows:

$$\frac{\Delta t}{L} = \frac{n_1}{n_2}\left(\frac{n_1 - n_2}{c}\right) \qquad (5.32)$$

As mentioned earlier, the term $\Delta t/L$ is referred to as *solidus multipath time dispersion* of the optical fibre. Multipath time dispersion can be minimized or even eliminated using a graded-index optical fibre. Figure 5.9 shows how multipath dispersion can be reduced to a negligible level using a parabolic refractive index profile.

5.7.3 Losses Due to Bending of Optical Fibre

Optical fibres radiate the propagating power if they are bent. Generally, two types of bends are encountered: bends with radii much larger than the fibre diameter, called *macrobends*, and smaller bends, called *microbends*. Random microbends of the fibre axis can arise due to faulty cabling.

Electric field distribution for any guided core has a tail extending into the cladding. This tail is called the *evanescent field* and decays exponentially with distance from the core. Thus, a portion of the net energy contained in the wave travels in the cladding. At the macrobend, the tail on the far side of the centre of curvature has to move faster to keep pace with the field within the core. At a certain distance r_c, called the *critical distance*, the field tail needs to move faster than the speed of light in the cladding material, in order to keep pace with the core field. Since this is not possible, the optical energy contained in the tail beyond the critical distance is lost in the form of radiation. This effect is demonstrated schematically in Fig. 5.14.

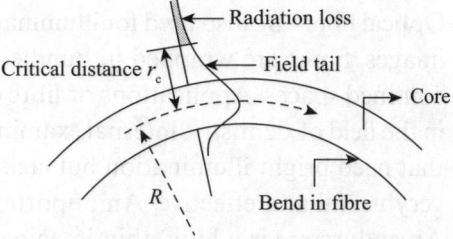

Fig. 5.14 Radiation loss at bends

5.7.4 Attenuation Curve

Light attenuates as it passes through an optical fibre. Two primary factors are responsible for this attenuation, namely, absorption and scattering. Light is absorbed by molecules in the glass and gets converted to heat. One of the

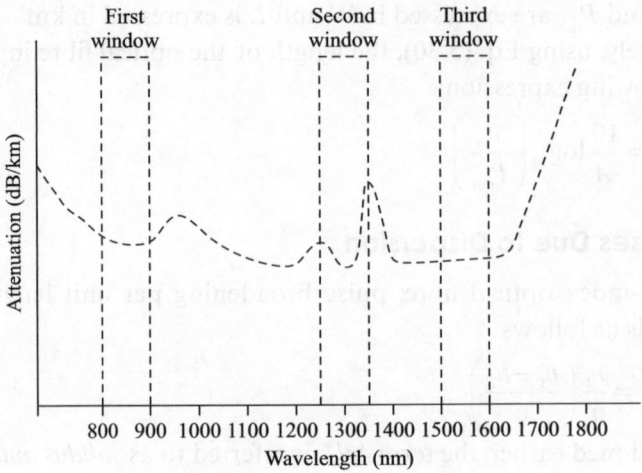

Fig. 5.15 Attenuation curve

primary absorbers is residual OH^+. Absorption of light is wavelength-dependent. Light also undergoes scattering as it collides with individual atoms in the glass. Scattering is wavelength-dependent and is inversely proportional to the fourth power of wavelength. Thus, if the wavelength of light is doubled, scattering losses are reduced by a factor of 16. On the other hand, absorption occurs strongly at select bands. For example, absorption due to residual OH^+ peaks around 1000, 1400, and above 1600 nm. A typical attenuation curve is shown in Fig. 5.15.

The attenuation curve clearly shows three transmission windows, at 800–900, 1260–1360, and 1500–1600 nm. The corresponding operating wavelengths are 850, 1310, and 1550 nm, respectively. Commonly used sources are LEDs and LDs operating at these infrared wavelengths.

5.7.5 Illumination and Image Transfer

Optical fibres are also used for illumination and image transfer. For transferring images, fibres are wrapped in bundles; these are useful in viewing images of confined spaces. Applications of fibre optic-based illuminations are prevalent in the field of dentistry, internal examination, and photography. In applications that need bright illumination but lack a clear line-of-sight path, fibres can be very useful and effective. An important device that uses fibres is endoscope. An endoscope is a long, thin imaging device used to view objects through a small hole. It uses a coherent bundle of fibres sometimes with a lens. Medical endoscopes assist during surgical procedures and diagnostics. Industrial endoscopes are used to inspect complex objects like jet engine interiors. Optical fibres also find use in spectroscopy. Sometimes, substances need to be analysed for their composition, even if they cannot be placed inside a spectrometer. In such situations, optical fibres carry light to and from the object under analysis.

Example 5.1   A step-index fibre has a core of refractive index 1.5 and a cladding of refractive index 1.47. The diameter of the core of the fibre is 100 μm, and the medium surrounding the fibre is air. Determine (a) NA of the fibre, and (b) angles α_m, ϕ_m, and θ_c.

Solution   The calculations are as follows:

(a) $$NA = (n_1^2 - n_2^2)^{\frac{1}{2}} \tag{5.33}$$

Substitution of the given values of n_1 and n_2 in Eq. (5.33) leads to the following relation: $NA = \sqrt{(1.5)^2 - (1.47)^2}$

(b) or $$NA = \sqrt{2.25 - 2.16} = 0.3 \tag{5.34}$$

(i)   We have

$$NA = n_0 \sin \alpha_m \tag{5.35}$$

Using Eq. (5.34) and the refractive index value for air in Eq. (5.35), we get the following relation: $0.3 = 1 \times \sin \alpha_m$

or $$\alpha_m = 17.46° \tag{5.36}$$

(ii)   From Eq. (5.5),

$$n_0 \sin \alpha_m = n_1 \sin \phi_m \tag{5.37}$$

The use of Eq. (5.36) and known values of n_0 and n_1 in Eq. (5.37) yields the following relation: $0.3 = 1.5 \sin \phi_m$

or $$\phi_m = \sin^{-1}\left(\frac{0.3}{1.5}\right) = 11.54°$$

(iii)   The critical angle θ_c is given by the following relation:

$$\theta_c = \sin^{-1}\left(\frac{n_2}{n_1}\right) \tag{5.38}$$

Substitution of the given values of n_1 and n_2 in Eq. (5.38) yields the following result:

$$\theta_c = \sin^{-1}\left(\frac{1.47}{1.5}\right) = 78.52°$$

Example 5.2   For the optical fibre of Example 5.1, calculate the pulse broadening per unit length due to multipath dispersion.

Solution   From Eq. (5.21), we get the following relation:

$$\frac{\Delta t}{L} = \frac{n_1}{n_2}\left(\frac{n_1 - n_2}{c}\right) \tag{5.39}$$

Using the given values of n_1 and n_2 and the known value of c in Eq. (5.39), the following result can be obtained:

$$\frac{\Delta t}{L} = \frac{1.5}{1.47}\left(\frac{1.5 - 1.47}{3 \times 10^8}\right)$$

or $$\frac{\Delta t}{L} = \frac{1.5 \times 0.03}{1.47 \times 3 \times 10^8} = 10.2 \times 10^{-11} \text{ s/m}$$

Example 5.3   Use the details of the optical fibre described in Example 5.1, and calculate the minimum and maximum number of total internal reflections per metre for the rays travelling through the optical fibre.

Solution From Fig. 5.7, we can write the following relation:

$$\tan\phi_m = \frac{a}{L} \qquad (5.40)$$

Using the value of ϕ_m from Example 5.1, we get the following expression:

$$\tan(11.54°) = \frac{a}{L}$$

or $\qquad L = \dfrac{a}{\tan(11.54°)} \qquad (5.41)$

The use of the values of a and $\tan(11.54°)$ in Eq. (5.41) leads to the following result:

$$L = \frac{0.5}{0.2042} \times 10^{-4}\,\text{m}$$

$$= 2.45 \times 10^{-4}\,\text{m} \qquad (5.42)$$

The rays travelling with $\alpha = 0$ suffer no internal reflection; therefore, the minimum number of reflections per metre is zero. For rays travelling with $\alpha \approx \alpha_m$, one internal reflection takes place for a transverse distance of $2L$. Thus, the total number of internal reflections for 1 m is as follows: $\dfrac{1}{2L} = \dfrac{1}{2 \times 2.45 \times 10^{-4}} = 2040/\text{m}$

5.8 FIBRE-OPTIC COMMUNICATION SYSTEMS

A simplified block diagram of a fibre-optic communication system is shown in Fig. 5.16.

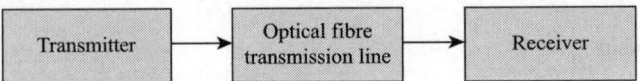

Fig. 5.16 Simplified block diagram of fibre-optic communication system

The communication system consists of a transmitter that converts the information to be transmitted, which is in the form of an electrical signal, into an optical signal. The optical signal then travels through the optical fibre transmission line to a receiver, which converts the optical signal back to the original electrical form. A more detailed block diagram is shown in Fig. 5.17.

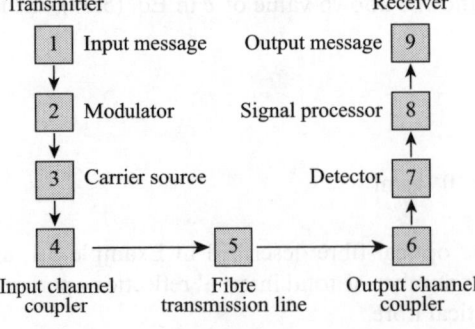

Fig. 5.17 Detailed block diagram of optical fibre-based communication system

5.9 ADVANTAGES OF FIBRE-OPTIC COMMUNICATION SYSTEM

Fibre-optic communication systems are becoming popular due to the several advantages these offer. The following are some of the more important advantages.

Immunity to electromagnetic interference   Flow of electrons through a conductor generates a magnetic field. Any change in the pattern of flow of electrons leads to a change in the generated magnetic field. This, in turn, leads to the generation of an electrical current. This phenomenon of the generation of an electrical current from a changing magnetic field is called electromagnetic interference or EMI. EMI is a common form of noise in communication systems involving coaxial cables. Fibre optic-based systems are, however, immune to EMI, since in these systems signals are transmitted as light instead of current.

Data security   Current-carrying conductors have magnetic fields associated with these currents. In spite of the most effective shielding, signal leakage can always take place. These leakages can result in tapping and therefore affect data security. In case of optical fibres, electromagnetic fields are confined within the fibre. Therefore, transmitted signals cannot be tapped. This leads to a high degree of data security.

Non-conductive cables   Electromagnetic design requires a uniform ground potential. The nominal ground potential can, however, differ by several volts if signal-carrying cables are spread over large distances. In circuits involving semiconductor devices, any difference in the ground potential leads to problems such as ground loop and noise. In extreme cases, these effects can also damage sensitive components of the circuit. Fibre-optic cables are fabricated using non-conducting materials such as *glass* and *plastic*, and are, therefore, much more immune to ground loop and noise-related problems associated with conducting materials.

Elimination of spark hazards   Signals being carried by an electric current can result in small sparks. These sparks can be hazardous, causing accidental fire in environments such as chemical plants or oil refineries. The risk associated with sparks is not present in fibre optic-based systems, since no current flow is involved in such systems.

Ease of installation   As the transmission capacity increases, coaxial cables can become extremely thick and rigid. Installation of such thick cables within a complex structure of a building is a big problem. Fibre cables are, however, leaner and more flexible, leading to a considerable ease of installation.

Higher bandwidth and distances   The information-carrying capacity of fibre-optic cables is much more than that of coaxial cables, that is, the bandwidth parameter of a fibre-optic cable is higher than that of a coaxial cable. Coaxial cables have a typical bandwidth parameter of a few MHz/km, whereas fibre-optic cables have a bandwidth in the region of 400 MHz/km.

Thus, systems based on fibre-optic cables can carry high-speed signals over longer distances without the need of repeaters. The information-carrying capacity of fibre-optic cables increases with frequency.

5.10 LIGHT SOURCES

Fibre-optic systems require a light source at the transmitter end. This light source converts electrical signals into optical signals. Generally, two types of sources are used: (a) incoherent optoelectronic sources like LEDs and (b) coherent optoelectronic sources like injection laser diodes (ILDs).

5.10.1 Light-emitting Diodes

When a p–n junction is forward biased, injection of carriers takes place across the junction. As the injected carriers travel, they undergo recombination in the space-charge region and in the neutral regions close to the junction. Figure 5.18 shows a schematic diagram of a p–n junction with no applied bias and with an applied forward bias. The corresponding band diagram showing injection of carriers across the junction is also shown in Fig. 5.18.

In the case of indirect-gap semiconductors such as Si and Ge, this recombination releases heat to the lattice. In direct-bandgap materials, on the other hand, the released energy can be given off as light emitted from the junction. This process is called *injection electroluminescence* and is the fundamental principle underlying the working of LEDs. LEDs find applications in digital

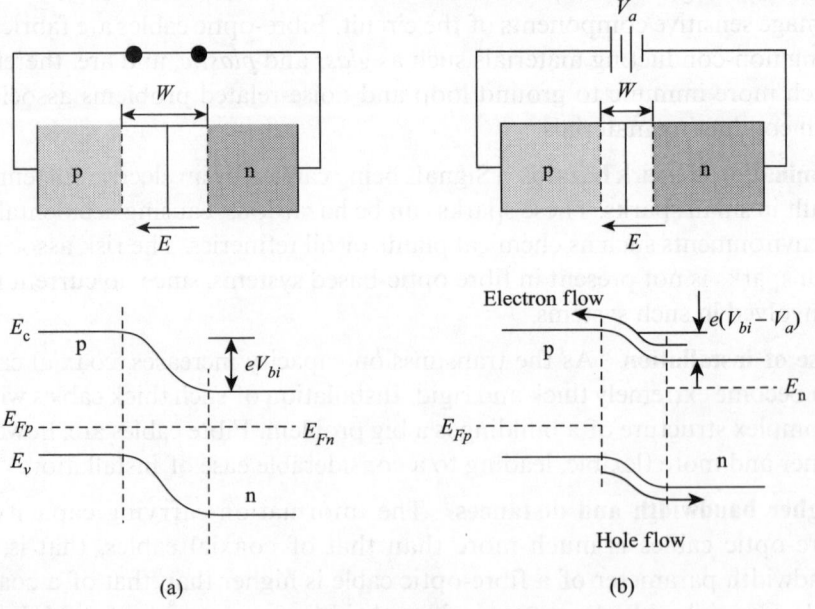

Fig. 5.18 Schematic diagram of p–n junction with its energy band (a) Zero bias (b) Forward bias

displays and communication systems. The wavelength of the emitted light is governed by the following equation:

$$\lambda = \frac{hc}{E_g} = \frac{1.24}{E_g} \, \mu m \qquad (5.43)$$

where E_g is the bandgap in eV.

5.10.2 Laser Diodes

The photons produced in an LED are the result of transitions of electrons from the conduction band to the valence band. Light emission is spontaneous because each band-to-band transition is, in itself, an independent event. The spectral output of the LED, however, has a fairly wide bandwidth. With suitable modifications in the structure and operating conditions of the LED, it is possible to operate the device in a mode that produces a coherent output with a wavelength bandwidth of < 0.1 nm. This special-mode-operated device is then called a *laser diode*.

5.11 DETECTORS

Photodetectors are devices that convert optical signals into electrical signals. Photons give their energies to valence-band electrons and excite them to the conduction band. The resultant electron–hole pairs increase the conductivity of the material. All photodetectors use these excess carriers in some way or the other to detect the incident radiation. Photoconductors, photodiodes, and phototransistors are some examples of such devices.

5.11.1 Photoconductors

A photoconductor consists of a bar of semiconductor material with ohmic contacts at the two ends. Figure 5.19 shows a semiconductor bar (n-type) of length L and cross-sectional area A, with illumination on the face having a larger area.

The photocurrent I_L generated by length L of the bar can be shown to be as follows:

$$I_L = eG_L \left(\frac{\tau_p}{t_n} \right) \left(1 + \frac{\mu_p}{\mu_n} \right) AL \qquad (5.44)$$

Where G_L is the photogeneration rate of excess carriers, τ_p the excess minority carrier lifetime, μ_p the mobility of the holes, and μ_n the mobility of the electrons.

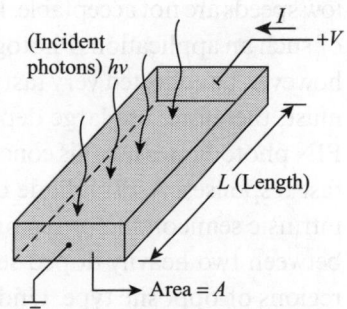

Fig. 5.19 Semiconductor bar under illumination

5.11.2 Photodiodes

A p–n junction can be used to separate the photogenerated excess carriers. A photodiode uses a reverse-biased p–n junction diode to detect photons.

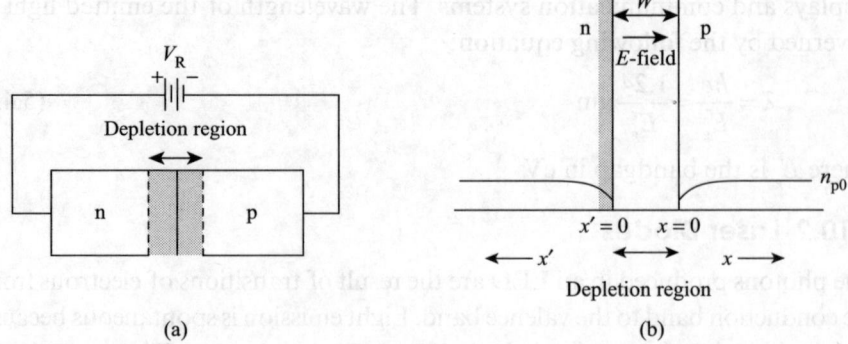

Fig. 5.20 Biasing and carrier profile of a long diode (a) Schematic diagram of reverse-biased long diode (b) Minority carrier concentration profile

Figure 5.20 shows a schematic diagram of a long reverse-biased diode along with the associated minority carrier concentration profile. Photocurrent is generated from three different sources. There are excess carriers in the space-charge region, neutral p-region, and neutral n-region. The total photocurrent density for a long diode is written as follows:

$$J_L = eG_L(W + L_n + L_p) \tag{5.45}$$

Here, G_L represents the photogenerated excess carrier generation rate, L_n is the diffusion length of the electrons, L_p is the diffusion length of holes, and W is the space-charge width.

5.11.3 PIN Photodiode

The photodiode discussed in the previous section is inherently slow due to the dominance of diffusion-based currents. In many photodetector applications such low speeds are not acceptable. Light-based communication systems are examples of such an application. Photogenerated carriers in the space-charge region can, however, be collected very fast due to the built-in electric field. Fast photodiodes must, therefore, use large depletion widths to enhance the speed of response. PIN photodiodes use this concept to achieve fast responses. A PIN diode consists of an intrinsic semiconductor region sandwiched between two heavily doped semiconductor regions of opposite type. Under an applied reverse bias, the complete intrinsic region can be swept off and the entire intrinsic region constitutes the space charge region. A schematic representation of a typical PIN photodiode with a suitable biasing arrangement is shown in Fig. 5.21.

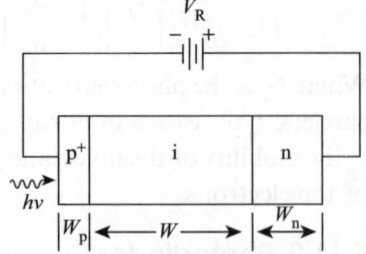

Fig. 5.21 Schematic representation of PIN photodiode along with its biasing arrangement

The photocurrent density J_L is as follows:

$$J_L = e\phi_i(1 - e^{-\alpha W}) \tag{5.46}$$

where ϕ_i represents the incident flux, α is the absorption coefficient, and W is the width of the ith layer.

5.11.4 Avalanche Photodiode

Under certain conditions, the reverse bias applied across a photodiode reaches the value that is sufficient to create electron–hole pairs by impact ionization. Excess carriers initially generated due to the absorption of photons create additional electron–hole pairs through the process of impact ionization. This avalanche process leads to huge current gains in photodiodes.

5.11.5 Phototransistors

A schematic representation of an n–p–n bipolar transistor is shown in Fig. 5.22.

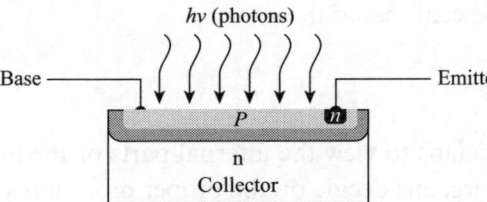

Fig. 5.22 Schematic representation of n–p–n phototransistor

The base terminal is generally kept open when a bipolar junction transistor is operated as a phototransistor. The incident photons create electron–hole pairs in the large-area base–collector junction. The built-in electric field in the space-charge region sweeps out these charge carriers to produce a photocurrent. The holes created are swept away into the p-type base to make the base positive with respect to the emitter. This results in forward biasing of the base–emitter junction, which causes the electrons to get emitted from the emitter region towards the base region, which leads to the usual transistor action.

5.11.6 Fibre-optic Sensors—Temperature and Displacement

A device that uses light guided within an optical fibre for detection of an external physical, chemical, or biomedical parameter is called a *fibre-optic sensor* (FOS). Figure 5.23 is a schematic diagram showing the general configuration of an FOS.

The sensor consists of a source, the output of which is fed into the modulating element using optical fibres. The measured parameter, which could be displacement or temperature, modulates the intensity of the light propagating through the optical fibre. The modulated light is fed through an optical fibre to a detector system. The signal processor then processes and calibrates the modulated light before it is displayed in the readout.

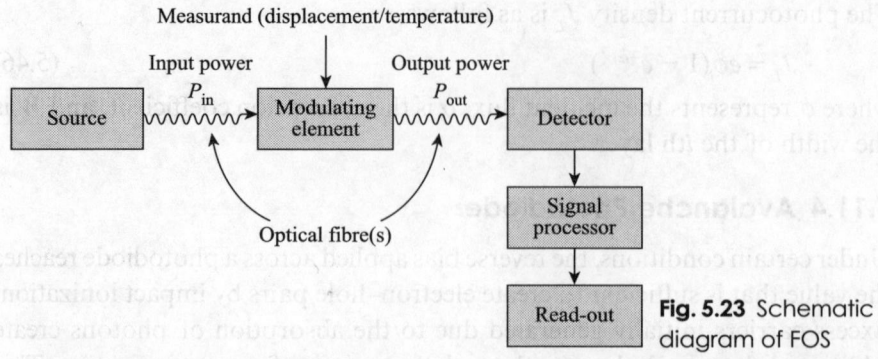

Fig. 5.23 Schematic diagram of FOS

A temperature sensor can also be designed based on the phenomenon of fluorescence. In this phenomenon, a material emits light after absorption of a suitable electromagnetic radiation. This material is called a *phosphor*. The intensity of fluorescence for most phosphors is temperature-dependent. A temperature sensor for the requisite temperature range can be designed by choosing an appropriate fluorescent material.

5.12 ENDOSCOPE

An endoscope is used by physicians to view the internal parts of the human body. This imaging can help surgeons decide on the proper procedures to be followed for treating patients. A typical endoscopy system is shown schematically in Fig. 5.24.

The endoscopy tube consists of two optical fibres. The outer fibre carries light from the source and projects it on the viewed object. The inner fibre is used to pick up the light reflected from the object surface and feeds it into a suitable imaging system.

Note: LEDs used in fibre-optic communication are generally fabricated using gallium arsenide (GaAs) or gallium arsenide phosphide (GaAsP). GaAsP LEDs operate at $1.3\,\mu m$, whereas those fabricated using GaAs operate at 0.81–$0.87\,\mu m$. The output spectrum of GaAsP LED is wider than that of GaAs LED by a factor of around 1.7. This leads to higher fibre dispersion in case of GaAsP LEDs. LEDs are increasingly being replaced by vertical cavity surface emitting laser (VCSEL) devices. These devices offer better speed, power,

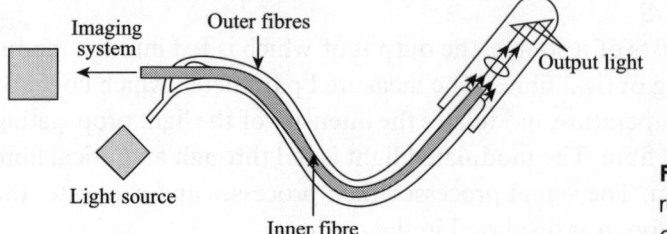

Fig. 5.24 Schematic representation of endoscope

and spectral properties in comparison to LEDs. VCSEL are semiconductor LDs that emit beams perpendicular to the top surface instead of the more conventional edge-emitting semiconducting lasers.

Example 5.4 A step-index fibre has a core of refractive index 1.5. If the NA of the fibre is 0.26, calculate the refractive index of the cladding material.

Solution For a step-index fibre,

$$NA = (n_1^2 - n_2^2)^{\frac{1}{2}} \tag{5.47}$$

Substituting the given value of NA in Eq. (5.47), we get the following relation:

$$0.26 = (n_1^2 - n_2^2)^{\frac{1}{2}}$$

yielding, $n_1^2 - n_2^2 = (0.26)^2 \cong 0.068$

resulting in $n_2^2 = n_1^2 - 0.068$

The use of the given value of n_1 yields the following result:

$$n_2^2 = (1.5)^2 - 0.068$$
$$= 2.25 - 0.068 = 2.182$$

giving $n_2 = \sqrt{2.182} = 1.48$

Example 5.5 Calculate α_m and ϕ_m for the fibre indicated in Example 5.4 if the medium surrounding the fibre is air.

Solution We have

$$NA = n_0 \sin \alpha_m \tag{5.48}$$

For a fibre surrounded by air, using the calculated value of NA in Eq. (5.48), we get the following relation: $0.26 = \sin \alpha_m$

resulting in $\alpha_m = \sin^{-1} 0.26 = 15.07°$

In addition, we have

$$n_0 \sin \alpha_m = n_1 \sin \phi_m \tag{5.49}$$

Substitution of the given values in Eq. (5.49) results in the following expression:

$$0.26 = 1.5 \sin \phi_m$$

giving $\sin \phi_m = \dfrac{0.26}{1.5} \cong 0.17$

resulting in $\phi_m \cong \sin^{-1}(0.17) \cong 9.79°$

Example 5.6 The cladding of a step-index fibre has a refractive index of 1.40. If NA of the fibre is 0.25, calculate the refractive index of the core material.

Solution We have

$$NA = (n_1^2 - n_2^2)^{\frac{1}{2}} \tag{5.50}$$

Using the given value of NA in Eq. (5.50), we get the following expression:

$$0.25 = (n_1^2 - n_2^2)^{\frac{1}{2}}$$

which, after solving, results in $n_1^2 - n_2^2 = (0.25)^2 = 0.0625$

yielding $n_1^2 = n_2^2 + 0.0625$ (5.51)

Substitution of the given value of n_2 in Eq (5.51) gives the following equation:

$$n_1^2 = (1.40)^2 + 0.0625$$

yielding $n_1^2 = 1.96 + 0.0625 = 2.0225$

resulting in $n_1 = \sqrt{2.0225} = 1.42$

Example 5.7 The pulse broadening per unit length for an optical fibre is 12×10^{-11} s/m. If the refractive index of the core is 1.5, calculate the refractive index of the cladding material.

Solution Using Eq. (5.21), we can write, for pulse broadening per unit length, that

$$\frac{\Delta t}{L} = \frac{n_1}{n_2}\left(\frac{n_1 - n_2}{C}\right) \tag{5.52}$$

Substituting the given values in Eq. (5.52), we get the following relation:

$$12 \times 10^{-11} = \frac{1.5}{n_2}\left(\frac{1.5 - n_2}{C}\right)$$

which results in $12 \times 10^{-11} \times 3 \times 10^8 = \frac{1.5}{n_2}(1.5 - n_2)$

giving $36 \times 10^{-3} = \frac{(1.5)^2}{n_2} - 1.5$

yielding $\frac{(1.5)^2}{n_2} = 36 \times 10^{-3} + 1.5 = 1.536$

giving us the following result: $n_2 = \frac{(1.5)^2}{1.536} \cong 1.46$

Example 5.8 The cladding material for the fibre indicated in Example 5.7 is changed keeping the core material unchanged. The pulse broadening per unit length changes to 20×10^{-11} s/m. Calculate the change in the refractive index of the cladding material.

Solution From Eq. (5.21), pulse broadening per unit length is expressed as follows:

$$\frac{\Delta t}{L} = \frac{n_1}{n_2}\left(\frac{n_1 - n_2}{C}\right)$$

Substituting these values and simplifying, we get the following relation:

$$20 \times 10^{-11} \times 3 \times 10^8 = \frac{(1.5)^2}{n_2} - 1.5$$

yielding $n_2 = \frac{(1.5)^2}{(0.06 + 1.5)} = 1.44$

Thus, $\Delta n_2 = 1.46 - 1.44 = 0.02$.

IMPORTANT CONCEPTS

1. A ray of light travelling from a medium of refractive index n_1 to a medium of refractive index n_2 undergoes refraction at the interface.

2. Optical fibres are classified into three categories based on the following criteria: (a) raw material of the fibre, (b) number of modes of propagation, and (c) refractive index profile.
3. A fibre in which the refractive index of the core varies with the radius helps minimize multipath time dispersion.
4. As the optical signal propagates through an optical fibre, it undergoes loss in signal strength.
5. LEDs utilize the process of injection electroluminescence.
6. PIN photodiodes have extremely high speed of response.

APPLICATIONS

1. An optical fibre consists of a cylindrical glass core surrounded by a cladding layer of a slightly lower refractive index.
2. Optical fibres are generally made from glass or plastic.
3. Optical fibres are of two types: single-mode and multimode.
4. Based on the refractive index profile, optical fibres can be classified into two types: step-index optical fibres and graded-index optical fibres.
5. Optical fibres are drawn using the double-crucible technique.
6. The process of forming a permanent joint between two optical fibres is called splicing.
7. Power loss in optical fibres can take place due to attenuation, dispersion, and bending.
8. Fibre optic-based communication systems are immune to EMI and are secure.
9. LEDs and LDs are two commonly used light sources.
10. Photoconductors, photodiodes, and phototransistors are commonly used photodetectors.
11. Fibre optic sensors can be used to monitor temperature and displacement.
12. An endoscope uses optical fibres to view the internal parts of the human body.

IMPORTANT FORMULAE

1. $\sin i / \sin r = n_2/n_1$

2. $\sin \theta_C = n_2/n_1$

3. $NA = (n_1^2 - n_2^2)^{1/2}$

4. $\Delta t/L = n_1/n_2 (n_1 - n_2)/c$

5. $L = 10/A \log_{10}(P_{in}/P_{out})$

6. $\lambda = hc/E_g = 1.24/E_g$ (μm)

7. $I_L = eG_L(\tau_p/t_n)(1 + \mu_p/\mu_n)AL$

8. $I_L = eG_L(W + L_n + L_p)$

9. $J_L = e\Phi_i(1 - \exp(-\alpha W))$

10. $NA = \left(n_1^2 - n_2^2\right)^{1/2}$

11. $\dfrac{\Delta t}{L} = \dfrac{n_1}{n_2}\left(\dfrac{n_1 - n_2}{c}\right)$

12. $A = \dfrac{10}{L} \log_{10} \dfrac{P_{in}}{P_{out}}$

13. $I_n = eG_n \left(\dfrac{\tau_p}{t_n}\right)\left(1 + \dfrac{\mu_p}{\mu_n}\right)^{AL}$

SELF-ASSESSMENT

Multiple-choice Questions

5.1 The correct form of Snell's law is
   (a) $\sin i / \sin r = n_2 / n_1$
   (b) $\sin r / \sin i = n_2 / n_1$
   (c) $\sin i \sin r = n_2 / n_1$
   (d) $\sin i = n_2 / n_1$

5.2 NA is given by
   (a) $NA = (n_1^2 - n_2^2)$
   (b) $NA = (n_1^2 - n_2^2)^{1/2}$
   (c) $(n_1^2 - n_2^2)^3$
   (d) $1/(n_1^2 - n_2^2)$

5.3 Typical multimode links have data rates of
   (a) 10–20 Mbit/s
   (b) 10–100 Mbit/s
   (c) 10 Mbit/s–10 Gbit/s
   (d) (1 – 100) Mbit/s

5.4 Multiple time dispersion of optical fibre is expressed as
   (a) $\Delta t/L = n_1 n_2 (n_1 - n_2)/c$
   (b) $\Delta t/L = n_1/n_2[c/(n_1 - n_2)$
   (c) $\Delta t/L = n_1 c/[n_2(n_1 - n_2)]$
   (d) $\Delta t/L = n_1/n_2[(n_1 - n_2)/c]$

5.5 Light travelling in a graded-index fibre follows a
   (a) helical path
   (b) zigzag path
   (c) circular path
   (d) straight-line path

5.6 Attenuation loss, A, is given by the relation
   (a) $A = 10L \log_{10} (P_{in}/P_{out})$
   (b) $A = (10/L) \log_{10} (P_{in}/P_{out})$
   (c) $A = (L/10) \log_{10} (P_{in}/P_{out})$
   (d) $A = \log_{10} (P_{in}/P_{out})$

5.7 Solidus multipath time dispersion of optical fibre refers to
   (a) $(\Delta t)L$
   (b) $L/\Delta t$
   (c) $\Delta t/L$
   (d) $(\Delta tL)^2$

5.8 Fibre-optic communication is immune to EMI since it uses
   (a) plastic
   (b) LEDs
   (c) photodiodes
   (d) light

5.9 Bandwidth parameter has the units
   (a) MHz/km
   (b) MHz-km
   (c) $(MHz)^2$ km
   (d) dB/km

5.10 Wavelength of light emitted by an LED is governed by the equation
   (a) h/cE_g
   (b) hc/E_g
   (c) E_g/hc
   (d) $E_g h/c$

5.11 LEDS use
   (a) indirect-gap semiconductors
   (b) direct-gap semiconductors
   (c) both indirect- and direct-gap semiconductors
   (d) none of these

5.12 The correct choice for fabricating an LED is
   (a) Si
   (b) Ge
   (c) GaAs
   (d) carbon

5.13 Spectral output of an LED is
   (a) narrow
   (b) wide
   (c) sharp
   (d) negligible

5.14 Wavelength bandwidth of an LD is typically
   (a) >1 nm
   (b) >10 nm
   (c) = 1 nm
   (d) < 0.1 nm

5.15 Choose the one that is not a photodetector:
   (a) Solar cell
   (b) Photoconductor
   (d) Photodiode
   (d) Phototransistor

5.16 Photocurrent generated in a photoconductor is proportional to
   (a) τ_p^2
   (b) τ_p
   (c) $\dfrac{1}{\tau_p^2}$
   (d) $\dfrac{1}{\tau_p}$

5.17 A photodiode uses a p–n junction that is
   (a) forward biased
   (b) not biased
   (c) reverse biased
   (d) none of these

Review Questions

5.1 Explain the phenomenon of total internal reflection.

5.2 What is an optical fibre?

5.3 Express Snell's law mathematically.

5.4 How many types of optical fibres are commonly used in fibre-optic communication?

5.5 Draw a schematic layout of a step-index optical fibre.

5.6 What are graded-index optical fibres?

5.7 Demonstrate, with a schematic diagram, refractive index dependence on radial distance for a graded-index optical fibre.

5.8 Write an expression for the numerical aperture of a graded-index fibre for $r = a$.

5.9 Provide a detailed description of an optical fibre-based communication system using a block diagram.

5.10 Explain the principle of operation of optical fibres.

5.11 What is the angle of acceptance for an optical fibre?

5.12 Define and explain numerical aperture for an optical fibre.

5.13 Derive the expression for NA.

5.14 Write an expression for $n(r)$ for a graded-index fibre.

5.15 Write expressions for the following parameters, for a graded-index fibre:
(a) $(NA)_{axis}$ and (b) $(NA)_{r=a}$

5.16 What is demodulation?

5.17 What are the advantages of using optical fibres in communication systems?

5.18 Give some medical applications of optical fibres.

5.19 State and explain Snell's law.

5.20 Show the process of total internal reflection using a suitable diagram.

5.21 What is a cladding layer?

5.22 What is the role of the core in an optical fibre?

5.23 What is fractional refractive index?

5.24 Can we have a negative fractional refractive index?

5.25 What is the significance of angle of acceptance of an optical fibre?

5.26 What is the relationship between the angle ϕ_m and the critical angle θ_c?

5.27 What is the significance of numerical aperture of an optical fibre?

5.28 What is the maximum possible value of numerical aperture?

5.29 What are the two materials generally used for fabricating optical fibres?

5.30 Give one disadvantage of plastic optical fibres.

5.31 What are meridional rays?

5.32 What are skew rays?

5.33 Give the typical cladding diameter of a single-mode optical fibre.

5.34 Describe step function dependence of refractive index with a suitable sketch.

5.35 Draw a typical attenuation curve for an optical fibre.

Numerical Problems

5.1 A step-index fibre has a core of refractive index 1.5 and a cladding of refractive index 1.48. Calculate the numerical aperture of the fibre.

$$\left[Hint: NA = (n_1^2 - n_2^2)^{\frac{1}{2}} \right]$$

5.2   Determine the angles α_m and ϕ_m for the optical fibre indicated in Problem 5.1.

$$\left[\text{Hint: NA} = n_0 \sin \alpha_m \text{ and } n_0 \sin \alpha_m = n_1 \sin \phi_m \right]$$

5.3   Determine the pulse broadening per unit length due to multipath dispersion for the optical fibre mentioned in Problem 5.1. $\left[Hint: \dfrac{\Delta t}{L} = \dfrac{n_1}{n_2}\left(\dfrac{n_1 - n_2}{c} \right) \right]$

5.4   The core of the optical fibre mentioned in Problem 5.1 is replaced with a material of refractive index 1.52. Calculate the percentage change in pulse broadening per unit length. $\left[Hint: \dfrac{\Delta t}{L} = \dfrac{n_1}{n_2}\left(\dfrac{n_1 - n_2}{c} \right) \right]$

5.5   Calculate the critical angle for the optical fibre indicated in Problem 5.1. $\left[Hint: \theta_c = \sin^{-1}\left(\dfrac{n_2}{n_1} \right) \right]$

5.6   The optical fibre indicated in Problem 5.1 has a core diameter of 100 μm. Calculate the maximum number of total internal reflections per metre for the rays travelling through the optical fibre. $\left[Hint: \tan \phi_m = \dfrac{a}{L} \right]$

Waves and Oscillations

Learning Objectives

After studying this chapter, students will be able to

- understand the potential energy function and restoring force of a system
- derive the equation of motions of linear harmonic oscillator
- understand simple pendulum
- comprehend the damped harmonic motion
- explain quality factor Q
- elucidate coupled oscillations

List of Symbols

$U(x)$ = Potential energy

$F(x)$ = Force acting on a particle

m = Mass of a particle

A = Amplitude of oscillation

ω = Angular frequency of oscillation

g = Gravitational force

T = Tension of massless string

τ = Restoring torque for an

angular displacement θ

K = Linear restoring force constant

ϑ = Frequency of a physical pendulum

L = Equivalent length of a physical pendulum

γ = A positive constant, called the damping coefficient

$V(t)$ = Velocity of a particle

ω_0 = Natural frequency of a system

$\langle E \rangle$ = Average energy of a system

$\langle P \rangle$ = Average power dissipation

$\langle E \rangle$ = Average kinetic energy

$\langle U \rangle$ = Average potential energy

Q = Quality factor

6.1 INTRODUCTION

The natural tendency of a particle or system is to attain the position of minimum energy or what is technically termed as the position of stable equilibrium. A particle or system, when disturbed from this position of minimum energy, tends to vibrate or oscillate about this mean position. This oscillatory or periodic behaviour of a particle/system may damp out after a certain time or can be sustained by application of a certain force. In this chapter, we will deal with the analytical treatment of periodic motion under ideal, damped, and forced conditions.

The periodic motion of a particle, as it oscillates under the influence of a restoring force, leads to the most common form of particle motion, that is, wave-like nature. In most natural occurrences, it is observed that as a particle oscillates, it traverses one complete wave or vice versa. In the sections that follow, the detailed discussion of the nature of oscillations of a particle, the conventional terms associated with waves, and reflection at boundaries have been discussed for a comprehensive understanding of the topic on waves and oscillations.

6.2 OSCILLATORY SYSTEMS

The term *oscillation* indicates a back and forth motion of any system or particle. An oscillatory motion is usually about any point of stable equilibrium of the system. In an oscillatory motion, the time period may not necessarily be constant for a given motion. Some of the common examples of oscillatory motion in real life are atoms in a molecule or a solid lattice, the strings of a sitar, a mass attached to a spring, the pendulum of a clock, and all such motions that repeat themselves periodically, as shown in Figs 6.1 and 6.2. (In fig. 6.1, the mass attached to a spring introduces oscillatory motion. In the real situation, the air-resistance acts as the frictional force and damps the motion). Visible light waves, radio waves, and microwaves are all examples of oscillatory electric and magnetic field vectors. Motions of the planets, stars, and satellites are all periodic in nature.

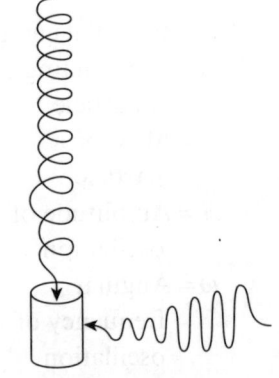

Fig. 6.1 Mass attached to spring

The *periodic or harmonic motion* repeats itself in equal intervals of time. Every system or particle possesses potential energy. Potential energy of a system or particle depends on its position. As a particle is displaced from its position of minimum energy or equilibrium position, it executes harmonic oscillation. Mathematically, a periodic motion is represented by the trigonometric functions of sine and cosine. The frictional force acts as a dissipative

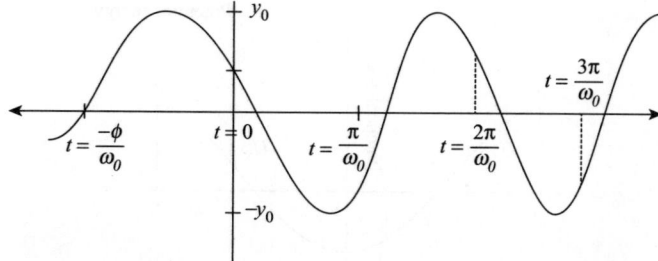

Fig. 6.2 Plot of oscillatory wave (sine/cosine wave) given by function $y = y_0 \sin(\omega_0 t + \phi)$ vs t

force to stop the regular or periodic oscillation of a particle or system. Such periodic motions are then termed as damped motions.

Forces that act on a particle to induce harmonic motion are usually general in nature and constant with respect to time. To overcome the dissipative forces, we have to introduce time-dependent forces for sustained oscillatory behaviour of a system or particle.

To understand the oscillatory motion of a system/particle, concepts of the fundamental parameters such as potential energy and restoring force need to be dwelled upon.

6.3 POTENTIAL ENERGY FUNCTION AND RESTORING FORCE

The potential energy of any general system can be represented as shown in Fig. 6.3. In this figure, the potential energy $U(x)$ has been shown as one-dimensional function of x. At points A, B, and O, $U(x)$ has a maximum or minimum value. At these points, the slope of the curve is zero such that the force is zero, that is, $F(x) = -\dfrac{dU(x)}{dx} = 0$

If a particle is placed at any of the position A or B, it will tend to move away from this position. The first derivative of the potential energy $U(x)$ with respect to space co-ordinate x is greater than zero to the left of x_1 and less than zero to the right of x_1. A small displacement from x_1, therefore, results in a force tending to increase the displacement. Such points are referred to as the points of *unstable equilibrium*. Point O is a point of *stable equilibrium* as the energy value is minimum there. The first derivative of the potential energy $U(x)$ with respect to the space co-ordinate x is less than zero to the left of O and greater than zero to the right of O. Hence, if the particle is displaced from this position, it will oscillate about this point or, in other words, try to restore itself to this position as the force is tending to decrease the displacement. We are interested in the oscillatory behaviour of the particle. Thus, we tend to displace the particle from its point of stable equilibrium. There is an infinitesimal change in the co-ordinates as the particle at the point x_0 changes

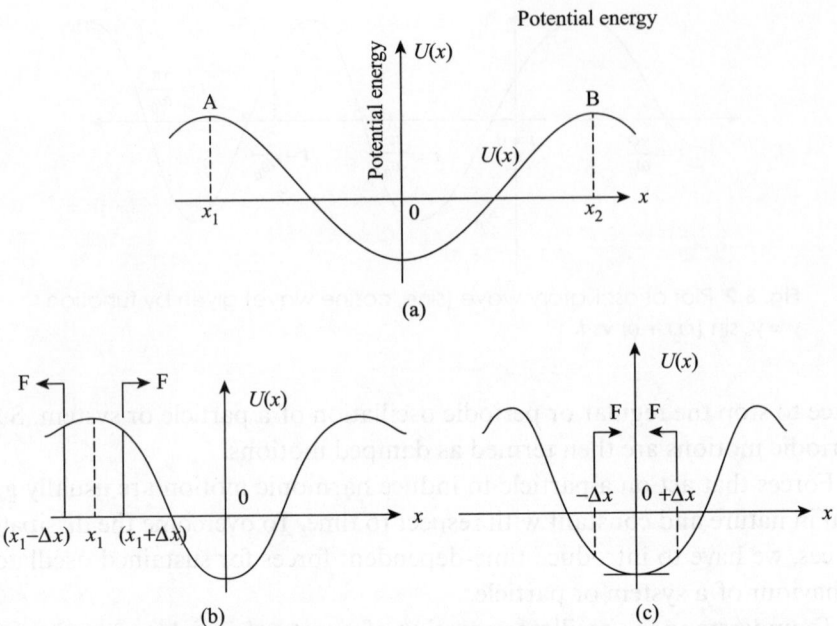

Fig. 6.3 Plot of one-dimensional potential energy function $u(x)$ vs x (a) At points $x = x_1$, 0 and $x = x_2$, $du/dx = 0$, that is, force F zero (points of equilibrium, but not necessarily stable) (b) At point $(x_1 - \Delta x)$, $du/dx > 0 \Rightarrow F < 0$ (to left) and at point $(x_1 + \Delta x)$, $du/dx < 0 \Rightarrow F > 0$ (to right). (Small displacement from x_1 results in force tending to increase displacement; x_1 is point of unstable equilibrium) (c) At $x = -\Delta x$, $du/dx < 0 \Rightarrow F > 0$ (to the right) and at $x = +\Delta x$, $du/dx > 0 \Rightarrow F < 0$ (to the left) (Small displacement from 'O' results in force tending to bring the point back to O; thus, 'O' is point of stable equilibrium)

to $(x_0 + \partial x)$ after displacement. The potential energy of the particle at $(x_0 + \partial x)$ can be expanded in the following Taylor series form:

$$U\left(x_0 + \partial x\right) = U\left(x_0\right) + \partial x \frac{\partial U}{\partial x}\bigg|x_0 + \frac{(\partial x)^2}{2!} \frac{\partial^2 U}{\partial x^2}\bigg| + \dots \qquad (6.1)$$

Let us fix the origin of the co-ordinates at x_0, that is, $x_0 = 0$ and $U(x_0 = 0)$. In addition, let K_1, K_2, and K_3 be constants for the first, second, and third derivatives of the potential energy, respectively, with respect to space co-ordinate x.

Then $\dfrac{\partial U}{\partial x}\bigg|x_0 = K_1; \dfrac{\partial^2 U}{\partial x^2}\bigg|x_0 = K_2; \dfrac{\partial^3 U}{\partial x^3}| x_0 = K_3; \dots$

Thus, using these assumptions and replacing ∂x by x, Eq. (6.1) can be rewritten as follows: $\Rightarrow U\left(x\right) = \dfrac{K_2}{2} x^2 + \dfrac{K_3}{3!} x^3 + \dots$

From the definition of force as the gradient of potential, we can write that

$$F\left(x\right) = -\frac{\partial U(x)}{\partial x} = -K_2 x - \frac{K_3}{2} x^2 + \dots$$

Neglecting the higher order terms, we get the following equation:

$$F(x) = -K_2 x$$

$$U(x) = \frac{1}{2} K_2 x^2 \tag{6.2}$$

Thus, any object that is displaced by a small distance from its position of stable equilibrium experiences a linear restoring force.

Example 6.1 In the potential energy curve shown in Fig. 6.4, enunciate the points of stable and unstable equilibrium.

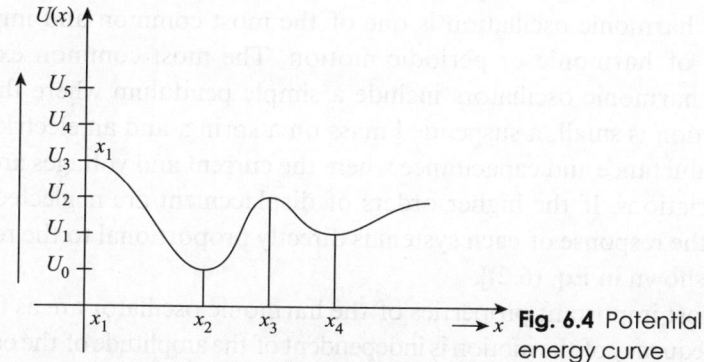

Fig. 6.4 Potential energy curve

Solution Points x_1 and x_3 are points of unstable equilibrium, as the potential energy curve is showing a maximum for these points. Points x_2 and x_4 are points of stable equilibrium, as the potential energy is minimum at these points.

Example 6.2 The potential energy function for a particle is given as $U(x) = A - Cx - kx^2$, where all the constants are positive.
(a) What is the restoring force for the system?
(b) Find the point for which the restoring force vanishes.
(c) Establish the force constant, if the point is of stable equilibrium.

Solution (a) Given the potential energy function, the restoring force is given as follows [from Eq. (6.2)]:

$$F(x) = -\frac{dU(x)}{dx} = -\frac{d}{dx}(A - Cx - kx^2) = -(-C - 2kx)$$

Restoring force for the given system $= C + 2kx$
(b) The restoring force will vanish when

$$F(x) = -\frac{dU(x)}{dx} = 0 \quad \Rightarrow \quad = C + 2kx = 0$$

This implies that $x = -\dfrac{C}{2k}$

(c) For stable equilibrium, the double derivative of the potential function with respect to space must be positive. Thus,

$$\frac{d^2U(x)}{dx^2} = \frac{d}{dx} F(x) = \frac{d}{dx}(C + 2kx) = 2k$$

Since the constants are positive, the double derivative is also evaluated as positive. Thus, the point $[x = -(C/2k)]$ is also a point of minimum potential energy and will therefore be a point of stable equilibrium, where $2k$ is the force constant.

6.4 LINEAR HARMONIC OSCILLATIONS

The preceding discussion on potential energy and restoring force has enabled us to define linear harmonic oscillations. If we consider the potential energy to be a quadratic function of displacement (neglecting the higher orders of displacement), then a particle experiences a *linear restoring force* and hence is said to be executing *linear harmonic oscillations*.

Linear harmonic oscillation is one of the most common and important examples of harmonic or periodic motion. The most common examples of linear harmonic oscillators include a simple pendulum where the angle of oscillation is small, a suspended mass on a spring, and an electric circuit having inductance and capacitance where the current and voltages are giving linear variations. If the higher orders of displacement are neglected for all systems, the response of each system is directly proportional to the restoring force [as shown in Eq. (6.2)].

The most important properties of the harmonic oscillator are as follows:
1. The frequency of the motion is independent of the amplitude of the oscillator.
2. The effects of the several driving forces may be superposed linearly.

6.4.1 Equation of Motion of Linear Harmonic Oscillators

The equation of motion of a linear harmonic oscillator subjected to a linear restoring force in one dimension can be written as follows:

$$m \cdot \frac{d^2 x}{dt^2} = -Kx \tag{6.3}$$

where m is the mass of the particle, x the displacement of the object from its equilibrium position, and K the linear restoring force constant.

Let us assume the solution of the double differential equation to be as follows:

$$x = ae^{\alpha t} \tag{6.4}$$

Equation (6.3) can be rewritten as follows:

$$\frac{d^2 x}{dt^2} = \frac{-K}{m} x = -\omega^2 x \tag{6.5}$$

where it is assumed that $\omega^2 = (-K/m)$. Substituting the value of x in the preceding equation, one gets the following relation:

$$\alpha^2 x = -\omega^2 x \Rightarrow \left(\alpha^2 + \omega^2\right)x = 0$$

Since $x \neq 0$

Thus, $\left(\alpha^2 + \omega^2\right)x = 0$

$$\alpha^2 = -\omega^2$$

$$\Rightarrow \alpha = \pm i\omega \tag{6.6}$$

Substituting the value of α in Eq. (6.4), one gets the following expression:

$$x_1 = a_1 e^{i\omega t}; \; x_2 = a_2 e^{-i\omega t} \tag{6.7}$$

The most general solution is as follows:

$$x = x_1 + x_2 \tag{6.8}$$

$$= a_1 e^{i\omega t} + a_2 e^{-i\omega t} = a_1 \left(\cos \omega t + i\sin \omega t\right) + a_2 \left(\cos \omega t - i\sin \omega t\right)$$

$$= \left(a_1 + a_2\right)\cos\omega t + i\left(a_1 - a_2\right)\sin\omega t \tag{6.9}$$

Assume that

$$\left(a_1 + a_2\right) = A\sin\theta$$

$$i\left(a_1 - a_2\right) = A\cos\theta$$

$$x = A\sin\theta\cos\omega t + A\cos\theta\sin\omega t$$

$$x = A\sin(\omega t + \theta) \tag{6.10}$$

Equation (6.10) indicates the displacement of a particle experiencing a linear restoring force. The amplitude of oscillation is indicated by A, angular frequency by ω, phase of the oscillator by θ, and time period of the oscillation by T (where $T = 2\pi/\omega$). The sinusoidal trigonometric function indicates that the displacement is oscillatory/periodic in nature.

Note: Now we are in a position to correlate how the trajectory of a particle can be determined when the force acting on the particle/system is ascertained.

Example 6.3 Which of the following equations represent harmonic and non-harmonic oscillations?

(a) $\dfrac{d^2x}{dt^2} = -\dfrac{3}{2}x$ (b) $\dfrac{d^2x}{dt^2} = -\dfrac{3}{2}x^2$ (c) $\dfrac{d^2x}{dt^2} = -4x$

In addition, ascertain the values of the restoring force constant (k) and mass (m) of the particle in harmonic oscillation.

Solution The explanations are as follows:

(a) $\dfrac{d^2x}{dt^2} = -\dfrac{3}{2}x$

This is a linear harmonic oscillation, with the restoring force constant being equal to 3 and the mass of the particle equal to 2.

(b) $\dfrac{d^2x}{dt^2} = -\dfrac{3}{2}x^2$

This is an anharmonic oscillation, as a non-linear term of displacement is present in the expression.

(c) $\dfrac{d^2x}{dt^2} = -4x$

This is a linear harmonic oscillation with a natural frequency equal to 2. This implies that the restoring force constant is 4 and the mass of the particle is 1.

Example 6.4   A simple harmonic motion is represented by $x = 0.1\sin(1000t + 0.10)$ where x and t are in metres. Find the following quantities: (a) amplitude, (b) angular frequency, (c) period of oscillation, (d) initial phase, and (e) frequency.

Solution   When we compare this equation with the simple harmonic equation, we can write the following equation:

$$x = A\sin(\omega t + \theta)\,;\, x = 0.1\sin(1000t + 0.10)$$

Thus,

(a) Amplitude $= A = 0.10$ m

(b) Angular frequency $= \omega = 1000$ rad/s

(c) Period of oscillation $= T = \dfrac{2\pi}{\omega} = \dfrac{2\pi}{1000} = \dfrac{\pi}{500}$ s

(d) Initial phase $= \theta = 0.10$ rad

(e) Frequency $= v = \dfrac{1}{T} = \dfrac{500}{\pi}$ per second

Example 6.5   For the vibrational motion shown in the graph in Fig. 6.5, find the (a) amplitude, (b) frequency, (c) time period, and (d) write the numerical equation of the form $x = A\sin(\omega t + \theta)$, where the symbols are as per the conventions used in the chapter.

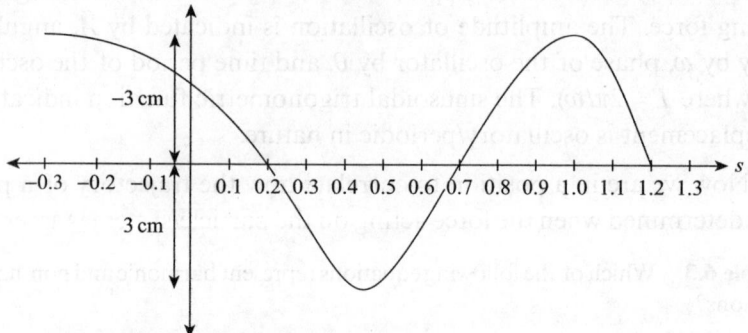

Fig. 6.5 Cosine wave with y-axis as 3 cm and time period as 1.2 s

Solution   From the graph in Fig. 6.5:

(a) Amplitude: 3.0 cm

(b) Frequency: One cycle completes in 1.2 s, so the frequency is $u = (1/1.2) = 0.83$ Hz.

(c) Time period: 1.2 s

(d) Since, $x = x_0$, the phase angle is $(\pi/2)$ for $t = 0$. So,

$$x = 3.0\sin\left[2\pi(0.83\text{ Hz})t + \left(\frac{\pi}{2}\right)\right]\text{cm} \Rightarrow x = 3.0\sin\left(5.2t + \frac{\pi}{2}\right)\text{cm}$$

The two common examples of a linear harmonic oscillator are a simple pendulum and a physical pendulum, which are discussed in the following subsections.

6.4.2 Simple Pendulum

A simple pendulum consists of a point mass hanging from a massless inextensible thread or string. The point mass is disturbed from its position of equilibrium and released. The pendulum oscillates in a vertical plane under the influence of gravity about its equilibrium position in a periodic and oscillatory manner.

Let us consider a pendulum of length l and a particle of mass m making an angle θ with the vertical (as shown in Fig. 6.6). The forces acting on the mass m are (mg), where g is the gravitational force, and T, the tension in the massless string. The force (mg) is resolved into two components: the radial component $(mg \cos\theta)$ and the tangential component $(mg \sin\theta)$. The radial component gives the neces-

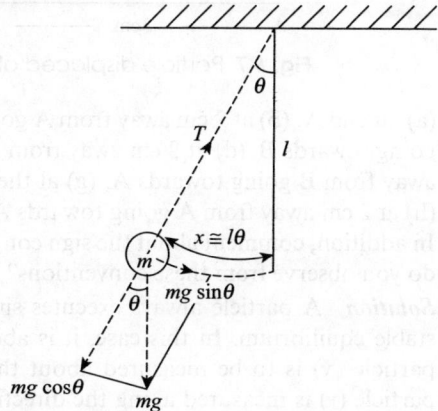

Fig. 6.6 Forces acting on simple pendulum

sary centripetal acceleration for the particle to have an oscillatory motion, whereas the tangential component acts as the restoring force that tends to bring the particle back to the equilibrium position.

$$F = -mg(\sin\theta) \qquad (6.11)$$

If the angle θ is very small, then one can assume that $\sin\theta \approx \theta$, where the higher orders of expansion of $\sin\theta = \theta - \dfrac{\theta^3}{3!} + \dfrac{\theta^5}{5!} - \dots$ Then,

$$F = -mg\theta = -mg\left(\frac{x}{l}\right) \quad (\text{since } x \cong l\theta)$$

$$= -(mg/l)x$$

$$= -kx \qquad (6.12)$$

For small displacements, therefore, the restoring force is proportional to the displacement and is oppositely directed. Now for the time period of the motion of a simple pendulum, we know that

$$T = 2\pi/\omega$$

$$T = 2\pi\sqrt{m/k} \qquad [\text{from Eq. (6.5)}]$$

$$T = 2\pi\sqrt{[m/(mg/l)]} \qquad [\text{from Eq. (6.12)}]$$

$$T = 2\pi\sqrt{(l/g)} \qquad (6.13)$$

The time period of a simple pendulum is independent of the amplitude of oscillation and is dependent on length of the string and the acceleration due to gravity at that point. A simple pendulum provides a simple and convenient method for measuring the value of g, acceleration due to gravity.

Example 6.6 A particle executes linear simple harmonic motion with the extreme positions A and B (Fig. 6.7), 6 cm apart. The positive direction is taken from left to right. Give the sign convention of the displacement x, velocity v, and acceleration a of the particle in each of the following cases:

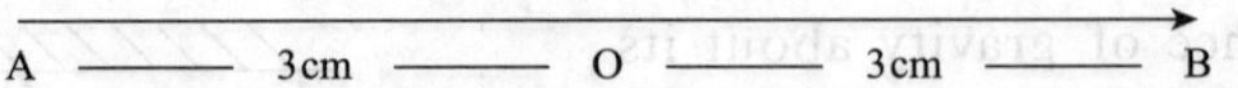

A ──────── 3cm ──────── O ──────── 3cm ──────── B

Fig. 6.7 Particle displaced along line AOB in +ve x-direction

(a) at end A, (b) at 2 cm away from A going towards B, (c) at the mid-point O of AB going towards B, (d) at 2 cm away from B going towards B, (e) at end B, (f) at 2 cm away from B going towards A, (g) at the mid-point O of AB going towards A, and (h) at 2 cm away from A going towards A.

In addition, comment about the sign convention of force acting on the particle. What do you observe from these conventions?

Solution A particle always executes simple harmonic motion about the point of stable equilibrium. In this case, it is about the mid-point O. Displacement of the particle (x) is to be measured about the equilibrium position O. Velocity of the particle (v) is measured along the direction of motion conventionally. Acceleration of the particle (a) is always directed towards the equilibrium position O in a simple harmonic motion. Thus,

(a) At end A: $x = -\text{ve}$; $v = 0$; $a = -\text{ve}$
(b) At 2 cm away from A going towards B: $x = -\text{ve}$; $v = +\text{ve}$; $a = +\text{ve}$
(c) At midpoint O of AB going towards B: $x = 0$; $v = +\text{ve}$; $a = 0$
(d) At 2 cm away from B going towards B: $x = +\text{ve}$; $v = +\text{ve}$; $a = -\text{ve}$
(e) At end B: $x = +\text{ve}$; $v = 0$; $a = -\text{ve}$
(f) At 2 cm away from B going towards A: $x = +\text{ve}$; $v = -\text{ve}$; $a = +\text{ve}$
(g) At mid-point O of AB going towards A: $x = 0$; $v = -\text{ve}$; $a = 0$
(h) At 2 cm away from A going towards A: $x = -\text{ve}$; $v = -\text{ve}$; $a = +\text{ve}$

Example 6.7 An 80 cm long simple pendulum has a 100 g bob attached to its end. The pendulum is pulled aside by 10° and released. If the timing clock is started just as the ball moves through its lowest position, write the equation of motion of the pendulum in terms of the pendulum angle in degrees.

Solution The amplitude of the simple pendulum is 10°. From the equation for time period of a simple pendulum,

$$T = 2\pi \sqrt{l/g}$$

$$2\pi 9 = \sqrt{(g/l)} = \sqrt{(980/80)} = \sqrt{12.25} = 3.5/s$$

The initial position of the pendulum indicates that the initial amplitude is zero at time $t = 0$. This indicates that the simple pendulum so formed is oscillating in the sine wave pattern. Thus, $\theta = 10 \sin(3.5t)$ degrees.

6.4.3 Physical Pendulum

A physical or compound pendulum is a body of irregular shape capable of oscillating about a horizontal frictionless axis about its equilibrium position. The equilibrium position of the physical body lies vertically below the (suspension) pivotal point P (Fig. 6.8). Let the distance from point P to the centre of mass of the body be l. Let m be the mass of the body and I the rotational inertia of the body about an axis through the point of suspension. The restoring torque for an angular displacement θ is given as follows:

$$\tau = -mgl\sin\theta \tag{6.14}$$

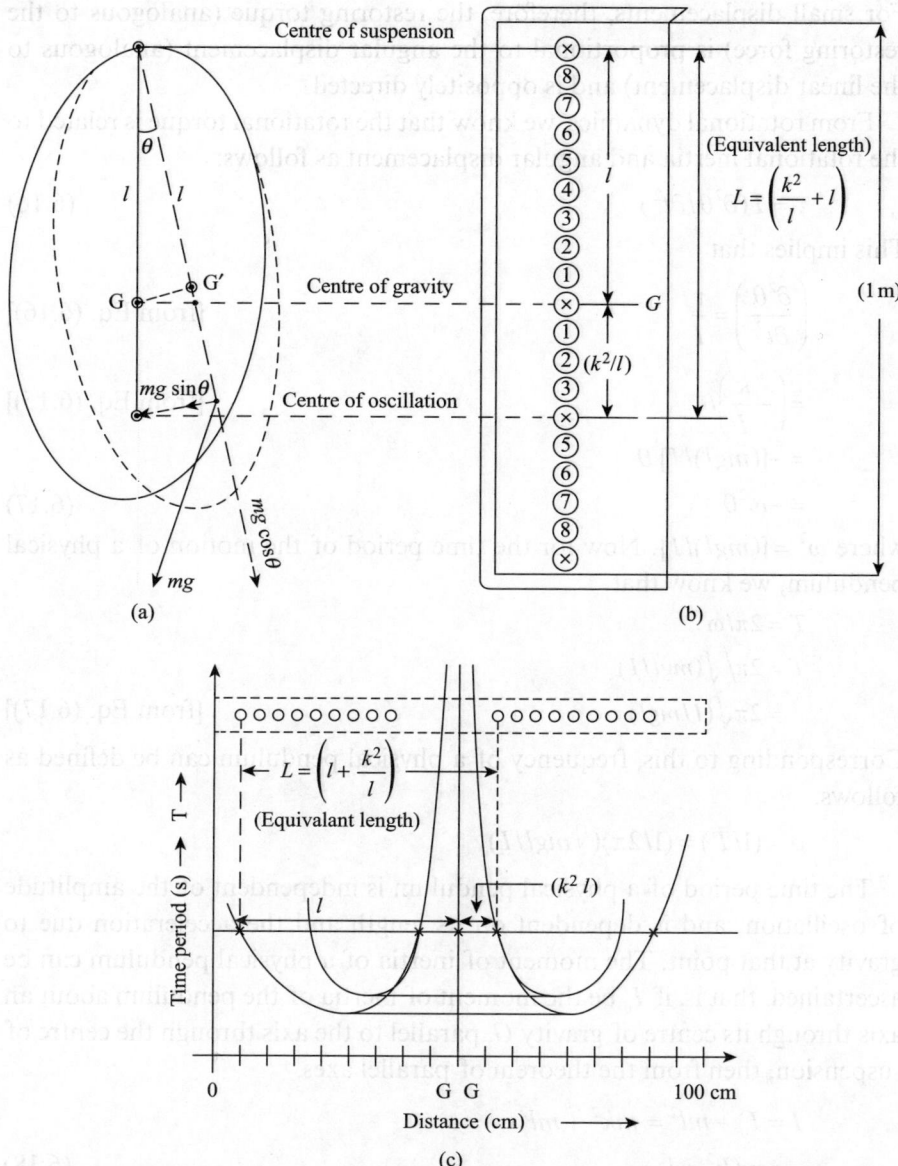

Fig. 6.8 Physical pendulum (a) Force diagram of rigid body when given swing θ (b) Conventional laboratory design of bar pendulum (c) Graph of time period vs distance, as measured in laboratory method

If angle θ is very small, then one can assume that $\sin\theta \approx \theta$ (where the higher orders of expansion of $\sin\theta = \theta - \dfrac{\theta^3}{3!} + \dfrac{\theta^5}{5!} - \ldots$). Hence,

$$\tau = -mgl\theta$$

$$\tau = -k\theta \qquad \text{where } k = (mgl) \quad \text{~} \qquad (6.15)$$

For small displacements, therefore, the restoring torque (analogous to the restoring force) is proportional to the angular displacement (analogous to the linear displacement) and is oppositely directed.

From rotational dynamics, we know that the rotational torque is related to the rotational inertia and angular displacement as follows:

$$\tau = I(\partial^2\theta/\partial t^2) \qquad (6.16)$$

This implies that

$$\left(\frac{\partial^2\theta}{\partial t^2}\right) = \frac{\tau}{I} \qquad \text{[from Eq. (6.16)]}$$

$$= \left(-\frac{k}{I}\right)\theta \qquad \text{[from Eq. (6.15)]}$$

$$= -[(mgl)/I]\,\theta$$

$$= -\omega^2\theta \qquad (6.17)$$

where $\omega^2 = [(mgl)/I]$. Now for the time period of the motion of a physical pendulum, we know that

$$T = 2\pi/\omega$$

$$T = 2\pi/\sqrt{(mgl/I)}$$

$$= 2\pi\sqrt{(I/mgl)} \qquad \text{[from Eq. (6.17)]}$$

Corresponding to this, frequency of a physical pendulum can be defined as follows:

$$\upsilon = (1/T) = (1/2\pi)(\sqrt{mgl/I})$$

The time period of a physical pendulum is independent of the amplitude of oscillation, and is dependent on its length and the acceleration due to gravity at that point. The moment of inertia of a physical pendulum can be ascertained, that is, if I_0 be the moment of inertia of the pendulum about an axis through its centre of gravity G, parallel to the axis through the centre of suspension, then from the theorem of parallel axes,

$$I = I_0 + ml^2 = mk^2 + ml^2$$

$$= m(k^2 + l^2) \qquad (6.18)$$

Thus, substituting the value of Eq. (6.18) in Eq. (6.17), we get the following expression:

$$T = 2\pi\sqrt{\frac{k^2 + l^2}{gl}}$$

$$T = 2\pi\sqrt{\left(\frac{k^2}{l} + l\right)/g}$$

$$T = 2\pi\sqrt{\left(\frac{L}{g}\right)} \qquad (6.19)$$

where L is now defined as the equivalent length of the physical pendulum corresponding to the simple pendulum and is mathematically defined as follows:

$$L = \left(\frac{k^2}{l} + l \right)$$

6.4.4 Interchangeability of Centre of Oscillation and Point of Suspension

As a rigid body is suspended about a knife edge, this point is referred to as the *point of suspension*. The rigid body is made to oscillate about this point by a small displacement θ. As a rigid body oscillates, at two points on either side of the centre of gravity of the body, the time period of oscillation becomes the same. The *centre of oscillation* is located at exactly a distance of (k^2/l) from the centre of gravity. Thus, as the point of suspension is inverted on a rigid body about the centre of gravity, the point of oscillation consequently gets changed. The point of suspension and the point of oscillation are interchangeable about the centre of gravity of a rigid body (Fig. 6.8).

6.4.5 Experimental Determination of Equivalent Length for Physical Pendulum

In most of the experiments on oscillations performed in laboratories, bar pendulums or Kater's pendulums are used as the learning devices. As a pendulum is displaced about its position of stable equilibrium, it oscillates, and the time period for 15 or 20 oscillations is determined using a stop watch. A graph is then plotted between the time period and the distance of the point of suspension from the centre of gravity. Thus, the equivalent length for the pendulum is determined by drawing a line on the symmetrical graph, for a given time period (the line is drawn in the bell region of the graph). From this equivalent length and for the given time period, the acceleration due to gravity is ascertained using the physical pendulum at that point.

Example 6.8 Find the frequency of oscillations when a meter stick is hung as a compound pendulum with the pivot at its 90 cm mark.

Solution The frequency of a physical pendulum can be defined as follows:

$$\vartheta = (1/T) = (1/2\pi)\left(\frac{\sqrt{mgl}}{I} \right)$$

where L is the distance from the axis of rotation to the centre of mass $= (90\ \text{cm} - 50\ \text{cm})$ $= 40\ \text{cm} = 0.40\ \text{m}$.

I is the moment of inertia about the axis of rotation

$$= I_{cm} + Mr^2 \qquad R = 100\ \text{cm} = 1\ \text{m},\ r = 0.40\ \text{m}$$

$$= \left(\frac{MR^2}{12} \right) + Mr^2 = M\left\{ \frac{R^2}{12} + r^2 \right\} = M\left\{ \frac{1}{12} + (0.40)^2 \right\}$$

$$= M[(0.08) + (0.16)] = M(0.24)$$

where M is the mass of the meter stick. Thus, the frequency is calculated to be as follows:

$$\vartheta = (1/2\pi)\left(\frac{\sqrt{mgl}}{I}\right) = \left(\frac{1}{2\pi}\right)\left(\frac{M \times 9.8 \times 0.40}{M(0.24)}\right) \text{MKS units} = 0.64 \text{ Hz}$$

6.5 DAMPED HARMONIC OSCILLATIONS

As we observe linear harmonic oscillators in real life, we find that frictional forces play a critical role in the ideal oscillatory motion of a particle. Frictional forces tend to dampen the oscillatory behaviour of the particle as the motion proceeds in time such that the oscillation decays gradually.

Thus, to explain the effect of friction on an ideal linear harmonic oscillator, let us first discuss the effect of frictional force.

6.5.1 Equation of Motion for Frictional Force

The effect of friction is to damp the free motion of an oscillator. The most commonly observed frictional forces are air resistance and internal forces. Consideration of the frictional force provides a realistic solution to a linear harmonic oscillator.

Frictional force is retarding in nature. The oscillatory motion of a particle is exhibited by its velocity during the motion. The simplest form of frictional force must oppose the velocity of the particle. Thus, the frictional force can be expressed as follows:

$$F_f \propto \frac{dx}{dt} \tag{6.20}$$

$$F_f = -\gamma \frac{dx}{dt} \tag{6.21}$$

where γ is a positive constant, called the damping coefficient, and (dx/dt) represents the velocity of the particle.

The equation of motion of a particle moving under the action of the frictional force is given as follows:

$$m\frac{d^2x}{dt^2} = -\gamma \frac{dx}{dt} \tag{6.22}$$

$$\frac{d^2x}{dt^2} = -\frac{\gamma}{m}\frac{dx}{dt} \tag{6.23}$$

where $(\gamma/m) = (1/\tau)$ is defined as the relaxation time and has the dimensions of time. In practical terms, the relaxation time indicates the time in which oscillations of an oscillator die out; for example, if $\tau = 5$ s, it indicates that the oscillator will have pendular oscillations with a decaying amplitude for about 5 s.

$$\frac{d^2x}{dt^2} + \frac{1}{\tau}\frac{dx}{dt} = 0 \tag{6.24}$$

This equation can be written in terms of velocity as follows:

$$\frac{dv}{dt} + \frac{1}{\tau}v = 0 \tag{6.25}$$

$$\frac{dv}{v} = -\frac{dt}{\tau}$$

Integrating both sides with respect to time, we get the following equations:

$$\int_{v_0}^{v} \frac{dv}{v} = -\frac{1}{\tau}\int_{0}^{t} dt \tag{6.26}$$

$$\log v - \log v_0 = -\frac{t}{\tau} \tag{6.27}$$

$$\log \frac{v}{v_0} = -\frac{t}{\tau} \tag{6.28}$$

Thus, expressing the equation exponentially, we get the following relation:

$$v(t) = v_0 e^{-t/\tau} \tag{6.29}$$

The effect of the frictional force is that the velocity decreases exponentially with time. The velocity is damped by a time constant τ. The relaxation time is the time interval in which the velocity reduces to $(1/e)$ times its initial value (Fig. 6.9).

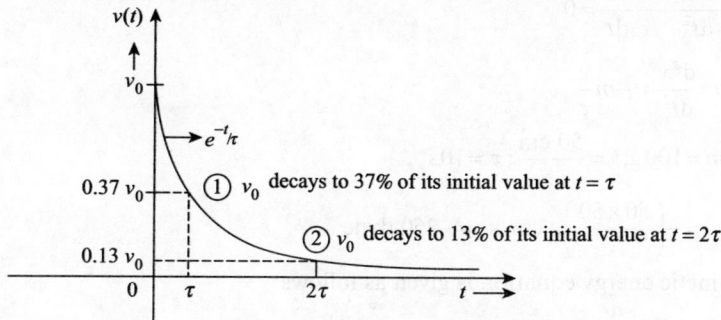

Fig. 6.9 Velocity as exponentially decaying function of time t on x-axis indicating time (relaxation) when velocity amplitude reduces to $(1/e)$

Note: Students must observe the exponential decay of a function. In the velocity–time graph, it is evident that velocity decays exponentially to 37% of its initial value when one decay constant has passed (i.e., $t = \tau$). Velocity decays to 13% when the second decay constant passes (i.e., $t = 2\tau$). In general, for an exponentially decaying function, this behaviour is universal. This can be verified for the energy–time curve as well.

Decay of kinetic energy due to the frictional force can be expressed as follows:

$$K = \frac{1}{2}mv^2 = \frac{1}{2}m v_0^2 e^{-2t/\tau}$$

$$K = K_0 e^{-2t/\tau} \tag{6.30}$$

Differentiating kinetic energy with time, one gets the following expression:

$$K = -\frac{2}{\tau}K \tag{6.31}$$

The effective relaxation time for kinetic energy is one-half of that for velocity.

Example 6.9 Velocity of a 100 g particle is decreased from 100 to 10 cm/s in 23 s. Calculate the (a) relaxation time, (b) damping force when its velocity is 50 cm/s, (c) time in which its kinetic energy is reduced to one-tenth of its initial value, and (d) total distance travelled when its initial velocity is 10 cm/s.

Solution We know that

(a) For the relaxation time,

$$v(t) = v_0 e^{-t/\tau}$$

where $v_0 = 100 \dfrac{cm}{s}$; $v = \dfrac{10 \ cm}{s}$; $t = 23 s$

$$10 = 100 e^{-23/\tau} \quad \Rightarrow \quad \frac{10}{100} = e^{-23/\tau} \quad \Rightarrow \quad \frac{1}{10} = e^{-23/\tau} \quad \Rightarrow \quad 10 = e^{23/\tau}$$

Taking natural log on both the sides,

$$2.303 = \frac{23}{\tau} \quad \Rightarrow \quad \tau = \frac{23}{2.303} s$$

$$\tau = 10 s$$

(b) The equation for the damping force is as follows:

$$\frac{d^2 x}{dt^2} + \frac{1}{\tau}\frac{dx}{dt} = 0$$

$$m\frac{d^2 x}{dt^2} = -m\frac{v}{\tau}$$

where $m = 100 \ g$; $v = \dfrac{50 \ cm}{s}$; $\tau = 10 s$

$$= -\left(\frac{50 \times 50}{10}\right) g \ cm/s^2 = -250 \ dyne$$

(c) The kinetic energy equation is given as follows:

$$K = K_0 e^{-2t/\tau}$$

Now, $(K/K_0) = (1/10)$; $\tau = 10$ s

$$\frac{1}{10} = e^{-2t/10} \quad \Rightarrow \quad 10 = e^{t/5}$$

Taking natural log on both the sides,

$$\log_e 10 = t_1/5 \quad \Rightarrow \quad t_1 = 2.303 \times 5 = 11.5 s$$

(d) The total distance can be evaluated by integrating the velocity equation, as follows:

$$v(t) = v_0 e^{-t/10}$$

$$x(t) = (-10) v_0 e^{-t/10} + c$$

The initial conditions indicate that at $t = 0$; $x = 0 \Rightarrow c = 10 v_0$

$$x = (-10) v_0 e^{-t/10} + 10 v_0$$

$$= 10 v_0 \left(1 - e^{-t/10}\right) \quad \text{(as for } t \to \infty, \text{ exponential term} \to 0)$$

$$= 10 \times 100 = 1000 \ cm = 10 \ m$$

6.5.2 Damped Harmonic Motion

As has been observed in Section 6.5.1, as the frictional force comes into play, the amplitude of oscillations gradually decays with time. Ideally, a linear harmonic oscillator should continue its oscillations indefinitely. Actually, in real life, the amplitude of oscillations gradually decreases to zero as a result of friction. The motion is said to be damped by friction and is called *damped harmonic motion*.

The equation of motion of a damped linear harmonic oscillator is given by Newton's second law of motion, that is, $F = ma$, where F is the sum of the restoring forces (kx) and the damping force $\left(-\gamma \dfrac{dx}{dt} \right)$, where γ is a positive constant.

$$m\frac{d^2x}{dt^2} = -kx - \gamma \frac{dx}{dt} \tag{6.32}$$

$$m\frac{d^2x}{dt^2} + \gamma \frac{dx}{dt} + kx = 0$$

$$\frac{d^2x}{dt^2} + \frac{\gamma}{m}\frac{dx}{dt} + \frac{k}{m}x = 0 \tag{6.33}$$

$$\frac{d^2x}{dt^2} + \frac{1}{\tau}\frac{dx}{dt} + \omega_0^2 \frac{k}{m}x = 0 \tag{6.34}$$

where $\dfrac{1}{\tau} = \dfrac{\gamma}{m}$; here, τ is the relaxation time.

$$\omega_0^2 = \frac{k}{m}$$

Here, ω_0 is the natural frequency of the system.

The solution for the double differential Eq. (6.34) can be assumed to be as follows:

$$x = x_0 e^{-\beta t} \sin \omega t \tag{6.35}$$

$$\frac{dx}{dt} = -\beta x_0 e^{-\beta t} \sin \omega t + \omega x_0 e^{-\beta t} \cos \omega t \tag{6.36}$$

Differentiating Eq. (6.36) with respect to time, one gets the following expression:

$$\frac{d}{dt}\left(\frac{dx}{dt} \right) = -\left(\beta^2 \right)x_0 e^{-\beta t}\sin \omega t - \beta x_0 e^{-\beta t}\omega \cos \omega t$$

$$-\omega^2 x_0 e^{-\beta t} \sin \omega t - \beta \omega x_0 e^{-\beta t} \cos \omega t \tag{6.37}$$

Thus, substituting Eqs (6.35)–(6.37) in Eq. (6.34), one gets the following expression:

$$(\beta^2 x_0 e^{-\beta t}\sin \omega t - 2\omega\beta x_0 e^{-\beta t}\cos \omega t - \omega^2 x_0 e^{-\beta t}\sin \omega t$$

$$+ \frac{1}{\tau}\left(-\beta x_0 e^{-\beta t}\sin \omega t + \omega x_0 e^{-\beta t}\cos \omega t \right) + \omega_0^2 x_0 e^{-\beta t}\sin \omega t = 0$$

$$\left(\beta^2 - \omega^2 + \omega_0^2 - \frac{\beta}{\tau} \right)x_0 e^{-\beta t}\sin \omega t + \left(-2\omega\beta + \frac{\omega}{\tau} \right)x_0 e^{-\beta t}\cos \omega t = 0 \tag{6.38}$$

For the right-hand side to be satisfied, the coefficients of $\sin\omega t$ and $\cos\omega t$ must go to zero individually. Thus, we get the following relations:

Either $\left(\beta^2 - \omega^2 + \omega_0^2 - \dfrac{\beta}{\tau}\right) = 0$ (6.39)

Or $\left(-2\omega\beta + \dfrac{\omega}{\tau}\right) = 0$ (6.40)

Considering Eq. (6.40), one gets the following relation:

$$\left(\dfrac{\omega}{\tau}\right) = 2\omega\beta \Rightarrow \beta = \dfrac{1}{2\tau}$$ (6.41)

Substituting this value in Eq. (6.39), one gets the following relation:

$$\omega^2 = \omega_0^2 + \dfrac{1}{\tau^2}\left(\dfrac{1}{4} - \dfrac{1}{2}\right) \quad \Rightarrow \quad \omega^2 = \omega_0^2 - \left(\dfrac{1}{2\tau}\right)^2$$

$$\omega = \left(\omega_0^2 - \left(\dfrac{1}{2\tau}\right)^2\right)^{1/2} = \omega_0\left(1 - \left(\dfrac{1}{2\omega_0\tau}\right)^2\right)^{1/2}$$ (6.42)

Thus, we observe that the frequency is lowered and the time period becomes longer when the frictional forces come into play. Frictional forces slow down the motion of a particle. When the relaxation time is infinite, that is, $\tau \to \infty$, a situation of no damping, then $\omega = \omega_0$, where ω_0 is the natural frequency of the system.

When the frictional forces come into play, as discussed earlier, the complete solution of the equation, after substituting Eq. (6.42) and (6.41) in Eq. (6.35), can be written as follows:

$$x = x_0 e^{-t/2\tau}\sin\left(\omega_0\left[\left(1 - \dfrac{1}{2\omega_0\tau}\right)^2\right]^{1/2}\right)$$ (6.43)

Displacement of a particle in a damped harmonic oscillator exhibits that the amplitude is dependent on the time and the frequency on the relaxation time. Thus, depending on the interplay of the restoring and frictional forces, damped harmonic oscillators have been classified as follows:

6.5.2.1 Under-damping

The natural frequency of a system is much greater than the frictional forces. Hence, $\omega_0^2 > \gamma^2$

that is, restoring force > frictional force

that is, $\omega_0\tau \gg 1$ limit of low damping

$$x \cong x_0 e^{-t/2\tau}\sin\omega_0 t$$ (6.44)

$$x \cong A(t)\sin\omega_0 t$$ (6.45)

where $A(t) = x_0 e^{-t/2\tau}$ (6.46)

where ω_0 is the natural frequency of the undamped oscillation. In the presence of friction, the amplitude, as seen in Eqs (6.45) and (6.46), is time dependent. As can be observed from Eq. (6.44), it varies exponentially with time. For the time $t = 2\tau = \dfrac{1}{\gamma}$, the amplitude becomes $(1/e)$ of its initial value; this value is referred to as the decay constant of the oscillator, that is,

$$x(t) = x_0 e^{-t/2\tau}$$

At $\qquad\qquad t = 2\tau,$

$$x(t) = \frac{x_0}{e} \qquad\qquad\qquad\qquad (6.47)$$

where x_0 is the initial displacement at the initial time $(t = 0)$.

Logarithmic decrement Damping is measured in terms of the 'amplitude' (during one oscillation), that is, how much the amplitude decreases as the particle completes one oscillation in time period T.

Let us assume that the amplitude at time t_1 be $A_1(t_1)$. Then the amplitude at time $(t_1 + T)$ after a time period T is given as $A_1(t_1 + T)$. From Eq. (6.46), the amplitudes can be mathematically represented as follows:

$$A_1(t_1) = x_0 e^{-t_1/2\tau} \quad\Rightarrow\quad A_2(t_1 + T) = x_0 e^{-(t_1+T)/2\tau}$$

Dividing these two equations for amplitude, we get the following relations:

$$\frac{A_1(t_1)}{A_2(t_1 + T)} = \frac{x_0 e^{\left(-t_1/2\tau\right)}}{x_0 e^{-\left((t_1+T)/2\tau\right)}} \qquad\qquad\qquad (6.48)$$

$$\frac{A_1(t_1)}{A_2(t_1 + T)} = \frac{e^{-\left(t_1/2\tau\right)}}{e^{-\left((t_1)/2\tau\right)}} e^{T/2\tau} = e^{T/2\tau}$$

$$= e^{\theta} \qquad\qquad\qquad\qquad (6.49)$$

where $\quad \theta = \dfrac{T}{2\tau} = \dfrac{T\gamma}{2M} \qquad\qquad\qquad\qquad (6.50)$

Thus, the ratio of the amplitudes for a time variation of one time period is defined by Eq. (6.49). If we take the natural logarithm of the amplitudes, we get the following equation:

$$\theta = \frac{T}{2\tau} = \log_e \frac{A_0}{A_1} = \log_e \frac{A_1}{A_2} \qquad\qquad\qquad (6.51)$$

Logarithmic decrement is defined as the natural logarithmic of the ratio of two successive maximum amplitudes, which are separated by one time period. Displacement of a particle in a damped harmonic oscillator can be written as follows [from Eq. (6.46)]:

$$x_1(t_1) = x_0 e^{-t_1/2\tau} \text{ for time } t_1$$

$$x_{n+1}(t_1 + nT) = x_0 e^{-(t_1+nT)/2\tau} \text{ for time } (t_1 + nT)$$

The decrease in amplitude during n complete oscillations is given as.

$$\frac{x_{n+1}(t_1+nT)}{x_1(t_1)} = e^{\left(-(t_1+nT)-(-t_1)/2\tau\right)} = e^{-nT/2\tau}$$

$$= e^{-n\theta} \tag{6.52}$$

For $n\theta = 1; \; x_{n+1} = \dfrac{x_1}{e}$ \hfill (6.53)

Hence, for n oscillations, the logarithmic decrement θ is defined as follows:

$$\theta = \frac{1}{n}\log_e \frac{A_1}{A_2} \tag{6.54}$$

Thus, we observe that for a damped harmonic oscillator, the under-damping case is a physically meaningful example that is observed in daily life. One observes that in laboratory conditions, the motion of a physical pendulum damps after 25–30 oscillations due to air friction.

6.5.2.2 Over-damping

The natural frequency of the system is much less than the frictional forces. Therefore,

$$\omega_0^2 \ll \left(\frac{1}{2\tau}\right)^2$$

$$\omega_0\tau \ll 1$$

Substituting this value in Eq. (6.42), $\sin\omega t \sim 1$

The displacement function can then be written as follows:

$$x \cong x_0 e^{-\beta t} \tag{6.55}$$

Thus, we observe that there is an exponential decay of the amplitude with time. This is logically true, as the frictional forces will overcome the natural frequency of the damped oscillator with time. Thus, one observes that in the case of an over-damped oscillator, the amplitude decays exponentially with time and finally dies out. The motion is not periodic at all. The particle just returns to its equilibrium position when released from its initial displacement.

Critical damping   The natural frequency of a system is nearly equal to the frictional forces. Then $\omega_0^2 = \left(\dfrac{1}{2\tau}\right)^2$

Substituting this value in Eq. (6.42), we get the following relations:

$$\omega \sim 0 \quad \Rightarrow \quad \sin\omega t \sim 1$$

The displacement function can then be written as follows:

$$x \cong x_0 e^{-\beta t} \tag{6.56}$$

In the critical damping case also, there is an exponential decay of the amplitude.

Having discussed the different classifications of damped harmonic oscillators depending on the interplay of the restoring and frictional forces, the most meaningful case is that of an under-damped oscillator.

Thus, we observe that only in the case of the *low damping* (under-damped harmonic oscillators), the amplitude of a system decays exponentially in an oscillatory manner. A further study of the power dissipation and quality factor analysis is performed for the under-damped low damping case.

Power dissipation   From the previous discussion, for a low damping case (under-damped harmonic case), $\omega_0 \tau \gg 1$

This implies that the frequency of the oscillator is nearly the same as the natural frequency of the system. Thus, $\omega \approx \omega_0$

The average power dissipated by a system is the negative of the rate of change of energy.

Hence, we first ascertain the kinetic and the potential energies for a low-damping under-damped harmonic oscillator.

The kinetic energy is defined as follows:

$$K = \frac{1}{2} m \left(\frac{dx}{dt} \right)^2 \tag{6.57}$$

$$\frac{dx}{dt} = \frac{d}{dt} (x_0 e^{-t/2\tau} \sin \omega_0 t) \tag{6.58}$$

$$= -\frac{1}{2\tau} x_0 e^{-t/2\tau} \sin \omega_0 t + \omega_0 x_0 e^{-t/2\tau} \cos \omega_0 t$$

Squaring Eq. (6.58), one gets the following relation:

$$\left(\frac{dx}{dt} \right)^2 = -\frac{1}{2\tau} x_0 e^{-t/2\tau} \sin \omega_0 t + \omega_0 x_0 e^{-t/2\tau} \cos \omega_0 t$$

$$- \left(\frac{\omega_0}{\tau} \right) x_0^2 e^{-t/\tau} \sin \omega_0 t \cos \omega_0 t \tag{6.59}$$

Since we are considering the case of a low-damping under-damped harmonic oscillator, that is, $\omega_0 \tau \gg 1$, to a good approximation, the factor $e^{-t/\tau}$ is considered to be independent of time and is hence taken outside the angle brackets denoting the time average. We assume that the amplitude of oscillation $x_0 e^{-t/\tau}$ does not change much in one cycle of motion. The time average is denoted as $\frac{1}{t} \int_0^T dt$ and is valid for the kinetic energy term. In Eq. (6.59), the time average of the trigonometric functions can be evaluated as follows:

$$< \cos^2 \theta > = < \sin^2 \theta > = \frac{1}{2}$$

$$< \cos \theta \sin \theta > = 0$$

since $< \cos \theta \sin \theta > = \dfrac{< \sin 2\theta >}{2} = 0$

Kinetic energy time-averaged over the time of one cycle is denoted as follows:

$$< K >_T \cong \frac{1}{2} m \left[\left(\frac{1}{2\tau} \right)^2 < \sin^2 \omega_0 t > + \omega_0^2 < \cos^2 \omega_0 t >\right.$$

$$\left. - \left(\frac{\omega_0}{\tau} \right) < \sin \omega_0 t \cos \omega_0 t > \right] x_0^2 e^{-t/\tau}$$

$$\cong \frac{1}{4} m \left[\left(\frac{1}{2\tau} \right)^2 + \omega_0^2 \right] x_0^2 e^{-t/\tau} \tag{6.60}$$

Since we are working for the low-damping under-damped harmonic oscillator

case, in Eq. (6.60), $\left(\dfrac{1}{2\tau} \right)^2 \ll \omega_0^2$. Hence, Eq. (6.60), representing the *average*

kinetic energy, can be written as follows:

$$\cong \frac{1}{4} m \omega_0^2 x_0^2 e^{-t/\tau}$$

$$\cong E_0 e^{-t/\tau} \tag{6.61}$$

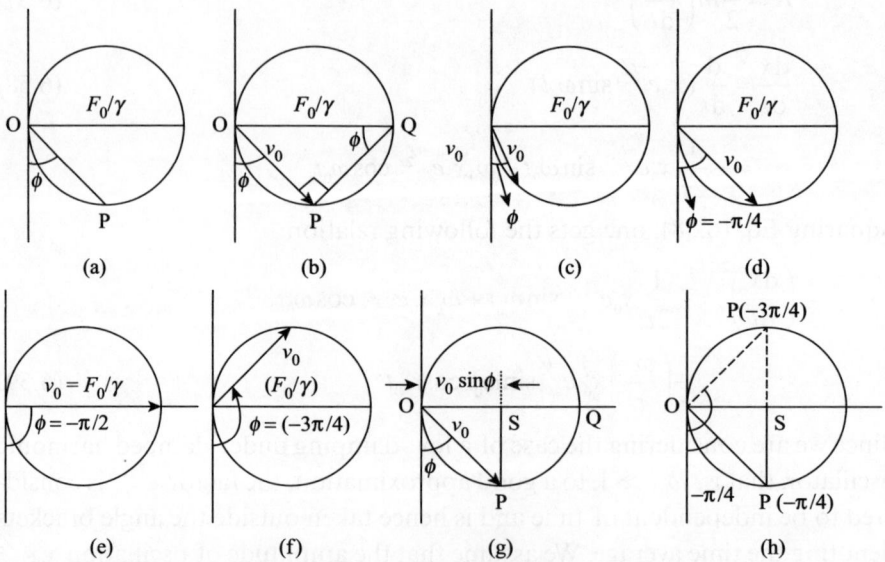

Fig. 6.10 Polar plot or graphical representation of driven harmonic oscillator
(a) Circle with diameter F_0/γ and line segment OP making angle ϕ with ordinate
(b) For any ϕ, triangle OPQ is a right triangle; line segment OP = $\omega x_0 = v_0$,
the amplitude of velocity (c) For $\omega \ll \omega_0$, $\phi \approx 0$ and $v_0 \ll F_0/\gamma$; response small in
this limit (d) As ω increases γ increases and so does v_0; at $S = -\dfrac{\pi}{4}, v_0 = \dfrac{F_0}{\sqrt{2}\gamma}$ (e) At
$\phi = -\pi/2$, $\omega = \omega_0$ and $v_0 = F_0/\gamma$; velocity amplitude is maximum at resonance (f) For
$\omega > \omega_0$, v_0 decreases; at $S = -3\pi/4$; $v_0 = F_0/\sqrt{2}\gamma$ (g) Power absorbed proportional to
OS (h) At phase angles $\phi = -\pi/4$ and $\phi = -3\pi/4$ line segment OS $= \dfrac{1}{2}(OS)_{max}$; half
maximum power absorption occurs at these points

In a similar manner, the average potential energy can be represented as follows:

$$< U >_T = \frac{1}{2} m \omega_0^2 x_0^2 \; < e^{-t/\tau} \sin^2 \omega_0 t >_T$$

$$\sim \frac{1}{4} m \omega_0^2 x_0^2 \, e^{-t/\tau} \tag{6.62}$$

From Eqs (6.61) and (6.62), we observe that the energy of an oscillator decays to ($1/e$) of its initial value in the time τ. In damped harmonic motion, energy of an oscillator is gradually dissipated by friction and falls to zero in time.

Figure 6.10 presents a graphical representation of a damped harmonic oscillator.

We know that the average power dissipation is the negative sum of the rate of change of kinetic and potential energy. Hence,

$$< P >= -\frac{d < E >}{dt} = \frac{d}{dt}[< K > + < U >] \tag{6.63}$$

Substituting the value of kinetic and potential energies, respectively, from Eqs (6.61) and (6.62) and differentiating with respect to time, we get the following expression:

$$= -\frac{1}{\tau} \left[\frac{1}{2} m \omega_0^2 x_0^2 e^{-t/\tau} \right] \tag{6.64}$$

Thus, the average power dissipated is given as follows: $< P(t) >= \dfrac{< E(t) >}{\tau}$

Physically, the average power dissipated represents the negative of the average work done by the frictional force.

6.5.3 Quality Factor Q

The quality factor is mathematically defined as 2π times the ratio of the energy stored to the average energy loss per time period. The quality factor is a dimensionless quantity. Thus,

$$Q = 2\pi \frac{\text{Energy stored}}{< \text{Energy loss in one time period} >} = 2\pi \frac{E}{(P/\vartheta)} \tag{6.65}$$

The time period is defined as ($1/v$) and also $2\pi v = \omega$. The inverse of angular frequency ($1/\omega$) is the time of 1 radian of motion. Thus,

$$Q = \frac{E}{(P/\omega)} \tag{6.66}$$

As we are discussing the case of an underdamped harmonic oscillator, $\omega_0 \tau \gg 1$. Then,

$$Q \cong \frac{E}{E/\omega\tau} \cong \omega_0 \tau \tag{6.67}$$

For a low-damping case, the frequency of the system is nearly equal to the natural frequency. Thus, physically the value of the quality factor is indicative of a lack of damping present in the system [we can recall that $\left(\dfrac{1}{\tau} = \dfrac{\gamma}{m} \right)$].

A high quality factor Q or $\omega_0\tau$ means that the oscillator is lightly damped. One can see from Eq. (6.67) that the oscillator performs $(\omega_0\tau/2\pi)$ oscillations in the time τ.

Having discussed the problem of a damped harmonic oscillator in detail, we now seek to overcome the frictional force such that we can get ideal harmonic oscillations.

Example 6.10   The amplitude of a damped harmonic oscillator reduces from 20 to 2.0 cm after 100 oscillations, each of period 2.3 s. Calculate the logarithmic decrement of the system ($\log_e 10 = 2.30$).

Solution   The logarithmic decrement is defined as follows:

$$\theta = \frac{1}{n}\log_e\frac{a_1}{a_2} = \frac{1}{100}\log_e\frac{20}{2.0} = 0.01 \times 2.303 = 0.023$$

Example 6.11   The quality factor in a damped oscillator is $10e^{-4}$. After what time, will its energy fall to (1/10) of its initial value? The frequency of the oscillator is 200 Hz.

Solution   The quality factor is defined as follows: $Q = \omega_0\tau$

where $Q = 10e^{-4}$; $\omega = 2\pi\vartheta = 2\pi \times 200 = 400\pi/s$

Substituting these values in the preceding equation, one gets the following relations:

$$10e^{-4} = 400\pi\tau \quad \Rightarrow \quad \tau = (10e^{-4})/400\pi$$

The following is the energy equation: $E \cong E_0 e^{-t/\tau}$

Let t be the time in which the energy falls to (1/10)th of its initial value. Then,

$$E_0/10 \cong E_0 e^{-t/\tau} \quad \Rightarrow \quad \left(\frac{1}{10}\right) = e^{-t/\tau} \quad \Rightarrow \quad 10 = e^{t/\tau}$$

Taking natural log on both the sides,

$$\log_e(10) = \left[\frac{t' \times 400\pi}{10e^{-4}}\right] \quad \Rightarrow \quad t' = \left((2.303 \times 10e - 4) \times \frac{1}{400} \times \frac{7}{22}\right)$$

$$t' = 0.0018 \times 10e^{-4}\,s = 1.8 \times 10^{-8}\,s$$

Example 6.12   A pendulum oscillates 100 times in a second. The quality factor of the pendulum is 1000. Calculate the time in which the amplitude of the pendulum decays to $(1/e^4)$ of its initial value.

Solution   The quality factor is defined as follows:

$$Q = \omega_0\tau \quad 1000 = 2 \times 3.14 \times 100 \times \tau \quad \Rightarrow \quad \tau = 1.59\ s$$

The amplitude equation for a damped oscillation is as follows, where x_0 represents the initial amplitude at time ($t = 0$):

$$A(t) = x_0\, e^{-t/2\tau} \ ; \ \frac{x_0}{e^4} = x_0 e^{-t/2\tau} \quad \Rightarrow \quad e^4 = e^{-t/2\tau} \quad \Rightarrow \quad 4 = \frac{t}{2\tau}$$

$$t = 4 \times 2 \times 1.59\ s = 12.72\ s$$

Example 6.13   Considering the quality factor of a sonometer wire of frequency 260 Hz as 2000, calculate the average power dissipated, when the initial amplitude of 2 cm (of a 2 g particle) damps to $(1/e)$ of its value.

Solution   The quality factor is defined as $Q = \omega_0\tau$

$$2000 = 2 \times 3.14 \times 260 \times \tau \quad \Rightarrow \quad \tau = 1.225s$$

The average power dissipated is defined as follows:

$$= \frac{1}{\tau}\left[\frac{1}{2}m\omega_0^2 x_0^2 e^{-t/\tau}\right] = \frac{1}{1.225} \times (0.5 \times 2 \times (2 \times 3.14 \times 260)^2 \times e^{-1} \cong 8.00e5 \,\mathrm{g\,cm^2/s^3}$$

6.6 FORCED VIBRATIONS

In Sections 6.4 and 6.5, we have discussed the cases where a particle is executing oscillations when it is released from its equilibrium position. The particle executes natural oscillatory motion about its equilibrium position, which is damped with time as frictional forces come into play.

In forced vibrations, the particle is subject to an oscillatory external force (which may or may not be time dependent). The most common example of forced vibration is a bridge oscillating due to the force of marching soldiers, a tuning fork oscillating due to the force of sound waves, a child sitting on a swing oscillating due to the force applied, and so on.

The problem of forced vibrations is a general one. Its solution is applicable to acoustic systems, mechanics, atomic physics, and alternating current circuits.

When a system is subject to an external force, the oscillations that result are *forced oscillations*. Forced oscillations have the frequency of the external force applied. The system takes a certain time to adjust to the external frequency applied by the external force. This is referred to as the *transient period*. In this time, free oscillations of the system die out and the system starts performing harmonic motion of constant amplitude with frequency ω of the external force. The time in which the free oscillations die out and steady-state motion is established is called the *transient period*. The steady-state motion is the state when a system/particle performs vibrations or oscillations with exactly the frequency of the external force.

While studying forced harmonic oscillators, we look at the steady-state response of a system or particle on which the external force is applied.

Thus, the *steady-state response* of a forced harmonic oscillator is precisely at the *driving frequency*. The *response* of an oscillator is exhibited as either the *displacement* (x) function or the *velocity* function (dx/dt) of the particle. For the *steady-state solution*, the initial conditions are irrelevant.

Thus, we now establish the equation of motion of a particle under the influence of an external time-dependent force.

6.6.1 Equation of Motion of Driven Harmonic Oscillator

The equation of motion of a driven oscillator is given by the second law of motion; in addition to the restoring force ($-kx$) and the damping force $\left(-\gamma\dfrac{dx}{dt}\right)$, there is an additional time-dependent oscillating external force $F(t)$. Thus,

$$m\frac{d^2 x}{dt^2} = -kx - \gamma\frac{dx}{dt} + F(t) \tag{6.68}$$

$$m\frac{d^2x}{dt^2} + kx + \gamma\frac{dx}{dt} = F(t)$$

$$\frac{d^2x}{dt^2} + \frac{k}{m}x + \frac{\gamma}{m}\frac{dx}{dt} = \frac{F(t)}{m} \tag{6.69}$$

$$\frac{d^2x}{dt^2} + \frac{1}{\tau}\frac{dx}{dt} + \omega_0^2 x = \frac{F(t)}{m} \tag{6.70}$$

$\frac{1}{\tau} = \frac{\gamma}{m}$, where τ is the relaxation time $\omega_0^2 = \frac{k}{m}$

where ω_0 is the natural frequency of the system in the absence of frictional forces.

The time-dependent force applied to the particle is, for convenience, assumed to be a complex function. The real part of the time-dependent force is the cosine function of frequency and time:

$$F(t) = F_0 e^{i\omega t} \Rightarrow F_0 \cos \omega t = ReF(t) \tag{6.71}$$

The solution for the double differential Eq. (6.70) is, for convenience, assumed to be a complex function for displacement. Thus,

$$x(t) = A_0 e^{i\alpha t}$$

$$x(t) = Re(Ae^{i\alpha t})$$

$$\frac{d^2}{dt^2}\left(Ae^{i\alpha t}\right) + \frac{1}{\tau}\frac{d}{dt}\left(Ae^{i\alpha t}\right) + \omega_0^2 Ae^{i\alpha t} = \frac{F_0}{m}e^{i\omega t}$$

$$-\alpha^2\left(Ae^{i\alpha t}\right) + \frac{1}{\tau}i\alpha\left(Ae^{i\alpha t}\right) + \omega_0^2 Ae^{i\alpha t} = \frac{F_0}{m}e^{i\omega t}$$

$$\left(-\alpha^2 + \frac{1}{\tau}i\alpha + \omega_0^2\right)Ae^{i\alpha t} = \frac{F_0}{m}e^{i\omega t}$$

Comparing the two sides,

$$e^{i\alpha t} = e^{i\omega t}$$

$$\Rightarrow \alpha = \omega \Rightarrow x = Ae^{i\omega t} \tag{6.72}$$

$$A = \frac{F_0}{m}\frac{1}{\left(-\omega^2 + \frac{1}{\tau}(i\omega) + \omega_0^2\right)}$$

$$A = \frac{F_0}{m}\frac{1}{\left((\omega_0^2 - \omega^2) + (i\omega/\tau)\right)}$$

Rationalizing the denominator,

$$A = \frac{F_0}{m}\frac{(\omega_0^2 - \omega^2) - \left(\dfrac{i\omega}{\tau}\right)}{(\omega_0^2 - \omega^2)^2 - (i\omega/\tau)^2}$$

$$A = \frac{F_0}{m} \frac{\left(\omega_0^2 - \omega^2\right) - \left(\dfrac{i\omega}{\tau}\right)}{\left(\left(\omega_0^2 - \omega^2\right)^2 - (\omega/\tau)^2\right)}$$

$$A = A_0 e^{-i\varphi} \tag{6.73}$$

$$A = \frac{F_0}{m} \frac{\left(\omega_0^2 - \omega^2\right) - \left(\dfrac{i\omega}{\tau}\right)}{\left(\left(\omega_0^2 - \omega^2\right)^2 - (\omega/\tau)^2\right)}$$

where

$$\mathrm{Re}(A) = A_0 \cos\varphi = \frac{F_0}{m} \frac{\left(\omega_0^2 - \omega^2\right)}{\left(\left(\omega_0^2 - \omega^2\right)^2 + (\omega/\tau)^2\right)} \tag{6.74}$$

$$\mathrm{Im}(A) = A_0 \sin\varphi = \frac{F_0}{m} \frac{(-\omega/\tau)}{\left(\left(\omega_0^2 - \omega^2\right)^2 + (\omega/\tau)^2\right)} \tag{6.75}$$

Squaring and adding,

$$A_0^2 (\cos^2\varphi + \sin^2\varphi) = \left(\frac{F_0}{m}\right)^2 \left[\frac{\left(\omega_0^2 - \omega^2\right)}{\left[\left(\omega_0^2 - \omega^2\right)^2 + (\omega/\tau)^2\right]^2} \right.$$

$$\left. + \frac{(\omega/\tau)^2}{\left[\left(\omega_0^2 - \omega^2\right)^2 + (\omega/\tau)^2\right]^2} \right]$$

$$A_0^2 = \left(\frac{F_0}{m}\right)^2 \frac{1}{\left(\omega_0^2 - \omega^2\right)^2 + (\omega/\tau)^2}$$

$$A_0 = \left(\frac{F_0}{m}\right) \frac{1}{\sqrt{\left(\left(\omega_0^2 - \omega^2\right)^2 + (\omega/\tau)^2\right)}} \tag{6.76}$$

In addition, dividing the preceding equations,

$$\tan\varphi = \frac{(\omega/\tau)}{\left(\omega_0^2 - \omega^2\right)} = \frac{(\gamma\omega/m)}{\left(\omega_0^2 - \omega^2\right)}$$

where, from Eqs (6.74) and (6.75), it is obvious that the values of the sine and cosine functions can be written as follows:

$$\cos\varphi = \frac{F_0}{m} \frac{\left(\omega_0^2 - \omega^2\right)}{\left(\left(\omega_0^2 - \omega^2\right)^2 + (\omega/\tau)^2\right)} \tag{6.77}$$

$$\sin\varphi = \frac{F_0}{m} \frac{(-\omega/\tau)}{\left[\left(\omega_0^2 - \omega^2\right)^2 + (\omega/\tau)^2\right]} \tag{6.78}$$

The displacement function can be seen, from Eq. (6.72), as follows: $x = A e^{i\omega t}$

From Eqs (6.72) and (6.73), the displacement can be written as

$$x = A_0 \, e^{-i\varphi} e^{i\omega t} \qquad (6.79)$$

$$x = A_0 \, e^{i(\omega t - \varphi)}$$

Considering only the real part of the displacement function:

$$x = A_0 \cos(\omega t - \varphi)$$

$$= \frac{(F_0/m)}{\sqrt{\left(\omega_0^2 - \omega^2\right)^2 + (\omega/\tau)^2}} \cos\left[\omega t - \tan^{-1}\left(\frac{(\omega/\tau)}{(\omega_0^2 - \omega^2)}\right)\right] \qquad (6.80)$$

This is the solution for the displacement of the particle under the influence of a time-dependent oscillatory force. Both the driving force and the displacement oscillate with a simple harmonic motion. The cycle from one maximum to another takes 360° or 2π.

The phase φ tells us by what angle the displacement (x) differs/leads from the time-dependent driving force [$F(t)$].

Driven harmonic oscillators can also be classified in terms of the relation between the natural frequency (ω_0) and the frequency of the driving force (ω).

Note: The phase factor φ in a driven harmonic oscillator has quite a different meaning than what we have studied in an undriven, undamped harmonic oscillator, where φ is related to the initial conditions. In the steady state of a driven oscillator, the initial conditions are irrelevant.

6.6.1.1 Low Driving Frequency—$\omega \ll \omega_0$

From Eq. (6.76), the maximum displacement can be written as follows:

$$A_0 = \left(\frac{F_0}{m}\right) \frac{1}{\sqrt{\left(\omega_0^2 - \omega^2\right)^2 + (\omega/\tau)^2}} \cong \frac{F_0}{m\omega_0^2}$$

$$\cong \frac{F_0}{m} \frac{m}{k} = \frac{F_0}{k} \qquad (6.81)$$

Thus, from the amplitude of the oscillator, it is evident that the spring constant (k) controls the response for the driven harmonic oscillator, in the low driving frequency case.

In this case, the external force acts as a constant static force F_0, which pushes the particle to a maximum distance x_{max}, where the restoring force balances F_0:

$$k x_{max} \cong F_0 \qquad (6.82)$$

From Eq. (6.80), the displacement function is obtained as follows:

$$x = A_0 \cos(\omega t - \varphi) \qquad (6.83)$$

$$\frac{dx}{dt} = \omega\left[-A_0 \sin(\omega t - \varphi)\right] \qquad (6.84)$$

$$\frac{d^2 x}{dt^2} = -\omega^2 A_0 \cos(\omega t - \varphi) = -\omega^2 x \qquad (6.85)$$

Thus, the equation of motion can be written as follows:

$$\frac{d^2}{dt^2}\left(A_0\cos(\omega t-\varphi)\right)+\frac{1}{\tau}\frac{d}{dt}\left(A_0\cos(\omega t-\varphi)\right)+\omega_0^2 A_0\cos(\omega t-\varphi)$$

$$=\frac{F_0}{m}(\cos\omega t+i\sin\omega t)-\omega^2 x+\frac{1}{\tau}\left[-\omega A_0\left(1-\frac{x^2}{A_0^2}\right)\right]+\omega_0^2 x=\frac{F_0}{m}\cos\omega t$$

$$\Rightarrow x\cong\frac{F_0}{m\omega_0^2}\cos\omega t \qquad\qquad (6.86)$$

$$x\cong\frac{F_0}{k}\cos\omega t \qquad\qquad (6.87)$$

In the case of a low driving frequency, there is a static interplay between elastic force and external force $(F(t))$ that is changing with time.

$$\tan\varphi=\frac{(\omega/\tau)}{\omega_0^2-\omega^2}=\frac{\sin\varphi}{\cos\varphi}$$

We know that for a low driving frequency, the value of the driven frequency of the time-dependent force is much less than the natural frequency of the oscillator, as

$$\sin\varphi\to 0\ \ \text{and}\ \ \cos\varphi\to 1\Rightarrow\varphi\to 0 \qquad\qquad (6.88)$$

Thus, the value of the phase tending to zero indicates that the response at the low driving frequency is said to be in phase with the driving force.

6.6.1.2 High Driving Frequency—$\omega \gg \omega_0$

From Eq. (6.76), the maximum displacement can be written as follows:

$$A_0=\left(\frac{F_0}{m}\right)\frac{1}{\sqrt{\left(\left(\omega_0^2-\omega^2\right)^2+(\omega/\tau)^2\right)}}$$

$$A_0\cong\frac{F_0}{m\omega^2} \qquad\qquad (6.89)$$

Thus, we observe that the amplitude decreases with the frequency of the driven oscillator. From Eq. (6.80), the displacement function is written as follows:

$$x=A_0\cos(\omega t-\varphi) \qquad\qquad (6.90)$$

$$\frac{dx}{dt}=\omega\left[-A_0\sin(\omega t-\varphi)\right] \qquad\qquad (6.91)$$

$$\frac{d^2x}{dt^2}=-\omega^2 A_0\cos(\omega t-\varphi)=-\omega^2 x \qquad\qquad (6.92)$$

Thus, the equation of motion can be written as follows:

$$\frac{d^2}{dt^2}\left(A_0\cos(\omega t-\varphi)\right)+\frac{1}{\tau}\frac{d}{dt}\left(A_0\cos(\omega t-\varphi)\right)$$

$$+\omega_0^2 A_0\cos(\omega t-\varphi)=\frac{F_0}{m}(\cos\omega t+i\sin\omega t)$$

$$-\omega^2 x + \frac{1}{\tau}\left(-\omega A_0\left(1 - \frac{x^2}{A^2}\right)\right)^{1/2} + \omega_0^2 x = \frac{F_0}{m}\cos\omega t$$

$$x \cong -\frac{F_0}{m\omega^2}\cos\omega t \tag{6.93}$$

Thus, the external force controls the displacement of the particle as if there is no damping force and no elastic force. The response decreases as $1/\omega^2$, that is, inertia of the mass controls the response in the high frequency limit.

For the high frequency limit, the phase factor can be analysed as follows:

$$\cos\varphi \to -1 \quad \text{and} \quad \sin\varphi \to 0 \Rightarrow \varphi \to -\pi$$

Thus, the value of the phase tending $(-\pi)$ indicates that the response at the high driving frequency is said to be $-\pi$ out of phase with the driving force.

6.6.1.3 Resonance Frequency—$\omega \cong \omega_0$ 🔊

When the driving frequency ω is nearly equal to the natural frequency ω_0, the resonance response of the system is expected to be very large. The amplitude response function is given as follows: $A_0 = \left(\dfrac{F_0}{m}\right)\dfrac{1}{\sqrt{\left((\omega_0^2 - \omega^2)^2 + (\omega/\tau)^2\right)}}$

To know the maximum of the amplitude function mathematically, the derivative of the denominator needs to be equal to zero:

$$\frac{d}{d\omega}\left(\sqrt{\left((\omega_0^2 - \omega^2)^2 + (\omega/\tau)^2\right)}\right) = 0$$

$$2(\omega_0^2 - \omega^2)(-2\omega) + (2\omega/\tau^2) = 0; \quad (\omega_0^2 - \omega^2) = (1/2\tau^2)$$

$$\omega^2 = \omega_R^2 = \left(\omega_0^2 - (1/2\tau^2)\right) \tag{6.94}$$

Thus, this is the position of the maximum response of the curve of the amplitude (A) versus ω. If $\omega_0 \gg 1$, then the maximum is very close to $\omega = \omega_0$. Thus, the resonance frequency is given as follows:

$$\Rightarrow \omega = \omega_R = \sqrt{\left(\omega_0^2 - (1/2\tau^2)\right)}$$

$$(A_0)_{max} = \left(\frac{F_0}{m}\right)\frac{1}{\sqrt{\left(\omega_0^2 - (1/2\tau^2)\right)}} \quad \text{as } \omega_R \ll \omega$$

Hence, $(A_0)_{max} = \dfrac{(F_0/m)\tau}{\omega_0}$ \hfill (6.95)

Thus, one observes that the lower the damping, the higher the values of τ and A_0. The amplitude–frequency response function is as shown in Fig. 6.11. The following observations can be drawn from the amplitude–frequency curve:

1. For a low driving frequency, the amplitude is nearly the same for all the values of damping.

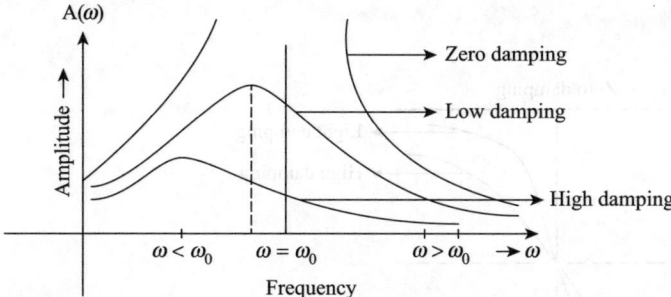

Fig. 6.11 Amplitude–Frequency response; low damping gives sharper resonance and larger damping flatter response

2. As the frequency increases, the amplitude increases depending on the damping present in the system, that is, for zero damping, amplitude is infinite at $\omega = \omega_0$. For low damping, the peak of the amplitude is higher, but the maximum value is shifted towards the left of ω_0. For higher damping, the peak of the amplitude reduces further and is shifted to the left of the resonant frequency.
3. As the driving frequency increases beyond the resonant frequency, the amplitude tends to reduce and attain a minimum value for all the damping.
4. The amplitude curve falls rapidly for low damping than for high damping.

Note: The lower the damping for a system, the sharper the resonance, and the larger the damping for a system, the flatter the resonance.

For the high-frequency limit, the phase factor can be analysed as follows:

$$\cos\varphi \to \pm 0 \quad \text{and} \quad \sin\varphi \to -1 \Rightarrow \varphi \to -\pi/2$$

Thus, the value of the phase tending to $-\pi/2$ indicates that the displacement response at the resonance frequency is said to be $\pi/2$ or $90°$ out of phase with the driving force. In other words, this indicates that the maximum response will be obtained when the velocity is exactly in phase with the driving forces. Thus, for resonance, with force and velocity in phase, we must have the force $90°$ ahead of the displacement.

6.6.2 Phase Characteristics of Driven Oscillators

The phase represents the lag between the force and the displacement. This is shown in Fig. 6.12. The following can be derived from the graph in Fig. 6.12:

1. When damping is zero, the curves for $\omega < \omega_0$ run from 0 to ω_0 for $\varphi = 0$ and is parallel to x-axis for $\omega > 0$ for $\varphi = \pi$.
2. Phase lag increases from 0 to $\pi/2$ as driving frequency increases from 0 to ω_0 and it approaches π as ω increases beyond ω_0. Thus, damping is present.
3. The resonance frequency is nearly ω_0. Thus for $\varphi = \pi/2$ at $\omega = \omega_0$, the particle experiences the same amplitude.
4. The rate of change of phase angle is faster when the damping is lower than when it is high.

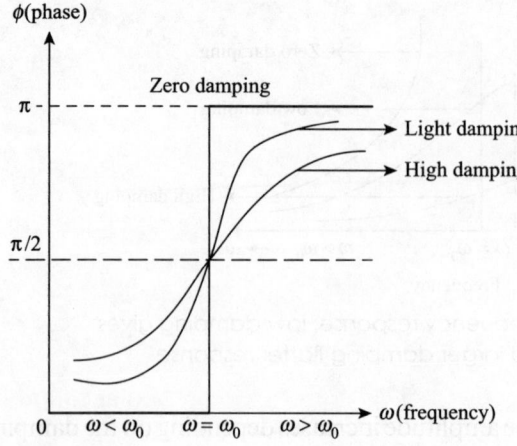

Fig. 6.12 Phase–Frequency response

From the analysis of a driven oscillator, one observes that the driving force F starts from zero at low (damping) frequencies, passes through $-\pi/2$ at resonance, and attains $-\pi$ at high frequencies, that is *displacement always lags behind the driving force.*

6.6.2.1 Low Driving Frequency—$\omega < \omega_0$

The phase angle φ has a small value and is positive. The displacement lags behind the force by a very small amount.

6.6.2.2 Resonance Frequency—$\omega = \omega_0$

The displacement lags behind the force by $-\pi/2$, that is, the largest displacement is obtained when the velocity is exactly in phase with the time-dependent driving force. The velocity of an oscillator leads its displacement by exactly 90°. Thus,

$$F(t) = F_0 \cos\omega t$$

$$x = A_0 \cos(\omega t - \pi/2)$$

$$(\mathrm{d}x/\mathrm{d}t) = -A_0\omega\sin(\omega t - \pi/2) = A_0\omega\cos\omega t$$

Power absorbed does not depend on the phase between the time-dependent driving force and displacement, but rather on the phase between time-dependent driving force and velocity.

6.6.2.3 High Driving Frequency—$\omega > \omega_0$, $\varphi = \pi$

The displacement is π out of phase with force. This implies that when the force is maximum in one direction, the displacement is maximum in the opposite direction.

6.6.3 Quality Factor

The quality factor measures the sharpness of tuning for a driven oscillator. Mathematically, the expression for quality factor is given as follows:

$$Q = \omega_0 \tau = \omega_0 / 2(\Delta\omega)_{1/2} = \frac{\text{Frequency at resonance}}{\text{Full width at half-maximum power}} \quad (6.96)$$

The quality factor is a measure of the lack of damping of an oscillator, that is, the larger the value of quality factor Q, the lower the damping. Thus, from the resonance case, the frequency has been ascertained as follows:

$$\omega = \omega_R = \sqrt{\left(\omega_0^2 - \left(1/2\tau^2\right)\right)} \text{ also } \omega_R < \omega_0$$

In terms of the quality factor, the frequency can be expressed as follows:

$$\omega = \omega_R = \omega_0\sqrt{1 - 1/2Q^2} \quad (6.97)$$

$$(A_0)_{\max} = \frac{(F_0/m)}{\sqrt{\left(\omega_0^2 - \left(1/2\tau\right)^2\right)}} \text{ as } \omega_R \ll \omega \quad (6.98)$$

In addition

$$(A_0)_{\max} = \frac{F_0/m}{\omega_0\sqrt{\left(1 - \left(1/2Q^2\right)\right)}} \text{ as } \omega_R \ll \omega \quad (6.99)$$

Furthermore,

$$\tan\varphi = \frac{\omega\omega_0^2/Q}{\left(\omega^2 - \omega_0^2\right)}$$

$$\omega_R \cong \omega_0 (A_0)_{\max} = \frac{F_0}{K}$$

$$\Rightarrow Q = \frac{(A_0)_{\max}}{(A_0)_{\text{static}}} \quad (6.100)$$

Equation (6.100) relates the quality factor to the maximum and static amplitudes, as discussed in the previous sections.

6.6.3.1 Power Absorbed by Driven Oscillators

The power absorbed by a driven oscillator is the time average of the work done per unit time on the oscillating system by the driving force. It is given as follows:

$$P = [F(\mathrm{d}v/\mathrm{d}t)] \quad (6.101)$$

Thus, we know from the previous discussion that

$$F(t) = F_0 \cos\omega t$$

$$x = A_0 \cos(\omega t - \varphi)$$

$$\frac{\mathrm{d}x}{\mathrm{d}t} = -A_0\omega \sin(\omega t - \varphi)$$

Substituting these values in Eq. (6.101), one gets the following relation:

$$P = -\frac{m\omega(F_0/m)^2}{\sqrt{(\omega_0^2 - \omega^2)^2 + (\omega/\tau)^2}} <\cos\omega t \, \sin(\omega t - \varphi)> \qquad (6.102)$$

Using the trigonometric identity,

$$\sin(\omega t - \varphi) = \sin\omega t \cos\varphi - \cos\omega t \sin\varphi \qquad (6.103)$$

We have the time-averaged term as follows:

$$<\cos\omega t(\sin\omega t \cos\varphi - \cos\omega t \sin\varphi)>$$
$$= -\sin\varphi <\cos^2\omega t> = -(1/2)\sin\varphi \qquad (6.104)$$

Thus, the phase factor is an important term in Eq. (6.104). Thus, substituting the relation for $\sin\varphi$ from Eq. (6.78) into Eq. (6.102), we get the following relation:

$$P = \frac{1}{2}m\frac{(F_0/m)^2\left(\dfrac{\omega^2}{\tau}\right)}{\left((\omega_0^2 - \omega^2)^2 + (\omega/\tau)^2\right)} \qquad (6.105)$$

The resonance power at $\omega = \omega_0$, from Eq. (6.105), is expressed as follows:

$$P = \frac{1}{2}m\tau\left(\frac{F_0}{m}\right)^2 \qquad (6.106)$$

Power absorption, as shown in Eq. (6.105), is reduced to one-half of the value at resonance when ω is changed by $(\pm\Delta\omega)_{1/2}$, such that

$$\frac{\omega}{\tau} = (\omega_0^2 - \omega^2) \equiv ((\omega_0 + \omega)(\omega_0 - \omega)) \cong 2\omega_0(\Delta\omega)_{1/2} \qquad (6.107)$$

Thus, from Eq. (6.107), one can draw the inference that the full width $(2\Delta\omega)_{1/2}$ of the resonance at half-maximum power is equal to $(1/\tau)$. The quality factor has been defined earlier in Eq. (6.96). The detailed graph can be seen in Figs 6.13(a) and (b).

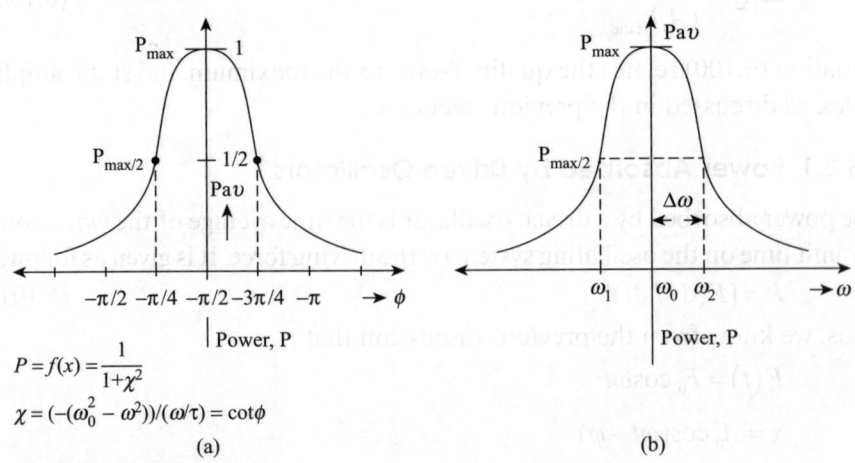

$$P = f(x) = \frac{1}{1+\chi^2}$$
$$\chi = (-(\omega_0^2 - \omega^2))/(\omega/\tau) = \cot\phi$$

(a)　　　　　　　　　(b)

Fig. 6.13 Response function for driven oscillator (a) Power–Phase response (b) Power–Frequency response

Note: Maximum power absorption occurs at phase angle $\varphi = (\pi/2)$, which is at the resonance position.

Example 6.14 A harmonic oscillator of quality factor 20 is subjected to a sinusoidal applied force of frequency one and a half times the natural frequency of the oscillator. Calculate the (a) amplitude of the forced oscillation in terms of its maximum amplitude and (b) angle θ by which the oscillator will be out of phase with the driving force. Assume that the damping factor is small.

Solution The following are the calculations involved:
(a) The quality factor is defined as follows:

$$Q = \omega_0 \tau = 20 \quad \Rightarrow \tau = (20/\omega_0)$$

In addition, $= (1/2)\omega_0 \quad \Rightarrow (\omega/\omega_0) = (1/2)$

The amplitude of the forced oscillation is given as follows:

$$A_0 = \frac{(F_0/m)}{\sqrt{\left((\omega_0^2 - \omega^2)^2 + (\omega/\tau)^2\right)}} \quad \Rightarrow A_0 = \frac{(F_0/m)}{\sqrt{\left((\omega_0^2 - \omega^2)^2 + (\omega/\tau)^2\right)}}$$

$$A_0 = \frac{(F_0/m)}{\sqrt{\left((\omega_0^2 - 0.5\omega_0^2)^2 + \left(0.5\omega_0^2/((20)^2 \times \omega_0^2\right)\right)}}$$

$$A_0 = \frac{(F_0/m)}{\omega_0^2 \sqrt{(9/16) + (1/1600)}} = \left(\frac{F_0}{m}\right)\frac{1}{\omega_0^2}\frac{40}{\sqrt{901}}$$

The maximum amplitude is defined as follows:

$$(A_0)_{max} = \frac{\left(\dfrac{F_0}{m}\right)\tau}{\omega_0} = \frac{\left(\dfrac{F_0}{m}\right) \times 20}{\omega_0 \times \omega_0}$$

Thus, the amplitude of the lightly damped forced oscillator in terms of its maximum amplitude is calculated as follows:

$$\frac{A_0}{(A_0)_{max}} = \left(\frac{F_0}{m}\right)\frac{1}{\omega_0^2}\frac{40}{\sqrt{901}}\frac{m}{F_0}\frac{\omega_0^2}{20} = \frac{20}{30.016} = 0.6663 \quad \Rightarrow A_0 = 0.663 A_{0max}$$

(b) The phase factor is defined as follows:

$$\tan\varphi = \frac{(\omega/\tau)}{(\omega_0^2 - \omega^2)} = \frac{(0.5 \times \omega_0^2)}{20 \times \omega_0^2 \times (1 - 0.5)} = \frac{1}{20}$$

$$\varphi = \tan^{-1}(0.05) = 2° - 52'$$

6.7 CHAOS

As discussed in Section 6.5, the time-dependent oscillating driving force, when applied to a damped linear harmonic oscillator, results in forced oscillations having the same angular frequency as that of the driving force. This results in an ideal harmonic oscillator.

If we apply a time-dependent force on a non-harmonic oscillator, that is, an oscillator that is executing oscillations dependent on the force equation that is dependent on the second, third, fourth, or higher powers of displacement, then the resultant oscillations will be non-harmonic in nature. Apparently, the resultant motion appears to be completely random and unpredictable, yet it follows a non-harmonic pattern of periodicity. Such a seemingly random motion is called a *chaotic motion*, and its behaviour is called *chaos*. Many physical systems in nature and technology display chaotic behaviour. The most common examples of chaotic motion are turbulent river flow, a mass oscillating on the top of a vertically oscillating piston, water dripping from a hand faucet, etc. Chaos is also visible in many economic and biological phenomena. It is an active field of research for those dealing with the perturbation theory in physics.

6.8 COUPLED OSCILLATIONS

In our discussion so far, we have discussed the one-body analysis of an oscillator. In nature, however, we find two-body oscillating systems, in which masses of both the bodies are almost comparable to each other, for example, diatomic molecules such as as H_2, CO_2, HCl, and many more.

We can redefine terms and equations for two-body oscillations by introducing the new concept of *reduced mass*. Let us assume that two bodies m_1 and m_2 be connected by a massless spring of force constant k, such that the system is free to oscillate on a frictionless horizontal surface (refer to Fig. 6.14). The two ends of the spring are fixed by the co-ordinates $x_1(t)$ and $x_2(t)$ such that the length of the spring at any instant is given by $(x_1 - x_2)$. Let l be the normal, unstressed length of the spring, then the change in length of the spring $x(t)$ is given by the following equation:

$$x = (x_1 - x_2) - l$$

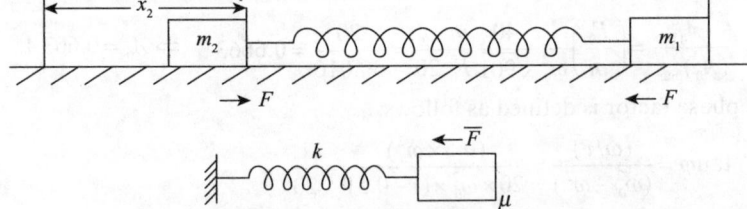

Fig. 6.14 Coupled oscillator

where if x is positive, then the spring is stretched; if $x = 0$, then the spring has its normal length; and if x is negative, then the spring is compressed.

Let a force F be exerted by the spring on m_1 and the force $-F$ be exerted on m_2. These forces are equal and opposite, and have the common magnitude $F = kx$. Using Newton's second law, $F = ma$, to masses m_1 and m_2, one gets the following expressions:

$$m_1 \frac{d^2 x_1}{dt^2} = -kx \tag{6.108}$$

$$m_2 \frac{d^2 x_2}{dt^2} = -kx \tag{6.109}$$

Multiplying the first equation by m_2 and the second equation by m_1, and subtracting one gets the following relation:

$$m_1 m_2 \frac{d^2 x_1}{dt^2} - m_1 m_2 \frac{d^2 x_2}{dt^2} = -m_2 kx - m_1 kx \tag{6.110}$$

This equation can further be written as follows:

$$\frac{m_1 m_2}{m_1 + m_2} \frac{d^2(x_1 - x_2)}{dt^2} = -kx \tag{6.111}$$

$\mu = \dfrac{m_1 m_2}{m_1 + m_2}$ is referred to as the reduced mass of the system.

The mass μ is always less than m_1 and m_2. Since l is a constant, $(x_1 - x_2) = x$, where x is the relative displacement of the two masses from their equilibrium positions.

Equation (6.111) can be rewritten as follows:

$$m \frac{d^2 x_1}{dt^2} + \frac{k}{\mu} x = 0 \tag{6.112}$$

Thus, Eq. (6.112) is identical to the equation formulated for single-body oscillation, where the displacement has been replaced by the relative displacement and individual masses by the reduced mass.

Let us discuss some basic definitions associated with the study of waveforms, specially wave features and nomenclature.

6.9 WAVES

The oscillatory behaviour of a particle due to perturbation/disturbance results in the formation of a wave. The conventional nomenclature of the waves is discussed in the following subsections.

Angular frequency, ω The unit for ω is radian/s. The angular frequency ω is defined as the rate of change of phase with time.

$$\omega = (2\pi/T) \tag{6.113}$$

The angular frequency of an oscillator plays the same role as the angular velocity of a pin.

Time period, T It is the time needed for one oscillation, that is, one complete cycle:

$$T = (2\pi/\omega) \tag{6.114}$$

that is, ω times the time period is one cycle of a cosine or sine wave.

Frequency, ϑ   Frequency is defined as the number of complete vibrations per unit time. It is the inverse of time period and is independent of the amplitude of motion. Mathematically, it can be represented as follows:

$$\vartheta = (\omega/2\pi) \tag{6.115}$$

$$c = \vartheta\lambda \tag{6.116}$$

where c is the velocity of light.

The wave function can essentially be discussed as a function of time t or a function of distance r. As a function of r or t, a wave is oscillatory in nature. One can define another quantity that is commonly used in this regard—space period λ.

Space period/wavelength, λ   A wavelength is the distance covered by one complete cycle, or in other words, it is the period in space. We mathematically define a wavelength as follows:

$$\lambda = \frac{2\pi}{k} \tag{6.117}$$

$$\lambda k = 2\pi \tag{6.118}$$

In terms of dimensions also, one understands wavelengths as (radians/metre) times (metre), which should give us one complete cycle, which is nothing but 2π. This relation is analogous to the relation that we have been studying always, which is as follows:

$$\omega t_0 = 2\pi \tag{6.119}$$

where t_0 is the initial time.

Wave number, k   This is defined as the rate of change of phase with distance (radians/metre). In other words, as we move in space at a fixed time, the phase changes, that is,

$$\frac{\partial\varphi}{\partial r} = k \quad \text{where} \quad \varphi = (\omega t - kx) \tag{6.120}$$

The wave number is also referred to as the *propagation wave vector*.

Phase, φ   The phase of an oscillation determines the initial position and velocity of a particle. The phase parameter is denoted by a general formula that incorporates time and displacement parameters. It can be mathematically represented as follows:

$$\varphi = (\omega t - k \cdot r) \tag{6.121}$$

Thus, we conclude from the aforementioned definitions that the frequency, wave number, and phase of a waveform are all very simply co-related quantities.

In general, the displacement $y(x, t)$ in mathematical waveform can be defined as follows:

$$y(x,t) = a\sin(\omega t \pm kx) \tag{6.122}$$

We observe that a wave described by a cosine or sine wave function is periodic in nature, that is, it repeats itself after a given cycle, as shown in Fig. 6.15.

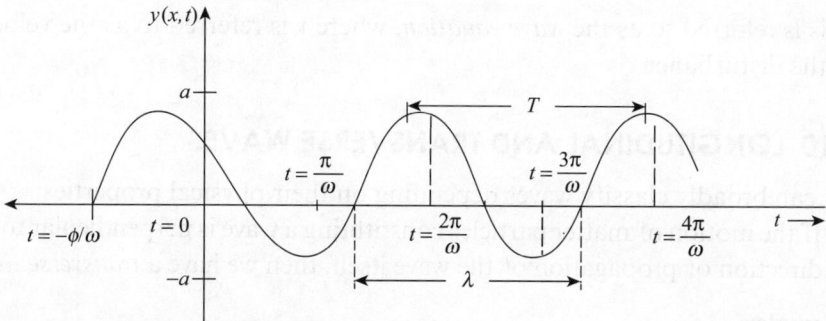

Fig. 6.15 Function $y(x, t) = a \sin(\omega t + kx)$ is plotted vs t. After one full oscillation $(2\pi/\omega)$, function repeats itself; a represents amplitude of motion and ϕ phase of motion

6.9.1 Wave Equation

Based on the discussion on wave functions, the periodic displacement is defined as follows: $y(x,t) = A\sin(\omega t \pm kx)$

Differentiating this displacement equation with respect to time, the velocity of the wave disturbance is defined as follows:

$$\frac{\partial y(x,t)}{\partial t} = \omega A \cos(\omega t - kx) \tag{6.123}$$

Thus, the maximum particle speed is ωA; this can be greater than, less than, or equal to the wave speed v, depending on the amplitude and frequency of the wave.

Acceleration of a wave disturbance can be evaluated by differentiating Eq. (6.123) with respect to t, as follows:

$$\frac{\partial^2 y(x,t)}{\partial t^2} = -\omega^2 A \sin(\omega t - kx) = -\omega^2 y(x,t) \tag{6.124}$$

Acceleration of a wave disturbance equals $-\omega^2$ times its displacement, which is an indication that the disturbance is simple harmonic in nature. If we double differentiate the displacement equation with respect to x, we get the following equation:

$$\frac{\partial^2 y(x,t)}{\partial x^2} = -k^2 A \sin(\omega t - kx) = -k^2 y(x,t) \tag{6.125}$$

Dividing Eq. (6.124) by (6.125), we get the following relation:

$$\frac{\dfrac{\partial^2 y(x,t)}{\partial t^2}}{\dfrac{\partial^2 y(x,t)}{\partial x^2}} = \frac{\omega^2}{k^2} = v^2 \tag{6.126}$$

Thus, we get

$$\frac{d^2 y(x,t)}{dx^2} = \frac{1}{v^2} \frac{d^2 y(x,t)}{dt^2} \tag{6.127}$$

This is referred to as the *wave equation*, where *v* is referred to as the velocity of the disturbance.

6.10 LONGITUDINAL AND TRANSVERSE WAVES

We can broadly classify waves depending on their physical properties:

1. If the motion of matter particles constituting a wave is perpendicular to the direction of propagation of the wave itself, then we have a *transverse wave*.

Examples

(a) A vertical string under tension is set into oscillation (Fig. 6.16). A transverse wave travels down the string where the direction of propagation of the wave is along the string, but the string particles vibrate at right angles to the direction of propagation of the oscillation.

(b) Light waves are transverse waves, as the electric and magnetic fields are perpendicular to the direction of propagation of light waves.

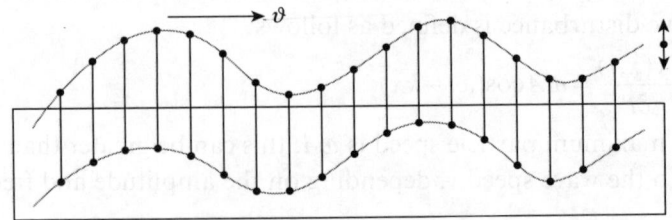

Fig. 6.16 Transverse wave

2. If the motion of the particles constituting a wave is along the direction of propagation of the wave, then we have a *longitudinal wave*.

Examples

(a) A vertical spring under tension is set oscillating up and down at one end. Coils of the spring vibrate back and forth in the direction in which the disturbance travels along the spring.

(b) Sound waves travelling in a gas are longitudinal waves.

6.11 STANDING WAVES

Having studied the superposition of waves in optics and with a reasonably good knowledge of waves, we are now in a position to understand the nature and behaviour of standing waves.

Considering the case of one-dimensional space, two waves having the same frequency, wavelength, and amplitude, and travelling in opposite directions when superimposed, will produce a standing or stationary wave, as shown in Fig. 6.17.

The most common example that one encounters in daily life is that of a wave travelling along a string. A wave travelling to the right along a stretched

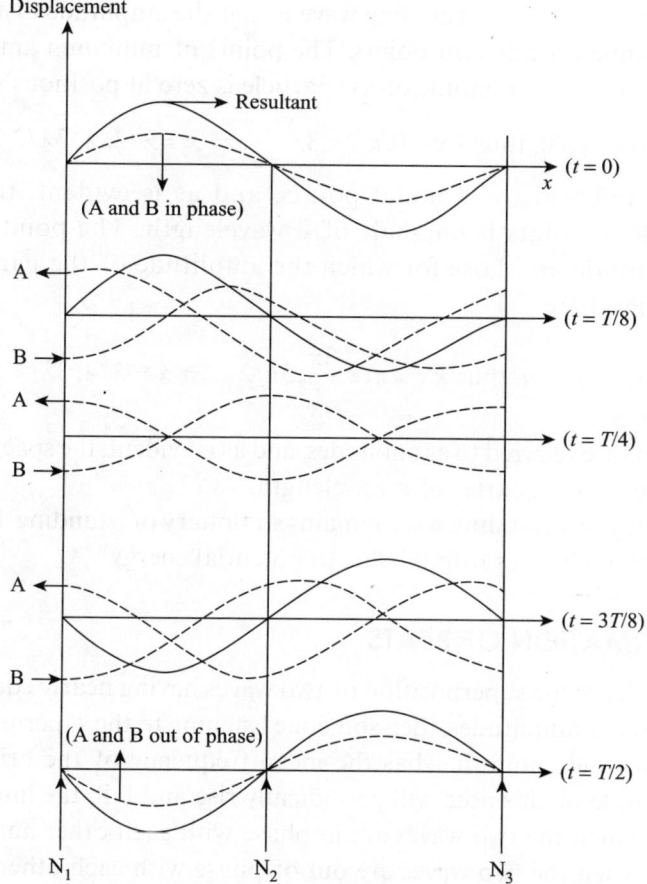

Fig. 6.17 Two waves producing stationary wave

string strikes the end (in the form of an obstacle) and is reflected back in the opposite direction along the string; the two waves superpose to produce a stationary wave. Mathematically, this is represented as follows:

$$y_1 = a\sin(\omega t - kx)$$

$$y_2 = a\sin(\omega t + kx)$$

$$y_R = y_1 + y_2$$

$$y_R = a\sin(\omega t - kx) + a\sin(\omega t + kx)$$

Using the trigonometric relation, $\sin A + \sin B = 2\sin\dfrac{(A+B)}{2}\dfrac{\cos(B-A)}{2}$, this relation reduces to the following form:

$$y_R = 2a\sin(kx)\cos(\omega t)$$

This is the equation of the standing wave.

The characteristic of a standing wave is that the amplitude of the particle is not the same for different points. The points of minimum amplitude are those for which the amplitude of the particle is zero at positions where

$$\sin(kx) = 0, \text{ thus } kx = 0, \pi, 2\pi, 3\pi, \ldots \Rightarrow x = \lambda/2, \lambda, 3\lambda/2, 2\lambda, \ldots$$

These are referred to as nodal points, and as is evident, the spacing between these points is one-half of a wavelength. The points of maximum amplitude are those for which the amplitude of the particles is $2a$ at positions where

$$\sin(kx) = 2a, \text{ thus } kx = \pi/2, \frac{3\pi}{2}, 5\pi/2 \ldots \Rightarrow x = \lambda/4, 3\lambda/4, 5\lambda/4, \ldots$$

These points are referred to as antinodes, and as is evident, the spacing between these points is one-quarter of a wavelength.

The energy in a standing wave remains stationary or 'standing' in the string even though it changes from kinetic to potential energy.

6.12 FORMATION OF BEATS

If one considers the superposition of two waves having nearly equal frequencies and similar amplitudes, then someone listening to the superimposed wave will hear a single note that has the mean frequency of the original notes. The amplitude of this note will periodically rise and fall, the intensity being maximum when the two waves are in phase with each other and going to a minimum when the two waves are out of phase with each other. The variations in amplitude are called beats. The number of times the sound reaches the maximum intensity in 1 s is called the *beat frequency*. The most common example that one encounters in daily life is the symphony of an orchestra or the melodious chorus of an enthralling group song.

Considering the case of one-dimensional space, two waves having slightly different frequencies ϑ_1 and ϑ_2, having the same amplitude α, and travelling along the positive x direction when superimposed will produce beats, as shown in Fig. 6.18.

For ease in mathematics, we assume that the waves are in phase at $t = 0$ without invalidating the result. Mathematically,

$$y_1 = a\sin(2\pi\vartheta_1 t)$$

$$y_2 = a\sin(2\pi\vartheta_2 t)$$

If the two waves are superimposed, the resultant displacement, y_R, is given by the following relations:

$$y_R = y_1 + y_2$$

$$y_R = a\sin(2\pi\vartheta_1 t) + a\sin(2\pi\vartheta_2 t)$$

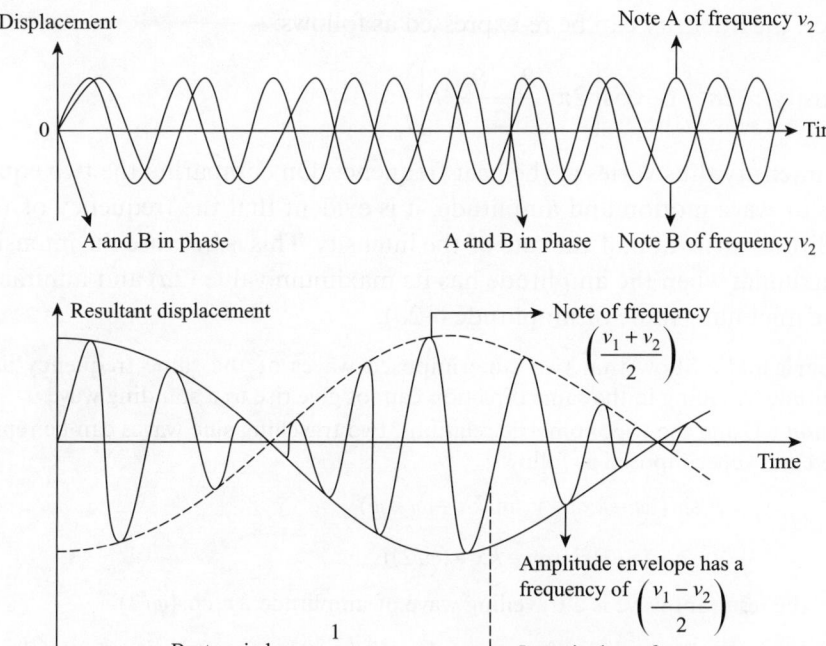

Displacement

Note A of frequency v_2

0 — Time

A and B in phase

A and B in phase   Note B of frequency v_2

Resultant displacement

Note of frequency $\left(\dfrac{v_1 + v_2}{2}\right)$

Time

Amplitude envelope has a frequency of $\left(\dfrac{v_1 - v_2}{2}\right)$

Beat period $= \dfrac{1}{(v_1 - v_2)}$

Intensity has a frequency of $(v_1 - v_2)$

Fig. 6.18 Formation of beats

Using the trigonometric relation $\sin A + \sin B = 2\sin((A+B)/2)\cos((B-A)/2)$, the preceding relation reduces to the following form:

$$y_R = 2a\sin\left(2\pi\left(\frac{\vartheta_1 + \vartheta_2}{2}\right)t\right)\cos\left(2\pi\left(\frac{\vartheta_1 - \vartheta_2}{2}\right)t\right)$$

This can further be written as follows:

$$y_R = A\sin\left(2\pi\left(\frac{\vartheta_1 + \vartheta_2}{2}\right)t\right) \tag{6.128}$$

where $\quad A = 2a\cos\left(2\pi\left(\frac{\vartheta_1 - \vartheta_2}{2}\right)t\right) \tag{6.129}$

The resultant equation is that of a wave having frequency $(\vartheta_1 + \vartheta_2)/2$. The amplitude of this wave is variable and is given by Eq. (6.129). The intensity of the wave motion is proportional to the square of its amplitude. Thus,

$$\text{Intensity} \propto 4a^2\cos^2\left[2\pi\left(\frac{\vartheta_1 - \vartheta_2}{2}\right)t\right]$$

From the trigonometric relation, $2\cos^2\theta = 1 + \cos 2\theta$

Hence, the intensity can be re-expressed as follows:

$$\text{Intensity} \propto 2a^2 \left(1 + \cos \left[2\pi \left(\frac{\vartheta_1 - \vartheta_2}{2} \right) t \right] \right)$$

The intensity thus varies at the beat frequency. On comparing the two equations of wave motion and amplitude, it is evident that the frequency of the amplitude varies at half the rate of the intensity. This is because the intensity is maximum when the amplitude has its maximum value ($2a$) and minimum at the minimum value of amplitude ($-2a$).

Example 6.15 Show that two superimposed waves of the same frequency and amplitude travelling in the same direction cannot give rise to a standing wave.

Solution Using the trigonometric relations, two travelling sine waves can be represented and superimposed as follows:

$$y_p = y_0 \sin(\omega t - kx) + y_0 \sin(kx - \omega t + \varphi)$$

$$= 2y_0 \cos(\varphi/2) \sin(\omega t - kx + (\varphi/2))$$

Thus, the resultant wave is a travelling wave of amplitude $2y_0 \cos(\varphi/2)$.

Example 6.16 The equation for a particular standing wave on a string is given as follows:

$$y = 0.10(\sin 4x \cos 400t)\ \text{m}$$

Find the (a) amplitude of vibration at the antinode, (b) distance between the nodes, (c) wavelength, (d) frequency, and (e) speed of the wave.

Solution The standard equation for a standing wave is given as follows:

$$y_p = A\sin(kx)\cos(\omega t)$$

By comparison we have the following results:
(a) Amplitude of vibration at the antinode: $A = 0.10$ m
(b) When the phase of the sine $= 0, 2\pi, \ldots.$, we have nodes. Thus, if $x = 0$ is a node, then the next node is at $4x = \pi \Rightarrow x = \pi/4 = 0.785$m.
(c) The wavelength is twice x, that is, $\lambda = 2(\pi/4) = 1.57$ m.
(d) The angular frequency $= \omega = 2 \times 3.14 \times \vartheta = 400 \Rightarrow \vartheta = (400/(2 \times 3.14)) = 63.6$ Hz
(e) Speed of the wave $v = \lambda\vartheta = (1.57 \times 63.6)$m/s $= 99.85$m/s

Example 6.17 A uniform string of length L, linear density μ, and tension F is vibrating with amplitude A_n in its nth mode. Find its total energy of oscillation.

Solution The displacement of the string is given by the following relation:

$$y(x,t) = A_n \sin \frac{n\pi x}{L} \cos (2\pi \vartheta_n t + \delta)$$

Hence, the transverse velocity is given as follows:

$$\frac{dy}{dt} = -2\pi \vartheta_n A_n \sin \frac{n\pi x}{L} \sin (2\pi \vartheta_n t + \delta)$$

The string's total energy of oscillation is equal to the maximum kinetic energy. All the points on the string achieve their maximum kinetic energy at the same time, that is, when $y = 0$ for all x.

In addition, $dm = \mu dx$, since μ is the linear density.

Thus,
$$E = K_{max} = \max\left[\frac{1}{2}\int dm\left(\frac{dy}{dt}\right)^2\right] = \max\left[\frac{1}{2}\int_0^L \mu\left(\frac{dy}{dt}\right)^2 dx\right]$$

The maximum value occurs when $\sin^2(2\pi\vartheta_0 nt + \delta) = 1$, so we find that

$$E = \frac{\mu}{2}(2\pi\vartheta_0 nA_n)^2 \int_0^L \sin^2\frac{n\pi x}{L} dx = 2\pi^2\mu\vartheta_n^2 A_n^2 (L/2)$$

where the average of the integral is $(1/2)$.

6.13 STRING MOTION

The simplest way to observe oscillatory particle motion is to fix the ends of a stretched string and then cause it to vibrate. The fixed ends of the string are displacement nodes. The three simplest modes of vibration that satisfy the condition in the case of a string of length L are shown in Figs 6.19(a)–(c). The simplest mode of vibration [Fig. 6.19(a)] is called the fundamental, and the frequency at which it vibrates is called the fundamental frequency. Higher frequencies, as depicted in Figs 6.19(b) and (c), are called overtones. The first overtone is referred to as the second harmonic. Wavelengths of the first, second, and third harmonics can be expressed in terms of the length of the string (L) as follows:

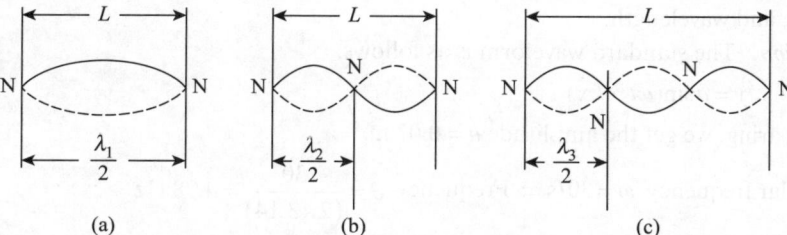

(a)   (b)   (c)

Fig. 6.19 Modes of vibration for stretched string (a) First harmonic (fundamental) (b) Second harmonic (first overtone) (c) Third harmonic (second overtone)

First harmonic $\dfrac{\lambda_1}{2} = L$

Second harmonic $\dfrac{\lambda_2}{2} = \dfrac{L}{2}$

Third harmonic $\dfrac{\lambda_3}{2} = \dfrac{L}{3}$

nth harmonic $\dfrac{\lambda_n}{2} = \dfrac{L}{n}$

The frequency associated with the nth harmonic is given by the following equation: $\vartheta_n = \dfrac{nv}{2L}$

It can easily be observed that the frequency of the nth harmonic can be expressed in terms of the fundamental frequency as follows: $\vartheta_n = n\vartheta_1$

Frequencies of the various overtones are whole number multiples of the fundamental frequency.

Note: The velocity of a wave remains the same for all the wavelengths. In addition, the velocity v represents the velocity of any of the progressive waves that have produced the stationary wave.

If the tension in the string is T and the mass per unit length of the string is μ, then the frequency of the string is given by the following equation:

$$\vartheta_n = \frac{n}{2L}\sqrt{\frac{T}{\mu}} \text{ where } n = 1, 2, 3, \ldots$$

When a string instrument, for example, guitar and piano, is struck, transverse waves travel along the string and are reflected on reaching the ends. Frequencies of the wave travelling along the string are those that satisfy the preceding equation; all the other frequencies die out. All the whole numbers of n give rise to new frequencies that are present on the instrument simultaneously. Among the experiments to verify the preceding formula for frequency, the most common are the Melde's experiment, sonometer, and frequency of the AC current mains.

Example 6.18   A wave along a string is represented by the following equation (x in metres and t in seconds): $y = 0.02 \sin(30t - 4.0x)$ m. Find the amplitude, frequency, speed, and wavelength.

Solution   The standard waveform is as follows:

$$y = a\sin(\omega t \pm kx)$$

Comparing, we get the amplitude $a = 0.02$ m.

Angular frequency $\omega = 30/s \Rightarrow$ Frequency $\vartheta = \dfrac{30}{(2\times 3.14)} = 4.78$ Hz

$k = 4.0/m \Rightarrow \left(\dfrac{2\times 3.14}{\lambda}\right) = 4.0/m \Rightarrow$ Wavelength $\lambda = \dfrac{2\times 3.14}{4.0} = 1.57\,m$

Velocity of the wave is $v = \vartheta\lambda = 4.78 \times 1.57\,m/s = 7.50$ m/s

Example 6.19   A wave travelling along a string has a frequency of 30 Hz and a wavelength of 60 cm. Its amplitude is 2 mm. Write the equation of the wave in SI units.

Solution   The given travelling wave has a frequency of 30 Hz.
$\Rightarrow$ Angular frequency $= 2 \times 3.14 \times 30 = 188.4/s$
Wavelength $= 60$ cm $= 0.60$ m
Propagation wave vector $= ((2\times 3.14)/0.60) = 10.46$; Amplitude $= 2$ mm $= 0.002$ m
Thus, the equation for the travelling wave is $y = 0.002 \sin(188.4t - 10.46x)$ m

Example 6.20   Prove that the average power transmitted by a sinusoidal wave in a string is $P_{av} = \dfrac{1}{2}\mu\vartheta\omega^2 A^2$ where μ is the mass per unit length and v the velocity of the wave.

Solution   If we assume that the infinitesimal length (dx) of the string executes a simple harmonic motion of amplitude A and angular frequency ω, then

Mass of the element $= \mu \, dx$

Energy of the string for the mass element $= \dfrac{1}{2}(\mu dx)\omega^2 A^2$

The energy has to be transmitted to all elements across the boundary, which corresponds to the length of the string $L = ct$; therefore,

Total energy of the string $= \dfrac{1}{2}(\mu ct)\omega^2 A^2$

Average power transmitted $= E/t = \dfrac{1}{2}(\mu c)\omega^2 A^2$

Example 6.21 A 100 Hz wave on a string has an amplitude of 0.100 mm. How much energy exists in 100 g of the string? Assume that one wavelength of the string has a mass much smaller than 100 g.

Solution The energy of the string for the mass element (μdx) is given as follows:

$$E = \dfrac{1}{2}(\mu dx)\omega^2 A^2$$

where amplitude $A = 0.100$ mm $= 1.0e-4$ m, angular frequency $\omega = 2 \times 3.14 \times 100/s$, Let mass of the string for one wavelength be $= 80$ g $= 0.080$ kg

Hence, $E = \dfrac{1}{2}(0.080)(2 \times 3.14 \times 100)^2 (1.0e-4)^2$ kg m^2/s^2

$= 157e-3 = 0.157$ mJ

Example 6.22 For the wave drawn in Fig. 6.20,
(a) Write the equation for the wave as it travels along the +ve x-axis if the initial position at $t = 0$ is as shown in the figure. Find the (i) amplitude, (ii) wavelength, and (iii) frequency, if its speed is 100 m/s.
(b) If the frequency of the wave on the string is 150 Hz, what is its speed?
(c) If the string has a mass of 0.20 g/m, how much energy is sent down the string each second for a 150 Hz wave?
From Fig. 6.20, the amplitude of the sine wave is 0.06 m and the displacement $x = 20$ cm.

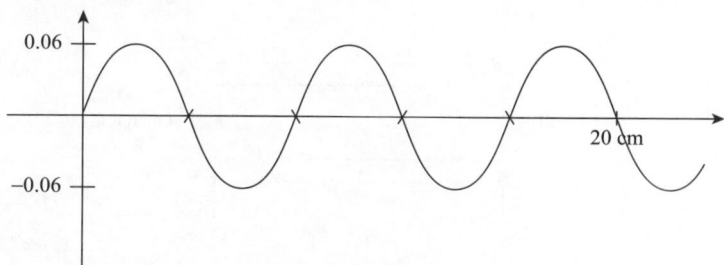

Fig. 6.20 Sine wave

Solution The following are the answers:
(a) (i) Amplitude $= 0.06$ m

(ii) From the figure, $\dfrac{5}{2}\lambda = 20$ cm; $\lambda = 0.080$ m

Propagation wave vector $k = \dfrac{2 \times 3.14}{0.080} = 78.50$

(iii) The frequency of the wave $\vartheta = (v/\lambda) = (100/0.080)/s = 1250$ Hz

Angular frequency $= 2 \times 3.14 \times 1250 = 7850/\text{s}$

Thus, the wave equation is as follows:

$$y = 0.06 \sin(78.5x - 7850t)\text{m}$$

(b) The speed of the wave $v = \vartheta\lambda = 150 \times 0.080 = 12 \text{ m}/\text{s}$

(c) The energy sent down the string each second is the average power. Hence,

Average power transmitted $P = (E/t) = \dfrac{1}{2}(\mu\vartheta)\omega^2 A^2$

$$= \frac{1}{2}(0.00020 \times 12)(2 \times 3.14 \times 150)^2 (0.06)^2 = 3.84 \text{ kW}$$

6.14 WAVES IN PIPES

As discussed in the earlier section just as a stretched string is capable of producing a frequency, so also air can be blown in a closed or an open pipe producing compressions and rarefactions. The production of compressions and rarefactions is the principle of the production of sound waves, which are longitudinal in nature. Vibrations produced in an organ pipe are due to reflection at the far end, producing stationary waves of the same speed, frequency, and amplitude.

6.14.1 Closed or Stopped Pipes

As can be seen in Fig. 6.21, the air at the far end Y cannot move and the pipe is open at the X end. The end X is the point of a displacement antinode, as vibrations are triggered at this point. The three simplest modes of vibrations that satisfy the conditions for a pipe of length L are shown in Figs 6.22(a)–(c).

The distance between a node and an antinode is a quarter of a wavelength. The wavelengths of the

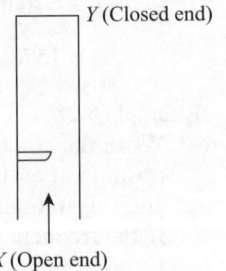

Fig. 6.21 View of closed pipe

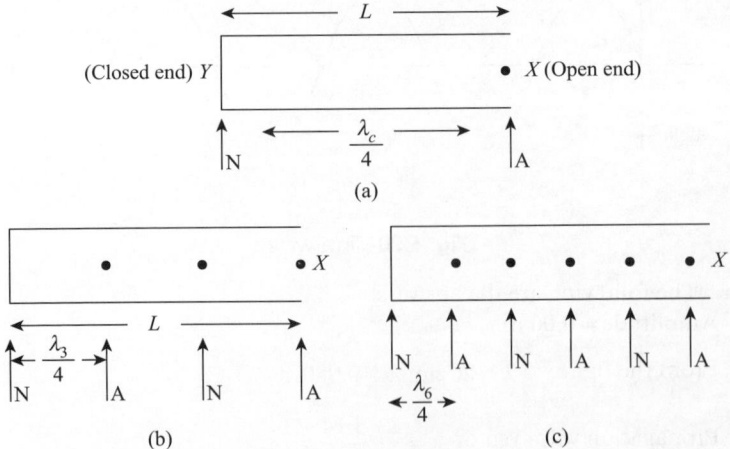

Fig. 6.22 Modes of vibration for closed pipe (a) Fundamental harmonic (b) First overtone (third harmonic) (c) Second overtone (fifth harmonic)

consecutive harmonics represented as λ_a, λ_b, λ_c are given in terms of the length (L) of the pipe as follows:

$$\frac{\lambda_a}{4} = L; \frac{\lambda_b}{4} = \frac{L}{3}; \frac{\lambda_c}{4} = \frac{L}{5}$$

The velocity of a sound wave in air is independent of its wavelength and hence has the same value for each of the consecutive overtones. The respective frequencies of the consecutive harmonics ϑ_a, ϑ_b, ϑ_c in terms of the velocity are given as follows:

$$\vartheta_a = \frac{v}{\lambda_a}; \vartheta_b = \frac{v}{\lambda_b}; \vartheta_c = \frac{v}{\lambda_c}$$

$$\vartheta_a = \frac{v}{4L}; \vartheta_b = \frac{3v}{4L}; \vartheta_c = \frac{5v}{4L}$$

We can observe that higher harmonics can be expressed in terms of the fundamental frequency. In addition, it is worth noting that a *closed pipe can produce only odd harmonics*. Thus, in general, the frequency of the nth harmonic for a closed pipe can be represented as follows:

$$\vartheta_n = \frac{nv}{4L} \ (n = 1,3,5, \ldots)$$

1. Fundamental (harmonic)
2. First overtone (third harmonic)
3. Second overtone (fifth harmonic)

6.14.2 Open Pipes

As can be seen in Fig. 6.23, air at the far end Y is free in the case of an open pipe. A longitudinal wave is produced by a vibration at the other open end X. In the case of an open pipe, air is free to move at both X and Y ends; thus, both the ends act as displacement antinodes. The three simplest modes of vibrations that satisfy the conditions for a pipe of length L are as shown in Figs 6.24(a)–(c).

The distance between a node and an antinode is a quarter of a wavelength. Wavelengths of the consecutive harmonics represented as λ_a, λ_b, λ_c are given in terms of the length (L) of the pipe as follows:

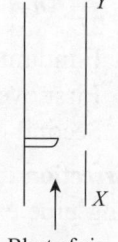

Fig. 6.23 Representation of open pipe; Y end is its open end

$$\frac{\lambda_a}{4} = \frac{L}{2}; \frac{\lambda_b}{4} = \frac{L}{4}; \frac{\lambda_c}{4} = \frac{L}{6}$$

The velocity of a sound wave in air is independent of its wavelength and hence has the same value for each of the consecutive overtones. The respective frequencies of the consecutive harmonics ϑ_a, ϑ_b, ϑ_c in terms of the velocity are given as follows:

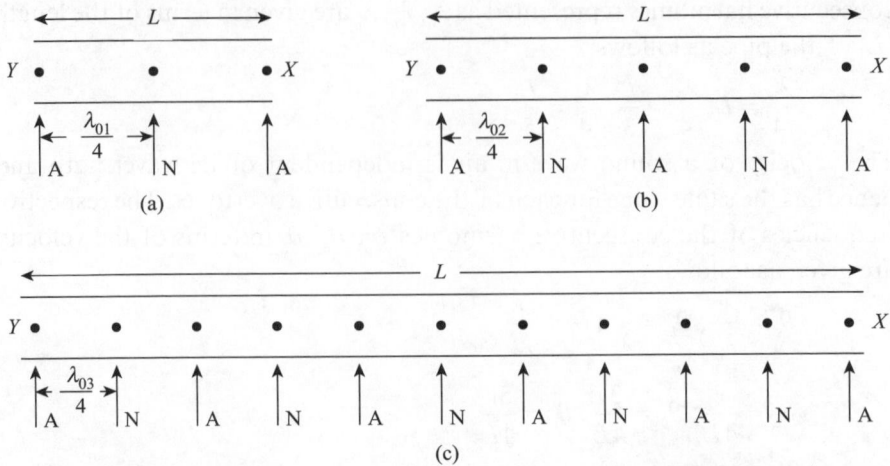

Fig. 6.24 Modes of vibration in open pipe (a) Fundamental (first harmonic) (b) First overtone (second harmonic) (c) Second overtone (third harmonic)

$$\vartheta_a = \frac{v}{\lambda_a}; \; \vartheta_b = \frac{v}{\lambda_b}; \; \vartheta_c = \frac{v}{\lambda_c}$$

$$\vartheta_a = \frac{v}{2L}; \; \vartheta_b = \frac{2v}{2L}; \; \vartheta_c = \frac{3v}{2L}$$

We can observe that higher harmonics can be expressed in terms of the fundamental frequency. In addition, it is of interest that an *open pipe can produce both even and odd harmonics*. Thus, in general, the frequency of the n*th* harmonic for an open pipe can be represented as follows:

$$\vartheta_n = \frac{nv}{2L} \; (n = 1, 2, 3, 4, 5, \ldots)$$

1.   Fundamental (harmonic)
2.   First overtone (second harmonic)
3.   Second overtone (third harmonic)

End correction in pipes   In actual practice, vibrations at an open end of a sounding pipe extend into the free air just outside. Thus, the displacement antinode gets reallocated to a further distance, d, beyond the end of the tube of length L. The effective length of the closed pipe becomes $(L + d)$ and that of an open pipe becomes $(L + 2d)$. Using geometry, one can show that $d = 0.6r$ approximately where r is the radius of the pipe

Thus, for pipes with increasing radius or wide pipes, the end correction becomes significantly meaningful.

6.15 REFLECTION AND TRANSMISSION AT BOUNDARY

As a wave travels along a string, it can encounter two situations; in one, both ends of the string are fixed, and in the other, one end is fixed and the other end is free.

6.15.1 Both Ends of String Fixed

As a pulse travels along a stretched string and strikes a fixed end, it exerts a force on the support. By virtue of Newton's third law of motion, there is an equal but oppositely directed force on the string, which generates a wave in the opposite direction. Thus, the incident signal is reflected from the fixed end of the string. The incident and the reflected waves superimpose destructively to produce a point of zero displacement or nodal point at the fixed end. Mathematically, it can be represented as follows:

For $x = 0$, the string is fixed, that is,

$$y = 2a \sin(k \cdot 0) \cos(\omega t) \text{ at } x = 0 \,; \quad y = 2a \sin(k \cdot L) \cos(\omega t) \text{ at } x = L$$

Thus, the values of wavelength can be ascertained as follows:

$$\sin(kL) = 0 = \sin(n\pi)$$

Or $\quad \sin((2\pi/\lambda) \cdot L) = 0 = \sin(n\pi)$

$$\lambda = \lambda_n = (2L/n), \, n = 1, 2, 3, \ldots$$

Thus, the corresponding frequencies are given as follows:

$$\vartheta_n = v/\lambda_n = ((n\vartheta)/2L) \, n = 1, 2, 3 \ldots$$

Thus, a string that is fixed at both ends can vibrate with certain well-defined frequencies. The fundamental mode is defined by $n = 1$, wherein the value of the wavelength is defined as $2L$. Subsequent frequencies or wavelengths are defined by the first harmonic ($\lambda = 2L/2$), second harmonic ($\lambda = 2L/3$), and so on.

Hence, a reflection at a fixed end generates a reflected wave that undergoes a phase change of π. The two fixed ends of a string are nodal points.

6.15.2 One End of String Fixed

As a pulse travels along a stretched string, it passes through the free end, which can be represented by a light ring or a pulley. Owing to the force exerted by the pulse on the ring or pulley, it experiences an acceleration, which in turn introduces a reaction force on the string. This gives rise to a pulse that travels in a direction opposite to the incident wave in the string.

Hence, there is a reflection at the free end as well, but it generates a reflected wave/pulse that does not undergo a phase change of π. The fixed end of the string has a node, whereas the free end has an antinode.

APPLICATIONS

1. Applications of the wave oscillations are manifold. One of the most common examples is the standard time piece machine that we have in our daily lives—a clock.

2. The planetary motion of the celestial bodies represents the oscillatory behaviour or periodic motion, where the masses involved are heavy in nature.
3. A major application by the understanding of oscillations is the precise calculation of the orbital motion of manmade satellites and observatory robot missions to the various planets; the most remarkable being the Mars Orbiter Mission by India.
4. Musical instruments such as sitar, flute, jal tarang, guitar, saxophone, and many others generate the tunes using the principles of standing waves, closed pipes, open pipes.
5. The idea of chaos has been used to make many random motion toys.

IMPORTANT FORMULAE

1. Time period for a simple pendulum:

$$T = 2\pi \sqrt{(l/g)}$$

2. Velocity decay exponentially in the presence of a frictional force:

$$v(t) = v_0 e^{-t/\tau}$$

3. Logarithmic decrement for a damped oscillator:

$$\theta = \frac{1}{n} \log_e \frac{a_1}{a_2}$$

4. Average power dissipated:

$$< P(t) > = \frac{< E(t) >}{\tau}$$

5. $Q = 2\pi \dfrac{\text{Energy stored}}{<\text{Energy loss in one time period}>}$

$$= 2\pi \frac{E}{(P/\vartheta)}, \text{ quality factor}$$

6. Quality factor for a driven oscillator:

$$Q = \frac{(A_0)_{max}}{(A_0)_{static}}$$

7. Power of a driven oscillator:

$$P = \frac{1}{2} m\tau \left(\frac{F_0}{m}\right)^2$$

8. Reduced mass for coupled oscillations:

$$\mu = \frac{m_1 m_2}{m_1 + m_2}$$

9. Frequency associated with the nth harmonic for an open pipe:

$$\vartheta_n = \frac{nv}{2L}$$

10. Frequency for a string:

$$\vartheta_n = \frac{n}{2L} \sqrt{\frac{T}{\mu}}$$

11. Frequency associated with the nth harmonic for a closed pipe:

$$\vartheta_n = \frac{nv}{4L}$$

12. Wave equation:

$$\frac{d^2 y(x,t)}{dx^2} = \frac{1}{v^2} \frac{d^2 y(x,t)}{dt^2}$$

13. Equation for a standing wave:

$$y_R = 2a \sin(kx) \cos(\omega t)$$

SELF-ASSESSMENT

Multiple-choice Questions

6.1 Which of the following waves does not carry energy?
    (a) Longitudinal vibration
    (b) Stationary waves
    (c) Electromagnetic waves
    (d) Transverse progressive waves

6.2 For the points of equilibrium on the potential energy curve,

(a) $\dfrac{dU}{dx} = 0$ (c) $\dfrac{dU}{dx} < 0$

(b) $\dfrac{dU}{dx} > 0$ (d) none of these

6.3 A small displacement from the point of equilibrium results in

(a) $\dfrac{dU}{dx} > 0$ (c) both (a) and (b)

(b) $\dfrac{dU}{dx} < 0$ (d) $\dfrac{dU}{dx} = 0$

6.4 Acceleration of a particle is maximum when its displacement is maximum—this statement is true for
(a) a simple harmonic motion   (c) a linear motion
(b) an oscillatory motion   (d) all of these

6.5 The total energy of a particle executing a simple harmonic motion is directly proportional to the
(a) inverse of the amplitude   (c) square of the amplitude
(b) square root of the amplitude   (d) amplitude

6.6 For a particle to execute simple harmonic motion the force should be
(a) constant
(b) proportional and opposite to the displacement
(c) opposite to the displacement
(d) proportional to its displacement

6.7 The total energy of a simple harmonic oscillator is
(a) minimum at the extreme positions
(b) minimum at the mean positions
(c) maximum at a point where potential energy is equal to the kinetic energy
(d) the same at every point

6.8 If a particle is made to oscillate along the diameter of the earth in a cavity, it
(a) oscillates between the two ends of the cavity
(b) comes to a stop at the extreme end of the cavity
(c) comes to rest at the centre of the earth
(d) comes out of the extreme end of the cavity

6.9 For a linear harmonic oscillator, the driving forces
(a) are superposed linearly   (c) cannot be superposed
(b) are superposed quadratically   (d) are neglected

6.10 In a simple pendulum,
(a) the radial component gives the restoring force, and the tangential component gives the centripetal acceleration
(b) the radial component gives the centripetal acceleration, and the tangential component gives the restoring force
(c) the oscillatory behaviour cannot be resolved into two components
(d) the oscillatory behaviour is only due to the back and forth motion of the system

6.11 For small displacements in a physical pendulum,
(a) the restoring torque is proportional to the angular displacement and is oppositely directed
(b) the restoring torque is indirectly proportional to the angular displacement and is oppositely directed

(c) the restoring torque is not related to the angular displacement

(d) dead beat formation takes place

6.12 The principal for a compound pendulum is interchangeability of the

(a) centre of gravity and the centre of suspension

(b) centre of gravity and the centre of oscillation

(c) centre of oscillation and the centre of suspension

(d) centre of suspension and the centre of oscillation

6.13 The fundamental note produced in an open pipe is represented by

(a) $\lambda/2$ (b) $\lambda/4$ (c) $\lambda/8$ (d) 2λ

6.14 Superposition of two simple harmonic motions of equal time period and equal amplitude with a phase difference $\varphi = \pi/2$ forms

(a) a circle   (c) a parabola

(b) an ellipse   (d) a straight line

6.15 The time for which the amplitude becomes A/e in the case of a damped oscillation is

(a) $t = 2m/b$ (c) $t = b^2/2m$

(b) $t = b/2m$ (d) $t = 2m/b^2$

6.16 In a simple harmonic motion, the particle is

(a) always accelerated

(b) alternately accelerated and retarded

(c) always retarded

(d) neither accelerated nor retarded

6.17 The pressure at the nodes in a stationary wave in air

(a) varies between the maximum and the minimum

(b) is initially maximum and then minimum

(c) is maximum

(d) is minimum

6.18 In a stationary wave, the physical characteristic that changes at the antinodes is

(a) pressure

(b) neither pressure nor density

(c) density

(d) pressure and density both

6.19 The effect of the frictional force on an oscillatory motion is that the velocity

(a) increases exponentially with time

(b) decreases exponentially with time

(c) is unaffected by the passage of time

(d) first increases and then decreases as time passes

6.20 The average power dissipated is

(a) negative of the average work done by the frictional force

(b) negative of the average energy for the system

(c) the rate of change of energy of the system

(d) the rate of change of work done

6.21 The transient period is the time for which the

(a) force oscillations die out and the steady-state motion is established

(b) force oscillations and the steady-state motion take place together

(c) force oscillations phase out

(d) steady-state motion phases out

6.22 For a higher damping factor, the resonance

(a) is very high   (c) is flatter

(b) is unaffected   (d) varies linearly

Review Questions

6.1 Define the following terms: (a) Periodic motion (b) Oscillatory motion (c) Linear harmonic oscillations (d) Time period (e) Phase (f) Angular frequency (g) Amplitude (h) Resonance (i) Damped and undamped oscillations (j) Forced oscillations

6.2 Explain the characteristics of a periodic function. Illustrate with examples.

6.3 How is periodic motion related to the potential energy of a particle?

6.4 What are points of stable and unstable equilibrium?

6.5 Explain the difference between a harmonic and an anharmonic oscillator.

6.6 Establish the equation of motion for a linear harmonic oscillator.

6.7 Give an example of the periodic motion that is not oscillatory in nature.

6.8 Is it necessary that all simple harmonic motions be periodic in nature and vice versa? Illustrate with examples.

6.9 Explain damping. On what factors does it depend?

6.10 How is the natural frequency affected by damping?

6.11 Derive the relation between displacement and frequency of a particle executing simple harmonic motion.

6.12 Show that the average kinetic energy and average potential energy of an oscillator are equal. In addition, show that the total energy of an oscillator is proportional to the square of its amplitude.

6.13 Derive the relation between the restoring force and potential energy of a system. Further, enunciate how a harmonic oscillator is different from an anharmonic oscillator.

6.14 Derive an expression for the total energy of a particle executing simple harmonic motion.

6.15 Establish the equation of motion for an oscillator experiencing the frictional force only. Further, derive that the frictional force decreases the velocity of the oscillating particle exponentially.

6.16 How can a pendulum be used to trace a sinusoidal curve?

6.17 Establish the equation of motion of a two-body oscillatory system using the concept of reduced mass.

6.18 Derive the equation for standing waves.

6.19 Discuss the various types of waves. Describe the propagation mechanism of transverse and longitudinal waves.

6.20 Derive the wave equation of motion of a transverse wave on a string.

6.21 Show that the total energy for a plane progressive wave is independent of both x and t.

6.22 Derive the wave equation and explain its physical significance.

Numerical Problems

6.1 The differential equation of an oscillator is

$$\frac{d^2x}{dt^2} + 2\gamma\frac{dx}{dt} + \omega_0^2 x = 0$$

If we assume that the $\omega_0 \gg \gamma$, then calculate the time in which (a) the amplitude becomes $(1/e)$ of its initial value, (b) energy becomes $(1/e)$ of the initial value, and (c) energy becomes $(1/e^4)$ of its initial value.

6.2 Establish that the frequency of a damped oscillator having quality factor Q reduces by $12.5Q^2\%$.

$$\left[Hint:\ \omega^2 = \left(\omega_0^2 - \left(\frac{1}{4\tau^2} \right) \right) \right]$$

6.3 A pendulum oscillates 150 times in a second. The quality factor of the pendulum is 1500. Calculate the time in which the amplitude of the pendulum decays to $(1/e^6)$ of its initial value. $\left[Hint:\ Q = \omega_0 \tau \right]$

6.4 The amplitude of a damped harmonic oscillator reduces from 10 to 1.0 cm after 50 oscillations, each of period 2.3 s. Calculate the logarithmic decrement of the system ($\log_e 10 = 2.30$).

$$\left[Hint:\ \theta = \frac{1}{n} \log_e \frac{a_1}{a_2} \right]$$

6.5 The velocity of a 50 g particle is decreased from 50 to 10 cm/s in 23 s. Calculate the (a) relaxation time (b) damping force when its velocity is 20 cm/s (c) time in which its kinetic energy is reduced to one-fifth of its initial value (d) total distance travelled when its initial velocity is 10 cm/s.

$$\left[Hint:\ v(t) = v_0 e^{-t/\tau};\ \frac{d^2x}{dt^2} + \frac{1}{\tau}\frac{dx}{dt} = 0 \right]$$

6.6 A wave along a string has the following equation (x in metres and t in seconds):
$y = 0.01 \sin(20t - 2.0x)$ m
Find the amplitude, frequency, speed, and wavelength.

$$\left[Hint:\ y = a \sin(\omega t \pm kx) \right]$$

6.7 A travelling wave on a string has a frequency of 20 Hz and a wavelength of 30 cm. Its amplitude is 1 mm. Write the equation of the wave in SI units.

$$\left[Hint: = a \sin(\omega t \pm kx) \right]$$

6.8 A harmonic oscillator of quality factor 10 is subjected to a sinusoidal applied force of frequency 3 and half times the natural frequency of the oscillator. Calculate the (a) amplitude of the forced oscillation in terms of its maximum amplitude and (b) the angle θ by which the oscillator will be out of phase with the driving force. Assume that the damping factor is small.

$$\left[Hint:\ Q = \omega_0 \tau;\ A_0 = \frac{(F_0/m)}{\sqrt{((\omega_0^2 - \omega^2)^2 + (\omega/\tau)^2}};\ \tan\varphi = \frac{(\omega/\tau)}{(\omega_0^2 - \omega^2)} \right]$$

6.9 A pendulum oscillates 150 times in a second. The quality factor of the pendulum is 10. Calculate the time in which the amplitude of the pendulum decays to $(1/e^2)$ of its initial value. $\left[Hint:\ Q = \omega_0 \tau;\ A(t) = x_0 e^{-t/2\tau} \right]$

Architectural Acoustics

Learning Objectives

After studying this chapter, students will be able to

- understand the characteristics of different types of sounds
- realize the characteristics of musical sounds
- comprehend the concept of intensity of sound
- define the Weber–Fechner law
- understand the different aspects of reverberation
- elucidate Sabine's formula
- derive the expression for reverberation time
- explain room acoustics in detail
- define absorption coefficient and describe its experimental determination
- appreciate the different aspects of Acoustics quieting

List of Symbols

y = Displacement	E = Energy	α = Absorption
λ = Wavelength	I = Intensity	coefficient
a = Amplitude	L = Loudness	P = Power
ρ = Density	S = Stimulus	t = Time
m = Mass	r = Radius	T = Reverberation time
v = Velocity	θ = Angle	
υ = Frequency	$d\Omega$ = Solid angle	

7.1 INTRODUCTION

Like many other physical phenomena, sound can have positive and negative impacts on life. Pleasant sounds, such as hymns and music, can improve the quality of life, whereas noise can have extremely unpleasant consequences. Very often, the audio experience is dependent upon the acoustic design of the auditorium or building. Thus, architectural acoustics is very important for improving the quality of the audio experience. This chapter deals with architectural acoustics. The treatment would begin with the classification of sound. Any person who is fond of music would know the difference between sound that is only loud and sound that is loud and clear. Like light, intensity of sound is an important parameter that can be used to express the loudness of sound using Weber–Fechner law, discussed later in this chapter. All of us have experienced the presence of sound even after the source emitting it has been turned off. Although this phenomenon may be pleasant in more serene environments like hill stations, it can be extremely irritating and can interrupt effective communication. The phenomenon of reverberation that is responsible for this effect will be discussed in this chapter. We will also discuss the methods to control it.

7.2 CLASSIFICATION OF SOUND

Commonly, sound is classified into two types: *ordered* and *disordered*. The distribution of instantaneous pressure and frequency is different for ordered and disordered sounds.

7.2.1 Ordered Sound

The instantaneous sound follows a regular pattern in ordered sound. A frequency analysis of an ordered sound reveals a definite structure. Ordered sound generally consists of a fundamental frequency, along with a series of overtones whose frequencies are integral multiples of the fundamental frequency. These overtones are referred to as *harmonics*. Figure 7.1 shows the waveform and spectrum of the sound produced by a violin, a string instrument.

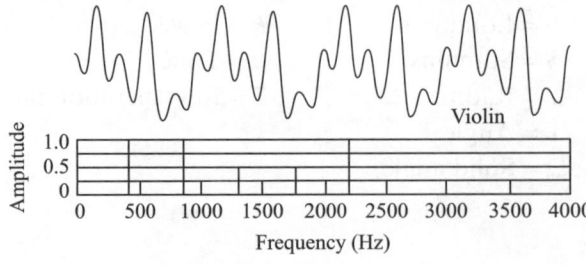

Fig. 7.1 Waveform and spectrum of sound produced by violin

7.2.2 Disordered Sound

Disordered sound has a random pattern of instantaneous pressure along with a non-definite frequency spectrum. A frequency analysis of disordered sound will have contributions from the entire range of audible frequencies.

7.3 CHARACTERISTICS OF MUSICAL SOUNDS

A musical sound has the following important characteristics:
1. Pitch
2. Loudness
3. Quality

7.3.1 Pitch

As discussed in Section 7.2.1, ordered sound consists of a definite structure. In a graphical representation, this would appear as a periodic repetition of a form, as shown in Fig. 7.1. Frequency is the number of cycles of this form per unit time. The sensation that a sound wave produces in the brain is called its pitch. Frequency of the sound wave is the cause that produces the specific sensation or pitch in the brain. As the frequency of the musical sound increases, the pitch of the sound also increases. Figure 7.2 shows a typical musical scale. In the musical scale shown, the pitch of sound also increases as we go towards higher notes. This sensation is registered by a trained brain.

Note	C (sa)	D (re)	E (ga)	F (ma)	G (pa)	A (dha)	B (ni)	C (sa)
Frequency (Hz)	256	288	320	341.3	384	426	480	512

Fig. 7.2 Typical musical scale

7.3.2 Loudness

The magnitude of an auditory sensation that is produced in the ear is the loudness of sound. The loudness of sound depends upon the frequency of the sound. A quantitative assessment of loudness is indicated by a parameter called *loudness level*. Loudness level is defined in terms of the sound pressure and frequency of a pure tone, using a 1000 Hz tone as a reference. It is measured in units of *phons*. Loudness level of a sound is numerically equal to the sound-pressure level, in decibels, of the 1000 Hz reference tone, which produces the same sensation in a listener. Usually, loudness is graphically indicated by drawing contours of equal loudness, also called Fletcher–Munson contours. Some typical Fletcher–Munson contours are shown in Fig. 7.3.

Loudness is related to the intensity of sound waves. This relationship is called the Weber–Fechner law and will be dealt with in Section 7.5.

From Fig. 7.3, one can conclude that a 500 Hz tone with a sound level of 25 dB produces the same sensation of loudness as a 50 Hz tone with a sound level of 64 dB. In fact, both the tones have a loudness level of 20 phons.

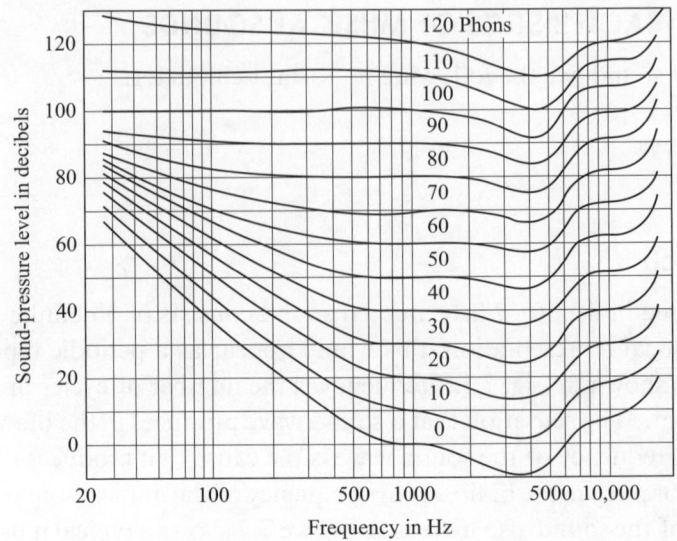

Fig. 7.3 Fletcher–Munson curves

7.3.3 Quality

Pitch and loudness do not completely define the sensation produced by sound. Sounds having the same pitch and loudness can produce different sensations. The characteristic of sound that enables us to differentiate between sounds of the same pitch and loudness is its *quality*. Two musical instruments playing the same note with identical pitch and loudness sound different due to a difference in quality. *The number of overtones present along with the relative intensity and pitch determines the quality of sound.*

7.4 INTENSITY OF SOUND

Suppose a sound wave is travelling in a particular direction. The travelling sound wave is carrying acoustic energy. Let us place an obstruction of unit area normal to the direction of propagation of sound. *Intensity of sound* at any point is then the average rate of flow of acoustic energy through this unit area. Intensity of sound varies inversely with the square of the distance from the source.

Sound can mathematically be expressed as a progressive wave. Such a progressive wave travelling in the x-direction can be represented by the following equation:

$$y = a\sin\frac{2\pi}{\lambda}(\upsilon t - x) \tag{7.1}$$

where y represents the displacement, a the amplitude, and υ the velocity of the wave. The particles within the medium are subjected to the travelling sound wave. The velocity of a particle in the medium can be written as follows:

$$\frac{dy}{dt} = a\cos\frac{2\pi}{\lambda}(\upsilon t - x)\cdot\left[\frac{2\pi\upsilon}{\lambda}\right] \tag{7.2}$$

Suppose the density of the medium is ρ. Mass dm of an elemental layer of unit area and thickness dx is then given by the following relation:

$$dm = \rho \, dx \tag{7.3}$$

This elemental mass has a kinetic energy associated with it. This is given by the following equation:

$$d(\text{K.E.}) = \frac{1}{2} dm \left(\frac{dy}{dt} \right)^2 \tag{7.4}$$

Substituting $\dfrac{dy}{dt}$ from Eq. (7.2) and dm from Eq. (7.3) into Eq. (7.4), we get the following expression:

$$d(\text{K.E.}) = \frac{1}{2} \rho a^2 \left[\frac{2\pi \upsilon}{\lambda} \right]^2 \cos^2 \frac{2\pi}{\lambda} (\upsilon t - x) dx \tag{7.5}$$

The frequency v of the sound wave is given as follows:

$$v = \frac{\upsilon}{\lambda} \tag{7.6}$$

Substitution of Eq. (7.6) into Eq. (7.5) yields the following relation:

$$d(\text{K.E.}) = 2\pi^2 a^2 \rho \upsilon^2 \cos^2 \frac{2\pi}{\lambda} (vt - x) dx \tag{7.7}$$

Maximum value of $(K \cdot E)$ of the elemental layer is expressed as follows:

$$[d(\text{K.E.})]_{\text{max}} = 2\pi^2 a^2 \rho \upsilon^2 dx \tag{7.8}$$

The total energy E of the elemental layer is the sum of kinetic energy and potential energy associated with this layer. The potential energy values must reach zero as the kinetic energy reaches its maximum value. Thus, the maximum kinetic energy also represents the total energy associated with the elemental layer. The total energy dE associated with the elemental layer is given by the following equation:

$$dE = [d(\text{K.E.})]_{\text{max}} = 2\pi^2 a^2 \upsilon^2 \rho dx \tag{7.9}$$

Energy contained in length v can be obtained by integrating Eq. (7.9) between suitable limits. Thus,

$$E = \int_o^v dE = \int_o^v 2\pi^2 a^2 \upsilon^2 \rho dx \tag{7.10}$$

resulting in

$$E = 2\pi^2 a^2 \upsilon^2 \rho \int_o^v dx \tag{7.11}$$

yielding $E = 2\pi^2 a^2 \upsilon^2 \rho v$ (7.12)

The progressive wave represented by Eq. (7.1) covers a distance v in 1 s. Thus, expression (7.12) represents the energy flowing per second across a unit area held perpendicular to the direction of propagation. Expression (7.12) is the intensity of the sound wave. Intensity I can then be written in the following form:

$$I = 2\pi^2 a^2 \rho v^2 \upsilon \tag{7.13}$$

The following conclusions can be drawn from Eq. (7.13):
1. Intensity of sound waves is proportional to the square of the frequency.
2. Intensity is proportional to the square of the amplitude.
3. Intensity is directly proportional to the density of the medium.
4. Intensity is directly proportional to the velocity.

Intensity, which is energy per second per unit area, is expressed in units of W/m². It is, however, a common practice to use relative intensity instead of absolute intensity. In relative terms, the intensity level is expressed in terms of the ratio of absolute intensity to standard reference intensity. Mathematically, this can be expressed as follows:

$$\text{Intensity level } (I) = \frac{\text{Intensity } (I_1)}{\text{Reference intensity } (I_o) \text{ standard}} \tag{7.14}$$

Generally, the reference intensity I_0 is taken to be 10^{-12} W/m² at a frequency of 1000 Hz, which is also the threshold of audibility.

It is common to express the intensity level in the logarithmic scale and write it in the units of *bel*. Thus,

$$I(\text{in bel}) = \log_{10} \frac{I}{I_0} \tag{7.15}$$

If, $I/I_0 = 10$, then Eq. (7.15) yields the following value:

$$I = \log_{10} 10 = 1 \text{ bel} \tag{7.16}$$

For practical purposes, bel is too large a unit to use regularly. It is common to use (1/10) of a bel, also called 1 decibel, to describe the intensity. It is clear that

$$1 \text{ bel} = 10 \text{ decibels} = 10 \text{ dB} \tag{7.17}$$

Example 7.1 The intensity of a source of sound is increased by a factor of 30. Calculate the increase in the intensity level in decibel unit.

Solution Suppose I and L_1 represent the initial intensity and sound level, respectively. Let L_2 represent the final sound level.
We can then write that

$$L_1 = 10 \log_{10} (I/I_0) \tag{7.18}$$

and

$$L_2 = 10 \log_{10} \left(\frac{30 I}{I_0} \right) \tag{7.19}$$

Using Eqs (7.18) and (7.19), we get the following expression:

$$L_2 - L_1 = 10 \log_{10} \left(\frac{30 I}{I_0} \right) - 10 \log_{10} \left(\frac{I}{I_0} \right)$$

leading to

$$L_2 - L_1 = 10 \log_{10} (30) + 10 \log_{10} \left(\frac{I}{I_0} \right) - 10 \log_{10} \left(\frac{I}{I_0} \right)$$

which results in

$$L_2 - L_1 = 10 \log_{10} 30 = 14.8 \, dB$$

Example 7.2 Determine the percentage change in intensity represented by 3 dB.
Solution Sound level, L, can be expressed as follows:

$$L = 10 \log_{10}(I/I_0) \text{ decibel} \tag{7.20}$$

For a loudness increase of 3 dB, we can write the following equation using Eq. (7.20):

$$3 = 10 \log_{10}(I/I_0)$$

leading to $\log_{10}(I/I_0) = \dfrac{3}{10} = 0.3$

yielding $(I/I_0) = \text{Antilog}(0.3) = 2$

Thus, $I = 2I_0$

A 3 dB increase in loudness, therefore, represents a 100% increase in intensity.

Example 7.3 For a given intensity I, $I/I_0 = 2$, where I_0 represents a reference intensity. Calculate the intensity in bel.
Solution The intensity is calculated as follows:

$$I(\text{bel}) = \log_{10} I/I_0 \tag{7.21}$$

Substituting the given value of I/I_0 in Eq. (7.21), we get the following relation:

$$I(\text{bel}) = \log_{10} 2 = 0.30 \, \text{bel}$$

Example 7.4 For the problem given in Example 7.3, calculate the intensity in decibels. In addition, calculate the intensity in decibels if $I/I_0 = 4$.
Solution The intensity is calculated as follows:

$$I(\text{dB}) = 10 \log_{10} I/I_0 \tag{7.22}$$

Using the given value of I/I_0 in Eq. (7.22), we get the following expression:

$$I(\text{dB}) = 10 \log_{10} 2 = 3 \, dB$$

For $I/I_0 = 4$, we get the following value: $I(\text{dB}) = 10 \log_{10} 4 = 6 \, dB$

Notice that the intensity doubles every 3 dB.

Example 7.5 Guess the ratio I/I_0 if the intensity is 12 dB.
Solution As indicated in Example 7.4, I/I_0 doubles every 3 dB.

Thus, for intensity to be 12 dB, $I/I_0 = 2 \times 2 \times 2 \times 2 = 16$.

Example 7.6 Express a 240% increase in intensity in units of decibel.
Solution Sound level L is represented as follows:

$$L = 10 \log_{10}(I/I_0) \tag{7.23}$$

For the given situation, we have $L = 10 \log_{10}\left(\dfrac{3.4 I_0}{I_0}\right)$

leading to $L = 10 \log_{10}(3.4) = 5.3 \, dB$

7.5 WEBER–FECHNER LAW

According to the Weber–Fechner law '*The magnitude of any sensation is proportional to the logarithm of the physical stimulus which produces the sensation.*'

Thus, mathematically,

$$L = k \log I \qquad (7.24)$$

where L is the loudness, I is the intensity of sound, and k is a constant of proportionality. Thus, the Weber–Fechner law relates the loudness of a sound wave to its intensity.

This law has its origin in a *psychophysical law* that dates back to the 19th century, which relates the perceived psychological magnitude to the logarithm of the physical stimulus. A more accurate relationship between stimulus and judgement is given by the Steven's law.

According to Steven's law, sensory perceptions correspond more closely to the powers of stimulus. Thus,

$$R = k\,(S - S_0) \qquad (7.25)$$

where S represents the stimulus, R the response, and S_0 the threshold of stimulus.

7.6 REVERBERATION

Suppose a source produces sound in a building. The sound travels at a particular velocity and reaches the listener to produce a sensation. There are, however, a lot of objects such as walls, windows, ceilings, and floors in the building. All these objects reflect sound and thus become secondary sources of sound, though of diminishing intensity. The intensity diminishes because a part of the energy is lost at every reflection. *The persistence of audible sound even after the source stops emitting sound energy is referred to as reverberation.* As the intensity of reflected sound falls below the audibility limit, the reflected sound does not produce any further sensation. The time taken for the sound to fall to a level below the minimum audibility limit is called the *time of reverberation*. Reverberation and time of reverberation would obviously be functions of frequency of a sound wave. Pioneering work in this area was done by Wallace Sabine, Harvard, in around 1900. Sabine used an organ pipe with a frequency of 512 vibrations/s to carry out his experiments.

He found the inaudibility limit to be one-millionth of the initial intensity and thus *defined reverberation time as the time taken for the sound intensity to fall to one-millionth of its initial intensity*. The time when the source of sound was switched off was taken to be zero.

The reverberation time was found to be 1–1.5 s for a hall of around 50,000 cubic feet volume, and 1.5–2 s for 400,000 cubic feet.

7.7 SABINE'S FORMULA

Reverberation time is an important factor that is used in determining the acoustic characteristics of a system. A quantitative estimate is very often an important requirement for designing an effective acoustic system. This can be accomplished using the Sabine's formula, which will be discussed in this section

Let us assume a uniform distribution of sound energy within a room. In Fig. 7.4, AB represents a plane wall and ds is a small element within the wall.

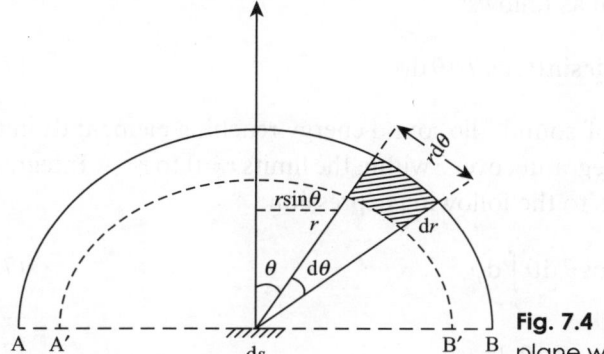

Fig. 7.4 Elemental area of plane wall

Figure 7.4 also shows two semi-circles of radii r and $(r + dr)$ drawn from the centre of the element ds. The circles have been drawn such that they lie in a plane that contains the normal to the element ds. The shaded portion in Fig. 7.4 is an elemental area between the angles θ and $(\theta + d\theta)$ lying within the two circles. The variable θ represents the angle between the radii and the direction of the normal to the element ds. The area dA of the shaded portion is expressed as follows:

$$dA = r\, d\theta\, dr \tag{7.26}$$

Suppose the area element is rotated about the normal through an angle dϕ, with the radius of rotating figure being kept at $r \sin \theta$. The distance travelled by the element area will be $r \sin \phi\, d\phi$; thus, the volume dV traced out by the elemental area would be as follows:

$$dV = r\, d\theta\, dr(r \sin\theta\, d\phi) \tag{7.27}$$

yielding $dV = r^2 \sin\theta\, d\theta\, dr\, d\phi$ \hfill (7.28)

If E represents the sound energy per unit volume, then the sound energy contained in the volume dV is given by the following expression:

$$dE = Er^2 \sin\theta\, d\theta\, dr\, d\phi \tag{7.29}$$

The sound energy per unit solid angle in any direction, dE', is given in the following form:

$$dE' = \frac{Er^2 \sin\theta\, d\theta\, dr\, d\phi}{4\pi} \tag{7.30}$$

The solid angle $d\Omega$ subtended by ds at the elemental volume dV is expressed as follows:

$$d\Omega \frac{ds\cos\theta}{r^2} \tag{7.31}$$

Using Eqs (7.30) and (7.31), the sound energy in the elemental volume that is travelling towards ds can be expressed in the following form:

$$(dE)'' = \frac{Er^2\sin\theta\,d\theta\,dr\,d\phi}{4\pi}\frac{ds\cos\theta}{r^2} \tag{7.32}$$

which can be rewritten as follows:

$$(dE)'' = \frac{Eds}{4\pi}dr\sin\theta\cos\theta\,d\theta\,d\phi \tag{7.33}$$

If v is the velocity of sound, the sound energy reaching element ds in unit time is obtained by integrating over r within the limits $r=0$ to $r=v$. Integrating Eq. (7.33) over ϕ leads to the following expression:

$$\frac{Eds}{4\pi}dr\sin\theta\cos\theta\,d\theta\int_0^{2\pi}d\phi \tag{7.34}$$

yielding $$\frac{Eds}{4\pi}dr\sin\theta\cos\theta\,d\theta.2\pi \tag{7.35}$$

implying that

$$\frac{Eds}{2}dr\sin\theta\cos\theta\,d\theta \tag{7.36}$$

Carrying out an integration of Eq. (7.36) over θ, the energy received at ds can be written as follows:

$$\frac{Eds\,dr}{2}\int_0^{\pi/2}\sin\theta\,\cos\theta\,d\theta \tag{7.37}$$

giving

$$\frac{Eds\,dr}{4}\int_0^{\pi/2}\sin 2\theta\,d\theta = \frac{Eds\,dr}{4} \tag{7.38}$$

Energy received by element ds per unit time can be obtained by integrating Eq. (7.38) between the limits $r = 0$ and $r = v$. This yields the following expression:

$$\frac{Eds}{4}\int_0^v dr = \frac{Eds\,v}{4} \tag{7.39}$$

Every surface absorbs a percentage of the incident power. Suppose that α_1 represents the absorption coefficient of the wall. Energy absorbed by ds per unit time is then given by the following equation:

$$\frac{1}{4}Ev\,\alpha_1\,ds \tag{7.40}$$

Total absorption rate at the surface of the wall is given as follows:

$$\frac{1}{4} Ev \sum \alpha_1 ds \tag{7.41}$$

where the summation has to be carried out over the complete surface of the wall or the absorbing surface.

If V is the total volume of a room, the total energy contained in the room is then EV. The rate of growth of energy is given as follows:

$$\frac{d}{dt}(EV) = V \frac{dE}{dt} \tag{7.42}$$

since the total volume of the room is independent of time.

In addition,

Rate of growth of energy in space = (Rate of supply of energy from the source) – (Rate of absorption of energy by all surfaces)

$$\tag{7.43}$$

If P represents the power from the source, then Eq. (7.43), using Eq. (7.42), can be written in the following form:

$$\frac{VdE}{dt} = P - \frac{Ev\alpha_T}{4} \tag{7.44}$$

where $\alpha_T = \Sigma \alpha_1 ds$ in Eq. (7.41) and represents the total absorption on all surfaces on which the sound energy is incident. Total absorption is expressed in units of sabin.

In steady state, $\frac{dE}{dt} = 0$, and Eq. (7.44) yields the following expression:

$$0 = P - \frac{E_m v\alpha_T}{4} \tag{7.45}$$

where E_m represents the steady-state energy density. From Eq. (7.45), we get the following relation: $P = \frac{E_m v\alpha_T}{4}$

yielding $E_m = \frac{4P}{v\alpha_T}$ $\tag{7.46}$

Using Eq. (7.44), we can write the following expression:

$$\frac{dE}{dt} = \frac{P}{V} - \frac{v\alpha_T}{4V} E \tag{7.47}$$

Assuming that $\frac{v\alpha_T}{4V} = A$, we can rewrite Eq. (7.47) in the following form:

$$\frac{dE}{dt} = \frac{4A}{v\alpha_T} P - AE \tag{7.48}$$

Multiplying both sides of Eq. (7.48) with e^{At} and rearranging, we get the following expression:

$$\left(\frac{dE}{dt} + AE \right) e^{At} = \frac{4PA}{v\alpha_T} e^{At} \tag{7.49}$$

which can be rewritten in the following form:

$$\frac{d}{dt}(Ee^{At}) = \frac{4P}{v\alpha_T}Ae^{At} \tag{7.50}$$

Integration of Eq. (7.50) yields the following equation:

$$Ee^{At} = \frac{4P}{v\alpha_T}e^{At} + \kappa \tag{7.51}$$

where K is a constant of integration. Equation (7.51) gives the expression for time dependence of energy density of sound.

7.7.1 Growth of Energy Density

Let us assume that $E = 0$ at $t = 0$. Use of this boundary condition in Eq. (7.51) results in the following relation:

$$\frac{4P}{v\alpha_T} + K = 0 \tag{7.52}$$

yielding $K = \dfrac{-4P}{v\alpha_T}$ \hfill (7.53)

Substitution of the value of K in Eq. (7.51) leads to the following expression:

$$Ee^{At} = \frac{4P}{v\alpha_T}e^{At} - \frac{4P}{v\alpha_T}$$

giving $E = \dfrac{4P}{v\alpha_T} - \dfrac{4P}{v\alpha_T}e^{-At}$

resulting in

$$E = \frac{4P}{v\alpha_T}(1 - e^{-At}) \tag{7.54}$$

From Eq. (7.46), we know that

$$E_m = \frac{4P}{v\alpha_T} \tag{7.55}$$

Substitution of Eq. (7.55) into Eq. (7.54) yields the following expression:

$$E = E_m\left(1 - e^{-At}\right) \tag{7.56}$$

A plot of Eq. (7.56) is shown schematically in Fig. 7.5.
From Fig 7.5, it is clear that the energy density increases exponentially with time and finally saturates at $E = E_m$ as $t \to \infty$.

7.7.2 Decay of Energy Density

Suppose the source producing the sound is switched off when E reaches its maximum value E_m. Let us assume that the corresponding time is $t = 0$. From Eq. (7.51), we get the following expression:

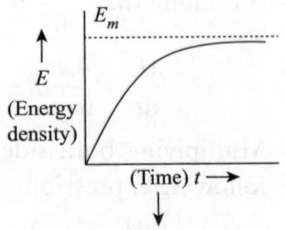

Fig. 7.5 Plot of energy density vs time

$$E_m = K \tag{7.57}$$

Since the source producing the sound has been switched off, we can put $P = 0$ in Eq. (7.51). Thus, using Eq. (7.57) in Eq. (7.51), we get the following relation:

$$E = E_m e^{-At} \tag{7.58}$$

Figure 7.6 shows a schematic representation of decay of E with time.

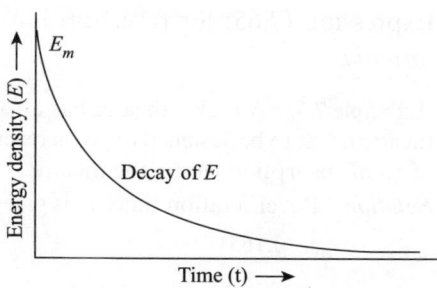

Decay of E

Fig. 7.6 Decay of energy density with time

7.7.3 Reverberation Time

Reverberation time, T, is defined as the time taken for the energy density to fall to the minimum audible value from an initial value that is 10^6 times higher. From Eq. (7.58), we can thus write that

$$\frac{E}{E_m} = 10^{-6} = e^{-AT} \tag{7.59}$$

leading to

$$e^{AT} = 10^6 \tag{7.60}$$

Logarithm of both the sides of Eq. (7.60) results in the following expression:

$$AT = 6\log_e 10 = 2.3026 \times 6 \tag{7.61}$$

Use of $A = \dfrac{v\alpha_T}{4V}$ in Eq. (7.61) leads to the following equation:

$$\frac{v\alpha_T}{4V} T = 2.3026 \times 6$$

which gives

$$T = \frac{4 \times 2.3026 \times 6}{v} \frac{V}{\alpha_T} \tag{7.62}$$

Substituting $v = 340\,\text{m/s}$ in Eq. (7.62), one gets the following expression:

$$T = \frac{4 \times 2.3026 \times 6}{340} \frac{V}{\alpha_T} \tag{7.63}$$

yielding $T = 0.165 \dfrac{V}{\alpha_T}$ (7.64)

Using $\alpha_T = \Sigma\alpha_1\, ds$ in Eq. (7.64), we can write the following expression for reverberation time T:

$$T = 0.165 \frac{V}{\sum \alpha_1 ds} \tag{7.65}$$

Expression (7.65) for reverberation time is also referred to as the *Sabine's formula*.

Example 7.7   A movie theatre has a total volume of $8000\,m^3$. The acoustics of the theatre needs to be designed to give a reverberation time of 2 s. Calculate the magnitude of total absorption within the theatre.

Solution   Reverberation time, T, is given by the following expression:

$$T = \frac{0.165V}{\alpha_T} \qquad (7.66)$$

which can be rewritten in the following form:

$$\alpha_T = \frac{0.165V}{T} \qquad (7.67)$$

Substitution of the given values in Eq. (7.67) results in the following expression:

$$\alpha_T = \frac{0.165 \times 8000}{2}$$

leading to $\alpha_T = 660$ sabin

Example 7.8   A room has a volume of $1000\,m^3$. The wall area of the room is $200\,m^2$, the total floor area is $100\,m^2$, and the total ceiling area is $100\,m^2$. The average sound absorption coefficient for the (a) walls is 0.02, (b) ceiling is 0.8, and (c) floor is 0.05. Determine the average absorption coefficient and the reverberation time.

Solution   Average absorption coefficient $\bar{\alpha}$ is given by the following expression:

$$\bar{\alpha} = \frac{\alpha_1 S_1 + \alpha_2 S_2 + \alpha_3 S_3}{S_1 + S_2 + S_3} \qquad (7.68)$$

Substituting the given values in Eq. (7.68), we get the following equation:

$$\bar{\alpha} = \frac{0.02 \times 200 + 0.8 \times 100 + 0.05 \times 100}{200 + 100 + 100}$$

yielding $\bar{\alpha} = \dfrac{89}{400} \approx 0.22 \qquad (7.69)$

Total absorption α_T can be obtained from the following expression:

$$\alpha_T = 0.22 \times 400 = 88 \text{ sabin}$$

Finally, the reverberation time, T, is given as follows:

$$T = \frac{0.165V}{\alpha_T} \qquad (7.70)$$

Using the given value of V and the calculated value of α_T in Eq. (7.70), we get the following result: $T = \dfrac{0.165 \times 1000}{88} = 1.88 \text{ s}$

Example 7.9   A conference room has a total volume of $2000\,m^3$. The magnitude of total absorption within the conference room is 100 sabin. Calculate the reverberation time.

Solution   Reverberation time T is given by the following equation:

$$T = \frac{0.165V}{\alpha_T} \qquad (7.71)$$

Using these values, we get the following result: $T = \dfrac{0.165 \times 2000}{100} = 3.3 \text{ s}$

7.8 ROOM ACOUSTICS

One of the important applications of sound is in the area of verbal communication. Apart from the characteristics of the source producing the sound, the environment in which the sound is created and heard also plays an important role in deciding the ultimate quality of sound. In this section, we will discuss some important parameters that can be used to judge the quality of the acoustics of a room, a hall, or an auditorium.

7.8.1 Clarity

Every note comprising the sound must arrive at the receiver in a clear, crisp, and unobscured manner to qualify as a clear sound. Any obstruction in the path of the sound waves as they move towards the listener can lead to distortion and therefore produce lack of clarity. Thus, the sound emanating from the source should be strong and must travel unobstructed towards the listener. This is often achieved by raising the stage level, bringing the listeners close, and making seating arrangements on a sloping floor or in balconies. Reverberation control also plays an important role in achieving clarity of sound in a room.

7.8.2 Uniformity

Listeners can be located anywhere in a given room or auditorium. A room with good acoustics would ensure that listeners hear the same quality of sound irrespective of their specific location. For example, if speakers in an auditorium are not evenly placed, the listening experience would vary greatly with the listener's location. Some common measures that can ensure uniformity are as follows:
1. There should not be any dead spots in the room.
2. A direct line-of-sight seating arrangement should be used.
3. Concave walls should be avoided since they focus sound instead of diffusing it.
4. Rectangular rooms with plain flat walls should be avoided because such situations allow sound waves to bounce back and forth repeatedly over the same path.
5. Rooms should be of irregular shape having nonparallel walls, convex surface, and several protruding edges to ensure thorough mixing of the sound energy.
6. Balancing fronts, open beams, chandeliers, etc., should be used to help diffuse sound to ensure a uniform distribution of sound energy. We must, however, remember that the critical dimensions of a structure need to be comparable to the wavelength of sound for effective diffusion of sound.

7.8.3 Envelopment

In a room with good acoustic quality, each listener must be surrounded by sound from all the sides. At the same time, the listener must be able to identify the original source of sound. To ensure this, early reflections must arrive from front, back, and side walls and the ceiling of the room. Sides and ceilings should, therefore, not be flat but must include appropriate structures to induce a large number of early reflections. This can be achieved by using surfaces that are wavy, irregular, or having contours.

7.8.4 Smoothness

Smoothness of sound refers to the absence of unpleasant roughness in the sound received by the listeners. Poorly placed concave surfaces and large flat surfaces can produce strong reflections that reach the listener more than 100 ms after the direct sound. These reflections produce distinct *echoes*. In fact, if a time interval of around 1/7th of a second exists between the times of the direct and the reflected sound reaching the listener, a distinct echo is registered by the listener. If the delay between the direct and reflected sounds is more than 30–40 ms, an unpleasant roughness can result in the sound received by the listener. If the time interval between the two is kept below 30 ms, the reflected sound appears to strengthen and lengthen the direct sound. Sound travels at a speed of 0.34 m/ms. Thus, the path length for the first reflection should not be longer than the direct path by 10 m. Proper designing must ensure that the path length for the first reflection is kept below 10 m with reference to the direct path of travel of sound for each and every listener in the room. Gaps between consecutive reflections (first and second, second and third, etc.) should also be less than 30 ms.

7.8.5 Reverberation

Reverberation is the persistence of sound in the room even after the source of sound has been switched off. In Section 7.6, we have discussed the topic of reverberation in detail. In a room with good acoustics, reverberation must be of appropriate loudness relative to the original sound and the rate of decay should also be pleasing. This can be achieved through a control of the size of the auditorium and relative amount of absorption on reflection of sound by the materials comprising the walls, ceiling, and floor. The preferred value of reverberation time is called the *optimum reverberation time*. The optimum reverberation time depends on the size of the room, loudness of sound, and the type of sound or music that is providing the stimulus to the ear.

7.8.6 Resonance

Absorption of sound energy by window panes, wooden portions, and walls lacking rigidity can lead to vibration produced in these structures. These vibrations lead to new sounds, and this phenomenon is called *resonance*. Sometimes, the frequency of the new sound may, in fact, be exactly equal to that of the original sound, thus fully justifying the name resonance given to this phenomenon.

If the original source consists of a wide spectrum of frequencies, a preferred reinforcement of certain frequencies may also take place due to this phenomenon. These effects produce distortions in the original sound produced and should therefore be adequately taken care of by suitable damping procedures.

7.8.7 Noise

Any unwanted sound is called noise. Noise produces a displeasing effect on the ear of the listener. Noise can result from sources outside a room as well as due to structures present within the room. There are three basic types of noise:
1. Air-borne noise
2. Noise due to structure or structure-borne noise
3. Inside noise
We shall now discuss each type of noise along with some suggestion regarding their control.

Air-borne noise Air-borne noise is noise that travels through air to reach the listener. Usually, the source of air-borne noise is located outside the room. Air-borne noise can be controlled in several ways, some of which are as follows:
1. Double-wall-suspended ceiling should be used.
2. Windows, doors, and partitions should be air tight.
3. Doors and windows should be mounted on separate frames, with suitable insulating material being used in between them and the mounting surface.
4. Windows and doors should be held in a mechanically strong manner.
5. Access points for pipes and ventilators should be avoided wherever possible. Otherwise, good insulation must be provided.

Structure-borne noise Noise that originates from mechanical impacts or vibration of parts of machinery is called structure-borne noise or solid-borne noise. Some common sources of such noise are drilling, cutting, hammering, and machining. This type of noise can be minimized using one or a combination of some of the following measures:
1. Use vibration isolation mounts for machinery.
2. Use rubber mats, carpets, and rubber mounts as cushions.
3. Use acoustic insulators to absorb travelling sound waves.
4. Reduce vibration level of equipment with moving parts, such as compressors and motors.

Inside noise If the source of noise is located inside the room, the noise produced is called *inside noise*. Some common sources of inside noise are printers, typewriters, air conditioners, etc. Several methods are possible to reduce or eliminate inside noise; the following are some of these methods:
1. Walls, floors, and ceilings should have sound-absorbing materials.
2. Floors should be covered by carpets.
3. All machinery should have suitable padding of wood or foam.
4. Equipment such as printers and typewriters should be placed on sound-absorbing pads or mats.

7.9 ABSORPTION COEFFICIENT

Every surface absorbs a part of the incident sound energy. The extent of absorption is determined by the absorption coefficient of the surface. *Absorption coefficient of a material is defined as the ratio of absorbed sound energy to the total incident sound energy.* Mathematically, this can be expressed in the following form:

$$Absorption\ coefficient(\alpha) = \frac{Absorbed\ sound\ energy}{Incident\ sound\ energy} \qquad (7.72)$$

Sound energy falling on an open window can pass through it completely. Therefore, an open window can be assumed to behave like a perfect absorber. *A unit area of open window is therefore taken as a standard unit of absorption.* Absorption of sound by any material can then be measured with respect to this standard.

Thus, *absorption coefficient of a material is defined to be the ratio of sound energy absorbed by a certain surface area of the material to the sound energy absorbed by an open window of the same area.*

A little reflection would tell us that the amount of sound energy absorbed by any material would always be less than that by an open window. Therefore, *absorption coefficient of a material is also defined as the reciprocal of the surface area of the material that absorbs the same sound energy as that absorbed by unit area of an open window.* For example, if 20 m² of a wooden surface absorbs the same amount of sound energy as that absorbed by 1 m² of an open window, then the absorption coefficient of the particular wooden surface is $\frac{1}{20} = 0.05$. Absorption coefficient is measured in terms of *open window unit* and expressed as OWU or *sabin*.

Table 7.1 gives the absorption coefficient values of some common materials for sound of medium pitch or frequency.

Table 7.1 Absorption coefficient of some common materials

Material	Absorption coefficient (OWU)
Brick wall (30 cm thick)	0.03
Painted brick wall	0.01
Carpets	0.15–0.30
Human body	4.3–4.7
Hair felt (2.5 cm thick)	0.58
Glass	0.02
Ordinary chair	0.17
Marble	0.01
Wooden floor	0.06

7.10 MEASUREMENT OF ABSORPTION COEFFICIENT

Absorption coefficient can be measured using two different methods. We will discuss these methods in this section.

7.10.1 Technique Based on Reverberation

In this technique, reverberation times are determined inside a test chamber. Two different measurements are carried out, one without the material under test and the other with a large sample of the material under text. Suppose T_A and T_B represent the reverberation times without and with the test material, respectively.

Using Eq. (7.65), we can write the following expressions for T_A and T_B:

$$\frac{1}{T_A} = \frac{\sum \alpha\,ds}{0.165V} \tag{7.73}$$

and

$$\frac{1}{T_B} = \frac{\sum \alpha\,ds + \alpha_t\,S_t}{0.165V} \tag{7.74}$$

Here, α represents the absorption coefficient of the empty chambers, α_t represents the absorption coefficient of test sample, and S_t is the surface area of the test sample.

From Eqs (7.73) and (7.74), we get the following relation:

$$\frac{1}{T_B} - \frac{1}{T_A} = \frac{\alpha_t S_t}{0.165V}$$

which leads to the following expression:

$$\alpha_1 = \frac{0.165V}{S_t}\left[\frac{1}{T_B} - \frac{1}{T_A}\right] \tag{7.75}$$

Equation (7.75) can then be used to determine the absorption coefficient.

7.10.2 Technique Based on Decay of Energy Density

This technique is based on the determination of decay of sound energy density at two different power levels for the test material.

Using Eq. (7.58), the decay of energy density is given as follows:

$$E = E_m e^{-At} \tag{7.76}$$

where

$$E_m = \frac{4P}{V\alpha_T} \tag{7.77}$$

and

$$A = \frac{v\alpha_T}{4V}$$

Assume that E_0 represents the minimum audible energy density. Suppose that two sources of power P_A and P_B are used, and the respective times of decay for the energy density to reach E_0 be t_A and t_B.

Then using Eqs (7.76) and (7.77), we can write that

$$E_O = \frac{4P_A}{v\alpha_T}e^{-At_A} \tag{7.78}$$

and $$E_O = \frac{4P_B}{v\alpha_T}e^{-At_B} \tag{7.79}$$

From Eqs (7.78) and (7.79), we get the following expression:

$$\frac{P_A}{P_B} = e^{A(t_A - t_B)} \tag{7.80}$$

resulting in

$$\log_e \frac{P_A}{P_B} = A(t_A - t_B) = \frac{v\alpha_T}{4V}(t_A - t_B) \tag{7.81}$$

leading to

$$\alpha_T = \frac{4V \log_e(P_A / P_B)}{v(t_A - t_B)} \tag{7.82}$$

We also know that

$$\alpha_T = \alpha_t S \tag{7.83}$$

where α_t is the absorption coefficient and S is the area of the test sample. Substitution of Eq. (7.83) into Eq. (7.82) yields the following expression:

$$\alpha_t S = \frac{4V \log_e(P_A / P_B)}{v(t_A - t_B)}$$

or $$\alpha_t = \frac{4V}{Sv}\frac{\log_e(P_A / P_B)}{(t_A - t_B)} \tag{7.84}$$

Equation (7.84) can then be used to determine the absorption coefficient of the material.

7.11 ACOUSTIC QUIETING

Most of us use several machines in our daily life. This includes refrigerators, air conditioners, washing machines, mixers, etc. They are useful in modern-day living; however, as they all contain mechanically moving parts, they can produce undesirable sound. The vibrating sound of an unbalanced washing machine and the irritating rattle of an ill-mounted air conditioner are some examples of unwanted sound encountered in our daily life. The process of controlling these is called acoustic quieting. As is true about many applications, initial developments in the field of acoustic quieting also have their origin in military applications. Submarines play a major role in naval strategy, and acoustic quieting knowledge was employed to make it difficult for sonar to detect submarines. As the field developed, it was realized that vibrations could be set up in machinery and these could in turn lead to sound waves in air, hydro-acoustic waves through water, and mechanical stresses in solid-material objects. Some of the common strategies involved in implementing acoustic quieting include absorption of vibrational energy, control of the vibration of the source generating sound waves, and redirection of the unwanted sound waves away from the observer. The entire sequence from generation to sensing of unwanted sound waves has to be thoroughly analysed before suitable measures can be initiated to implement acoustic quieting. Very often, effective acoustic quieting can be achieved by more than one method, and a cost–benefit analysis is then carried out to implement the optimum solution.

7.12 METHODS OF QUIETING

We will discuss some common methods of quieting in this section.

7.12.1 Quieting for Specific Observers

Special quieting techniques are targeted towards specific observers. These special quieting techniques are employed, in addition to general quieting techniques, to make the quieting more effective. Let us discuss a submarine as a specific example. A submarine can operate below a particular depth called the *sound channel axis*. At depths below the sound channel axis, speed of sound in water can reach the lowest values, thereby helping the submarine in evading detection. Another special technique used by submarines is refraction to hide its acoustic signature from surface-based vessels. The same principle can be used to prevent noise from reaching ground-based observers. For example, it is known that observers close to the ground will receive sound waves refracted towards them when the ground is cooler than the ambient air, but sound waves would be refracted away from them if the ground is hotter than the surrounding air. Many sources of sound do not emit sound uniformly in all directions. For example, a jet engine emits the highest-energy sound waves directly in line with the jet's exhaust. Observers, who could be the crew and passengers travelling in the aircraft, can be made to sit away from the maximum-energy locations to avoid unpleasant noise levels. Another common quieting technique is the use of ear plugs in areas such as firing range and airport.

7.12.2 Mufflers

We have all experienced the noise pollution caused by vehicles. In fact, apart from polluting gases, noise pollution from vehicles is a cause of great concern, as it can lead to health problems such as hearing loss, depression, and high blood pressure. Mufflers are devices that are used to reduce the magnitude of noise emitted by the exhaust of internal combustion engines. They are also called silencers. Mufflers were invented by Million O. Reeves. Mufflers reduce the exhaust noise through the process of absorption. Exhaust fumes are routed through a series of chambers and passages where noise reduction takes place due to sound absorption by fiberglass wool, destructive interference in resonating chambers, and other similar measures. Catalytic converters installed in vehicles also serve as mufflers. Usually mufflers in cars are located underneath the base of the car and blow the fumes backwards. Motorcycles usually incorporate exhaust/muffler in the form of long straight cylindrical barrels that finally merge into a pipe. It is also important to remember that mufflers are useful from the point of view of noise regulation and are not merely cosmetic contraptions.

7.12.3 Soundproofing

The means adopted to minimize unwanted sound and improve the quality of desirable sound is called soundproofing. Unwanted sounds can originate from the leakage of sound from adjacent rooms or external environment, reflections that cause echoes, resonances that cause reverberation, etc. Several measures are adopted to accomplish effective soundproofing. Some common techniques adopted are *distance*, *damping*, and *absorption*.

We are aware that energy density of sound waves decreases with distance from the source of sound. This fact can be utilized to create effective sound-proofing, by placing listeners away from the sources of unwanted sound. Unwanted sound can also be reduced by damping, which may involve pro-cedures such as redirection of sound or elimination of resonance by suitable designing of the room and use of obstructions. Unwanted sound can also be reduced by using suitable absorbing materials. Some common categories of sound-absorbing materials are as follows:

Porous absorbents   A porous material absorbs a large percentage of the incident sound energy. One of the main reasons for this absorption is the existence of pores in the material. The absorbed sound energy ultimately gets converted to heat. Some common porous absorbents are fibre boards, soft plasters, wood wool, asbestos fibre spray, etc.

Cavity resonators   A cavity resonator is a suitably designed chamber that has a small opening, through which sound waves can enter the cavity. However, once these waves enter the cavity, they cannot escape easily and undergo multiple reflections within the cavity, leading to their absorption. Resonators work best when they are operated at one particular frequency, and this is also their drawback. Reduction of noise due to individual sources can, however, be achieved with high efficiency using cavity resonators.

Panel absorbers   In this type of absorber, an absorbent material is fixed on a frame and the frame is then fixed on the wall, while maintaining a finite air space between the frame and the wall. As sound waves strike the panel, they result in flexural vibrations of the panel. These vibrations lead to the absorption of energy, which is ultimately converted to heat. Some common examples of panel absorbers are gypsum boards, hard board panels, wood panels, plastic boards, etc.

Composite absorbers   Composite absorbers utilize a combination of the principles of porous absorbers, cavity resonators, and panel absorbers into a combined absorber. Composite absorbers are thus designed in the form of perforated panels fixed on walls with an air space that is filled with porous absorbers. The panels may be fabricated using metal, wood, plywood, hard board, or plaster board. The striking sound waves pass through the panel and get damped by resonance of the air column in the cavity.

IMPORTANT CONCEPTS

1. Ordered sound consists of a fundamental frequency along with a series of overtones whose frequencies are integral multiples of the fundamental frequency.
2. A frequency analysis of disordered sound will have contributions from the entire range of audible frequencies.
3. A musical sound is characterized using the following parameters:
   (a) Pitch
   (b) Loudness
   (c) Quality
4. Intensity of sound varies inversely with the square of the distance from the source.
5. Intensity of sound waves is proportional to the following quantities:
   (a) Square of the frequency
   (b) Square of the amplitude.
   (c) Density of the medium.
   (d) Velocity
6. The Weber–Fechner law relates the loudness of a sound wave with its intensity.
7. The following are the three basic types of noise:
   (a) Air-borne noise
   (b) Noise due to structure or structure-borne noise
   (c) Inside noise
8. The process of controlling unwanted sound is called acoustic quieting.

APPLICATIONS

1. A young, healthy individual can hear sounds in the frequency range of 20 Hz–20 kHz, with frequency resolution of around 2%. Thus, the individual is capable of differentiating between a tone of 1000 Hz and that of 1020 Hz. In terms of mechanical movement, it is interesting to note that a sound of 1 kHz displaces the eardrum by less than 1 Å, which is much lower than the diameter of a hydrogen atom. In terms of intensity, the human ear is capable of hearing sounds over an intensity range of approximately 100,000,000–1. Thus, it can sense a soft whisper as well as the sound of a jet fighter taking off.
2. The Boston Symphony Hall designed by Wallace C. Sabine was the first scientifically designed music hall. The structure has a high, textured ceiling and two balconies extending along three walls. The hall has a reverberation time of around 1.8 s in 500–1000 Hz frequency range.
3. The most recent advancement in the field of noise control is active noise control. The principle behind this technique is to first duplicate the noise, but with a 180° phase difference. The offending noise and the duplicated out-of-phase noise are then combined for noise cancellation.

IMPORTANT FORMULAE

1. Progressive wave travelling in the x-direction is represented as follows:

$$y = a\sin\frac{2\pi}{\lambda}(vt - x)$$

2. Intensity, $I = 2\pi^2 a^2 \rho \upsilon^2 v$.
3. I (in bel) $= \log_{10} I/I_0$
4. 1 bel = 10 dB
5. $L = k \log I$
6. $E = E_m(1 - e^{-At})$
7. $E = E_m e^{-At}$

8. Reverberation time, $T = 0.165\dfrac{V}{\sum \alpha_1 ds}$

9. Absorption coefficient

$$\alpha = \frac{\text{Absorbed sound energy}}{\text{Incident sound energy}}$$

10. The Sabine's formula

$$T = 0.165\frac{V}{\sum \alpha_1 ds}$$

SELF-ASSESSMENT

Multiple-choice Questions

7.1 Pitch of a musical note is closely related to
   (a) amplitude
   (b) frequency
   (c) wave velocity
   (d) all of these

7.2 A progressive wave travelling in the +x-direction can be represented by
   (a) $y = a\sin\dfrac{2\pi}{\lambda}(vt + x)$
   (b) $y = a\sin\dfrac{2\pi}{\lambda}(vt - x)$
   (c) $x = a\sin\dfrac{2\pi}{\lambda}(vt - y)$
   (d) $x = a\sin\dfrac{2\pi}{\lambda}(vt - y)$

7.3 Intensity of sound wave is proportional to
   (a) v^{-1} (b) v^{-2} (c) v^2 (d) v

7.4 The Weber–Fechner law is represented as
   (a) $L = k \log I^2$
   (b) $L = k \log (I)^{-1}$
   (c) $L = k \log I$
   (d) $L = k \log (I)^{-2}$

7.5 Growth of sound energy density in represented by
   (a) $E = E_m e^{At}$
   (b) $E = E_m e^{-At}$
   (c) $E = E_m (1 - e^{At})$
   (d) $E = E_m (1 - e^{-At})$

7.6 The preferred value of reverberation time is
   (a) zero
   (b) infinity
   (c) optimum reverberation time
   (d) none of these

7.7 Absorption coefficient has
   (a) units of OWU
   (b) units of sabin
   (c) no units
   (d) units of cm^{-1}

7.8 Speed of sound below the sound channel axis is
   (a) highest
   (b) lowest
   (c) zero
   (d) ∞

7.9 Cavity resonators use the phenomenon of
   (a) absorption
   (b) multiple reflections
   (c) refraction
   (d) amplification

7.10 Intensity of a source of sound is increased by a factor of 100. Increase in the intensity level in decibels is
   (a) 2 dB   (b) 10 dB   (c) 20 dB   (d) 40 dB

7.11 Ordered sound consists of
(a) fundamental frequency
(b) harmonics and fundamental frequency
(c) only harmonics
(d) noise

7.12 Which of the following does not produce ordered sound?
(a) Violin
(b) Sitar
(c) Air conditioner
(d) Tabla

7.13 If the fundamental frequency is n, then which of the following is not a harmonic?
(a) $2n$
(b) $3n$
(c) $4n$
(d) $n/2$

7.14 Disordered sound has
(a) fundamental frequency
(b) harmonics
(c) harmonics and fundamental frequency
(d) entire range of audible frequencies

7.15 Which of the following is not a characteristic of musical sound?
(a) Frequency
(b) Pitch
(c) Loudness
(d) Quality

7.16 Frequency of the note C (Sa) is
(a) 240 Hz
(b) 256 Hz
(c) 270 Hz
(d) 200 Hz

7.17 Fletcher–Munson contours are contours of equal
(a) intensity
(b) pitch
(c) quality
(d) loudness

7.18 Quality of sound is decided by
(a) loudness
(b) number of overtones
(c) intensity
(d) frequency

7.19 Intensity of sound depends
(a) directly on the distance from the source
(b) directly on the square of the distance from the source
(c) inversely on the square of the distance from the source
(d) inversely on the distance from the source

7.20 Frequency v and wavelength λ are related through the expression
(a) $v = \upsilon\lambda$
(b) $v = \dfrac{\lambda}{\upsilon}$
(c) $v = \dfrac{\lambda^2}{\upsilon}$
(d) $v = \dfrac{\upsilon}{\lambda}$

7.21 Intensity of a sound wave is proportional to its
(a) amplitude
(b) $\dfrac{1}{\text{amplitude}}$
(c) $(\text{amplitude})^2$
(d) $\dfrac{1}{(\text{amplitude})^2}$

7.22 Intensity of a sound wave is proportional to its
(a) density
(b) $\dfrac{1}{\text{density}}$
(c) $(\text{density})^2$
(d) $\dfrac{1}{(\text{density})^2}$

7.23 Intensity of a sound wave has units of
(a) W
(b) W/m^2
(c) W–m^2
(d) W/m^3

7.24 Relative intensity has units of
(a) W
(b) W/m^2
(c) no unit
(d) W–s

7.25 Reference intensity is usually taken as
(a) 10^{-10} W/m^2
(b) 10^{-9} W/m^2
(c) 10^{-15} W/m^2
(d) 10^{-12} W/m^2

7.26 Reference intensity is taken to be a frequency of
    (a) 100 Hz   (c) 200 Hz
    (b) 1000 Hz   (d) 500 Hz

7.27 Intensity in bel is expressed as
    (a) $\log_{20} I/I_0$ (b) $\log_{20} I_0/I$ (c) $\log_{10} I_0/I$ (d) $\log_{10} I/I_0$

7.28 If $I/I_0 = 10$, then I in units of bel is
    (a) 1   (b) 10   (c) 2   (d) 20

7.29 1 bel is equal to
    (a) 20 dB   (b) 10 dB   (c) 0.1 dB   (d) 0.01 dB

7.30 Steven's law states that
    (a) $R = k(S - S_0)^2$ (c) $R = k(S - S_0)$
    (b) $R = k^2(S - S_0)^2$ (d) $R = \dfrac{k}{(S - S_0)}$

7.31 Inaudibility limit is around
    (a) one-hundredth of the initial intensity
    (b) one-tenth of the initial intensity
    (c) one-thousandth of the initial intensity
    (d) one-millionth of the initial intensity

7.32 Total absorption is expressed in units of
    (a) $(\text{sabines})^2$ (c) 1/sabines
    (b) sabines   (d) $1/(\text{sabines})^2$

7.33 To avoid focussing of sound, one should not use
    (a) concave walls   (c) plane walls
    (b) convex walls   (d) straight walls

7.34 Thorough mixing of sound energy can be ensured by using a
    (a) concave surface   (c) plane surface
    (b) convex surface   (d) flat surface

7.35 Effective diffusion of sound needs critical dimensions to be comparable with
    (a) λ^2 (b) $\dfrac{1}{\lambda}$ (c) λ (d) $\dfrac{1}{\lambda^2}$

Review Questions

7.1 Differentiate between noise, music, and speech.
7.2 Explain the difference between ordered sound and disordered sound.
7.3 Explain the main characteristics of a musical sound.
7.4 In what units is loudness represented?
7.5 Describe an expression for the intensity of sound waves.
7.6 Explain the relationship between the units bel and decibel.
7.7 State and explain Weber–Fechner law.
7.8 Derive Sabine's formula for reverberation time.
7.9 Demonstrate the difference between noise and music graphically.
7.10 What is noise?
7.11 What is the correlation between pitch and frequency?
7.12 How does the intensity of a sound wave change if its frequency is doubled?
7.13 State and explain Steven's law.
7.14 Derive an expression for the growth of energy density of a sound wave.
7.15 Explain the terms envelopment and resonance.
7.16 Suggest some methods for controlling air-borne noise.
7.17 Define reverberation time.

7.18 Define sound intensity.
7.19 Discuss Sabine's formula.
7.20 State four factors that affect acoustics of buildings.
7.21 List and explain the characteristics of musical sound.
7.22 Explain intensity level in terms of dB.
7.23 Classify sound based on frequency.
7.24 Suggest remedies to improve acoustics of buildings.
7.25 Suggest a method for measuring absorption coefficient.
7.26 Write a note on noise pollution.

Numerical Problems

7.1 Intensity of sound waves produced by a source is increased by a factor of 20. Determine the increase in the sound level in decibels. [*Hint*: $L = 10 \log_{10} (I/I_0)$]

7.2 A 6 dB increase in the sound level is to be achieved. Calculate the factor by which the intensity of sound has to be increased. [*Hint*: $L = 6 \log (I_2/I_1)$]

7.3 The loudness of sound waves is increased by 9 dB. Calculate the percentage change in intensity. [*Hint*: $L = 10 \log_{10} (I/I_0)$]

7.4 The intensity of a sound wave is increased by 300% of its original value. Calculate the corresponding change in loudness in decibels. [*Hint*: $L = 10 \log_{10} (I/I_0)$]

7.5 The acoustics of a lecture hall have to be so designed that the resultant reverberation time is 3 s. If the total volume of the hall is 9000 m³, calculate the magnitude of total absorption. $\left[Hint: T = \dfrac{0.165\,V}{\alpha_T} \right]$

7.6 A presentation room has a total absorption magnitude of 400 sabin and its volume is 10,000 m³. Calculate the resultant reverberation time. $\left[Hint: T = \dfrac{0.165\,V}{\alpha_T} \right]$

7.7 The total absorption for a movie theatre is 300 sabin. The reverberation time has been measured to be 3.5 s. Calculate the total volume from this information. $\left[Hint: V = \dfrac{\alpha_T T}{0.165} \right]$

7.8 An amphitheatre has the following important specifications: volume = 500 m³; wall area = 100 m²; floor area = 50 m²; ceiling area = 50 m²; and the average sound absorption coefficient for (i) wall = 0.01; (ii) ceiling = 0.4; and (iii) floor = 0.03. Calculate the (a) average absorption coefficient and (b) reverberation time. $\left[Hint: \text{(a) } \bar{\alpha} = \dfrac{\alpha_1 S_1 + \alpha_2 S_2 + \alpha_3 S_3}{S_1 + S_2 + S_3} \text{ (b) } T = \dfrac{0.165V}{\alpha_T} \right]$

7.9 For a given intensity, I, $I/I_0 = 3$ where I_0 represents a reference intensity. Calculate the intensity in bel. [*Hint*: $I = \log_{10} I/I_0$ (bel)]

7.10 For Problem 7.9, calculate the intensity in decibels. In addition, calculate the intensity in decibels if $I/I_0 = 8$. [*Hint*: I (dB) = $10 \log_{10} I/I_0$]

7.11 Guess the ratio I/I_0 if the intensity is 15 dB. [*Hint*: I/I_0 doubles every 3 dB]

7.12 Express a 350% increase in intensity in units of decibel. [*Hint*: $L = 10 \log_{10}(I/I_0)$]

7.13 A seminar room has a total volume of 1500 m³. The magnitude of total absorption within the conference room is 80 sabin. Calculate the reverberation time. $\left[Hint: T = \dfrac{0.165\,V}{\alpha_T} \right]$

Ultrasonics

Learning Objectives

After studying this chapter, students will be able to

- explain the different methods of production of ultrasonic waves
- elucidate the different techniques for detection of ultrasonic waves
- demonstrate the properties of ultrasonic waves
- describe the phenomenon of cavitation
- realize the concept of acoustic grating
- demonstrate the important industrial applications of ultrasonic waves
- comprehend the principle of operation of SONAR
- explain the different techniques of non-destructive testing
- demonstrate the important medical applications of ultrasonic waves

List of Symbols

f = Frequency of alternating current ρ = Density
v = Natural frequency λ = Wavelength
E = Young's modulus V = Velocity

8.1 INTRODUCTION

Human ears are capable of hearing frequencies in the range of 20 Hz–20 kHz. This frequency range is called the *audible range*. Sound waves having frequencies beyond 20 kHz are termed as *ultrasonic waves*. Ultrasonic waves are used in sound navigation and ranging (SONAR) applications to locate submerged objects such as icebergs and submarines. Targets and obstructions can be located, and route guidance obtained with a great degree of precision. Since ultrasonic waves possess higher energy, they have higher penetrating power. This characteristic is used in the medical field to study the functioning of internal organs of the human body using ultrasonic imaging. This chapter discusses some important aspects of ultrasonic waves.

To begin with, the different methods of production of ultrasonic waves are discussed. The relative merits and demerits of these techniques are discussed next. The chapter then elaborates on the important properties of ultrasonic waves. Finally, the various industrial and medical applications of ultrasonic waves are discussed.

8.2 PRODUCTION OF ULTRASONIC WAVES

Sound waves in the audible range are produced by the vibrating diaphragm of a loudspeaker that has been fed a suitable alternating voltage. Inductance of the loudspeaker coil offers inductive reactance to any alternating current flowing through it. As the frequency of the impressed signal increases, the reactance offered through the inductance of this coil also increases. Thus, the current flowing through the coil reduces with increasing frequency. In addition to this, with increasing frequencies, the diaphragm of a loudspeaker also loses its ability to vibrate efficiently. Both these effects prevent the loudspeaker from functioning as a useful source of ultrasonic waves. Two important methods for producing ultrasonic waves are as follows:
1. Magnetostriction method
2. Piezoelectric method

The magnetostriction method is generally used for the production of ultrasonics of up to around 100 kHz, whereas generators based on piezoelectricity are generally used for producing higher frequencies. These methods are discussed in the following sections.

8.2.1 Magnetostriction Effect

When a ferromagnetic material such as iron, nickel, or cobalt is placed in a magnetic field, it undergoes a change in dimension. This phenomenon is known as the *magnetostriction effect*. The extent of change is independent of the polarity of the magnetic field, but is dependent on its magnitude and the nature of the material. Figure 8.1 shows a schematic representation of the magnetostriction effect. A permanent magnet is used to create the magnetic field that brings about a change in length.

Suppose a ferromagnetic rod is placed inside the magnetic field of a coil carrying an alternating current. This alternating current would give rise to a time-varying field, and the rod will be put into vibration. The amplitude of this vibration is generally small. If the frequency of the alternating signal is the same as the natural frequency of the rod, a resonance will occur. This resonance will reinforce the vibrations that would result in the rod. The ends

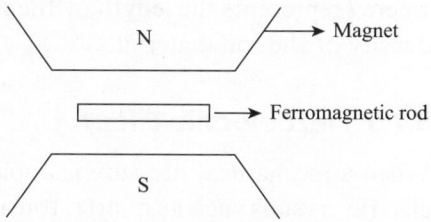

Fig. 8.1 Magnetostriction effect

of the rod would then emit sound waves. A sufficiently high applied frequency leads to the production of ultrasonic waves.

8.2.2 Magnetostriction Generator

Figure 8.2 exhibits a schematic circuit diagram of a set-up for producing ultrasonic waves using the magnetostriction method. A short nickel rod is clamped at the centre. The rod is magnetized by passing a direct current through a coil wrapped around the rod. Two other coils L_1 and L_2 are wrapped around the rod, as shown in Fig. 8.2.

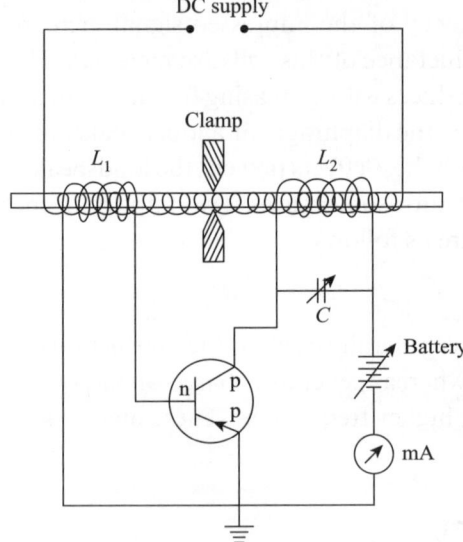

Fig. 8.2 Circuit diagram of magnetostriction method

Coil L_2 is a part of the collector circuit of the transistor circuit. Suitable biasing is provided to the transistor such that it is not reflected (see Fig. 8.2). Coil L_1 forms a part of the base circuit. Frequency (f) of the alternating current set up by the resonant circuit L_2C is given by the following equation: $f = \dfrac{1}{2\pi\sqrt{L_2C}}$

The natural frequency v of the vibrating rod is expressed as follows:

$$v = \frac{1}{2l}\sqrt{\frac{E}{\rho}}$$

where l represents the length of the rod, E the Young's modulus, and ρ is the density of the rod material.

8.2.3 Piezoelectric Effect

When a mechanical pressure is applied along the mechanical axis of piezoelectric crystals such as quartz, tourmaline, and Rochelle salt, equal amount of opposite electric charges are developed at the two end faces along the perpendicular direction, that is, along the electrical axis, as shown in Fig. 8.3. This phenomenon is known as the piezoelectric effect.

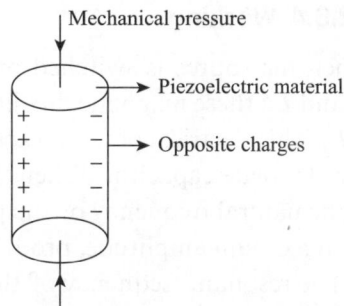

8.2.3.1 Inverse Piezoelectric Effect

When an emf is applied along the electrical axis of a *piezoelectric crystal*, the dimension of the crystal changes along the mechanical axis. This phenomenon is known as the *inverse piezoelectric effect*.

Fig. 8.3 Piezoelectric effect caused due to mechanical pressure applied along mechanical axis of piezoelectric crystals

8.2.3.2 Principle

Inverse piezoelectric effect is the principle behind the production of ultrasonics using piezoelectric crystals. In this method, ultrasound waves with high amplitudes are produced when the natural frequency of the piezoelectric crystal equals the frequency of the oscillatory circuit.

8.2.3.3 Construction

A piezoelectric oscillator circuit consists of a primary and a secondary circuit. In the primary circuit, L_1 and L_2 are two inductances connected with a p–n–p transistor through a variable source, as shown in Fig. 8.4. C_1 is a variable condenser connected parallel to L_1. This combination of L_1 and C_1 is known as a *tank circuit*. L_1 and L_2 are connected inductively with the secondary circuit.

The secondary circuit consists of an inductance L_3 and a parallel plate condenser. The quartz crystal is kept between the plates of the parallel-plate condenser.

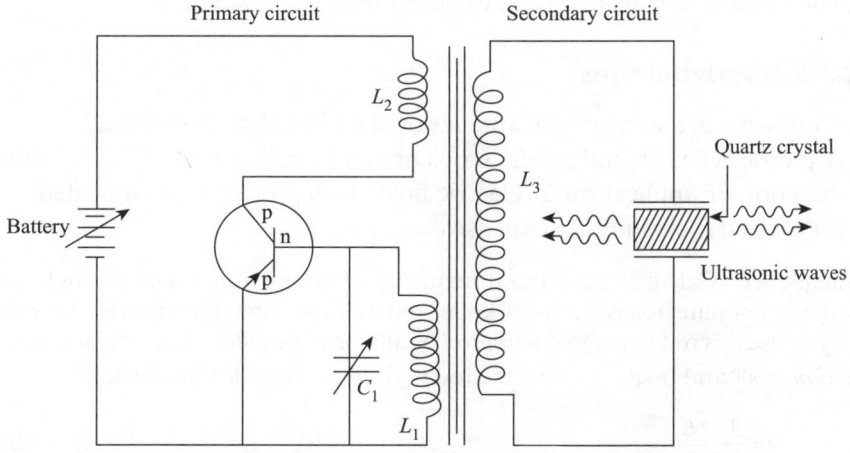

Fig. 8.4 Piezoelectric oscillator circuit

8.2.3.4 Working

When the source is switched on, magnetic flux lines are produced by coils L_1 and L_2; these magnetic flux lines induce a current in the opposite direction in L_3. This electrical potential acts on the piezoelectric crystal placed inside the parallel-plate capacitor. When the frequency of the oscillator circuit is equal to the natural frequency of the piezoelectric crystal, the crystal vibrates with the maximum amplitude, producing ultrasonic waves.

The resonant frequency of the circuit is expressed as follows:

$$v = \frac{1}{2\pi\sqrt{L_1 C_1}} \tag{8.1}$$

The natural frequency of the crystal is given by the following equation:

$$v = \frac{p}{2l}\sqrt{\frac{E}{\rho}} \tag{8.2}$$

where $p = 1, 2, 3, \ldots$ for the fundamental frequency, first overtone, second overtone, etc.; l is the length of the crystal; E represents the Young's modulus of the crystal; and ρ is the density of the crystal.

Thus, the frequency of the ultrasound produced is given by the following expression:

$$v = \frac{p}{2l}\sqrt{\frac{E}{\rho}} \tag{8.3}$$

8.2.3.5 Advantages

The following are some of the advantages of the piezoelectric method:
1. This method can produce ultrasonic waves of frequencies up to 500 MHz.
2. A piezoelectric crystal is not affected by temperature and moisture. Therefore, the ultrasound produced is of a constant frequency.
3. The method can also be used to detect ultrasonic waves.

8.2.3.6 Disadvantages

The following are some disadvantages of the piezoelectric method:
1. A piezoelectric crystal needs to be cut in a direction perpendicular to the direction of application of electric field. This process is complicated.
2. Piezoelectric crystals are expensive.

Example 8.1   Calculate the natural frequency of a pure iron rod of 40 mm length. The density of pure iron is 7.25×10^3 kg/m³ and its Young's modulus is 115×10^9 N/m². Can you use this rod in magnetostriction oscillator to produce ultrasonic waves?

Solution   Natural frequency v of the rod is given by the following relation:

$$v = \frac{1}{2l}\sqrt{\frac{E}{\rho}} \tag{8.4}$$

where l represents the length of the rod, E the Young's modulus, and ρ the density of iron. Substituting the given values in Eq. (8.4) leads to the following result:

$$v = \frac{1}{2 \times 40 \times 10^{-3}} \left[\frac{115 \times 10^9}{7.25 \times 10^3} \right]^{\frac{1}{2}}$$

Or $v = (12.5) \, (15.86 \times 10^6)^{1/2} = 49,780 \, \text{Hz} = 49.78 \, \text{kHz}$

Yes, the rod can be used for producing ultrasonic waves because its frequency lies in the ultrasonic range.

Example 8.2 A quartz crystal of length 1 mm is vibrating at resonance. Calculate its fundamental frequency. (Assume that for quartz, $E = 7.9 \times 10^{10} \, \text{N/m}^2$ and $\rho = 2650 \, \text{kg/m}^3$.)

Solution   The frequency v of vibration is given by the following equation:

$$v = \frac{p}{2l} \sqrt{\frac{E}{\rho}} \tag{8.5}$$

For the fundamental mode, $p = 1$, giving

$$v = \frac{1}{2l} \sqrt{\frac{E}{\rho}} \tag{8.6}$$

The use of the given values in Eq. (8.6) leads to the following result:

$$v = \frac{1}{2 \times 0.001} \sqrt{\frac{7.9 \times 10^{10}}{2650}} = 2.73 \times 10^6 \, \text{Hz}$$

Example 8.3   Calculate the natural frequency of a 50 mm long ferromagnetic rod whose density is 7250 kg/m³; Young's modulus of the material is $11.5 \times 10^{10} \, \text{N/m}^2$.

Solution   Length of the rod, $l = 50 \times 10^{-3} \, \text{m}$, density of the rod, $\rho = 7250 \, \text{kg/m}^3$, Young's modulus of the material, $E = 11.5 \times 10^{10} \, \text{N/m}^2$

The natural frequency of the rod is calculated as follows:

$$v = \frac{1}{2l} \left[\frac{E}{\rho} \right]^{\frac{1}{2}} = \frac{1}{2 \times 50 \times 10^{-3}} \left[\frac{11.5 \times 10^{10}}{50 \times 10^{-3}} \right]^{\frac{1}{2}} \quad v = 39.83 \, \text{kHz}$$

Example 8.4   A quartz crystal of length 2 mm is vibrating at the resonance frequency. The density of quartz is 2650 kg/m³ and its Young's modulus is $7.9 \times 10^{10} \, \text{N/m}^2$. Calculate the frequency of the ultrasonic waves produced by the piezoelectric method.

Solution   Length of the crystal, $l = 2 \times 10^{-3} \, \text{m}$, density of the crystal, $\rho = 2650 \, \text{kg/m}^3$, Young's modulus of the crystal, $E = 7.9 \times 10^{10} \, \text{N/m}^2$

The frequency of the crystal is calculated as follows:

$$v = \frac{p}{2l} \left[\frac{E}{\rho} \right]^{\frac{1}{2}}$$

Here $p = 1$, so

$$v = \frac{1}{2 \times 2 \times 10^{-3}} \left[\frac{7.9 \times 10^{10}}{2650} \right]^{\frac{1}{2}} \quad \text{or } v = 1.365 \, \text{MHz}$$

Example 8.5   A quartz crystal of length 3 mm is vibrating at the resonance frequency. The density of the material is 2500 kg/m³ and its Young's modulus is 8×10^{10}

N/m^2. Calculate the frequency of the ultrasound produced by the piezoelectric method.

Solution   Length of the crystal, $l = 3 \times 10^{-3}$ m; density of the crystal, $\rho = 2500\,kg/m^3$; Young's modulus of the crystal, $E = 8 \times 10^{10}\,N/m^2$

The frequency of the crystal is calculated as follows:

$$v = \frac{p}{2l}\left[\frac{E}{\rho}\right]^{1/2}$$

Here $p = 1$, so

$$v = \frac{1}{2 \times 3 \times 10^{-3}}\left[\frac{8 \times 10^{10}}{2500}\right]^{1/2} \quad \text{or } v = 942.80\,kHz$$

Example 8.6   A quartz crystal of length 1.5 mm is vibrating at the resonance frequency. The density of the material is $2650\,kg/m^3$ and its Young's modulus is $7.9 \times 10^{10}\,N/m^2$. Calculate the frequency of the ultrasonic waves produced by the piezoelectric method.

Solution   Length of the crystal, $l = 1.5 \times 10^{-3}$ m; density of the crystal, $\rho = 2650\,kg/m^3$; Young's modulus of the crystal, $E = 7.9 \times 10^{10}\,N/m^2$

The frequency of the crystal is calculated as follows:

$$v = \frac{p}{2l}\left[\frac{E}{\rho}\right]^{1/2}$$

Here $p = 1$, so

$$v = \frac{1}{2 \times 1.5 \times 10^{-3}}\left[\frac{7.9 \times 10^{10}}{2650}\right]^{1/2} \quad \text{or } v = 1.82 \text{ MHz}$$

Example 8.7   A ferromagnetic rod has a length of 40 mm, and the density of the material is $7250\,kg/m^3$. Evaluate the natural frequency of the rod if Young's modulus of the material is $11.5 \times 10^{10}\,N/m^2$.

Solution   Length of the rod, $l = 40 \times 10^{-3}$ m, density of the rod, $\rho = 7250\,kg/m^3$, Young's modulus of the material, $E = 11.5 \times 10^{10}\,N/m^2$

The natural frequency of the rod, v, is given by the following relation:

$$v = \frac{p}{2l}\left[\frac{E}{\rho}\right]^{1/2}$$

which yields the following result:

$$v = \frac{1}{2 \times 40 \times 10^{-3}}\left[\frac{11.5 \times 10^{10}}{7250}\right]^{1/2} \quad \text{or } v = 49.78\,kHz$$

8.3 PROPERTIES OF ULTRASONICS

Ultrasonic waves have many characteristic properties. Some of the properties of ultrasonic waves are as follows:

1. Ultrasonic waves have extremely high energy content. This is because of their high frequency. The high energy of these waves can be used in applications such as drilling and cutting.

2. Ultrasonic waves display all the characteristics of sound waves including reflection, refraction, and absorption. These properties can, in turn, be used to design systems based on ultrasonic waves. Like sound waves, reflection is possible if the obstructing surface has dimensions much larger than the wavelength.

3. Ultrasonic waves can be transmitted over long distances with extremely low loss of energy. This is possible because of their small wavelengths, which result in negligible diffraction effects.

4. Ultrasonic waves can produce intense heating as they pass through materials. This is due to absorption of the energy content of the ultrasonic waves by the medium.

5. The speed of propagation of ultrasonic waves is frequency dependent, increasing with an increase in frequency. This dependence is used in developing basic ultrasonic-wave detection systems.

6. Ultrasonic waves passing through liquids lead to the formation of bubbles.

8.4 CAVITATION

Microscopic bubbles with diameters in the range of 10^{-9}–10^{-8} m are generally present in a liquid. Reduction of pressure in regions around these bubbles leads to evaporation, resulting in the growth of the bubbles. This growth, however, is not unlimited. Ultimately, it leads to the collapse of the bubbles. All these happen within a very short span of time, just a few milliseconds. The process of collapse of the bubbles results in the generation of shock waves, and the temperature in the region of the collapse increases manifold. Ultrasonic waves passing through a liquid induce alternate regions of rarefaction and compression. Rarefaction regions are local negative-pressure regions and result in the process of bubble growth and collapse. This phenomenon is called *cavitation*. As a result of the collapse of bubbles, local pressures can reach thousands of atmospheres and local temperatures increase by as much as 10,000°C.

The phenomenon of cavitation can be used for the following applications:
1. Ultrasonic cleaning
2. Exploration of minerals and oil deposits
3. Speeding up of chemical reactions
4. Emulsification
5. Formation of stoichiometric alloys and compounds

8.5 DETECTION OF ULTRASONICS

Human ears do not respond to ultrasonic waves, unlike that of some animals (e.g., bat). We, therefore, need special methods to detect ultrasonic waves. These methods are discussed in detail in this section.

8.5.1 Piezoelectric Detectors

Piezoelectric crystals have the ability to develop an electric potential when a stress is applied across certain faces of the crystal. This phenomenon can be used to detect ultrasonic waves. One pair of faces of a quartz crystal (piezoelectric material) is subjected to ultrasonic waves, as shown in Fig. 8.5.

An alternating potential then develops across the perpendicular faces. This potential can be amplified and measured to detect the presence of ultrasonic waves.

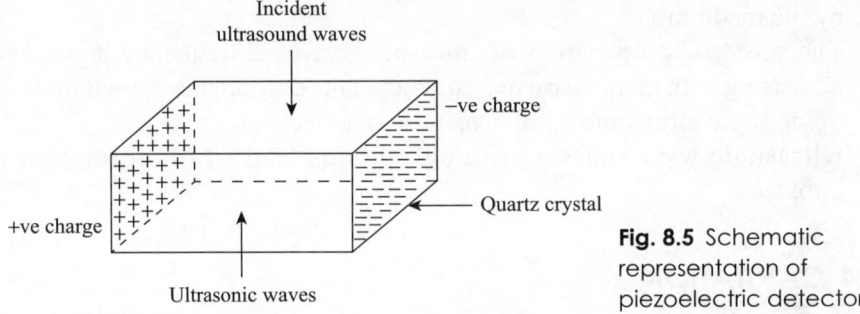

Fig. 8.5 Schematic representation of piezoelectric detector

8.5.2 Kundt's Tube Method

Kundt devised an experimental technique in 1889 to study the transmission of sound in different materials. This technique is based on the Kundt's tube shown schematically in Fig. 8.6.

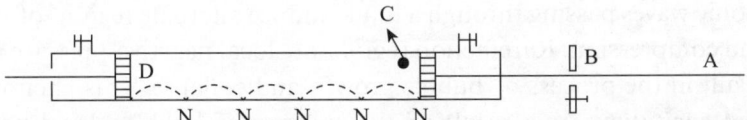

Fig. 8.6 Kundt's tube set-up for detection of ultrasonic waves

A Kundt's tube consists of a horizontal glass tube that is about 1 m long and 5 cm in diameter. One end of the tube has an adjustable piston D, and the other end has a loosely fitted cardboard cap C that is firmly fixed to a metal rod CA. The metal rod is clamped in the middle at B on a horizontal table to ensure minimum disturbance during the use of the Kundt's tube. A small amount of lycopodium powder is scattered in the portion CD of the tube. When ultrasonic waves are incident on the tube and pass through it, the lycopodium powder is collected in the form of heaps at the nodal points and blown off at the antinodes. The distance between subsequent nodes is then equal to half the magnitude of the wavelength of ultrasonic waves. This information can then be used to determine the frequency of the waves.

8.5.3 Sensitive Flame Method

The formation of node and antinodes in the presence of ultrasonic waves can be exploited in another interesting way to detect and determine the frequency of the waves. If a narrow, sensitive flame is moved through a medium that carries ultrasonic waves, the flame remains stationary at antinodes and tends to flicker at nodes. The frequency of the ultrasonic wave can be found by equating the distance between subsequent nodes or antinodes to half the wavelength.

8.5.4 Thermal Detector Method

Whenever an ultrasonic wave propagates through a medium, it causes alternate compressions and rarefactions in the medium. Due to these compressions and rarefactions, the temperature of the medium changes at the nodes, while remaining almost constant at antinodes. A thermal detector uses a fine platinum wire as one of the arms of a sensitive bridge arrangement. Using this bridge arrangement, changes in the resistance of the platinum wire at the nodes can be measured as a function of time. These measurements can then be used to determine the frequency of ultrasonic waves. As the detector element is moved through the medium, the bridge remains balanced at antinodes but gets off-balance at nodes.

8.6 ACOUSTIC GRATING

When ultrasonic waves are made to propagate through a liquid, its density varies from layer to layer due to the periodic variation of pressure in the liquid. If a monochromatic light is made to pass through the liquid at right angles to the direction of propagation of ultrasonic waves, the liquid then behaves like a diffraction grating. Since the grating action has been created with the help of acoustic waves, the grating is also referred to as *acoustic grating*. This set-up can be used to determine the wavelength and velocity of ultrasonic waves.

8.6.1 Velocity Measurement

Figure 8.7 shows a schematic diagram of the experimental set-up of an acoustic grating.

Light originating from a monochromatic source S is focused on a narrow slit O using a focusing lens L_1. The second lens L_2 is used to produce a parallel beam of light. This parallel beam of light is then made to pass through an ultrasonic cell that consists of a tank containing the test liquid. A piezoelectric crystal forming a part of the oscillator circuit is dipped inside the liquid. Ultrasonic waves generated by the crystal travel through the liquid and get reflected from the walls of the container of the ultrasonic cell.

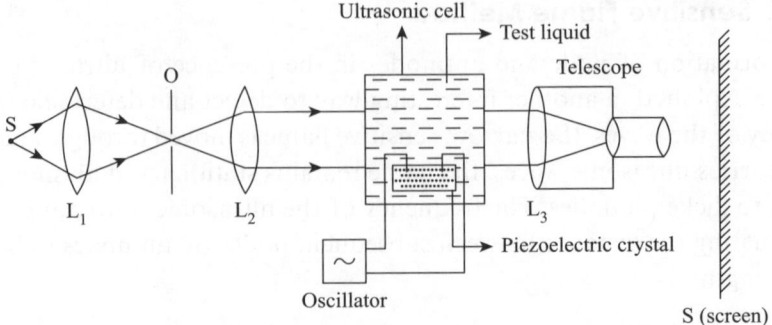

Fig. 8.7 Experimental set-up for acoustic grating

The position of the crystal is adjusted to ensure that the resultant station-ary waves are formed in the direction of propagation of ultrasonic waves. The light emerging out of the ultrasonic cell is focused onto a telescope with the help of the lens L_3.

Initially, no power is fed to the piezoelectric crystal and thus no ultrasonic waves are generated. A single image of the slit is then observed through the telescope. As the piezoelectric crystal starts generating ultrasonic waves, a complete diffraction pattern is formed on the screen S. The angular separa-tion θ between the direct image of the slit and the maxima of the diffraction fringe of any order n is then determined. If λ_A represents the wavelength of the generated ultrasonic waves and λ the wavelength of the incident mono-chromatic light, then applying the theory of diffraction grating, we can write the following equation:

$$\lambda_A \sin \theta_n = n\lambda \qquad (8.7)$$

The grating element for the acoustic grating is clearly the wavelength of the generated ultrasonic waves.

Equation (8.7) can be rewritten in the following form:

$$\lambda_A = \frac{n\lambda}{\sin\theta_n} \qquad (8.8)$$

Equation (8.8) can be used to calculate the wavelength of the generated ultrasonic wave λ_A. If v represents the frequency of the generated ultrasonic waves, then the velocity V_L of ultrasonic waves in the liquid medium is given by the following relation:

$$V_L = v\lambda_A \qquad (8.9)$$

Example 8.8   An ultrasonic interferometer is used to measure the velocity of ultra-sonic waves in sea water. If the distance between two consecutive antinodes is 0.55 mm, compute the velocity of the wave in sea water. The frequency of the crystal is 1.45 MHz. **Solution**   The distance between two antinodes is given by $\lambda/2$, where λ is the wave-length of the ultrasonic waves. Thus,

$$\lambda/2 = 0.55 \times 10^{-3}$$

or $\qquad \lambda = 1.1 \times 10^{-3}\,\text{m}$ (8.10)

In addition, $\upsilon = \nu\lambda$ (8.11)

where υ represents the velocity of the ultrasonic waves and ν the frequency. The use of Eq. (8.10) in Eq. (8.11) leads to the following result:

$$\upsilon = 1.5 \times 10^{6} \times 1.1 \times 10^{-3} = 1650\,\text{m/s}$$

Example 8.9   An ultrasonic interferometer-based system is used to measure the velocity of ultrasonic waves in sea water. The distance between two consecutive antinodes is found to be 0.4 mm. Calculate the velocity of the waves in sea water. Frequency of the waves generated by the crystal is 1.5 MHz.

Solution   The distance between two antinodes gives $\lambda/2$, where λ is the wavelength of the ultrasonic waves. Thus,

$$\lambda/2 = 0.4 \times 10^{-3}\,\text{m}$$ (8.12)

or $\qquad \lambda = 0.8 \times 10^{-3}\,\text{m}$

We know that

$$\upsilon = \nu\lambda$$ (8.13)

where υ represents the velocity of the ultrasonic waves and ν the frequency. Use of Eq. (8.12) in Eq. (8.13) yields the following result: $\upsilon = 1.5 \times 10^{6} \times 0.8 \times 10^{-3} = 1200\,\text{m/s}$

8.7 APPLICATIONS OF ULTRASONICS

Ultrasonics can be used in a variety of ways. We will discuss the different application areas of ultrasonics in this section.

8.7.1 Industrial Applications

Ultrasonic waves can be used in a variety of industrial applications. We will discuss some important industrial applications in this section.

8.7.1.1 Drilling

Ultrasonics can be used to drill holes in hard materials such as glass and diamond. A schematic diagram of an ultrasonic drilling system is shown in Fig. 8.8.

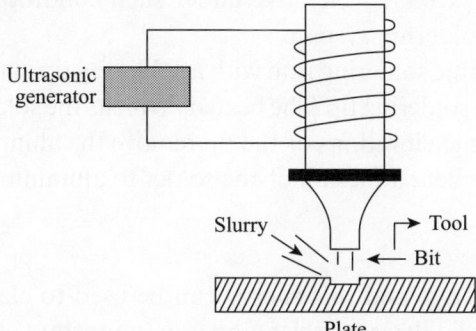

Fig. 8.8 Ultrasonic drilling system

The system consists of a tool bit connected to an ultrasonic generator. The tool bit carries out a vertical up–down motion due to the generated ultrasonic waves. A slurry [a thin paste of carborundum, i.e., silicon carbide (CSi) powder and water] flows in the region between the plate of the material to be drilled and the tool bit. As the tool bit undergoes a vertical motion, the slurry removes material from the plate. Holes with very good control of dimensions can be obtained using this technique.

8.7.1.2 Welding

Welding is the process of joining metals. Ultrasonics can be used to carry out welding. Figure 8.9 is a schematic representation of an ultrasonic welding system.

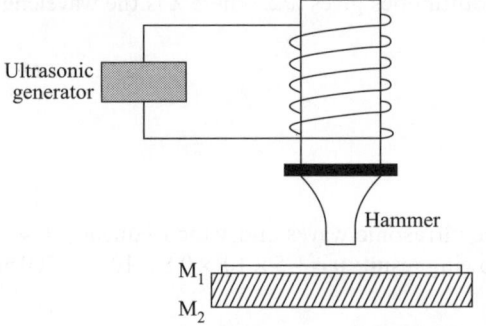

Fig. 8.9 Ultrasonic welding system

The set-up consists of a hammer connected to an ultrasonic generator. M1 and M2 represent two metal sheets that are to be welded together. The ultrasonic generator makes the hammer vibrate vertically at ultrasonic frequencies, generating pressure on the metal surfaces and causing the molecules of the metals to diffuse into each other. This results in welding of the two metal parts without the need for heating the plates to high temperatures. This process of welding is, therefore, also called *cold welding*.

8.7.1.3 Soldering

Aluminium has diverse industrial applications. However, using the conventional soldering technique, aluminium cannot be soldered without the use of fluxes. Ultrasonic soldering is extremely effective under such conditions. Figure 8.10 shows an ultrasonic soldering system.

The set-up consists of an ultrasonic soldering iron with a soldering tip at the end. Provision exists for heating the soldering tip. The heated tip melts the solder placed on aluminium, and ultrasonic vibrations of the tip remove the aluminium oxide layer. This results in excellent adhesion of the solder to aluminium.

8.7.1.4 Ultrasonic Cleaning

Ultrasonic waves possess high energy, and this energy can be used to clean utensils, clothes, machine parts, etc. Ultrasonic cleaning is an important step

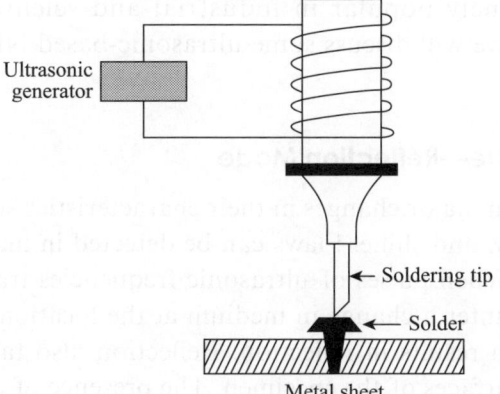

Ultrasonic
generator

Soldering tip

Solder

Metal sheet

Fig. 8.10 Ultrasonic soldering system

in the processing of semiconductor wafers to realize integrated circuits and devices.

A schematic diagram of a typical ultrasonic cleaning system is shown in Fig. 8.11. The set-up consists of a transducer that converts electrical energy to mechanical energy. The ultrasonic waves so generated are coupled to the container vessel. This vessel contains the component to be cleaned, in a suitable cleaning solution. The high energy of the ultrasonic waves acts on the contaminants to loosen and remove them, thus cleaning the components.

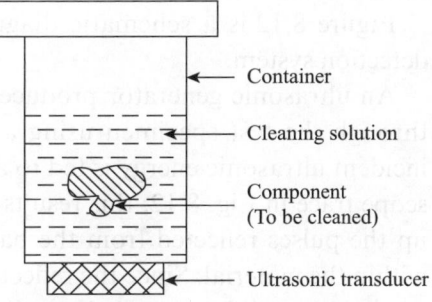

Container

Cleaning solution

Component
(To be cleaned)

Ultrasonic transducer

Fig. 8.11 Ultrasonic cleaning system

8.7.2 SONAR

SONAR is an acronym for sound navigation and ranging. This technique is primarily used for maritime purposes only. In this system, sharp beams of ultrasonic frequency are sent out in various directions. Reflections from icebergs, shipwrecks, etc., are picked up, amplified, and displayed on suitable screens. The time lag between the incident and the reflected pulse is used to determine the location of the object that resulted in reflection. If the object producing reflection is moving, the reflected signal also incorporates a change in frequency due to the *Doppler effect*. The change in frequency is then used to determine the direction and magnitude of the velocity of the object.

8.7.3 Non-destructive Testing Using Ultrasonics

Testing techniques that do not cause any harm or damage to the component being tested are referred to as non-destructive testing (NDT) techniques.

These techniques are extremely popular in industrial and scientific applications. In this section, we will discuss some ultrasonic-based NDT techniques.

8.7.3.1 Pulse Echo Technique—Reflection Mode

Flaws in materials can result in major changes in their characteristics such as yield strength, conductivity, and shine. Flaws can be detected in materials using ultrasonic waves. When pulses of ultrasonic frequencies travel through materials, they encounter a change in medium at the location of flaws. This change in medium results in reflection. Reflection also takes place at extremities or back surfaces of the specimen. The presence of any reflection, other than that from the back surface, is indicative of flaws in the materials.

Figure 8.12 is a schematic diagram of an ultrasonic pulse-based flaw detection system.

An ultrasonic generator produces ultrasonic pulses that are transmitted through the test specimen using a suitable transducer. A portion of the incident ultrasonic energy is fed to an oscilloscope. As shown in the oscilloscope trace in Fig. 8.12, this results in peak 1. A receiving transducer picks up the pulses reflected from the back surfaces and from any flaw existing within the material. Since the reflected signals are extremely low in strength, a suitable amplifier is used to amplify them. Peaks 2 and 3 represent the signals reflected from a flaw and from the back surface, respectively. The flaw detection system should also have a suitable arrangement to scan the entire spectrum for possible flaws.

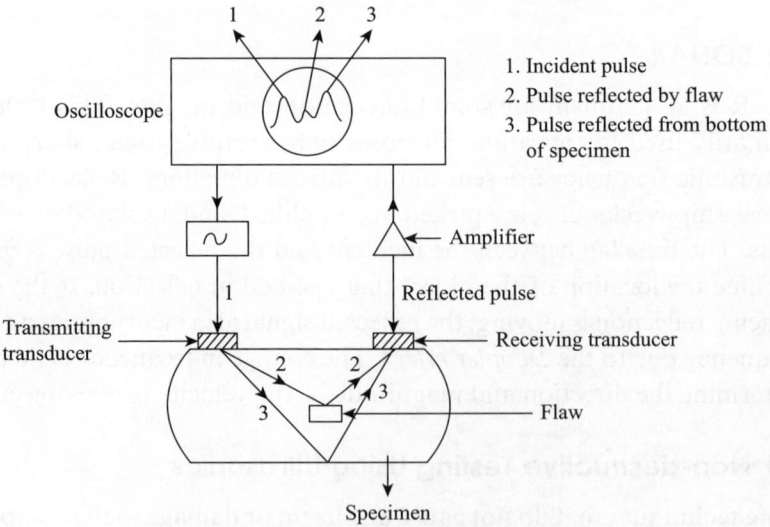

Fig. 8.12 Ultrasonic flaw detection system

8.7.3.2 Transmission Technique

In this technique, ultrasonic pulses are made to travel through the specimen and are received as they come out. A schematic diagram of the system is shown in Fig. 8.13.

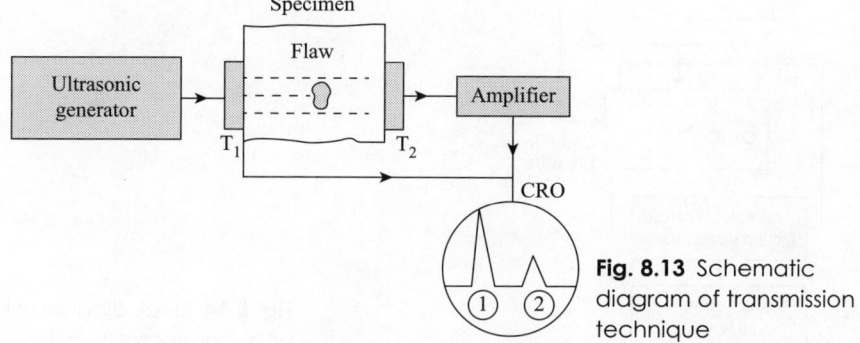

Fig. 8.13 Schematic diagram of transmission technique

The system consists of two transducers T_1 and T_2. T_1 converts high-frequency waves into ultrasonic pulses for transmission through the sample. A part of the input electrical signal is fed to a cathode ray oscilloscope (CRO). This is indicated by pulse 1, as shown in Fig. 8.13. The signal received after transmission is converted into an electrical signal by transducer T_2 and fed to the CRO. This transmitted signal appears as pulse 2 shown in Fig. 8.13. In the presence of flaws, the transmitted signal is of lower amplitude than the direct signal, because of absorption of ultrasonic energy by the flaws. The exact size and location of a flaw are, however, difficult to predict using this technique.

8.7.3.3 Resonance Technique

In this technique, the thickness of a specimen is an integral multiple of half the wavelength of the ultrasonic pulse. This is achieved by varying the frequency of the ultrasonic waves till standing waves are formed within the sample. The resonance frequency is found to change at the location of flaws in the specimen. This change is used to detect flaws within the specimen.

8.7.4 Medical Applications

Some important medical applications of ultrasonic waves are discussed in this section.

8.7.4.1 Echocardiogram/Sonogram

An echocardiogram records the movement of the valves and other structures of the heart as a function of time. A block diagram of the basic echocardio-graph unit is shown in Fig. 8.14.

A rate generator produces high-frequency pulses that are fed to a transmitter. The transmitter output is applied to the transducer probe, as shown in Fig. 8.14.

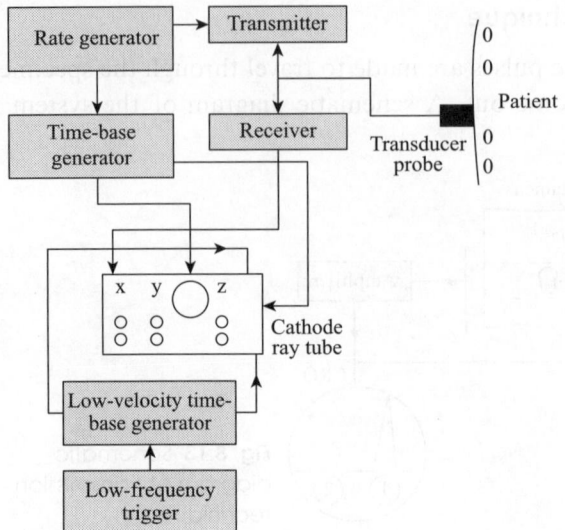

Fig. 8.14 Block diagram of echocardiograph unit

The reflected echo received by the probe forms the z-axis input for the cathode ray tube (CRT). Pulsed output from the rate generator is also fed to the time-base generator whose output is given to the x-plates of the CRT. The signal applied between the x-plates is used to deflect the electron beam of the CRT. The vertical deflecting plates (y-plates) are fed the output of a low-frequency time-base generator. This input makes a bright spot on the screen of the CRT that executes vertical motion at a slow speed.

An expansion or a contraction of the heart leads to the horizontal movement of the bright spot coupled with the vertical movement of the spot. This results in a zigzag path of the spot on the CRT. A typical zigzag path of the spot is shown in Fig. 8.15.

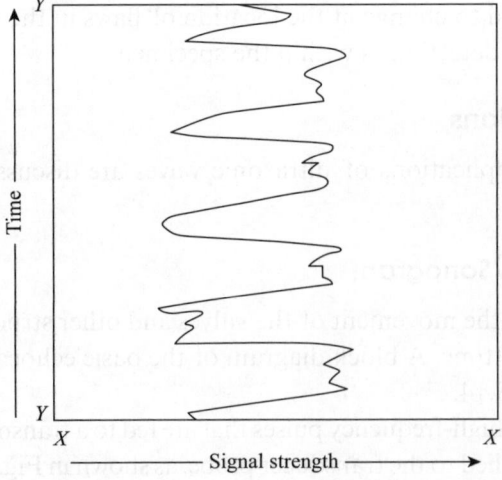

Fig. 8.15 Typical zigzag movement of CRT spot

8.7.4.2 Ultrasonic Imaging—Scan Displays

When an ultrasonic wave passes from one medium into another, reflection and refraction take place. The extent of reflection depends on the change in wave speed. A higher change in wave speed increases the amount of wave energy that is reflected back. An ultrasound imaging device fires a pulse of ultrasound into the body. The waves get reflected back as they meet a bone or an organ. The intensity of the reflection and the time delay from the instant when the pulse was emitted are characteristic of the nature of the reflecting medium. An image of the bone or organ is then built based on the information of the intensity and time delay. Velocity of sound in air is much less than that in human skin. Thus, air and skin are said to be acoustically mismatched. When the ultrasound transducer that forms a part of any ultrasound imaging system is placed against the skin, a high proportion of the sound energy is immediately reflected back, if there is any air gap. To eliminate the air gap, a gel is placed on the skin. The velocity of sound in the gel is midway between that in air and the skin. This helps in reducing reflection from the skin. In an A-scan, information regarding the strength of the reflected signal and the time at which the sound pulse was sent is used to display a graph of the distance that sound waves travelled into the body (as calculated from the time delay) versus the strength of reflection. In a B-scan, a sequence of dots is placed on the screen along a line, the distance between them being proportional to the distance that sound waves travelled into the body. The strength of reflection decides the relative brightness of the dots. As the transducer is moved about, many such lines are obtained, which are then assembled into an image that represents a section of the body. Figure 8.16 shows a schematic representation of A-scan and B-scan displays.

Figure 8.17 shows an image formed from a sequence of B-scan lines.

C-scan display In the C-scan display, energy or frequency of the ultrasonic pulse from a probe is adjusted such that the ultrasonic pulse can reach a

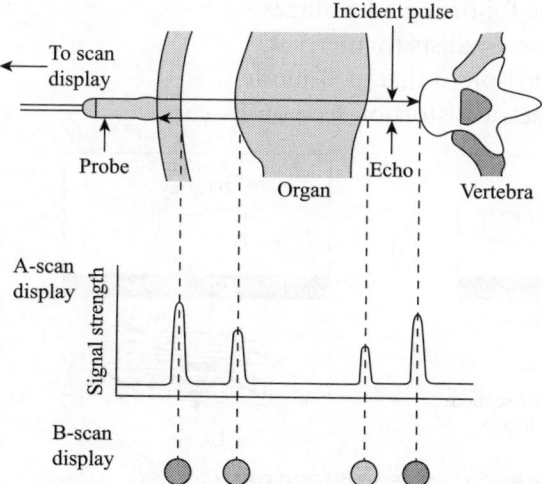

Fig. 8.16 Schematic representation of A-scan and B-scan displays

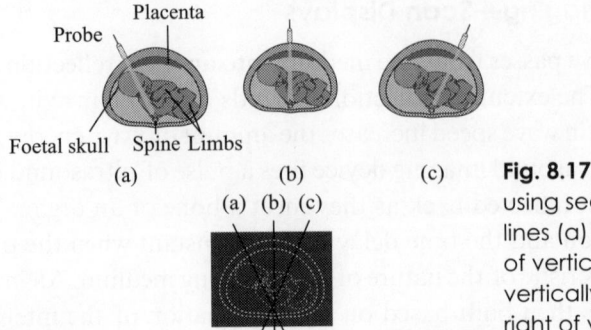

Fig. 8.17 Formation of image using sequence of B-scan lines (a) Probe placed left of vertical (b) Probe placed vertically (c) Probe placed right of vertical

particular depth from the surface of the specimen, and the cross section of the specimen at that depth is scanned. The depth information of the scan is not recorded. For this purpose, an ultrasonic probe, which is connected to an x–y plotter, is moved over the surface of the specimen either in a zigzag manner or in a manner similar to the parallel-line pattern. The intensity of the echo received from the section of the specimen is recorded as either 'variation in line shading' or 'shading with blank spaces corresponding to defect regions'. We can get both the position and the cross-sectional area of the defect across the section studied. The depth of the defect is, in general, not recorded in this method. However, with a number of 2D observations being made, as explained, it may be possible to get a 3D image of the defect. Normally, the C-scan procedure is adopted to carry out automatic testing (Figs 8.18 and 8.19).

Advantages of the C-scan procedure are as follows:
1. The testing method is automatic.
2. Position and cross-sectional area of the defect are recorded.

The following are the disadvantages of the C-scan procedure:
1. Cost of the technique is very high.
2. Depth information is not obtained.

Conventional ultrasound imaging uses the B-mode. The C-mode ultrasound imaging, however, offers the following advantages:
1. It is much easier for non-specialists to interpret.
2. Its cost is lower in comparison to that of B-mode.
3. C-scan is free from geometric distortion seen on B-scan.

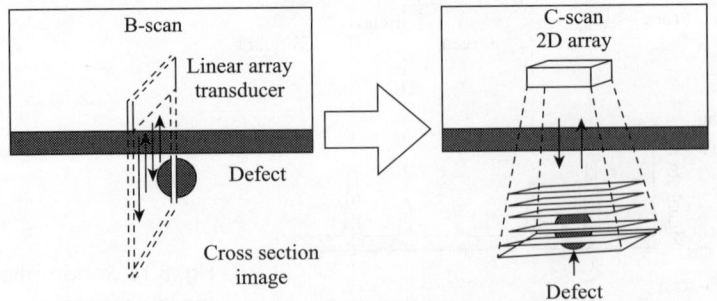

Fig. 8.18 Difference between B-scan and C-scan

4. C-scan has far greater spatial resolution.
5. It takes multiple round trips to generate one B-scan image. Although image quality can be improved by sending more and more pulses, it sacrifices the refresh rate. C-scan requires only one round trip to generate a full field image.
6. C-scan, therefore, does not include the shadows that normally streak across B-scans.
7. Structures in C-scan appear to reflect ultrasound waves the way they would reflect light.

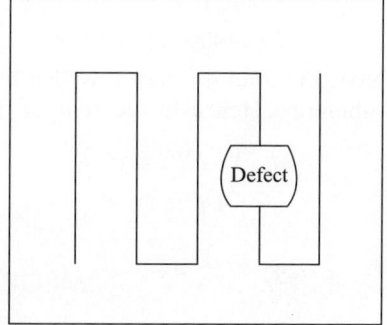

Fig. 8.19 Top view of scans or plane view

Example 8.10 Given that the velocity of ultrasonic waves in sea water is equal to 1440 m/s, find the depth of a submerged submarine, if an ultrasonic pulse reflected from the submarine is received 0.33 s after being sent out.

Solution Distance d travelled by the ultrasonic wave is given by the following equation:
$$d = v \times t \tag{8.14}$$
where v is the velocity of the ultrasonic waves and t is the time lapsed.

Substituting the given values in Eq. (8.14), one gets the following result:
$$d = 1440 \times 0.33 = 475.2 \, \text{m} \tag{8.15}$$
Now, the total distance travelled by the ultrasonic waves is twice the depth of the submarine. Hence, the depth d_1 of the submarine is calculated as follows:
$$d_1 = \frac{d}{2} = \frac{475.2}{2} = 237.6 \, \text{m}$$

Example 8.11 An ultrasonic source generating waves of frequency 800 kHz is used to find the depth of the sea. The velocity v of sound in sea water is 1440 m/s, and the time t taken by the sound to reach the source after reflection from the sea bed is 0.95 s. Calculate the depth of the sea.

Solution Depth of the sea is given by half of the total distance d traversed by the ultrasonic pulse. Therefore,
$$d = vt = 1440 \times 0.95 \, \text{m}$$
The depth of the sea is calculated as follows: $= \dfrac{1440 \times 0.95}{2} \, \text{m} = 684 \, \text{m}$

Example 8.12 Find the depth of a submerged submarine if the ultrasound pulse reflected from the submarine reaches the source after 0.83 s and the velocity of sound in sea water is 1400 m/s.

Solution Total distance d travelled by sound in sea $= v \times t$. Hence, the depth at which the submarine is detected is calculated as follows: $1162/2 = 581$ m

Example 8.13 The velocity of ultrasonic waves in sea water is equal to 1440 m/s. Find the depth of a submerged submarine if an ultrasonic pulsed reflected from the submarine is received 0.5 s after being sent out.

Solution Distance d travelled by the ultrasonic waves is given by the following relation:
$$d = v \times t \tag{8.16}$$

where v is the velocity of the ultrasonic waves and t is the time lapsed. Substituting the given values in Eq. (8.16), one gets the following result:

$$d = 1440 \times 0.5 = 720 \, \text{m} \tag{8.17}$$

Now, the total distance travelled by the ultrasonic waves is twice the depth of the submarine. Hence, the depth d_1 of the submarine is calculated as follows:

$$d_1 = \frac{d}{2} = \frac{720}{2} = 360 \, \text{m}$$

IMPORTANT CONCEPTS

1. Sound waves having frequencies beyond 20 kHz are called ultrasonic waves.
2. When a ferromagnetic material is placed in a magnetic field, it undergoes a change in dimension. This phenomenon is known as the magnetostriction effect.
3. When mechanical pressure is applied along the mechanical axis of piezoelectric crystals, equal amount of opposite charges are developed at the two end faces along the perpendicular direction.
4. The process of growth and collapse of bubbles in a liquid due to the passage of ultrasonic waves is called cavitation.
5. A liquid through which ultrasonic waves pass behaves like an acoustic grating.
6. SONAR is an acronym for sound navigation and ranging.

APPLICATIONS

1. Ultrasonics are used to disperse and deagglomerize solids into liquids. The high shear forces generated during cavitation are used to break particle agglomerates into uniformly dispersed particles. This process is used to develop products such as paint, ink, shampoo, and beverages.
2. Very-high-intensity ultrasound waves are used in ultrasonic emulsification. This process leads to dispersion of two or more immiscible liquids. This technique is used to prepare cosmetics, skin lotions, pharmaceutical ointments, varnishes, lubricants, and fuels.
3. High-power ultrasonic waves are used to disintegrate fibrous, cellulosic materials into fine particles. They are also used to break the cell walls. This process is used to release starch or sugar into liquids, and also used in fermentation, digestion and conversion of organic matter
4. Ultrasound can also be used to modify chemical reactions and processes. This branch of science is called sonochemistry. The use of ultrasound leads to increases in the reaction speed, reaction output, and the energy usage efficiency.

IMPORTANT FORMULAE

1. The frequency f of the alternating current set up by a resonant LC circuit is given as follows: $f = \dfrac{1}{2\pi\sqrt{LC}}$

2. The natural frequency v of a vibrating rod is given by the following relation:
$$v = \frac{1}{2l}\sqrt{\frac{E}{\rho}}$$

3. An *x*-cut crystal plate of length *l* of a piezoelectric crystal has the following frequency: $v = \dfrac{p}{2l}\sqrt{\dfrac{E}{\rho}}$

4. For an acoustic grating, $\lambda_A \sin \theta_n = n\lambda$

Multiple-choice Questions

8.1 Ultrasonic waves have frequencies
   (a) <20 kHz   (b) >20 kHz   (c) <30 kHz   (d) >40 kHz

8.2 Magnetostriction effect requires the presence of
   (a) an electric field   (c) a magnetic field
   (b) an electromagnetic wave   (d) a gravitational field

8.3 An LC resonant circuit has a resonant frequency given by
   (a) $f = 2\pi LC$ (c) $f = \dfrac{1}{2\pi\sqrt{LC}}$
   (b) $f = \dfrac{1}{2\pi}\dfrac{L}{C}$ (d) $f = 2\pi\sqrt{LC}$

8.4 Inverse piezoelectric effect results in the development of
   (a) stress   (b) strain   (d) a field   (d) a voltage

8.5 The natural frequency of a piezoelectric crystal is given by
   (a) $v = \dfrac{1}{2pl}\sqrt{\dfrac{E}{\rho}}$ (c) $v = \dfrac{p}{2l}\sqrt{\dfrac{E}{\rho}}$
   (b) $v = \dfrac{p}{2l}\sqrt{\dfrac{\rho}{E}}$ (d) $v = \dfrac{l}{2p}\sqrt{\dfrac{E}{\rho}}$

8.6 During collapse of bubbles, local pressures can reach
   (a) a few atmospheric pressure
   (b) hundreds of atmospheric pressure
   (c) thousands of atmospheric pressure
   (d) less than an atmospheric pressure

8.7 Ultrasonic detectors use the principle of
   (a) inverse piezoelectric effect   (c) cavitation effect
   (b) piezoelectric effect   (d) thermoelectric effect

8.8 Ultrasonic waves do not show
   (a) reflection   (c) absorption
   (b) refraction   (d) polarization

8.9 Ultrasonic welding is extremely effective for welding
   (a) gold   (c) iron
   (b) steel   (d) aluminium

8.10 In the resonance technique, the thickness of the specimen is an integral multiple of
   (a) λ (b) $\lambda/2$ (c) $\lambda/4$ (d) $\lambda/3$

Review Questions

8.1 What are the properties of ultrasonics?
8.2 What are the various methods of producing ultrasonic waves?

8.3   What is meant by the magnetostriction effect?

8.4   What is meant by the piezoelectric effect?

8.5   What are the methods used for detection of ultrasonics?

8.6   What is meant by SONAR?

8.7   Mention some applications of ultrasonic waves.

8.8   What is Doppler effect?

8.9   Mention a few medical applications of ultrasonic waves.

8.10   What are the industrial applications of ultrasonic waves?

8.11   Give an expression for the natural frequency of vibration of a rod.

8.12   Explain the Kundt's tube method for determination of frequency of sound waves.

8.13   Explain the sensitive flame method for the determination of frequency of ultrasonic waves.

8.14   Describe the phenomenon of cavitation.

8.15   Give some applications of the phenomenon of cavitation.

8.16   What is an acoustic grating?

8.17   How can one use an acoustic grating to determine the velocity of an ultrasonic wave?

8.18   Differentiate between A-scan and B-scan.

Numerical Problems

8.1   Velocity of ultrasonic waves in sea water is 1440 m/s. Find the depth of a submerged submarine if a reflected ultrasonic pulse is received 0.25 s after being sent out.   [*Hint*: $d = v \times t$]

8.2   The submarine indicated in Problem 8.1 relocates itself to a depth of 200 m. What will be the new time interval between the received and the sent ultrasonic pulses?   [*Hint*: $d = v \times t$]

8.3   Calculate the natural frequency of a 30 mm long pure iron rod, given that the density of pure iron is 7.25×10^3 kg/m^3 and its Young's modulus is 115×10^9 N/m^2.

$$\left[Hint: v = \frac{1}{2l}\sqrt{\frac{E}{\rho}} \right]$$

8.4   The length of rod indicated in Problem 8.3 is changed to 33 mm. Calculate the change in natural frequency. $\left[Hint: v = \dfrac{1}{2l}\sqrt{\dfrac{E}{\rho}} \right]$

8.5   A quartz crystal of length 0.8 mm is vibrating at resonance. Determine the fundamental frequency. (Assume that E for quartz $= 7.9 \times 10^{10}$ N/m^2 and ρ for quartz $= 2650$ kg/m^3.) $\left[Hint: v = \dfrac{1}{2l}\sqrt{\dfrac{E}{\rho}} \right]$

8.6   What should be the length of the quartz crystal that resonates at 6 MHz. Use the data given in Problem 8.5. $\left[Hint: v = \dfrac{1}{2l}\sqrt{\dfrac{E}{\rho}} \right]$

8.7   An ultrasonic interferometer is being used to measure the velocity of sound waves in sea water. If the distance between two consecutive antinodes is 0.6 mm, compute the velocity of the waves in sea water. The frequency of the crystal is 2 MHz.   [*Hint*: $v = v\lambda$]

PART II
MECHANICS

Classical Mechanics

Learning Objectives

After studying this chapter, students will be able to

- understand and apply important properties of vectors
- learn how to handle mechanics of system of particles
- comprehend important elastic properties of materials
- demonstrate important characteristics of rotational motion
- elucidate important topics related to fluids in motion
- solve numerical problems based on vector analysis, rotational motion, and fluid dynamics

List of Symbols

$\hat{i}, \hat{j}, \hat{k}$ = Unit vectors in x, y, and z directions

F = Force

r = Position vector

r_{cm} = Position vector of the centre of mass

τ = Torque

L = Angular momentum

v = Velocity

m = Mass

p = Linear momentum

A = Area

η = Coefficient of viscosity

ρ = Density

g = Acceleration due to gravity

v_t = Terminal velocity

D = Diameter

Re = Reynolds number

h = Height

9.1 INTRODUCTION

Any meaningful study of physics needs some basic understanding of some mechanics. Mechanics deals with the motion of a single or a group of particles under the influence of forces. Due to the significance of this field, a brief treatment of elastic properties of matter would be presented in this chapter. The presentation will include some important characteristics related to rigid bodies in motion. Fluids in motion is an area which has several interesting applications. This chapter will present some basic ideas about fluid dynamics. In short, this chapter will discuss some basic tools of mechanics that would prove useful as we learn about the different areas of physics in the chapters to follow.

9.2 VECTOR ANALYSIS

Physical quantities that have both magnitude and direction are called vectors or vector quantities, for example, displacement, velocity, acceleration, force, weight, and torque. A vector (see Fig. 9.1) is represented by either a single letter with an arrowhead on it, as $\vec{a}$ or as a line segment $\overrightarrow{OP}$.

The magnitude of the vector is represented as $|\vec{a}|$ or a, $|\overrightarrow{OP}|$ or OP.

9.2.1 Important Definitions

In the following subsections, some important definitions have been discussed.

9.2.1.1 Unit Vector

A unit vector of a given vector $\vec{a}$ is a vector of unit magnitude that has the same direction as the given vector $\vec{a}$. It is represented as $\hat{a}$. The magnitude of a unit vector is always 1.

$$\hat{a} = \frac{\vec{a}}{|\vec{a}|} = \frac{\vec{a}}{a} \tag{9.1}$$

$$\therefore \quad \vec{a} = a \, \hat{a} \tag{9.2}$$

In Cartesian coordinates, in three dimensions, the unit vectors along the x, y, and z axes are, $\hat{i}$, $\hat{j}$, and $\hat{k}$ respectively.

9.2.1.2 Equal Vectors

Two vectors $\vec{A}$ and $\vec{B}$ are said to be equal if they have the same magnitude as well as the same direction (see Fig. 9.2).

9.2.1.3 Co-initial Vectors

Two vectors are said to be co-initial if their initial point is common. Vectors $\vec{A}$ and $\vec{B}$ are co-initial as they both originate at O, that is, the point of origin of co-initial vector is always the same (see Fig. 9.3).

Fig. 9.1 Representation of vector   **Fig. 9.2** Equal vectors

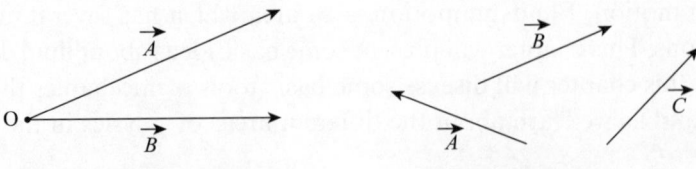

Fig. 9.3 Co-initial vectors   **Fig. 9.4** Coplanar vectors

9.2.1.4 Collinear Vectors

Vectors that act along the same straight line in a plane are called collinear vectors. If several weights are hanging from the same rope, the vectors representing the force exerted by each weight are collinear vectors.

9.2.1.5 Coplanar Vectors

As evident from the name, vectors that act in the same plane are called coplanar vectors. Vectors $\vec{A}$, $\vec{B}$, and $\vec{C}$ in Fig. 9.4 all lie on the plane of the paper and are coplanar.

9.2.2 Properties of Vectors

Vectors have certain properties associated with them based on which their treatment is different from scalar and other quantities. We will discuss some important properties of vectors in this section, which influence the way we operate, analyze, and simplify situations involving vector quantities.

9.2.2.1 Multiplication of Vector by Real Number

Multiplication of a vector by a real number n gives another vector nA whose direction is the same or opposite to that of A, depending on whether n is positive or negative (see Fig. 9.5).

Thus,

$$n(A) = nA \qquad (9.3)$$

and

$$-n(A) = -nA \qquad (9.4)$$

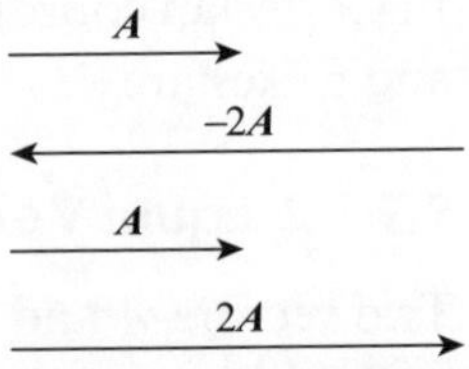

The units and dimensions of nA are the same as those of A.

Fig. 9.5 Multiplication of vector by real number

9.2.2.2 Multiplication of Vector by Scalar

When a vector A is multiplied by a scalar S, it becomes a vector SA whose magnitude is S times the magnitude of A and acts along the direction of A (see Fig. 9.6).

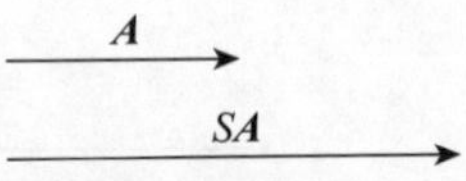

The units and dimensions of SA may be different from those of A, for example, if a is acceleration

Fig. 9.6 Multiplication of vector by scalar

[dimension (L/T^2)] and M is a mass (scalar), then Ma represents the force F [dimension (ML/T^2)] in the direction of a.

9.2.2.3 Addition of Vectors

Vectors can be added in the following ways:

1. When two vectors A and B act along the same direction,

$$R = A + B \qquad (9.5)$$

where R is called the resultant vector (see Fig. 9.7).

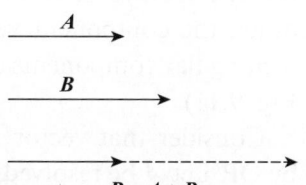

Fig. 9.7 Resultant of two vectors acting along same direction

2. When two vectors A and B act in opposite directions, the resultant is given as follows:

$$R = A + (-B) \qquad (9.6)$$

where R is the resultant vector (see Fig. 9.8).

3. When two vectors A and B are inclined at an angle θ to each other, the resultant vector is obtained using the *parallelogram law of vector addition* or the *triangle law of vector addition*.

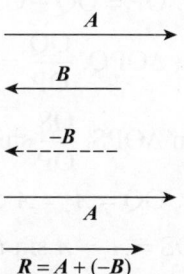

Fig. 9.8 Resultant of two vectors in opposite directions

Parallelogram law of vector addition   *If two vector quantities are represented by two adjacent sides of a parallelogram, then the diagonal of the parallelogram will be equal to the resultant of these two vectors.*

Thus, this law states that if two vectors acting simultaneously at a point are represented in both magnitude and direction by two sides of a parallelogram drawn from that point, their resultant vector is represented in both magnitude and direction by the diagonal of the parallelogram drawn from the same point (see Fig. 9.9).

Triangle law of vector addition   This law states that *if two vectors acting simultaneously at a point are represented in magnitude and direction by the two sides of a triangle taken in one order, their resultant is represented in magnitude and direction by the third side of the triangle taken in opposite order* (see Fig. 9.10).

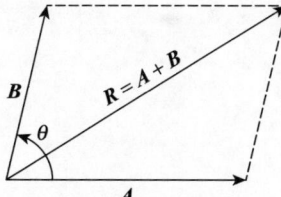

Fig. 9.9 Parallelogram law of vector addition

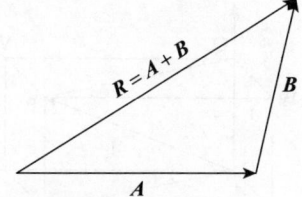

Fig. 9.10 Triangle law of vector addition

9.2.2.4 Rectangular Components of Vectors in Planes

When a vector is resolved into two component vectors at right angles to each other, the component vectors are called rectangular components of the vector (see Fig. 9.11).

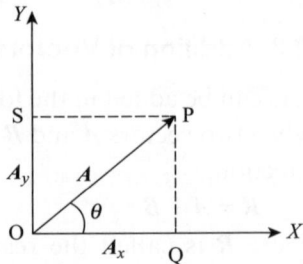

Consider that vector A is represented by OP. Let A be resolved into rectangular components parallel to the x and y axes; OQ = A_x and OS = A_y are the rectangular components of the vector. Therefore,

Fig. 9.11 Rectangular components of vector in plane

$$\therefore \text{OP} = \text{OQ} + \text{OS or } A = A_x + A_y \tag{9.7}$$

$$\text{In } \Delta \text{OPQ}, \frac{\text{OQ}}{\text{OP}} = \cos\theta \text{ or } \text{OQ} = \text{OP}\cos\theta \tag{9.8}$$

$$\text{In } \Delta \text{OPS}, \frac{\text{OS}}{\text{OP}}\sin\theta \text{ or } \text{OS} = \text{OP}\sin\theta \tag{9.9}$$

$$\therefore \text{OQ} = A_x = A\cos\theta$$

$$\text{OS} = A_y = A\sin\theta$$

Let $\hat{\imath}$ and $\hat{\jmath}$ be unit vectors along the x and y directions, respectively. Therefore,

$$A_x = A_x\hat{\imath} \text{ and } A_y = A_y\hat{\jmath}$$

Or $\quad A = A_x\hat{\imath} + A_y\hat{\jmath}$

In addition, $\tan\theta = \dfrac{\text{PQ}}{\text{OQ}} = \dfrac{\text{OS}}{\text{OQ}} = \dfrac{A_y}{A_x}$

A vector can be resolved into rectangular components in an infinite number of ways. However, if the vector is to be resolved into rectangular components along specified x and y axes, the components are unique.

9.2.2.5 Rectangular Components of Vector in Three Dimension

Consider a vector A represented by OP in Fig. 9.12, with O as the origin; also consider x, y, and z axes form a right-handed coordinate system.

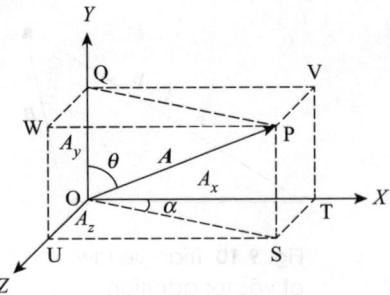

Fig. 9.12 Rectangular components in three dimension

To obtain the rectangular components A_x, A_y, and A_z of vector A along the x, y, and z axes, consider the plane OSPQ containing the vector A. Let OP make an angle θ with the y-axis, and let the plane OSPQ of vector A make an angle α with the xy plane. In the plane OSPQ,

$$OP = OS = OQ \tag{9.10}$$

where S is the foot of the perpendicular PS drawn from P onto the xz plane, and Q is the foot of the perpendicular drawn from P on the yz plane.

$$\therefore OQ = OP \cos \theta, \ OS = OP \sin \theta \tag{9.11}$$

However, $OQ = A_y$, the component of A along the y-axis. Therefore,

$$A_y = A \cos \theta \tag{9.12}$$

In the xz plane, OTSU, let ST be drawn perpendicular to the y-axis and SU be drawn perpendicular to the z-axis, and $\angle SOT = \alpha$

Then,

$$OS = OT + OU \tag{9.13}$$

$$OT = OS \cos\alpha = OP \sin\theta \cos\alpha = A \sin\theta \cos\alpha \tag{9.14}$$

$$OU = OS \sin\alpha = OP \sin\theta \sin\alpha = A \sin\theta \sin\alpha \tag{9.15}$$

However, OP represents the body diagonal of the cuboid OTVQWUS and

$$OP^2 = OT^2 + OQ^2 + OU^2 \tag{9.16}$$

Or

$$OP = OT + OQ + OU \tag{9.17}$$

Therefore, OP is the component of A along the x-axis and OU the component of A along the z-axis.

$$\therefore OT = A_x, \ OU = A_z \tag{9.18}$$

Using Eqs (9.14) and (9.15), we get the following expressions:

$$A_x = A \sin\theta \cos\alpha \tag{9.19}$$

$$A_z = A \sin\theta \sin\alpha \tag{9.20}$$

From Eqs (9.12), (9.17), (9.19), and (9.20), we have the following expressions:

$$A = A_x + A_y + A_z$$

Or, if $\hat{\imath}$, $\hat{\jmath}$, and $\hat{k}$ represent unit vectors along the x, y, and z axes, respectively, then $A = A_x\hat{\imath} + A_y\hat{\jmath} + A_z\hat{k}$

Using Eqs (9.12), (9.19), and (9.20), it can be seen that

$$A^2 = A_x^2 + A_y^2 + A_z^2 \tag{9.21}$$

Using the rectangular components of vectors, we may simplify the calculation of sum or difference of two vectors as follows:

Let $\quad A = A_x\hat{\imath} + A_y\hat{\jmath} + A_z\,\hat{k}$

$$B = B_x \hat{i} + B_y \hat{j} + B_z \hat{k}$$

$$A + B = \left(A_x + B_x\right)\hat{i} + \left(A_y + B_y\right)\hat{j} + \left(A_z + B_z\right)\hat{k} \qquad (9.22a)$$

and

$$A - B = \left(A_x - B_x\right)\hat{i} + \left(A_y - B_y\right)\hat{j} + \left(A_z - B_z\right)\hat{k} \qquad (9.22b)$$

9.2.2.6 Multiplication of Vectors

The product of two vectors can be of two types:

Scalar product or dot product The scalar or dot product of two vectors A and B, represented by $A \cdot B$, is a *scalar*, which is equal to the product of the magnitudes of A and B and the smaller angle between their directions. If θ is the smaller angle between A and B, then from Fig. 9.13,

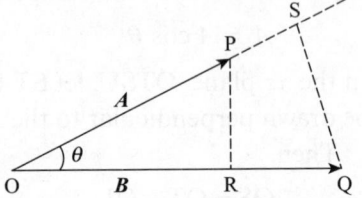

Fig. 9.13 Scalar product of vectors

$$A \cdot B = AB \cos\theta$$

Thus, $A \cdot B = A(B \cos\theta) = A(OS) = (OP)\ (OS)$ or
$$A \cdot B = B(A \cos\theta) = B(OR) = (OQ)\ (OR)$$

Thus, the scalar product or dot product of two vectors is defined as the product of the magnitude of one vector and the magnitude of a component of the other vector (projection of the other vector) in the direction of the first vector.

Special cases

1. When vectors A and B are parallel, $\theta = 0°$

   $\cos\theta = 1$

   $\therefore \quad A \cdot B = AB$

Similarly, $\hat{i} \cdot \hat{i} = 1,\ \hat{j} \cdot \hat{j} = 1,\ \hat{k} \cdot \hat{k} = 1$

2. When A and B are perpendicular to each other, $\theta = 90°$, $\cos\theta = 0$.

   $A \cdot B = AB\ (0) = 0$

   $\therefore \quad \hat{i} \cdot \hat{j} = \hat{j} \cdot \hat{k} = \hat{k} \cdot \hat{i} = 0$

3. When A and B are anti-parallel,

   $\theta = 180°$, $\cos\theta = -1$

   $AB = AB(-1) = -AB$

4. $A \cdot A = A\ A \cos\theta = A^2$

Vector product or cross product The vector product of two vectors A and B, represented by $A \times B$, is a vector C whose magnitude is equal to the product of the magnitudes of the two vectors and the sine of the smaller angle between to them, and whose direction is perpendicular to the plane containing *A* and *B*, according to the right-hand screw rule (see Fig. 9.14).

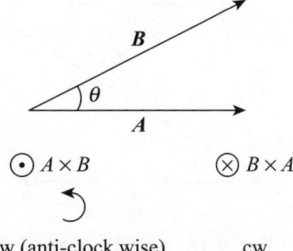

∴ $A \times B = (AB \sin\theta)\,\hat{C} = C$

where the direction of *C* is given by the right-hand screw rule, that is, direction in which a right-handed screw would move if rotated from *A* towards *B*.

⊙ $A \times B$ ⊗ $B \times A$

acw (anti-clock wise)   cw

Fig. 9.14 Vector cross product

The symbol '·' is used to indicate a vector coming out of the page; '×' is used to indicate a vector going into the page.

Special cases

1. When the vectors *A* and *B* are parallel or anti-parallel:

$$\theta = 0° \text{ or } \theta = 180°$$

∴ $\sin\theta = 0$

∴ $|A \times B| = AB \sin\theta = 0$

For unit vectors, $\hat{\imath} \times \hat{\imath} = 0,\ \hat{\jmath} \times \hat{\jmath} = 0,\ \hat{k} \times \hat{k} = 0$

2. When vectors *A* and *B* are perpendicular to each other:

$$\theta = 90°,\ \sin\theta = 1$$

$$|A \times B| = AB \sin\theta = AB$$

For unit vectors, $\hat{\imath} \times \hat{\jmath} = \hat{k},\ \hat{\jmath} \times \hat{k} = \hat{\imath},\ \hat{k} \times \hat{\imath} = \hat{\jmath}$

$\hat{\jmath} \times \hat{\imath} = -\hat{k},\ \hat{k} \times \hat{\jmath} = -\hat{\imath},\ \hat{\imath} \times \hat{k} = -\hat{\jmath}$

Example 9.1 $|\vec{P}| = 4$ N and $|\vec{Q}| = 3$ N, where $\vec{P}$ and $\vec{Q}$ are two mutually perpendicular vectors. Find $\vec{P} + \vec{Q}$.

Solution   The calculations are as follows:

$$|\vec{P} + \vec{Q}| = \sqrt{P^2 + Q^2} = \sqrt{16 + 9} = 5 \text{ N}$$

Let $(\vec{P} + \vec{Q})$ make an angle of α with respect to $\vec{P}$. Hence,

$$\tan\alpha = \frac{|\vec{Q}|}{|\vec{P}|} = \frac{3}{4}$$

or $\alpha = \tan^{-1}\dfrac{3}{4}$ (see Fig. 9.15)

Example 9.2   Rain is falling vertically with a speed of 30 m/s. If a wind is blowing with a speed of 10 m/s in the direction from north to south. In which direction would a boy hold his umbrella? (See Fig. 9.16.)

Solution   Let θ be the angle to the vertical to which the boy should hold the umbrella. Hence, $\tan\theta = \dfrac{10}{30} = \dfrac{1}{3}$

The boy should hold his umbrella at an angle θ where θ is given by

$$\theta = \tan^{-1}\frac{1}{3} = (18°26')$$

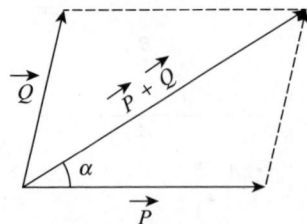

Fig. 9.15 Addition of
vectors $\vec{P}$ and $\vec{Q}$

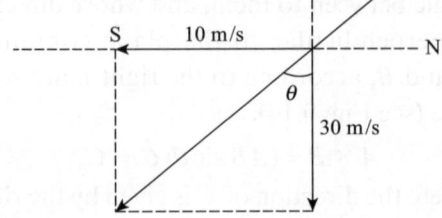

Fig. 9.16 Direction for umbrella
in rain

Example 9.3 A 52 kg man is riding a bicycle. The crank arm of the bicycle pedal
is 165 cm long. The man puts all his weight on one pedal. Calculate the torque produced when the crank is horizontal.

Solution When the crank arm is horizontal, the schematic representation is shown
in Fig. 9.17.

Angle between the man's weight and the crank arm is 90°. Torque τ is then given
by the following relation:

$$\tau = r \times F = rF \sin 90° \tag{9.23}$$

Magnitude of torque is, therefore, $\tau = 840.8$ N m

Example 9.4 $|\vec{P}| = 5$ N and $|\vec{Q}| = 4.6$ N, where the vectors $\vec{P}$ and $\vec{Q}$ are mutually
perpendicular. Find the angle that the resultant makes with the vector $\vec{P}$.

Solution Figure 9.18 shows the schematic of vector addition of the vectors $\vec{P}$ and $\vec{Q}$.

From Fig. 9.18, we can write that $\tan\alpha = \dfrac{|\vec{Q}|}{|\vec{P}|} = \dfrac{4.6}{5} = 0.92$

leading to $\alpha = \tan^{-1}0.92 = 42.6°$

Example 9.5 $|\vec{P}|$ and $|\vec{Q}|$ are two mutually perpendicular vectors. The resultant of
the two vectors makes an angle of 40° with the vector $\vec{P}$. If the magnitude of $\vec{P}$ is 4 N,
find the magnitude of $\vec{Q}$.

Solution Figure 9.19 is a schematic representation of addition of vectors $\vec{P}$ and $\vec{Q}$.

From Fig. 9.19, we get the following relation: $\tan\alpha = \dfrac{|\vec{Q}|}{|\vec{P}|} = \tan 40° = 0.84$

yielding $|\vec{Q}| = |\vec{P}| \times 0.84 = 4 \times 0.84 = 3.36$ N

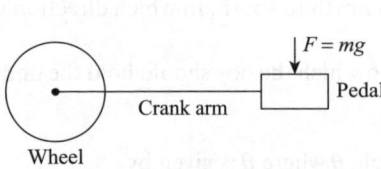

Fig. 9.17 Man pedalling bicycle
with horizontal crank

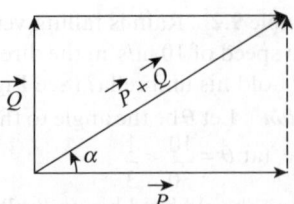

Fig. 9.18 Addition of
vectors $\vec{P}$ and $\vec{Q}$

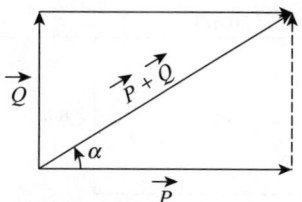

Fig. 9.19 Addition of vectors $\vec{P}$ and $\vec{Q}$

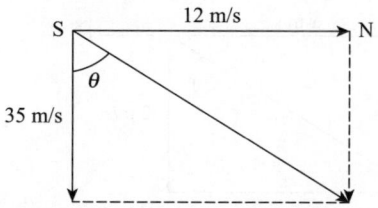

Fig. 9.20 Direction of umbrella in rain

Example 9.6 $\left|\vec{P}\right|$ and $\left|\vec{Q}\right|$ are two mutually perpendicular vectors. The resultant of the two vectors makes an angle of 38° with the vector $\vec{P}$. If the magnitude of $\left|\vec{Q}\right|$ is 4 N, find the magnitude of $\left|\vec{P}\right|$.

Solution The expression for tan α is as follows:

$$\tan\alpha = \frac{\left|\vec{Q}\right|}{\left|\vec{P}\right|} \tag{9.24}$$

For the given problem, $\tan\alpha = \tan 38 = 0.78 = \dfrac{\left|\vec{Q}\right|}{\left|\vec{P}\right|}$

leading to $\left|\vec{P}\right| = \dfrac{\left|\vec{Q}\right|}{\tan 38} = \dfrac{4}{0.78} = 5.13 \text{N}$

Example 9.7 Rain is falling vertically with a speed of 35 m/s. Wind is blowing with a speed of 12 m/s from south to north. In what direction should a boy hold his umbrella to protect himself?

Solution The situation is shown schematically in Fig. 9.20.

Let θ be the angle with respect to the vertical that the boy should make while holding the umbrella. Then,

$$\tan\theta = \frac{12}{35} = 0.34$$

yielding $\theta = \tan^{-1} 0.34 = 18°80'$

Example 9.8 Rain is falling vertically with a speed of 32 m/s and wind is blowing from north to south. A man wanting to protect himself has to hold an umbrella at an angle of 20° with respect to the vertical. Calculate the magnitude of wind speed.

Solution Let x be the magnitude of wind speed. The situation is shown in Fig. 9.21.

From Fig. 9.21, we get the following relation: $\tan\theta = \tan 20 = \dfrac{x}{32}$

leading to $x = 32\tan 20 = 11.6 \text{m/s}$

Example 9.9 Rain is falling vertically. Wind is blowing at a speed of 20 m/s from north to south. A man wanting to protect himself holds the umbrella at an angle of 25° w.r.t. the vertical. Calculate the speed at which rain is falling vertically.

Solution The situation is depicted schematically in Fig. 9.22.

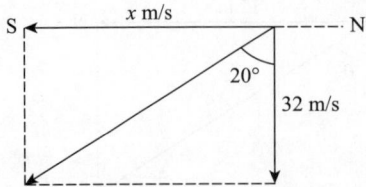

Fig. 9.21 Direction of umbrella in rain

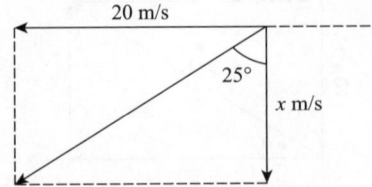

Fig. 9.22 Direction of umbrella in rain

From Fig. 9.22, we get the following expression: $\tan\theta = \tan 25 = \dfrac{20}{x}$

leading to $x = \dfrac{20}{0.47} = 42.6\,\text{m/s}$

Example 9.10 A man is riding a bicycle. The crank arm of the bicycle pedal is 160 cm long. The man puts all his weight on one pedal. The torque produced when the crank is horizontal is 800 N m. Calculate the weight of the man.

Solution The schematic representation when the crank arm is horizontal is shown in Fig. 9.23.

The angle between the man's weight and the crank arm is 90°.

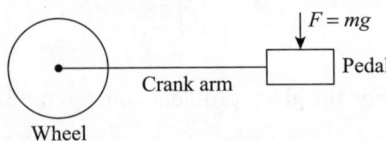

Fig. 9.23 Man pedalling bicycle with horizontal crank

Torque is given as follows:

$\tau = r \times F = 800 = 1.6 \times m \times g$

leading to $m = \dfrac{800}{1.6 \times 9.8} = 51\,\text{kg}$

Example 9.11 A 40 kg man is riding a bicycle. The man puts all his weight on one pedal. The torque produced when the crank is horizontal is 700 Nm. Calculate the length of the crank arm.

Solution When the crank arm is horizontal, it is schematically represented in Fig. 9.24.

The angle between the man's weight and crank arm is 90°.

Torque is then given by the following relation: $\tau = rF = 700$

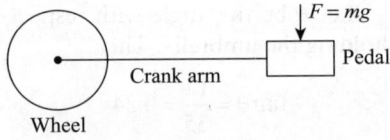

Fig. 9.24 Man pedalling bicycle with horizontal crank

leading to $r = \dfrac{700}{40 \times 9.8} = 1.79\,\text{m}$

9.3 CENTRAL FORCE

A force whose magnitude depends only on the distance r of the object from the origin and acts along the line joining them is called a *central force*, as shown in Fig. 9.25. In mathematical form, we express a central force as follows:

$$F(r) = F\hat{r} \tag{9.25}$$

The unit vector $\hat{r}$ can be represented as follows:

$$\hat{r} = \frac{r}{|r|} \tag{9.26}$$

A central force is conservative in nature and can be expressed in terms of potential as its negative gradient, that is,

$$F(r) = -\nabla V(r) \tag{9.27}$$

In a conservative force field, the total mechanical energy is conserved, that is,

$$E = \frac{1}{2} m |r|^2 + V(r) = \text{constant} \tag{9.28}$$

The angular momentum is also conserved in a central force field, that is,

$$L = r \times m \; dr/dt = \text{constant}$$

This is because the net torque exerted by the force is zero.

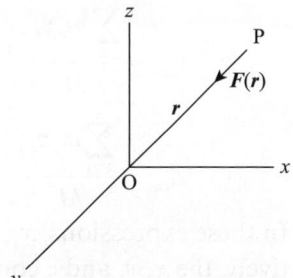

Fig. 9.25 Schematic representation of general central force

Two important examples of central forces are gravitational force and Coulomb's force. For both these examples, $F(r)$ is proportional to $\frac{1}{r^2}$.

9.4 MECHANICS OF SYSTEM OF PARTICLES

How do we handle a system consisting of a large number of particles? For such a system, a point exists such that the entire mass of the system is concentrated at that point: The motion of this point is predictable assuming that the force acting on this system is acting at this point. This point is called the *centre of mass* of this system.

Suppose there are n particles in a system with masses $m_1, m_2, m_3, ..., m_n$ with position vectors $r_1, r_2, r_3, ..., r_n$, respectively. Then the position vector r_{cm} of the centre of mass is given by the following equation:

$$r_{cm} = \frac{m_1 r_1 + m_2 r_2 + m_3 r_3 + \cdots + m_n r_n}{\left(m_1 + m_2 + m_3 + \cdots m_n \right)} \tag{9.29}$$

which can also be written as follows:

$$r_{cm} = \frac{\sum\limits_{k=1}^{n} m_k r_k}{\sum\limits_{k=1}^{n} m_k} \tag{9.30}$$

If M represents the total mass of the system, then the expression becomes as follows:

$$r_{cm} = \frac{\sum\limits_{k=1}^{n} m_k r_k}{M} \tag{9.31}$$

X_{cm}, Y_{cm}, and Z_{cm} of the centre of mass are then given by the following equations:

$$X_{cm} = \frac{\sum\limits_{k=1}^{n} m_k x_k}{M} \tag{9.32}$$

$$Y_{cm} = \frac{\sum\limits_{k=1}^{n} m_k y_k}{M} \tag{9.33}$$

$$Z_{cm} = \frac{\sum\limits_{k=1}^{n} m_k z_k}{M} \tag{9.34}$$

In these expressions, $x_1, x_2, ..., x_n; y_1, y_2, ..., y_n;$ and $z_1, z_2, ..., z_n$ are, respectively, the x, y, and z coordinates for the n particles. For a continuous distribution of mass, the corresponding expressions become,

$$X_{cm} = \frac{\int x\,dm}{M} \tag{9.35}$$

$$Y_{cm} = \frac{\int y\,dm}{M} \tag{9.36}$$

$$Z_{cm} = \frac{\int z\,dm}{M} \tag{9.37}$$

Let v_k represent the velocity vector of the kth particle; now, we can write that

$$M\frac{d}{dt}\mathbf{r}_{cm} = \sum_{k} m_k \mathbf{v}_k \tag{9.38}$$

Differentiation of Eq. (9.38) leads to the following expression:

$$M\frac{d^2}{dt^2}\frac{\text{Deforming force}}{\text{Area}}\mathbf{r}_{cm} = \sum_{k} m_k \frac{d^2 r_k}{dt^2} \tag{9.39}$$

or,

$$Ma_{cm} = \sum_{k} m_k \,\boldsymbol{a}_k \tag{9.40}$$

where $\boldsymbol{a}_{cm}$ represents acceleration of the centre of mass and $\boldsymbol{a}_k$ represents that of the kth particle. Newton's second law of motion can then be written in the following form:

$$M\,a_{cm} = \sum_{k} F_k = F_{ext} \tag{9.41}$$

where F_{ext} represents the net external unbalanced force.

9.5 ELASTIC PROPERTIES

Every object tries to oppose any effort/force trying to change its shape and size. The extent of this opposition depends upon its elastic properties. A change in size or shape of a solid body requires the application of an external force. *Any force resulting in a change in shape or size of a body is called a deforming force.* This force can thus change the length, volume, or merely the shape of a body. Such a force tries to produce a change in the normal equilibrium position of the atoms or molecules constituting the body. The body responds by generating an internal restoring force that resists any change in its shape or size. *The internal restoring force per unit area of a deformed body is called the stress in the body.*

Thus,

$$\text{Stress} = \frac{\text{Internal restoring force}}{\text{Area}} \tag{9.42}$$

Two possibilities exist at this stage. First, on removal of the deforming force, the restoring force brings the body back to its original shape or size. Second, the body does not regain its original shape or size on removal of the deforming force. Bodies of the first type are called *elastic bodies*, that is, bodies that regain their original shape and size upon removal of the external deforming force. Bodies of the second type are called *plastic bodies*, that is, bodies that cannot regain their original shape or size even after the external force is removed. For perfectly elastic bodies, the restoring force has got to be equal to the deforming force and expression (9.42) reduces to the following form:

$$\text{Stress} = \frac{\text{Deforming force}}{\text{Area}} \tag{9.43}$$

The SI unit of stress is N/m^2.

A deforming force acting normal to the area of a body generates *normal stress* and that acting tangential to the area generates *tangential stress*. Normal stress can be of two types, namely, *tensile stress* and *compressive stress*.

Tensile stress   The restoring force per unit area of a body whose length has been increased in the direction of the deforming force is called tensile stress. This type of stress results from extension produced in a body. A spring gets extended when a mass is at one of its ends. A restoring force generates tensile stress.

Compressive stress   The restoring force per unit area of a body whose length has decreased under the application of a deforming force is called compressive stress. Springs in the shock absorbers of vehicles experience compressive stress.

Tangential stress   The restoring force per unit area of a body whose shape changes due to the application of a tangential force is called tangential stress.

Let us now turn our attention to the consequences of stresses acting on a body. The deforming force acting on a body leads to a change in shape or size of the body. The change produced in the dimensions of a body is reflected in the strain. *Strain is the ratio of change in dimension to the original dimension.* Since it is a ratio it is dimensionless. Thus,

$$\text{Strain} = \frac{\text{Change in dimension}}{\text{Original dimension}} \tag{9.44}$$

There are three types of strain:

1. Linear strain $= \dfrac{\text{Change in length}}{\text{Original length}} = \dfrac{\Delta l}{l}$ 　　　　　(9.45)

2. Volume strain $= \dfrac{\text{Change in volume}}{\text{Original volume}} = \dfrac{\Delta v}{V}$ 　　　　　(9.46)

3. Shape strain or shear = Angular deformation in radians 　　　　　(9.47)

Linear strain and volume strain are easy to visualize. The extension produced in a stretched rubber band is an example of linear strain. The increasing radius of a balloon as it is inflated is an example of volume strain. To visualize shear strain, let us consider a cubical body of side l, as shown in Fig. 9.26. A tangential force F is shown to be applied to the face PQRS. An angular deformation θ is produced in the process, as shown in the figure.

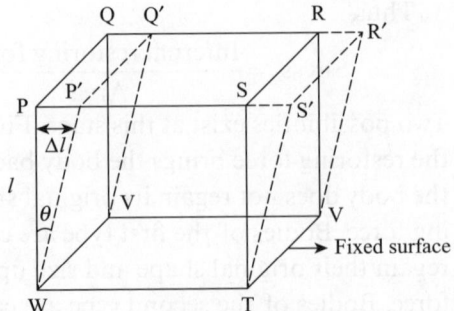

Fig. 9.26 Angular deformation produced by tangential force

The surface PQRS gets shifted to P'Q'R'S' under the influence of this force. Then,

$$\text{Shearing strain} = \theta = \frac{\Delta l}{l} \tag{9.48}$$

Thus, shearing strain is the angle through which a surface perpendicular to the fixed surface gets angularly displaced.

Suppose we carry out the following experiment. A wire of a uniform cross section is hung vertically from a rigid support. The free end of the wire has a pan on which weights can be kept to subject the wire to different stress levels. Another similar wire is kept close to the first wire. A small weight is attached to the free end of the second wire, and a fixed weight and a Vernier scale is attached to the wire carrying the pan with variable weights. Such a set-up is shown in Fig. 9.27.

Wire A in the figure carries the main scale and wire B the Vernier scale. We now vary the stress by keeping different weights on the pan, and note the corresponding extensions produced using the Vernier and main scale combination. What is the type of curve that we obtain if we plot strain as a function of stress? A typical plot is shown in Fig. 9.28.

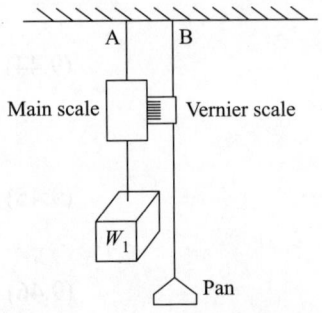

Fig. 9.27 Experimental set-up to study stress–strain relationship

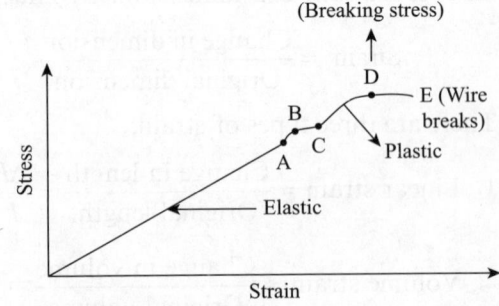

Fig. 9.28 Stress–Strain relationship for a wire

Important points in the curve are clearly identified. Let us now understand the curve in detail. Up to point A, stress and strain are linearly related. In this region, stress is directly proportional to strain. Point A is called the *proportional limit* of the material. After this point, non-linearity sets in for the stress–strain relationship. Up to point B, the wire would return to its original length if the deforming force is removed. B thus represents the *elastic limit*. Beyond point B, the material does not return to its original length on removal of the deforming force. In fact, a residual strain or *permanent strain* remains on removing the force that caused the wire to extend. The material is said to have acquired a *permanent set*. This is the beginning of the plastic region, which continues up to point E. Point C, slightly ahead of point B, is called the yield point of the wire. On increasing the stress further, the strain increases rapidly due to a greatly increased extension of the wire. Point D in the figure represents the maximum of the breaking stress. The wire finally breaks or fractures at point E.

Materials that elongate considerably and go through a plastic deformation region before breaking are called ductile materials. Wrought iron, lead, and copper are common examples of ductile materials. *Materials that break just beyond the elastic limit are called brittle materials.* High-carbon steel and glass are some examples. There exists another category of material that does not display any region in their stress–strain curve, where stress is directly proportional to strain. These materials, however, show elastic behaviour for a large stress range. Thus, the materials return to its original length on removal of the deforming force; such materials that can be stressed to large strain values are called *elastomers*. We have all handled rubber bands and experienced that they can withstand a large amount of extension. Rubber is a well-known elastomer. The elastic tissue of the large vessel carrying blood from the heart (aorta) is also an elastomer.

9.6 ROTATIONAL MOTION

The quantity measuring the effect of a force that causes rotation is called the *moment of the force* or *torque*. It is defined as *the product of the applied force and the perpendicular distance between the line of action of the force and the point about which the body rotates freely.* A symbol τ is used to represent torque. Torque τ about a point is then defined as follows:

$$\tau = rF \sin\theta \tag{9.49}$$

where r represents the magnitude of displacement from the axis to the point of application of the force F. The angle between the direction of force and r is θ. In vector notation, Eq. (9.49) can be rewritten as follows:

$$\tau = r \times F \tag{9.50}$$

We can easily draw the following conclusions from Eq. (9.50):

1. $\theta = 90°$ leads to the maximum torque
2. $\theta = 0$ produces zero torque

There is another physical interpretation of the definition of torque. Suppose $r \sin \theta$ is the lever arm. The torque τ can then be seen as the product of the force and the lever arm. Thus, applying a torque is equivalent to applying a force perpendicular to a lever, provided the lever arm is defined as the perpendicular distance from the axis of rotation to the line of action of the force.

Another useful concept in rotational motion is angular momentum, L, which is defined as follows:

$$L = r \times p = mr \times v \tag{9.51}$$

where r represents the distance vector and p the linear momentum. Equation (9.51) can be rewritten as follows:

$$L = (r \sin \theta)p \tag{9.52}$$

Thus, angular momentum of a particle/body about a point/axis is the product of the linear momentum and the perpendicular distance between the point/axis and the line of action of linear momentum.

According to the principle of conservation of angular momentum, the net angular momentum of a system does not change with time in the absence of a net external torque.

Example 9.12   A stone attached to a string is whirled around in a horizontal circle, as shown in Fig. 9.29. The stone is moving at a rate of 0.5 rad/s. Determine the angular velocity of the stone if the radius of the circumscribed circle is halved.

The radius of the circle formed by the rotational motion of the stone is reduced by applying a force F, as shown in Fig. 9.29. This force acts along the axis of the stone's motion and, therefore, does not result in any applied torque. The angular momentum is, therefore, conserved, according to the principle of conservation of angular momentum.

Let w_2 and r_2 represent the initial angular velocity and radius vectors, respectively; the corresponding final quantities are represented by w_1 and r_1. Conservation of angular momentum implies that

$$mr_1^2 w_1 = mr_2^2 w_2 \tag{9.53}$$

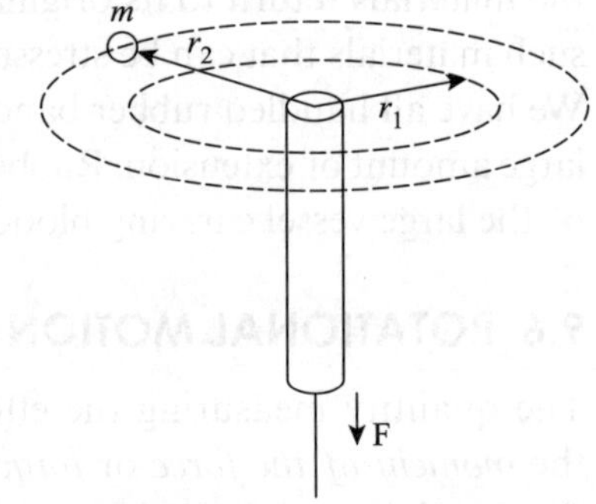

Fig. 9.29 Whirling of stone in horizontal circle

which leads to

$$w_1 = (r_2/r_1)^2 w_2 \tag{9.54}$$

Substituting the given value of w_2 and the given ratio $\frac{r_2}{r_1}$ in Eq. (9.54), the following relation is obtained:

$$w_1 = \left(\frac{1}{0.5}\right)^2 \times (0.5)$$

which leads to the following result:

$$w_1 = 2 \text{ rad/s}$$

Thus, a reduction in the radius results in an increase in the angular velocity.

Example 9.13 A stone attached to a string is whirled around in a horizontal circle, as shown in Fig. 9.30. The stone is moving at a rate of 0.42 rad/s. Calculate the angular velocity of the stone if the radius of the circumscribed circle is reduced by a factor of 1/1.6.

Solution The radius of the rotational motion of stone is reduced by applying a force, as shown in Fig. 9.30. This force does not result in any applied torque. The angular momentum is therefore conserved.

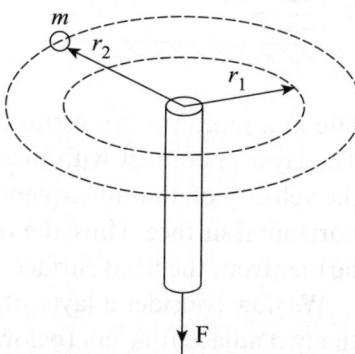

Let w_2 and r_2 represent the initial angular velocity and radius, respectively. The corresponding final quantities are w_1 and r_1. Conservation of angular momentum leads to the following relation: $mr_1^2 w_1 = mr_2^2 w_2$ yielding

$$w_1 = w_2 \left(\frac{r_2}{r_1} \right)^2 \qquad (9.55)$$

Fig. 9.30 Stone being whirled around in horizontal circle

In the present situation,

$$r_1 = \frac{r_2}{1.6} \qquad (9.56)$$

Use of Eq. (9.56) in Eq. (9.55) results in the following relation:

$$w_1 = w_2 (1.6)^2 \text{ that is, } w_1 = (0.42)(1.6)^2 = 1.08 \, \text{rad/s}$$

Example 9.14 The angular velocity of the stone indicated in Example 9.13 is changed to 0.46 rad/s and the radius is increased by a factor of 1.4. Calculate the new angular velocity of the stone.

Solution Once again, we have

$$w_1 = \frac{r_2^2}{r_1^2} w_2 \qquad (9.57)$$

In addition,

$$r_1 = 1.4 r_2 \qquad (9.58)$$

Using Eq. (9.58) in Eq. (9.57), we get $w_1 = \dfrac{0.46}{(1.4)^2} = 0.23 \, \text{rad/s}$

9.7 FLUID DYNAMICS

In this section, we shall discuss some important topics related to fluids (i.e., liquids and gases) in motion. These subjects are important in handling and analyzing situations where liquids and gases are in a dynamic flow under a variety of conditions. Important material properties that decide fluid flow would be discussed in this section.

9.7.1 Viscosity

Let us assume that a liquid is flowing over a fixed horizontal surface. Such a situation would be prevalent when water is flowing steadily at a low enough

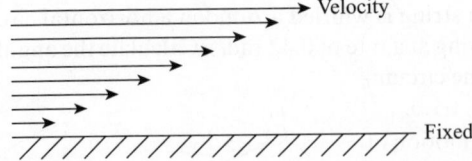

Fig. 9.31 Velocity increases with distance from fixed surface

rate in a pipe. We can assume the liquid to be made up of different layers. The layer in contact with the horizontal surface would then be at rest, and the velocity of the subsequent layers would increase with distance from the horizontal surface. Thus, the velocity of the layers would increase as one goes farther from the fixed surface, as shown in Fig. 9.31.

We now consider a layer of the liquid and see the effect of the motion of the two adjacent layers (below and above) on the given layer's motion. The layer above the chosen layer is moving faster, whereas the one below is moving slower; therefore, our chosen layer's motion is retarded. *There is a tendency on the part of adjacent layers to destroy their relative motion. This is as if a backward dragging force is acting tangentially on the layers of the liquid. The property of a liquid by virtue of which it opposes relative motion between its constituent layers is called viscosity.*

The quantitative nature of the viscous force was elaborated by Newton. He showed that the viscous force F acting tangentially on any layer of a liquid is

1. directly proportional to surface area A of the layer,
2. directly proportional to velocity V of the layer, and
3. inversely proportional to the distance x of the layer from the stationary layer.

Thus we have,

$$F \, \alpha \, \frac{-Av}{x} \tag{9.59}$$

where the negative sign indicates that the direction of F is opposite to that of v. The proportionality constant is called the coefficient of viscosity and it is denoted by the symbol η. We then have the following expression:

$$F = -\eta \frac{Av}{x} \tag{9.60}$$

Suppose that we have two layers separated by a distance dx, and a relative velocity dv exists between them. We can then write Eq. (9.60) in the following form:

$$F = -\eta A \frac{dv}{dx} \tag{9.61}$$

Equation (9.61) is sometimes referred to as the *Newton's law of viscous force.* We should keep in mind at this stage that Eq. (9.61) is valid only for the case of a liquid in streamline motion. If we put $A = 1$ m^2 and $\frac{dv}{dx} = 1$ in Eq. (9.61), we get the following expression: $F = -\eta$

Thus, the coefficient of viscosity is the tangential force required to maintain a unit velocity gradient when the surface area of the liquid is unity. The SI unit of coefficient of viscosity is Poiseuille (Pl) or Pa s (Pascal second).

9.7.2 Stokes' Law and Terminal Velocity

A body moving through a fluid always experiences a *viscous drag* that opposes its motion. Such opposing forces are felt by cars, airplanes, and projectiles. They result in wasteful expenditure of energy. Sir George Stokes (1819–1903) showed in 1845 that the viscous force F felt by a spherical body of radius r moving in a medium of coefficient of viscosity η with a velocity v is given as follows:

$$F = 6\pi\eta r v \tag{9.62}$$

Equation (9.62) is known as the *Stroke's law* and is valid only for a body moving so slowly that no eddy currents or waves are set up.

Let us assume that a sphere is falling through a viscous medium, as shown in Fig. 9.32.

The sphere has a radius r and is made of a material of density ρ. The medium has density ρ' and its coefficient of viscosity is η. At any given instant of time, three forces would act on the sphere, which are as follows:

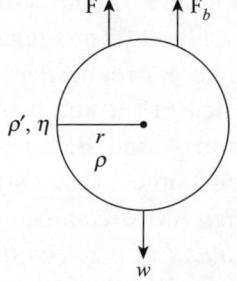

1. Downwards force W due to the weight of the sphere
2. Buoyant force F_b acting upwards
3. Viscous drag F acting upwards

Fig. 9.32 Sphere falling through viscous medium

The downward force due to weight and upward force due to buoyancy are constant, whereas the viscous drag increases as the sphere is under acceleration. A stage is finally reached when the sphere feels no net force and hence a zero acceleration. *The constant velocity v_t, of the sphere when the net force on it is zero is called the terminal velocity.*

From Fig. 9.32, we can see that this condition implies that $W = F_b + F$

Using Eq. (9.62), we can rewrite the preceding equation in the following form:

$$\frac{4\pi}{3}r^3\rho g = \frac{4\pi}{3}r^3\rho' g + 6\pi\eta r v_t \tag{9.63}$$

leading to

$$v_t = \frac{2r^3 g}{9\eta}(\rho - \rho') \tag{9.64}$$

From Eq. (9.64), for the terminal velocity of a sphere falling through a viscous medium, we can draw the following conclusions:

1. Terminal velocity varies directly with the square of the sphere's radius and inversely with the coefficient of viscosity of the medium.
2. For $\rho < \rho'$, the terminal velocity is negative. Gas bubbles rising up in a soda water bottle is an example of this case.

3. Equation (9.64) can be used to determine η if the other terms in it are known.

9.7.3 Streamline Flow

In the simplest type of flow of a liquid, the velocity at every point in it remains constant in magnitude as well as in direction. In this case, the energy being supplied to maintain the flow of the liquid is used up in overcoming the *viscous drag* between the different layers constituting the liquid. Such a steady, orderly flow is called a streamline flow. Thus, *any particle of a liquid in a streamline flow follows exactly the same path and velocity as its predecessor*. Let us consider a particle within the liquid and trace the line along which the particle moves. Such a line is called a streamline. The direction of the line at any point then gives the direction of velocity of the particle at that point. Thus, a streamline is a curve, the tangent to which at any point gives the direction of flow of the liquid at that point.

Figure 9.33 shows some streamlines drawn for a liquid from a region A to a region B.

These streamlines would represent any particle of the liquid as it passes between the region A and B. Streamlines display some important characteristics, as follows:

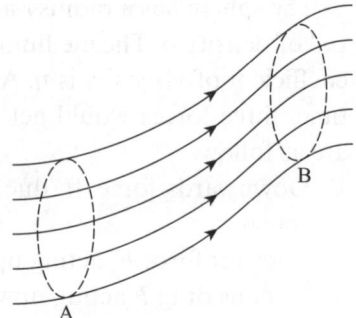

1. *No two streamlines cross each other for a liquid in a streamline flow*. This characteristic follows from the fact that if this is not true then the particles of the liquid will have multiple directions at the points of intersection of the streamlines. This would then be contrary to the condition of a steady flow.

Fig. 9.33 Streamlines of liquid between two regions

2. *Density of streamlines decides the magnitude of velocity of a liquid in a streamline flow*. Thus, the velocity of the liquid is higher in regions showing greater crowding of streamlines.

9.7.4 Equation of Continuity

Let us consider an incompressible liquid flowing through a tube of non-uniform cross section, as shown in Fig. 9.34.

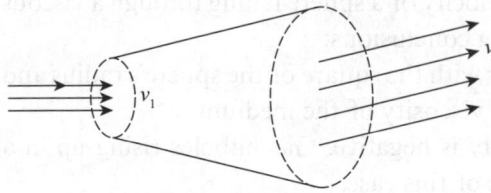

Fig. 9.34 Flow of incompressible liquid through tube of non-uniform cross section

We will also assume that a streamline flow exists. A_1 represents a perpendicular area of cross section in the narrower region and A_2 is a perpendicular area of cross section in the wider region. Let ρ represent the density of the flowing liquid. Since the liquid is incompressible, the amount of liquid that leaves any region towards the right per unit time must be the same as that entering the region from the left per unit time. In addition, the volume of liquid that flows into the tube across cross section A_1 in a given time interval Δt is given by the following expression:

$$V_1 = A_1 v_1 \Delta t \tag{9.65}$$

where v_1 is the velocity of the liquid at A_1. Thus, the mass m_1 entering the tube at A_1 is given by the following equation:

$$m_1 = V_1 \rho = A_1 v_1 \rho \Delta t \tag{9.66}$$

Similarly, mass m_2 leaving the tube through A_2 is given as follows:

$$m_2 = A_2 v_2 \rho \Delta t \tag{9.67}$$

where v_2 is the velocity of the liquid at A_2.

Keeping in mind the incompressible nature of the liquid, the following relation must hold: $m_1 = m_2$

leading to $A_1 v_1 \rho \Delta t = A_2 v_2 \rho \Delta t$

which further yields the following relation:

$$v_1 A_1 = v_2 A_2 \tag{9.68}$$

Equation (9.68) is called the *equation of continuity*. Another way of expressing Eq. (9.68) is to say that

$$vA = \text{constant} \tag{9.69}$$

The product of velocity and area of cross section is thus a constant, according to the equation of continuity.

9.7.5 Reynolds Number

Liquids are not always in a streamline flow. Beyond a certain velocity, called the critical velocity, particles constituting a liquid move in a haphazard, disorderly manner. These particles then do not follow the path of the preceding particle. The path and velocity of the particles of the liquid thus change with position as well as time. Such a flow is called a *turbulent flow*. An Irish-born engineer Osborne Reynolds (1842–1912) discovered this velocity dependence of the transition from a streamline flow to a turbulent flow in the year 1883. Reynolds number Re is very useful as a parameter for characterizing the type of flow of a fluid. It is given as follows:

$$\text{Re} = \frac{\rho v L}{\eta} \tag{9.70}$$

where ρ and η are density and viscosity of the fluid, respectively. Velocity of the fluid is represented by v and L is a characteristic length. For a liquid flowing through a tube, the Reynolds number becomes as follows:

$$\text{Re} = \frac{\rho v D}{\eta} \tag{9.71}$$

where D is the diameter of the tube. The value of the Reynolds number, a dimensionless constant, determines whether the flow of a liquid is turbulent or streamline. Experimental observations for flows through a pipe have shown that Re values lower than 2000 correspond to streamline flows, whereas values greater than 3000 result in turbulent flows. Values between 2000 and 3000 result in flows that are unstable, changing back and forth between streamlined and turbulent flows.

The origin of the concept of the Reynolds number lies in the interplay of two forces, namely, the inertial resistive force and the viscous drag force. In fact, *the ratio of inertial force to viscous force is the Reynolds number.*

Let us discuss this in detail for a fluid flowing through a tube. We will assume the area of cross section of the tube to be A, and the flowing fluid would be assumed to be having a velocity v and density ρ. Mass Δm flowing through the tube per second is given by the following equation:

$$\Delta m = A v \rho \tag{9.72}$$

This flowing mass leads to a rate of change of linear momentum $\dfrac{\Delta p}{\Delta t}$, given as follows:

$$\frac{\Delta p}{\Delta t} = (\Delta m)v = A v^2 \rho \tag{9.73}$$

By definition, Eq. (9.73) also gives the inertial force, F_{inertial}; thus,

$$F_{\text{inertial}} = A v^2 \rho$$

The inertial force per unit area is given by the following expression:

$$\frac{F_{\text{inertial}}}{A} = v^2 \rho \tag{9.74}$$

Using the expression for magnitude of viscous force $F_{\text{viscous,}}$ we have the following relation:

$$\frac{F_{\text{viscous}}}{A} = \eta \frac{dv}{dx} = \frac{\eta v}{D} \tag{9.75}$$

where D is the diameter of the tube. Dividing Eq. (9.74) by Eq. (9.75), we get the following relation:

$$\frac{F_{\text{inertial}}}{F_{\text{viscous}}} = \frac{v^2 \rho D}{\eta v} = \frac{\rho v D}{\eta}$$

which is the Reynolds number Re, as given by Eq. (9.71).

9.7.6 Bernoulli's Theorem

Let us consider a non-viscous, incompressible fluid flowing through a tube, as shown in Fig. 9.35.

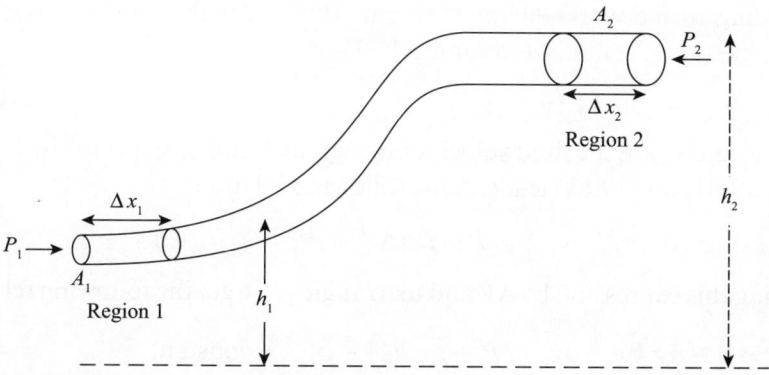

Fig. 9.35 Non-viscous, incompressible fluid flowing through tube

The pressure at the lower end of the tube is P_1 and at the upper end P_2. A_1 and A_2 represent the respective areas of cross section. The centre of the lower end of the tube is at a height h_1 from some arbitrary reference level, whereas a similar height of the upper end of the tube is h_2. The work done W_1 to move a small element of the fluid through a distance Δx_1 in region 1 is given as follows:

$$W_1 = P_1 A_1 \Delta x_1 \qquad (9.76)$$

The quantity $A_1 \Delta x_1$ represents volume ΔV of the fluid moving in this process. Thus, Eq. (9.76) can be rewritten as follows:

$$W_1 = P_1 \Delta V \qquad (9.77)$$

Since the fluid is assumed to be incompressible, the same volume of fluid moves in region 2 of the tube. The distance moved in region 2 is Δx_2. We thus have the following relation for the work done W_2 in moving this elemental fluid:

$$W_2 = -P_2 A_2 \Delta x_2 \qquad (9.78)$$

The negative sign indicates that the liquid element moves in the direction opposite to that of the applied pressure. Since the liquid is incompressible, we must have the following relation:

$$A_1 \Delta x_1 = A_2 \Delta x_2 = \Delta V$$

resulting in

$$W_2 = -P_2 \Delta V \qquad (9.79)$$

The mass m of an element of the moving fluid is given as follows: $m = \rho \Delta V$
where ρ is the density of the fluid.

The work done W_g against gravity in moving this mass of fluid from a height h_1 to a height h_2 is then given by the following expression:

$$W_g = \rho \Delta V g (h_2 - h_1) \qquad (9.80)$$

The net work done W_{net} can, thus be expressed as follows:

$$W_{net} = (W_1 + W_2) - W_g$$

which yields the following relation:

$$W_{net} = (P_1 - P_2) \Delta v - \rho \Delta V g (h_2 - h_1) \qquad (9.81)$$

According to the work–energy theorem, W_{net} is equal to the net change in kinetic energy of a fluid of volume ΔV. Thus,

$$W_{net} = \frac{1}{2}\rho \, \Delta V \left(v_2^2 - v_1^2\right) \tag{9.82}$$

where v_1 and v_2 are the fluid velocities in regions 1 and 2, respectively. The use of Eqs (9.81) and (9.82) leads to the following relation:

$$\frac{1}{2}\rho \, \Delta V \left(v_2^2 - v_1^2\right) = \left(P_1 - P_2\right)\Delta V - \rho \, \Delta V \, g\left(h_2 - h_1\right)$$

Dividing this expression by ΔV and rearranging, we get the following relation:

$$P_1 + \rho g \, h_1 + \frac{1}{2}\rho v_1^2 = P_2 + \rho g \, h_2 + \frac{1}{2}\rho v_2^2 = \text{constant} \tag{9.83}$$

Equation (9.83) expresses the *Bernoulli's theorem*. Thus, according to the Bernoulli's theorem, *for an ideal fluid (incompressible and non-viscous) in a streamline flow, the sum of pressure per unit volume (P), potential energy per unit volume (ρgh), and kinetic energy per unit volume* $\left(\dfrac{1}{2}\rho v^2\right)$ *is a constant independent of the cross section of the fluid flow.*

Example 9.15   Determine the terminal speed of an oil drop of radius 2×10^{-5} m in Millikan's soil drop experiment. Density of the oil used is 1.2×10^3 kg/m³. Assume the coefficient of velocity of oil at the temperature of experiment to be 1.8×10^{-5} N s/m². Neglect the buoyant force on the drop due to air. In addition, calculate the viscous force at the terminal velocity.

Solution   The calculations are as follows:

$$\text{Terminal velocity}\, v_t = \frac{2r^3(\rho - \rho')g}{9\eta}$$

where r is the radius, ρ the density of oil, ρ' the density of air, and η the coefficient of viscosity of air. If the buoyancy due to air is neglected, this expression gets simplified to the following form:

$$v_t = \frac{2}{9}\frac{2\times\left(2\times 10^{-5}\right)^2 \times 1.2\times 10^3 \times 9.8}{9\times 1.8\times 10^{-5}} \tag{9.84}$$

leading to $v_t = 0.058 = 5.8 \times 10^{-2}$ m/s^{-1}

Viscous force F on the drop can be obtained from the following relation: $F = 6\pi\eta r v$

Use of the given values in this aforementioned expression leads to the following expression:

$$F = 6\times\frac{22}{7}\times\left(1.8\times 10^{-5}\right)\times\left(2\times 10^{-5}\right)\left(5.8\times 10^{-2}\right)$$

yielding $F = 3.94 \times 10^{-10}$ N

Example 9.16   Water is being carried by a horizontal pipe of 25 cm² cross section at a velocity of 3 m/s. The supply feeds into a small-bore pipe of 15 cm² cross section. Determine the velocity of water in the pipe of smaller cross section.

Solution   Let v_1 and A_1 represent the velocity and area of cross section of the larger pipe, and v_2 and A_2 are the corresponding quantities for the smaller pipe. From the equation of continuity, we have the following relation: $v_2 A_2 = v_1 A_1$

leading to $v_2 = \dfrac{3 \times 25}{15} = 5$ m/s

Note: One of the most important applications of the Bernoulli's theorem is in the aerodynamics of a flying airplane. The wings of an aircraft are given a special shape to produce a lifting force. This shape is referred to as an airfoil. Air (fluid) flowing over an airfoil travels a longer distance in comparison to the air flowing under the bottom side of the airfoil. The air, therefore, travels with a higher velocity over the top surface to maintain a streamline flow. The shape of wings also results in a crowding of streamlines over the surface; according to the Bernoulli's theorem, the region below the wing experiences a higher pressure relative to the region above it. This produces a lifting force.

IMPORTANT CONCEPTS

1. Vectors have both magnitude and direction.
2. Two vectors are said to be co-initial if their initial point is common.
3. Vectors can be added following the parallelogram law of vector addition and triangle law of vector addition.
4. Scalar product and vector cross product are the two ways in which vectors can be multiplied.
5. Central forces can be represented as follows: $F = F(r) = F(|r|) \, \hat{r}$
6. The entire mass of a multi-particle system can be assumed to be concentrated at the centre of mass.
7. Stress $= \dfrac{\text{Internal restoring force}}{\text{area}}$
8. Beyond the elastic limit, a material does not return to its original length on removal of the deforming force.
9. The net angular momentum of a system does not change with time in the absence of a net external torque.
10. The property of a liquid by virtue of which it opposes relative motion between its constituent layers is called its viscosity.
11. Stokes' law is given by the following relation: $F = 6\pi \eta \gamma v$
12. Any particle of a liquid in a streamline flow follows exactly the same path and velocity as its predecessor.

    The expression $vA = $ constant is called the equation of continuity.
13. Reynolds number, Re, is given by the following equation: $\text{Re} = \dfrac{\rho v L}{\eta}$
14. According to the Bernoulli's theorem, $P + \rho g h + \dfrac{1}{2}\rho v^2 = $ constant

APPLICATIONS

1. The tensile strength of bone fibres is due to the protein collagen. The compressive strength of bones is due to the presence of inorganic salt crystals. Bones have an ultimate strength (both in tension and in compression) that is greater than

that of concrete. The ultimate strength of materials is indicated in multiples of 10^8 N/m^2. For concrete under tensile stress, the ultimate strength is 0.02, for bone it is 1.3, and for cast iron it is 1.7. The comparison is difficult to accept because the volume of concrete used in most situations is much greater than the volume of any bone in the body.

2. Very thick blood can damage blood vessels and increase the risk of having a heart attack. Physicists have shown that electric and magnetic fields can be used to reduce the viscosity of blood by up to 30%. This is because red blood cells contain iron. A magnetic field polarizes the red blood cells, causing them to form short chains. These chains are larger than single blood cells, which lead to reduced friction against the walls of the blood vessels. The conventional method of reducing the viscosity of blood is to use drugs like aspirin, but with significant side effects.

3. Flying of kites demonstrates the Bernoulli's principle. The air above the kite moves faster than the air below, resulting in a pressure difference that leads to its flight. Perfume atomizers, chimneys, and sprays all work based on the Bernoulli's theorem. Carburettors present in older vehicles also work on the Bernoulli's theorem.

IMPORTANT FORMULAE

1. $r_{cm} = \dfrac{m_1 r_1 + m_2 r_2 + m_3 r_3 + \cdots + m_n r_n}{(m_1 + m_1 + m_3 + \cdots + m_n)}$ can be used to determine the position of the centre of mass.

2. $\tau = r \times F$ can be used to determine torque.

3. $L = r \times p$ can be used to determine angular momentum.

4. $v_t = \dfrac{2r^2 g}{9\eta}(\rho - \rho')$ can be used to determine terminal velocity.

SELF-ASSESSMENT

Multiple-choice Questions

9.1 The vectors $\hat{\imath}$, $\hat{\jmath}$, and $\hat{k}$ have a magnitude of
   (a) 0   (b) 1   (c) ∞ (d) 2

9.2 Co-initial vectors have the same
   (a) initial point
   (b) mid-point
   (c) end point
   (d) both initial and end points

9.3 Scalar product of two vectors has a magnitude that is proportional to
   (a) $\dfrac{1}{\cos\theta}$ (b) $\sin\theta$ (c) $\dfrac{1}{\sin\theta}$ (d) $\cos\theta$

9.4 Which one of the following is not a central force?
   (a) Coulomb force
   (b) Frictional force
   (c) Gravitational force
   (d) Viscous force

9.5 Stress is proportional to
   (a) area   (b) $\dfrac{1}{(\text{area})}$ (c) $(\text{area})^2$ (d) $\dfrac{1}{(\text{area})^2}$

9.6 Shear strain is measured in units of
(a) radian   (b) centimetre   (c) gram   (d) second

9.7 Identify the brittle material:
(a) Iron   (b) Lead   (c) Copper   (d) Glass

9.8 Angular momentum is proportional to
(a) $\dfrac{1}{p}$ (b) p^2 (c) p (d) $\dfrac{1}{p^2}$

9.9 The SI unit of coefficient of viscosity is
(a) Pa-s   (b) Pa   (c) Pa-m   (d) $\dfrac{1}{Pa}$

9.10 Terminal velocity is negative if
(a) $\rho > \rho'$ (b) $\rho = \rho'$ (c) $\rho < \rho'$ (d) $\rho = 2\rho'$

9.11 Select the correct option:
(a) $\mathbf{a} = \dfrac{a}{\hat{a}}$ (b) $\mathbf{a} = a\hat{a}$ (c) $\mathbf{a} = \dfrac{\hat{a}}{a}$ (d) $\mathbf{a} = a^2\hat{a}$

9.12 Equal vectors have
(a) equal magnitudes
(b) equal directions
(c) equal magnitudes and directions
(d) equal magnitudes but different directions

9.13 Collinear vectors
(a) lie in the same plane
(b) act along the same straight line
(c) lie in different planes
(d) lie in the same plane and act along the same straight line

9.14 Coplanar vectors
(a) lie in the same plane   (c) coexist or exist together
(b) act along the same line   (d) cooperate with each other

9.15 Multiplication of a vector with a real number results in
(a) a scalar   (c) either a scalar or a vector
(b) a vector   (d) a real number

9.16 On multiplication by a scalar, the units of a vector
(a) do not change   (c) sometimes change
(b) always change   (d) disappear

9.17 $A + B$ can be obtained by
(a) using the triangle law of addition
(b) using the parallelogram law of addition
(c) just adding the magnitudes
(d) both (a) and (b)

9.18 $\dfrac{A_y}{A_x}$ is given by
(a) $\tan\theta$ (b) $\dfrac{1}{\tan\theta}$ (c) $\tan^2\theta$ (d) $\dfrac{1}{\tan^2\theta}$

9.19 For $\theta = 0$, $A \cdot B$ is
(a) 0   (b) AB (c) A/B (d) B/A

9.20 For $\theta = 90°$, $A \cdot B$ is
(a) AB (b) 90   (c) 0   (d) A/B

9.21 For $\theta = 180°$, $A \cdot B$ is
(a) AB (b) 0   (c) BA (d) $-AB$

9.22 For $\theta = 0°$, $|\mathbf{A} \times \mathbf{B}|$ is
    (a) 0   (b) AB (c) A/B (d) B/A

9.23 For $\theta = 180°$, $|\mathbf{A} \times \mathbf{B}|$ is
    (a) AB (b) 0   (c) A/B (d) B/A

9.24 $\hat{i} \times \hat{i}$ is
    (a) 1   (b) 2   (c) 0   (d) −1

9.25 For $\theta = 90°$, $|\mathbf{A} \times \mathbf{B}|$ is
    (a) 0   (b) A/B (c) $-AB$ (d) AB

9.26 Choose the correct expression for angular momentum:

    (a) $\mathbf{L} = \mathbf{r} \times \dfrac{m\,d\mathbf{r}}{dt}$ (c) $\mathbf{L} = \dfrac{m\,d\mathbf{r}}{dt} \times \hat{r}$

    (b) $\mathbf{L} = \mathbf{r} \times \dfrac{d\mathbf{r}}{dt}$ (d) $\mathbf{L} = \mathbf{r} \times m\mathbf{r}$

9.27 In a central force field,
    (a) energy is not conserved
    (b) energy is conserved
    (c) angular momentum is not conserved
    (d) net torque is not zero

9.28 Choose the correct option:
    (a) $Ma_{cm} < f_{ext}$ (c) $Ma_{cm} = f_{ext}$
    (b) $Ma_{cm} > f_{ext}$ (d) $Ma_{cm} \geq f_{ext}$

9.29 Linear strain
    (a) has dimension of cm   (c) has dimensions of cm^2
    (b) has dimensions of cm^{-1} (d) is dimensionless

9.30 Permanent set exists in a material
    (a) beyond the elastic limit   (c) at the elastic limit
    (b) below the elastic limit   (d) both (a) and (b)

Review Questions

9.1 What are vectors?

9.2 What is a unit vector?

9.3 When are two vectors considered equal?

9.4 What are co-initial vectors?

9.5 What are collinear vectors?

9.6 What are coplanar vectors?

9.7 Explain the parallelogram law of vector addition using a suitable diagram.

9.8 Explain the triangle law of vector addition using a suitable diagram.

9.9 What do you understand by the scalar product of two vectors?

9.10 Give examples of the cross product of two vectors.

9.11 What is a central force?

9.12 Give an expression for the position vector of the centre of mass of a multi-particle system.

9.13 Define stress and give its units.

9.14 Differentiate between tensile and compressive stress.

9.15 What is tangential stress?

9.16 Write an expression for torque.

9.17 State and explain the Newton's law of viscous force.

9.18 State the SI unit of coefficient of viscosity.

9.19 State and explain Stoke's law.

9.20 Derive an expression for the terminal velocity of a spherical body falling through a viscous medium.

9.21 What is a streamline flow?

9.22 State and explain the equation of continuity.

9.23 Write an expression for Reynolds number. What is its significance?

9.24 Derive the Bernoulli's theorem.

9.25 What is an ideal fluid?

Numerical Problems

9.1 $\vec{Q}$ makes an angle of 60° with $\vec{P}$ where $|\vec{Q}| = 8$ N and $|\vec{P}| = 8$ N. Find $\vec{P} + \vec{Q}$.

$$\left[\text{Hint: } |\vec{P} + \vec{Q}| = \sqrt{P^2 + Q^2 + 2PQ \cos 60^0} \right]$$

9.2 A man can swim with a speed of 4 km/h in still water. How long does he take to cross a river of 1 km width, if the river flows steadily at 3 km/h and he swims normal to the flow.

$$\left[\text{Hint: } v' = \sqrt{16 + 9} \right]$$

9.3 For problem 9.2, how far down the river does the man reach the opposite bank?

[Hint: Distance = Time × Velocity]

9.4 Add the following vector and find the magnitude: $\vec{P} = 3\hat{\imath} + 4\hat{\jmath}$ and $\vec{Q} = 8\hat{\imath} + 3\hat{\jmath}$

[Hint: $\vec{P} + \vec{Q} = (P_x + Q_x)\hat{\imath} + (P_y + Q_y)\hat{\jmath}$]

9.5 What should be the angle between two vectors of equal magnitude so that their resultant has a magnitude equal to that of either vector?

[Hint: $P^2 = P^2 + P^2 + 2P^2 \cos \theta$]

9.6 A stone attached to a string is whirled around in a horizontal circle. The stone is moving at a rate of 1 rad/s. Determine the angular velocity of the stone if the radius of the circumscribed circle is made one-fourth the original value.

[Hint: $mr_1^2 w_1 = mr_2^2 w_2$]

9.7 A glass bottle contains glycerine at a temperature of 20°C. An aluminium ball of radius 1 mm is dropped into it. Determine the terminal velocity of the ball. Assume the density of glycerine to be 1.26×10^3 kg/m³ and that of aluminium 2.7×10^3 kg/m³. The coefficient of viscosity for glycerine is 1.47/Pas.

$$\left[\text{Hint: } \frac{P}{\rho} + gh + \frac{1}{2}v_1^2 = \frac{P_2}{\rho} + gh + \frac{1}{2}v_2^2 \right]$$

9.8 A model of an aeroplane is being tested in a wind tunnel experiment. The flow speeds on the lower and upper surfaces of the wing are 63 and 70 m/s, respectively. The area of the wing is 2.5 m². Calculate the lift produced, assuming the density of air to be 1.3 kg/m³.

$$\left[\text{Hint: } \frac{P}{\rho} + gh + \frac{1}{2}v_1^2 = \frac{P_2}{\rho} + gh + \frac{1}{2}v_2^2 \right]$$

9.9 $|\vec{P}| = 4.8$ N and $|\vec{Q}| = 4.2$ N where the vectors $\vec{P}$ and $\vec{Q}$ are mutually perpendicular. Find the angle that the resultant makes with the vector $\vec{P}$.

$$\left[\text{Hint: } \tan \alpha = \frac{|\vec{Q}|}{|\vec{P}|} \right]$$

9.10 $\vec{P}$ and $\vec{Q}$ are two mutually perpendicular vectors. The resultant of the two vectors makes an angle of 36° with the vector $\vec{P}$. If the magnitude of $\vec{P}$ is 4.3 N, find the magnitude of $\vec{Q}$. $\left[\text{Hint: } \tan\alpha = \dfrac{|\vec{Q}|}{|\vec{P}|}\right]$

9.11 $\vec{P}$ and $\vec{Q}$ are two mutually perpendicular vectors. The resultant of the two vectors makes an angle of 32° with the vector $\vec{P}$. If the magnitude of $\vec{Q} = 4.3\text{N}$, find the magnitude of $\vec{P}$. $\left[\text{Hint: } \tan\alpha = \dfrac{|\vec{Q}|}{|\vec{P}|}\right]$

9.12 Rain is falling vertically with a speed of 30.6 m/s. If wind is blowing with a speed of 11.2 m/s from to north, in what direction should a girl hold her umbrella to protect herself? $\left[\text{Hint: } \tan\theta = \dfrac{11.2}{30.6}\right]$

9.13 Rain is falling vertically with a speed of 26.2 m/s and wind is blowing from north to south. A lady wanting to protect herself from rain has to hold an umbrella at an angle of 22.3° with respect to the vertical. Calculate the magnitude of wind speed. $\left[\text{Hint: } \tan 22.3 = \dfrac{x}{26.2}\right]$

9.14 Rain is falling vertically. Wind is blowing at a speed of 28.3 m/s from north to south. A girl wanting to protect herself from rain holds the umbrella at an angle of 26.2° w.r.t the vertical. Calculate the speed at which rain is falling vertically. $\left[\text{Hint: } \tan\theta = \dfrac{28.3}{x}\right]$

9.15 A woman is riding a bicycle. The crank arm of the bicycle pedal is 162.5 cm long. The woman puts all her weight on one pedal. The torque produced when the crank is horizontal is 727.8 N m. Calculate the weight of the woman.
[Hint: $\tau = \mathbf{r} \times \mathbf{f}$]

9.16 A 42.8 kg woman is riding a bicycle. The woman puts all her weight on one pedal. The torque produced when the crank is horizontal is 690.2 N m. Calculate the length of the crank arm. [Hint: $\tau = rF$]

9.17 A stone attached to a string is whirled around in a horizontal circle, as shown in Fig. 9.29. The stone is moving at a rate of 0.48 rad/s. Calculate the angular velocity of the stone if the radius of the circumscribed circle is reduced by a factor of 1/1.4. $\left[\text{Hint: } mr_1^2 w_1 = mr_2^2 w_2\right]$

9.18 The angular velocity of a stone indicated in Problem 9.17 is changed to 0.52 rad/s and the radius is increased by a factor of 1.2. Calculate the new angular velocity of the stone. $\left[\text{Hint: } w_1 = \dfrac{r_2^2}{r_1^2} w_2\right]$

Relativistic Mechanics

Learning Objectives

After studying this chapter, students will be able to

- understand the meaning of inertial and non-inertial reference frames
- explain Galilean transformations
- comprehend the details of Michelson–Morley experiment
- describe the postulates of special theory of relativity
- elucidate Lorentz transformations
- describe length contraction and time dilation
- demonstrate relativistic addition of velocities
- explain how mass changes with velocity
- understand the equivalence of mass and energy
- solve numerical problems based on length contraction, time dilation, relativistic addition of velocities, change of mass with velocity, and mass–energy equivalence

List of Symbols

x, y, z = Coordinate system
v = Velocity of reference frame
t = Time
v_x, v_y, v_z = Components of velocity

i, j, k = Unit vectors in x, y, and z directions
u = Particle velocity
a = Particle acceleration
c = Velocity of light
Δt = Time interval
λ = Wavelength

δ = Optical path difference
l = Length
m = Mass
F = Force
k = Kinetic energy
m_0 = Rest mass
E = Total energy

10.1 INTRODUCTION

Albert Einstein realized the importance of a reference frame in the observation of physical phenomena. Just as Newton laid the foundation of classical mechanics, Einstein's theory of relativity laid the foundation of modern physics. He presented his theory of relativity in the year 1905, which forced physicists to revise their theories, especially in the context of velocities approaching or comparable to the velocity of light. Even a quantity as simple as mass was found to vary with velocity. Moreover, mass and energy were found to be interchangeable, and this led to some profound consequences. We will study some important aspects of relativistic mechanics in this chapter. This chapter will introduce important concepts such as inertial and non-inertial reference frames. Michelson–Morley experiment would be discussed in detail and its inferences would be presented. This chapter will also discuss the principle of mass–energy equivalence.

It is important to remember that several events in our cosmos involve moving reference frames and velocities approaching the velocity of light. The study of these phenomena, which involve stars, black holes, galaxies, comets, and particle physics, benefits a lot from an understanding of relativity. The dependence of mass on velocity has, in fact, been experimentally verified. The equivalence of mass and energy has resulted in exploitation of nuclear energy for both generation of power and destructive purposes.

10.2 INERTIAL AND NON-INERTIAL FRAMES

The motion of any object can be analysed with respect to a particular choice of coordinate system. This chosen coordinate system is called the reference frame or frame of reference. This reference frame is then used to understand the motion of bodies under the influence of external forces. The frames of reference used to analyse the motion of an object are of two types: inertial reference frames and non-inertial reference frames.

10.2.1 Inertial Reference Frames

The three coordinates x, y, and z, and the time t define a reference frame. All physical phenomena need a reference frame for understanding and analysis. If the reference frame is either at rest or moving with a uniform velocity, then it is called an *inertial reference frame*. Suppose a person is standing on the bank of a river and another person is standing on a boat that is moving through a river at a constant velocity V. The reference frame used by the person on the bank to observe physical phenomena and the reference frame used by the person on the boat are both inertial reference frames. All inertial reference frames are equivalent in terms of physical laws. Specific values of measured physical quantities may vary between specific choices of reference frames, but the concluded physical laws do not change. To take a specific example—suppose that the person on the bank observes a newly developed model of an aircraft and the same is observed by the person on the boat. The observation of the behaviour of the aircraft would then not be dependent upon the two reference frames.

10.2.2 Non-inertial Reference Frames

An accelerating reference frame is called a non-inertial reference. The inertial reference frame discussed in Section 10.2.1 has one defining feature. A body not acted upon by an external force does not undergo any acceleration in such a reference frame. On the contrary, there are reference frames in which a body can experience acceleration even in the absence of any external force. Such reference frames are called non-inertial reference frames. In non-inertial reference frames, bodies do not obey the Newton's laws. If the boat in the situation discussed in Section 10.2.1 did not move with a uniform velocity, then the reference frame used by the person standing on the boat would cease to be an inertial reference frame.

10.3 GALILEAN TRANSFORMATIONS

Different inertial reference frames may have different values for physical quantities, but the laws of physics do not depend upon any specific choice of inertial reference frame. Galilean transformations are used for transformation of position coordinates and time from one inertial reference frame to another. Figure 10.1 is a representation of the two reference frames.

The two reference frames are represented by S_1 and S_2 and the coordinate systems by x_1, y_1, z_1 and x_2, y_2, z_2, respectively. Reference frame S_2 is moving with a uniform velocity v with respect to reference frame S_1. The point P is the location in space where a particular event is taking place. The coordinates of P in reference frame S_1 are x_1, y_1, z_1, t_1 and in reference frame S_2 are x_2, y_2, z_2, t_2. To simplify matters, the x, y, z axes of the two coordinate systems are assumed to be parallel to each other. Let us start counting time from the moment when the origins O_1 and O_2 coincide.

From Fig. 10.1, we can write the following equations:

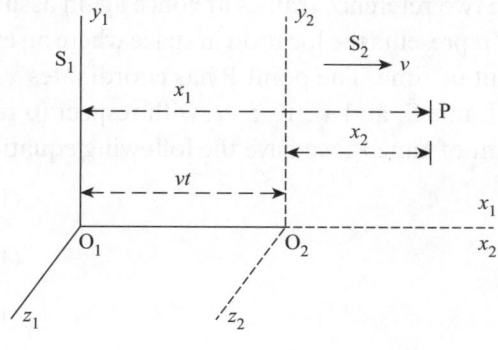

Fig. 10.1 Schematic representation of two reference frames

$$x_2 = x_1 - vt_1 \tag{10.1}$$

$$y_2 = y_1 \tag{10.2}$$

$$z_2 = z_1 \tag{10.3}$$

In addition,

$$t_2 = t_1 \tag{10.4}$$

Equations (10.1)–(10.4) represent *the Galilean transformation.*

In deriving Eqs (10.1)–(10.4), it was assumed that the corresponding axes in the two reference frames are parallel to one another. This may not always be true. Figure 10.2 shows such a situation. Reference frame S_2 moves with respect to S_1 along a straight line in any direction.

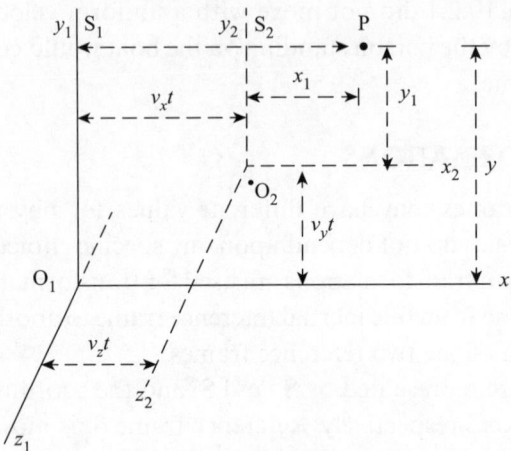

Fig. 10.2 Relative movement of reference frames

Let us assume that reference frame S_2 moves w.r.t. frame S_1 with a velocity v, given as follows:

$$V = iv_x + jv_y + kv_z \tag{10.5}$$

where v_x, v_y, and v_z are components of velocity along x_1, y_1, and z_1 axes, respectively.

The origins O_1 and O_2 of the two reference frames are once again assumed to coincide at $t = 0$. The point P represents the location in space where an event is taking place at a given instant of time. The point P has coordinates x_1, y_1, z_1, t_1 with respect to reference frame S_1 and x_2, y_2, z_2, t_2 with respect to reference frame S_2. At a given instant of time t_1, we have the following equations:

$$x_2 = x_1 - v_x t_1 \tag{10.6}$$

$$y_2 = y_1 - v_y t_1 \tag{10.7}$$

$$z_2 = z_1 - v_z t_1 \tag{10.8}$$

$$t_2 = t_1 \tag{10.9}$$

Equations (10.6)–(10.9) represent the *Galilean transformation for position and time.* Let us now derive an expression for Galilean transformation for velocity. Differentiation of Eqs (10.6)–(10.9) yields the following expressions:

$$dx_2 = dx_1 - v_x dt_1 \tag{10.10}$$

$$dy_2 = dy_1 - v_y dt_1 \tag{10.11}$$

$$dz_2 = dz_1 - v_z dt_1 \tag{10.12}$$

$$dt_2 = dt_1 \tag{10.13}$$

Equations (10.10)–(10.12) assume v_x, v_y, and v_z to be constant quantities.

Equations (10.10)–(10.12) lead to the following relations:

$$\frac{dx_2}{dt_2} = \frac{dx_1}{dt_1} - v_x \tag{10.14}$$

$$\frac{dy_2}{dt_2} = \frac{dy_1}{dt_1} - v_y \tag{10.15}$$

and

$$\frac{dz_2}{dt_2} = \frac{dz_1}{dt_1} - v_z \tag{10.16}$$

Implying that

$$u_{x_2} = u_{x_1} - v_x \tag{10.17}$$

$$u_{y_2} = u_{y_1} - v_y \tag{10.18}$$

$$u_{z_2} = u_{z_1} - v_z \tag{10.19}$$

where u_{x_1}, u_{y_1}, and u_{z_1} represent the particle velocities in reference frame S_1 and u_{x_2}, u_{y_2}, and u_{z_2} represent the corresponding velocities in reference frame S_2.

In terms of unit vector i, j, and k along the three axes, we can then write, using Eqs (10.17)–(10.19), the following relations:

$$(iu_{x_2} + ju_{y_2} + ku_{z_2}) = (iu_{x_1} + ju_{y_1} + ku_{z_1}) - (iv_x + jv_y + kv_z) \tag{10.20}$$

Giving,

$$u_2 = u_1 - v \tag{10.21}$$

Equation (10.21) represents the *Galilean transformation for velocity of a particle*.

Differentiating Eq. (10.20), we get the following expression:

$$du_2 = du_1 \tag{10.22}$$

Here, we assume velocity v to be a constant quantity. From Eq. (10.22), we can conclude that *a change in velocity observed in reference frame S_2 is equal to the change observed in reference frame S_1.*

From Eq. (10.22), we can also infer that

$$a_2 = \frac{du_2}{dt_2} = \frac{du_1}{dt_1} = a_1 \tag{10.23}$$

Thus, *the particle P, possesses the same acceleration in the two reference frames, S_1 and S_2.*

We must, however, remember that the equal nature of acceleration in the two reference frames S_1 and S_2 has been shown only for reference frames that move with respect to each other with a uniform velocity.

10.4 MICHELSON–MORLEY EXPERIMENT

In the early days of wave theory of light, it was thought that space is completely filled with a material called *ether*. Ether was supposed to have the following properties: being invisible, massless, perfectly transparent, perfectly non-resistive, and at absolute rest. Earth was supposed to move through ether without producing any disturbance. Thus, the presence of ether would enable the determination of absolute velocity of the earth with respect to the medium ether. An experiment to determine this absolute velocity of the earth was carried out by Michelson and Morley.

The experimental arrangement used by Michelson and Morley is shown in Fig. 10.3.

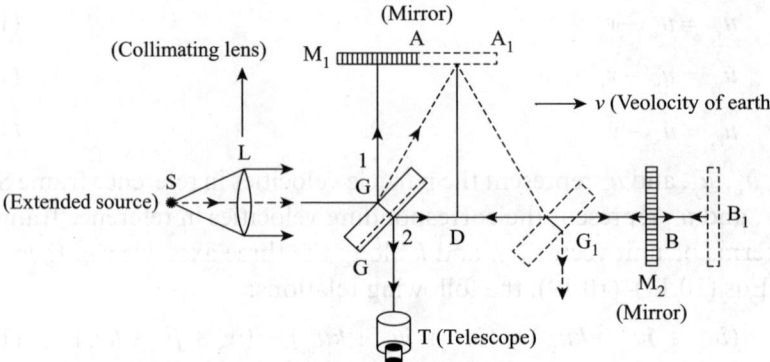

Fig. 10.3 Experimental set-up used by Michelson and Morley

The arrangement consists of an extended source S, which emits a monochromatic beam. This monochromatic beam is rendered parallel due to the collimating lens L. This parallel beam is then incident on the semi-silvered glass plate G, which is placed at an angle of 45° with respect to the incident beam. This splits the beam into two parts. One is reflected from the semi-silvered surface of the glass plate G and travels towards mirror M_1. The other part gets transmitted through the glass plate G and moves towards mirror M_2. Mirrors M_1 and M_2 are held perpendicular to the respective incident beams, which then retrace their original paths, as shown in Fig. 10.3. The reflected rays then meet at the semi-silvered surface of the plate G and finally enter the telescope, where an interference pattern is formed. A compensating plate is used to ensure that the optical paths of rays from glass plate G to mirrors M_1 and M_2 are made equal.

If the apparatus is assumed to be at rest with respect to ether, then the two reflected rays would take equal time to return to plate G. The apparatus is, however, moving along with the earth. Let us assume that the direction of motion of the earth is along the direction of the initial beam. The optical paths traversed by the two rays are not now equal due to the motion of the earth. The actual reflection from the two mirrors takes place at A_1 and B_1 instead of at A and B, as shown in Fig. 10.3.

Let us assume that the two mirrors M_1 and M_2 are at an equal distance d from glass plate G. In addition, assume that c is the velocity of light and v the velocity of the earth. Obviously, v is also the speed of the apparatus. The total path followed by ray 1 is then from G to A_1 and back to G_1. The law of reflection implies that

$$GG_1 = 2GD = 2AA_1 \tag{10.24}$$

Thus,

$$GA_1G_1 = GA_1 + A_1G_1 = 2GA_1 \tag{10.25}$$

In addition,

$$\left(GA_1\right)^2 = \left(GD\right)^2 + \left(A_1D\right)^2 \tag{10.26}$$

Use of Eq. (10.24) in Eq. (10.26) yields the following relation:

$$\left(GA_1\right)^2 = \left(AA_1\right)^2 + \left(A_1D\right)^2 \tag{10.27}$$

Suppose that t represents the time taken by the ray to travel from G to A_1; then using Eq. (10.27), one gets the following relation:

$$\left(ct\right)^2 = \left(vt\right)^2 + \left(d\right)^2$$

yielding $(c^2 - v^2)\, t^2 = d^2$

resulting in,

$$t = \frac{d}{\sqrt{c^2 - v^2}} \tag{10.28}$$

Suppose that t_1 represents the time taken by the ray to travel the path GA_1G_1; then we have the following relation:

$$t_1 = 2t = \frac{2d}{\sqrt{c^2 - v^2}}$$

which can be rewritten in the following form:

$$t_1 = \frac{2d}{c\left(1 - \dfrac{v^2}{c^2}\right)^{\frac{1}{2}}}$$

or,

$$t_1 = \frac{2d}{c}\left(1 - \frac{v^2}{c^2}\right)^{-\frac{1}{2}} \tag{10.29}$$

We also know that the binomial series expansion of $(1 + x)^{-m}$ is given by the following equation:

$$(1+x)^{-m} = 1 - mx + \frac{m(m+1)x^2}{2!} - \frac{m(m+1)(m+2)x^3}{3!} + \cdots \qquad (10.30)$$

Since $\frac{v}{c} \ll 1$, taking only the first term of the binomial series, Eq. (10.29) leads to the following form:

$$t_1 = \frac{2d}{c}\left(1 + \frac{v^2}{2c^2}\right) \qquad (10.31)$$

Transmitted ray 2 moves longitudinally towards mirror M_2. This ray has a velocity $(c - v)$ with respect to the apparatus as it moves from G towards B, and a velocity $(c + v)$ during the opposite journey. The total time t_2, taken by ray 2 to reach G_1 is given as follows:

$$t_2 = \frac{d}{(c-v)} + \frac{d}{(c+v)} \qquad (\because \text{GB} = G_1 B_1 = d) \qquad (10.32)$$

which leads to the following form:

$$t_2 = \frac{d(c+v) + d(c-v)}{(c^2 - v^2)} \qquad (10.33)$$

yielding

$$t_2 = \frac{2dc}{(c^2 - v^2)} = \frac{2d}{c}\left[1 - \frac{v^2}{c^2}\right]^{-1} \qquad (10.34)$$

Using the binomial series expansion [Eq. (10.30)] and keeping only the first two terms, we get the following expression:

$$t_2 = \frac{2d}{c}\left[1 - \frac{v^2}{c^2}\right] \qquad (10.35)$$

The difference between the travel times of the two rays, using Eqs. (10.31) and (10.35), is expressed as follows:

$$\Delta t = t_2 - t_1 = \frac{2d}{c}\left[1 + \frac{v^2}{c^2}\right] - \frac{2d}{c}\left[1 + \frac{v^2}{2c^2}\right] \qquad (10.36)$$

Simplifying this equation, we get the following relation:

$$\Delta t = \frac{2d}{c} \cdot \frac{v^2}{2c^2} = \frac{dv^2}{c^3} \qquad (10.37)$$

The optical path difference between the two rays is then expressed as follows:

$$\delta = \frac{\text{Path difference}}{\text{Wavelength}} = \frac{c(\Delta t)}{\lambda} \qquad (10.38)$$

Using Eq. (10.37) in Eq. (10.38), we get the following expression:

$$\delta = \frac{c}{\lambda}\left(\frac{dv^2}{c^3}\right) = \frac{dv^2}{\lambda c^2} \qquad (10.39)$$

Michelson and Morley experiment was conducted in two steps. The configuration depicted in Fig. 10.3 was first used to obtain an interference fringe pattern. The experimental set-up was then turned through 90°. This changes the position of the two mirrors, and the new optical path difference becomes as follows:

$$\delta_1 = \frac{-dv^2}{\lambda c^2} \qquad (10.40)$$

The difference between Eqs (10.39) and (10.40) is given by the following equation: $\Delta\delta = \delta - \delta_1 = \dfrac{dv^2}{\lambda c^2} - \left(\dfrac{-dv^2}{\lambda c^2}\right)$

resulting in

$$\Delta\delta = \frac{2dv^2}{\lambda c^2} \qquad (10.41)$$

Expression (10.41) also gives the fringe shift in the observed interference pattern. For the Michelson and Morley experiments, the following data has to be used:

$$d = 1.0 \times 10^3 \, \text{cm}, \; \lambda = 5 \times 10^{-5} \, \text{cm}, \; v = 3 \times 10^6 \, \text{cm}, \; \text{and} \; c = 3 \times 10^{10} \, \text{cm/s}$$

Use of these values in Eq. (10.41) leads to the following expression:

$$\Delta\delta = \frac{2 \times 1.0 \times 10^3 \times (3 \times 10^6)^2}{5 \times 10^{-5} \times (3 \times 10^{10})^2}$$

yielding $\Delta\delta = 0.4$ fringe

In the actual experiment, Michelson and Morley could observe a shift of about 0.01 fringe. Moreover, a shift of 0.01 fringe was within the limits of error of observation. The experiment was repeated by Michelson and Morley at different points on the surface of the earth and in different seasons of a year. However, no measurable shift in the interference pattern was observed by them. *The extensive experimentation thus failed to observe the expected shift of 0.4 fringe.*

The negative result thus conclusively proved the fallacy of the assumption of the all-pervading ether. Ether as a fixed frame of reference was thus discarded. Hence, it was concluded that the speed of light in vacuum is the same in all the reference frames that are in a uniform relative motion.

10.5 POSTULATES OF SPECIAL THEORY OF RELATIVITY

Experiments conducted by Michelson and Morley and others conclusively rejected the concept of ether. These results were used by Albert Einstein to offer two fundamental postulates in the year 1905, which form the cornerstones of the Einstein's theory of relativity. These postulates are discussed in the following subsections:

10.5.1 Einstein's First Postulate

'All physical laws are the same in all inertial reference frames that are moving relative to each other with constant velocity.'

Thus, according to this postulate, it is not possible to demonstrate *absolute motion* by any means. This postulate also makes it impossible to demonstrate absolute position. The concept of ether is, therefore, invalid. All motions of bodies can be visualized relative to one another. Einstein reasoned that if an observer was located far out in space, far removed from all the stars and planets so that he/she could not use anything as a reference point to determine his/her speed, then he/she can never distinguish between the state of rest and the state of uniform motion. A reference object is, therefore, a must to define the state of motion of a body.

10.5.2 Einstein's Second Postulate

'The speed of light in free space has the same value 'c' in all inertial reference frames.'

Thus, this postulate places an upper limit on the velocity of any object. It also implies that the speed of light does not change even when any change takes place in the direction of measurement. It also does not matter whether the measurement is being carried out by an observer who is stationary, or moving towards or away from the source of light. All the measurements reveal the same value for the velocity of light.

Einstein's theory of relativity based on these two postulates is also called the *special theory of relativity.*

10.6 LORENTZ TRANSFORMATION

Lorentz transformations are a set of equations that help us transform position and time coordinates from one inertial reference frame to another. These equations are derived assuming the constant nature of the velocity of light (Einstein's second postulate). These transformations are, therefore, consistent with the results of Michelson and Morley experiment.

Figure 10.4 shows two inertial reference frames S_1 and S_2. Reference frame S_2 moves at a uniform velocity v with respect to reference frame S_1.

The axes of the two coordinate systems are assumed to coincide at $t_1 = t_2 = 0$. Suppose a pulse of light gets generated at $t_1 = t_2 = 0$ at the common origin $O_1 = O_2$. Let us consider the instant at which the pulse reaches point P. Let us assume that there are two observers, one at O_1 and the other at O_2 in

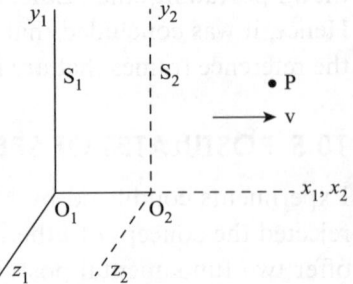

Fig. 10.4 Relative motion of reference frame S_2 w.r.t. S_1

the respective reference frames S_1 and S_2. Coordinates of point P, as observed by the observers at O_1 and O_2, are given by (x_1, y_1, z_1, t_1) and (x_2, y_2, z_2, t_2), respectively.

In reference frame S_1, we have the following equation:

$$C = \text{velocity of light} = \frac{(x_1^2 + y_1^2 + z_1^2)^{\frac{1}{2}}}{t_1} \tag{10.42}$$

yielding

$$x_1^2 + y_1^2 + z_1^2 - c^2 t_1^2 = 0 \tag{10.43}$$

whereas in reference frame S_2, we get the following expression:

$$x_2^2 + y_2^2 + z_2^2 - c^2 t_2^2 = 0 \tag{10.44}$$

Since $y_1 = y_2$ and $z_1 = z_2$, Eqs (10.43) and (10.44) lead to the following expression:

$$x_1^2 - c^2 t_1^2 = x_2^2 - c^2 t_2^2 \tag{10.45}$$

Transformation between x_1 and x_2 can be represented as follows:

$$x_2 = \lambda_1 (x_1 - vt_1) \tag{10.46}$$

where λ_1 is independent of x_1 and t_1.

With respect to reference frame S_2, reference frame S_1 can be assumed to move with a velocity $-v$ along the $+x$ direction. The transformation between x_1 and x_2 can, therefore, also be written in the following form:

$$x_1 = \lambda_2 (x_2 + vt_2) \tag{10.47}$$

Substitution of the expression for x_2 from Eq. (10.46) in Eq. (10.47) yields the following expression:

$$x_1 = \lambda_2 \left[\lambda_1 (x_1 - vt_1) + vt_2 \right] \tag{10.48}$$

yielding $\dfrac{x_1}{\lambda_2} = \lambda_1 (x_1 - vt_1) + vt_2$

leading to $vt_2 = -\lambda_1 (x_1 - vt_1) + \dfrac{x_1}{\lambda_2}$

implying that

$$t_2 = \lambda_1 \left[t_1 - \frac{x_1}{v} \left(1 - \frac{1}{\lambda_1 \lambda_2} \right) \right] \tag{10.49}$$

Substitution of t_2 from Eq. (10.49) and x_2 from Eq. (10.46) in Eq. (10.45) leads to the following expression:

$$x_1^2 - c^2 t_1^2 = \lambda_1^2 (x_1 - vt_1)^2 - c^2 \lambda_1^2 \left[t_1 - \frac{x_1}{v} \left(1 - \frac{1}{\lambda_1 \lambda_2} \right) \right]^2$$

yielding $x_1^2 - c^2 t_1^2 - \lambda_1^2 (x_1 - vt_1)^2 + c^2 \lambda_1^2 \left[t_1 - \frac{x_1}{v} \left(1 - \frac{1}{\lambda_1 \lambda_2} \right) \right]^2 = 0$

giving

$$x_1^2 - c^2 t_1^2 - \lambda_1^2 \left(x_1 - 2x_1 vt_1 + v^2 t_1^2 \right)$$

$$+ c^2 \lambda_1^2 \left[t_1^2 - \frac{2x_1 t_1}{v} \left(1 - \frac{1}{\lambda_1 \lambda_2} \right) + \frac{x_1^2}{v^2} \left(1 - \frac{1}{\lambda_1 \lambda_2} \right) \right]^2 = 0 \qquad (10.50)$$

In Eq. (10.50), the coefficients of various powers of x_1 and t_1 must vanish separately; thus, for the coefficient of t_1^2, we can write the following equation:

$$-c^2 - \lambda_1^2 v^2 + c^2 \lambda_1^2 = 0$$

giving

$$\lambda_1^2 \left(c^2 - v^2 \right) = c^2 \qquad (10.51)$$

yielding

$$\lambda_1^2 = \frac{c^2}{\left(c^2 - v^2 \right)} = \frac{1}{1 - \dfrac{v^2}{c^2}} \qquad (10.52)$$

resulting in

$$\lambda_1 = \frac{1}{\sqrt{\left(1 - \dfrac{v^2}{c^2} \right)}} \qquad (10.53)$$

Use of the coefficients of $x_1 t_1$ from Eq. (10.50) leads to the following expression:

$$2\lambda_1^2 v + c^2 \lambda_1^2 \left[\frac{-2}{v} \left(1 - \frac{1}{\lambda_1 \lambda_2} \right) \right] = 0$$

implying that $2\lambda_1^2 v + c^2 \lambda_1^2 \left[\dfrac{-2}{v} + \dfrac{2}{v \lambda_1 \lambda_2} \right] = 0$

yielding $-\lambda_1^2 v + \dfrac{2c^2 \lambda_1^2}{v} + \dfrac{2c^2 \lambda_1^2}{v \lambda_1 \lambda_2} = 0$

which, on simplification, results in the following expression:

$$\lambda_1 \lambda_2 \left(c^2 - v^2 \right) = c^2 \qquad (10.54)$$

A comparison of Eqs (10.51) and (10.54) implies that

$$\lambda_1 = \lambda_2 = \lambda \qquad (10.55)$$

substituting the value of λ_1 from Eq. (10.53) into Eq. (10.46), one gets the following expression:

$$x_2 = \frac{x_1 - vt_1}{\sqrt{1 - \left(\dfrac{v^2}{c^2} \right)}} \qquad (10.56)$$

In addition, substitution of the value of λ_1 from Eq. (10.53) in Eq. (10.49) results in the following relation: $t_2 = \dfrac{1}{\sqrt{1-\dfrac{v^2}{c^2}}}\left[t_1 - \dfrac{x_1}{v}\left(1 - \dfrac{\left(c^2 - v^2\right)}{c^2}\right)\right]$

which can be rewritten as follows: $t_2 = \dfrac{1}{\sqrt{1-\dfrac{v^2}{c^2}}}\left[t_1 - \dfrac{x_1}{v}\dfrac{v^2}{c^2}\right]$

giving

$$t_2 = \frac{t_1 - \dfrac{v x_1}{c^2}}{\sqrt{1-\dfrac{v^2}{c^2}}} \tag{10.57}$$

Equations (10.56) and (10.57) are called *Lorentz transformations for space and time,* respectively.

With respect to reference frame S_2, reference frame S_1 is moving with a velocity of $-v$ along the $+x$ direction. The corresponding Lorentz transformation can be written as follows:

$$x_1 = \frac{x_2 + v t_2}{\sqrt{1-\dfrac{v^2}{c^2}}}; \; y_1 = y_2; \; z_1 = z_2 \tag{10.58}$$

and

$$t_1 = \frac{t_2 + \left(\dfrac{v x_2}{c^2}\right)}{\sqrt{1-\dfrac{v^2}{c^2}}} \tag{10.59}$$

Equations (10.58) and (10.59) are called *inverse Lorentz transformation for space and time,* respectively.

Example 10.1 A light pulse is emitted from the origin of reference frame S_2 at $t_2 = 0$. The distance covered by the light pulse in time t_2 is given by $x_2^2 = c^2 t_2^2$. Use Lorentz transformation to express this equation in terms of x_1 and t_1 and show that $x_1^2 = c^2 t_1^2$.

Solution The given equation is as follows:

$$x_2^2 = c^2 t_2^2 \tag{10.60}$$

From Eqs (10.56) and (10.57), Lorentz transformation can be written as follows:

$$x_2 = \frac{x_1 - v t_1}{\sqrt{1-\dfrac{v^2}{c^2}}} \tag{10.61}$$

and

$$t_2 = \frac{t_1 - \left(\dfrac{v x_1}{c^2}\right)}{\sqrt{1-\dfrac{v^2}{c^2}}} \tag{10.62}$$

Substitution of x_2, t_2 from Eqs (10.61) and (10.62) in Eq. (10.60) leads to the following form:

$$\frac{(x_1 - vt_1)^2}{\left[1 - \dfrac{v^2}{c^2}\right]} = \frac{c^2\left[t_1 - \left(\dfrac{vx_1}{c^2}\right)\right]^2}{\left[1 - \dfrac{v^2}{c^2}\right]}$$

resulting in $(x_1 - vt_1)^2 = c^2\left[t_1 - \left(\dfrac{vx_1}{c^2}\right)\right]^2$

yielding $x_1^2 + v^2 t_1^2 - 2x_1 vt_1 = c^2 t_1^2 + \dfrac{v^2 x_1^2}{c^2} - 2v x_1 t_1$

giving $\left(x_1^2 - c^2 t_1^2\right) - \dfrac{v^2}{c^2}\left(x_1^2 - c^2 t_1^2\right) = 0$

resulting in

$$\left(x_1^2 - c^2 t_1^2\right)\left[1 - \frac{v^2}{c^2}\right] = 0 \tag{10.63}$$

Since $v \neq c$, $\left[1 - \dfrac{v^2}{c^2}\right] \neq 0$; thus, Eq. (10.63) implies that

$$x_1^2 - c^2 t_1^2$$

or

$$x_1^2 = c^2 t_1^2 \tag{10.64}$$

Equation (10.64) also implies that the velocity of light is an absolute constant, that is, independent of the choice of reference frame.

Example 10.2   Show that the equation $x_1^2 + y_1^2 + z_1^2 = c^2 t_1^2$ is invariant under Lorentz transformation.

Solution   From Eqs (10.58) and (10.59), we can write the following expression:

$$x_1 = \frac{x_2 + vt_2}{\sqrt{1 - \dfrac{v^2}{c^2}}}; \; y_1 = y_2; \; z_1 = z_2 \tag{10.65}$$

and

$$t_1 = \frac{t_2 + \left(\dfrac{vx_2}{c^2}\right)}{\sqrt{1 - \dfrac{v^2}{c^2}}} \tag{10.66}$$

Substituting Eqs (10.65) and (10.66) in the given equation, we get the following

expression: $\dfrac{(x_2 + vt_2)^2}{\left[1 - \dfrac{v^2}{c^2}\right]} + y_2^2 + z_2^2 = \dfrac{c^2\left[t_2 + \left(\dfrac{vx_2}{c^2}\right)\right]^2}{\left[1 - \dfrac{v^2}{c^2}\right]}$

leading to $\quad y_2{}^2 + z_2{}^2 = \dfrac{1}{\left[1 - \dfrac{v^2}{c^2}\right]}\left[c^2 t_2{}^2 + \dfrac{v^2 x_2{}^2}{c^2} - x_2{}^2 - v^2 t_2{}^2\right]$

resulting in $\quad y_2{}^2 + z_2{}^2 = \dfrac{1}{\left[1 - \dfrac{v^2}{c^2}\right]}\left[\left(c^2 t_2{}^2 - x_2{}^2\right)\left(1 - \dfrac{v^2}{c^2}\right)\right]$

giving $\quad y_2{}^2 + z_2{}^2 = c^2 t_2{}^2 - x_2{}^2$

yielding $\quad x_2{}^2 + y_2{}^2 + z_2{}^2 = c^2 t_2{}^2$

Thus, the given equation is invariant under Lorentz transformation.

10.7 LENGTH CONTRACTION

The length of an object in a moving reference frame is found to reduce as observed from a stationary reference frame. This phenomenon is called 'length contraction'.

Let us once again consider two reference frames S_1 and S_2. Reference frame S_2 is assumed to move with a velocity v with respect to reference frame S_1, along the positive x direction. Suppose a rod is placed along the x-axis, with the ends of the rod being at coordinates x_2' and x_1' in frame S_2. The length of the rod l_2', in reference frame S_2 is then given as follows:

$$l_2 = x_2' - x_1' \tag{10.67}$$

For an observer in reference frame S_1, the length l_1 of the rod is given as follows:

$$l_1 = x_2 - x_1 \tag{10.68}$$

where x_1 and x_2 represent coordinates of the ends of the rod in reference frame S_1.

Using Lorentz transformation represented by Eq. (10.56), we can write that

$$x_2' = \dfrac{x_2 - vt_1}{\sqrt{1 - \left(\dfrac{v^2}{c^2}\right)}} \tag{10.69}$$

and

$$x_1' = \dfrac{x_1 - vt_1}{\sqrt{1 - \left(\dfrac{v^2}{c^2}\right)}} \tag{10.70}$$

Use of Eqs (10.69) and (10.70) in Eq. (10.67) yields the following expression:

$$l_2 = \dfrac{\left(x_2 - vt_1\right)}{\sqrt{1 - \left(\dfrac{v^2}{c^2}\right)}} - \dfrac{\left(x_1 - vt_1\right)}{\sqrt{1 - \left(\dfrac{v^2}{c^2}\right)}}$$

giving

$$l_2 = \frac{(x_2 - x_1)}{\sqrt{1 - \left(\frac{v^2}{c^2}\right)}}$$

(10.71)

Use of Eq. (10.68) in Eq. (10.71) results in the following expression:

$$l_2 = \frac{l_1}{\sqrt{1 - \left(\frac{v^2}{c^2}\right)}}$$

(10.72)

or,

$$l_1 = l_2 \sqrt{1 - \left(\frac{v^2}{c^2}\right)}$$

(10.73)

Thus, the observer in the stationary reference frame S_1 observes a shorter length. This phenomenon is called length contraction and the contraction

factor is $\sqrt{1 - \left(\frac{v^2}{c^2}\right)}$ in the direction of motion. The length contraction

observed is called the *Lorentz–Fitzerald contraction*.

Some special situations deserve a separate mention:

1. When $v \ll c$, the quantity $\frac{v^2}{c^2}$ becomes a negligible quantity and Eq. (10.73) reduces to the following form: $l_2 = l_1$
2. When v is comparable to c, the quantity $\sqrt{1 - \frac{v^2}{c^2}}$ is less than unity. Thus, $l_1 < l_2$ and the length of rod appears less than that measured by an observer at rest with respect to the rod.
3. For $v = c$, $l_1 = 0$. This is impossible; therefore, it is not possible for any material object to attain the velocity of light.
4. Length contraction is observed only along the direction of motion. No length contraction is observed in the perpendicular direction.

Example 10.3   A meter scale is kept in a satellite moving with velocity $0.6c$ with respect to the earth. Calculate the length of the rod as measured by an observer on the earth.

Solution   Length l_1 of the meter scale measured by the observer on earth is given by the following relation:

$$l_1 = l_2 \sqrt{1 - \frac{v^2}{c^2}}$$

(10.74)

Using the given values in Eq. (10.74), we get the following expression:

$$l_1 = 100 \sqrt{1 - \left(\frac{0.6c}{c}\right)^2}$$

yielding $l_1 = 100 \times \sqrt{0.64} = 80$ cm

Example 10.4 A box of length 50 cm is kept in a spaceship travelling with speed 0.7c. What is the length of the box as observed by an observer on the earth?

Solution The length l_1 of the box measured by the observer on earth is given by the following relation:

$$l_1 = l_2\sqrt{1 - \frac{v^2}{c^2}} \tag{10.75}$$

Using the given values in Eq. (10.75), we get the following expression:

$$l_1 = 50\sqrt{1 - \left(\frac{0.7c}{c}\right)^2}$$

which can be rewritten as follows: $l_1 = 50\sqrt{1 - 0.49} = 50\sqrt{0.51} = 35.5$ cm

Example 10.5 A rod has a length of 90 cm. It is kept in a satellite moving with a velocity 0.5c with respect to the earth. Calculate the length of the rod as measured by an observer (a) in the satellite and (b) on the surface of earth.

Solution The calculations are as follows:

(a) When the observer is in the satellite, the rod is at rest with respect to the observer. Therefore, the length of the rod measured by the observer is 90 cm.

(b) Length l_1 of the rod measured by the observer on earth is given as follows:

$$l_1 = l_2\sqrt{1 - \frac{v^2}{c^2}} \tag{10.76}$$

Use of the given values in Eq. (10.76) leads to the following relation:

$$l_1 = 90\sqrt{1 - \left(\frac{0.5c}{c}\right)^2}$$

yielding $l_1 = 90 \times (0.75)^{1/2} = 77.9$ cm

Example 10.6 A rocket measures 90 m in length on the ground. During flight, its length is measured to be 88 m by an observer on the ground. Calculate the speed of the rocket.

Solution We know that

$$l_1 = l_2\sqrt{1 - \frac{v^2}{c^2}} \tag{10.77}$$

Using the given values in Eq. (10.77), one gets the following expression:

$$(88) = (90)\sqrt{1 - \frac{v^2}{c^2}}$$

giving $\sqrt{1 - \frac{v^2}{c^2}} = \frac{(88)}{(90)} = 0.98$

yielding $1 - \frac{v^2}{c^2} = (0.98)^2 = 0.96$

Thus, $\frac{v^2}{c^2} = 1 - 0.96 = 0.04$

resulting in $v = 3 \times 10^{10} \times \sqrt{0.04} = 3 \times 10^{10} \times 0.2 = 0.6 \times 10^{10}$ cm/s

Example 10.7 The rod of length 90 cm indicated in Example 10.5 slows down to 0.4c. Find the percentage increase in length of the rod as observed by an observer on the earth.

Solution Length l_1 of the rod measured by an observer on earth is given by the following equation:

$$l_1 = l_2\sqrt{1 - \frac{v^2}{c^2}} \tag{10.78}$$

Using $v = 0.4c$ in Eq. (10.78), we get the following expression: $l_1 = 90\sqrt{1 - \left(\frac{0.4c}{c}\right)^2}$

yielding

$$l_1 = 90\sqrt{1 - 0.16} = 82.5 \text{ cm} \tag{10.79}$$

The percentage increase in length is, therefore, $\frac{(82.5 - 77.9)}{77.9} \times 100 = 5.9\%$

Example 10.8 The rod of length 90 cm indicated in Example 10.5 increases its velocity to 0.6c. Find the percentage reduction in length of the rod as observed by an observer on the earth.

Solution The length l_1 of the rod measured by an observer on earth is given as follows:

$$l_1 = l_2\sqrt{1 - \frac{v^2}{c^2}} \tag{10.80}$$

Using $v = 0.6c$ in Eq. (10.80), we get the following expression: $l_1 = 90\sqrt{1 - \left(\frac{0.6c}{c}\right)^2}$

yielding $l_1 = 90\sqrt{1 - 0.36} = 90\sqrt{0.64}$

resulting in $l_1 = 72 \text{ cm}$

$$\% \text{ decrease in length} \frac{(77.9 - 72)}{77.9} \times 100 = 7.6\%$$

Example 10.9 A moving object of length 100 cm is measured to be 90 cm long by an observer on the ground. Calculate the speed of the object.

Solution We know that

$$l_1 = l_2\sqrt{1 - \frac{v^2}{c^2}} \tag{10.81}$$

Using the given values in Eq. (10.81) leads to the following expression:

$$90 = 100\sqrt{1 - \frac{v^2}{c^2}}$$

giving $\sqrt{1 - \frac{v^2}{c^2}} = 0.90$

yielding $1 - \frac{v^2}{c^2} = (0.9)^2 = 0.81$

implying that $\frac{v^2}{c^2} = 1 - 0.81 = 0.19$

resulting in $v = 3 \times 10^{10}\sqrt{0.19} = 1.3 \times 10^{10}$ cm/s

Example 10.10 At what speed will an object of length 100 cm be measured as 50 cm by an observer at rest?

Solution We know that

$$l_1 = l_2 \sqrt{1 - \frac{v^2}{c^2}} \qquad (10.82)$$

Substituting the given values in Eq. (10.82), we get the following expression:

$$50 = 100 \sqrt{1 - \frac{v^2}{c^2}}$$

leading to $\dfrac{v^2}{c^2} = 1 - (0.5)^2 = 1 - 0.25 = 0.75$

yielding $v = 3 \times 10^{10} \sqrt{0.75} = 2.61 \times 10^{10}$ cm/s

Example 10.11 Find the speed at which a rocket appears to be 75% of its actual length, to an observer at rest.

Solution We know that

$$l_1 = l_2 \sqrt{1 - \frac{v^2}{c^2}} \qquad (10.83)$$

Using the given values in Eq. (10.83), we get the following expression:

$$\frac{l_1}{l_2} = 0.75 = \sqrt{1 - \frac{v^2}{c^2}}$$

yielding $v = 3 \times 10^{10} \sqrt{0.44} = 1.98 \times 10^{10}$ cm/s

Example 10.12 Length of a moving space ship is found to be 80% of its actual length to an observer at rest. Calculate the speed of the space ship.

Solution We know that

$$l_1 = l_2 \sqrt{1 - \frac{v^2}{c^2}} \qquad (10.84)$$

Substituting the given values in Eq. (10.84), we get the following expression:

$$\frac{l_1}{l_2} = 0.8 = \sqrt{1 - \frac{v^2}{c^2}}$$

yielding $1 - \dfrac{v^2}{c^2} = 0.64 \Rightarrow \dfrac{v^2}{c^2} = 1 - 0.64 = 0.36$

leading to $v = 3 \times 10^{10} \sqrt{0.36} = 3 \times 10^{10} \times 0.6 = 1.8 \times 10^{10}$ cm/s

10.8 TIME DILATION

The time taken for an event to take place increases in a moving reference frame as compared to the time taken for the same event in a stationary reference frame. This phenomenon is called 'time dilation'.

Let us once again assume two reference frames S_1 and S_2. Reference frame S_2 is moving with respect to reference frame S_1 with a velocity v along the $+x$ direction.

Suppose a clock is located at position x_1 with respect to reference frame S_1. A particular time interval $(\Delta t)_1$ in reference frame S_1 is defined as follows:

$$(\Delta t)_1 = t_{B_1} - t_{A_1} \tag{10.85}$$

The same time interval is observed to be $(\Delta t)_2$ by an observer in reference frame S_2. Thus,

$$(\Delta t)_2 = t_{B_2} - t_{A_2} \tag{10.86}$$

Utilizing Lorentz transformation for time, using Eq. (10.57) in Eq. (10.86), we get the following expressions:

$$t_{B_2} = \frac{t_{B_1} - \left(\dfrac{v x_1}{c^2}\right)}{\sqrt{1 - \dfrac{v^2}{c^2}}} \tag{10.87}$$

and

$$t_{A_2} = \frac{t_{A_1} - \left(\dfrac{v x_1}{c^2}\right)}{\sqrt{1 - \dfrac{v^2}{c^2}}} \tag{10.88}$$

Substitution of t_{B_2} and t_{A_2} in Eq. (10.86) leads to the following expression:

$$(\Delta t)_2 = \frac{t_{B_1} - \left(\dfrac{v x_1}{c^2}\right)}{\sqrt{1 - \dfrac{v^2}{c^2}}} - \frac{t_{A_1} - \left(\dfrac{v x_1}{c^2}\right)}{\sqrt{1 - \dfrac{v^2}{c^2}}}$$

which can be simplified to the following form:

$$(\Delta t)_2 = \frac{t_{B_1} - t_{A_1}}{\sqrt{1 - \frac{v^2}{c^2}}} = \frac{(\Delta t)_1}{\sqrt{1 - \frac{v^2}{c^2}}} \tag{10.89}$$

Equation (10.89) implies that $(\Delta t)_2 > (\Delta t)_1$. Thus, the time interval as observed by the observer from reference frame S_2 is greater than that observed in reference frame S_1. Some interesting conclusions can be drawn from Eq. (10.89), which are as follows:

1. For $v \ll c$, $\left(\dfrac{v^2}{c^2}\right)$ is a negligible quantity; therefore,

$$(\Delta t)_2 = (\Delta t)_1 \tag{10.90}$$

Thus, the time interval observed by a moving clock equals that observed by a stationary clock.

2. As v becomes comparable to c, the term $\dfrac{1}{\sqrt{1 - \frac{v^2}{c^2}}}$ becomes less than unity and Eq. (10.89) leads to the following form:

$$(\Delta t)_2 > (\Delta t)_1 \tag{10.91}$$

Thus, the time interval observed by the moving clock is greater than that observed by the stationery clock.

3. When $v = c$, $(\Delta t)_2 = \infty$ and thus not possible. This leads to the conclusion that a material object can never attain the velocity of light.

The concept of time dilation has been verified experimentally. Mesons are nuclear particles that exist as π mesons and μ mesons. A π^+ meson decays into μ mesons and neutrons. Half-life of π mesons is 1.8×10^{-8} s. In a laboratory experiment, π^+ mesons were produced by bombarding a target with high-energy particles generated by a synchrotron. These π^+ mesons in turn travel with a speed of $0.99c$, where c is the velocity of light. Flux of these π^+ mesons was measured at two places separated by a distance of 30 m. The corresponding time interval $(\Delta t)_2$ can be obtained as follows:

$$(\Delta t)_2 = \frac{30}{0.99 \times 3 \times 10^8 \, \text{m/s}} = 10 \times 10^{-8} \, \text{s} \tag{10.92}$$

The magnitude of $(\Delta t)_2$ is around 5.6 times the half-life of 1.8×10^{-8} s for the π mesons. The flux of π^+ mesons should, therefore, decrease to a value of $(2)^{-5.6}$ or less than 2% of the original flux. Experimentally, a flux reduction to 60% of the original value is only observed. According to Eq. (10.89),

$$(\Delta t)_2 = \frac{(\Delta t)_1}{\sqrt{1 - \left(\frac{v^2}{c^2}\right)}}$$

which yields

$$(\Delta t)_1 = (\Delta t)_2 \sqrt{1 - \left(\frac{v^2}{c^2}\right)} \tag{10.93}$$

Use of Eq. (10.92) in Eq. (10.93) results in the following expression:

$$(\Delta t)_1 = 10 \times 10^{-8} \sqrt{1 - \left(0.99 \frac{c}{c}\right)^2} \tag{10.94}$$

giving

$$(\Delta t)_1 = 1.4 \times 10^{-8} \, \text{s} \tag{10.95}$$

Thus, $(\Delta t)_1$ is 0.78 times the half-life of 1.8×10^{-8} s of π^+ meson. The flux should, therefore, correspondingly fall to $(2)^{-0.78} \approx 60\%$.

The experimentally observed reduction in flux of π^+ mesons can, therefore, be easily explained using the concept of time dilation. From Eqs (10.94) and (10.95), one can conclude that π^+ mesons take a time of 10×10^{-8} s in the reference frame of the laboratory and a time of 1.4×10^{-8} s in their own frame of reference. A seven-fold time dilation is thus observed in this case.

Example 10.13 A process needs 1 μs to complete in an atom at rest in a laboratory. Calculate the time required for this process to complete with respect to an observer in the laboratory, if the atom is moving at a speed of 4×10^9 cm/s.

Solution From Eq. (10.89), we have the following relation:

$$(\Delta t)_2 = \frac{(\Delta t)_1}{\sqrt{1 - \frac{v^2}{c^2}}} \tag{10.96}$$

Substitution of the given values in Eq. (10.96) yields the following expression:

$$(\Delta t)_2 = \frac{10^{-6}}{\sqrt{1 - \left(\frac{4\times10^{+9}}{3\times10^{10}}\right)^2}}$$

giving $(\Delta t)_2 = \frac{10^{-6}}{\sqrt{1 - \left(\frac{16}{900}\right)}} = \frac{10^{-6}}{\sqrt{0.982}} = 1.01\times10^{-6}\,\text{s}$

Example 10.14 A particle with a mean life time of $3\,\mu\text{s}$ is moving at a speed of $0.8c$ in a laboratory set-up. Calculate its life time as measured by an observer in the laboratory.

Solution From Eq. (10.89), we have the following relation:

$$(\Delta t)_2 = \frac{(\Delta t)_1}{\sqrt{1 - \frac{v^2}{c^2}}} \tag{10.97}$$

Using the given values in Eq. (10.97), we get the following expression:

$$(\Delta t)_2 = \frac{3\times10^{-6}}{\sqrt{1 - (0.8)^2}}$$

yielding $(\Delta t)_2 = \frac{3\times10^{-6}}{\sqrt{0.36}} = \frac{3\times10^{-6}}{0.6} = 5\times10^{-6}\,\text{s}$

Example 10.15 The proper life of some π^- mesons is 3×10^{-8} s. Calculate the velocity of these π mesons if their observed mean life is 3×10^{-7} s.

Solution From Eq. (10.89), we have the following relation:

$$(\Delta t)_2 = \frac{(\Delta t)_1}{\sqrt{1 - \frac{v^2}{c^2}}} \tag{10.98}$$

Substitution of the given values in Eq. (10.98) yields the following expression:

$$3\times10^{-7} = \frac{3\times10^{-8}}{\sqrt{1 - \frac{v^2}{c^2}}}$$

resulting in, $10 = \dfrac{1}{\sqrt{1 - \dfrac{v^2}{c^2}}}$

giving, $100 = \dfrac{1}{\left(1 - \dfrac{v^2}{c^2}\right)} \Rightarrow 100 - 100\left(\dfrac{v^2}{c^2}\right) = 1$

or $\dfrac{99}{100} = \dfrac{v^2}{c^2}$ or $v = \sqrt{\dfrac{99}{100}}\ c = 0.995c$

Example 10.16 An atomic process needs $2\,\mu s$ to complete in the laboratory. If the atom starts moving with a speed of 8×10^9 cm/s, calculate the time required to complete the same process with respect to an observer in the laboratory.

Solution From Eq. (10.89), we have the following relation:

$$(\Delta t)_2 = \frac{(\Delta t)_1}{\sqrt{1 - \dfrac{v^2}{c^2}}} \tag{10.99}$$

Substituting the given values in Eq. (10.99), we get the following expression:

$$(\Delta t)_2 = \frac{2 \times 10^{-6}}{\sqrt{1 - \left(\dfrac{8 \times 10^9}{3 \times 10^{10}}\right)^2}}$$

yielding $(\Delta t)_2 = \dfrac{2 \times 10^{-6}}{\sqrt{1 - \left(\dfrac{64}{900}\right)}} = \dfrac{2 \times 10^{-6}}{0.96} = 2.08 \times 10^{-6}$ s

Example 10.17 A process takes $1\,\mu s$ to complete in a rocket at rest. Calculate the time required to complete the same process with respect to an observer on the earth when the rocket is moving with a speed of 6×10^9 cm/s.

Solution From Eq. (10.89), we have the following relation:

$$(\Delta t)_2 = \frac{(\Delta t)_1}{\sqrt{1 - \dfrac{v^2}{c^2}}} \tag{10.100}$$

Using the values in Eq. (10.100), we get the following expression:

$$(\Delta t)_2 = \frac{1 \times 10^{-6}}{\sqrt{1 - \left(\dfrac{6 \times 10^9}{3 \times 10^{10}}\right)^2}}$$

yielding $(\Delta t)_2 = \dfrac{1 \times 10^{-6}}{\sqrt{1 - 0.04}} = 1.02 \times 10^{-6}$ s

Example 10.18 A spaceship is moving at a speed of 5×10^9 cm/s with respect to earth. The time required for an event to take place on the spaceship, as observed by an observer on the earth, is $2\,\mu s$. What will be the time required for the same event if the observer is travelling on the spaceship?

Solution From Eq. (10.89), we have the following relation:

$$(\Delta t)_1 = (\Delta t)_2 \sqrt{1 - \dfrac{v^2}{c^2}} \tag{10.101}$$

Using the given values in Eq. (10.101), we get the following relation:

$$(\Delta t)_1 = 2 \times 10^{-6} \sqrt{1 - \left(\dfrac{5 \times 10^9}{3 \times 10^{10}}\right)^2}$$

yielding $(\Delta t)_1 = 2 \times 10^{-6} \sqrt{1 - \dfrac{25}{900}} = 2 \times 10^{-6} \sqrt{0.97}$

resulting in $(\Delta t)_1 = 1.96 \times 10^{-6}$ s

Example 10.19   A rocket is travelling at a speed of 2×10^9 cm/s with respect to the earth. The time required for a fuse on the rocket to blow is $1 \,\mu s$, as observed by an observer on the earth. Calculate the actual time taken for the fuse to blow.

Solution   From Eq. (10.89), we have the following relation:

$$(\Delta t)_1 = (\Delta t)_2 \sqrt{1 - \frac{v^2}{c^2}} \qquad (10.102)$$

Substituting the given values in Eq. (10.102), we get the following expression:

$$(\Delta t)_1 = 1 \times 10^{-6} \sqrt{1 - \left(\frac{2 \times 10^9}{3 \times 10^{10}}\right)^2}$$

yielding $(\Delta t)_1 = 1 \times 10^{-6} \sqrt{1 - \left(\dfrac{4}{900}\right)}$

resulting in $(\Delta t)_1 = 1 \times 10^{-6} \sqrt{0.996} = 0.998 \times 10^{-6}$ s

Example 10.20   The particle indicated in Example 10.14 speeds up to $0.9c$. What is its new lifetime, as measured by an observer in the laboratory?

Solution   From Eq. (10.89), we have the following relation:

$$(\Delta t)_2 = \frac{(\Delta t)_1}{\sqrt{1 - \dfrac{v^2}{c^2}}} \qquad (10.103)$$

Substitution of the given values in Eq. (10.103) leads to the following expression:

$$(\Delta t)_2 = \frac{3 \times 10^{-6}}{\sqrt{1 - (0.9)^2}}$$

giving $(\Delta t)_2 = \dfrac{3 \times 10^{-6}}{\sqrt{0.19}} = 6.8 \times 10^{-6}$ s

Example 10.21   The particle indicated in Example 10.14 slows down to $0.7c$. Calculate its life time, as measured by an observer in the laboratory.

Solution   From Eq. (10.89), we have the following relation:

$$(\Delta t)_2 = \frac{(\Delta t)_1}{\sqrt{1 - \dfrac{v^2}{c^2}}} \qquad (10.104)$$

Using the given values, we get the following result:

$$(\Delta t)_2 = \frac{3 \times 10^{-6}}{\sqrt{1 - (0.7)^2}} = \frac{3 \times 10^{-6}}{\sqrt{0.51}} = 4.2 \times 10^{-6} \text{ s}$$

Example 10.22   A particle at rest has a lifetime of $2 \,\mu s$. When it speeds up, its lifetime is measured to be $4 \,\mu s$ by an observer in the laboratory. Calculate the speed of the particle.

Solution From Eq. (10.87), we have the following relation:

$$(\Delta t)_2 = \frac{(\Delta t)_1}{\sqrt{1 - \dfrac{v^2}{c^2}}} \qquad\qquad (10.105)$$

Substituting the given values in Eq. (10.105), we get the following expression:

$$4 \times 10^{-6} = \frac{2 \times 10^{-6}}{\sqrt{1 - \dfrac{v^2}{c^2}}}$$

yielding $\sqrt{1 - \dfrac{v^2}{c^2}} = \dfrac{2}{4} = \dfrac{1}{2} \Rightarrow 1 - \dfrac{v^2}{c^2} = \dfrac{1}{4} \Rightarrow v = \sqrt{0.75}\,c = 2.6 \times 10^{10}$ cm/s

Example 10.23 The ratio of the lifetime of a particle at rest and that when the same particle is moving is 1/3, as measured by an observer in the laboratory. Find the speed of the particle.

Solution From Eq. (10.87), we have the following relation:

$$(\Delta t)_2 = \frac{(\Delta t)_1}{\sqrt{1 - \dfrac{v^2}{c^2}}} \qquad\qquad (10.106)$$

Using the given values in Eq. (10.106), we get the following expression:

$$\frac{(\Delta t)_1}{(\Delta t)_2} = \frac{1}{3} = \sqrt{1 - \frac{v^2}{c^2}}$$

leading to $1 - \dfrac{v^2}{c^2} = \dfrac{1}{9}$

implying that $v = \sqrt{\dfrac{8}{9}}\,c = \sqrt{0.89}\,c = 2.82 \times 10^{10}$ cm/s

Example 10.24 The proper life of some π mesons is found to be 3×10^{-8} s. Calculate the velocity of these π mesons if their observed mean life is 5×10^{-7} s.

Solution From Eq. (10.87), we have the following relation:

$$(\Delta t)_2 = \frac{(\Delta t)_1}{\sqrt{1 - \dfrac{v^2}{c^2}}} \qquad\qquad (10.107)$$

Substituting the given values in Eq. (10.107), we get the following expression:

$$5 \times 10^{-7} = \frac{3 \times 10^{-8}}{\sqrt{1 - \dfrac{v^2}{c^2}}}$$

yielding $\dfrac{50}{3} = \dfrac{1}{\sqrt{1 - \dfrac{v^2}{c^2}}}$

resulting in $\dfrac{v^2}{c^2} = \dfrac{2491}{2500}$

giving $v = 0.998c$

Example 10.25 The proper life of a fundamental particle is 1×10^{-8} s. Calculate the observed mean life of this particle when it moves with speed $0.9c$.

Solution From Eq. (10.87), we have the following relation:

$$(\Delta t)_2 = \frac{(\Delta t)_1}{\sqrt{1 - \dfrac{v^2}{c^2}}} \tag{10.108}$$

Using the given values in Eq. (10.108), we get the following expression:

$$(\Delta t)_2 = \frac{1 \times 10^{-8}}{\sqrt{1 - (0.9)^2}}$$

leading to $(\Delta t)_2 = \dfrac{1 \times 10^{-8}}{\sqrt{0.44}} = 2.27 \times 10^{-8}$ s

Example 10.26 The observed mean life of a fundamental particle is 5×10^{-8} s when the particle is moving with speed $0.95c$. Calculate the proper life of the particle.
Solution From Eq. (10.87), we have the following expression:

$$(\Delta t)_2 = \frac{(\Delta t)_1}{\sqrt{1 - \dfrac{v^2}{c^2}}} \tag{10.109}$$

yielding

$$(\Delta t)_1 = (\Delta t)_2 \sqrt{1 - \frac{v^2}{c^2}} \tag{10.110}$$

Using the given values in Eq. (10.110), we get the following expression:

$$(\Delta t)_1 = 5 \times 10^{-8} \sqrt{1 - (0.95)^2}$$

giving $(\Delta t)_1 = 5 \times 10^{-8} \sqrt{0.1} = 5 \times 10^{-8} \times 0.32 = 1.6 \times 10^{-8}$ s

Example 10.27 The ratio of the proper life to the mean life of a moving fundamental particle is 1/5. Calculate the speed of the fundamental particle.
Solution From Eq. (10.87), we have the following relation:

$$(\Delta t)_2 = \frac{(\Delta t)_1}{\sqrt{1 - \dfrac{v^2}{c^2}}} \tag{10.111}$$

leading to $\sqrt{1 - \dfrac{v^2}{c^2}} = \dfrac{(\Delta t)_1}{(\Delta t)_2} = \dfrac{1}{5}$

Solving it, we get the following expression: $\dfrac{v^2}{c^2} = \dfrac{24}{25}$

or $\qquad v = 0.98c$

10.9 ADDITION OF VELOCITIES

We know how velocities add, as long as the numerical magnitude of the velocities is very small in comparison to the velocity of light. Things, however, change drastically as velocities approach the velocity of light. In this section, we shall discuss the addition of velocities as they approach the velocity of light.

Let us consider two reference frames S_1 and S_2; reference frame S_2 is assumed to move with velocity v with respect to reference frame S_1 along the $+x$ direction. Suppose that a body covers a distance of $(dx)_1$ in time $(dt)_1$ in reference frame S_1. The corresponding distance in reference frame S_2 is $(dx)_2$ in time $(dt)_2$. We can then write that

$$\frac{(dx)_1}{(dt)_1} = u_1 \tag{10.112}$$

and

$$\frac{(dx)_2}{(dt)_2} = u_2 \tag{10.113}$$

Lorentz transformation for position and time are given by the following equations:

$$x_1 = \lambda(x_2 + vt_2) \tag{10.114}$$

and

$$t_1 = \lambda\left(t_2 + \frac{vx_2}{c^2}\right) \tag{10.115}$$

Differentiation of Eqs (10.114) and (10.115) result in the following expressions:

$$(dx)_1 = \lambda\left[(dx)_2 + v(dt)_2\right] \tag{10.116}$$

and,

$$(dt)_1 = \lambda\left[(dt)_2 + \frac{v(dx)_2}{c^2}\right] \tag{10.117}$$

Dividing Eq. (10.116) by Eq. (10.117), one obtains the following relation:

$$\frac{(dx)_1}{(dt)_1} = \frac{\lambda\left[(dx)_2 + v(dt)_2\right]}{\lambda\left[(dt)_2 + \frac{v(dx)_2}{c^2}\right]}$$

which, on simplification, gives the following equation:

$$\frac{(dx)_1}{(dt)_1} = \frac{\frac{(dx)_2}{(dt)_2} + v}{1 + \left(\frac{v}{c^2}\right)\frac{(dx_2)_2}{(dt)_2}} \tag{10.118}$$

Use of Eqs (10.112) and (10.113) in Eq. (10.118) yields the following expression:

$$(u)_1 = \frac{(u)_2 + v}{1 + \left[\frac{(u)_2 v}{c^2}\right]} \tag{10.119}$$

Equation (10.119) represents the *relativistic law of addition of velocities*.

Some special cases deserve a special mention, which are as follows:

Case 1 $(u)_2$ and v are much smaller than the velocity of light, c

Equation (10.119) then simplifies to the following form:

$$(u)_1 = (u)_2 + v \tag{10.120}$$

Equation (10.120) is the classical law of addition of velocities.

Case 2 $u = c$

Equation (10.119) then leads to the following expression:

$$(u)_1 = \frac{(u)_2 + c}{1 + \left[\dfrac{(u)_2 c}{c^2}\right]} = \left[\frac{(u)_2 + c}{c + (u)_2}\right] c = c \tag{10.121}$$

Thus, if an object moves with a velocity c with respect to another object, the relative velocity is also c, since the velocity of light cannot be exceeded.

Case 3 $(u)_2 = v = c$

Using this, Eq. (10.119) leads to the following expression:

$$(u)_1 = \frac{c + c}{1 + 1} = c \tag{10.122}$$

Thus, addition of the velocity of light to the velocity of light also leads to the velocity of light.

Example 10.28 Two particles are travelling towards each other with a speed of $0.85c$ with respect to an observer in a laboratory. Calculate their relative speed.

Solution From Eq. (10.119),

$$(u)_1 = \frac{(u)_2 + v}{1 + \left[\dfrac{(u)_2 v}{c^2}\right]} \tag{10.123}$$

In the present problem, $(u)_2 = v = 0.85c$

Thus, Eq. (10.123) leads to the following expression:

$$(u)_1 = \frac{0.85c + 0.85c}{1 + \dfrac{0.85c \times 0.85c}{c^2}}$$

yielding $(u)_1 = \dfrac{1.7c}{1 + 0.72} = \dfrac{1.7c}{1.72} = 0.988c$

Example 10.29 A missile is chasing an enemy aircraft. From earth, it is found that the speed of the missile is 2×10^{10} cm/s, while that of the enemy aircraft is 1×10^{10} cm/s. Calculate the relative velocity of the enemy aircraft with respect to the missile.

Solution From Eq. (10.119), we have the following relation:

$$(u)_1 = \frac{(u)_2 + v}{1 + \left[\dfrac{(u)_2 v}{c^2}\right]} \tag{10.124}$$

In the present problem, we have

$$(u)_2 = 1 \times 10^{10} \text{ and } v = -2 \times 10^{10}$$

Use of these values in Eq. (10.124) leads to the following expression:

$$(u)_1 = \frac{1 \times 10^{10} - 2 \times 10^{10}}{1 - \left[\dfrac{1 \times 10^{10} \times 2 \times 10^{10}}{(3 \times 10^{10})^2}\right]}$$

which yields the following result:

$$(u)_1 = \frac{-1 \times 10^{10}}{1 - \dfrac{2}{9}} = \frac{-1 \times 10^{10} \times 9}{7} = -1.29 \times 10^{10} \, \text{cm/s}$$

Thus, the missile finds the enemy aircraft moving towards it at a speed of 1.29×10^{10} cm/s.

Example 10.30 Two objects are travelling towards each other with speed $0.8c$ with respect to an observer in a laboratory. Calculate their relative speed.

Solution From Eq. (10.119), we have the following relation:

$$(u)_1 = \frac{(u)_2 + v}{1 + \left[\dfrac{(u)_2 v}{c^2}\right]} \tag{10.125}$$

In the present problem, $(u)_2 = v = 0.8c$. Thus, Eq. (10.125) leads to the following expression:

$$(u)_1 = \frac{0.8c + 0.8c}{1 + \dfrac{0.8c \times 0.8c}{c^2}}$$

yielding $(u)_1 = \dfrac{1.6c}{1 + (0.8)^2} = \dfrac{1.6}{1.64}c = 0.976c$

Example 10.31 Two objects are travelling towards each other with speed $0.9c$ with respect to an observer in a laboratory. Calculate their relative speed.

Solution From Eq. (10.119), one gets the following relation:

$$(u)_1 = \frac{(u)_2 + v}{1 + \left[\dfrac{(u)_2 v}{c^2}\right]} \tag{10.126}$$

In the given situation, $(u)_2 = v = 0.9c$. Thus, Eq. (10.126) leads to the following expression:

$$(u)_1 = \frac{0.9c + 0.9c}{1 + \dfrac{0.9c \times 0.9c}{c^2}} = \frac{1.8c}{1 + 0.81} = 0.99c$$

10.10 VARIATION OF MASS WITH VELOCITY

One of the most profound differences between Newtonian mechanics and the Einstein's theory of relativity is the way the mass of a body behaves with respect to its velocity. According to the Newtonian mechanics, the force being felt by a body does not depend on its velocity. This is because the mass of the body does not change with a change in its velocity. The conditions change dramatically as the velocity of the body approaches the velocity of light, and effects pertaining to the special theory of relativity have to be taken into account.

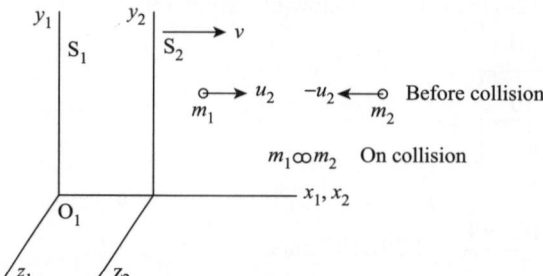

Fig. 10.5 Collision in stationary and moving reference frames

Consider two reference frames S_1 and S_2, with frame S_2 moving with velocity v with respect to frame S_1 in the positive x direction, as shown in Fig. 10.5.

Two masses m_1 and m_2 travel towards each other with velocities u_2 and $-u_2$ parallel to the x-axis in reference frame S_2. The two masses collide and coalesce into one body, as shown in Fig. 10.5. The definition of momentum and conservation of momentum and mass are valid even at relativistic speeds. Thus, the mass of the combined body after coalescing is $(m_1 + m_2)$. In addition, as per the law of conservation of linear momentum, the combined mass $(m_1 + m_2)$ must come to rest, since the net momentum was zero before collision.

For an observer in reference frame S_1, the two masses m_1 and m_2 have velocities u_{11} and u_{12}, respectively.

Using Eq. (10.119), we can write the following expressions:

$$u_{11} = \frac{u_2 + v}{1 + \dfrac{u_2 v}{c^2}} \tag{10.127}$$

and

$$u_{12} = \frac{-u_2 + v}{1 - \dfrac{u_2 v}{c^2}} \tag{10.128}$$

Conservation of momentum in reference frame S_1 yields the following expression:

$$m_1 u_{11} + m_2 u_{12} = (m_1 + m_2) v \tag{10.129}$$

Substitution of u_{11} and u_{12} from Eqs (10.127) and (10.128) in Eq. (10.129) leads to the following relation:

$$m_1 \left[\frac{u_2 + v}{1 + \dfrac{u_2 v}{c^2}} \right] + m_2 \left[\frac{-u_2 + v}{1 - \dfrac{u_2 v}{c^2}} \right] = (m_1 + m_2) v$$

giving

$$m_1 \left[\frac{u_2 + v}{1 + \dfrac{u_2 v}{c^2}} - v \right] = m_2 \left[v - \frac{(-u_2 + v)}{1 - \dfrac{u_2 v}{c^2}} \right]$$

which can be simplified to the following expression:

$$m_1 \left[\frac{u_2 + v - v - \dfrac{u_2 v}{c^2}^2}{\left(1 + \dfrac{u_2 v}{c^2}\right)} \right] = m_2 \left[\frac{v - \dfrac{u_2 v}{c^2}^2 + u_2 - v}{\left(1 - \dfrac{u_2 v}{c^2}\right)} \right]$$

resulting in

$$\frac{m_1}{m_2} = \frac{\left(1 + \dfrac{u_2 v}{c^2}\right)}{\left(1 - \dfrac{u_2 v}{c^2}\right)} \tag{10.130}$$

Equations (10.127) and (10.128) can be used to show that

$$\frac{u_2 v}{c^2} = \frac{2c^2 - u_{11}^2 - u_{12}^2 - 2\sqrt{\left(c^2 - u_{11}^2\right)\left(c^2 - u_{12}^2\right)}}{u_{11}^2 - u_{12}^2} \tag{10.131}$$

Substituting $\dfrac{u_2 v}{c^2}$ from Eq. (10.131) in Eq. (10.130) and simplifying, we get the following relation:

$$\frac{m_1}{m_2} = \frac{\left(1 - \dfrac{u_{12}^2}{c^2}\right)^{\frac{1}{2}}}{\left(1 - \dfrac{u_{11}^2}{c^2}\right)^{\frac{1}{2}}} \tag{10.132}$$

Let us now assume that mass m_2 is at rest in reference frame S_1 before collision. Substitution of $u_{12} = 0$ in Eq. (10.132) leads to the following expression:

$$\frac{m_1}{m_2} = \frac{1}{\sqrt{1 - \left(\dfrac{u_{11}}{c}\right)^2}} \tag{10.133}$$

Let the mass m_2 at rest be represented by m_0 and let m represent mass m_1 at rest. Using this notation, Eq. (10.133) can rewritten in the following form:

$$\frac{m_1}{m_0} = \frac{1}{\sqrt{1 - \left(\dfrac{u_{11}}{c}\right)^2}} \tag{10.134}$$

Representing u_{11} by a general term for velocity, Eq. (10.134) can be expressed as follows:

$$m = \frac{m_0}{\sqrt{1 - \left(\dfrac{u^2}{c^2}\right)}} \tag{10.135}$$

Equation (10.135) expresses the dependence of mass on velocity. In Eq. (10.135), mass m_0 is called the rest mass and m the effective mass.

Some special cases deserve a special mention, which are as follows:

Case 1 Velocity of the object is much less than that of light: $u \ll c$

Under this condition, $\dfrac{u^2}{c^2}$ can be neglected with respect to 1 and Eq. (10.135) reduces to the following form:

$$m = m_0 \tag{10.136}$$

Thus, at velocities that are much smaller than the velocity of light, the effective mass equals the rest mass.

Case 2 When u is comparable to c, the term $\sqrt{1 - \frac{v^2}{c^2}}$ is less than unity, implying that

$$m > m_0 \tag{10.137}$$

Thus, as the velocity of a body approaches the velocity of light, the effective mass becomes greater than the rest mass.

Case 3 Velocity of object is equal to the velocity of light, $u = c$

Equation (10.135) then leads to the following form:

$$m = \infty \tag{10.138}$$

This is absurd and implausible, and therefore, it is not possible for any physical body to attain a velocity that equals the velocity of light.

Example 10.32 Calculate the speed at which the mass of an object becomes double its rest mass.

Solution From Eq. (10.135), we have the following expression:

$$m = \frac{m_0}{\sqrt{1 - \dfrac{u^2}{c^2}}} \tag{10.139}$$

Since $m = 2m_0$ in the present problem, Eq. (10.139) yields the following expression:

$$2m_0 = \frac{m_0}{\sqrt{1 - \dfrac{u^2}{c^2}}}$$

giving $\sqrt{1 - \dfrac{u^2}{c^2}} = \dfrac{1}{2}$

that is, $1 - \dfrac{u^2}{c^2} = \dfrac{1}{4}$

implying that $\dfrac{u^2}{c^2} = 1 - \dfrac{1}{4} = \dfrac{3}{4}$

Thus, $u = \sqrt{\left(\dfrac{3}{4}\right)} \times 3 \times 10^{10} = 2.6 \times 10^{10}$ cm/s

Example 10.33 The rest mass of electron is 9.1×10^{-28} g. Calculate the effective mass of the electron when it is travelling with a velocity $\left(\dfrac{1}{2}\right)$ the speed of light.

Solution   From Eq. (10.135), we have the following relation:

$$m = \frac{m_0}{\sqrt{1 - \dfrac{u^2}{c^2}}} \tag{10.140}$$

Using the given values in Eq. (10.140), we get the following expression:

$$m = \frac{9.1 \times 10^{-28}}{\sqrt{1 - (0.5)^2}} = \frac{9.1 \times 10^{-28}}{0.87}$$

giving $m = 10.46 \times 10^{-28}\,\text{g}$

Example 10.34   Calculate the fractional increase of mass for a particle travelling with a velocity of $0.2c$.

Solution   From Eq. (10.135), we have the following relation:

$$m = \frac{m_0}{\sqrt{1 - \dfrac{u^2}{c^2}}} \tag{10.141}$$

Using the given values in Eq. (10.141), we get the following expression:

$$\frac{m}{m_0} = \frac{1}{\sqrt{1 - (0.2)^2}} = \frac{1}{\sqrt{1 - 0.04}} = \frac{1}{\sqrt{0.96}}$$

yielding $\dfrac{m}{m_0} = \dfrac{1}{0.98} = 1.02$

resulting in $\dfrac{m}{m_0} - 1 = \dfrac{m - m_0}{m_0} = 0.02$

Example 10.35   Calculate the speed at which the mass of an object becomes 2.5 times its rest mass.

Solution   From Eq. (10.135), we have the following relation:

$$m = \frac{m_0}{\sqrt{1 - \dfrac{u^2}{c^2}}} \tag{10.142}$$

Since, $m = 2.5 m_0$, Eq. (10.142) reduces to the following form: $2.5 m_0 = \dfrac{m_0}{\sqrt{1 - \dfrac{u^2}{c^2}}}$

leading to $\dfrac{u^2}{c^2} = 1 - 0.16 = 0.84$

Thus, $u = \sqrt{0.84}\, c = 2.76 \times 10^{10}$ cm/s

Example 10.36   The ratio of mass of a moving object to its rest mass is 1.6. Calculate the speed of the object.

Solution   From Eq. (10.135), we have the following relation:

$$m = \frac{m_o}{\sqrt{1 - \dfrac{u^2}{c^2}}} \tag{10.143}$$

Since $\dfrac{m}{m_o} = 1.6$, we can write Eq. (10.143) in the following form: $1.6 = \dfrac{1}{\sqrt{1 - \dfrac{u^2}{c^2}}}$

resulting in $1 - \dfrac{u^2}{c^2} = \dfrac{1}{(1.6)^2} = 0.39$

leading to $u = \sqrt{0.61} \times 3 \times 10^{10} = 2.34 \times 10^{10}$ cm/s

Example 10.37 The rest mass of an electron is 9.1×10^{-28} g. Calculate the speed at which the effective mass of the electron becomes 10×10^{-28} g.

Solution From Eq. (10.135), we have the following relation:

$$m = \frac{m_o}{\sqrt{1 - \dfrac{u^2}{c^2}}} \tag{10.144}$$

For the given situation, we have the following expression: $\dfrac{10}{9.1} = \dfrac{1}{\sqrt{1 - \dfrac{u^2}{c^2}}}$

leading to $1 - \dfrac{u^2}{c^2} = (0.91)^2 = 0.83$

yielding $u = 0.41 \times 3 \times 10^{10} = 1.23 \times 10^{10}$ cm/s

10.11 MASS ENERGY EQUIVALENCE

Mass and energy are interrelated as one approaches relativistic speeds. Thus, a mass change in a reaction has the potential to generate energy. This equivalence has resulted in the production of nuclear energy for both power and destructive purposes. We will discuss mass–energy equivalence in this section.

Force is the rate of change of momentum, that is,

$$F = \frac{d}{dt}(mu) \tag{10.145}$$

At a relativistic speed, Eq. (10.145) can be written in the following expanded form:

$$F = m\frac{du}{dt} + u\frac{dm}{dt} \tag{10.146}$$

Suppose that a force F displaces a particle through a distance dx. The increase in the kinetic energy is then given as follows:

$$dk = Fdx \tag{10.147}$$

Substitution of F from Eq. (10.146) in Eq. (10.147) results in the following expression:

$$dk = m\frac{du}{dt}.dx + u\frac{dm}{dt}dx \tag{10.148}$$

Since $\dfrac{dx}{dt} = u$, Eq. (10.148) can be rewritten as follows:

$$dk = mu\,du + u^2\,dm \qquad (10.149)$$

Equation (10.135) becomes as follows:

$$m = \frac{m_0}{\sqrt{1 - \frac{v^2}{c^2}}}$$

Squaring of both sides of Eq. (10.135) yields the following relation:

$$m^2 = \frac{m_0^2 c^2}{\left(c^2 - u^2\right)}$$

resulting in $m^2 c^2 - m^2 u^2 = m_0^2 c^2$

Differentiating, we get the following equation:

$$c^2\,(2m)\,dm - u^2\,(2m)\,dm - m^2\,(2u)\,du = 0$$

yielding $c^2 dm - u^2 dm - mu\,du = 0$

that is,

$$c^2 dm = mu\,du + u^2 dm \qquad (10.150)$$

A comparison of Eqs (10.149) and (10.150) implies that

$$dk = c^2 dm \qquad (10.151)$$

Let us now assume that a body of initial mass m_0 at rest attains a velocity v and a final mass m. The total kinetic energy acquired by the body is given by the following equation:

$$\int dk = \int_{m_0}^{m} c^2 dm$$

yielding

$$k = c^2\left(m - m_0\right) \qquad (10.152)$$

Equation (10.152) gives the kinetic energy acquired by the body due to an increase in velocity from zero to v. Total energy E of the body can be expressed in the following form: $E = c^2\left(m - m_0\right) + m_0 c^2$

resulting in

$$E = mc^2 \qquad (10.153)$$

Equation (10.153) is a statement of the universal equivalence between mass and energy.

The equivalence of mass and energy is demonstrated in nature also. Collision of a particle with its antiparticle results in mutual annihilation, and the mass represented by the two particles is converted into radiant energy. This demonstrates convertibility of mass into energy. On the contrary, a radiation of appropriate energy interacting with a charged nucleus results in the production of particles and antiparticles. This demonstrates conversion of energy into mass.

Example 10.38 How much energy can be liberated by annihilation of 0.5 g of matter?

Solution From Eq. (10.153), we have the following expression:

$$E = mc^2 \tag{10.154}$$

Using the given values in Eq. (10.154), we get the following expression:

$$E = 0.5 \times 10^{-3} \times (3 \times 10^8)^2$$

yielding $E = \dfrac{1}{2} \times 10^{-3} \times 9 \times 10^{16} = 4.5 \times 10^{13} \, \text{J}$

We also know that $1 \text{eV} = 1.6 \times 10^{-19} \, \text{J}$

Thus, we have the following value: $E = \dfrac{4.5 \times 10^{13}}{1.6 \times 10^{-19}} = 2.81 \times 10^{32} \, \text{eV}$

Example 10.39 The total energy of a particle is exactly twice its rest energy. Calculate the velocity of the particle.

Solution From Eq. (10.153), we have the following expression:

$$E = mc^2 \tag{10.155}$$

We are given that $mc^2 = 2m_0c^2$

Thus,

$$m = 2m_0 \tag{10.156}$$

Using Eq. (10.156) in Eq. (10.135), we get the following expression:

$$2m_0 = \frac{m_0}{\sqrt{1 - \dfrac{u^2}{c^2}}}$$

yielding $\sqrt{1 - \dfrac{u^2}{c^2}} = \dfrac{1}{2}$

that is, $1 - \dfrac{u^2}{c^2} = \dfrac{1}{4}$

leading to $\dfrac{u^2}{c^2} = 1 - \dfrac{1}{4} = \dfrac{3}{4}$

implying that $u = \sqrt{\dfrac{3}{4}} \times c = \sqrt{\dfrac{3}{4}} \times 3 \times 10^{10} = 2.60 \times 10^{10} \, \text{cm/s}$

Example 10.40 Calculate the mass m and speed u of an electron with kinetic energy 1.2 MeV. Assume that the rest mass of electron is $m_0 = 9.11 \times 10^{-31}$ kg and velocity of light in vacuum $c = 3 \times 10^8$ m/s.

Solution Kinetic energy, $\text{KE} = (m - m_0)c^2$

Thus, $(1.2 \times 10^6)(1.6 \times 10^{-19}) = (m - 9.11 \times 10^{-31})(3 \times 10^8)^2$

implying that $m - 9.11 \times 10^{-31} = \dfrac{(1.2 \times 10^6) \times (1.6 \times 10^{-19})}{(3 \times 10^8)^2} = 2.13 \times 10^{-30}$

yielding

$$m = 9.11 \times 10^{-31} + 2.13 \times 10^{-30} = 30.41 \times 10^{-31} \text{kg} \tag{10.157}$$

Equation (10.135) now becomes as follows:

$$m = \frac{m_0}{\sqrt{1 - \dfrac{u^2}{c^2}}} \tag{10.158}$$

which can be simplified to the following form:

$$u = c\sqrt{1 - \left(\frac{m_0}{m}\right)^2} \tag{10.159}$$

Use of Eq. (10.158) in Eq. (10.159) yields the following expression:

$$u = 3 \times 10^8 \sqrt{1 - \left(\frac{9.11 \times 10^{-31}}{30.41 \times 10^{-31}}\right)^2}$$

giving $\quad u = 3 \times 10^8 \sqrt{1 - \left(\dfrac{9.11}{30.41}\right)^2} = 3 \times 10^8 \sqrt{0.91}$

or $\qquad u = 2.86 \times 10^8 \text{m/s}$

10.12 KINETIC ENERGY OF RELATIVISTIC PARTICLE

Kinetic energy of a relativistic particle is given by the following expression:

$$\text{KE} = \lambda mc^2 - mc \tag{10.160}$$

where λ = Lorentz factor = $\left(1 - u^2 / c^2\right)^{-\frac{1}{2}}$

The quantity λmc^2 is known as the total energy, E, and mc^2 as the rest energy of the particle. Thus,

$$\text{KE} = E - mc^2 \tag{10.161}$$

The quantity mc^2 is also called the mass energy. The binomial expansion for $(a + x)^n$ is as follows:

$$(a + x)^n = a^n + na^{n-1}x + n(n - 1)a^{n-2}x^2 / 2! + \cdots \tag{10.162}$$

Using the binomial expansion in the expression for λ and keeping only the first two terms, we get the following equation:

$$\left(1 - u^2 / c^2\right)^{-\frac{1}{2}} = 1 + u^2 / 2c^2 \tag{10.163}$$

Using Eq. (10.163) in Eq. (10.160), we get the following expression:

$$\text{KE} = \left(1 + u^2 / 2c^2\right)mc^2 - mc^2$$

yielding $\text{KE} = \dfrac{1}{2}mu^2$

which is the Newtonian approximation for kinetic energy.

10.13 ENERGY–MOMENTUM RELATION

The rest mass m_0, total energy E, and momentum p of a relativistic particle are related through the following equation:

$$E^2 = (pc)^2 + (m_0c^2)^2 \qquad (10.164)$$

The following are some special cases:

Case 1   For a massless particle, $m_0 = 0$, and we get the following relation:

$$E = pc \qquad (10.165)$$

Equation (10.165) is valid for a photon, since it has a zero rest mass.

Case 2   If speed $u \ll c$, then

$$E = m_0u^2/2 + m_0c^2 = KE + Rest\ energy$$

Case 3   If $u = 0$, then

$$E = E_0$$

and $m = m_0$

10.14 TWINS PARADOX

Relativity has several paradoxes associated with itself. An interesting paradox that is linked to relativity is called the *twins paradox*. Suppose there is a pair of twins, one of whom chooses to take a space flight at a speed approaching the speed of light. On returning to earth after the space flight, the person would find that the stay-at-home twin would have aged at a much faster rate. It is called a paradox because, relatively speaking, both the twins are travelling with respect to each other, whereas only the stay-at-home twin ages faster. The paradox is resolved by arguing that only one twin experiences acceleration and deceleration, and therefore ages at a slower rate.

IMPORTANT CONCEPTS

1. A reference frame that is either at rest or moving with a uniform velocity is called an inertial reference frame. A body experiences acceleration even in the absence of any external force in a non-inertial reference frame.
2. The Michelson and Morley experiment proved the fallacy of the assumption of the all-pervading ether.
3. Einstein's first postulate states that all physical laws are the same in all inertial reference frames that are moving relative to each other with a constant velocity.
4. Einstein's second postulate states that the speed of light has the same value in all inertial reference frames.
5. Lorentz transformations enable us to transform position and time from one inertial reference frame to another.
6. The length of an object in a moving reference frame is found to reduce when observed from a stationary reference frame.

7. Time taken for an event to take place increases in a moving reference frame as compared to the time taken for the same event in a stationary reference frame.
8. Mass starts depending upon the speed as speed approaches the speed of light.
9. Mass and energy are interrelated at relativistic speeds, through the following expression:

$$E = mc^2$$

APPLICATIONS

1. Atomic clocks are very accurate in recording time. Experiments have been carried out using two atomic clocks that are first synchronized. One of the clocks is then kept at rest and the other is flown in a high-speed aircraft. The clock on the plane is found to record marginally lower passage of time, as expected from theory. The difference is, however, not very high, since even the fastest aircraft cannot fly at a speed anywhere near the speed of light.
2. Ultra-short-lived muons, fundamental particles, can travel at 99.9% the speed of light. These fast-moving muons have been found to live 25 times longer and travel up to 25 times further than they are theoretically expected to do. The particle accelerator at CERN, Switzerland, produces particles that travel at speeds approaching 99.99% speed of light. These high-speed particles also experience time dilation with a Lorentz factor of approximately 5000.

IMPORTANT FORMULAE

1. Galilean transformations are represented by the following expressions:

$$x_2 = x_1 - vt_1$$
$$y_2 = y_1$$
$$z_1 = z_1$$
$$t_2 = t_1$$

2. Galilean transformation for velocity is given as $u_2 = u_1 - v$
3. Acceleration does not change while going from an inertial to a non-inertial reference frame.
4. Lorentz transformation for space and time are as follows:

$$x_2 = \frac{x_1 - vt_1}{\sqrt{1 - \dfrac{v^2}{c^2}}} ; \quad y_2 = y_1; z_2 = z_1$$

and,

$$t_2 = \frac{t_1 - \left(\dfrac{vx_1}{c^2}\right)}{\sqrt{1 - \dfrac{v^2}{c^2}}}$$

5. Corresponding inverse Lorentz transformation are as follows:

$$x_1 = \frac{x_2 + vt_2}{\sqrt{1 - \dfrac{v^2}{c^2}}} ; \quad y_1 = y_2; z_1 = z_2$$

and,

$$t_1 = \frac{t_2 + \left(\dfrac{vx_2}{c^2}\right)}{\sqrt{1 - \dfrac{v^2}{c^2}}}$$

6. Lorentz–Fitzerald length contraction is given by the following equation:

$$l_1 = l_2 \sqrt{1 - \left(\dfrac{v^2}{c^2}\right)}$$

7. Time dilation is given as follows:

$$(\Delta t)_2 = \frac{(\Delta t)_1}{\sqrt{1 - \dfrac{v^2}{c^2}}}$$

8. The relativistic law of addition of velocities is as follows:

$$(u)_1 = \frac{(u)_2 + v}{1 + \left[\dfrac{(u)_2 v}{c^2}\right]}$$

9. The variation of mass with velocity can be expressed as follows:

$$m = \frac{m_0}{\sqrt{1 - \left(\dfrac{u^2}{c^2}\right)}}$$

10. The universal equivalence between mass and energy is mathematically expressed in the following form:

$$E = mc^2$$

SELF-ASSESSMENT

Multiple-choice Questions

10.1 Mass energy equivalence was proposed by
   (a) Newton
   (b) Madam curie
   (c) Lorentz
   (d) Einstein.

10.2 The quantity e/m of electrons
   (a) increases with speed
   (b) decreases with speed
   (c) does not change with speed
   (d) increases and then decreases

10.3 Two identical clocks A and B are used to record an event. With respect to the observer, clock A is at rest, whereas clock B moves with a uniform velocity. Then clock A will be
   (a) faster than B
   (b) slower than B
   (c) showing equal time as B
   (d) showing time that depends on the velocity of B

10.4 The theory of relativity proves
   (a) the existence of ether
   (b) that the velocity of light is independent of observer's motion
   (c) variation of mass with velocity
   (d) relative nature of time

10.5 A body in a non-inertial reference frame experiences acceleration
   (a) in presence of force
   (b) in absence of force
   (c) both (a) and (b)
   (d) depending on the magnitude of force

10.6 According to Galilean transformation,
   (a) change of velocity is independent of the reference frame
   (b) change of velocity is dependent on the reference frame
   (c) acceleration depends on the reference frame
   (d) velocity does not depend on the choice of the reference frame

10.7 According to mass–energy equivalence,
   (a) mass and energy are inter-convertible
   (b) mass can never be converted to energy
   (c) mass and energy are two different entities
   (d) annihilation of mass cannot produce energy

10.8 Michelson–Morley experiment was carried out to
   (a) measure speed of light
   (b) measure speed of earth
   (c) prove existence of ether
   (d) test isotropy of space

10.9　The speed at which the mass of a particle becomes double its rest mass is
(a)　2.60×10^8 m/s
(b)　6.20×10^8 m/s
(c)　2.60×10^{10} m/s
(d)　1.30×10^8 m/s

10.10　When 1 g mass is converted into energy, the amount of energy released in kilowatt-hour is
(a)　0.25×10^9
(b)　0.25×10^8
(c)　0.25×10^{10}
(d)　0.25×10^{11}

10.11　Twins paradox exits due to
(a)　length contraction
(b)　time dilation
(c)　both (a) and (b)
(d)　none of these

10.12　According to the twins paradox, a space traveller
(a)　ages faster
(b)　does not grow old at all
(c)　ages slower
(d)　ages at the same rate as a person at rest

10.13　Pick up the correct energy–momentum relation:
(a)　$E = pc + m_0 c^2$
(b)　$E^2 = pc + (m_0 c^2)^2$
(c)　$E^2 = (pc)^2 + m_0 c^2$
(d)　$E^2 = (pc)^2 + (m_0 c)^2$

10.14　When $u \ll c$, we have
(a)　$E = KE + \text{Rest energy}$
(b)　$E = KE$
(c)　$E = \text{Rest energy}$
(d)　$E = E_0 + KE + \text{Rest energy}$

10.15　Total energy is given by the relation:
(a)　mc^2
(b)　λmc^2
(c)　$\lambda^2 mc^2$
(d)　$\lambda + mc^2$

Review Questions

10.1　What is an inertial reference frame?
10.2　Define a non-inertial reference frame.
10.3　Write and explain Galilean transformation for space and time.
10.4　Does change in velocity depend on the choice of a reference frame?
10.5　Explain the Michelson–Morley experiment.
10.6　Give the two postulates of the special theory of relativity.
10.7　State and explain Lorentz transformations for space and time.
10.8　Derive an expression for Lorentz–Fitzerald contraction.
10.9　Derive an expression for time dilation at relativistic speed.
10.10　Deduce the relativistic law of addition of velocities.
10.11　Derive an expression for velocity dependence of mass.
10.12　Justify the statement of universal equivalence between mass and energy.

Numerical Problems

10.1　A rod having a length of 80 cm is kept in a satellite moving with a velocity $0.5c$ with respect to the earth. Calculate the length of the rod as measured by an observer (a) in the satellite and (b) on the surface of earth

$$\left[Hint : l_1 = l_2 \sqrt{1 - \frac{v^2}{c^2}} \right]$$

10.2 A rocket measures $85 \, \text{m}$ in length on the ground. During flight, its length is measured to be $83 \, \text{m}$ by an observer on the ground. Determine the speed of

the rocket.

$$\left[Hint : l_1 = l_2 \sqrt{1 - \frac{v^2}{c^2}} \right]$$

10.3 A process requires $3 \, \mu\text{s}$ to complete inside an atom at rest in a laboratory. Determine the time required to complete this process with respect to an observer in the laboratory when the atom is moving at a speed of $3 \times 10^9 \, \text{m/s}$.

$$\left[Hint : (\Delta t)_2 = \frac{(\Delta t)_1}{\sqrt{1 - \frac{v^2}{c^2}}} \right]$$

10.4 A particle having a mean life time of $2 \, \mu\text{s}$ is moving at a speed of $0.9c$ in a laboratory set-up. Determine its lifetime, as calculated by an observer in the

laboratory.

$$\left[Hint : (\Delta t)_2 = \frac{(\Delta t)_1}{\sqrt{1 - \frac{v^2}{c^2}}} \right]$$

10.5 The proper life of π^- mesons is $2 \times 10^{-8} \, \text{s}$. Determine the velocity of these π

mesons if their observed mean life is $2 \times 10^{-7} \, \text{s}$.

$$\left[Hint : (\Delta t)_2 = \frac{(\Delta t)_1}{\sqrt{1 - \frac{v^2}{c^2}}} \right]$$

10.6 Two particles are travelling towards each other with speed $0.91c$ with respect to an observer in a laboratory. Calculate their relative speed.

$$\left[Hint : (u)_1 = \frac{(u)_2 + v}{1 + \left[\frac{(u)_2 v}{c^2} \right]} \right]$$

10.7 A missile chases an enemy aircraft. For an observer on the earth, the speed of the missile is $1 \times 10^{10} \, \text{cm/s}$ and that of the enemy aircraft is $0.7 \times 10^{10} \, \text{cm/s}$. Determine the relative velocity of the enemy aircraft with respect to the missile.

$$\left[Hint : (u)_1 = \frac{(u)_2 + v}{1 + \left[\frac{(u)_2 v}{c^2} \right]} \right]$$

10.8 Determine the speed at which the mass of a given object becomes thrice its

rest mass.

$$\left[Hint : m = \frac{m_0}{\sqrt{1 - \frac{u^2}{c^2}}} \right]$$

10.9 The rest mass of electron is 9.1×10^{-28} g. Determine the effective mass of the electron when it is travelling with velocity $0.8c$.

$$\left[Hint : m = \frac{m_0}{\sqrt{1 - \dfrac{u^2}{c^2}}} \right]$$

10.10 Determine the fractional increase of mass for a particle travelling with a velocity of $0.4c$.

$$\left[Hint : m = \frac{m_0}{\sqrt{1 - \dfrac{u^2}{c^2}}} \right]$$

10.11 Calculate the energy liberated by annihilation of 0.6 g of matter.

$$\left[Hint : E = mc^2 \right]$$

10.12 The total energy of a particle is exactly 1.5 times its rest energy. Determine the velocity of the particle.

$$\left[Hint : E = mc^2 = \frac{m_0}{\sqrt{1 - \dfrac{u^2}{c^2}}} c^2 \right]$$

10.13 Determine the mass m and speed u of an electron with kinetic energy 1.3 MeV. Assume that the rest mass of electron is $m_0 = 9.11 \times 10^{-31}$ kg and the velocity of light in vacuum, $c = 3 \times 10^8$ m/s. $\left[Hint : KE = (m - m_0)c^2 \right]$

10.14 A meter scale is kept in a satellite moving with velocity $0.65c$ with respect to the earth. Calculate the length of the rod as measured by an observer on the earth.

$$\left[Hint : l_1 = l_2 \sqrt{1 - \dfrac{v^2}{c^2}} \right]$$

10.15 A box of length 60 cm is kept in a spaceship travelling with speed $0.72c$. What is the length of the box, as observed by an observer on the earth?

$$\left[Hint : l_1 = l_2 \sqrt{1 - \dfrac{v^2}{c^2}} \right]$$

10.16 The rod of length 90 cm indicated in Example 10.5 slows down to $0.35c$. Find the percentage increase in the length of the rod as observed by an observer on the earth?

$$\left[Hint : l_1 = l_2 \sqrt{1 - \dfrac{v^2}{c^2}} \right]$$

10.17 The rod of length 90 cm indicated in Example 10.5 increases its velocity to $0.66c$. Find the percentage reduction in the length of the rod as observed by an observer on the earth.

$$\left[Hint : l_1 = l_2 \sqrt{1 - \dfrac{v^2}{c^2}} \right]$$

10.18 A moving object of length 95 cm is measured to be 86 cm long by an observer on the ground. Calculate the speed of the object.

$$\left[Hint : l_1 = l_2 \sqrt{1 - \dfrac{v^2}{c^2}} \right]$$

10.19 At what speed will an object of length 92 cm move for it to be measured as 40 cm long by an observer at rest? $\left[Hint : l_1 = l_2 \sqrt{1 - \dfrac{v^2}{c^2}} \right]$

10.20 Find the speed at which a rocket appears to be 82% of its actual length, to an observer at rest. $\left[Hint : l_1 = l_2 \sqrt{1 - \dfrac{v^2}{c^2}} \right]$

10.21 The length of a moving ship is found to be 76% of its actual length to an observer at rest. Calculate the speed of the space ship. $\left[Hint : l_1 = l_2 \sqrt{1 - \dfrac{v^2}{c^2}} \right]$

10.22 An atomic process needs 1.8 µs to complete in the laboratory. If the atom starts moving at a speed of 8.2×10^9 cm/s, calculate the time required to complete the same process with respect to an observer in the laboratory.

$$\left[Hint : (\Delta t)_2 = \dfrac{(\Delta t)_1}{\sqrt{1 - \dfrac{v^2}{c^2}}} \right]$$

10.23 A process takes 0.8 µs to complete in a rocket at rest. Calculate the time required to complete the same process with respect to an observer on the earth when the rocket is moving at a speed of 5×10^9 cm/s.

$$\left[Hint : (\Delta t)_2 = \dfrac{(\Delta t)_1}{\sqrt{1 - \dfrac{v^2}{c^2}}} \right]$$

10.24 A spaceship is moving at a speed of 4.6×10^9 cm/s with respect to the earth. The time required for an event to take place on the spaceship, as observed by an observer on the earth, is 1.2 µs. What will be the time required for the same event if the observer is travelling on the spaceship?

$$\left[Hint : (\Delta t)_1 = (\Delta t)_2 \sqrt{1 - \dfrac{v^2}{c^2}} \right]$$

10.25 A rocket is travelling at a speed of 8.2×10^9 cm/s with respect to the earth. The time required for a fuse on the rocket to blow is 5.1 µs, as observed by an observer on the earth. Calculate the actual time taken for the fuse to blow.

$$\left[Hint : (\Delta t)_1 = (\Delta t)_2 \sqrt{1 - \dfrac{v^2}{c^2}} \right]$$

10.26 The particle indicated in Example 10.14 speeds up to 0.93c. What is its new lifetime, as measured by an observer in the laboratory?

$$\left[Hint : (\Delta t)_2 = \dfrac{(\Delta t)_1}{\sqrt{1 - \dfrac{v^2}{c^2}}} \right]$$

10.27 The particle indicated in Example 10.14 slows down to 0.53c. Calculate its lifetime, as measured by an observer in the laboratory.

$$\left[Hint : (\Delta t)_2 = \frac{(\Delta t)_1}{\sqrt{1 - \dfrac{v^2}{c^2}}} \right]$$

10.28 A particle at rest has a lifetime of 2.4 μs. When it speeds up, its lifetime is measured to be 3.6 μs by an observer in the laboratory. Calculate the speed of the particle.

$$\left[Hint :(\Delta t)_2 = \frac{(\Delta t)_1}{\sqrt{1 - \dfrac{v^2}{c^2}}} \right]$$

10.29 The ratio of the lifetime of a particle at rest to that when the same particle is moving is 1/2.2, as measured by an observer in the laboratory. Find the speed of the particle.

$$\left[Hint : (\Delta t)_2 = \frac{(\Delta t)_1}{\sqrt{1 - \dfrac{v^2}{c^2}}} \right]$$

10.30 The proper life of a fundamental particle is found to be 1.2×10^{-8} s. Calculate the velocity of this particle if the observed mean life is 6.8×10^{-7} s.

$$\left[Hint : (\Delta t)_2 = \frac{(\Delta t)_1}{\sqrt{1 - \dfrac{v^2}{c^2}}} \right]$$

10.31 The proper life of a fundamental particle is 2.8×10^{-8} s. Calculate the observed mean life of this particle when it moves at a speed of 0.86c.

$$\left[Hint : (\Delta t)_2 = \frac{(\Delta t)_1}{\sqrt{1 - \dfrac{v^2}{c^2}}} \right]$$

10.32 The observed mean life of a fundamental particle is 6.8×10^{-8} s when the particle is moving with speed 0.89c. Calculate the proper life of the particle.

$$\left[Hint : (\Delta t)_2 = \frac{(\Delta t)_1}{\sqrt{1 - \dfrac{v^2}{c^2}}} \right]$$

10.33 The ratio of the proper life to the mean life of a moving fundamental particle is 1/6.1. Calculate the speed of the fundamental particle.

$$\left[Hint : (\Delta t)_2 = \frac{(\Delta t)_1}{\sqrt{1 - \dfrac{v^2}{c^2}}} \right]$$

11

Concepts of Quantum Mechanics

Learning Objectives

After studying this chapter, students will be able to

- understand the photoelectric effect which explains the photon as an energy packet of light
- realize De Broglie's concept that elucidates the dual nature of light
- comprehend the concept of phase and group velocity
- apply the knowledge of X-rays to study crystal structure highlighting inter-planar spacing
- describe the two-slit interference experiment together with the Heisenberg uncertainty principle, which leads to the conceptualization of quantum mechanics
- explain the Schrodinger equation, which acquaints the learner with the operator formalism and the wave function concept
- elucidate the quantum phenomena such as free electron states, quantum tunnelling, and bound states that help realize the concept of discreteness

List of Symbols

h = Planck's constant

ϑ = Frequency of radiation

e = Charge on an electron

V_0 = Threshold voltage

φ_0 = Photoelectric work function

λ = Wavelength of function

m = Mass of an electron

B = Uniform magnetic field

ω = Angular frequency

ω_C = Energy lost by collision

V_c = Contact potential

V_S = Retarding potential

d = Inter-planar spacing

φ = Wave function

$\bar{x}$ = Average position

$\langle U(x) \rangle$ = Expectation value of the function $U(x)$

$\widehat{p_x}$ = Momentum operator

$\hat{E}$ = Energy operator

$\hat{H}$ = Hamiltonian operator

T = Transmission coefficient

11.1 INTRODUCTION

As the various domains of physics extended from the classical approach to the introduction of the basic concepts of what is now referred to as quantum mechanics, the basic properties of light have been studied in detail by the following most salient branches. Maxwell's equation suggested that electrical and magnetic fields coexist. The electromagnetic theory suggested the idea that light was formed of electromagnetic waves. Blackbody radiations were extensively studied by Planck, and quantization of energy was highlighted by the Plank's law. Magnetic theories have been established using the concepts of spin and orbital moments coming from the quantum nature of atomic particles. Discovery of the photoelectric effect strongly contradicted the idea of light consisting of waves, but suggested that light was formed of photons.

These thoughts towards the discrete behaviour of particles led to a reasoning among the then scientific fraternity to introduce a new mathematical approach to the problems of physics, which should be able to explain both the particle and the wave nature. Heisenberg pioneered this thought by introducing the concept of probabilistic nature in the exact determination of the space, time coordinates of a particle. The work of Schrodinger gave physics its new found theory of wave mechanics, wherein each problem went beyond the realms of classical, Newtonian physics. The solutions of Schrodinger established a new theory, which reduced to the classical solutions as the Planck's constant ceases its significance in the mathematical calculations. In developing this chapter of wave mechanics, care has been taken to introduce the reader to the initial concepts that led to its development.

11.2 PHOTOELECTRIC EFFECT

Photoelectric emission is the phenomenon in which a good number of substances, chiefly metals, under the influence of radiations such as γ-rays, X-rays, ultraviolet rays, and even visible light, emit electrons. It was first discovered by W. Smith and Hertz, and later established through experiments by Hallwach, Elster, and Geitel. A fairly simple set-up explained most of the fundamental characteristics of photoelectricity (Fig. 11.1).

Two zinc plates A and B are placed in an evacuated quartz bulb and connected to a battery E and a galvanometer G. When an ultraviolet light fell on the plate A, which was connected to the negative terminal of the battery, a current was found to flow, as indicated by the galvanometer (Fig. 11.1). However, when the light fell on the positive

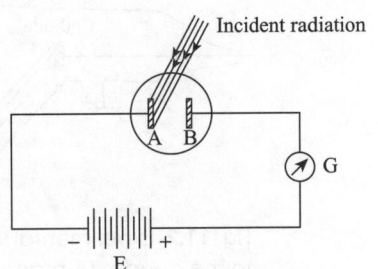

Fig. 11.1 Simple electronic set-up to understand photoelectric effect

plate B, there was no flow of current. This proves that not only a negatively charged body loses its charge when irradiated with an ultraviolet light, while no such effect occurs with a positively charged body, but also the ions emitted from bodies that lose charge under the action of light must be negatively charged; the observed current can only be expressed by the fact that the negative ions emitted by the negative plate are attracted towards the positive plate and this cause the flow of electricity. Since this effect is produced by light, it is called the *photoelectric effect*, and the negative ions are called *photoelectrons* and the resulting current the *photoelectric current*.

When plates of different materials were used, Zn, Mg, Li, Na, K, and Rb were found to respond photoelectrically to the ultraviolet light in an increasing order of sensitivity. This shows that the *photoelectric emission depends upon both the nature of the emitter and the quality of light*. It was also established that the photoelectric effect takes place even in the highest vacuum, which shows that the gas surrounding the emitter does not play a role in the effect.

11.2.1 Experimental Study

To study the various effects of photoelectric emission, an apparatus was set up by Lenard. The results are postulated in the following subsections and are referred to as the properties/laws of photoelectric emission.

11.2.1.1 Nature of Negative Ions Emitted

The experimental set-up shown in Fig. 11.2 is constructed. When rays from the light source S fall on the cathode, which is negatively charged, particles are liberated and accelerated towards the anode, after being repelled by the cathode. The central hole marked A in the anode sends the particle through to plate P_1, which is connected with electrometer 1. If a suitable magnetic field is applied at right angle to the plane of the figure and directed towards

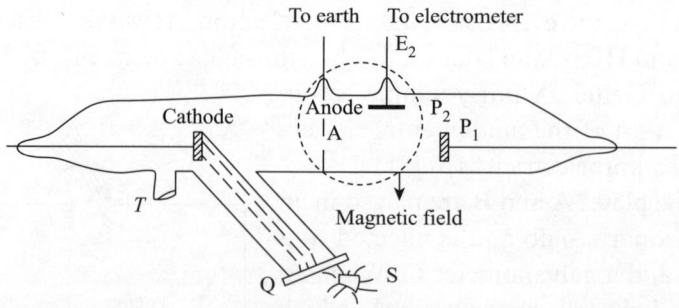

Fig. 11.2 Experimental set-up wherein cathode sends particles to the anode. (A magnetic field is applied to deviate path of particles so as to determine their charge–mass ratio. As photoelectric equation is established, experiment shows that particles emitted from cathode are electrons.)

the reader, in the region between A and P_1, the ions will be deflected upwards in a circular path and strike plate P_2, as indicated by the electrometer E_2. In order to eliminate the effect due to velocity imparted to the ions, an initial potential is given to the cathode so that the photoelectric current just starts to flow. Thus, when the particles arrive at A they possess a velocity v solely due to the potential between A and C. The ions, on reaching A, will have a kinetic energy given by the following expression:

$$eV = \frac{1}{2}mv^2 \tag{11.1}$$

If ions are subject to a uniform magnetic field after leaving the anode, the circular path traversed by them is given by the following equation:

$$Bev = \frac{mv^2}{r} \tag{11.2}$$

or, $$v = \frac{Ber}{m} \tag{11.3}$$

Substituting the value of v in Eq. (11.1), we get the following expression:

$$eV = \frac{1}{2}m\left(\frac{Ber}{m}\right)^2 = \frac{1}{2}\frac{B^2 e^2 r^2}{m} \tag{11.4}$$

Thus,

$$\frac{e}{m} = \frac{2V}{B^2 r^2} \tag{11.5}$$

The ratio of (e/m) is found to be in agreement with that obtained for electrons. Thus, *the ions emitted in the photoelectric emission are identical to the electrons.*

11.2.1.2 Dependence of Photoelectric Current on Intensity of Illumination

In the absence of any significant change in the spectral quality of the light causing the emission of the photoelectrons, the photoelectric current is directly proportional to the intensity of illumination on the emitting surface. The photoelectric current, however, attains a saturation current value for a given potential difference between the irradiated plate and an auxiliary electrode that is used to clear away electrons as they are ejected.

11.2.1.3 Velocity and Energy of Photoelectrons

The Richardson and Compton experiment was performed to exhibit the relations existing between photoemission and the wavelength, frequency, and energy of electrons.

1. Richardson and Compton first studied the relation between the photoelectric current I and the retarding potential V. The cathode C is illuminated with a monochromatic light of a given intensity, and the potential on the

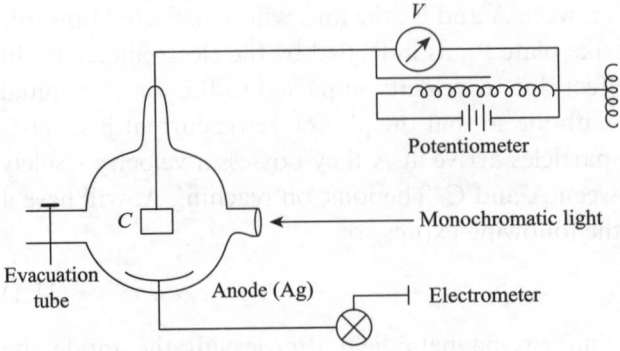

Fig. 11.3 Monochromatic light is incident on cathode with varying voltage and intensity to observe variation in current

cathode is varied from a few positive volts to zero and a few negative volts, as indicated in Fig. 11.3. The strength of the photoelectric current I at different values of V was measured with an electrometer, and observations were taken by doubling and tripling the intensities of illumination.

At potentials just less than the positive V_0, a small current is observed, and as the potential is decreased to zero, the current rises to a maximum value at $V = 0$; then no further increase is observed as V becomes negative. For different intensities, the critical potential V_0 remains the same, but the corresponding current attains a different maximum value.

Thus, it is seen that while the maximum current (I_m) is proportional to the intensity of illumination, the *critical potential V_0 is independent of the intensity.*

2. The rate of emission of photoelectrons at any retarding potential V is given by (I/e), where I is the photoelectric current and e is the electronic charge (Fig. 11.4). The slope of the experimental curve at any point V_0 is 0 and is proportional to the number of photoelectrons possessing energy corresponding to V at that point. A curve plotted (Fig. 11.5) between these slopes as ordinates and the corresponding value of eV as abscissae gives the *energy distribution* curve of photoelectrons. Thus, in the stream of photoelectrons emerging from the emitter, there are electrons possessing energies from 0 to a maximum value eV_0.

3. Using several different monochromatic radiations of wavelengths $\lambda_1, \lambda_2, \lambda_3, \ldots$ to irradiate the emitter successively, in each case, the photoelectric

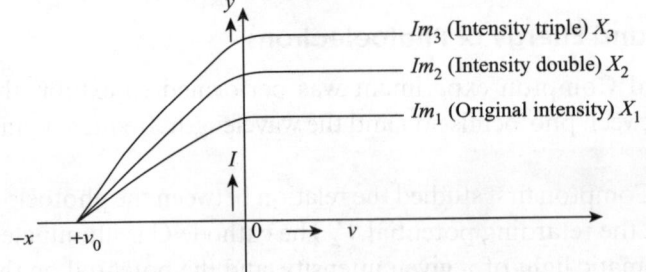

Im_3 (Intensity triple) X_3
Im_2 (Intensity double) X_2
Im_1 (Original intensity) X_1

Fig. 11.4 Retarding potential curve showing variation in current as intensity of incident radiation is varied

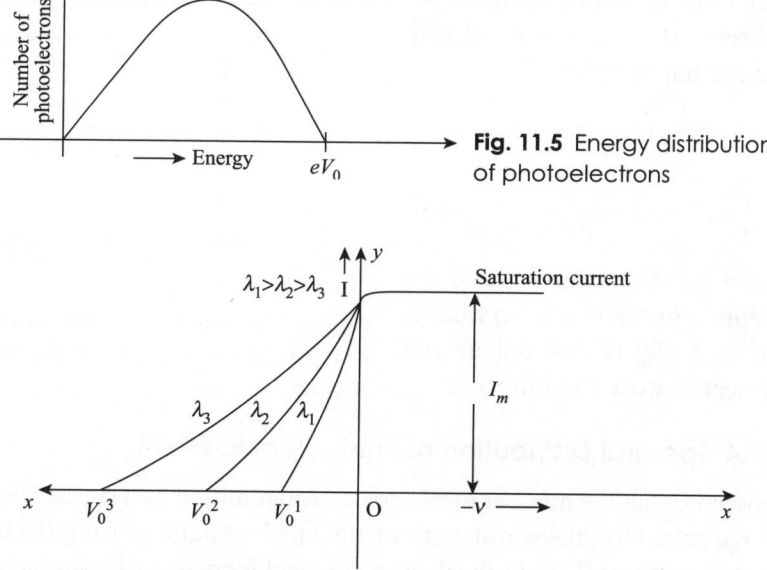

Fig. 11.5 Energy distribution of photoelectrons

Fig. 11.6 Frequency variation curves (frequency versus current)

current I for different values of voltages applied to the emitter was determined. It is seen from the curves that if $\lambda_1 > \lambda_2 > \lambda_3$, the corresponding critical voltage values at which the current starts in the three cases are found to be $V_0^1 < V_0^2 < V_0^3$, as shown in Fig. 11.6. Hence, as the wavelength of light increases, the critical retarding potential decreases. This means that the maximum kinetic energy of photoelectrons is given by the following relation:

$$\frac{1}{2}mv_m^2 = eV_0 \tag{11.6}$$

Kinetic energy increases with the increasing frequency of the light that causes the emission. Since we already know that V_0 is independent of the intensity of illumination, we conclude that the velocity and kinetic energy of photoelectrons are independent of the intensity of illumination, but dependent on the frequency of light.

If the value of the maximum energy given by eV_0 is plotted against the corresponding frequency ϑ of the light, a straight line is obtained, which has an intercept ϑ_0 on the frequency axis. Thus, it implies that a frequency of light of less than ϑ_0 cannot cause photoelectric emission from the emitter concerned. The quantity ϑ_0 is characteristic of the emitter, varying from substance to substance, and is referred to as the *threshold frequency* as it represents the beginning of the photoelectric activity of the emitter. Equation of the straight line curve is represented as follows:

$$eV_0 = h(\vartheta - \vartheta_0) \tag{11.7}$$

where h is the slope of the line (represented in Fig. 11.7). From Eq. (11.6), we can observe that the left-hand side can be written as follows:

$$eV_0 = \frac{1}{2}mv_m^2 = h(\vartheta - \vartheta_0) \quad (11.8)$$

$$\frac{1}{2}mv_m^2 = h\vartheta - \varphi_0 \quad (11.9)$$

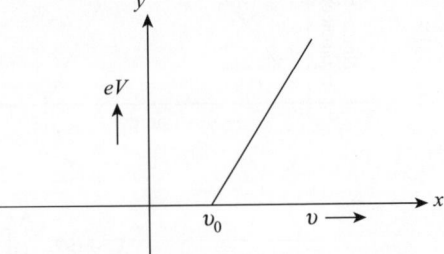

Fig. 11.7 Variation of eV with frequency is a straight line such that slope of this line is h

Equation (11.9) is referred to as the Einstein's photoelectric equation, where $\varphi_0 = h\vartheta_0$ is defined as the photoelectric work function.

11.2.1.4 Spectral Distribution of Photoelectric Effect

It is observed that the number of photoelectrons emitted by a surface depends, to a large extent, on the wavelength of the incident light. During the normal photoelectric effect [Fig. 11.8(a)], electron yield increases with an increase in the frequency of light or a decrease in the wavelength.

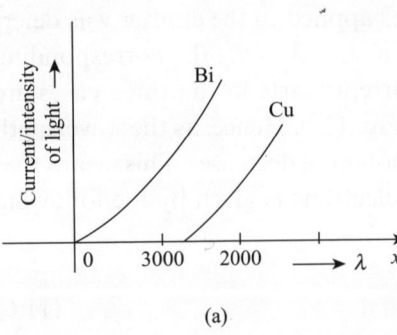

(a)

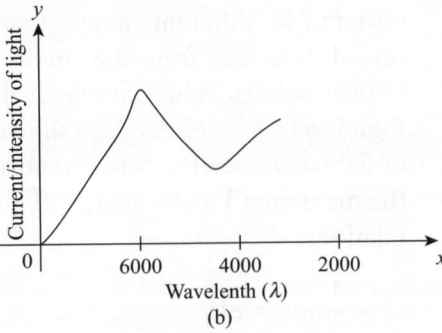

(b)

Fig. 11.8 Dependence of photoelectric effect on the emitter surface (a) Normal photoelectric effect (b) Selective photoelectric effect

Selective photoelectric effect Production of electrons exhibits a maximum at a particular frequency. It generally occurs in alkali and alkaline earth metals. It is exhibited only when the electric vector in the incident light has a component normal to the surface of the emitter. The position and height of the maxima are greatly influenced by the structure and condition of the surface. It was later shown that this effect arises due to contamination of the surface of the emitter either by adsorbed gases or by impurity deposits.

11.2.1.5 Dependence of Photoelectric Effect on Nature of Metal Used as Emitter

Photoelectric emission is greatly affected by the surface of the emitter. The presence of layers of gases on the surface or absorbed inside the emitter is of great importance, as it not only increases the yield considerably (selective effect), but also shifts the threshold very evidently towards the long wavelength side.

11.2.1.6 Photoelectric Fatigue

It has been observed that the photoelectric sensitivity of a pure metal diminishes with time. This is due to slow oxidation of the surface, which results in an increase of the photoelectric work function with a corresponding decrease in the yield of photoelectrons or the *ionic bombardment of the surface* (for impure metals) by the positive ions created in the surrounding gas by the photoelectrons that produce changes in the temperature at the surface and thereby alters the nature of the surface.

11.2.1.7 Photoelectric Effects in Non-metals

Photoelectric effect is seen in non-metals, liquids, vapours, and gases.

11.2.1.8 Photochemical Reactions

Photochemical reactions occur due to photoelectric emission. A photographic plate is a practical application of these reactions.

11.2.1.9 Inner Photoelectric Effect

Photoelectric effect is observed in an inner volume emission of electrons if the incident radiation is capable of penetrating into the interior of the body. This liberation of electrons inside a dielectric (insulator) or a body with a high resistance (semiconductor) results in an increase in the specific inductive capacity or conductivity.

11.3 EINSTEIN'S PHOTOELECTRIC EQUATION

Radiation is regarded as a shower of photons, each of energy $h\vartheta$, moving in space with the velocity of light. We assume that the collision between a photon and an atom results in the absorption of the whole photon by the atom and the consequent emission of a photoelectron with practically the total energy of the photon:

$$h\vartheta = \frac{1}{2}mv^2 + \omega_0 \tag{11.10}$$

where the term $h\vartheta$ is the energy of the photon, the first term on the right-hand side is the kinetic energy of the photoelectron, and the second term on the right-hand side is the energy spent in extracting the electron from the emitter to which its bound.

The following are the consequences of the preceding relation:

1. The velocity of a photoelectron is directly proportional to the frequency of radiation.
2. The velocity of a photoelectron is independent of the intensity of radiation.
3. There is a threshold frequency that varies with the nature of the emitter.
4. The process of photoemission is instantaneous.

If the atom that is involved in the process lies on the surface of the metal, the photoelectron escapes into the surrounding space with the amount of kinetic energy as estimated using the Einstein's equation; otherwise, it may lose part of its energy in passing through a layer of matter. This explains the existence of photoelectrons whose energies vary from a maximum value down to zero. If the incident radiation penetrates a distance equal to many atomic diameters in a metal, photoelectrons can originate at different depths. Now in order to reach the surface, they have to pass by many atoms and collisions would be probable, at each of which they might lose some of the energy ($h\vartheta$) and hence would reach the surface with an energy less than $h\vartheta$. At the surface, they have to give up a further amount ω_0 to overcome the surface barrier and finally emerge with the following amount of energy:

$$\frac{1}{2}mv^2 = h\vartheta - \omega_c - \omega_0 \tag{11.11}$$

where ω_c is the energy lost by collision, and varies from electron to electron depending on the depth of their place of origin and the subsequent collisions encountered. Some electrons thus emerge with zero velocity after having lost all their initial energy ($h\vartheta$) in collisions and overcoming the surface barrier. The maximum kinetic energy is given by the Einstein's equation.

11.4 EXPERIMENTAL VERIFICATION OF EINSTEIN'S PHOTOELECTRIC EQUATION

Several experiments were performed to testify the Einstein's photoelectric equation, which can be broadly classified into two types: one for the low frequencies for which the Milikan's experiment is taken into consideration and the other for very high frequencies for which the De Broglie's experiment may be cited as an example.

11.4.1 Milikan's Experiment for Visible and Ultraviolet Rays

The following subsections describe this experiment.

11.4.1.1 Principle

The experiment is based on the idea of 'stopping potential'. Einstein's equation states that

$$\frac{1}{2}mv^2 = h\vartheta - \omega_0 \tag{11.12}$$

that is, the energy of a photoelectron leaving the surface of a metal receiving light of frequency (ϑ) is equal to $(h\vartheta - \omega_0)$, which implies that

$$eV = \frac{1}{2}mv^2 \qquad (11.13)$$

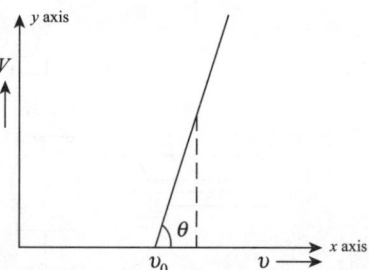

Fig. 11.9 Frequency versus voltage curve showing straight line; intercept on frequency axis gives threshold frequency

where V is the potential difference that is applied between the emitting surface and a collecting electrode in order to prevent the emitting photoelectrons from just leaving the emitter, the emitter being at a positive potential with respect to the collector. Consequently, V is called the 'stopping potential' as it just stops the electrons from leaving the surface.

$$eV = \frac{1}{2}mv^2 = (h\vartheta - \omega_0) \qquad (11.14)$$

$$V = \left(\frac{h}{e}\right)\vartheta - \frac{\omega_0}{e} \qquad (11.15)$$

$$V \propto \vartheta \qquad (11.16)$$

where h, e, and ω_0 are constants. Figure 11.9 suggests that if Einstein's equation is correct then for different values of V versus ϑ, the graph should be a straight line. Further, the intercept on the ϑ – axis should give the threshold frequency ϑ_0 for the given emitter from which $\omega_0 = h\vartheta_0$ can be calculated. In addition, the slope of the line given by $\tan\theta$ gives the coefficient (h/e) of ϑ, which offers a method to measure the value of the Planck's constant (h), which will verify the Einstein's equation and hence the validity of the quantum theory of radiation.

11.4.1.2 Experimental Arrangement

Alkali metals were employed as emitters. The stopping potential of the liberated photoelectrons was measured by raising the emitting surface to a positive potential just sufficient to prevent any of the electrons from reaching the collector, which was a Faraday cylinder FF of oxidized copper gauge, not photosensitive to the light used, placed opposite to the emitting surface, as shown in Fig. 11.10. This cylinder was connected to an electrometer and so acted as a detector of any electronic charge that might reach it. The stopping potential is the positive potential applied to the emitter, which corresponds to zero current in the electrometer.

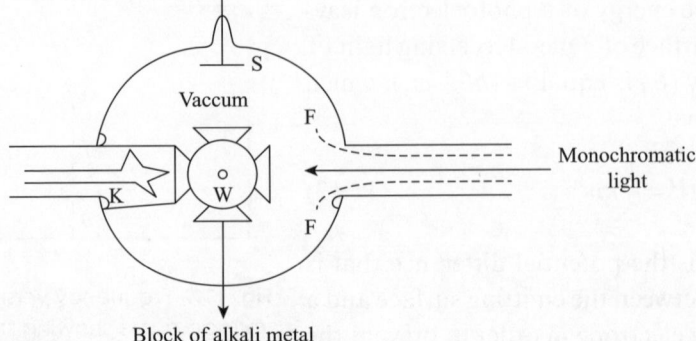

Fig. 11.10 Monochromatic light made incident on alkali metal to observe flow of photoelectrons

11.4.1.3 Result

The experiment was conducted with different lights from red to violet, and in each case the stopping potential was measured:

$$V = V_c + V_S \tag{11.17}$$

where V_c is the contact potential difference as the emitter and the collector are different metals. V_s is the actual retarding potential required to stop the electrons.

$$e(V_c + V_S) = h\vartheta - \omega_0 \tag{11.18}$$

The potential difference was measured between the metal surface and the test plate S, which was of the same material as the Faraday cylinder and had a similarly treated surface.

$$V = \frac{h}{e}(\vartheta - \vartheta_0) \tag{11.19}$$

A straight line was plotted in a graph of V vs $(V_c + V_S)$, and the value of (h/e) was calculated from the slope of the straight line and the potential difference was measured directly by a Kelvin's potentiometer. The value of the photo-electric work function is easily evaluated. Then substituting the value of e, h was calculated and was found to be $(6.57 \pm 0.01) \times 10^{-27}$ (CGS units), which agrees well with the value of h obtained by other methods. Thus, the Einstein's photoelectric equation is based on the quantum theory and has been verified for low frequencies.

Example 11.1 Find the maximum kinetic energy of photoelectrons ejected from a potassium surface, for which $\phi_0 = 2.1 \text{eV}$, by photons of wavelengths 2000 and 5000 Å. What is the threshold frequency ϑ_0 and the corresponding wavelength?

Solution Energy of the photon is given as follows:

$$E = h\vartheta = \frac{hc}{\lambda} = \frac{1}{\lambda} \frac{(6.626 \times 10^{-34} \text{ J.s}) \left(\frac{3 \times 10^8 \text{ m}}{\text{s}} \right)}{\left(\frac{1.602 \times 10^{-19} \text{ J}}{\text{eV}} \right)} = \frac{1}{\lambda} (1.240 \times 10^{-6} \text{ eV.m})$$

$$E_{p2000A} = \frac{1}{2000 \times 10^{-10} \text{ m}} \left(1.240 \times 10^{-6} \text{ eV.m}\right) = 6.204 \text{ eV}$$

$$KE_{max} = h\vartheta - \phi = 6.204 \text{ eV} - 2.1 \text{ eV} = 4.104 \text{ eV}$$

$$KE_{max} = 4.104 \text{ eV} = h\vartheta - h\vartheta_0 = 6.204 \text{ eV} - h\vartheta_0$$

$$\vartheta_0 = \frac{6.204 \text{ eV} - 4.104 \text{ eV}}{4.136 \times 10^{-15} \text{ eV.s}} = 5.08 \times 10^{14} \text{ Hz}$$

$$\lambda = \frac{c}{\vartheta_0} = \frac{3 \times 10^8 \text{ m/s}}{5.08 \times 10^{14} \text{ Hz}} = 5900 \text{ Å}$$

Similarly, $E_{p5000A} = \dfrac{1}{5000 \times 10^{-10} \text{ m}} (1.240 \times 10^{-6} \text{ eV.m}) = 2.50 \text{ eV}$

$$KE_{max} = h\vartheta - \phi = 2.50 \text{ eV} - 2.1 \text{ eV} = 0.4 \text{ eV}$$

Example 11.2　An experiment on the photoelectric effect of caesium shows that the stopping potential for $\lambda = 4358$ and 5461 Å are, respectively, 0.95 and 0.38 eV. From this data, find h, the threshold frequency, and the work function of caesium.

Solution　The work function calculation is as follows:

$$\text{Work function} = \frac{hc}{\lambda_1} - eV_{s1} = \frac{1.240 \times 10^{-6} \text{ eV.m}}{4358 \times 10^{-10} \text{ m}} - 0.95 \text{ eV}$$

$$= 2.84 \text{ eV} - 0.95 \text{ eV} = 1.9 \text{ eV}$$

$$h\vartheta_0 = \text{Work function}$$

$$\vartheta_0 = \frac{\text{Work function}}{h} = \frac{1.9 \text{ eV}}{4.136 \times 10^{-15} \text{ eV.s}} = 0.459 \times 10^{15} \text{ Hz} = 459 \text{ THz}$$

Example 11.3　Using the concept of photoelectric effect, find the wavelength corresponding to a photon of energy 3.5 eV capable of breaking a chemical bond in the molecules of human skin, causing sunburn.

Solution　The corresponding wavelength can be calculated as follows:

$$\lambda = \frac{h}{E_{\text{photon energy}}} \Rightarrow \lambda = \frac{1240 \text{ eV} \cdot \text{nm}}{3.5 \text{ eV}} = 354 \text{ nm}$$

This corresponds to the wavelength of the ultraviolet radiation. Thus, UV protection has become mandatory for skin protection creams.

Example 11.4　A source is emitting 150 W of red light at a wavelength of 600 nm. How many photons per second are emerging from the source?

Solution　Energy emitted $= (150 \text{ W})(1 \text{s}) = 150 \text{ J}$

$$\text{Photon flux} = N = \frac{\text{Energy emitted}}{h\vartheta} = \frac{(\text{Energy emitted})\lambda}{hc}$$

$$= \frac{(150 \text{ J})(600 \times 10^{-9} \text{ m})}{(6.6 \times 10^{-34} \text{ J.s}) \left(\dfrac{3 \times 10^8 \text{ m}}{\text{s}}\right)} = 45 \times 10^{19} \text{ photons/s}$$

11.5 DE BROGLIE'S EXPERIMENT

De Broglie verified the Einstein's equation in the case of high-frequency radiations, such as X-rays.

11.5.1 Principle

The method is known as the magnetic spectrograph method, since it involves the measurement of energies of emitted photoelectrons by making them traverse a circular path under the action of a suitably applied magnetic field, as well as 'focusing' them so that those leaving the emitter at slightly different directions but moving with the same velocity are made to strike at a common point on a photographic plate placed to receive the photoelectrons at the end of the semicircular journey.

11.5.2 Experimental Arrangement

When a fine beam of X-rays falls on the target T, photoelectrons are expelled from T in all directions and with different velocities. The whole apparatus is placed in a uniform magnetic field acting at right angles to the plane of Fig. 11.11. Some of the photoelectrons pass through the narrow magnetic slit S and are deflected by the magnetic field to describe a semicircular path and eventually strike the photographic plate PP placed horizontally on the metal block D. If the electrons leaving T have velocities $v_1, v_2, v_3, \ldots$ they will move in circles of radii $r_1, r_2, r_3, \ldots$ given by the following relation:

$$BeV = \frac{mv^2}{r} \qquad (11.20)$$

and will strike the plate at K, L, ... points. By suitably adjusting the position of the plate and the width of the slit S, it is possible to get a certain focusing effect. Electrons of the same velocity but slightly different initial directions are made to strike at the same point (such as at K, L, ...), and the diameters of the respective circular paths are the distances between S and K, S and L, ... When the plate is developed, linear traces are found, each of which was been made by photoelectrons of a definite velocity. A monochromatic beam of

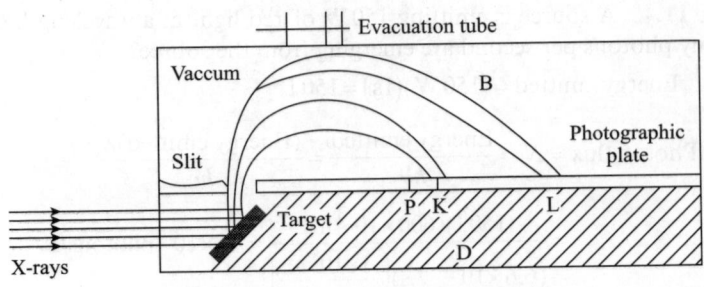

Fig. 11.11 Magnetic spectrograph

X-rays is obtained by intensifying one particular wavelength by filtering the beam through a material of a particular absorption coefficient.

11.5.3 Results

The linear traces on the photographic plate can be traced easily because the radius r of the circular path followed by the photoelectrons that made each one of the traces can be readily found and also the value of B is known:

$$v = Br \frac{e}{m} \tag{11.21}$$

Thus, the kinetic energy of the photoelectrons is deduced. The relation of kinetic energy is given by the theory of relativity, as we are dealing with high-frequency radiations whose velocities are comparable with the velocity of light. Having thus determined the kinetic energy of the photoelectrons ejected from the different energy levels in the atom and knowing $h\vartheta$, the relations $\frac{1}{2}mv_k^2 = h\vartheta - \omega_k$ and $\frac{1}{2}mv_L^2 = h\vartheta - \omega_L$ can be verified. The values of ω_k, ω_L, ... determined agree well with those obtained by spectroscopic methods.

This establishes the validity of the Einstein's equation for very-high-frequency radiations also.

These experiments indicate that energy is exchanged only in certain 'discrete packets', unlike the electromagnetic theory. The photon theory (photoelectric effect) was a revival of the corpuscular theory. Thus, during transmission of energy, light behaves more like particles than like waves. Hence, the *dualism of light behaviour* became highlighted. The phenomena of interference and diffraction could be explained only on the basis of the wave theory of light. Thus, a number of phenomena were explained to strengthen the *dualism of light*.

11.6 ORIGIN OF DE-BROGLIE'S CONCEPT OF MATTER WAVES

According to this principle, the two fundamental forms, matter and energy, in which nature manifests herself must be mutually symmetrical. Since radiant energy shows the dual nature of wave and particle, particles of matter must also possess the same dual nature ; that is, particles of matter and in particular electrons must possess certain oscillatory characteristics, so that they too might exhibit a dual nature.

De Broglie associated photons with waves and referred to them as 'phase waves'. A material particle carries energy; the basic idea of the quantum theory is the impossibility of imagining an isolated quantity of energy without associating it with a certain frequency. Therefore, material particles should be similar to photons to be accompanied by phase waves of some type, and these waves, under suitable circumstances, should give rise to interference effects. Waves associated with a particle should have a frequency equal to the energy of the particle divided by the Planck's constant (h). We assume that any material particle of mass m, moving at speed v, has associated with it waves of frequency ϑ given by the following relation:

$$h\vartheta = \frac{1}{2}mv^2 + V = E \tag{11.22}$$

Where h is the Planck's constant and V is the potential energy of the particle.

11.7 PHASE AND GROUP VELOCITIES

The wavelength to be expected for matter waves can be discovered by the fact that individual waves cannot stay with the particle permanently since u and v are usually different. Whenever the wave velocity (or phase velocity) varies with frequency, a finite group of waves moves with a velocity different from the phase velocity. Let us assume that the group velocity of matter waves is equal to the particle velocity so that a group of waves can accompany the particle in its motion. An ideal wave travelling in the x-direction can be represented by the following equation:

$$\varphi = A\cos 2\pi\left(\vartheta t - \frac{x}{\lambda}\right)$$

$$= A\cos(\omega t - kx) \tag{11.23}$$

where A represents the amplitude of the wave, $\omega = 2\pi\vartheta$ is the angular velocity, $k = \frac{2\pi}{\lambda}$ is the propagation number, and phase velocity $u = \vartheta\lambda = \frac{\omega}{k}$.

When the phase velocity varies with the wavelength, a finite group of waves and energy are propagated at the group velocity (v). We consider two components with angular frequencies ($\omega - \Delta\omega$) and ($\omega + \Delta\omega$). Let the components have the same amplitude A and are given by the following expressions:

$$\varphi_1 = A\cos(\omega - \Delta\omega)t - (k - \Delta k)x \tag{11.24}$$

$$\varphi_2 = A\cos(\omega + \Delta\omega)t - (k + \Delta k)x \tag{11.25}$$

$$\varphi = \varphi_1 + \varphi_2 = A(\cos(\omega - \Delta\omega)t - (k - \Delta k)x$$

$$+\cos(\omega + \Delta\omega)t - (k + \Delta k)x$$

$$= 2A\cos(\omega t - kx)\cos(\Delta\omega t - \Delta kx) \tag{11.26}$$

Let $\quad \theta = \omega t - kx; \quad \varphi = \Delta\omega t - \Delta kx$

Using the trigonometric identity, Eq. (11.26) reduces to the following form:

$$\varphi = 2A\cos\theta\cos\varphi \tag{11.27}$$

The two cosine terms show the presence of beats in the resulting wave. For a constant phase point,

$$\omega t - kx = \text{constt.} \tag{11.28}$$

Differentiating this equation, we get the following equation:

$$\frac{dx}{dt} = \frac{\omega}{k} = \vartheta\lambda = u \tag{11.29}$$

where u represents the phase velocity.

$$\Delta\omega t + \Delta k x = \text{constt} \tag{11.30}$$

Differentiating this equation, we get the following formula:

$$\frac{dx}{dt} = \frac{\Delta\omega}{\Delta k} = \Delta f \cdot \Delta\lambda \tag{11.31}$$

Equation (11.31) represents the velocity of the wave envelope; the group velocity (v) is given by the following relation:

$$v = \lim_{\Delta k \to 0} \frac{\Delta\omega}{\Delta k} = \frac{\partial\omega}{\partial k} = \frac{\partial\vartheta}{\partial(1/\lambda)} = -\lambda^2 \frac{\partial\vartheta}{\partial\lambda} \tag{11.32}$$

Since $\omega = ku$,

$$v = \frac{\partial\omega}{\partial k} = \frac{\partial(ku)}{\partial k} = u + k\frac{\partial u}{\partial k} = u - \lambda\frac{\partial u}{\partial\lambda} \tag{11.33}$$

Differentiating Eq. (11.33) and taking the potential energy term (V) to be constant, we get the following equation:

$$h.\frac{\partial\vartheta}{\partial\lambda} = mv\frac{\partial v}{\partial\lambda} \tag{11.34}$$

Substituting from Eq. (11.32), we get the following expressions:

$$h \cdot \left(-\frac{v}{\lambda^2}\right) = mv\frac{\partial v}{\partial\lambda} \tag{11.35}$$

$$\frac{\partial v}{\partial\lambda} = -\frac{h}{m} \cdot \frac{1}{\lambda^2} \tag{11.36}$$

Integrating Eq. (11.36), we get the following relation:

$$v = \frac{h}{m} \cdot \frac{1}{\lambda} + c \tag{11.37}$$

where c is the constant of integration. Value of c cannot be determined uniquely; hence, it is arbitrarily taken to be zero.

$$\lambda = \frac{h}{mv} \tag{11.38}$$

Equation (11.38) represents the De Broglie wavelength, where mv represents the momentum of the particle.

The De Broglie equation holds for photons as well as for material particles, as the momentum of a photon is represented as $\frac{h\vartheta}{c} = \frac{h}{\lambda}$.

The De Broglie equation combines corpuscular (particle) and undulatory (wave) concepts in a very intimate way, as λ (the wavelength) has a clear meaning only with the wave theory and p (the momentum) is associated with a moving particle.

For an electron starting from rest and being accelerated through a small potential difference (V), the kinetic energy (KE) is given by the following relation:

$$KE = eV = \frac{1}{2}mv^2 \tag{11.39}$$

$$mv = \sqrt{2meV} = p \tag{11.40}$$

Thus, $\quad \lambda = \frac{h}{p} = \frac{h}{\sqrt{2meV}} \tag{11.41}$

In addition, based on the concept of a De Broglie wave being associated with a particle, we now replace the idea of a single wave with a wave packet.

The associated velocity of the De Broglie wave can be ascertained. The De Broglie wavelength is represented by Eq. (11.38), and the energy possessed by a photon is $E = h\vartheta$. Using the aforementioned equations, we get the following expression:

$$v = \vartheta\lambda = \frac{h}{mv} \cdot \frac{E}{h} = \frac{mc^2}{mv} = \frac{c^2}{v} > c \tag{11.42}$$

Equation (11.42) indicates that the phase velocity of a wave is greater than the velocity of light.

Considering Eq. (11.33), we have the following relation:

$$v = u - \lambda \frac{\partial u}{\partial \lambda} \tag{11.43}$$

Differentiating this equation, one gets the following expression:

$$v = \lambda^2 \left(\frac{u}{\lambda^2} - \frac{1}{\lambda}\frac{\partial u}{\partial \lambda} \right) = -\lambda^2 \frac{\partial}{\partial \lambda}\left(\frac{u}{\lambda} \right) = -\lambda^2 \frac{\partial \vartheta}{\partial \lambda} \tag{11.44}$$

where ϑ is the frequency of the wave. Now,

$$\frac{1}{v} = -\frac{1}{\lambda^2}\frac{\partial \lambda}{\partial \vartheta} = \frac{\partial}{\partial \vartheta}\left(\frac{1}{\lambda} \right) \tag{11.45}$$

If E is the kinetic energy and V is the potential energy, then we can write the following equations:

$$\frac{1}{2}mu^2 = E - V \tag{11.46}$$

$$u = \left(\frac{2(E-V)}{m} \right)^{\frac{1}{2}} \tag{11.47}$$

Substituting the value of u from Eq. (11.47) in Eq. (11.46), one gets the following relation:

$$\frac{1}{v} = \frac{\partial}{\partial \vartheta}\left(\frac{mu}{h} \right) = \frac{\partial}{\partial \vartheta}\left(\frac{1}{h}\left(2m(E-V)^{\frac{1}{2}}\right) \right) \tag{11.48}$$

Substituting the energy of a photon as $E = h\vartheta$, one gets the following expressions:

$$\frac{1}{v} = \frac{1}{h} \cdot \frac{\partial}{\partial \vartheta} \left(2m(h\vartheta - V)\right)^{\frac{1}{2}} \tag{11.49}$$

$$= \frac{1}{h} \cdot \frac{1}{2} \left(2m(h\vartheta - V)\right)^{-\frac{1}{2}} \cdot 2mh$$

$$\frac{1}{v} = \frac{m}{\left(2m(h\vartheta - V)\right)^{\frac{1}{2}}} = \frac{1}{u} \tag{11.50}$$

That is, the group velocity (v) is equal to the phase velocity (u).

Thus, we conclude that a material particle in motion is equivalent to a group of waves or a *wave packet* as it is termed. The wave packet thus formed by the superposition of a number of waves and travelling with the velocity of the particle behaves like a single corpuscular unit (analogy with the corpuscular theory) (Fig. 11.12).

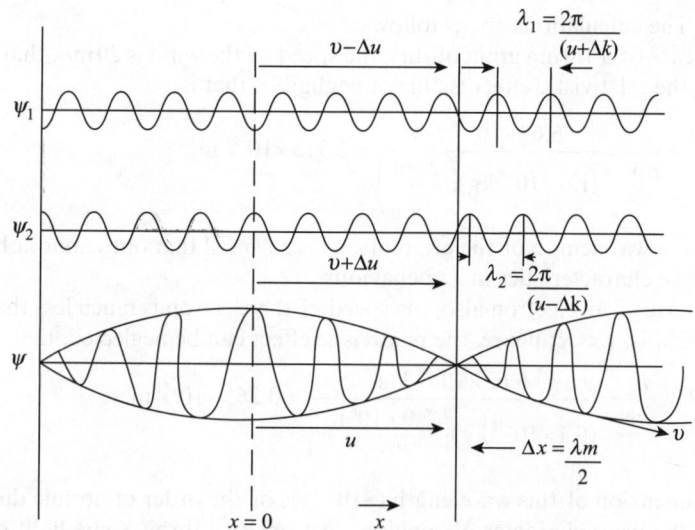

Fig. 11.12 Concept of phase and group velocity

For wavelengths of atomic waves, molecule waves, and even waves of larger masses, Eq. (11.49) is used. Thus, for a given velocity v, the larger the mass (m), the shorter the wavelength (λ). Hence, one observes that it is not easy to measure the wavelengths of heavy particles.

Example 11.5 An electron has a De Broglie wavelength of 4.00 pm. Find (a) its kinetic energy and (b) the phase and group velocities of its De Broglie waves.

Solution Kinetic energy of the electron in terms of its relativistic components and the rest energy can be represented as follows: Kinetic energy $= E - E_0 = \sqrt{E_0^2 + (pc)^2} - E_0$

Rest energy of the electron is $E_0 = 511 \text{keV}$

$$pc = \frac{hc}{\lambda} = \frac{\left(4.136 \times 10^{-15} \, eV \cdot s\right)\left(\frac{3 \times 10^8 \, m}{s}\right)}{4.00 \times 10^{-12} \, m} = 3.102 \times 10^5 \, eV$$

$$\text{Kinetic energy} = \sqrt{(511 keV)^2 + (310.2 keV)^2} - 511 keV = 597.8 keV$$

(b) Electron velocity can be found using the following relativistic energy relation:

$$E = \frac{E_0}{\sqrt{1 - v^2 / c^2}}$$

$$v = c\sqrt{1 - \frac{E_0^2}{E^2}} = c\sqrt{1 - \frac{(511 keV)^2}{(598 keV)^2}} = 0.519c$$

$$\text{Phase velocity} = v_p = \frac{c^2}{v} = \frac{c^2}{0.519c} = 1.926c$$

$$\text{Group velocity} = v_g = v = 0.519c$$

Example 11.6 Find the De Broglie wavelength of (a) a 1.0 mg grain of dirt blown by a wind of speed 20 m/s and (b) an electron whose speed is 2.0 × 10⁸ m/s

Solution The calculations are as follows:

(a) In the case of a 1.0 mg grain of dirt, the speed of the wind is 20 m/s, that is, $v \ll c$; hence, the relativistic effect is almost negligible, that is,

$$\lambda = \frac{h}{mv} = \frac{6.63 \times 10^{-34} \, J \cdot s}{(1.0 \times 10^{-6} \, kg)\left(\frac{20 \, m}{s}\right)} = 3.315 \times 10^{-29} \, m$$

Thus, the wavelength of the grain of dirt is so small that one cannot observe the wave-like characteristics in its behaviour.

(b) In the case of an electron also, the speed of the electron is much less than that of light, that is, $v \ll c$; hence, the relativistic effect can be neglected, that is,

$$\lambda = \frac{h}{mv} = \frac{6.63 \times 10^{-34} \, J \cdot s}{(9.1 \times 10^{-31} \, kg)\left(\frac{2.0 \times 10^8 \, m}{s}\right)} = 0.365 \times 10^{-11} \, m$$

The dimension of this wavelength ($\sim 10^{-11}$) is of the order of atomic dimensions. Thus, the wave character of moving electrons can throw some light on atomic structures.

Example 11.7 Find the kinetic energy (in eV) of a neutron with a De Broglie wavelength of 0.30 nm. Is a relativistic calculation needed?

Solution The De Broglie relation is given as follows: $\lambda = \dfrac{h}{mv}$

Thus, $\lambda = 0.30 \, nm = 0.30 \times 10^{-9} \, m$, $h = 6.634 \times 10^{-34} \, J \cdot s$, and $m = 1.67 \times 10^{-27} \, kg$

$$0.30 \times 10^{-9} \, m = \frac{6.634 \times 10^{-34} \, J \cdot s}{1.67 \times 10^{-27} \, kg \times v}$$

$$v = \frac{6.634 \times 10^{-34} \, J \cdot s}{1.67 \times 10^{-27} \times 0.30 \times 10^{-9} \, kg \cdot m} = 13.24 \times 10^2 \, m/s$$

$$KE = \frac{1}{2}mv^2 = \frac{1}{2} \times 1.67 \times 10^{-27} \times (13.24 \times 10^2)^2 \, J$$

$$= 146.37 \times 10^{-23} \, J = 0.00915 \, eV$$

There is no need for relativistic calculation in this case, as the kinetic energy is not comparable to the rest energy of neutron (of the order of GeV).

Note: It is commonly observed that the atomic spacing is of the order of the wavelength of particles such as neutrons having a wavelength of 0.30 nm. They are used for studying crystal structures.

Example 11.8   Green light has a wavelength of about 5461 Å. Through what potential difference must an electron be accelerated to have this wavelength?

Solution   From our discussion, we know that the kinetic energy gained by the electron $\left(\frac{1}{2}mv^2\right)$ equals the electric potential energy lost (eV). Thus,

$$eV = \frac{1}{2}mv^2$$

$$V = \frac{1}{2e}mv^2 = \left(\frac{1}{2e}\right)m\left(\frac{h}{m\lambda}\right)^2 = \left(\frac{1}{2e}\right)\left(\frac{h^2}{m\lambda^2}\right)$$

$$V = \frac{1}{2 \times 1.6 \times 10^{-19}} \frac{(6.634 \times 10^{-34})^2}{9.1 \times 10^{-31} \times (5461 \times 10^{-10})^2} = 5.06 \times 10^{-6} \, V = 5.0 \, \mu V$$

Example 11.9   (a) Assume that a photon ($m = 0$) and a particle have the same wavelength. Compare the linear momenta of the two particles.
(b) Both an electron ($m = 0.511$ MeV/c^2) and a photon ($m = 0$) have momenta of 5.00 MeV/c. Compare their total energies.

Solution   (a) Linear momentum for a photon is given as follows: $E = h\vartheta = pc$

$$\Rightarrow p = \frac{h\vartheta}{c} = \frac{h}{\lambda}$$

Linear momentum of a particle of mass m and velocity v is given as follows:

$$\lambda = \frac{h}{p} \Rightarrow p = \frac{h}{\lambda}$$

Since the wavelength is same for the photon and the particle, the linear momenta will be the same.

(b) Electron's total energy $= \sqrt{m^2c^4 + p^2c^2} = \sqrt{\left(\frac{0.511 \, \text{MeV}}{c^2}\right)^2 c^4 + \left(\frac{5.00 \, \text{MeV}}{c}\right)^2 c^2}$

$$= 5.026 \, \text{MeV}$$

Photon's total energy $= E = pc = 5.00 \dfrac{\text{MeV}}{c} c = 5.00 \, \text{MeV}$

Thus, the total energy of the photon is less than that of the particle, assuming that their momenta are the same.

The existence of De Broglie wavelength as a concept has been verified through diffraction experiments.

11.8 EXPERIMENTAL STUDY OF MATTER WAVES

Electrons are used for experimental verification of matter waves, as they are easily produced in fairly intense beams with a definite velocity. The mass of an electron is very small and hence its wavelengths are of very high magnitudes, which can be measured using X-rays.

11.8.1 Davisson and Germer Experiment

Electron waves predicted by De Broglie were first observed experimentally by Davisson and Germer, who also succeeded in measuring the De Broglie wavelengths for slow electrons by diffraction methods.

They were studying the reflection of electrons from a nickel target and accidentally subjected a target to heat treatment, which transformed it into a group of large crystals. As a result, reflection of the electron beam became anomalous, that is, instead of decreasing continuously from the angular position of the maximum reflection, the reflected intensity showed striking maxima and minima. This result suggested that, like X-rays, a beam of electrons might be diffracted by crystals, implying that electrons behave like waves under certain circumstances.

11.8.1.1 Apparatus

An electron beam is produced from the electron gun (G), which contains a tungsten filament (F) that is heated to dull red and electrons are emitted by thermionic action. These electrons are then accelerated in an electric field of known potential difference and made to hit a target T, which is capable of rotating about an axis parallel to the axis of the incident beam. The electrons are reflected from the crystal in different directions, the angular distribution being measured with a Faraday cylinder called the collector (C), which is connected to a sensitive galvanometer and can rotate about a graduated circular scale. A retarding potential is applied across the insulated walls C (inner) and D (outer) of the collector so that only the fastest electrons, that is, those possessing nearly the incident velocity but not the secondary slow electrons excited by collisions with atoms, may enter the collector and be detected by the galvanometer. The whole apparatus (Fig. 11.13) was completely enclosed, highly evacuated, and degassed.

The nickel crystal, a face-centred cube, is cut so as to have a smooth reflecting surface parallel to the lattice plane (1, 1, 1), that is, perpendicular to one of the diagonals of the cube. By rotating the crystal about the axis, any azimuth of the crystal can be presented to the plane defined by the incident beam and the beam entering the collector.

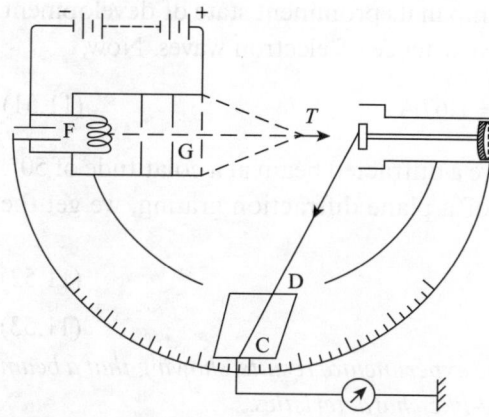

Fig. 11.13 Apparatus used for Davisson and Germer experiment for diffraction of electrons

11.8.1.2 Experimental Procedure

The experimental procedure consists of the following steps:

Beam of electrons at normal incidence on surface of crystal For each azimuth of the crystal, a beam of low-voltage electrons was made to fall normally on the surface of the crystal, the collector was moved to various positions on scale 'S', and the galvanometer current at each position was noted. The current (a measure of the intensity of the diffracted beam) was plotted against the angle between the incident beam and the beam entering the collector, known as the 'colatitude'.

It is observed that a bump begins to appear in the curve for 44 V electrons [Figs 11.14(a–g)]. With increasing voltage, the bump moves outwards and attains its greatest development in the curve for 54 V at a colatitude of 50°. At higher voltages, the bump gradually diminishes, with there being hardly

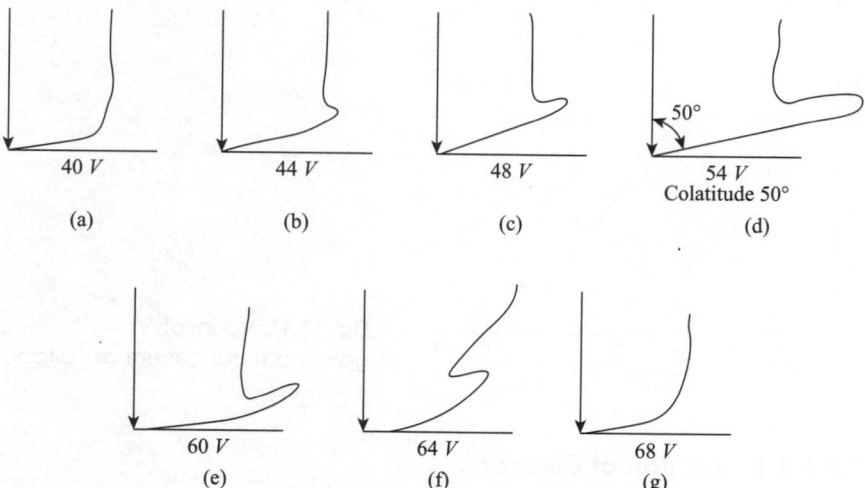

Fig. 11.14 Diffraction graphs depicting position of lattice points as voltage is varied

any trace of it at about 68 V. The bump in its prominent state of development offers a convincing evidence for the existence of electron waves. Now,

$$\lambda = \frac{h}{mv} = \frac{h}{\sqrt{2meV}} = \frac{12.27}{\sqrt{54}} = 1.67 \text{ Å} \qquad (11.51)$$

According to the experiment we have a diffracted beam at a colatitude of 50°. Applying the well-known relation of a plane diffraction grating, we get the following equations:

$$n\lambda = d\sin\theta \qquad (11.52)$$

$$\lambda = 2.15\sin 50° = 1.65 \text{ Å} \qquad (11.53)$$

This is in excellent agreement with the experimental results, showing that a beam of electrons does really possess wave-like characteristics.

Beam of electrons falling 'obliquely' upon crystal   This phenomenon occurs when diffraction from a space lattice, that is, from successive parallel layers of atoms in a crystal, which is analogous to Bragg's X-ray diffraction, occurs, as indicated by the existence of 'regular selective reflections' depending upon the velocity of the incident electrons. If the electrons are simple particles then this phenomenon cannot be explained, but if the electrons are considered to have wave characteristics associated with them, their wavelengths can be calculated using the Bragg's formula. The values so found agree with those calculated by the De Broglie relation.

   If the collector and gun are fixed, the glancing angles of incidence and reflection can be kept constant; then by varying the electron velocity, the galvanometer current is measured for each value of velocity. Plotting the values of current against the corresponding electron velocities or accelerating voltages, a curve with several maxima (sharp) are obtained. The different maxima correspond to the various orders in the diffraction pattern (Fig. 11.15).

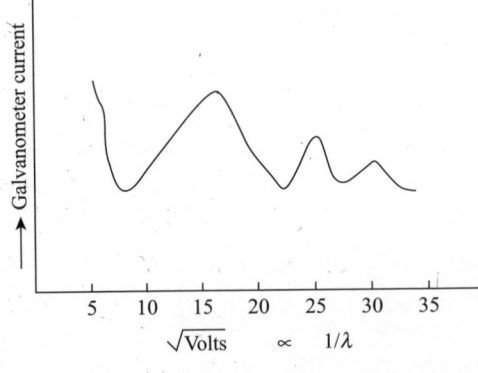

Fig. 11.15 Variation in galvanometer current as voltage is varied

11.8.1.3 Refraction of Electrons

In the case of the oblique incidence, the phenomenon of refraction of a beam of electrons is also observed. It was found by Davisson and Germer that the values of λ obtained from the Bragg's law ($n\lambda = 2d \sin \theta$) and from the De Broglie's theoretical equation were not perfect, the former being systematically less than the latter.

An explanation of this discrepancy was that the electron beam is refracted as it enters the crystal. In the case of X-rays, such an effect takes place, which can be explained by modifying the Bragg's relation as follows:

$$n\lambda = 2d\sqrt{\mu^2 - \cos^2 \theta} \tag{11.54}$$

where μ is the refractive index of the X-rays entering the crystal. This relation holds for electron waves. To determine μ and the phenomenon of refraction arising in electron waves, an experiment was performed by *Eckart* and *Bethe*. During thermionic emission, work is done by electrons while escaping the metal surface. Hence, we say that a metallic crystal is at a positive potential (ΔV) above that of the surroundings. Consequently, if a pencil of V volt-electrons is incident at an angle i on the surface of a crystal, then on entering it, it is further accelerated through (ΔV) volts and is refracted. Thus,

$$\frac{1}{2}mv_1^2 = eV \tag{11.55}$$

$$\frac{1}{2}mv_2^2 = e(V + \Delta V) \tag{11.56}$$

where v_1 and v_2 are velocities of an electron outside and inside the crystal, respectively, and r is the angle of refraction. Since the electric field parallel to the surface does not change, components of the velocities parallel to the surface are equal. Hence, we can say that

$$v_1 \sin i = v_2 \sin r \tag{11.57}$$

$$\frac{\sin i}{\sin r} = \frac{v_2}{v_1} = \mu = \sqrt{\frac{V + \Delta V}{V}} \tag{11.58}$$

Example 11.10 A beam of 54 eV electrons is directed at a nickel target. The potential energy of an electron that enters the target changes by 25 eV.
(a) Compare electron speeds outside and inside the target.
(b) Compare respective De Broglie wavelengths.

Solution The calculations involved are as follows:
(a) (i) Following the Davisson and Germer experiment, the electron momentum is given as follows:

$$mv = \sqrt{2m(KE)}$$

$$= \sqrt{2 \times (9.1 \times 10^{-31} \, kg) \times (54 \, eV) \left(\frac{1.6 \times 10^{-19} \, J}{eV} \right)}$$

$$mv = 4.0 \times x10^{-24} \, kg \cdot m/s$$

$$v = \frac{4.0 \times 10^{-24} \, \text{kg} \cdot \text{m/s}}{9.1 \times 10^{-31} \, \text{kg}} = 0.439 \times 10^{7} \, \text{m/s} \Rightarrow v = 4.39 \times 10^{6} \, \text{m/s}$$

This is the velocity of the electron outside the nickel target.

(ii) As the potential energy changes by 25 eV, the total energy would then be $(54 + 25) = 79$ eV; hence,

$$mv = \sqrt{2mKE}$$

$$= \sqrt{2 \times (9.1 \times 10^{-31} \, \text{kg})(79 \, \text{eV})\left(\frac{1.6 \times 10^{-19} \, \text{J}}{\text{eV}}\right)} = \sqrt{2300.48 \times 10^{-50}}$$

$$mv = 47.9633 \times 10^{-25} \, \text{kg} \cdot \frac{\text{m}}{\text{s}}$$

$$v = \frac{47.9633 \times 10^{-25} \, \text{kg} \cdot \text{m/s}}{9.1 \times 10^{-31} \, \text{kg}} = 5.270 \times 10^{6} \, \text{m/s}$$

Thus, the velocity inside the target is 5.270×10^{6} m/s.

(b) The De Broglie wavelength is dependent on the momentum. Hence, inside the target:

$$\lambda = \frac{h}{mv} = \frac{6.63 \times 10^{-34} \, \text{J} \cdot \text{s}}{4.0 \times 10^{-24} \, \text{kg} \cdot \text{m/s}} = 1.66 \times 10^{-10} \, \text{m} = 0.166 \, \text{nm}$$

Outside the target:

$$\lambda = \frac{h}{mv} = \frac{6.63 \times 10^{-34} \, \text{J} \cdot \text{s}}{47.9633 \times 10^{-25} \, \text{kg} \cdot \text{m/s}} = 0.138 \times 10^{-9} \, \text{m} = 0.138 \, \text{nm}$$

11.9 G.P. THOMSON EXPERIMENT

The G.P. Thomson experiment extended the research on electron waves of high-speed electrons ranging from 10,000 to 50,000 V diffracted by very thin films.

11.9.1 Experimental Apparatus

A beam of cathode rays is produced in a discharge tube AC by means of an induction coil. The rays are passed through a diaphragm tube A to obtain a fine pencil of electrons, which is then allowed to fall upon a very thin metallic film F of gold (Au) or aluminium (Al) (thickness of the film is nearly of the order of 10^{-6} cm) [Figs 11.16(a) and (b)]. P is a photographic plate that can be slid down into a position to receive the pencil of electrons after it has traversed the film. S is a fluorescent screen that can be used instead of the photographic plate for the visual examination of the result obtained after the passage of electrons through the foil. The camera part FP of the apparatus is exhausted to a high vacuum, while air is allowed to leak into the discharge tube section through a needle valve. Since the only connection between the camera and the discharge tube is through diaphragm A, it is possible to have the camera at a low pressure and yet maintain the discharge tube sufficiently soft to give a

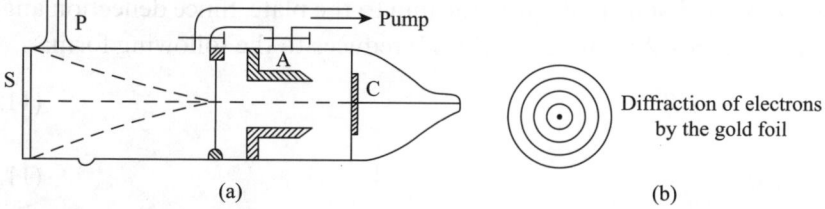

Fig. 11.16 Diffraction of electrons (a) G.P. Thomson apparatus for diffraction of electrons (b) Diffraction of electrons by gold foil

beam of the required voltage. The current from the induction coil is rectified by a kenatron, and several smoothing condensers are connected in parallel with the discharge tube.

11.9.2 Experimental Procedure

When a pencil of electrons of known velocity falls on the photographic plate after traversing the thin foil, after developing the plate, a symmetrical pattern consisting of concentric rings surrounding a central spot is obtained, which is very similar to that produced by X-rays. To confirm whether the pattern is produced by secondary X-rays generated by the electrons during their passage through the foil or not, the cathode rays in the discharge tube are deflected by means of a magnetic field; this causes a shift in the whole pattern observed on the fluorescent screen 'S' (refer to Fig. 11.16), which cannot happen if X-rays are responsible for the pattern. Further, on removing the film, the diffraction pattern disappears. This shows that the presence of a thin film is essential. The Thomson experiment highlights the fact that the electron pencil behaves like waves, since diffraction is a phenomenon that is characteristic of wave-like behaviour.

Next, quantitative verification of the De Broglie wave equation follows.

In a polycrystalline film, some crystals will be set at the correct angle to produce Bragg reflection. If enough crystals are distributed at random, such reflections will result in a series of rings arising from the intersection of the cones of diffraction with the photographic plate. Let AB be the incident beam passing through the film at B and let BE be a beam that has suffered a Bragg reflection in some small crystal in the film at B and falls at point E on the photographic plate at some distance R from the central point C (Fig. 11.17). From Bragg's reflection:

$$n\lambda = 2d\sin\theta \tag{11.59}$$

$$R = L\tan 2\theta \tag{11.60}$$

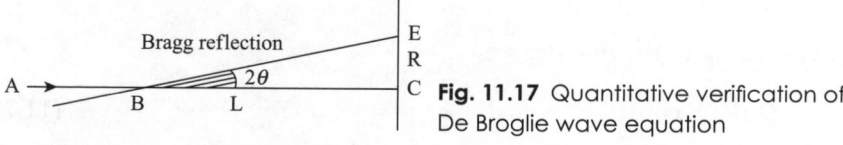

Fig. 11.17 Quantitative verification of De Broglie wave equation

where L is the distance BC from the film to the plate. Since deflection angle θ is small, $\tan 2\theta \cong 2\theta$. Thus, Eq. (11.60) reduces to the following form:

$$n\lambda = 2d\theta \tag{11.61}$$

$$\theta = \frac{n\lambda}{2d} \tag{11.62}$$

Substituting Eq. (11.62) in Eq. (11.60), one gets the following expression:

$$R = \frac{Ln\lambda}{d} \tag{11.63}$$

In Eq. (11.63), the De Broglie's value of λ can be substituted after duly deriving it from the following equation:

$$\frac{1}{2}mv^2 = \frac{eV}{300} \tag{11.64}$$

$$m^2v^2 = \frac{meV}{300} \times 2$$

$$mv = \sqrt{\frac{meV}{150}} \tag{11.65}$$

For the De Broglie wave:

$$\lambda = \frac{h}{mv} = h\left(\frac{150}{meV}\right)^{\frac{1}{2}} \tag{11.66}$$

Applying the relativistic corrections,

$$\lambda = \sqrt{\frac{150}{V}}\left(1 + \frac{\alpha}{2}\right)^{-\frac{1}{2}} \quad \text{where } \alpha = \frac{eV}{300\,m_0c^2} \tag{11.67}$$

Relativistic corrections are included due to the high speed of electrons. Thus, substituting Eq. (11.67) in Eq. (11.63), one gets the following relations:

$$R = \frac{nL}{d}\sqrt{\frac{150}{V}}\left(1 + \frac{\alpha}{2}\right)^{-\frac{1}{2}} \tag{11.68}$$

$$RV^{\frac{1}{2}}\left(\frac{\alpha}{2}\right)^{\frac{1}{2}} = \frac{nL\sqrt{150}}{d} = \text{constt} \tag{11.69}$$

$$\frac{D}{2}V^{\frac{1}{2}}\left(1 + \frac{\alpha}{2}\right)^{\frac{1}{2}} = \text{constt} \tag{11.70}$$

Approximating to the first degree,

$$D\sqrt{V} = \text{constt} \tag{11.71}$$

This relation was verified by Thomson and was found to be true.

Alternately, grating spaces (d) can be calculated from other readily measured quantities according to the following relation:

$$d = \frac{nL}{R}\sqrt{\frac{150}{V\left(1+\dfrac{\alpha}{2}\right)}} \tag{11.72}$$

The grating spaces were verified in the case of various metals and comparing them with those found by means of X-rays, the De Broglie law was verified again within the limits of experimental error.

G.P. Thomson also obtained diffraction patterns using fast electrons by reflection from both single crystal and polycrystalline surfaces at small angles. In general, the pattern obtained with a single crystal showed a number of separate spots, whereas the polycrystalline surfaces showed a ring pattern that is similar to that obtained by transmission of electrons through a thin foil.

The idea of electron waves has been established on a firm footing and is used for determining structures of unknown compositions owing to their advantages over X-rays, which are as follows:

1. They penetrate less deeply and their interaction with an atom of the body under test is more intimate.

2. The practical advantage of electron diffraction is the high intensity of the diffraction pattern so that only very short exposure times are necessary for photographic purposes.

Polarization of electron waves appears doubtful, but investigations are still in the initial stages. Since polarization of light waves is attributed to the spin of a photon, the spin of an electron should also cause polarization of electron waves.

The De Broglie's concept of waves can be applied to any material particle. Experiments designed to show the wave characteristics of free atoms and molecules have been tried by several workers; specific difficulties are encountered in producing diffraction effects with atoms and molecules, which do not arise while working with electrons. These difficulties are as follows:

1. Atoms and molecules are much smaller; hence, it is difficult to detect the diffraction effects.

2. Free atoms or molecules possess a Maxwellian velocity, whereas for diffraction experiments a beam of uniform velocity is desirable.

3. Neutral particles (i.e., atoms or molecules) are much harder to detect than charged particles.

Nevertheless, Dempster, Estermann, et al. succeeded in obtaining clear diffraction effects with hydrogen and helium by suitably reflecting them from crystal gratings, thus lending full support to the De Broglie's general concept of matter waves.

11.10 TWO-SLIT INTERFERENCE EXPERIMENT

This experiment was performed to emphasize the common behaviour of electron waves and photons of the same wavelength. Let us imagine a pair of slits through which we can alternatively send a beam of photons and that of electrons of comparable wavelength from either an electron gun or a photon source. Wave aspects of the electrons and photons were observed on a photographic plate. In terms of the classical theory, the picture of electrons would consist simply of spots at P_1 and P_2, corresponding to electrons that would pass directly through the slits and impinge on the detecting plate. Instead, we find photons and electrons exhibiting the same wave properties.

We should obtain identical distribution of intensities if we use a little scintillator that moves along the plate [shown in Fig. 11.18(a)] and flashes each time an electron or photon strikes it. Individual scintillations would provide clear evidence that the photons or electrons had particle characteristics, since classical waves would produce a uniform glow proportional to the intensity, and not individual flashes of light.

The similarity of behaviour of electrons and photons is further illustrated when we consider what happens if one of the slits is closed. If only one slit is open, for either electrons or photons, the observed intensity pattern corresponds to curve I of Fig. 11.18(c). Similarly, if slit 1 is closed and slit 2 is open, the observed pattern resembles curve I_2. The pattern with both slits open is not the sum of the patterns of the two individual slits. Even when one counts the arrival of individual particles by a scintillation detector, the pattern obtained resembles the curve of Fig. 11.18(b) and not the sum of the curves of Fig. 11.18(c).

We cannot explain the aforementioned experiment based on the true classical view that the 'electrons are particles'. 'Particle' is a convenient name of an

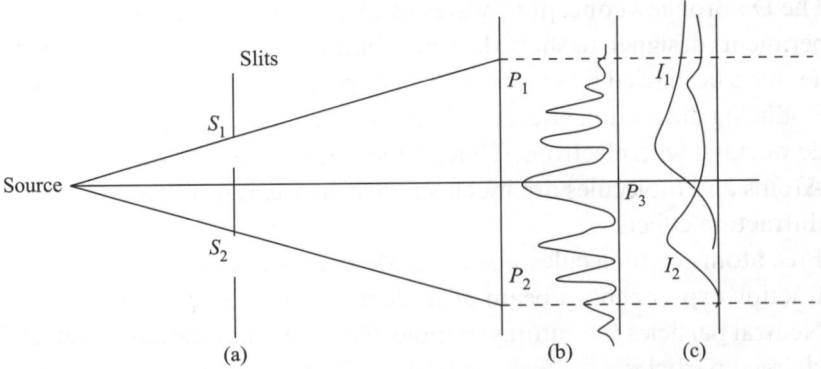

| (a) | (b) | (c) |

Fig. 11.18 Electron interference experiment (a) If electrons behave like classical particles, pattern expected for electrons passing through two slits would be two spots at P_1 and P_2 (b) Actual pattern observed on plate, same as what is expected for classical waves (c) Patterns expected from waves through each slit if other slit is closed

object that can 'localize', but we use another name to describe its behaviour in a diffraction experiment. An intriguing fact is that an entity that can be located in space in some experiments can exhibit wave properties in the others. There is a continuum between the two where neither property is well defined. The classical theory explains the interference pattern shown in Fig. 11.18(b) satisfactorily. Since electrons exhibit an analogous pattern, one can explain the electron 'interference' pattern in analogy with that of electromagnetic waves. When two waves interfere, the resultant intensity is not the sum of their intensities; rather, the resulting displacement is the resultant of the individual displacements. Two electromagnetic waves with electric vectors E1 and E2 can be given as follows:

$$E_1 = A_1 \cos 2\pi \left(\frac{x}{\lambda} - vt \right) \tag{11.73}$$

$$E_2 = A_2 \cos 2\pi \left(\frac{x}{\lambda} - vt - \varphi \right) \tag{11.74}$$

The electric vectors interfere to give a resultant displacement as follows:

$$E = E_1 + E_2 = A_1 \cos 2\pi \left(\frac{x}{\lambda} - vt \right) + A_2 \cos 2\pi \left(\frac{x}{\lambda} - vt - \varphi \right) \tag{11.75}$$

$$= A_1 \cos 2\pi \left(\frac{x}{\lambda} - vt \right) + A_2 \cos 2\pi \left(\frac{x}{\lambda} - vt \right) \cos \varphi$$

$$+ A_2 \cos 2\pi \left(\frac{x}{\lambda} - vt \right) \sin \varphi \tag{11.76}$$

$$= \sqrt{A_1^2 + A_2^2 + 2A_1 A_2 \cos \varphi} \; \cos \left(2\pi \left(\frac{x}{\lambda} - vt + \delta \right) \right) \tag{11.77}$$

where $\delta = \dfrac{A_2 \sin \varphi}{(A_1 + A_2 \cos \varphi)}$

The corresponding intensity of wave 1 alone is proportional to A_1^2 and that of wave 2 is proportional to A_2^2, and the time average of the intensity over a full cycle is proportional to $A_1^2 + A_2^2 + 2A_1 A_2 \cos \varphi$. We observe that the resulting intensity differs from the sum of individual intensities by $2A_1 A_2 \cos \varphi$, which is referred to as the 'interference term'.

In free space, the intensity (I) of a beam with a single frequency (ϑ) for an electromagnetic wave is given as follows:

$$I = C \epsilon_0 E^2 = Nh\vartheta \tag{11.78}$$

where c is the speed of light, ϵ_0 is the permittivity of the medium, and h is the Planck's constant.

We make an analogy between photon and electron beams by assuming that the electrons in a beam travelling in the x-direction with constant momentum

p_x are controlled by 'pilot waves' with a wave function φ that plays the role for electrons that the electric intensity plays for photons. Therefore,

$$\varphi = A e^{2\pi i\left(\frac{x}{\lambda}-vt\right)} = A e^{i(p_x x - Et)/h} \tag{11.79}$$

Here the wavelength (λ) is given by the De Broglie's concept (h/p) and the energy (E) is $h\vartheta$.

For an electron beam, the wave function φ is a complex number, whereas the electric intensity (E) is an observable quantity and is real. There are differences in electron waves and photons. An electron has a rest mass, charge, and a spin different from that of a photon. Referring to Fig.11.18(b), we observe that Eq. (11.79) is applicable to electrons. At any point on the screen, there is a contribution φ_1 from the first slit and a contribution φ_2 from the second slit, so that the resulting wave function probability amplitude becomes as follows:

$$\varphi = \varphi_1 + \varphi_2 \tag{11.80}$$

The probability that an electron arrives at this point on the screen is proportional to $\varphi^*\varphi$. If there are places where $\varphi_1 = -\varphi_2$; $\varphi^*\varphi$ goes to zero and an electron interference pattern is expected.

There are other cases as in the photoelectric effect in which one wishes to follow the flight of a single electron or other particles; then φ is taken to refer to a single particle. Usually, in a given case, values of φ appreciably different from zero occur only within some finite regions. A solution of the wave equation of this latter type is sometimes called a 'wave packet'. One can raise the question of the position of a particle with respect to the 'wave packet'. It is usually considered that the position of the particle is not defined any more closely than is indicated by the values of φ At any given instant, the particle is found at any point where φ is different from zero. The term $\left|\varphi^2\right|$ represents the 'probability density' at any point; the probability of finding the particle within any element of volume $dxdydz$ is as follows:

$$\left|\varphi^2\right| = \varphi^*\varphi\, dxdydz \tag{11.81}$$

The wave scalar φ itself is sometimes called the probability amplitude for the position of the particle. The probability amplitude, as explained earlier, necessarily imposes a condition, that is, the probability of finding the particle must be unity, which can be mathematically expressed as follows:

$$\iiint \varphi^*\varphi\, dxdydz = 1 \tag{11.82}$$

The triple integral indicates that the integration is extending over all the three dimensions x, y, and z. A wave function φ that satisfies this requirement is said to be normalized.

Suppose that the particle is in free space with a potential equal to zero ($V = 0$) such that the energy of the particle is equal to $p^2/2m$. Such a particle cannot be represented by Eq. (11.82), which denotes an infinite train of waves, since the probability amplitude becomes as follows:

$$\iiint \varphi^* \varphi \, dxdydz = \infty$$

unless the amplitude (A) of the wave function becomes zero. Thus, we conclude that it is not possible for a particle to have a perfectly definite momentum or velocity. The value of the wave function may be so chosen that it defines the momentum nearly definitely over as large a region as desired, going to zero outside the boundary of this region. The position of the particle is not well defined, since $\varphi^* \varphi$ is finite over a considerable volume.

11.11 HEISENBERG UNCERTAINTY PRINCIPLE

De Broglie proposed that the motion of a particle with a velocity v is controlled by pilot waves with a group velocity u. He further suggested that *there is a limit beyond which we cannot determine simultaneously both the momentum and the position of the particle.* Consider that the two waves shown in Fig. 11.19 represent the pilot waves for a particle; the particle is somewhere in the shaded half of the envelope. Then the uncertainty in position (Δx) corresponds to ($\lambda_m/2$) the following relation:

$$\Delta x = \frac{\lambda_m}{2} = \frac{2\pi}{\Delta k.2} = \frac{\pi}{\Delta k} \tag{11.83}$$

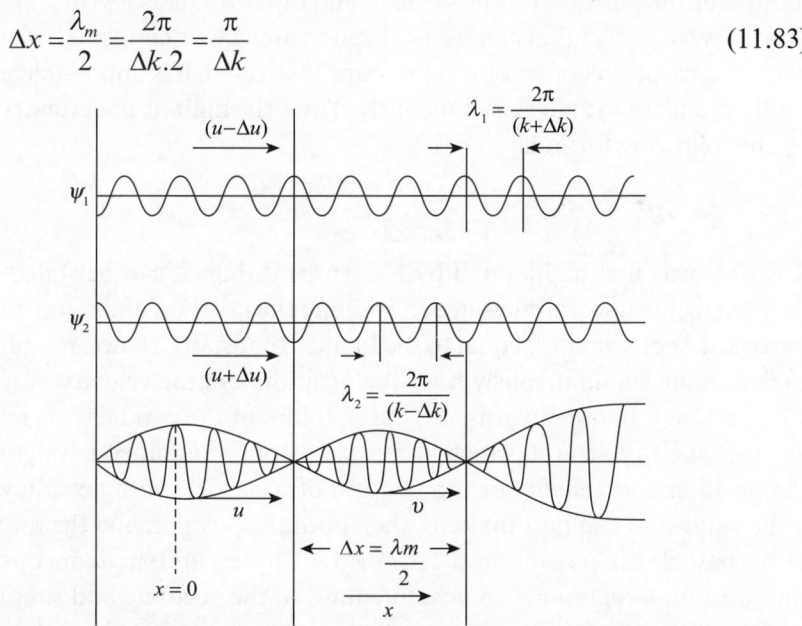

Fig. 11.19 Wave travelling with phase velocity (u), wavelength (l), and frequency (ϑ). (Modulation envelope (or wave packet) has group velocity (v), and the number of beats per unit time is the frequency difference ($2\Delta\vartheta$), so that modulation wave frequency becomes ($\Delta\vartheta$). Consequently, wavelength (λ_m) of modulation wave is $\left(\dfrac{v}{\Delta\vartheta}\right)$ and modulation wave number is $k_m = \left(\dfrac{2\pi}{\lambda_m}\right)$ is just Δk. Thus, $\lambda_m = \dfrac{2\pi}{\Delta k}$)

The propagation wave vector can be redefined using De Broglie's definition of wavelength as follows:

$$k = \frac{2\pi}{\lambda} = \frac{2\pi p}{h} \tag{11.84}$$

$$\Delta k = \frac{2\pi \Delta p}{h} \tag{11.85}$$

The propagation wave number differs by $2\Delta k$ so that the uncertainty in the x component of the momentum is given as follows:

$$\Delta p_x = 2 . \frac{h \Delta k}{2\pi} \tag{11.86}$$

$$\Delta x \, \Delta p_x = \frac{\pi}{\Delta k} \frac{2h\Delta k}{2\pi} \tag{11.87}$$

Thus,

$$\Delta x \Delta p_x \approx h \tag{11.88}$$

It is not surprising that our choice of two beating waves of slightly varying frequency is far from our optimum way of choosing the pilot waves for the motion of the particle. It is possible to add other frequencies (or k's) and thus make a wave packet that can be localized to any Δx we choose. However, even when a wave packet of an optimum shape is selected, it is impossible to reduce both Δx and Δp to zero simultaneously. Thus, the limit of uncertainty is given by the following formula:

$$\Delta x \, \Delta p_x \geq \frac{h}{4\pi} \tag{11.89}$$

This fact was first highlighted by Heisenberg; thus, it can be stated that the product of the uncertainties in determining the position and momentum of a particle is approximately equal to the Planck's constant. Hence, it is impossible to determine simultaneously both the position and the velocity of a particle with accuracy. In other words, it is the existence of the parameter h, referred to as the Planck's constant, which prevents us from simultaneously determining the position and velocity or momentum of a particle with accuracy. Hence, if the value of h can become zero, then both the position and the momentum of the particle can possibly be determined accurately and simultaneously. Thus, the constant h represents an absolute limit to the accurate and simultaneous measurement of the position and momentum of the particle.

Note: The essential point in the uncertainty principle is not simultaneity of determination of the two quantities, but the accuracy involved in the simultaneous determination.

The indefiniteness that has been found to exist in the values of certain mechanical magnitudes associated with a particle, such as its position and momentum, is a fundamental feature of wave mechanics. The indefiniteness in position can be minimized by making the 'wave packet' very small, but this will lead to a broad range of propagation numbers, thereby leading to an inaccuracy in the measurement of momentum and velocity. On the other hand, if the 'wave packet' is made large in order to fix the momentum or velocity of the particle, there is large indefiniteness in the measurement of position.

The Heisenberg uncertainty principle applies to any pair of variables that are canonically conjugate in the Hamiltonian formulation of mechanics. Canonically conjugate pairs are those pairs whose exact simultaneous measurements are not possible. In addition to $\Delta x \Delta p_x$, we have $\Delta y \Delta p_y$, $\Delta z \Delta p_z$; $\Delta E \Delta t$, where ΔE is the uncertainity in energy and Δt that in time; $\Delta \theta \Delta A_\theta$, where ΔA_θ is the uncertainty associated with the angular momentum; and $\Delta \theta$ is the indefiniteness in the corresponding angle.

Physical significance of the uncertainty principle is at the microscopic level and not at the macroscopic level. In classical mechanics, this very small amount of indeterminacy is overlooked. As we discuss microscopic details of elementary particles, this minute indeterminacy becomes significant and starts playing a critical role in the behaviour of the particle. This has been quantified with precise measuring devices and, when used in technology, has bought about a revolutionary change in the outlook of modern devices.

11.11.1 Experimental Verification of Uncertainty Principle

The following sections discuss this.

11.11.1.1 Determination of Position of Particle by Microscope

An attempt is made to determine accurately the position of an electron using a microscope of high resolution. The lower limit of resolution depends upon the wavelength of light used to illuminate the particle; $\gamma-rays$ are used for accuracy as it has the shortest wavelength. Employment of the $\gamma-rays$ involves the Compton effect so that the electron experiences a recoil. Thus, the instant at which we locate the position of the electron by irradiating it with $\gamma-rays$ and observe the scattered rays, its momentum undergoes a discontinuous change. Furthermore, there is indeterminateness about the magnitude of this change, for it will vary according to the direction in which the scattered $\gamma - rays$ leave the point of impact. Scattering of the $\gamma-rays$ is dependent on the angular aperture of the microscope. Thus, one cannot limit closely the range of possible directions of the scattered $\gamma-rays$ without a serious loss of resolving power. As there is a limitation on the resolving power, the position of the electron would be defined less accurately.

Quantitative analysis　We know that

$$\Delta x = \frac{\lambda}{2 \sin \alpha} \tag{11.90}$$

Here, Δx is the distance between two points that can just be resolved, λ is the wavelength of light, and α is the angular aperture of the microscope [Fig. 11.20].

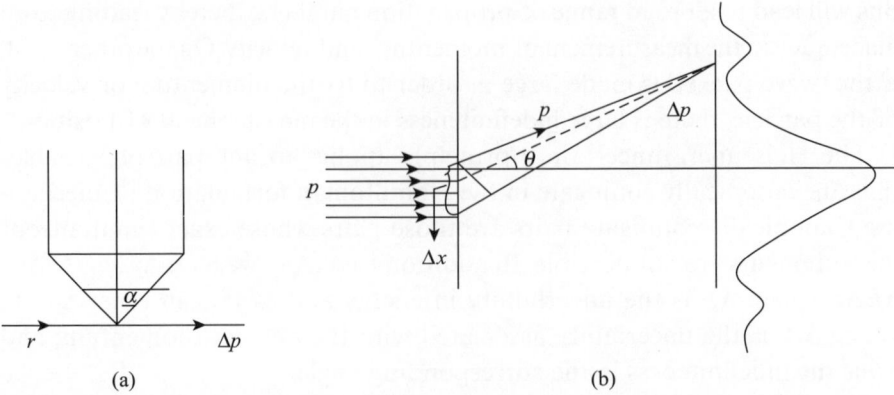

Fig. 11.20 Heisenberg's uncertainty principle (a) Uncertainty in measurement of momentum (Δp) (or position r) of electron owing to resolution limit, as there is angular aperture α at the tip of microscope (b) Diffraction of electrons through narrow slit

The minimum amount of light that can be used for irradiation is a single quantum ($h\vartheta$). The electron can actually be seen only when it scatters the quantum into the microscope. During this process, the electron suffers a Compton recoil of the order of magnitude $\left(\dfrac{h\vartheta}{c}\right)$, the direction of which is indeterminate to the same extent as the direction of the scattered quantum. Since the scattered quantum can enter the microscope anywhere within its angular aperture or the angle, the uncertainty in the direction of both the scattered quantum and the recoil electron is given by α. Hence, the component of momentum of the electron in a direction perpendicular to the axis of the microscope uncertain to the amount Δp is given as follows:

$$\Delta p \approx \frac{2h\vartheta}{c}\sin\alpha \approx \frac{2h}{\lambda}\sin\alpha \tag{11.91}$$

Thus, uncertainty in the simultaneous determination of the position and momentum is given as follows:

$$\Delta x \Delta p \approx \frac{\lambda}{2\sin\alpha} \times \frac{2h}{\lambda}\sin\alpha \tag{11.92}$$

$$\Delta x \Delta p \approx h \tag{11.93}$$

This is the same as the Heisenberg uncertainty principle.

11.12 DIFFRACTION OF ELECTRON BEAM THROUGH NARROW SLIT

Let us consider the phenomenon of diffraction of electrons. The phenomenon of diffraction occurs due to deflection of individual electrons at the slit (either upwards or downwards). The diffraction pattern recorded on a photographic plate is obtained from the statistical superposition of the electrons in the beam. Every electron that is registered on the plate must have passed through the slit, but it is indefinite which position in the slit it has passed. If Δx is the width of the slit, the uncertainty in the specification of position of the electron perpendicular to the direction of flight is given by Δx. Since the electron is deflected at the slit, it acquires an additional component of momentum perpendicular to the original direction of flight. If p is the momentum of the electron, the component perpendicular to the initial direction is $p \sin \theta$, where θ is the angle of deflection. Since the electron may be anywhere within the diffraction pattern, the uncertainty in the aforementioned component is given as follows:

$$\Delta p \approx 2p \sin \theta \tag{11.94}$$

From the wave theory of diffraction,

$$\sin \theta = \frac{\left(\dfrac{\lambda}{2}\right)}{\Delta x} \tag{11.95}$$

$$\Delta x = \frac{\lambda}{2 \sin \theta} \tag{11.96}$$

From Eqs (11.95) and (11.96), we get the following relations:

$$\Delta p \Delta x \approx \frac{\lambda}{2 \sin \theta} . 2p \sin \theta \tag{11.97}$$

$$\Delta p \Delta x \approx \lambda p \tag{11.98}$$

$$\Delta p \Delta x \approx h \tag{11.99}$$

Thus, the uncertainty principle is verified.

Example 11.11   If the uncertainty in the time during which an electron remains in the excited state is 10^{-6} s, what is the least uncertainty in the energy of the excited state?

Solution   If E represents the energy of the excited state, then $\Delta E \Delta t \geq \dfrac{h}{4\pi}$

$$(\Delta E)(10^{-6}) \geq \frac{6.63 \times 10^{-34} \times 7}{4 \times 22} \geq 0.527 \times 10^{-28} \text{ J}$$

Example 11.12   The position and momentum of a 1.0 keV electron are simultaneously measured. If the position is located to an accuracy of 0.2 nm, what is the percentage uncertainty in its momentum?

Solution   It is given to us that $\Delta x = 0.2 \times 10^{-9}$ m

$$E = 1.0 \text{ keV} = 1000 \times 1.6 \times 10^{-19} \text{ J} = 1.6 \times 10^{-16} \text{ J}$$

$$\Delta x \Delta p = \frac{h}{4\pi}$$

$$p = \sqrt{2mE} = \sqrt{2 \times 9.1 \times 10^{-31} \times 1.6 \times 10^{-16}} = \frac{1.71 \times 10^{-23} \text{ kg m}}{\text{s}}$$

$$\Delta p = \frac{h}{2\times 2\pi \times \Delta x} = \frac{7\times 6.63\times 10^{-34}}{2\times 2\times 22\times 0.2\times 10^{-9}} = \frac{2.63\times 10^{-25}\ \text{kg m}}{\text{s}}$$

The percentage of uncertainty in momentum: $\dfrac{\Delta p}{p}\times 100 = \dfrac{2.63\times 10^{-25}}{1.71\times 10^{-23}} = 1.5\%$

Example 11.13   Compare the uncertainties in velocity of a proton and an electron contained in a $10\,\text{Å}$ box.

Solution   The following is the dimension of the box: $\Delta x = 10\times 10^{-10}\ \text{m} = 10^{-9}\ \text{m}$

$$\Delta p \Delta x = \frac{h}{4\pi}$$

$$\Delta p = \frac{h}{4\pi \Delta x} = \Delta(mv) = m\Delta v$$

$$\Delta v_p = \frac{\Delta p_p}{m_p};\ \Delta v_e = \frac{\Delta p_e}{m_e}$$

$$\frac{\Delta v_p}{\Delta v_e} = \frac{\Delta p_p}{\Delta p_e}\frac{m_e}{m_p} = \frac{m_e}{m_p}$$

$$\left(\text{as the change in momentum for the electron and proton is equal}\right)$$

$$\frac{\Delta v_p}{\Delta v_e} = \frac{m_e}{m_p} = \frac{9.1\times 10^{-31}}{1.67\times 10^{-27}} = 5.45\times 10^{-4}$$

Example 11.14   Explain, using the uncertainty principle, the non-existence of electrons in the nucleus. Assume that the maximum possible kinetic energy of an electron emitted by a radioactive nuclei is of the order of $4\,\text{MeV}$; the radius of the nucleus is $10^{-14}\ \text{m}$.

Solution   The uncertainty principle states that $\Delta x \Delta p = \dfrac{h}{4\pi}$

$$\Delta p = \frac{h}{4\pi \Delta x} = \frac{7\times 6.634\times 10^{-34}}{4\times 22\times 10^{-14}} = 5.275\times 10^{-21}\ \text{kg m/s}$$

$$KE = \frac{p^2}{2m} = \frac{(5.275\times 10^{-21})^2}{2\times 9.1\times 10^{-31}} = 1.528\times 10^{-11}\ \text{J} = 95.55\times 10^{6}\ \text{eV} \approx 96\,\text{MeV}$$

Comparing the kinetic energy of $96\,\text{MeV}$ with the kinetic energy of the radioactive nucleus of the order of $4\,\text{MeV}$, it is evident that such an electron cannot exist inside the nucleus. Hence, the uncertainty principle emphasizes the fact that the electrons do not reside inside a nucleus.

Example 11.15   The speed of an electron is measured to be $\dfrac{4.0\times 10^{3}\ \text{m}}{\text{s}}$ to an accuracy of 0.002%. Find the uncertainty in determining the position of this electron.

Solution   The speed of the electron $= v = 4.0\times 10^{3}\ \text{m/s}$

$$\Delta v = v\times \frac{0.002}{100} = 4.0\times 10^{3}\times \frac{0.002}{100} = 0.08\,\text{m/s}$$

$$\Delta x \Delta p = \Delta x \,(m\Delta v) = \frac{h}{4\pi}$$

$$\Delta x = \frac{1}{(m\Delta v)}\frac{h}{4\pi} = \frac{1}{(9.1\times10^{-31}\times0.08)}\frac{6.63\times10^{-34}\times7}{4\times22} = 0.724\times10^{-3}\,\text{m}$$

Example 11.16 If an excited state of hydrogen atom has a life time of the order of 10^{-14} s, what is the minimum error with which the energy of this state can be measured?

Solution The uncertainty in time is given as follows: $\Delta t = 10^{-14}$ s

$$\Delta E\,\Delta t = \frac{h}{4\pi}$$

$$\Delta E = \frac{h}{4\pi}\Delta t = \frac{6.63\times10^{-34}\text{ J.s}\times7}{4\times22}\times\frac{1}{10^{-14}\text{ s}} = 0.527\times10^{-20}\,\text{J}$$

Example 11.17 An excited atom has an average life time of the order 10^{-8} s. What is the minimum uncertainty in the frequency of this photon?

Solution The uncertainty in time is given as follows: $\Delta t = 10^{-8}$ s

The energy for a photon is given as follows: $E = h\vartheta \Rightarrow \Delta E = \Delta(h\vartheta) = h\Delta\vartheta$

In addition, the uncertainty principle states that $\Delta E \Delta t = \frac{h}{4\pi} \Rightarrow h\Delta\vartheta\Delta t = \frac{h}{4\pi}$

Thus, $\Delta\vartheta = \frac{1}{4\pi}\frac{1}{\Delta t} = \frac{7}{4\times22}\frac{1}{10^{-8}} = 0.079\times10^8\,\text{s} = 7.9\times10^8\,\text{s}$

Example 11.18 The radius of a hydrogen atom is of the order of half an angstrom. Estimate the minimum energy that an electron can have in this hydrogen atom using the uncertainty principle.

Solution The uncertainty in position for the hydrogen atom is given by the following expression: $\Delta x = 0.5\,\text{Å}$

$$\Delta x\,\Delta p = \frac{h}{4\pi}$$

$$\Rightarrow \Delta p = \frac{h}{4\pi}\frac{1}{\Delta x} = \frac{6.634\times10^{-34}\times7}{4\times22}\frac{1}{0.5\times10^{-10}} = \frac{1.056\times10^{-24}\,\text{kg m}}{\text{s}}$$

$$(KE)_{min} = \frac{(\Delta p)^2}{2m} = \frac{(1.056\times10^{-24})^2}{2\times9.1\times10^{-31}} = 0.061\times10^{-17}\,\text{J}$$

Example 11.19 Show that for a particle undergoing circular motion, the uncertainty principle takes the following form: $\Delta L\Delta\theta \geq h/4\pi$

Solution The relation between linear momentum and angular momentum is as follows: $L = mvR$

$$\Rightarrow \Delta p = \Delta L/R$$

We know that a circular motion is characterized by the arc length as the displacement function: $\theta = s/R$

$$\Delta s = R\Delta\theta$$

Thus, the displacement function is replaced by the arc function, and the momentum function has been redefined in terms of the angular momentum, as follows:

$$\Delta p \Delta x = \Delta p \Delta s = \left(\frac{\Delta L}{R} \right) (R\Delta\theta) = \Delta L\Delta\theta \geq (h/4\pi)$$

11.13 PHYSICAL SIGNIFICANCE OF WAVE FUNCTION

The wave function φ was considered to be merely an auxiliary mathematical quantity employed to facilitate computations relative to experimental results. Max Born suggested an interpretation for wave function. Schrodinger based his interpretation of φ on charge density, where the square of the absolute value of wave function φ is a measure of the particle density at a given point in space at a given instant.

The restrictions put on the value of φ are as follows:

1. The wave function should be physically acceptable.
2. It should be such that $\varphi \rightarrow 0$ as the particle coordinates tend to infinity, that is, the wave function should be normalizable.
3. The wave function φ and its first derivatives, that is, $\dfrac{\partial\varphi}{\partial x}$, and so on, should be single valued and continuous.
4. $\Sigma\varphi^2 dV$ must be unity, that is, $\int \varphi^2 dV = 1$, where dV is any point in space at any given instant of time.

The probability density $|\varphi^2|$, for a complex φ, is thus represented by the product $\varphi^*\varphi$, where φ^* represents the complex conjugate of φ. Let the wave function be represented as follows:

$$\varphi = C + iD \tag{11.100}$$

$$\text{Complex conjugate } \varphi^* = C - iD \tag{11.101}$$

$$|\varphi|^2 = \varphi^*\varphi = C^2 - i^2 D^2 = C^2 + D^2 \tag{11.102}$$

where we observe that $|\varphi|^2$ is always a positive real quantity. The probability density function represents an absolute quantity that gives us the probability of finding a body or particle at any specific position in space at a specific time.

The normalized wave function ensures that the particle or body is definitely present or definitely exists in the region defined, which can be mathematically expressed as follows:

$$\int_{-\infty}^{\infty} |\varphi|^2 \, dV = 1 \tag{11.103}$$

Thus, a wave function that is physically acceptable is normalizable.

The expectation value of a wave function $\varphi(x, t)$ is the value of x that one would obtain if one measured the positions of many particles described by the same wave function at some instant t and then averaged these results. The average position $\bar{x}$ of a particle using the statistics is given as follows:

$$\bar{x} = \frac{\sum N_i x_i}{\sum N_i} \tag{11.104}$$

Changing the summations to integrals, the expectation value of the position of the single particle can be given as follows:

$$\langle x \rangle = \frac{\int_{-\infty}^{\infty} x |\varphi|^2 \, dx}{\int_{-\infty}^{\infty} |\varphi|^2 \, dx} \tag{11.105}$$

If the wave function is normalized to 1, then the denominator vanishes.

Thus, in general, the expectation value of a function dependent on x is given as follows:

$$U(x) = \frac{\int_{-\infty}^{\infty} U(x) |\varphi|^2 \, dx}{\int_{-\infty}^{\infty} |\varphi|^2 \, dx} \tag{11.106}$$

Note: The expectation value for momentum and energy will be explained after defining the concept of operators in quantum mechanics.

Example 11.20 A particle limited to the x-axis has the wave function $\varphi(x) = bx^2$ between $x = 0$ and $x = 2$; the wave function $\varphi(x) = 0$ elsewhere.
(a) Find the probability that the particle can be found between $x = 1.0$ and $x = 1.5$.
(b) Find the expectation value $\langle x \rangle$ of the particles' position.
Solution The following are the calculations involved:
(a) The probability density function is calculated as follows:

$$\int_{x_1}^{x_2} |\varphi|^2 \, dx = b^2 \int_{1.0}^{1.5} x^4 dx = b^2 \left[\frac{x^5}{5} \right] = b^2 \left[\frac{1.5^5}{5} - \frac{1.0^5}{5} \right]$$

$$= b^2 \left[\frac{7.594}{5} - \frac{1}{5} \right] = b^2 [1.5188 - 1.0] = b^2 0.5188$$

(b) The expectation value is calculated as follows: $\langle x \rangle = \dfrac{b^2 \int_{1.0}^{1.5} x.x^4 dx}{b^2 \int_{1.0}^{1.5} x^4 dx}$

$$= \frac{b^2 \left[\dfrac{1.5^6}{6} - \dfrac{1.0^6}{6} \right]}{b^2 \, (0.5188)} = \frac{\left[\dfrac{11.390}{6} - \dfrac{1.0}{6} \right]}{0.5188} = \frac{[1.8985 - 0.167]}{0.5188}$$

$$= \frac{1.7315}{0.5188} = 3.337$$

Example 11.21 A particle limited to the z-axis has the wave function $\varphi(z) = bz$ between $z = 0$ and $z = 2$; the wave function $\varphi(z) = 0$ elsewhere.
(a) Find the probability that the particle can be found between $z = 0$ and $z = 0.5$.
(b) Find the expectation value $\langle z \rangle$ of the particles' position.

Solution   The following are the calculations.

(a) The probability density function is given as follows:

$$\int_{z_1}^{z_2} |\varphi|^2 \, dz = b^2 \int_0^{0.5} z^2 \, dz = b^2 \left[\frac{z^3}{3} \right] = b^2 \left[\frac{0.5^3}{3} \right]$$

$$= b^2 \left[\frac{0.125}{3} \right] = b^2 [0.042] = 0.042 b^2$$

(b) The expectation value is as follows: $\langle x \rangle = \dfrac{b^2 \int_0^{0.5} z \cdot z^2 \, dz}{b^2 \int_0^{0.5} z^2 \, dz}$

$$= \frac{b^2 \left[\dfrac{0.5^4}{4} \right]}{b^2 \, (0.042)} = \frac{\left[\dfrac{0.625}{4} \right]}{0.042} = \frac{[0.156]}{0.042} = 3.72$$

11.14 SCHRODINGER WAVE EQUATION

If a particle has wave properties, as De Broglie proposed, it is expected that there should be a wave equation that describes the behaviour of the wave function.

The Schrodinger wave equation was introduced as a concept, which was subsequently able to explain the new thoughts of De Broglie and Heisenberg. The wave equation could reduce to the classical mechanics, where the values of the Planck's constant became insignificant or negligible. In this section, how the various concepts introduced bring about the mathematical basis of the wave equation is explained.

Let us assume a particle in the xy plane. Let the displacement (y) originate at the point O. Then,

$$y = F(t) \tag{11.107}$$

The disturbance at the point P at a distance x from the point 'O' is given as follows:

$$P(t) = O\left(t - \frac{x}{v} \right) \tag{11.108}$$

where t is the time and v is the velocity of propagation of the wave.

Thus, at the same instant of time, the simultaneous disturbance at points O and P, respectively, is given by the following relations:

$$y_0 = A F(t) \tag{11.109}$$

$$y_p = A F\left(t - \frac{x}{v} \right) \tag{11.110}$$

Thus, the most general form of the equation is as follows:

$$y_p = A F\left(t - \frac{x}{v} \right) \tag{11.111}$$

Let us seek the solution of the wave equation. Differentiating Eq. (11.111) with respect to the x-coordinate, we get the following relations:

$$\frac{\partial y_p}{\partial x} = AF'\left(t - \frac{x}{v}\right) \times \left(-\frac{1}{v}\right) \tag{11.112}$$

$$\frac{\partial^2 y_p}{\partial x^2} = AF''\left(t - \frac{x}{v}\right) \times \left(-\frac{1}{v^2}\right) \tag{11.113}$$

Differentiating Eq. (11.111) with respect to time, one gets the following expressions:

$$\frac{\partial y_p}{\partial t} = AF'\left(t - \frac{x}{v}\right) \tag{11.114}$$

$$\frac{\partial^2 y_p}{\partial t^2} = AF''\left(t - \frac{x}{v}\right) \tag{11.115}$$

Substituting differential Eqs (11.112)–(11.115) in the form of the wave equation, one gets the following expression:

$$\frac{\partial^2 y_p}{\partial x^2} = \frac{1}{v^2} \frac{\partial^2 y}{\partial t^2} \tag{11.116}$$

This is the differential form of the wave equation in one dimension. The general solutions of the preceding equations are given as follows:

$$y_p = a \sin \omega\left(t - \frac{x}{v}\right) \tag{11.117}$$

or $\quad y_p = a \cos \omega\left(t - \frac{x}{v}\right) \tag{11.118}$

Combining the two solutions (11.117) and (11.118), the simplest and general form of the wave equation is obtained, which is as follows:

$$y = a \exp\left[\pm i\omega\left(t - \frac{x}{v}\right)\right] \tag{11.119}$$

Schrodinger suggested that the wave function can be expressed in the form of a simple wave equation, as acquired from Eq. (11.119):

$$\varphi(x,t) = A \exp\left[-i\omega\left(t - \frac{x}{v}\right)\right] \tag{11.120}$$

Energy of the particle can be represented by the following equations:

$$E = \frac{h}{2\pi}\omega \tag{11.121}$$

$$2\pi E = h\omega \tag{11.122}$$

The wavelength of the particle, based on the De Broglie's concept, is given as follows:

$$\lambda = \frac{h}{p} \tag{11.123}$$

$$\frac{2\pi p}{h} = \frac{2\pi}{\lambda} \tag{11.124}$$

The frequency–wavelength relation is given as follows:

$$v = \vartheta\lambda \tag{11.125}$$

$$\frac{2\pi}{\lambda} = \frac{2\pi\vartheta}{v} = \frac{\omega}{v} \tag{11.126}$$

Thus, from Eqs (11.121)–(11.126), we get the following relations:

$$\varphi(x,t) = A \exp\left[-i\left(\frac{2\pi E}{h}t - \frac{2\pi p}{h}x\right)\right] \tag{11.127}$$

$$\varphi(x,t) = A \exp\left(\frac{2\pi i}{h}(p_x x - Et)\right) \tag{11.128}$$

This form of the wave function is the simplest form. The corresponding differential equations for the wave function (φ) are as follows:

$$\frac{\partial\varphi}{\partial t} = -\frac{2\pi i}{h}E\varphi \tag{11.129}$$

$$E\varphi = \frac{ih}{2\pi}\frac{\partial}{\partial t}\varphi \tag{11.130}$$

$$\hat{E} = \frac{ih}{2\pi}\frac{\partial}{\partial t} \tag{11.131}$$

The variable $\hat{E}$ is defined mathematically as the energy operator in wave mechanics formalism.

$$\frac{\partial\varphi}{\partial x} = \frac{2\pi i}{h}p_x\varphi \tag{11.132}$$

$$p_x\varphi = -\frac{ih}{2\pi}\frac{\partial}{\partial x}\varphi \tag{11.133}$$

$$\widehat{p_x} = -\frac{ih}{2\pi}\frac{\partial}{\partial x} \tag{11.134}$$

The variable $\widehat{p_x}$ is defined mathematically as the momentum operator in wave mechanics formalism.

The expectation value for momentum is defined as follows:

$$\langle p \rangle = \int_{-\infty}^{\infty} \varphi^* p\varphi\, dx = \int_{-\infty}^{\infty} \varphi^*\left(\frac{h}{2\pi i}\frac{\partial}{\partial x}\right)\varphi\, dx = \frac{h}{2\pi i}\int_{-\infty}^{\infty} \varphi^*\frac{\partial\varphi}{\partial x}\, dx \tag{11.135}$$

The total energy of a system is defined by the sum total of the kinetic and potential energy:

$$\text{Total energy} = \text{kinetic energy} + \text{potential energy} \tag{11.136}$$

$$E = \frac{p^2}{2m} + V \tag{11.137}$$

The expectation value for energy is defined as follows:

$$\langle E \rangle = \int_{-\infty}^{\infty} \varphi^* E \varphi \, dx = \int_{-\infty}^{\infty} \varphi^* \left(\frac{ih}{2\pi} \frac{\partial}{\partial t} \right) \varphi \, dx = \frac{ih}{2\pi} \int_{-\infty}^{\infty} \varphi^* \frac{\partial \varphi}{\partial t} \, dx \tag{11.138}$$

In the wave mechanics formalism, the momentum operator is defined as follows:

$$p^2 = p.p = \left(-\frac{ih}{2\pi} \frac{\partial}{\partial x} \right) . \left(-\frac{ih}{2\pi} \frac{\partial}{\partial x} \right) \tag{11.139}$$

$$= -\left(\frac{h}{2\pi} \right)^2 \frac{\partial^2}{\partial x^2} \tag{11.140}$$

Thus, Eq. (11.136) can be written in the following form:

$$\hat{H} = -\left(\frac{h}{2\pi} \right)^2 \frac{1}{2m} \frac{\partial^2}{\partial x^2} + V \tag{11.141}$$

Equation (11.141) defines the total energy in the wave mechanics formalism, as defined by the Hamiltonian operator. Now,

$$\hat{H} = -\left(\frac{h}{2\pi} \right)^2 \frac{1}{2m} \nabla^2 + V \tag{11.142}$$

Equation (11.142) represents the Hamiltonian operator in three dimension.

$$\hat{H}\varphi = \widehat{E}\varphi \tag{11.143}$$

$$-\left(\frac{h}{2\pi} \right)^2 \frac{1}{2m} \nabla^2 \varphi + V\varphi = i\left(\frac{h}{2\pi} \right) \frac{\partial \varphi}{\partial t} \tag{11.144}$$

Equation (11.144) is referred to as the *time-dependent Schrodinger wave equation* for the steady state of a system. The agreement of results deduced from these equations justifies the belief that the equation is valid for matter waves associated with a particle of mass m as long as the relativistic effects can be neglected. The equation resembles some other mechanical equations except for the factor of the imaginary 'iota,' which distinctly differentiates it from all the other equations established so far.

Let us consider the case in which E, P_x, and V are all constants satisfying the Hamiltonian equation and A is a constant representing the wave amplitude.

1. If $E > 0; p > 0$; The solution of the Schrodinger wave equation represents the plane waves of φ travelling towards the positive x-direction; for at a

point moving with velocity $\left(\dfrac{E}{p_x}\right)$, the value of φ remains; $\vartheta = \dfrac{E}{h}; \lambda = \dfrac{h}{p_x}$

in accordance with the previous conclusions.

2. If $p_x < 0; E > 0$; both the particle and the waves are moving towards negative x-direction.

3. If $E < 0$, then V is sufficiently negative; the phase waves and the particle travel in opposite directions.

When the potential energy V is a function of x only and the total energy E is constant, we can seek the solution of the second-order double-differential Schrodinger equation (11.144) by the method of separation of variables:

$$\varphi(x, t) = \varphi(x)\phi(t) \tag{11.145}$$

$$\frac{\partial^2 \varphi}{\partial x^2} = \varphi(t)\varphi''(x) \tag{11.146}$$

$$\frac{\partial \varphi}{\partial t} = \phi'(t)\varphi(x) \tag{11.147}$$

Substituting Eqs (11.145)–(11.147) in Eq. (11.144), one gets the following relation:

$$-\left(\frac{h}{2\pi}\right)^2 \frac{1}{2m}\phi(t)\varphi''(x) + V\varphi(x)\phi(t) = i\left(\frac{h}{2\pi}\right)\varphi(x)\frac{\partial \phi}{\partial t} \tag{11.148}$$

Dividing Eq. (11.148) by $\varphi(x)\phi(t)$ throughout, we get the following relation:

$$-\left(\frac{h}{2\pi}\right)^2 \frac{1}{2m} \frac{1}{\varphi(x)}\varphi''(x) + V = i\left(\frac{h}{2\pi}\right)\frac{1}{\phi(t)}\frac{\partial \phi}{\partial t} \tag{11.149}$$

where the potential energy is assumed to be time independent. Thus, the left-hand side of Eq. (11.149) is dependent only on the space coordinates, and the right-hand side is dependent on the time coordinates. Hence, using the technique of separation of variables, the time-dependent Schrodinger wave equation can be solved as follows:

$$i\left(\frac{h}{2\pi}\right)\frac{1}{\phi(t)}\frac{\partial \phi}{\partial t} = E, \quad \text{where } E \text{ is an arbitrary constant} \tag{11.150}$$

$$\frac{\partial \phi(t)}{\phi(t)} = \frac{E}{i}\frac{2\pi}{h} \tag{11.151}$$

$$\phi(t) = A\exp\left(-\frac{2\pi}{h}iEt\right) \tag{11.152}$$

Equation (11.152) explains the time dependence of the wave function $\varphi(x, t)$. The space-dependent Schrodinger wave equation can be solved as follows:

$$-\left(\frac{h}{2\pi}\right)^2 \frac{1}{2m}\frac{\varphi''(x)}{\varphi(x)} + V = E \tag{11.153}$$

This equation can be rewritten in the following form:

$$\frac{\partial^2 \varphi}{\partial x^2} + \left(\frac{2\pi}{h}\right)^2 2m(E-V)\varphi(x) = 0 \tag{11.154}$$

This is the one-dimensional space-dependent Schrodinger wave equation.

$$\nabla^2 \varphi(x) + \left(\frac{2\pi}{h}\right)^2 2m(E-V)\varphi(x) = 0 \tag{11.155}$$

This is the three-dimensional space-dependent Schrodinger wave equation.

Thus, combining Eqs (11.154) and (11.155), we get the solution of the wave equation $\varphi(x, t)$. The solution of the Schrodinger equation can be represented as follows:

$$\varphi(x, t) = Ae^{-(iEt.2\pi/h)} \varphi(x) \tag{11.156}$$

Example 11.22 Show that $y = Ae^{-i\omega(t-x/v)}$ is a solution of the wave equation.

Solution Taking the partial derivative of y with respect to x, one gets the following relation:

$$\frac{\partial y}{\partial x} = \frac{\partial}{\partial x} Ae^{-i\omega(t-x/v)} = \frac{i\omega}{v} y$$

Taking the second partial derivative of this equation with respect to x, one obtains the following expression:

$$\frac{\partial^2 y}{\partial x^2} = \frac{i\omega}{v} \frac{\partial y}{\partial x} = \frac{i\omega}{v} \frac{i\omega}{v} y = -\frac{\omega^2}{v^2} y$$

The differential of the displacement equation with respect to t is given by the following relation:

$$\frac{\partial^2 y}{\partial t^2} = i^2 \omega^2 y = -\omega^2 y$$

Thus, substituting the two equations using the wave equation, the following equation is obtained:

$$\frac{\partial^2 y}{\partial x^2} = \frac{1}{v^2} \frac{\partial^2 y}{\partial t^2} - \frac{\omega^2}{v^2} y = \frac{1}{v^2} (-\omega^2 y)$$

Hence, we observe that the displacement equation is the solution of the wave equation.

Example 11.23 Explain the concept of an operator, eigen function, and eigen value operating $\dfrac{\partial^2}{\partial x^2}$ on the wave function $\varphi(x) = e^{4x}$.

Solution Operating the operator on the eigen function (x), we get the following relations:

$$\frac{\partial^2}{\partial x^2} e^{4x} = 4\frac{\partial}{\partial x} e^{4x} = 4 \times 4 e^{4x} = 16 e^{4x}$$

$$\frac{\partial^2}{\partial x^2} e^{4x} = 16 e^{4x}$$

Thus, in this equation $\dfrac{\partial^2}{\partial x^2}$ is the operator, e^{4x} is the eigen function, and 16 is referred to as the eigen value of the given eigen function.

Example 11.24 An eigen function of the operator $\dfrac{\partial^2}{\partial x^2}$ is $\cos nx$, where $n = 1, 2, 3, \ldots$ Find the corresponding eigen values.

Solution Operating the operator on the eigen function (x), we get the following equations:

$$\frac{\partial^2}{\partial x^2}\cos nx = n\frac{\partial}{\partial x}\sin nx = n^2\cos nx$$

$$\frac{\partial^2}{\partial x^2}\cos nx = n^2\cos nx$$

Since, the values of $n = 1, 2, 3, \ldots$, the corresponding eigen values are 1, 4, 9, $\ldots$

Note: The eigen value corresponding to a eigen function (operator) is the characteristic or proper energy value for that equation, thus defining the concept of discreteness of a particle or a system in quantum mechanics.

11.15 ELECTRON BEAM IN FIELD-FREE SPACE

An electron, as it moves in space, acts as a free particle. Having understood the discrete behaviour of an electron during its existence in an atom, it would be interesting to study its motion using the Schrodinger equation. Schrodinger has been able to solve the time-dependent equations completely [Eqs (11.150)–(11.152)]. The time-independent part involves the space potential [Eq. (11.155)] that an electron or a particle encounters as it travels in space. Thus, the solution for this simple problem will set the basis for the credibility of the Schrodinger equation and the essence of quantum mechanics. The space potential barriers can classically be visualized as the force acting on a particle or an electron moving from its state of uniform motion or rest.

Consider a region in which the potential energy V of an electron is constant, so that no forces are acting on it. Let us consider that a uniform beam of electrons is travelling in the x-direction, each electron has momentum p_x, and the total energy $E = (p_x^2/2m) + V$. For this beam, the space-dependent Schrodinger wave equation is written as follows:

$$-\frac{1}{2m}\left(\frac{h}{2\pi}\right)^2\frac{\partial^2\varphi}{\partial x^2} + V_0\varphi = E\varphi \tag{11.157}$$

where $\varphi(x)$ is a wave function such that the probability of finding the electron between x and $(x + dx)$ is $\varphi^*(x)\varphi(x)\,dx$. Hence, we have

$$-\frac{1}{2m}\left(\frac{h}{2\pi}\right)^2\frac{\partial^2\varphi}{\partial x^2} + (V_0 - E)\varphi = 0 \tag{11.158}$$

$$\frac{\partial^2 \varphi}{\partial x^2} + 2m\left(\frac{2\pi}{h}\right)^2 (E - V_0)\varphi = 0 \qquad (11.159)$$

This is a simple harmonic wave equation, since the coefficient of the wave function $\varphi(x)$ is a positive constant. Let us consider that $k = \sqrt{2m(E - V_0)}$, then the general solution of Eq. (11.159) together with the time-dependent solution [as in Eq. (11.152)] is given as follows:

$$\varphi(x,t) = Ae^{2\pi i(kx - Et)/h} + Be^{2\pi i(k(-x) - Et)/h} \qquad (11.160)$$

Equation (11.160) represents the complete solution for the space-dependent Schrodinger wave equation. Physically, the second term of the equation represents the wave motion to the left; while invoking the problem we have postulated that the beam associated with $\varphi(x)$ is moving in the +ve x-direction. Therefore, the constant B is zero as it is representing the electron beam travelling in the −ve x-direction. Hence, the physically meaningful solution of the wave function $\varphi(x,t)$ is as follows:

$$\varphi(x,t) = Ae^{2\pi i(kx - Et)/h}$$

$$= Ae^{2\pi i(k_x x - Et)/h}$$

$$= A\cos\left(\frac{2\pi}{h}(k_x x - Et)\right) + iA\sin\left(\frac{2\pi}{h}(k_x x - Et)\right) \qquad (11.161)$$

Equation (11.161) resembles the initial Schrodinger equation and represents a propagative solution in the positive x-direction.

The probability of finding an electron between x and $(x + dx)$ is given in the one-dimensional form by the probability amplitude as follows:

$$\varphi(x)\varphi^*(x)dx = Ae^{2\pi i(k_x x - Et)/h} Ae^{-2\pi i(k_x x - Et)/h} dx \qquad (11.162)$$

$$= A^2 dx \qquad (11.163)$$

As can be seen from Eq. (11.163), the probability density function is not a function of x. The constant A is related to the intensity of the beam. As the particle is moving freely in space, the probability density function indicates that the probability of finding the particle at any point in space is equally probable.

The wave function can be normalized. If we assume that there are n such electrons in the length L, then one can normalize the wave function as follows:

$$\int_x^{x+L} \varphi(x)\varphi^*(x)dx = \int_x^{x+L} A^2 dx = A^2 L = n \qquad (11.164)$$

$$A = \sqrt{\frac{n}{L}} \qquad (11.165)$$

Thus, using the concept of normalization, the arbitrary constant A is defined as per the specification of the problem.

Example 11.25 Using the concept of normalization of a wave function, find the constant A for the wave function $\varphi(x) = Axe^{-x^2/2}$ in the region varying from 0 to ∞.

Solution Using the concept of normalization for one particle, we get the following result:

$$\int_0^\infty Axe^{-x^2/2}Axe^{-x^2/2} = 1 \Rightarrow \int_0^\infty A^2x^2e^{-x^2} = 1 \Rightarrow A^2 = 4/\sqrt{\pi} \Rightarrow A = 2/\pi^{1/4}$$

11.16 ELECTRON BEAM IN FREE STATE OF STEP BARRIER

Let us consider a uniform beam of electrons of energy E coming from the left, as shown in Fig. 11.21. Let the potential energy $V(x)$ be 0 to the left of the origin and a constant V_0 to the right. Such a rectangular barrier does not exist in nature, but it is a reasonable approximation to assume the instantaneous potential difference between the 'Dees' of the cyclotron or the potential barrier at the surface of a metal to be a step barrier potential. The wave function of such a system is ascertained as follows:

$$\Psi(x, t) = \varphi(x)\phi(t) \tag{11.166}$$

where the time-dependent part of the wave function has been completely solved as follows:

$$\phi(t) = A\exp\left(-i\frac{2\pi}{h}Et\right) \tag{11.167}$$

The spatial wave function $\varphi(x)$ is solved for the given step barrier potential function as follows:

$$\left(\frac{h}{2\pi}\right)^2 \frac{1}{2m}\frac{\partial^2\varphi(x)}{\partial x^2} + V(x)\varphi(x) = E\varphi(x) \tag{11.168}$$

Observing Fig. 11.21, the potential function $V(x)$ can be defined as follows:

$$\text{Region I } V(x) = 0 \quad \text{for } -\infty \le x < 0 \tag{11.169}$$

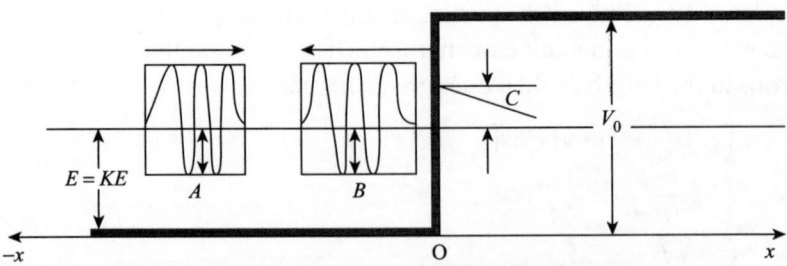

Fig. 11.21 When beam of free electrons with kinetic energy E is incident on ideal potential barrier of height $V_0 > E$, beam is reflected (inset rectangles depict real part of incident and reflected waves)

Region II $V(x) = V_0$ for $0 \le x < \infty$ (11.170)

The general solution for region I is as follows:

$$\varphi_1(x) = A \exp\left(i \frac{2\pi\sqrt{2mE}}{h} x \right) + B \exp\left(-i \frac{2\pi\sqrt{2mE}}{h} x \right) \quad (11.171)$$

$$= A e^{ik_1 x} + B e^{-ik_1 x} \quad (11.172)$$

where the constant $k_1 = \frac{2\pi}{h}\sqrt{2mE}$. The constants A and B are to be determined later, as the constraints on the wave function are discussed. It can be observed evidently that the first term in Eq. (11.172) indicates the incident beam of particles or electrons with A being the constant quantifies the incident electrons. The second term in Eq. (11.172) represents the particles or electrons moving to the left, which arise due to reflection at the barrier.

The general solution for region II can be distinguished into two situations, based on whether $KE = E - V_0$ can take a negative or positive value, depending on the kinetic and potential energy. If we consider the concept of classical physics, then electrons for which the value of $KE = E - V_0$ is a positive quantity would enter region II and pass freely through this region. On the contrary, classically, particles having a negative value of $KE = E - V_0$ would be forbidden to go to region II. Wave mechanics predicts differently, as we will observe in the mathematical formulations that follow, that there is a finite probability for the existence of the particle in region II in both the situations.

11.16.1 Situation I: $(E - V_0) < 0$, that is, It Is Negative Quantity

This situation arises physically, as the height of the potential barrier is greater than the total energy (E) and the kinetic energy becomes negative. Thus, although we do not foresee a situation like this classically, using the wave mechanics mathematics it has been observed that there is a finite probability of φ that is not zero in region II, which is expressed by the following mathematical equations:

$$-\left(\frac{h}{2\pi}\right)^2 \frac{1}{2m} \frac{\partial^2 \varphi_{II}(x)}{\partial x^2} + V_0(x)\varphi(x) = E\varphi(x) \quad (11.173)$$

$$-\left(\frac{h}{2\pi}\right)^2 \frac{1}{2m} \frac{\partial^2 \varphi_{II}(x)}{\partial x^2} + (V_0(x) - E)\varphi(x) = 0 \quad (11.174)$$

The solution for this double differential equation (11.174) is as follows:

$$\varphi_{II}(x) = C \exp\left(-\frac{2\pi}{h}\sqrt{2m(V_0 - E)}x \right) + D \exp\left(\frac{2\pi}{h}\sqrt{2m(V_0 - E)}x \right) \quad (11.175)$$

$$= Ce^{-\alpha x} + De^{\alpha x} \tag{11.176}$$

where $\quad \alpha = \dfrac{2\pi}{h}\sqrt{2m(V_0 - E)}x$

The constants C and D define the space-dependent wave function in region II for this specific barrier potential. As discussed earlier in Section 11.14, there are certain physical conditions that have to be placed on the wave function and its derivatives in order that the probability function is finite and single valued:

1. The wave function $\varphi(x)$ is finite, continuous, and single valued.
2. The first derivative (or gradient in three dimension) of $\varphi(x)$ is finite, continuous, and single valued at all points where the potential energy (V) is infinite. For an infinite potential energy anywhere, an appropriate boundary condition is found by starting with a finite V and taking the limit as V goes to infinity.

Thus, for the space-dependent wave function $\varphi(x)$ to be finite, the constant D should go to zero as it comes with an exponentially positive term. Thus,

$$\varphi_{II}(x) = Ce^{-\alpha x} \tag{11.177}$$

As the wave function should be continuous everywhere, considering Eqs (11.175) and (11.177), one can write that

$$Ae^{ik_1 x} + Be^{-ik_1 x} = Ce^{-\alpha x} \tag{11.178}$$

The first boundary is at $x = 0$; thus,

$$A + B = C \tag{11.179}$$

While for the first derivative of the wave function to be continuous at $x = 0$, the Eq. (11.178) yields the following relation:

$$ik_1(A - B) = -\alpha C \tag{11.180}$$

Here, as defined earlier, A is the amplitude of the incident wave, B is the amplitude of the reflected wave, and C represents the amplitude of the transmitted waves. For the determination of these constants, the simultaneous equations become as follows:

$$\alpha B - \alpha C = -\alpha A \tag{11.181}$$

$$ik_1(-B) + \alpha C = -ik_1 A \tag{11.182}$$

$$B(\alpha - ik_1) = -A(\alpha + ik_1)$$

$$B = -\dfrac{A(\alpha + ik_1)}{(\alpha - ik_1)} \tag{11.183}$$

Substituting the value of B from Eq. (11.183), one gets the following relations:

$$A + A.\dfrac{(\alpha + ik_1)}{(\alpha - ik_1)} = C \tag{11.184}$$

$$C = \frac{2ik_1}{ik_1 - \alpha} A \tag{11.185}$$

The constants B and C are determined in terms of the constant A for incident waves. Equation (11.178) determines the waves entering region II, which are exponentially damped, although there is a finite probability of finding electrons in the classically forbidden region, that is, there is a steady transmission of particles in this region. In the steady state, all the particles are ultimately reflected. We use the polar form of complex variables to highlight the latter statement.

Let us consider that

$$k_1 + i\alpha_1 = \rho e^{\pm i\delta} \tag{11.186}$$

Then,

$$\rho^2 = k_1^2 + \alpha^2 \tag{11.187}$$

$$\tan \delta = \frac{\alpha}{k_1} \tag{11.188}$$

$$B = \frac{\rho e^{-i\delta}}{\rho e^{i\delta}} = e^{-2i\delta} A \tag{11.189}$$

where $\quad \delta = \tan^{-1} \left(\frac{V_0 - E}{E} \right)^{\frac{1}{2}} \tag{11.190}$

Substituting Eq. (11.189) into Eq. (11.178), one gets the following expression:

$$\varphi_1 = A \left(e^{ik_1 x} + e^{-2i\delta} e^{-ik_1 x} \right) = A e^{-i\delta} \cos \left(\sqrt{\frac{2\pi}{h} 2mE} \, x + \delta \right)$$

$$= A e^{-i\delta} \cos \left(\sqrt{\frac{2\pi}{h} 2mE} \, x + \delta \right) \tag{11.191}$$

Equation (11.191) represents a standing wave in which equal numbers of particles are going to the left and right directions. This is similar to the classical behaviour of particles at the metallic surface.

11.16.2 Situation II: $(E - V_0) > 0$, That Is, It Is Positive Quantity

When the total energy exceeds the potential barrier, the kinetic energy is a positive quantity and region II is classically accessible to the electrons or particles. In this case, the general solution for Eq. (11.168) is given by the following relation:

$$\varphi_{II} = G e^{ik_2 x} + H e^{-ik_2 x} \tag{11.192}$$

where $\quad k_2 = \frac{2\pi \sqrt{2m(E - V_0)}}{h}.$

The first term in Eq. (11.192) represents a wave travelling to the right and the second term a wave travelling to the left. Since we have assumed that the initial electrons are incident from the left, we have $H = 0$.

Using the boundary conditions for the wave function, that is, at $x = 0$ both the wave function φ and its first derivatives are continuous, the following relation is obtained:

$$A + B = G \tag{11.193}$$

$$k_1 A - k_1 B = k_2 G \tag{11.194}$$

From Eqs (11.193) and (11.194), the amplitudes of the reflected and transmitted waves can be calculated in terms of that of the incident wave (Fig. 11.22). Multiplying Eq. (11.194) with k_1, the following relations can be obtained:

$$k_1 A + k_1 B = k_1 G \tag{11.195}$$

$$k_1 A - k_1 B = k_2 G \tag{11.196}$$

$$2k_1 A = (k_1 + k_2) G \tag{11.197}$$

$$G = \left(\frac{2k_1}{k_1 + k_2}\right) A \tag{11.198}$$

$$A + B = G \tag{11.199}$$

Substituting the value of G from Eq. (11.198), one gets the following expressions:

$$A + B = \frac{2k_1}{(k_1 + k_2)} A \tag{11.200}$$

$$B = \frac{(k_1 - k_2)}{(k_1 + k_2)} A \tag{11.201}$$

From the physical significance of the constants, it can be easily inferred that

The number of reflected electrons or particles $\propto B^2$

The number of incident electrons or particles $\propto A^2$

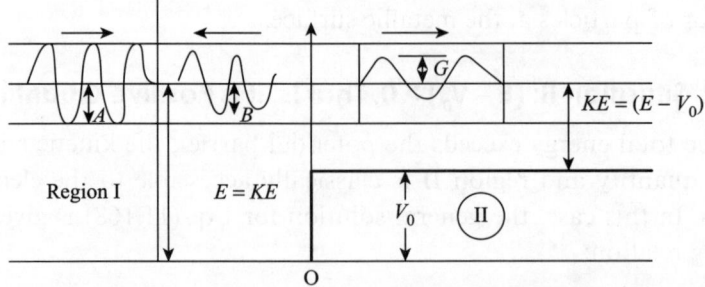

Fig. 11.22 When beam of free electrons or particles with kinetic energy $KE = E$ is incident on potential barrier of height $V_0 < E$, part of beam is reflected and part is transmitted (small sections of beam are shown as reflected and transmitted waves)

Fraction of incident electrons reflected (reflection coefficient R)

$$R = \frac{|B^2|}{|A^2|} = \frac{(k_1 - k_2)^2}{(k_1 + k_2)^2} \tag{11.202}$$

The reflection coefficient (R) is intimately related to the height of the potential barrier V_0. The situation is analogous to that in optics, where, as the light is incident on the interface, optical properties change.

The transmission coefficient (T) is the fraction of the incident electrons transmitted across the barrier. In analogy to optics, the transmission coefficient can be defined as follows:

$$T = \sqrt{\frac{E - V_0}{E}} \frac{|G^2|}{|A^2|} = \frac{k_2}{k_1} \frac{4k_1^2}{(k_1 + k_2)^2} = \frac{4k_1 k_2}{(k_1 + k_2)^2} \tag{11.203}$$

From the conservation of particles, one can observe that

$$T + R = 1 \tag{11.204}$$

Note: Thus, the fact that is highlighted here is that in quantum mechanics all the regions are accessible to the particle irrespective of its energy, unlike the concept of classical mechanics.

Example 11.26 A 20 eV electron is incident on a barrier of height 30 eV. What is the probability that the electron will tunnel through the barrier if its width is (a) 1.0 nm and (b) 0.20 nm.

Solution The parameter that matters most in tunnelling is the difference in energy:

$$(V_0 - E) = (30\,\text{eV} - 20\,\text{eV}) = 10\,\text{eV} = 10 \times 1.6 \times 10^{-19}\,\text{J} = 1.6 \times 10^{-18}\,\text{J}$$

Considering the wave equation from the previous discussion, the tunnelling probability is represented by $e^{-2\alpha L}$. Thus,

$$\alpha = \frac{1}{h}\sqrt{2m(V_0 - E)} = \frac{1}{1.054 \times 10^{-34}} \sqrt{2 \times 9.1 \times 10^{-31} \times 1.6 \times 10^{-18}}$$

$$= \frac{1}{1.054 \times 10^{-34}} \sqrt{2 \times 9.1 \times 1.6 \times 10^{-49}} = 195.36 \times 10^{-10}$$

(a) The tunnelling probability $= e^{-2\alpha L} = e^{-2 \times 195.36 \times 10^{10} \times 1.0 \times 10^{-9}} \approx 10^{-15}$

Thus, the electron has about one chance in 10^{14} times of tunnelling through the barrier.

(b) $L = 0.20\,\text{nm};\quad 2\alpha L = 3.24$

The tunnelling probability $\approx e^{-2\alpha L} = e^{-3.24} = 0.039$

Now, one can observe that the tunnelling probability of the electron is much higher, that is, 3.9%.

11.17 QUANTUM TUNNELLING

Having studied the problem of a step barrier in Section 11.16, it is evident that there is a finite probability of finding the electron in a cross boundary, where its energy is less than the potential barrier. This idea was then extended to a thin barrier like a potential. The probability to see the electron after having crossed the thin barrier would open an era of new technological advancement. As we see in the following section, the quantum mathematics predicts a finite probability for the electron to cross this thin barrier potential referred to as quantum mechanical tunnelling. Applications of this phenomenon are manifold, ranging from transistors to scanning tunnelling microscopes (STMs).

A beam of free particles or electrons is coming from the left and strikes a potential function $V(x)$, which is given by the following relations:

Region I $V(x) = 0$ for $-\infty \leq x \leq 0$

Region II $V(x) = V_0$ for $0 \leq x \leq L$

Region III $V(x) = 0$ for $L \leq x \leq \infty$ (11.205)

Region II acts as a thin barrier for this beam of electrons or particles. Classically, only particles possessing energy greater than the potential barrier will be seen in region III. Quantum mechanically, though, there is a finite probability for a particle to penetrate into a region where its total energy (E) is less than its potential energy (V_0), leading to tunnelling of particles through thin barriers, is not possible classically. Such an effect is referred to as the quantum mechanical tunnelling of particles (Fig. 11.23).

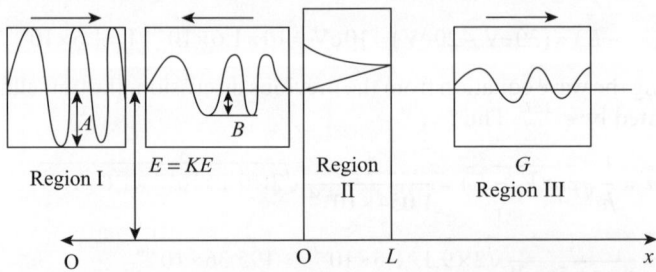

Fig. 11.23 When beam of free electrons or particles with kinetic energy $KE = E$ is incident on thin ideal barrier of height $V_0 > E$, beam is transmitted through the barrier (small sections of real part of incident, reflected, and transmitted beams are shown in rectangles)

The space-dependent wave equations for the potential barrier given in Eq. (11.205) for the given regions can be represented as follows:

Region I $-\left(\dfrac{h}{2\pi}\right)^2 \dfrac{1}{2m} \dfrac{\partial^2 \varphi(x)}{\partial x^2} + 0 = E\varphi(x)$ for $-\infty \leq x \leq 0$ (11.206)

Region II $-\left(\dfrac{h}{2\pi}\right)^2 \dfrac{1}{2m} \dfrac{\partial^2 \varphi(x)}{\partial x^2} + V_0 = E\varphi(x)$

$$-\left(\frac{h}{2\pi}\right)^2 \frac{1}{2m} \frac{\partial^2 \varphi(x)}{\partial x^2} + (V_0 - E)\varphi(x) = 0 \text{ for } 0 \le x \le L \qquad (11.207)$$

$$\text{Region III } -\left(\frac{h}{2\pi}\right)^2 \frac{1}{2m} \frac{\partial^2 \varphi(x)}{\partial x^2} + 0 = E\varphi(x) \text{ for } L \le x \le \infty \qquad (11.208)$$

The solution for the space-dependent wave function $\varphi(x)$ can be determined for each of the regions. The solution for this case can be solved for both the situations where $(E - V_0) < 0$, that is, it is a negative quantity and for $(E - V_0) > 0$, that is, it is a positive quantity.

The quantum mechanical case is $(E - V_0) < 0$, that is, it is a negative quantity. Thus, the general solution is given as follows:

$$\varphi_1(x) = A \exp i \frac{2\pi}{h}\sqrt{2mE}x + B \exp\left(-i\frac{2\pi}{h}\sqrt{2mE}x\right) \text{ for } -\infty \le x \le 0 \qquad (11.209)$$

$$\varphi_2(x) = C \exp i \frac{2\pi}{h}\sqrt{2m(V_0 - E)}x$$

$$+ D \exp\left(-i\frac{2\pi}{h}\sqrt{2m(V_0 - E)}x\right) \text{ for } 0 \le x \le L \qquad (11.210)$$

$$\varphi_3(x) = G \exp\left(i\frac{2\pi}{h}\sqrt{2mE}x\right) + H \exp\left(-i\frac{2\pi}{h}\sqrt{2mE}x\right) \text{ for } L \le x \le \infty \qquad (11.211)$$

Let us consider that

$$k_1 = \frac{2\pi}{h}\sqrt{2mE}; \ \alpha = \frac{2\pi}{h}\sqrt{2m(V_0 - E)}$$

Thus,

$$\varphi_1(x) = A \exp(ik_1 x) + B \exp(-ik_1 x) \quad \text{for } -\infty \le x \le 0 \qquad (11.212)$$

$$\varphi_2(x) = C \exp(i\alpha x) + D \exp(-i\alpha x) \quad \text{for } 0 \le x \le L \qquad (11.213)$$

$$\varphi_3(x) = G \exp(ik_1 x) + H \exp(-ik_1 x) \quad \text{for } L \le x \le \infty \qquad (11.214)$$

The constants are to be determined on the basis of the physical conditions defined by the problem. The constant H appearing in Eq. (11.214) represents the particles or electrons propagating from the right. Since there is no incident beam of particles or electrons from the right, $H = 0$.

The boundary conditions are applied to Eqs (11.212)–(11.214). The value of the space-dependent wave function is continuous at the boundaries, that is, at $x = 0$, $x = L$; thus,

$$|\varphi_1(x)| = |\varphi_2(x)| \text{ at } x = 0; \quad A + B = C + D \qquad (11.215)$$

$$\left|\frac{\partial \varphi_1}{\partial x}\right| = \left|\frac{\partial \varphi_2}{\partial x}\right| \text{ at } x = 0; \quad ik_1 A - ik_1 B = \alpha C - \alpha D \qquad (11.216)$$

$$|\varphi_2(x)| = |\varphi_3(x)| \text{ at } x = L; \quad Ce^{\alpha L} + De^{-\alpha L} = Ge^{ik_1 L} \tag{11.217}$$

$$\left|\frac{\partial \varphi_2}{\partial x}\right| = \left|\frac{\partial \varphi_3}{\partial x}\right| \text{ at } x = L; \quad \alpha Ce^{\alpha L} - \alpha De^{-\alpha L} = ik_1 Ge^{ik_1 L} \tag{11.218}$$

Evaluating the value of A from Eqs (11.215)–(11.217), we have the following relation:

$$A + B = C + D \Rightarrow B = C + D - A \tag{11.219}$$

Substituing the value of B in Eq.(11.216):

$$ik_1 A - ik_1 (C + D - A) = \alpha C - \alpha D \tag{11.220}$$

$$2ik_1 A = (ik_1 + \alpha)C + (ik_1 - \alpha)D \tag{11.221}$$

Now from Eqs (11.217) and (11.218), one gets the following relations:

$$\alpha(Ce^{\alpha L} + De^{-\alpha L}) = \alpha Ge^{ik_1 L} \tag{11.222}$$

$$\alpha(Ce^{\alpha L} - De^{-\alpha L}) = ik_1 Ge^{ik_1 L} \tag{11.223}$$

Subtracting and adding the two, respectively, one gets the following relations:

$$C = \frac{(ik_1 + \alpha)}{2\alpha} e^{-\alpha L} Ge^{ik_1 L} \tag{11.224}$$

$$D = \frac{(ik_1 - \alpha)}{-2\alpha} e^{\alpha L} Ge^{ik_1 L} \tag{11.225}$$

Substituting Eqs (11.224) and (11.225) into Eq. (11.218), one gets the following expressions:

$$2ik_1 A = \left(\frac{(ik_1 + \alpha)^2}{2\alpha} e^{-\alpha L} + \frac{(ik_1 - \alpha)^2}{-2\alpha} e^{\alpha L}\right) Ge^{ik_1 L} \tag{11.226}$$

$$A = \left(\frac{(ik_1 + \alpha)^2}{4ik_1 \alpha} e^{-\alpha L} + \frac{(ik_1 - \alpha)^2}{(-4ik_1 \alpha)} e^{\alpha L}\right) Ge^{ik_1 L} \tag{11.227}$$

$$= \left(\frac{1}{2}(e^{\alpha L} + e^{-\alpha L}) + \frac{i\alpha}{4k_1}(e^{\alpha L} - e^{-\alpha L}) - \frac{ik_1}{4\alpha}(e^{\alpha L} - e^{-\alpha L})\right) Ge^{ik_1 L} \tag{11.228}$$

$$= \left(\cosh \alpha L + \frac{i\alpha}{2k_1} \sinh \alpha L - \frac{ik_1}{2\alpha} \sinh \alpha L\right) Ge^{ik_1 L} \tag{11.229}$$

$$= \left(\cosh \alpha L + \frac{i}{2}\left(\frac{\alpha}{k_1} - \frac{k_1}{\alpha}\right) \sinh \alpha L\right) Ge^{ik_1 L} \tag{11.230}$$

Thus, the transmission coefficient is defined as follows:

$$\frac{1}{T} = \frac{\left|A^2\right|}{\left|G^2\right|} = 1 + \frac{1}{4}\left(\frac{\alpha}{k_1} + \frac{k_1}{\alpha}\right)^2 \sinh^2 \alpha L \qquad (11.231)$$

$$= 1 + \frac{1}{4}\left(\frac{\alpha^2 + k_1^2}{\alpha k_1}\right)^2 \sinh^2 \alpha L$$

$$= 1 + \frac{1}{4}\left(\frac{\left(2m(V_0 - E) + 2mE\right)^2}{2m(V_0 - E).2mE}\right)\sinh^2 \frac{2\pi}{h}\sqrt{2m(V_0 - E)}L$$

$$\frac{1}{T} = 1 + \frac{V_0^2}{4(V_0 - E)E}\sinh^2 \frac{2\pi}{h}\sqrt{2m(V_0 - E)}\,L \qquad (11.232)$$

The transmission coefficient T increases rapidly, with the square of the hyperbolic sine. The argument of the hyperbolic sine function has the potential barrier height V_0 and the width of the barrier L, which plays a sensitive role. The transmission coefficient is strongly dependent on the potential barrier height V_0.

Note: Reducing the width of the barrier by only one order of magnitude increases the probability of tunnelling by about 12 orders of magnitude.

Example 11.27 A beam of electrons is incident on a barrier of 6.00 eV height and 0.200 nm width. Find the energy level they should have if 1.0% of them are to get through the barrier.

Solution Tunnelling probability $\approx e^{-2\alpha L} \approx 1.0\% \approx 0.01$

$$\log_e e^{-2\alpha L} = \log_e(0.01) \Rightarrow \alpha L = 2.3026$$

$$\frac{1}{h}\sqrt{2m(V_0 - E)}\,L = 2.3026 \Rightarrow V_0 = 6.00\,\text{eV}; \; L = 0.2 \times 10^{-9}\,\text{m}$$

$$h = 6.634 \times 10^{-34}\,\text{J.s} \Rightarrow E = 0.96\,\text{eV}$$

The major applications of the tunnelling effect are as follows:

1. Alpha decay: It is the emission of alpha-particles by unstable, heavy nuclei. For an alpha-particle to escape from the nucleus, it penetrates through the potential barrier created by the combination of the nuclear force and the Coulomb repulsion (between the alpha particle and the rest of the nucleus).
2. STMs.

11.17.1 Scanning Tunnelling Microscope

The discovery of the scanning probes is as recent as the late 19th century. Gerd Binnig and Heinrich Rohrer developed the first working STM in the year 1981 while working at IBM Zurich Research Laboratories in Switzerland. They were later awarded the Nobel Prize for this in the year 1986.

11.17.1.1 Principle

The microscope uses three basic principles of physics to probe a surface. They are as follows:

Quantum mechanical tunnelling   One can observe the surface of the system under detection using quantum mechanical tunnelling. The resolution of the measurement of the dimensions of the surface improved multifold by observing the exponentially decaying wave function.

Piezoelectric effect   The piezoelectric effect is used to design a tube wherein the scanning tip is placed. The scanning tip is usually made up of tungsten. A piezoelectric tube helps us scan the tip precisely with an angstrom-level control $(4-7\,\text{Å})$.

Feedback loop   This helps us monitor the tunnelling current, and coordinates the current and the positioning of the scanning tip.

11.17.1.2 Working

A sharp tip made up of a conducting material is brought within about $(4-7\,\text{Å})$ distance of the surface that has to be analysed. The surface is a conducting material. Vacuum is to be created between the sample and the tip. A certain voltage is applied to the surface due to which the Fermi levels experience a shift. The wave functions of an electron in the tip overlap those of the sample surface (Fig. 11.24). Thus, the electrons tunnel from one surface to another of lower potential. The tunnelling current so observed is dependent on the applied

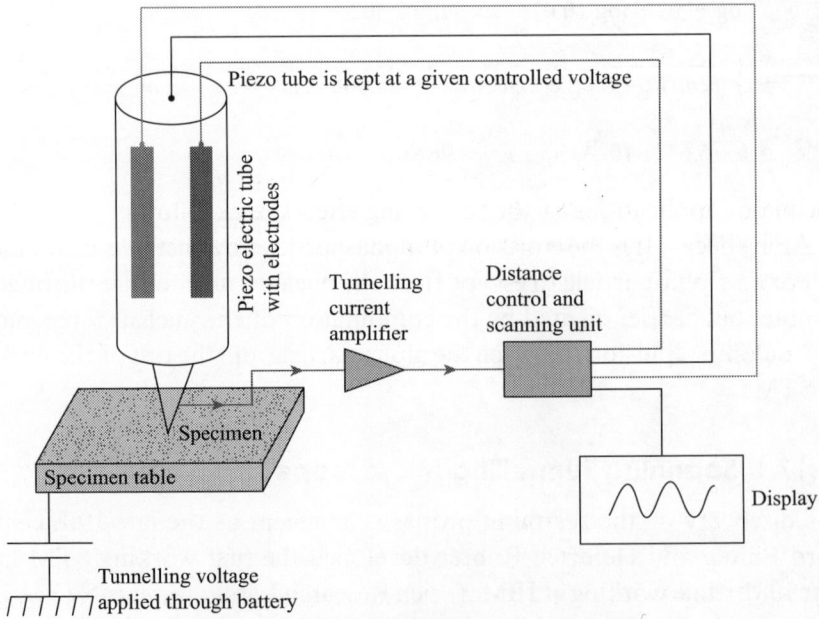

Fig. 11.24. Scanning Tunnelling Microscope

voltage, tip voltage, and local density of states of the sample. An image is taken of the current as the tip scans the entire surface of the conductor sample.

11.17.1.3 Advantages

The following are the advantages of an STM:

1. The atomic resolution of an STM is several orders higher than that of an electron microscope. The depth resolution is about 0.01 nm and the lateral resolution is about 0.01 nm.
2. An STM is capable of performing localized, non-destructive measurements.
3. The images of an STM give a good resolution for the plane wave functions direction of the standing-wave patterns in the local density of states of any given conducting sample.

11.17.1.4 Disadvantages

The following are the disadvantages of an STM:

1. The major disadvantage of an STM is that it can scan only the conductive surfaces or thin non-conductive films and small objects deposited on conductive substrates. The scanning tip does not work on non-conductive materials such as rocks and glass.
2. The fast kinetics of the electrochemical processes require precision in the temporal coherence of the instrument. In an STM, the spatial resolution is taken care of by introducing the concept of piezoelectric effect in the designing of the scanning tip, but the temporal coherence is not as good. Thus, the images of an STM give a good resolution for the plane wave functions, unlike the kinetic processes.

11.17.1.5 Applications

The following are the applications of an STM:

1. The STM technique can be used for atomic deposition of metals such as gold and silver. Any desired pre-programmed pattern can be used.
2. An STM has recently been used as a molecular switch. The STM tip rotates the individual bonds within the single molecules, resulting in the change of resistance of the molecule such that it can be used as a switch.
3. An STM is used to tunnel electrons into a layer of electron beam photo-resist on a sample in order to perform lithography. This technique is more advantageous than the traditional electron beam, as it offers more control of the exposure.

11.17.2 Atomic or Scanning Force Microscope

A scanning force microscope is a modification of the STM, and its first experimental implementation was made by Binning, Quate, and Gerber in 1986.

The resolution limit for this microscope is of the order of fractions of a nanometre. This microscope provides one of the foremost tools for imaging, measuring, and manipulating matters at the nanoscale.

11.17.2.1 Principle

A scanning force microscope works on the following principles:

1. The specimen surface is scanned using a cantilever that has a sharp tip, of the dimension of a nanometre.
2. Forces that act between the tip and the sample lead to deflection of the cantilever, as defined by the Hooke's law. Depending on the specimen, the forces that can be measured are the mechanical contact force, Van der Waals force, capillary forces, electrostatic forces, magnetic forces, and others.

11.17.2.2 Main Components and Their Functions

The main components of a scanning force microscope and their functions are described in the following subsections:

Laser diode   A laser diode is used to illuminate a specimen.

Cantilever   A cantilever spring deflects as the probe tip scans the sample surface. The atomic forces in the specimen cause the deflection of the cantilever.

Position-sensitive photo-detector   A mirror placed normal to the specimen makes the incident radiation impinge on the position-sensitive photo-detector that measures the deflection of the cantilever. Three types of sampling can be done in the specimen:

Contact mode imaging   Here, frictional and adhesive forces damage the specimen and distort the data.

Non-contact imaging   It provides low resolution as the pollutants in the specimen interfere with the oscillation.

Tapping mode imaging   The probe contacts the surface of the specimen, intermittently contacting the surface to a sufficient extent and oscillating it with sufficient amplitude, to prevent the tip from being trapped by the adhesive meniscus forces from the pollutant layers.

Feedback loop   The error between the actual signal and the fixed point is measured as the error. The feedback loop controls the third dimension of the specimen. The piezoelectric scanner that positions the specimen is connected to the feedback loop. The computer controls the entire system, in addition to performing data acquisition, display, and analysis of the data.

11.17.2.3 Advantages over Optical and Electron Microscopes

A scanning force microscope has the following advantages over optical and electron microscopes:

1. The vertical dimension measurement, that is, the height or the depth of the sample, can be measured, in addition to the two-dimensional features of a specimen (surface).
2. The magnification factor is at least hundred times higher than those of electron and optical microscopes.
3. It can study samples that are non-conductive, particularly protein molecules.

11.17.3 Scanning Electron Microscope

The scanning electron microscope (SEM) was developed in the early 1950s. The invention of SEM has been an evolving process, having contributions from Manfred von Ardenne (1937) in magnification and removal of chromatic aberration; the Cambridge groups later took it to the commercial stages. An SEM is a microscope that uses electrons to form an image, unlike an optical microscope that uses light. The resolution for an SEM is of nanometre dimensions. Observation of a specimen can be done under various conditions such as high vacuum, low vacuum, wet conditions, and a wide range of cryogenic or high temperatures.

11.17.3.1 Principle

An SEM works on the following principles:

1. The lenses formed in an SEM are made of electromagnets formed using high voltages.
2. Electrons are made to strike the specimen vertically after being guided through these lenses and electromagnetic fields.
3. The signal produced by an SEM depends upon the interaction of the incident radiation with the specimen surface. The signals include those from secondary electrons, back-scattered electrons, characteristic X-rays, specimen current, and transmitted electrons.
4. Different signals are capable of providing different information about the specimens that are analysed using detectors. For example, the intensity of the back-scattered electrons reflected from the specimen by elastic scattering is related to the atomic number (of the specimen), and hence provides information about the composition of the various elements in the sample.

11.17.3.2 Working

An SEM works under the following conditions:

1. An SEM works under vacuum conditions and uses electrons to form an image.

2. The specimen to be analysed needs to be prepared with care, removing water completely from it. Conductive samples, particularly metals, require no preparation before being used.
3. All non-metals have to be made conductive, which is done by covering the sample with a thin layer of conductive material, using a device called a 'sputter coater'.
4. The sputter coater uses an electric field and argon gas. The specimen is placed in a small chamber that is at a vacuum. The electric field causes the electron to be removed from the argon gas, making the atoms positively charged. The positively charged argon ions are attracted to the negatively charged gold foil or any other metal whose coating is required. Argon ions knock gold atoms from the surface of the gold foil. These gold atoms fall and settle onto the surface of the specimen, producing a thin gold coating.

11.17.3.3 Radiation Safety Standards

Back-scattered electrons and X-rays produced from the specimen raise the concern for safety standards for people involved in taking measurements and maintenance of the SEM. An SEM usually comes with a well-covered metal shielding that absorbs all the radiation; nonetheless, safety measures have to be taken and proper inventories need to be performed. State Departments are responsible for the registration of applications and maintenance of radiation safety standards.

Shielding of radiation is maintained for an exposure rate of $< 0.5 \dfrac{\text{mrem}}{\text{hr}}$ at 5 cm.

11.18 STATIONARY OR QUANTUM STATES OR BOUND STATES OF PARTICLES

Bound or stationary states of a particle are those for which a particle is restricted within the boundaries of a given potential. Wave mechanics predicts that the particle then picks up certain definite energy states that are governed by the bound potential state of the particle. These definite energy states are the discrete value of the energy states. As one deals with the bound state problems in wave mechanics, the concept of discrete energy states or quantum energy states becomes significant.

11.18.1 Particle in Rectangular Box

The box represents a three-dimensional problem—the simplest three-dimensional problem in wave mechanics. A rectangular box with sides of length L_x, L_y, L_z is considered. The potential energy inside the box is zero and that outside the box is infinity. Mathematically,

Region I $V(x) = 0$ for $0 \leq x \leq L_x$ (11.233)

$$V(y) = 0 \quad \text{for} \quad 0 \le y \le L_y \tag{11.234}$$

$$V(z) = 0 \quad \text{for} \quad 0 \le z \le L_z \tag{11.235}$$

This indicates the region inside the box.

Region II $V(x) = \infty \quad \text{for} \quad 0 \le x \le L_x \tag{11.236}$

$$V(y) = \infty \quad \text{for} \quad 0 \le y \le L_y \tag{11.237}$$

$$V(z) = \infty \quad \text{for} \quad 0 \le z \le L_z \tag{11.238}$$

This indicates the region outside the box. For the potential value to be infinite, the wave function has to be zero, that is, $\varphi = 0$ outside the box (Fig. 11.25).

The wave equation inside the box for the given potential for which $V = 0$ is given as follows:

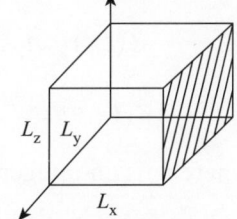

Fig. 11.25 Dimensions of the 3D box

$$E\varphi = -\left(\frac{h}{2\pi}\right)^2 \frac{1}{2m} \nabla^2 \varphi + 0 \tag{11.239}$$

$$E\varphi = -\left(\frac{h}{2\pi}\right)^2 \frac{1}{2m} \left(\frac{\partial^2 \varphi}{\partial x^2} + \frac{\partial^2 \varphi}{\partial y^2} + \frac{\partial^2 \varphi}{\partial z^2}\right) \tag{11.240}$$

$$-\left(\frac{2\pi}{h}\right)^2 (2mE)\varphi = \left(\frac{\partial^2 \varphi}{\partial x^2} + \frac{\partial^2 \varphi}{\partial y^2} + \frac{\partial^2 \varphi}{\partial z^2}\right) \tag{11.241}$$

Let us define that $k^2 = \left(\frac{2\pi}{h}\right)^2 (2mE)$,

Using the technique of separation of variables, we have that

$$\varphi(x, y, z) = X(x)Y(y)Z(z)$$

$$\frac{1}{X(x)}\frac{d^2 X}{dx^2} + \frac{1}{Y(y)}\frac{d^2 Y}{dy^2} + \frac{1}{Z(z)}\frac{d^2 Z}{dz^2} = -k^2 \tag{11.242}$$

Assuming that

$$k^2 = k_x^2 + k_y^2 + k_z^2 \tag{11.243}$$

$$\frac{1}{X(x)}\frac{d^2 X}{dx^2} = -k_x^2 \tag{11.244}$$

$$\frac{1}{Y(y)}\frac{d^2 Y}{dy^2} = -k_y^2 \tag{11.245}$$

$$\frac{1}{Z(z)}\frac{d^2 Z}{dz^2} = -k_z^2 \tag{11.246}$$

Equations (11.243)–(11.246) indicate simple harmonic motion equations. Thus, the general solution for these equations is given as follows:

$$X(x) = A_x \sin k_x x + B_x \cos k_x x \tag{11.247}$$

Using the boundary conditions, at $x = 0$, the continuity of φ requires that $X(0) = 0$ and so in Eq. (11.247) the constant $B_x = 0$, such that

$$X(x) = A_x \sin k_x x \tag{11.248}$$

The subsequent boundary condition indicates that $X(L_x) = 0$, so that

$$X(L_x) = A_x \sin k_x L_x = 0 \tag{11.249}$$

$$k_x L_x = n_x \pi \tag{11.250}$$

where n_x is an integer

$$X(x) = A_x \sin \frac{n_x \pi}{L_x} x \tag{11.251}$$

Similarly,

$$Y(y) = A_y \sin \frac{n_y \pi}{L_y} y \tag{11.252}$$

$$Z(z) = A_z \sin \frac{n_z \pi}{L_z} z \tag{11.253}$$

$$\varphi(x, y, z) = X(x) Y(y) Z(z)$$

$$= A_x \sin \frac{n_x \pi}{L_x} x A_y \sin \frac{n_y \pi}{L_y} y A_z \sin \frac{n_z \pi}{L_z} z \tag{11.254}$$

$$\varphi(x, y, z) = A_x A_y A_z \sin \frac{n_x \pi}{L_x} x \sin \frac{n_y \pi}{L_y} y \sin \frac{n_z \pi}{L_z} z \tag{11.255}$$

$$k^2 = k_x^2 + k_y^2 + k_z^2 \tag{11.256}$$

Thus, the wave function is defined in terms of the quantum numbers n_x, n_y, n_z. This further endorses the thought that the particle trapped inside a potential well has discrete energy states.

Example 11.28 Find the probability that a particle trapped in a box of width L can be found between $0.2L$ and $0.4L$ for the ground and the first excited states, respectively.
Solution The wave function of the particle in a box is given as follows:

$$\varphi_n(x) = A_x \sin \frac{n_x \pi}{L_x} x$$

Probability density function $= \int_{x_1}^{x_2} |\varphi_n(x)|^2 \, dx = \int_{0.2L}^{0.4L} \sin^2 \frac{n_x \pi}{L} x \, dx$

$$= \left[\frac{x}{L} - \frac{1}{2n\pi} \sin\left(\frac{2n\pi x}{L} \right) \right]$$

For the ground state, $n = 1$.

$$= \left[\frac{0.4}{L} - \frac{0.2}{L} \right] - \left[\frac{1}{2\pi} \left(\sin\left(\frac{2\pi(0.4)}{L} \right) - \sin\left(\frac{2\pi(0.2)}{L} \right) \right) \right] = \frac{0.2}{L}$$

Thus, the probability of finding the particle accurately in the box of length L is 20% of its length.

Example 11.29 Find the expectation value $\langle x \rangle$ of the position of a particle trapped in a box of width L.

Solution The expectation value for position has been defined earlier in the text; we only have to replace the wave function by that for a box of width L:

$$\langle x \rangle = \frac{2}{L} \int_0^L x \sin^2 \frac{n\pi x}{L} dx$$

$$= \frac{2}{L} \left[\frac{x^2}{4} - \frac{x \sin\left(\frac{2n\pi x}{L} \right)}{4n\pi/L} - \frac{\cos\left(\frac{2n\pi x}{L} \right)}{8\left(\frac{n\pi}{L} \right)^2} \right]$$

As $\sin n\pi = 0$, $\cos(2n\pi) = 1$, $\cos(0) = 1$, substituting it in this equation one gets the following relation:

$$x = \frac{2}{L} \left(\frac{L^2}{4} \right) = \frac{L}{2}$$

Thus, as expected, the most probable position of finding the particle in the box is at the centre of its dimensions ($L/2$).

Example 11.30 An electron is confined between two impenetrable walls 0.200 nm apart. Determine the energy levels for the states $n = 1, 2$, and 3.

Solution The discussion on a particle in infinite dimensions of an infinite well gives the discrete values for energy, as follows:

$$E_n = \frac{n^2 h^2}{8 m_e L^2} = \frac{(6.634 \times 10^{-34} \text{ J.s})^2}{8 \times (9.1 \times 10^{-31}) \times (2.0 \times 10^{-10})^2} = 1.51 \times 10^{-18} \text{ J} = 9.37 \text{ eV}$$

Consequently, $n = 2$; $E_2 = 2^2 E_1 = 4 \times 9.37 \text{ eV} = 37.5 \text{ eV}$

$$n = 3; \quad E_2 = 3^2 E_1 = 9 \times 9.37 \text{ eV} = 84.33 \text{ eV}$$

Example 11.31 Determine the energy of the lowest three levels for an electron in a square well of width 3 Å.

Solution The discrete energy levels of the particle in a box are given by the following expression:

$$E_n = \frac{n^2 h^2}{8 m L^2}$$

For $n = 1$, $E_1 = \frac{1^2 \times (6.63 \times 10^{-34})^2}{8 \times 9.1 \times 10^{-31} (3 \times 10^{-10})^2} = 0.067 \times 10^{-17} \text{ J}$

For $n = 2$, $E_2 = 2^2 \times E_1 = 4 \times 0.067 \times 10^{-17} \text{ J} = 0.268 \times 10^{-17} \text{ J}$

For $n = 3$, $E_3 = 3^2 \times E_1 = 9 \times 0.067 \times 10^{-17} \text{ J} = 0.603 \times 10^{-17} \text{ J}$

Thus, the lowest three energy levels are defined by these values of energy, the ground state being defined by the energy level for $n = 1$.

Example 11.32   Evaluate the lowest energy that a neutron possesses while it is confined inside the nucleus of an atom.

Solution   The conventional dimensions of the diameter of a nucleus are as follows:

$$L = 10^{-14} \text{ m, mass of the neutron} = 1.67 \times 10^{-27} \text{ kg, and } h = 6.634 \times 10^{-34} \text{ J.s}$$

The lowest energy of the confined nucleus is defined by the discrete energy levels of a particle confined in a box, which is as follows:

$$E_n = \frac{n^2 h^2}{8mL^2} = \frac{1^2 \times \left(6.634 \times 10^{-34}\right)^2}{8 \times 1.67 \times 10^{-27} \times \left(10^{-14}\right)^2} = 3.294 \times 10^{-13} \text{ J}$$

Example 11.33   Evaluate the momentum and energy of an electron confined in a box of length $1.0\,\text{Å}$ for the ground and the first excited states. Find the corresponding De Broglie wavelength.

Solution   The discrete energy states for a particle in a box are related to the momentum p_n follows:

$$E_n = \frac{n^2 h^2}{8mL^2} = \frac{p_n^2}{2m} \Rightarrow p_n = \frac{nh}{2L}$$

$$L = 1.0 \times 10^{-10} \text{ m}$$

For the ground state $n = 1$, $p_1 = \dfrac{1 \times 6.634 \times 10^{-34} \text{ J.s}}{2 \times 1.0 \times 10^{-10} \text{ m}} = 3.317 \times 10^{-24}$ kg m/s

$$E_1 = \frac{p_1^2}{2m} = \frac{(3.317 \times 10^{-24})^2}{2 \times 9.1 \times 10^{-31}} = 0.604 \times 10^{-17} \text{ J}$$

$$\lambda_1 = \frac{h}{p_n} = \frac{6.634 \times 10^{-34} \text{ J.s} \times 2 \times 1.0 \times 10^{-10} \text{ m}}{1 \times 6.634 \times 10^{-34} \text{ J.s}} = 2.0 \times 10^{-10} \text{ m} = 2.0\,\text{Å}$$

For the first excited state $n = 2$, $p_2 = \dfrac{nh}{2L} = \dfrac{2 \times 6.634 \times 10^{-34}}{2 \times 1.0 \times 10^{-10}} = 6.634 \times 10^{-24}$ kg m/s

$$E_2 = \frac{p_2^2}{2m} = \frac{(6.634 \times 10^{-24})^2}{2 \times 9.1 \times 10^{-31}} = 2.418 \times 10^{-17} \text{ kg m/s}$$

$$\lambda_2 = \frac{h}{p_n} = \frac{6.634 \times 10^{-34} \text{ J.s} \times 2 \times 1.0 \times 10^{-10} \text{ m}}{2 \times 6.634 \times 10^{-34} \text{ J.s}} = 1.0 \times 10^{-10} \text{ m} = 1.0\,\text{Å}$$

Thus, one observes that, as stated, the De Broglie wavelength decreases as we go higher up in the energy states.

Example 11.34   Compare the ground state energy eigen value of an electron confined in a box of length $1.0\,\text{Å}$ to that of a $50\,\text{g}$ golf ball confined in a box of length $50\,\text{cm}$. Hence, infer the significance of nano dimensions in quantum mechanics.

Solution Energy eigen values are computed using the following relation: $E_n = \dfrac{n^2 h^2}{8mL^2}$

For an electron, $m = 9.1 \times 10^{-31}$ kg; $L = 1.0 \times 10^{-10}$ m

$$E_1 = \frac{1^2 (6.634 \times 10^{-34} \text{ J.s})^2}{8 \times 9.1 \times 10^{-31} \times (1.0 \times 10^{-10})^2} = 0.604 \times 10^{-17} \text{ J}$$

For a golf ball, $m = 50$ g; $L = 50$ cm

$$E_1 = \frac{1^2 (6.634 \times 10^{-34} \text{ J.s})^2}{8 \times 50 \times 10^{-3} \times (50.0 \times 10^{-2})^2} = 4.5 \times 10^{-66} \text{ J}$$

Thus, it is evident from the energy eigen value that, in nano dimensions, the order of 10^{-17} is measurable and significant. In the case of macroscopic dimensions like that of a golf ball, the order of the energy eigen value 10^{-66} is insignificant and not measurable (almost tending to zero).

Example 11.35 A particle is in a cubical box with infinitely hard walls whose edges are L long. The wave functions of the particle are given as follows:

$$\varphi(x, y, z) = A_x A_y A_z \sin\frac{n_x \pi x}{L} \sin\frac{n_y \pi y}{L} \sin\frac{n_z \pi z}{L}$$

Find the value of the normalization constant $A = A_x A_y A_z$.

Solution The wave function has to be normalized for one particle in a box, as follows:

$$= \iiint_0^{L_{x,y,z}} \varphi^*(x, y, z) \varphi(x, y, z) \, dx \, dy \, dz$$

$$= \iiint_0^{L_{x,y,z}} A_x A_y A_z \sin\frac{n_x \pi x}{L} \sin\frac{n_y \pi y}{L} \sin\frac{n_z \pi z}{L} A_x A_y A_z$$

$$\sin\frac{n_x \pi x}{L} \sin\frac{n_y \pi y}{L} \sin\frac{n_z \pi z}{L} \, dx \, dy \, dz$$

Thus, considering each of the dimensions in the integral individually,

$$\int_0^{L_x} A_x^2 \sin^2\frac{n_x \pi x}{L} \, dx$$

Using the trigonometric relation,

$$= \frac{A_x^2}{2} \left[\int_0^{L_x} dx - \int_0^{L_x} \cos\left(\frac{2n_x \pi x}{L}\right) dx \right]$$

$$= \frac{A_x^2}{2} \left[L_x - \frac{L_x}{2n\pi} \sin\frac{2n_x \pi x}{L} \right] = \frac{A_x^2}{2} [L_x]$$

$$\Rightarrow A_x^2 = \frac{2}{L_x} \Rightarrow A_x = \sqrt{\frac{2}{L_x}}$$

Similarly, $A_y = \sqrt{\dfrac{2}{L_y}}$; $A_z = \sqrt{\dfrac{2}{L_z}}$

$$A_x A_y A_z = \sqrt{\frac{2}{L_x}} \sqrt{\frac{2}{L_y}} \sqrt{\frac{2}{L_z}}$$

In this special case, the edges of the cube are equal to L; thus, $A_x A_y A_z = A = \left[\dfrac{2}{L}\right]^{3/2}$

IMPORTANT CONCEPTS

1. Photoelectric emission is the phenomenon of emission of visible light electrons when metals are exposed to radiations such as γ-rays, X-rays, and ultraviolet radiations. The photoelectric equation is given as follows:

$$\frac{1}{2}mv_m^2 = h\vartheta - \varphi_0$$

2. De Broglie postulated the dual, that is, wave and particle, behaviour of material particles.

   The waves associated with a particle should have a frequency equal to the energy of the particle divided by the Planck's constant (h). We assume that any material particle of mass m, moving at speed v, has associated with it waves of frequency ϑ given by the following relation:

$$h\vartheta = \frac{1}{2}mv^2 + V = E$$

   The existence of the parameter h, referred to as the Planck's constant, prevents us from simultaneously measuring any two conjugate physical parameters (e.g., position or velocity or momentum of the particle) with accuracy. That is,

$$\Delta x \, \Delta p_x \geq \frac{h}{4\pi}$$

3. The wave function should be physically acceptable.
4. The wave function should be such that $\varphi \to 0$ as the particle coordinates tend to infinity, that is, the wave function should be normalizable.
5. The wave function φ and its first derivatives, that is, $\dfrac{\partial \varphi}{\partial x}$, should be single valued and continuous.

   The quantity $\sum \varphi^2 dV$ must be unity, that is, $\int \varphi^2 dV = 1$, where dV is any point in space at any given instant of time.

   The Hamiltonian operator defining total energy is explicitly given as follows:

$$\widehat{H} = -\left(\frac{h}{2\pi}\right)^2 \frac{1}{2m}\frac{\partial^2}{\partial x^2} + V$$

   The Schrodinger wave equation is mathematically represented as follows:

$$\hat{H}\varphi = \widehat{E}\varphi$$

$$-\left(\frac{h}{2\pi}\right)^2 \frac{1}{2m}\nabla^2\varphi + V\varphi = i\left(\frac{h}{2\pi}\right)\frac{\partial \varphi}{\partial t}$$

   This equation is referred to as the 'time-dependent Schrodinger wave equation' for the steady state of a system.

$$\nabla^2\varphi(x) + \left(\frac{2\pi}{h}\right)^2 2m(E - V)\varphi(x) = 0$$

   This equation is referred to as the 'space-dependent Schrodinger wave equation' for the steady state of a system.

The expectation value for momentum is defined as follows:

$$\langle p \rangle = \int_{-\infty}^{\infty} \varphi^* p \varphi \, dx = \int_{-\infty}^{\infty} \varphi^* \left(\frac{h}{2\pi i} \frac{\partial}{\partial x} \right) \varphi \, dx = \frac{h}{2\pi i} \int_{-\infty}^{\infty} \varphi^* \frac{\partial \varphi}{\partial x} \, dx$$

The expectation value for energy is defined as follows:

$$\langle E \rangle = \int_{-\infty}^{\infty} \varphi^* E \varphi \, dx = \int_{-\infty}^{\infty} \varphi^* \left(\frac{ih}{2\pi} \frac{\partial}{\partial t} \right) \varphi \, dx = \frac{ih}{2\pi} \int_{-\infty}^{\infty} \varphi^* \frac{\partial \varphi}{\partial t} \, dx$$

6. The eigen value corresponding to a eigen function (operator) is the characteristic or proper energy value for that equation, thus defining the concept of discreteness of a system in quantum mechanics.

7. The electron beam in a field-free space is given using the Schrodinger wave equation as follows:

$$\varphi(x, t) = A e^{2\pi i (kx - Et)/h}$$

The step barrier space-dependent potential barrier akin to an electron facing a potential in the cyclotron D gives the fraction of incident electrons reflected (refection coefficient R)

$$= \frac{|B^2|}{|A^2|} = \frac{(k_1 - k_2)^2}{(k_1 + k_2)^2}$$

The transmission coefficient is defined as follows:

$$T = \sqrt{\frac{E - V_0}{E}} \frac{|G^2|}{|A^2|} = \frac{k_2}{k_1} \frac{4k_1^2}{(k_1 + k_2)^2} = \frac{4k_1 k_2}{(k_1 + k_2)^2}$$

From the conservation of particles, one can observe that $T + R = 1$

The quantum tunnelling of particles/electrons, owing to the concept of quantum mechanics, predicts a definite probability for transmission. The transmission coefficient whose probability is negligible in classical physics is shown to have a definite probability in quantum mechanics:

$$\frac{1}{T} = 1 + \frac{V_0^2}{4(V_0 - E)E} \sinh^2 \frac{2\pi}{h} \sqrt{2m(V_0 - E)} L$$

The concept of bound state of a particle discussed with only a three-dimensional example of a particle in a box leads to the following wave function:

$$\varphi(x, y, z) = A_x A_y A_z \sin \frac{n_x \pi}{L_x} x \sin \frac{n_y \pi}{L_y} y \sin \frac{n_z \pi}{L_z} z$$

APPLICATIONS

1. The Schrodinger equation is used to determine the wave function for various potential problems resulting in the explanation of discreteness of energy or concept of energy packet.

   The electron beam in a field-free space is given using the Schrodinger wave equation: $\varphi(x, t) = A e^{2\pi i (kx - Et)/h}$

The step barrier space-dependent potential barrier akin to the electron facing a potential in the cyclotron D, gives the fraction of incident electrons reflected (reflection coefficient R)

$$= \frac{|B^2|}{|A^2|} = \frac{(k_1 - k_2)^2}{(k_1 + k_2)^2}$$

The transmission coefficient is defined as follows:

$$T = \sqrt{\frac{E - V_0}{E}} \frac{|G^2|}{|A^2|} = \frac{k_2}{k_1} \frac{4k_1^2}{(k_1 + k_2)^2} = \frac{4k_1 k_2}{(k_1 + k_2)^2}$$

From the conservation of particles, one can observe that $T + R = 1$
The quantum tunnelling of particles/electrons owing to the concept of quantum mechanics, the transmission coefficient whose probability is negligible in classical physics is shown to have a definite probability in quantum mechanics:

$$\frac{1}{T} = 1 + \frac{V_0^2}{4(V_0 - E)E} \sinh^2 \frac{2\pi}{h} \sqrt{2m(V_0 - E)} L$$

2. The major applications of the tunnelling effect are as follows:
   (a) Alpha decay is the emission of alpha-particles by unstable, heavy nuclei. For an alpha-particle to escape from the nucleus, it penetrates through the potential barrier created by the combination of the nuclear force and the Coulomb repulsion (between the alpha particle and the rest of the nucleus)
   (b) This effect can be seen in scanning tunnelling microscopes (STMs).
3. This definite transmission probability over a barrier in quantum mechanics led to a boom in the technology of fabrication of electronic goods. Transistors and subsequently diodes (discussed in Chapter 19) introduced a new era of electronic goods.
4. Quantum mechanics as a subject has wide and varied applications, the latest being the development of superfast quantum computers using the bound space potential and discrete behaviour of particles.

IMPORTANT FORMULAE

1. Photoelectric equation:

$$\frac{1}{2} m v_m^2 = h\vartheta - \varphi_0$$

2. De Broglie material particle of energy E, moving with velocity v, having waves of frequency ϑ:

$$h\vartheta = \frac{1}{2} mv^2 + V = E$$

3. Kinetic energy $= E - E_0$

$$= \sqrt{E_0^2 + (pc)^2} - E_0$$

4. Energy operator:

$$\hat{E} = \frac{ih}{2\pi} \frac{\partial}{\partial t}$$

5. Momentum operator:

$$\widehat{p_x} = -\frac{ih}{2\pi} \frac{\partial}{\partial x}$$

6. Uncertainty principle for the position and momentum conjugate parameters:

$$\Delta x \, \Delta p_x \geq \frac{h}{4\pi}$$

7. For a wave function to be normalized,

$$\int \varphi^2 \, dV = \text{the total number of particles}$$

in the volume dV

8. Hamiltonian operator defining total energy:

$$\hat{H} = -\left(\frac{h}{2\pi}\right)^2 \frac{1}{2m} \frac{\partial^2}{\partial x^2} + V$$

9. Schrodinger wave equation:

$$\hat{H}\varphi = \widehat{E}\varphi$$

Time-dependent Schrodinger wave equation:

$$-\left(\frac{h}{2\pi}\right)^2 \frac{1}{2m} \nabla^2 \varphi + V\varphi = i\left(\frac{h}{2\pi}\right)\frac{\partial \varphi}{\partial t}$$

Space-dependent Schrodinger wave equation:

$$\nabla^2 \varphi(x) + \left(\frac{2\pi}{h}\right)^2 2m(E-V)\varphi(x) = 0$$

10. Expectation value for momentum:

$$\langle p \rangle = \int_{-\infty}^{\infty} \varphi^* p\varphi \, dx = \int_{-\infty}^{\infty} \varphi^* \left(\frac{h}{2\pi i}\frac{\partial}{\partial x}\right)$$

$$\varphi \, dx = \frac{h}{2\pi i} \int_{-\infty}^{\infty} \varphi^* \frac{\partial \varphi}{\partial x} dx$$

11. Expectation value for energy:

$$\langle E \rangle = \int_{-\infty}^{\infty} \varphi^* E\varphi \, dx = \int_{-\infty}^{\infty} \varphi^* \left(\frac{ih}{2\pi}\frac{\partial}{\partial t}\right)\varphi \, dx$$

$$= \frac{ih}{2\pi} \int_{-\infty}^{\infty} \varphi^* \frac{\partial \varphi}{\partial t} dx$$

12. Step barrier space-dependent potential barrier akin to the electron facing a potential in the cyclotron D, giving the fraction of incident electrons reflected (reflection coefficient R):

$$= \frac{|B^2|}{|A^2|} = \frac{(k_1 - k_2)^2}{(k_1 + k_2)^2}$$

13. Transmission coefficient:

$$T = \sqrt{\frac{E - V_0}{E}} \frac{|G^2|}{|A^2|} = \frac{k_2}{k_1} \frac{4k_1^2}{(k_1 + k_2)^2}$$

$$= \frac{4k_1 k_2}{(k_1 + k_2)^2}$$

From the conservation of particles, one can observe that $T + R = 1$

14. The quantum tunnelling of particles predicting a definite probability for the transmission of the electrons through a thin barrier:

$$\frac{1}{T} = 1 + \frac{V_0^2}{4(V_0 - E)E}$$

$$\sin h^2 \frac{2\pi}{h} \sqrt{2m(V_0 - E)} L$$

15. Concept of bound state of a particle discussed with only a three-dimensional example of a particle in a box:

$$\varphi(x, y, z) = A_x A_y A_z \sin\frac{n_x \pi}{L_x}x$$

$$\sin\frac{n_y \pi}{L_y}y \sin\frac{n_z \pi}{L_z}z$$

SELF-ASSESSMENT

Multiple-choice Questions

11.1 The photoelectric effect highlights the phenomenon of
   (a) emission of quanta of energy 'photon'
   (b) absorption of quanta of energy 'photon'
   (c) both (a) and (b)
   (d) none of these

11.2 The wavelength of the λ-ray energy of 10^{19} eV is
   (a) 1.240×10^{-25} m
   (b) 12.40×10^{-25} m
   (c) 0.1240×10^{-25} m
   (d) 0.01240×10^{-25} m

11.3 Which of the following is the momentum–energy relation?
(a) $E^2 = (pc)^2 + (m_0c^2)^2$
(b) $E = pc$
(c) both (a) and (b)
(d) $E = m_0c^2$

11.4 Sunburn of skin is associated with the breakage of chemical bonds. The wavelength corresponding to photon energy of $3.5\,eV$ is
(a) the visible light
(b) the ultraviolet light
(c) the green light
(d) microwaves

11.5 A stream of photons impinging on a completely absorbing screen in vacuum exerts pressure P given by
(a) $P = \dfrac{1}{c}$
(b) $P = \Delta E/\Delta t$
(c) $P = \Delta p/\Delta t$
(d) none of these

11.6 The largest momentum for a microwave photon is
(a) $\dfrac{6.6\times10^{-34}\,kg\,m}{s}$
(b) $\dfrac{6.6\times10^{-30}\,kg\,m}{s}$
(c) $\dfrac{6.6\times10^{-31}\,kg\,m}{s}$
(d) $6.6\times10^{-28}\,kg\,m/s$

11.7 The rest mass of a photon is
(a) 0
(b) 1
(c) negligible
(d) all of these

11.8 The De Broglie wavelength of a particle of mass m and kinetic energy E_k is given by
(a) $\lambda = \dfrac{hc}{\sqrt{E_k(E_k + 2mc^2)}}$
(b) $\lambda = \dfrac{hc}{\sqrt{E_k(E_k + mc^2)}}$
(c) $\lambda = \dfrac{hc}{\sqrt{2E_k(E_k + 2mc^2)}}$
(d) $\lambda = \dfrac{hc}{\sqrt{2E_k(E_k + 2mc^2)}}$

11.9 The De Broglie wavelength (λ) of a particle having potential V is
(a) $\lambda = \dfrac{h}{\sqrt{2Vqm}}$
(b) $\lambda = \sqrt{\dfrac{mh}{2Vq}}$
(c) $\lambda = \sqrt{\dfrac{hV}{2qm}}$
(d) $\lambda = \sqrt{\dfrac{qh}{2m}}$

11.10 The uncertainty principle states that
(a) $\Delta x \Delta p \geq \dfrac{h}{2\pi}$
(b) $\Delta x \Delta p \geq \dfrac{h}{4\pi}$
(c) $\Delta x \Delta p \leq \dfrac{h}{2\pi}$
(d) $\Delta x \Delta p \leq \dfrac{h}{\pi}$

11.11 Heisenberg uncertainty principle holds for
(a) microscopic and macroscopic particles
(b) only microscopic particles
(c) only macroscopic particles
(d) all of these

11.12 To be a solution of the Schrodinger equation, the wave function should be
(a) single valued, continuous, and continuous first derivatives
(b) double valued, continuous, and discontinuous firs derivatives
(c) single valued, discontinuous, and continuous first derivatives
(d) none of these

11.13 Quantum mathematics is popularly known as
(a) wave formalism
(b) operator formalism
(c) particle formalism
(d) Schrodinger formalism

11.14 The Hamiltonian operator defines the
(a) total energy of the system
(b) kinetic energy of the system
(c) potential energy of the system
(d) all of these

11.15 Quantum mechanics is different from classical mechanics in that
(a) there is a probabilistic approach
(b) a particle without the energy to pass over a potential barrier may still tunnel through
(c) there is a wave function approach
(d) all of these

11.16 The eigen value of an eigen function is its
(a) probabilistic value
(b) characteristic value
(c) wave value
(d) potential energy value

11.17 The operator $\dfrac{d^2}{dx^2}$ when operated on $\varphi = e^{2x}$ gives an eigen value of
(a) 1
(b) 2
(c) 3
(d) 4

11.18 The transmission probability in the tunnel effect is given by
(a) $T = e^{-2k_2 l}$, where k_2 is the propagation constant, L is barrier width
(b) $T = e^{-k_2 l}$
(c) $T = e^{-L}$
(d) $T = e^{-2L}$

11.19 The expectation value of a physical quantity in quantum formalism is
(a) the definite value of the physical quantity
(b) the probabilistic value of the physical quantity
(c) the most probabilistic value of the physical quantity
(d) none of these

11.20 The expectation value of the position of a particle in a box is given by
(a) $\langle x \rangle = L/2$
(b) $\langle x \rangle = L/4$
(c) $\langle x \rangle = 2L$
(d) $\langle x \rangle = L/6$

11.21 The eigen energy value of a particle confined in a box is
(a) $E_n = \dfrac{n^2 h^2}{8mL^2}$
(b) $E_n = \dfrac{nh^2}{2mL^2}$
(c) $E_n = \dfrac{nh}{4mL}$
(d) $E_n = \dfrac{nh}{6mL}$

11.22 The salient features of an STM are
(a) quantum mechanical tunnelling
(b) piezoelectric effect
(c) feedback loop
(d) all of these

11.23 Which of the following particles are used as detecting agents in an SEM?
(a) Electrons
(b) Protons
(c) Neutrons
(d) Light

11.24 An atomic force microscope uses
   (a) the atomic force in the specimen as the detection device as the probe tip scans
   (b) the cantilever deflection due to the various internal forces of the specimen as the probe tip scans
   (c) the feedback loop to study the specimen
   (d) none of these

Review Questions

11.1 Underline the thought process that contemporary phenomena such as photo-electric effect, De Broglie wave, uncertainty principle lead to the understanding of quantum mechanics.

11.2 Explain how the idea of De Broglie wavelength led to the conceptualization of the Heisenberg uncertainty principle.

11.3 What is the difference between the normalized and orthogonal wave functions?

11.4 Explain one major feature of the wave function for which the wave function and its first derivative needs to be continuous everywhere.

11.5 Enunciate the physical significance of the wave function.

11.6 Is there a distinct possibility of transmission of a particle having energy less than the potential barrier height? Explain.

11.7 Conceptualize mathematically the term 'energy eigen value'.

11.8 Conceptualize mathematically the term 'degeneracy'.

11.9 Explain in detail the meaning of the term work function and hence the pho-toelectric effect.

11.10 Explain the uncertainty principle.

11.11 Write two major differences between classical mechanics and quantum mechanics.

11.12 Explain briefly the basis of a normalized function.

11.13 Derive the Schrodinger wave equation and hence differentiate between the time-dependent and time-independent Schrodinger equations.

11.14 Explain, enunciating the mathematical formalism, the physical significance of wave function.

11.15 Obtain the Schrodinger wave equation for a step potential barrier. Hence, show that there is a distinct possibility of transmission of a particle having energy less than the potential barrier height.

11.16 Explain the phenomenon of quantum tunnelling across a thin barrier of width L using the Schrodinger wave equation.

11.17 Establish the one-dimensional Schrodinger wave equation for a particle con-fined in a box, and hence give the discrete energy levels that are available to the particle.

11.18 Establish the three-dimensional Schrodinger wave equation for a particle confined in a box. Explain the concept of energy eigen value and degeneracy using this problem.

Numerical Problems

11.1 Find the work function of sodium metal whose photoelectric threshold wave-length is 600 nm.

$$\left[Hint: W_{min} = \left(\frac{hc}{\lambda_{max}} \right) = 2.067 \text{ eV} \right]$$

11.2   Find the maximum kinetic energy of the photoelectrons ejected from a sodium surface, for which $\varphi_0 = 2.1\,\text{eV}$, by photons of wavelengths 3000 and 4000 Å. What is the threshold frequency ϑ_0 and the corresponding wavelength?

$$\left[Hint: \text{KE}_{max} = h\vartheta - \varphi\right]$$

11.3   An experiment on the photoelectric effect of caesium gives the results that the stopping potentials for $\lambda = 4000$ and 5000 Å are, respectively, 0.95 and 0.38 eV. From these data, find h, the threshold frequency, and the work function of caesium.

$$\left[Hint: \text{Work function} = \left(\frac{hc}{\lambda}\right) - e\text{Vs}_1;\ h\vartheta_0 = \text{work function}\right]$$

11.4   A source is emitting 100 W of green light at a wavelength of 540 nm. How many photons per second are emerging from the source?

$$\left[Hint: \text{Photon flux} = (\text{energy emitted})\left(\frac{(\lambda)}{(hc)}\right)\right]$$

11.5   An electron has a De Broglie wavelength of 2.00 pm. Find its kinetic energy and the phase and group velocities of its De Broglie waves.

$$\left[Hint: \text{Kinetic energy} = E - E_0 = \sqrt{E_0^2 + (pc)^2} - E_0;\right.$$

$$\left.E = \frac{E_0}{\sqrt{1 - v^2/c^2}};\text{phase velocity} = v_p = \frac{c^2}{v};\ \text{group velocity} = v_g = v\right]$$

11.6   Find the De Broglie wavelength of (a) a 0.5 mg grain of dirt blown by a wind of speed 10 m/s and (b) an electron whose speed is $\dfrac{1.0 \times 10^8\,\text{m}}{\text{s}}$.

$$\left[Hint: \lambda = \frac{h}{mv}\right]$$

11.7   Find the De Broglie wavelength of a 2.00 MeV proton. By what percentage will the non-relativistic calculation be in error?

$$\left[Hint: pc = \frac{hc}{\lambda}\right]$$

11.8   A beam of 54 eV electrons is directed at a nickel target. The potential energy of an electron that enters the target changes by 12 eV.
(a)   Compare the electron speeds outside and inside the target.
(b)   Compare the respective De Broglie wavelengths.

$$\left[Hint: mv = \sqrt{2m(\text{KE})}; \lambda = \frac{h}{mv}\right]$$

11.9   If the uncertainty in the time during which an electron remains in the excited state is 10^{-5} s, what is the least uncertainty in the energy of the excited state?

$$\left[Hint: \Delta E \Delta t \geq \frac{h}{4\pi}\right]$$

11.10   Compare the uncertainties in velocity of a proton and an electron contained in a 5 Å box.

$$\left[Hint: \Delta E \Delta t \geq \frac{h}{4\pi}\right]$$

11.11   The speed of an electron is measured to be $\dfrac{2.0 \times 10^3 \, \text{m}}{\text{s}}$ to an accuracy of 0.001%. Find the uncertainty in determining the position of this electron.

$$\left[\text{Hint}: \Delta x \Delta p = \Delta x \, (m \Delta v) = \frac{h}{4\pi} \right]$$

11.12   A particle limited to the x-axis has the wave function $\varphi(x) = bx^2$ between $x = 0$ and $x = 1$; the wave function $\varphi(x) = 0$ elsewhere.
   (a)   Find the probability that the particle can be found between $x = 0.5$ and $x = 1.0$.
   (b)   Find the expectation value $\langle x \rangle$ of the particles' position.

$$\left[\text{Hint}: \int_{z_1}^{z_2} |\varphi|^2 \, dz = b^2 \int_0^{1.0} z^2 dz = b^2 \left[\frac{z^3}{3} \right]; \langle x \rangle = \frac{b^2 \int_0^{1.0} z \cdot z^2 dz}{b^2 \int_0^{1.0} z^2 dz} \right]$$

11.13   A particle limited to the z-axis has the wave function $\varphi(z) = bz$ between $z = 0$ and $z = 1$; the wave function $\varphi(z) = 0$ elsewhere.
   (a)   Find the probability that the particle can be found between $z = 0$ and $z = 0.5$.
   (b)   Find the expectation value $\langle z \rangle$ of the particles' position.

$$\left[\text{Hint}: \int_{z_1}^{z_2} |\varphi|^2 \, dz = b^2 \int_0^{1.0} z^2 dz = b^2 \left[\frac{z^3}{3} \right]; \langle x \rangle = \frac{b^2 \int_0^{1.0} z \cdot z^2 dz}{b^2 \int_0^{1.0} z^2 dz} \right]$$

11.14   A particle limited to the x-axis has the wave function $\varphi(x) = bx^2$ between $x = 0$ and $x = 2$; the wave function $\varphi(x) = 0$ elsewhere.
   (a)   Find the probability that the particle can be found between $x = 0.5$ and $x = 1.0$.
   (b)   Find the expectation value $\langle x \rangle$ of the particles' position.

$$\left[\text{Hint}: \int_{x_1}^{x_2} |\varphi|^2 \, dx = b^2 \int_{0.5}^{1.0} x^4 dx = b^2 \left[\frac{x^5}{5} \right] \right]$$

11.15   Explain the concept of operator and eigen function, and eigen value operating

$$\dfrac{\partial^2}{\partial x^2} \text{ on the wave function } \varphi(x) = e^{2x}. \qquad \left[\text{Hint}: \dfrac{\partial^2}{\partial x^2} e^{2x} \right]$$

11.16   A 10 eV electron is incident on a barrier of height 20 eV. What is the probability that the electron will tunnel through the barrier if its width is (a) 0.5 nm and (b) 0.10 nm.

$$\left[\text{Hint}: \alpha = \frac{1}{h} \sqrt{2m(V_0 - E)}; \text{ Tunnelling probability} = e^{-2\alpha L} \right]$$

11.17   Find the probability that a particle trapped in a box of width L can be found between $0.1L$ and $0.2L$ for the ground and the first excited states.

$$\left[\text{Hint}: \text{Probability density function} = \int_{x_1}^{x_2} |\varphi_n(x)|^2 \, dx = \int_{0.1L}^{0.2L} \sin^2 \frac{n_x \pi}{L} x \, dx \right]$$

11.18 Determine the energy of the lowest three levels for an electron in a square well of width $2\,\text{Å}$.

$$\left[Hint: E_n = \frac{n^2 h^2}{8mL^2} \right]$$

11.19 Compare the ground state energy eigen value of an electron confined in a box of length $2.0\,\text{Å}$ to that of a $1\,\text{mg}$ marble confined in a box of length $20\,\text{cm}$.

$$\left[Hint: E_n = \frac{n^2 h^2}{8mL^2} \right]$$

CHAPTER 12

Statistical Mechanics

Learning Objectives

After studying this chapter, students will be able to

- understand the formation of energy levels, energy states, and degeneracy
- use thermodynamic probability in defining the three distinguished statistical distributions: Bose–Einstein, Fermi–Dirac, and Maxwell–Boltzmann
- understand the classical radiation theory
- comprehend how the radiation laws follow the Kirchhoff's Law, Stefan–Boltzmann Law, and Wien's Displacement Law
- appreciate the quantum theory of radiation, that is, Planck's Law of blackbody radiation is holistic for radiation laws and reduces to the Wien's and Rayleigh–Jeans law for special conditions
- understand Stefan's Law

List of Symbols

ρ = Density

N_{Av} = Avogadro's number

k_B = Boltzmann constant

C_p = Specific heat capacity at constant pressure.

C_v = Specific heat capacity at constant volume

γ = Degree of freedom

ϵ_j = Energy level

p_j = Magnitude of momentum

g_j = Degenerate energy level

W_k = Thermodynamic probability or the statistical count

12.1 INTRODUCTION

This branch of statistics deals with the average behaviour of particles in a gas. Statistical mechanics dwells on the overall behaviour of particles, which is governed by the average value of a given thermodynamic variable in a system. Statistical mechanics is not concerned about the actual motions or interactions or detailed considerations of the individual particles. It takes into account the fact that molecules and particles are numerous, thus helping us calculate the average property of a large number of molecules and particles. Some of the physical quantities that can be defined using this simple mathematical technique of averaging are pressure, temperature, specific heat, and internal energy at the atomic level. This is usually dealt with in the kinetic theory of gases and is the classical aspect of statistics. Many scientists have contributed to the kinetic theory of gases—Robert Boyle, Daniel Bernoulli, James Joule, A. Kronig, Rudolph Clausius, and Clark Maxwell being some of them.

The quantum mechanical aspect of this branch has primarily been developed by J. Willard Gibbs and Ludwig Boltzmann in the form of statistical mechanics, with landmark contributions from Paul A.M. Dirac, S.N. Bose, and Planck. Statistical mechanics uses the modern concepts of quantum mechanics to explain the behaviour of particles in a system of more than one body.

12.1.1 Ideal Gas Behaviour—Molecular Energies

For an ideal gas of mass m confined in a container of volume V, and the density ρ of the gas being $\dfrac{m}{V}$ in a state of thermal equilibrium, the *Boyles* and *Charles* laws together give the following equation:

$$\frac{PV}{T} = C \quad \text{(a constant, for a fixed mass of the gas)} \tag{12.1}$$

The constant on the right-hand side for a fixed mass of gas is given as (μR) where μ is the molar gas constant and R is a constant that is determined for each gas by experiment.

$$R = 8.314 \frac{\text{J}}{\text{mole}} \text{K} \tag{12.2}$$

$$PV = \mu RT \tag{12.3}$$

This is referred to as the *equation of state* of an ideal gas. This equation explains the macroscopic behaviour of an ideal gas.

In the microscopic view of an ideal gas, the product of the number of particles (N) and mass (m) per unit volume gives the density (ρ). The average value of the velocity of these N particles is given as follows:

$$\overline{v}_x = \frac{n_1 v_{x1} + n_2 v_{x2} + \ldots + n_N v_{xN}}{n_1 + n_2 + n_3 + \ldots + n_N}$$

$$\overline{v_x^2} = \frac{\left(v_{x1}^2 + v_{x2}^2 + \ldots\right)}{N} \tag{12.4}$$

where $v_{x1}, v_{x2}, \ldots$ are the velocities of the particles. Each particle is moving randomly in all the directions, and the motion in each direction is equally probable such that the average values of v_x^2, v_y^2, v_z^2 are equal and the value of each is exactly one-third of the average value of v^2, such that the velocity along any one dimension, let us say x, is given as follows:

$$\overline{v_x^2} = \frac{1}{3}\overline{v^2} \tag{12.5}$$

As the motions of particles are random in nature, elastic collisions occur between identical particles, that is, for every molecule that collides with a particle of momentum mv_x, there is a corresponding particle that has collided with another particle of the same momentum. This process is dynamic and on-going. The pressure generated due to this process of random collision is given as follows:

$$p = \rho\overline{v_x^2} = \rho\left(\frac{1}{3}\overline{v^2}\right) \tag{12.6}$$

The root mean square speed of the molecules is inferred as follows:

$$v_{rms} = \sqrt{\overline{v^2}} = \sqrt{\frac{3P}{\rho}} \tag{12.7}$$

In Eq. (12.7), a physically macroscopic quantity (pressure) has been related to the average value of a microscopic quantity (the velocity of the particle). Multiplying Eq. (12.6) by V we get the following equation:

$$pV = \frac{1}{3}\rho V\overline{v^2} \tag{12.8}$$

where (ρV) is the total mass (M) of the gas. If M is the molecular weight of the gas and μ is the molar mass of the gas, then

$$pV = \frac{1}{3}\mu\frac{M}{V}.V.\overline{v^2} = \frac{1}{3}\mu M\overline{v^2} \tag{12.9}$$

The quantity on the right-hand side actually represents two-third of the total kinetic energy of translation of molecules, that is,

$$pV = \frac{2}{3}\left(\frac{1}{2}\mu M\overline{v^2}\right) \tag{12.10}$$

Comparing this with the gas Eq. (12.3), one gets the following relation:

$$\frac{1}{2}M\overline{v^2} = \frac{3}{2}RT \tag{12.11}$$

Hence, it is evident from Eq. (12.11) that the total translational kinetic energy per mole of an ideal gas is proportional to its temperature.

If N_{Av} represents the Avogadro's number (i.e., the number of molecules present in one mole of a gas), then $\left(\dfrac{M}{N_0}\right) = \mu$ represents the mass of a single molecule; hence, we have the following equation:

$$\frac{1}{2}\left(\frac{M}{N_0}\right)\overline{v^2} = \frac{1}{2}\mu\overline{v^2} = \frac{3}{2}\left(\frac{R}{N_0}\right).T \tag{12.12}$$

where $\dfrac{1}{3}\mu\overline{v^2}$ represents the average translational kinetic energy per molecule.

The ratio $\left(\dfrac{R}{N_0}\right)$ is denoted as k_B, the Boltzmann constant that can be physically defined as the gas constant per molecule

$$\frac{1}{3}\mu\overline{v^2} = \frac{3}{2}k_B T \tag{12.13}$$

where $\quad k_B = \dfrac{R}{N_0} = \dfrac{8.317\,\text{J}\,\dfrac{\text{K}}{\text{mole}}}{6.023\times10^{23}\ \text{molecules/mole}}$ $\tag{12.14}$

$$= 1.38\times10^{23}\ \text{J/molecule K}$$

$$T = \frac{2}{3k_B}\frac{\mu\overline{v^2}}{3} \tag{12.15}$$

$$T = \frac{2}{3k_B}\frac{m_1\overline{v_1^2}}{2} = \frac{2}{3k_B}\frac{m_1\overline{v_2^2}}{2} \tag{12.16}$$

$$\sqrt{\frac{\overline{v_1^2}}{\overline{v_2^2}}} = \frac{v_{1\text{rms}}}{v_{2\text{rms}}} = \sqrt{\frac{m_2}{m_1}} \tag{12.17}$$

It is obviously clear from Eq. (12.17) that the gas having a heavier mass will diffuse more slowly than the gas having lower mass, whose molecules move more rapidly on an average.

A mixture of gases of different atomic weights can be separated using the principle of Eq. (12.17) for diffusion of gases. A porous barrier is created in an evacuated space. The lighter gas particles with a heavier average speed diffuse through the barrier faster. The heavier gas particles that are remaining in the chamber will now be in abundance, due to their low average speed (Fig. 12.1). A term called the 'ideal separation factor' is defined as follows:

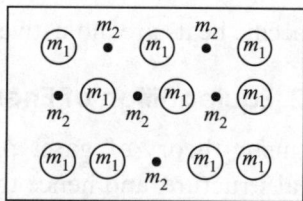

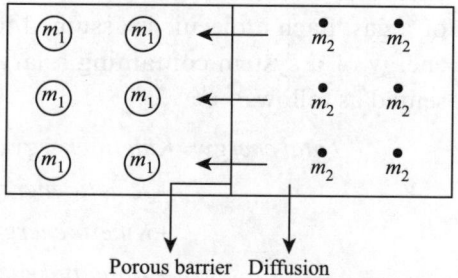

Fig. 12.1 Assembly of particles with masses m_1 and m_2: $m_2 > m_1$

$$\alpha = \sqrt{m_2/m_1} \tag{12.18}$$

where m_1 is the molecular weight of the lighter gas and m_2 that of the heavier gas.

From the aforementioned discussion, we can infer that according to the kinetic theory model, particles or molecules of an ideal gas are hard elastic spheres, possessing an average translational kinetic energy per molecule, which is $\frac{3}{2}kT$, such that the *internal energy* U of an ideal gas containing N molecules is expressed as follows:

$$U = \frac{3}{2}NkT = \frac{3}{2}\mu RT \tag{12.19}$$

Thus, the internal energy is proportional to the Kelvin temperature, independent of the pressure and volume. It is noteworthy that translational kinetic energy exists for a monatomic gas for which rotational and vibrational energies are not possible.

The specific heat of a substance can be defined as the heat required per unit mass per unit temperature change. The molar heat capacity (C) corresponds to the specific heat, being C_v for constant volume and C_p for constant pressure. The molar gas constant can be redefined as the difference between the molar heat capacity at a constant pressure and that at a constant volume:

$$(C_p - C_v) = R \tag{12.20}$$

$$\gamma = \frac{C_p}{C_v} \tag{12.21}$$

The specific heats are indicative of the absorbing energy.

12.1.2 Equipartition of Energy

The kinetic theory of gases proposes that a particle or molecule has no internal structure, and hence the total energy that it possesses is the kinetic energy of translation. Let us assume a new model for the internal energy of a gas. Each molecule is assumed to have an internal structure. The total energy of a system containing a large number of molecules can be represented as follows:

Total energy = *Kinetic energy of translation*

+ Kinetic energy of rotation

+ Kinetic energy of vibration

+ Potential energy of vibration

Therefore,

$$\text{Total energy} = \frac{1}{2}mv_x^2 + \frac{1}{2}I\omega_x^2 + \frac{1}{2}\mu v^2 + \frac{1}{2}kx^2 \tag{12.22}$$

Clark Maxwell suggested the concept of equipartition of energy, that is, the total available energy distributes itself in equal shares to each of the aforementioned ways of absorbing energy in independent ways. Each such independent mode of energy absorption is called a degree of freedom. The total available energy depends on the temperature intricately.

$$\text{Kinetic energy of translation per mole} = \frac{1}{2}m\overline{v_x^2} + \frac{1}{2}m\overline{v_y^2} + \frac{1}{2}m\overline{v_z^2}$$

(12.23)

The theorem of equipartition states that each term contributes the same amount to the total energy per mole, that is, $\frac{1}{2}RT$ per degree of freedom

For a monatomic gas, the translational degree of freedom is accounted for, that is,

$$U = \frac{3}{2}\mu RT \tag{12.24}$$

$$C_v = \frac{dU}{\mu\,dT} = \frac{d}{\mu\,dT}\left[\frac{3}{2}\mu RT\right] = \frac{3}{2}R \tag{12.25}$$

$$C_p = \frac{3}{2}R + R = \frac{5}{2}R \tag{12.26}$$

$$\gamma = \frac{C_p}{C_v} = \frac{5}{3} = 1.67 \tag{12.27}$$

For a diatomic gas, the two molecules form a dumb-bell shape that is joined by a rigid rod. The molecules can rotate about any one of the three mutually perpendicular axes. The rotational motion is accounted for along the two axes perpendicular to the axis of motion. The translational degree of freedom is accounted for in all the three directions; hence,

$$U = 3\mu\left(\frac{1}{2}RT\right) + 2\mu\left(\frac{1}{2}RT\right) = \frac{5}{2}\mu RT \tag{12.28}$$

$$C_v = \frac{dU}{\mu\,dT} = \frac{d}{\mu\,dT}\left[\frac{5}{2}\mu RT\right] = \frac{5}{2}R \tag{12.29}$$

$$C_p = \frac{5}{2}R + R = \frac{7}{2}R \tag{12.30}$$

$$\gamma = \frac{C_p}{C_v} = \frac{7}{5} = 1.40 \tag{12.31}$$

For a polyatomic gas, the molecules are more than two and hence are capable of rotating about each of the axes. Thus, the rotational and translational degrees of freedom are accounted for equally along each of the perpendicular axes; hence,

$$U = 3\mu\left(\frac{1}{2}RT\right) + 3\mu\left(\frac{1}{2}RT\right) = 3\mu RT \tag{12.32}$$

$$C_v = \frac{dU}{\mu\,dT} = \frac{d}{\mu\,dT}[3\mu RT] = 3R \tag{12.33}$$

$$C_p = 4R \tag{12.34}$$

$$\gamma = \frac{C_p}{C_v} = \frac{4}{3} = 1.33 \tag{12.35}$$

From the preceding discussion, we are able to see how the specific heat of a gas depends on the different modes of vibration. As temperature is increased, gas molecules start changing their motion from translational to vibrational. For different gases, the temperature at which this occurs may vary. As the internal energy changes with temperature, the theory of equipartition of energy predicts that it will be equally distributed among the translational, rotational, and vibrational modes of a gas. The kinetic theory follows the laws of Newtonian mechanics for a large assembly of particles.

Classical behaviour of a system ceases as its velocity approaches the velocity of light; similarly, the kinetic theory of gases takes the form of statistical mechanics under special circumstances. Predictions of quantum mechanics help us understand the atomic structure of elements. In the subsequent section, we will briefly recapitulate the relevant quantum mechanics predictions to proceed to the newly found branch of statistical mechanics.

Example 12.1   There are five types of molecules (n_i = 6, 5, 7, 6, and 6) with velocities 5, 10, 15, 20, and 25, respectively. Calculate the (a) average speed, (b) root mean square speed, and (c) the most probable speed.

Solution   the calculations are explained here.

(a) The average speed is given as follows:

$$\bar{v} = \frac{n_1 v_{x1} + n_2 v_{x2} + \ldots + n_N v_{xN}}{n_1 + n_2 + n_3 + \ldots + n_N} = \frac{(6 \times 5) + (5 \times 10) + (4 \times 15) + (6 \times 20) + (6 \times 25)}{6 + 5 + 7 + 6 + 6}$$

$$= \frac{30 + 50 + 60 + 120 + 150}{27} = \frac{410}{30} = 13.67\,\text{m/s}$$

(b) The root mean square speed is calculated as follows:

$$v_{\text{rms}} = \sqrt{\frac{n_1 v_1^2 + n_1 v_2^2 + n_1 v_3^2 + n_1 v_4^2 + n_1 v_5^2 + n_1 v_6^2}{n_1 + n_2 + n_3 + n_4 + n_5 + n_6}}$$

$$= \sqrt{\frac{6 \times 5^2 + 5 \times 10^2 + 4 \times 15^2 + 6 \times 20^2 + 6 \times 25^2}{30}}$$

$$= \sqrt{\frac{150 + 500 + 900 + 2400 + 3750}{30}}$$

$$v_{rms} = \sqrt{\frac{7700}{30}} = \sqrt{256.67} = 16.02 \, m/s$$

(c) The most probable speed is the maximum speed possessed by the maximum number of particles. In this case, it is 15 m/s.

Example 12.2 There are three molecules per cm^3 on an average in a very dilute gas. The temperature is 6 K. What is the pressure of this very dilute gas $\left(k_B = 1.38 \times 10^{-23} \, J/K\right)$?

Solution We know that

$$pV = \mu RT = \mu\left(k_B N_A\right)T = \left(\mu N_A\right)k_B T = N k_B T$$

Hence, $p = \dfrac{N}{V}k_B T = 3 \times 10^{-6} \times 1.38 \times 10^{-23} \times 6 = 24.84 \times 10^{-29} \, J - \dfrac{K}{Km^3} = \dfrac{24.84 \times 10^{-29} \, N}{m^2}$

will be the calculated pressure of the given dilute gas.

Example 12.3 The density of air at NTP is 1.29 kg/m^3. Find the rms velocity of air molecules at NTP.

Solution Let us consider the following equation:

$$p = \frac{1}{3}\rho\overline{v^2} \Rightarrow \overline{v^2} = \frac{3p}{\rho} \quad \text{Also,} \quad v_{rms} = \sqrt{\overline{v^2}} = \sqrt{\frac{3P}{\rho}}$$

At NTP, Pressure $= p = 76 cmHg = \dfrac{76 \times 10^{-2} \times 13.6 \times 10^3 \times 9.8 \, N}{m^2}$

$$v_{rms} = \sqrt{\frac{3 \times 76 \times 10^{-2} \times 13.6 \times 10^3 \times 9.8}{1.29}} = 485.35 \, m/s$$

Example 12.4 An astronaut takes a 20 L cylinder containing nitrogen gas at a temperature of 27°C and pressure of 50 atm. The astronaut makes a hole of 1 cm^2 area in this cylinder and places it in an open space. Estimate the time it would take for the cylinder to become empty $\left(k_B = 1.38 \times 10^{-23} \, J/K\right)$.

Solution We know that

$$v_{rms} = \sqrt{\frac{3RT}{M}} = \sqrt{\frac{3N_A k_B T}{M}}$$

$$v_{rms} = \sqrt{\frac{3 \times \left(6.023 \times 10^{23}\right) \times \left(1.38 \times 10^{-23}\right) \times (300)}{28 \times 10^{-3}}} = \frac{516.87 \, m}{s}$$

As the gas escapes from the cylinder, the temperature decreases such that the root mean square speed also decreases thus going to zero. The average root mean square speed is given as:

$$v = \frac{0 + 516.87}{2} = 258.25 \frac{m}{s}$$

Volume of the gas escaping per second = Area of hole $\times$ v

$$= 1 \times 10^{-4} \times 258.25 \, m/s$$

$$\text{Time taken} = \frac{\text{Total volume}}{\text{Volume of the gas escaping per second}}$$

$$= \frac{10 \times 10^{-3}}{258.25 \times 10^{-4}} = 0.387 \, s$$

Example 12.5 Estimate the total number of air molecules (where the air is constituted of particles of oxygen, nitrogen, water vapour, and other constituents) in a room of capacity 30 m³ at a temperature of 27°C and 1 atm pressure $\left(k_B = 1.38 \times 10^{-23} \, \text{J/K}\right)$.
Solution We know that $pV = Nk_BT$

$$\Rightarrow N = \frac{pV}{k_BT} = \frac{(1.01 \times 10^5) \times (30)}{(1.38 \times 10^{-23} \times 300)} = 7.32 \times 10^{26} \text{ number of air molecules}$$

Example 12.6 The temperature of the gas is –73°C. To what temperature should it be heated so that
(a) the average kinetic energy of the molecules be doubled, and
(b) the rms velocity of the molecules be doubled
Solution The calculations are as follows:

(a) $E_1 = \frac{3}{2} k_B T_1$; and $E_2 = \frac{3}{2} k_B T_2 \Rightarrow \frac{E_1}{E_2} = \frac{T_1}{T_2}$

$$\frac{E_1}{2E_1} = \frac{(273 - 73)}{T_2} = \frac{200}{T_2}$$

$$T_2 = 200 \times 2 = 400 \, K$$

(b) $v_{rms} = \sqrt{\frac{3RT}{M}} = \sqrt{\frac{3N_A k_B T}{M}}$

$$\frac{v_{rms}}{2v_{rms}} = \sqrt{\frac{T_1}{T_2}}$$

$$\frac{1}{2} = \sqrt{\frac{(273 - 73)}{T_2}} = \sqrt{\frac{200}{T_2}}$$

$$\Rightarrow T_2 = 200 x 4 = 800 K \text{ or alternatively } T_2 = 800K - 273K = 527K$$

Example 12.7 Two gases (A and B) having an equal volume diffuse at 27°C through a porous partition in 4 and 8s respectively. Molecular weight of A is 6. Find the root mean square speed of B $(R = 8.31 \text{J/mol}^{-1}/\text{K}^{-1})$
Solution If the volume of gases A and B diffused $= V$ ml

Rate of diffusion (r_A) of A $= \frac{V}{4}$ and Rate of diffusion (r_A) of B $= \frac{V}{8}$

Molecular weight of A $= M_A = 6$ and Molecular weight of B $= M_B$

Thus, $\frac{r_A}{r_B} = \sqrt{\frac{M_B}{M_A}} \Rightarrow \frac{V/4}{V/8} = \sqrt{\frac{M_B}{6}} \Rightarrow M_B = 24$

Thus, the molecular weight of B is 24.

$$v_{\text{rmsB}} = \sqrt{\frac{3RT_B}{M_B}} = \sqrt{\frac{3 \times 8.31 \times 300}{24}} = 17.65 \, \text{m/s}$$

12.2 ENERGY LEVELS, ENERGY STATES, AND DEGENERACY

Statistical mechanics tells us the state of each particle. *A particle can exist only in one of the states having discrete energies.* These discrete energy states are quantized for a given system. The *state of the particle* is defined by the quantum numbers n_x, n_y, n_z (in rectangular coordinates). Energies corresponding to the different values of n_j^2 are called 'energy levels' (ϵ_j). These energy levels depend only on the values of n_j^2 and not on the individual values of the quantum numbers, that is, the energy depends only on the magnitude of the momentum p_j and not on its direction. If a number of different states (corresponding to different directions of the momentum) have the same energy, then the energy levels are referred to as 'degenerate energy levels' (represented by g_j).

Let us consider the example of a standing wave associated with the De Broglie wavelength. The momentum associated with this wavelength is given as follows:

$$p = \frac{h}{\lambda} \tag{12.36}$$

For a standing wave,

$$\lambda_j = \frac{1}{n_j} . 2L \tag{12.37}$$

where L is the length of the string.

$$n_x = \frac{2L}{\lambda} = 1, 2, 3, \ldots = \text{Number of wavelengths in the } x\text{-direction}$$

$$n_y = \frac{2L}{\lambda} = 1, 2, 3, \ldots = \text{Number of wavelengths in the } y\text{-direction}$$

$$n_z = \frac{2L}{\lambda} = 1, 2, 3, \ldots = \text{Number of wavelengths in the } z\text{-direction}$$

The standing waves in any direction in a cubical cavity of length L is given as follows:

$$n_x^2 + n_y^2 + n_z^2 = \left(\frac{2L}{\lambda}\right)^2 \quad \text{where } n_x, n_y, n_z = 0, 1, 2, \ldots$$

The vectors n_x, n_y, n_z are drawn in a spherical shell as its coordinate axes. The permissible values of n_x, n_y, n_z represent the standing waves in this spherical cavity. For the total number of standing waves that lie between λ and $(\lambda + d\lambda)$, we count the number of permissible sets of n_x, n_y, n_z that give this wavelength in this interval. The total number of wavelengths between λ and $(\lambda + d\lambda)$ is the same as the number of points in the n-space whose distances from the origin lie between n and $(n + dn)$. In addition, for each standing wave counted in this way, there are two perpendicular directions of polarization (which constitute the multiplicative factor of 2). Thus, the number of independent standing waves in a spherical cavity is given as follows:

$$g(n)\,dn = (2).\left(\frac{1}{8}\right)\left(4\pi n^2\,dn\right)$$

$$= \pi n^2\,dn$$

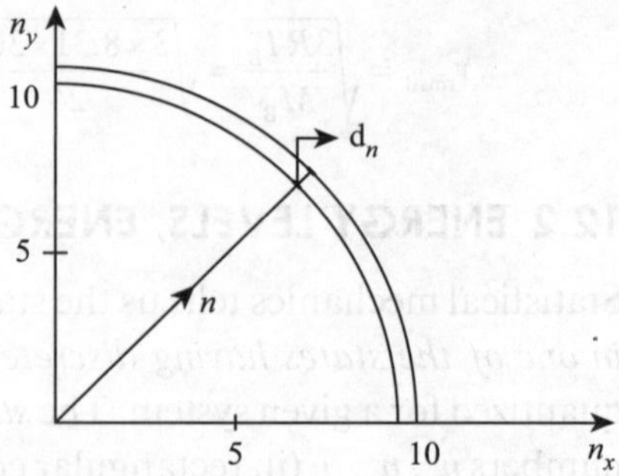

Figure 12.2 represents the *n*-space in which standing waves can be found.

Thus, by substituting Eq. (12.37) into Eq. (12.36), one gets the following relation:

Fig. 12.2 Each point in the *n* space corresponding to one possible standing wave

$$p_j = n_j\frac{h}{2L} \tag{12.38}$$

This can explicitly be expressed as follows:

$$p_x = n_x\frac{h}{2L} \tag{12.39}$$

$$p_y = n_y\frac{h}{2L} \tag{12.40}$$

$$p_z = n_z\frac{h}{2L} \tag{12.41}$$

The momentum vectors p_x, p_y, p_z are presented in Fig. 12.3, where n_x, n_y, n_z are referred to as the principal quantum numbers that take values 1, 2, 3, ... Each set of quantum numbers corresponds to a certain direction of the momentum. The resultant momentum corresponding to a set of quantum numbers n_x, n_y, n_z, is given as follows:

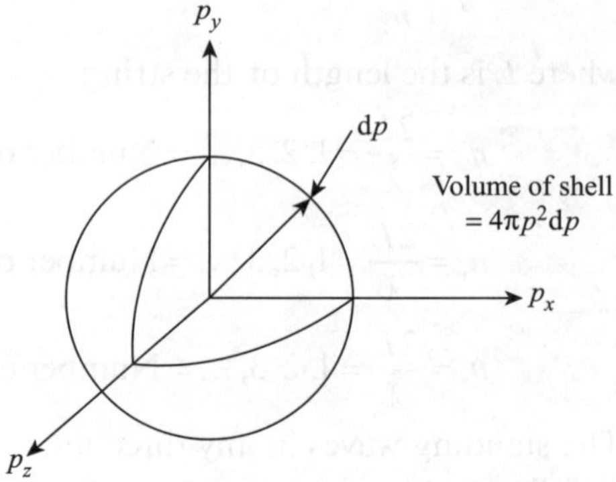

Fig. 12.3 Spherical shell representation of momentum vectors

$$p_j^2 = p_x^2 + p_y^2 + p_z^2 \tag{12.42}$$

$$p_j^2 = (n_x^2 + n_y^2 + n_z^2).\frac{h^2}{4L^2} \tag{12.43}$$

The kinetic energy can be expressed in terms of momentum as follows:

$$\epsilon_j = \frac{p_j^2}{2m} = n_j^2\frac{h^2}{8mL^2} \tag{12.44}$$

Here, for $j = 1$,

$$\epsilon_1 = n_1^2 \frac{h^2}{8mL^2} \tag{12.45}$$

This implies that

$$n_1^2 = \left(1^2 + 1^2 + 1^2\right) = 3 \tag{12.46}$$

From Eq. (12.46) it can be said only one state has energy of this magnitude $\left(h^2/8mL^2\right)$, that is, the lowest energy state is non-degenerate.

For $j = 2$,

$$\epsilon_2 = n_2^2 \frac{h^2}{8mL^2} \tag{12.47}$$

This implies that

$n_x = 1; n_y = 1; n_z = 2$ state (1–1–2)

$n_x = 2; n_y = 1; n_z = 1$ state (2–1–1)

$n_x = 1; n_y = 2; n_z = 1$ state (1–2–1)

$$n_2^2 = \left(1^2 + 1^2 + 2^2\right) = 6 \tag{12.48}$$

$$g_2 = 3$$

$$\epsilon_2 = 6 \frac{h^2}{8mL^2} \tag{12.49}$$

Thus, the second energy state has degeneracy of 3, with six times the magnitude of the lowest energy state.

For $j = 3$,

$n_x = 2; n_y = 1; n_z = 2$ state (2–1–2)

$n_x = 1; n_y = 2; n_z = 2$ state (1–2–2)

$n_x = 2; n_y = 2; n_z = 1$ state (2–2–1)

$$n_3^2 = \left(2^2 + 2^2 + 1^2\right) = 9 \tag{12.50}$$

$$g_3 = 3$$

Thus, degeneracy of the third state is 3, that is, it is triply degenerate.

For $j = 4$,

$n_x = 3; n_y = 1; n_z = 1$ state (3–1–1)

$n_x = 1; n_y = 1; n_z = 3$ state (1–1–3)

$n_x = 1; n_y = 3; n_z = 1$ state (1–3–1)

$$n_3^2 = \left(3^2 + 1^2 + 1^2\right) = 11 \tag{12.51}$$

$$g_4 = 3$$

Thus, degeneracy of the fourth state is 3.

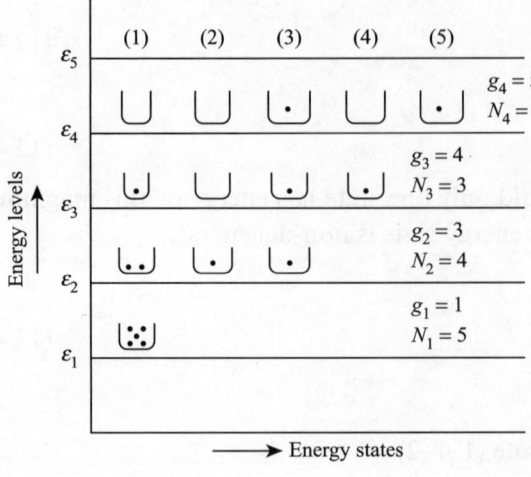

Fig. 12.4 Representation of energy levels, their degeneracies (g), and occupation number (N_i)

For $j = 5$,

$$n_x = 2n_y = 2n_z = 2 \quad \text{state (3–1–1)}$$

$$n_5^2 = \left(2^2 + 2^2 + 2^2\right) = 12 \tag{12.52}$$

$$g_4 = 1$$

It can be observed that one can work out the degeneracy of a given system using the characteristic energy value corresponding to that energy state. In the rigorous treatment of quantum mechanical problems, Schrodinger wave equation can be used to work out the characteristic energy state and value.

The aforementioned problems have been detailed to highlight the fact that energy levels are like shelves (elevated as the height varies), an energy state comprises a set of boxes on each of these shelves, and degeneracy of a level is the number of boxes on the corresponding shelf (Fig. 12.4).

12.3 SALIENT FEATURES DEFINING STATISTICAL DISTRIBUTION FUNCTIONS

Some important salient features on the basis of which statistical distribution functions are defined are as follows:

1. The total number of particles in each of the states (i.e., boxes) at any level j is called the 'occupation number' $\left(N_j\right)$ of the level. The sum of the occupation numbers N_j of a level equals the total number of particles N, that is,

$$\sum_{j=1}^{N} N_j = N \tag{12.53}$$

2. Since the particles present in one state included in any level j have the same energy ϵ_j, the total energy of the particle in level j is $\epsilon_j N_j$ and the total energy is calculated as follows:

$$\sum_{j=1}^{N} \epsilon_j N_j = E \tag{12.54}$$

In addition, if the potential energy is zero, then the total energy is equal to the internal energy, that is,

$$\sum_{j=1}^{N}\epsilon_j N_j = U \tag{12.55}$$

3. Entropy of a system is related to the probability of occurrence of an event, which in this case is the filling up of an energy state, that is,

$$S = k \ln W_k \tag{12.56}$$

where W_k is the thermodynamic probability or the statistical count. W_k is defined as *the number of equally probable microstates that can correspond to a given macrostate k.*

The total number of microstates, that is, $\left(\sum_k W_k\right)$ is equal to the thermodynamic probability of each macrostate, that is,

$$\sum_k W_k = \Omega \tag{12.57}$$

W_k is different for different statistical distribution functions.

These salient features are now consolidated by using Lagrange's method of undetermined multipliers, and from the principle of conservation of particles and energy, the following equations are obtained:

$$\alpha\left(\sum_{j=1}^{N}\delta N_j - N\right) = 0 \tag{12.58}$$

This equation indicates that the number of particles in a system is conserved.

$$\beta\left\{\left(\sum_{j=1}^{N}\epsilon_j \delta N_j\right) - E\right\} = 0 \tag{12.59}$$

This equation indicates that the total energy of the system is conserved.

$$\Delta S = 0 \tag{12.60}$$

$$\Delta k \ln W_k = 0 \tag{12.61}$$

From Eqs (12.60) and (12.61), it can be stated that *any system tends to attain the maximum entropy, that is, in thermodynamics, any equilibrium configuration is of maximum entropy.*

Combining Eqs (12.58), (12.59), and (12.61), one gets the following relation:

$$\Delta k \ln W_k + \alpha\sum_{j=1}^{N}\delta N_j + \beta\sum_{j=1}^{N}\epsilon_j \delta N_j = 0 \tag{12.62}$$

Depending on the values of α and β calculated by the Lagrange's method, we can solve Eq. (12.62).

The value of W_k varies for different distributions, that is, the thermodynamic probability W_k of a macrostate of an assembly depends on the particular statistics obeyed by the assembly. There exist three distinct ways of filling the

energy levels and states for different assembly of particles, which are cited in the following quantum statistical distribution functions:
1. The Bose–Einstein statistics
2. The Fermi–Dirac statistics
3. The Maxwell–Boltzmann statistics

12.4 BOSE–EINSTEIN STATISTICS

The Bose–Einstein statistics is obeyed by particles that have zero or integral spin (the term spin is defined conventionally, as per the atomic theory). The occupation number W_k is defined by the following criteria:
1. Particles are indistinguishable (although one can say that they are identical). In quantum mechanics, the physical state of the particle is represented by the wave function that it represents. Thus, for indistinguishable particles, wave functions must overlap.
2. There is no restriction on the number of particles that can occupy any energy state. The energy states are, however, distinguishable.
   For example, one can find the possible distinct ways in which two particles can be placed in three energy states using these two postulates (Fig. 12.5), that is,

(1)	(2)	(3)	States
× ×			$N_j = 2$ (No. of particles)
	× ×		$g_j = 3$ (states)
		× ×	Number of possible ways
×	×		$w_j = \dfrac{(g_j + N_j - 1)!}{(g_j - 1)!\, N_j!}$
×		×	
	×	×	$= \dfrac{4!}{2!2!} = 6$

Fig. 12.5 Number of ways in which two particles can be placed in three energy states using Bose–Einstein statistics

$$\sum_{j=1}^{N=2} N_j = 2 \qquad\qquad (12.63)$$

$$\sum_{j=1}^{N=3} g_j = 3 \qquad\qquad (12.64)$$

This feature of a particle, in which it is free to occupy any of the energy states, leads to the degeneracy of the energy level and is represented by the symbol g_j

Thus, the number of different distributions for the jth level, on the basis of the Bose–Einstein postulates, is given as follows:

$$W_j = \frac{(g_j + N_j - 1)!}{(g_j - 1)!\, N_j!} = \frac{(3 + 2 - 1)!}{(3 - 1)!(2)!} = \frac{4!}{2!2!} = \frac{4.3.2.1}{2.1.2.1} = 6 \qquad (12.65)$$

The occupation number W_k is determined or defined based on the way the energy states are filled. For the Bose–Einstein statistics, it is defined as follows:

$$W_{BE} = W_k = \prod_j W_j = \prod_j \frac{(g_j + N_j - 1)!}{(g_j - 1)! N_j !} \tag{12.66}$$

Note: $\prod$ is a greek symbol that symbolizes multiplication of all numbers for which j is varying.

3. A system of two particles, 1 and 2, one of which is in state a and the other in state b, is defined by the wave function φ as follows:

$$\varphi_I = \varphi_a(1)\varphi_b(2) \tag{12.67}$$

$$\varphi_{II} = \varphi_a(2)\varphi_b(1) \tag{12.68}$$

In these equations, both the states are equally probable. The wave function representing the particles obeying the Bose–Einstein statistics (φ_B) is given as the sum of Eqs (12.67) and (12.68), with a normalizing factor of $1/\sqrt{2}$ as given in Eq. (12.69). This is because the particles are not distinguishable and can occupy the states in either way, that is, both the possibilities are valid.

$$\varphi_B = \frac{1}{\sqrt{2}}\{\varphi_a(1)\varphi_b(2) + \varphi_a(2)\varphi_b(1)\} \tag{12.69}$$

4. Particles having zero or integral spin obey the Bose–Einstein statistics.
5. Particles that obey the Bose–Einstein statistics are called *bosons*.
6. Examples of particles that obey the Bose–Einstein statistics are photons, phonons, pions, and α-particles. This statistics is also used for studying the frequency spectrum of blackbody radiation (refer to applications of statistical mechanics at the end of the chapter).
7. The occupation number W_k is used to define the Bose–Einstein statistical distribution function:

$$f_{BE}(\epsilon_j) = \frac{1}{e^\alpha e^{\beta \epsilon_j} - 1} \tag{12.70}$$

where $\beta = \dfrac{1}{k_B T}$

Using the value of β in Eq. (12.70), we get the following expression:

$$f_{BE}(\epsilon_j) = \frac{1}{e^\alpha e^{\epsilon_j / kT} - 1} \tag{12.71}$$

12.4.1 Bose–Einstein Condensation

We are interested in the low-temperature behaviour of phonons. The Bose–Einstein statistics does not obey the exclusion principle, that is, more than one phonon can occupy the same energy state, as shown in Fig 12.6(a). This property

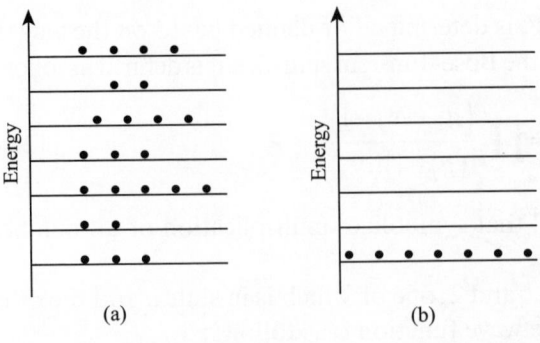

Fig. 12.6 Phonons at same energy levels (a) Phonon distribution at normal temperature (b) Bose–Einstein condensation

of bosons helps them to occupy the lowest energy state or the ground state as the temperature is lowered. This is shown in Fig. 12.6; bosons behave as a single particle, as they all possess the same wave function and the same velocity.

In the discussion that follows, we evaluate the temperature for which the atoms in a gas tend to form a conglomeration in its ground state. This conglomeration of atoms in the ground state at a particular temperature is termed as the 'Bose–Einstein condensate'.

The number of atoms in the energy state ϵ, with a density of state $g(\epsilon)$, is given as follows:

$$n(\epsilon)d\epsilon = \frac{g(\epsilon)d\epsilon}{\left(\dfrac{\exp(\epsilon - \alpha)}{k_B T} - 1\right)} \tag{12.72}$$

As we are dealing with the atoms of a gas, the density of state is taken for an ideal gas, that is,

$$g(\epsilon) = \frac{4m\pi V}{h^3}(2m\epsilon)^{\frac{1}{2}} \tag{12.73}$$

The Lagrange's multiplier α, in this case in Eq. (12.72), is taken to be the chemical potential μ. The chemical potential for an atom in the ground state at low temperatures is taken to be zero, that is, $\mu = 0$. As the energy of the total number of particles (N) ranges from the ground state to higher energy states, the limit of integration ranges from 0 to ∞.

$$N = \int_0^\infty \frac{g(\epsilon)d\epsilon}{\exp(\epsilon / k_B T) - 1} \tag{12.74}$$

Equation (12.74) represents the number of atoms in the excited state N_{ex} of an ideal gas, following the Bose–Einstein statistics. This number is very large, approximately of the order of a mole, that is, $\sim 10^{24}$ atoms. Using the standard integral tables, one can estimate N_{ex} from Eq. (12.74) as follows:

$$N_{ex} = \left(\frac{2\pi mk_B T}{h^2}\right)^{\frac{3}{2}} 2.612\, V \tag{12.75}$$

As the temperature decreases, atoms in the excited state start disappearing. In fact, the integral includes the atoms above the ground state $\epsilon = 0$. Thus, *the missing atoms are in fact going back to the ground state.* Thus, the number of atoms in the excited state N_{ex} starts becoming less than that in the original ground state (N). Thus, $(N - N_{ex})$ atoms start going back to the ground state. The condensation takes place when $N = N_{ex}$. Thus, from Eq. (12.75), the temperature T_{BE} can be estimated to be the following:

$$T_{BE} = \frac{h^2}{2\pi m k_B}\left(\frac{N}{2.612V}\right)^{2/3} \tag{12.76}$$

Above T_{BE}, there are very few atoms in the ground state; below T_{BE}, this number increases until all the atoms fall into the ground state. When this state is achieved, it is referred to as the *Bose–Einstein condensation.*

This peculiar phenomenon, which is a characteristic of bosons, has successfully explained the phenomenon of superconductivity.

The phenomenon of Bose—Einstein condensation was first observed by Cornell and Weirmann for rubidium gas with a condensation temperature of 1.7×10^{-7} K. This is far below the temperature that can be reached using liquid helium (melting point 4.2 K.)

Note: Bosons are categorized into two types: virtual and real bosons. Virtual bosons are the phonons and photons that disappear as the object is heated or cooled. Real bosons are the atoms that possess the integral spin number.

12.4.2 Specific Heat Capacity of Solids

We know that the atoms in a solid are fixed at the edge and vibrate. The quantum of vibration energy is called a 'phonon'. As the phonons obey the Bose–Einstein statistics, more than one phonon can occupy the same energy state, as shown in Fig. 12.7. The number of atoms in the energy state ϵ is given as follows:

$$N(\epsilon)\,d\epsilon = \frac{g(\epsilon)\,d\epsilon}{\exp(\epsilon/k_B T) - 1}$$

Hence, $\epsilon = \hbar\omega$ for a phonon.

Atoms in a solid behave as simple harmonic oscillators. The phonon mode is simply vibration of the atoms coordinated in a particular way. The corresponding energy takes the following form:

$$E_n = \left(n + \frac{1}{2}\right)\hbar\omega$$

Thus, phonon distribution takes the following form:

$$N(\omega)\,d\omega = \frac{g(\omega)\,d\omega}{\exp(\hbar\omega/k_B T) - 1}$$

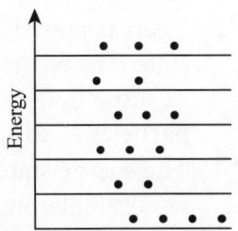

Fig. 12.7 Phonons can occupy same energy states

$$\hbar\omega N(\omega)\,d\omega = \frac{g(\omega)\,d\omega}{\exp(\hbar\omega/k_BT)-1}$$

The total energy is then given as follows:

$$U = \int_0^{\omega_D} \hbar\omega N(\omega)\,d\omega = \int_0^{\omega_D} \frac{\hbar\omega g(\omega)\,d\omega}{\exp(\hbar\omega/k_BT)-1}$$

The density of states is given as follows: $g(\omega) = 3 \times \dfrac{V\omega^2}{2\pi^2 \vartheta^3}$

The factor of 3 corresponds to a wave in a solid that can have two transverse polarizations and one longitudinal polarization. In addition, the three different directions of vibrations are in fact different states with the same wave vector (n_x, n_y, n_z).

Substituting this value of density of states in the preceding integral for the total energy, the following expression is obtained:

$$U = \frac{3Vh}{2\pi^2\vartheta^3} \int_0^{\omega_D} \frac{\omega^3\,d\omega}{(\exp(\hbar\omega/k_BT)-1)}$$

At a low temperature, the specific heat capacity $C \propto T^3$, as the total energy is proportional to T^4. At high temperatures, the heat capacity tends to be $3Nk_B$.

These results have experimentally been proved by measurements of the heat capacity for copper at various, particularly high temperatures.

12.5 FERMI–DIRAC STATISTICS

The Fermi–Dirac statistics is obeyed by particles having half-integral spins (the term spin is defined conventionally, as per the atomic theory). The occupation number W_k is defined by the following criteria:

1. The particles are indistinguishable (although one can say that they are identical). Quantum mechanics identifies the physical state of a particle with the help of the wave function. Thus, for indistinguishable particles, the wave functions must overlap.
2. There is a restriction on the number of particles that can occupy any energy state. The particles obey the Pauli exclusion principle, that is, there cannot be more than one particle in each energy state. Evidently, the number of particles N_j in any level cannot exceed the number of states g_j in that level.
3. The energy states are however distinguishable.

   For example, one can find the possible distinct ways in which two particles can be placed in three energy states following the Fermi–Dirac postulate (Fig. 12.8):

$$\sum_{j=1}^{N=2} N_j = 2 \tag{12.77}$$

$$\sum_{j=1}^{N=2} g_j = 3 \tag{12.78}$$

$$W_j = \frac{(g_j)!}{(g_j - N_j)! N_j!} = \frac{(3)!}{(3-2)!(2)!}$$

$$= \frac{3!}{1!2!} = \frac{4.3.2.1}{2.1.2.1} = 3 \tag{12.79}$$

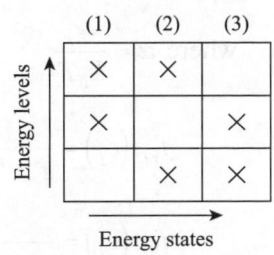

Fig. 12.8 Number of ways in which two particles can be placed in three energy states using Fermi–Dirac statistics

The occupation number W_k is determined or defined by the way the energy states are filled. For the Fermi–Dirac statistics, it is defined as follows:

$$W_{\text{FD}} = W_k = \prod_j W_j = \prod_j \frac{(g_j)!}{(g_j - N_j)! N_j!} \tag{12.80}$$

4. For particles having half integral spins, the wave function is anti-symmetric. In the wave function of a system of fermions, if there is an exchange of state for any pair of fermions, the wave function will change its sign because each state is unique. Hence, for a system of two particles 1 and 2, one of which is in state a and the other in state b, the following relation holds:

$$\varphi_{\text{I}} = \varphi_a(1)\varphi_b(2) \tag{12.81}$$

$$\varphi_{\text{II}} = \varphi_a(2)\varphi_b(1) \tag{12.82}$$

In these equations both the states are equally probable. The wave function representing the particles obeying Fermi–Dirac statistics (φ_F) is given as the sum of Eqs (12.81) and (12.82), with a normalizing factor of $1/\sqrt{2}$. This is because the particles are not distinguishable and can occupy the states only discretely, obeying the exclusion principle. Thus, for fermions,

$$\varphi_F = \frac{1}{\sqrt{2}} \{ \varphi_a(1)\varphi_b(2) - \varphi_a(2)\varphi_b(1) \} \tag{12.83}$$

5. Particles that have odd half-integral spins obey such statistics.
6. Particles that obey the Fermi–Dirac statistics are called fermions.
7. Examples of particles that obey the Fermi–Dirac statistics are electrons, muons, protons, neutrons, and in general all particles of one-half integral spins.
8. The occupation number W_k is used to define the Fermi-Dirac statistical distribution function:

$$f_{\text{FD}}(\epsilon_j) = \frac{1}{e^\alpha e^{\beta \epsilon_j} + 1} \tag{12.84}$$

where $\alpha = \dfrac{-\epsilon_F}{k_B T}$

$$f_{\text{FD}}\left(\epsilon_j\right) = \frac{1}{e^{-\epsilon_F/k_B T} e^{\epsilon_j/k_B T} + 1} \tag{12.85}$$

$$f_{\text{FD}}\left(\epsilon_j\right) = \frac{1}{e^{(\epsilon_j - \epsilon_F)/k_B T} + 1} \tag{12.86}$$

From the frequency distribution function for Fermi–Dirac statistics, it is evident that the distribution function is simply the fraction of g_j states that are filled. Thus, $f_{\text{FD}}\left(\epsilon_j\right)$ is the probability that a state of energy ϵ_j is filled. This is owing to the fact that the energy levels have been filled obeying Pauli's' principle.

For $\epsilon_j = \epsilon_f$,

$$f_{\text{FD}}\left(\epsilon_j\right) = \frac{1}{1+1} = \frac{1}{2} \tag{12.87}$$

that is, the Fermi–Dirac statistics predict that if the jth energy level is the Fermi level, then the probability of occupancy of the energy levels is one-half. The term Fermi energy is an important term/quantity in a system of fermions such as electron gas in a metal. It is conventionally referred to mark the uppermost filled energy band (or level) for electron gases, in particular generalizing to assemblies with $\left(\dfrac{1}{2}, \dfrac{3}{2}, \dfrac{5}{2}, \ldots\right)$ or half integral spins. Let us further understand this by considering a system of fermions at an absolute temperature $T = 0$ and investigate the occupancy of states whose energies are less than and greater than ϵ_F. We find that

1. When $T = 0$ and $\epsilon < \epsilon_F$, then

$$f_{\text{FD}}\left(\epsilon_j\right) = \frac{1}{e^{(\epsilon_j - \epsilon_F)/k_B T} + 1} \tag{12.88}$$

$$f_{\text{FD}}\left(\epsilon_j\right) = \frac{1}{e^{-\infty} + 1} = 1 \tag{12.89}$$

2. When $T = 0$ and $\epsilon > \epsilon_F$, then

$$f_{\text{FD}}\left(\epsilon_j\right) = \frac{1}{e^{(\epsilon_j - \epsilon_F)/k_B T} + 1} \tag{12.90}$$

$$f_{\text{FD}}\left(\epsilon_j\right) = \frac{1}{e^{\infty} + 1} = 0 \tag{12.91}$$

This implies that all the filled energy levels/states lie below the Fermi energy level. Above the Fermi energy level, all the states are unoccupied. The Fermi energy level is akin to the water-table mark. For the electron filled energy levels the Fermi energy marks the uppermost level. As the water bodies are brought together, the water-table mark balances, that is, the water flows from the higher level to the lower level, so also for the electron gas assemblies.

12.5.1 Fermi Dirac Statistics as Applied to Free Electrons

The Fermi Dirac statistics with the Pauli exclusion principle at its core made it possible to understand the behaviour of metals. The previous theories of Einstein and Debye could not explain the specific heats for metals, as they both consider particles as bosons, unlike the Femi–Dirac model that treats metals as electron gases. The inherent property of metals to have free electrons contributes to the electron gas picture. Thus, metals are akin to a sea of electrons that are the fermions.

From the Fermi–Dirac statistics, the average occupancy of a state of energy ϵ for fermions is given as follows:

$$f_{\text{FD}}(\epsilon) = \frac{1}{e^{(\epsilon - \epsilon_F)/k_B T} + 1} \tag{12.92}$$

Fermions, being half-integral particles, possess a state of spin that is referred to as the spin up and the spin down. This corresponds to our earlier discussion in the chapter where the standing waves in a spherical cavity have two states of polarization. Thus, to evaluate the number of quantum states available to electrons with energies ϵ and $(\epsilon + d\epsilon)$ the expression for $g(\epsilon)d\epsilon$ should be identical to the expression discussed in the section for standing waves in a cubical cavity L, that is,

$$g(\epsilon) \, d\epsilon = g(n) \, dn = \pi n^2 \, dn \tag{12.93}$$

If we assume that an electron or a fermion possesses a De Broglie wavelength, then $\lambda = h/p$ such that

$$n = \frac{2L}{\lambda} = \frac{2Lp}{h} = \frac{2L\sqrt{2m\epsilon}}{h} \tag{12.94}$$

$$dn = \frac{L}{h}\sqrt{\frac{2m}{\epsilon}} \, d\epsilon \tag{12.95}$$

Substituting Eqs (12.94) and (12.95) in Eq. (12.93), we get the following expression:

$$g(\epsilon) \, d\epsilon = \pi \left(\frac{2L\sqrt{2m\epsilon}}{h}\right)^2 \frac{L}{h}\sqrt{\frac{2m}{\epsilon}} \, d\epsilon = \frac{8\sqrt{2}\pi L^3 m^{3/2}}{h^3}\sqrt{\epsilon} \, d\epsilon \tag{12.96}$$

$$g(\epsilon) \, d\epsilon = \frac{8\sqrt{2}\,\pi V m^{3/2}}{h^3}\sqrt{\epsilon} \, d\epsilon \tag{12.97}$$

Where V represents the volume of a cavity where the electron gas is present.

The energy levels in an electron gas are filled from the lowest energy level at 0 to the highest energy level ϵ_F. In the case of an electron gas or a Fermion system, the number of energy states is equal to the number of electrons or fermions present in the system. The total number of electrons present is calculated as follows:

$$N = \int_0^{\epsilon_F} g(\epsilon)\, d\epsilon = \frac{8\sqrt{2}\,\pi V m^{3/2}}{h^3} \int_0^{\epsilon_F} \sqrt{\epsilon}\, d\epsilon = \frac{16\sqrt{2}\,\pi V\, m^{3/2}}{3h^3} \epsilon_F^{3/2} \quad (12.98)$$

Thus, the Fermi energy for an electron gas can be formulated as follows:

$$\epsilon_F = \frac{h^2}{2m}\left(\frac{3N}{8\pi V}\right)^{2/3} \quad (12.99)$$

The Fermi energy expression clearly indicates that it is keenly dependent on the density of the system. For a monovalent metal, that is, an ion giving one electron, the typical metallic density is $\sim 10^{29}$ m^{-3}:

$$\epsilon_F = \frac{(6.634\times 10^{-34}\ \text{J s})^2}{2\times\left(\dfrac{9.11\times 10^{-31}\text{kg}}{\text{electron}}\right)}\left(\frac{3}{8\pi}\times\frac{10^{29}\,\text{electrons}}{\text{m}^3}\right)^{2/3} \approx 8\times 10^{-19}\ \text{J} \approx 5\,\text{eV}$$

This energy is extremely large, roughly about 200 times more than the energy of the molecule of air at room temperature. This is the picture of an electron gas at the absolute zero.

The average energy of electrons in an electron gas can also be evaluated and is seen to be different from $k_B T$. The number of electrons that have energies between ϵ and $\epsilon + d\epsilon$ is calculated as follows:

$$N(\epsilon)\, d\epsilon = g(\epsilon) f(\epsilon)\, d\epsilon$$

$$= \frac{8\sqrt{2}\,\pi V m^{3/2}}{h^3}\sqrt{\epsilon}\, d\epsilon\,\frac{1}{e^{(\epsilon-\epsilon_F)/k_B T}+1} = \frac{3N}{2}\left(\epsilon_F\right)^{-3/2}\frac{\sqrt{\epsilon}\, d\epsilon}{e^{(\epsilon-\epsilon_F)/k_B T}+1}$$

$$(12.100)$$

The plot of energy versus the number density is shown in Fig. 12.9. The total energy at an absolute temperature is given as follows:

$$E_T = \int_0^{\epsilon_F} \epsilon N(\epsilon)\, d\epsilon = \frac{3N}{2}\left(\epsilon_F\right)^{-3/2}\int_0^{\epsilon_F} \epsilon\,\frac{\sqrt{\epsilon}\, d\epsilon}{e^{(\epsilon-\epsilon_F)/k_B T}+1} \quad (12.101)$$

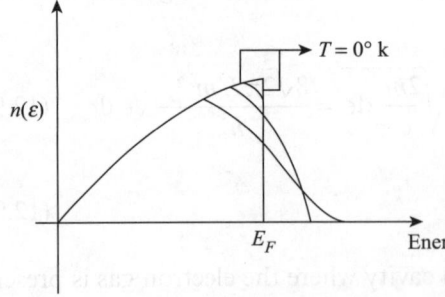

Fig. 12.9 Translational energy distribution of free electrons changes with temperature

As the exponential term in the denominator tends to zero for $\epsilon = \epsilon_F$, one gets the following expression:

$$E_T = \frac{3N}{2}\epsilon_F^{-3/2} \int_0^{\epsilon_F} \epsilon^{3/2} d\epsilon = \frac{3}{5}N\epsilon_F \tag{12.102}$$

where the average electron energy at $T = 0\,K$ is given as follows:

$$E_T = \frac{3}{5}\epsilon_F \tag{12.103}$$

Electron gas energy at the absolute zero has already been established; we now look at this system as the temperature is increased from the absolute zero. In the case of a metal, as the temperature is increased from the absolute zero, electrons gain an energy of the order of $k_B T$ (this is expected classically). Actually, only those electrons within an energy range $k_B T$ of the Fermi level are excited thermally. Thus, out of the total number of electrons N, only NT/T_F electrons will get thermally excited. Each of these $\left(\dfrac{NT}{T_F}\right)$ electrons has a thermal energy of the order of $k_B T$. The same can be observed from the graph shown in Fig. 12.9. Thus,

$$U \approx N\left(\frac{T}{T_F}\right)k_B T \tag{12.104}$$

$$C_{vE} = \frac{dU}{dT} \cong Nk_B\left(\frac{T}{T_F}\right) \tag{12.105}$$

Here, the new term T_F referred to as the Fermi temperature is actually defined as follows:

$$T_F = \frac{E_F}{k_B} \tag{12.106}$$

Thus, one observes that metals remain at room temperature even though their Fermi energy levels are very high at the absolute zero. At room temperature, the specific heat ranges from 0.016 for cesium to 0.0021 for aluminium. At low temperatures, the specific heat is proportional to T^3. At high temperatures, the contribution of the specific heat starts exhibiting itself to the overall value of the specific heat.

12.5.2 Fermi Energies and Free Electron Gas

Field emission is a phenomenon in which the electrons can tunnel through a barrier in the presence of a large electric field gradient. Metals with a very high work function or very high Fermi energies are preferred. Filed emission is commonly used in modern-day technology; examples include filed emission displays and flash memory (a type of read-only memory commonly used in memory cards, digital cameras, mobile phones, and so on).

Field emission microscopy is used instead of a cathode ray tube (CRT) electron gun; it uses a large array of tubes (carbon nanotubes) placed behind the phosphor dots. As they undergo field emission, the liberated electrons illuminate the screen, as was done earlier by the solid-state CRT.

Example 12.8 There are three assemblies of a gas of classical molecules, bosons, and fermions at the same temperature. Which of the assemblies exerts the maximum and the least pressures? Justify your answer.

Solution In an assembly of classical molecules, the energy possessed by the molecules by virtue of the Maxwell–Boltzmann distribution function is about $\dfrac{3}{2}k_BT$.

The energy possessed by the molecules in an assembly of bosons will be a large number of particles in their ground state or the lowest possible energy state. Thus, the assembly of bosons will exert the least pressure.

The energy possessed by the molecules in an assembly of fermions will have electrons stacked at the most two in an energy level such that average electron energy is about $\dfrac{3}{5}\epsilon_F$. Fermi levels usually have very high energies compared to those of bosons or classical molecules. Thus, the assembly of fermions exert the maximum pressure.

Example 12.9 Find the Fermi energy of copper, given that its density is $7.13\,\text{g/cm}^3$ and its atomic mass is $65.4u$. (The copper atom possesses two 4s electrons.) The effective mass of an electron in copper is $0.85\,m_e$.

Solution The electron density (N/V) in zinc is given as follows:

$$\frac{N}{V}=\frac{\text{Mass/cm}^3}{\text{Mass/atom}}=\frac{7130\,\text{kg/m}^3}{(65.4u)\times\left(\dfrac{1.66\times10^{-27}\,\text{kg}}{u}\right)}=6.5675\times10^{28}\ \text{electrons/m}^3$$

The Fermi energy is given as follows:

$$\epsilon_F=\frac{(6.634\times10^{-34}\,\text{Js})^2}{2\times0.85\times9.11\times10^{-31}\,\text{kg/electron}}\left[\frac{7\times3\times6.5675\times10^{28}}{8\times22}\right]^{\frac{2}{3}}$$

$$\epsilon_F=2.841\times10^{-37}\times\left(0.783\times10^{28}\right)^{\frac{2}{3}}=2.841\times10^{-37}\times(0.614\times10^{56})^{\frac{1}{3}}$$

$$\epsilon_F=2.841\times10^{-37}\times3.94\times10^{18}=11.207\times10^{-19}\,J=7.00\,\text{eV}$$

Thus, the electrons present in copper have a Fermi energy of 7.00 eV but one must remember that this is at the absolute zero.

Example 12.10 As calculated in the previous example, the Fermi energy in copper is 7.00 eV.
(a) Calculate the average energy of free electrons present in copper at room temperature if they follow the Maxwell–Boltzmann statistics.
(b) Calculate the energy states below the Fermi energy level for a volume of 1 cm³.

Solution The following are the calculations:
(a) For room temperature, let $T = 20°C = 293\,\text{K}$.

$$\text{Hence, } k_BT = 1.381\times\frac{10^{-23}\,J}{K}\times293\,\text{K} = 4.04\times10^{-21}\,J = 0.025\,\text{eV}$$

Average energy for a free electron following the Maxwell–Boltzmann statistics is calculated as follows:

$$= \frac{3}{2} k_B T = \frac{3}{2} \times 0.025 \, \text{eV} = 0.0375 \, \text{eV}$$

(b) The energy states below the Fermi Energy level are as follows:

$$g(\epsilon) d = \frac{8\sqrt{2} \, \pi V m^{3/2}}{h^3} \sqrt{\epsilon} \approx \frac{8\sqrt{2} \times 22 \times 10^{-6} \, \text{m}^3 (9.1 \times 10^{-31} \, \text{kg})^{3/2}}{7 \times (6.634 \times 10^{-34} \, \text{Js})^3}$$

$$\approx 2 \times 10^{22} \, \text{states/eV}$$

Thus, the number of available energy states below the Fermi energy level is nearly equal to that of particles present. Hence, the system of an electron gas or fermions is a nearly continuous system.

Example 12.11 Calculate the following parameters:
(a) Average energy of free electrons in silver at 0 K
(b) Speed of the electron possessing this energy
(c) Temperature for which particles of silver possess this average molecular energy in an ideal gas
(Assume the Fermi energy of silver to be 5.51 eV.)

Solution The parameter are calculated as follows:

(a) Average energy of the free electrons of silver $= \frac{3}{5} \epsilon_F = \frac{3}{5} \times 5.51 \, \text{eV} = 3.306 \, \text{eV}$

(b) Speed of the electron $= \frac{1}{2} m_e v_f^2 = \frac{1}{2} \times 9.1 \times 10^{-31} \, \text{kg} \times v_f^2 = 3.306 \, \text{eV}$

$$\Rightarrow v_f = \sqrt{\frac{2 \times 3.306 \times 1.6 \times 10^{-19} \, \text{J}}{9.1 \times 10^{-31} \, \text{kg}}} = 1.08 \times 10^6 \, \text{m/s}$$

(c) Average molecular energy $= \frac{3}{2} k_B T = 3.306 \, \text{eV}$

$$\Rightarrow T = 3.306 \times 1.6 \times 10^{-19} \, \text{J} \times \frac{2}{3} \times \frac{1}{1.380 \times 10^{-23} \, \text{J/K}} = 2.56 \times 10^4 \, \text{K}$$

12.6 MAXWELL–BOLTZMANN STATISTICS

Classical particles obey the Maxwell–Boltzmann statistics, for example, gas molecules. The occupation number W_k is defined by the following criteria:

1. The particles are at a considerable distance from each other, for example, the molecules of a gas. Such particles are said to be distinguishable, although they are identical.
2. There is a restriction on the number of particles that can occupy the same energy state; for example, the possible distinct ways in which two particles can be placed in three energy states are as follows:

$$\sum_{j=1}^{N=2} N_j = 2 \tag{12.107}$$

$$\sum_{j=1}^{N=2} g_j = 3 \tag{12.108}$$

As we know that the particles are distinguishable, let us denote the two particles distinctly as 'a' and 'b', as shown in Fig. 12.10. For the particles to follow the Maxwell–Boltzmann postulates, the following relation applies:

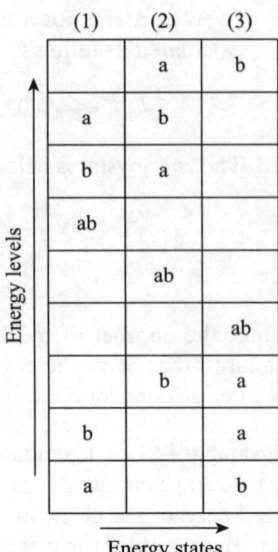

$$W_j = \frac{(N)!}{\Pi_j N_j!} \Pi_j g_j^{N_j}$$

$$= \frac{2!}{\Pi_j 2!} \Pi_j 3^2 = 9 \tag{12.109}$$

There are nine distinct ways to arrange two particles as per the Maxwell–Boltzmann statistics.

The occupation number W_k is determined or defined by the way the energy states are filled. For the Maxwell–Boltzmann statistics, it is defined as follows:

Fig. 12.10 Ways in which two particles can be placed in three energy states using Maxwell–Boltzmann statistics

$$W_j = \frac{(N)!}{\Pi_j N_j!} \Pi_j g_j^{N_j}$$

$$= N! \Pi_j \frac{g_j^{N_j}}{N_j!} \tag{12.110}$$

3. The Maxwell–Boltzmann distribution function is a purely exponentially decreasing function, decreasing by a factor of $(1/e)$ for every increase in ϵ by a factor of $(k_B T)$. (The details of the Boltzmann constant are given in Section 12.1.) Figure 12.11 shows a graph representing the Maxwell–Boltzmann energy distribution:

$$f_{MB}(\epsilon_j) = A e^{-\epsilon_j / k_B T} \tag{12.111}$$

If we compare the two energy levels, then we get the following relation:

$$\frac{f(\epsilon_i)}{f(\epsilon_j)} = e^{-(\epsilon_i - \epsilon_j)/k_B T} \tag{12.112}$$

$$= e^{-\Delta\epsilon / k_B T} \tag{12.113}$$

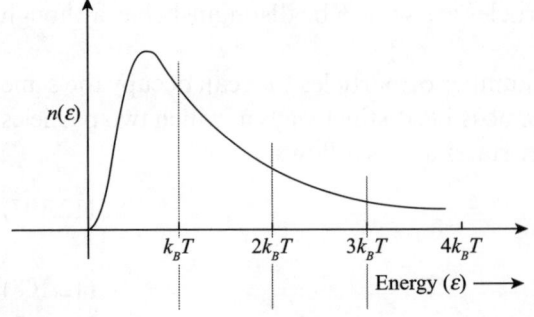

Fig. 12.11 Maxwell–Boltzmann energy distribution

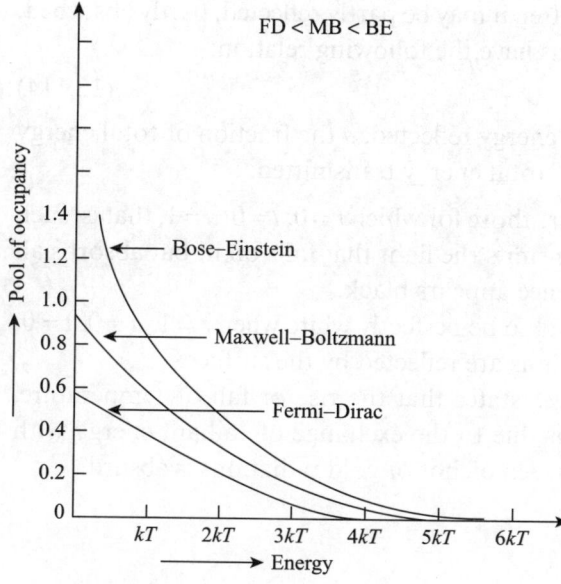

Fig. 12.12 Various statistical functions give the probability of occupancy of a state of energy (ϵ) at an absolute temperature (T)

4. In quantum terms, the wave functions of particles (distinguishable particles) overlap to a negligible extent, that is, there is no overlapping of wave functions.
5. Particles that obey these statistics are generally classical particles, for example, particles or molecules of a gas.
   Having studied the three statistical distributions in detail, a comparative graph for the individual distribution functions can be plotted (Fig. 12.12).

12.7 CLASSICAL THEORY OF RADIATION

Loss of heat from a body in the form of quanta of energy is referred to as the heat loss by radiation. The radiant energy has same properties and obeys the same laws as light. These properties and laws are as follows:

1. Radiant energy (or radiation) travels in straight lines and forms shadows when obstructed, according to the laws of geometric optics. It can travel through vacuum.
2. The intensity of radiant energy follows the law of inverse square.
3. Radiant energy obeys the basic laws of reflection of light.
4. It suffers from refraction. A spectrum is obtained by refraction of radiation energy through a prism.
5. Radiation energy or thermal radiation can be polarized as light by transmission through a tourmaline crystal or Nicol prism.
6. Radiation travels very quickly, like light as when observed that there is obstruction of the radiant energy coming from the sun by clouds or other means, it is immediately accompanied by a fall in temperature.

Substances that allow the passage of the radiant energy are referred to as *diathermanous* and objects obstructing the passage are called *athermanous*.

When radiation falls on the matter, it may be partly reflected, partly absorbed, and partly transmitted. We thus have the following relation:

$$r + a + t = 1 \tag{12.114}$$

where r is the fraction of total energy reflected, a the fraction of total energy absorbed, and t the fraction of total energy transmitted.

Notes: *Perfectly black bodies* are those for which $r = 0$, $t = 0$, $a = 1$, that is, such a body neither reflects nor transmits the light that falls on it, but absorbs all the incoming radiation and hence appears black.

Perfectly white bodies are said to be perfectly white when $r = 1$, $a = 0$, $t = 0$, that is, all the incoming radiations are reflected by the surface.

Prevost's theory of exchanges states that the rise or fall of temperature, which is observed in a body, is due to the exchange of radiant energy with surrounding bodies. Thus, the idea of hot or cold radiations is absurd.

12.7.1 Kirchoff's Law

Let us try to understand the exchange of energy between radiation and matter with the following example. We take an enclosed space such that the walls are opaque to radiations of all wavelengths. The walls are first subjected to a uniform temperature. Then the walls are thermally insulated from the surroundings. The whole enclosed space is filled with the radiations emitted by the walls. Now, if the body is of a heterogeneous composition, then different parts of the body will have different emissive and absorptive powers. When the body is in a state of equilibrium, the total energy absorbed by the body will be equal to the energy it emits. It is also known that the energy emitted by a body always remains the same, irrespective of the orientation or position of the body with respect to the enclosure. As the different surfaces of the body have different coefficients of absorption, the total energy absorbed by the body remains the same. This condition holds only if the radiation travelling inside the enclosure in different directions is identical in quantity and quality (i.e., isotropic).

The following statements are derived from the Kirchhoff's law:
1. *The radiation inside a hollow enclosure is independent of the nature or the geometrical shape of the walls of the enclosure or a body placed inside it.*
2. *At any temperature, the ratio of the emissive power of a substance to its absorptive power is constant and equal to the emissive power of a perfect black body.*

Mathematical interpretation of the Kirchhoff's law can be understood as follows. Let us consider an enclosure filled with radiation having wavelengths between λ and $(\lambda + \Delta\lambda)$, and let a body be placed in it. Then from the preceding discussion, we can say that

Energy emitted by the body = Energy absorbed by the body

that is,

$$e_\lambda \, \mathrm{d}\lambda = a_\lambda \, \mathrm{d}Q \tag{12.115}$$

where e_λ is the emissive power of the body, a_λ is the absorptive power of the body, and dQ is the energy of the incident radiation.

$$\frac{e_\lambda}{a_\lambda} = \frac{dQ}{d\lambda} \tag{12.116}$$

In the case of a perfect black body with emissivity E_λ (as $a_\lambda = 1$), we get the following relations:

$$E_\lambda d\lambda = 1.dQ \tag{12.117}$$

$$\frac{dQ}{d\lambda} = E_\lambda \tag{12.118}$$

Equating Eqs (12.116)–(12.118), we have the following expression:

$$\frac{e_\lambda}{a_\lambda} = E_\lambda \tag{12.119}$$

Thus, the emissivity of a perfect black body is the ratio of the emissive power of the body to its absorptive power.

We can observe that the Kirchhoff's law embodies two distinct relations— qualitative and quantitative. *Qualitatively, it implies that if a body is capable of emitting certain radiations, it will absorb them when they fall on it. Quantitatively, it signifies that the ratio is the same for all the bodies.*

The most important conclusion given by Kirchhoff's laws is about the solar spectrum. Fraunhofer, a German scientist, observed the solar spectrum and found that it was not a continuous spectrum but consisted of certain 'dark bands', a phenomenon that he was not able to explain at that time. When Kirchhoff observed the spectrum, with the help of his laws, he explained that the sunlight has a continuous spectrum; however, as the rays pass through a cooler atmosphere surrounding the central mass, which contains Na, Cu, etc., in gaseous form, these atoms or substances absorb rays of specific frequencies, resulting in the appearance of dark lines.

Thus, *for the first time it was asserted that every different type of atom, when it is properly excited, emits a wavelength that is characteristic of that atom.*

12.7.2 Stefan–Boltzman Law

The Stefan–Boltzman law is a mathematical representation of heat transfer by radiation. The rate at which heat is emitted by the electromagnetic radiation of an object of surface area A and absolute temperature T_e is given as follows:

$$P_e = \epsilon \, \sigma \, A T_e^4 \quad \text{(emission)} \tag{12.120}$$

Similarly, the same object when placed in an enclosure with walls at an absolute temperature T_a will absorb radiation from the walls at the following rate:

$$P_a = \epsilon \, \sigma \, A T_a^4 \quad \text{(absorption)} \tag{12.121}$$

Thus, if $T_e > T_a$, that is, the object is hotter than the walls of the enclosure, then there will be a net flow of energy from the object to the walls at the following rate:

$$P = P_e - P_a = \epsilon \, \sigma \, A \left(T_e^4 - T_a^4 \right) \quad \text{(net loss)} \tag{12.122}$$

where ϵ is a dimensionless parameter called the emissivity or absorptivity, which varies from 0 to 1 depending on the nature of the surface; A is the surface area of the object; σ is the Stefan–Boltzmann constant whose value is $5.67 \times 10^{-8} \, \dfrac{\text{W}}{\text{m}^2} \, \text{K}^4$.

Now let us consider another situation. If we consider radiation in a black body chamber and apply the thermodynamic laws of radiation, what will happen? To understand this, let us assume that u is the energy density of radiation inside the enclosure, V its total volume, and p the pressure of radiation. Then the total energy U of the radiation is expressed as follows:

$$U = uV \tag{12.123}$$

From the laws of thermodynamics, we know that

$$\left(\frac{\partial U}{\partial V} \right)_T = T \left(\frac{\partial p}{\partial T} \right)_V - p \tag{12.124}$$

As the radiation is diffuse,

$$p = \frac{u}{3}; \frac{\partial u}{\partial V} = u; \tag{12.125}$$

The right-hand side of Eq. (12.124) takes the following form:

$$T \left(\frac{\partial \left(\dfrac{u}{3} \right)}{\partial T} \right)_V - \frac{u}{3} = \frac{T}{3} \frac{\partial u}{\partial T} - \frac{u}{3} \tag{12.126}$$

From Eq. (12.124), the right-hand side of Eq. (12.126) takes the following form:

$$\left(\frac{\partial (uV)}{\partial V} \right)_T = u \tag{12.127}$$

Hence, Eq. (12.124) can be written as follows:

$$u = \frac{T}{3} \frac{\partial u}{\partial T} - \frac{u}{3} \tag{12.128}$$

$$\frac{4u}{3} = \frac{T}{3} \frac{\partial u}{\partial T} \tag{12.129}$$

$$\frac{\partial u}{u} = 4 \frac{dT}{T} \tag{12.130}$$

Integrating on both the sides, we get the following equation:

$$u = aT^4 \tag{12.131}$$

where a is a constant independent of the properties of the body. Equation (12.131) gives us the density of radiation in the enclosure.

The amount of energy lost through an aperture of unit area per second is given by the following equation:

$$u = \frac{4\pi K}{C} \tag{12.132}$$

where K is the specific intensity of radiation, that is, the amount of radiation proceeding in a particular direction per solid angle per unit area per unit time.

From Eq. (12.132),

$$K = \frac{Cu}{4\pi} = \frac{CaT^4}{4\pi} \tag{12.133}$$

The rate of energy/heat lost is given by the following equations:

$$P = \pi \epsilon \, A K = \pi \epsilon \, A \frac{CaT^4}{4\pi} = \epsilon \, A \frac{aC}{4} T^4 \tag{12.134}$$

$$P = \epsilon \, A \sigma T^4 \tag{12.135}$$

where σ is referred to as the Stefan's constant and its value is given as follows:

$$\sigma = \frac{5.67 \times 10^{-8} \text{ W}}{\text{m}^2} . \text{K}^4 \quad \text{or} \quad 5.67 \times 10^{-5} \frac{\text{ergs}}{\text{s cm}^2} \text{K}^4$$

12.8 ADIABATIC EXPANSION OF RADIATION OR WIEN'S DISPLACEMENT LAW

We have observed from our previous discussion that the radiation emitted by a body is not confined to a single wavelength but is spread over a continuous spectrum. In this section, we consider how energy is spread or distributed over different wavelengths. Wien's law states that the amount of energy contained in the spectral region within the wavelengths λ and $\lambda + \Delta\lambda$ emitted by a black body at different temperatures takes the following form:

$$E_\lambda \, d\lambda = \frac{A}{\lambda^5} f(\lambda T) \, d\lambda \tag{12.136}$$

where A is a constant and $f(\lambda T)$ is a function of the product λT.

Let us consider a spherical enclosure having perfectly reflecting walls, which are made of an elastic material and are capable of moving outwards slowly. A body is kept inside the enclosure maintained at a temperature T. After some time, the radiation inside the enclosure attains equilibrium with the black body. The black body is then taken out so that the enclosure remains filled with diffused radiation at T.

Now, we allow the black body radiation within the enclosure to expand adiabatically, that is, the walls begin to move outwards slowly with a uniform velocity v $(v \ll c)$ such that the volume expands to V, the temperature falls to T', and the quality of radiation may also change as the radiation becomes black radiation, which is characteristic of the lower temperature T'. We further calculate the change in frequency due to Doppler's effect, which every wave suffers on reflection from the walls as they are moving outwards. Waves inside the enclosure are incident on the wall at all possible angles. In case of normal incidence, a particular wave of frequency ϑ changes to $\vartheta\left(1 - \dfrac{v}{c}\right)$ on reaching the wall, where the resultant frequency is as follows:

$$\vartheta + d\vartheta = \frac{\vartheta\left(1 - \dfrac{v}{c}\right)}{\left(1 + \dfrac{v}{c}\right)} = \vartheta\left(1 - \frac{2v}{c}\right) \tag{12.137}$$

$$\vartheta + d\vartheta = \vartheta - \frac{2\vartheta v}{c} \tag{12.138}$$

$$\frac{d\vartheta}{\vartheta} = -\frac{2v}{c} \tag{12.139}$$

Thus,

$$\frac{d\lambda}{\lambda} = \frac{2v}{c} \tag{12.140}$$

When the incident ray falls obliquely on the walls, then

$$d\lambda = \frac{2v\cos\theta}{c} \cdot \lambda \tag{12.141}$$

Therefore, between successive reflections, the wave travels a distance of $2r\cos\theta$, and hence the number of reflections per second is $\left(\dfrac{c}{2r\cos\theta}\right)$. The change of wavelength per second is given by the following expression:

$$d\lambda = \frac{c}{2r\cos\theta} \frac{2\vartheta\cos\theta}{c}\lambda = \frac{\vartheta\lambda}{r} \tag{12.142}$$

$$\frac{d\lambda}{\lambda} = \frac{v}{r} = \frac{\delta r}{r} \tag{12.143}$$

where $v = \delta r$ is the distance traversed in 1 s.

The temperature of the enclosure changes, since the energy is distributed over a larger volume and a part of it is spent doing work. As the process is an adiabatic one, we have the following relation:

$$dQ = dU + PdV = 0 \tag{12.144}$$

where u is the density of radiation, V the volume of the sphere, dU the total change in internal energy, and PdV the work done by the radiant energy.

Substituting $U = uV$ and $p = \dfrac{u}{3}$ in Eq. (12.124), one gets the following expressions:

$$d(uV) + \frac{u}{3}dV = 0 \qquad (12.145)$$

$$\frac{du}{u} + \frac{4}{3}\frac{dV}{V} = 0 \qquad (12.146)$$

Integrating both the sides,

$$uV^{\left(\frac{4}{3}\right)} = \text{const} \qquad (12.147)$$

Volume of sphere, $V = \dfrac{4\pi r^3}{3}$ and from Stefan's law $U = aT^4$.

Substituting these values in Eq. (12.147), we get the following relations:

$$aT^4 \left(\frac{4\pi r^3}{3}\right)^{\frac{4}{3}} = C \qquad (12.148)$$

$$aT^4 \left(\frac{4\pi}{3}\right)^{\frac{4}{3}} r^4 = C' \qquad (12.149)$$

$$a\left(\frac{4\pi}{3}\right)^{\frac{4}{3}} T^4 r^4 = C'' \qquad (12.150)$$

$$rT = C''' \qquad (12.151)$$

$$\frac{dT}{T} = -\frac{dr}{r} \qquad (12.152)$$

Equating Eqs (12.143) in (12.152), we get the following expression:

$$\frac{d\lambda}{\lambda} = -\frac{dT}{T} \qquad (12.153)$$

Thus,

$$\lambda T = C'''' \qquad (12.154)$$

Thus, if radiation of a particular wavelength at a certain temperature is adiabatically altered to another wavelength, then the temperature changes in the inverse ratio.

We now further prove that $U_\lambda \lambda^5 = \text{constt}$, that is, the density of radiation is changed inversely with λ^5 in the aforementioned process. Let us suppose that waves between λ and $(\lambda + \Delta\lambda)$ are in a spherical chamber. These are then subjected to an adiabatic expansion. The work done in an adiabatic expansion is $\dfrac{1}{3}U_\lambda\, d\lambda\, \Delta V$, and this should be equal to the decrease in the total energy content, that is, $-\Delta(U_\lambda \vartheta\, d\lambda)$.

C, C′, C″, C‴, and C‴″ refer to constants.

Hence,

$$\frac{1}{3}U_\lambda \; d\lambda \, \Delta V = -\Delta(U_\lambda \vartheta \, d\lambda) \qquad (12.155)$$

$$\frac{1}{3}U_\lambda \; d\lambda \, \Delta V = -\Delta(U_\lambda \vartheta \, d\lambda) \qquad (12.156)$$

$$\frac{1}{3}U_\lambda \; d\lambda \Delta V = -\Delta U_\lambda \vartheta \, d\lambda - U_\lambda \Delta \vartheta \, d\lambda - U_\lambda \vartheta \Delta \, d\lambda \qquad (12.157)$$

Dividing by $U_\lambda \vartheta \, d\lambda$, we get the following expression:

$$\frac{1}{3}\frac{\Delta V}{V} = -\frac{\Delta U_\lambda}{U_\lambda} - \frac{\Delta d\lambda}{d\lambda} - \frac{\Delta V}{V} \qquad (12.158)$$

Since $d\lambda$ changes in the same way as λ, we use $\dfrac{\Delta V}{V} = \dfrac{3\Delta\lambda}{\lambda}$. In addition, V is proportional to λ. Therefore,

$$\frac{1}{3}\frac{3\Delta\lambda}{\lambda} = -\frac{\Delta U_\lambda}{U_\lambda} - \frac{\Delta d\lambda}{d\lambda} - \frac{3\Delta\lambda}{\lambda} \qquad (12.159)$$

$$\frac{5\Delta\lambda}{\lambda} = -\frac{\Delta U_\lambda}{U_\lambda} \qquad (12.160)$$

$$\frac{5\Delta\lambda}{\lambda} + \frac{\Delta U_\lambda}{U_\lambda} = 0 \qquad (12.161)$$

$$U_\lambda \lambda^5 = \text{constt} = U'_\lambda \lambda'^5 \qquad (12.162)$$

where U'_λ is the density of the radiation λ' which is the transformed λ as a result of expansion. U_λ must be a function of T, and hence the constant term in Eq. (12.162) must involve the temperature T. However, the constant should not vary throughout such adiabatic alteration of wavelengths. We further observe that λT is a constant for the process. Clearly, therefore, this constant is a function of the product of λT. Hence, we have the following expression:

$$U_\lambda \, d\lambda = \frac{A}{\lambda^5} f(\lambda T) d\lambda \qquad (12.163)$$

If the frequencies ϑ and $(\vartheta + \Delta\vartheta)$ correspond to the wavelength λ and $(\lambda + \Delta\lambda)$ and $U_\vartheta \, d\vartheta$ represents the energy in this frequency interval, then we have the following expression:

$$U_\vartheta \, d\vartheta = U_\lambda \, d\lambda \qquad (12.164)$$

Since $\quad \dfrac{c}{\vartheta} = \lambda; d\lambda = -\dfrac{c}{\vartheta^2} d\vartheta$

$$U_\vartheta \, d\vartheta = -\frac{A\vartheta^5}{c^5} f\left(\frac{c}{\vartheta} \cdot T\right) - \frac{c}{\vartheta^2} \, d\vartheta \qquad (12.165)$$

C'''' refers to constant.

$$U_\vartheta \, d\vartheta = -\frac{A\vartheta^3}{c^4} \, f\left(\frac{c}{\vartheta}.T\right) d\vartheta \tag{12.166}$$

$$U_\vartheta \, d\vartheta = B\vartheta^3 \varphi\left(\frac{\vartheta}{T}\right) d\vartheta \tag{12.167}$$

We can also write

$$U_\lambda \, d\lambda = AT^5 F\left(\lambda T\right) d\lambda \tag{12.168}$$

Where

$$F\left(\lambda T\right) = \left(\lambda T\right)^{-5} f\left(\lambda T\right) \tag{12.169}$$

The emission E_λ is proportional to the energy density U_λ. It follows that

$$\frac{E_\lambda}{E_\lambda'} = \left(\frac{T}{T'}\right)^5 \tag{12.170}$$

It is now apparent that if the distribution function of E_λ be plotted against λ (on the *x*-axis) for a given temperature, then the distribution at a second higher temperature T' can be obtained graphically by shortening the abscissa in the ratio of $\left(\dfrac{T}{T'}\right)$, and this increases the corresponding ordinate in the ratio $(T'/T)^5$. The wave interval $d\lambda$ becomes shortened in the ratio (T'/T). The curve (Fig. 12.13) now becomes higher and more closed up, and the total area that represents the intensity is changed in the ratio $(T'/T)^4$ corresponding to point P of maximum emission in the first curve. There is a point P′ of maximum emission in the second curve such that

$$\lambda_m T = \lambda_m' T' = b = \text{C}'''' \tag{12.171}$$

where λ_m is the wavelength corresponding to the maximum emission. The constant b is the root of the following equation:

$$\frac{dE_\lambda}{d\lambda} = 0 \tag{12.172}$$

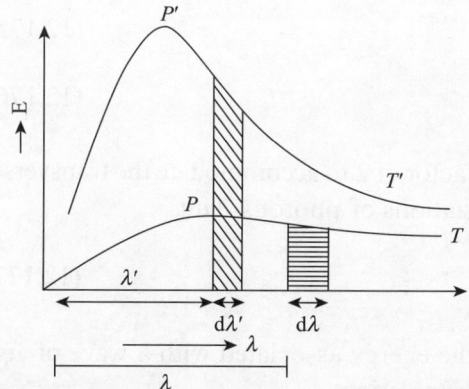

Fig. 12.13 Correspondence of black body curves for two temperatures

---

C′′′′ refers to constant.

12.9 QUANTUM THEORY OF RADIATION—PLANCK'S LAW OF BLACKBODY RADIATION

Radiant energy is regarded as atomic in structure, consisting of discrete particles called 'photons' that retain their total stock of energy unimpaired throughout their flight in space and may only be broken up during their encounter with matter. In a blackbody chamber, photons may have energy between 0 and ∞, and may be moving in all possible directions. Planck's law of black body radiation describes the following three specific properties of photons:

1. Photons behave, in many ways, like particles of spin 1. Hence, photons obey the Bose–Einstein statistics.
2. Photons do not interact with each other, that is, a photon gas is a perfect gas.
3. The process that produces thermal equilibrium between the radiation in a cavity and the enclosure consists of continual emission and absorption of photons by the atoms of the surrounding wall maintained at temperature T. This implies that the number of photons in a cavity is not a constant.

Thus, we observe that a photon gas consists of bosons and hence obeys the Bose–Einstein statistics. In addition, since photons do not obey the laws of conservation of numbers, we have the following expression:

$$\sum_j \alpha \, \delta N_j = 0 \tag{12.173}$$

From Eq. (12.173), it is evident that since the total number of particles cannot be zero, the constant α should be zero. Since photons are zero-spin particles, α is equal to zero.

The energy distribution function is given as follows:

$$dn(E) = \frac{g(E)\,dE}{e^{\epsilon/kT} - 1} \tag{12.174}$$

The energy function is related to the momentum function as follows:

$$g(E)\,dE = g(p)\,dp \tag{12.175}$$

$$g(p)\,dp = V \cdot \frac{4\pi p^2 \, dp}{h^3} \tag{12.176}$$

Equation (12.176) is multiplied by a factor of 2 to accommodate the transverse, clockwise, and anticlockwise polarizations of photons. Thus,

$$g(p)\,dp = 2V \cdot \frac{4\pi p^2 \, dp}{h^3} \tag{12.177}$$

We are familiar with the fact that the energy associated with a wave of frequency ϑ is given by the following expression:

$$E = h\vartheta \tag{12.178}$$

Therefore, $p = \dfrac{h\vartheta}{c}$ (12.179)

$$dp = \frac{h}{c}d\vartheta$$ (12.180)

Now,

$$g(E)dE = 2V \cdot \frac{4\pi\vartheta^2}{h^3}dp = 2V \cdot \frac{4\pi\vartheta^2}{h^3} \cdot \frac{h}{c}d\vartheta$$

$$g(E)dE = \frac{8\pi V\vartheta^2 d\vartheta}{c^3}$$ (12.181)

Thus, substituting Eq. (12.181) into Eq. (12.174), one gets the following relations:

$$dn(E) = \frac{8\pi V\vartheta^2}{c^3\left(e^{(\epsilon/kT)}-1\right)}d\vartheta$$ (12.182)

$$dn(E) = \frac{8\pi V\vartheta^{2^-}}{c^3\left(e^{(h\vartheta/kT)}-1\right)}d\vartheta$$ (12.183)

Here, $dn(E)$ defines the number of particles in the frequency range ϑ and $(\vartheta + d\vartheta)$. Each photon has an energy $h\vartheta$; therefore, the frequency of these photons is in the following frequency range:

$$E_\vartheta = h\vartheta \, dn(E)$$ (12.184)

$$E_\vartheta = h\vartheta \frac{8\pi V\vartheta^2}{c^3\left(e^{(h\vartheta/kT)}-1\right)}d\vartheta$$ (12.185)

The energy density U_ϑ is given by the following expressions:

$$U_\vartheta = \frac{E\vartheta}{V} = \frac{h\vartheta}{V}\frac{8\pi V\vartheta^2}{c^3\left(e^{(h\vartheta/kT)}-1\right)}d\vartheta$$ (12.186)

$$U_\vartheta = \frac{8\pi h}{c^3}\frac{\vartheta^3 d\vartheta}{e^{h\vartheta/kT}-1}$$ (12.187)

Equation (12.187) is referred to as the Planck's radiation formula and is the energy density of black body radiation in the frequency range ϑ to $(\vartheta + d\vartheta)$. We further consider the wavelength of the photons in the range λ and $(\lambda + \Delta\lambda)$. We also know that

$$\vartheta = \frac{c}{\lambda}$$ (12.188)

$$d\vartheta = -\frac{c}{\lambda^2}d\lambda$$ (12.189)

Taking the absolute value of frequency from Eq. (12.189):

$$d\vartheta = \frac{c}{\lambda^2} d\lambda \qquad (12.190)$$

Substituting Eqs (12.189) and (12.190) into Eq. (12.142), one gets the following expression:

$$U_\lambda = \frac{8\pi hc}{\lambda^5} \frac{d\lambda}{\left(e^{\frac{hc}{\lambda kT}} - 1\right)} \qquad (12.191)$$

Equation (12.191) gives the energy density for wavelength λ in the spectrum of a black body radiation.

Planck, for the first time, proposed a law that was true for both higher and lower wavelengths.

Higher wavelengths or low frequencies:

$$\lambda kT \gg\gg ch \qquad (12.192)$$

or $\qquad \dfrac{ch}{\lambda kT} \ll 1 \qquad (12.193)$

or $\qquad \left(e^{\frac{ch}{\lambda kT}} - 1\right) \cong 1 + \dfrac{ch}{\lambda kT} + \ldots - 1 \cong \dfrac{ch}{\lambda kT} \qquad (12.194)$

(where the higher orders are neglected)

Therefore,

$$U_\lambda = \frac{8\pi ch}{\lambda^5} \frac{d\lambda}{ch} \lambda kT = \frac{8\pi k(\lambda T)d\lambda}{\lambda^5} \qquad (12.195)$$

Or $\qquad U_\lambda \propto \dfrac{1}{\lambda^4} \qquad (12.196)$

Equation (12.196) represents the *Rayleigh–Jeans Law*. *This equation shows that, in this limit, the mean energy per photon (or per normal mode if we think of the radiation as a superposition of normal modes) is kT.* We can see that the result holds only for the low frequencies, but its energy distribution is predicted for all frequencies by classical physics.

According to the classical theorem of equipartition, the energy of a harmonic oscillator is kT and the equation of motion describing a normal mode is the same as that of a harmonic oscillator. Equations (12.195) and (12.196) were first classically derived by Rayleigh. Rayleigh observed that when we calculate the total energy density by integrating the equation over all the frequencies, the result is divergent because of the contributions from high frequencies. This was given the name of *ultraviolet catastrophe.* The result obtained by Rayleigh was only partially correct for low frequencies or high wavelengths.

Lower wavelengths or higher frequencies:

$$ch \gg \lambda kT \qquad (12.197)$$

Or $\qquad \dfrac{ch}{\lambda kT} \gg 1 \qquad (12.198)$

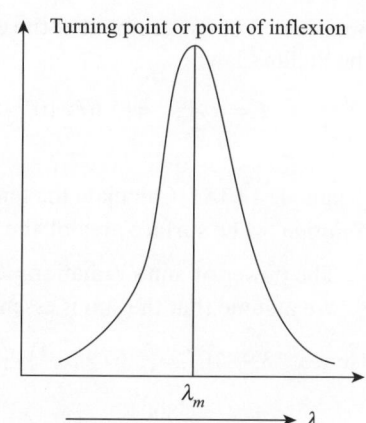

Turning point or point of inflexion

u_λ

λ_m

λ

Fig. 12.14 $U\lambda$ vs λ curve of Planck's radiation law

Therefore, the value of $e^{ch/\lambda kT}$ is very large with respect to 1. Thus, one cannot neglect 1 with respect to $e^{ch/\lambda kT}$. Therefore,

$$U_\lambda = \dfrac{8\pi ch}{\lambda^5} e^{ch/\lambda kT} \, d\lambda \qquad (12.199)$$

Equation (12.199) represents the Wein's displacement law, that is,

$$U_\lambda \propto \dfrac{1}{\lambda^5} \qquad (12.200)$$

Planck's law reduces to the Rayleigh law for higher wavelengths and to the Wein's displacement law for lower wavelengths. It also explains the turning point or the point of inflexion (Fig. 12.14). According to Planck's law, as the wavelength decreases, the first term increases but the second term decreases. It compensates for the increase in the first term, avoiding the ultraviolet catastrophe. For a certain value of wavelength, the $\left(1/\lambda^5\right)$ term is completely balanced by the term $\left(1/e^{ch/\lambda kT} - 1\right)$.

This term is the point of inflexion, and this wavelength is referred to as the characteristic wavelength. This occurs for a maximum number, and so it is called the minimum characteristic wavelength:

$$U_\lambda = \dfrac{8\pi ch}{\lambda^5} \qquad (12.201)$$

Example 12.12   A sphere of 5 cm radius acts like a black body. It is in equilibrium with its surroundings and absorbs 50 kW of power radiated to it from the surroundings. What is the temperature of the sphere?

Solution   The power absorbed by a black body is given as follows:

$$P_a = \left(50 \times 10^3 \, \text{W}\right) = \sigma \epsilon \, A T_a^4 = \left(\dfrac{5.67 \times x10^{-8} \, \text{W}}{\text{m}^2} . \text{K}^4\right)\left(4\pi\left(0.05 \, \text{m}\right)^2\right) T_a^4$$

$$T_a^4 = 5.611 \times 10^{13} \;\Rightarrow T_a \cong 2700 \, \text{K}$$

As the body is in equilibrium with the surroundings, it is also at a temperature of 2700 K.

Example 12.13   There is a small hole of area 100 mm² in an electric furnace used for treating metals. Find the amount of power that should travel through the hole to maintain the metal at 1000°C.

Solution Let us assume that the electric furnace is nearly a black body. Thus, from the Stefan's law:

$$P = \sigma AT^4 = (5.67 \times 10^{-8}) \times (10^{-4})(1273)^4 = 14.89 \text{ W}$$

Example 12.14 Calculate the temperature of the sun using the Stefan's law.

Solution The surface area of the sun $= 6.1 \times 10^{18} \text{ m}^2$

The power of sun's radiation $= 3.9 \times 10^{26} \text{ W}$

We assume that the sun is essentially emissive, that is, $\epsilon = 1$.

Hence, $3.9 \times 10^{26} = (5.67 \times 10^{-8}) \times (6.1 \times 10^{18}) T_a^4$

$$\Rightarrow T_a = 5800 \text{ K}$$

Example 12.15 If a surface has an absorption factor of 0.4, calculate the power radiated per square metre by a filament at a temperature of 1727°C.

Solution Using Stefan's law,

$$P = 0.4 \times (5.67 \times 10^{-8})(2000)^4 = 0.36 \text{ MW/m}^2$$

Example 12.16 A black body is at a temperature of 427°C. To what value must its temperature be increased so that the black body radiates thrice as much energy per second?

Solution As $P \propto T^4$, $3^{1/4} (700 \text{ K}) \cong 922 \text{ K} = 649°C$

$$\Rightarrow \text{Temperature increase} = 649 - 427 = 222°C$$

Example 12.17 The sun's surface temperature is about 6000 K. The sun's radiation is maximum at wavelength of 0.5 µm. A light bulb filament emits maximum radiation at a wavelength of 1 µm. If the surfaces of both the sun and the filament have the same emissive characteristics, what is the temperature of the filament?

Solution Form the Wein's displacement law,

$$\lambda_m T = \text{constant}$$

$$1 . T_{\text{bulb}} = 0.5 \times 6000 \text{ K} = 3000 \text{ K}$$

Example 12.18 If the increase in the temperature of a black body is from 7°C to 337°C, how much is the increase in the radiated power?

Solution Using Stefan's law, the increase in radiated power is as follows:

$$\frac{P(610)}{P(280)} = \left(\frac{610}{280}\right)^4 \cong 23$$

IMPORTANT CONCEPTS

1. Quantum numbers n_x, n_y, n_z (in rectangular coordinates) define the *state of the particle*.

2. Energies corresponding to the different possible values of n_j^2 are the possible 'energy levels' $\left(\epsilon_j\right)$.

3. When a number of different states [corresponding to different directions of the physical quantity (momentum)] have the same energy, the energy level is then referred to as the 'degenerate energy level' (represented by g_j).

4. The postulates of statistical mechanics, on the basis of which the statistical distribution functions, are as follows:

$$\sum_{j=1}^{N} N_j = N$$

$$\sum_{j=1}^{N} \epsilon_j N_j = E$$

$$\sum_{j=1}^{N} \epsilon_j N_j = U$$

5. Entropy of a system is linked with the probability of happening, that is, $S = k \ln W_k$ where W_k is the thermodynamic probability or the statistical count.

6. W_k is defined as the number of equally probable microstates that can correspond to a given macrostate k.

   The total number of microstates, that is, $\sum_k W_k$, is equal to the thermodynamic probability of each macrostate, that is, $\sum_k W_k = \Omega$

   W_k is different for different statistical distribution functions.

7. Particles obeying the Bose–Einstein statistics have zero integral spin and obey the following frequency distribution function:

$$f_{\mathrm{BE}}\left(\epsilon_j\right) = \frac{1}{e^{\alpha} e^{\beta \epsilon_j} - 1}$$

   where $\beta = \dfrac{1}{k_B T}$

$$f_{\mathrm{BE}}\left(\epsilon_j\right) = \frac{1}{e^{\alpha} e^{\epsilon_j / kT} - 1}$$

8. Particles obeying the Fermi–Dirac statistics are generally particles of one-half integral spins:

$$f_{\mathrm{FD}}\left(\epsilon_j\right) = \frac{1}{e^{\alpha} e^{\beta \epsilon_j} + 1} \qquad \alpha = \frac{-\epsilon_F}{k_B T}$$

$$f_{\mathrm{FD}}\left(\epsilon_j\right) = \frac{1}{e^{-\epsilon_F / k_B T} e^{\epsilon_j / k_B T} + 1}$$

$$f_{\mathrm{FD}}\left(\epsilon_j\right) = \frac{1}{e^{(\epsilon_j - \epsilon_F)/k_B T} + 1}$$

9. The Fermi–Dirac statistics predicts that if the jth energy level is the Fermi level, then the probability of occupancy of the energy levels is one-half. The Fermi energy is an important term/quantity in a system of fermions, for example, the electron gas in a metal.

10. The Maxwell–Boltzmann distribution function is a purely exponentially decreasing function, decreasing by a factor of $(1/e)$ for every increase in ϵ by kT.

$$f_{MB}(\epsilon_j) = A e^{-\epsilon_j/k_B T}$$

11. *Perfect black bodies* are those for which $r = 0$, $t = 0$, $a = 1$, that is, such a body neither reflects nor transmits the light that falls on it, but absorbs all the incoming radiation and hence appears black.

12. *Perfectly white bodies* are said to be perfectly white when $r = 1$, $a = 0$, $t = 0$, that is, all the incoming radiations are reflected by the surface.

13. *The Prevost's theory of exchanges* states that the rise or fall of temperature observed in a body is due to the exchange of radiant energy with the surrounding bodies.

14. Kirchhoff's law is the exchange of energy between radiation and matter. The following are some postulates of Kirchhoff's law:

    (a) Radiation inside a hollow enclosure is independent of the nature or the geometrical shape of the walls of the enclosure or any of the body placed inside it.

    (b) At any temperature the ratio of the emissive power to the absorptive power of a substance is constant and equal to the emissive power of a perfectly black body. $\dfrac{e_\lambda}{a_\lambda} = E_\lambda$

    Thus, the emissivity of a perfectly black body is the ratio of the emissive power of the body to its absorptive power.

15. The Kirchhoff's law for the first time asserted that every different type of atom, when properly excited, emits a wavelength which is characteristic of that atom.

16. According to Stefan's law, the rate of energy lost is given as follows:

    $$P = \epsilon\, A \sigma\, T^4$$

    where σ is referred to as the Stefan's constant and its value is calculated as follows:

    $$\sigma = \frac{5.67 \times 10^{-8}\,\text{W}}{\text{m}^2} \text{K}^4 \quad \text{or} \quad 5.67 \times 10^{-5}\,\frac{\text{ergs}}{\text{s.cm}^2}\text{K}^4$$

17. Wien's law states that the amount of energy contained in the spectral region included within the wavelengths λ and $\lambda + \Delta\lambda$ emitted by a blackbody at different temperatures is of the following form:

    $$E_\lambda d\lambda = \frac{A}{\lambda^5} f(\lambda T) d\lambda$$

    $$\lambda T = C''''$$

    $$U_\lambda \lambda^5 = C'''' = U'_\lambda \lambda'^5$$

APPLICATIONS

1. Statistical mechanics uses distribution functions to explain various unknown assemblies of particles. This helps understand the distribution of energy levels, states, and degeneracies. This knowledge is helpful for research purposes.

2. Direct application of statistical mechanics is observed in the study of metals and semiconductors as electron assemblies using the Fermi–Dirac statistics.

3. The application of the Bose–Einstein statistics has helped us understand the concept of Bose–Einstein condensate.

4. The field of cryogenics (especially space-related discoveries) is endowed to us by the Bose–Einstein statistics.

5. The phenomenon of superconductivity can be explained using the Bose–Einstein statistics.
6. The Maxwell–Boltzmann statistics explains the phenomenon of all ordinary gases.
7. The radiation laws help us understand the stellar world and its intricacies.

IMPORTANT FORMULAE

1. Degenerate energy levels:

$$\epsilon_j = \frac{p_j^2}{2m} = n_j^2 \frac{h^2}{8mL^2}$$

2. Total number of particles present in an assembly: $\sum_{j=1}^{N} N_j = N$

3. Total energy of a system: $\sum_{j=1}^{N} \epsilon_j N_j = E$

4. Internal energy of a system:

$$\sum_{j=1}^{N} \epsilon_j N_j = U$$

5. Entropy of a system: $S = k \ln W_k$
6. Occupation number of a system:

$$\sum_{k} W_k = \Omega$$

7. Frequency distribution of the Bose–Einstein distribution:

$$f_{BE}\left(\epsilon_j\right) = \frac{1}{e^{\alpha} e^{\epsilon_j / kT} - 1}$$

8. Frequency distribution of the Fermi–Dirac distribution:

$$f_{FD}\left(\epsilon_j\right) = \frac{1}{e^{(\epsilon_j - \epsilon_F)/k_B T} + 1}$$

9. Frequency distribution of the Maxwell–Boltzmann distribution:

$$f_{MB}\left(\epsilon_j\right) = A e^{-\epsilon_j / k_B T}$$

10. Emissivity of a body: $\dfrac{e_\lambda}{a_\lambda} = E_\lambda$

11. According to Stefan's law,
$$P = \epsilon A \sigma T^4$$

12. According to Wein's law,

$$E_\lambda d\lambda = \frac{A}{\lambda^5} f\left(\lambda T\right) d\lambda$$

where $\lambda T = \text{constt}$

13. According to Rayleigh–Jeans law,
$$U_\lambda \lambda^5 = \text{constt} = U_\lambda' \lambda'^5$$

SELF-ASSESSMENT

Multiple-choice Questions

12.1 The quantum numbers n_x, n_y, n_z (in rectangular coordinates) define the
(a) state of the particle
(b) energy of the particle
(c) occupation number of the system
(d) all of these

12.2 Energies corresponding to the different possible values of n_j^2 are the
(a) possible energy states
(b) possible energy levels $\left(\epsilon_j\right)$
(c) possible degenerate energy levels
(d) possible occupation numbers

12.3 A number of different states corresponding to different directions of the physical quantity have the same energy. The energy level is then referred to as

(a) an energy state

(b) an occupation energy level

(c) a eigen energy level

(d) a degenerate energy level (represented by g_j)

12.4 The postulates of statistical mechanics, on the basis of which the statistical distribution functions are defined, are

(a) $\sum_{j=1}^{N} N_j = N$

(d) $S = k \ln W_k$

(b) $\sum_{j=1}^{N} \epsilon_j N_j = E$

(e) all of these

(c) $\sum_{j=1}^{N} \epsilon_j N_j = U$

12.5 The thermodynamic probability of each macrostate is

(a) S (b) W_k (c) U (d) N

12.6 W_k is different for different statistical distribution functions. The statement is

(a) true

(b) false

(c) incomplete

(d) none of these

12.7 Particles obeying the Bose–Einstein statistics have

(a) zero integral spin

(b) frequency distribution function: $f_{BE}(\epsilon_j) = \dfrac{1}{e^{\alpha} e^{\beta \epsilon_j} - 1}$

(c) (a) and (b)

(d) all of these

12.8 Particles obeying the Fermi–Dirac statistics are generally

(a) particles of one-half integral spins

(b) particles that follow the frequency distribution function:

$$f_{FD}(\epsilon_j) = \frac{1}{e^{(\epsilon_j - \epsilon_F)/k_B T} + 1}$$

(c) (a) and (b)

(d) all of these.

12.9 The Fermi–Dirac statistics predicts that if the *j*th energy level is the Fermi level, then

(a) the probability of occupancy of the energy level is one-half

(b) the probability of occupancy of the energy level is full

(c) the probability of occupancy of the energy level is unknown

(d) the probability of occupancy of the energy level is incomplete

12.10 Which of the following frequency distribution functions defines the Maxwell–Boltzmann distribution function:

(a) $f_{MB}(\epsilon_j) = A e^{-\epsilon_j / k_B T}$

(b) $f_{MB}(\epsilon_j) = A e^{\epsilon_j / k_B T}$

(c) $f_{MB}(\epsilon_j) = 1 / \left(A e^{\epsilon_j / k_B T} + 1 \right)$

(d) $f_{MB}(\epsilon_j) = 1 / \left(A e^{\epsilon_j / k_B T} + 1 \right)$

12.11 Perfect black bodies are those for which
    (a) $r = 0; t = 0; a = 1$
    (c) $r = 0; t = 1; a = 1$
    (b) $r = 1; t = 0; a = 1$
    (d) $r = 0; t = 0; a = 1$

12.12 Perfectly white bodies are said to be perfectly white when
    (a) $r = 0; t = 0; a = 1$
    (c) $r = 0; t = 1; a = 1$
    (b) $r = 1; t = 0; a = 0$
    (d) $r = 1; t = 1; a = 1$

12.13 The Prevost's theory of exchanges states that the rise or fall of temperature, which is observed in a body, is due to
    (a) exchange of radiant energy with surrounding bodies
    (b) conservation of energy
    (c) radiation energy
    (d) convection flow

12.14 The Kirchoff's law states that
    (a) radiation inside a hollow enclosure is independent of the nature or the geometrical shape of the walls of the enclosure or any of the body placed inside it
    (b) at any temperature, the ratio of the emissive power to the absorptive power of a substance is constant and equal to the emissive power of a perfect blackbody, that is, $\dfrac{e_\lambda}{a_\lambda} = E_\lambda$
    (c) (a) and (b)
    (d) all of these

12.15 Stefan's law states that the rate of energy lost is given by
    (a) $P = A\sigma T^4$
    (c) $P = \epsilon A\sigma T^2$
    (b) $P = \epsilon A T^4$
    (d) $P = \epsilon A\sigma T^4$

12.16 Wien's law stated that the amount of energy contained in the spectral region included within the wavelengths λ and $\lambda + \Delta\lambda$ emitted by a black body at different temperatures is of the form
    (a) $E_\lambda d\lambda = \dfrac{A}{\lambda^5} f(\lambda T) d\lambda$
    (c) $U_\lambda \lambda^5 = \text{constt} = U'_\lambda \lambda'^5$
    (b) $\lambda T = \text{constt}$
    (d) all of these

12.17 Which of the following is the Planck's radiation formula?
    (a) $U_\vartheta = \dfrac{8\pi h}{c^3} \dfrac{\vartheta^3 d\vartheta}{e^{h\vartheta/kT} - 1}$
    (c) $U_\vartheta = \dfrac{8\pi h}{c^3} \dfrac{\vartheta d\vartheta}{e^{h\vartheta/kT} - 1}$
    (b) $U_\vartheta = \dfrac{8\pi h}{c^3} \dfrac{\vartheta^3 d\vartheta}{e^{h\vartheta/kT} + 1}$
    (d) $U_\vartheta = \dfrac{8\pi h}{c^3} \dfrac{\vartheta^2 d\vartheta}{e^{\frac{h\vartheta}{kT}} - 1}$

12.18 The Bose–Einstein condensation has
    (a) a characteristic ground state
    (c) a characteristic energy level
    (b) a characteristic excited state
    (d) all of these

12.19 An electron gas follows
    (a) the Fermi–Dirac statistics
    (d) Maxwell–Boltzmann statistics
    (b) the Bose–Einstein statistics
    (d) none of these

12.20 Fermions contribute to the specific heat at
    (a) very low temperatures
    (c) room temperature
    (b) very high temperatures
    (d) both (a) and (B)

12.21   Black body radiation follows
   (a)   the Fermi–Dirac statistics
   (b)   the Bose–Einstein statistics
   (c)   the Maxwell–Boltzmann statistics
   (d)   none of these

Review Questions

12.1   Explain how statistical distribution is followed by an electron, a Cooper pair, a photon, and a molecule of oxygen.

12.2   Write six salient points of quantum mechanics and statistical mechanics, each of which has a direct impact in the field of engineering.

12.3   Give two distinct features that necessitate the need of studying statistical mechanics.

12.4   Considering that a standing wave is associated to the De Broglie wavelength, explain
   (a)   the concept of quantum numbers;
   (b)   the occupation number N_j, j being a dummy index representing the energy level;
   (c)   the degeneracy of an energy level; and
   (d)   the energy state, energy level, and degeneracy using a graphical (pictorial) representation.

12.5   Considering that a particle is confined in a three-dimensional box, explain
   (a)   the concept of quantum numbers;
   (b)   the occupation number N_j, j being a dummy index representing the energy level;
   (c)   the degeneracy of an energy level; and
   (d)   the energy state, energy level, and degeneracy using a graphical (pictorial) representation.

12.6   Explain how the Planck's radiation law has used the quantum concept to arrive at a general radiation law.

12.7   Distinguish between the Maxwell–Boltzmann, Bose–Einstein, and Fermi–Dirac statistical distributions qualitatively.

12.8   What is an electron gas? Discuss its basic characteristics and the statistics that it follows.

12.9   Differentiate between Maxwell–Boltzmann, Bose–Einstein, and Fermi–Dirac statistical distribution functions. Give a graphical comparison for the three distinct statistical distribution functions.

12.10   Give the salient postulates of statistical mechanics on the basis of which the statistical distribution functions are defined. In addition, state clearly the differences in evaluation of the thermodynamic probability for each of the distribution functions.

12.11   Give the wave function representations of the Bose–Einstein, Fermi–Dirac, and Maxwell–Boltzmann statistical distribution functions.

12.12   Enunciate with an example the differences in the occupation number of the three statistical distribution functions, that is, the Maxwell–Boltzmann, Bose–Einstein, and Fermi–Dirac statistical distribution functions.

12.13   Explain Kirchhoff's law of radiation.

12.14   Derive the Stefan–Boltzmann law.

12.15 Postulate the Planck's radiation law. Show how it limits to the Wein's and the Rayleigh–Jeans law in their specific limits.

12.16 Electrons pair up to settle in the ground state for a system following the Bose–Einstein statistics. Justify.

12.17 Electrons in a conducting metal do not contribute to the specific heat only at room temperature. Justify.

12.18 Bosons and fermions follow different quantum statistics. Explain.

Numerical Problems

12.1 There are five types of molecules (n_i = 3, 4, 5, 7, 8) with velocities 2, 4, 6, 8, and 10, respectively. Calculate the (a) average speed, (b) root mean square speed, and (c) the most probable speed.

$$\left[Hint: \bar{v} = \frac{n_1 v_{x1} + n_2 v_{x2} + \ldots + n_N v_{xN}}{n_1 + n_2 + n_3 + \ldots + n_N}; \right.$$

$$\left. v_{rms} = \sqrt{\frac{n_1 v_1^2 + n_1 v_2^2 + n_1 v_3^2 + n_1 v_4^2 + n_1 v_5^2 + n_1 v_6^2}{n_1 + n_2 + n_3 + n_4 + n_5 + n_6}} \right]$$

12.2 There are five molecules per cm^3 on an average in a very dilute gas. The temperature is 10 K. What is the pressure of this very dilute gas $\left(k_B = 1.38 \times 10^{-23} \text{ J/K} \right)$?

$$\left[Hint: pV = \mu RT = \mu(k_B N_A)T = (\mu N_A)k_B T = Nk_B T \right]$$

12.3 There are four molecules per cm^3 on an average in a very dilute gas. The temperature is 20 K. What is the pressure of this very dilute gas $\left(k_B = 1.38 \times 10^{-23} \text{ J/K} \right)$?

$$\left[Hint: pV = \mu RT = \mu(k_B N_A)T = (\mu N_A)k_B T = Nk_B T \right]$$

12.4 The density of air at NTP is 1.29kg/m^3. Find the rms velocity of air molecules at NTP.

$$\left[Hint: p = \frac{1}{3}\rho \bar{v^2} \Rightarrow \bar{v^2} = \frac{3p}{\rho} \text{ Also, } v_{rms} = \sqrt{\bar{v^2}} = \sqrt{\frac{3P}{\rho}} \right]$$

12.5 An astronaut takes a 25 L cylinder containing nitrogen gas at a temperature of 25°C and a pressure of 50 atm. The astronaut makes a hole of 1 cm² area in this cylinder and places it in an open space. Estimate the time it would take for the cylinder to become empty $\left(k_B = 1.38 \times 10^{-23} \text{ J/K} \right)$.

$$\left[Hint: v_{rms} = \sqrt{\frac{3RT}{M}} = \sqrt{\frac{3N_A k_B T}{M}} \right]$$

12.6 Estimate the total number of air molecules (where the air is constituted of particles of oxygen, nitrogen, water vapour, and other constituents) in a room of capacity 20m^3 at a temperature of 25°C and 1 atm pressure ($k_B = 1.38 \times 10^{-23} \text{ J/K}$).

$$\left[Hint: pV = Nk_B T \right]$$

12.7 The temperature of a gas is $-37°C$. To what temperature should it be heated so that
(a) the average kinetic energy of the molecules is doubled, and
(b) the rms velocity of the molecules is doubled?

$$\left[Hint: \text{(i)}\ \frac{E_1}{E_2} = \frac{T_1}{T_2}\ \text{(ii)}\ \frac{v_{rms}}{2v_{rms}} = \sqrt{\frac{T_1}{T_2}} \right]$$

12.8 Two gases (A and B) having equal volume diffuse at $25°C$ through a porous partition in 3 and 6 s respectively. The molecular weight of A is 4; find the root mean square speed of B $(R = 8.31\,J/mol/K)$

$$\left[Hint: \frac{r_A}{r_B} = \sqrt{\frac{M_B}{M_A}} \implies \frac{V/3}{V/6} = \sqrt{\frac{M_B}{4}}; v_{rmsB} = \sqrt{\frac{3RT_B}{M_B}} \right]$$

12.9 A sphere of 6 cm radius acts like a black body. It is in equilibrium with its surroundings and absorbs 20 kW of power radiated to it from the surroundings. What is the temperature of the sphere?

$$\left[Hint: P_a = \sigma \epsilon A T_a^4 \right]$$

12.10 There is a small hole of area $25\,mm^2$ in an electric furnace used for treating metals. Determine the amount of power that travels through the hole, in order to maintain the metal at $1000°C$.

$$\left[Hint: P = \sigma A T^4 \right]$$

12.11 There is a small hole of area $10\,mm^2$ in an electric furnace used for treating metals. Calculate the power that should travel through the hole to maintain the metal at $500°C$.

$$\left[Hint: P = \sigma A T^4 \right]$$

12.12 If a surface has an absorption factor of 0.3 calculate the power radiated per square metre by a filament at a temperature of $1427°C$.

$$\left[Hint: P = \sigma A T^4 \right]$$

12.13 Find the temperature of a filament, if the power radiated by it per square metre is $1.50\,MW/m^2$ and the surface has an absorption factor of 0.25.

$$\left[Hint: P = \sigma A T^4 \right]$$

PART III
ELECTROMAGNETISM

Dielectric Properties of Materials

Learning Objectives

After studying this chapter, students will be able to

- understand the fundamentals of electric dipole, dipole moment, and dielectric constant
- comprehend the concept of polarizability and hence displacement vector
- realize the various types of polarization mechanisms in dielectric materials
- calculate local fields
- explain the various applications of dielectric materials in the field of engineering
- understand the fundamentals of ferroelectric materials

List of Symbols

q = Magnitude of charge

δ = Distance between charges

N = Number of atoms per unit volume

P = Polarization vector

ϵ_0 = Absolute permittivity

ϵ_κ = Relative permittivity

C = Capacitance

V = Applied voltage

A = Area of capacitor plate

σ = Surface charge density

σ_{pol} = Polarized surface charge density

σ_{free} = Free-surface charge density

E = Electric field

χ = Electric susceptibility

D = Displacement vector

ϵ = Permittivity of medium

α_e = Electronic polarizability

F = Restoring force

β = Restoring force constant

M_n = Mass of negative ion

M_p = Mass of positive ion

ω_0 = Angular frequency

α_i = Ionic polarizability

B = Magnetic field

k_B = Boltzmann constant

T = Absolute temperature

τ = Relaxation time

P_L = Power loss

v = Frequency of the signal

X_L = Capacitive reactance

13.1 INTRODUCTION

The term 'dielectric' (combination of 'dia' and 'electric') was coined by William Whewell. In a dielectric material, electric charges do not flow, as is the case in a conductor, but get slightly shifted from their equilibrium positions, on being placed in an external electric field. Thus, dielectric materials do not belong to the class of insulators or conductors, but are materials that can be easily polarized and hence are referred to as electric insulators. This property of easy polarization of a dielectric places it in the class of energy-storing materials. The forbidden energy gap in dielectrics is large. In an ideal dielectric, electric conductivity is also zero. Common examples of dielectric materials are polymers, glass, paper, mica, and oil.

The following are the properties of dielectric materials:
1. Dielectric materials are insulators having high resistivity.
2. Forbidden energy gap for dielectric materials is more than 3eV.
3. All the electrons in dielectric materials are localized, that is, engaged in bonding.
4. Electrical conductivity of dielectric materials is very low.
5. These materials have negative temperature coefficients of resistance.

13.2 ELECTRIC DIPOLE

In a neutral atom, the positive charge rests on the nucleus, which is surrounded by negative electrons. As an external electric field is applied, the nucleus and the electrons align themselves according to the polarity. Wave patterns of the electrons are distorted, that is, the centre of the negative charges no longer falls on the centre of the positive charge of the nucleus. Hence, we now observe two charges or 'dia' electric charges (Fig. 13.1), leading to the formation of a dipole. In a dielectric material, many such dipoles exist.

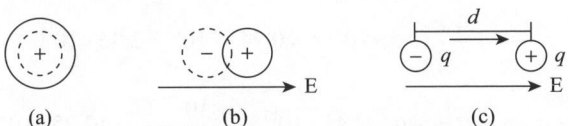

(a)　　　　　(b)　　　　　(c)

Fig. 13.1 Electric dipole, symbol *q*, representing the magnitude of charge and *d* the distance between the charges (a) No electric field (b) Electric field (just applied) (c) Electric field introducing separation of charges

The extent of displacement of charges from their equilibrium positions is dependent on the strength of the external electric field applied.

13.3 DIPOLE MOMENT

Each atom in a dielectric material possesses a little electric dipole. If each of the charges *q* is separated by a distance δ, then the 'dipole moment' per atom is given by $q\delta$. The product of the magnitude of the charge and the distance between the two charges is called 'dipole moment'. The unit is coulomb-metre.

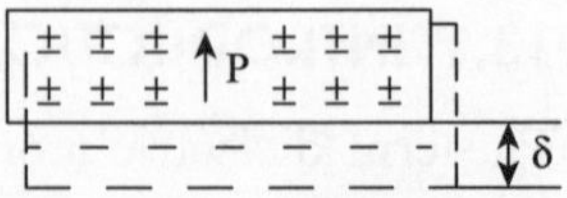

Fig. 13.2 Dielectric slab in uniform electric field

$$\mu = q\delta \qquad (13.1)$$

If there are N number of atoms per unit volume in a given material, then the dipole moment per unit volume is equal to $Nq\delta$. *This quantity is defined as the 'polarization vector' for a given dielectric material, where the magnitude is given by $Nq\delta$ and the direction is defined by the displacement vector δ of the charges:*

$$P = Nq\delta \qquad (13.2)$$

The unit of polarization vector is coulomb/m^2. In terms of dipole moment, the polarization vector can be defined as the average dipole moment and is mathematically represented as follows:

$$P = N\mu \qquad (13.3)$$

The polarization vector is proportional to the electric field, but it varies from one position to another in the dielectric (Fig. 13.2).

Example 13.1   A water molecule has a dipole moment of 5×10^{-25} C-cm, where all the molecular dipoles point in the same direction. If the radius of a water drop is 0.5 mm, find the polarization in the water drop.

Solution   Molecular mass of water $= (2 \times 1) + (8 \times 2) = 18\,g$

Hence, 18 g of H_2O contains 6.023×10^{23} molecules.

That is, $\dfrac{18}{10^3}\,m^3$ of H_2O contains 6.023×10^{26} molecules.

Volume of the water drop $= \dfrac{4\pi}{3}(0.5 \times 10^{-3})^3\,m^3$

Number of molecules in the water drop

$$= \frac{6.023 \times 10^{26}}{18} \times \frac{4\pi}{3}\,(0.5 \times 10^{-3})^3 = 0.245 \times 10^{17} = 2.45 \times 10^{16}$$

$$\text{Polarization} = P = N\mu = 2.45 \times 10^{16} \times \frac{5 \times 10^{-25}\,C}{cm^2} = 12.25 \times 10^{-9}\ C/cm^2$$

Example 13.2   If the induced dipole moment of each atom due to a gas at NTP is 7.27×10^{-36} C-m, determine its polarization vector.

Solution   From Eq. (13.3), the polarization vector is given as follows:

$$P = N\mu$$

$$P = \frac{6.023 \times 10^{23}}{22.4 \times 10^{-3}\,m^3}\left(7.27 \times 10^{-36}\right) C\text{-m} = 1.954 \times 10^{-10}\ \frac{C}{m^2}$$

Example 13.3   The polarization vector of a gas at NTP is 4.0×10^{-10} C/m^2; find the induced dipole moment due to each atom.

Solution   The polarization vector is given as follows: $P = N\mu$

$$4.0 \times 10^{-10} \frac{C}{m^2} = \frac{6.023 \times 10^{23}}{22.4 \times 10^{-3} m^3} \mu$$

$$\Rightarrow \mu = \frac{22.4 \times 10^{-3} m^3}{6.023 \times 10^{23}} \times 4.0 \times 10^{-10} \frac{C}{m^2} \mu = 14.87 \times 10^{-36} \text{ C-m}$$

13.4 DIELECTRIC CONSTANT

For understanding dielectrics, the picture of a capacitor has to be recapitulated in mind (Fig 13.3). The charge on the capacitor is defined as follows:

$$Q = CV \tag{13.4}$$

where V is the voltage; capacitance is defined in terms of area of each plate (A) and the spacing between the plates (d) of the capacitor, as follows:

$$C = \frac{\epsilon_0 A}{d} \tag{13.5}$$

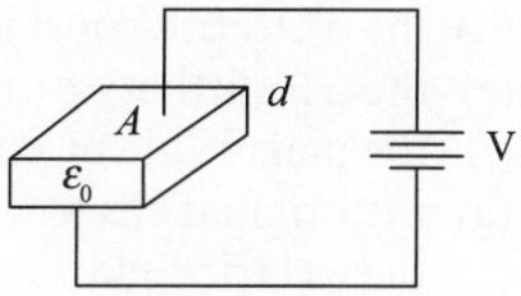

where ϵ_0 is the absolute permittivity, which is equal to $8.854187817 \times 10^{-12}$ F/m in vacuum or free space.

Fig. 13.3 Voltage (V) applied across capacitor of area A and plate separation d

Assume now that the capacitor is filled with a dielectric medium; with the charge remaining the same, the following relation holds:

$$Q = C_\kappa V \tag{13.6}$$

where $$C_\kappa = \frac{\epsilon_\kappa A}{d} \tag{13.7}$$

Here, ϵ_κ is the relative permittivity.

Note: Permittivity is physically understood as the measure of resistance that is encountered when forming an electric field in a medium. The permittivity of a medium gives the amount of electric flux produced per unit charge in that medium. Thus, the higher the permittivity of a material, the higher the degree to which it can be polarized. We can correlate permittivity to electric susceptibility as follows: $\epsilon = \epsilon_\kappa \epsilon_0 = (1 + \chi)\epsilon_0$

Thus, it was observed that the capacitance of a capacitor is strongly related to the medium that fills the space between its plates. *When a capacitor is filled with a dielectric material or an insulating material, its capacitance alters by a factor κ, termed as the dielectric constant.* The dielectric constant is a property of the dielectric medium. Mathematically,

$$\kappa = \frac{C_\kappa}{C_0} = \frac{\epsilon_\kappa}{\epsilon_0} = \frac{V_0}{V_\kappa} \tag{13.8}$$

The properties of a dielectric constant are as follows:
1. Dielectric constant is a dimensionless quantity. It is a pure number.
2. It is assumed to be equal to 1 for free space or vacuum.

3. Since the capacitance of a capacitor always increases after filling the dielectric medium, the value of the dielectric constant κ is always greater than unity.

Example 13.4   The parallel plates of a capacitor have an area of $1.00 \times 10^{-1} \, \text{m}^2$ each and are $1.00 \times 10^{-2} \, \text{m}$ apart. The capacitor is connected to a power supply of $2.00 \, \text{kV}$. A thin insulating plastic sheet is inserted between the capacitor plates and it fills the space between the plates completely. The potential difference between the plates drops to $1.00 \, \text{kV}$, whereas the charge on each capacitor plate remains constant. Calculate
(a) the original capacitance C_0;
(b) the magnitude of charge Q on each plate;
(c) the capacitance C after the dielectric is inserted;
(d) the dielectric constant κ of the dielectric;
(e) the permittivity ε of the dielectric;
(f) the magnitude of charge Q_i on each face of the dielectric;
(g) the original electric field E_0 between the plates; and
(h) the electric field E after the dielectric is inserted.
Solution   The calculations are as follows:
(a) The original capacitance C_0 is given by Eq. (13.5) as follows:

$$C_0 = \epsilon_0 \frac{A}{d} = \left(\frac{8.85 \times 10^{-12} \, \text{F}}{\text{m}} \right) \frac{1.00 \times 10^{-1} \, \text{m}^2}{1.00 \times 10^{-2} \, \text{m}} = 8.85 \, \text{pF}$$

(b) $Q = C_0 V_0 = 8.85 \times 10^{-12} \, \text{F} \times 2.0 \times 10^3 \, \text{V} = 17.7 \times 10^{-9} \, \text{C}$
(c) The capacitance C_κ after the dielectric has been introduced from Eq. (13.6) is given as follows: $C_\kappa = \dfrac{Q}{V} = \dfrac{17.7 \times 10^{-9} \, \text{C}}{1.0 \times 10^3 \, \text{V}} = 17.7 \, \text{pF}$

(d) The dielectric constant is given as follows: $\kappa = \dfrac{C_\kappa}{C_0} = \dfrac{17.7 \, \text{pF}}{8.85 \, \text{pF}} = 2$

Or $\kappa = \dfrac{V_0}{V_\kappa} = \dfrac{2.0 \times 10^3 \, \text{V}}{1.0 \times 10^3 \, \text{V}} = 2$

(e) The permittivity of the dielectric is calculated as follows:

$$\epsilon = \kappa \epsilon_0 = \left(\frac{2.00 \times 8.85 \times 10^{-12} \, \text{C}^2}{\text{Nm}^2} \right) = 17.7 \times 10^{-12} \, \text{C}^2/\text{Nm}^2$$

(f) The magnitude of induced charge on each face of the dielectric is calculated as follows: $Q_i = Q \left(1 - \dfrac{1}{\kappa} \right) = \left(17.7 \times 10^{-9} \right) C \left(1 - \dfrac{1}{2} \right) = 8.85 \times 10^{-9} \, \text{C}$

(g) The following is the original electric field E_0 between the plates:

$$E_0 = \frac{V_0}{d} = \frac{2.0 \times 10^3 \, \text{V}}{1.00 \times 10^{-2} \, \text{m}} = \frac{2.00 \times 10^5 \, \text{V}}{\text{m}}$$

(h) The following is the value of the electric field after the dielectric is inserted:

$$E = \frac{Q}{\epsilon A} = \frac{E_0}{\kappa} = \frac{V}{d} = \frac{1.0 \times 10^3 \, \text{V}}{1.00 \times 10^{-2} \, \text{m}} = 10^5 \, \text{V/m}$$

Thus, this numerical problem highlights how capacitance, dielectric constant, and electric field are intricately related.

13.5 POLARIZABILITY

A dipole moment is formed when an external electric field is applied to a dielectric. This has been explained earlier in terms of the polarization vector (P). For a parallel-plate capacitor, P is uniform, such that we have a surface polarization charge, where effectively the negative charges move out a distance d corresponding to the positive charges (Fig. 13.2). If A is the area of a plate and N is the number of atoms per unit volume, the magnitude of the surface charge density is given as follows:

$$\sigma_{pol} = Nq_e\delta = P \tag{13.9}$$

where, q_e is the electronic charge.

The surface density of charge is equal to the polarization inside the material. An important feature of a dielectric is that one of the surfaces is positive and the other negative.

Polarized surface charge is a result of the free charge existing on the surface of a parallel-plate capacitor due to its charging. As the capacitor discharges, the free charge decreases, correspondingly decreasing the polarized charge (Fig. 13.4).

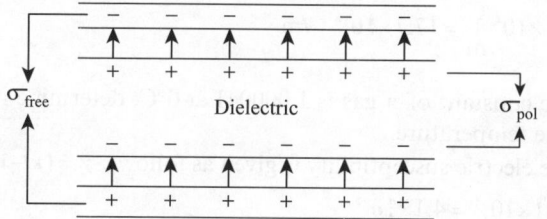

Fig. 13.4 Parallel-plate capacitor with dielectric filled inside it

Having discussed the surface charge density of a dielectric, it is easy to correlate it to the electric field inside the dielectric using the Gauss's law:

$$E = \frac{\sigma_{free} - \sigma_{pol}}{\epsilon_0} \tag{13.10}$$

$$E = \frac{\sigma_{free} - P}{\epsilon_0} \tag{13.11}$$

Thus, one can observe that the electric field is dependent on the polarization vector, or, in other words, the vectors are directly proportional to each other, that is,

$$P \propto E \tag{13.12}$$

$$P = \chi\epsilon_0 E \tag{13.13}$$

where electric susceptibility χ and the absolute permittivity ϵ_0 are constants of proportionality.

$$\Rightarrow \epsilon_0 E = \sigma_{free} - \chi\epsilon_0 E \tag{13.14}$$

$$(1 + \chi)\epsilon_0 E = \sigma_{free} \tag{13.15}$$

$$E = \frac{\sigma_{\text{free}}}{(1+\chi)\epsilon_0} = \frac{\sigma_{\text{free}}}{\epsilon_0} \frac{1}{(1+\chi)} \tag{13.16}$$

Thus, physically, the electric field is reduced by a factor of $1/(1+\chi)$ due to the introduction of the dielectric inside the parallel plates of the capacitor. The dielectric constant can now be redefined in terms of the electric susceptibility as follows:

$$\kappa = (1+\chi) \tag{13.17}$$

Equation (13.17) can be inferred from Eqs (13.5) and (13.6).

$$\Rightarrow \chi = (\kappa - 1) \tag{13.18}$$

$$P = (\kappa - 1)\epsilon_0 E \tag{13.19}$$

Thus, the polarization vector is defined in terms of the dielectric constant.

Example 13.5 If the electric field strength inside two parallel plates of a capacitor is 10^4 V/m due to a dielectric constant of 3, find the polarization vector.

Solution From Eq. (13.19), the polarization vector is given as follows: $P = (\kappa - 1)\epsilon_0 E$

$$E = 10^4 \text{ V/m}; \kappa = 3; \epsilon_0 = 8.85 \times 10^{-12} \text{ C}^2/\text{N/m}^2$$

$$P = (3-1) \times 8.85 \times 10^{-12} \times 10^4 \frac{C}{m} = 17.7 \times 10^{-8} \text{ C/m}$$

Example 13.6 If the dielectric constant of a gas is 1.000041 at 0°C, determine its electric susceptibility at the same temperature.

Solution From Eq. (13.18), the electric susceptibility is given as follows: $\chi = (\kappa - 1)$

$$\chi = (1.000041 - 1) = 0.41 \times 10^{-4} = 4.1 \times 10^{-5}$$

Example 13.7 A thick sheet of polythene of dimension 0.5mm possessing a relative dielectric constant of 2.0, is subjected to 110 V. Calculate the polarization.

Solution We know that

$$P = \epsilon_0 (\kappa - 1) E$$

Hence, $\kappa = 2.0; E = \dfrac{V}{d} = \dfrac{110 \text{ V}}{0.5 \times 10^{-3} \text{ m}} = 220 \times 10^3 \text{ V/m}$

$$\epsilon_0 = 8.85 \times \frac{10^{-12} \text{ C}}{} . \text{m}$$

$$P = 8.85 \times 10^{-12} \text{ C} . \frac{m^2}{V} (2.0 - 1) \times \frac{220 \times 10^3 \text{ V}}{m}$$

$$P = 1947 \times 10^{-9} \text{ C-m} = 1.947 \times 10^{-6} \text{ C-m}$$

13.6 ELECTRIC DISPLACEMENT VECTOR

The concept of displacement vector came into existence owing to the existence of the polarization vector and hence the polarization charge. Thus, a new vector called the electric displacement vector was introduced, which is a linear combination of electric vector E and polarization vector P, and is mathematically defined as follows:

$$D = \epsilon_0 E + P \tag{13.20}$$

From the definition of the polarization vector in Section 13.5 [Eq. (13.13)], we get the following relation: $P = \chi \epsilon_0 E$

Thus, $\quad D = \epsilon_0 E + \chi \epsilon_0 E = \epsilon_0 E (1 + \chi) = \kappa \epsilon_0 E$

$$D = \epsilon E \tag{13.21}$$

The variable ε has been earlier defined as the permittivity of the medium and is mathematically represented as follows:

$$\epsilon = \kappa \epsilon_0 = (1 + \chi) \epsilon_0 \tag{13.22}$$

Note: The electric displacement vector is taken to be proportional to the electric field vector, but it actually depends on the substance being considered. In addition, the proportionality constant is dependent on the variation of electric field with time.

Example 13.8 Two parallel plates having equal and opposite charges are separated by a 4 cm thick slab that has a dielectric constant of 4. If the electric field is 10^4 N/C, calculate the polarization and displacement vectors.

Solution We have the following values:

$$E = \frac{10^4 \text{ N}}{\text{C}} = 10^4 \text{ V/m}; \quad \kappa = 4; \quad \epsilon_0 = 8.85 \times 10^{-12} \text{ C}^2/\text{N/m}^2$$

From Eq. (13.22), the displacement vector is given as follows: $D = \epsilon_0 \kappa E = \epsilon_0 E + P$

$$D = 8.85 \times 10^{-12} \times 4 \times 10^4 \frac{\text{C}}{\text{m}^2} = 35.4 \times 10^{-8} \text{ C/m}^2$$

$$P = D - \epsilon_0 E = 35.4 \times 10^{-8} \frac{\text{C}}{\text{m}^2} - 8.85 \times 10^{-12} \text{ C}^2/\text{N/m} \times 10^4 \text{ N/C}$$

Prior to discussing polarization in dielectrics, it is noteworthy to mention that in a polar molecule the centre of gravity for the positive charge and that for the negative charge do not coincide; as a result, the molecule inherently possesses a net dipole moment. On the contrary, for the non-polar molecule the centre of gravity for the positive charge and that for the negative charge coincide with each other such that there is no net dipole moment (refer to Section 13.2 for dipole moment).

13.7 DIFFERENT POLARIZATIONS IN DIELECTRICS

Polarization in dielectric materials occurs due to several microscopic mechanisms that take place at the atomic or molecular level. The following are the various possible polarizations in dielectrics:

1. Electronic polarization
2. Ionic polarization
3. Orientational polarization
4. Space-charge polarization

13.7.1 Electronic Polarization

Electronic polarization occurs due to the displacement of positively charged nucleus and negatively charged electrons of an atom in the opposite directions when an external electric field is applied. *This results in the creation of a dipole moment in the dielectric.* We know from our previous discussion that the dipole moment μ is proportional to the external electric field (E); therefore.

$$\mu \propto E \tag{13.23}$$

$$\mu = \alpha_e E \tag{13.24}$$

where α_e is electronic polarizability. Electronic polarization is a common feature of almost all dielectrics.

13.7.1.1 Calculation of Electronic Polarizability (α_e)

Without electric field   Consider an atom of a dielectric material of nuclear charge Ze, where Z is the atomic number. The electrons of charge $(-Ze)$ are distributed uniformly throughout the sphere of radius R, as shown in Fig. 13.5.

The centres of the electron cloud and positive nucleus are at the same point, and hence there is no dipole moment. The negative charge density of an atom of radius R is given as follows:

$$\rho = \frac{\text{Total negative charge}}{\text{Volume of the atom}} = \frac{(-Ze)}{\frac{4}{3}\pi R^3} \tag{13.25}$$

$$\rho = \frac{3}{4}\frac{(-Ze)}{\pi R^3} \tag{13.26}$$

With electric field   As the atom of dielectric is placed in a DC electric field of strength E, the following two phenomena occur:

1. *Lorentz force* (due to an electric field) will tend to move the nucleus and electron cloud of that atom from their equilibrium positions. The positive nucleus will move towards the external electric field and the electron cloud will move in the opposite direction of the field applied (Fig. 13.5). The Lorentz force between the nucleus and the electron is given as follows:

$$F_L = \text{Charge} \times \text{Electric field} = (Ze) \times E \tag{13.27}$$

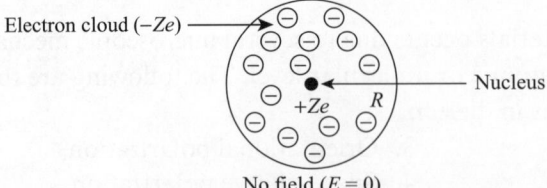

Fig. 13.5 Atom without any electric field

2. After separation, an attractive Coulomb force arises between the nucleus and the electron cloud, which tends to maintain the original equilibrium position. Let the separation between the electron cloud and the nucleus be x. This attractive Coulomb force is mathematically represented as follows:

$$F_C = \frac{1}{4\pi\epsilon_0}\frac{q_p q_e}{x^2} \tag{13.28}$$

When these two forces are equal and opposite, a new equilibrium will be established between the nucleus and the electron cloud of the atom.

The total number of negative charges enclosed in the sphere of radius x

$$= \left(\text{Charge density}\,(\rho) \times \text{Volume of the sphere of radius } x\right)$$

$$= \frac{3}{4}\frac{(-Ze)}{\pi R^3} \times \frac{4\pi x^3}{3} = (-Ze)\frac{x^3}{R^3} \tag{13.29}$$

The total positive charge of an atom present in the sphere of radius $x = (Ze)$ $\tag{13.30}$

Substituting this in the Coulomb force equation, we get the following relation:

$$F_c = \frac{1}{4\pi\epsilon_0}\frac{(Ze)\left(-Ze\,\dfrac{x^3}{R^3}\right)}{x^2} = -\frac{Z^2 e^2 x}{4\pi\epsilon_0 R^3} \tag{13.31}$$

At equilibrium, the two forces (Coulomb and Lorentz) must be equal and opposite. Hence,

$$F_L = -F_C \tag{13.32}$$

$$(Ze)\times E = \frac{Z^2 e^2 x}{4\pi\epsilon_0 R^3} \tag{13.33}$$

$$E = \frac{Zex}{4\pi\epsilon_0 R^3} \tag{13.34}$$

This equation represents the electric field corresponding to which the charge centres are displaced from their equilibrium position, giving rise to a dipole moment. Now, we can write that

$$x = \frac{4\pi\epsilon_0 R^3 E}{Ze} \tag{13.35}$$

This induced dipole moment (μ_{ind}) due to the applied external electric field can be mathematically represented as follows:

$$\mu_{\text{ind}} = \text{Magnitude of charge} \times \text{Displacement} \tag{13.36}$$

$$\mu_{\text{ind}} = (Ze)\times x = \alpha_e E \tag{13.37}$$

$$\mu_{\text{ind}} = 4\pi\epsilon_0 R^3 = \alpha_e \tag{13.38}$$

where α_e represents *electronic polarizability*. Thus, the salient features of electronic polarizability are as follows:

1. Electronic polarization is independent of temperature.
2. It is proportional to the volume of atoms in the material.
3. Electronic polarization occurs in all materials.

Example 13.9 An isotropic material has a volume $1.0\,\mathrm{m}^3$ and magnitude of polarization $1.0\times10^{-4}\,\mathrm{C/m}^2$, which introduces an electric field of $\dfrac{10^4\,\mathrm{N}}{\mathrm{C}}$. Find the electronic polarizability of the material.

Solution The dipole moment for the isotropic slab is given as follows:

$$\mu = |\text{Polarization vector}| \times \text{Volume} = \frac{1.0\times10^{-4}\,\mathrm{C}}{\mathrm{m}^2}\times1.0\,\mathrm{m}^3$$

$$\mu = 10^{-4}\,\mathrm{C\text{-}m}$$

$$\alpha_e = \frac{\mu}{E} = \frac{10^{-4}\,\mathrm{C\text{-}m}}{10^4\,\mathrm{N/C}} = 10^{-8}\,\mathrm{C}^2\mathrm{m/N}$$

Example 13.10 If an electric field of magnitude 10^2 N/C introduces a displacement of 1.0×10^{-6} m between the electron cloud and the nucleus, what is the electronic polarizability thus introduced in the material having an atomic number Z?

Solution The induced dipole moment is given as follows:

$$\mu_{\mathrm{ind}} = (\text{Magnitude of charge})\times \text{Displacement}$$

$$\mu_{\mathrm{ind}} = (Ze)\times x = \alpha_e E$$

$$\mu_{\mathrm{ind}} = \left(Z\times1.6\times10^{-19}\,\mathrm{C}\right)\times\left(1.0\times10^{-6}\,\mathrm{m}\right) = \alpha_e\left(\frac{10^2\,\mathrm{N}}{\mathrm{C}}\right)$$

$$\alpha_e = \left(Z\times1.6\times10^{-27}\right)\mathrm{N}^{-1}\mathrm{C}^2\mathrm{m}$$

The electronic polarizability has been calculated, and it depends on the atomic number of the material.

Example 13.11 Calculate the polarizability of an argon atom if the relative permittivity of argon at NTP is 1.000435.

Solution The polarization vector is defined [from Eq. (13.19)] as follows:

$$P = (\kappa - 1)\epsilon_0 E$$

In addition, electronic polarization is defined as follows: $P = N\alpha_e E$

Hence, $P = (\kappa - 1)\epsilon_0 E = N\alpha_e E$

$$(\kappa - 1)\epsilon_0 = N\alpha_e$$

$$\alpha_e = \frac{(\kappa - 1)}{N}\epsilon_0$$

The number of atoms of argon at NTP is $= \dfrac{6.023\times10^{23}\times10^3}{22.4} = 0.2688\times10^{26}$ $= 2.688\times10^{25}$. Hence,

$$\alpha_e = \frac{(1.000435 - 1.000)}{2.688 \times 10^{25}} \left(8.85 \times 10^{-12} \text{ C}^2/\text{N/m}^2\right)$$

$$= \frac{0.000435 \times 8.85 \times 10^{-12}}{2.688 \times 10^{25}} \text{C}^2/\text{N/m}^2$$

$$\alpha_e = 0.001432 \times 10^{-37} \text{ C}^2/\text{N/m}^2 = 1.432 \times 10^{-40} \text{ C}^2/\text{N/m}^2$$

Example 13.12 If the electronic polarizability of Kr atom is 2.18×10^{-40} C^2/N^{-1}/m^2, calculate the dielectric constant of Kr at NTP.

Solution $\alpha_e = 2.18 \times 10^{-40}$ C^2/N/m^2

The number of atoms of argon at NTP is $= \dfrac{6.023 \times 10^{23} \times 10^3}{22.4} = 0.2688 \times 10^{26} = 2.688 \times 10^{25}$.

Thus,

$$\alpha_e = 2.18 \times 10^{-40} \text{ C}^2/\text{N/m}^2 = \frac{(\kappa - 1)}{2.688 \times 10^{25}} \times 8.85 \times 10^{-12} \text{ C}^2/\text{N/m}^2$$

$$2.18 \times 10^{-40} \times 2.688 \times 10^{25} \times \frac{1}{8.85 \times 10^{-12}} = 0.66212 \times 10^{-3} = (\kappa - 1)$$

$$\kappa = 1 + 0.66212 \times 10^{-3} = 1.00066212$$

13.7.2 Ionic Polarization

Ionic polarization occurs due to the displacement of cations (positive ions) and anions (negative ions) in opposite directions. This occurs in ionic dielectrics (e.g., NaCl crystal) by the influence of an external electric field (Fig. 13.6). When an electric field (E) is applied to an ionic dielectric, there is a shift of one ion with respect to another. Positive ions move in the direction of the applied electric field through x_1 units of distance, whereas neg-ative ions move in the opposite direction through x_2 units of distance (Fig. 13.6).

We assume that there is one cation and one anion in each unit cell of that ionic crystal. Hence, the net distance between two ions is given as follows:

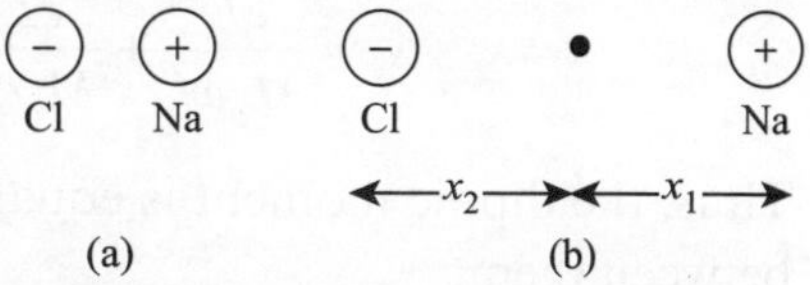

Fig. 13.6 Ionic polarization (a) NaCl crystal without electric field (b) NaCl crystal with electric field

$$x = x_1 + x_2 \tag{13.39}$$

When the ions are displaced from their mean positions, in their respective direc-tions a restoring force appears, which tends to bring the ions back to their mean position. The restoring force produced is proportional to the displacement.

For the positive ion,

$$\text{Restoring force} \left(F_p\right) \propto x_1 \tag{13.40}$$

$$\text{Restoring force} \left(F_p\right) = \beta_1 \, x_1 \tag{13.41}$$

For the negative ion,

$$\text{Restoring force} \left(F_n\right) \propto x_2 \tag{13.42}$$

$$\text{Restoring force} \left(F_n \right) = \beta_2 \, x_2 \tag{13.43}$$

Here, β_1 and β_2 are restoring force constants, which depend upon the masses of ions and angular frequency of the molecule in which ions are present. If M_p is the mass of a positive ion, M_n the mass of a negative ion, and ω_0 the angular frequency, then

$$\beta_1 = M_p \omega_0^2 \tag{13.44}$$

$$\beta_2 = M_n \omega_0^2 \tag{13.45}$$

Substituting for β_1 in the force equation, the restoring force for positive ion can be rewritten as follows:

$$F_p = M_p \omega_0^2 x_1 \tag{13.46}$$

If the force is represented by the electric force, then

$$eE = M_p \omega_0^2 x_1 \tag{13.47}$$

$$x_1 = \frac{eE}{M_p \omega_0^2} \tag{13.48}$$

Similarly, for the negative ion, we can write the following expression:

$$x_2 = \frac{eE}{M_n \omega_0^2} \tag{13.49}$$

The final displacement is represented as the sum of displacements of the positive and the negative ions:

$$x = x_1 + x_2 = \frac{eE}{M_p \omega_0^2} + \frac{eE}{M_n \omega_0^2} = \frac{eE}{\omega_0^2} \left(\frac{1}{M_p} + \frac{1}{M_n} \right) \tag{13.50}$$

Thus, the dipole moment is equal to the product of charges and separation between them:

$$\mu = e \times x \tag{13.51}$$

$$\mu = e \frac{eE}{\omega_0^2} \left(\frac{1}{M_p} + \frac{1}{M_n} \right) = \frac{e^2 E}{\omega_0^2} \left(\frac{1}{M_p} + \frac{1}{M_n} \right) \tag{13.52}$$

$$\mu \propto E \tag{13.53}$$

$$\mu \propto \alpha_i \tag{13.54}$$

$$\mu = \alpha_i \, E \tag{13.55}$$

where α_i is the ionic polarizability of the dielectric material and can be represented in terms of physical parameters as follows:

$$\alpha_i = \frac{e^2}{\omega_0^2} \left(\frac{1}{M_p} + \frac{1}{M_n} \right) \tag{13.56}$$

The following are the salient features of ionic polarizability:
1. Ionic polarizability (α_i) is inversely proportional to the square of angular frequency of the ionic molecule and directly proportional to its reduced mass, given by $\left(\dfrac{1}{M_p} + \dfrac{1}{M_n}\right)$.
2. It is temperature independent.

13.7.3 Orientational Polarization

Orientational polarization takes place only in *polar dielectrics*. Polar dielectrics have molecules with permanent dipole moments even in the absence of an electric field, as shown in Fig. 13.7. When polar dielectrics are subjected to an electric field, molecular dipoles tend to get oriented in the direction of the electric field. Orientational polarization occurs only in polar dielectrics such as H_2O, HCl, and nitrobenzene. *The polarization that arises due to the orientation of molecular dipoles is called orientational polarization.*

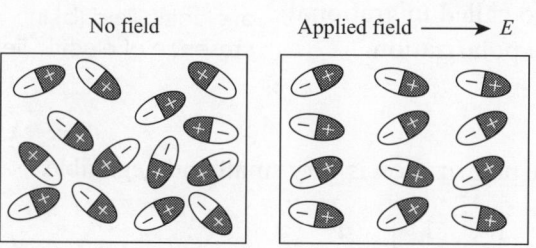

Fig. 13.7 Orientational polarization

Orientational polarization depends on temperature. When temperature is increased, thermal energy tends to disturb the alignment.

From Langevin's theory of paramagnetism,

$$\text{Net intensity of magnetization} = \frac{N\mu^2 B}{3kT} \tag{13.57}$$

Since the same principle can be used for the application of an electric field to dielectrics, we may write that

$$\text{Orientation polarization } P_0 = \frac{N\mu^2 E}{3kT} \tag{13.58}$$

As orientation polarization is proportional to the applied field ($\vec{E}$), it can be written that

$$P_0 = N\alpha_0 E = \frac{N\mu^2 E}{3kT} \tag{13.59}$$

Thus, orientation polarization can be written as follows:

$$\alpha_0 = \frac{\mu^2}{3kT} \tag{13.60}$$

Therefore, orientational polarization is inversely proportional to the absolute temperature of a material.

13.7.4 Space-charge Polarization

Space-charge polarization occurs due to the accumulation of charges at the electrodes or at the interfaces of multiphase dielectric materials (two or more phases). Space-charge polarization occurs in ferrites and semiconductors. When such materials are subjected to an electric field at a high temperature, charges get accumulated at the interfaces (Fig. 13.8). These charges create dipoles, leading to polarization. This type of polarization is known as space-charge polarization. Space-charge polarization is very small compared to other polarization mechanisms, and it is uncommon in most dielectrics. Space-charge polarization is also called migrational polarization or interfacial polarization.

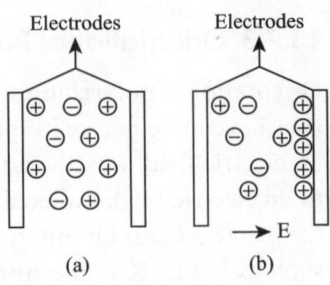

Fig. 13.8 Space-charge polarization (a) In absence of electric field (b) In presence of electric field

Total polarization:

$$\alpha = \alpha_e + \alpha_i + \alpha_0 \tag{13.61}$$

The value of space-charge polarization is very small and negligible.

$$\alpha = 4\pi\epsilon_0 R^3 + \frac{e^2}{\omega_0^2}\left(\frac{1}{M_p} + \frac{1}{M_n}\right) + \frac{\mu^2}{3kT} \tag{13.62}$$

Total polarization $= P = N\alpha E = N\alpha\epsilon_0 E \tag{13.63}$

$$P = NE\left(4\pi\epsilon_0 R^3 + \frac{e^2}{\omega_0^2}\left(\frac{1}{M_p} + \frac{1}{M_n}\right) + \frac{\mu^2}{3kT}\right) \tag{13.64}$$

This equation is referred to as the *Langevin–Debye equation.*

13.8 FREQUENCY AND TEMPERATURE DEPENDENCE OF POLARIZATION

When an alternating field is applied across a material, polarization occurs as a function of time. Polarization as a function of time t is given by the following equation:

$$P(t) = P\left(1 - e^{-t/\tau}\right) \tag{13.65}$$

where P is the maximum polarization that occurs due to the application of a static field for a long time; τ is the relaxation time (time taken for a particular polarization process to take place).

13.8.1 Frequency Dependence

Electric polarization is the fastest polarization, which will be complete at the instant the field is applied, the reason being that electrons are lighter elementary particles than ions. Therefore, even when a very-high-frequency electric field is applied [in the optical range ($\approx 10^{15}$ Hz)], electric polarization occurs during every cycle of the applied field. Ionic polarization is a little slower than electronic polarization. As ions are heavier than an electron cloud, the time taken for their displacement is larger. In addition, the frequency of the applied electric field, by which the ions will be displaced, is equal to the frequency of lattice vibration ($\sim 10^{13}$ Hz). This means that, if the frequency of the applied electric field is in the optical frequency range, ions do not respond, as the time required for lattice vibration to occur is nearly 100 times more than the period of applied voltage at the optical frequency. Hence, at optical frequencies, there is no ionic polarization. If the frequency of the applied field is $<10^{13}$ Hz (i.e., infrared range), the ions have enough time to respond during each cycle of the applied field. Orientational polarization is even slower than ionic polarization. This type of polarization occurs only in the frequency (audio and radio frequencies) range ($\approx 10^6$ Hz), as shown in Fig. 13.9. Space-charge polarization is the slowest because ions have to diffuse (jump) over several atomic distances. This process occurs at a very low frequency [i.e., at power frequencies (10^2 Hz)], as shown in Fig. 13.9. This figure illustrates the four types of polarization at different frequency ranges. At optical frequencies ($\sim 10^{15}$ Hz), electronic polarization alone is present. At $\sim 10^{13}$ Hz frequency range, ionic polarization occurs in addition to electronic polarization. In the 10^6–10^{10} Hz frequency range, contribution due to orientation polarization gets added, whereas at 10^2 Hz frequency range space-charge polarization also contributes (Table 13.1).

From Fig. 13.9, we observe that at lower frequencies all the four types of polarizations occur and total polarization is very high (maximum). Total polarization decreases with the increase in frequency and becomes minimum

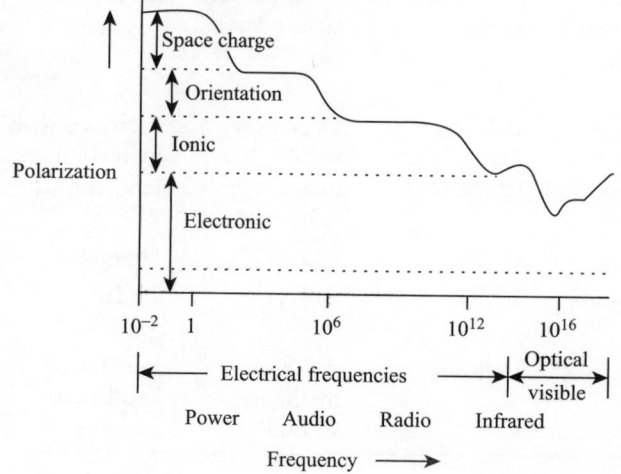

Fig. 13.9 Frequency dependence of polarization

Table 13.1 Polarization and frequency ranges

Frequency range in Hz	Type of polarization
$\sim 10^{15}$	Electronic
$\sim 10^{13}$	Ionic + Electronic
$\sim 10^6 - 10^{10}$	Orientation + Ionic + Electronic
$\sim 10^2$	Space charge + Orientation + Ionic + Electronic

in the optical frequency range. Hence, at high frequencies, the value of polarization is very small (minimum).

13.8.2 Temperature Dependence

When a dielectric material is subjected to an increase in temperature, electronic distribution in the constituent molecules is not affected. Hence, there will be no influence of temperature on electronic and ionic polarization mechanisms. Therefore, electronic and ionic polarizations are practically independent of temperature. An increase in temperature may bring about a higher degree of randomness in molecular orientation of the material. This will affect the tendency of permanent dipoles to align along the direction of the field. Hence, orientation polarization decreases with the increase in temperature. However, in space-charge polarization, an increase in temperature helps in the movement of ions by diffusion. As a result, it will increase polarization. Thus, both orientational and space-charge polarization mechanisms are strongly temperature dependent (Table 13.2).

Table 13.2 Comparison of various types of polarization mechanisms

Factor	Electronic polarization	Ionic polarization	Orientational polarization	Space-charge polarization
Definition	Electron clouds are shifted with respect to nucleus	Cations and anions are shifted	Alignment of random molecules takes place	Ion diffusion takes place
Temperature dependence	Independent of temperature	Independent of temperature	Decreases with increase in temperature	Increases with increase in temperature
Power loss	Low	High	Higher	Highest
Frequency range	10^{15} Hz and above	10^{14} Hz	10^{12} Hz	10^5 Hz
Examples	Inert gases	Ionic crystals	Alcohol, methane, CH_3Cl	Ferrites, semiconductors

13.9 POLARIZABILITY AND INTERNAL FIELD

When a dielectric material is placed in an external electric field, it exerts a dipole moment. There is an interplay between the following two fields that are developed:
1. A macroscopic electric field due to the external field
2. A field due to the electric dipole moment

This long-range Coulomb field that is created due to dipoles is known as an internal field or a local field. This internal field is responsible for polarization of each atom or molecule in a solid.

13.9.1 Lorentz Method for Finding Internal Field

A dielectric material is uniformly polarized by placing it in between two plates of a parallel-plate capacitor (uniform electric field), as shown in Fig. 13.10.

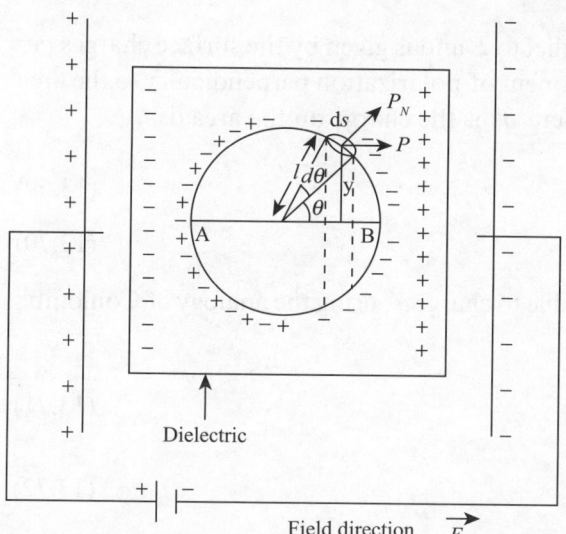

Dielectric

Field direction $\overrightarrow{E}$

Fig. 13.10 Calculation of local field

Assume an imaginary small spherical cavity around an atom for which the internal field must be calculated at its centre. The internal field (E_{int}) at the atom site can be considered to be made up of the following four components: E_1, E_2, E_3, and E_4. Thus,

$$E_{int} = E_1 + E_2 + E_3 + E_4 \tag{13.66}$$

where E_1 is the electric field due to charges on the plates of the capacitor (without dielectric), E_2 is the electric field due to polarized charges (induced charges) on the plane surface of the dielectric, E_3 is the electric field due to polarized charges induced on the surface of the imaginary spherical cavity (to be calculated), and E_4 is the electric field due to permanent dipoles of atoms inside the spherical cavity considered.

Macroscopically, we can assume that the total internal field is the sum of the field externally applied (E_1) and the field induced on the plane surface of the dielectric (E_2), and can be represented as E:

$$E = E_1 + E_2 \tag{13.67}$$

If we consider a dielectric that is highly symmetric, the field due to dipoles present inside the imaginary cavity will cancel out each other. Therefore, the electric field due to permanent dipoles $E_4 = 0$. Hence, the internal field can be written as follows:

$$E_{int} = E + E_3 \tag{13.68}$$

13.9.1.1 Finding E_3

Let us consider a small area ds taken at an angle θ with the direction of the field E, on the surface of a spherical cavity. Let this small area ds be confined within an angle $d\theta$.

The polarization (P) is parallel to E and is given by the surface charges per unit area. Let P_N be the component of polarization perpendicular to the area ds, as shown in Fig. 13.10, where q' is the charge on the area ds.

$$P_N = P\cos\theta = \frac{q'}{ds} \tag{13.69}$$

$$\Rightarrow q' = P\cos\theta\, ds \tag{13.70}$$

The electric field intensity at C due to charge q', using the analogy of Coulomb's law, is given as follows:

$$E = \frac{q'}{4\pi\epsilon_0 r^2} \tag{13.71}$$

$$= \frac{P\cos\theta\, ds}{4\pi\epsilon_0 r^2} \tag{13.72}$$

This field intensity is along the radius r and can be resolved into two components, as shown in Fig. 13.11.

The component of intensity *parallel* to the field direction:

$$E = E\cos\theta \tag{13.73}$$

$$E = \frac{P\cos\theta\, ds}{4\pi\epsilon_0 r^2}\cos\theta \tag{13.74}$$

The component of intensity *perpendicular* to the field direction:

$$E_y = E\sin\theta \tag{13.75}$$

Since the perpendicular components are in opposite directions (Fig. 13.11), they cancel each other out. Hence, only the parallel components are taken into consideration.

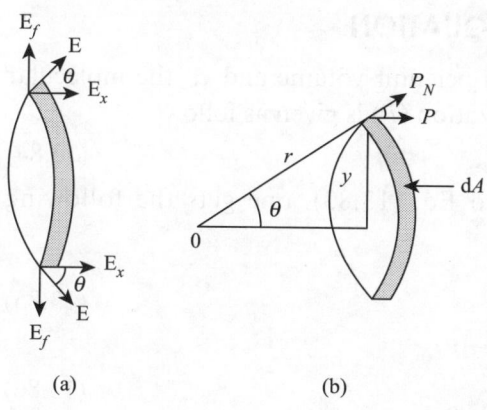

Fig. 13.11 Resolution of field intensity (a) Resolution of field intensity E along x- and y-axis. Field vector E is inclined at an angle θ to the x-axis. (b) Polarization vector P is parellel to the electric field vector E. Normal polarization vector makes angle θ with respect to the horizontal axis.

(a)   (b)

Now, consider a ring area dA, which is obtained by revolving ds about AB (Fig. 13.11), such that

$$\text{Ring area } dA = \text{Circumference} \times \text{Thickness} \qquad (13.76)$$

$$dA = 2\pi y \times r\,d\theta \qquad (13.77)$$

(since $q \sin \theta = y/r$)

$$dA = 2\pi r \sin\theta \times r\,d\theta \qquad (13.78)$$

$$dA = 2\pi r^2 \sin\theta\,d\theta \qquad (13.79)$$

Substitution of the area of cross section in the electric field gives the following relation:

$$E = \frac{P\cos^2\theta\left(2\pi r^2 \sin\theta\,d\theta\right)}{4\pi\epsilon_0 r^2} \qquad (13.80)$$

$$E = \frac{P\cos^2\theta \sin\theta\,d\theta}{2\epsilon_0} \qquad (13.81)$$

The electric field intensity due to charge present in the whole sphere is obtained by integrating Eq. (13.81) over the limits from 0 to π. This represents the field E_3:

$$E_3 = \int_0^\pi \frac{P\cos^2\theta \sin\theta\,d\theta}{2\epsilon_0} \quad \left(\text{as } \int_0^\pi \cos^2\theta \sin\theta = \frac{2}{3}\right)$$

$$E_3 = \frac{P}{2\epsilon_0}\frac{2}{3} = \frac{P}{3\epsilon_0} \qquad (13.82)$$

Thus, the value of the total internal field or Lorentz field is given as follows:

$$E_{\text{int}} = E + \frac{P}{3\epsilon_0} \qquad (13.83)$$

Thus, the internal local electric field is different from E. The local intensity E_{int} is larger than the macroscopic intensity E. Hence, the molecules are more effectively polarized.

13.10 CLAUSIUS–MOSSOTTI EQUATION

Let N be the number of molecules per unit volume and α the molecular polarizability. Then the total polarization (P) is given as follows:

$$P = N\alpha E_{int} \tag{13.84}$$

Substituting the value of E_{int} into Eq. (13.83), one gets the following expressions:

$$P = N\alpha\epsilon_0\left(E + \frac{P}{3\epsilon_0}\right) \tag{13.85}$$

$$P = N\alpha\epsilon_0 E + N\alpha\frac{P}{3} \tag{13.86}$$

$$P\left(1 - \frac{N\alpha}{3}\right) = N\alpha\epsilon_0 E \tag{13.87}$$

$$P = \frac{N\alpha}{(1 - (N\alpha/3))}\epsilon_0 E = (\kappa - 1)\epsilon_0 E \tag{13.88}$$

The polarizability is related to the dielectric constant, as derived in Eq. (13.88). Now,

$$(\kappa - 1) = \frac{N\alpha}{(1 - (N\alpha/3))} = \left(\frac{\epsilon}{\epsilon_0} - 1\right) \tag{13.89}$$

In this equation, the dielectric constant of a liquid is related to the atomic polarizability (α). This is referred to as the Clausius–Mossotti equation. The physical significance of this equation is that it is relating a macroscopic quantity κ, the dielectric constant, to a microscopic quantity α, the polarizability. To a good approximation, the equation can be reduced to a simple form if the value of $N\alpha$ is small compared to unity, that is, N is small; hence,

$$(\kappa - 1) = N\alpha \tag{13.90}$$

Note: This equation is valid only for electronic polarization and does not hold for polar molecules.

Example 13.13 If the electronic polarization of a material is $(Z \times 10^{-23})\,C^2/N^{-1}/m$ at NTP, what is the dielectric constant of the material?
Solution Using the Clausius–Mossotti equation,

$$(\kappa - 1) = N\alpha_e$$

$$\kappa = (6.023 \times 10^{23})(Z \times 10^{-23})\,C^2/N/m$$

$$\kappa = 6.023 \times Z$$

13.11 SOLID DIELECTRICS

In solid crystals, due to the near permanent locations of atoms, there is a possibility of permanent polarization, which is appreciable in the absence of an external electric field. On the basis of how dielectric materials are used, we can categorize them as follows:

1. Active dielectrics, that is, ferroelectrics, piezoelectrics, and pyroelectrics. These materials are used to generate, amplify, modulate, and convert electrical signals. These materials store electrical signals.
2. Ferroelectric materials possess electric polarization in the absence of the applied external electric field.

Barium titanate ($BaTiO_3$), potassium dihydrogen phosphate (KH_2PO_4), ammonium dihydrogen phosphate ($NH_4H_2PO_4$), lithium niobate ($LiNbO_3$), and Rochelle salt ($NaKC_4H_4O_6 \cdot 4H_2O$) are a few typical examples of ferroelectric materials.

The following are some salient features of ferroelectric materials:

1. Ferroelectric materials can easily be polarized even by very weak electric fields.
2. They exhibit dielectric hysteresis. Lagging of polarization behind the applied electric field is called dielectric hysteresis. Ferroelectricity is a result of dielectric hysteresis (Fig. 13.12).
3. Ferroelectric materials possess spontaneous polarization, which is the polarization that persists even when the applied field is zero.
4. They possess permanent electric dipoles and an internal electric field, which develop spontaneous polarization at the ferroelectric Curie temperature.
5. Ferroelectric materials exhibit ferroelectricity when the temperature $T \le T_c$, where T_c = ferroelectric Curie temperature. When $T > T_c$, they are converted into paraelectric materials.
6. Ferroelectric materials exhibit piezoelectricity, which means the creation of electric polarization by mechanical stress.
7. *Piezoelectric materials* are used to make pressure transducers, ultrasonic transducers, and microphones. Examples include the following:
   (a) Quartz, lithium niobate, and barium titanate among the crystalline materials
   (b) Lead zirconium titanate, calcium barium titanate, and lead barium niobate among the ceramic materials
8. Electrets are ferroelectric materials and electrostatic analogues of permanent magnets. Electrets possess a gross permanent electric dipole moment. They are manufactured from certain types of waxes, plastics, and ceramics. Electrets are used in capacitor microphones and gas filters to capture submicron particles by electrostatic attraction. Further, electret bondages are used over fractured bones to speed up the healing process. The following are examples of some important ferroelectric materials:
   (a) Barium titanate: It is a very useful ferroelectric material and an active dielectric. It is used in ultrasonic transducers, harmonic generators, electrets, etc.

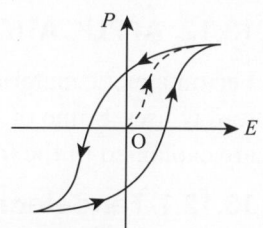

Fig. 13.12 Dielectric hysteresis in ferroelectric materials

(b) Lithium niobate:
(i) It is an artificial (or) synthetic ferroelectric material.
(ii) It has the highest Curie temperature of 1210°C. Therefore, this is a suitable piezoelectric material for space applications.
(iii) It has high electric polarizability and low-loss optical transmission. It is a uniaxial crystal with two refractive indices. It exhibits nonlinear optical characteristics.
(iv) Even though it is a dielectric material, at high temperatures it becomes electrically conductive due to lattice distortion and ionic conductivity.
(v) It is a high-frequency single-crystal piezoelectric material. Due to its high electromechanical coupling and low attenuation for elastic waves up to 2 GHz, it is primarily used in surface acoustic waves (SAW) devices as delay lines, and in narrow-band and ultra high frequency (UHF) filters.
(vi) Its dielectric constant is roughly about 44 (range 27–84). A rapid change of temperature can result in a highly strong field on the surface, leading to a breakdown of the surface.
(c) Lithium tantalite ($LiTaO_3$): It is similar to lithium niobate in many physical properties, and its Curie point is 660°C and melting point is 1650°C.

The domain size is much smaller than $LiNbO_3$ domains. Further, the domains are antiparallel polar domains. It has a low temperature coefficient of resonance, and so it can be used in resonators.

It is a more efficient electro-optic modulator material than $LiNbO_3$. If we take X cut plates of $LiNbO_3$ and $LiTaO_3$, $LiNbO_3$ has the advantage of combining higher coupling factors and lower mechanical impedance values.

Pyroelectricity means the creation of electric polarization by thermal stress. *Pyroelectric materials* are used to make highly sensitive infrared detectors. Some ferroelectric semiconductors, such as $BaTiO_3$–$SrTiO_3$, $BaTiO_3$–$PbTiO_3$, and $SrTiO_3$–$PbTiO_3$ are used to measure and control temperature in thermistors. Examples include barium titanate, triglycine sulphate, lithium niobate, lithium tantalate, and polyvinyl fluoride. These are passive dielectric materials, that is, electrical insulating materials, and hence obstruct the flow of electric current.

13.12 APPLICATIONS OF FERROELECTRIC MATERIALS

Ferromagnetic materials often find applications in various industrial and day-to-day uses. Some of the important applications of ferromagnetic materials are explained in the following subsections.

13.12.1 Ferroelectric Energy Converter

Ferroelectric crystals exhibit the pyroelectric effect, that is, a change in the temperature of the crystal produces a change in its polarization. Using this effect, one can convert heat energy into useful electrical energy.

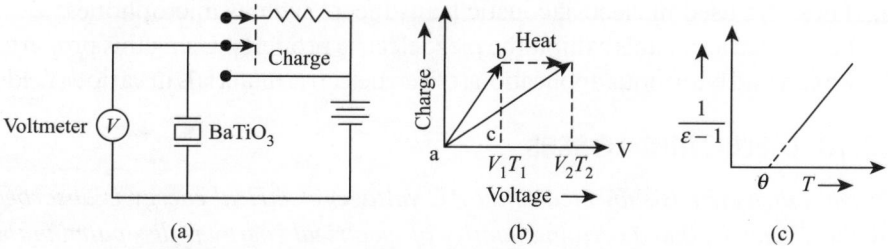

Fig. 13.13 Ferroelectric energy conversion

First, the capacitor with $BaTiO_3$ is charged to a voltage of V_1 at a temperature T_1 just below the Curie temperature θ (Fig. 13.13).

$$Q = C_1 V_1 \tag{13.91}$$

where C_1 is the capacitance of the capacitor.
From Fig. 13.13, it is evident that

$$\text{Initial energy} = \frac{1}{2} Q V_1 = \text{Area of the triangle ABC} \tag{13.92}$$

Now, the capacitor is isolated from the battery and it is heated up to temperature T_2. Therefore, the dielectric constant decreases. For example, in $BaTiO_3$, when temperature is increased from 15°C to 30°C the dielectric constant is reduced about five times. Hence, one can represent this mathematically as follows:

$$\frac{\text{Value of dielectric constant at 15°C}}{\text{Value of dielectric constant at 30°C}} = \frac{\kappa 15}{\kappa 30} = 5 \tag{13.93}$$

Since the value of the dielectric constant decreases, the value of the capacitance correspondingly decreases, and at temperature T_2, the capacitance of the capacitor $C_2 < C_1$. Since the charge Q in the capacitor is constant and $C_2 < C_1$,

$$Q = C_1 V_1 = C_2 V_2 \tag{13.94}$$

As $\quad V_2 > V_1$,

$$V_2 = \frac{V_1 C_1}{C_2} = \frac{V_1 \kappa_1}{\kappa_2} \tag{13.95}$$

$$\text{Increase in energy} (\Delta W) = \frac{1}{2} Q (V_2 - V_1) \tag{13.96}$$

The capacitor is now discharged through a load resistor R and a rechargeable battery of voltage V_1 to get electrical power. The capacitor is cooled to temperature T_1 to complete the cycle. The following are some important uses of ferromagnetic materials:

1. The high dielectric constant of ferroelectric crystals is also useful for storing energy in small-sized capacitors in electrical circuits.
2. In optical communication, the ferroelectric crystals are used for optical modulation.

3. These are used in electroacoustic transducers such as microphones.
4. Ferroelectric crystals exhibit the piezoelectric property. Using this property, we can find enormous applications of ferroelectric materials in various fields.

13.13 DIELECTRIC LOSSES

When a dielectric is subjected to an AC voltage, electrical energy is absorbed by the dielectric, and a certain quantity of electrical energy is dissipated in the form of heat energy. This dissipation of electrical energy in the dielectric is called dielectric loss.

Dielectric loss can occur in both direct and alternating voltages. It is less in direct voltage than that in alternating voltage.

13.13.1 Expression for Dielectric Loss (or Loss Tangent)

When an AC voltage is applied to a perfect insulator like vacuum or a purified gas, it does not absorb electrical energy and there is no loss of electrical energy [Fig. 13.14(a)].

Polarization of the dielectric is in phase with the voltage. In such a case, the charging current leads the applied voltage by an angle of 90°, as shown in Fig. 13.14.

We know that power loss is given as follows:

$$P_L = VI \cos\theta \tag{13.97}$$

When $\theta = 90°$; $P_L = VI \cos 90°$;

$$P_L = 0 \tag{13.98}$$

This means that there is no power loss in the insulator.

However, a practical dielectric always has some loss of electrical energy. In this case, the leakage current does not lead the applied voltage by exactly 90°. The phase angle is always less than 90°, as shown in Fig. 13.14.

The current leads the voltage by $(90° - \delta)$. This shows that there is some loss in electrical energy. The symbol δ represents dielectric loss angle, which is a measure of the power dissipated in each cycle.

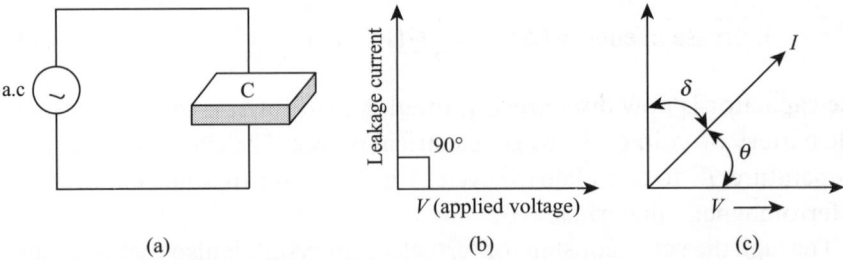

Fig. 13.14 Dielectric loss graph (a) ac source used to charge capacitor (b) Lag of 90° between applied voltage and leakage current (c) In practical dielectric, current leads voltage by 90° δ where δ represents dielectric loss angle

The power loss for a dielectric having a capacitance C and a voltage V applied to it at a frequency of ϑ Hz is given by Eq. (13.97) as follows:

$$P_L = VI \cos\theta$$

Since $\theta = (90 - \delta)$, we have

$$P_L = VI \cos(90 - \delta) \tag{13.99}$$

$$P_L = VI \sin\delta \tag{13.100}$$

From Ohm's law, we know that $V = IR$ or $I = \dfrac{V}{R}$.

If the capacitive reactance is X_C, we can write that

$$I = \frac{V}{X_C} \tag{13.101}$$

We know that frequency $= \dfrac{1}{2\pi RC}$.

Comparing Eq. (13.101) with the Ohm's law equation (as $R = X_C$), we get the following expression:

$$\vartheta = \frac{1}{2\pi X_C C}$$

$$X_C = \frac{1}{2\pi\vartheta C} \tag{13.102}$$

Substituting Eq. (13.102) into Eq. 13.101), we have the following relation:

$$I = \frac{V}{X_C} = \frac{V}{\dfrac{1}{2\pi\vartheta C}} = V(2\pi\vartheta C) \tag{13.103}$$

Using Eq. (13.103) in Eq. (13.100), we have the following expression:

$$P_L = 2\pi\vartheta C V^2 \sin\delta \tag{13.104}$$

In most dielectrics, the angle δ is very small. Thus, the sine argument can be easily translated to the tangent argument, that is, $\sin\delta = \tan\delta$. Hence, Eq. (13.104) takes the following form:

$$P_L = 2\pi\vartheta C V^2 \tan\delta \tag{13.105}$$

Thus, the dielectric power loss is dependent on the $\tan\delta$ factor as long as other factors such as voltage, frequency, and capacitance are constant. The power factor for a dielectric is $\tan\delta$. Power loss changes with frequency. Its value is high in the electrical frequency range and low in the optical frequency range (Fig. 13.15).

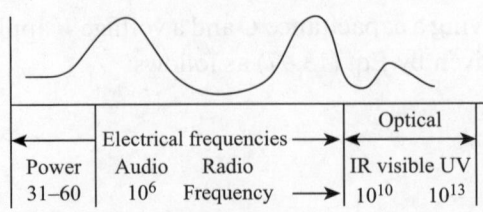

Power | Audio | Radio | IR visible UV
31–60 | 10^6 | Frequency ⟶ | 10^{10} 10^{13}

Electrical frequencies ⟶

Optical

Fig. 13.15 Variation of power loss with frequency

13.13.2 Factors Affecting Dielectric Loss

Dielectric loss may increase due to the following factors:
1. High frequency of the applied voltage
2. High value of the applied voltage
3. Presence of humidity
4. Temperature rise

13.14 DIELECTRIC BREAKDOWN

When the strength of the electric field applied to a dielectric exceeds a critical value, a very large current flows through it. The dielectric loses its insulating property and becomes a conductor. This phenomenon is known as a dielectric breakdown.

13.14.1 Dielectric Strength

The electric field strength at which a dielectric breakdown occurs is known as dielectric strength. It is the breakdown voltage per unit thickness of the material.

$$\text{Dielectric strength} = \frac{\text{Dielectric breakdown voltage}}{\text{Thickness of dielectric}}$$

In practical applications, the failure or breakdown of a dielectric material is of great concern to engineers. There are different mechanisms by which a dielectric breakdown takes place.

Example 13.14 Calculate the maximum potential gradient to which a 0.5 mm thick mica sheet can be subjected. The dielectric strength for mica is $100 \times \dfrac{10^6 \, \text{V}}{\text{m}}$.

Solution The calculations are as follows:

$$\text{Dielectric strength} = \frac{\text{Breakdown voltage}}{\text{Thickness of the sheet}}$$

$$100 \times 10^6 \, \frac{\text{V}}{\text{m}} = \frac{\text{Breakdown voltage}}{0.5 \times 10^{-3} \, \text{m}}$$

Breakdown voltage $= 0.5 \times 10^{-3} \times 100 \times 10^6 \, \text{V} = 5.0 \times 10^4 \, \text{V}$

This is the reason that mica is widely used for insulation purposes, as it can sustain high voltages even for a thin sheet.

Example 13.15 What should be the thickness of a porcelain sheet that can be inserted in an instrument that is to be subjected to a breakdown voltage of the order of 10^6 V? The dielectric strength of porcelain is $6*\dfrac{10^6 V}{m}$.

Solution The thickness of the porcelain sheet is calculated as follows

$$\text{Dielectric strength} = \frac{\text{Breakdown voltage}}{\text{Thickness of the sheet}}$$

$$6 \times 10^6 \, \frac{V}{m} = \frac{10^6 \, V}{d} \;\Rightarrow\; d = \frac{1}{6} \, m = 0.167 \times 100 \, cm$$

The thickness of the porcelain plate is 16.7 cm.

13.14.2 Types of Dielectric Breakdown

Some important types of dielectric breakdown are as follows:
1. Intrinsic breakdown
2. Avalanche breakdown
3. Thermal breakdown
4. Chemical and electrochemical breakdowns
5. Discharge breakdown
6. Defect breakdown

13.14.2.1 Intrinsic Breakdown

For a dielectric, charge displacement increases with increasing field strength. Beyond a critical value of field strength, there is an electrical breakdown due to physical deterioration in the dielectric material (Fig. 13.16).

When the applied electrical field is large, some of the electrons in the valence band cross over to the conduction band across the large forbidden energy gap. They become conduction electrons, producing a large conduction current. As a result, a large current flows through the dielectric, and a breakdown occurs. This type of breakdown is called an *intrinsic breakdown*.

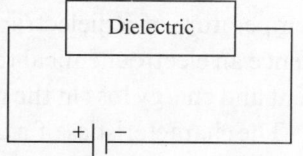

Fig. 13.16 Intrinsic breakdown

13.14.2.2 Avalanche Breakdown

Conduction electrons are accelerated to very high velocity and energy on further application of an electrical field. They collide with valence electrons in the covalent bond. Valence electrons, on acquiring this energy, jump from the valence band to the conduction band. This process continues as more and more valence electrons jump to the conduction band by breaking the covalent bond. As a result, a large current flows through the dielectric, and a breakdown occurs. This type of breakdown is called an *avalanche breakdown*.

The following are the characteristics of an avalanche breakdown:
1. Intrinsic and avalanche breakdowns require large electrical fields.
2. They occur at low temperatures (at about room temperature or lower temperatures).
3. They do not depend on the size and shape of the sample and the configuration of electrodes.
4. They can occur in thin samples.
5. They occur within a short span of time (microseconds).

13.14.2.3 Thermal Breakdown

When an electric field is applied to a dielectric material, some amount of heat is produced. This heat must dissipate from the material. In some cases, the amount of heat produced is very large when compared to the heat dissipated. Due to this excess heat, the temperature inside the dielectric increases, which may produce local heating in the dielectric material. During this process, a large amount of current flows through the material and causes the dielectric to break down. This type of breakdown is known as a *thermal breakdown*.

The following are the characteristics of a thermal breakdown:
1. It occurs only at high temperatures.
2. The strength of the electric field to create a dielectric breakdown depends on the size and shape of the dielectric sample.
3. The breakdown time is of the order of a few milliseconds.
4. It requires moderate electric fields.

13.14.2.4 Chemical and Electrochemical Breakdowns

An electrochemical breakdown is similar to a thermal breakdown. When the temperature of a dielectric material increases, mobility of ions increases, and hence an electrochemical reaction may take place. This results in leakage current and energy loss in the material, and finally a dielectric breakdown occurs.

The characteristics of a chemical/electrochemical breakdown are as follows:
1. It occurs only at low temperatures.
2. It depends on the concentration of ions and magnitude of the leakage current.

13.14.2.5 Discharge Breakdown

Discharge breakdown occurs when the dielectric contains occluded gas bubbles (Fig. 13.17). When this type of dielectric is subjected to an electric field, the gas present in the material will easily ionize, producing a large ionization current.

Gaseous ions bombard the solid dielectric. This causes electrical deterioration and leads to a dielectric breakdown. This is known as a *discharge breakdown*.

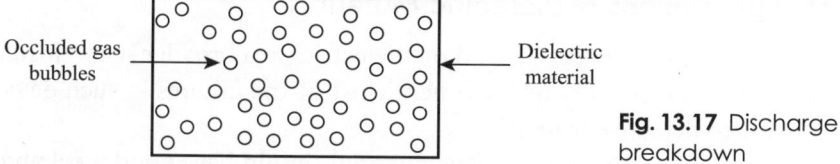

Fig. 13.17 Discharge breakdown

The characteristics of a discharge breakdown are as follows:
1. It occurs at low voltages.
2. It occurs in the dielectric material that has a large number of occluded gas bubbles.
3. When discharge takes place at a point, the surrounding places are burnt and hence their electrical properties are affected. Thus, the life of an insulating material depends on the number of discharges taking place inside the material, that is, it depends upon the frequency of applied voltage.

13.14.2.6 Defect Breakdown

The surface of a dielectric material may have defects such as cracks, porosity, and blow holes (Fig. 13.18). Impurities such as dust or moisture may collect at these discontinuities (defects), leading to a breakdown.

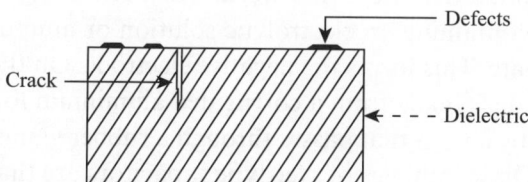

Fig. 13.18 Defect breakdown

Remedies for dielectric breakdown   To avoid a breakdown, the insulating material should have the following properties:
1. It should have high resistivity to reduce the leakage current.
2. It should have high dielectric strength to withstand higher voltages.
3. It should have a smaller dielectric loss.
4. It should have sufficient mechanical strength.
5. It should be resistant to oils, liquids, gas fumes, acids, and alkalis.
6. It should have a small thermal expansion to prevent mechanical damage.
7. It should be fire proof.

13.14.3 Characterization of Dielectric Materials

Dielectric materials are classified into the following types based on their physical state:
1. Solid dielectric materials
2. Liquid dielectric materials
3. Gaseous dielectric materials

13.15 Applications of Dielectric Materials

Almost all electrical devices depend on dielectric (in a broader sense, insulating) materials in some way or the other. Most of the failures in such devices may be due to defective insulation.

In general, an insulating/dielectric material should have good mechanical and electrical strengths and acceptable thermal and chemical stabilities. Insulating materials are used in power and distribution transformers, rotating machines, capacitors, cables, transmission equipment, switchgear, and electronic appliances. Some examples of these materials have been discussed in the subsequent sections.

13.15.1 Capacitors

A capacitor is made up of conducting plates and a dielectric. The dielectric may be air, a solid insulating material, or a liquid, depending on the functions of the capacitor and the electronic circuitry.

Capacitors are used in many electrical utilities. Figure 13.19 shows the different types of capacitors. Power capacitors use tissue paper and/or polypropylene films as dielectrics. Such an insulation system is impregnated with mineral oil or synthetic liquids.

Electrolytic capacitors are produced in two forms: wet and dry. The wet type consists of an aluminium can containing an electrolytic solution of ammonium borate or sodium phosphate. This forms the negative electrode. On the other hand, the positive electrode is made from a corrugated aluminium foil held in the electrolyte solution by a stem that passes through a rubber gland. They are used for DC supply. Disadvantages of wet-type capacitors are that they must be placed vertically; however, this type of alignment causes the electrolyte to evaporate.

A dry-type capacitor is made from two very thin aluminium foil strips separated by two layers of circulating papers saturated with an electrolytic paste of glycol and ammonium tetraborate.

Many types of capacitors are now used specially for DC applications. They include paper capacitors (paper with impregnates such as castor oil, synthetic oils, mineral oil, and polyesters), electrolytic capacitors (aluminium

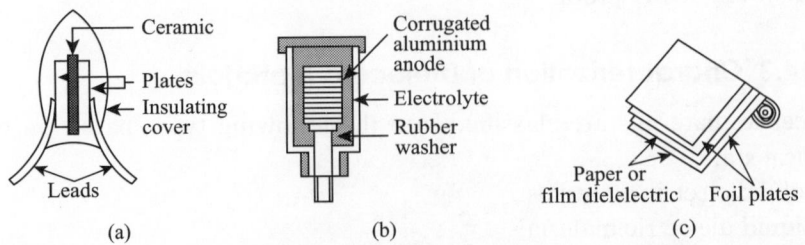

Fig. 13.19 Different types of capacitors (a) Ceramic (b) Electrolytic (c) Power

or tantalum, and ceramic capacitors) with a variety of ceramic bodies (tubular capacitor, disc capacitor, blocking capacitor, etc.), and film dielectric (polystyrene, Teflon, and polyethylene) capacitors.

13.15.2 Power and Distribution Transformers

The insulating materials used in power and distribution transformers must possess the following properties.
1. Good resistance to withstand power frequency voltages and impulse over voltages
2. Good strength to withstand fabrication and handling during manufacture, and electromagnetic forces during overload, short circuit, and normal operating conditions
3. Good thermal stability and low ageing effect

Insulating materials for different machine parts are listed in Table13.3:

Table 13.3 Insulating materials used in power and distribution transformers

Instrument	Dielectric material
Low-voltage coil to ground; high-voltage coil to low voltage coil	Thick radial spacers or tubes made up of pressboard, paper, glass fabric, paper filled plastic laminates, or porcelain
Turn to turn	Organic enamel, paper, glass tapes
Layer to layer and coil to coil	Craft paper, glass fabric, pressboard, varnished paper
Fluid (used as insulating and cooling media)	Mineral oil, air, nitrogen
Bushings	Porcelain, phenolic bonded tubes

IMPORTANT CONCEPTS

1. Dielectric materials belong to the group of insulating materials in which the dipoles can be produced by applying an external electric potential.
2. The product of the magnitude of a charge and the distance between the two charges is called the 'dipole moment'.
3. Permittivity represents the dielectric property of a material. It indicates the polarizability nature of the material/medium.
4. Dielectric constant (ε_r) or relative permittivity is the ratio of the absolute permittivity of the medium (ε) to the permittivity of free space (ε_0)
5. Polarization is the process of producing electric dipoles inside a dielectric material by the application of an external electric field.
6. Polarizability is defined as the ratio of average dipole moment to the electric field applied.
7. Polarization vector is directly proportional to the external electric field applied to the dielectric material.

8. Electronic polarization is due to the displacement of positively charged nucleus and negatively charged electrons of an atom in opposite directions by the application of an electric field.

9. Ionic polarization is due to the displacement of cations (positive ions) and anions (negative ions) in opposite directions.

10. Polarization due to orientation of molecular dipoles is called orientation polarization.

11. Space-charge polarization occurs due to the accumulation of charges at the electrodes or at the interfaces of multiphase dielectric materials.

12. Electric polarization is the fastest polarization, which will be complete at the instant the field is applied.

13. Electronic and ionic polarizations are practically independent of temperature.

14. Local field is the actual field experienced by an atom/ion/molecule inside the dielectric material in the presence of an external electric field.

15. When a dielectric is subjected to an AC voltage, electrical energy is absorbed by the dielectric, and a certain quantity of electrical energy is dissipated in the form of heat energy. This dissipation of electrical energy in the dielectric is called dielectric loss.

16. When the strength of the electric field applied to a dielectric exceeds a critical value, a very large current flows through it; this phenomenon is termed as a dielectric breakdown.

17. Materials that exhibit electric polarization even in the absence of an applied electric field are known as ferroelectric materials.

18. Examples of ferroelectric materials are barium titanate ($BaTiO_3$), potassium di-hydrogen phosphate (KH_2PO_4), ammonium di-hydrogen phosphate ($NH_4H_2PO_4$), lithium niobate ($LiNbO_3$), and Rochelle salt ($NaKC_4H_4O_6 \cdot (4H_2O)$).

19. When a dielectric is subjected to an AC voltage, electrical energy is absorbed by the dielectric, and a certain quantity of electrical energy is dissipated in the form of heat energy. This dissipation of electrical energy in the dielectric is called dielectric loss.

20. As an electric potential is applied to a dielectric (insulating material), a dipole moment is produced, which is the product of the magnitude of the charge and the distance between the two charges.

APPLICATIONS

1. Almost all electrical devices depend on dielectric (in a broader sense, insulating) materials in some way or the other.

2. An insulating/dielectric material has good mechanical and electrical strengths and acceptable thermal and chemical stability. Insulating materials are used in power and distribution transformers, rotating machines, capacitors, cables, transmission equipment, switchgear, and electronic appliances.

3. Capacitors are used in many electrical utilities. There are different types of capacitors (refer to Table 13.2).

4. Power and distribution transformers use various dielectric/insulating materials.

IMPORTANT FORMULAE

1. The polarization vector is mathematically derived using the following formula: $P = Nq\delta = N\mu = (\kappa - 1)\epsilon_0 E$

2. The dielectric constant is governed by the following relation: $\kappa = \dfrac{C_\kappa}{C_0} = \dfrac{\epsilon_\kappa}{\epsilon_0} = \dfrac{V_0}{V_\kappa}$

3. The displacement vector is given as follows: $D = \epsilon_0 E + P$

4. Electronic polarizability is given as follows: $\mu_{ind} = 4\pi\epsilon_0 R^3 = \alpha_e$

5. Ionic polarizability is given as follows:

$$\alpha_i = \frac{e^2}{\omega_0^2}\left(\frac{1}{M_p} + \frac{1}{M_n}\right)$$

6. Orientation polarizability is given as follows: $\alpha_0 = \dfrac{\mu^2}{3kT}$

7. The Langevin–Debye equation is written as follows:

$$P = NE\left(4\pi\epsilon_0 R^3 + \frac{e^2}{\omega_0^2}\left(\frac{1}{M_p} + \frac{1}{M_n}\right) + \frac{\mu^2}{3kT}\right)$$

8. Local field is the actual field experienced by an atom/ion/molecule inside the dielectric material in the presence of the external electric field:

$$E_{int} = E + \frac{P}{3\epsilon_0}$$

9. The Clausius–Mossotti equation can be memorized as follows:

$$P = \frac{N\alpha}{(1 - (N\alpha/3))}\epsilon_0 E = (\kappa - 1)\epsilon_0 E$$

SELF-ASSESSMENT

Multiple-choice Questions

13.1 What happens to the capacitance of a capacitor as a dielectric is introduced in between its plates:
  (a) Increases
  (b) Decreases
  (c) No change
  (d) None of these

13.2 Insulating materials that can be polarized by the application of an electric field are called
  (a) dielectrics
  (b) insulating materials
  (c) polarized materials
  (d) all of these

13.3 The polarization vector P comes into play when an external electric field induces the
  (a) formation of dipoles inside the material
  (b) formation of holes inside the material
  (c) formation of electrons inside the material
  (d) formation of ions inside the material

13.4 The magnitude of the surface charge density is equal to the magnitude of
  (a) displacement vector
  (b) polarization vector
  (c) electric vector
  (d) none of these

13.5 From the Gauss law of electrostatics, the polarization charge density is given as
  (a) $\rho_{pol} = \nabla.P$
  (b) $\rho_{pol} = -\nabla.D$
  (c) $\rho_{pol} = -\nabla.E$
  (d) $\rho_{pol} = -\nabla.E$

13.6 Dimensions of atomic polarizability in SI units is
  (a) /cm^2
  (b) CV m
  (c) C/V/m^2
  (d) none of these

13.7 The displacement vector is defined as

(a) $D = \epsilon_0 E + P$

(c) $D = \epsilon_0 \left(\dfrac{1}{E} + P \right)$

(b) $D = \epsilon_0 (E + P)$

(d) $D = (\epsilon_0 E + 1/P)$

13.8 The dielectric strength is
(a) directly proportional to the dielectric breakdown voltage
(b) inversely proportional to the thickness of the dielectric
(c) both (a) and (b)
(d) none of these

13.9 Electronic polarization is directly proportional to the
(a) induced dipole moment
(b) induced magnetic moment
(c) induced electric field
(d) induced electric moment

13.10 Dielectric constants of different gases depend on the
(a) density of the gas
(b) frequency of its optical absorption
(c) both (a) and (b)
(d) none of these

13.11 Electronic polarizability is related to the dielectric constant as
(a) $(\kappa - 2) = N\alpha$
(b) $(\kappa - 1) = N\alpha$
(c) $(\kappa - 1) = (N + 1)\alpha$
(d) $(\kappa - 3) = N\alpha$

13.12 Solid materials that have permanent polarization even when the field is removed are called
(a) electrets
(b) magnates
(c) dielectrics
(d) none of these

13.13 The electrical analogue of a magnet is
(a) electra   (b) dipole   (c) electret   (d) electre

13.14 Ferroelectricity is the property of a
(a) special class of crystals possessing permanent moment
(b) special class of crystals possessing non-permanent magnetic moment
(c) special class of crystals possessing no electric moment
(d) special class of crystals possessing no permanent moment

13.15 Pyroelectric materials possess moment due to
(a) heat transfer
(b) mechanical stress
(c) thermal stress
(d) none of these

13.16 Dielectric constant is the property of the
(a) dielectric medium
(b) polarizing medium
(c) charging medium
(d) resistive medium

13.17 Understanding permittivity physically means
(a) resistance encountered when forming a magnetic field in a medium
(b) resistance encountered when forming an electric field in a medium
(c) resistance encountered when forming an electromagnetic field in a medium
(d) resistance encountered when forming a polarizing medium

13.18 Polarized surface charge exists due to
(a) dense charge existing on the surface of a parallel-plate capacitor due to charging
(b) polarized charge existing on the surface of a parallel-plate capacitor due to charging

(c) free charge existing on the surface of a parallel-plate capacitor due to charging

(d) free charge existing on the surface of a parallel-plate capacitor due to discharging

13.19 Ionic polarizability is directly proportional to the
   (a) reduced mass
   (b) frequency
   (c) square of angular frequency
   (d) centre of mass

13.20 Space-charge polarization is observed in
   (a) ferrites and semiconductors
   (b) diamagnetics
   (c) paramagnetics
   (d) all magnetic materials

13.21 The fastest polarization is
   (a) ionic polarization
   (b) space-charge polarization
   (c) orientation polarization
   (d) electric polarization

13.22 Polarization is
   (a) space-charge dependent
   (b) charge dependent
   (c) density dependent
   (d) frequency dependent

13.23 The dielectric constant is given as

   (a) $\kappa = \dfrac{C_\kappa}{C_o} = \dfrac{\epsilon_\kappa}{\epsilon_o} = \dfrac{V_o}{V_\kappa}$

   (b) $\kappa = \dfrac{C_\kappa}{C_o} = \dfrac{\epsilon_o}{\epsilon_\kappa} = \dfrac{V_o}{V_\kappa}$

   (c) $\kappa = \dfrac{C_\kappa}{C_o} = \dfrac{\epsilon_\kappa}{\epsilon_o} = \dfrac{V_\kappa}{V_o}$

   (d) $\kappa = \dfrac{C_\kappa}{C_o} = \dfrac{\epsilon_o}{\epsilon_\kappa} = \dfrac{V_\kappa}{V_o}$

13.24 Dielectric constant is
   (a) always > 1
   (b) always < 1
   (c) always equal to 1
   (d) none of these

13.25 Orientation polarization is called so due to the orientation of the
   (a) pre-existing dipoles
   (b) electrons
   (c) protons
   (d) nucleus

Review Questions

13.1 How can we differentiate between the local field and the applied field in the context of a dielectric medium?

13.2 Why does the capacitance of a capacitor increase as it is filled with a dielectric medium?

13.3 How does polarization change the definition of the electric field vector?

13.4 Using the definition of a dielectric constant, show that it is infinity for a metal.

13.5 Is there a relation between the dielectric constant and electronic polarizability? Explain why.

13.6 What happens to the electric field of a dielectric due to polarization?

13.7 Can a ferroelectric material be used as a potential source of green energy?

13.8 Why is a capacitor known to be an energy-storing device?

13.9 Using the definition of a dielectric constant, show that it is necessarily greater than one.

13.10 Define dielectric constant. Is it a dimensionless quantity?

13.11 Explain the phenomenon of polarization of dielectric materials.

13.12 Write the essential difference between a conductor, an insulator, and a dielectric material.

13.13 What is meant by a local field in the context of a dielectric medium?

13.14 Explain the term polarization. How does the concept of a polarization vector change the definition of an electric field vector?

13.15   What is a dielectric breakdown?

13.16   Explain the difference between a ferroelectric, a pyroelectric, and a piezoelectric material.

13.17   What do you understand by dielectric loss?

13.18   Write the Clausius–Mossotti equation highlighting its physical significance.

13.19   Discuss, with proper diagrams, the formation of dipoles in a dielectric material.

13.20   State and prove the Gauss law in dielectrics.

13.21   Derive the relation $D = \epsilon_0 E + P$.

13.22   Explain the phenomenon of polarization. Derive the expressions for electronic and ionic polarizations in dielectrics.

13.23   Derive the dielectric constant of a liquid in terms of electronic polarizability.

13.24   Explain how temperature affects the dielectric constant of a material.

13.25   Derive the relation between electric susceptibility and the dielectric constant of a material.

13.26   Derive the relation for dielectric loss.

13.27   Explain the physical significance of the term δ in the expression for dielectric loss.

13.28   What do you understand by the term dielectric strength? How is it related to a dielectric breakdown?

13.29   Explain the different types of solid dielectric materials.

Numerical Problems

13.1   The parallel plates of a capacitor have an area of $2.00 \times 10^{-1}\,\mathrm{m}^2$ and are $1.00 \times 10^{-2}\,\mathrm{m}$ apart. The capacitor is connected to a 4.00 kV power supply. A thin insulating plastic sheet is inserted between the capacitor plates, which fills the space between the plates completely. The potential difference between the plates drops to 2.00 kV, while the charge on each capacitor plate remains constant. Calculate the following:

   (a)   The original capacitance C_0
   (b)   The magnitude of charge Q on each plate
   (c)   The capacitance C after the dielectric is inserted
   (d)   The dielectric constant κ of the dielectric
   (e)   The permittivity ε of the dielectric
   (f)   The magnitude of charge Q_i on each face of the dielectric
   (g)   The original electric field E_0 between the plates and the electric field E after the dielectric is inserted

$$\left[Hint: C_0 = \epsilon_0 \frac{A}{d}; Q = C_0 V_0; C_\kappa = \frac{Q}{V}; \kappa = \frac{C_\kappa}{C_0}; \right.$$

$$\left. \epsilon = \kappa \epsilon_0; Q_i = Q\left(1 - \frac{1}{\kappa}\right); E_0 = \frac{V_0}{d} \right]$$

13.2   If the electric field strength inside two parallel plates of a capacitor is 10^2 V/m due to a dielectric constant of 2, find the polarization vector.

$$\left[Hint: P = (\kappa - 1)\epsilon_0 E \right]$$

13.3   If the dielectric constant of a gas is 1.000039 at 0°C, determine its electric susceptibility at the same temperature.

$$\left[Hint: \chi = (\kappa - 1) \right]$$

13.4 If the induced dipole moment of each atom is 7.27×10^{-26} C-m at NTP, give its polarization vector.

$$\left[\text{Hint}: P = N\mu\right]$$

13.5 The polarization vector of a gas at NTP is given by 2.0×10^{-10} C/m^2; find the induced dipole moment of each atom.

$$\left[\text{Hint}: P = N\mu\right]$$

13.6 Two parallel plates having equal and opposite charges are separated by a 2 cm thick slab that has a dielectric constant of 2. If the electric field is 10^2 N/C, calculate the polarization and the displacement vector.

$$\left[\text{Hint}: D = \epsilon_0 \kappa E = \epsilon_0 E + P\right]$$

13.7 A water molecule has a dipole moment of 7×10^{-28} C-cm, where all the molecular dipoles point in the same direction. If the radius of the water drop is 0.21 mm, find the polarization in the water drop.

$$\left[\text{Hint}: P = N\mu\right]$$

13.8 The parallel plates of a capacitor have an area of 2.00×10^{-1} m^2 and are 1.50×10^{-2} m apart. The capacitor is connected to a power supply of 3.00 kV. A thin insulating plastic sheet is inserted between the capacitor plates and fills the plates completely. The potential difference between the plates drops to 1.50 kV while the charge on each capacitor plate remains constant. Calculate the following:
   (a) The original capacitance C_0
   (b) The magnitude of charge Q on each plate
   (c) The capacitance C after the dielectric is inserted
   (d) The dielectric constant κ of the dielectric
   (e) The permittivity ε of the dielectric
   (f) The magnitude of charge Q_i on each face of the dielectric
   (g) The original electric field E_0 between the plates
   (h) The electric field E after the dielectric is inserted

$$\left[\begin{array}{l} \text{Hint}: C_0 = \epsilon_0 \dfrac{A}{d}; Q = C_0 V_0; C_\kappa = \dfrac{Q}{V}; \kappa = \dfrac{C_\kappa}{C_0}; \\[2mm] \epsilon = \kappa \epsilon_0; Q_i = Q\left(1 - \dfrac{1}{\kappa}\right); E_0 = \dfrac{V_0}{d} \end{array}\right]$$

13.9 The polarization vector of a gas at NTP is given by 4.0×10^{-10} C-m; find the induced dipole moment due to each atom.

$$\left[\text{Hint}: P = N\mu\right]$$

13.10 Calculate the electronic polarizability of a silicon atom if the relative permittivity of silicon at NTP is 11.68.

$$\left[\text{Hint}: P = N\alpha_e E; \alpha_e = \dfrac{(\kappa - 1)}{N}\epsilon_0\right]$$

13.11 Calculate the electronic polarizability of a mica atom if the relative permittivity of mica at NTP is 6.0.

$$\left[\text{Hint}: P = N\alpha_e E; \alpha_e = \dfrac{(\kappa - 1)}{N}\epsilon_0\right]$$

13.12 A thick sheet of polythene of dimension 0.45 mm possessing a relative dielectric constant of 2.0 is subjected to 220 V. Calculate the polarization.

$$\left[Hint: P = \epsilon_0 (\kappa - 1) E \right]$$

13.13 Calculate the maximum potential gradient to which a 0.25 mm thick mica sheet can be subjected. The dielectric strength for mica is $100 \times \dfrac{10^6 \text{ V}}{\text{m}}$.

$$\left[Hint: \text{Dielectric strength} = \frac{\text{Breakdown voltage}}{\text{Thickness of the sheet}} \right]$$

13.14 Calculate the maximum potential gradient to which a 0.4 mm thick bakelite sheet can be subjected. The dielectric strength for bakelite is $15 \times \dfrac{10^6 \text{ V}}{\text{m}}$.

$$\left[Hint: \text{Dielectric strength} = \frac{\text{Breakdown voltage}}{\text{Thickness of the sheet}} \right]$$

13.15 What should be the thickness of the porcelain sheet that can be inserted in an instrument that is to be subject to breakdown voltages of the order of 2×10^6 V. The dielectric strength of porcelain is $6 \times \dfrac{10^6 \text{ V}}{\text{m}}$.

$$\left[Hint: \text{Dielectric strength} = \frac{\text{Breakdown voltage}}{\text{Thickness of the sheet}} \right]$$

13.16 What should be the thickness of the silica sheet that can be inserted in an instrument that is to be subjected to breakdown voltages of the order of 3×10^6 V. The dielectric strength of porcelain is $10 \times \dfrac{10^6 \text{ V}}{\text{m}}$.

$$\left[Hint: \text{Dielectric strength} = \frac{\text{Breakdown voltage}}{\text{Thickness of the sheet}} \right]$$

13.17 Two parallel plates having equal and opposite charges are separated by a 1 cm thick slab that has a dielectric constant of 2. If the electric field is 2×10^2 N/C, calculate the polarization and the displacement vector.

$$\left[Hint: D = \epsilon_0 \kappa E = \epsilon_0 E + P; P = D - \epsilon_0 E \right]$$

13.18 If the relative permittivity of Kr atom is 1.0006612 $C^2/N/m^2$, calculate the electronic polarizability of Kr at NTP.

$$\left[Hint: P = (\kappa - 1) \epsilon_0 E \right]$$

13.19 If the electronic polarizability of argon atom is 1.432×10^{-40} $C^2/N/m^2$, calculate the relative permittivity of argon at NTP.

$$\left[Hint: P = (\kappa - 1) \epsilon_0 E = N \alpha_e E; \alpha_e = \frac{(\kappa - 1)}{N} \epsilon_0 \right]$$

13.20 A carbon dioxide molecule has a dipole moment of 5×10^{-25} C-cm, where all the molecular dipoles point in the same direction. If the radius of the molecule drop is 0.5 mm, find the polarization in the carbon dioxide drop.

$$\left[Hint: P = N \mu \right]$$

13.21 A molecule has a dipole moment of 3×10^{-25} C-cm, where all the molecular dipoles point in the same direction. If the radius of the molecule as a drop is 0.25 mm, find the polarization in the molecule drop.

$$[Hint: P = N\mu]$$

Magnetic Properties of Materials

Learning Objectives

After studying this chapter, students will be able to

- understand the origin of magnetic moment
- realize the concept of Bohr magneton
- explain the phenomenon of diamagnetism, paramagnetism, and ferromagnetism
- elucidate the phenomenon of hysteresis in a magnetic material
- comprehend the concept of ferromagnetic and anti-ferromagnetic materials
- demonstrate some important applications of magnetic materials

List of Symbols

e = Charge on the electron

r = Radius of the orbit

I = Current in loop

T = Time period

V = Linear velocity

A = Area of the orbit

M = Magnetic moment associated with a current loop

ω = Angular velocity

h = Planck's constant

n = Number of the orbit

μ_B = Bohr magneton

χ_m = Magnetic susceptibility

μ_r = Relative permeability

B = Magnetic induction

H = Magnetic intensity

μ_m = Magnetic permeability

14.1 INTRODUCTION

Different conventional forms of matter, such as free atoms, ions, molecules, condensates, and the like, exhibit magnetic behaviour. The core of the earth possesses magnetic properties. Magnetism is exhibited by many naturally occurring materials. These magnetic properties have been traced to their unique orbital motion of electrons. Alignment of the orbital motion of electrons to the external applied magnetic field renders a characteristic behaviour to each of these materials. Some of the atoms or ions may possess an effectively permanent magnetic dipole moment. Knowledge of the structure of atoms and their further constituents lead to the better understanding of their alignments. Spin angular momenta, in addition to the orbital motion of electrons, lead to permanent atomic moments, which are the aligning atomic moments. Magnetic materials can be categorized as diamagnetic, paramagnetic, and ferromagnetic materials. Although the explanation of magnetic properties of materials by Faraday, Curie, and also Langevin seems to be based on the classical theory, these had their roots in quantum physics.

These characteristic magnetic properties are used by technologists for applications in several areas, such as magnetic recording, storage of data, and subsequent retrieval of recorded and stored information. In this chapter, we will discuss these unique magnetic materials and their magnetic properties.

14.1.1 Basic Definitions

Magnetic field   A magnetic field (B) is usually generated externally and is a relative quantity. In general, the flow of current in any material will lead to the polarization of orbitals and spin moments of its electrons. Thus, the magnitude of the current applied to a material will define the magnitude of the magnetic field or vice versa.

Magnetization   Magnetization (M) for a bulk sample is defined as the magnetic dipole moment with respect to either its unit mass or its unit volume.

Magnetic susceptibility   The term magnetic susceptibility per unit mass is defined as magnetization per unit mass. The following is the mathematical definition: $\chi = \mu_0 \dfrac{M}{B}$

where B is the initial magnetic field and μ_0 the vaccum permeability.

14.2 ORIGIN OF MAGNETIC MOMENT

The moment of a particle is associated with its circular or rotational motion. In the case of an atom, an electron possesses an inherent spin motion in addition to its orbital motion about the nucleus. These motions together constitute the magnetic moment; we will study in detail how this happens in the following sections. Suppose that an electron is revolving around a nucleus. Let the charge on the electron be e and the radius of its orbit be r (Fig. 14.1). The revolving electron is akin to a loop of current. The direction of current in this loop is opposite to the direction of revolution of the electron.

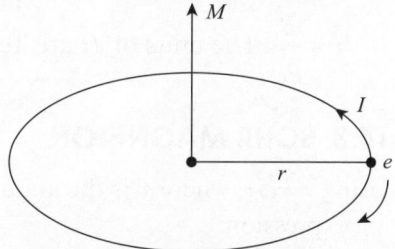

Fig. 14.1 Electron revolving around nucleus

Thus, if the electron revolves in the clockwise direction, it constitutes an anticlockwise current, and vice versa. The equivalent current I is given by the following equation:

$$I = \frac{\text{Charge}}{\text{Time}} = \frac{e}{T} \tag{14.1}$$

where T is the time period of revolution of the electron. If v is the linear velocity of the electron, then we can write the following relation:

$$T = \frac{2\pi r}{v} \tag{14.2}$$

In addition, the area enclosed by the orbit is given by the following equation:

$$A = \pi r^2 \tag{14.3}$$

The magnetic moment M associated with a current loop is given by the following expression:

$$M = I A \tag{14.4}$$

Using Eqs (14.1)–(14.3) in Eq. (14.4), we get the following expression:

$$M = \frac{ev}{2\pi r} \times \pi r^2 = \frac{evr}{2} \tag{14.5}$$

The magnetic dipole moment of a revolving electron is thus half the product of its charge, linear velocity, and the radius of its orbit. The dipole moment vector of a revolving electron points upwards for an electron revolving in the clockwise direction and downwards for an electron revolving in the anti-clockwise direction.

Note: Physical quantities B, H, and M are interdependent. The magnetic dipole moment (M) per unit volume or magnetization or magnetic polarization is the sum of the dipole moments due to orbital motion of electrons, their (electronic) spin, and the nuclear spin. In terms of magnetic susceptibility χ_m, M is given as follows: $M = \chi_m H$

Essentially, in a given medium, M is a part of B that depends on the microscopic properties of the medium; this is unlike H, which is a part of B that is independent of the microscopic properties of the medium.

For a free space,

1. $M = 0$. The units for M are $\dfrac{A}{m}$ or $\dfrac{J}{T}.m^3$.

2. $H = \dfrac{B}{\mu_0}$. The units of H are Tesla (T).

14.3 BOHR MAGNETON

Using $v = \omega r$, where ω is the angular velocity, in Eq. (14.5), we get the following expression:

$$M = \frac{1}{2} e \omega r^2 \tag{14.6}$$

The physical significance of this equation is that, in it, magnetization has been defined in terms of fundamental atomic parameters. We can explicitly observe from this equation how the spin and orbital momenta of an electron are involved in this expression of magnetization.

Electrons in an atom obey the Bohr's theory. According to the Bohr's theory, *an electron in an atom can revolve only in certain stationary orbits.* In these stationary orbits, the angular momentum of an electron takes values that are integral multiples of $h/2\pi$, where h is Planck's constant. This implies that

$$mvr = \frac{nh}{2\pi} \tag{14.7}$$

where n denotes the number of the orbit and can take values $n = 1, 2, 3, \ldots$ We also know that

$$v = r\omega \tag{14.8}$$

Using Eq. (14.8) in Eq. (14.7), we get the following expression:

$$m(r\omega)r = \frac{nh}{2\pi} \tag{14.9}$$

which leads to

$$r^2\omega = \frac{nh}{2\pi m} \tag{14.10}$$

Using Eq. (14.10) in Eq. (14.6), we get, for magnet moment M, the following relation:

$$M = \frac{1}{2}e\frac{nh}{2\pi m} = \frac{neh}{4\pi m} \tag{14.11}$$

The quantity $(eh)/(4\pi m)$ is called *Bohr magneton* and is represented by the symbol μ_B. Bohr magneton can be defined as *the magnetic dipole moment associated with an atom due to the orbital motion of an electron in the first orbit of a hydrogen atom.* In terms of μ_B, Eq. (14.11) can be rewritten as follows:

$$M = n\mu_B \tag{14.12}$$

The magnetic moment of an atom is thus quantized in multiples of Bohr magneton. Substituting appropriate values in the expression for Bohr magneton, we get the following expression:

$$\mu_B = \frac{eh}{4\pi m} = \frac{1.6\times 10^{-19}\times 6.6\times 10^{-34}}{4\times \dfrac{22}{7}\times 9\times 10^{-31}} = 9.27\times 10^{-24} \text{ Am}^2 \tag{14.13}$$

Example 14.1 A plane loop is carrying a current of 12 A. The loop is of irregular shape and encloses an area of 7.5×10^{-4} m^2. To an observer, the current appears to flow in the clockwise direction. Determine the magnitude and direction of the magnetic moment associated with the current loop.

Solution The magnetic moment M associated with the loop is given by the following relation: $M = IA$

Using the given values, we get the following result: $M = 12 \times 7.5 \times 10^{-4} = 9 \times 10^{-3} \, \text{Am}^2$

M is directed away from the observer and is perpendicular to the plane of the loop.

Example 14.2 An electron in a hydrogen atom moves in a circular orbit of radius $0.5 \, \text{Å}$. The electron performs 10^{16} revolutions per second. Determine the magnetic moment associated with the orbital motion of the electron.

Solution Radius r of the orbit is calculated as follows: $r = 0.5 \, \text{Å} = 0.5 \times 10^{-10} \, \text{m}$

The charge e on the electron is the following: $e = 1.6 \times 10^{-19} \, C$

The motion represents a current I given by the following relation: $I = \dfrac{e}{T} = e \vartheta$

where ϑ is the frequency of revolution of the electron and $\vartheta = 1016 \, \text{rps}$.

Putting the values of e, we get $I = 1.6 \times 10^{-19} \times 10^{16} = 1.6 \times 10^{-3} \, A$

The area A enclosed by the orbiting electron is given by the following relation:

$$A = \pi \, (0.5 \times 10^{-10})^2 = 7.85 \times 10^{-21} \, \text{m}^2$$

The magnetic moment M associated with the motion of the electron is given by the following equation: $M = IA$

Substituting for I and A, we get the following value:

$$M = (1.6 \times 10^{-3})(7.85 \times 10^{-21}) = 1.26 \times 10^{-23} \, \text{Am}^2$$

Example 14.3 Express the smallest allowed magnitude of the atomic dipole moment in units of $\dfrac{\text{J}}{\text{T}}$.

Solution The smallest allowed magnitude of the atomic dipole moment is the Bohr magneton. It can be calculated as follows:

$$\mu_B = \frac{e h}{4 \pi m} = \frac{1.6 \times 10^{-19} C \times 6.6 \times 10^{-34} \, \text{J.s}}{4 \times \dfrac{22}{7} \times 9 \times 10^{-31} \, \text{kg}}$$

$$= 9.27 \times 10^{-24} \, \text{C.J.} \frac{\text{s}}{\text{kg}} = 9.27 \times 10^{-24} \, \text{J/T}$$

Example 14.4 The saturation magnetic induction of nickel is $0.65 \, \text{Wb/m}^2$. If the density of nickel is $8906 \, \text{kg/m}^3$ and its atomic weight is 514.7, calculate the magnetic moment of nickel in Bohr magneton.

Solution Magnetic induction of nickel, $B = 0.65 \, \text{Wb/m}^2$

Density of nickel, $d = 8906 \, \text{kg/m}^3$

Atomic weight $M = 514.7$

We know that $B = N \mu_0 M$ and $N = d \dfrac{N_A}{M}$ where $\mu_0 = 4\pi \times 10^{-7}$

N is the number of atoms per unit volume and N_A is the Avogadro constant. Substituting these values, we get the following results:

$$N = \frac{(8906 \times 6.023 \times 10^{26})}{58.7} = 9.14 \times 10^{28} \, \text{atoms/m}^3$$

$$\mu_r = (B/N\mu_0) = \frac{0.65}{9.14 \times 10^{28} \times 4 \times 10^{-7}} = 5.66 \times 10^{-24} \text{ A/m}^2$$

$$M = \frac{5.66 \times 10^{-24}}{9.27 \times 10^{-24}} \mu_B = 0.61 \mu_B$$

14.4 DIAMAGNETISM, PARAMAGNETISM, AND FERROMAGNETISM

Magnetic phenomenon for an empty space is defined by the magnetic induction B, which we know to be a vector quantity accounting for the force acting on a moving charge. The current I, a macroscopic flow of charge in a medium, leads us to the Ampere's law that defines a new vector quantity H, which is magnetic field strength or magnetic intensity. The inherent orbital motion of an electron also introduces a new vector quantity M (discussed in Sections 14.1–14.3). These quantities are simply related for an isotropic medium (i.e., B, H, and M are all in the same direction) as follows: $B = \mu_0(H + M)$

For free space: $M = 0 \Rightarrow B = \mu_0 H$

Else, $B = \mu_0(H + M) = \mu_0(H + \chi_m H) = \mu H$

Having established the relation of the quantity B (magnetic induction or magnetic flux density) with H (magnetizing field or magnetic intensity) and M (magnetization or magnetic polarization), we will now study the response of a magnetic material to any external magnetic field. Faraday was the first to classify magnetic materials into different categories depending on their response.

All magnetic materials fall into one of the following three categories: diamagnetic, paramagnetic, and ferromagnetic

Let us now undertake a detailed discussion of these materials with examples for each category.

14.4.1 Diamagnetism

In diamagnetic materials, the individual atoms/molecules/ions do not possess any net magnetic moment of their own. A diamagnetic material, when placed in a magnetic field, acquires magnetism in a direction opposite to that of the applied magnetic field.

Suppose a sample of a diamagnetic material is kept in an external magnetic field of induction B. Each atom/molecule/ion then acquires a small magnetic moment, which points in a direction opposite to the applied magnetic field. Some examples of diamagnetic materials are bismuth, antimony, copper, gold, quartz, mercury, water, alcohol, air, hydrogen, etc.

Some important properties of diamagnetic materials are as follows:
1. If a diamagnetic material is suspended in a uniform magnetic field, it comes to rest in a position such that the longest axis is perpendicular

to the direction of the magnetic field. The shortest axis is along the direction of the field. Figure 14.2 illustrates this situation schematically.

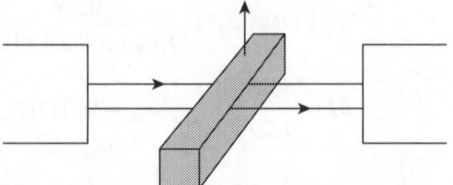

Fig. 14.2 Orientation of diamagnetic material under the influence of external magnetic field

2. If a diamagnetic material is kept in a non-uniform magnetic field, it moves away from stronger-field regions to weaker-field regions. Suppose a diamagnetic liquid is put in a watch glass placed on the two pole pieces of an electromagnet. On switching the current through the electromagnet, the liquid meniscus takes the shape shown in Fig. 14.3.

Two characteristics of the liquid meniscus are significant. The liquid accumulates on the sides as the magnetic field is weaker in that region. A depression is created in the central portion, where the magnetic field is strongest. The

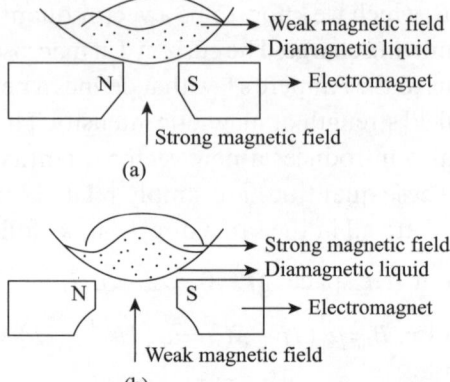

Fig. 14.3 Diamagnetic liquid in magnetic field

situation prevails as long as the distance between the pole pieces is small. As the distance between the pole pieces increases, the magnetic field starts becoming stronger near the poles than the field between the poles. The liquid then prefers to collect over the region between the pole pieces and thus acquires a shape opposite to the one shown in Fig. 14.3.

3. If a diamagnetic liquid is filled in a U-tube and one of the arms of the U-tube is kept between the pole pieces of a magnet, level of the liquid in the tube gets depressed. This is shown in Fig. 14.4.

Arrows in the figure indicate the direction of motion of the liquid in the U-tube when one of its columns is brought under the influence of an external magnetic field.

4. The magnetic lines of force get expelled by a diamagnetic substance. They prefer not to pass through the diamagnetic specimen. This is shown in Fig. 14.5.

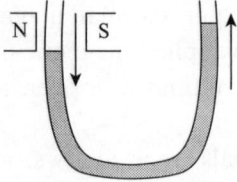

Fig. 14.4 Motion of diamagnetic liquid in magnetic field

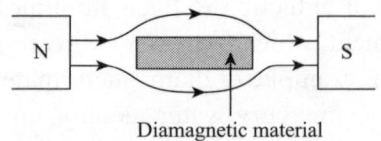

Diamagnetic material

Fig. 14.5 Expulsion of magnetic lines of force by diamagnetic material

Due to this property, the magnetic field *B* inside a diamagnetic material gets reduced. The magnetic permeability of diamagnetic substances is thus always less than unity.

5. The magnetic susceptibility χ_m of diamagnetic substances is always negative. This follows from the fact that the relative permeability, μ_r, and the magnetic susceptibility are related to each other through the following expression:

$$\mu_r = \frac{\mu}{\mu_0} = \left(1 + \chi_m\right) \tag{14.14}$$

Since μ_r for diamagnetic substance is less than unity, χ_m must take only negative values.

6. The magnetic susceptibility χ_m of diamagnetic materials does not change with temperature. Bismuth violates this property at low temperatures.

14.4.2 Paramagnetism

Paramagnetic materials acquire feeble magnetism in the direction of the magnetic field in which they are placed. In paramagnetic materials, each individual atom/molecule/ion has a net magnetic moment of its own.

In the absence of an external magnetic field, these individual magnetic dipoles are randomly aligned due to thermal agitation. If a paramagnetic material is placed in an external magnetic field *B*, the individual dipole moments try to align themselves along the direction of the external field. This is because a torque is generated among the dipole moments such that there is a net magnetization in a direction parallel to the magnetic field. If the magnetic field is strong enough, the material acquires a net average magnetic dipole moment density in the direction of *B*.

Some examples of paramagnetic material are aluminium, platinum, chromium, manganese, copper sulphate, crown glass, salt solutions of iron and nickel, oxygen, etc.

Some important properties of paramagnetic materials are as follows:

1. If a paramagnetic material is freely suspended in a uniform magnetic field, it comes to rest in a position such that the longest axis is in the direction of the external magnetic field. As explained earlier, the external magnetic field exerts a torque on all the dipole moments such that they align parallel to the external applied field, as shown in Fig. 14.6.

2. If a paramagnetic material is placed in a non-uniform magnetic field, it moves away from regions of weaker magnetic field to those of stronger fields. Suppose a watch glass containing a paramagnetic material is placed on the pole pieces of an electromagnet. On passing a current, the liquid takes the shape shown in Fig. 14.7.

The salient feature of the meniscus is the accumulation of the liquid at the centre, where the magnetic field is strongest. This meniscus is observed as long as the distance between the pole pieces is kept small. If the distance is

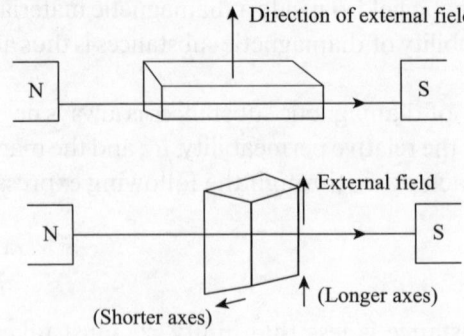

Fig. 14.6 Paramagnetic material in magnetic field

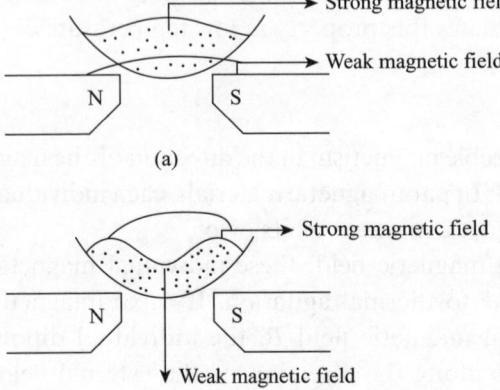

Fig. 14.7 Paramagnetic liquid in non-uniform field
(a) Paramagnetic material in non-uniform magnetic field
(b) Tranformation of shape of liquid as current is passed

increased, the magnetic field becomes weaker in the region between the pole pieces and stronger near the two poles. The paramagnetic material's meniscus then acquires a shape opposite to the one shown in Fig. 14.7.

3. If one arm of a U-tube filled with a paramagnetic liquid is kept between the poles of a magnet, the liquid level in that arm rises, as shown in Fig.14.8. Arrows indicate the direction of movement of the liquid when an arm of the U-tube is brought between the two poles. This effect is also due to the property of paramagnetic materials moving to regions of stronger magnetic field.

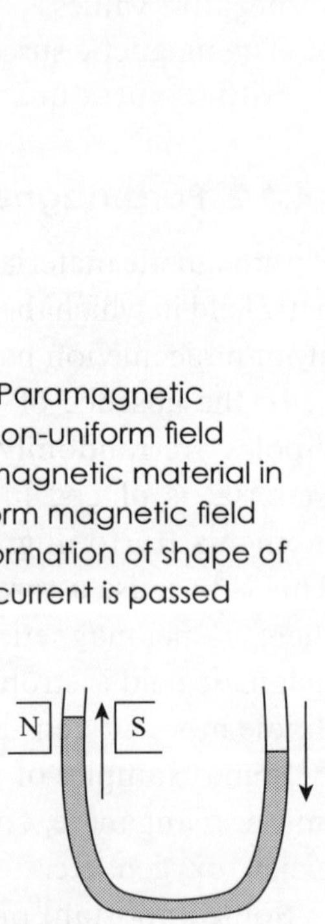

Fig. 14.8 U-tube filled with paramagnetic liquid kept in magnetic field

4. If a paramagnetic material is kept in an external magnetic field, the lines of force associated with this field show a preference for this material. This results in concentration of the magnetic lines of force in the paramagnetic material compared to the surrounding non-paramagnetic air. This phenomenon is illustrated in Fig. 14.9.

The magnetic induction B inside a paramagnetic material is, therefore, higher than the magnetic intensity H. Thus,

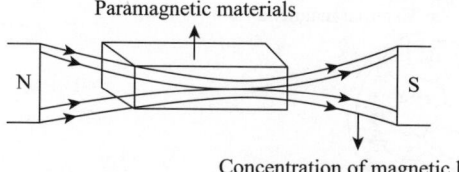

Paramagnetic materials

N   S

Concentration of magnetic lines
of force along the material

Fig. 14.9 Concentration of magnetic lines due to paramagnetic material

$$\frac{B}{H} = \mu > 1 \tag{14.15}$$

which implies that $\mu > 1$ for a paramagnetic material. The magnetic permeability μ_m of a paramagnetic material is thus greater than unity.

5. The magnetic susceptibility χ_m of a paramagnetic material is positive. This can be concluded from Eqs (14.14) and (14.15).

6. The magnetic susceptibility χ_m of a paramagnetic material varies inversely with temperature, that is,

$$\chi_m \propto \frac{1}{T} \tag{14.16}$$

This is called Curie's law and leads to the following relation:

$$\chi_m = \frac{C}{T} \tag{14.17}$$

where C is the Curie constant.

This effect leads to loss of magnetism of a paramagnetic material with increasing temperature.

14.4.3 Ferromagnetism—Domain Theory

Individual atoms/molecules/ions of a ferromagnetic material also possess non-zero magnetic moments, similar to paramagnetic materials. The difference here is that the magnetic moments of the individual atoms interact with each other and align themselves spontaneously along a common direction over macroscopic volumes due to exchange of coupling forces. This happens even in the absence of any external magnetic field.

The macroscopic volumes over which an alignment of the individual magnetic moments of the atoms takes place are called *domains*. Every domain has a net magnetic moment. Magnetic moments of the different domains are, however, randomly distributed. This can be compared to a situation where small groups in a big class are discussing different topics. All the groups are focused and, therefore, oriented like the domains. There is, however, no net focus or alignment of the class. In quite a similar manner, there is no bulk magnetic moment of a ferromagnetic material in the absence of an external magnetic field.

If a ferromagnetic material is kept in an external magnetic field, all the magnetic moments of the different domains align themselves, and the ferromagnetic

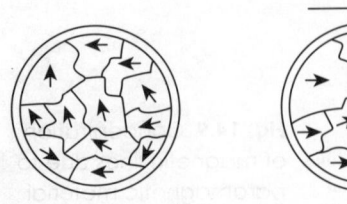

No magnetic field applied   Magnetic field is applied

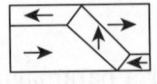

No external field   Weak applied field   Strong applied field

Fig. 14.10 Ferromagnetic domains in presence and absence of magnetic fields

material gets strongly magnetized in the direction of the magnetic induction *B*, as shown in Fig. 14.10. Iron, cobalt, nickel, and a large number of their alloys are some examples of ferromagnetic materials found in nature.

Some important properties of ferromagnetic materials are as follows:

1. Ferromagnetic materials have large values of magnetic permeability. Their permeability can be of the order of hundreds and thousands.
2. Ferromagnetic materials also have large values of magnetic susceptibility.
3. The susceptibility of ferromagnetic materials decreases with the rise in temperature.
4. The domain structure disintegrates with temperature. The disappearance of magnetization with temperature is a gradual process. It is a phase transition, reminding us of the process of melting of a solid crystal.

Ferromagnetic materials lose their ferromagnetism above a certain transition temperature. This transition temperature is called the *Curie temperature*. Above the Curie temperature, ferromagnetic materials are paramagnetic. The Curie temperature of iron is approximately 1000 K. Curie temperatures T_C of some ferromagnetic materials are given in Table 14.1.

Table 14.1 Curie temperature T_C of common ferromagnetic materials

Material	T_C(K)
Cobalt	1394
Iron	1043
Fe_2O_3	893
Nickel	631
Gadolinium	317

14.4.4 Langevin's Theory of Magnetism

In the year 1905, to explain the magnetic behaviour of atoms, ions, and molecules, Langevin made the simple assumption that each atom of a gas carried a permanent moment μ. This thought was purely classical at that time, but with shades of quantum theory due to the existence of an individual permanent moment. As an external magnetic field (B) is applied, these permanent moments orient themselves, though the thermal motion of the particles hinders this process. These N atoms are oriented in the direction of the applied field. Thus, the moment inclined at an angle θ to the applied field B gives a magnetic contribution, $-\mu B\cos\theta$, to the total free energy of the atom. The Maxwell–Boltzmann distribution gives the number of atoms, with their moments inclined at an angle θ, using the following relation:

$$dN = A\exp\left(-\frac{\mu B\cos\theta}{k_B T}\right)d\omega \tag{14.18}$$

where the constant is A and $d\omega$ refers to an element of solid angle bounded by the cones θ and $(\theta + d\theta)$. Let us assume that

$$\alpha = \frac{\mu B}{k_B T} \tag{14.19}$$

$$\Rightarrow dN = A e^{-\alpha\cos\theta}\,d\omega \tag{14.20}$$

$$N = 2\pi\int_0^\pi A e^{\alpha\cos\theta}\sin\theta\,d\theta = \frac{4\pi}{\alpha}A\sinh\alpha \tag{14.21}$$

The total resolved moment in the field direction is as follows:

$$M = N<\mu>=\int_0^\pi \mu\cos\theta\,dN \tag{14.22}$$

$$= 2\pi A\mu\int_0^\pi e^{\alpha\cos\theta}\sin\theta\,d\theta \tag{14.23}$$

Using the standard integrals,

$$M = N<\mu>= N\mu\left(\coth\alpha - \frac{1}{\alpha}\right) \tag{14.24}$$

Where $L(\alpha)=\left(\coth\alpha - \frac{1}{\alpha}\right)$ $\qquad$ (14.25)

is referred to as the Langevin function.

The equation gives the magnetization for a classical magnetic gas. Magnetization is dependent on the field strength and temperature. At room temperature, magnetization remains linearly dependent on the external field applied. However, as the temperature rises, to about $100T$, the moment starts to saturate and the linear dependence starts to give way to a saturated graph. The constant is dependent on temperature, and so at room temperature $\mu B \ll k_B T$, such that α is a small quantity $(\alpha < 10^{-2})$; thus,

$L(\alpha)\approx\frac{1}{3}\alpha$ so that Eqs (14.24) and (14.25) can be reduced to the following expression:

$$\frac{<\mu>}{\mu} = \frac{\alpha}{3} = \frac{\mu B}{3k_B T} \tag{14.26}$$

$$M = N <\mu> = \frac{N\mu^2 B}{3k_B T} \qquad (14.27)$$

The Curie law states that for paramagnetic non-metals,

$$\chi = \frac{C}{T} \qquad (14.28)$$

$$\chi = \frac{\mu_0 M}{B} = \frac{\mu_0 N \mu^2}{3k_B T} = \frac{C}{T}$$

The molar susceptibility can be defined in terms of the molar moment, as for one gram molecule of a gas N becomes the Avogadro's number:

$$\chi = \frac{\mu_0 N^2 \mu^2}{3RT} = \frac{\mu_0 \Sigma^2}{3RT} \qquad (14.29)$$

where $\Sigma = N\mu$ is defined as the molar moment

Thus, it can be seen that

$$C_{\text{mol}} = \chi T = \frac{\mu_0 \Sigma^2}{3R} \qquad (14.30)$$

Hence, the Langevin's approach helps us understand the susceptibility of a paramagnetic gas, which in turn helps us comprehend atomic moments.

14.5 HYSTERESIS—SOFT AND HARD MAGNETIC MATERIALS

Suppose a magnetic material is being magnetized by keeping it in an external magnetic field. Let us see how the intensity of magnetization changes with changes in the magnetizing field H. The intensity of magnetization is defined as the magnetic moment per unit volume of a material. It is represented by the symbol I. The variation of the intensity of magnetization with the magnetizing field is shown in Fig. 14.11.

In Fig. 14.11, the magnetizing field H is shown along the X-axis and the intensity of magnetization along the Y-axis. As the magnetic intensity increases

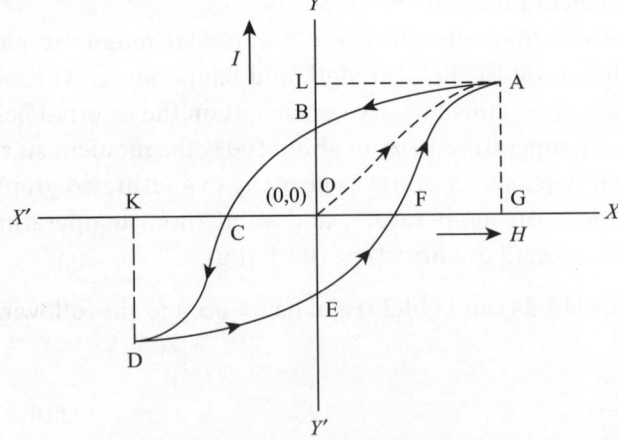

Fig. 14.11 *I–H* curve

from O to G, the intensity of magnetization can be seen to follow the path OA. Suppose that, at this point, the magnetic intensity is reduced to zero and then allowed to continue to grow in the opposite direction. Suppose K represents (Fig. 14.11) the point up to which the magnetic intensity H is allowed to grow in the opposite direction. The intensity of magnetization then follows the curve ABCD. The maximum value of the intensity of magnetization corresponds to the maximum magnetic intensity. These are indicated by the points L and G, respectively, in Fig. 14.11.

Another interesting feature of the *I–H* curve is that the intensity of magnetization does not become zero when the magnetic intensity becomes zero.

The value of the intensity of magnetization I, when the magnetic intensity H has been reduced to zero is called the *retentivity* of magnetic materials (OB, OE, Fig. 14.11). The corresponding magnetic induction is called the *residual magnetism, B_r,* of the material.

A magnetizing field, of the magnitude represented by the point C, has to be applied in the opposite direction in order to reduce the intensity of magnetization to zero. The value of this magnetizing field applied in the opposite direction is called the *coercivity* (H_c) of magnetic materials. As can be observed, OC, OF (refer to Fig. 14.11) lie along the *x*-axis such that the intensity of magnetization is zero.

If the magnetic intensity is again reduced to zero and then allowed to grow in the original direction, the portion DEFA of the curve is obtained.

The complete curve ABCDEFA (Fig. 14.11) is called the *hysteresis curve* or the *cycle of magnetization*. A similar curve is also obtained for the variation of magnetic induction B with magnetic intensity H. Such a curve is called the *B–H curve*. The hysteresis curve indicates that the intensity of magnetization I, or the magnetic induction B, lags behind the magnetic intensity H.

Thus, hysteresis is the phenomenon of lagging of I or B behind H as a magnetic sample is taken through its cycle of magnetization.

As the magnetic sample is taken through the *I–H* hysteresis curve, some energy is dissipated in the form of heat. Magnetic dipoles of individual atoms/molecules/ions are made to go through a process of orientation and reorientation. This results in the production of heat.

Energy loss per unit volume of a magnetic substance per cycle of magnetization is equal to the area of the *B–H* loop. The area of the *B–H* loop depends on its shape and size. Different materials exhibit different shapes of their *B–H* loops. Hysteresis loops for soft iron and steel are shown in Fig. 14.12. The *I–H* curve for soft

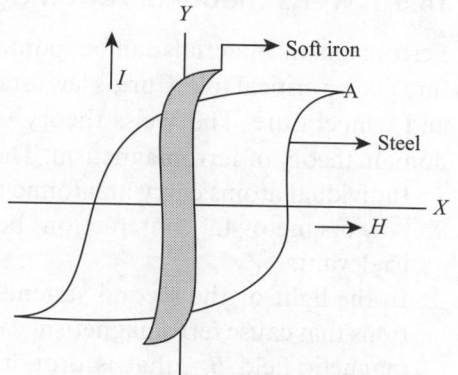

Fig. 14.12 *I–H* curve for soft iron steel

iron can be seen to be narrow and large. The loop for steel, on the other hand, is wide and short. These aspects of a hysteresis loop are taken into account while deciding the suitability of magnetic materials for various applications.

Depending on the area of the hysteresis loop, magnetic materials are categorized into two types: soft magnetic materials and hard magnetic materials. Materials with narrow hysteresis loops are called *soft magnetic materials* and those with a large area within their hysteresis loop are called *hard magnetic materials*. Soft magnetic materials are used to make cores of rotating electric machines. Hard magnetic materials are used to make permanent magnets, magnetic tapes, and disks.

Generally, soft magnetic materials are classified into the following categories:

Heavy-duty flux multipliers   These materials have a high working flux density, and are usually found in the cores of transformers, generators, and motors. These materials are subjected to alternating and rotating magnetic fields. Minimization of the energy loss per cycle is a design objective. Electrical steel and iron are usually the materials of choice owing to their low cost and lighter weight (as they are required in large sizes their weight should be less).

Light-duty flux multipliers   These materials have a low working flux density and are usually required in electronic equipment used particularly for communication purpose. Thus, the requirement for the frequencies that the magnetic materials can take is particularly in the range of megahertz. Thus, these are found in the cores of small, special-purpose transformers and inductors. Soft ferrites and nickel–iron alloys belong to this class.

Square loop materials   These are used in computers, magnetic amplifiers, and other suitable core devices. They also include soft ferrites and nickel–iron alloys.

Microwave system components   These comprise soft ferrites and garnets. Various soft magnetic materials such as Fe–Si or Fe–Ni alloys, and ferrimagnetic oxides like ferrites belong to this class of material.

14.5.1 Weiss Theory of Ferromagnetism

Ferromagnetic materials can be spontaneously magnetized, and the temperature plays a critical role. Curie's law establishes a relation between susceptibility and temperature. The Weiss theory used the simplest model to explain the domain theory of ferromagnetism. The assumptions of Weiss were as follows:

1. Individual atoms carry an atomic moment.
2. Any magneto-static interactions between atomic moments are completely irrelevant.
3. In the light of the second statement, Weiss stated that the true interactions that cause ferromagnetism can be represented by an inner molecular magnetic field B_{mw} that is proportional to spontaneous magnetization:
$$B_{mw} = \lambda M$$

where λ is referred to as the molecular coefficient and the field B_{mw} is the molecular field. The constant λ includes μ_0. As the external field (B) is now applied to the system, the total effective field now becomes as follows:

$$B = B_{mw} + \lambda M$$

Behaviour of ferromagnetic materials in small regions follows that of paramagnetic materials. Thus, following the magnetization equation from Section 14.4.4, we know that

$$M = N <\mu> = \frac{N\mu^2 B}{3k_B T} = \frac{C}{\mu_0 T} B \tag{14.31}$$

$$M = \frac{C}{\mu_0 T}(B_{mw} + \lambda M) \tag{14.32}$$

$$M = \frac{C}{\mu_0 T} B_{mw} + \frac{C}{\mu_0 T} \lambda M \tag{14.33}$$

$$M\left(1 - \frac{\lambda C}{\mu_0 T}\right) = \frac{C}{\mu_0 T} B_{mw} \tag{14.34}$$

$$M\left(1 - \frac{T_c}{T}\right) = \frac{C}{\mu_0 T} B_{mw} \tag{14.35}$$

$$M = \frac{C}{\mu_0 T} B_{mw} \frac{T}{(T - T_c)} = \frac{C}{\mu_0 (T - T_c)} B_{mw} \tag{14.36}$$

Thus, susceptibility is defined as follows:

$$\chi = \frac{\mu_0 M}{B_{mw}} = \frac{\mu_0}{B_{mw}} \frac{C}{\mu_0 (T - T_c)} B_{mw} = \frac{C}{(T - T_c)} \tag{14.37}$$

This is referred to as the Curie–Weiss law that once again expresses the inherent dependence of susceptibility on the temperature. Temperature T_c is the Curie temperature and is dependent on the inherent nature of the material. Thus, for temperatures $T > T_c$, ferromagnetic materials behave like paramagnetic materials; for temperatures $T < T_c$, the equation for magnetization [Eq. (14.31)] can be seen to be dependent on the orbital angular momentum (J) such that the atomic magnetic moment (μ) is $g\mu_B J$. Thus, the saturation magnetization M_s is temperature dependent.

Example 14.5 Iron has a relative permeability of 5000. Calculate its magnetic susceptibility.

Solution Magnetic susceptibility and relative permeability are related through the following equation: $\mu_r = 1 + \chi_m$

This can be rewritten as follows, to calculate the magnetic susceptibility:

$$\chi_m = \mu_r - 1 = 5000 - 1 = 4999$$

Example 14.6 A magnetic field of 1800 A/m produces a magnetic flux of 3×10^{-5} Wb in an iron bar of cross-sectional area $0.2 \, \text{cm}^2$. Calculate the permeability.

Solution Magnetizing field $H = 1800 \, \text{A/m}$, magnetic flux $\Pi = 3 \times 10^{-5}$ Wb, area of cross section of the iron rod, $A = 0.2 \times 10^{-4} \, \text{m}^2$

We know that the magnetic flux density can be written as follows: $B = \dfrac{\varphi}{A}$

which gives $B = \dfrac{3 \times 10^{-5} \, \text{Wb}}{0.2 \times 10^{-4} \, \text{m}^2} = 1.5 \, \text{Wb/m}^2$

Permeability is given as follows: $\mu = \left(\dfrac{B}{H} \right) = \dfrac{1.5 \, \text{H}}{1800 \, \text{m}} = 8.3 \times 10^{-4} \, \text{H/m}$

Example 14.7 A sample of carbon steel has a permeability of 0.01 T.m/A when the magnetic intensity is $75 \dfrac{\text{A}}{\text{m}}$.

(a) Find the magnetic field in the sample at this value of H.

(b) Find the field (in air) at this value of H.

Solution The calculations are as follows:

(a) The magnetic field is given as follows: $B = \mu H = \left(0.01 \, \text{T.} \dfrac{\text{m}}{\text{A}} \right) \times \dfrac{75 \, \text{A}}{\text{m}} = 0.75 \, \text{T}$

(b) $B = \mu_0 H = \left(4\pi \times 10^{-7} \, \text{T.m/A} \right) \times (0.75 \, \text{T}) = 9.42 \times 10^{-7} \, \text{T}$

Example 14.8 An ion-core solenoid is wound with 20 turns/cm carrying a current of 0.1 A.

(a) Find H and B inside the solenoid.

(b) An ion core whose permeability is 6×10^{-3} T.m/A is inserted in the solenoid. Find H and B now.

Solution The calculations are as follows:

(a) $H = nI = \left(20 \times 100 \, \text{m}^{-1} \right) (0.1 \, \text{A}) = 200 \dfrac{\text{A}}{\text{m}}$

$B = \mu_0 H = \left(4\pi \times 10^{-7} \, \text{T.m/A} \right) \times (200 \, \text{A/m}) = 2.514 \times 10^{-4} \, \text{T}$

(b) $H = nI = (20 \times 100/\text{m})(0.1 \, \text{A}) = 200 \dfrac{\text{A}}{\text{m}}$

$B = \mu H = \left(6 \times 10^{-3} \, \text{T.} \dfrac{\text{m}}{\text{A}} \right) (200 \, \text{A/m}) = 1.2 \, \text{T}$

14.6 ANTI-FERROMAGNETIC MATERIALS

Some materials have neighbouring dipoles lined up in opposition to one another in the presence of a magnetic field. Thus, although the strength of each dipole is very high, these materials have zero magnetization. Such materials are called *anti-ferromagnetic materials*. Some examples of such materials are manganese, chromium, MnO, NiO, CoO, and $MnCl_2$. This effect is illustrated for MnO in Fig. 14.13.

The magnetic susceptibility of anti-ferromagnetic materials is positive and small.

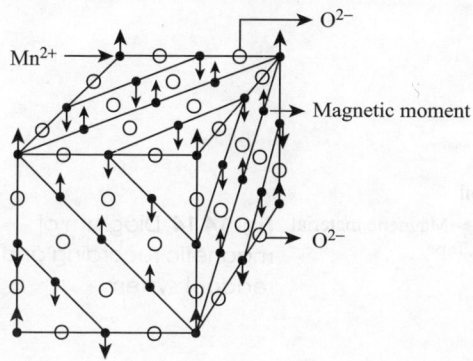

(Has a dipole moment) Ferromagnetic material

Anti ferromagnetic substance (net dipole moment is zero)

Fig. 14.13 Anti-ferromagnetism in MnO

14.7 FERRITES

Different ions have different magnetic moments in ceramic materials. For example, dipoles of a cation A may align along the field, whereas those of another cation B may oppose the field. Since the strength or number of dipoles of either type of material is in general not the same, these materials have a net magnetization. The group of ceramics displaying this behaviour is called *ferrites*. Such materials show a large and magnetic field-dependent magnetic susceptibility. Such materials are also called *ferromagnetic* materials. Since most of the ferromagnetic materials are insulators of electricity, the corresponding electrical losses due to eddy current can be made very small in comparison with metallic ferromagnetic materials. Ferrites are, therefore, useful in high-frequency applications such as wireless communication.

14.8 APPLICATIONS OF MAGNETIC MATERIALS

Some important applications of magnetic materials are mentioned in the following subsections.

14.8.1 Magnetic Recording and Readout

Magnetic materials can be used for recording information as well as for reading out the recorded information. The recording head (e.g., a tape recorder's recording head) consists of a laminated electromagnet, fabricated using

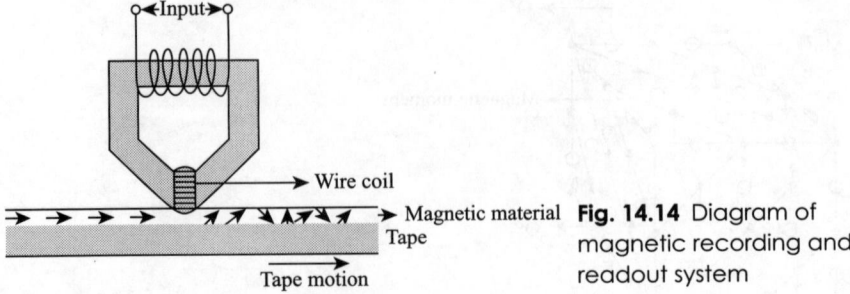

Fig. 14.14 Diagram of magnetic recording and readout system

permalloy or a soft ferrite with an air gap. A wire coil is wound around the magnetic material core. A schematic diagram of a typical set-up is shown in Fig. 14.14.

As a magnetic tape or disc is passed along an electromagnet, the electrical signal fed to the coil generates a magnetic field within the gap. The magnetization of this portion of the tape is proportional to the current passing through the coil. The magnetic material coated on the tape is such that magnetization is retained even after the removal of the field. The signal thus gets stored in the tape. The moving tape induces an emf in the wire coil. To read the tape, the induced alternating emf is amplified through a suitable amplifier and fed into an output device.

14.8.2 Storage of Magnetic Data—Tapes, Floppy, and Magnetic Disc Drives

Magnetic data can be stored in storage devices such as tapes, floppies, and magnetic discs. Magnetic materials used for information storage must have a square-shaped hysteresis loop and a low coercive field. This permits a rapid transmission of information. Magnetic tapes for audio or video applications are fabricated by plating, sputtering, or evaporating magnetic materials such as Fe_2O_3 or CrO_2 on a polyester tape. Magnetic oxide materials are dispersed in a binder to ensure strong bonding. The oxide particles are needle-like, with a size of around $0.1–0.5\,mm^2$. Coercivity of Fe_2O_3 is around $20–28\,kA/m$. The elongated particles are aligned using a field during manufacturing so that the long axes become parallel to the length of the tape or the track.

Floppy discs and hard discs used in computer data storage are also fabricated in a similar manner. Magnetic particles used in a hard disc, for example, are embedded in a polymer film on a flat substrate of aluminium. The domains can rotate very fast in a magnetic field due to the presence of small particles in the polymer matrix. Typical magnetic recording materials and their important properties are listed in Table 14.2.

Table 14.2 Important properties of magnetic recording materials

	Particle length	Aspect ratio	Magnetization B_r		Coercivity H_c		Surface area	Curie temp.
	(µm)		(Wb/m²)	(emu/cc)	(kA/m)	(Oe)	(m²/g)	(Tc) (°C)
Co–γFe$_2$O$_3$	0.20	6:1	0.48	380	30–75	940	20–35	700
CrO$_2$	0.20	10:1	0.50	400	30–75	950	18–55	125
Fe	0.15	10:1	1.40	1100	56–176	2200	20–60	770
γ–Fe$_2$O$_3$	0.20	5:1	0.44	350	22–34	420	15–30	600

IMPORTANT CONCEPTS

1. Magnetic moment M associated with a current loop is given as $M = IA$.
2. The quantity $(eh)/(4\pi m)$, called *Bohr magneton*, is the minimum packet of magnetization and is represented by the symbol μ_B.
3. The quantities B, H, and M are all inter-related. For an isotropic medium (i.e., B, H, and M are all in the same direction) the following relation holds: $B = \mu_0 (H + M)$
4. All magnetic materials fall into the category of diamagnetic, paramagnetic, or ferromagnetic materials, depending on their response to the external applied field.
5. Hysteresis is the phenomenon that tells us that magnetic materials do not completely lose their magnetization when the applied magnetizing field is removed.
6. Magnetic induction B is constituted by magnetization (M) and magnetic intensity (H).
7. Bohr magneton is the minimum packet of atomic dipole moment.
8. Diamagnetic, paramagnetic, and ferromagnetic materials are categorized on the basis of their response to the external applied field.
9. Materials having zero magnetization are called the anti-ferromagnetic materials.

APPLICATIONS

1. Magnetic materials can be used for recording information as well as for reading out the recorded information; examples include the recording heads of tape recorder. Although tape recorders is an obsolete technique now.
2. Magnetic data can be stored in storage devices such as tapes, floppies, and magnetic discs. Magnetic materials used for information storage must have a square-shaped hysteresis loop and a low coercive field. This permits a rapid transmission of information.
3. Magnetic materials are used for communication purpose, as in electronic devices. Typical examples include nickel–iron alloys and soft ferrites.

IMPORTANT FORMULAE

1. Magnetization: $M = \dfrac{evr}{2}$
2. Bohr magneton: $\mu_B = \dfrac{eh}{4\pi m}$

3. $\mu_r = (1 + \chi_m)$

5. Susceptibility: $\chi_m \propto \dfrac{1}{T}$

4. $\dfrac{B}{H} > 1$ for paramagnetic materials

SELF-ASSESSMENT

Multiple-choice Questions

14.1 The Bohr magneton is defined as

(a) $\mu_B = \dfrac{eh^2}{4\pi m}$ (b) $\mu_B = \dfrac{eh}{4\pi m}$ (c) $\mu_B = \dfrac{e^2 h^2}{4\pi m}$ (d) $\mu_B = \dfrac{eh^2}{4\pi m^2}$

14.2 Magnetic moment is defined as

(a) $M = \dfrac{evr}{2}$ (b) $M = \dfrac{ev^2 r}{2\pi r}$ (c) $M = \dfrac{ev}{2\pi r}$ (d) $M = \dfrac{ev}{2\pi}$

14.3 The magnetic susceptibility μ_m of a diamagnetic substance is always
(a) negative   (c) neutral
(b) positive   (d) none of these

14.4 The magnetic permeability of diamagnetic substances is always
(a) less than unity   (c) equal to unity
(b) greater than unity   (d) not equal to unity

14.5 The magnetic susceptibility μ_m of a diamagnetic material
(a) changes with temperature
(b) changes inversely with temperature
(c) does not change with temperature
(d) none of these

14.6 The magnetic permeability μ_m of a paramagnetic material
(a) is greater than unity   (c) is equal to unity
(b) is less than unity   (d) none of these

14.7 Magnetization is defined as the
(a) magnetic moment per unit area of the material
(b) magnetic moment per unit volume of the material
(c) magnetic material per unit length of the material
(d) none of these

14.8 Materials showing large and magnetic-field-dependent magnetic susceptibility
are called
(a) diamagnetic materials   (c) ferromagnetic materials
(b) paramagnetic materials   (d) none of these

14.9 Materials possessing zero magnetization are
(a) ferromagnetic materials   (c) diamagnetic materials
(b) anti-ferromagnetic materials (d) paramagnetic materials

14.10 Hysteresis curve or the B–H curve is the variation of
(a) magnetic induction B with magnetic intensity H
(b) magnetic field with dipole moment
(c) magnetization with magnetic susceptibility
(d) magnetic susceptibility with dipole moment

14.11 Curie temperature is the transition temperature at which
    (a) diamagnetic materials lose their magnetic properties
    (b) paramagnetic materials lose their magnetic properties
    (c) ferromagnetic materials lose their ferromagnetism
    (d) all of these

14.12 Magnetic susceptibility μ_m of a paramagnetic material varies inversely with temperature. This is stated by the
    (a) Curie's Law   (c) Lenz's law
    (b) Magnetic law   (d) none of these

14.13 The Langevin's theory was based on the principle that the
    (a) magnetic moment is quantized
    (b) magnetic moment is classical
    (c) each atom of a gas carries a permanent moment μ
    (d) none of these

14.14 The Langevin's theory gives the
    (a) magnetization for a classical magnetic gas
    (b) magnetization for a quantum gas
    (c) magnetization for a dipole
    (d) all of these

14.15 The Langevin function for high temperatures is given as
    (a) $L(\alpha) = \left(\coth \alpha - \dfrac{1}{\alpha} \right)$ (c) both (a) and (b)

    (b) $L(\alpha) = \dfrac{1}{3}\alpha$ (d) none of these

14.16 Molar susceptibility, as defined by the Langevin's theory, is
    (a) $C_{\text{mol}} = \chi T = \dfrac{\mu_0 \Sigma^2}{3R}$ (c) $C_{\text{mol}} = \chi T^2 = \dfrac{\mu_0 \Sigma^2}{3R}$

    (b) $C_{\text{mol}} = \chi / T = \dfrac{\mu_0 \Sigma^2}{3R}$ (d) $C_{\text{mol}} = \chi / T^2 = \dfrac{\mu_0 \Sigma^2}{3R}$

14.17 The Weiss theory states that
    (a) the inner molecular magnetic field has no relation to the spontaneous magnetization
    (b) the inner molecular magnetic field is not proportional to the spontaneous magnetization
    (c) the inner molecular magnetic field is proportional to the spontaneous magnetization
    (d) the inner molecular magnetic field is by the virtue of dipole moment

14.18 The Curie–Weiss law gives the inherent dependence of susceptibility and
    (a) temperature   (c) magnetization
    (b) dipole moment   (d) molecular magnetic field

14.19 When an external magnetic field is applied to a magnetic material,
    (a) a torque acts on the individual dipole moments
    (b) no force is experienced by the dipole moments
    (c) the particles are in random motion
    (d) there is no change in the dipole moments of the molecules

14.20 Heavy-duty flux multipliers are materials having
    (a) a low working flux density   (c) zero working flux density
    (b) a high working flux density.   (d) no magnetic properties

14.21   Soft magnetic material possesses
    (a)   low coercivity
    (b)   high saturation flux density
    (c)   narrow hysteresis loop
    (d)   all of these

14.22   Hard/Permanent magnets possess
    (a)   high coercivity
    (b)   high remanence
    (c)   bigger hysteresis loop
    (d)   all of these

14.23   The relative permeability of vacuum is
    (a)   0   (b)   1   (c) ∞ (d)   none of these

14.24   The relative permeability of paramagnetic, diamagnetic, and ferromagnetic materials is in the ratio of
    (a) $0.9:1.001:10^4$
    (c) $0.2:0.9:10^4$
    (b) $0.5:1.0:10^{-2}$
    (d) $1.0:1.001:10^4$

Review Questions

14.1   Does a circulating electron generate a magnetic moment? Justify.

14.2   For paramagnetic materials, which of the following is true?
    (a) $\mu_r > 1$
    (c) $\mu_r = 1$
    (b) $\mu_r < 1$
    (d)   Justify your answer.

14.3   How does one assign direction to a dipole?

14.4   Which materials are high-frequency applications attributed to?

14.5   Magnetic materials do not completely lose their magnetization when a magnetizing field is applied and then removed. Justify.

14.6   The terms retentivity and coercivity are attributed to magnetization. Explain.

14.7   Magnetic materials can be used as good storage devices. How? Devise a new simple magnetic storage device.

14.8   Define magnetic moment. What is Bohr magneton? Give its units.

14.9   What are diamagnetic materials? Give some examples.

14.10   What are paramagnetic materials? Give some examples.

14.11   Write three important properties of diamagnetic materials.

14.12   State the relationship between μ_r and μ_m.

14.13   What is Curie's law?

14.14   What is ferromagnetism?

14.15   What are the properties of ferromagnetic materials?

14.16   What is the domain theory of ferromagnetism?

14.17   What are ferrites?

14.18   What are the applications of ferrites?

14.19   Distinguish between soft and hard magnetic materials.

14.20   Mention a few magnetic storage devices.

14.21   What are anti-ferromagnetic materials?

14.22   Explain the working of a magnetic recording and readout system with a suitable diagram.

14.23   On the basis of their magnetic properties, name the materials that are preferred in (a) a permanent magnet, (b) the core of an electromagnet, and (c) the core of an electric transformer.

Numerical Problems

14.1 A plane loop is carrying a current of 10 A. The loop is of irregular shape and encloses an area of $10^{-3}\,m^2$. To an observer, the current appears to flow in the clockwise direction. Determine the magnitude and direction of the magnetic moment associated with the current loop.

$$\left[\text{Hint}: M = IA = 10^{-2}\,A\,m^2; \text{perpendicular to the plane of the loop} \atop \text{and directed away from the observer}\right]$$

14.2 A plane loop is carrying a current of 8 A. The loop is of irregular shape and encloses an area of $5 \times 10^{-4}\,m^2$. To an observer, the current appears to flow in the anticlockwise direction. Determine the magnitude and direction of the magnetic moment associated with the current loop.

$$[\text{Hint}: M = IA]$$

14.3 The electron in a hydrogen atom moves in a circular orbit of radius 0.5 Å. The electron performs 700 revolutions per second. Determine the magnetic moment associated with the orbital motion of the electron.

$$\left[\text{Hint}: I = \frac{e}{T} = e\vartheta\right]$$

14.4 The saturation magnetic induction of nickel is $0.50\,Wb/m^2$. If the density of nickel is $8906\,kg/m^3$ and the atomic weight is 58.7, calculate the magnetic moment of nickel in terms of Bohr magneton.

$$[\text{Hint}: B = N\mu_0 M]$$

14.5 A magnetic field of 800 A/m produces a magnetic flux of 2×10^{-5} Wb in an iron bar of cross-sectional area $0.2\,cm^2$. Calculate the permeability.

$$\left[\text{Hint}: B = \frac{\varphi}{A}; \mu = \left(\frac{B}{H}\right)\right]$$

14.6 The current flowing through the loop in Problem 14.1 is doubled and the enclosed area is made five times the value in Problem 14.1. The current direction is also reversed. Calculate the magnitude and direction of the magnetic moment under the changed circumstances.

$$\left[\text{Hint}: M = IA = 10^{-1}\,A\,m^2; \text{perpendicular to the plane} \atop \text{of the loop and directed towards the observer.}\right]$$

14.7 An electron is circulating in an orbit of radius 1 Å. The electron performs 10^{10} revolutions per second. Calculate the magnetic moment associated with the orbital motion of the electron.

$$\left[\text{Hint}: I = \frac{e}{T} = e\vartheta; M = IA\right]$$

14.8 A given material has a magnetic susceptibility of 900. Determine the relative permeability of the material.

$$[\text{Hint}: \mu_r = 1 + \chi_m]$$

14.9 A given material has a relative permeability of 2000. Calculate its magnetic susceptibility.

$$[\text{Hint}: \chi_m = \mu_r - 1]$$

14.10   A sample of carbon steel has permeability of 0.02 T.m/A when the magnetic intensity is 80 $\frac{A}{m}$. Find the following parameters:

(a)   Magnetic field in the sample at this value of H

(b)   Field (in air) at this value of H

$$\left[Hint: = \mu H, B = \mu_0 H \right]$$

14.11   A magnetic field of 1500 A/m produces a magnetic flux of 1×10^{-5} Wb in an iron bar of cross-sectional area 0.1 cm^2. Calculate the permeability.

$$\left[Hint: B = \frac{\varphi}{A}; \mu = \left(\frac{B}{H} \right) \right]$$

14.12   The magnetizing field for a set of materials is varying from 0.48 to 1.40 Wb/m^2 and the coercivity varies from 22 to 176 $\frac{m}{kA}$. For the construction of which computer devices can these materials be used?

[*Hint*: Refer to Table 14.2]

14.13   If the coercivity of a material fall in the range of 6.0–110 m/kA and the remanence ranges from 0.36 to 1.35/T, these materials can be categorized to be hard or soft magnets. Justify.

[*Hint*: Refer to Table 14.2]

14.14   Draw the hysteresis loop for materials having coercivity 43.0 and 58 m/kA, respectively, and a remanence of 0.65–1.35 / T. Categorize these materials as soft or hard magnets.

Electromagnetic Field Theory

Learning Objectives

After studying this chapter, students will be able to

- use the concepts of gradient, divergence, and curl from vector calculus to physically relate them to the flow of energy for electric and magnetic fields
- employ the concept of electromagnetic field by the introduction of Maxwell's equations
- explain the energy vector, that is, Poynting vector, for conservation of energy through the Poynting theorem
- elucidate the use of Maxwell's equations in recognizing the electric and magnetic vectors for a dielectric
- demonstrate the use of Maxwell's equations in understanding the electric and magnetic vectors for a conducting medium
- use the concept of penetration depth of an electromagnetic wave in a conducting medium

List of Symbols

λ = Wavelength $\qquad$ τ = Relaxation $\qquad$ ϵ = Permittivity

β = Attenuation $\qquad\quad$ constant $\qquad$ σ = Mobility

$\quad$ constant $\qquad$ $\int$ = Integration $\qquad$ ω = Angular frequency

κ = Dielectric $\qquad$ μ = Permeability $\qquad$ ρ = Charge density

15.1 INTRODUCTION

The behaviour of a light wave as it travels through various media has always been intriguing physicists. A light wave is constituted of both the electric and the magnetic vector. As the idea of electricity and magnetism has been dealt with at great length, it has been conjectured that a light wave is an electromagnetic radiation or wave.

The phenomenon of electricity and magnetism has been dealt with extensively independently. Maxwell thought that the two phenomena must be complementary to each other. This thought lead to a new unified field of physics, called electromagnetism or electromagnetic field theory.

In this chapter, new mathematical techniques of gradient, divergence, and curl have been discussed that help us understand the new Maxwell's equations of electromagnetism. Further discussion helps us understand how the simple concept of Maxwell's equations leads us to understand the guided light waves through any medium, opening up an era of new technological developments.

15.2 VECTOR CALCULUS—GRADIENT AND DIVERGENCE

We are introducing some elementary mathematical operations in vector calculus to understand the concepts of electromagnetism better.

As per the convention, any position vector in translational co-ordinates is defined as follows:

$$r = x\hat{i} + y\hat{j} + z\hat{k} \qquad (15.1)$$

where r is conventionally defined as the position vector, where x, y, z are the rectangular co-ordinates and $\hat{i}, \hat{j}, \hat{k}$ are the corresponding unit vectors in the specific direction [Fig. 15.1(a)].

The vector differential operator 'del', written as ∇, is defined as follows:

$$\nabla = \frac{\partial}{\partial x}\hat{i} + \frac{\partial}{\partial y}\hat{j} + \frac{\partial}{\partial z}\hat{k} \qquad (15.2a)$$

$$= \hat{i}\frac{\partial}{\partial x} + \hat{j}\frac{\partial}{\partial y} + \hat{k}\frac{\partial}{\partial z} \qquad (15.2b)$$

Thus, the del operator helps us understand the variation of a physical parameter in space. This can further be explained from the concepts of gradient and divergence that follow.

15.2.1 Gradient

A gradient represents a directional derivative. In simple physical terms, it is the rate of change of a function in a specified direction, that is, with respect to space. The gradient of a scalar function is a vector quantity. The magnitude of the vector quantity is the maximum directional derivative at the point being considered, and its direction is given by the directional derivative at that point.

Mathematically, if $\phi(x, y, z)$ is differentiable at each point (x, y, z) in a certain region of space, then

$$\nabla\phi(x, y, z) = \left(\frac{\partial}{\partial x}\hat{i} + \frac{\partial}{\partial y}\hat{j} + \frac{\partial}{\partial z}\hat{k}\right)\phi(x, y, z)$$

$$= \left(\frac{\partial\phi}{\partial x}\hat{i} + \frac{\partial\phi}{\partial y}\hat{j} + \frac{\partial\phi}{\partial z}\hat{k}\right) \qquad (15.3)$$

(Cartesian co-ordinates)

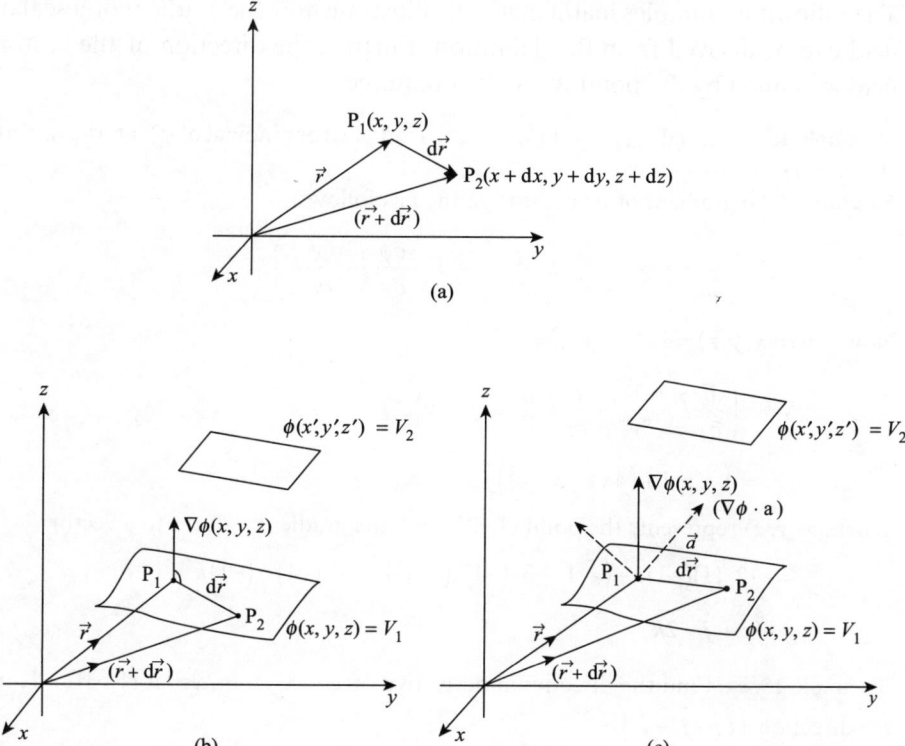

Fig. 15.1 Pictorial representation of the del operator (a) Representation of position vector **r** at point P_1 in translational co-ordinates. P_2 represents position vector $(r + dr)$ at point P_2 at an infinitesimal distance dr from vector r; del operator helps us quantify this infinitesimal change (b) Gradient vector $\nabla\varphi(x,y,z)$ normal to tangent dr drawn between two infinitesimally close points P_1 and P_2 on an equipotential surface $\varphi(x,y,z) = V_1$ (c) Directional derivative $\bar{\nabla}\varphi \cdot \bar{a}$ and quantity $\nabla\varphi(x,y,z)$ representing maximum directional derivative being perpendicular to tangent dr

It is evident from this equation that $\nabla\phi$ represents the vector field. The component of $\nabla\phi$ in the direction of a unit vector $\bar{a}$ is given by $\nabla\phi \cdot \bar{a}$ and is called the directional derivative of the scalar field ϕ in $\bar{a}$. Physically, it is the rate of change of the physical parameter (or scalar field) $\phi(x, y, z)$ at the co-ordinate $P_1(x, y, z)$ in the direction $\bar{a}$. The direction of the vector field has the direction normal to the level surface of $\phi(x, y, z)$ through the point being considered [Figs 15.1(b) and (c)]. In other words, the gradient of a potential function is a vector field that is normal to the equipotential surfaces.

$$\nabla\varphi(x,y,z) = \frac{\partial\phi}{\partial x}r + \frac{\partial\phi}{r\partial\varphi}\varphi + \frac{\partial\phi}{\partial z}z \qquad \text{(cylindrical co-ordinates)}$$

$$\nabla\varphi(x,y,z) = \frac{\partial\phi}{\partial x}r + \frac{\partial\phi}{r\partial\theta}\theta + \frac{\partial\phi}{r\sin\theta\partial\varphi}\varphi \qquad \text{(spherical co-ordinates)}$$

The following examples mathematically illustrate how the gradient of a scalar field can be derived from the definition. Further, the direction of the vector field is defined by the point where it is required.

Example 15.1 If $\varphi(x,y,z) = 4x^2y - y^3z^2$, find $\nabla\varphi$ (or gradient of φ) at the point $(1, -1, -1)$.

Solution The gradient of a vector is defined as follows:

$$\nabla\phi = \left(\frac{\partial}{\partial x}\hat{i} + \frac{\partial}{\partial y}\hat{j} + \frac{\partial}{\partial z}\hat{k}\right)\phi = \left(\frac{\partial\phi}{\partial x}\hat{i} + \frac{\partial\phi}{\partial y}\hat{j} + \frac{\partial\phi}{\partial z}\hat{k}\right)$$

Now, $\varphi(x,y,z) = 4x^2y - y^3z^2$

$$\nabla\phi = \left(\frac{\partial}{\partial x}\hat{i} + \frac{\partial}{\partial y}\hat{j} + \frac{\partial}{\partial z}\hat{k}\right)(4x^2y - y^3z^2)$$

$$= 4\cdot 2xy\hat{i} + \left(4x^2 - 3y^2z^2\right)\hat{j} - 2y^3z\hat{k}$$

When (x, y, z) represents the point $(1, -1, -1)$, the gradient reduces to a vector:

$$= 4\cdot 2\cdot(1)(-1)\hat{i} + \left(4\cdot 1^2 - 3\cdot(-1)^2(-1)^2\right)\hat{j} - 2\cdot(-1)^3\cdot(-1)\hat{k}$$

$$= -8\hat{i} + \hat{j} - 2\hat{k}$$

Example 15.2 Find the directional derivative of $\varphi = x^2yz + 2xz^2$ at $(1, -1, -1)$ in the direction $\left(\hat{i}, -\hat{j}, -\hat{k}\right)$

Solution The gradient is defined as follows:

$$\nabla\phi = \left(\frac{\partial}{\partial x}\hat{i} + \frac{\partial}{\partial y}\hat{j} + \frac{\partial}{\partial z}\hat{k}\right)\phi = \left(\frac{\partial\phi}{\partial x}\hat{i} + \frac{\partial\phi}{\partial y}\hat{j} + \frac{\partial\phi}{\partial z}\hat{k}\right)$$

$$\varphi = x^2yz + 2xz^2$$

$$\nabla\phi = \left(\frac{\partial}{\partial x}\hat{i} + \frac{\partial}{\partial y}\hat{j} + \frac{\partial}{\partial z}\hat{k}\right)(x^2yz + 2xz^2)$$

$$= \left(2xyz + 2z^2\right)\hat{i} + \left(x^2z\right)\hat{j} + \left(x^2y + 4xz\right)\hat{k}$$

Substituting the co-ordinates (x, y, z) as $(1, -1, -1)$,

$$\nabla\varphi = \left(2\cdot(1)\cdot(-1)(-1) + 2\cdot(-1)^2\right)\hat{i} + \left(1^2\cdot(-1)\right)\hat{j} + (1^2\cdot(-1) + 4\cdot 1\cdot(-1))\hat{k}$$

$$= 4\hat{i} - \hat{j} - 5\hat{k}$$

The unit vector along the direction $(\hat{i} - \hat{j} - \hat{k})$ is given as follows: $= \frac{1}{\sqrt{3}}(\hat{i} - \hat{j} - \hat{k})$

The directional derivative is expressed as follows:

$$\nabla\varphi\cdot a = \left(4\hat{i} - \hat{j} - 5\hat{k}\right)\cdot\frac{1}{\sqrt{3}}(\hat{i} - \hat{j} - \hat{k}) = \frac{10}{\sqrt{3}}$$

Example 15.3 Using the concept of gradient co-relate the electric field intensity (E) to the potential function (V).

Solution The potential of a point A with respect to point B is defined as the work done (W) in moving a unit positive charge, Q, from B to A: $V_{AB} = \frac{W}{Q} = -\int_B^A E\cdot dl \quad \left(\frac{J}{C}\right)$ or V

Translating this definition in the differential form, one gets the following equation:

$$dV_{AB} = -E \cdot dI$$

From the definition of the gradient from Eq. (15.3), one gets the following expression:

$$dV = \nabla V \cdot dr$$

where dr is the arbitrary small displacement; it can also be equated as the small change in the displacement dr. Thus,

$$E = -\nabla V$$

Hence, the electric field intensity (E) is the negative gradient of the potential function (V). The physical significance of the negative sign is that the electric field is directed from the higher to the lower levels of potential.

15.2.2 Divergence

The divergence of a vector field represents the net amount of flux of the vector field E diverging per unit area. The vector field $V(x, y, z)$ be defined as follows:

$$V(x, y, z) = V_1 \hat{i} + V_2 \hat{j} + V_3 \hat{k} \tag{15.4}$$

and is assumed to be differentiable at point (x, y, z) in a certain region of co-ordinate space. Let the edges of the cube be Δx, Δy, and Δz parallel to the x, y, and z axes, respectively. The surface integral of the vector field $V(x, y, z)$ over this cube must cover all the six faces with dS as the outward drawn normal on each face. Considering the surface integral for the x-component, we get the following relations:

$$\iint_S^{\text{left face}} V(x, y, z) \cdot dS = -V_x(x) \Delta y \Delta z \tag{15.4a}$$

$$\iint_S^{\text{right face}} V(x, y, z) \cdot dS = V_x(x + \Delta x) \Delta y \Delta z$$

$$= \left[V_x(x) + \frac{\partial V_x}{\partial x} \Delta x \right] \Delta y \Delta z \tag{15.4b}$$

$$\iint_S^{\text{left face}} V(x, y, z) \cdot dS + \iint_S^{\text{right face}} V(x, y, z) \cdot dS$$

$$= -V_x(x) \Delta y \Delta z + \left[V_x(x) + \frac{\partial V_x}{\partial x} \Delta x \right] \Delta y \Delta z = \frac{\partial V_x}{\partial x} \Delta x \Delta y \Delta z \tag{15.4c}$$

Similarly, one can simplify for the y and z components. As the volume ($\Delta x \, \Delta y \, \Delta z$) of the cube is assumed to be approaching zero, it is a negligible quantity and can be eliminated from the left and right hand side of the equation by dividing throughout:

$$\nabla \cdot V = \left(\frac{\partial}{\partial x} \hat{i} + \frac{\partial}{\partial y} \hat{j} + \frac{\partial}{\partial z} \hat{k} \right) (V_1 \hat{i} + V_2 \hat{j} + V_3 \hat{k})$$

$$= \left(\frac{\partial V_1}{\partial x} \hat{i} + \frac{\partial V_2}{\partial y} \hat{j} + \frac{\partial V_3}{\partial z} \hat{k} \right) \tag{15.5}$$

This definition of divergence in Eq. (15.5) is specific to the Cartesian co-ordinates. Mathematically, divergence can be defined as the dot product of the del operator and the given vector field. For completeness, expressions for divergence in the cylindrical (r, φ, z) and spherical co-ordinates are given as follows:

$$\nabla \cdot V = \frac{1}{r} \frac{\partial}{\partial r} (rV_r) + \frac{1}{r} \frac{\partial V_\varphi}{\partial \varphi} + \frac{\partial V_z}{\partial z}$$

(Divergence in cylindrical co-ordinates)

$$\nabla \cdot V = \frac{1}{r^2} \frac{\partial}{\partial r} (r^2 V_r) + \frac{1}{r \sin\theta} \frac{\partial}{\partial \theta} (V_\theta \sin\theta) + \frac{1}{r \sin\theta} \frac{\partial V_\varphi}{\partial \varphi}$$

(Divergence in spherical co-ordinates)

The following mathematical steps show that the order of the del operator and the vector field in a mathematical operation are of great significance while considering the definition of divergence:

$$\nabla \cdot V = \left(\frac{\partial}{\partial x} \hat{i} + \frac{\partial}{\partial y} \hat{j} + \frac{\partial}{\partial z} \hat{k} \right) (V_1 \hat{i} + V_2 \hat{j} + V_3 \hat{k})$$

$$= \left(\frac{\partial V_1}{\partial x} \hat{i} + \frac{\partial V_2}{\partial y} \hat{j} + \frac{\partial V_3}{\partial z} \hat{k} \right)$$

$$V \cdot \nabla = (V_1 \hat{i} + V_2 \hat{j} + V_3 \hat{k}) \cdot \left(\frac{\partial}{\partial x} \hat{i} + \frac{\partial}{\partial y} \hat{j} + \frac{\partial}{\partial z} \hat{k} \right)$$

$$= \left(V_1 \frac{\partial}{\partial x} \hat{i} + V_2 \frac{\partial}{\partial y} \hat{j} + V_3 \frac{\partial}{\partial z} \hat{k} \right)$$

$$\Rightarrow \nabla \cdot V \neq V \cdot \nabla \tag{15.6}$$

Physically, the divergence of a field represents the gain or loss of that field in the specified direction. If the divergence is a positive quantity, it indicates that, at that point, the vector quantity is expanding and its density is falling with time, or the point is a 'source' of flux. On the other hand, if the divergence is a negative quantity, it indicates that, at that point, the field is contracting and its density is rising, or the point is a 'negative source' or 'sink'. If the divergence of a vector field is zero, then there is no 'source' or 'sink'; else, the density of the field will be a constant or there is no gain of field anywhere. Such a vector field whose divergence is zero is called 'solenoidal'. This also represents the 'equation of continuity' for an incompressible fluid (Fig. 15.2). Usually, this concept is useful where fluid dynamics is involved, and hence the vector field can also be referred to as a fluid in many instances.

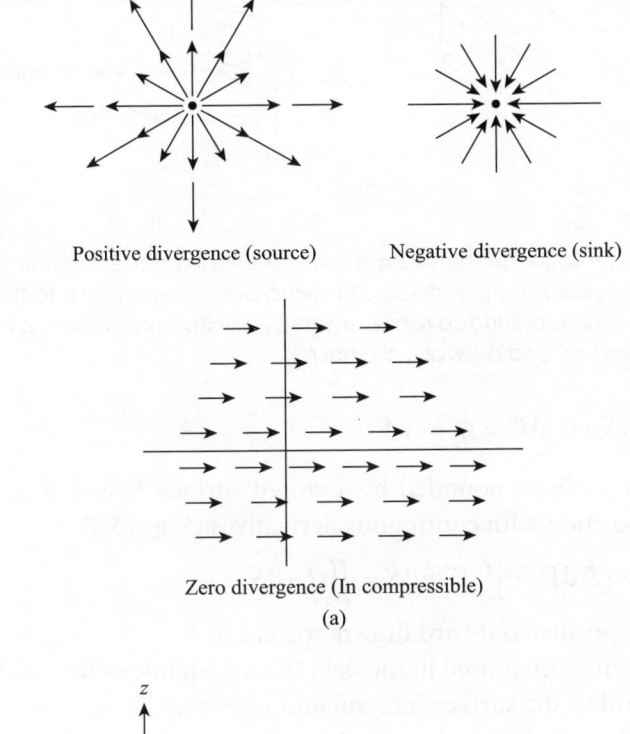

Positive divergence (source)   Negative divergence (sink)

Zero divergence (In compressible)

(a)

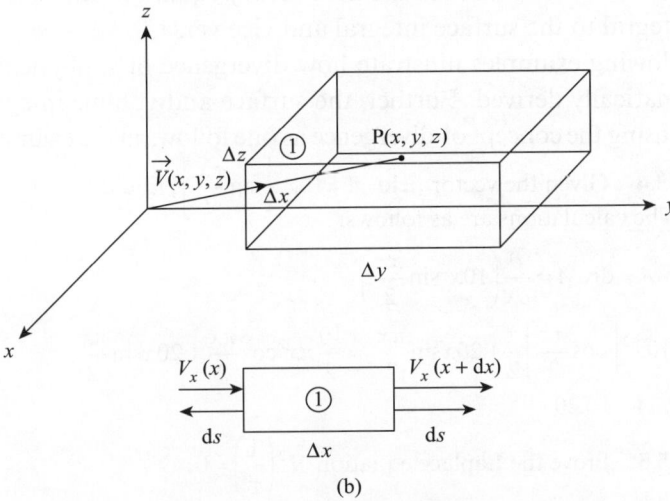

(b)

Fig. 15.2 Representation of Divergence (a) Divergence is flux of vector field lines from given point (b) Divergence of vector field as calculated using concept of cube with Cartesian co-ordinates

15.3 GAUSS DIVERGENCE THEOREM

The divergence theorem states that *the volume integral of the divergence of a vector field over any volume V is equal to the surface integral of that field taken over the closed surface enclosing the volume V; that is,*

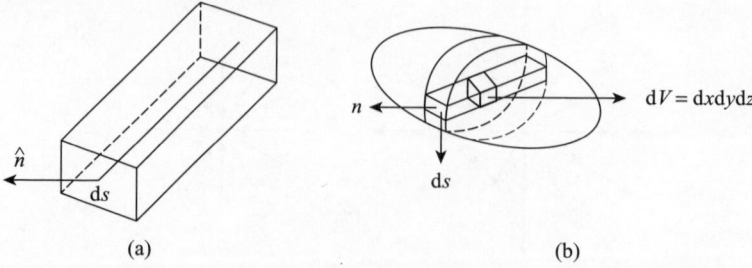

Fig. 15.3 Representation of the volume and spherical surface depicting the normal 'n' (a) Cuboid h with surface dS having outward normal n to this surface (b) Volume of spherical surface represented by infinitesimal volume ΔV = dxdydz with surface area dS and outward normal n

$$\iiint_V (\nabla \cdot E) dV = \oint_S E \cdot dS \qquad (15.7)$$

where V is the volume bounded by a closed surface S, and E is the vector function of position with continuous derivatives (Fig. 15.3).

$$\iiint_V \nabla \cdot E \, dV = \iint_S E \cdot n \, dS = \iint_S E \cdot dS \qquad (15.8)$$

where n is the positive outward drawn normal to S.

This theorem is often used in the field theory equations for translating the volume integral to the surface integral and vice versa.

The following examples illustrate how divergence of a physical field can be mathematically derived. Further, the surface and volume integral can be co-related using the concept of divergence as one follows in the examples below:

Example 15.4 Given the vector field $A = 10x^2 \left[\sin \dfrac{\pi x}{2} \right]$, find div A at $x = 1$.

Solution The calculations are as follows:

$$\nabla \cdot A = \text{div } A = \frac{\partial}{\partial x}\left(10x^2 \sin\frac{\pi x}{2} \right)$$

$$= 10x^2 \left[\cos\frac{\pi x}{2} \right]\frac{\pi}{2} + 20x\sin\frac{\pi x}{2} = \frac{10}{2}\pi x^2 \cos\frac{\pi x}{2} + 20x\sin\frac{\pi x}{2}$$

$$\left| \nabla \cdot A \right|_{x=1} = 20$$

Example 15.5 Prove the Laplace equation $\nabla^2\left(\dfrac{1}{r}\right) = 0$.

Solution Using the concept of divergence, the Laplace equation can be expressed as follows:

$$\nabla^2\left(\frac{1}{r}\right) = \left(\frac{\partial^2}{\partial x^2} + \frac{\partial^2}{\partial y^2} + \frac{\partial^2}{\partial z^2} \right)\left(\frac{1}{\sqrt{x^2 + y^2 + z^2}} \right)$$

$$\frac{\partial}{\partial x}\left(\frac{1}{\sqrt{x^2 + y^2 + z^2}} \right) = -x(x^2 + y^2 + z^2)^{-3/2}$$

$$\frac{\partial^2}{\partial x^2}\left(\frac{1}{\sqrt{x^2 + y^2 + z^2}} \right) = 3x^2(x^2 + y^2 + z^2)^{-5/2} - (x^2 + y^2 + z^2)^{-3/2}$$

$$= \frac{2x^2 - y^2 - z^2}{(x^2 + y^2 + z^2)^{5/2}}$$

$$\frac{\partial^2}{\partial y^2}\left(\frac{1}{\sqrt{x^2 + y^2 + z^2}}\right) = \frac{2y^2 - z^2 - x^2}{(x^2 + y^2 + z^2)^{5/2}} \Rightarrow \frac{\partial^2}{\partial z^2}\left(\frac{1}{\sqrt{x^2 + y^2 + z^2}}\right) = \frac{2z^2 - x^2 - y^2}{(x^2 + y^2 + z^2)^{5/2}}$$

Thus, by addition, $\nabla^2\left(\dfrac{1}{r}\right) = \left(\dfrac{\partial^2}{\partial x^2} + \dfrac{\partial^2}{\partial y^2} + \dfrac{\partial^2}{\partial z^2}\right)\left(\dfrac{1}{\sqrt{x^2 + y^2 + z^2}}\right) = 0$

The equation $\nabla^2\left(\dfrac{1}{r}\right) = 0$ is called the Laplace equation, where, now $\dfrac{1}{r}$, is a solution of this equation.

Example 15.6   Show that the electric and displacement vectors have a divergence of zero in any isotropic charge-free region.

Solution   From the definition of divergence,

$$\lim_{\Delta v \to 0} \frac{\iint_S D \cdot dS}{\Delta V} = \text{div}\, D = \lim_{\Delta v \to 0} \frac{Q}{\Delta V} = \rho$$

that is, $\nabla \cdot D = \rho$ and $\nabla \cdot E = \dfrac{\rho}{\epsilon}$

where ρ is the charge density. For an isotropic charge-free region, ϵ is constant throughout; thus, the divergence is equal to zero.

Example 15.7   A cube of 4 m edge is centred at the origin, the edges being parallel to the axes. Verify the divergence theorem for a vector $V(x,y,z) = 2x^2\hat{i}\ C/m^2$.

Solution   The divergence theorem states that

$$\iint_S V(x,y,z) \cdot dS = \iiint_V (\nabla \cdot V(x,y,z))\, dV$$

Considering the left-hand side of the divergence theorem:

$$\iint_S V(x,y,z) \cdot dS = \iint_{-2}^{2}(2x^2\hat{i}) \cdot dydz\,\hat{i} + \iint_{-2}^{2}(2x^2\hat{i})dydz(-\hat{i})$$

$$= \iint_{-2}^{2}(2 \cdot 2^2)\hat{i} \cdot dydz\,\hat{i} + \iint_{-2}^{2}(2 \cdot 2^2)\hat{i}\,dydz\,(-\hat{i}) = 0$$

where the limits of the integral vary from (-2) to (2), as the cube is centred about the origin. In addition, the vector field has only the x-component; thus, $(V(x,y,z) \cdot dS)$ is zero on the y and z faces.

The right-hand side of the divergence theorem:

$$\nabla \cdot V(x,y,z) = 4x$$

$$= \iiint_{-2}^{2}(4x)dxdydz = \iint_{-2}^{2} 4 \cdot \left.\frac{x^2}{2}\right|_{-2}^{2} dydz = 0$$

Hence, the divergence theorem is verified.

Example 15.8   Verify the divergence theorem physically (you can take the help of Fig.15.2).

Solution   The divergence theorem physically relates to the flow of a fluid at any given point. Thus, considering the left-hand side of the divergence theorem, let $V(x, y, z)$ represent the velocity of a moving fluid in a cylinder of base dS in time Δt such that $(V \cdot \Delta t)$ is the slant height of the cylinder; then

Volume of the fluid crossing dS in Δt seconds $= (V \cdot \Delta t) \cdot n \, dS = V \cdot n \, dS \, \Delta t$

Volume of the fluid crossing dS per seconds $= V \cdot n \, dS = V \cdot n \, dS$

Total volume of the fluid crossing dS per second $= \iint_S V \cdot n \, dS$

For the right-hand side of the divergence theorem,

The volume per second of the fluid emerging from volume element $dV = \nabla \cdot V \, dV$

Total volume per second of the fluid emerging from all volume elements in $S = \iiint_V \nabla \cdot V \, dV$

LHS = RHS

$$\iint_S V \cdot n \, dS = \iiint_V \nabla \cdot V \, dV$$

15.4 CURL

The curl of a vector field $V(x, y, z)$ is a vector field indicating the rotational properties of a field. The vector field $V(x, y, z)$ be defined as follows:

$$V(x, y, z) = V_1 \hat{i} + V_2 \hat{j} + V_3 \hat{k} \tag{15.9}$$

and is assumed to be differentiable at point P(x, y, z) in a certain region of space. The point P(x, y, z) lies in a plane area ΔS bounded by a closed curve C traversed in a manner such that the enclosed area is to the left or in other words the contour is drawn in the clockwise direction to the point. The direction of the normal component (n) to the contour C is defined by the right-hand rule. The component of the curl of $V(x, y, z)$ in the direction of the normal is given as follows:

$$\left(\text{curl} V(x, y, z)\right) \cdot n = \lim_{\Delta S \to 0} \frac{\oint V(x, y, z) \cdot dr}{\Delta S}$$

This expression in the Cartesian co-ordinates for one of the components, let us say x, is given as follows:

$$\left(\text{curl } V(x, y, z)\right) \cdot \hat{i} = \lim_{\Delta y \Delta z \to 0} \frac{\oint V(x, y, z) \cdot dr}{\Delta y \, \Delta z}$$

The vector field $V(x, y, z) = V_x \hat{i} + V_y \hat{j} + V_z \hat{k}$ is at the corner of the area ΔS closest to the origin (point A). The contour C defined for the x-component is a rectangular plane shown in Fig. 15.4. Hence, the contour integration in the denominator for the x-component plane can be explicitly written in the following terms:

$$\oint V(x, y, z) dr = \oint_A^B V(x, y, z) dr + \oint_B^C V(x, y, z) dr$$
$$+ \oint_C^D V(x, y, z) dr + \oint_D^A V(x, y, z) dr$$

$$V_y \Delta y = V_y \Delta y + \left(V_z + \frac{\partial V_z}{\partial y} \Delta y\right) \Delta z + \left(V_y + \frac{\partial V_y}{\partial z} \Delta z\right)(-\Delta y) + V_z(-\Delta z)$$

$$V_y \Delta y = \left(\frac{\partial V_z}{\partial y} - \frac{\partial V_y}{\partial z}\right) \Delta y \Delta z$$

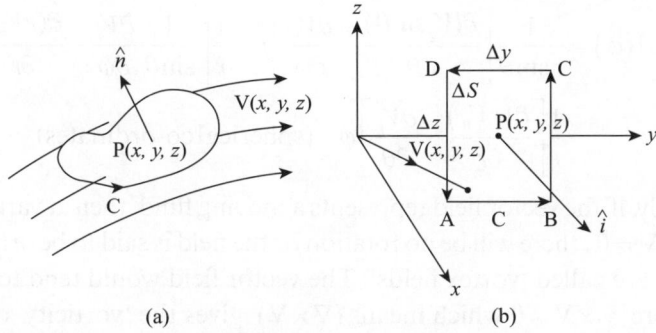

Fig. 15.4 Diagrammatic representation of curl (a) Curl of vector field as shown in potential surface $V(x, y, z)$ with outward normal n (b) Contour C defined by a square in x = constant plane through P

$$(\text{curl}V) \cdot \hat{\boldsymbol{i}} = \left(\frac{\partial V_z}{\partial y} - \frac{\partial V_y}{\partial z} \right)$$

Similarly, the corresponding y and z components may be determined such that

$$(\text{curl}V) = \left(\frac{\partial V_z}{\partial y} - \frac{\partial V_y}{\partial z} \right)\hat{\boldsymbol{i}} + \left(\frac{\partial V_x}{\partial z} - \frac{\partial V_z}{\partial x} \right)\hat{\boldsymbol{j}} + \left(\frac{\partial V_y}{\partial x} - \frac{\partial V_x}{\partial y} \right)\hat{\boldsymbol{k}}$$

$$\nabla \times V(x, y, z) = \left(\frac{\partial}{\partial x}\hat{\boldsymbol{i}} + \frac{\partial}{\partial y}\hat{\boldsymbol{j}} + \frac{\partial}{\partial z}\hat{\boldsymbol{k}} \right) \times (V_x\hat{\boldsymbol{i}} + V_y\hat{\boldsymbol{j}} + V_z\hat{\boldsymbol{k}}) \qquad (15.10)$$

Mathematically, $\text{curl}V = \begin{vmatrix} \hat{\boldsymbol{i}} & \hat{\boldsymbol{j}} & \hat{\boldsymbol{k}} \\ \dfrac{\partial}{\partial x} & \dfrac{\partial}{\partial y} & \dfrac{\partial}{\partial z} \\ V_x & V_y & V_z \end{vmatrix}$

This equation represents the mathematical expression for the curl of a vector field in Cartesian co-ordinates (Fig. 15.4).

It is evident from these mathematical steps that $\Rightarrow \nabla \times V \neq V \times \nabla$

Students can verify this in simple mathematical steps.

Notes:
1. The divergence of a curl of the vector field $V(x, y, z)$ is the scalar zero, that is, $\nabla \cdot (\nabla \times V(x, y, z)) = 0$
2. The curl of the gradient of a vector field is the zero vector, that is, $\nabla \times (\nabla \phi) = 0$

The expression for the curl of a vector in the spherical and cylindrical co-ordinates is given for completeness:

$$\text{curl}(\varphi) = \left[\frac{1}{r}\frac{\partial V_z}{\partial \varphi} - \frac{\partial V_\varphi}{\partial z} \right]\hat{\boldsymbol{r}} + \left[\frac{\partial V_r}{\partial z} - \frac{\partial V_z}{\partial r} \right]\hat{\varphi} + \frac{1}{r}\left[\frac{\partial(rV_\varphi)}{\partial r} - \frac{\partial V_r}{\partial \varphi} \right]\hat{z}$$

(cylindrical co-ordinates)

$$\text{curl}(\varphi) = \frac{1}{r\sin\theta}\left[\frac{\partial(V_\varphi\sin\theta)}{\partial\theta} - \frac{\partial V_\theta}{\partial\varphi}\right]\hat{r} + \frac{1}{r}\left[\frac{1}{\sin\theta}\frac{\partial V_r}{\partial\varphi} - \frac{\partial(rV_\varphi)}{\partial r}\right]\hat{\theta}$$

$$+ \frac{1}{r}\left[\frac{\partial(rV_\theta)}{\partial r} - \frac{\partial V_r}{\partial\theta}\right]\hat{\varphi} \quad \text{(spherical co-ordinates)}$$

Physically, if the vector field represents a moving fluid, then at various points where $V \times V = 0$, there will be no rotation or the field is said to be 'irrotational'. Such fields are called 'vortex fields'. The vector field would tend to rotate for points where $\nabla \times V \neq 0$ which means $(\nabla \times V)$ gives the 'vorticity' of the field at the point (x, y, z). For example, the electrostatic field is an irrotational field. Figure 15.5 represents the curl of a vector.

The mathematical concepts of vector calculus relate the curl of a vector to its line integration, as seen in the following section.

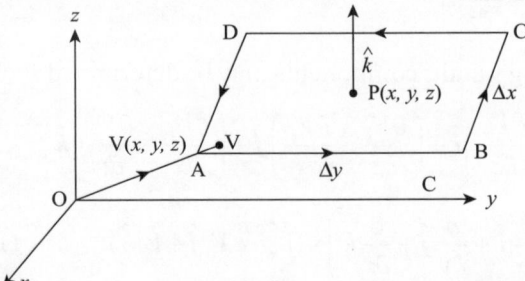

Fig. 15.5 Curl (z-component) of vector field

15.4.1 Stoke's Theorem

The Stoke's theorem states that the integral of the tangential component of the vector field $V(x, y, z)$ around the contour C is equal to the integral of the normal component of the curl of the vector field. It can mathematically be represented as follows:

$$\oint V \cdot d\boldsymbol{r} = \iint_S (\nabla \times V) \cdot n \, dS \tag{15.11}$$

where S is an open, two-sided surface bounded by a closed, non-intersecting curve C (simple closed curve), the vector field $V(x, y, z)$ has continuous derivatives, and the contour C is traversed in the positive direction.

It can be observed from the definition of the Stokes theorem that if the tangential component of a vector field is known, then it can be simply translated to the integral of the normal component of the curl of the vector field. Often in physics, it becomes meaningful to co-relate these components of electric and magnetic vectors to understand better the physical significance of an equation or a concept, particularly the Maxwell's equations.

Example 15.9 Show using the Stokes theorem that if the vector field represents the magnetic potential, then the contour integration would result in the magnetic flux.

Solution The vector magnetic potential A is defined as follows: $\nabla \times A = B$ where the unit of A is Weber/m or Tesla.

From the Stokes theorem, $\oint A \cdot dr = \iint_S B \cdot dS = \phi$

where integration of the magnetic field over the surface results in a magnetic flux through that surface.

Example 15.10 From the conventional definition of the curl of a vector field, find the z-component of the curl of a vector field.

Solution From the conventional definition of the curl of a vector field, the contour C is traversed in the positive direction, that is,

$$\oint_C V(x,y,z) \cdot dr = \oint_A^B V \cdot dr + \oint_B^C V \cdot dr + \oint_C^D V \cdot dr + \oint_D^A V \cdot dr$$

$$V_y \Delta y = V_y \Delta y - \left(V_x + \frac{\partial V_x}{\partial y} \Delta y \right) \Delta x - \left(V_y + \frac{\partial V_y}{\partial x}(-\Delta x) \right)(\Delta y) + V_x(\Delta x)$$

$$V_y \Delta y = \left(\frac{\partial V_y}{\partial x} - \frac{\partial V_x}{\partial y} \right) \Delta x \Delta y$$

Thus, the z-component of the curl of a vector field is given as follows:

$$(\text{curl}V) \cdot k = \left(\frac{\partial V_y}{\partial x} - \frac{\partial V_x}{\partial y} \right)$$

Example 15.11 Show that $\iiint_V (\nabla \times B) dV = 0$, if the vector B is always normal to the given closed surface S.

Solution Using the Stokes theorem, $\oint_C B \cdot dS = \oint_C B dS \cos\theta = 0$

This is because B is normal to the closed surface S enclosing the volume V. From the Stokes theorem, $\oint_C B \cdot dS = \iiint_V (\nabla \times B) dV = 0$

Example 15.12 Show that the curl of $V(x,y,z) = \dfrac{x\hat{i} + y\hat{j} + z\hat{k}}{(x^2 + y^2 + z^2)^{3/2}}$ is zero.

Solution The curl of a vector is conventionally defined as follows:

$$\text{curl}V = \begin{vmatrix} \hat{i} & \hat{j} & \hat{k} \\ \dfrac{\partial}{\partial x} & \dfrac{\partial}{\partial y} & \dfrac{\partial}{\partial z} \\ V_x & V_y & V_z \end{vmatrix} = \frac{1}{(x^2 + y^2 + z^2)^{3/2}} \begin{vmatrix} \hat{i} & \hat{j} & \hat{k} \\ \dfrac{\partial}{\partial x} & \dfrac{\partial}{\partial y} & \dfrac{\partial}{\partial z} \\ x & y & z \end{vmatrix}$$

$$(\text{curl}V) = \frac{1}{(x^2 + y^2 + z^2)^{3/2}} \left(\left(\frac{\partial z}{\partial y} - \frac{\partial y}{\partial z} \right) \hat{i} + \left(\frac{\partial x}{\partial z} - \frac{\partial z}{\partial x} \right) \hat{j} + \left(\frac{\partial y}{\partial x} - \frac{\partial x}{\partial y} \right) \hat{k} \right)$$

$$\text{curl}V = 0$$

Having assimilated the vector calculus concepts of gradient, divergence, and curl, we are now in a position to understand the Maxwell's equations of electromagnetism.

15.5 DERIVATION OF MAXWELL'S EQUATIONS

James Clerk Maxwell, belonging to the Scottish province, was a mathematician with an astute knowledge of physics. Maxwell linked the electric and magnetic fields as complementary identities, and gave the classical ideas of electric and magnetic fields a new meaning by introducing the concept of displacement current in the Amperes law. The contribution of Maxwell came in the form of providing a mathematical form to the existing physical equations of Gauss, Ampere, and Faraday. He then unified these equations by capturing their physical ideas. The four Maxwell's equations unified together formed the new branch of 'electromagnetic field theory', which could discuss light as an electromagnetic wave moving through various media.

Maxwell's equations have now taken a new form as the basis of all modern circuit theories, electrical engineering, and other electromagnetic technologies. The general force equation of Lorentz together with the four Maxwell's equations gives us a complete insight into the creation of charges and currents, and also how the fields can be induced by changing other fields.

The following section helps us understand how the previous existing facts on electric and magnetic fields were brought together by Maxwell, a clever mathematician and then linked together to form the new electromagnetic theory.

15.5.1 Derivation of Maxwell's First Equation

$$\nabla \cdot \boldsymbol{D} = \rho \tag{15.12}$$

Let us assume an arbitrary surface 'S' bounding an arbitrary volume V in a dielectric medium. In a dielectric medium, the total charge density present is the sum of the free charge density (ρ_f) and the polarized charge density (ρ_p). Mathematically,

$$\left(\rho_f + \rho_p\right) = \rho \tag{15.13}$$

The Gauss law of electrostatics states that the total outward electric flux is equal to the total charge enclosed in that volume (Fig. 15.6).

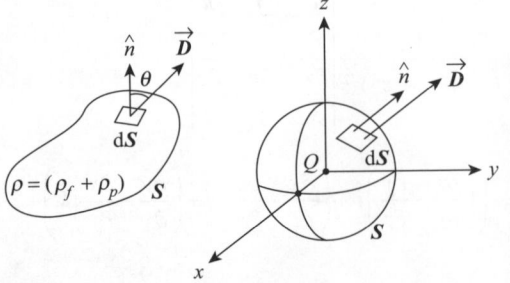

Fig. 15.6 Net outward flux of electric displacement vector (**D**) through surface enclosing volume V equal to net charge density contained in that volume (charge density corresponds to net charge Q)

Physical interpretation The net outward flux of the electric displacement vector through the surface enclosing a volume V is equal to the net charge density contained within that volume.

$$\iint E.dS = \frac{1}{\epsilon_0} \iiint_V (\rho + \rho_p) dV \tag{15.14}$$

where $q = \rho V$.

The polarization charge is represented as the divergence of the polarization vector:

$$\rho_p = -\nabla \cdot P \tag{15.15}$$

Hence, Eq. (15.14) can be written as follows:

$$\int_S E \cdot dS = \frac{1}{\epsilon_0} \iiint_V (\rho - \nabla \cdot P) dV \tag{15.16}$$

$$\iint_S \epsilon_0 E \cdot dS = \iiint_V \rho dV - \iiint_V (\nabla \cdot P) dV$$

The Gauss divergence theorem can be used to convert the surface integral to the volume integral:

$$\iiint \nabla \cdot (\epsilon_0 E) dV = \iiint_V \rho dV - \iiint_V (\nabla \cdot P) dV$$

$$\iiint \nabla \cdot (\epsilon_0 E + P) dV = \iiint_V \rho dV \tag{15.17}$$

The electric displacement vector is defined as follows:

$$D = \epsilon_0 E + P \tag{15.18}$$

Thus, Eq. (15.17) reduces as follows:

$$\iiint \operatorname{div} D \, dV = \iiint \rho dV$$

$$\iiint (\operatorname{div} D - \rho) dV = 0$$

Since the volume of the integral is arbitrary, the integrand becomes zero, that is,

$$\operatorname{div} D = \rho$$

$$\nabla \cdot D = \rho \tag{15.19}$$

Example 15.13   Using Maxwell's first equation, derive the Poisson's equation. In addition, give their physical significance.

Solution   The Maxwell's equation states that $\nabla \cdot D = \rho$

We know that $\epsilon E = D$, also $-\nabla V = E$; thus, one gets the following relation:

$$\nabla \cdot \epsilon (-\nabla V) = \rho$$

If the medium is homogeneous throughout, then ϵ may be extracted out of the differential operations:

$$\nabla \cdot \nabla V = -\frac{\rho}{\epsilon} \Rightarrow \nabla^2 V = -\frac{\rho}{\epsilon}$$

This is the Poisson's equation. The physical significance of the Poisson's equation is that it helps us know the potential function of a surface whose charge distribution (ρ) is known.

When the surface is charge free and has a uniform permittivity, the Poisson's equation reduces to the Laplace equation: $\nabla^2 V = 0$

The Laplace equation helps us in knowing the potential of a function throughout the surface, provided the conditions on the bounding conductors are known.

15.5.2 Derivation of Maxwell's Second Equation

$$\nabla \cdot \boldsymbol{B} = 0 \tag{15.20}$$

This is the Gauss's law for magnetism. The divergence of a magnetic field on the left-hand side in this equation physically indicates the presence of lines of force associated with this magnetic field. The occurrence of an isolated magnetic pole or a magnetic current due to the virtue of the pole has no physical significance. We know that magnetic force lines are closed curves, that is, arising in the North Pole and converging to the South Pole; else, they go off to infinity. Thus, it can be physically inferred that the flux of a magnetic field across any closed surface is zero. The analogous equation indicating the magnetic induction $\boldsymbol{B}$ can be written as follows:

$$\iint_S \boldsymbol{B} \cdot d\boldsymbol{S} = 0 \tag{15.21}$$

The Gauss divergence equation helps us change the surface integral to the volume integral; thus, we get the following equation:

$$\iiint_V \nabla \cdot \boldsymbol{B} \, dV = 0 \tag{15.22}$$

As the surface bounding the volume is arbitrary, the integrand is equated to the right-hand side, that is,

$$\nabla \cdot \boldsymbol{B} = 0 \tag{15.23}$$

Hence, Eq. (15.23) derived from the original concept of magnetic poles is concurrent with the equation given by the Gauss's law for magnetism.

Physical interpretation   The net outward flux of the magnetic vector through a surface enclosing a volume is equal to zero (Fig. 15.7).

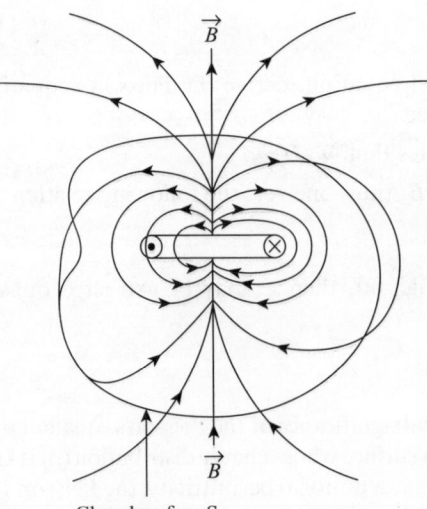

Closed surface S

Fig. 15.7 Net outward flux through surface enclosing a volume is equal to zero

15.5.3 Derivation of Maxwell's Third Equation

$$\nabla \times \boldsymbol{E} = -\frac{\mathrm{d}\boldsymbol{B}}{\mathrm{d}t} \tag{15.24}$$

The electromotive force induced in a closed loop (Fig. 15.8) is the negative rate of change of the magnetic flux, according to the Faraday's law of electromagnetic induction:

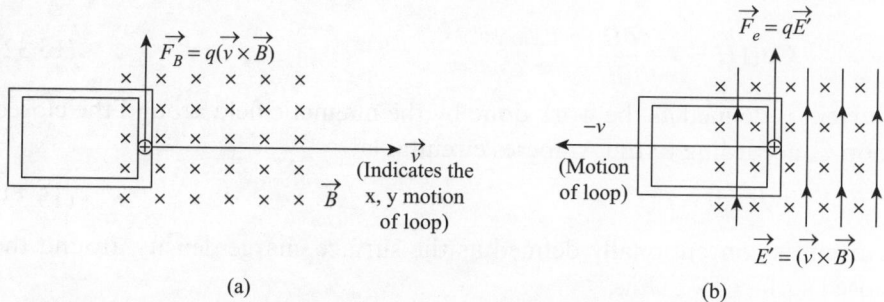

(a)   (b)

Fig. 15.8 Maxwell's third equation (Faraday's law of electromagnetic motion) (a) Conducting loop moving with velocity $\vec{v}$ into a magnetic field, producing a magnetic force $\vec{F}_B = q(\vec{v} \times \vec{B})$, induced e.m.f (**E**) = vLB in rest frame (S) (b) Magnetic field exerting no force on charges, but the electric field does, thus, $\vec{F}_E = q\,\vec{E} = q\,(\vec{v} \times \vec{B}) = (qvB, \text{upward})$ in frame S

Physical interpretation The electromotive force around a closed path is equal to the negative rate of change of the magnetic flux.

$$e = -\frac{\mathrm{d}\varphi}{\mathrm{d}t} \tag{15.25}$$

The magnetic flux is defined as follows:

$$\varphi = \iint_S \boldsymbol{B} \cdot \mathrm{d}\boldsymbol{S} \tag{15.26}$$

Substituting the value of φ from Eq. (15.26) into Eq. (15.25), we get the following expression:

$$e = -\frac{\mathrm{d}}{\mathrm{d}t} \iint_S \boldsymbol{B} \cdot \mathrm{d}\boldsymbol{S} = -\iint_S \frac{\mathrm{d}\boldsymbol{B}}{\mathrm{d}t} \cdot \mathrm{d}\boldsymbol{S} \tag{15.27}$$

The electromotive force is the work done in carrying a unit charge round the closed loop C. If E is defined as the electric field intensity and dl the element of loop, then

$$e = \int_C \boldsymbol{E} \cdot \mathrm{d}l \tag{15.28}$$

Substituting Eq. (15.28) into Eq. (15.27), one gets the following relation:

$$\int_C \boldsymbol{E} \cdot \mathrm{d}l = -\iint_S \frac{\mathrm{d}\boldsymbol{B}}{\mathrm{d}t} \cdot \mathrm{d}\boldsymbol{S} \tag{15.29}$$

On the left-hand side of the equation is contour integration, which can be translated to surface integration using the Stokes theorem. Thus,

$$\iint_S (\nabla \times \boldsymbol{E}) \cdot \mathrm{d}\boldsymbol{S} = -\iint_S \frac{\mathrm{d}\vec{\boldsymbol{B}}}{\mathrm{d}t} \cdot \mathrm{d}\boldsymbol{S} \tag{15.30}$$

$$\iint_S \left(\nabla \times E + \frac{dB}{dt} \right) \cdot dS = 0$$

Since the surface is arbitrary, the integrand is set equal to zero.

$$\nabla \times E = -\frac{dB}{dt} \tag{15.31}$$

15.5.4 Derivation of Maxwell's Fourth Equation

$$\text{Curl}\, H = J + \frac{\partial D}{\partial t} \tag{15.32}$$

Current is defined as the work done by the magnetic field around the closed loop C, according to the Amperes circuital law:

$$\oint H \cdot dl = I \tag{15.33}$$

Current is conventionally defined as the surface charge density around the closed surface '*S*'. Thus,

$$I = \oint J \cdot dS \tag{15.34}$$

Thus, one can equate Eqs (15.33) and (15.34):

$$\oint H \cdot dl = \oint J \cdot dS \tag{15.35}$$

The knowledge of the Stoke's theorem helps us convert the contour integration on the left-hand side equation to a surface integration. Thus,

$$\oint (\nabla \times H) \cdot dS = \oint J \cdot dS \tag{15.36}$$

$$\oint ((\nabla \times H) - J) \cdot dS = 0$$

This implies that the surface charge current is equivalent to the cross-product of the magnetic field:

$$\nabla \times H = J \tag{15.37}$$

The validity of Eq. (15.37) needs to be verified for time-varying fields only, as it seems to hold for the steady state. Mathematically, the divergence of the curl of a quantity is always zero; thus,

$$\text{div}(\nabla \times H) = \text{div}(J) = 0$$

$$\text{div}(J) = 0 \tag{15.38}$$

The equation of continuity states that for the conservation of charge to be true, the divergence of the surface charge current should be equal to the time rate of change of the charge density, that is,

$$\text{div}(J) + \frac{\partial \rho}{\partial t} = 0 \tag{15.39}$$

$$\text{div}(J) = -\frac{\partial \rho}{\partial t} \tag{15.40}$$

Thus, Eq. (15.40) indicates that the rate of change of the surface charge density is equal to the divergence of the current density. This can be valid only if the rate of change of charge density with time goes to zero, that is, the charge density needs to be static in nature or, in other words, be independent of time. Thus, the conclusion that can be drawn from the aforementioned mathematical analysis is that *the Ampere's circuital law holds only for the steady-state condition* and is not adequately defined for cases where the field (or the charge) varies with time. The aspect of the variation of charge (or field) with time needs to be included in the Ampere's law. Thus, this anomaly in the Ampere's law led to the contribution of Maxwell, thus modifying the Ampere's Circuital law. The modification by Maxwell was such that the time-varying fields could now be included.

Table 15.1 gives the details of Maxwell's equations in their differential and integral forms.

Table 15.1 Maxwell's Equations for Charge Density(s)

S. No	Name of the Law	Equation		Description of Concept
		Point Form	**Integral Form**	
1.	Gauss's Law of Electricity	$\vec{\nabla}\cdot\vec{D}=\rho$	$\oint_S \vec{D}\cdot d\vec{S}$ $=\int_V \rho\,dV$	Enclosed charge density and Electric field.
2.	Gauss's Law of Magnetism	$\vec{\nabla}\cdot\vec{B}=0$	$\oint \vec{B}\cdot d\vec{S}=0$	Magnetic field. Non-existence of monopole.
3.	Faraday's Law	$\vec{\nabla}\times\vec{E}=\dfrac{-\partial\vec{B}}{\partial t}$	$\oint \vec{E}\cdot d\vec{l}$ $=\int_S \left(\dfrac{-\partial\vec{B}}{\partial t}\right)d\vec{S}$	Electric effect of a changing magnetic field. Also changing magnetic field creates an electric field.
4.	Ampere's Law or Modified Ampere's Law by Maxwell	$\vec{\nabla}\times\vec{H}$ $=\vec{J}_c+\dfrac{\partial\vec{D}}{\partial t}$	$\oint \vec{H}\cdot d\vec{l}$ $=\int_S \left(\vec{J}_c+\dfrac{\partial\vec{D}}{\partial t}\right)\cdot d\vec{S}$	The magnetic effect of a changing electric field or of a current. New concept of electric flux akin to magnetic flux.

15.6 MODIFICATION OF AMPERE'S CIRCUITAL LAW

In Section 15.5, on derivation of Maxwell's fourth equation, it has been shown aptly how there exists a missing link in the understanding of the Ampere's circuital law. The law holds only for static fields; hence, for its general application Maxwell thought of a modification to the current density. Let the total current density be defined as the sum of the current density (J) and another quantity (J_d) such that the fourth equation may be rewritten as follows:

$$(\nabla \times H) = (J + J_d) \qquad (15.41)$$

Taking divergence on both the sides,

$$\nabla \cdot (\nabla \times H) = \nabla \cdot (J + J_d) \qquad (15.42)$$

Using the vector identity where the divergence of the curl of a quantity is zero,

$$\text{div}(J + J_d) = 0 \qquad (15.43)$$

$$\text{div}\, J = -\text{div}\, J_d$$

Now, from the equation of continuity:

$$\text{div}\, J_d = \frac{\partial \rho}{\partial t} \qquad (15.44)$$

In addition, from the Maxwell's first equation:

$$\nabla \cdot D = \rho$$

$$\text{div}\, J_d = \frac{\partial(\nabla \cdot D)}{\partial t}$$

$$\text{div}\, J_d = \text{div}\left(\frac{\partial D}{\partial t}\right)$$

$$J_d = \frac{\partial D}{\partial t} \qquad (15.45)$$

This implies that the fourth equation is now modified to the following form:

$$\nabla \times H = J + J_d$$

$$\nabla \times H = J + \frac{\partial D}{\partial t} \qquad (15.46)$$

Maxwell thus incorporated the time-varying fields, which corroborated with the thought of conservation of charge.

15.7 DISPLACEMENT CURRENT

The electric displacement vector (D) (defined as the sum of the electric field vector and the polarization vector) when differentiated with respect to time leads to the displacement current. Maxwell established that the displacement current is as important as the conventional conduction current. Maxwell applied the Ampere's law to a charging capacitor; this led to the discovery of the displacement current.

We will see how the simple process of charging a capacitor explains the significance and importance of the existence of the displacement current, which had been missed due to the conventional classical approach in understanding this concept earlier.

Let us observe a parallel-plate capacitor being charged by a current I from the left. This current moves as per the convention, carrying the positive

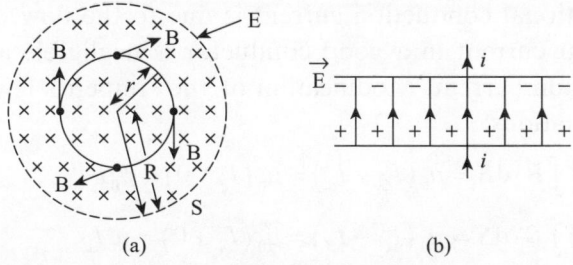

(a)   (b)

Fig. 15.9 Diagrammatic repersentation of magnetic and electric fields (a) Generation of magnetic field *B* at different points due to changing electric field (b) Charging of parallel-plate capacitor (circuit example of changing electric field)

charges to the left plate and negative charges to the right plate, the magnetic field being ascertained by the right-hand rule. The surface S_C is bounded by the closed curve C around the current-carrying wire from the left (Fig. 15.9). According to Ampere's law,

$$\iint_S \boldsymbol{B} \cdot \boldsymbol{dS} = \mu_0 I \tag{15.47}$$

The magnetic field is evaluated correctly for a current-carrying wire. If now a surface S_d bounding a closed curve C is made around the current-carrying wire from the right-hand side and the Ampere's law is applied, then we get the following expression:

$$\iint_S \boldsymbol{B} \cdot \boldsymbol{dS} = \mu_0 I = 0 \tag{15.48}$$

This equation is incorrect, as physically the current exhibits itself by the means of charge accumulation on the other plate of the capacitor. Hence, there was a need to modify the Ampere's law. Maxwell introduced the concept of *electric flux*. The electric flux between the two capacitor plates of surface area A is given as follows:

$$\varphi_e = EA \tag{15.49}$$

where E is the electric field $[E = (Q/\epsilon_0 A)]$ and A the area of the capacitor plates.

$$\varphi_e = (Q/\epsilon_0) \tag{15.50}$$

The rate at which the electric flux is changing is directly proportional to the charging current I. Thus,

$$\frac{d\varphi e}{dt} = \frac{1}{\epsilon_0} \frac{dQ}{dt} = \frac{I}{\epsilon_0}$$

$$I_d = \epsilon \frac{d\varphi_e}{dt} \tag{15.51}$$

Here, I_d is defined as the displacement current. The displacement current is due to the changing electric flux with time generating a magnetic field.

The conventional conduction current is due to the flow of charges. The displacement current in a good conductor is negligible as compared to the conduction current. Modification of the Ampere's law now gives the following relations:

$$S_c: \iint B \cdot dS = \mu_0 \left(I_C + I_d \right) = \mu_0 \left(I_c + 0 \right) = \mu_0 I_c \qquad (15.52a)$$

$$S_d: \iint B \cdot dS = \mu_0 \left(I_C + I_d \right) = \mu_0 \left(I_d + 0 \right) = \mu_0 I_d \qquad (15.52b)$$

Thus, the introduction of the displacement current serves the purpose of making the total current continuous across boundaries. The modification of the Ampere's law set the basis for the new electromagnetic theory.

Thus, the charging of a capacitor is a phenomenon worth observing (Fig. 15.10). In this context, it is worthwhile to ponder on the classification of materials as good conductors, perfect dielectric materials, or neither good conductors nor perfect dielectric materials/media. As we discuss the concepts of mobility, permittivity, and frequency in the sections that follow, these will become clearer. The total current density for such materials is defined as follows: $J_t = J_C + J_D$

where $J_C = \sigma E$; $J_D = \dfrac{\partial D}{\partial t} = \dfrac{\partial(\epsilon E)}{\partial t} = \dfrac{\partial(\epsilon E e^{(i\omega t)})}{\partial t} = i\omega \epsilon E$

Thus, $\dfrac{J_C}{J_D} = \left| \dfrac{\sigma}{\omega \epsilon} \right|$

The ratio of the conductivity (σ) to the frequency (ω) and permittivity (ϵ) plays a critical role in classifying the material as a good conductor, a perfect dielectric, or an intermediate conductor. The frequency of the radiation plays a crucial role in the existence of the displacement current, that is, as the frequency of the electromagnetic wave increases, the concept of displacement current becomes increasingly important.

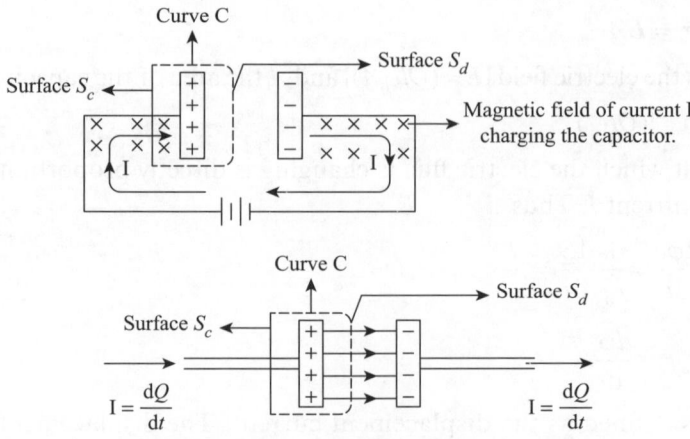

Fig. 15.10 Charging of parallel-plate capacitor

It would be appropriate here to draw a simile between the parameters of Planck's constant (h) in quantum mechanics and that of the displacement current in the electromagnetic theory. As the order of h is insignificant in calculations when the Newtonian laws of mechanics hold good, the displacement current is also insignificant when the characteristics of the electric and magnetic fields are well defined. When the parameter h becomes significant, the quantum theory starts taking lead over the classical theory. Similarly, as the concept of displacement current becomes significant, particularly in the time-dependent electric and magnetic fields, the Maxwell's theory of electromagnetics starts taking lead over the classical laws of electrostatics and magnetostatics.

The following examples illustrate in detail how the concept of displacement current can be mathematically utilized in circuit analysis.

Example 15.14 For the circuit shown in Fig. 15.11,
(a) use the circuit analysis to show that $\hat{i}_C = \hat{i}_D$ in the circuit, and
(b) use the Stokes theorem to show that $\hat{i}_C = \hat{i}_D$ in the circuit.

Here, $\hat{i}_C$ represents the conventional current and $\hat{i}_D$ the displacement current in the circuit.

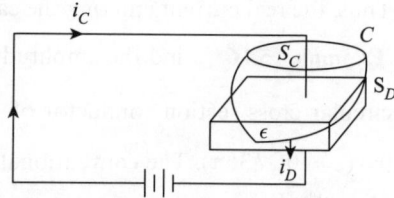

Fig. 15.11 Charging of a parallel plate capacitor of separation 'd' and plate area A.

Solution The following is the explanation:
(a) The conventional definition of capacitance of a parallel-plate capacitor of separation d and plate area A is given as follows: $C = \dfrac{\epsilon A}{d}$

The conventional current $\hat{i}_C$ is defined as follows: $\hat{i}_C = C\dfrac{dV}{dt} = \dfrac{\epsilon A}{d}\dfrac{dV}{dt}$

The displacement current $\hat{i}_D$ is defined as follows: $\hat{i}_D = \int_S \boldsymbol{J}_D \cdot d\boldsymbol{S} = \int_S \dfrac{\partial \boldsymbol{D}}{\partial t} d\boldsymbol{S}$

From the conventional definition of the displacement vector D,

$$\boldsymbol{D} = \epsilon \boldsymbol{E} = \epsilon \frac{V}{d} \Rightarrow \frac{\partial \boldsymbol{D}}{\partial t} = \frac{\epsilon}{d}\frac{dV}{dt}$$

$$\hat{i}_D = \int_S \frac{\epsilon}{d}\frac{dV}{dt} d\boldsymbol{S} = \frac{\epsilon A}{d}\frac{dV}{dt} = \hat{i}_C$$

(b) The contour C is common to both the surfaces S_c and S_d. The Stokes theorem can be written as follows:

$$\oint_C \boldsymbol{H} \cdot d\boldsymbol{l} = \iint_{S_c} (\nabla \times \boldsymbol{H}) \cdot d\boldsymbol{S} = \iint_{S_d} (\nabla \times \boldsymbol{H}) \cdot d\boldsymbol{S}$$

$$\iint_{S_c} \left(\boldsymbol{J}_C + \frac{\partial \boldsymbol{D}}{\partial t} \right) \cdot d\boldsymbol{S} = \iint_{S_d} \left(\boldsymbol{J}_C + \frac{\partial \boldsymbol{D}}{\partial t} \right) \cdot d\boldsymbol{S}$$

For the surface S_c, $\boldsymbol{D} = 0 \Rightarrow \dfrac{\partial \boldsymbol{D}}{\partial t} = 0$; for the surface S_d, $\boldsymbol{J}_C = 0$. Thus,

$$\iint_{S_c} \boldsymbol{J}_C \cdot d\boldsymbol{S} = \iint_{S_d} \frac{\partial \boldsymbol{D}}{\partial t} \cdot d\boldsymbol{S} \Rightarrow \hat{i}_C = \hat{i}_D$$

Example 15.15　The charge of a parallel-plate capacitor is $q = q_0 \sin 2\pi \vartheta t$. The plates are very large and close together, with an area of cross section A and separation d. Find the displacement current through the capacitor and show that it equals the current entering the capacitor.

Solution　The displacement current is defined as follows: $I_d = \epsilon \dfrac{d\varphi_e}{dt}$

Let us assume that all the electric field lines passing through that area are circular in nature and lie between the two plates of the capacitor. The electric field is uniform and perpendicular between the plates. Thus, electric field is given as follows: $E = \dfrac{q}{(A\epsilon)}$

Thus, the electric flux through cross section area A is written as follows: $EA = \dfrac{q}{\epsilon}$

The displacement current can then be mathematically expressed as follows:

$$\epsilon \frac{\partial}{\partial t}\left(\frac{q}{\epsilon}\right) = \frac{dq}{dt} = I_d \Rightarrow I_d = \left[2\pi \vartheta q_0 \sin(2\pi \vartheta t)\right]$$

Thus, the real current entering the capacitor is the displacement current.

Example 15.16　Find the amplitude of the displacement current density (J_D) for a circular cross section conductor of radius 2.0 mm, $\sigma = \dfrac{61.7 MS}{m} = \dfrac{6.17 \times 10^7 S}{m}; \epsilon_r = 1$ $(\Rightarrow (\epsilon = 10^{-9}/36\pi)$. The conventional current flowing through the wire is $I_C = 4.0 \sin\left(2 \times 10^{10} t\right)(\mu A)$.

Solution　We know that $\dfrac{J_C}{J_D} = \dfrac{\sigma}{\omega \epsilon} = \dfrac{6.17 \times 10^7}{(2 \times 10^{10})\left(\dfrac{10^{-9}}{36\pi}\right)}$

Therefore, $J_D = J_c \cdot \dfrac{(2 \times 10^{10})\left(\dfrac{10^{-9}}{36\pi}\right)}{6.17 \times 10^7} = \dfrac{I_{oC}}{(\pi r^2)} \dfrac{(2 \times 10^{10})\left(\dfrac{10^{-9}}{36\pi}\right)}{6.17 \times 10^7}$

$$= \frac{4.0 \times 10^{-6}}{\pi(2.0 \times 10^{-3})^2} \frac{(2 \times 10^{10})\left(\dfrac{10^{-9}}{36\pi}\right)}{6.17 \times 10^7} = 9.1 \times 10^{-4} \mu A / m^2$$

Note: The value of permittivity is taken to be ϵ for any general space and ϵ_0 for free space.

15.8 MAXWELL'S EQUATIONS IN INTEGRAL AND DIFFERENTIAL/ POINT FORMS

Maxwell aimed at linking the knowledge of electric and magnetic fields, introducing the electromagnetic field theory. The speed of light can be inferred from the Maxwell's equations. This led to the understanding of wave guides and electromagnetic technology, the present-day engineering marvel. The Maxwell's theory of electromagnetism is represented by four equations that incorporate the entire domain of electrostatics and magnetostatics. The modification made by Maxwell takes us a step further wherein the concept of electromagnetic energy has also been incorporated for various media.

Maxwell's equations can be represented in the differential/point form as follows:

$$\nabla \cdot \boldsymbol{D} = \rho \quad \text{Gauss law in electrostatics} \tag{15.53a}$$

$$\nabla \cdot \boldsymbol{B} = 0 \quad \text{Gauss law in magnetostatics} \tag{15.53b}$$

$$\nabla \times \boldsymbol{E} = -\frac{\partial \boldsymbol{B}}{\partial t} \quad \text{Faraday's law of electromagnetic induction} \tag{15.53c}$$

$$\nabla \times \boldsymbol{H} = \boldsymbol{J} + \frac{\partial \boldsymbol{D}}{\partial t} \tag{15.53d}$$

Maxwell's modification of the Ampere's law

For a linear, isotropic, and homogeneous medium, the constitutive relations are given as follows:

$$\boldsymbol{D} = \epsilon \boldsymbol{E} \tag{15.54a}$$

$$\boldsymbol{B} = \mu \boldsymbol{H} \tag{15.54b}$$

$$\boldsymbol{J} = \sigma \boldsymbol{E} \tag{15.54c}$$

where $\boldsymbol{E}$, $\boldsymbol{D}$, $\boldsymbol{B}$, and $\boldsymbol{H}$ represent the electric field, electric displacement, magnetic induction, and magnetic field, respectively; ϵ, μ, and σ denote the dielectric permittivity, magnetic permeability, and conductivity of the medium, respectively. Maxwell's equations can be solved using the constitutive relations for respective media.

The integral form of the Maxwell's equations can be represented as follows:

$$\iint_S \vec{\boldsymbol{D}} \cdot \mathrm{d}\vec{\boldsymbol{S}} = \iiint_V \rho \mathrm{d}V \tag{15.55a}$$

$$\iint_S \vec{\boldsymbol{B}} \cdot \mathrm{d}\vec{\boldsymbol{S}} = 0 \tag{15.55b}$$

$$\oint \vec{\boldsymbol{E}} \cdot \mathrm{d}l = \iint_S -\left(\frac{\partial \boldsymbol{B}}{\partial t}\right) \cdot \mathrm{d}\vec{\boldsymbol{S}} \tag{15.55c}$$

$$\oint \vec{\boldsymbol{H}} \cdot \mathrm{d}l = \iint_S \left(\boldsymbol{J}_c + \frac{\partial \vec{\boldsymbol{D}}}{\partial t}\right) \cdot \mathrm{d}\vec{\boldsymbol{S}} \tag{15.55d}$$

The following examples illustrate how the Maxwell's equations help us derive the electric and magnetic fields.

Example 15.17 The E field of a harmonic plane electromagnetic wave has the following form: $E_z(y,t) = E_{0z}\sin\left[\omega\left(t - \frac{y}{c}\right) + \theta\right]$

Determine the corresponding magnetic field.

Solution Maxwell's equations yield the following relation:

$$\frac{\partial E_z}{\partial y} = -\frac{\partial B_x}{\partial t} = \frac{\omega}{c} E_{0z}\cos\left[\omega\left(t - \frac{y}{c}\right) + \theta\right]$$

Integration with respect to t gives the following expression:

$$B_x(y,t) = \frac{1}{c} E_{0z}\left[\omega\left(t - \frac{y}{c}\right) + \theta\right] = \frac{1}{c} E_z(y,t)$$

The constant of integration, if introduced, will indicate a steady magnetic field. The electric and magnetic fields are orthogonal; their magnitudes are related by the velocity of light. Both the fields are normal to the propagation direction.

Example 15.18   Imagine an electromagnetic plane wave in vacuum whose E field (in SI units) is as follows: $E_z = 10^2 \sin\pi(3 \times 10^6 z - 9 \times 10^{14} t), E_y = 0, E_z = 0$
(a) Determine the speed, frequency, wavelength, period, initial phase, and E-field amplitude and polarization.
(b) Determine the magnetic field associated with the corresponding electric field wave.

Solution   The calculations are as follows:
(a) The electric field vector resembles the following equation: $E_z(z,t) = E_{0z} \sin \hat{k}(z - vt)$
Thus, on comparing, one gets the following relation:

$$\hat{k} = 3 \times 10^6 \pi / m; \ v = 3 \times 10^8 \ m/s; \ \hat{k} = \frac{2\pi}{\lambda}, \ \lambda = 666 \ nm$$

$$\vartheta = \frac{v}{\lambda} = \frac{3 \times 10^8}{\frac{2}{3} \times 10^{-6}} = 4.5 \times 10^{14} \ Hz; \ T = \frac{1}{\vartheta} = 2.2 \times 10^{-15} \ s$$

The initial phase of the wave is zero. The wave is linearly polarized in the x-direction and is propagating along the z-direction. The wavelength matches that of the visible red light.
(b) The magnetic field is perpendicular to both the direction of the electric vector and the direction of propagation. The plane of the electric vector is the xz plane. The magnetic field vector is thus confined to the yz plane, that is,

$$B_x = B_z = 0$$

$$B = B_y(z,t)\hat{j} = (E_x / c)$$

$$B_y(z,t) = \frac{10^2}{3 \times 10^8} \sin\pi(3 \times 10^6 z - 9 \times 10^{14} t) T$$

$$B_y(z,t) = 0.33 \times 10^{-6} \sin\pi(3 \times 10^6 z - 9 \times 10^{14} t) T$$

Example 15.19   A plane electromagnetic harmonic wave of frequency 500×10^{12} Hz, propagating in the positive x-direction in vacuum, has an electric field amplitude of 42.42 V/m. The wave is linearly polarized such that the plane of vibration of the electric field is at 45° to the xz plane. Express the electric and magnetic fields.

Solution   The amplitude of the electric field is given as follows: $E_0 = (E_{0y}^2 + E_{0z}^2)^{\frac{1}{2}}$

Since the plane of vibration of the electric field is at an angle of 45°,

$$E_{0y} = E_{0z}$$

$$E_0 = 42.42 = (1^2 E_{0y}^2 + 1^2 E_{0z}^2)^{\frac{1}{2}} = \sqrt{2} E_{0y}$$

$$E_{0y} = E_{0z} = 30 \ V/m$$

The electric and magnetic field can be expressed as follows:

$$E_y(z,t) = E_z(z,t) = E_{0z} \sin\left[2\pi\vartheta\left(t - \frac{x}{c}\right)\right]$$

$$E_x = 0, E_y = E_z = 30\sin\left[2\pi600 \times 10^{12}\left(t - \frac{x}{3 \times 10^8}\right)\right]$$

$$E = E_y\hat{j} + E_z\hat{k} = E_y(\hat{j} + \hat{k})$$

$$B_x = 0, \quad B_z = -B_y = 10^{-7} \sin\left[2\pi600 \times 10^{12}\left(t - \frac{x}{3 \times 10^8}\right)\right]$$

$$B = B_y\hat{j} + B_z\hat{k} = B_z(-\hat{j} + \hat{k})$$

Note that

$$E \cdot B = E_y B_z(-1+1) = 0, \, E \text{ is perpendicular to } B$$

$$E \times B = E_y B_z(\hat{i} + \hat{i}) = \left(\frac{E_y^2}{c}\right)2i, \text{ along the positive } x\text{-axis}$$

15.9 ELECTROMAGNETIC ENERGY DENSITY

Electric and magnetic fields are associated with a certain flow of energy. Classical physics had ascertained the potential energy associated with the electric and magnetic fields as the electromagnetic potential energy. Now under the Maxwell's theory, both the electric energy and the magnetic energy together contribute towards the flow of energy in a medium. Hence, we now define the new concept of electromagnetic energy density and check whether it is able to reduce to its classical counterparts.

Electromagnetic potential energy is defined as follows:

$$U_C = \frac{1}{2}\iiint E \cdot B \, dV \tag{15.56}$$

The magnetic field is associated with an energy represented as follows:

$$U_m = \frac{1}{2}\iiint H \cdot B \, dV \tag{15.57}$$

The magnetic and electric fields individually represent static or time-independent situations. Electromagnetic energy can now be studied for non-static situations. We consider the differential form of Maxwell's equations.

Let us consider the following equations:

$$\text{Curl } E = -\frac{\partial B}{\partial t} \tag{15.58}$$

$$\text{Curl } H = J + \frac{\partial D}{\partial t} \tag{15.59}$$

Taking the scalar products of Eq. (15.58) and *H* and of Eq. (15.59) and *E*, we get the following expressions:

$$H \cdot \text{curl } E = -H\frac{\partial B}{\partial t} \tag{15.60}$$

$$E \cdot \text{curl } H = E \cdot J + E \cdot \frac{\partial D}{\partial t} \tag{15.61}$$

Subtracting Eq. (15.61) from Eq. (15.60), one gets the following expression:

$$H \cdot \text{curl } E - E \cdot \text{curl } H = -H \cdot \frac{\partial B}{\partial t} - E \cdot \frac{\partial D}{\partial t} - E \cdot J$$

$$= -\left(H \cdot \frac{\partial B}{\partial t} + E \cdot \frac{\partial D}{\partial t}\right) - E \cdot J \tag{15.62}$$

Using vector identity,

$$\operatorname{div}(E \times H) = -\left(H \cdot \frac{\partial B}{\partial t} + E \cdot \frac{\partial D}{\partial t} \right) - E \cdot J \tag{15.63}$$

For a medium that is linear, the following relations are used:

$B = \mu H$ and $D = \epsilon E$

One can write these relations as follows:

$$H \cdot \frac{\partial B}{\partial t} = H \cdot \frac{\partial}{\partial t}(\mu H) = \frac{1}{2}\mu\frac{\partial}{\partial t}(H)^2 = \frac{\partial}{\partial t}\left(\frac{1}{2}H \cdot B\right) \tag{15.64}$$

$$E \cdot \frac{\partial D}{\partial t} = E \cdot \frac{\partial}{\partial t}(\epsilon E) = \frac{1}{2}\epsilon\frac{\partial}{\partial t}(E)^2 = \frac{\partial}{\partial t}\left(\frac{1}{2}E \cdot D\right) \tag{15.65}$$

Having used this in vector identity (15.63), one gets the following expression:

$$\operatorname{div}(E \times H) = -\frac{\partial}{\partial t}\left[\frac{1}{2}(E \cdot D + H \cdot B)\right] - J \cdot E \tag{15.66}$$

This can further be written as follows:

$$\operatorname{div} S + \frac{\partial U}{\partial t} = -J \cdot E \tag{15.67}$$

Equation (15.67) resembles the equation of continuity. The new term S is defined as the Poynting vector (discussed in detail in Section 15.11) and

$$U = \frac{1}{2}(E \cdot D + H \cdot B) \tag{15.68}$$

represents the electrostatic and the magnetic energy, respectively.

Hence, there is continuity and conservation in the flow of the electromagnetic energy, as can be observed from the preceding equation. The term on the right-hand side, $J \cdot E$, represents the kinetic (or heat) energy of charged particles, similar to the Joule loss. This can be inferred from the following, already familiar, equations of work done for a charged particle (q) under the influence of an electromagnetic field. We know that the Lorentz equation defines the force acting on a particle and ds is the distance traversed by the particle; thus, the equation of motion can be written as follows:

$$F \cdot \frac{\mathrm{d}s}{\mathrm{d}t} = F \cdot v \tag{15.69}$$

$$= \left[qE + q(v \times B) \right] \cdot v$$

$$= q E \cdot v \tag{15.70}$$

If there are N charged particles per unit volume, each carrying a charge q, then the work done per unit volume would be given as follows:

$$N qv \cdot E = J \cdot E \tag{15.71}$$

where J is the current density.

The volume integration of Eq. (15.71) gives the following relation:

$$\int \operatorname{div} S \, \mathrm{d}V + \frac{\mathrm{d}}{\mathrm{d}t}\int U \, \mathrm{d}V = -\int J \cdot E \, \mathrm{d}V \tag{15.72}$$

Using Gauss divergence theorem, the volume integral can be changed to a surface integral:

$$\oint_s S \cdot dS + \int J \cdot E \, dV = -\frac{d}{dt} \int U \, dV \tag{15.73}$$

The left-hand side of the equation represents the sum of the net flow out of energy plus the Joule loss for a given volume enclosing that surface. The right-hand side of the equation represents the rate of decrease of the total energy. Hence, this equation represents the equation of continuity or the conservation of electromagnetic energy density.

15.10 INTENSITY OF ELECTROMAGNETIC WAVES

Conventionally, the intensity of an electromagnetic wave is the power transmitted per second by the wave that impinges on an area A. Intensity can also be defined as the average energy transfer. Mathematically, the intensity of an electromagnetic wave can be obtained from the product of the average energy of the radiation and the wave (radiation) velocity. Thus, the intensity can mathematically be given as follows:

$$I = \frac{1}{2} \in v E_0^2 \tag{15.74}$$

The velocity of propagation of a wave is given as follows:

$$v = \frac{1}{\sqrt{\in \mu}} \tag{15.75}$$

$$I = \frac{1}{2} \sqrt{\frac{\epsilon}{\mu}} E_0^2 \tag{15.76}$$

In free space,

$$I = \frac{1}{2} \epsilon_0 c E_0^2 \tag{15.77}$$

Substituting the value of the permittivity and velocity of sound, one gets the value of intensity as follows:

$$I = \frac{1}{2} \left(8.854 \times \frac{10^{-12} \, C^2}{N} \, m^2 \right) \times \left(3 \times 10^8 \, m/s \right) E_0^2$$

$$= \left(1.33 \times 10^{-3} \, W/V^2 \right) E_0^2 \tag{15.78}$$

Thus, the intensity of a radiation can be ascertained for a given distance if its power is known. Using this knowledge, one can compare the electric fields produced due to an ordinary light source with that due to a laser beam.

The following examples illustrate how the electromagnetic energy and intensity can be evaluated mathematically.

Example 15.20 Find the electric field produced by a 50 W ordinary bulb at a distance of 5 m from any given surface.

Solution The intensity is defined as the power per unit area and is given as follows:

$$I = \frac{P}{4\pi r^2} = \frac{50}{4\pi(5)^2} = 0.159 \, \text{W/m}^2$$

In addition, we know that $\epsilon_0 = \dfrac{8.85 \times 10^{-12} \, \text{C}^2}{\text{N m}^2}; c = 3 \times 10^8 \, \text{m/s}$

$$E_0 = \sqrt{\frac{2I}{\epsilon_0 c}} \Rightarrow E_0^2 = \frac{I}{1.33 \times 10^{-3}}$$

$$E_0^2 = \left[\frac{0.159}{1.33 \times 10^{-3}}\right] = 119.5 \, \text{N/C} \Rightarrow E_0 = \sqrt{119.5} = 10.93 \, \text{N/C}$$

Example 15.21 Find the electric field produced by a 10^2 W laser beam focused by a lens of cross section area $10^{-6} \, \text{cm}^2$.

Solution The intensity for a laser beam can be found as follows:

$$I = \frac{P}{4\pi r^2} = \frac{10^2}{10^{-10}} = 10^{12} \, \text{W/m}^2 \Rightarrow E_0^2 = \left[\frac{10^{12}}{1.33 \times 10^{-3}}\right] = 7.5 \times 10^{14}$$

$$E_0 = \sqrt{7.5 \times 10^{14}} = 2.74 \times 10^7$$

Thus, one can observe from the previous examples that the electric field intensity for a low-watt laser beam is also high enough, which, when converted to heat energy, can produce high temperatures.

15.11 POYNTING VECTOR

Poynting vector represents the directional energy flux density [the rate of energy transfer per unit area, in units of watts per square metre (W/m^2)] of an electromagnetic field. Mathematically, the Poynting vector (S) is represented as follows:

$$S = E \times H \tag{15.79}$$

where E is the electric field vector and H the magnetic field vector.

The physical significance of the Poynting vector is that it gives us the direction of flow of an electromagnetic radiation. The rate at which any source is emitting electromagnetic energy can be ascertained by integrating the normal (perpendicular) component of the Poynting vector (Fig. 15.12) over the closed surface drawn enclosing it (source). For the source where the electric and magnetic fields are varying with time, the Poynting vector gives

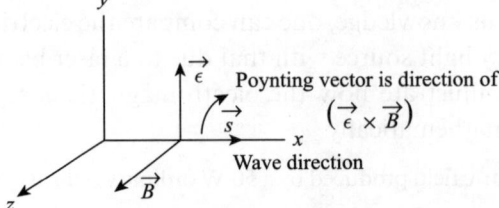

Poynting vector is direction of
$$(\vec{\epsilon} \times \vec{B})$$
Wave direction

Fig. 15.12 Poynting vector

the instantaneous value of the electromagnetic energy. Further, the quantity $S.da$ may be interpreted as the electromagnetic energy crossing the area da per unit time.

Note: The radiant flux density is the average value of the Poynting vector over a given time interval. When energy is incident on a surface, the flux density is called irradiance ($I \equiv < S >$).

The following examples illustrate how the average energy density and intensity can be evaluated.

Example 15.22 Show that the average energy density in an electromagnetic plane wave is $< \rho_e > = \dfrac{< S >}{c}$, where S is the Poynting vector. What is the momentum density in the wave?

Solution The average energy density in a harmonic electric field and a magnetic field is given as follows:

$$< E_{av} > = \frac{1}{4}\epsilon_0 E_0^2; \quad < B_{av} > = \frac{1}{4}\frac{B_0^2}{\mu_0}$$

We know that $B_0 = \dfrac{E_0}{c}$; thus,

$$< \rho_e > = \frac{1}{4}\epsilon_0 E_0^2 + \frac{1}{4}\frac{E_0^2}{c^2 \mu_0} = \frac{1}{2}\epsilon_0 E_0^2; \quad < S > = \frac{c\epsilon_0 E_0^2}{2} = \frac{< \rho_e >}{c}$$

The radiation carries momentum and hence exerts pressure. Therefore,

$$\frac{< S >}{c^2} = \frac{< \rho_e >}{c} = < \rho_m >$$

The analogy can be seen from the quantum theory also:

$$E = h\vartheta = \frac{hc}{\lambda} = pc \implies < \rho_e > = c < \rho_m >$$

Example 15.23 A plane electromagnetic wave moving through free space has an E field given by the following expression:

$$E_x = 0; \quad E_y = 0; \quad E_z = 100\sin\left[8\pi \times 10^{14}\left(t - \frac{x}{3 \times 10^8}\right)\right]$$

Calculate the corresponding flux density.

Solution The flux density is given as follows:

$$I = \frac{c\epsilon_0}{2}E_0^2; \epsilon_0 = 8.8542 \times 10^{-12}\, C^2 / N / m^2; E_0 = 100\, V / m$$

$$I = \frac{(3 \times 10^8)(8.85 \times 10^{-12})100^2}{2} = 13.3\, W / m^2$$

Example 15.24 Considering the problem in Example 15.23, calculate the magnetic field B, energy density, and rate of flow of energy per unit area.

Solution The electric field has been evaluated to be equal to $\dfrac{100\, V}{m}$ or 100 N/C; hence, the magnetic field is defined as follows: $B = \dfrac{E}{c} = \dfrac{100\, V / m}{3 \times 10^8\, m / s} = 3.33 \times 10^{-7}\, T$

The energy density is defined as follows:

$$u_{em} = \epsilon_0 E^2 = \frac{8.85 \times 10^{-12}\,\mathrm{C^2}}{\mathrm{Nm^2 x}} \left(\frac{100\,\mathrm{N}}{\mathrm{C}}\right)^2 = 8.85 \times 10^{-8}\,\mathrm{N/m^2}$$

The rate of flow of energy per unit area = Magnitude of the Poynting vector

$$S = \frac{EB}{\mu_0} = \frac{\dfrac{100\,\mathrm{V}}{\mathrm{m}} \times 3.33 \times 10^{-7}\,\mathrm{T}}{4\pi \times 10^{-7}\,\mathrm{T.m/A}} = \frac{2.65\,\mathrm{W}}{\mathrm{m^2}}$$

15.12 ELECTROMAGNETIC WAVE EQUATION FOR ELECTRIC VECTOR (E) AND MAGNETIC VECTOR (B)

This section now highlights the Maxwell's equations as a tool to understand the behaviour of the electric and magnetic fields, and hence the electromagnetic wave. The first and foremost check of the Maxwell's equations is to establish the wave equation to understand the electromagnetic phenomenon. Maxwell's equations (Eqs 15.53a–d) written in the compact form can explicitly be expressed as follows for better understanding:

$$\frac{\partial D_x}{\partial x} + \frac{\partial D_y}{\partial y} + \frac{\partial D_z}{\partial z} = \rho \tag{15.80a}$$

$$\frac{\partial B_x}{\partial x} + \frac{\partial B_y}{\partial y} + \frac{\partial B_z}{\partial z} = 0 \tag{15.80b}$$

$$\frac{\partial E_z}{\partial y} - \frac{\partial E_y}{\partial z} = -\frac{\partial B_x}{\partial t}$$

$$\frac{\partial E_x}{\partial z} - \frac{\partial E_z}{\partial x} = -\frac{\partial B_y}{\partial t}$$

$$\frac{\partial E_y}{\partial x} - \frac{\partial E_x}{\partial y} = -\frac{\partial B_z}{\partial t} \tag{15.80c}$$

$$\frac{\partial H_z}{\partial y} - \frac{\partial H_y}{\partial z} = J_x + \frac{\partial D_x}{\partial t}$$

$$\frac{\partial H_x}{\partial z} - \frac{\partial H_z}{\partial x} = J_y + \frac{\partial D_y}{\partial t}$$

$$\frac{\partial H_y}{\partial x} - \frac{\partial H_x}{\partial y} = J_z + \frac{\partial D_z}{\partial t} \tag{15.80d}$$

where the symbols are as per conventions.

Let us derive the equations for electromagnetic waves using the Maxwell's equations for a non-charged ($\rho = 0$) medium. Thus, the Maxwell's equations for this specific case in a compact form can be given as follows:

$$\operatorname{div} E = 0 \quad (\text{as } \rho = 0) \tag{15.81a}$$

$$\operatorname{div} H = 0 \tag{15.81b}$$

$$\operatorname{curl} E = -\mu \frac{\partial H}{\partial t} \tag{15.81c}$$

$$\operatorname{curl} H = \epsilon \frac{\partial E}{\partial t} \quad (\text{as } J = 0) \tag{15.81d}$$

Taking curl of Eqs (15.81c) and (15.81d), we get the following relation:

$$\operatorname{curl} \operatorname{curl} E = -\mu \frac{\partial}{\partial t} \operatorname{curl} H \tag{15.82}$$

Substituting curl H from Eq. (15.81d), we get the following expression:

$$\operatorname{curl} \operatorname{curl} E = -\mu \frac{\partial}{\partial t} \left(\sigma E + \epsilon \frac{\partial E}{\partial t} \right) \tag{15.83}$$

Using the following vector identity in Eq. (15.83) helps us to reduce it to Eq. (15.84):

$$\operatorname{curl} \operatorname{curl} E = \operatorname{grad} \operatorname{div} E - \nabla^2 E \tag{15.84}$$

The right-hand side of Eq. (15.83) now becomes as follows:

$$\operatorname{grad} \operatorname{div} E - \nabla^2 E = -\sigma\mu \frac{\partial E}{\partial t} - \epsilon\mu \frac{\partial^2 E}{\partial t^2} \tag{15.85}$$

On applying Maxwell's equations (Eqs 15.53a–d), the first term becomes zero; hence, Eq. (15.85) takes the following form:

$$-\nabla^2 E = -\sigma\mu \frac{\partial E}{\partial t} - \epsilon\mu \frac{\partial^2 E}{\partial t^2} \tag{15.86}$$

$$\nabla^2 E - \sigma\mu \frac{\partial E}{\partial t} - \epsilon\mu \frac{\partial^2 E}{\partial t^2} = 0 \tag{15.87}$$

Similarly, the magnetic field equation can be derived by analogy with Eq. (15.81):

$$\nabla^2 H - \sigma\mu \frac{\partial H}{\partial t} - \epsilon\mu \frac{\partial^2 H}{\partial t^2} = 0 \tag{15.88}$$

Equations (15.87) and (15.88) together represent the wave equations that give the electromagnetic field in a homogeneous, linear medium, in which the charge density is zero. The electric and magnetic vectors will be oscillatory in nature. The complex form of the electric and magnetic field vectors is preferred for ease in mathematics than their sinusoidal or cosine form. Thus, the electric field vector is of the following form:

$$E(r,t) = E_0 \exp(\hat{i}(\hat{k} \cdot r - \omega t) \tag{15.89}$$

$$H(r,t) = H_0 \exp(\hat{i}(\hat{k} \cdot r - \omega t) \tag{15.90}$$

where $\hat{k}$ is the propagation wave vector, r is the position vector, and ω is the angular velocity such that it is directly proportional to the speed of propagation $[v = (\omega / \hat{k})]$.

15.13 PLANE WAVES IN DIELECTRIC MEDIUM

Having understood how to use the Maxwell's equations as a tool to analyze the behaviour of an electromagnetic wave in different media, we now redefine the medium as per our specifications. The plane wave in a dielectric medium is shown in Fig. 15.13. It is propagating along the z direction, which is coming out of the paper.

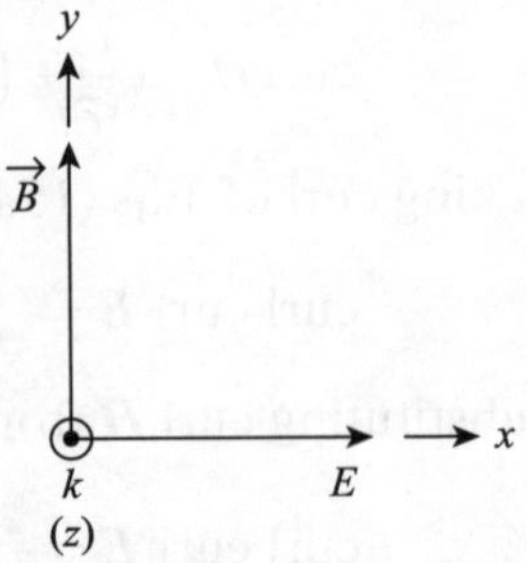

Let us consider a non-charged, current-free dielectric. The physical parameters of charge density ($\rho = 0$) and current density ($J = 0$) go to zero. When these specific parameters are substituted in the Maxwell's equations, the equations take the following form:

$$D = \epsilon E \tag{15.91a}$$

$$B = \mu H \tag{15.91b}$$

$$J = 0 \tag{15.91c}$$

$$\frac{\partial E_x}{\partial x} + \frac{\partial E_y}{\partial y} + \frac{\partial E_z}{\partial z} = 0 \tag{15.92a}$$

$$\frac{\partial H_x}{\partial x} + \frac{\partial H_y}{\partial y} + \frac{\partial H_z}{\partial z} = 0 \tag{15.92b}$$

$$\frac{\partial E_z}{\partial y} - \frac{\partial E_y}{\partial z} = -\mu \frac{\partial H_x}{\partial t}$$

$$\frac{\partial E_x}{\partial z} - \frac{\partial E_z}{\partial x} = -\mu \frac{\partial H_y}{\partial t}$$

$$\frac{\partial E_y}{\partial x} - \frac{\partial E_x}{\partial y} = -\mu \frac{\partial H_z}{\partial t} \tag{15.92c}$$

$$\frac{\partial H_z}{\partial y} - \frac{\partial H_y}{\partial z} = 0 + \epsilon \frac{\partial E_x}{\partial t}$$

$$\frac{\partial H_x}{\partial z} - \frac{\partial H_z}{\partial x} = 0 + \epsilon \frac{\partial E_y}{\partial t}$$

$$\frac{\partial H_y}{\partial x} - \frac{\partial H_x}{\partial y} = 0 + \epsilon \frac{\partial E_z}{\partial t} \tag{15.92d}$$

Fig. 15.13 Plane waves in dielectric medium

In a compact form, Maxwell's equations can be written as follows:

$$\operatorname{div} E = 0 \quad (\text{as } \rho = 0) \tag{15.93a}$$

$$\operatorname{div} H = 0 \tag{15.93b}$$

$$\operatorname{curl} E = -\mu \frac{\partial H}{\partial t} \tag{15.93c}$$

$$\operatorname{curl} H = \epsilon \frac{\partial E}{\partial t} \quad (\text{as } J = 0) \tag{15.93d}$$

Taking curl of Eqs (15.93c) and (15.93d), we get the following expression:

$$\operatorname{curl} \operatorname{curl} E = -\mu \frac{\partial}{\partial t} \operatorname{curl} H \tag{15.94}$$

Substituting curl H from Eq. (15.93d), we get the following expression:

$$\text{curl curl } E = -\mu \frac{\partial}{\partial t}\left(\epsilon \frac{\partial E}{\partial t}\right) \tag{15.95}$$

On applying vector identity,

$$\text{curl curl } E = \text{grad div } E - \nabla^2 E \tag{15.96}$$

The right-hand side of Eq. (15.95) now becomes as follows:

$$\text{grad div } E - \nabla^2 E = -\epsilon\mu \frac{\partial^2 E}{\partial t^2}$$

In Maxwell's equation (15.93a), the first term becomes zero. Hence,

$$-\nabla^2 E = -\epsilon\mu \frac{\partial^2 E}{\partial t^2}$$

$$\nabla^2 E - \epsilon\mu \frac{\partial^2 E}{\partial t^2} = 0 \tag{15.97a}$$

Similarly, the magnetic field equation can be derived by analogy with Eq. (15.93d):

$$\nabla^2 H - \epsilon\mu \frac{\partial^2 H}{\partial t^2} = 0 \tag{15.97b}$$

Equations (15.97a) and (15.97b) together represent the wave equations that give the electromagnetic field in an isotropic dielectric, in which the charge density is zero. The electric and magnetic vectors will be oscillatory in nature. The forms of the electric and magnetic vectors are as follows:

$$E(r,t) = E_0 \exp(\hat{i}(\hat{k}.r - \omega t) \tag{15.98a}$$

$$H(r,t) = H_0 \exp(\hat{i}(\hat{k}.r - \omega t) \tag{15.98b}$$

where $\hat{k}$ is the propagation wave vector, r is the position vector, and ω is the angular velocity such that it is directly proportional to the speed of propagation $[v = (\omega/\hat{k})]$. The velocity of an electromagnetic wave is given by analogy with the wave equation (refer to Chapter 10):

$$v = \frac{1}{\sqrt{\mu\epsilon}} = \frac{\omega}{\hat{k}} \tag{15.99}$$

Since the velocity of an electromagnetic wave has been physically verified by Eq. (15.99), the existence of an electromagnetic wave is proved. Let us now assume that plane waves are propagating in the +ve z-direction, so the electric and magnetic vectors indicated in Eqs (15.98a) and (15.98b) take the following forms:

$$E(r,t) = E_0 \exp(\hat{i}(\hat{k}z - \omega t) \tag{15.100a}$$

$$H(r,t) = H_0 \exp(\hat{i}(\hat{k}z - \omega t) \tag{15.100b}$$

As the electric and magnetic vectors are independent of the x and y co-ordinates, and are propagating along the z-direction, substituting this fact in Eqs (15.92a) and (15.92b), one gets the following relations:

$$\frac{\partial E_z}{\partial z} = 0 \tag{15.101a}$$

$$\cdot \frac{\partial H_z}{\partial z} = 0 \tag{15.101b}$$

Thus, there is no variation of the electric and magnetic vectors along the z-direction. Now, the vectors are either constant or zero. Electric and magnetic vectors are plane waves shown in Eq. (15.99), so they are not taken to be constant. Thus, one infers from Eqs (15.101a) and (15.101b) that the solution for these vectors would be as follows:

$$E_z = 0 \tag{15.102a}$$

$$H_z = 0 \tag{15.102b}$$

Thus, the longitudinal components are zero; hence, the waves would be transversal in nature. The electric vector is taken along the x-axis (arbitrarily, it can be taken along the y-axis also without any loss of generality following the Maxwell's equations) such that

$$E_y = 0 \tag{15.103}$$

Substituting this specific solution in Maxwell's equations (15.92a–d):

$$0 = -\mu \frac{\partial}{\partial t} \left(H_{0x} e^{i(\hat{k}z - \omega t)} \right) \tag{15.104a}$$

$$\frac{\partial}{\partial z} \left(E_{0x} e^{i(\hat{k}z - \omega t)} \right) - 0 = -\mu \frac{\partial}{\partial t} \left(H_{0y} e^{i(\hat{k}z - \omega t)} \right) \tag{15.104b}$$

$$0 - \frac{\partial}{\partial z} \left(H_{0y} e^{i(\hat{k}z - \omega t)} \right) = \epsilon \frac{\partial}{\partial t} \left(E_{0x} e^{i(\hat{k}z - \omega t)} \right) \tag{15.104c}$$

$$\frac{\partial}{\partial z} \left(H_{0x} e^{i(\hat{k}z - \omega t)} \right) = 0 \tag{15.104d}$$

Thus, further simplifying Eqs (15.104a)–(15.104d), one gets the following expressions:

$$0 = \hat{i} \omega \mu H_{0x} \tag{15.105a}$$

$$\hat{i}\hat{k} E_{0x} = \hat{i} \omega \mu H_{0y} \tag{15.105b}$$

$$-\hat{i}\hat{k} H_{0y} = -\hat{i} \omega \epsilon \epsilon E_{0x} \tag{15.105c}$$

$$\hat{i}\hat{k} H_{0x} = 0 \tag{15.105d}$$

where

$$E_0 = \hat{i} E_{0x} + \hat{j} E_{0y} + \hat{k} E_{0z} \tag{15.106a}$$

$$H_0 = \hat{i} H_{0x} + \hat{j} H_{0y} + \hat{k} H_{0z} \tag{15.106b}$$

Thus, from Eqs (15.105a)–(15.105d), we get the following relations:

$$H_{0x} = 0 \tag{15.107a}$$

$$\frac{H_{0y}}{E_{0x}} = \frac{\hat{k}}{\omega \mu} = \frac{\omega \epsilon}{\hat{k}} \tag{15.107b}$$

$$\hat{k}^2 = \omega^2 \epsilon \mu \tag{15.107c}$$

Electromagnetic plane waves are propagating along the z-direction, with the electric vector along the x-axis; consequently, the magnetic vector is along the y-axis (although there is no compulsion of the vectors to be along the said direction). Thus, the electric and magnetic vectors are expressed as follows:

$$E(z,t) = \hat{i}\, E_0 e^{i\left(kz - \omega t\right)} \tag{15.108a}$$

$$H(z,t) = \hat{j}\, H_0 e^{i\left(kz - \omega t\right)} \tag{15.108b}$$

where $\quad H_0 = \dfrac{\hat{k}}{\omega\mu} E_0 \tag{15.108c}$

From Eq. (15.108b), the expression for magnetic induction is obtained as follows:

$$B = \hat{j} B_0 e^{i(kz - \omega t)} \tag{15.109a}$$

From Eqs (15.91b) and (15.108c),

$$B_0 = \frac{\hat{k}}{\omega} E_0 = \sqrt{\epsilon\mu}\, E_0 = \frac{1}{v} E_0 \tag{15.109b}$$

Thus, Maxwell's equations have been used to deduce the nature of plane electric and magnetic vectors in a non-charged, current-free dielectric.

The following inferences are deduced from the Maxwell's equations for plane waves in a dielectric medium:

1. Electromagnetic plane waves are transverse in nature. The electric, magnetic, and propagation vectors form an orthogonal basis.

    Mathematically, it can be illustrated as follows:

    Maxwell's first two equations are as follows:

    $\nabla.E = 0$

    $\nabla.H = 0$

In explicit form, these two equations take the following forms:

$$\nabla.E = \left(\hat{i}\frac{\partial}{\partial x} + \hat{j}\frac{\partial}{\partial y} + \hat{k}\frac{\partial}{\partial z} \right) \cdot E_0\, e^{(i\hat{k}.r - \omega t)} \tag{15.110a}$$

$$\nabla.H = \left(\hat{i}\frac{\partial}{\partial x} + \hat{j}\frac{\partial}{\partial y} + \hat{k}\frac{\partial}{\partial z} \right) \cdot H_0 e^{(i\hat{k}.r - \omega t)} \tag{15.110b}$$

$$k = \hat{k}_x \hat{i} + \hat{k}_y \hat{j} + \hat{k}_z \hat{k}$$

$$r = x\hat{i} + y\hat{j} + z\hat{k}$$

$$E_0 = E_{0x}\hat{i} + E_{0y}\hat{j} + E_{0z}\hat{k}$$

$$k.r = \left(\hat{k}_x \hat{i} + \hat{k}_y \hat{j} + \hat{k}_z \hat{k} \right) \cdot \left(x\hat{i} + y\hat{j} + z\hat{k} \right)$$

$$= (\hat{k}_x x + \hat{k}_y y + \hat{k}_z z) \tag{15.110c}$$

$$\nabla.E = \left(\hat{i} \frac{\partial}{\partial x} + \hat{j} \frac{\partial}{\partial y} + \hat{k} \frac{\partial}{\partial z} \right) \cdot (E_{0x} \hat{i} + E_{0y} \hat{j} + E_{0z} \hat{k}) e^{\left(k_x x + k_y y + k_z z \right) - \omega t}$$

$$= \hat{i} (\hat{i} \hat{k}_x + \hat{j} \hat{k}_y + \hat{k} \hat{k}_z) \cdot (E_{0x} \hat{i} + E_{0y} \hat{j} + E_{0z} \hat{k}) e^{\left(k_x x + k_y y + k_z z \right) - \omega t}$$

$$= \hat{i} \hat{k}.E_0 e^{i(\hat{k}.r - \omega t)} \tag{15.110d}$$

$$= \hat{i} \hat{k}.E$$

$$\nabla.E = \hat{i} \hat{k}.E = 0 \tag{15.110e}$$

Similarly, $\nabla.H = \left(\hat{i} \frac{\partial}{\partial x} + \hat{j} \frac{\partial}{\partial y} + \hat{k} \frac{\partial}{\partial z} \right) \cdot \left(H_{0x} \hat{i} + H_{0y} \hat{j} + H_{0z} \hat{k} \right) e^{\left(\hat{i} k_x x + \hat{k}_y y + \hat{k}_z z \right) - \omega t}$

$$= \hat{i} \hat{k}.H$$

$$\nabla.H = \hat{i} \hat{k}.H = 0 \tag{15.110f}$$

Thus, if the dot product of any two vectors is zero, it implies that they are at a right angle to each other. Hence, both electric and magnetic vectors are perpendicular to the direction of propagation.

Now, Maxwell's equations can further be explicitly expressed mathematically as follows:

$$\nabla \times E = -\mu \frac{\partial H}{\partial t}$$

$$\begin{vmatrix} \hat{i} & \hat{j} & \hat{k} \\ \dfrac{\partial}{\partial x} & \dfrac{\partial}{\partial y} & \dfrac{\partial}{\partial z} \\ E_{0x} e^{i(\hat{k}.r - \omega t)} & E_{0y} e^{i(\hat{k}.r - \omega t)} & E_{0z} e^{i(\hat{k}.r - \omega t)} \end{vmatrix} = -\mu \frac{\partial}{\partial t} H_0 e^{(\hat{i}\hat{k}.r - \omega t)}$$

$$\hat{i} \left(\hat{k} \times E \right) = -\mu (-\hat{i} \omega H) \quad \Rightarrow \quad \hat{k} \times E = \mu \omega H$$

$$H = \frac{1}{\mu \omega} \left(\hat{k} \times E \right) = \frac{\hat{k}}{\mu \omega} (n \times E) \tag{15.110g}$$

where *n* is a unit vector in the direction of wave propagation. Thus, $= \hat{k} n$; also $\dfrac{\omega}{k} = c$. Hence,

$$H = \frac{1}{\mu v} (n \times E)$$

$$= \sqrt{\frac{\epsilon}{\mu}} (n \times E) \tag{15.110h}$$

This implies that the magnetic vector is perpendicular to the propagation vector $\hat{k}$ and the electric vector *E*. Thus,

$$\nabla \times H = \epsilon \frac{\partial E}{\partial t}$$

$$\hat{i}\left(\hat{k} \times H\right) = \epsilon(-\hat{i}\omega E)$$

$$\hat{k} \times H = -\epsilon\omega E \tag{15.110i}$$

This implies that the electric vector is perpendicular to the propagation vector $\hat{k}$ and the magnetic vector H.

Considering the cross and dot products of the electric, magnetic, and propagation vectors, it can be concluded and has been mathematically proved that they are all mutually perpendicular. Thus, electromagnetic plane waves are transverse in nature, and $(E, H, \hat{k})$ together form a right-handed orthogonal basis.

2. The ratio of the electric field component to the magnetic field component is a real number [Eq. (15.110g)]. This quantity is represented as wave impedance (the unit being ohm) and is mathematically represented as follows:

$$Z = \left|\frac{E}{H}\right| = \frac{E_0}{H_0} = \sqrt{\frac{\mu}{\epsilon}} \tag{15.110j}$$

This indicates that the electric and magnetic field vectors are in phase, that is, there is a relative magnitude at all points at all times. Thus, the electric and magnetic vectors can be written as follows:

$$E_x(z,t) = \hat{i}E_0 e^{i\left(\hat{k}z - \omega t\right)} \tag{15.110k}$$

$$H_y(z,\hat{t}) = \hat{j}\frac{E_0}{z} e^{i\left(\hat{k}z - \omega t\right)} \tag{15.110l}$$

3. The speed of propagation of an electromagnetic wave (v) has been ascertained to be as follows:

$$v = \left(\frac{\omega}{\hat{k}}\right) = \left(\frac{1}{\sqrt{\mu\epsilon}}\right) \tag{15.110m}$$

4. The refractive index of a dielectric can be represented as follows:

$$n = \frac{c}{v} = \sqrt{\frac{\epsilon\mu}{\epsilon_0\mu_0}} \tag{15.110n}$$

Most cases: $\mu = \mu_0$

$$n = \frac{c}{v} = \sqrt{\frac{\epsilon}{\epsilon_0}} = \sqrt{\kappa} \tag{15.110o}$$

where κ is refered to as the dielectric constant of the medium.

5. The most important inference of the Maxwell's equations is that the electric and magnetic vectors are interdependent. The changing electric field vector produces the magnetic field vector and vice versa.

6. Plane waves are plane polarized, as the electric and magnetic vectors are along the fixed directions x and y, respectively.

7. The Poynting vector (i.e., energy flow per unit area per unit time) for a plane electromagnetic wave is given by the following relation:

$$S = E \times H$$

$$= E \times \sqrt{\frac{\epsilon}{\mu}}(n \times E) = \frac{1}{Z}E \times (n \times E) = \frac{1}{Z}\left[(E \cdot E)n - (E \cdot n)E\right]$$

$$= \frac{1}{Z}E^2 n$$

$$= \frac{E^2}{Z}n \qquad\qquad (15.110\text{p})$$

Thus, the flow of energy of an electromagnetic wave is along the direction of propagation of the wave.

8. The electrostatic and the magnetostatic energy densities in an electromagnetic wave field are given as follows:

$$u_e = \frac{1}{2}\epsilon E^2 \qquad\qquad (15.110\text{q})$$

$$u_m = \frac{1}{2}\mu H^2 \qquad\qquad (15.110\text{r})$$

The total electromagnetic energy density is given as follows:

$$u_e + u_m = 2u_e = \epsilon E^2 \qquad\qquad (15.110\text{s})$$

Thus, the ratio of the intensities is given as follows:

$$\frac{u_e}{u_m} = \frac{\epsilon}{\mu}\frac{E^2}{H^2} = \frac{\epsilon}{\mu}Z^2 = \frac{\epsilon}{\mu}\cdot\frac{\mu}{\epsilon} = 1 \qquad\qquad (15.110\text{t})$$

That is, the electrostatic energy density is equal to the magnetostatic energy density.

9. The energy flux or power flow for an electromagnetic wave is given by the following expression:

Energy flux = $v \times$ energy density

$$<S> = v \times <u> n \qquad\qquad (15.110\text{u})$$

Thus, if the energy is flowing with a velocity v (which happens to be the phase velocity of the electromagnetic wave) in the direction of propagation of the wave, then the energy density of the wave also propagates in the direction of the wave.

10. The wavelength (λ) corresponding to an electromagnetic wave travelling in a dielectric medium is given as follows:

$$\lambda = \frac{2\pi}{\hat{k}} = \frac{2\pi}{\omega\sqrt{\mu\epsilon}} \qquad\qquad (15.110\text{v})$$

Note: Free space acts like a perfect dielectric, for which

$$\mu = \mu_0 = 4\pi \times 10^{-7}\text{H}/\text{m}; \quad \epsilon = \epsilon_0 = \frac{8.854 \times 10^{-12}\text{ F}}{\text{m}} \approx \frac{10^{-9}}{36\pi}\text{F}/\text{m}$$

$$Z = Z_0 \approx 120\pi\,\Omega; \quad u = c \approx 3 \times 10^8\text{ m}/\text{s}$$

Note: It is also important here to understand that when a travelling wave reaches an interface between two regions, it is partly reflected and partly transmitted; the magnitudes of the two parts are determined by the constants of the two regions.

Thus, for a normal incidence, E and H are entirely tangential to the interface and, thus, are continuous across it, that is,

$$E_0^i + E_0^r = E_0^t$$

$$H_0^i + H_0^r = H_0^t$$

where E^i and E^r are in region 1 with E^t in region 2.

The intrinsic impedence in either region is equal to the following relations:

$$\frac{E_0^i}{H_0^i} = Z_1; \quad \frac{E_0^r}{H_0^r} = -Z; \quad \frac{E_0^t}{H_0^t} = Z_2$$

The following examples illustrate the mathematical derivation of the concepts learnt in this section.

Example 15.25 Travelling E and H waves in free space (region 1) are normally incident on the interface with a perfect dielectric (region 2) for which $\epsilon_r = 3.0$. Compare the magnitudes of the incident, reflected, and transmitted E and H waves at the interface.

Solution The impedance for the two regions is defined as follows:

$$Z_1 = Z_0 = 120\pi\Omega; \qquad Z_2 = \sqrt{\frac{\mu}{\epsilon}} = \frac{120\pi}{\sqrt{\epsilon_r}} = 217.7\Omega$$

$$\frac{E_0^r}{E_0^i} = \frac{Z_2 - Z_1}{Z_1 + Z_2} = -0.268 ; \qquad \frac{H_0^r}{H_0^i} = \frac{Z_1 - Z_2}{Z_1 + Z_2} = 0.268$$

$$\frac{E_0^t}{E_0^i} = \frac{2Z_2}{Z_1 + Z_2} = 0.732; \qquad \frac{H_0^t}{H_0^i} = \frac{2Z_1}{Z_1 + Z_2} = 1.268$$

Note: This numerical has been introduced to highlight the behaviour of electric and magnetic fields at the boundary/interfaces. It is of interest to see how the concept of Maxwell's equations helps us understand electromagnetic waves at interfaces, similar to the light phenomenon in optics.

Example 15.26 In free space $E(z,t) = 10^3 \sin(\omega t - \alpha z)\,\hat{j}\left(\dfrac{V}{m}\right)$.

(a) Obtain the magnetic field vector $H(z, t)$.

(b) Determine the propagation constant k of the wave with a frequency $\vartheta = 95.5$ Hz.

Solution The phase factor in the electric field vector shows that the direction of propagation is along the positive z-direction. This implies that the Poynting vector $(S = E \times H)$ is along the positive z-direction. Thus, the magnetic field vector is along the negative x-axis. The impedance for such field vectors is represented as follows:

$$\frac{E_y}{-H_x} = Z_0 = 120\pi\Omega$$

(a) The magnetic field vector is given as follows:

$$H_x = -\frac{10^3 \sin(\omega t - \alpha z)}{120\pi} \frac{V}{\Omega m} \quad \Rightarrow \quad H_x = -\frac{10^3 \sin(\omega t - \alpha z)}{120\pi} \frac{A}{m}$$

(b) In general, the propagation wave vector is defined as follows:

$$\hat{k} = \sqrt{\hat{i}\omega\mu(\sigma + \hat{i}\omega\epsilon)} \text{ (refer to the subsequent section)}$$

In free space, $\sigma = 0$, so that

$$\hat{k} = \hat{i}\omega\sqrt{\mu_0\epsilon_0} = \hat{i}\left(\frac{2\pi\vartheta}{c}\right) = \hat{i}\frac{2\pi(95.5\times10^6)}{3\times10^8} \frac{s}{m.s}$$

$$= \hat{i}(200\times10^{-8+6}) = \hat{i}2.0/m$$

Thus, the space attenuation factor is zero, as it is free space. In addition, in general, the phase shift constant/the propagation wave vector is given as follows:

$$\hat{k} = \alpha + \hat{i}\beta = \alpha = 2.0 \text{ rad/m}.$$

Example 15.27 Examine the following electric field vector:

$$E(z,t) = 5\sin(\omega t + \alpha z)\hat{i} + 5\cos(\omega t + \alpha z)\hat{j}$$

for $\quad \omega t = 0, \dfrac{\pi}{4}, \dfrac{\pi}{2}, \dfrac{3\pi}{4}$, π in the $z = 0$ plane.

Solution Considering the components of the electric field vector at various angles, one observes that

$$E_x = 5\sin(\omega t); \quad E_y = 5\cos(\omega t); \quad E = E_x\hat{i} + E_y\hat{j}$$

For $\quad \omega t = 0$,

$$E_x = 5\sin(0) = 0; \quad E_y = 5\cos(0) = 5; \quad E = 0\hat{i} + 5\hat{j} = 5\hat{j}$$

For $\quad \omega t = \dfrac{\pi}{4}$,

$$E_x = 5\sin\left(\frac{\pi}{4}\right); \quad E_y = 5\cos\left(\frac{\pi}{4}\right) \Rightarrow E = \frac{5}{\sqrt{2}}\hat{i} + \frac{5}{\sqrt{2}}\hat{j}$$

For $\quad \omega t = \dfrac{\pi}{2}$,

$$E_x = 5\sin\left(\frac{\pi}{2}\right) = 5; \quad E_y = 5\cos\left(\frac{\pi}{2}\right) = 0 \Rightarrow E = 5\hat{i} + 0\hat{j} = 5\hat{i}$$

For $\quad \omega t = \dfrac{3\pi}{4}$,

$$E_x = 5\sin\left(\frac{3\pi}{4}\right) = \frac{5}{\sqrt{2}}; \quad E_y = 5\cos\left(\frac{3\pi}{4}\right) = -\frac{5}{\sqrt{2}} \Rightarrow E = \frac{5}{\sqrt{2}}\hat{i} - \frac{5}{\sqrt{2}}\hat{j} = 5\left(\frac{\hat{i}+\hat{j}}{\sqrt{2}}\right)$$

For $\quad \omega t = \pi$,

$$E_x = 5\sin(\pi) = 0; \quad E_y = 5\cos(\pi) = -5 \Rightarrow E = 0\hat{i} - 5\hat{j} = -5\hat{j}$$

Thus, the electric field vector is circularly polarized. It is evident from the change in the phase factor that the wave is travelling along the negative x-direction.

Thus, as the phase of a wave changes, it affects the state of the field (i.e., it gets polarized), and its direction and magnitude.

Note: Students should now be able to deduce plane electromagnetic waves for free space using the Maxwell's equations, where the physical parameters are as follows: $\rho = 0$, $\sigma = 0$, $J = 0$, $\mu = \mu_0$, ϵ.

Example 15.28 Evaluate the speed of light using the concept of Maxwell's electromagnetic theory.

Solution The speed of propagation in a non-charged, current-free dielectric is defined as follows:

$$v = \frac{\omega}{\hat{k}} = \sqrt{\frac{1}{\epsilon\mu}}$$

For free space, $c = \sqrt{\dfrac{1}{\epsilon_0\mu_0}}$

$\epsilon = \epsilon_0 = 8.8542 \times 10^{-12} \, C^2/Nm^2$; $\mu = \mu_0 = 4\pi \times 10^{-7} \, N - s^2 C^2$

Substituting the values, one gets the following result:

$$c = \sqrt{\frac{1}{8.8542 \times 10^{-12} \times 4\pi \times 10^{-7}}} = 2.99794 \times 10^8 \text{ m/s}$$

15.14 PLANE ELECTROMAGNETIC WAVES IN CONDUCTING MEDIUM

Maxwell's equations are to be used to understand and explain the behaviour of electromagnetic waves in a conducting medium. Characteristics of the conducting medium are defined as follows.

Let us consider a linear and isotropic medium. The medium is characterized by the physical parameters of permittivity (ϵ), permeability (μ), and conductivity (σ). The charge density (ρ) for the medium is zero.

When these specific parameters are substituted in the Maxwell's equations, the equations take the following forms:

$$D = \epsilon E$$

$$B = \mu H$$

$$J = \sigma E$$

$$\rho = 0 \tag{15.111a}$$

$$\frac{\partial E_x}{\partial x} + \frac{\partial E_y}{\partial y} + \frac{\partial E_z}{\partial z} = 0 \tag{15.111b}$$

$$\frac{\partial H_x}{\partial x} + \frac{\partial H_y}{\partial y} + \frac{\partial H_z}{\partial z} = 0 \tag{15.111c}$$

$$\frac{\partial E_z}{\partial y} - \frac{\partial E_y}{\partial z} = \mu \frac{\partial H_x}{\partial t}$$

$$\frac{\partial E_x}{\partial z} - \frac{\partial E_z}{\partial x} = -\mu \frac{\partial H_y}{\partial t}$$

$$\frac{\partial E_y}{\partial x} - \frac{\partial E_x}{\partial y} = -\mu \frac{\partial H_z}{\partial t} \tag{15.111d}$$

$$\frac{\partial H_z}{\partial y} - \frac{\partial H_y}{\partial z} = \sigma E + \epsilon \frac{\partial E_x}{\partial t}$$

$$\frac{\partial H_x}{\partial z} - \frac{\partial H_z}{\partial x} = \sigma E + \epsilon \frac{\partial E_y}{\partial t}$$

$$\frac{\partial H_y}{\partial x} - \frac{\partial H_x}{\partial y} = \sigma E + \epsilon \frac{\partial E_z}{\partial t} \tag{15.111e}$$

In a compact form, Maxwell's equations are represented as follows:

$$\text{div } E = 0 \quad (\text{as } \rho = 0) \tag{15.112a}$$

$$\text{div } H = 0 \tag{15.112b}$$

$$\text{curl } E = -\mu \frac{\partial H}{\partial t} \tag{15.112c}$$

$$\text{curl } H = \sigma E + \epsilon \frac{\partial E}{\partial t} \tag{15.112d}$$

Taking curl of Eqs (15.112c) and (15.112d), we get the following expression:

$$\text{curl curl } E = -\mu \frac{\partial}{\partial t} \text{curl } H \tag{15.113}$$

Substituting curl H from Eq. (15.112d), we get the following expression:

$$\text{curl curl } E = -\mu \frac{\partial}{\partial t}\left(\epsilon \frac{\partial E}{\partial t}\right) \tag{15.114}$$

If we use the vector identity, then $\text{curl curl } E = \text{grad div } E - \nabla^2 E$

Equation (15.114) takes the following form:

$$\text{grad div } E - \nabla^2 E = -\mu\sigma \frac{\partial E}{\partial t} - \epsilon\mu \frac{\partial^2 E}{\partial t^2} \tag{15.115}$$

In Maxwell's equations (Eqs 15.111a–d), the first term becomes zero; hence,

$$\nabla^2 E - \mu\sigma \frac{\partial E}{\partial t} - \epsilon\mu \frac{\partial^2 E}{\partial t^2} = 0 \tag{15.116a}$$

Similarly, the magnetic field equation can be derived by analogy with Eqs (15.112a)–(15.112d):

$$\nabla^2 H - \mu\sigma \frac{\partial E}{\partial t} - \epsilon\mu \frac{\partial^2 H}{\partial t^2} = 0 \tag{15.116b}$$

Equations (15.116a) and (15.116b) together represent the wave equations that give the electromagnetic field in a conducting medium in which the charge density is zero. The electric and magnetic vectors will be oscillatory in nature. The complex form is preferred for ease in mathematics; thus, the electric field vector is of the following form:

$$E(r, t) = E_0 \exp(\hat{i}(\hat{k}.r - \omega t)) \tag{15.117a}$$

$$H(r, t) = H_0 \exp(\hat{i}(\hat{k}.r - \omega t)) \tag{15.117b}$$

where $\hat{k}$ is the propagation wave vector, r the position vector, and ω the angular velocity.

Thus, electromagnetic waves exist, as the wave equation has been established. Let us assume that plane waves are propagating in the +ve z-direction; then the electric and magnetic vectors indicated in Eqs (15.117a) and (15.117b) now take the following form:

$$E(r, t) = E_0 \exp(\hat{i}(\hat{k}z - \omega t)) \tag{15.118a}$$

$$H(r, t) = H_0 \exp(\hat{i}(\hat{k}z - \omega t)) \tag{15.118b}$$

$$\left[\left(\frac{\partial^2}{\partial x^2} + \frac{\partial^2}{\partial y^2} + \frac{\partial^2}{\partial z^2} \right) E_0 \exp(\hat{i}(\hat{k}z - \omega t)) - \mu\sigma \frac{\partial}{\partial t} E_0 \exp(\hat{i}(\hat{k}z - \omega t)) \right.$$

$$\left. - \epsilon\mu \frac{\partial}{\partial t^2} E_0 \exp(\hat{i}(\hat{k}z - \omega t)) \right] = 0$$

$$(-\hat{k}^2 + \hat{i}\omega\mu\sigma + \omega^2\epsilon\mu) E_0 \exp(\hat{i}(\hat{k}z - \omega t)) = 0 \tag{15.119}$$

Thus, this equation will hold only if

$$\left(-\hat{k}^2 + \hat{i}\omega\mu\sigma + \omega^2\epsilon\mu \right) = 0 \tag{15.120}$$

This means that the propagation wave vector, $\hat{k}$, is complex. Let us represent this complex propagation vector in the following form:

$$\hat{k} = \alpha + \hat{i}\beta \tag{15.121a}$$

Substituting Eq. (15.121a) in Eq. (15.120), we get the following form:

$$-\left(\alpha^2 + 2\hat{i}\alpha\beta - \beta^2 \right) + \hat{i}\omega\mu\sigma + \omega^2\epsilon\mu = 0 \tag{15.121b}$$

Thus, equating the real and imaginary parts, we get the following expression:

$$\alpha^2 - \beta^2 = \omega^2\epsilon\mu \tag{15.122a}$$

$$\beta = \frac{\omega\mu\sigma}{2\alpha} \tag{15.122b}$$

Substituting the value of β in Eq. (15.122a) and solving, we get the following expression:

$$\alpha^2 = \omega^2\epsilon\mu + \left[\frac{\omega\mu\sigma}{2\alpha} \right]^2 \tag{15.123a}$$

$$\alpha^4 - \mu\epsilon\omega^2\alpha^2 - \frac{\mu^2\omega^2\sigma^2}{4} = 0 \tag{15.123b}$$

Solving this quadratic equation in α^2, one gets the following relation:

$$\alpha^2 = \frac{\mu\epsilon\omega^2 \pm \sqrt{(\mu^2\epsilon^2\omega^4 \pm \mu^2\omega^2\sigma^2)}}{2} \tag{15.123c}$$

$$\alpha = \omega\sqrt{\frac{\mu\epsilon}{2}} \left[\sqrt{1 + \frac{\sigma^2}{\omega^2\epsilon^2}} + 1 \right] \tag{15.124a}$$

$$\beta = \omega\sqrt{\frac{\mu\epsilon}{2}\left[\sqrt{1+\frac{\sigma^2}{\omega^2\epsilon^2}}-1\right]} \tag{15.124b}$$

Substituting, the value of $\hat{k}$ in Eq. (15.118a) for the electric vector and (15.118b) for the magnetic vector, we get the following relation:

$$E(r,t) = E_0\exp(\hat{i}((\alpha+\hat{i}\beta)z - \omega t)$$

$$= E_0\exp(-\beta z)\exp\left(\hat{i}(\alpha z - \omega t)\right) \tag{15.125a}$$

$$H(r,t) = H_0\exp(\hat{i}((\alpha+\hat{i}\beta)z - \omega t)$$

$$= H_0\exp(-\beta z)\exp\left(\hat{i}(\alpha z - \omega t)\right) \tag{15.125b}$$

From these equations, it is evident that the electric and magnetic vectors travelling in a conducting medium have exponentially decaying terms in the amplitude. These exponentially decaying terms indicate that although the electromagnetic waves show oscillatory behaviour, they attenuate spatially due to the presence of this term. The spatial attenuation is an important parameter in transmission of signals in wave guides.

The quantity β is a measure of attenuation and is referred to as the absorption coefficient. Attenuation in an electromagnetic wave is due to the Joule loss.

15.15 WAVES FOR PARTIALLY CONDUCTING MEDIA

In the previous sections, the media for which Maxwell's equations have been solved were either conducting or non-conducting. We now look for a solution of the waves for a region in which there is some conductivity but not much, for example, moist earth and sea water, using the Maxwell's equations as in the previous sections, gives us the detailed solutions for the electric and the magnetic vectors. The solutions of the electric vector and the corresponding magnetic vector are given as follows:

$$E = \hat{i}E_0\exp(-\beta z)\exp\left(\hat{i}(\alpha z - \omega t)\right) \tag{15.126a}$$

$$H = \hat{j}\sqrt{\frac{\sigma+\hat{i}\omega\epsilon}{\hat{i}\omega\mu}}E_0\exp(-\beta z)\exp\left(\hat{i}(\alpha z - \omega t)\right) \tag{15.126b}$$

This expression indicates that there is attenuation in space unless the conductivity (σ) becomes zero. Thus, the impedance of the medium is given as follows:

$$Z = \sqrt{\frac{\hat{i}\omega\mu}{\sigma+\hat{i}\omega\epsilon}} \tag{15.127}$$

where, in the polar form, the magnitude and angle of the impedance vector are given as follows:

$$|Z| = \frac{\sqrt{\frac{\mu}{\epsilon}}}{\sqrt[4]{1+\frac{\sigma^2}{\omega^2\epsilon^2}}} \tag{15.128a}$$

$$\tan 2\theta = \frac{\sigma}{\omega\epsilon} \qquad 0° < \theta < 45° \tag{15.128b}$$

The electric and magnetic vectors can be represented in the phase form as follows:

$$E = \hat{i}E_0 \exp(-\beta z)\exp\left(\hat{i}(\alpha z - \omega t)\right) \tag{15.129a}$$

$$H = \hat{j}\sqrt{\frac{\sigma + \hat{i}\omega\epsilon}{\hat{i}\omega\mu}}E_0 \exp(-\beta z)\exp\left(\hat{i}(\alpha z - \omega t - \theta)\right) \tag{15.129b}$$

The speed of propagation $[v = (\omega/\hat{k})]$ of the electromagnetic wave is given as follows:

$$v = \frac{\omega}{\hat{k}} = \frac{\omega}{\alpha} = \frac{1}{\sqrt{\frac{\mu\epsilon}{2}\left[\sqrt{1 + \frac{\sigma^2}{\omega^2\epsilon^2}} + 1\right]}} \tag{15.130}$$

$$\lambda = \frac{2\pi}{\hat{k}} = \frac{2\pi}{\alpha} = 2\pi / \left(\omega\sqrt{\frac{\mu\epsilon}{2}\left[\sqrt{1 + \frac{\sigma^2}{\omega^2\epsilon^2}} + 1\right]} \right) \tag{15.131}$$

The conducting medium is categorized as good or bad depending upon the ratio of mobility to the product of the angular frequency and permittivity of the medium. The following are the characteristics of good and bad conductors.

For a good conductor,

$$\frac{\sigma}{\omega\epsilon} \gg 1 \quad \text{or} \quad \sigma \gg \omega\epsilon \tag{15.132}$$

The expression for the propagation constant can be determined for this special case using Eqs (15.124a) and (15.124b). Thus, one observes that the value of α and β are approximately equal, that is,

$$\alpha = \beta = \mu\epsilon\omega\sqrt{\frac{\left(\frac{\sigma}{\omega\epsilon}\right)}{2}} = \sqrt{\frac{\mu\sigma\omega}{2}} \tag{15.133}$$

The propagation vector for a good conducting medium is given in the following form:

$$\hat{k} = \alpha + \hat{i}\beta = \sqrt{\frac{\mu\sigma\omega}{2}} \tag{15.134}$$

This physically means that in a conductor, the electric and magnetic vectors are spatially attenuated. The rate of spatial attenuation is very rapid.

The impedance of a good conductor can be derived from Eqs (15.128a) and (15.128b). The impedance for a good conductor is numerically given as follows:

$$Z = \sqrt{\frac{\omega\mu}{\sigma}}\angle 45° \tag{15.135}$$

This indicates that the electric and magnetic vectors will be out of phase by a factor of $\frac{\pi}{4}$ or 45°. Thus, assuming the direction of propagation to be along the z-axis, the direction of electric and magnetic vectors will be along the x- and *y*-axis, respectively. Thus, incorporating the conclusions of Eq. (15.135) in the electric and magnetic vectors, we get the following expressions:

$$E(z, t) = E_0 \exp(-\beta z)\exp\left(\hat{i}(\alpha z - \omega t)\right)\hat{i} \tag{15.136a}$$

$$H(z, t) = \frac{E_0}{|Z|_0} \exp(-\beta z)\exp\left(\hat{i}\left(\alpha z - \omega t - \frac{\pi}{4}\right)\right)\hat{j} \tag{15.136b}$$

The velocity of an electromagnetic wave in a good conductor is given as follows:

$$v = \frac{\omega}{\alpha} = \sqrt{\frac{2\omega}{\mu\sigma}} = \omega\delta \tag{15.137}$$

The wavelength of an electromagnetic wave in a good conducting medium is given as follows:

$$\lambda = \frac{2\pi}{\alpha} = \frac{2\pi}{\sqrt{\pi\vartheta\mu\sigma}} = 2\pi\delta \tag{15.138}$$

Example 15.29 Calculate the wavelength and wave velocity of an electromagnetic wave in a good conductor, given that its frequency is $\vartheta = 100\,\text{MHz}$; $\mu = 1$; $\sigma = \dfrac{5.8\times10^7 \text{ mhos}}{\text{m}}$

Solution The calculations are as follows:

$$\lambda = \frac{2\pi}{\sqrt{\pi\vartheta\mu\sigma}} = \frac{2\pi}{\sqrt{\pi\times100\times10^6\times1\times5.8\times10^7}} = 4.6\times10^{-8}\,\text{m}$$

$$v = \frac{2\pi\vartheta}{\sqrt{\pi\vartheta\mu\sigma}}$$

15.16 SKIN DEPTH/PENETRATION DEPTH

In understanding the behaviour of electromagnetic waves using the Maxwell's equations, it has come to light that as a wave traverses through different media it follows different paths. As observed in the case of a conducting medium, the amplitude contains an exponentially decaying term. Spatial damping is directly proportional to attenuation β, that is, greater the value of β, the greater the attenuation. The term 'penetration depth' is coined for the parameter δ, which measures the depth/distance at which the electromagnetic wave entering a conductor is damped to $1/e$, that is,

$$\delta = \frac{1}{\beta} = \frac{1}{\omega\sqrt{\mu\epsilon}} \left[\frac{\sqrt{\left[1+\dfrac{\sigma}{\omega\epsilon}\right]^2} - 1}{2} \right]^{-\frac{1}{2}}$$

$$= \sqrt{\frac{2}{\mu\epsilon\omega}}$$

$$= \sqrt{\frac{1}{\pi\vartheta\mu\sigma}} \tag{15.139}$$

The significance of the penetration depth is that it can measure the depth to which an electromagnetic wave can penetrate a conducting medium. Thus, knowing the skin depth for a particular conducting medium, it can be made thicker or thinner, depending on the application. Electromagnetic shields used by police are made of conducting materials. Once the penetration depth of a given conducting material is ascertained, its dimension can be made in accordance such that the radiation is not able to penetrate and the person using it is protected.

We are now in a position where, on the basis of the frequency, permittivity and mobility of a surface can be categorized as follows.

For a poor conductor: $\dfrac{\sigma}{\omega\epsilon} \ll 1$

From Eqs (15.124a)–(15.124c), the values of α and β can be inferred as follows:

$$\alpha = \sqrt{\mu\epsilon}\,\omega$$

$$\beta = \sqrt{\dfrac{\mu}{\epsilon}}\,\dfrac{\sigma}{2}$$

The propagation vector for a bad conducting medium takes the following

form: $\hat{k} = \alpha + \hat{i}\beta = \sqrt{\mu\epsilon}\,\omega + \hat{i}\dfrac{\sigma}{2}\sqrt{\dfrac{\mu}{\epsilon}}$

In the case of a poor conductor, it can thus be inferred that the value of α is much greater than the value of β.

For a quasi-conductor: $0.01 \leq \dfrac{\sigma}{\omega\epsilon} \leq 100$

Hence, a quasi-conductor lies in between a poor and a good conductor.

The following examples make us understand the concept of space attenuation and penetration depth.

Example 15.30 An electric field vector $E = 1.0e^{-\beta z}e^{i(\omega t - \alpha z)}\hat{i}$ V/m, with $\vartheta = 100$ MHz, at the surface of a copper conductor, $\sigma = 58$ MS/m, is located at $z > 0$. Determine the space attenuation constant as the wave propagates into the conductor.

Solution The magnitude of the electric field vector is calculated as follows:

$$|E| = 1.0e^{-\beta z} = 1.0e^{-\frac{z}{\delta}}; \ \delta = \dfrac{1}{\sqrt{\pi\vartheta\mu\sigma}} = 6.61\,\mu m$$

Thus, after 6.61 μm, the field is attenuated in space to ($1/e$) of its initial value. Further, at 5δ, the field is attenuated to 33 μm.

Example 15.31 Find the frequency for which the skin depth δ in aluminium is 0.01 mm, where $\sigma = 3.54 \times 10^7$ mho/m, $\epsilon_r = 1$, and $\mu_r = 1$. Find the wave velocity (v) and the propagation constant ($\hat{k}$).

Solution The skin depth is defined as follows:

$$\delta = \dfrac{1}{\sqrt{\pi\vartheta\mu\sigma}} = \dfrac{1}{\sqrt{3.14 \times \vartheta \times 3.54 \times 10^7}} = 0.01 \times 10^{-3}\,m$$

Example 15.32 Find the depth of penetration, δ, of an electromagnetic wave in copper at $\vartheta = 60$ Hz and $\vartheta = 100$ Hz. For copper, $\sigma = \dfrac{5.8 \times 10^7 \text{mho}}{\text{m}}$, $\mu_r = 1$, $\epsilon_r = 1$.

Solution For copper at $\vartheta = 60$ Hz, $\dfrac{\sigma}{\omega\epsilon} = \dfrac{5.8 \times 10^7}{2\pi \times 60 \times 8.854 \times 10^{-12}} = 174 \times 10^{14}$

This indicates that copper acts as a very good conductor. The depth of penetration can be calculated using the following formula:

$$\delta = \frac{1}{\beta} = \sqrt{\frac{2}{\omega\mu\sigma}} = \sqrt{\frac{2}{2\pi \times 60 \times 4\pi \times 10^{-7} \times 5.8 \times 10^7}} = 8.53 \times 10^{-3}\,\text{m}$$

Note: The frequency plays a critical role in determining whether a particular material can behave as a dielectric or a conductor.

Example 15.33 Is water a perfect dielectric at discrete frequencies? Assume that $\sigma = 10^{-3}\,\text{S/m}$, $\mu_r = 1$, $\dfrac{\epsilon}{\epsilon_0} = 80$.

Solution We know that $\dfrac{\sigma}{\epsilon} = \dfrac{10^{-3}}{80 \times 8.85 \times 10^{-12}} = 1.4 \times 10^6\,/\text{s}$

For $\omega = 2\pi \times 10/\text{s}$; $\dfrac{\sigma}{\omega\epsilon} = 2 \times 10^4$

For $\omega = 2\pi \times 10^{10}$; $\dfrac{\sigma}{\omega\epsilon} = 2 \times 10^{-5}$

Thus, it is evident from these calculations that water behaves as a perfect dielectric for discrete frequencies, that is, $\vartheta \geq 10^7\,/\text{s}$.

It is also evident that it behaves as a good conductor for $\vartheta \leq 10^3\,/\text{s}$.

Note: Using the Maxwell's equations, students should now be able to deduce plane electromagnetic waves for free space where the physical parameters are $\rho = 0$, $\sigma = 0$, $J = 0$, $\mu = \mu_0$, $\epsilon = \epsilon_0$.

IMPORTANT CONCEPTS

1. Mathematical understanding of the concepts of vector calculus, that is, gradient, divergence, and curl, helps us understand the flow of energy for electric and magnetic fields in space, their spread over a given angle, and their rotation about a given point, respectively.
2. Maxwell's equations integrate the laws of pure electric and magnetic fields to give a new field called 'electromagnetics'. Maxwell's equations are a basis for understanding the flow of electromagnetic energy (simply said 'light energy') through different media.
3. The electromagnetic energy density equation indicates conservation of energy, that is, the rate of decrease of the total energy is equal to the sum of the net flow out of energy and the Joule loss for a given volume enclosing that surface.
4. The Poynting vector gives the direction of flow of an electromagnetic radiation. The radiant flux density is the average value of the Poynting vector over a given time interval.

5. The electromagnetic wave equation for electric and magnetic fields indicates that both the fields are oscillatory in nature.

6. Plane waves in a dielectric medium show the following characteristics:
    (a) The electric and magnetic vectors are both perpendicular to the direction of propagation.
    (b) The electric and magnetic vectors are in phase.
    (c) The Poynting vector indicates that fields are interdependent, and the flow of energy of the electromagnetic wave is along the direction of propagation of the wave.
    (d) Electrostatic energy is equal to the magnetostatic energy density.

7. Plane electromagnetic waves in a conducting medium are oscillatory in nature but attenuate spatially. This spatial attenuation is known as the absorption coefficient, which is akin to the Joule loss.

8. A conducting medium is categorized as good or bad depending on the ratio of the mobility to the angular velocity and permittivity of the medium.

9. The penetration depth is the depth to which an electromagnetic wave can penetrate a conducting medium.

10. The concept of displacement current gives a modification to the already existing definitions of electric and magnetic fields. This is represented as follows: $I_d = \epsilon \dfrac{\partial \varphi_e}{\partial t}$

11. The equation of continuity is the conservation of energy equation represented as follows: $\text{div}(\boldsymbol{J}) = 0$

12. The Poynting theorem or the equation representing energy conservation, while energy is flowing from one medium to another, is given as follows:

$$\text{div}(\boldsymbol{S}) + \frac{\partial U}{\partial t} = -(\boldsymbol{J} . \boldsymbol{E})$$

    where the Poynting vector is defined as $\boldsymbol{S} = \boldsymbol{E} \times \boldsymbol{H}$.

13. The solution for the wave equation for various media predicts the electric and magnetic vectors to be perpendicular to each other and the direction of propagation.

14. The skin depth or penetration depth for a conducting medium is given as follows:

$$\delta = \sqrt{\frac{2}{\mu \epsilon \omega}} = \sqrt{\frac{1}{\pi \vartheta \mu \sigma}}$$

APPLICATIONS

1. The idea of the Maxwell's displacement current brought about the electromagnetism theory in a unifying way, wherein the propagation of waves in various media could be studied.

2. The Maxwell's electromagnetic theory is a tool to understand how electromagnetic waves travel through different media. The concept of penetration depth for a conducting medium has far-reaching applications in defence arsenal, particularly electromagnetic shields. Transmission of light through fibre optics, which has introduced an electronic revolution, follows the Maxwell's equations.

IMPORTANT FORMULAE

1. $\nabla . D = \rho$

2. $\nabla . B = 0$

3. $\nabla \times E = -\dfrac{\partial B}{\partial t}$

4. $\nabla \times H = J + \dfrac{\partial D}{\partial t}$

5. $\operatorname{div}(J) = 0$

6. $E(r, t) = E_0 \exp(\hat{i}(\hat{k}z - \omega t))$

7. $H(r, t) = H_0 \exp(\hat{i}(\hat{k}z - \omega t))$

8. $S = E \times H$

9. $\operatorname{div}(S) + \dfrac{\partial U}{\partial t} = -(J.E)$

10. $\delta = \sqrt{\dfrac{2}{\mu \epsilon \omega}} = \sqrt{\dfrac{1}{\pi \vartheta \mu \sigma}}$

SELF-ASSESSMENT

Multiple-choice Questions

15.1 The gradient operator is defined by

(a) $\bar{\nabla} = \hat{i}\dfrac{\partial}{\partial x} + \hat{j}\dfrac{\partial}{\partial y} + \hat{k}\dfrac{\partial}{\partial z}$

(b) $\bar{\nabla} = \hat{i}\dfrac{\partial}{\partial x^2} + \hat{j}\dfrac{\partial}{\partial y^2} + \hat{k}\dfrac{\partial}{\partial z^2}$

(c) $\nabla = \dfrac{\partial}{\partial x} + \dfrac{\partial}{\partial y} + \dfrac{\partial}{\partial z}$

(d) $\bar{\nabla} = \dfrac{\partial}{\partial x^2} + \dfrac{\partial}{\partial y^2} + \dfrac{\partial}{\partial z^2}$

15.2 The mathematical operation of a gradient on a physical quantity gives the
   (a) derivative of that quantity
   (b) directional derivative of the physical quantity in space
   (c) derivative of the physical quantity in direction
   (d) derivative of the physical quantity $\bar{\nabla} = \hat{i}\dfrac{\partial}{\partial x} + \hat{j}\dfrac{\partial}{\partial y} + \hat{k}\dfrac{\partial}{\partial z}$

15.3 The divergence of a physical field vector gives
   (a) the flow of a fluid from the source to the sink
   (b) gain or loss of a fluid per unit volume per unit time in a given parallelepiped
   (c) the directional derivative of the physical quantity
   (d) the vector perpendicular to the surface given by the physical quantity

15.4 The curl of a physical quantity highlights its
   (a) directional properties
   (b) derivative properties
   (c) diverging properties
   (d) rotational properties

15.5 Maxwell's equations give
   (a) the variation of the electric and magnetic fields in the classical domain
   (b) the variation of the electric and magnetic fields in the quantum domain
   (c) the unified approach called electromagnetic theory explaining the variation of static and time-varying electric and magnetic fields
   (d) the variation of only the electric fields

15.6 The oscillating or time-varying mathematical expression is valid for both the electric and magnetic fields because they are able to satisfy
   (a) the wave equation
   (b) the wave nature of light
   (c) the interference effect of light
   (d) the diffraction effect of light

15.7   The equation of continuity explains
   (a)   non-conservative nature of charge
   (b)   conservation of charge for a static electric field
   (c)   conservation of charge for a non-static electric field
   (d)   non-destructive nature of charge

15.8   The electromagnetic energy density equation indicates
   (a)   conservation of energy, that is, the rate of decrease of the total energy is equal to the sum of the net flow out of energy and the Joule loss for a given volume enclosing that surface
   (b)   non-conservation of energy for static and time-varying fields
   (c)   flow of energy in a cavity
   (d)   destruction of energy in a cavity

15.9   The Poynting vector indicates that the fields are
   (a)   interdependent and the flow of energy of the electromagnetic wave is along the direction of propagation of the wave
   (b)   independent and the flow of energy of the electromagnetic wave is along the direction of propagation of the wave
   (c)   interdependent and the flow of energy of the electromagnetic wave is not along the direction of propagation of the wave
   (d)   not dependent and the flow of energy of the electromagnetic wave is not along the direction of propagation of the wave

15.10   Plane waves in a dielectric medium show that
   (a)   the electric and magnetic vectors are both perpendicular to the direction of propagation
   (b)   the electric and magnetic vectors are in phase
   (c)   both (a) and (b)
   (d)   none of these

15.11   Plane electromagnetic waves in a conducting medium are
   (a)   oscillatory in nature but attenuate spatially
   (b)   non-oscillatory and non-attenuating in nature
   (c)   oscillatory but attenuating in nature
   (d)   oscillatory and exponentially increasing spatially

15.12   A conducting medium is categorized as
   (a)   good or bad depending on the ratio of the conductivity and permittivity of the medium
   (b)   good or bad depending on the ratio of the mobility to the angular velocity and permittivity of the medium
   (c)   good or bad depending on the ratio of the mobility to the permittivity of the medium and angular velocity
   (d)   good or bad depending on the ratio of the mobility to the angular velocity

15.13   The penetration depth is the depth to which an electromagnetic wave
   (a)   can penetrate a conducting medium
   (b)   can penetrate a non-conducting medium
   (c)   cannot penetrate a conducting medium
   (d)   can penetrate a dielectric medium

15.14   Ampere's circuital law is valid for
   (a)   varying current only
   (b)   steady current only
   (c)   alternative current only
   (d)   none of these

15.15 Displacement current through an ideal capacitor
    (a) is greater than the conduction current
    (b) is less than the conduction current
    (c) is equal to the conduction current
    (d) none of these

15.16 $\nabla.B = 0$ signifies that
    (a) B is a conservative field
    (b) magnetic monopole does not exist
    (c) $B = 0$
    (d) there exists a magnetic monopole

15.17 If a time-dependent electromagnetic wave is represented as $e^{-i\omega t}$, a wave represented as $f(z) = Ae^{-i\beta z}$ is termed as a
    (a) standing wave   (c) backward travelling wave
    (b) forward travelling wave   (d) longitudinal wave

15.18 The energy stored per unit volume of a field at any point in an electromagnetic cavity resonator is given as
    (a) $U = \dfrac{1}{2}\epsilon_0 E^2$ (c) $U = \dfrac{1}{2}\epsilon_0 E^2 + \dfrac{1}{2\mu_0} B^2$
    (b) $U = \dfrac{1}{2\mu_0} B^2$ (d) $U = \dfrac{1}{2}\epsilon_0 E^2 - \dfrac{1}{2\mu_0} B^2$

15.19 If one connects an AC source between two plates of a parallel plate capacitor,
    (a) conduction current flows due to the actual motion of charge between plates
    (b) displacement as well as conduction current flows
    (c) displacement current appears without the actual motion of charge carriers and conduction current flows due to actual motion outside the capacitor plates
    (d) the capacitor is capable of blocking any type of current

15.20 In a dielectric medium, the phase difference between E and B is
    (a) zero   (c) $\pi/2$
    (b) π (d) any non-zero value

15.21 The ratio of the electric field to the magnetic field in free space is given as
    (a) the velocity of light
    (b) the negative velocity of light
    (c) the product of permeability and permittivity in free space
    (d) the ratio of permeability to permittivity in free space

Review Questions

15.1 Show that $\nabla\varphi$ is a vector perpendicular to the surface $\varphi(x, y, z) = $ const.

15.2 Show that the speed of electromagnetic waves in an isotropic dielectric is less than the speed of the electromagnetic waves in free space.

15.3 Explain why radio communication with submerged submarines becomes increasingly difficult at several skin depths.

15.4 Show that the penetration depth decreases with increases in frequency for a conducting material like copper.

15.5 Give the physical significance of the equation of continuity.

15.6 A fluid moves so that its velocity at any points is $v(x, y, z)$. Show that the gain of a fluid per unit volume per unit time in a small parallelepiped, having centre at P(x, y, z), edges parallel to the co-ordinate axes, and magnitude $\Delta x, \Delta y, \Delta z$, respectively, is given approximately by $\mathrm{div}\, v = \nabla.v$.

15.7 Illustrate mathematically how and under what conditions the Ampere's circuital law fails. How did Maxwell modify the Ampere's law to make it consistent under all conditions? Give mathematical justification to prove its consistency.

15.8 Obtain the equation for the propagation of plane electromagnetic waves in an isotropic dielectric medium. Discuss the following for these two cases:
(a) Relative direction of E and H
(b) Phase of E and H, and impedance
(c) Poynting vector

15.9 State and prove the Poynting theorem. Give the physical significance of the Poynting vector.

15.10 Derive the relation for skin depth. How is this concept used to classify a material/medium as a good conductor or a good dielectric?

15.11 Show that the electromagnetic energy density is conserved for an electromagnetic wave.

15.12 Derive the formula for the intensity of an electromagnetic wave.

Numerical Problems

15.1 Find $\nabla\varphi$ if (a) $\varphi = \ln(r)$ and (b) $\varphi = \dfrac{1}{r}$.

15.2 Deduce that $\nabla r^n = nr^{n-2}r$.

$$\left[\text{Hint: } \nabla r^n = \nabla(\sqrt{x^2 + y^2 + z^2})^n = \nabla\left(\sqrt{x^2 + y^2 + z^2}\right)^{n/2} \right]$$

15.3 Prove that $\nabla \cdot \left(\dfrac{r}{r^3}\right) = 0$.

15.4 Determine the constant a so that the following vector V is solenoidal:
$$V = (x+3y)\hat{i} + (y-2z)\hat{j} + (x+az)\hat{k} \qquad [\text{Hint: } \nabla.V = 0]$$

15.5 If $v = \omega xr$, prove that $\omega = \dfrac{1}{2}\text{curl}\, v$, where ω is a constant vector.

15.6 Evaluate $<\rho_e>$ and $<\rho_m>$ for sunlight, assuming that the average energy flux in sunlight at the earth's surface is $<S> = 1.0\ \text{kW/m}^2$.

$$\left[\text{Hint: } <S> = \frac{<\rho_e>}{c}; \quad \frac{<S>}{c^2} = \frac{<\rho_e>}{c} = <\rho_m> \right]$$

15.7 When the amplitude of the magnetic field in a plane wave is 2 A/m,
(a) determine the magnitude of the electric field when the wave propagates in a medium characterized by $\sigma = 0$, $\mu = \mu_0$ and $\epsilon = 4\epsilon_0$
(b) determine the magnitude of the electric field for the plane wave in free space

15.8 Is earth a perfect dielectric at discrete frequencies? Assume that $\sigma = 5.0 \times 10^{-3}$ S / m, $\mu_r = 1$, $\epsilon_r = 8$. How does α vary at these frequencies?

$$\left[\text{Hint: } \delta = \frac{1}{\sqrt{\pi \vartheta \mu \sigma}}; \ \mu = \mu_0 \right]$$

15.9 Ocean water can be assumed to be a non-magnetic dielectric with $\kappa = \left(\dfrac{\epsilon}{\epsilon_0}\right) = 80$

and $\sigma = 4.3$ mho/m.
(a) Calculate the frequencies at which the penetration depth will be 5 and 10 cm.
(b) Show that for frequencies less than 10^8/s, it can be considered a good

conductor.
$$\left[Hint: \delta = \frac{1}{\sqrt{\pi \vartheta \mu \sigma}}; \mu = \mu_0\right]$$

15.10 What is the magnetic field amplitude of an electromagnetic wave whose electric field amplitude is 500 V/m? What is the intensity of the wave?

$$\left[Hint: B = \frac{E}{c}; I = \frac{1}{2}\epsilon_0 c E_{max}^2\right]$$

15.11 A 1000 W carbon dioxide laser emits light of wavelength 10 µm into a 3.0 mm diameter laser beam. What force does the laser beam exert on a completely

absorbing target?
$$\left[Hint: I = \frac{P}{4\pi r^2}; E_0 = \sqrt{\frac{2I}{\epsilon_0 c}} \Rightarrow E_0^2 = \frac{I}{1.33 \times 10^{-3}}\right]$$

15.12 A helium–neon laser, commonly used for laboratory demonstrations, emits a 1.0 mm diameter laser beam with a power of 1.0 mW. What is the amplitude of the oscillating electric field of the laser beam?

$$\left[Hint: I = \frac{P}{4\pi r^2}; E_0 = \sqrt{\frac{2I}{\epsilon_0 c}} \Rightarrow E_0^2 = \frac{I}{1.33 \times 10^{-3}}\right]$$

15.13 Show that for frequencies $\leq 10^8$/s, a sample of silicon will act as a good conductor. For silicon, one may assume that $\dfrac{\epsilon}{\epsilon_0} = 12$ and $\sigma = \dfrac{2\,\text{mho}}{\text{m}}$. In addition,
calculate the penetration depth for this sample at $\vartheta = 10^6$/s.

$$\left[Hint: \frac{\sigma}{\omega \epsilon} \gg 1; \delta = \frac{1}{\sqrt{\pi \vartheta \mu \sigma}} \text{ where } \mu = 1\right]$$

15.14 The electric field components are $E_x = E_y = 0$ and $E_z = E_0 \cos(\hat{k} x)\sin(\omega t)$ in free space. Calculate the magnetic field components for the same.

$$\left[Hint: \nabla \times E = -\frac{\partial B}{\partial t}\right]$$

15.15 Find the average density of a wave travelling in free space with an average
Poynting's vector of magnitude $\dfrac{6\,J}{m^2 s}$.

$$\left[Hint: S = EH; \text{Energy density} = \epsilon_0 E^2 = \mu_0 H^2\right]$$

15.16 Show that electric and magnetic densities in a plane travelling wave are equal. In addition, prove that the total energy density $= \epsilon_0 E^2 = \mu_0 H^2$.

[*Hint*: Refer to Section 15.9.]

15.17 Estimate the signal in free space of an FM radio channel broadcasting its signal at 106.4 MHz.
$$\left[\text{Hint: } \lambda_{\text{free space}} = \frac{c}{v}\right]$$

15.18 Give the physical significance of the term 'frequency'. You can highlight your answer in the context that the power-line frequency is about 50 Hz and that of an FM radio signal is in the range of 100 MHz.
$$\left[\text{Hint } \lambda = \frac{c}{\vartheta}\right]$$

15.19 Determine the propagation constant for a wave travelling in free space with a frequency of 90 MHz.
$$\left[\begin{array}{l}\text{Hint: } k = \alpha + i\beta; k = \sqrt{i\omega\mu(\sigma + i\omega\epsilon)};\\ \text{in free space } \sigma = 0; \text{ so that } k = i\omega\sqrt{\mu_0\epsilon_0})\end{array}\right]$$

15.20 At a frequency $\vartheta = 2\,\text{GHz}$ a material has a $\sigma = \dfrac{25\,\text{S}}{\text{m}}; \epsilon_r = 80$. Explain why this material will act as a conductor at this frequency.
$$\left[\text{Hint: } \frac{\sigma}{\omega\epsilon} \gg 1 \text{ or } \sigma \gg \omega\epsilon\right]$$

15.21 Find the skin depth δ at a frequency of 1.6 MHz in aluminium, where $\sigma = 38.2\dfrac{\text{MS}}{\text{m}}$ and $\mu_r = 1, \epsilon_r = 1$.
$$\left[\text{Hint: } \delta = \frac{1}{\sqrt{\pi\vartheta\mu\sigma}}\right]$$

15.22 Travelling E and H waves in free space (region 1) are normally incident on the interface with a perfect dielectric (region 2) for which $\epsilon_r = 2.0$. Compare the magnitudes of the incident, reflected, and transmitted E and H waves at the interface.

15.23 In free space $E(z,t) = 10^3 \sin(\omega t - \alpha z)\,\hat{j}\left(\dfrac{V}{m}\right)$.

   (a) Obtain the magnetic field vector $H(z, t)$.
   (b) Determine the propagation constant $\hat{k}$ of the wave with a frequency $\vartheta = 100$ Hz.
$$\left[\text{Hint: } H_x = -\frac{10^3 \sin(\omega t - \alpha z)}{120\pi}\frac{\text{A}}{\text{m}}; k = \sqrt{i\omega\mu(\sigma + i\omega\epsilon)}\right]$$

15.24 Find the electric field produced by a 2×10^2 W laser beam focused by a lens of cross section area 2×10^{-6} cm².
$$\left[\text{Hint: } I = \frac{P}{4\pi r^2}; E_0 = \sqrt{\frac{2I}{\epsilon_0 c}} \Rightarrow E_0^2 = \frac{I}{1.33 \times 10^{-3}}\right]$$

15.25 Find the amplitude of the electric and magnetic fields in a beam in vacuum if the laser has a power of 1.5 mW and a diameter of 2 mm.
$$\left[\text{Hint: } I = \frac{P}{4\pi r^2}; E_0 = \sqrt{\frac{2I}{\epsilon_0 c}} \Rightarrow E_0^2 = \frac{I}{1.33 \times 10^{-3}}\right]$$

PART IV

SOLID STATE PHYSICS

X-rays

Learning Objectives

After studying this chapter, students will be able to

- understand the lattice configuration of various crystals from the diffraction phenomenon exhibited by crystals, obtained using X-rays
- demonstrate crystal structure using Bragg's law
- understand large crystals opaque to X-rays, based on the Van Laue back reflection technique
- explain the Debye–Scherrer fine powder method used for crystals having relatively microscopic inter-planar spacing
- demonstrate the production of X-rays through various phenomena of fluorescence, Bremsstrahlung, and synchrotron radiation
- elucidate practical applications of X-rays, particularly in the areas of medicine, radiograph, computed tomography, angiography, fluoroscopy, radiotherapy, and crystallography

List of Symbols

d = Inter-planar spacing

θ = Diffraction angle

n = Integral number or order of diffraction

λ = Wavelength of the incident beam

k = Wave vector

R_B = Position vector of atom B with respect to A

φ = Amplitude of the incident beam

f = Atomic form factor

Δ = Phase difference caused by the displaced position of atom B relative to atom A

α, β, γ = Angles between the scattering normal (S) and the a, b, and c axes of the crystal, respectively

16.1 INTRODUCTION

X-rays are waves of electromagnetic radiation. The wavelength varies from 0.1 to 100 Å. As the X-ray are incident on a crystal the electrons present in the atoms act as scatterers. These produce secondary spherical waves pertaining to their placement in the atomic structure. The atoms are usually placed in a regular manner in the crystal. This scattering is also elastic in nature.

William Rontgen gave them the name 'X-rays', though many referred to these as 'Röntgen rays' (and the associated X-ray radiograms as 'Rontgenograms'). Actually, the name 'X-rays' implies that these rays could not be placed in the category of optical waves or any other rays at that time; it was only later that their origin was known.

These rays opened up a new era in our understanding of structures of crystals and bone structures.

16.2 DIFFRACTION OF X-RAYS

Crystal structures are determined by diffraction of X-rays. For this, one needs to know the physical basis of crystal diffraction, the scattering power of an atom, and the effect of lattice geometry.

Let us assume that a beam of X-rays is incident at an angle θ on a plane of atoms. The X-rays penetrate into the solids. A small fraction of X-rays is reflected back (obeying the laws of reflection) from the various parallel planes present in the atom. These reflected X-rays correspond to certain specific angles of incidence for which the phases of all the reflected waves are same (Fig. 16.1). The relation between the spacing of atomic planes in a crystal and the angles of incidence at which the crystal planes produce the most intense reflection of electromagnetic waves like X-ray is referred to as the Bragg's law and is mathematically expressed as follows:

$$2d \sin\theta = n\lambda \tag{16.1}$$

where d is the inter-planar spacing, θ is the diffraction angle, n is an integral number or order of diffraction, and λ is the wavelength of the X-ray beam.

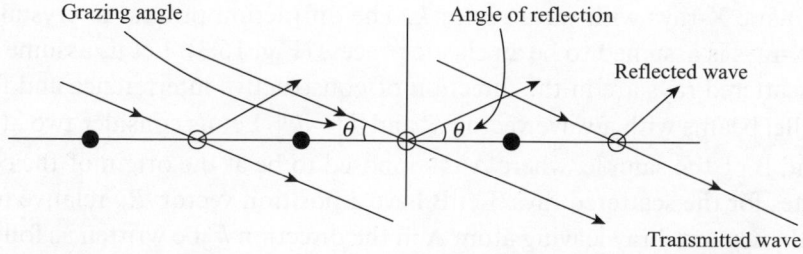

Fig. 16.1 X-rays reflect from parallel atomic planes of crystal; reflected X-rays interfere constructively, giving maximum intensities for certain angles of incidence

Hence, the wavelength of X-rays is chosen to be around 0.1–100 Å, which is comparable to the distance (d) between atoms in a crystal. As an X-ray photon is impinged on a crystal, secondary X-rays are produced, which have the same wavelength and phase as the incident X-ray radiation. The specific direction of the secondary X-rays appears in the form of a diffraction pattern. Thus, X-ray diffraction experiments provide the spacing between crystals accurately, aiding in understanding crystal structures with precision. This phenomenon of diffraction of X-rays establishes the wave nature of these rays. Later in this chapter, the dual behaviour of X-rays has been proved experimentally as well.

16.2.1 Rigorous Approach to Bragg's Law

X-rays are incident on lattice points A and B. Scattering of the incident plane wave (k) takes place from two atoms situated at positions A and B. The scattered wave has a component with wave vector k'. Scattering of X-rays by atoms may be elastic, that is, $|k| = |k'|$ or inelastic, that is, $|k| \neq |k'|$. The diffraction process is characterized by elastic scattering (Fig. 16.2). The vector R_B is the position vector of atom B with respect to atom A, which is assumed to be at the origin.

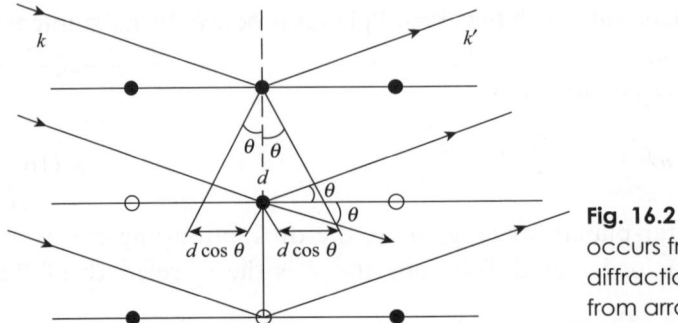

Fig. 16.2 Bragg reflection occurs from planar array; diffraction takes place from array of point scatters

A given sample of the crystal is irradiated by a collimated beam of monochromatic X-rays with wave vector k. The diffraction process in crystals by the X-rays is assumed to be an elastic process (Fig. 16.3). Let us assume that the scattered rays are in the direction of constructive interference and form parallel beams with a wave vector k' and $|k| = |k'|$. Let us consider two atoms A and B of the sample, where A is supposed to be at the origin of the coordinates for the scattered rays. Let B have a position vector R_B relative to A. Let the scattered ray leaving atom A in the direction k' be written as follows:

$$\frac{\phi}{r'} f e^{i(k'r' - \omega t)}$$

(16.2)

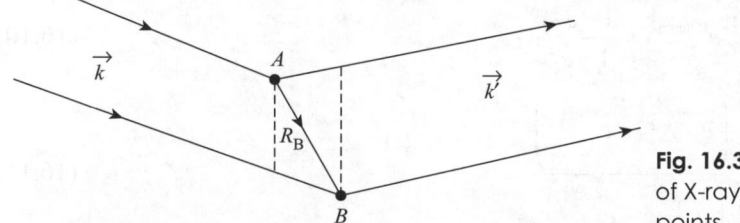

Fig. 16.3 Diffraction of X-rays from lattice points

where ϕ is the amplitude of the incident beam, f the atomic form factor, and r' the distance to the detector measured from A. The scattered amplitude from atom B can be written in the following form:

$$\frac{\phi}{r_B} f e^{i(k'r'-\omega t+\Delta)} \tag{16.3}$$

where $\Delta = R_B \cdot (k'-k) = R_B \cdot \Delta k$ $\tag{16.4}$

A phase difference is caused by the displacement of atom B relative to the position of atom A. If we consider any atom j in the sample at R_{Bj} relative to atom A and at a distance r_j from the detector, then this atom produces a signal amplitude at the detector and in the direction k', which can be written as follows:

$$\frac{\phi}{r_j} f e^{i(k'r'-\omega t+R_{Bj} \cdot \Delta k)} \tag{16.5}$$

Thus, the signal amplitude at the detector and in the direction k' arising from the total sample can be written as follows:

$$\sum_{\text{all atoms}} \frac{\phi}{r_j} f e^{i(k'r'-\omega t)} e^{(iR_{Bj} \cdot \Delta k)} \tag{16.6}$$

The term that contributes to the diffraction effects is given by the following expression:

$$\sum_{\text{all atoms}} e^{(iR_{Bj} \cdot \Delta k)} \tag{16.7}$$

which arises because atoms in different positions produce scattered waves with different phases. Let us suppose that the sample has atoms D, E, F such that the rectangular coordinates are expressed as follows:

$$R_{Bj} = da + eb + fc \tag{16.8}$$

Thus, we get the following expression:

$$\sum_{0}^{(d-1)(e-1)(f-1)} e^{i(da+eb+fc) \cdot \Delta k} \tag{16.9}$$

Considering one of the Cartesian coordinates, it is evident that this equation represents a geometric progression. Hence,

$$\sum_{0}^{(d-1)} e^{i(da)\cdot\Delta k} = \frac{1-e^{ida\cdot\Delta k}}{1-e^{ia\cdot\Delta k}} \tag{16.10}$$

$$\frac{e^{ida\cdot\frac{\Delta k}{2}}\left(e^{-ida\cdot\frac{\Delta k}{2}} - e^{ida\cdot\frac{\Delta k}{2}}\right)}{e^{ia\cdot\frac{\Delta k}{2}}\left(e^{-ia\cdot\frac{\Delta k}{2}} - e^{ia\cdot\frac{\Delta k}{2}}\right)} \tag{16.11}$$

The magnitude of this term is given as follows:

$$\frac{\sin\frac{1}{2}da\cdot\Delta k}{\sin\frac{1}{2}a\cdot\Delta k} \tag{16.12}$$

This term is the same as that we have observed in the case of a diffraction grating, where the maximum is given as follows:

$$\sin\frac{1}{2}a\cdot\Delta k = 0, \text{ that is, } \Delta k = n_1 A \text{ where } n_1 \text{ is an integral and}$$

the vector A is defined as

$$A = 2\pi\frac{b\times c}{a\cdot(b\times c)} \tag{16.13}$$

The aforementioned definition of Δk is a special case where the vector A has been defined in terms of the primitive vectors a, b, and c. This equation can be derived for the remaining ordinates as well by taking the general definition for Δk.

$$\Delta k = n_1 A + n_2 B + n_3 C \tag{16.14}$$

where n_1, n_2, n_3 are integral multiples.

Note: This vector can take the form of the reciprocal lattice which can be defined as follows:

$$G_{hkl} = hA + kB + lC \tag{16.15}$$

$$\Delta k = G_{hkl} = hA + kB + lC \tag{16.16}$$

As we expect that each of the diffracted beams is associated with a particular family of planes, and if 2θ represents the angle separating the incident and diffracted beams, then

$$|\Delta k| = 2|k|\sin\theta = |G_{hkl}| = \frac{2\pi}{d_{hkl}} \tag{16.17}$$

$$\Rightarrow 2d_{hkl}\sin\theta = \lambda \tag{16.18}$$

This is referred to as the Bragg's law. Thus, in terms of Bragg reflection, this is the result of constructive interference of rays diffracted from successive planes of a given family of planes (*hkl*), where in general, one can write the following equation:

$$n\lambda = 2d_{hkl}\sin\theta \tag{16.19}$$

where *n* is the order of diffraction.

Note: If we study the Miller indices, then one can introduce a triplet of lowest indices. The planes (100) are real planes, and all the planes of the form (*n*00), (*nn*0), and (*nnn*), *n* > 1, are not real and represent only hypothetical planes. Physically meaningful planes will be those that pass through the lattice points.

The planes (*nh nk nl*) have inter-planar spacing (d_{hkl}/n), so *n*th-order diffraction in planes (*h k l*) is equivalent to first-order diffraction in planes (*nh nk nl*). For all practical purposes, we observe that lattices (or crystals) always diffract X-rays, neutrons, electrons, or any other particles in the first order.

16.2.2 Experimental Verification of Bragg's Law—Van Laue Method

Von Laue first thought of using crystals as diffraction gratings. He illustrated effectively that X-rays, because these have shorter wavelengths, could be scattered from different atoms, producing well-defined spots forming a diffraction pattern. This provided the experimental verification for Bragg's law.

We assume a one-dimensional lattice, in which atoms are arranged in a line and act as diffraction centres. Let us assume that there are two identical scattering centres S and P, separated by a distance *r*. Let i_0 be the unit vector in the direction of the incident beam and d_0 be the unit vector in any arbitrary scattering direction, as shown in Fig.16.4. Two perpendiculars are dropped from the points S and P, respectively, on the incident and the diffracted (reflected) beam at the points I and D. The path difference between the radiation scattered at S and that scattered at P can be given as follows:

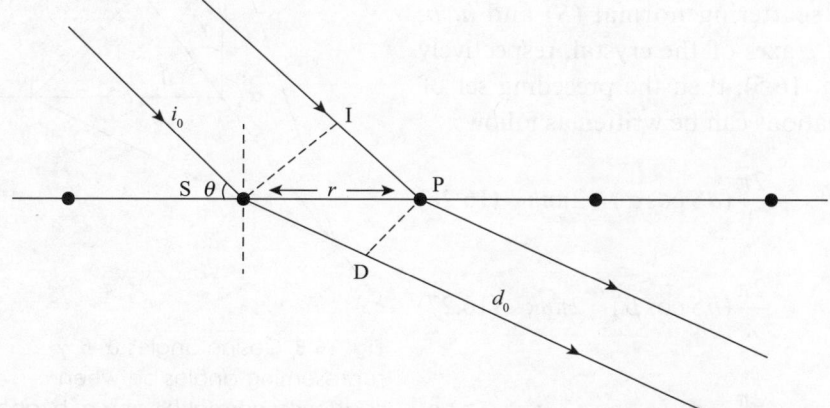

Fig. 16.4 Scattering of X-rays from one-dimensional lattice

$$PI - SD = r \cdot i_0 - r \cdot d_0 = r \cdot (i_0 - d_0) = r \cdot S \tag{16.20}$$

where $S = (i_0 - d_0)$ physically represents the direction of the normal to a plane that reflects the incident direction into the scattering direction. The Bragg's law defines this plane as a reflection plane. The mathematical relation between the vector S and the angle of incidence θ is given as follows:

$$|S| = 2 \sin \theta \tag{16.21}$$

The phase difference, as we know, is related to the path difference as follows:

$$\varphi_r = \frac{2\pi}{\lambda} (r \cdot S) \tag{16.22}$$

This equation represents the phase difference between the radiation scattered by the two points S and P in the crystal. We know, from our knowledge of diffraction analysis, that for a diffraction maximum, the phase difference should be an integral multiple of 2π. The primitive translation distances are a, b, and c for the respective nearest neighbour atoms. Hence,

$$\frac{2\pi}{\lambda} (a \cdot S) = 2\pi n h \tag{16.23}$$

$$\frac{2\pi}{\lambda} (b \cdot S) = 2\pi n k \tag{16.24}$$

$$\frac{2\pi}{\lambda} (c \cdot S) = 2\pi n l \tag{16.25}$$

where h, k, and l represent the three smallest integers identical to the Miller indices.

If α, β, γ represent the angles between the scattering normal (S) and a, b, and c axes of the crystal, respectively (Fig. 16.5), then the preceding set of equations can be written as follows:

$$\frac{2\pi}{\lambda} (aS \cos \alpha) = 2\pi n h \tag{16.26}$$

$$\frac{2\pi}{\lambda} (bS \cos \beta) = 2\pi n k \tag{16.27}$$

$$\frac{2\pi}{\lambda} (cS \cos \gamma) = 2\pi n l \tag{16.28}$$

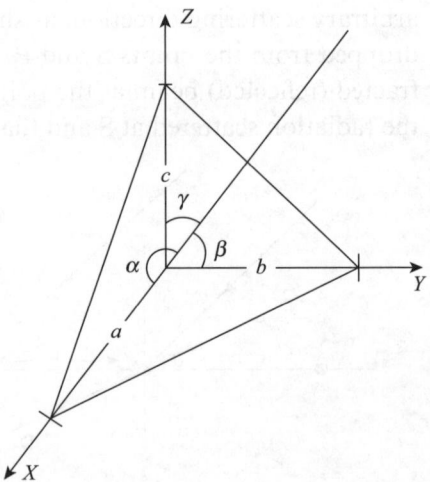

Fig. 16.5 Cosine angles α, β, γ representing angles between scattering normal (S) and a, b, and c axes of crystal, respectively

These equations can further be simplified to the following forms:

$$(2a\sin\theta\cos\alpha) = nh\lambda \tag{16.29}$$

$$(2b\sin\theta\cos\beta) = nk\lambda \tag{16.30}$$

$$(2c\sin\theta\cos\gamma) = nl\lambda \tag{16.31}$$

These three equations together are referred to as the Laue equations. The importance of these equations is that they define the unique value for the diffraction direction θ and the scattering normal S. The angles or directional cosines, $\cos\alpha$, $\cos\beta$, $\cos\gamma$, in a orthogonal coordinate system are related as follows:

$$\cos^2\alpha + \cos^2\beta + \cos^2\gamma = 1 \tag{16.32}$$

These direction cosines of the normal to the (h, k, l) family are proportional to $\left(\dfrac{n}{a}, \dfrac{k}{b}, \dfrac{l}{c}\right)$. Thus, one can easily derive that the scattering normal S is identical to the normal to the (h, k, l) planes. Thus, the (h, k, l) planes are the reflecting planes in the Bragg's treatment also. Hence, the preceding equations can be modified to represent the inter-planar spacing of a family of (h, k, l) planes that are related to the direction cosines as follows:

$$d = \frac{a}{h}\cos\alpha = \frac{a}{k}\cos\beta = \frac{a}{l}\cos\gamma \tag{16.33}$$

Hence, $2d\sin\theta = n\lambda$ (16.34)

which is the Bragg's equation. Thus, the Laue equations provided the experimental basis for the Bragg's equation. The integer n is called the order of diffraction. The diffraction order indicates the reflection plane, which is represented as (h, k, l); for example, for $n = 1$, reflection from the (111) planes is referred to as the (111) reflection; for $n = 2$, for the (111) plane, it is the (222) reflection; for $n = 3$, for the (111) plane, it is the (333) reflection; and so on.

16.3 BRAGG'S SPECTROMETER

It was the prevailing thought of solid-state physicist Max Von Laue that crystals should diffract the incident radiation, thus depicting both wave and particle nature that propelled William Bragg to design an instrument for the measurement of this angle (Fig. 16.6).Bragg's spectrometer has two fine slits S_1 and S_2 cut into two lead plates L_1 and L_2. The crystal is mounted on a turntable (TT), which is capable of rotation read from the circular table (S). The X-ray beam enters the ionization chamber, which is connected to an electrometer (E). The ionization curve is plotted against the incident angle θ.

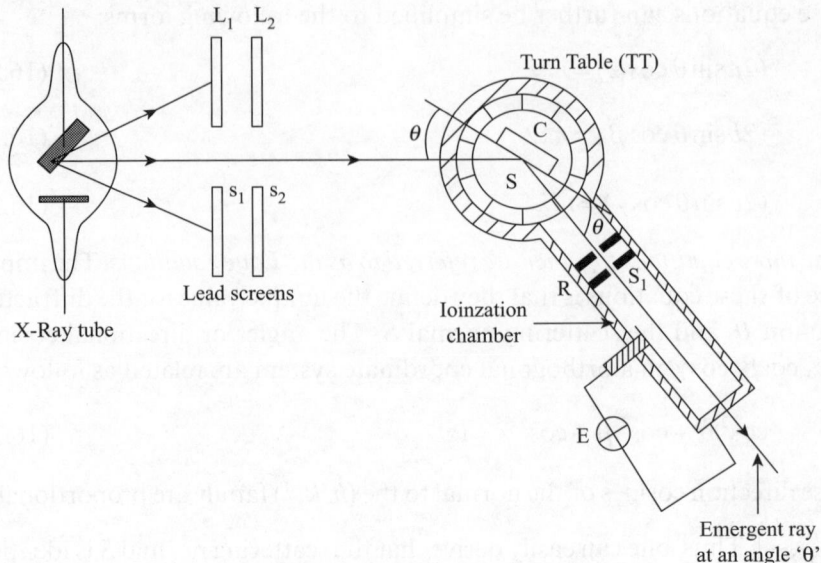

Fig.16.6 Bragg's X-ray spectrometer

16.3.1 Principle

Bragg's spectrometer was designed on the principle that if the incident radiation has a wavelength of the order of the inter-atomic spacing of the crystal (under study), then the angle of diffraction due to the crystal lattice sites is observable.

16.3.2 Working

The following steps explain how a Bragg's spectrometer works:
1. X-rays possess the wavelength that is comparable to the inter-atomic spacing of a crystal lattice site.
2. The X-ray tube is operated by an induction coil that provides high voltage. The X-ray source is an anti-cathode or a target made of a heavy metal (platinum, palladium, rhodium, or osmium). The X-ray tube was later modified according to the design of a Coolidge tube.
3. The X-ray target is bombarded with ions of the gas present in the X-ray tube.
4. X-rays pass through a hole in the lead screen and strike the crystal mounted on the graduated turntable.
5. X-rays reflected from the crystal pass through a slit of variable width into an ionization chamber. Intensity of the reflected radiation is recorded.
6. The intensity so recorded then explains the crystal structure.

Example 16.1 An X-ray radiation of wavelength 2.0Å is incident on a common salt crystal with a *d* spacing of 3.0 Å. What is the highest order that the crystal can diffract?

Solution   The Bragg's law states that

$$\sin\theta = \frac{n\lambda}{2d_{hkl}}$$

One can observe that the value of $\sin\theta$ cannot exceed 1; hence, we can write the following statement:

$$1 = \frac{n\cdot(2\times10^{-8}\,\text{cm})}{2\times(3\times10^{-8}\,\text{cm})} = \frac{n}{3} \Rightarrow n = 3$$

The highest order for which the crystal can diffract is 3.

Example 16.2   An X-radiation of wavelength 1.0 Å strikes a crystal of d spacing 2.0 Å. Estimate the diffraction angle for the second order.

Solution   The diffraction angle can be calculated from the following equation:

$$\sin\theta = \frac{n\lambda}{2d_{hkl}} = \frac{(2\times1)}{(2\times2)} = \frac{1}{2} \Rightarrow \theta = 33°$$

Example 16.3   Find the wavelength of the first-order Bragg reflection if the glancing angle is nearly 50°, assuming that the grating spacing is 8.0 Å.

Solution   The Bragg's law is given as follows:

$$2d_{hkl}\sin\theta = n\lambda \quad (2\times8.0)\sin50° = 16\times0.8 = 1\times\lambda$$

$$\lambda = 12.8°\text{Å}$$

Example 16.4   If the potential difference applied across an X-ray tube is 4 kV and the current flowing through it is 4 mA, calculate the following parameters:
(a) The number of electrons striking the target per second
(b) The speed with which they strike

Solution   The calculations are as follows:
(a) $I = ne$, where n is the number of electrons striking the anode per second.

$$n = \frac{I}{e} = \frac{4\times10^{-3}}{1.6\times10^{-19}} = 2.5\times10^{16}\,\text{electrons}$$

(b) The speed v is the velocity of the striking electron and is determined as follows:

$$\frac{1}{2}mv^2 = eV$$

$$v = \sqrt{\frac{2eV}{m}} = \sqrt{\frac{2\times1.6\times10^{-19}\times4\times10^3}{16.1\times10^{-31}}} = 3.75\times10^7\,\text{m/s}$$

Example 16.5   Find the Duane–Hunt wavelength (minimum wavelength) of X-rays generated from a Coolidge tube that operates at 40 kV.

Solution   As we know,

$$\frac{hc}{\lambda_{min}} = eV$$

$$\Rightarrow \lambda_{min} = \frac{hc}{eV} = \frac{6.62 \times 10^{-34} \text{ joules-s } 3 \times 10^8 \text{ m/s}}{1.6 \times 10^{-19} \text{°C}} \frac{1}{V} = \frac{12.41 \times 10^{-7-3+3} \text{ m}}{V}$$

$$\lambda_{min} = \frac{12,400}{V} \text{ Å} = \frac{12,400}{40 \times 10^3} \text{ Å} = 310 \times 10^{-3} = 0.31 \text{Å}$$

Example 16.6 Calculate the longest wavelength that can be analysed by a sodium chloride crystal of spacing $d = 2.80$ Å in the first order.

Solution According to the Bragg's equation,

$$2d \sin\theta = n\lambda$$

For the longest wavelength, $(\sin \theta)_{max} = 1$. Hence,

$$\lambda_{max} = \frac{2d}{n} = \frac{2 \times 2.80 \times 10^{-10} \text{ m}}{1} = 5.60 \times 10^{-10} \text{ m} = 5.60 \text{Å}$$

Example 16.7 X-ray analysis of a crystal is made with monochromatic X-rays of wavelength 0.58Å. Bragg reflections are obtained at angles of (a) 0.1126 rad (b) 0.1596 rad, and (c) 0.2268 rad. Calculate the inter-planar spacing of the crystal.

Solution According to the Bragg's equation,

$$2d \sin\theta = n\lambda \Rightarrow \frac{d}{n} = \frac{\lambda}{2 \sin\theta}$$

(a) $\dfrac{d}{n} = \dfrac{0.58 \text{Å}}{2 \sin(0.1126)} = \dfrac{0.58 \text{Å}}{2 \times 0.1126} = 2.568 \text{Å}$

(b) $\dfrac{d}{n} = \dfrac{0.58 \text{Å}}{2 \sin(0.1596)} = \dfrac{0.58 \text{Å}}{2 \times 0.1596} = 1.817 \text{Å}$

(c) $\dfrac{d}{n} = \dfrac{0.58 \text{Å}}{2 \sin(0.2268)} = \dfrac{0.58 \text{Å}}{2 \times 0.2268} = 1.288 \text{Å}$

16.4 POWDER METHOD

Some crystals are micro-crystalline in dimensions and referred to as micro-crystals. The Bragg's law helps us understand the structure of the micro-crystal. The principle of a Bragg's spectrometer is to observe the diffraction of X-rays from a single crystal (free of any distortions) at lattice planes oriented to satisfy the Bragg's law.

The powder crystal method [Fig.16.7(a)] is a standard technique used to study the structure of a micro-crystal or the powder form of the crystal. In the experimental arrangement, a fine pencil of monochromatic X-rays is generated, by passing the X-rays produced in an X-ray tube through a filter F that absorbs all the wavelengths except one. This monochromatic X-ray is collimated by passing through two fine slits S_1 and S_2 cut in two lead plates [Fig.16.7(b)]. This monochromatic beam of X-rays is made to fall on the powdered crystal specimen.

The powdered crystal is coated on a glass fibre. These finely powdered crystals have all the possible random orientations. There are many micro-crystals in the powdered crystal. Some of these micro-crystals have their crystal planes oriented to satisfy the Bragg's law. Reflection of the incident monochromatic radiation takes place from all the [$h\,k\,l$] planes, satisfying the following equation:

$$\theta_{hkl} = \sin^{-1}\left(\frac{\lambda}{2d_{hkl}}\right) \le \frac{1}{2}\pi \tag{16.35}$$

As the incident X-ray beam is impinged on an axis of symmetry, each set of planes produces a cone of diffracted rays. The cones intercept the film in a series of concentric rings, with a circular spot at the centre corresponding to the undeflected beam.

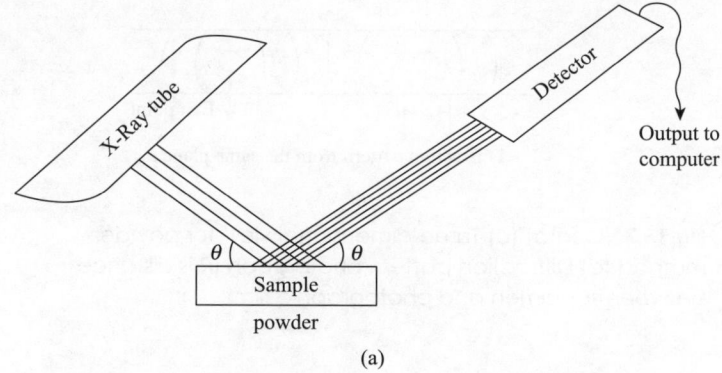

(a)

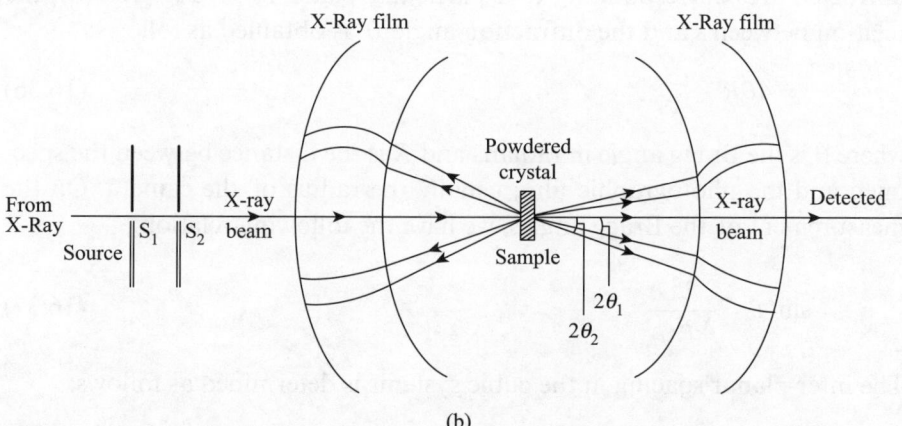

(b)

Fig.16.7 X-ray powder diffraction (a) Powder method (b) Experimental X-ray powder diffraction apparatus (*contd*)

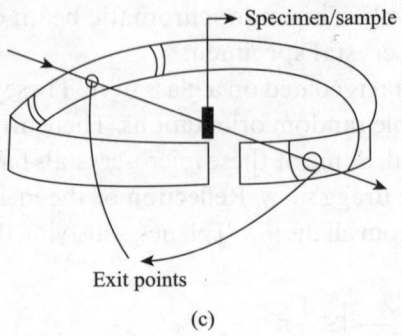

Specimen/sample

Exit points

(c)

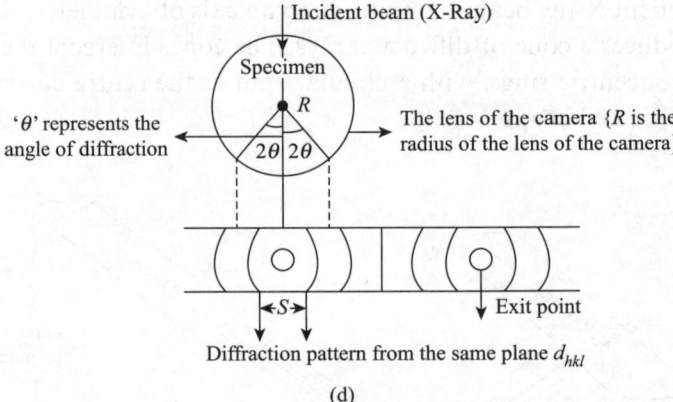

Incident beam (X-Ray)

Specimen

R

'θ' represents the
angle of diffraction

2θ 2θ

The lens of the camera {R is the
radius of the lens of the camera}

←S→

Exit point

Diffraction pattern from the same plane d_{hkl}

(d)

Fig.16.7 (Contd) (c) Three-dimensional view for powder
method (d) Diffraction pattern photograph (R is distance
between specimen and photographic film)

In the diffraction pattern, let s be the distance on the film between the diffracted arcs corresponding to a particular plane. From Fig. 16.7(d), the relation between s and the diffraction angle θ is obtained as follows:

$$s = 4\theta R \qquad (16.36)$$

where θ is the Bragg angle in radians and R is the distance between the specimen and the photographic film, usually the radius of the camera. On the measurement of the Bragg angles, we have the following relation:

$$\sin^2 \theta = \frac{\lambda^2}{4d^2_{hkl}} \qquad (16.37)$$

The inter-planar spacing in the cubic systems is determined as follows:

$$\sin^2 \theta = \frac{(h^2 + k^2 + l^2)\lambda^2}{4a^2} \qquad (16.38)$$

The values of $(h^2 + k^2 + l^2)$ depend on the crystal structure, a list of the mea-sured $\sin^2\theta$, and $(\lambda^2/4a^2)$, which allow a to be evaluated.

The Debye–Scherrer powder method has the following uses:

1. It can determine the lattice spacing with an accuracy of about one in 10^4.
2. It adapts to a temperature range that varies from low to high values.
3. This method helps in the identification of alloy phases and determination of phase boundaries.
4. It helps in the studies of phase transformations.
5. Structure determination of simple cases can be done using this method.

16.5 PRODUCTION OF X-RAYS

A high voltage is applied to a cathode to release high-velocity electrons in a vacuum tube, which strike a metallic target, an anode, thus producing X-rays (Fig. 16.8). The maximum energy that can be produced by an X-ray photon is governed by the incident electron. Energy of the X-rays can be ascertained

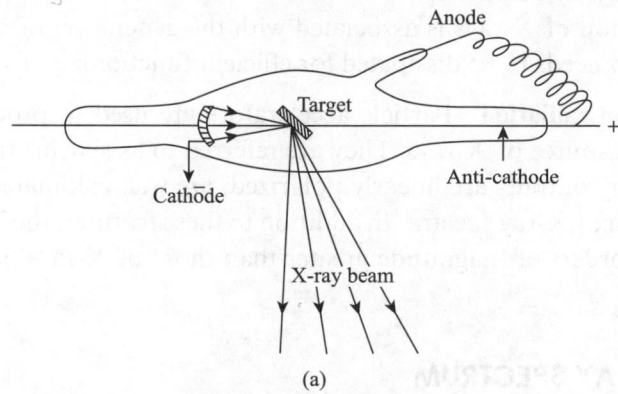

(a)

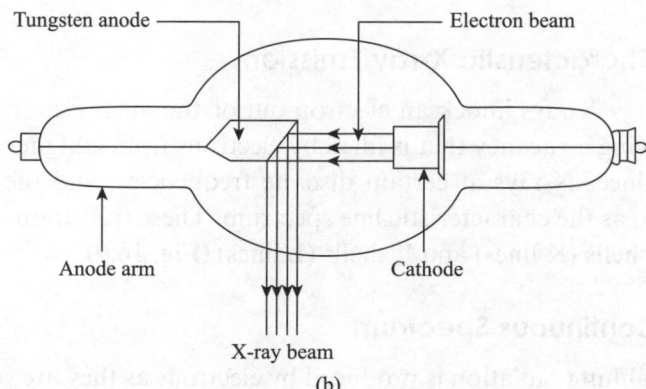

(b)

Fig.16.8 Production of X-rays (a) Using gas-filled tube (b) Using typical Coolidge tube

by the voltage times the electronic charge mentioned on the X-ray tube; for example, for a 50 kV tube, X-rays produced are of the energy of 50 keV.

There are three major methods for producing X-rays, which are discussed in the following sections.

X-ray fluorescence method   In the process of fluorescence, an electron having sufficient energy is knocked out of the innermost shell of an atom, resulting in the production of X-ray photons at discrete frequencies corresponding to the transitions, wherein electrons from a higher energy level fall to a lower energy level to fill up the vacancy that has been created. These characteristic discrete frequencies are referred to as K lines (corresponding to transitions from upper shells to K shells), L lines (corresponding to transitions from upper shells to L shell), and so on. These characteristic lines are different for different anode elements.

Bremsstrahlung method   Elements possessing a high proton number tend to scatter electrons due to the generation of a strong electric field near their nuclei. The X-rays so produced are continuous, with their intensity increasing linearly as the frequency decreases.

Production of X-rays is associated with the generation of a lot of waste heat, which needs to be dissipated for efficient functioning of the tube.

Synchrotron radiation   Particle accelerators are used to produce a highly specialized source of X-rays. They are referred to as synchrotron radiation. These X-ray outputs are linearly polarized, are well collimated, and have a wide region of X-ray spectra. In addition to these features, the X-ray outputs are many orders of magnitude greater than those of X-rays produced from tubes.

16.6 X-RAY SPECTRUM

The characteristics of X-rays are discussed in the following sections.

16.6.1 Characteristic X-ray Emission

High-energy X-rays knock an electron out of the inner electron shell of an atom, leaving a vacancy that is filled by electrons from a higher energy level. This produces X-rays at certain discrete frequencies, and the spectrum is referred to as the characteristic line spectrum. These transitions are from the upper K shells (K lines) and L shells (L lines) (Fig. 16.9).

16.6.2 Continuous Spectrum

Bremsstrahlung radiation is produced by electrons as they are scattered by a strong electric field near the high-Z proton number nuclei. These are continuous in nature (Fig. 16.10).

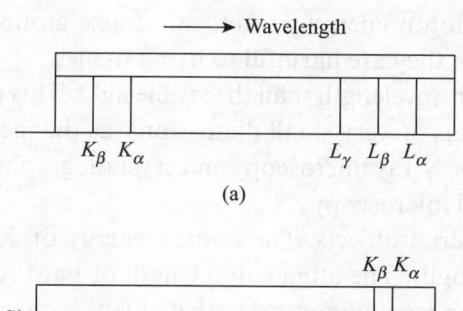

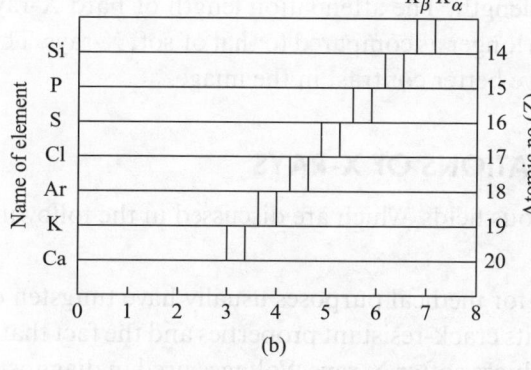

Fig.16.9 Characteristic spectrum for X-rays (a) Characteristic X-ray spectrum of given element (b) K-series of neighbouring elements in periodic table

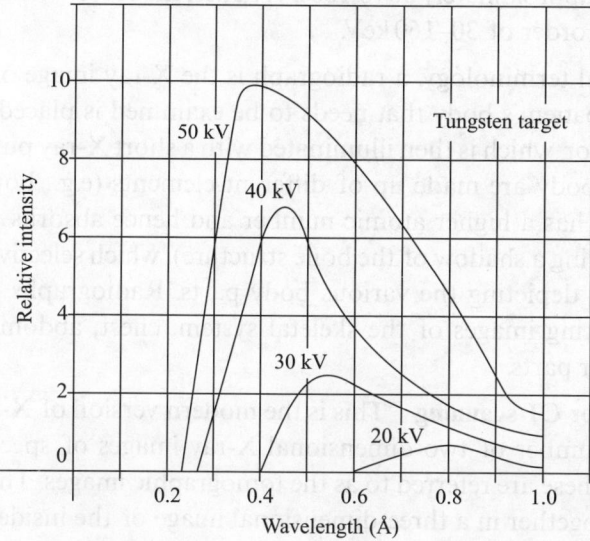

Fig. 16.10 Continuous X-ray spectra at different voltages for tungsten target

16.6.3 Properties of X-rays

The following are the properties of X-rays:

1. X-rays possess photon energies in the range of $100\,\text{eV}-100\,\text{keV}$ corresponding to wavelengths in the range of $0.01-10\,\text{nm}$, with frequencies in the range from 30 petahertz ($\sim10^{16}\,\text{Hz}$) to exahertz ($\sim10^{19}\,\text{Hz}$). X-rays with energy in the range of $5-10\,\text{keV}$ are called hard rays and those with energy $100\,\text{eV}-1\,\text{keV}$ are called soft rays.

2. X-rays, by virtue of their photon energy packet, can ionize atoms and break molecular bonds. Thus, they are harmful to living tissues.

3. X-rays possess a much shorter wavelength than the visible light. This property helps us to study materials of very small dimensions, as the incident radiation used is X-rays. Thus, X-ray microscopy and crystallography give better resolution than optical microscopy.

4. X-rays are absorbed by material objects. The photon energy of X-rays determines its attenuation length. The attenuation length of hard X-rays is four orders of magnitude longer as compared to that of soft X-rays. This property helps us to observe better contrast in the image.

16.7 PRACTICAL APPLICATIONS OF X-RAYS

X-rays have been used in various fields, which are discussed in the following subsections.

Medicines   X-ray tubes used for medical purposes usually have tungsten or molybdenum as the target for its crack-resistant properties and the fact that it (especially molybdenum) produces softer X-rays. Voltages used in diagnostic X-ray tubes range from approximately 30 to 160 kV; hence, the corresponding photon energy is of the order of 30–160 keV.

Radiograph   In medical terminology, a radiograph is the X-ray image of a patient. A part of the patient's body that needs to be examined is placed in front of an X-ray detector, which is then illuminated with a short X-ray pulse. Different parts of the body are made up of different elements (e.g., bones possess calcium, which has a higher atomic number and hence absorbs the incident radiation, creating a shadow of the bone structure), which selectively absorb the X-radiation depicting the various body parts. Radiographs are commonly used for taking images of the skeletal system, chest, abdomen, tooth cavities, and other parts.

Computed tomography or CT scanning   This is the modern version of X-ray radiography. A large number of two-dimensional X-ray images of specific body parts are taken. These are referred to as the tomographic images. These images are combined together in a three-dimensional image of the inside of the body, which is referred to as the CT scan of that particular body part.

Angiography   Angiography is basically an image of the cardiovascular system that includes arteries and veins. During this process, a first image is taken of the area of interest. In the second image, an iodinated contrast agent is injected into the blood vessels of the same area. These two images are then digitally subtracted, producing an image in which the iodinated contrast outlines the blood vessels. These images are then compared to the normal anatomical images to determine if there has been any damage or blockade in the artery.

Fluoroscopy In a fluoroscope, there is an X-ray source and a fluorescent screen. In the updated versions, there may be an X-ray image intensifier or a CCD video camera to produce real-time moving images of the patients' specific body part. Some common applications of fluoroscopy are in cardiac catheterization, which is used to examine coronary artery blockages, and in barium swallow, which is used to examine oesophageal disorders.

Radiotherapy It is a common modern-day treatment for the management of cancerous cells in specific body parts. Cancerous cells should be irradiated with high-energy radiations. However, exposure to this high-energy radiation can result in many adverse effects on the body. X-rays have been declared carcinogenic to the human body. Thus, radiation therapy is performed strictly under medical advice.

Crystallography Each dot, called a reflection, in this diffraction pattern is a result of the constructive interference of scattered X-rays passing through a crystal. These data can be used to determine the crystalline structure.

16.8 BIOLOGICAL APPLICATIONS OF X-RAYS

Biological effects of radiation depend upon how the energy is absorbed by the body and how tissues react to the different forms of radiation.

Exposure is a measure of the ionizing ability of X-rays. The various units associated with X-rays are as follows:

1. Coulomb per kilogram (C/kg) is the SI unit of the ionizing radiation exposure, and it is the amount of radiation required to create 1 C of charge of each polarity in 1 kg of matter.
2. Roentgen (R) is an obsolete traditional unit of exposure, which represents the amount of radiation required to create one electrostatic unit of charge of each polarity in one cubic centimetre of dry air. $1 R = 2.58 \times 10^{-4} °C/kg$.

The effect of ionizing radiation on matter (especially living tissues) is more closely related to the amount of energy deposited into them, than to the charge generated. The measure of the energy absorbed is called the absorbed dose; the units used to measure this parameter are as follows:

1. Gray (Gy), which has units of joules/kilogram, is the SI unit of the absorbed dose, and it is the amount of radiation required to deposit 1 J of energy in 1 kg of any kind of matter. Thus, 1 gray $= 1 Gy = \dfrac{1.00}{kg}$ of absorbed energy
2. Rad is the (obsolete) corresponding traditional unit, equal to 10 mJ of energy deposited per kilogram. 100 rad = 1 Gy.

The absorbed dose depends only on the energy absorbed, irrespective of the type of radiation or absorbing material. It is also observed that the biological consequences of 1 Gy dose of gamma rays are quite different from those of 1 Gy dose of alpha rays. The term relative biological effectiveness

(RBE) is defined as the biological effect of a given dose relative to the biological effect of an equal dose of X-rays.

3. Roentgen equivalent man (rem) is the traditional unit of the equivalent dose. For X-rays, it is equal to the rad or, in other words, 10 mJ of energy deposited per kilogram. 100 rem = 1 Sv.

4. Sievert (Sv) is the SI unit of the equivalent dose, and also of the effective dose. For X-rays, the 'equivalent dose' is defined as follows:

Dose equivalent in Sv = Absorbed dose in Gy × RBE

One Sievert of all types of radiations produces the same biological damage, regardless of the type of radiation. The exposure to the X-ray radiation varies with the medical diagnostic technique (refer to Table 16.1). The safe dose of radiations is a topic of concern and debate, as far as the human body is concerned.

Table 16.1 Typical X-ray exposure for medical diagnostic techniques

Radiation source	Typical exposure (mrem)
CT scan	1000
Natural background (1 year)	300
Mammogram X-ray	80
Chest X-ray	30
Dental X-ray	03

16.9 COMPTON EFFECT

The classical theory was unable to explain X-ray scattering. Predictions of the classical theory are as follows:

1. A scattered X-ray should have the same wavelength as that of the incident one.

2. The scattering coefficient (σ) should be independent of the wavelength of the incident radiation, having a constant value of 0.2.

3. The scattered radiation should be symmetrical with regard to the distribution of intensity.

Observed experimental results highlighted the following facts:

1. The scattered radiation was found to possess a greater wavelength than that of the incident radiation.

2. The scattering coefficient (σ) varied with the wavelength of the incident radiation, diminishing as the wavelength decreased.

3. Distribution of the scattered radiations was not symmetrical, scattering taking place practically in the forward direction, that is, the same direction as that of the incident radiation.

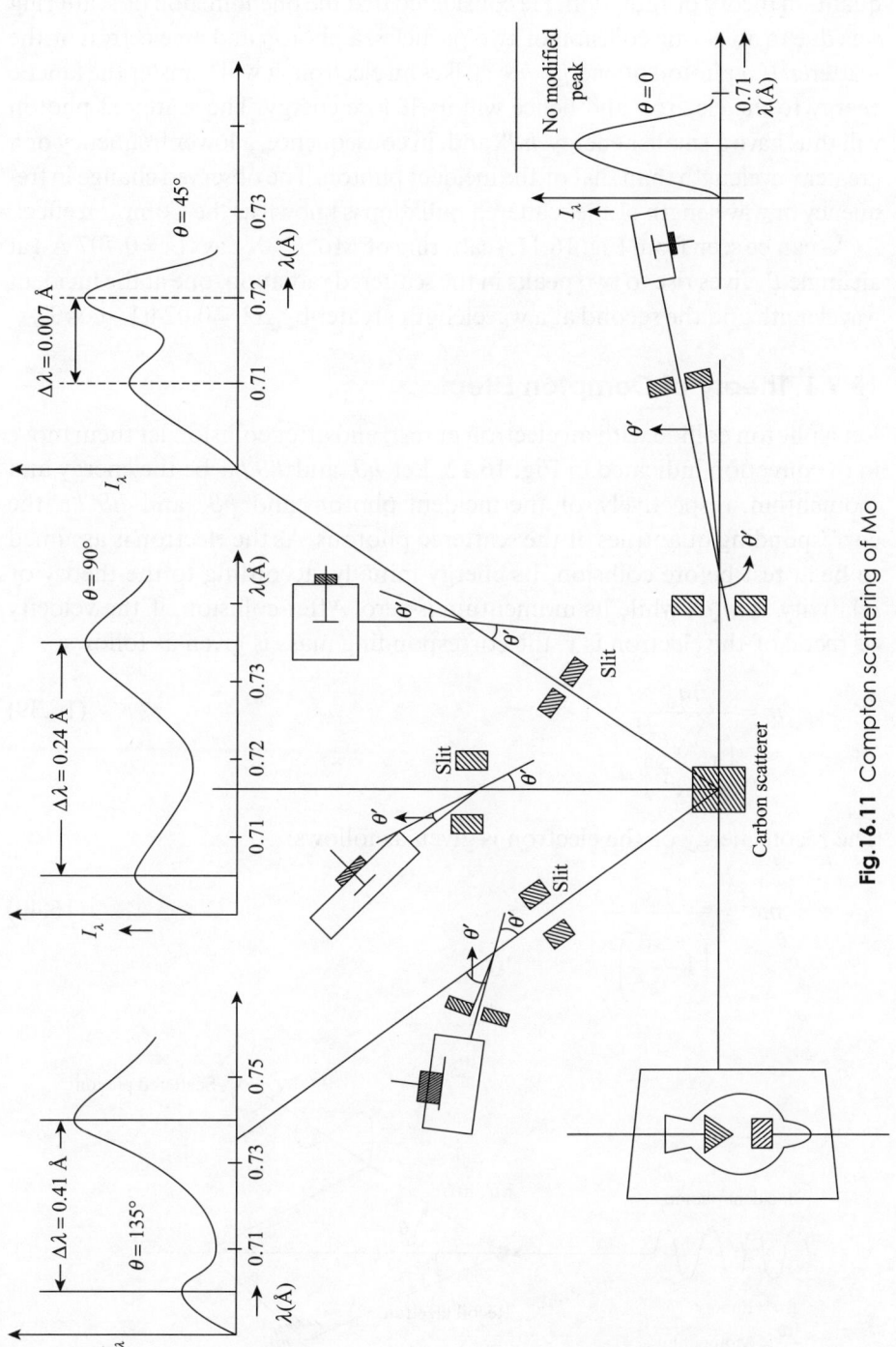

Fig.16.11 Compton scattering of Mo

Compton proposed an adequate explanation of these facts on the basis of the quantum theory of radiation. He considered that the phenomenon of scattering was due to an elastic collision of two particles, a photon and an electron of the scatterer. If a photon of energy $h\vartheta$ strikes an electron, it will transfer the kinetic energy to the electron and hence will itself lose energy. The scattered photon will thus have a smaller energy $h\vartheta'$ and, in consequence, a lower frequency or a greater wavelength than that of the incident photon. The observed change in frequency or wavelength of the scattered radiation is known as the Compton effect.

As can be seen from Fig. 16.11, scattering of Mo K_α X-rays ($\lambda = 0.707$ Å) at an angle θ gives rise to two peaks in the scattered radiation, one at the incident wavelength and the second at a wavelength greater by $\Delta\lambda = 0.024(1 - \cos\theta)$Å.

16.9.1 Theory of Compton Effect

Let a photon collide with an electron at rest, and after collision let them move in the direction indicated in Fig. 16.12. Let $h\vartheta$ and $h\vartheta/c$ be the energy and momentum, respectively, of the incident photon, and $h\vartheta'$ and $h\vartheta'/c$ the corresponding quantities of the scattered photons. As the electron is assumed to be at rest before collision, its energy initially, according to the theory of relativity, is m_0c^2, while its momentum is zero. After collision, if the velocity of recoil of the electron is v, the corresponding mass is given as follows:

$$m = \frac{m_0}{\left(1 - \dfrac{v^2}{c^2}\right)^{\frac{1}{2}}} \tag{16.39}$$

The recoil energy of the electron is given as follows:

$$mc^2 = \frac{m_0c^2}{\left(1 - \dfrac{v^2}{c^2}\right)^{\frac{1}{2}}} \tag{16.40}$$

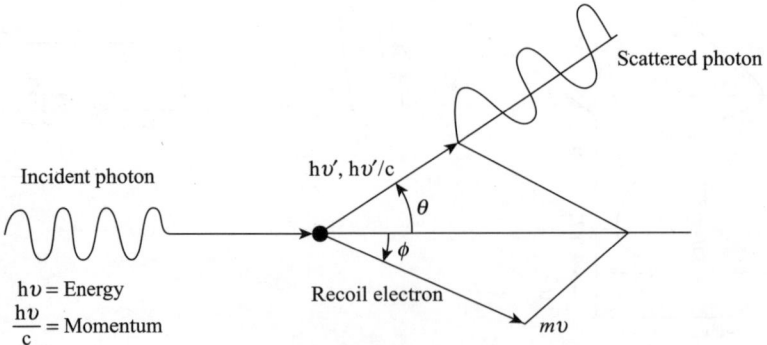

Fig.16.12 Photon of energy $h\vartheta$ and momentum $h\vartheta/c$ colliding with electron at rest

The momentum associated with the electron is expressed as follows:

$$mv = \frac{m_0 v}{\left(1 - \dfrac{v^2}{c^2}\right)^{\frac{1}{2}}} \tag{16.41}$$

Let θ and φ be the angles that the scattered photon and recoil electron make, respectively, with the direction of the incident photon. Applying the principles of conservation of energy [Eq. (16.42)] and momentum [along x and y components, respectively; Eqs (16.43) and (16.44)], we get the following expressions:

$$h\vartheta + m_0 c^2 = h\vartheta' + mc^2 \tag{16.42}$$

$$\frac{h\vartheta}{c} + 0 = \frac{h\vartheta'}{c} + \cos\theta + mv\cos\varphi \tag{16.43}$$

$$0 = \frac{h\vartheta}{c}\sin\theta - mv\sin\varphi \tag{16.44}$$

Eliminating φ from Eqs (16.43) and (16.44), respectively,

$$mvc\cos\varphi = h(\vartheta - \vartheta'\cos\theta) \tag{16.45}$$

$$mvc\sin\varphi = h\vartheta'\sin\theta \tag{16.46}$$

Squaring and adding these two equations, one gets the following expressions:

$$m^2 v^2 c^2 = h^2(\vartheta^2 - 2\vartheta\vartheta'\cos\theta + \vartheta'^2\cos^2\theta + \vartheta^2\sin^2\theta) \tag{16.47}$$

$$m^2 v^2 c^2 = h^2(\vartheta^2 - 2\vartheta\vartheta'\cos\theta + \vartheta'^2)$$

$$mc^2 = h^2(\vartheta^2 - 2\vartheta\vartheta'\cos\theta + \vartheta'^2) \tag{16.48}$$

$$mc^2 = h\left((\vartheta - \vartheta') + m_0 c^2\right)$$

$$m^2 c^4 = h^2(\vartheta^2 - 2\vartheta\vartheta' + \vartheta'^2) + 2h(\vartheta - \vartheta')m_0 c^2 + m_0^2 c^4 \tag{16.49}$$

Subtracting Eq. (16.48) from Eq. (16.49), one gets the following expressions:

$$m^2 c^2(c^2 - v^2) = -2h^2\vartheta\vartheta'(1 - \cos\theta) + 2h(\vartheta - \vartheta')m_0 c^2 + m_0^2 c^4 \tag{16.50}$$

$$= -2h^2\vartheta\vartheta'(1 - \cos\theta) + 2h(\vartheta - \vartheta')m_0 c^2 + m_0^2 c^4$$

$$m^2 c^2(c^2 - v^2) = \frac{m_0^2 c^2}{\left(1 - \dfrac{v^2}{c^2}\right)^{\frac{1}{2}}} \cdot \frac{1}{\left(1 - \dfrac{v^2}{c^2}\right)^{\frac{1}{2}}}(c^2 - v^2)$$

$$m_0^2 c^4 = -2h^2\vartheta\vartheta'(1 - \cos\theta) + 2h(\vartheta - \vartheta')m_0 c^2 + m_0^2 c^4$$

$$2h^2 \vartheta \vartheta'(1 - \cos\theta) = 2h(\vartheta - \vartheta')m_0 c^2$$

$$\frac{\vartheta - \vartheta'}{\vartheta \vartheta'} = \frac{h}{m_0 c^2}(1 - \cos\theta)$$

$$\frac{1}{\vartheta'} - \frac{1}{\vartheta} = \frac{h}{m_0 c^2}(1 - \cos\theta)$$

$$= \frac{m_0 c^2 + h\vartheta(1 - \cos\theta)}{\vartheta m_0 c^2}$$

$$\vartheta' = \frac{m_0 c^2 \vartheta}{m_0 c^2 + h\vartheta(1 - \cos\theta)}$$

$$\vartheta' = \frac{\vartheta}{1 + \dfrac{h\vartheta}{m_0 c^2}(1 - \cos\theta)} \tag{16.51}$$

Substituting the value of $\alpha = \dfrac{h\vartheta}{m_0 c^2}$; and $(1 - \cos\theta) = 2\sin^2(\theta/2)$, we get the following equation:

$$\vartheta' = \frac{\vartheta}{1 + 2\alpha \sin^2(\theta/2)} \tag{16.52}$$

That is, $\vartheta'^2 < \vartheta$.

The *change in wavelength* is obtained as follows:

$$\frac{1}{\vartheta'} - \frac{1}{\vartheta} = \frac{h}{m_0 c^2}(1 - \cos\theta)$$

$$\frac{c}{\vartheta'} - \frac{c}{\vartheta} = \frac{h}{m_0 c^2}(1 - \cos\theta) \tag{16.53}$$

$$\lambda' = \lambda + \frac{2h}{m_0 c^2}\sin^2(\theta/2) \tag{16.54}$$

That is, $\lambda'^2 < \lambda$.

Considering the complementary phenomenon of the recoil electron that, according to the initial assumption, is produced in the scattering process and acquires a kinetic energy equal to the amount of energy lost by the photon, for example, $(h\vartheta - h\vartheta')$, the expression for the 'energy of the recoil electron' is derived as follows.

From Eqs (16.43)–(16.44), one gets the following expressions:

$$(h\vartheta - h\vartheta' \cos\theta)\frac{1}{c} = mv\cos\varphi \tag{16.55}$$

$$\frac{h\vartheta'}{c}\sin\theta = mv\sin\varphi \qquad (16.56)$$

$$\tan\varphi = \frac{h\vartheta'\sin\theta}{h\vartheta - h\vartheta'\cos\theta}$$

Thus, $\tan\varphi = \dfrac{\vartheta'\sin\theta}{\vartheta - \vartheta'\cos\theta}$

Substituting from Eq. (16.50), we get the following relation:

$$\tan\varphi = \frac{\vartheta'\sin\theta}{\vartheta - \vartheta\,\cos\theta}$$

$$\tan\varphi = \frac{\vartheta\sin\theta / \left(1 + 2\alpha\sin^2\left(\theta/2\right)\right)}{\vartheta - \left(\dfrac{\vartheta\cos\theta}{1 + 2\alpha\sin^2\left(\theta/2\right)}\right)} = \frac{\sin\theta}{1 + 2\alpha\sin^2\left(\theta/2\right) - \cos\theta}$$

$$\tan\varphi = \frac{1}{(1+\alpha)\tan(\theta/2)} \qquad (16.57)$$

Thus, we get a relation between θ and φ, which shows that as θ varies from 0° to 180°, φ varies from 90° to 0°.

Thus, the energy of the recoil of the electron (W) is given by the following equation:

$$W = h\vartheta - h\vartheta' \qquad (16.58)$$

$$W = h\vartheta\left(1 - \frac{1}{1 + 2\alpha\sin^2\theta/2}\right)$$

$$W = \frac{h\vartheta \cdot 2\alpha(1+\alpha)^2\tan^2\varphi + 1}{\left((1+\alpha)^2\tan^2\varphi + 1\right)\left((1+\alpha)^2\tan^2\varphi + 2\alpha + 1\right)}$$

$$W = h\vartheta \cdot \frac{2\alpha}{(1+\alpha)^2\tan^2\varphi + 2\alpha + 1} \qquad (16.59)$$

Thus, $W \propto (h\vartheta)$ \hfill (16.60)

$$W \propto \frac{1}{\tan^2\varphi} \qquad (16.61)$$

This detailed theoretical analysis of the Compton effect gives the following results of the collision between an electron and the atom:

Looking at Eq. (16.52), $\vartheta' = \dfrac{\vartheta}{1 + 2\alpha \sin^2(\theta/2)}$, one observes that for a given

scattering angle θ when ϑ is small, $2\alpha \sin^2(\theta/2)$ is small. Since $\alpha = \dfrac{h\vartheta}{m_0 c^2}$

becomes small, $\sin^2(\theta/2)$ varies from 0 to 1 as θ varies from $0°$ to $180°$ and hence $\vartheta' \approx \vartheta$ in accordance with the classical theory, which holds for radiations of low frequencies; however, as ϑ increases and has an appreciable value, this approximation cannot be made and ϑ' will decrease, which can be explained by the classical theory:

Thus, for $\theta = 0°$, $\vartheta' = \vartheta$ $\hspace{4cm}$ (16.62)

$$\theta = 90°; \sin^2(\theta/2) = \frac{1}{2}; \vartheta' = \frac{\vartheta}{(1+\alpha)}; \vartheta' < \vartheta \hspace{2cm} (16.63)$$

$$\theta = 180°; \vartheta' = \frac{\vartheta}{1+2\alpha}; \vartheta' < \vartheta \hspace{3cm} (16.64)$$

Thus, as the scattering angle completes one cycle from $0°$ to $180°$, the frequency of the scattered photon changes from a maximum of (ϑ) to a minimum of $(\vartheta/(1+2\alpha))$

1. Looking at Eq. (16.48), $\lambda' = \lambda + \dfrac{2h}{m_0 c} \sin^2(\theta/2)$ it can be observed that the change in wavelength $(d\lambda)$ is independent of the wavelength of the incident photon as well as of the nature of the scatterer. Thus,

$$d\lambda \alpha \sin^2(\theta/2) \hspace{5cm} (16.65)$$

When $\theta = 0; \lambda' > \lambda; d\lambda = 0$ $\hspace{4cm}$ (16.66)

$$\theta = 90°; d\lambda = \frac{h}{m_0 c} \hspace{5cm} (16.67)$$

which is a constant quantity and can be calculated from the known values of h, m_0, and c. It is referred to as the 'Compton wavelength', which, according to the Einstein's law of equivalence, corresponds to an energy of 0.51 million volts—the same as the self-energy of an electron.

$$\theta = 180°; d\lambda = \frac{2h}{m_0 c} \hspace{5cm} (16.68)$$

which is twice the Compton wavelength. Hence, as θ varies from $0°$ to $180°$, the wavelength of the scattered photon varies from a minimum value of λ to a maximum value of $\lambda + \dfrac{2h}{m_0 c}$ provided that the wavelength of the incident radiation is sufficiently small; otherwise, there will be no change in the wavelength, as predicted by the classical theory.

2. *Angular distribution* of the intensity of the scattered radiation: In the Compton theory, the dissymmetry is evident. Scattering is all in the forward direction, that is, between 0° and 90°, whereas the backward scattering between 90° and 180° is very small or practically nil (Fig. 16.13). This dissymmetry in the distribution increases as $\alpha = h\vartheta/m_0 c^2$. As ϑ increases, the dissymmetry increases, which can clearly be observed in case of γ rays.

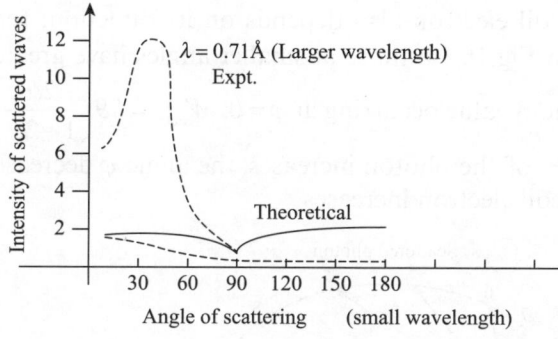

Fig.16.13 Plot of angle of scattering versus intensity of scattered waves

3. Coefficient of scattering: In the Compton effect we may distinguish two parts based on the coefficient of scattering; σ_s measures the reduction in intensity of the incident beam by scattering, σ_a measures the intensity absorbed during the production of the recoil electrons, and σ_0 is the classical coefficient of scattering. We can write that

$$\frac{\sigma}{\sigma_0} = \frac{1}{1+2\alpha}; \frac{\sigma_s}{\sigma_0} = \frac{1+\alpha}{(1+2\alpha)^2}; \frac{\sigma_a}{\sigma_0} = \frac{\alpha}{(1+2\alpha)^2} \qquad (16.69)$$

where $\alpha = \dfrac{h\vartheta}{m_0 c^2}$

When the frequency is small, the value of α is negligible; this results in $\sigma = \sigma_0$, $\sigma_s = \sigma_0$, and $\sigma_a = 0$. Hence, the Compton coefficient is reduced to classical scattering coefficient practically due to scattering, while the absorption coefficient σ_a is practically nil.

When the frequency increases, α becomes high and the total scattering coefficient (σ), becomes smaller, that is, has a lesser value than that predicted by the classical theory, and σ_0 decreases faster than σ_a that is, the true scattering effect will become less than the absorption effect.

4. Looking at Eq. (16.57), $\tan\varphi = \dfrac{1}{(1+\alpha)\tan(\theta/2)}$, one observes that it relates the direction of the recoil electron with that of the scattered photon. It shows that as θ varies from 0° to 180°, φ varies from 90° to 0°. This means that while the photon is scattered in all directions, the recoil electron is restricted to the forward direction, projected at angles less than 90°.

5. The equation $= h\vartheta \cdot \dfrac{2\alpha}{(1+\alpha)^2 \tan^2\varphi + 2\alpha + 1}$ shows that the energy of the recoil electron depends on the energy of the incident photon ($h\vartheta$). The ratio $w/h\vartheta$ always has a rather small value, even for very hard X-rays. Thus, in comparison with photoelectrons whose energy is nearly equal to the incident photon ($h\vartheta$), the energy of 'Compton electrons' is of a considerably small value.

The energy of the recoil electron also depends on its direction, that is, on φ, as evident from Fig.16.14; those at smaller angles have greater energies, with the maximum value occurring at $\varphi = 0$; $W_{max} = h\vartheta\left(\dfrac{2\alpha}{1+2\alpha}\right)$.

As the scattering angle θ of the photon increases, the angle φ decreases and the energy of the recoil electron increases.

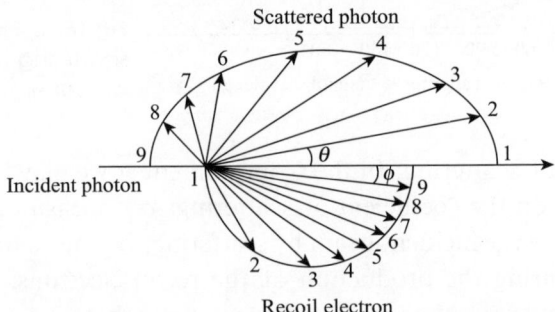

Fig.16.14 Debye's diagram[arrows indicate energy of scattered photon (top part) and recoil electrons (bottom part)]

The numbers in the figure correspond to a given scattered photon and its complementary recoil electrons, the sum of the lengths in each case representing the energy of the incident photon. For each photon scattered in a definite direction (θ), there is a simultaneous ejection of an electron at a definite direction (φ) and vice versa. The Compton experiment, by its successful explanation of the X-ray diffraction phenomenon, further endorsed the quantum nature of discreteness in the behaviour of atoms. Having established the quantum nature of X-rays, the dualism of these rays is also highlighted. The phenomena of interference and diffraction of X-rays can now be easily explained on the basis of the wave theory of light.

Example 16.8   A photon is Compton scattered by an electron through an angle of 90.° Find the energy of the scattered photon for incident photon energies of 10 keV, 0.511 MeV, and 10 MeV.

Solution   The Compton wavelength corresponding to the electron is given by the following relation:

$$\lambda' = \lambda + \frac{h}{m_0 c}(1-\cos\varphi) \Rightarrow \lambda' = \frac{hc}{E} + \frac{h}{m_0 c}(1-\cos\varphi)$$

$$\lambda' = \frac{1.240 \times 10^{-6} \text{eV} \cdot \text{m}}{10 \times 10^2 \text{eV}} + (2.43 \times 10^{-12})(1-0)$$

$$\lambda' = (1.240 \times 10^{-10} \text{m}) + (2.43 \times 10^{-12} \text{m}) = 1.2643 \times 10^{-10} \text{m}$$

Energy of the scattered photon $= \dfrac{hc}{\lambda'}$

$$= \frac{1.24 \times 10^{-6} \text{eV} \cdot \text{m}}{1.2643 \times 10^{-10} \text{m}} = 0.980 \times 10^4 \text{eV} = 0.0098 \text{ MeV}$$

$$\lambda' = \frac{1.240 \times 10^{-6} \text{eV} \cdot \text{m}}{0.511 \times 10^{-6} \text{eV}} + (2.43 \times 10^{-12} \text{m})(1-0)$$

$$\lambda' = (2.4266 \times 10^{-12} \text{m}) + (2.43 \times 10^{-12} \text{m}) = 4.856 \times 10^{-12} \text{m}$$

Energy of the scattered photon $= \dfrac{hc}{\lambda'}$

$$= \frac{1.24 \times 10^{-6} \text{eV} \cdot \text{m}}{4.856 \times 10^{-12} \text{m}} = 0.255 \times 10^6 \text{eV} = 0.256 \text{ MeV}$$

$$\lambda' = \frac{1.240 \times 10^{-6} \text{eV} \cdot \text{m}}{10 \times 10^{-6} \text{eV}} + (2.43 \times 10^{-12} \text{ m})(1-0)$$

$$\lambda' = (0.1240 \times 10^{-12} \text{m}) + (2.43 \times 10^{-12} \text{m}) = 2.554 \times 10^{-12} \text{m}$$

Energy of the scattered photon $= \dfrac{hc}{\lambda'}$

$$= \frac{1.24 \times 10^{-6} \text{eV} \cdot \text{m}}{2.554 \times 10^{-12} \text{m}} = 0.486 \times 10^{-6} \text{eV} = 0.486 \text{ MeV}$$

Example 16.9 During a diagnostic X-ray examination, 1.5 kg portion of a broken leg receives an equivalent dose of 0.50 mSv.
(a) What is the equivalent dose in mrem?
(b) What is the absorbed dose in mrad and mGy?
(c) If the X-ray energy is 50 keV, how many X-ray photons are absorbed?
(d) Solve the same problem, if the incident radiation is of α-particles.

Solution The conversion of units gives the following value:

$$100 \text{ rem} = 1 \text{ Sv} \Rightarrow 1 \text{rem} = 0.01 \text{ Sv}$$

(a) Thus, the equivalent dose in mrem is calculated as follows:

$$\frac{0.50 \text{ mSv}}{0.01 \text{ Sv/rem}} = 50 \text{ mrem}$$

(b) For X-rays, RBE $= 1$ rem/rad or $\dfrac{1 \text{ Sv}}{\text{Gy}}$, so the absorbed dose is calculated as follows:

$$\frac{50 \text{ mrem}}{1 \text{ rem/rad}} = 50 \text{ mrad}$$

$$\frac{0.50 \text{ mSv}}{1 \text{ Sv/Gy}} = 0.50 \text{ mGy} = 5.0 \times 10^{-4} \text{ J/kg}$$

(c) The total energy absorbed is calculated as follows:

$$\left(\frac{5.0\times10^{-4}\,\text{J}}{\text{kg}}\right)(1.5\,\text{kg})=7.5\times10^{-4}\,\text{J}=4.69\times10^{15}\;\text{eV}$$

Energy of the X-ray photon $50\,\text{keV}=5.0\times10^{4}\,\text{eV}$.

Thus, the number of X-ray photons is calculated as follows:

$$\frac{4.69\times10^{15}\,\text{eV}}{5.0\times10^{4}\;\text{eV/photon}}=0.94\times10^{11}=0.94\times10^{10}\;\text{photons}$$

(d) If the incident radiation is a beam of α-particles, then for this

$$\text{RBE}=20$$

The absorbed dose needed for an equivalent dose of $0.50\,\text{mSv}$ would be as follows:

$$0.50\;\text{mSv/20 Sv/Gy}=0.025\;\text{mGy}=2.5\times10^{-5}\;\text{J/kg}$$

The total absorbed energy is calculated as follows:

$$\left(\frac{2.5\times10^{-5}\,\text{J}}{\text{kg}}\right)(1.5\,\text{kg})=3.75\times10^{-5}\,\text{J}=2.34\times10^{-14}\;\text{eV}$$

Example 16.10 An alpha particle beam passes through flesh and deposits. $0.4\,\text{J}$ of energy in each kilogram of flesh. The RBE for these particles is $\dfrac{14\;\text{rem}}{\text{rad}}$ Find the dose in rad and in rem.

Solution The calculations are as follows:

$$\text{Dose in rad}=\left(\frac{\text{Absorbed energy}}{\text{Mass}}\right)\times\left(\frac{100}{\text{J/kg}}\right)=\left(\frac{0.4\,\text{J}}{\text{kg}}\right)\times\left(\frac{100\;\text{rad}}{\text{J/kg}}\right)=40\;\text{rad}$$

$$\text{Dose in rem}=(\text{RBE})\times(\text{Dose in rad})=(14)(20)=280\;\text{rem}$$

Example 16.11 A beam of gamma rays has a cross-sectional area of $1\,\text{cm}^2$ and carries 5×10^{8} photons through the cross section each second. Each photon has an energy of $1.5\,\text{MeV}$. The beam passes through a $0.5\,\text{cm}$ thickness of flesh $\left(\rho=\dfrac{0.95\;\text{g}}{\text{cm}^3}\right)$ and loses 2% of its intensity in the process. What is the average dose (in rad) applied to the flesh each second?

Solution The number of gamma rays absorbed per second is calculated as follows:

$$(5\times10^{8}\,\text{/s})(0.02)=0.1\times10^{8}\,\text{/s}=1.0\times10^{7}\,\text{/s}$$

The mass of the flesh in which this energy was absorbed is given as follows:

$$\text{Mass}=\rho V=\left(\frac{0.95}{\text{cm}^3}\right)(1\,\text{cm}^2)(0.5\,\text{cm})=0.475\;\text{g}$$

Dose is measured as the energy absorbed per kilogram of flesh. Hence,

$$\frac{\text{Dose}}{s}=\frac{\text{Energy/s}}{\text{Mass}}$$

Energy absorbed per second $=(1.0\times10^{7}\,\text{/s})\times(1.5\,\text{MeV})=1.5\times10^{7}\;\text{MeV/s}$

$$\frac{\text{Dose}}{s}=\frac{(1.5\times10^{7}\;\text{MeV/s})(1.6\times10^{-13}\,\text{/MeV})}{(0.475\times10^{-3}\;\text{kg})}\times\frac{100\;\text{rad}}{1\;\text{J/kg}}=0.5053\;\text{rad/s}$$

Example 16.12 In a radiation therapy experiment, an equivalent dose of 0.80 rem is given by 0.80 MeV protons to a localized area of tissue with mass 0.20 kg.
(a) What is the absorbed dose in rad?
(b) How many protons are absorbed by the tissue?
(c) How many α-particles of the same energy of 0.80 MeV are required to deliver the same equivalent dose of 0.80 rem?

Solution The calculations are as follows:
(a) The equivalent dose of 0.80 rem is given in rad as follows:

$$\frac{0.80 \text{ rem}}{10 \text{ rem/rad}} = 0.080 \text{ rad}$$

As for protons RBE $= \dfrac{10 \text{ rem}}{\text{rad}}$, or $\dfrac{1 \text{ Sv}}{\text{Gy}}$ In addition, we know that 100 rad $= 1$ Gy

$$\Rightarrow 0.080 \text{ rad} = 8.0 \times 10^{-4} \text{ Gy} = 8.0 \times 10^{-4} \text{ J/kg}$$

(b) The total energy absorbed is calculated as follows:

$$\left(\frac{8.0 \times 10^{-4}}{\text{kg}}\right)(0.20 \text{ kg}) = 1.6 \times 10^{-4} \text{ J} = 1.0 \times 10^{15} \text{ eV}$$

Energy per proton $= 0.80 \text{ MeV} = 0.80 \times 10^6 \text{ eV}$

The number of protons absorbed is calculated as follows:

$$\frac{1.0 \times 10^{15} \text{ eV}}{0.80 \times 10^6 \text{ eV / protons}} = 1.25 \times 10^9 \text{ protons}$$

(c) If the incident radiation is a beam of α-particles, then

$$\text{RBE} = 20 \text{ rem/rad}$$

The absorbed dose needed for an equivalent dose of 0.80 rem would be calculated as follows:

$$\frac{(0.80 \text{ rem})}{(20 \text{ rem/rad})} = 0.04 \text{ rad} = 4.0 \times 10^{-4} \text{ J/kg}$$

The total absorbed energy is calculated as follows:

$$\left(\frac{4.0 \times 10^{-4}}{\text{kg}}\right)(0.2 \text{ kg}) = 0.8 \times 10^{-4} \text{ J} = 0.5 \times 10^{15} \text{ eV}$$

Energy per α-particle $= 0.80 \text{ MeV} = 0.80 \times 10^6 \text{ eV}$
The number of protons absorbed is calculated as follows:

$$\frac{0.5 \times 10^{15} \text{ eV}}{0.80 \times 10^6 \text{ eV / } \alpha \text{-particle}} = 0.625 \times 10^9 \, \alpha \text{-particles}$$

Example 16.13 In a diagnostic X-ray procedure, 6.50×10^{10} photons are absorbed by a tissue of mass 0.60 kg. The X-ray wavelength is 0.020 nm.
(a) What is the total energy absorbed by the tissue?
(b) What is the equivalent dose in rem?

Solution

(a) The energy of the X-ray can be found as follows:

$$E = \frac{hc}{\lambda} = \frac{(6.63 \times 10^{-34}\,\text{Js})\left(\dfrac{3 \times 10^{8}\,\text{m}}{\text{s}}\right)\left(\dfrac{1.6 \times 10^{-19}}{\text{eV}}\right)}{(0.02\,\text{nm})}$$

$$= \frac{(1240\,\text{eV} \cdot \text{nm})}{(0.02\,\text{nm})} = 620.00\,\text{eV}$$

The number of photons can be found as follows:

$$6.5 \times 10^{10} = (\text{energy eV}) / 620\,\text{eV}$$

$$\Rightarrow \text{Energy (eV)} = 6.5 \times 10^{10} \times 620\,\text{eV} = 4030 \times 10^{10}\,\text{eV}$$

$$= 4030 \times 10^{10}\,\text{eV} \times 1.6 \times 10^{-19}\,\text{J} = 6.49 \times 10^{-6}\,\text{J}$$

Total energy absorbed by the tissue = $6.49 \times 10^{-6}\,\text{J}$.

(b) The total energy absorbed is calculated as follows:

$$\frac{x\text{J}}{\text{kg}}(0.6\,\text{kg}) = 6.49 \times 10^{-6}\,\text{Js}$$

$$\frac{6.49 \times 10^{-6}\,\text{J}}{0.60\,\text{kg}} = \frac{10.816 \times 10^{-6}\,\text{J}}{\text{kg}} = 0.108\,\text{rem}$$

IMPORTANT CONCEPTS

1. X-ray diffraction is actually strong X-ray reflections at certain specific angles. This is referred to as the Bragg's law:

   $$2d \sin\theta = n\lambda$$

   where d is the inter-planar spacing, θ is the diffraction angle, n is an integral number or order of diffraction, and λ is the wavelength of the X-ray beam.

2. $\varphi_r = \dfrac{2\pi}{\lambda}(r \cdot S)$

   This equation represents the phase difference between the radiations scattered by any two arbitrary points S and P in the crystal. The primitive translation distances are a, b, and c for the respective nearest neighbour atoms. Hence,

   $$\frac{2\pi}{\lambda}(a \cdot S) = 2\pi nh$$

   $$\frac{2\pi}{\lambda}(b \cdot S) = 2\pi nk$$

   $$\frac{2\pi}{\lambda}(c \cdot S) = 2\pi nl$$

where $h, k,$ and l represent the three smallest integers identical to the Miller indices. If α, β, γ represent the angles between the scattering normal (S) and the $a, b,$ and c axes of the crystal, respectively, then the aforementioned equations can be written as follows:

$$\frac{2\pi}{\lambda}(aS\cos\alpha) = 2\pi nh$$

$$\frac{2\pi}{\lambda}(bS\cos\beta) = 2\pi nk$$

$$\frac{2\pi}{\lambda}(cS\cos\gamma) = 2\pi nl$$

These equations can be further simplified to the following forms:

$$(2a\sin\theta\cos\alpha) = nh\lambda$$

$$(2b\sin\theta\cos\beta) = nk\lambda$$

$$(2c\sin\theta\cos\gamma) = nl\lambda$$

These three equations together are referred to as the Laue equations.

3. The basic principle of the powder method is that micro-crystals have their crystal planes oriented to satisfy the Bragg's law. Reflection of the incident monochromatic radiation takes place from all the [hkl] planes, satisfying the following equation:

$$\theta_{hkl} = \sin^{-1}\left(\frac{\lambda}{2d_{hkl}}\right) \le \frac{1}{2}\pi$$

4. There are three methods of producing X-ray emission:
   (a) X-ray fluorescence
   (b) Bremsstrahlung
   (c) Synchrotron radiation
5. Crystal structures are determined by the diffraction of X-rays.
6. Bragg reflection is the result of the constructive interference of rays diffracted from the successive planes of a given family of planes [hkl] such that $n\lambda = 2d_{hkl}\sin\theta$
7. Van Laue illustrated that X-rays having shorter wavelengths can be scattered from different atoms, producing well-defined spots referred to as the diffraction pattern.

APPLICATIONS

Applications of X-rays have been exhaustively dealt within the realms of this chapter.

IMPORTANT FORMULAE

1. Diffraction law:

$$n\lambda = 2d_{hkl}\sin\theta$$

2. Energy of the recoil electron:

$$W = h\vartheta \cdot \frac{2\alpha}{(1+\alpha)^2\tan^2\varphi + 2\alpha + 1}$$

SELF-ASSESSMENT

Multiple-choice Questions

16.1 X-rays are
   (a) electromagnetic radiation
   (b) light radiation
   (c) ultraviolet radiation
   (d) none of these

16.2 X-rays were discovered by
   (a) William Rontgen
   (b) William Chadwick
   (c) William Rutherford
   (d) William Wallace

16.3 X-ray radiograms are referred to as
   (a) radiographs
   (b) nuclear magnetic spectroscopes
   (c) Rontgenograms
   (d) none of these

16.4 Crystal structure can be determined by
   (a) interference of light
   (b) diffraction of light
   (c) polarization of light
   (d) diffraction of X-rays

16.5 X-ray diffraction is
   (a) strong X-ray deflection at certain specific angles
   (b) strong X-ray reflection at certain specific angles
   (c) strong X-ray deflection at all scattered angles
   (d) strong X-ray deflection at no specific angle

16.6 The Bragg's law states that (symbols having conventional meaning)
   (a) $2d \sin \theta = \lambda$
   (b) $2d_{hkl} \sin \theta = n\lambda$
   (c) $2n \sin \theta = \lambda$
   (d) $2 \sin \theta = \lambda$

16.7 The thought that crystals can be used as diffraction gratings was perpetuated by
   (a) Von Frank
   (b) Von Laue
   (c) Von Rontgen
   (d) William Rontgen

16.8 The significance of the Laue equations is that
   (a) the unique value for the diffraction direction θ and the scattering normal S is not defined by these equations
   (b) they do not define the unique value for the diffraction direction θ and the scattering normal S
   (c) they define the arbitrary diffraction direction θ and the scattering normal S
   (d) they define the unique value for the diffraction direction θ and the scattering normal S

16.9 The primitive translation distances a, b, and c are for the respective
   (a) farthest neighbour atoms
   (b) nearest neighbour atoms
   (c) diagonally placed neighbour atoms
   (d) diagonally placed farthest neighbour atoms

16.10 The $[h, k, l]$ planes are
   (a) refracting planes
   (b) diffracting planes
   (c) reflecting planes
   (d) interfering planes

16.11 The Laue equations provide the
    (a) order of diffraction
    (b) order of interference
    (c) theoretical basis for the Bragg's equation
    (d) experimental basis for the Bragg's equation

16.12 The diffraction order n indicates the
    (a) diffraction plane   (c) refraction plane
    (b) polarization plane   (d) reflection plane

16.13 The crystal powder method is recommended for the study of structure of
    (a) macro dimensions of crystals
    (b) micro dimensions of crystals
    (c) hexagonally packed crystals
    (d) face-centred crystals

16.14 X-ray fluorescence is the process of
    (a) knocking an electron out of the outermost shell of an atom, resulting in the production of X-ray photons at discrete frequencies
    (b) knocking an electron out of the innermost shell of an atom, resulting in the production of X-ray photons at discrete frequencies
    (c) knocking an electron out of the nucleus of an atom, resulting in the production of X-ray photons at discrete frequencies.
    (d) none of these

16.15 Bremsstrahlung is the characteristic of elements
    (a) possessing a high proton number
    (b) possessing a low proton number
    (c) possessing an equal number of protons and electrons
    (d) that are unstable in nature

16.16 The energy of the Compton electrons
    (a) is nearly equal to that of the incident photoelectrons
    (b) is greater than that of the incident photoelectrons
    (c) is less than that of the incident photoelectrons
    (d) has no correlation with that of the incident photoelectrons

16.17 The Compton effect highlights the fact that
    (a) energy can be exchanged only in 'discrete packets'
    (b) energy cannot be exchanged only in 'discrete packets'
    (c) energy is not exchanged in this process
    (d) none of these

16.18 Relative biological effectiveness is defined as the biological effect of a given dose relative to the biological effect of
    (a) an equal dose of X-rays   (c) an equal dose of β-rays
    (b) an equal dose of γ-rays   (d) an equal dose of α-rays

Review Questions

16.1 Is the diffraction in X-rays characterized by elastic or inelastic scattering?
16.2 Why do reflected X-rays correspond to certain specific angles of incidence?
16.3 Give a characteristic property of X-rays that helps us determine the structure of crystals.
16.4 Is it possible to use crystals as diffraction gratings? Explain briefly.

16.5 Argue for the physical significance of the Laue equations using mathematical expressions.

16.6 The experiment of Compton highlights the dual nature of light. Justify.

16.7 Highlight the difference in the production of X-rays through the different phenomena of X-ray fluorescence, Bremsstrahlung, and synchrotron radiation.

16.8 Enunciate the fundamental principle underlying the biological applications of radiations (particularly X-rays).

16.9 Explain in detail the phenomenon of diffraction of X-rays.

16.10 Discuss how the rigorous approach to the derivation of the Bragg's law helps us understand diffraction in crystal structure.

16.11 Derive the Bragg's law in crystals.

16.12 Derive the Laue equations.

16.13 Explain the physical significance of the directional cosines in a crystal. Support your answer with proper ray diagrams.

16.14 Explain how the powder method is able to make us understand microscopic crystal structure.

16.15 Give the salient assumptions of the Compton method and hence derive the Compton equations.

16.16 Explain how the Compton effect corroborated the existence of dualism in the behaviour of light.

16.17 Give the salient features of the experimental results of the Compton effect that directly indicated the existence of dualism of light. Hence, highlight the experimental evidence that indicated that classical concepts needed to be redefined.

16.18 Explain in detail how the discovery of X-rays led to the better understanding of the crystal structure.

16.19 Discuss in detail the biomedical applications of X-rays.

16.20 Define the SI unit of radiation measurement. How is it quantified in comparison to the other units?

16.21 Discuss in detail the advantages of radiation in biological and industrial applications.

Numerical Problems

16.1 An X-radiation of wavelength 1.0Å is incident on a common salt crystal with a d-spacing of 4.0Å. What is the highest order that the crystal can diffract?

$$\left[Hint: \sin\theta = \frac{n\lambda}{2d_{hkl}} \right]$$

16.2 An X-radiation of wavelength 0.5Å strikes a crystal of d spacing 1.0Å. Estimate the diffraction angle for the second order.

16.3 Find the wavelength of the first-order Bragg reflection if the glancing angle is nearly 50°, assuming that the grating spacing is 6.0Å.

$$\left[Hint: 2d_{hkl}\sin\theta = n\lambda \right]$$

16.4 If the potential difference applied across an X-ray tube is 2 kV and the current flowing through it is 2 mA, calculate the following:
(a) The number of electrons striking the target per second
(b) The speed with which they strike

$$\left[Hint: I = ne, \, v = \sqrt{\frac{2eV}{m}} \right]$$

16.5   Find the Duane–Hunt wavelength (minimum wavelength) of X-rays generated from a Coolidge tube that operates at 30 kV.

$$\left[Hint: \frac{hc}{\lambda_{min}} = eV \right]$$

16.6   Calculate the longest wavelength that can be analysed by a sodium chloride crystal of spacing $d = 1.80\text{Å}$ in the first order.

$$\left[Hint: 2d\sin\theta = n\lambda \right]$$

16.7   A photon is Compton scattered by an electron through an angle of 90°. Find the energy of the scattered photon for incident photon energies of 5 keV, 0.400 MeV, and 8 MeV.

$$\left[Hint: \lambda' = \frac{hc}{E} + \frac{h}{m_0 c}(1 - \cos\varphi) \right]$$

16.8   During a diagnostic X-ray examination, 1.0 kg portion of a broken leg receives an equivalent dose of 0.50 mSv.
   (a)   What is the equivalent dose in mrem?
   (b)   What is the absorbed dose in mrad and mGy?
   (c)   If the X-ray energy is 50 keV, how many X-ray photons are absorbed?
   (d)   Solve the same problem, if the incident radiation is of α-particles.

16.9   An alpha-particle beam passes through flesh and deposits 0.2 J of energy in each kilogram of flesh. The RBE for these particles is $\dfrac{14\,\text{rem}}{\text{rad}}$ Find the dose in rad and in rem.

$$\left[Hint: \text{Dose in rad} = \left(\frac{\text{absorbed energy}}{\text{mass}} \right) \times \left(\frac{100\,\text{rad}}{\dfrac{J}{kg}} \right) \right]$$

16.10   A beam of gamma rays has a cross-sectional area of 1 cm² and carries 2×10^8 photons through the cross-section each second. Each photon has an energy of 1.5 MeV. The beam passes through a 0.5 cm thickness of flesh $\rho\,\dfrac{0.95\,g}{cm^3}$ and loses 2% of its intensity in the process. What is the average dose (in rad) applied to the flesh each second?

$$\left[Hint: \frac{\text{Dose}}{s} = \frac{\text{energy}/s}{\text{mass}} \right]$$

16.11   In a radiation therapy experiment, an equivalent dose of 0.80 rem is given by 0.80 MeV protons to a localized area of tissue of mass 0.10 kg.
   (a)   What is the absorbed dose in rad?
   (b)   How many protons are absorbed by the tissue?
   (c)   How many α-particles of the same energy of 0.80 MeV are required to deliver the same equivalent dose of 0.80 rem?

16.12   In a diagnostic X-ray procedure, 6.50×10^{10} photons are absorbed by a tissue of mass 0.50 kg. The X-ray wavelength is 0.020 nm.
   (a)   What is the total energy absorbed by the tissue?
   (b)   What is the equivalent dose in rem?

$$\left[Hint: E = \frac{hc}{\lambda}; \text{Total energy absorbed} = \left(\frac{xJ}{kg} \right)(mkg) \right]$$

Crystal Physics

Learning Objectives

After studying this chapter, students will be able to

- understand the concepts of lattice and unit cell
- explain the various types of Bravais lattices
- comprehend the concept of lattice planes and the method of determining the Miller indices
- calculate the number of atoms per unit cell and the atomic radius
- understand the concepts of coordination number and packing factor
- describe some important crystal structures
- differentiate between polymorphism and allotropy
- elucidate the different types of crystal defects and the concept of Burger vector

List of Symbols

r_1 = Displacement vector

a, b, c = Basis vectors

h, k, l = Miller indices

d_{hkl} = Interplanar spacing

a_0 = Lattice parameter

r = Atomic radius

f = Packing factor

V_1 = Atomic volume per unit cell

V = Volume of unit cell

n_1 = Equilibrium concentration of interstitial atom

E_{FI} = Formation energy of interstitial atom

H = Number of interstitial sites in a given volume

B = Burger vector

17.1 INTRODUCTION

Solid-state materials can exist in crystalline, polycrystalline, and amorphous forms. Atoms in a crystalline material are arranged in a periodic arrangement. This periodic arrangement is called a *lattice*. We will learn some properties of the lattice in this chapter. The study of crystal physics needs a representation to define the different planes and directions in a crystal. Miller indices offer such a representation, and a complete understanding of this concept helps us utilize the immense potential of a variety of crystals. One must, however, remember that Miller indices are at best a representation that helps us exploit crystal structures, but this is not the only way to represent planes and directions in a crystal. Many common materials can be classified into well-defined crystal structures. This classification is useful in predicting some common characteristics of these materials. A detailed study of one member of any of these groups can help us understand some characteristics of other members of the group. A perfect crystal is not realizable in the real world. Real crystals invariably have defects that can have a defining effect on their properties. In this chapter, we would be looking at some of the more common defects in crystals. This can help us undertake measures to minimize these defects, if not eliminate; understand the cause of these defects; and design devices that do not show degradation due to the presence of these defects.

17.2 CLASSIFICATION OF SOLIDS

In general, solids can be classified into three types, depending upon the arrangement of the constituent atoms. These three types are *crystalline* or *single crystal*, *polycrystalline*, and *amorphous crystal solids*. Atoms are arranged in a regular manner in a crystalline solid. A three-dimensional repetition of a certain basic pattern constitutes the solid. In a single crystal, periodicity of the basic pattern extends throughout the material. This periodicity of the structure gets interrupted at certain boundaries in a polycrystalline material. These so-called boundaries are called grain boundaries. Specific grain sizes may vary from several Angstroms to macroscopic dimensions. An interesting situation is one in which the size of the grains within the grain boundaries becomes comparable to the size of the basic pattern. Periodicity is completely absent at this stage, and the material is then called amorphous. Figure 17.1 shows the three types of solids schematically in a two-dimensional representation.

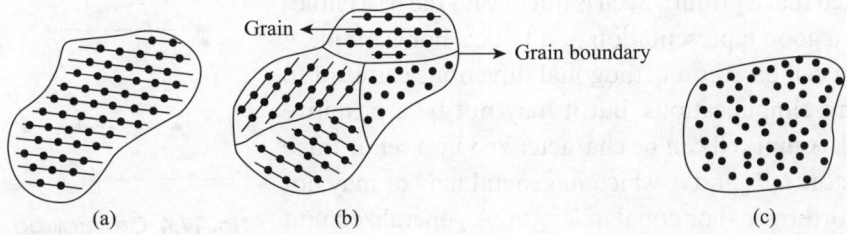

Fig. 17.1 Three types of solids (a) Crystalline (b) Polycrystalline (c) Amorphous

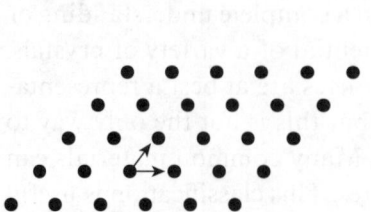

Fig. 17.2 Two-dimensional single crystal lattice

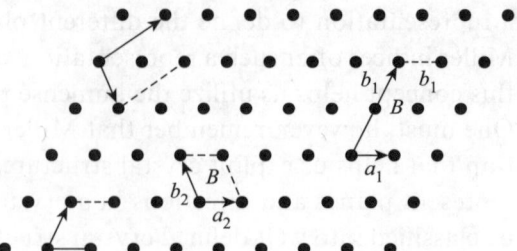

Fig. 17.3 Two-dimensional single crystal lattice showing various possible unit cells

17.3 LATTICE—UNIT CELL

In a single crystal, a basic pattern, which may consist of a single atom or a group of atoms, is repeated at regular intervals in all the three dimensions. This periodic arrangement of atoms in a crystal is called a *lattice*. Due to this periodic arrangement of atoms, the environment around any point within a crystal appears to be the same. A lattice may also be defined as an array of points in space such that the environment around each point is the same. The distance between constituent atoms and the relative orientation of these atoms can have different magnitudes. The fundamental unit that undergoes a regular repetition to create the lattice is called a *unit cell*, which is the basic building block of the entire crystal.

Figure 17.2 is a schematic representation of an infinite two-dimensional array of lattice atoms. In this representation, an atom is denoted by a dot, referred to as a *lattice point*. Any lattice point may be translated through a distance *a* in one direction and by a distance *b* in a second noncollinear direction, to generate a two-dimensional lattice. The two translational directions can have any angle between them, but the angle is fixed for a particular crystal. A three-dimensional lattice can similarly be created by translating a lattice point in a third noncollinear direction through a distance *c*. A general three-dimensional lattice can be obtained by carrying out a periodic repetition of the unit cell. A unit cell is not unique to a particular lattice. As a demonstration of this feature, Fig. 17.3 shows several possible unit cells for a two-dimensional lattice.

Unit cells may vary in size; *the smallest unit cell that can generate a crystal is called a primitive cell*. It must, however, be remembered that a primitive cell is not always the best choice for a good representation of a lattice. For example, a unit cell based on orthogonal directions can lead to some simplifications, but it may not be a primitive cell. A unit cell can be characterized by a set of three vectors *a*, *b*, and *c*, which in general may or may not be orthogonal or equal in length. A generalized unit cell is shown in Fig. 17.4.

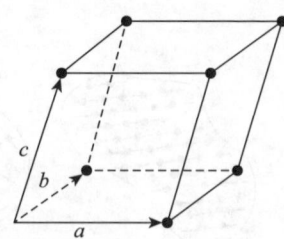

Fig. 17.4 Generalized unit cell

17.4 BRAVAIS LATTICE

Any lattice point is indistinguishable from another lattice point, as long as the displacement vector between the two lattice points can be represented by the following equation:

$$r = pa + qb + sc \tag{17.1}$$

where p, q, and s are integers, and the vectors a, b, and c are called the basis vectors.

Seven unique lattice point arrangements can be used to fill up a three-dimensional space. These seven unique arrangements are called *crystal systems*, which include cubic, tetragonal, orthorhombic, rhombohedral (or trigonal), hexagonal, monoclinic, and triclinic systems. There are a total of 14 distinct arrangements of lattice points, and these are called *Bravais lattices*, named after Auguste Bravais (1811–1863), a French crystallographer. Figure 17.5 shows schematic representations of these 14 types of Bravais lattices.

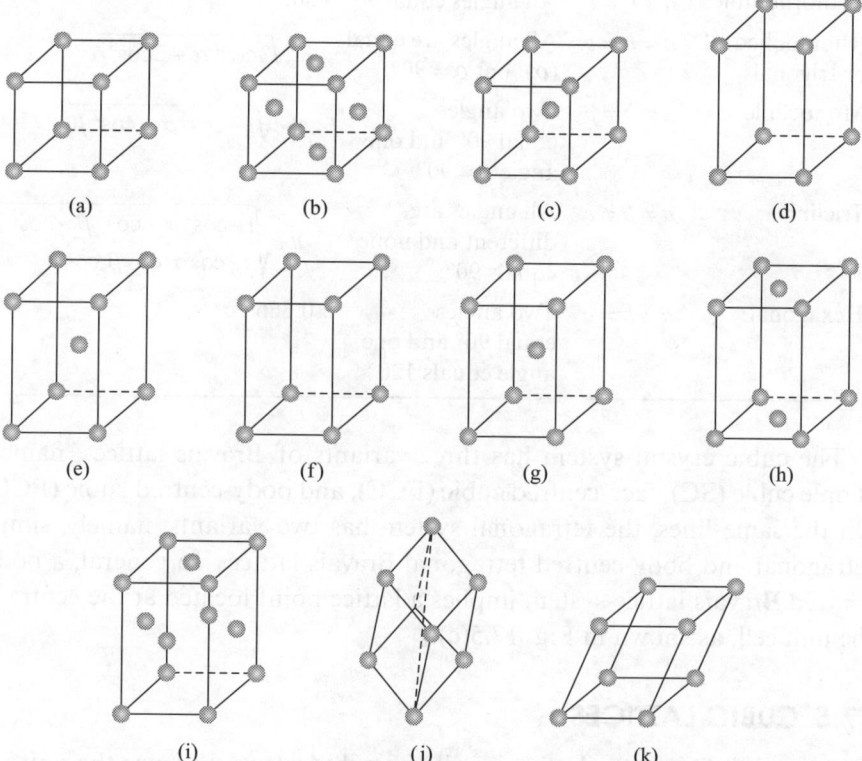

Fig. 17.5 Schematic representation of Bravais lattices (a) Simple cubic (b) Face-centred cubic (c) Body-centred cubic (d) Simple tetragonal (e) Body-centred tetragonal (f) Simple orthorhombic (g) Body-centred orthorhombic (h) Base-centred orthorhombic (i) Face-centred orthorhombic (j) Rhombohedral (k) Simple monoclinic

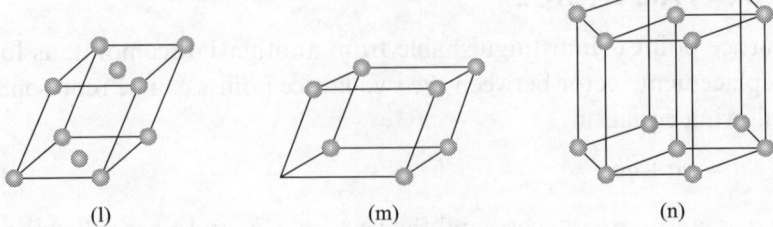

(l)   (m)   (n)

Fig. 17.5 (contd) (l) Base-centred monoclinic (m) Triclinic (n) Hexagonal

Table 17.1 lists some salient features of the seven crystal systems.

Table 17.1 Salient features of seven crystal systems

Structure	Basis vectors	Angles between basis vectors	Volume of the unit cell
Cubic	$a = b = c$	All angles equal 90°	a^3
Tetragonal	$a = b \neq c$	All angles equal 90°	a^2c
Orthorhombic	$a \neq b \neq c$	All angles equal 90°	abc
Rhombohedral or trigonal	$a = b = c$	All angles are equal (α), but $\alpha \neq 90°$	$a^3\sqrt{3\cos^2\alpha + 2\cos^3\alpha}$
Monoclinic	$a \neq b \neq c$	Two angles equal 90° and one angle $\beta \neq 90°$	$abc\sqrt{1 - \cos^2\alpha - \cos^2\beta}$
Triclinic	$a \neq b \neq c$	All angles are different and none equals 90°	$abc\sqrt{\begin{array}{c}1 - \cos^2\alpha - \cos^2\beta - \cos^2\gamma \\ + 2\cos\alpha\,\cos\beta\,\cos\gamma\end{array}}$
Hexagonal	$a = b \neq c$	Two angles equal 90° and one angle equals 120°	$0.866a^2c$

The cubic crystal system has three variants of Bravais lattices, namely, simple cubic (SC), face-centred cubic (FCC), and body-centred cubic (BCC). On the same lines, the tetragonal system has two variants, namely, simple tetragonal and body-centred tetragonal Bravais lattices. In general, a body-centred Bravais lattice system implies a lattice point located at the centre of the unit cell, as shown in Fig. 17.5(c).

17.5 CUBIC LATTICES

The simplest form of a lattice is called a cubic lattice, because the unit cell for such a lattice is a cubic volume. For a cubic lattice, the vectors a, b, and c are equal in length and perpendicular to each other. The three most common cubic lattices are simple cubic (sc), body-centred cubic (bcc), and face-centred cubic (fcc). These three lattice types are shown in Fig. 17.6.

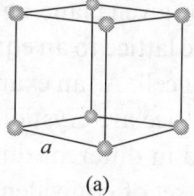 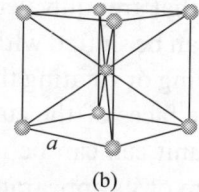 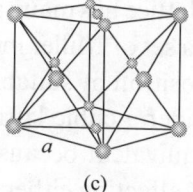

(a) (b) (c)

Fig. 17.6 Unit cells for cubic lattices (a) Simple cubic (b) Body-centred cubic (c) Face-centred cubic

The simple cubic structure consists of an atom located at each corner. The body-centred cubic structure has an additional atom at the centre of the cube. The face-centred cubic structure has additional atoms on each face of the cube.

17.6 MILLER INDICES—CRYSTAL PLANES AND DIRECTIONS

Single crystal materials are used in a variety of applications. Many of these applications rely on properties that are orientation- or direction-dependent. To remove any ambiguity, a standard method of indicating planes and directions in a crystalline material is used. In this method, a set of three integers h, k, and l, placed within parentheses [e.g., (hkl)] is generally used to indicate a particular crystalline plane. The set h,k,l is called the *Miller indices*. These integers are determined using the following procedure:

1. Determine the intercepts of the given plane on the crystal axes and express them as integral multiples of the respective basis vectors. A translation of the particular plane with respect to the origin is allowed as long as the basic direction of the plane is maintained.
2. Evaluate the reciprocal of the three integers found in Step 1 and reduce these reciprocals to the smallest set of integers, while maintaining their relationship. Designate these integers as h, k, and l.
3. Label the plane then as (hkl).

A few common lattice planes for a cubic crystal are shown in Fig. 17.7 (the shaded regions). The figure also indicates the Miller indices for these planes. One must, however, remember that the letters h, k, and l define a set of parallel planes inside a lattice, not just a single plane.

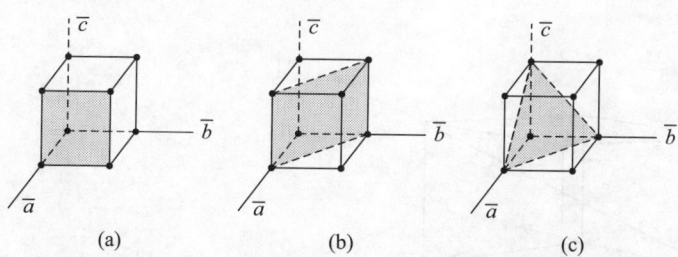

(a) (b) (c)

Fig. 17.7 Three lattice planes for cubic lattice (a) (100) plane (b) (110) plane (c) (111) plane

Any lattice has many planes that are equivalent. Any given plane, character-ized by a set of Miller indices can be shifted within the lattice to an equivalent plane position by suitably moving or rotating the unit cell. As an example, let us consider a cubic lattice. The faces of the cubic lattice are crystallograph-ically equivalent because the unit cell can be rotated in different directions, without affecting either its form or its appearance. A set of equivalent planes is represented by curly brackets {*hkl*}. Figure 17.8 shows the six equivalent faces {100} of a cubic lattice.

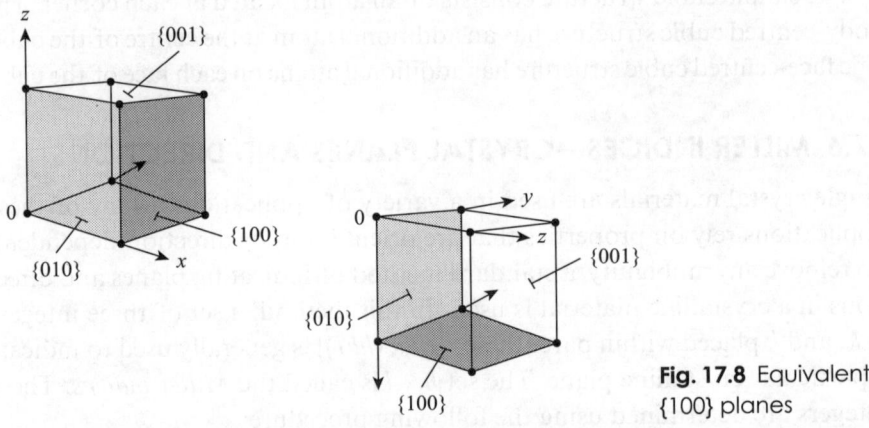

Fig. 17.8 Equivalent {100} planes

The following procedure is used to determine directions in a lattice:
1. Choose the axis vectors with respect to a suitable origin.
2. The vector components for a particular direction are then expressed in multiples of the basis vectors.
3. The three integers are then reduced to their smallest values, while retaining the relationship between them.
4. The given direction is then expressed within square brackets, for example, [*abc*].

Figure 17.9 depicts the [111] direction, which happens to be the body diagonal of a cubic lattice.

Like planes, many directions in a lattice are equivalent. These depend on the choice of the orientation of the axes. Equivalent directions are expressed

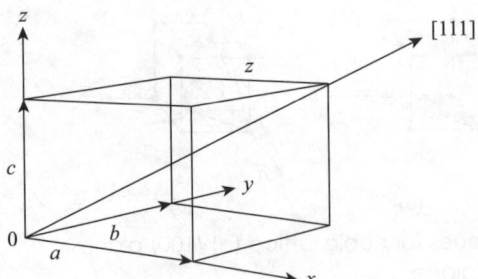

Fig. 17.9 Schematic diagram of [111] direction

within angular brackets (e.g., <*abc*>). A few equivalent <100> directions are shown schematically in Fig. 17.10.

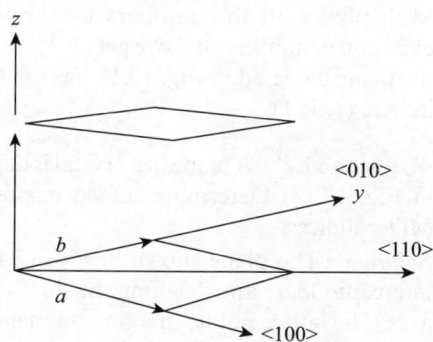

Fig. 17.10 Equivalent <100> directions

Example 17.1 A plane in a crystal is depicted in Fig. 17.11. Determine the corresponding Miller indices.

Solution The plane shown in Fig. 17.11 has intercepts a, $2b$, and $3c$ along the three crystal axes. We know that the lattice points in a three-dimensional lattice are given by the following expression:

$$r = pa + qb + sc \qquad (17.2)$$

From Fig. 17.11 and Eq. (17.2), we can conclude that $p = 1$, $q = 2$, and $s = 3$

Taking reciprocals, we get $\left[1, \dfrac{1}{2}, \dfrac{1}{3}\right]$

Multiplying all the numbers by the lowest common denominator 6, we get (6, 3, 2). Thus, the plane depicted in Fig. 17.11 has the Miller indices (6, 3, 2).

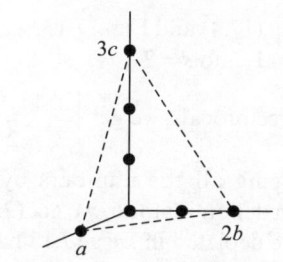

Fig. 17.11 Plane in crystal

Example 17.2 The lattice constant a of Si is 5.43×10^{-8} cm and its atomic weight is 28.1. Calculate the density of Si.

Solution We know that for Si,
Number of atoms/cell = 8, $a = 5.43 \times 10^{-8}$ cm
Therefore, the atomic concentration is

$$\frac{\text{No. of atoms/cell}}{\text{Cell volume}} = \frac{8}{a^3} = \frac{8}{(5.43 \times 10^{-8})^3} \text{ atoms/cm}^3 = 5 \times 10^{22} \text{ atoms/cm}^3$$

Hence, the density of Si is calculated as follows: $\text{Density} = \dfrac{5 \times 10^{22} \times 28.1}{6.02 \times 10^{23}} = 2.33 \text{ g/cm}^3$

where 6.02×10^{23} is the Avogadro number.

Example 17.3 A plane in a crystal is shown in Fig. 17.12. Calculate the corresponding Miller indices.

Solution The plane shown in Fig. 17.12 has intercepts $2a$, $2b$, and $2c$ along the three crystal axes. The lattice points in a three-dimensional lattice are given by the following expression:

$$r = pa + qb + sc \qquad (17.3)$$

From Fig. 17.12 and Eq. (17.3), we can conclude that $p = 2$, $q = 2$, and $s = 2$

Taking reciprocals, we get $\left[\dfrac{1}{2}, \dfrac{1}{2}, \dfrac{1}{2}\right]$

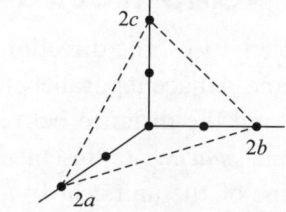

Fig. 17.12 Particular plane in a crystal

Multiplying all the numbers by the lowest common denominator 2, we get (1, 1, 1). Thus, the plane depicted in Fig. 17.12 has the Miller indices (1, 1, 1).

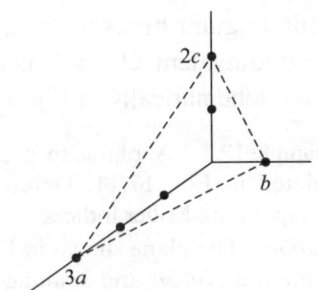

Example 17.4   A plane in a crystal is depicted in Fig. 17.13. Determine the corresponding Miller indices.

Solution   The plane shown in Fig. 17.13 has intercepts $3a$, b, and $2c$ along the three crystal axes. The lattice points in a three-dimensional lattice are given by the following expression:

$$r = pa + qb + sc \qquad (17.4)$$

Fig. 17.13 Particular plane in a crystal

From Eq. (17.4) and Fig. 17.13, we can see that $p = 3$, $q = 1$, and $s = 2$

Taking reciprocals, we get $\left[\dfrac{1}{3}, \dfrac{1}{1}, \dfrac{1}{2}\right]$

Multiplying all the numbers by the lowest common denominator 6, we get (2, 6, 3). Thus, the plane depicted in Fig. 17.13 has the Miller indices (2, 6, 3).

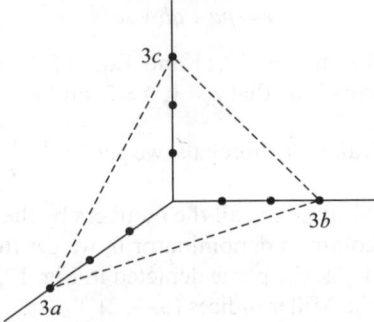

Example 17.5   A plane in a crystal is shown in Fig. 17.14. Calculate the corresponding Miller indices.

Fig. 17.14 Particular plane in a crystal

Solution   The plane shown in Fig. 17.14 has intercepts $3a$, $3b$, and $3c$ along the three crystal axes. We know that the lattice points in a three-dimensional lattice are given by the following expression:

$$r = pa + qb + sc \qquad (17.5)$$

From Fig. 17.14 and Eq. (17.5), we can conclude that $p = 3$, $q = 3$, and $s = 3$

Taking reciprocals, we get $\left[\dfrac{1}{3}, \dfrac{1}{3}, \dfrac{1}{3}\right]$

Multiplying all the numbers by lowest common denominator 3, we get (1, 1, 1). Thus, the plane depicted in Fig. 17.14 has the Miller indices (1, 1, 1).

Notice that Examples 17.4 and 17.6 have the same answers since the referred planes are parallel.

17.7 SPACING IN CUBIC LATTICES

For cubic lattices, the direction represented by [*hkl*] is perpendicular to the (*hkl*) plane. Adjacent parallel planes of atoms in a crystal have the same Miller indices, and the distance between two adjacent parallel planes is called the *interplanar spacing*, represented by d_{hkl}. Lattice parameters describe the size and shape of the unit cell. In a cubic system, the length of one of the sides of the cube completely describes the unit cell. This length is called the *lattice parameter* a_0. For cubic materials, the interplanar spacing is given by the following equation:

$$d_{hkl} = \frac{a_0}{\sqrt{h^2 + k^2 + l^2}} \tag{17.6}$$

where a_0 is the lattice parameter. For non-cubic materials, the interplanar spacing is given by more complex equations.

Example 17.6 The lattice constant of a unit cell of aluminium is 4.049 Å. Calculate the spacing of (220) planes.

Solution Let (hkl) be the Miller indices; then from Eq. (17.6),

$$d = \frac{a}{\sqrt{h^2 + k^2 + l^2)}}, \text{ with } a = 4.049\,\text{Å}$$

where $h = 2, k = 2, l = 0$

Thus, $d = \dfrac{4.049}{\sqrt{(2^2 + 2^2 + 0^2)}} = 1.432\,\text{Å}$

Example 17.7 The lattice constant of a crystalline material with cubic lattices is found to be 4.24 Å. Calculate the spacing of (110) planes.

Solution Let (hkl) be the Miller indices, then from Eq. (17.6)

$$d = \frac{a}{\sqrt{h^2 + k^2 + l^2}}, \text{ with } a = 4.24\,\text{Å}$$

where $h = 1, k = 1,$ and $l = 0$. Thus, $d = \dfrac{4.24}{\sqrt{1^2 + 1^2 + 0}} = \dfrac{4.24}{\sqrt{2}}$

leading to $d \cong \dfrac{4.24}{1.414} \cong 2.999\,\text{Å}$

Example 17.8 Spacing between (220) planes is found to be 1.41 Å. Determine the lattice constant.

Solution From Eq. (17.6), we get

$$d = \frac{a}{\sqrt{h^2 + k^2 + l^2}} \tag{17.7}$$

which can be rewritten as follows:

$$a = d\sqrt{h^2 + k^2 + l^2} \tag{17.8}$$

Substituting given values in Eq. (17.8) leads to $a = 1.41\sqrt{2^2 + 2^2 + 0^2}$

resulting in $a = 1.41\sqrt{8} = 3.988\,\text{Å}$

Example 17.9 The spacing between the planes (22x) is 1.85 Å where x is an unknown. If the lattice constant is 5.55 Å, calculate x.

Solution Equation (17.6) for spacing *d* is as follows:

$$d = \frac{a}{\sqrt{h^2 + k^2 + l^2}} \qquad (17.9)$$

This equation can be rewritten as follows:

$$\sqrt{h^2 + k^2 + l^2} = \frac{a}{d} \qquad (17.10)$$

Using the given values in Eq. (17.10), we get the following relation:

$$\sqrt{2^2 + 2^2 + x^2} = \frac{5.55}{1.85} = 3$$

which results in $x^2 = 9 - 8 = 1$
implying that $x = 1$.

Example 17.10 Determine the spacing between [100], [110], and [111] planes in an NaCl crystal having a lattice constant $a = 5.64$ Å.
Solution We know that

$$d = \frac{a}{\sqrt{(h^2 + k^2 + l^2)}}$$

For [100] planes, $h = 1$, $k = 0$, $l = 0$; hence, $d_{100} = \dfrac{5.64}{\sqrt{(1^2 + 0^2 + 0^2)}} = 5.64$ Å

For [110] planes, $h = 1$, $k = 1$, $l = 0$; hence, $d_{110} = \dfrac{5.64}{\sqrt{(1^2 + 1^2 + 0^2)}} = 4.00$ Å

For [111] planes, $h = 1$, $k = 1$, $l = 1$; hence, $d_{111} = \dfrac{5.64}{\sqrt{(1^2 + 1^2 + 1^2)}} = 3.26$ Å

17.8 NUMBER OF ATOMS PER UNIT CELL

Each unit cell is defined by a specific number of lattice points. A given lattice point can, however, be shared by more than one unit cell. Referring to Fig. 17.15, we can see that a lattice point at a corner of one unit cell is shared by seven adjacent unit cells. Thus, one lattice point is shared by eight unit cells. This implies that one-eighth of each corner lattice point belongs to any one particular unit cell. Each unit cell in turn has eight corners. Thus, the number

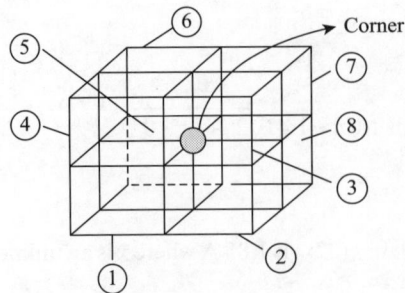

Fig. 17.15 Corner lattice point shared by eight unit cells

of lattice points contributed by all the corner positions in one unit cell is given by the following expression:

$$\frac{1}{8}\left(\begin{array}{c}\text{lattice point}\\ \text{per corner}\end{array}\right) \times 8\left(\begin{array}{c}\text{corner}\\ \text{per unit cell}\end{array}\right) = \frac{1 \text{ lattice point}}{\text{per unit cell}} \qquad (17.11)$$

Finally, the number of atoms per unit cell is given by the product of the number of atoms per lattice point and the number of lattice points per unit cell. For most metals, each lattice point has one atom.

For a simple cubic unit cell, lattice points are present only at the corners of the cube. Thus,

$$\text{Number of lattice points per unit cell} = (8 \text{ corners}) \times \left(\frac{1}{8}\right) = 1 \quad (17.12)$$

Since one atom is located per lattice point, the number of atoms per unit cell for a simple cubic unit cell is 1.

For a BCC unit cell, lattice points are present at each corner and at the centre of the cube. Thus, the number of lattice points per unit cell is as follows:

$$\text{Number of lattice points per unit cell} = (8 \text{ corners}) \times \frac{1}{8}$$

$$+ (1 \text{ centre}) \times 1 = 2 \quad (17.13)$$

The number of atoms per unit cell for a BCC unit cell is thus 2.

For an FCC unit cell, lattice points are present at all the corners and all the faces of the cube. Note that each face is shared by two unit cells. Hence, the number of lattice points per face is 1/2. The number of lattice points per unit cell is, therefore, given by the following expression:

$$\text{Number of lattice points per unit cell} = (8 \text{ corners}) \times \frac{1}{8}$$

$$+ (6 \text{ faces}) \times \frac{1}{2} = 4 \quad (17.14)$$

Thus, an FCC unit cell has four atoms.

Example 17.11 A crystalline material has a structure identical to silicon. If the atomic concentration of the material is 1×10^{22} atoms/cm³, calculate the lattice constant.

Solution Number of atoms/cell = 8

$$.\text{Atomic concentration} = \frac{\text{No. of atoms/cell}}{\text{Cell volume}} = \frac{8}{a^3} \qquad (17.15)$$

where a is the lattice constant.

From Eq. (17.15), we get $a^3 = \dfrac{8}{1 \times 10^{22}} = 8 \times 10^{-22}$

resulting in, $a = 9.28 \text{Å}$

Example 17.12 If the density of the material indicated in Example 17.11 is 1.55 g/cm³, calculate the atomic weight of the material.

Solution The atomic weight is calculated as follows:

$$\text{Density} = \frac{\text{Atomic concentration} \times \text{atomic weight}}{6.02 \times 10^{23}} \quad (17.16)$$

Equation (17.16) can be rewritten as follows:

$$\text{Atomic weight} = \frac{\text{Density} \times 6.02 \times 10^{23}}{\text{Atomic concentration}} \quad (17.17)$$

Substituting the values in Eq. (17.17), we get atomic weight $= \dfrac{1.55 \times 6.02 \times 10^{23}}{1 \times 10^{22}}$

yielding atomic weight $= 93.3$

17.9 ATOMIC RADIUS

Figure 17.16 shows the arrangement of atoms in a simple cubic structure. From this figure, we can conclude that

$$\text{Atomic radius} = r = \frac{a_0}{2} \quad (17.18)$$

where a_0 is the lattice parameter.

The arrangement of atoms in a BCC structure is shown in Fig. 17.17. From this figure we can see that $4r = \sqrt{3}\, a_0$

or $\quad r = \dfrac{\sqrt{3}}{4} a_0 \quad (17.19)$

The corresponding schematic arrangement of atoms for an FCC structure is shown in Fig. 17.18.

In this case, we can write that $4r = \sqrt{2}\, a_0$

or $\quad r = \dfrac{\sqrt{2}}{4} a_0 \quad (17.20)$

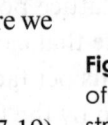

Fig. 17.16 Arrangement of atoms in simple cubic structure

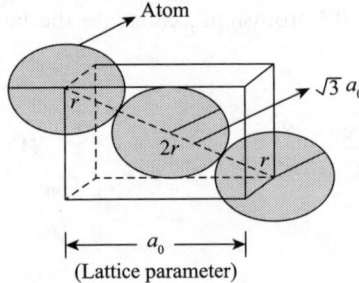

Fig. 17.17 Arrangement of atoms in BCC structure

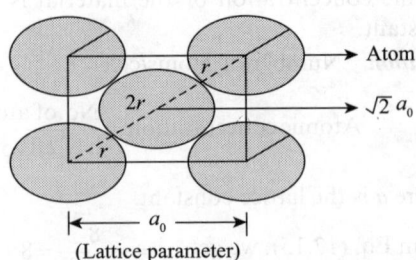

Fig. 17.18 Arrangement of atoms in FCC structure

17.10 COORDINATION NUMBER

The *coordination number* of an atom is the number of atoms touching this particular atom, or in other words, the number of nearest neighbours of the atom. Thus, the coordination number indicates how tightly and efficiently atoms are packed together in a lattice. In ionic solids, the coordination number of anions is the number of nearest cations. Similarly, the coordination number of cations is the number of nearest anions.

Figure 17.19 shows the nearest neighbours for a simple cubic lattice. It is clear from this figure that any atom in a simple cubic lattice has six nearest neighbours. The coordination number of an atom in a simple cubic lattice is thus 6.

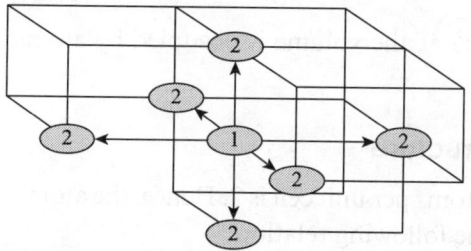

Fig. 17.19 Nearest neighbours for simple cubic lattice

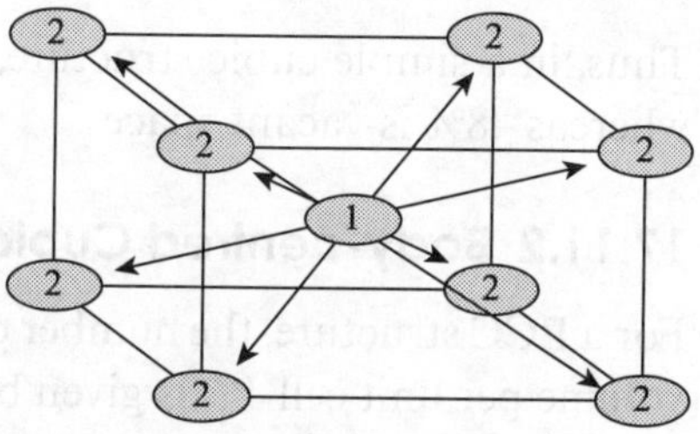

Fig. 17.20 Nearest neighbour configuration for BCC lattice

Figure 17.20 shows the nearest neighbour configuration for a BCC lattice. The coordination number for an atom in a BCC structure is 8. Similarly, the coordination number for an atom in an FCC structure is 12.

17.11 PACKING FACTOR

Atoms in a crystal can be visualized as hard spheres whose sizes are such that they touch their closest neighbours. Vacant spaces would exist in such a model of a crystalline material. In such a model of a crystal, the fraction of space occupied by atoms is called the *packing factor*. The packing factor f is defined as follows:

$$f = \frac{\text{number of atoms/unit cell} \times \text{volume of each atom}}{\text{volume of unit cell}} \quad (17.21)$$

The volume of each atom in this equation is calculated, assuming the atom to be spherical in shape. Packing factors for some important structures are evaluated in the following subsections.

17.11.1 Simple Cubic Structure

For a simple cubic structure, the number of atoms per unit cell is 1. Hence, the atomic volume per unit cell V_1, is given by the following equation:

$$V_1 = \frac{4\pi}{3} r^3 \quad (17.22)$$

where r represents the atomic radius. From Eq. (17.18), we know that

$$r = \frac{a_0}{2} \quad \text{or} \quad a_0 = 2r$$

Therefore, the volume V of a unit cell is given as follows:

$$V = a_0^3 = 8r^3 \tag{17.23}$$

Hence, the packing factor is given by the following equation:

$$f = \frac{V_1}{V} = \frac{\dfrac{4\pi}{3} r^3}{8r^3} = \frac{\pi}{6}$$

or $\quad f \cong 0.52 \tag{17.24}$

Thus, in a simple cubic structure, 52% of the volume is occupied by atoms, whereas 48% is vacant space.

17.11.2 Body-centred Cubic Structure

For a BCC structure, the number of atoms per unit cell is 2. Hence, the atomic volume per unit cell V_1, is given by the following relation:

$$V_1 = 2 \times \frac{4}{3} \pi r^3 \tag{17.25}$$

From Eq. (17.19), we know that

$$r = \frac{\sqrt{3}}{4} a_0 \quad \text{or} \quad a_0 = \frac{4r}{\sqrt{3}} \tag{17.26}$$

Therefore, the volume V of a unit cell is expressed as follows:

$$V = a_0^3 = \frac{64r^3}{3\sqrt{3}} \tag{17.27}$$

Hence, the packing factor is given by the following relation:

$$f = \frac{V_1}{V} = \frac{2 \times \dfrac{4\pi}{3} r^3}{\dfrac{64r^3}{3\sqrt{3}}} \tag{17.28}$$

$$= \frac{8\pi}{3} \times \frac{3\sqrt{3}}{64}$$

or $\quad f = \frac{\sqrt{3}}{8} \pi \approx 0.68 \tag{17.29}$

Thus, in a BCC structure 68% of the volume is occupied by atoms, whereas 32% is vacant space. A BCC structure is thus more densely packed than a simple cubic structure.

17.11.3 Face-centred Cubic Structure

For an FCC structure, the number of atoms per unit cell is 4. The volume V_1 is, therefore, given by the following relation:

$$V_1 = 4 \times \frac{4}{3} \pi r^3 \tag{17.30}$$

From Eq. (17.20), we know that

$$r = \frac{\sqrt{2}}{4} a_0 \quad \text{or} \quad a_0 = \frac{4r}{\sqrt{2}} \tag{17.31}$$

Therefore, the volume V of a unit cell is expressed as follows:

$$V = a^3_0 = \frac{64}{2\sqrt{2}} r^3 \tag{17.32}$$

Hence, the packing factor is given by the following relation:

$$f = \frac{V_1}{V} = \frac{\dfrac{16}{3} \pi r^3}{\dfrac{64 r^3}{2\sqrt{2}}}$$

$$= \frac{16\pi}{3} \times \frac{2\sqrt{2}}{64}$$

or $\qquad f = \frac{\pi\sqrt{2}}{6} \cong 0.74 \tag{17.33}$

Thus, in an FCC structure 74% of the volume is occupied by atoms, whereas 26% is vacant space. The FCC configuration of atoms thus represent the most densely-packed structure among the various cubic structures.

17.11.4 Hexagonal Close-packed Structure

A schematic diagram of a hexagonal close-packed (HCP) structure is shown in Fig. 17.21. The unit cell of this structure has one atom at each of the 12 corners of a hexagonal prism, an atom each at the centre of the two hexagonal faces, and three atoms that are symmetrically arranged within the body of the unit cell.

Each corner atom of the unit cell is shared by six other unit cells. Thus, the number of atoms per unit cell from the top hexagonal face is $(1/6) \times 6 = 1$. Similarly, the number of atoms per unit cell from the bottom hexagonal face is $(1/6) \times 6 = 1$. Each atom at the centre of the two hexagonal faces is shared by two unit cells. Thus, the number of atoms per unit cell, due to atoms at the centre of the top and bottom hexagonal faces is $(1/2) \times 2 = 1$. In addition, the number of atoms per unit cell due to the three atoms placed symmetrically within the unit cell is 3. Adding up all these contributions yields the following result:

$$\text{Number of atoms per unit cell} = 6 \tag{17.34}$$

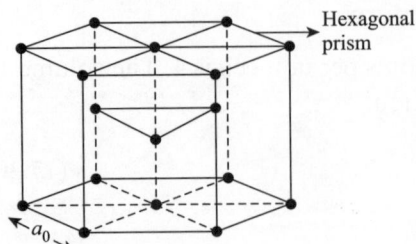

Fig. 17.21 Schematic diagram of HCP structure

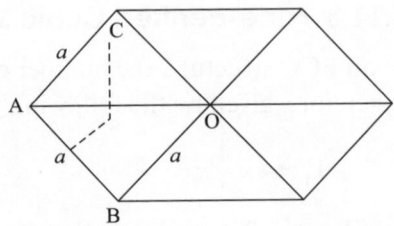

Fig. 17.22 Schematic diagram of bottom hexagonal face of HCP structure

The atomic radius r and the lattice parameter a for an HCP structure are related through the following expression:

$$r = \frac{a}{2} \tag{17.35}$$

Figure 17.22 is a schematic diagram of the bottom hexagonal face of an HCP structure. It can be seen from this figure that

$$\text{Area of the hexagonal base} = 6 \times \text{Area of triangle AOB} \tag{17.36}$$

In addition, from Fig. 17.23,

$$\text{Area of triangle AOB} = \frac{1}{2}(\text{BO}) \times (\text{AY})$$

In triangle ABY in Fig. 17.23,

$$\frac{\text{AY}}{\text{AB}} = \cos\ 30° = \frac{\sqrt{3}}{2} \tag{17.37}$$

or $\quad$ $$\text{AY} = \frac{\text{AB}\sqrt{3}}{2} \tag{17.38}$$

Since $\text{AB} = a$, this gives

$$\text{AY} = \frac{a\sqrt{3}}{2} \tag{17.39}$$

Hence,

$$\text{Area of triangle AOB} = \frac{1}{2} \times a \times \frac{a\sqrt{3}}{2} = \frac{a^2}{2} \times \frac{\sqrt{3}}{2} = \frac{\sqrt{3}a^2}{4} \tag{17.40}$$

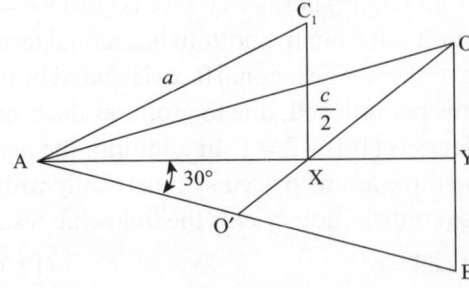

Fig. 17.23 Schematic arrangement of atoms in HCP structure

Using Eq. (17.36), we get the following relation:

$$\text{Area of the base} = 6 \times \frac{\sqrt{3}}{4}a^2 = \frac{3}{2}\sqrt{3}a^2 \tag{17.41}$$

Hence, the volume V of the unit cell is given by the following expression:

$$V = \text{Area} \times \text{Height} = \frac{3}{2}\sqrt{3}a^2 c \tag{17.42}$$

where c represents the height of the unit cell. Points A, B, and O in Fig. 17.22 represent lattice points. At a perpendicular distance $c/2$, the next atom lies at C_1, as shown in Fig. 17.23.

We have

$$AX = \frac{2}{3}AY \tag{17.43}$$

Substituting from Eq. (17.39), we get the following relation:

$$AX = \frac{2}{3}\frac{a\sqrt{3}}{2} = \frac{a}{\sqrt{3}} \tag{17.44}$$

In triangle AXC_1 in Fig. 17.23,

$$\left(AC_1\right)^2 = \left(AX\right)^2 + \left(C_1X\right)^2 \tag{17.45}$$

Using Eq. (17.44) and substituting values of AC_1 and C_1X in Eq. (17.45), one gets the following relation:

$$a^2 = \left(\frac{a}{\sqrt{3}}\right)^2 + \left(\frac{c}{2}\right)^2$$

or $$a^2 = \frac{a^2}{3} + \frac{c^2}{4}$$

or $$\frac{c^2}{4} = a^2 - \frac{a^2}{3} = \frac{2a^2}{3}$$

or $$\frac{c^2}{a^2} = \frac{8}{3}$$

or $$\frac{c}{a} = \sqrt{\frac{8}{3}} = 1.633 \tag{17.46}$$

We know that the lattice constant $a = 2r$ or $r = a/2$.

In addition, since the number of atoms per unit cell is 6, the volume V_1 of the atoms is given by the following relation:

$$V_1 = 6 \times \frac{4}{3}\pi r^3 = \frac{24}{3}\pi\left(\frac{a}{2}\right)^3 = \pi a^3 \tag{17.47}$$

Hence, from Eqs (17.47) and (17.42), the packing factor can be calculated as follows:

$$f = \frac{\pi a^3}{\frac{3}{2}\sqrt{3}a^2 c} \tag{17.48}$$

which in conjunction with Eq. (17.46) gives

$$f = \frac{\pi}{\frac{3}{2}\sqrt{3}}\left(\frac{3}{8}\right)^{1/2} = \frac{\pi}{3\sqrt{3}} = 0.74 \tag{17.49}$$

Thus, for an HCP structure, 74% of the volume is occupied by atoms, whereas 26% is vacant space.

Example 17.13 Calculate the surface density of atoms in the (111) plane of a BCC structure. Assume that the lattice constant $a = 5$ Å. In addition, assume the atoms to be hard spheres, with the closest atoms touching each other.

Solution Figure 17.24 shows a (111) plane for a BCC crystal.

To calculate the planar concentration n_{hkl} on a given (hkl) plane, only atoms whose centres lie on the bound area A need to be considered. The shaded area is an equilateral triangle defined by face diagonals of length $a\sqrt{2}$.

$$\text{Height of this triangle} = a\sqrt{\frac{3}{2}}$$

Thus, the area of the shaded triangular portion is given the following relation:

$$\frac{1}{2}a\sqrt{2} \times a\sqrt{\frac{3}{2}} = \frac{a^2\sqrt{3}}{2}$$

Every atom at the corner contributes 1/6 of its part to this area. Thus, planar concentration $n(111)$ is given by the following relation:

$$n_{(111)} = \frac{\frac{3}{6}}{\frac{a^2\sqrt{3}}{2}} = \frac{1}{a^2\sqrt{3}}$$

Substituting the given value of a, we get the following result:

$$n_{(111)} = \frac{1}{(5\times10^{-10})^2 \times \sqrt{3}} = 2.3\times10^{18} \text{ atoms/m}^2$$

(111)

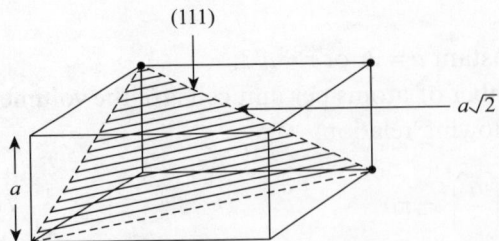

Fig. 17.24 BCC crystal showing (111) plane

Example 17.14 The distance d_{110} between (110) planes in a BCC crystal is 2.03 Å. What is the size of the unit cell?

Solution We know that

$$d_{110} = \frac{a}{\sqrt{(h^2 + k^2 + l^2)}}$$

Therefore, the size a of the unit cell is calculated as follows:

$$a = d_{110} \times \sqrt{(h^2 + k^2 + l^2)} \quad \text{or} \quad a = 2.03\,\text{Å} \times \sqrt{(1^2 + 1^2 + 0^2)} \quad \text{or} \quad a = 2.87\,\text{Å}$$

Example 17.15 An element has the HCP structure. If the radius r of the atom is 1.605 Å, find the volume of the unit cell.

Solution In the HCP structure, $a = 2r = 3.2 \times 10^{-10}$ m. We know that

$$\frac{c}{a} = \sqrt{\left(\frac{8}{3}\right)}$$

Hence, $c = a\sqrt{\left(\frac{8}{3}\right)} = 3.2 \times 10^{-10}\,\text{m} \times 1.66$

or $\qquad c = 5.24$ Å

The volume of the unit cell is calculated as follows:

$$V = \frac{3\sqrt{3}a^2c}{2}$$

$$= \frac{3\sqrt{3} \times (3.2 \times 10^{-10})^2 \times 5.24 \times 10^{-10}}{2}$$

Thus, the volume of the unit cell is $V = 1.4 \times 10^{-28}\,\text{m}^3$

Example 17.16 Surface density of atoms in the (111) plane of a BCC structure is 1.8×10^{18} atoms/m². Calculate the lattice constant.

Solution From Example 17.13, we know that

$$n_{(111)} = \frac{1}{a^2\sqrt{3}} \tag{17.50}$$

rewriting, we get

$$a^2 = \frac{1}{n_{(111)}\sqrt{3}} \tag{17.51}$$

Substitution of the values in Eq. (17.51) yields the following relation:

$$a^2 = \frac{1}{\sqrt{3} \times 1.8 \times 10^{18}} = 3.207 \times 10^{-19}$$

resulting in

$$a = \sqrt{3.207 \times 10^{-19}} = 5.66\,\text{Å}$$

17.12 CRYSTAL STRUCTURES

Crystalline materials can exist in a variety of structures. The characteristic properties of these materials are often dictated by the specific structure. We will be discussing some common crystal structures in this section.

17.12.1 NaCl

The sodium chloride crystal has a structure that consists of one Na^+ ion and one Cl^- ion associated with each lattice point in an FCC configuration. A schematic diagram of the NaCl lattice is shown in Fig. 17.25.

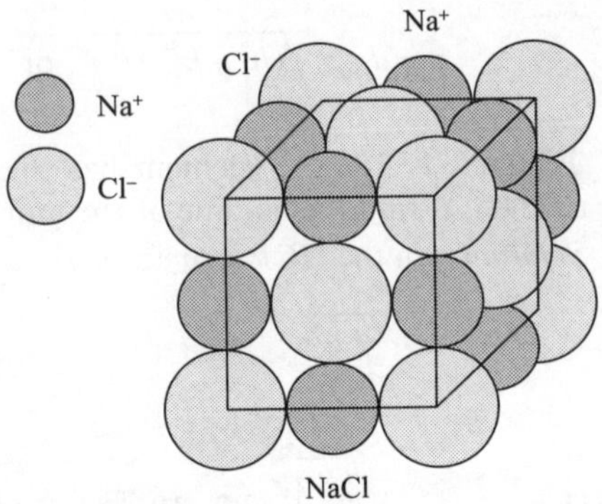

17.12.2 Diamond

The unit cell of a diamond lattice is shown in Fig. 17.26.

The diamond lattice is obtained by inserting one FCC lattice into another FCC lattice and displacing the same along the body diagonal by one-fourth of its length. This results in the formation of a tetrahedron within the diamond lattice, as shown in Fig. 17.27.

Each atom in the diamond lattice is surrounded by four nearest neighbours, which are located at the apexes of a tetrahedron having an edge of $a/2$. The basic lattice structure of many semiconductor materials used in the modern semiconductor industry, for example, Si and Ge, is a diamond lattice.

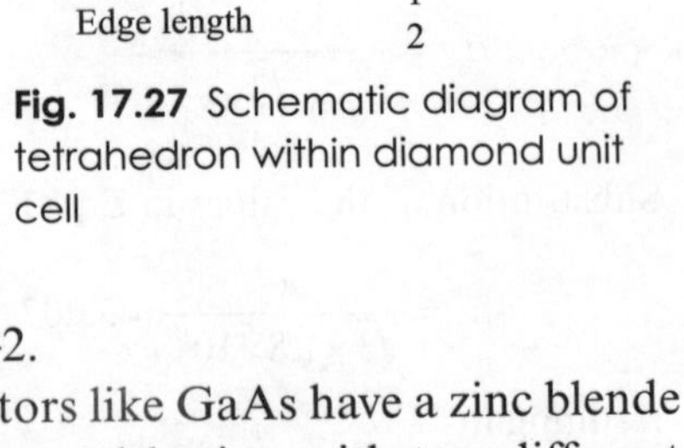

17.12.3 ZnS

The ZnS or zinc blende structure is shown in Fig. 17.28. Zn ions have a charge of +2 and S ions have a charge of −2.

Some common compound semiconductors like GaAs have a zinc blende lattice, which is closely related to the diamond lattice, with two different types of atoms in the lattice. For example, in a GaAs lattice, one sub-lattice

Fig. 17.25 Schematic diagram of NaCl lattice

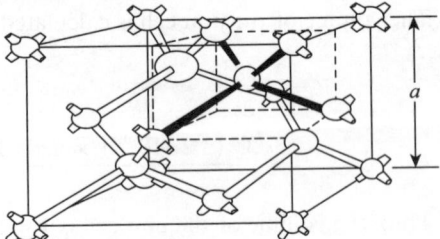

Fig. 17.26 Diamond unit cell

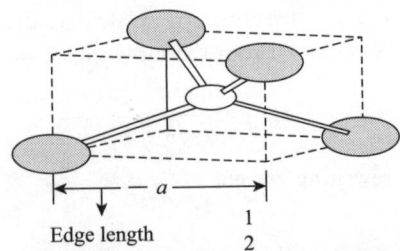

Fig. 17.27 Schematic diagram of tetrahedron within diamond unit cell

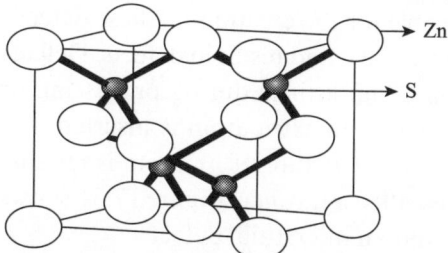

Fig. 17.28 Schematic diagram of ZnS structure

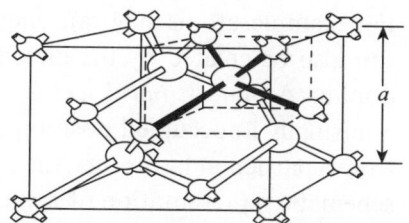

Fig. 17.29 Zinc blende (sphalerite) lattice of GaAs

is of Ga and the other is of As. Figure 17.29 shows the zinc blende lattice of GaAs.

17.12.4 Graphite

The graphite structure consists of carbon atoms arranged in regular hexagons in flat parallel layers. The bonding between the different parallel layers is not strong. The layers are, therefore, easily separable from each other. A schematic representation of the graphite structure is shown in Fig. 17.30.

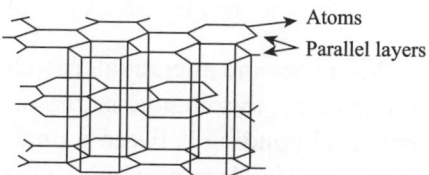

Fig. 17.30 Schematic representation of graphite structure

Weak bonding between the layers lends softness and lubricating property to graphite. Carbon atoms in the hexagonal layers are united by covalent or metallic bonds. The good electrical conductivity of graphite can be attributed to these metallic bonds.

17.13 POLYMORPHISM AND ALLOTROPY

Some materials can exist in more than one crystalline structure. This phenomenon is called *allotropy* or *polymorphism*. The term allotropy is generally used for pure elements, whereas polymorphism is used for compounds. Iron and titanium can exist in more than one crystal structure. For example, iron has a BCC structure at low temperatures, but it transforms to an FCC structure at higher temperatures. These transformations produce corresponding changes in the properties of such materials.

17.14 CRYSTAL DEFECTS

Localized disruptions in the otherwise perfect arrangement of atoms in a crystal structure are called *point defects*. A localized disruption affects not only one atom at a particular location, but also several atoms in the region around it. Several processes are responsible for the creation of these defects. For example, atoms may gain energy due to heat and result in defects. Defects can also be created by the introduction of impurities during processing or doping. An *interstitial defect* is formed when an extra atom is inserted into a normally unoccupied position in a crystal structure. If an atom is missing from a particular lattice site, the defect is called a *vacancy*. Figure 17.31 shows schematic representation of a vacancy and an interstitial defect.

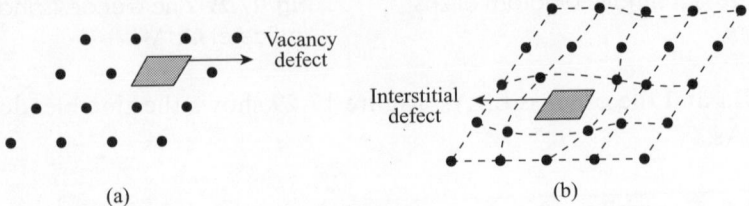

(a)   (b)

Fig. 17.31 Two-dimensional single-crystal lattice showing defects (a) Vacancy (b) Interstitial defect

Vacancies and interstitials can change the electrical properties of a material. Quite often, this change is due to the deviations produced in the nature of chemical bonding between atoms. Sometimes, a vacancy and an interstitial may occur in close proximity. One common way in which this may happen is when atoms move from their natural sites to interstitials, thereby creating a vacancy. This type of vacancy–interstitial defect is called a *Frenkel defect*. Frenkel defects produce effects that are characteristically different from those produced by simple vacancies or interstitials alone. Detailed calculations show that the equilibrium concentration n_I of interstitial atoms at a given temperature is as follows:

$$n_I = ANe^{-E_{FI}/kT} \tag{17.52}$$

where E_{FI} is the formation energy of the interstitial (generally several eV), N is the number of sites in the given volume, and A is an integer (generally, ~1) indicating the number of identical interstitial positions per lattice atom.

More complex defects can also occur in crystals. One such defect is a *Schottky defect*. In this type of defect, vacancies are not accompanied by a simultaneous transition of atoms to interstitials. If an entire row of atoms is missing from the normal lattice site, the defect is called a *line defect*. A line defect is also referred to as a line dislocation. A two-dimensional schematic diagram of a line dislocation is shown in Fig. 17.32.

If one atom in a crystal is replaced by a different atom in the normal lattice site, the defect created is called a *substitutional defect*. Figures 17.33(a) and (b) show two types of substitutional defects in a lattice.

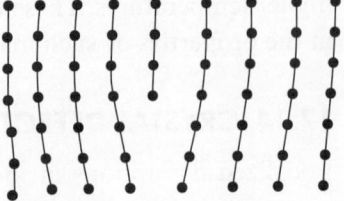

Fig. 17.32 Line dislocation in two-dimensional lattice

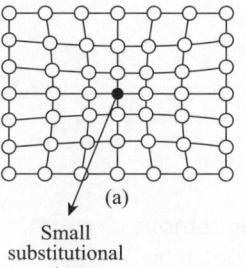

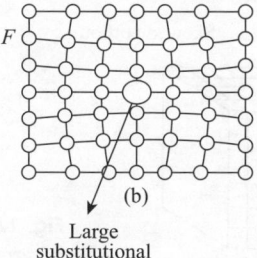

(a)

Small
substitutional
atom

(b)

Large
substitutional
atom

Fig. 17.33 Two types of substitutional atoms (a) Small substitutional atom (b) Large substitutional atom

Boundaries or planes that separate a material into different regions are called *surface defects*. The individual regions have the same crystal structure but differ in orientations. A material's surface results in abrupt disruption of the crystalline structure. Surface atoms suffer distortions in coordination number and atomic bonding. These properties often play an important role in the operation of microelectronic devices. Surfaces of materials can also exhibit defects such as roughness and notches, which make them more reactive than the bulk portion of the materials.

17.15 LINE AND VOLUME DEFECTS

A line discontinuity in a crystalline structure is called a dislocation. There are three basic types of dislocations, namely, edge dislocation, screw dislocation, and mixed dislocation. Insertion of extra half planes of unit cells leads to the development of an edge dislocation. The regions that surround the dislocation line are made up of perfect crystals. There is, however, a disruption in the crystal structure along the dislocation line. In a screw dislocation, the atomic planes do not exist separately from each other. The planes form a single surface like the threads of a screw and spiral from one end of the crystal to the other. In three-dimensional visualization, it appears like a helical structure and is not flat like a spiral. In mixed dislocations, both edge and screw dislocations are present together. Absence of a number of atoms within a crystal leads to the formation of volume defects or voids. Voids result in the formation of internal surfaces within the crystal and give rise to broken bonds at the surface. These broken bonds have properties similar to those of micro-cracks, which are areas where a solid is liable to fracture easily.

17.16 BURGER VECTOR

As mentioned earlier, line imperfections in a crystal are called dislocations. There are three types of dislocations, namely, screw dislocation, edge dislocation, and mixed dislocation. A screw dislocation is shown schematically in Fig. 17.34.

A crystal gets skewed by one atom spacing due to a screw dislocation. Suppose we start from the point x and follow a crystallographic plane for one revolution around the axis about which the crystal is skewed, and travel equal

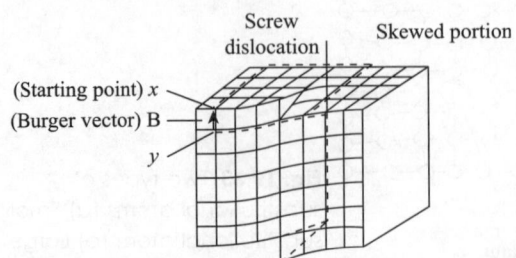

Fig. 17.34 Schematic diagram of screw dislocation

atom spacings in every direction. The complete revolution would then end at the point *y*, which is one atom spacing below the starting point *x*. The vector *B* needed to complete the loop and return to the starting point is called a *Burger vector*. Thus, a Burger vector can be seen to be parallel to the screw dislocation.

IMPORTANT CONCEPTS

1. Lattices are of two types: translational Bravais lattices and lattices with a basis.
2. A diamond lattice is an example of a three-dimensional lattice with a basis.
3. Some solids have two or more crystal structures, each of which is stable for a particular range of temperature and pressure.
4. Solids are of three types: crystalline, polycrystalline, and amorphous.
5. The periodic arrangement of atoms in a crystal is called a *lattice.*
6. A *unit cell* is the fundamental unit that undergoes regular repetition to create a lattice.
7. The smallest unit cell is called a *primitive cell.*
8. Displacement between two lattice points is represented by The following expression:

   $r = pa + qb + sc$

9. There are 14 Bravais lattices.
10. Simple cubic, body-centred cubic and face-centred cubic lattices are the three most common cubic lattices.
11. Equivalent planes in a crystal are characterized by a set of Miller indices.
12. The number of atoms touching a particular atom gives its *coordination number.*
13. The fraction of space occupied by atoms in a crystal is called the *packing* factor.
14. NaCl has an FCC configuration.
15. A diamond lattice is obtained by inserting one FCC lattice into another FCC lattice.
16. GaAs has a zinc blende lattice.
17. A graphite structure contains carbon atoms arranged in regular hexagons in flat parallel layers.
18. Allotropy or polymorphism is the phenomenon by which materials can exist in more than one crystalline structure.
19. Localized disruptions in a crystalline structure are called *point defects.*
20. An extra atom inserted into a normally unoccupied position in a crystal structure results in an *interstitial defect.*
21. A vacancy–interstitial defect is called a *Frenkel defect.*
22. In a Schottky defect, vacancies are not accompanied by a simultaneous transition of atoms to interstitials.

23. The absence of an entire row of atoms from normal lattice sites results is a *line defect*.
24. *Substitutional defect* is due to the replacement of an atom in a crystal by a different atom in the normal lattice site.
25. Boundaries or planes that separate a material into different regions are called *surface defects*.

APPLICATIONS

1. Knowledge of crystallographic planes and directions helps us in growing bulk and thin film crystals in preferred directions.
2. Knowledge of number and types of defects present in a crystalline material helps us decide its suitability for application in various devices.
3. Interplanar spacing can be verified using X-ray diffraction studies.
4. Humans express a huge number of proteins that are essential for life. Proteins are also expressed in other organisms like bacteria. Protein crystals are grown and studied by scientists to understand the related biochemistry. This knowledge is extremely useful towards the development of new drugs to fight diseases.
5. Silicon crystals form the backbone of the microelectronics industry. The fact that silicon can be grown, cut, polished, coated, and etched in a reproducible and accurate manner is due to the detailed knowledge of its crystalline structure.
6. The knowledge of single-crystal solids is used in the development of high-strength materials with low thermal creep. These materials are used in applications such as turbine blades and aircraft engines. Single crystals like sapphire are used for developing lasers and other components of nonlinear optics.

IMPORTANT FORMULAE

1. $r = pa + qb + sc$

2. $d_{hkl} = \dfrac{a_0}{\sqrt{h^2 + k^2 + l^2}}$

3. $f = \dfrac{\text{Number of atoms/unit cell} \times \text{Volume of each atom}}{\text{Volume of unit cell}}$

4. $n_I = ANe^{-E_{FI}/kT}$

SELF-ASSESSMENT

Multiple-choice Questions

17.1 Grain boundaries are present in
   (a) polycrystalline materials
   (b) crystalline materials
   (c) amorphous material
   (d) crystalline and amorphous materials

17.2 The fundamental unit that repeats to form a lattice is called a
   (a) cell   (b) unit cell   (c) crystal   (d) defect

17.3 The smallest unit cell is named
    (a) prime cell
    (b) simple cell
    (c) primitive cell
    (d) fundamental cell

17.4 The number of Bravais lattices is
    (a) 13   (b) 12   (c) 15   (d) 14

17.5 For a cubic structure
    (a) $a = b = c$
    (b) $a = b \neq c$
    (c) $a \neq b \neq c$
    (d) $a + b = c$

17.6 A set of equivalent planes is represented by
    (a) $[hkl]$ (b) $\{hkl\}$ (c) (hkl) (d) $<hkl>$

17.7 For a simple cubic lattice, d_{hkl} is given by
    (a) $\dfrac{a_0}{\left(h^2 + k^2 + l^2\right)}$
    (c) $\dfrac{a_0}{\sqrt{h^2 + k^2 + l^2}}$
    (b) $\dfrac{a_0{}^2}{\sqrt{h^2 + k^2 + l^2}}$
    (d) $\dfrac{a_0}{h + k + l}$

17.8 A unit cell of an FCC structure has
    (a) two atoms
    (b) one atom
    (c) three atoms
    (d) four atoms

17.9 For a simple cubic structure, f is
    (a) 0.52   (b) 0.52   (c) 0.51   (d) 0.49

17.10 The semiconductor GaAs has
    (a) an FCC lattice
    (b) a zinc blende lattice
    (c) a diamond-like lattice
    (d) a hexagonal lattice

17.11 Burger vector and screw dislocation
    (a) are anti-parallel
    (b) are parallel
    (c) are perpendicular
    (d) make an arbitrary angle

17.12 A vacancy–interstitial defect is called a
    (a) Schottky defect
    (b) Poole defect
    (c) Frenkel defect
    (d) line defect

17.13 Iron at low temperatures has
    (a) an FCC structure
    (b) a simple cubic structure
    (c) a hexagonal structure
    (d) a BCC structure

17.14 Zn ions in ZnS have a charge of
    (a) +2   (b) −2   (c) +1   (d) −1

17.15 For an HCP structure, the packing factor value is
    (a) 0.23   (b) 0.74   (c) 0.89   (d) 0.50

Review Questions

17.1 How many grains does a single crystal have?
17.2 Can one have a one-dimensional lattice?
17.3 What is the relation between a lattice point and a lattice?
17.4 Differentiate between orthorhombic and triclinic crystal systems.
17.5 What is the angle between (hkl) plane and $[hkl]$ direction?
17.6 Write the equivalent $<100>$ directions.
17.7 What is the number of lattice points per unit cell for a BCC unit cell?
17.8 What is the percentage of volume of vacant space in a simple cubic structure?

17.9 Name two common semiconductors that have a diamond lattice as the basic lattice structure.

17.10 Why does graphite display softness and lubricating property?

17.11 What is a space lattice?

17.12 What is a unit cell?

17.13 Define the coordination number.

17.14 Define the atomic packing factor.

17.15 What are the Miller indices?

17.16 What is a point defect?

17.17 What are line defects? What are the various types of line defects?

17.18 What is a Burger vector?

17.19 What is a lattice?

17.20 Define a unit cell.

17.21 What are basis vectors?

17.22 Explain the procedure followed to obtain the Miller indices.

17.23 How are directions represented in a lattice?

17.24 Sketch the unit cell of a diamond lattice.

17.25 Define packing factor.

17.26 Explain the meaning of coordination number.

17.27 Is the choice of unit cell unique for a given lattice?

17.28 What is a primitive cell?

17.29 Write an expression for a general displacement vector for a lattice.

17.30 What are Bravais lattices?

17.31 Explain the notations (a) (hkl) and (b) $\{hkl\}$.

17.32 Explain the expression for interplanar spacing for cubic materials.

17.33 Show the NaCl lattice schematically.

17.34 What is the basic assumption used for defining the packing factor f for a crystal?

17.35 Give the coordination numbers for (a) a BCC structure and (b) an FCC structure.

17.36 What is an interstitial defect?

17.37 What is a Schottky defect?

Numerical Problems

17.1 Hard spheres are packed in an FCC lattice with maximum packing density. Calculate the fraction of the cell filled.

$$\left[\begin{array}{l} Hint\text{: Atoms per unit cell} = 1 \text{ (corners)} + 3 \text{ (from faces)} \\ \qquad\qquad = 4; \text{ radius of each sphere} = \frac{1}{4}\left(a\sqrt{2}\right) \end{array}\right]$$

17.2 Assume that each atom is a hard sphere, with the surface of each atom being in contact with the surface of its nearest neighbour. Calculate the percentage of the total unit cell volume in a simple cubic lattice. [*Hint*: A cubic unit cell contains only one sphere and the edge length is exactly equal to the diameter of the sphere.]

17.3 Find the percentage of the total unit cell volume in a diamond lattice.
[*Hint*: There are eight atoms in one unit cell.]

17.4 Find the Miller indices for the plane shown in Fig. 17.35.

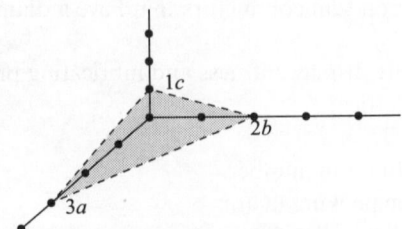

Fig. 17.35 Crystal lattice plane

17.5   Calculate the Miller indices for a plane that cuts intercepts $a = 1/2$, $b = 2$, and $c = 1/3$ along the x, y, and z axes, respectively.

[*Hint*: Lowest common denominator = 2]

17.6   Sketch the following principal planes in a cubic crystal:

(a) (100)   (b) (010)   (c) (001)

17.7   Sketch the planes (110), (101), and (011) in a cubic lattice.

17.8   Show the planes (111) and (212) in a cubic crystal.

17.9   Calculate the surface density of atoms on the (110) plane of a silicon crystal.

$$\left[\textit{Hint: (110) plane has } 4\times\frac{1}{4}+2\times\frac{1}{2}+2=4 \text{ atoms/cell}\right]$$

17.10   Evaluate the ratio $\dfrac{d_{110}}{d_{111}}$ for a cubic lattice.

17.11   A plane in a crystal is shown in Fig. 17.36. Calculate the Miller indices.

[*Hint*: $p = 3$, $q = 3$, $s = 3$]

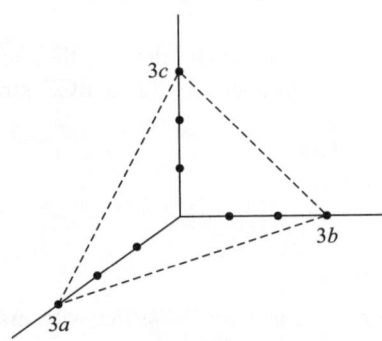

Fig. 17.36

17.12   A plane in a crystal is shown in Fig. 17.37. Determine the Miller indices for the plane.   [*Hint*: $p = 3$, $q = 3$, $s = 1$]

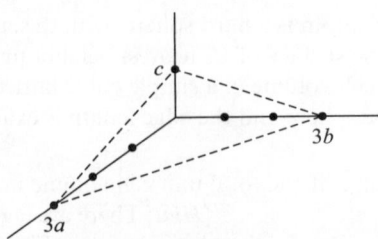

Fig. 17.37

17.13 A plane in a crystal is shown in Fig. 17.38. Calculate the Miller indices for the plane. [*Hint: $p = 4$, $q = 2$, $s = 2$*]

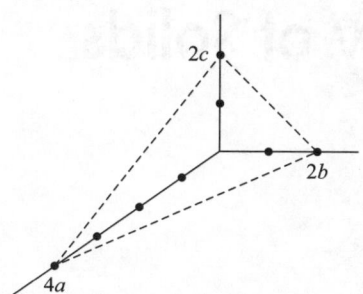

Fig. 17.38

17.14 The lattice constant of a crystalline material with cubic lattice is found to be 4.35 Å. Calculate the spacing of (110) planes.

$$\left[Hint: d = \frac{a}{\sqrt{h^2 + k^2 + l^2}} \right]$$

17.15 Spacing between (220) planes is found to be 1.38 Å. Calculate the lattice constant.

$$\left[Hint: d = \frac{a}{\sqrt{h^2 + k^2 + l^2}} \right]$$

17.16 The spacing between the planes $(22x)$ is 1.80 Å where x is an unknown. If the lattice constant is 5.4 Å, calculate x.

$$\left[Hint: d = \frac{a}{\sqrt{h^2 + k^2 + l^2}} \right]$$

17.17 A crystalline material has a structure identical to silicon. If the atomic concentration of the material is 6×10^{22} atoms/cm³, calculate the lattice constant.

17.18 If the density of the material indicated in Problem 17.17 is 1.68 g/cm³, calculate the atomic weight of the material.

$$\left[Hint: At.wt = \frac{Density \times 6.02 \times 10^{23}}{At.conc.} \right]$$

17.19 Surface density of atoms in the (111) plane of a BCC structure is 2.5×10^{18} atoms/m². Calculate the lattice constant.

$$\left[Hint: n_{(111)} = \frac{1}{a^2 \sqrt{3}} \right]$$

Band Theory of Solids

Learning Objectives

After studying this chapter, students will be able to

- elucidate the fundamentals of band theory of solids
- state the salient features of the Kronig–Penney model
- explain the behaviour of electrons in a periodic lattice
- demonstrate k-space diagrams
- solve numerical problems based on the conservation of energy
- solve numerical problems based on the band theory of solids
- point out salient features of the Fermi–Dirac distribution function
- explain the concept of work function
- solve numerical problems on effective mass
- solve numerical problems on density of states
- resolve numerical problems based on the Fermi–Dirac distribution function

List of Symbols

r_0 = Equilibrium interatomic distance

E_g = Bandgap

$\Psi(x, t)$ = Wave function (time dependent)

$V(x)$ = Potential function (time independent)

$j = \sqrt{-1}$

$\Psi(x)$ = Wave function (time independent)

$\Phi(x)$ = Time-dependent component of $\Psi(x, t)$

h = Planck's constant

$\hbar = h/2\Pi$

ω = Angular frequency

k = Wave number

E = Energy

b = Potential barrier width

p = Momentum

$f(E)$ = Fermi–Dirac distribution function

E_F = Fermi level

T = Absolute temperature

n_0 = Electron concentration in the conduction band

m^* = Band curvature effective mass

E_c = Conduction band edge

E_v = Valence band edge

$N(E) =$ Density of states

$N_c =$ Effective density of states at the conduction band edge

$m_n^* =$ Density-of-state effective mass of electrons

$p_0 =$ Concentration of holes in valence band

$m_l =$ Longitudinal effective mass

$m_t =$ Transverse effective mass

$N_v =$ Effective density of states in the valence band

$n_i =$ Intrinsic electron concentration

$p_i =$ Intrinsic hole concentration

$E_i =$ Position of Fermi level for an intrinsic semiconductor

$F =$ Force

$q =$ Charge on an electron

$a_1 =$ Acceleration

$W =$ Work done

$m_{\text{eff}} =$ Effective mass

18.1 INTRODUCTION

Solid-state materials play a dominant role in the modern-day world. Communication, entertainment, transportation, defence, and all other important aspects linked to our lives are critically dependent on the developments in this field. This chapter will deal with the theoretical concepts that determine the properties of solid-state materials. Semiconductors, metals, and insulators are important classes of solids with unique properties. These properties can be well understood by assuming the energy of the system that consists of allowed and forbidden bands. We will discuss the Kronig–Penny model that explains the formation of bands in solids. It must, however, always be remembered that in reality, no band exists as a physical entity. Bands are mathematical conclusions that result from some very basic assumptions. Electrons residing in solids are forced to come out when sufficient/threshold energy is supplied. This energy is called the work function of a material. We will learn the significance of this parameter in this chapter. As we know, mass of a particle decides the resistance it offers to any attempt to change its state of motion. Electrons and other charged particles moving inside a solid are not free, as they move with restrictions imposed on them by the presence of other atoms and charged particles. The parameter that now decides their resistance to any attempt to change of state is called the effective mass. In this chapter, we will see how this parameter is related to the band theory of solids.

18.2 BAND THEORY

Many important solids can be obtained in crystalline forms. As discussed in Chapter 17, a crystal consists of a collection of atoms. As atoms are brought together, various interactions take place between the neighbouring atoms. Some of these interactions result in bonding forces. A system consisting of a collection of atoms has properties that are different from individual atoms. One of the fundamental functions characterizing an electron in an atom is its

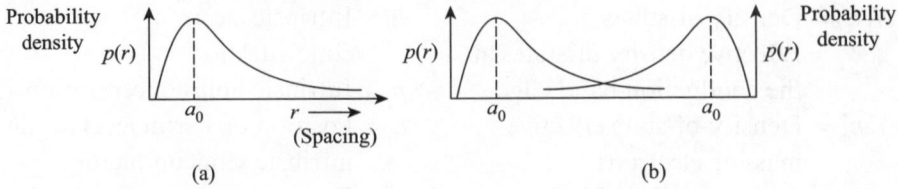

Fig. 18.1 Radial probability density functions (a) For one hydrogen atom (b) For two hydrogen atoms in close proximity

radial probability density, which is defined as the probability of finding the electron at a particular distance from the nucleus. The radial probability density function for the lowest electron energy state of a single, non-interacting hydrogen atom is shown in Fig. 18.1(a). The parameter a_0 is equal to the Bohr radius. Probability density functions for two atoms in close proximity are shown in Fig. 18.1(b).

One can easily infer from Fig. 18.1(b) that the probability functions of the two electrons overlap, indicating an interaction. Thus, the two electrons now form one system and, therefore, cannot have the same energy level, according to the *Pauli's exclusion principle*. The discrete quantized energy level of the individual atom, therefore, splits into two states, as shown in Fig. 18.2.

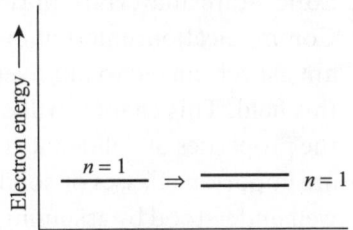

Fig. 18.2 Splitting of discrete quantized energy level

Suppose we now generalize the aforementioned discussion and bring a large number of hydrogen-type atoms in close proximity in a periodic arrangement. Consistent with the Pauli's exclusion principle, the initial quantized levels of individual atoms would now split into a band of discrete energy levels, as shown schematically in Fig. 18.3.

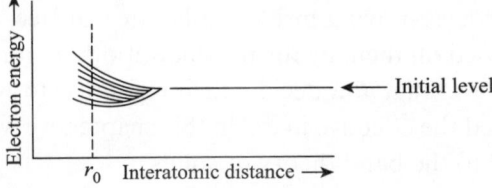

Fig. 18.3 Formation of bands of allowed energy

Let us consider that the parameter r_0 represents the equilibrium interatomic distance in the periodic arrangement of atoms.

In this example, we have considered a regular periodic arrangement of atoms having one electron. Suppose we now consider a periodic arrangement of atoms containing electrons up to the $n = 3$ energy level. When the atoms are separated by large distances, there is no interaction between individual atoms and each atom has its allowed discrete energy levels. When the atoms are brought together, the outermost electrons ($n = 3$ energy shell) start interacting.

This results in the discrete energy levels splitting into a band of allowed energies. When the atoms come even closer, electrons in the $n = 2$ shell start interacting and the corresponding energy levels also go through the same fate. When the atoms get sufficiently closer, the innermost electrons ($n = 1$ shell) also interact. The process of formation of allowed energy bands is shown in Fig. 18.4.

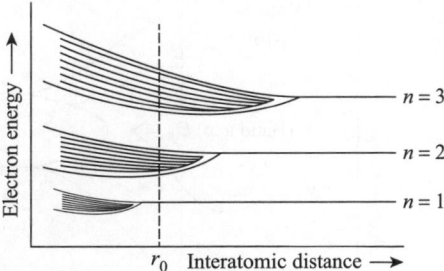

Fig. 18.4 Splitting of three energy states into allowed bands

The parameter r_0 represents the equilibrium interatomic distance. Some interesting characteristics of the schematic shown in Fig. 18.4 are as follows:

1. Discrete energy levels of individual atoms split into allowed energy bands.
2. There exist allowed energy bands corresponding to every energy level.
3. Forbidden bands exist, separating any two allowed energy bands.

Let us now look at the consequences of the band theory of solids for a semiconductor such as silicon. A silicon atom has 14 electrons, 10 of which occupy deep-lying energy levels close to the nucleus, as shown in Fig. 18.5.

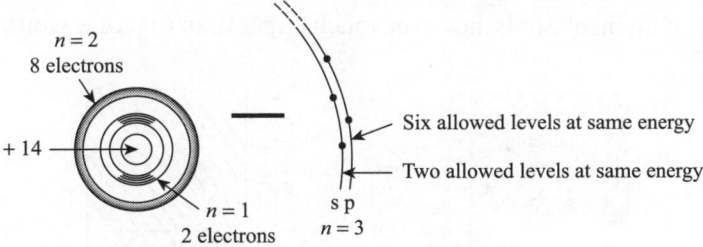

Fig. 18.5 Energy levels of isolated silicon atom

The remaining four valence electrons are relatively weakly bound and participate in chemical reactions. These also are shown in Fig. 18.5. The $n = 3$ energy level consists of the 3s and the 3p states. The 3s state corresponds to $n = 3$ and $l = 0$, and contains two quantum states per atom. At absolute zero temperature, that is, $T = 0$ K, the state contains two of the four valence electrons. The 3p state corresponds to $n = 3$ and $l = 1$, and contains six quantum states per atom. This state contains the remaining two electrons of an individual silicon atom. Let us now take a large number (N) of silicon atoms and arrange them in a periodic manner. When we bring the atoms closer, the 3s and 3p states go through band splitting, as shown in Fig. 18.6.

At the equilibrium interatomic distance a_0, we have four quantum states per atom in the lower band and four quantum states per atom in the upper band. At absolute zero temperature, all the valence electrons are in the lower band, and, for this reason, this band is often referred to as the *valence band*.

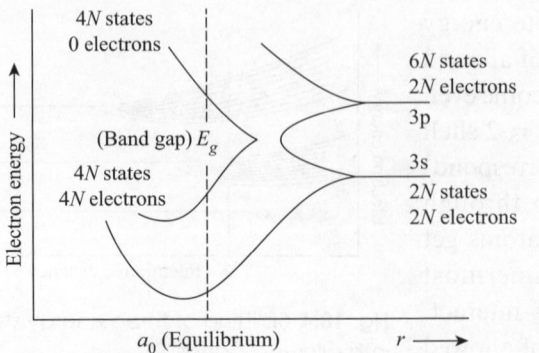

Fig. 18.6 Splitting of 3s and 3p states of silicon

The upper band is completely empty at absolute zero temperature and is called the *conduction band*. The gap between the top of the valence band and the bottom of the conduction band is called the bandgap energy E_g.

Formation of energy bands in a semiconductor plays an important role in deciding the electrical characteristics of semiconductors. The concept of energy bands can also be used to differentiate between the three kinds of solids, namely, insulators, semiconductors, and metals. Insulators have a bandgap separating a filled valence band and a conduction band. The band structure of insulators is thus quite similar to that of semiconductors, as shown in Fig. 18.7. The energy bandgap of an insulator is, however, much larger than that of a semiconductor.

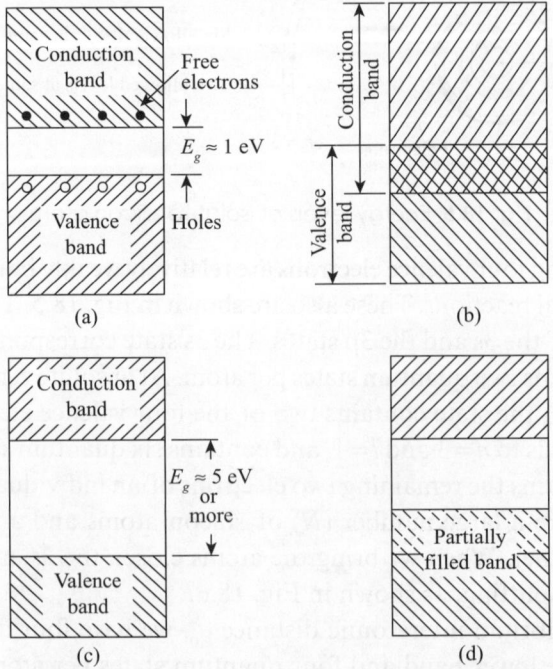

Fig. 18.7 Energy band diagrams (a) Semiconductor (b) Metal having overlapping bands (c) Insulator (d) Metal having partially filled bands

Typically speaking, the energy bandgap E_g of a semiconductor is of the order of 1 eV, whereas insulators have energy bandgaps in the range of 5–15 eV. Metals, on the other hand, present a completely different type of band structure. The valence band may merge into an empty band, resulting in overlapping bands. The last filled band may overlap with the first empty band, resulting in a partially filled band. The high conductivity of metals like Cu is due to their energy band structure.

18.3 KRONIG–PENNEY MODEL

In the last section, we have seen how individual atoms react to their grouping in a periodic arrangement of atoms. The presentation was, however, qualitative and therefore not rigorous. A more detailed theory involves quantum mechanics and needs the solution of Schrodinger's wave equation. This theory is called the Kronig–Penney model. The Kronig–Penney model is able to predict the detailed band structure and electron dynamics within actual crystal structures.

18.3.1 Schrodinger's Wave Equation

The origin of Schrodinger's wave equation lies in the fact that many particles exhibit wave–particle duality. You would recollect that the wave nature of particles such as electrons and protons was explained by de Broglie. In 1926, Schrodinger developed a theory called *wave mechanics* that considered both the concept of *quanta*, introduced by Planck, and the principle of wave–particle duality together. The one-dimensional, non-relativistic Schrodinger's wave equation is given as follows:

$$\frac{-\hbar^2}{2m}\frac{\partial^2 \psi(x,t)}{\partial x^2} + V(x)\psi(x,t) = \hat{i}\hbar\frac{\partial \psi(x,t)}{\partial t} \tag{18.1}$$

where $\Psi(x, t)$ represents the wave function, $V(x)$ is the time-independent potential function, and $\hat{i}$ is equal to $\sqrt{-1}$.

The wave function $\Psi(x, t)$ would have time-dependent and time-independent components. Separating the two, we can write that

$$\psi(x,t) = \psi(x)\phi(t) \tag{18.2}$$

where $\Psi(x)$ represents the time-independent, position-dependent component and $\Phi(t)$ represents the time-dependent component. Substitution of Eq. (18.2) in Eq. (18.1) leads to the following expression:

$$\frac{-\hbar^2}{2m}\phi(t)\frac{\partial^2 \psi(x)}{\partial x^2} + V(x)\psi(x)\phi(t) = \hat{i}\hbar\psi(x)\frac{\partial \phi(t)}{\partial t} \tag{18.3}$$

Dividing Eq. (18.3) by $\Psi(x, t)$, one gets the following equation:

$$\frac{-\hbar^2}{2m}\frac{1}{\psi(x)}\frac{\partial^2 \psi(x)}{\partial x^2} + V(x) = \hat{i}\hbar \cdot \frac{1}{\phi(t)}\frac{\partial \phi(t)}{\partial t} \tag{18.4}$$

The left-hand side of Eq. (18.4) is a function of only position, whereas the right-hand side is a function of only time. The two sides should, therefore, be equal to a constant. The time-dependent component of Eq. (18.4) can be written in the following form:

$$i\hbar \cdot \frac{1}{\phi(t)} \frac{\partial \phi(t)}{\partial t} = A_1 \tag{18.5}$$

where A_1 represents the separation of variables constant. Solution of Eq. (18.5) is of the following form:

$$\phi(t) = e^{-i\left(\frac{A_1}{\hbar}\right)t} \tag{18.6}$$

If $A_1 = E$, where E is the total energy of the particle, from Eq. (18.6), we can conclude that

$$\frac{A_1}{\hbar} = \frac{E}{\hbar} = \frac{hv}{\hbar} = \omega \tag{18.7}$$

leading to

$$\phi(t) = e^{-i\omega t} \tag{18.8}$$

Thus, this form of solution of Eq. (18.5) results in the classical exponential form of a sinusoidal wave form.

From Eq. (18.3), the time-independent form of Schrodinger's equation can be written in the following form:

$$\frac{-\hbar^2}{2m} \cdot \frac{1}{\psi(x)} \frac{\partial^2 \psi(x)}{\partial x^2} + V(x) = E$$

which can be rewritten as follows:

$$\frac{\partial^2 \psi(x)}{\partial x^2} + \frac{2m}{\hbar^2}[E - V(x)]\psi(x) = 0 \tag{18.9}$$

18.3.2 Electrons in Periodic Lattice

In the qualitative treatment of band formation, we have seen the effect of bringing a group of atoms closer on the energy levels of the system. Let us start with a one-electron atom. The potential function for such a non-interacting system is schematically shown in Fig. 18.8. The existence of discreet energy levels can be seen from the figure. The shape of the potential function is due

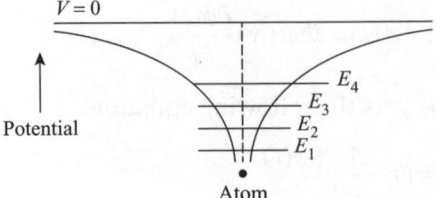

Fig. 18.8 Potential function for non-interacting system

to the $(1/r)$ dependence. The potential function has negative values and finally approaches zero.

As individual atoms are brought together, overlapping of potential functions of individual atoms takes place, as shown in Fig. 18.9(a).

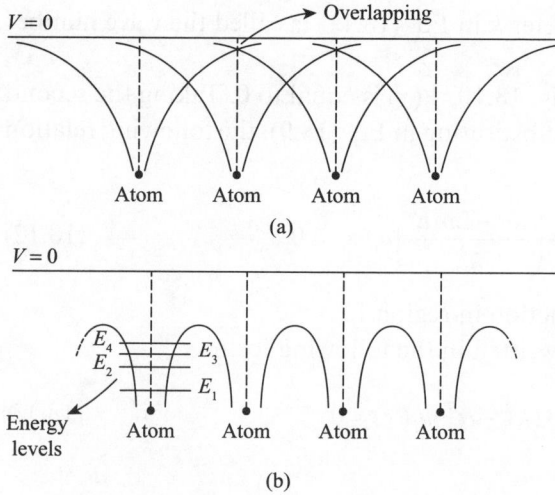

Fig. 18.9 Electron wave function (a) Overlapping potential function (b) Potential function of one-dimensional array of atoms

Figure 18.9(b) shows the potential function existing for a one-dimensional array of atoms. A solution of the Schrodinger's wave equation with a potential function represented by Fig. 18.9(b) is needed to model a one-dimensional lattice. The Kronig–Penney model simplifies the situation by assuming a simpler potential function, shown schematically in Fig. 18.10 as an approximation.

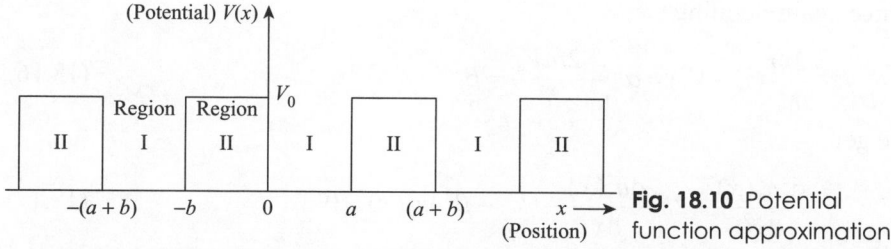

Fig. 18.10 Potential function approximation

To solve the Schrodinger's wave equation, we need to use the Bloch's theorem. According to the Bloch's theorem, *all one-electron wave functions that involve potential functions with periodic variation have the following form:*

$$\psi(x) = u(x)e^{ikx} \qquad (18.10)$$

where $u(x)$ represents a periodic function and k is a constant of motion. From Fig. 18.10, we can see that the period of $u(x)$ is $(a + b)$ for the given situation. Using Eq. (18.10) in Eq. (18.2), one gets the following relation:

$$\psi(x,t) = u(x)e^{ikx} \cdot e^{-i\left(E/\hbar\right)t}$$

which can be rewritten in the following form:

$$\psi(x,t) = u(x)e^{i\left[kx - \left(E/\hbar\right)t\right]}$$ (18.11)

Thus, the wave function of an electron moving in a single crystal is represented by a travelling wave. The amplitude of this travelling wave turns out to be a periodic function. The parameter k in Eq. (18.11) is called the wave number or wave vector.

In region I $(0 < x < a)$ of Fig. 18.10, $V(x)$ is equal to 0. Taking the second derivative of Eq. (18.10) and substituting in Eq. (18.9), the following relation is obtained:

$$\frac{d^2 u_1(x)}{dx^2} + 2ik \frac{du_1(x)}{dx} - \left(\frac{k^2 - 2mE}{\hbar^2} \right) u_1(x) = 0$$ (18.12)

where $u_1(x)$ is the periodic function in region I.

Equation (18.12) can be rewritten in the following form:

$$\frac{d^2 u_1(x)}{dx^2} + 2ik \frac{du_1(x)}{dx} - (k^2 - \alpha^2) u_1(x) = 0$$ (18.13)

where

$$\alpha^2 = \frac{2mE}{\hbar^2}$$ (18.14)

For region II, we have $-b < x < 0$ and $V(x) = V_0$. The time-independent Schrodinger's wave equation now leads to the following form:

$$\frac{d^2 u_2(x)}{dx^2} + 2ik \frac{du_2(x)}{dx} - \left(k^2 - \alpha^2 + \frac{2mV_0}{\hbar^2} \right) u_2(x) = 0$$ (18.15)

Once again defining

$$\frac{2m}{\hbar^2}(E - V_0) = \alpha^2 - \frac{2mV_0}{\hbar^2} = \beta^2$$ (18.16)

we get

$$\frac{d^2 u_2(x)}{dx^2} + 2ik \frac{du_2(x)}{dx} - (k^2 - \beta^2) u_2(x) = 0$$ (18.17)

From Eq. (18.16) we can note that parameter β is real for $E > V_0$ and imaginary for $E < V_0$. Equation (18.13) has solutions of the following form:

$$u_1(x) = A_2 e^{i(\alpha - k)x} + B_2 e^{-i(\alpha + k)x} \text{ for } (0 < x < a)$$ (18.18)

In addition, Eq. (18.17) has solutions of the following form:

$$u_2(x) = A_3 e^{i(\beta - k)x} + B_3 e^{-i(\beta + k)x} \text{ for } (-b < x < 0)$$ (18.19)

Potential function $V(x)$ being finite everywhere, the wave function $\psi(x)$ and its derivative $\partial \psi(x)/\partial t$ must be continuous. This also implies that the wave amplitude function $u(x)$ and the derivative $du(x)/dx$ are also continuous.

In Fig. 18.10, $x = 0$ is the boundary, and continuity at the boundary leads to the following relation:

$$u_1(0) = u_2(0) \tag{18.20}$$

and

$$\left.\frac{du_1}{dx}\right|_{x=0} = \left.\frac{du_2}{dx}\right|_{x=0} \tag{18.21}$$

Use of Eqs (18.18) and (18.19) in Eq. (18.20) yields the following expression:

$$A_2 + B_2 = A_3 + B_3$$

giving

$$A_2 + B_2 - A_3 - B_3 = 0 \tag{18.22}$$

Use of Eqs (18.18) and (18.19) in Eq. (18.21) yields the following relation:

$$(\alpha - k)A_2 - (\alpha + k)B_2 = (\beta - k)A_3 - (\beta + k)B_3$$

giving

$$(\alpha - k)A_2 - (\alpha + k)B_2 - (\beta - k)A_3 + (\beta + k)B_3 = 0 \tag{18.23}$$

Since $u(x)$ is continuous and periodic, we must have the following relation:

$$u_1(a) = u_2(-b) \tag{18.24}$$

Use of Eqs (18.18) and (18.19) in Eq. (18.24) leads to the following expression:

$$A_2 e^{i(\alpha-k)a} + B_2 e^{-i(\alpha+k)a} = A_3 e^{-i(\beta-k)b} - B_3 e^{i(\beta+k)b}$$

yielding

$$A_2 e^{i(\alpha-k)a} + B_2 e^{-i(\alpha+k)a} - A_3 e^{-i(\beta-k)b} - B_3 e^{i(\beta+k)b} = 0 \tag{18.25}$$

In addition,

$$\left.\frac{du_1}{dx}\right|_{x=a} = \left.\frac{du_2}{dx}\right|_{x=b} \tag{18.26}$$

Using Eqs (18.18) and (18.19) in Eq. (18.26), one gets the following equation:

$$A_2(i)(\alpha - k)e^{i(\alpha-k)a} + B_2(-i)(\alpha + k)e^{-i(\alpha+k)b}$$
$$= A_3(i)(\beta - k)e^{i(\beta-k)b} + B_3(-i)(\beta + k)e^{-i(\beta+k)b} \tag{18.27}$$

Equations (18.22), (18.23), (18.25), and (18.27) are a set of four homogeneous equations with four unknown quantities. A non-trivial solution exists only if the determinant of the coefficients is zero. It can be shown that this condition translates to the following equation:

$$\frac{-(\alpha^2 + \beta^2)}{2\alpha\beta}(\sin\alpha a)(\sin\beta b) + (\cos\alpha a)(\cos\beta b) = \cos k(a+b) \tag{18.28}$$

For an electron bound within the crystal, we have $E < V_0$. Equation (18.16) implies that for $E < V_0$, β is an imaginary quantity. Let us define β as follows:

$$\beta = j\gamma \tag{18.29}$$

where γ is a real quantity. Use of Eq. (18.29) in Eq. (18.28) results in the following relation:

$$\frac{\gamma^2 - \alpha^2}{2\alpha\gamma}(\sin\alpha a)(\sin j\gamma b) + \cos(\alpha a)\cos j(\gamma b) = \cos k(a+b) \tag{18.30}$$

At this stage, let us assume that the potential barrier width $b \to 0$ and the barrier height $V_0 \to \infty$, while keeping the product bV_0 finite. Under these conditions, Eq. (18.30) simplifies to the following form:

$$\left(\frac{mV_0 ba}{\hbar^2}\right)\frac{\sin\alpha a}{\alpha a} + \cos\alpha a = \cos ka \tag{18.31}$$

Let us define a parameter P_1 as follows:

$$P_1 = \frac{mV_0 ba}{\hbar^2} \tag{18.32}$$

Using Eq. (18.32), Eq. (18.31) becomes as follows:

$$P_1\frac{\sin\alpha a}{\alpha a} + \cos\alpha a = \cos ka \tag{18.33}$$

Equation (18.33) relates k, E, and potential barrier. It gives a condition under which the Schrodinger's wave equation has a solution.

18.3.3 Diagram for *k*-Space

The graphical representation of energy as a function of momentum or the wave number helps us understand the dynamic behaviour of electrons in a crystal lattice. This is an important result based on the Kronig–Penny model.

For a free particle there are no potential barriers. Thus, for a free particle, $V_0 = 0$. Using Eq. (18.32), this implies that $P_1 = 0$. Use of $P_1 = 0$ in Eq. (18.31) leads to the following relation:

$$\cos\alpha a = \cos ka \tag{18.34}$$

resulting in

$$\alpha = k \tag{18.35}$$

Use of Eq. (18.14) in Eq. (18.35) yields the following expression:

$$\alpha = \sqrt{\frac{2mE}{\hbar^2}} = k \tag{18.36}$$

Since the particle has no potential energy, the only form of energy is kinetic energy. Thus, Eq. (18.36) can be rewritten in the following form:

$$\alpha = \sqrt{\frac{2m\left(\frac{1}{2}mv^2\right)}{\hbar^2}} = \frac{p}{\hbar} = k \tag{18.37}$$

where p is the momentum of the free particle.

We can also write

$$E = \frac{p^2}{2m} = \frac{k^2\hbar^2}{2m} \qquad (18.38)$$

The parabolic relationship between energy and p (or k) as outlined by Eq. (18.38) is shown schematically in Fig. 18.11.

The parameter P_1 is related to the potential barrier. As the value of P_1 increases, the particle gets more and more tightly bound to the atom. The left-hand side of Eq. (18.33) is clearly a function of the quantity (αa). Thus,

$$f(\alpha a) = P_1 \frac{\sin \alpha a}{\alpha a} + \cos \alpha a \qquad (18.39)$$

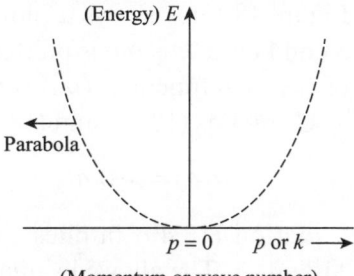

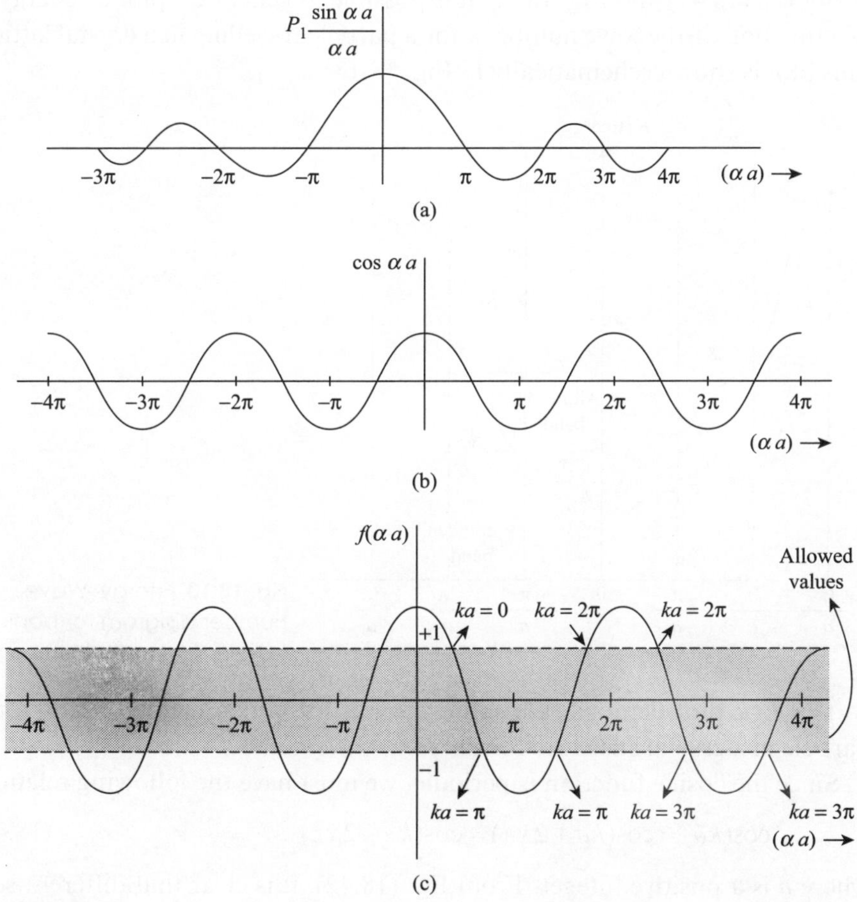

Fig. 18.11 Parabolic relationship between energy and momentum

Fig. 18.12 Plots of components of Eq. 18.39 (a) Plot of $(P_1 \sin \alpha a)/\alpha a$ vs αa (b) Plot of cos αa vs αa (c) Plot of $f(\alpha a)$ vs αa

Figure 18.12(a) shows a plot of $(P_1 \sin \alpha a/\alpha a)$ as a function of (αa). The second term, $\cos \alpha a$, is plotted as a function of (αa) in Fig. 18.12(b). A plot of the total function $f(\alpha a)$ versus (αa) is shown in Fig. 18.12(c).

Using Eqs (18.33) and (18.39), we can also write

$$f(\alpha a) = \cos ka \qquad (18.40)$$

Equation (18.40) implies that $f(\alpha a)$ must have values that lie between $+1$ and -1. The allowed values of $f(\alpha a)$ for (αa) falling in the range between $+1$ and -1 are shown as shaded portions in Fig. 18.12(c). The values of (ka) as per Eq. (18.40) corresponding to allowed values of $f(\alpha a)$ are also shown in Fig. 18.12(c).

From Eq. (18.36), we can write the following relation:

$$\alpha^2 = \frac{2mE}{\hbar^2} = k^2 \qquad (18.41)$$

Using Eq. (18.41) and Fig. 18.12, it is possible to generate a plot of energy E as a function of the wave number k for a particle travelling in a crystal lattice. This plot is shown schematically in Fig. 18.13.

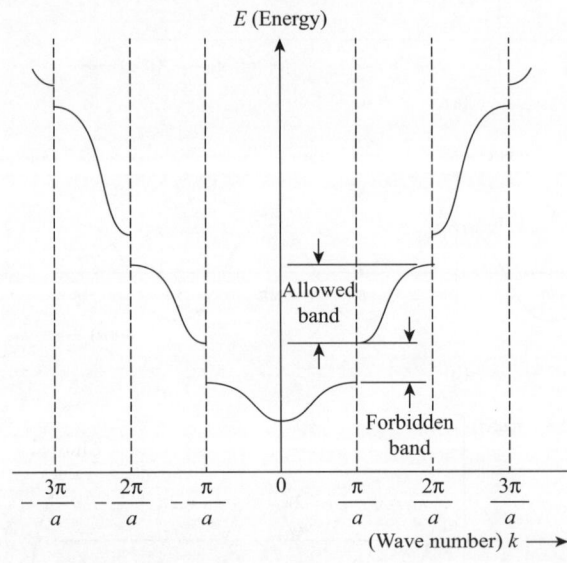

Fig. 18.13 Energy–Wave number diagram for particle travelling in crystal lattice

The E–k plot demonstrates allowed and forbidden energy bands for a particle in a crystal structure for different wave number values.

Since the cosine function is periodic, we must have the following relation:

$$\cos(ka) = \cos(ka + 2x\pi) = \cos(ka - 2x\pi) \qquad (18.42)$$

where n is a positive integer. From Eq. (18.42), it is clear that different segments of the plot shown in Fig. 18.13 can be displaced in any direction by a factor of 2π without disturbing their basic characteristics. Figure 18.14 shows

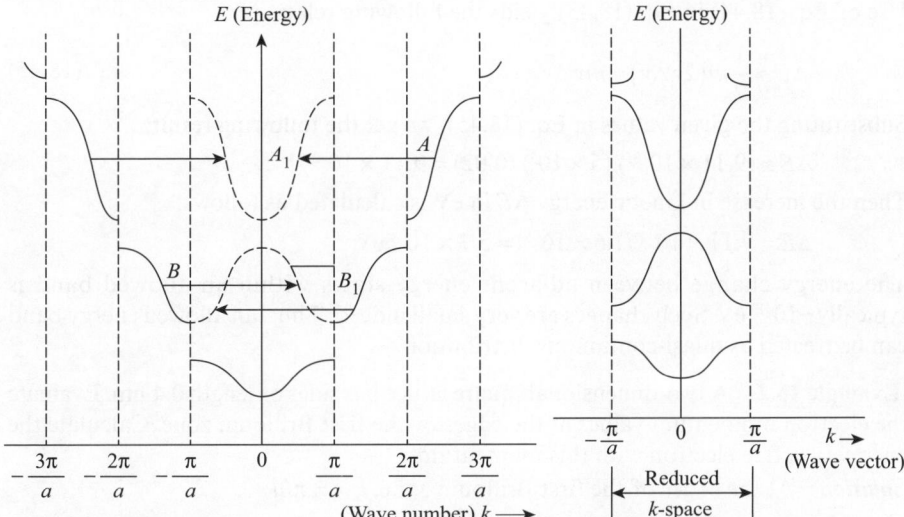

Fig. 18.14 Plot of energy vs wave number

Fig. 18.15 Energy-Wave number plot for region $-\pi/a < k < \pi/a$

a plot that has been reconstructed by displacing segments of Fig. 18.13 by 2π. The displaced portions are shown as dotted curves. The shifting of sections of the plot shown in Fig. 18.13 does not affect the magnitude of the energy gap between the allowed and forbidden bands, but makes the visualization of the bandgaps easier, as is clear from a comparison of Figs 18.13 and 18.14.

The entire E versus k plot can also be constructed within the region $-\pi/a < k < \pi/a$, as shown in Fig. 18.15.

The diagram shown in Fig. 18.15 is called a reduced k-space or a reduced – zero diagram. The different length segments of wave numbers in diagrams such as those shown in Fig. 18.14 are called Brillouin zones. The first segment $-\pi/a < k < \pi/a$ is the first Brillouin zone. The second Brillouin zone comprises the two segments $-\pi/a < k < -2\pi/a$ and $\pi/a < k < 2\pi/a$.

Example 18.1 The velocity of an electron initially travelling with a velocity of 5×10^7 cm/s increases by a value of 2 cm/s. Calculate the corresponding increase in the kinetic energy of the electron.

Solution The increase in kinetic energy, ΔE, is given by the following relation:

$$\Delta E = \frac{1}{2} m \left(v_f^2 - v_i^2 \right) \tag{18.43}$$

where v_f and v_i represent the final and initial velocities, respectively. Let us assume that

$$v_f = v_i + \Delta v$$

Then we get the following expression: $v_f^2 = (v_i + \Delta v)^2 = v_i^2 + 2v_i \Delta v + (\Delta v)^2$

Since Δv is a small quantity, the term $(\Delta v)^2$ can be neglected. This gives the following expression: $v_f^2 = v_i^2 + 2v_i \Delta v$

which implies that

$$v_f^2 - v_i^2 = 2v_i \Delta v \tag{18.44}$$

Use of Eq. (18.44) in Eq. (18.43) yields the following relation:

$$\Delta E \approx \frac{1}{2}m(2v_i\Delta v) = mv_i\Delta v \qquad (18.45)$$

Substituting the given values in Eq. (18.45), we get the following result:

$$\Delta E \approx (9.11 \times 10^{-31})\,(5 \times 10^5)\,(0.02) = 9.11 \times 10^{-27}\,\text{J}$$

Then the increase in kinetic energy ΔE in eV is calculated as follows:

$$\Delta E = 9.11 \times 10^{-27}/1.6 \times 10^{-19} = 5.7 \times 10^{-8}\,\text{eV}$$

The energy change between adjacent energy states within an allowed band is typically $\sim 10^{-19}$ eV. Such changes are very small indeed. Thus, an allowed energy band can be treated as quasi-continuous distribution.

Example 18.2 A two-dimensional square lattice has sides of length 0.4 nm. Evaluate the electron momentum values at the edges of the first Brillouin zone. Calculate the energy of a free electron with this momentum.

Solution At the edges of the first Brillouin zone, $k = \pm\,\pi/a$

Electron momentum is expressed as follows:

$$p = \frac{kh}{2\pi} = \frac{h}{2\pi}\left(\frac{\pi}{a}\right) = \frac{h}{2a} \qquad (18.46)$$

Substituting the given values in Eq. (18.46), one obtains the following result:

$$p = \frac{6.626 \times 10^{-34}}{2 \times 0.4 \times 10^{-9}} = 7.8 \times 10^{-25}\,\text{kg m/s} \qquad (18.47)$$

In addition, $E = p^2/2m$
which, on using Eq. (18.46), leads to the following relation:

$$E = \frac{1}{2m}\left(\frac{h}{2a}\right)^2 \qquad (18.48)$$

Use of the given values in Eq. (18.48) yields the following result:

$$E = (7.8 \times 10^{-25})^2/(2 \times 9.1 \times 10^{-31} \times 1.6 \times 10^{-19}) = 2.1\,\text{eV}$$

18.4 FERMI–DIRAC DISTRIBUTION FUNCTION

Electrons existing in solids obey the Fermi–Dirac statistics. The distribution of electrons over a range of allowed energy levels at thermal equilibrium is given by the *Fermi–Dirac distribution function*, $f(E)$, which is expressed as follows:

$$f(E) = \frac{1}{1 + e^{(E-E_F)/kT}} \qquad (18.49)$$

where k is Boltzmann's constant ($k = 8.62 \times 10^{-5}$ eV/K = 1.38×10^{-23} J/K). The Fermi–Dirac distribution function, $f(E)$, gives the probability that an available energy state at E will be occupied by an electron at absolute temperature T. The quantity E_F in the function $f(E)$ is called the *Fermi level*. To gain more insight into the physical meaning of E_F, let us use $E = E_F$ in Eq. (18.49). We get the following expression: $f(E_F) = [1 + \exp\,(E_F - E_F)/kT]^{-1}$

implying that

$$f(E_F) = \frac{1}{1+1} = \frac{1}{2} \qquad (18.50)$$

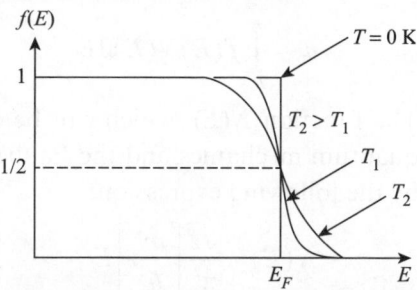

Fig. 18.16 Fermi–Dirac distribution function for different temperatures

Thus, the Fermi level E_F represents the energy that has a 50% probability of being occupied by an electron. A plot of $f(E)$ for some temperatures is shown in Fig. 18.16.

A few salient features of function $f(E)$ are as follows:

1. At 0 K, the distribution function $f(E)$ takes a simple rectangular shape. This is because at $T = 0$ K, $f(E) = 1/(1 + 0) = 1$ for $E < E_F$ and $f(E) = 1/(1 + \infty) = 0$ for $E > E_F$
   The rectangular shape implies that at $T = 0$ K, all states up to E_F are filled with electrons and states above E_F are empty.
2. At temperatures greater than 0 K, some probability exists for states above the Fermi level to get filled with electrons. At any temperature T_1, there is a probability $f(E)$ that states above E_F are filled. A corresponding probability $[1 - f(E)]$ exists that states below E_F are empty.
3. The Fermi function is symmetrical about E_F for all temperatures. Thus, the probability $f(E_F + \Delta E)$ that a state ΔE above E_F is filled is equal to the probability $[1 - f(E_F - \Delta E)]$ that a state ΔE below E_F is empty.
4. Owing to the symmetry of the function $f(E)$ about the Fermi level E_F, the Fermi level becomes a natural choice as a reference point for electron and hole concentrations in semiconductors.

Example 18.3 Calculate the probability that an energy level $2kT$ above the Fermi energy is occupied by an electron.

Solution Probability $f(E)$ of occupation of any level of energy E is given by the following relation:

$$f(E) = \frac{1}{1 + \exp\left(\dfrac{E - E_F}{kT}\right)} \qquad (18.51)$$

For the given situation, we have the following expression:

$$f(E) = 1/[1 + \exp(2kT/kT)] = 1/[1 + \exp(2)]$$

resulting in $f(E) = 0.119 = 11.9\%$

18.5 DENSITY OF STATES AND CARRIER CONCENTRATION

To evaluate the electron and hole concentrations in a semiconductor, we need to know the density of states in a particular energy range. Let $N(E)dE$ represent the density of states (in cm^{-3}) in the energy range dE. The electron concentration n_0 in the conduction band at equilibrium is then given by the following equation:

$$n_0 = \int\limits_{E_c}^{\infty} f(E) N(E) \, dE \qquad (18.52)$$

The function $N(E)$, which can be calculated using standard techniques of quantum mechanics and the Pauli's exclusion principle, can be represented by the following expression:

$$N(E) = \frac{\sqrt{2}}{\pi^2} \left[\frac{m^*}{\hbar^2} \right] E^{\frac{1}{2}} \qquad (18.53)$$

where m^* is the effective mass (we will understand this in Section 18.6). Since $N(E)$ is proportional to $E^{\frac{1}{2}}$, the density of states in the conduction band increases with electron energy. The Fermi function $f(E)$, however, as we have seen, has extremely small values for large energies. The product $f(E)N(E)$ in Eq. (18.52) thus decreases rapidly above the conduction band edge E_c. This is the reason why very few electrons are at energy levels far above the conduction band edge. A similar effect is also observed for holes in the valence band. The probability of finding a hole in the valence band $[1 - f(E)]$ decreases sharply below E_v, and therefore most holes exist near the top of the valence band. Figure 18.17 shows

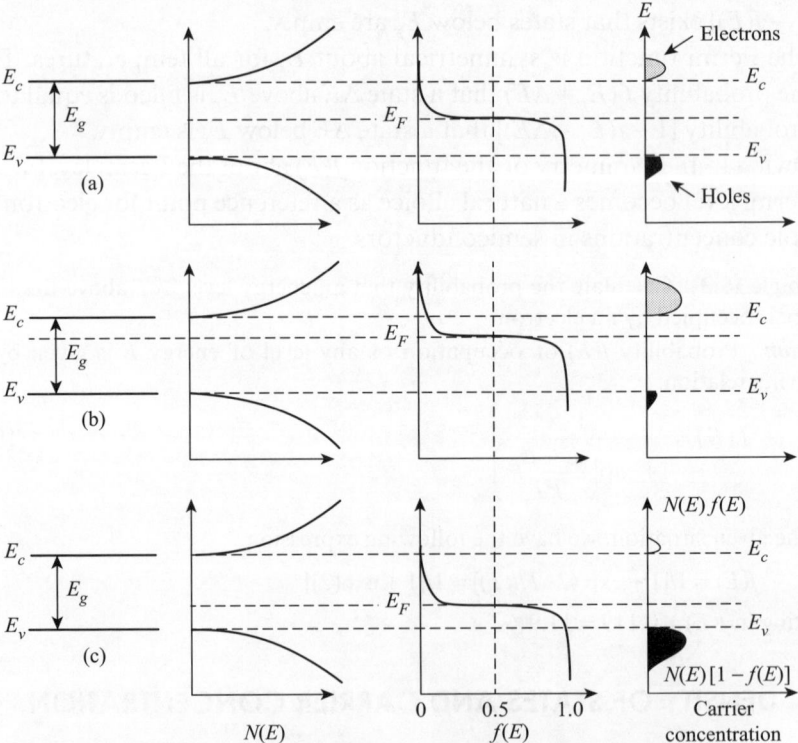

Fig. 18.17 Thermal equilibrium band diagram, density of states, Fermi–Dirac distribution, and carrier concentration for semiconductors at thermal equilibrium (a) Intrinsic semiconductor (b) n-Type semiconductor (c) p-Type semiconductor

the band diagram, density of states, Fermi–Dirac distribution function, and carrier concentrations for extrinsic, n-type, and p-type semiconductors at thermal equilibrium.

We can replace the density of states $N(E)$ in Eq. (18.52) with an effective density of states N_c located at the conduction band edge E_c. The conduction band electron concentration n_0 can, therefore, be written as follows:

$$n_0 = N_c f(E_c) \tag{18.54}$$

where $f(E_c)$ is the probability of occupancy at E_c. The Fermi function $f(E_c)$ is given as follows:

$$f(E_c) = 1/[1 + \exp (E_c - E_F)/kT]$$

If the Fermi level E_F is assumed to lie several kT below the conduction band, then this expression can be simplified as follows:

$$f(E_c) = \frac{1}{1 + e^{(E_c - E_F)/kT}} \approx e^{-(E_c - E_F)/kT} \tag{18.55}$$

Using Eq. (18.55) in Eq. (18.53), we get the following expression:

$$n_0 = N_c e^{-(E_c - E_F)/kT} \tag{18.56}$$

The effective density of states N_c can be given as follows:

$$N_c = 2\left[\frac{2\pi m_n^* kT}{h^2}\right]^{3/2} \tag{18.57}$$

where m_n^* is the density-of-state effective mass of electrons. This effective mass is different from the band curvature effective mass m^*. This is because in any particular direction in a crystal, there is often more than one equivalent conduction band minima. For silicon (Si), for example, there are six equivalent conduction band minima along the x-direction. Thus, we have more than one band curvature to deal with for arriving at the effective mass. There is a longitudinal effective mass along the major axis of the ellipsoid (see Fig. 18.18) and the transverse effective mass m_t along the two minor axes.

Using dimensional equivalence and adding contributions from all six valleys, we get the following relation:

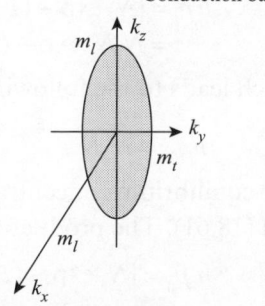

Conduction band

Fig. 18.18 Ellipsoidal constant energy surface for Si, near six conduction band minima along x-direction

$$\left(m_n^*\right)^{3/2} = 6\left(m_l m_t^2\right)^{1/2} \tag{18.58}$$

Thus, the density-of-state effective mass is the geometric mean of the effective masses. Similarly, the concentration of holes in the valence band, at equilibrium, is given as follows:

$$p_0 = N_v[1 - f(E_v)] \tag{18.59}$$

where N_v is the effective density of states in the valence band. The probability of finding an empty state (hole) at E_v is given by the following equation:

$$1 - f(E_v) = 1 - \frac{1}{1 + e^{(E_v - E_F)/kT}} \simeq e^{\frac{-(E_F - E_v)}{kT}} \tag{18.60}$$

if E_F is several kT larger than E_v. Using Eq. (18.60) in Eq. (18.59), we get the following relation:

$$p_0 = N_v e^{-(E_F - E_v)/kT} \tag{18.61}$$

The effective density of states in the valence band reduced to its band edge is given as follows:

$$N_v = 2 \left[\frac{2\pi m_p^* kT}{h^2} \right]^{3/2} \tag{18.62}$$

From Eq. (18.61), it is clear that the hole concentration increases as E_F moves closer to the valence band.

For an intrinsic semiconductor, the equilibrium electron and hole concentrations are represented by n_i and p_i, respectively. Using Eqs (18.56) and (18.61), we get the following relation:

$$n_i = N_c e^{-(E_c - E_i)/kT}, p_i = N_v e^{-(E_i - E_v)/kT} \tag{18.63}$$

where E_i is the position of the Fermi level E_F for an intrinsic semiconductor. E_i lies near the middle of the bandgap. Multiplying n_i and p_i, we have the following relation:

$$n_i p_i = [N_c \exp - (E_c - E_i)/kT) (N_v \exp - (E_i - E_v)/kT)]$$
$$= N_c N_v \exp - (E_c - E_v)/kT$$

which leads to the following expression:

$$n_i p_i = N_c N_v e^{-E_g/kT} \tag{18.64}$$

The equilibrium concentrations n_0 and p_0 are, in general, given by Eqs (18.56) and (18.61). The product of n_0 and p_0 results in the following expression:

$$n_0 p_0 = [N_c \exp - (E_c - E_F)/kT] [N_v \exp - (E_F - E_v)/kT]$$

which yields

$$n_0 p_0 = N_c N_v e^{-(E_c - E_v)/kT} = N_c N_v e^{-E_g/kT} \tag{18.65}$$

Thus, the product of electron and hole concentrations at equilibrium is a constant for a particular material at a particular temperature.

For an intrinsic semiconductor, $n_i = p_i$, implying that

$$n_0 p_0 = n_i^2 \tag{18.66}$$

Using Eqs (18.65) and (18.63), we can conclude that $E_c - E_i = E_g/2$ if the effective densities of states, N_c and N_v, are equal. In general, however, N_c and N_v are slightly different due to some difference between the effective mass for electrons and holes. The displacement of E_i from the middle of the bandgap is more for GaAs than for Si or Ge.

From Eq. (18.56), we get the following expression:

$$N_0 = N_c \exp-\left(E_c - E_F\right)/kT$$
$$= \left[N_c \exp-\left(E_c - E_i\right)/kT\right] \exp-\left(E_F - E_i\right)/kT \tag{18.67}$$

which on using Eq. (18.63) results in the following expression:

$$n_0 = n_i e^{-(E_F - E_i)/kT} \tag{18.68}$$

Similarly, we can write that

$$p_0 = n_i e^{(E_i - E_F)/kT} \tag{18.69}$$

Equations (18.68) and (18.69) lead to the following two important conclusions:
1. $n_0 = p_0 = n_i$ when $E_F = E_i$.
2. The equilibrium electron concentration n_0 increases exponentially as the Fermi level moves away from E_i towards the conduction band. The equilibrium hole concentration p_0 increases exponentially as E_F moves away from E_i towards the valence band.

Example 18.4 Determine the density-of-state effective mass of electrons in silicon. Assume that $m_l = 0.98m_0$ and $m_t = 0.19m_0$, where m_0 is the rest mass of electrons.
Solution From Eq. (18.58), we get the following relation:

$$\left(m_n^*\right)^{3/2} = 6\left(m_l m_t^2\right)^{1/2}$$

This gives the following expression:

$$m_n^* = 6^{2/3}\left(m_l m_t^2\right)^{1/3} \tag{18.70}$$

Substitution of the given values in Eq. (18.70) leads to the following result:

$$m_n^* = 6^{2/3}\left[0.98 \, (0.19)^2\right]^{1/3} m_0 = 1.1m_0$$

Example 18.5 Calculate the intrinsic carrier concentration in gallium arsenide at $T = 300$ and 400 K. Assume N_c and N_v values at 300 K to be 4.7×10^{17} and 7.0×10^{18}/cm³, respectively. Assume that the bandgap energy of gallium arsenide is 1.42 eV and it is independent of temperature for this temperature range.

Solution The intrinsic carrier concentration, n_i, is given as follows:

$$n_i = \sqrt{N_C N_V}\, e^{-E_g/2kT} \tag{18.71}$$

Using the given values and equating $T = 300$ K, we get the following expression:

$$n_i = \sqrt{\left(4.7 \times 10^{17} \times 7 \times 10^{18}\right)} \exp-\left(1.42/0.026 \times 2\right)$$

yielding $n_i = 2.5 \times 10^6$/cm³
At 400 K,

$$kT = (0.026)\left(\frac{400}{300}\right) = 0.035\,\text{eV} \tag{18.72}$$

Use of Eq. (18.72) in Eq. (18.71) leads to the following expression:

$$n_i = \sqrt{\left(\left(4.7 \times 10^{17}\right) \times \left(7 \times 10^{18}\right)\left(400/300\right)^3\right)} \exp\left(-1.42/0.035 \times 2\right)$$

resulting in $n_i = 2.79 \times 10^9/\text{cm}^3$

18.6 EFFECTIVE MASS

Mass is the property of a particle by virtue of which it resists any change in its state of motion. The relation between the momentum of a free electron and its wave vector k is given by the de Broglie formula:

$$p = \hbar k \tag{18.73}$$

where p represents the momentum and k represents the wave vector.

Thus, $E = (\hbar^2/2m)k^2$

The second derivative of energy E leads to the following expression:

$$(1/\hbar^2)d^2E/dk^2 = 1/m$$

that is, $m = \hbar^2/(d^2E/dk^2)$

So far, we have considered the electron to be free. It turns out that the expression for momentum for electrons moving in a periodic crystal is also given by the following equation:

$$p = \hbar k \tag{18.74}$$

Differentiating the expression for E once with respect to k, we get the following expression:

$$\frac{dE}{dk} = \frac{\hbar^2}{m}k \tag{18.75}$$

which leads to the following relation: $k = (m/\hbar^2)\,dE/dk$

Using this in Eq. (18.74), we get the following expression:

$$p = \frac{m}{\hbar}\frac{dE}{dk} \tag{18.76}$$

which yields (for velocity v) the following relation:

$$v = \frac{1}{\hbar}\frac{dE}{dk} \tag{18.77}$$

Suppose an external electric field E is set up in the crystal. The force F acting on the electron is then given as follows: $F = -qE$

The corresponding acceleration a_1 is then given as follows:

$$a_1 = \frac{dv}{dt} = \frac{1}{\hbar}\frac{d}{dt}\left[\frac{dE}{dk}\right] = \frac{1}{\hbar}\frac{d^2E}{dk^2}\frac{dk}{dt} \tag{18.78}$$

Work dW done by force F in a time interval dt is given as follows:

$$dW = Fv\,dt = (F/\hbar)\,(dE/dk)\,dt$$

This work done increases the electron's energy by an amount dE. Thus,

$$dE = (F/\hbar)\,(dE/dk)\,dt$$

which leads to the following expression:

$$\frac{F}{\hbar} = \frac{dk}{dt} \tag{18.79}$$

Use of Eq. (18.79) in Eq. (18.78) results in the following expression:

$$a_1 = \frac{F}{\hbar^2}\frac{d^2 E}{dk^2} = \frac{-qE}{\hbar^2}\frac{d^2 E}{dk^2} = \frac{-qE}{m_{\text{eff}}} \tag{18.80}$$

Equation (18.80) is a statement of Newton's second law. Thus, the electron acted upon by an external force moves in a periodic crystal field, on average, in the same manner as a free electron if its mass is given by the following relation:

$$m_{\text{eff}} = \frac{\hbar^2}{d^2 E / dk^2} \tag{18.81}$$

The mass m_{eff} is called the *effective mass of the electron*. The effective mass may be positive or negative, many times larger or many times smaller in magnitude than the electron's rest mass, m_0.

18.7 WORK FUNCTION

A metal lattice consisting of positive ions leads to a positive potential changing periodically along a straight line passing through the lattice sites. The situation is depicted schematically in Fig. 18.19(a). As an approximation, it is common to neglect this variation, assuming the potential to be constant. Let us assume this constant value to be V_0 everywhere on the metal. A free electron in the metal has a negative potential energy $U_0 = -eV_0$ due to this constant potential.

Figure 18.19(b) shows the change in the electron's potential energy as the electron moves from vacuum into the metal.

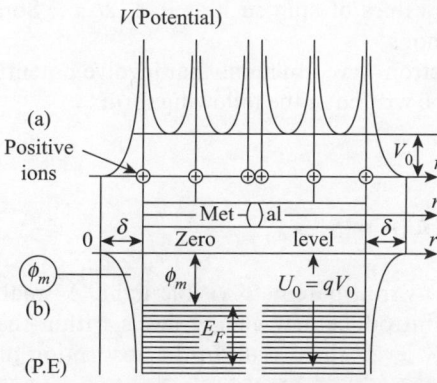

Fig. 18.19 Potential and potential energy variation along lattice (a) Potential variation (b) Potential energy variation

In reality, this change in potential energy occurs over a distance δ that is approximately equal to the lattice parameter. Thus, from Fig. 18.19(b), it is clear that a potential trough exists for the electrons in the metal. The work to be done to remove electrons from a metal is called its work function.

If electrons do not possess any kinetic energy, the work needed to liberate them from the metal equals the potential trough V_0. Electrons do, however, possess kinetic energy of translational motion. The energy needed to ensure that the electrons leave the metal is, therefore, generally less than V_0. The least amount of work is needed to be performed to liberate electrons that lie close to the Fermi level. The separation Φ_m of the Fermi level from the zero level is called the *thermodynamic work function*. For semiconductors, it can be shown theoretically that the average energy of electrons leaving a semiconductor surface equals the Fermi energy. The distance between the vacuum level and the Fermi level is also defined as the work function for semiconductors, although no electrons actually exist at the Fermi level. The work function is generally expressed in units of electron volts.

IMPORTANT CONCEPTS

1. The band theory of solids explains the formation of allowed and forbidden energy bands.
2. An expression that relates momentum and energy is called the *dispersion relation*. Thus, $E = p^2/2m$ is the dispersion relation for a free electron.
3. Very often, for carrying out a simplified analysis of the energy band structure of semiconductors, two parallel lines, one tangential to the bottom of the conduction band and the other tangential to the top of the valence band, are used. The separation between these lines is called the forbidden gap.
4. Monolayers of particular materials absorbed by the surface of solids can lead to drastic reduction in work function values. For example, adsorption of caesium on the surface of tungsten can lower the work function value of pure tungsten from 4.52 to 1.36 eV. This effect is widely used for manufacturing vacuum tube cathodes and photocathodes.
5. Fermions are elementary particles with half-odd integral spin values such as $\hbar/2, 3\hbar/2, \ldots$ Some common fermions are electrons, protons, and neutrons. Bosons are elementary particles with integral values of spin such as $0, \hbar, 2\hbar, \ldots$ Some common bosons are photons and phonons.
6. Bloch's theorem states that all one-electron wave functions that involve potential functions with periodic variation can be written in the following form:

$$\Psi(x) = u(x) \exp(jkx)$$

APPLICATIONS

1. Glass is an insulating material that is transparent to visible light. A small amount of impurity can, however, introduce new energy levels within the large bandgap of glass. These energy levels then result in the absorption of

certain colours of light, imparting colours to glass. This is true about other insulators as well. For example, the ruby mineral corundum, which is an aluminium oxide, gets a pink or red colour when around 0.05% of chromium is introduced. This doped ruby absorbs green and blue light, which results in the pink or red colour.

2. The simple concepts of the band theory are unable to explain the way electrons transport themselves within high-temperature superconductors. This is because of the existence of very strong electrical and magnetic interactions in these materials. A different approach, called the many-body theory, has to be used to explain the transport properties of high-temperature superconductors.

3. Energy band diagrams of semiconductor homo- and hetero-junctions can be used to analyse and predict the characteristics of novel concepts and devices. Perturbations such as interface states, traps, and recombination centres can be incorporated, and their effects can be understood using the concept of energy bands.

4. Lattice periodicity can be controlled by using modern sophisticated layer growth techniques such as molecular beam epitaxy and metal-organic chemical vapour deposition techniques. These form the basis of the very interesting and challenging field of bandgap engineering.

5. The fact that the effective mass of a charge carrier in a semiconductor depends upon its band structure is a very useful concept. This concept is used to develop devices like Gunn diode, where transfer of charge carriers from one band to another leads to a change in its effective mass.

IMPORTANT FORMULAE

1. Energy E of a particle:
$$E = p^2/2m = k^2\hbar^2/2m.$$

2. The Fermi–Dirac distribution function:
$$f(E) = 1/[1 + \exp(E - E_F)/kT].$$

3. The density-of-state effective mass for electrons m_n^*:
$$\left(m_n^*\right)^{3/2} = 6\left(m_l\, m_t^2\right)^{1/2}.$$

4. The product of equilibrium electron concentration n_0 and hole concentration p_0:
$$n_0 p_0 = n_i^2.$$

5. The intrinsic carrier concentration n_i:
$$n_i = \sqrt{(N_C N_V)} \exp\left(-E_g/2kT\right).$$

6. The effective mass m_{eff} of a particle:
$$m_{\text{eff}} = \hbar^2/(\mathrm{d}^2E/\mathrm{d}k^2).$$

7. One-dimensional non-relativistic Schrodinger's wave equation:
$$(-\hbar^2/2m)(\partial^2\psi(x, t)/\partial x^2) + V(x)\psi(x, t) = j\hbar(\partial\psi(x, t)/\partial t)$$

8. The time-independent form of Schrodinger's wave equation:
$$(\partial^2\psi(x)/\partial x^2) + (2m/\hbar^2) [E - V(x)]\psi(x) = 0$$

9. The first Brillouin zone:
$$-\pi/a < k < \pi/a.$$

10. The Fermi–Dirac distribution function, $f(E)$:
$$f(E) = 1/[1 + \exp(E - E_F)/kT].$$

11. At 0 K,
$$f(E) = 1 \text{ for } E < E_F$$
and
$$f(E) = 0 \text{ for } E > E_F$$

12. For silicon, the density-of-state effective mass, m_n^*:

$$\left(m_n^*\right)^{\frac{3}{2}} = 6\left(m_l \, m_t^2\right)^{\frac{1}{2}}.$$

SELF-ASSESSMENT

Multiple-choice Questions

18.1 Splitting of states is due to
  (a) Pauli's exclusion principle
  (b) quantum nature of light
  (c) Schrodinger's wave equation
  (d) Bloch's theorem

18.2 In silicon, band splitting takes place for the states
  (a) 3s and 2s
  (b) 3s and 3p
  (c) 2s and 2p
  (d) 2s and 3p

18.3 The typical range of bandgaps for insulators is
  (a) 2–4 eV
  (b) 1–2 eV
  (c) 5–15 eV
  (d) 20–40 eV

18.4 Potential function for a group of atoms has a dependence given by
  (a) $1/r^2$
  (b) $1/r^3$
  (c) r^2
  (d) $1/r$

18.5 Momentum p for a free particle is given by
  (a) $k\hbar$
  (b) $k^2\hbar$
  (c) $k^2\hbar^2$
  (d) $k\hbar^2$

18.6 First Brillouin zone is given by
  (a) $-2\pi/a < k < \pi/a$
  (b) $-\pi/a < k < \pi/a$
  (c) $-\pi/a < k < 2\pi/a$
  (d) $2\pi/a < k < 2\pi/a$

18.7 At $E = E_f$, $f(E)$ has the value
  (a) 0
  (b) 1
  (c) 1/2
  (d) 1/3

18.8 For an intrinsic semiconductor,
  (a) $n_0 p_0 > n_i^2$
  (b) $n_0 p_0 < n_i^2$
  (c) $n_0 p_0 = n_i$
  (d) $n_0 p_0 = n_i^2$

18.9 Effective mass m^* is given by
  (a) $\hbar^2/(d^2 E/dk^2)$
  (b) $\hbar/(d^2 E/dk^2)$
  (c) $\hbar^2/(dE/dk)$
  (d) $\hbar/(dE/dk)$

18.10 Work function is distance between
  (a) conduction band minimum and Fermi level
  (b) valence band maximum and Fermi level
  (c) vacuum level and Fermi level
  (d) vacuum level and conduction band minimum

18.11 Choose the correct expression for wave vector k:
  (a) $k = 2\pi\lambda$
  (b) $k = 2\pi/\lambda$
  (c) $k = \lambda/2\pi$
  (d) $k = \lambda^2 2\pi$

18.12 Identify the correct expression for n_0:
  (a) $n_i \, e^{E_f/kT}$
  (b) $n_i \, e^{E_i/kT}$
  (c) $n_i \, e^{(E_f - E_i)/kT}$
  (d) $n_i \, e^{(E_f + E_i)/kT}$

18.13 When $E_f = E_i$,
  (a) $n_0 > p_0$
  (b) $n_0 < p_0$
  (c) $n_0 = 1 = p_0$
  (d) $n_0 = p_0 = n_i$

18.14 Density of states in the valence band, N_V, is proportional to

(a) $T^{3/2}$ (b) $T^{-3/2}$ (c) $T^{1/2}$ (d) $T^{-1/2}$

18.15 Energy E of a free particle is given as

(a) $\dfrac{k\hbar}{2m}$ (b) $\dfrac{k^2\hbar^2}{2m}$ (c) $\dfrac{k^2\hbar^2}{m}$ (d) $\dfrac{k^2\hbar}{m}$

Review Questions

18.1 Why do individual discreet quantized energy levels of individual atoms split up?
18.2 What are forbidden bands?
18.3 What is the main difference between semiconductors and insulators?
18.4 What is the origin of Schrodinger's wave equation?
18.5 Sketch the shape of the potential function of a one-electron atom.
18.6 State and explain Bloch's theorem.
18.7 What is a reduced k-space diagram?
18.8 How does the hole concentration change as the Fermi level moves towards the valence band?
18.9 How does n_i vary with temperature?
18.10 What is implied by a negative effective mass?
18.11 Define radial probability density.
18.12 Explain Pauli's exclusion principle.
18.13 Use band theory to explain the existence of forbidden bands.
18.14 Define valence band.
18.15 Write and explain the non-relativistic Schrodinger's wave equation.
18.16 Write the Fermi–Dirac distribution function and explain the different terms.
18.17 Give one consequence of the symmetry of $f(E)$ about E_F.
18.18 Derive an expression for the effective mass for an electron.
18.19 Explain the concept of work function using a suitable schematic diagram.
18.20 Give a schematic representation of the radial probability density function for the lowest energy state of a single, non-interacting hydrogen atom.
18.21 Differentiate between semiconductors, insulators, and metals using band theory.
18.22 Indicate the first two Brillouin zones.
18.23 Show that $f(E_F) = \dfrac{1}{2}$ at all temperatures.
18.24 Derive an expression for intrinsic carrier concentration.

Numerical Problems

18.1 An electron is initially moving with a velocity of 6×10^7 cm/s. An accelerating voltage increases its velocity by 5 cm/s. Determine the increase in the kinetic energy of the electron in eV.
18.2 A two-dimensional square lattice has sides of length 0.5 nm. Calculate the electron momentum values at the edges of the first Brillouin zone. Determine the energy of a free electron of this momentum.

[*Hint: $p = h/2a$ and $E = p^2/2m$*]

18.3 The magnitude of electron momentum at the edges of the first Brillouin zone for a two-dimensional square lattice is 5×10^{-25} kg m/s. Calculate the length of the side of the lattice.

[*Hint: $a = h/2p$*]

18.4 Two possible conduction and valence bands are shown in the *E* versus *k* diagrams given in Fig. 18.20.

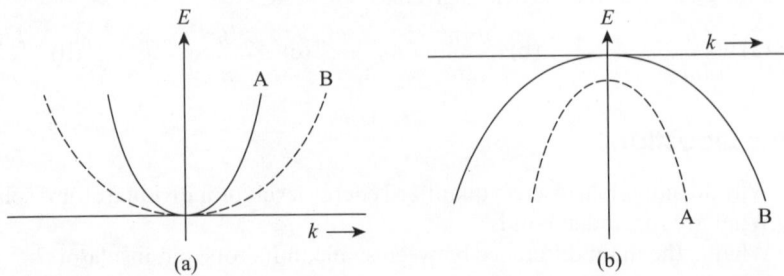

Fig. 18.20 Plot of *E* versus *k* (a) Overlapping plots (b) Non-overlapping plots

(a) Identify the particles for which each of these diagrams is valid.
(b) Which band will have heavier particle mass in each of these figures and why?

[*Hint*: $d^2E/dk^2 = \hbar^2/m$]

18.5 Figure 18.21 is an *E–k* diagram in a particular allowed energy band. Predict the sign of the effective mass for the positions shown in the figure.

[*Hint*: $m_{\text{eff}} = \hbar^2/(d^2E/dk^2)$]

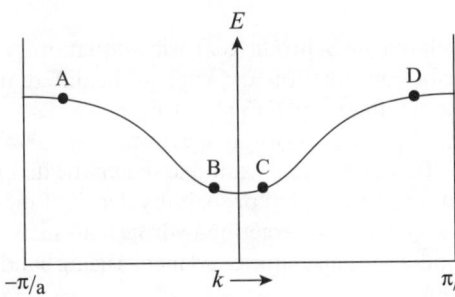

$-\pi/a$ $k \longrightarrow$ π/a **Fig. 18.21** *E–k* diagram

18.6 The energy near a valence band edge is given by the expression $E(k) = -2 \times 10^{-33}k^2$ J. Determine the magnitude of the effective mass of a hole.

[*Hint*: $m_{\text{eff}} = \hbar^2/(d^2E/dk^2)$]

18.7 The energy in the valence band of a simple cubic metal can be represented in the form $E = Ak^2 + B$, where $A = 5 \times 10^{-38}$ J m². Determine the magnitude of m_{eff}/m_0.

[*Hint*: $m_{\text{eff}} = \hbar^2/(d^2E/dk^2)$]

18.8 Determine the intrinsic carrier concentration in gallium arsenide at $T = 450$ K. Assume N_c and N_v values at 300 K to be $4.7 \times 10^{17}/\text{cm}^3$ and $7.0 \times 10^{18}/\text{cm}^3$, respectively. Assume that the bandgap energy of gallium arsenide is 1.42 eV and it is independent of temperature for this temperature range.

$$\left[Hint: n_i = \sqrt{(N_C N_V)} \exp\left(-E_g / 2kT\right) \right]$$

18.9 Determine the probability that an energy level $3kT/2$ above the Fermi energy is occupied by an electron.

[*Hint*: $f(E) = 1/[1 + \exp(E - E_F)/kT]$

18.10 The probability that an energy level is occupied by an electron is 10%. Locate the energy level with respect to the Fermi level.

$$[Hint: f(E) = 1/[1 + \exp{(E - E_F)/kT}]$$

18.11 An electron is travelling with an initial velocity of 2.8×10^7 cm/s. If the change in KE is 7.21×10^{-27} J, calculate the change in velocity.

$$[Hint: \Delta E \approx mv_i \Delta v]$$

18.12 The electron momentum at the edges of the first Brillouin zone of a two-dimensional square lattice is 6.2×10^{-25} kg m/s. Calculate the side length.

$$\left[Hint: p = \frac{h}{2a} \right]$$

18.13 The density-of-state effective mass of electrons in silicon is $1.2m_0$. If $m_l = 0.92m_0$, calculate m_t.

$$\left[Hint: m_n^* = 6^{\frac{2}{3}} \left[m_l m_t^2 \right]^{\frac{1}{3}} \right]$$

18.14 The intrinsic carrier concentration of a semiconductor at $T = 300$ K is found to be 4.2×10^6/cm³. Assume that $N_c = 3.9 \times 10^{17}$/cm³ and $N_v = 6.2 \times 10^{18}$/cm³ at 300 K. Calculate E_g assuming that it is independent of temperature.

$$\left[Hint: n_i = \sqrt{N_c N_v} e^{-Eg/2kT} \right]$$

18.15 The probability that an energy level above the Fermi level is occupied is 12.8%. Calculate $(E - E_f)$.

$$\left[Hint: f(E) = \cfrac{1}{1 + \exp\left(\cfrac{E - E_f}{kT} \right)} \right]$$

Semiconductor Physics

Learning Objectives

After studying this chapter, students will be able to

- explain the meaning of intrinsic semiconductors
- elucidate the expressions for carrier concentration
- comprehend the concept of Fermi level and its variation with temperature
- realize the concept of band gap of a semiconductor and its determination
- calculate carrier concentration in an extrinsic semiconductor
- elucidate the concept of Fermi level in extrinsic semiconductors
- understand compound semiconductors
- describe the Hall effect

List of Symbols

E_G = Forbidden energy gap

n = Concentration of electrons/cm^3

p = Concentration of holes/cm^3

n_i = Intrinsic carrier concentration

g_i = Generation rate of e–h pairs

r_i = Recombination rate

n_0 = Equilibrium concentration of electrons

p_0 = Equilibrium concentration of holes

$N(E)$ = Density of states

m_e^* = Effective mass

dE = Energy range

E_c = Conduction band edge

E_F = Fermi band edge

$F(E_c)$ = Probability of occupancy at E_c

m_n = Effective mass of electrons

m_l = Longitudinal effective mass

m_t = Transverse effective mass

N_v = Effective density of states in valence band

E_v = Valence band edge

m_p^* = Effective mass of hole

E_{Fi} = Position of Fermi band for an intrinsic semiconductor

k_B = Boltzmann constant

m_e = Mass of electron

E_n = Total energy of the electron in the nth orbit

τ_t = Lifetime of electrons

V_H = Hall voltage

V_d = Drift velocity

R_H = Hall coefficient

m_e = Mass of electron

N_n = Number of electrons in conduction band of n-region

N_p = Number of protons in p-region

19.1 INTRODUCTION

The band theory of solids introduced the concept of conduction band, valence band and an energy band gap separating the two bands. Materials have been classified on the basis of the separation between these energy bands. There is a class of materials that have their forbidden energy band gap, E_G, between that of a conductor and an insulator. In a semiconductor, the two valence and conduction bands play critical roles in the conductivity of electrons. At absolute zero temperature, that is, 0 K, the valence band is completely full and the conduction band is empty such that the semiconductor is similar to an insulator. As the temperature is increased, some of the electrons reach the conduction band, thereby leaving vacant places in the valence band. The semiconductor now starts exhibiting conducting properties. The conductivity of these materials can be modified by the introduction of certain impurities in controlled amount.

Semiconductor elements appear in Group IV of the periodic table, which include carbon, silicon, germanium, and tin. In addition, elements of Group III and Group V form semiconductors by combining with one another, for example, gallium–arsenide and indium–antimonide.

19.2 TYPES OF SEMICONDUCTORS

There are certain naturally occurring semiconductors. Electrical properties of a semiconductor can be controlled using certain elements that help enhance the recombination process, thereby helping in conduction. Depending on this, semiconductors have been classified as described in the following subsections.

19.2.1 Intrinsic Semiconductors

A perfect, pure semiconductor crystal containing no impurities or lattice defects is called an *intrinsic semiconductor*. At 0 K, the valence band of such a semiconductor is filled with electrons and the conduction band is empty. An intrinsic semiconductor, therefore, has no charge carriers at 0 K.

As the temperature is increased, valence band electrons acquire energy, are thermally excited across the forbidden energy band gap, and start moving to the conduction band, leaving behind holes. Thus, electron–hole (e–h) pairs are created, which are the only charge carriers in an intrinsic semiconductor. The physical mechanism underlying e–h pair creation can be understood with the help of the broken bond model of a Si crystal, as shown in Fig. 19.1.

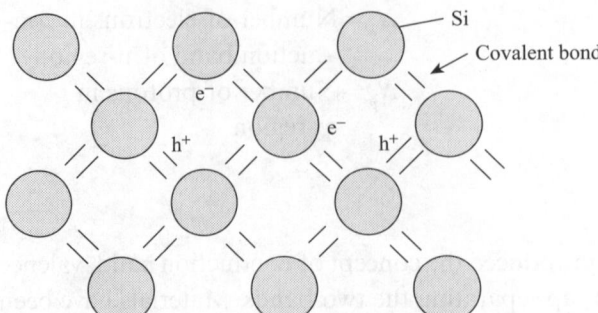

Fig. 19.1 Electron–hole pairs in broken bond model of Si crystal

At 0 K, all the covalent bonds are intact. As the temperature increases, some valence electrons acquire sufficient energy to break away from their position in the bonding structure and become free to move about in the lattice. This is how a conduction band electron is created and a hole (broken bond) is left behind in the valence band. The energy required by an electron to break free from a covalent bond is the forbidden energy gap E_g. It must, however, be emphasized that this picture is deceptively simplified and, at best, qualitative. This model seems to indicate that free electrons and holes remain localized at the lattice site. The actual picture is quite different, with electrons and holes being spaced out over several lattice spacing.

All the charge carriers created in an intrinsic semiconductor are solely by the process of e–h pair generation. Hence, the concentration n of electrons (electrons per cm³) in the conduction band is equal to the concentration p of holes (holes per cm³) in the valence band. The symbol n_i is used to represent each of these intrinsic carrier concentrations. Thus, for an intrinsic semiconductor,

$$n = p = n_i \tag{19.1}$$

At a given temperature, two competing processes decide the equilibrium concentration of e–h pairs: (a) generation of e–h pairs due to transition of valence electrons to the conduction band and (b) annihilation of these pairs due to *recombination*. Recombination occurs when a conduction band electron makes a transition to an empty state (hole) in the valence band. Such transitions can be direct or indirect. Suppose that g_i is the generation rate of e–h pairs (per cm³ per s) and r_i is the recombination rate. Then, at equilibrium, we must have the following expression:

$$r_i = g_i \tag{19.2}$$

Both g_i and r_i are functions of temperature.

If n_0 and p_0 represent the equilibrium concentrations of electrons and holes, respectively, then we can write the following equation:

$$r_i = \alpha_r n_0 p_0 = \alpha_r n_i^2 = g_i \tag{19.3}$$

where α_r is a constant of proportionality and is dependent on the particular recombination process.

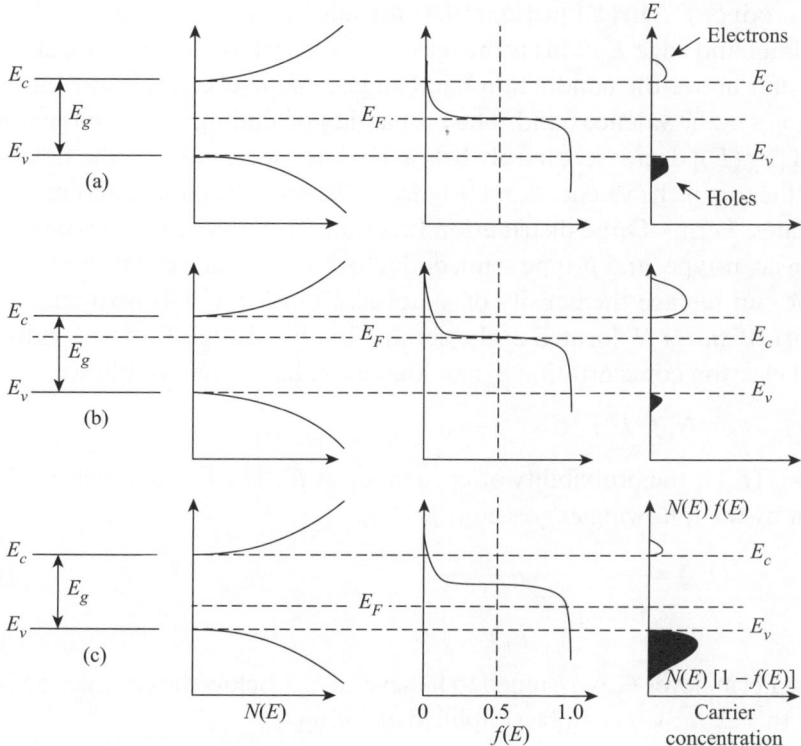

Fig. 19.2 Thermal equilibrium band diagram, density of states, Fermi–Dirac distribution function, and carrier concentrations for different semiconductors (a) Intrinsic (b) n-type (c) p-type

19.2.2 Carrier Concentration in an Intrinsic Semiconductor

To evaluate the electron and hole concentrations in a semiconductor, we need to know the density of states in a particular energy range. Let $N(E)dE$ represent the density of states (in per cm³) in the energy range d E. The electron concentration n_0 in the conduction band at equilibrium is then given by the following relation:

$$n_0 = \int_{E_c}^{\infty} f(E) N(E) dE \tag{19.4}$$

The function $N(E)$, calculated using standard techniques of quantum mechanics and the Pauli exclusion principle, can be shown to be given by the following expression:

$$N(E) = \frac{\sqrt{2}}{(\pi)^2} \left[\frac{m_e^*}{\hbar^2} \right]^{3/2} E^{1/2} \tag{19.5}$$

where m_e^* is the effective mass of electron. Since $N(E)$ is proportional to $E^{1/2}$, the density of states in the conduction band increases with electron energy. The Fermi function $f(E)$, however, has extremely small values for large energies.

The product $f(E)N(E)$ in Eq. (19.4) thus decreases rapidly above the conduction band edge E_c. This is the reason why very few electrons are at energy levels far above the conduction band edge. A similar effect is also observed for holes in the valence band. The probability of finding a hole in the valence band $[1-f(E)]$ decreases sharply below E_v. Therefore, most of the holes exist near the top of the valence band. Figure 19.2 shows the band diagram, density of states, Fermi–Dirac distribution function, and carrier concentrations for intrinsic, n-type, and p-type semiconductors at thermal equilibrium.

We can replace the density of states $N(E)$ in Eq. (19.4) with an *effective density of states* N_c located at the conduction band edge E_c. The conduction band electron concentration n_0 can, therefore, be written as follows:

$$n_0 = N_c f(E_c) \tag{19.6}$$

where $f(E_c)$ is the probability of occupancy at E_c. The Fermi function $f(E_c)$ is given by the following expression:

$$f(E_c) = \frac{1}{1 + \exp\left(\dfrac{E_c - E_F}{k_B T}\right)} \tag{19.7}$$

If the Fermi level E_F is assumed to lie several $k_B T$ below the conduction band, then this expression can be simplified as follows:

$$f(E_c) = \frac{1}{1 + \exp\left(\dfrac{E_c - E_F}{k_B T}\right)} \approx e^{-(E_c - E_F)/k_B T} \tag{19.8}$$

Substituting Eq. (19.8) into Eq. (19.6), one gets the following expression:

$$n_0 = N_c f(E_c) = N_c e^{-(E_c - E_F)/k_B T} \tag{19.9}$$

The effective density of states N_C can be expressed as follows:

$$N_c = 2\left[\frac{2\pi m_n^* k_B T}{h^2}\right]^{3/2} \tag{19.10}$$

where m_n^* is the density-of-states effective mass of electrons. This effective mass is different from the band curvature effective mass m_e^*. This is because in any particular direction in a crystal there is often more than one equivalent conduction band minimum. For Si, for example, there are six equivalent conduction band minima along the x-direction. Thus, we have more than one band curvature to deal with in arriving at the effective mass. There is a longitudinal effective mass m_l along the major axis of the ellipsoid (see Fig.19.3) and transverse effective mass m_t along the two minor axes.

By using dimensional equivalence and adding contributions from all the six valleys, we get the following equation:

$$(m_n^*)^{3/2} = 6(m_l m_t^2)^{1/2} \tag{19.11}$$

Thus, the density-of-states effective mass is the geometric mean of the longitudinal and transverse effective masses.

Similarly, the concentration of holes in the valence band at equilibrium is given by the following expression:

$$p_0 = N_v[1 - f(E_v)] \qquad (19.12)$$

where N_v is the effective density of states in the valence band. The probability of finding an empty state (hole) at E_v is given by the following equation:

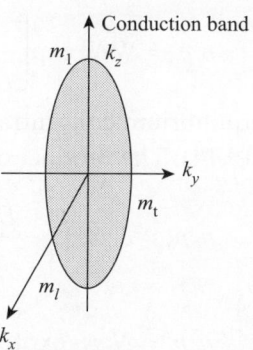

Fig. 19.3 Ellipsoidal constant-energy surface for Si near six conduction band minima along x-direction

$$(1 - f(E_v)) = 1 - 1 + \exp\left[\frac{E_v - E_F}{k_B T}\right] \approx \exp\left[-\frac{E_F - E_v}{k_B T}\right] \qquad (19.13)$$

where the effective density of states in the valence band is reduced to its band edge. If E_F is several $k_B T$ larger than E_v; substitution of Eq. (19.13) into Eq. (19.12) leads to the concentration of holes in the valence band at equilibrium:

$$p_0 = N_v \exp\left[-\frac{E_F - E_v}{k_B T}\right] \qquad (19.14)$$

where m_p^* is the density of states effective mass of a hole. We can write that

$$N_v = 2\left[\frac{2\pi m_p^* k_B T}{h^2}\right]^{3/2} \qquad (19.15)$$

From Eq. (19.14) it is clear that the hole concentration increases as E_F moves closer to the valence band. For an intrinsic semiconductor, equilibrium electron and hole concentrations are represented by n_i and p_i, respectively.

Using Eqs (19.8) and (19.14), we get the following relation:

$$n_i = N_C \exp\left[-\frac{E_C - E_{Fi}}{k_B T}\right] \qquad (19.16)$$

$$p_i = N_v \exp\left[-\frac{E_{Fi} - E_v}{k_B T}\right] \qquad (19.17)$$

where E_{Fi} is the position of the Fermi level E_F for an intrinsic semiconductor. E_{Fi} lies near the middle of the bandgap. Multiplying n_i and p_i, we have the following equations:

$$n_i p_i = N_c \exp\left[-\frac{E_c - E_{Fi}}{k_B T}\right] x N_v \exp\left[-\frac{E_{Fi} - E_v}{k_B T}\right] \qquad (19.18)$$

$$n_i p_i = N_c N_v \exp\left[-\frac{E_c - E_v}{k_B T}\right] = N_c N_v \exp\left[-\frac{E_g}{k_B T}\right] \tag{19.19}$$

The equilibrium concentrations n_0 and p_0 are in general given by Eqs (19.8) and (19.14). The product of n_0 and p_0 results in the following equations:

$$n_0 p_0 = N_c \exp\left[-\frac{E_c - E_{Fi}}{k_B T}\right] x N_v \exp\left[-\frac{E_{Fi} - E_v}{k_B T}\right] \tag{19.20}$$

$$n_0 p_0 = N_c N_v \exp\left[-\frac{E_c - E_v}{k_B T}\right] = N_c N_v \exp\left[-\frac{E_g}{k_B T}\right] \tag{19.21}$$

Thus, the product of electron and hole concentrations at equilibrium is a constant for a given material at a given temperature.

For an intrinsic semiconductor, $n_i = p_i$; hence,

$$\Rightarrow n_0 p_0 = n_i^2 \tag{19.22}$$

Substituting Eqs (19.22) in Eqs (19.20) and (19.21), we get the following relation:

$$n_i = \sqrt{N_c N_v} \exp\left[-\frac{E_g}{k_B T}\right] \tag{19.23}$$

From Eqs (19.23) and (19.16) we can conclude that $(E_c - E_{Fi}) = (E_g/2)$ if the effective densities of states N_c and N_v are equal. In general, however, N_c and N_v are slightly different due to some differences in the effective mass for electrons and holes. The displacement of E_{Fi} from the middle of the band gap is more for GaAs than for Si or Ge.

From Eq. (19.8), we get the following relations:

$$n_0 = N_c \exp\left[-\frac{(E_c - E_{Fi})}{k_B T}\right] \tag{19.24}$$

$$n_0 = N_c \exp\left[-\left(\frac{E_c - E}{k_B T}\right)\right] \exp\left[-\left(\frac{E_{Fi} - E}{k_B T}\right)\right] \tag{19.25}$$

which on using Eqs (19.20) and (19.21) results in the following relations:

$$n_0 = n_i \exp\left[-\frac{E_{Fi} - E}{k_B T}\right] \tag{19.26}$$

$$p_0 = n_i \exp\left[-\frac{E_{Fi} - E}{k_B T}\right] \tag{19.27}$$

Equations (19.24)–(19.27) lead to the following two important conclusions:
1. $n_0 = p_0 = n_i$ for $E_F = E_{Fi}$.

2. The equilibrium electron concentration n_0 increases exponentially as the Fermi level moves away from E_{Fi} towards the conduction band. The equilibrium hole concentration eases p_0 increases exponentially as E_F moves towards the valence band.

19.2.3 Fermi Level of Intrinsic Semiconductor

Electrons present in solids obey the Fermi–Dirac statistics. Distribution of electrons over a range of allowed energy levels at thermal equilibrium is given by the *Fermi–Dirac distribution function $f(E)$*, as follows:

$$f(E)\frac{1}{1+\exp\left[\dfrac{E_F-E}{k_BT}\right]} \tag{19.28}$$

where k_B is the Boltzmann constant $\left(k_B=8.62\times\dfrac{10^{-5}\,\text{eV}}{\text{K}}=1.38\times\dfrac{10^{-23}\,\text{J}}{\text{K}}\right)$. The Fermi–Dirac distribution function $f(E)$ gives the probability of an available energy state at E being occupied by an electron at absolute temperature T. The quantity E_F in the function $f(E)$ is called the *Fermi level*. To gain more insight into the physical meaning of E_F, substitute $E=E_F$ in Eq. (19.28). We get

$$f(E_F)=\frac{1}{1+\exp\left[\dfrac{E_F-E_F}{k_BT}\right]}=\frac{1}{1+1}=\frac{1}{2} \tag{19.29}$$

Thus, the Fermi level E_F represents that energy state that has 50% probability of being occupied by an electron. A plot of $f(E)$ for some temperatures is shown in Fig. 19.4.

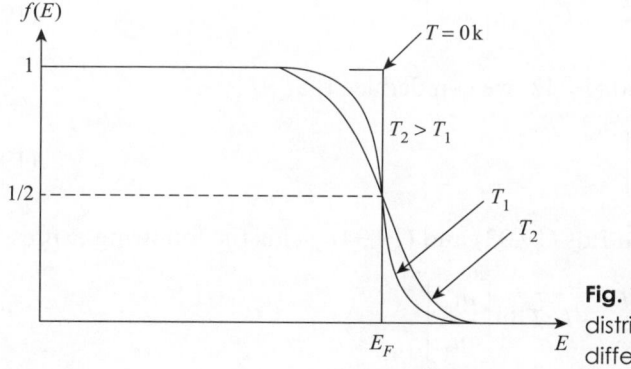

Fig. 19.4 Fermi-Dirac distribution functions at different temperatures

A few salient features of the function $f(E)$ are as follows:
1. At $0\,\text{K}$, the distribution function $f(E)$ takes a simple rectangular shape. This is because at $T=0\,\text{K}$,

$$f(E)=\frac{1}{1+0}=1,\ E<E_F \tag{19.30}$$

$$f(E) = \frac{1}{1+\infty} = 1, E > E_F \qquad (19.31)$$

The rectangular shape implies that at $T = 0\,\text{K}$, all states up to E_F are filled with electrons and all states above E_F are empty.

2. At temperatures greater than $0\,\text{K}$, some probability exists for states above the Fermi level to get filled with electrons. Thus, at any temperature T, there is a probability $f(E)$ that the states above E_F are filled. A corresponding probability $[1 - f(E)]$ exists that the states below E_F are empty.

3. The Fermi function is symmetrical about E_F for all temperatures. Thus, the probability $f(E_F + \delta E)f$ that a state δE above E_F is filled is equal to the probability $[1 - f(E_F - \delta E)]$ that a state δE below E_F is empty.

4. Owing to the symmetry of the function $f(E)$ about the Fermi level, E_F, the Fermi level becomes a natural choice as a reference point for electron and hole concentrations in semiconductors.

19.2.4 Variation of Fermi Level with Temperature for Intrinsic Semiconductor

For an intrinsic semiconductor, $n_0 = p_0$. Substituting Eqs (19.8) in (19.14), one gets the following relation:

$$N_c \exp\left[-\frac{E_c - E_F}{k_B T}\right] = N_v \exp\left[-\frac{E_F - E_v}{k_B T}\right] \qquad (19.32)$$

Simplification of Eq. (19.32) results in the following expressions:

$$E_F = \frac{E_c + E_v}{2} + \frac{k_B T}{2} \ln\frac{N_v}{N_c} \qquad (19.33)$$

$$E_F = \frac{E_c + E_v}{2} + k_B T \ln\left[\frac{N_v}{N_c}\right]^{1/2} \qquad (19.34)$$

From Eqs (19.9) and (19.32) we can deduce that

$$\frac{N_v}{N_c} = \left[\frac{m_p^*}{m_n^*}\right]^{3/2} \qquad (19.35)$$

Using Eq. (19.35) in Eqs (19.33) and (19.34) yields the following expression:

$$E_F = \frac{E_c + E_v}{2} + k_B T \ln\left[\frac{m_p^*}{m_n^*}\right]^{3/4} \qquad (19.36)$$

Equation (19.36) gives the variation of Fermi level with temperature for an intrinsic semiconductor. Figure 19.5 is a schematic plot of Eq. (19.36). We can draw the following conclusions from Fig. 19.5:

1. At $T = 0\,\text{K}$, the Fermi level lies in the middle of the forbidden energy gap.

2. As T increases, the Fermi level position shifts due to the difference between m_p^* and m_n^*. If $m_p^* > m_n^*$, the Fermi level shifts towards the bottom of the

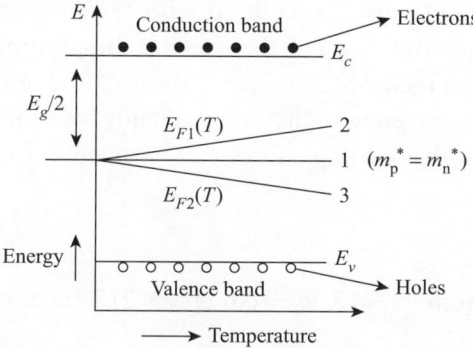

Fig. 19.5 Schematic plot of Eq. (19.36)

conduction band, as shown by curve 2 in Fig. 19.5. If $m_p^* < m_n^*$, the Fermi level shifts towards the top of the valence band, as depicted by curve 3 in Fig. 19.5.

3. The position of the Fermi level is linearly dependent on temperature.

19.2.5 Extrinsic Semiconductors

Properties of intrinsic semiconductors can be suitably modified by introducing controlled amounts of known impurities. This process is called *doping*. An *extrinsic semiconductor* is doped in such a way that the equilibrium carrier concentrations n_0 and p_0 become different from the intrinsic carrier concentration n_i. There are two types of doped semiconductors: n-type and p-type. Electrons are the majority charge carriers in an n-type semiconductor and holes are that in a p-type semiconductor. The additional carriers that come about in the system due to doping are not so tightly bound to the system.

Let us illustrate the concept of binding energy using donor impurities. The energy required to excite the fifth electron of a donor atom to the conduction band is called its *binding energy*. For the purpose of an approximate calculation of the binding energy, let us assume that the four covalent bonding electrons are tightly bound and the fifth extra electron is loosely bound to the atom. This situation can be approximated to the case of a loosely bound electron around a tightly bound core, in a hydrogen-like orbit. The total energy of the electron in the nth orbit, in accordance with the Bohr model, is given as follows:

$$E_n = \frac{m_e e^4}{2K^2 n^2 h^2} \tag{19.37}$$

where m_e is the mass of the electron and the constant K is given by $K = 4\pi \in_0 \in_r ;$, $\in_r$ being the relative permittivity of the semiconductor material. For the ground-state energy, $n = 1$, and we get the following expression:

$$E_n = \frac{m_e e^4}{2K^2 h^2} \tag{19.38}$$

The mass m_e in expression (19.38) must be replaced with the conductivity effective mass m_{ec}^* rather than the free mass of the electron. In addition, the radius of the orbit is modified to include the effective mass of the electron, where the relative permittivity of the semiconductor has already been included in the constant factor (K). Thus, the binding energy can be given as follows:

$$E_b = -\frac{13.6}{\epsilon_r^2}\left[\frac{m_e^*}{m_e}\right]\text{eV} \tag{19.39}$$

For the special case of germanium, $\epsilon_r = 16$, $m_e^* = 0.6\, m_e$, $r = 215 \times (r_h) = 215 \times (0.05)\,\text{nm} = 1.4\,\text{nm}$.

Putting these values in Eq. (19.39) we get the following result:

$$E_b = -(13.6 \times 0.6)/16^2 = 0.03\,\text{eV}$$

The dopants usually have an energy of about 0.01 eV, which is in addition to the preceding calculations. Thus, in the discussion that follows, one can observe the value of energy of the orders indicated earlier.

19.2.6 Fermi Level of Extrinsic Semiconductor

The process of doping can be understood using the band diagram (Fig. 19.6). Introduction of impurities or lattice defects in a perfect crystal leads to the creation of additional energy levels in the energy band structure, generally within the band gap. For example, if we introduce an impurity from column V of the periodic table (P, As, and Sb), the new energy level is created near the conduction band. This energy level is filled with electrons at 0 K in semiconductors such as Ge or Si. As the new level lies very close to the conduction band (Fig. 19.6), very little thermal energy is required to excite these electrons to the conduction band. At temperatures above 0 K, the electrons in these additional levels get excited to the conduction band. Such an impurity level is called a *donor* level and the corresponding impurities are called donor impurities, since these levels donate electrons to the conduction band. The donor level can contribute a significant number of electrons to the conduction band even at temperatures that are low for appreciable intrinsic e–h pair concentration.

As the temperature is varied from 0 to 50 K, the electrons in the donor energy level are excited to the conduction band. Thus, the thermal energy required for the process is low.

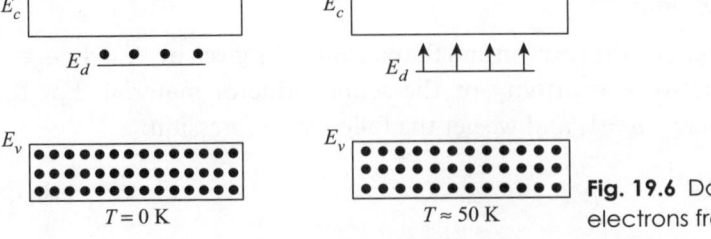

Fig. 19.6 Donation of electrons from donor level

For a semiconductor doped with a considerable amount of donor atoms, we have at room temperature,

$$n_0 \gg (n_i, p_0) \tag{19.40}$$

Such a doped semiconductor is thus an n-type semiconductor.

On the other hand, introduction of impurity atoms from column III of the periodic table (B, Al, Ga, and In) into Ge or Si results in the creation of impurity levels near the valence band. This is shown schematically in Fig.19.7.

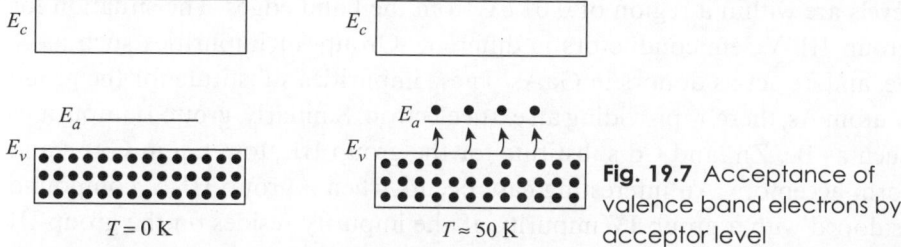

Fig. 19.7 Acceptance of valence band electrons by acceptor level

These energy levels are empty at 0 K. At other temperatures (which are still low enough), the available thermal energy is sufficient to excite the electrons from the valence band to the impurity levels. Such impurity levels are called *acceptor* levels, and the corresponding impurities are called acceptor impurities since they accept electrons from the valence band. Acceptor impurities can result in an equilibrium hole concentration p_0 much greater than the conduction band electron concentration n_0. Thus,

$$p_0 \gg (n_i, n_0) \tag{19.41}$$

Such an extrinsic semiconductor is called a p-type semiconductor. Figure 19.8 is a schematic representation showing the introduction of a donor and an acceptor atom in the Si lattice.

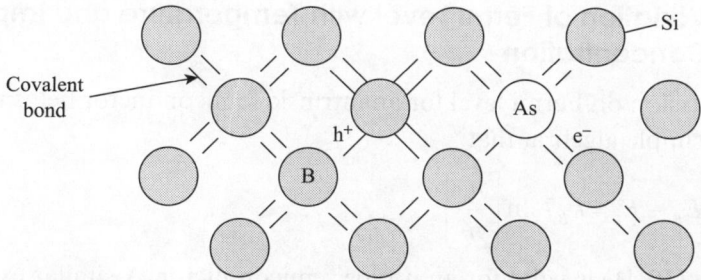

Fig. 19.8 Donor and acceptor atoms in covalent bond model

The As atom in the Si lattice has an extra electron after sharing four of its valence electrons with the neighbouring Si atoms. The fifth electron is loosely bound to the As atom since it does not fit into the bonding structure of the lattice. At 0 K, there is no thermal energy available, but at low enough temperatures, sufficient thermal energy is available to this extra electron to overcome its Columbic binding to the impurity As atom. This extra electron thus gets donated to the crystal and takes part in current conduction. The column III impurity B atom has only three valence electrons. These three electrons participate in covalent bonding with the neighbouring Si atoms, but one bond stays incomplete. At low enough temperatures, an electron from an adjacent neighbouring atom hops to the incomplete bond at the B site, creating an incomplete bond, or a hole, at its original location. This hole can also contribute to current conductions.

In Si, the donor and acceptor energy levels lie within a region of 0.03–0.06 eV from the band edges. For Ge, the corresponding donor and acceptor levels are within a region of 0.01 eV from the band edges. The situation for group III–V semiconductors is different. Group VI impurities such as S, Se, and Te act as donors in GaAs. These impurities substitute for the group V atom As, thereby providing an extra electron. Similarly, group II impurities such as Be, Zn, and Cd substitute for the group III atom Ga in GaAs and form acceptors. An interesting case occurs when a group III–V compound is doped with a group IV impurity. If the impurity resides on the group III sub-lattice of the crystal, it serves as a donor.

On the other hand, if it resides on the group V sub-lattice, it serves as an acceptor. Such impurities are called *amphoteric impurities*. Si and Ge are some examples of amphoteric impurities for group III–V compounds. In GaAs, Si usually occupies the Ga sites and thus serves as a donor. If, however, an excess of As vacancies arise due to processing or during growth, Si impurity atoms can occupy the As sites, thereby acting as acceptors.

With suitable doping, a semiconductor can be made n-type or p-type. In n-type semiconductors, the conduction band electrons far outnumber the holes. Therefore, in an n-type semiconductor, electrons are called majority carriers, whereas holes are called minority carriers. The corresponding majority carriers in a p-type semiconductor are holes and the minority carriers are electrons.

19.2.7 Variation of Fermi Level with Temperature and Impurity Concentration

The discussion on Fermi level for an intrinsic semiconductor helps us know, through simple algebra, that

$$E_F - E_i = k_B T \ln\left[\frac{n_0}{n_i}\right] \tag{19.42}$$

Equation (19.41) is valid for an n-type semiconductor. A similar expression for a p-type semiconductor can be written as follows:

$$E_i - E_F = k_B T \ln\left[\frac{p_0}{n_i}\right] \tag{19.43}$$

Equations (19.41) and (19.42) can be used to plot the variation of the Fermi level position with impurity concentration and temperature. Figure 19.9 shows the Fermi level position dependence on impurity concentration for Si at $T = 300$ K.

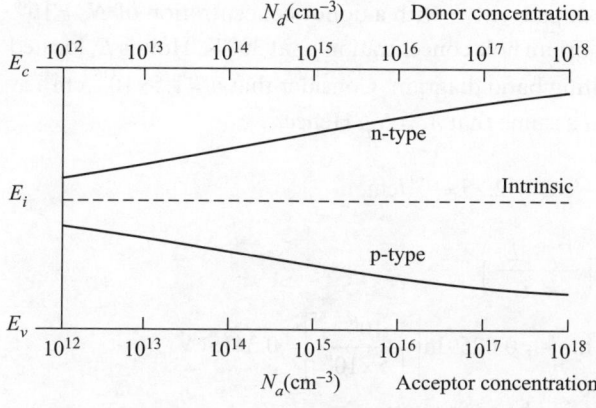

Fig. 19.9 Plot of Fermi level position vs impurity concentration for Si at $T = 300\,\text{K}$

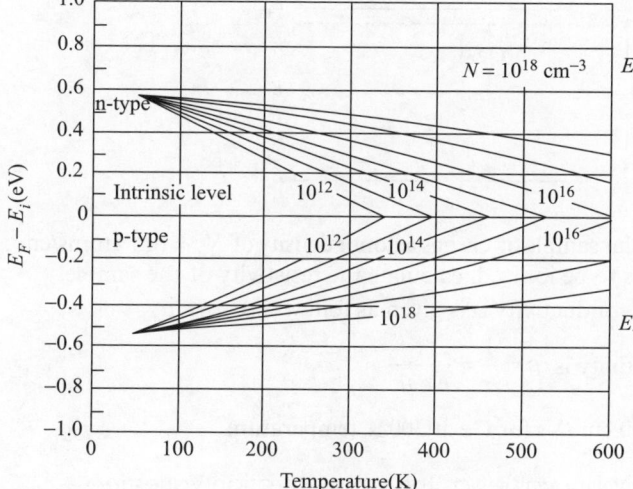

Fig. 19.10 Fermi level vs temperature for Si

Intrinsic carrier concentration is also a function of temperature. From Eqs (19.42) and (19.43) one can, therefore, conclude that the position of Fermi level changes with temperature also. As the temperature increases, n_i increases; as a result, the Fermi level moves closer to the intrinsic Fermi level. Thus, at high temperatures, semiconductors lose their extrinsic behaviour and start behaving like intrinsic semiconductors. The variation of Fermi level position as a function of temperature for Si is shown in Fig. 19.10.

Example 19.1 Evaluate the approximate donor binding energy for GaAs. Assume that $\epsilon_r = 13.2$, $m_{nc}^* = 0.0615 m_e$.

Solution The energy is defined as follows:

$$E = \frac{m_{nc}^* e^4}{8(\epsilon_0 \epsilon_r)^2 h^2} = \frac{0.0615 \times 9.11 \times 10^{-31} \times (1.6 \times 10^{-19})^4}{8 \times (8.85 \times 10^{-12} \times 13.2)^2 (6.63 \times 10^{-34})^2}$$

$$= 8.34 \times 10^{-22}\,\text{J} = 0.0052\,\text{eV}$$

Thus, 5.2 meV energy is required to excite the donor atom.

Example 19.2   A Si wafer is doped with P with a donor concentration of $N_d = 10^{16}$ atoms/cm³. Calculate the equilibrium hole concentration p_0 at 300 K. How is E_F located relative to E_i? Sketch the resulting band diagram. Consider that $n_i = 1.5 \times 10^{10}$/cm³.
Solution   For $N_d \gg n_i$, we can assume that $n_0 = N_d$. Hence,

$$p_0 = \frac{n_i^2}{n_0} = \frac{(1.5 \times 10^{10})^2}{10^{16}}/cm^3 = 2.25 \times 10^4/cm^3$$

We also know that $n_0 = n_i \exp\left(\dfrac{E_F - E_i}{k_B T}\right)$

This gives that $E_F - E_i = k_B T \ln\dfrac{n_0}{n_i} = 0.0259 \ln\left[\dfrac{10^{16}}{1.5 \times 10^{10}}\right] = 0.3415\,eV$

The resulting band diagram would be as follows:

Example 19.3   A particular sample of Ge has a donor density of $N_d = 10^{14}$ atoms/cm³. Assuming all donor atoms to be ionized, calculate the resistivity of the sample.
Solution   For $n \gg p$, the conductivity σ is given as follows: $\sigma = n e \mu_n$

This implies that the resistivity is $\rho = \dfrac{1}{\sigma} = \dfrac{1}{n e \mu_n}$

From Table 19.1, $\mu_n = 3900$ cm²/Vs for Ge at 300 K temperature.

Table 19.1 Typical mobility values as the concentration varies in common semiconductors

For Silicon	N	$\mu_n\left(cm^2/Vs\right)$	$\mu_p\left(cm^2/Vs\right)$
	$10^{15}\,cm^{-3}$	1362	462
	$10^{16}\,cm^{-3}$	1184	429
	$10^{17}\,cm^{-3}$	721	317
	$10^{18}\,cm^{-3}$	277	153
	$10^{19}\,cm^{-3}$	115	71
At T = 300K and low doping concentrations			
Silicon		1350	480
Gallium Arsenide		8500	400
Germanium		3900	1900

Thus, the resistivity is calculated as follows:

$$\rho = \frac{1}{10^{14} \times 1.6 \times 10^{-19} \times 3900} = 16.03\,\Omega\,cm$$

19.3 COMPOUND SEMICONDUCTORS

Elemental semiconductors are found in group IV of the periodic table and, as the name implies, are composed of a single species of atoms. Compound semiconductors, on the other hand, are made up of a combination of more than one species of atoms. Some common compound semiconductors are made up of either a combination of group III and group V atoms (GaAs is the most common example) or a combination of group II and group VI atoms (ZnS is a prominent example). Two group IV elements can also combine to form a compound semiconductor. Table 19.2 lists some elemental and compound semiconductors.

More complicated ternary ($Al_xGa_{1-x}As$) and quaternary (In GaAsP) compound semiconductors are also finding increasingly sophisticated applications.

19.4 EQUATION OF CONTINUITY

The continuity equation is essentially the conservation principle that highlights the interplay of the rate of generation or annihilation (destruction) of carriers and the drift of carriers into or out of a semiconducting element. In a semiconducting element, the minority carriers play a significant role which is physically exhibited as the drift current. The current constituted by the majority carriers is the diffusion current.

Table 19.2 Elemental and compound semiconductors

(a)	**II**	**III**	**IV**	**V**	**VI**
		B	C	N	
		Al	Si	P	S
	Zn	Ga	Ge	As	Se
	Cd	In		Sb	Te

(b)	**Elemental**	**IV components**	**Binary III–V compounds**	**Binary II–VI compounds**
	Si	SiC	AlP	ZnS
	Ge	SiGe	AlAs	ZnSe
			AlSb	ZnTe
			GaN	CdS
			GaP	CdSe
			GaAs	CdTe
			InP	
			InAs	
			InSb	

(a) Region of the periodic table where semiconductors are found
(b) Elemental and compound semiconductors

Consider a semiconducting element; the rate at which the minority carriers hole density changes (N_h) is equal to the rate at which the holes are generated (g_h) minus the rate at which they recombine (r_h) with the holes; that is,

$$\frac{dN_h}{dt} = (g_h - r_h) \tag{19.44}$$

The process of recombination is a dynamic process proportional to the product of the densities of the holes and the electrons:

$$r_h \propto r\,n\,p \tag{19.45}$$

$$r_h = r\,n\,p \tag{19.46}$$

where r is the constant of proportionality and is called the recombination coefficient.

As discussed in the previous section, generation of minority carriers is a thermal process, that is, the generation rate is dependent on the temperature of the semiconducting element. The dynamic equilibrium is established when the rate of generation will be equal to the rate of recombination, that is, there is no change in the density of holes with time:

$$g_h(T) = r\,n_0 p_0 = r\,n_i^2 \tag{19.47}$$

For the time $t = 0$, let us say that δp additional holes are injected into the semiconducting element owing to which there is a corresponding change in the electron concentration, that is, $\delta n = \delta p$. Therefore,

$$n = n_0 + \delta n \tag{19.48}$$

$$p = p_0 + \delta p \tag{19.49}$$

$$\frac{dp}{dt}\bigg|(j=0) = \frac{d(\delta p)}{dt}\bigg|(j=0) = g_h(T) - r_h(n_0 + \delta n)(p_0 + \delta p) \tag{19.50}$$

As the temperature is constant, we can use Eq. (19.47) to get the following result to a first degree of approximation:

$$\frac{d(\delta p)}{dt}\bigg|(j=0) = g_h(T) - rn_i^2 - r_h n_0 \delta p - rp_0 \delta n - r\delta n\delta p \tag{19.51}$$

$$\frac{d(\delta n)}{dt}\bigg|(j=0) = -rp_0 \delta n = -\frac{\delta n}{\tau_{lh}} \tag{19.52}$$

where is τ_l is the lifetime of electrons. Integrating this equation, one gets the following relation:

$$\delta n(t) = \delta n\,|(t=0)e^{-t/\tau_{lh}} \tag{19.53}$$

The left-hand side of Eq. (19.53) represents the change in hole density with time. As time passes, the right-hand side of the equation indicates that the hole carrier density decays exponentially and at $t = \tau_{lh}$, returns to its equilibrium value.

Hence, the time τ_{lh} is the average time for which the holes behave as free charge carriers before recombination. We can write that

$$\frac{d}{dt}(\delta n) = -\frac{\delta p}{\tau_{lh}} \tag{19.54}$$

Here, τ_{le} is the average lifetime of minority electrons.

19.4.1 Continuity Equation with Current Flow

Let us assume that an electron current J_e flows along the positive x-direction across a volume element of thickness is Δx and of a unit area perpendicular to the direction of current flow. The current density and electron concentration are related as follows:

$$n_{x+\delta x} = \frac{J_e(x+\delta x)}{ev_D} \tag{19.55}$$

where the left-hand side indicates the number of electrons that are entering the volume at $(x + \delta x)$. For dynamic equilibrium, the number of electrons leaving at x are given as follows:

$$n_x = \frac{J_e(x)}{ev_D} \tag{19.56}$$

Thus, the number of electrons that increase in the given volume element is given by the following relations:

$$n_{x+\delta x} - n_x = \frac{1}{ev_D}(J_e(x+\delta x) - J_e(x)) \tag{19.57}$$

$$= \frac{1}{ev_D}\left[J_e(x) + \frac{\partial J_e}{\partial x}\delta x - J_e(x)\right] = \frac{1}{ev_D}\frac{\partial J_e}{\partial x}\delta x \tag{19.58}$$

$$\frac{\partial}{\partial t}(\delta n) = \frac{1}{e}\frac{\partial J_e}{\partial x} \tag{19.59}$$

$$\frac{\partial}{\partial t}(\delta p) = \frac{1}{e}\frac{\partial J_p}{\partial x} \tag{19.60}$$

$$\frac{\partial}{\partial t}(\delta n) = -\frac{\delta n}{\tau_{Le}} + \frac{1}{e}\frac{\partial J_e}{\partial x} \tag{19.61}$$

This happens to be the continuity equation for minority electrons/holes when the current is flowing. Thus, the total rate of change of the minority electrons/holes is expressed as follows:

$$\frac{\partial}{\partial t}(\delta n) = -\frac{\delta n}{\tau_{Le}} + \frac{1}{e}\frac{\partial J_e}{\partial x} \tag{19.62}$$

$$\frac{\partial}{\partial t}(\delta p) = -\frac{\delta p}{\tau_{Lh}} + \frac{1}{e}\frac{\partial J_h}{\partial x} \tag{19.63}$$

We know that

$$J_{ex} = ne\mu_e E_x + eD_e \frac{\partial n}{\partial x} \tag{19.64}$$

Substituting the value of the current density in the equation of continuity and knowing that $n = n_0 + \delta n$, where n_0 is a constant, one gets the following expressions:

$$\frac{\partial}{\partial t}(\delta n) = -\frac{\delta n}{\tau_{Le}} + \frac{1}{e}\frac{\partial}{\partial x}\left(ne\mu_e E_x + eD_e \frac{\partial n}{\partial x}\right)$$

$$\frac{\partial}{\partial t}(\delta n) = -\frac{\delta n}{\tau_{Le}} + \mu_e E_x \frac{\partial}{\partial x}(\delta n) + D_e \frac{\partial^2}{\partial x^2}(\delta n) \tag{19.65}$$

$$\frac{\partial}{\partial t}(\delta p) = -\frac{\delta p}{\tau_{Lh}} - \mu_h E_x \frac{\partial}{\partial x}(\delta p) + D_h \frac{\partial^2}{\partial x^2}(\delta p) \tag{19.66}$$

Thus, the continuity equation enables us to know about the excess density of electron or holes in time and space. The continuity equation and current density equation together explain the flow of carriers in semiconducting devices.

Example 19.4 The intrinsic carrier density is 1.5×10^{16} /m³. If the electron and hole mobilities are 0.13 and 0.05 m²/Vs, respectively, calculate its electrical conductivity.
Solution The electrical conductivity is calculated as follows:

$$\sigma = n_i e(\mu_n + \mu_p) = 1.5 \times 10^{16} \times 1.6 \times 10^{-19}(0.13 + 0.05)$$

$$\sigma = 4.32 \times 10^{-4} / \Omega m$$

Example 19.5 Find the intrinsic resistivity of Ge at room temperature (300 K) if the carrier density is 2.15×10^{-13}/cm³. (The electron mobility $\mu_n = 3900$ cm²/Vs; hole mobility $\mu_p = 1900$ cm²/Vs.)
Solution The electron mobility is calculated as follows:

$$\sigma = n_i e(\mu_n + \mu_p)$$

$$\sigma = 2.15 \times 10^{-13} \times 1.6 \times 10^{-19} \times (3900 + 1900) = 2.32 \times 10^{-2} /\Omega cm$$

The intrinsic resistivity is calculated as follows:

$$\rho_i = \frac{1}{\sigma_i} = 43 \,\Omega cm$$

Example 19.6 In a p-type Ge semiconductor, the intrinsic carrier concentration is 2.1×10^{19}/m³ and the density of B is 4.5×10^{23} atoms/m³. The electron and hole mobilities are 0.4 and 0.2 m²/Vs, respectively. What is its electrical conductivity before and after the addition of B atoms?
Solution The electron mobility is calculated as follows:

$$\sigma = n_i e(\mu_n + \mu_p)$$

$$\sigma = 2.1 \times 10^{19} \times 1.6 \times 10^{-19} \times (0.40 + 0.20) = 2.016/\Omega m$$

Example 19.7 Calculate the intrinsic carrier concentration, intrinsic conductivity, and resistivity of Ge at 300 K using the following data:

$$\mu_n = \frac{0.4 \text{ m}^2}{\text{Vs}}, \mu_p = \frac{0.2 \text{ m}^2}{\text{Vs}}, E_g = 0.15 \text{ eV}, m_n^* = 0.55m_e, \text{ and } m_p^* = 0.315m_0.$$

Solution The intrinsic concentration is calculated as follows:

$$n_i = 2(2\pi k_B T/h^2)^{3/2}(m_n^* m_p^*)^{3/2} \exp(-E_g/2k_B T)$$

$$= 2(((2 \times 22/17) \times (1.38 \times 10^{-23}) \times 300)/(6.634 \times 10^{-34})^2)^{3/2}$$

$$(0.55 \times 9.1 \times 10^{-31} \times 0.315)^{3/4} \exp(-0.15/(2 \times 1.38062 \times 10^{-23} \times 300)$$

$$= 1.352 \times 10^{13}/\text{m}^3$$

Intrinsic conductivity is calculated as follows:

$$\sigma = n_i e(\mu_n + \mu_p) = (1.352 \times 10^{13})(1.6 \times 10^{-19})(0.4 + 0.2)$$

$$= 1.298 \times 10^{-6}/\Omega\text{m}$$

Intrinsic resistivity is calculated as follows:

$$\rho = 1/\sigma = \left(\frac{1}{1.298 \times 10^{-6}}\right) = 0.1515 \times 10^6 \,\Omega\text{m}$$

Example 19.8 An n-type Si wafer has been doped uniformly with Sb atoms, and the doped Si has a donor concentration of $10^{16}/\text{cm}^3$. Calculate the Fermi energy with respect to the Fermi energy in intrinsic Si. (For Si, $n_i = 1.45 \times 10^{10}/\text{cm}^3$.)

Solution Doping of Sb converts Si into n-type material with $N_d = 10^{16}/\text{cm}^3$
For Si, $n_i = 1.45 \times 10^{10}/\text{cm}^3$
As $N_d > n_i$, for intrinsic Si:

$$n_i = N_c \exp[-(E_c - E_i)/k_B T]$$

and for doped Si:

$$N_d = N_c \exp\left[-\frac{E_c - E_{Fd}}{k_B T}\right]$$

Dividing N_d by n_i, we get the following value:

$$E_{Fd} - E_i = k_B T \ln(N_d/n_i) = 0.348 \text{ eV}$$

The Fermi energy with respect to E_F in intrinsic Si = 0.348 eV

Example 19.9 Find the resistance of a 1 cm³ pure Si crystal. What is the resistance of the crystal when doped with As with a doping concentration of 1 in 10^9. Intrinsic resistivity is $2.3 \times 10^{-5}/\Omega\text{m}^3$.

Solution The resistance is calculated as follows:

$$R = L/\sigma A = 2.39 \times 10^{15} \,\Omega$$

If the crystal is doped with arsenic, then the donor concentration is calculated as follows:

$$N_d = N_{Si}/10^9 = 5 \times 10^9/\text{m}^3$$

The concentration of hole is calculated as follows:

$$P = n_i^2/N_d = \frac{2.3 \times 10^{-5}}{5 \times 10^{-19}} = 4.2 \times 10^6/m^3$$

Conductivity is given by the following relation:

$$\sigma = ne\mu_e$$

Therefore, the resistance R has the following value:

$$R = L/\sigma A = 5.26\,\Omega$$

Example 19.10   Mobilities of electrons and holes in a sample of intrinsic Ge at 300 K are 0.36 and $\dfrac{0.15\,m^2}{Vs}$, respectively. If the resistivity of the specimen is 2.12 Ωm, compute the forbidden energy gap for Ge.

Solution   We have the following values:
Temperature, $T = 300\,K$; resistivity, $\rho = 2.12\,\Omega m$; electron mobility, $\mu_n = 0.36\,m^2/Vs$; hole mobility, $\mu_p = 0.15\,m^2/Vs$

$$\sigma = 1/\rho = \left(\frac{1}{2.12}\right) = 0.415/\Omega m$$

We know that $\sigma = n_i e(\mu_n + \mu_p)$

$$\Rightarrow n_i = \sigma/(e(\mu_n + \mu_p)) = \left(\frac{0.415}{1.6 \times 10^{-19}(0.36 + 0.15)}\right) = 5.56 \times 10^{18}/m^3$$

The intrinsic concentration of carriers is also given by the following relation:

$$n_i = (N_c N_v)^{\frac{1}{2}} \exp(-E_g/2k_B T)$$

$$N_c = 2(2\pi k_B T/h^2)^{\frac{3}{2}}(m_n^*)^{\frac{3}{2}} = 8.85 \times 10^{24}/m^3$$

$$N_v = 2(2\pi k_B T/h^2)^{\frac{3}{2}}(m_p^*)^{\frac{3}{2}} = 5.634 \times 10^{24}/m^3$$

$$E_g = 2k_B T\left[\ln(N_c N_v)^{\frac{1}{2}}/n_i)\right] = 2k_B T \ln\left(\frac{8.852 \times 10^{24} \times (5.635 \times 10^{24})^{\frac{1}{2}}}{5.56 \times 10^{18}}\right)$$

$$= 2k_B T \times 14.054 = 0.15215\,eV$$

Example 19.11   The following data is given for an intrinsic Ge semiconductor at 300 K. Calculate the conductivity of the sample. Given that $n_i = 2.4 \times 10^{19}/m^3$, $\sigma_n = 0.39\,m^2/Vs$, $\sigma_n = 0.19\,m^2/V$.

Solution   The intrinsic conductivity is calculated as follows:

$$\sigma = n_i e(\mu_n + \mu_p) = 2.4 \times 10^{19} \times 1.6 \times 10^{-19}(0.39 + 0.19) = 2.22/\Omega m$$

Example 19.12   In an n-type semiconductor the Fermi level lies 0.3 eV below the conduction band at 300 K. If the temperature is increased to 330 K, find the new position of the Fermi level.

Solution The carrier concentration at 300 K is given by the following relation:

$$n_{i300} = N_c \exp\left[-\frac{E_c - E_{F300}}{k_B T}\right]$$

$$= N_c \exp[-(0.3 \times 1.6 \times 10^{-19})/300 k]$$

Taking $N_{i300} = N_{i330}$, from this equation we get the following expression:

$$\frac{0.3}{300} = \left(E_c - \frac{E_{F330}}{330}\right)$$

Hence,

$$E_c - E_{F330} = 0.333 \text{ eV}$$

At 330 K, the Fermi energy level lies 0.33 eV below the conduction band.

Example 19.13 The energy gap of Si is 1.1 eV. The average electron effective mass is $0.31 m_e$, where m_e is the free electron mass. Calculate the electron concentration in the conduction band of Si at room temperature ($T = 300$ K). Assume that $E_F = E_g/2$.

Solution For Si the intrinsic concentration is calculated as follows:

$$n_i = \left(\frac{2\pi k_B T m_e^*}{h^2}\right)^{3/2} \exp\left[-\frac{(E_c - E_v)}{2k_B T}\right]$$

$$= \left[\frac{2 \times \dfrac{22}{15} \times (1.38 \times 10^{-23} \times 300 \times 9.1 \times 10^{-31})}{[(6.634 \times 10^{-34})^2]^{3/2}}\right] \exp\left[-\frac{(1.1 \times 1.6 \times 10^{-19})}{2 \times 1.38 \times 10^{-23} \times 300}\right]$$

$$= 4.32 \times 10^{24} \exp\left[-\frac{(1.1 \times 1.6 \times 10^{-19})}{2 \times 1.38 \times 10^{-23} \times 300}\right]$$

$$= 2.54 \times 10^{15}/\text{m}^3$$

The intrinsic electron concentration of Si at 300 K is $2.54 \times 10^{15}/\text{m}^3$.

Example 19.14 Find the resistance of an intrinsic Ge rod (1 cm long, 1 mm wide, and 1 mm thick) at 300 K. For Ge, $n_i = 2.5 \times 10^{19}/\text{m}^3$; $\mu_n = 0.39 \dfrac{\text{m}^2}{\text{Vs}}$; $\mu_p = 0.19 \text{ m}^2/\text{Vs}$.

Solution We know that

$$\sigma = ne(\mu_n + \mu_p) = 2.5 \times 10^{19} \times 1.6 \times 10^{19}(0.39 + 0.19) = 2.32/\Omega\text{m}$$

The resistance is calculated as follows:

$$R = L/\sigma A = 1/(2.32 \times 100 \times 3.14 \times 10^{-1} \times 10^{-1}) = 0.13152 \Omega$$

Example 19.15 For an Si semiconductor with a band gap 1.12 eV, determine the position of the Fermi level at 300 K if $m_n^* = 0.12 m_e$, $m_p^* = 0.28 m_e$.

Solution We know that

$$E_F = \frac{E_g}{2} + \frac{3k_B T}{4}\left[\log_e \frac{m_p^*}{m_n^*}\right]$$

$$E_F = \frac{1.12}{2} + \frac{3 \times 1.38 \times 10^{-23} \times 300}{4 \times 1.6 \times 10^{-19}} \log\left[\frac{0.28m_e}{0.12m_e}\right]$$

$$= 0.56 + 15.14 \times 10^{-3} \text{ eV}$$

$$= 0.5615 \text{ eV}$$

Example 19.16 The intrinsic carrier density is $1.5 \times 10^{16}/\text{m}^3$. If the electron and hole mobilities are 0.13 and 0.05 m^2/Vs, respectively, calculate its electrical conductivity.

Solution The electrical conductivity is calculated as follows:

$$\sigma = n_i e(\mu_n + \mu_p) = 1.5 \times 10^{16} \times 1.6 \times 10^{-19}[0.13 + 0.05] = 4.32 \times 10^{-4}/\Omega\text{m}$$

Example 19.17 Find the intrinsic resistivity of Ge at room temperature (300 K), if the carrier density is 2.15×10^{-13}. Given that electron mobility $\mu_n = 3900 \text{ cm}^2/\text{Vs}$ hole mobility $= \mu_p = 1900 \text{ cm}^2/\text{Vs}$.

Solution We know that

$$\sigma = n_i e(\mu_n + \mu_p) = 2.15 \times 10^{-13} \times 1.6 \times 10^{-19}(3900 + 1900) = 2.32 \times 10^{-2}/\Omega\text{cm}$$

The intrinsic resistivity is calculated as follows:

$$\rho_i = \left(\frac{1}{\sigma_i}\right) = \frac{1}{2.32 \times 10^{-2}} = 43\ \Omega\text{cm}$$

Example 19.18 In a p-type Ge semiconductor, the intrinsic carrier concentration is $2.1 \times 10^{19}/\text{m}^3$ and the density of B is 4.5×10^{23} atoms/m³. The electron and hole mobilities are 0.4 and 0.2 m^2/Vs, respectively. What is its electrical conductivity before and after the addition of B atoms?

Solution We know that

$$\sigma = n_i e(\mu_n + \mu_p) = 2.1 \times 10^{19} \times 1.6 \times 10^{19}(0.4 + 0.2) = 2.016/\Omega\text{m}$$

19.5 HALL EFFECT

Moving charges experience forces in the presence of applied electric and magnetic fields. One interesting consequence of these forces is the *Hall effect*. Simply put, the Hall effect is a process that leads to the development of a voltage across one of the faces of a semiconductor slab when crossed electric and magnetic fields are applied across the other two pairs of faces of the slab. Hall effect can be used to distinguish between n- and p-type semiconductors, and to measure the majority carrier concentration and majority carrier mobility. The effect can also be used to develop a magnetic probe for circuit application.

Figure 19.11 shows a schematic diagram explaining the Hall effect. A current I_x flows through a semiconductor in the x-direction. A magnetic field B_z is applied in the z-direction. Charge carriers present in the semiconductor experience a force due to the magnetic field. This magnetic force is felt in the $-y$-direction by both electrons and holes. In an n-type semiconductor ($n_0 > p_0$), a build-up of electrons would take place on the $y = 0$ surface. For a p-type semiconductor ($p_0 > n_0$), the build-up would be of holes on the $y = 0$ surface.

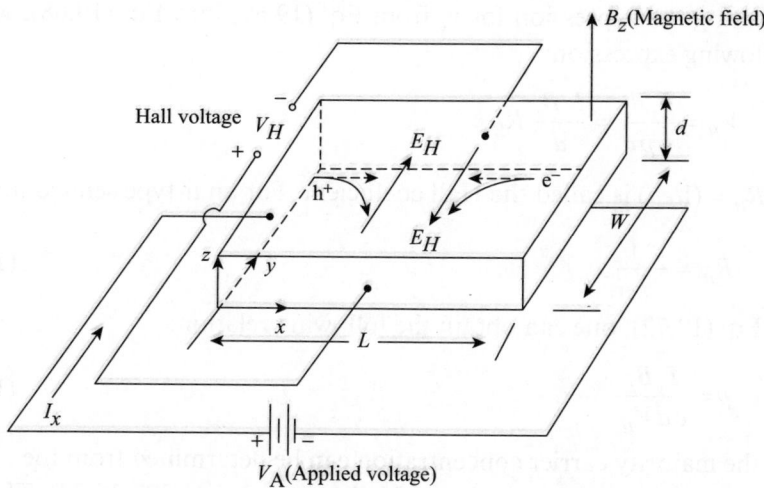

Fig. 19.11 Typical set-up for observing Hall effect

This charge build-up would result in an induced electric field in the y direction. In the steady state, the resulting force on the charge carriers due to this induced electric field would exactly be balanced by the magnetic force. Thus, at equilibrium we can write that

$$eE_y = ev_x B_z \tag{19.67}$$

where the induced electric field E_y is called the Hall field and the corresponding voltage is called the *Hall voltage*. Designating the Hall voltage as V_H and the Hall field as E_H, we can see from Fig. 19.11 that

$$V_H = E_H W = E_y W \tag{19.68}$$

where E_H is assumed to be positive in the $+y$-direction and V_H is considered positive if it has the polarity shown in Fig. 19.11. W is the specimen width.

For a p-type semiconductor, the Hall voltage would be positive; for an n-type semiconductor, it would be negative. Thus, the polarity of the Hall voltage can be used to determine the type of the extrinsic semiconductor.

Use of Eq. (19.67) in Eq. (19.68) leads to the following expression:

$$V_H = v_x W B_z \tag{19.69}$$

For a p-type semiconductor, we can express the current density J_x as follows:

$$J_x = epv_{dx} \tag{19.70}$$

(where $v_{dx} = v_x$)
which leads to

$$v_{dx} = \frac{J_x}{ep} = \frac{I_x}{epWd} \tag{19.71}$$

where v_{dx} is the drift velocity along the x-direction, the product Wd represents the relevant face area, and e is the magnitude of the electron charge.

Substituting the expression for v_x from Eq. (19.69) into Eq. (19.68), we get the following expression:

$$V_H = \frac{I_x B_z}{e p d} = \frac{I_x B_z}{d} R_H \tag{19.72}$$

Here $R_H = (1/ep)$ is called the Hall coefficient. For an n-type semiconductor,

$$R_H = -\frac{1}{e n} \tag{19.73}$$

From Eq. (19.72), one can obtain the following relation:

$$p = \frac{I_x B_z}{e d V_H} \tag{19.74}$$

Thus, the majority carrier concentration can be determined from the knowledge of current, magnetic field, and Hall voltage using Eq. (19.74). The corresponding expressions for an n-type semiconductor are

$$V_H = -\frac{I_x B_z}{n e d} \tag{19.75}$$

$$n = -\frac{I_x B_z}{e d V_H} \tag{19.76}$$

Since the Hall voltage is negative for electrons, the electron concentration obtained using Eq. (19.76) is positive, as expected. We will now see how the Hall effect can be utilized to determine low-field carrier mobility. The current density J_x can be expressed as follows:

$$J_x = e p \mu_p E_x \tag{19.77}$$

where μ_p is the low-field hole mobility. Equation (19.77) can be rewritten as follows:

$$\frac{I_x}{Wd} = \frac{e p \mu_p V_x}{L} \tag{19.78}$$

which leads to

$$\mu_p = \frac{I_x L}{e p V_x W d} \tag{19.79}$$

The corresponding expression for μ_n would be as follows:

$$\mu_n = \frac{I_x L}{e n V_x W d} \tag{19.80}$$

Equations (19.79) and (19.80) can be used to calculate the low-field carrier mobility.

Example 19.19   An n-type semiconductor specimen has a Hall coefficient $R_H = 3.66 \times 10^{-11}$ m³/As. The conductivity of the specimen is found to be 112×10^{-15} m³/As. Calculate the charge carrier density n and the electron mobility at room temperature.

Solution The Hall coefficient is given by the following expression:

$$R_H = \frac{1}{ne}$$

$$\Rightarrow n = \frac{1}{eR_H} = \frac{1}{1.6\times10^{-19}\times3.66\times10^{-11}} = 2\times10^{29}/m^3$$

For an *n*-type semiconductor, $\sigma = ne\mu_n$

$$\Rightarrow \mu_n = \sigma/ne = \frac{112\times10^{15}}{2\times10^{29}\times1.6\times10^{-19}} = 0.035\ m^2/As$$

Example 19.20 A current of 50 A is established in a Cu slab (0.2 cm thick and 2 cm wide). The slab is placed in a magnetic field *B* of 1.5 T. The magnetic field is perpendicular to the plane of the slab and to the current. The free electron concentration in Cu is $8.4\times10^{28}/m^3$. What will be the magnitude of the Hall voltage across the width of the slab?

Solution The Hall voltage is given by the following expression:

$$V_H = \frac{BI}{ned} = \frac{1.5\times50}{8.4\times10^{28}\times1.6\times10^{-19}\times2\times10^{-3}} = 2.159\times10^{-15}\ V$$

Example 19.21 A sample of Si is doped with As to a level of 5×10^{16} atoms/cm³. Calculate the resistivity of the sample. Determine the Hall coefficient and the Hall voltage if the thickness *d* of the sample is 2×10^{-2} cm, $I_x = 2$ mA, $B_z = 5$ kG $= \dfrac{5\times10^{-5}\ Wb}{cm^2}$.

Assume that $\mu_n = 800\dfrac{cm^2}{Vs}$.

Solution Resistivity is defined as follows:

$$\rho = \frac{1}{\sigma} = \frac{1}{e\mu_n n_0} = \frac{1}{1.6\times10^{-19}\times800\times5\times10^{16}} = 0.156\ \Omega\ cm$$

Hall coefficient R_H for an n-type semiconductor is given by the following expression:

$$R_H = -\frac{1}{en} = \frac{1}{1.6\times10^{-19}\times5\times10^{16}} = -125\ cm^3/C$$

The Hall voltage V_H is given by the following expression:

$$V_H = \frac{I_x B_z}{d}R_H = \frac{(2\times10^{-3})\times(5\times10^{-5})(-125)}{(2\times10^{-2})} = -62.5\times10^{-5}\ V$$

Example 19.22 Find the Hall coefficient and electron mobility of Ge for a given sample (length 1 cm, breadth 5 mm, and thickness 1 mm). A current of 5 mA flows from a 1.35 V supply and develops a Hall voltage of 20 mV across the specimen in a magnetic field of 0.45 Wb/m².

Solution The following is the equation for resistivity:

$$\rho = \frac{RA}{L}$$

We know that the resistance R is calculated as follows: $R = \dfrac{V}{I} = \dfrac{1.35}{5 \times 10^{-3}} \, \Omega$

The area can be calculated as follows: $A = 5 \times 10^{-3} \times 1 \times 10^{-3} \, \text{m}^2 = 5 \times 10^{-6} \, \text{m}^2$

Thus, $\rho = \dfrac{RA}{L} = \dfrac{1.35 \times 5 \times 10^{-6}}{5 \times 10^{-3}} = 0.135 \, \Omega$

The Hall field is calculated as follows: $E_y = \dfrac{v_y}{d} = \dfrac{20 \times 10^{-3}}{10^{-3}} = 20 \, \text{V/m}$

$$\text{Current density} = \frac{\text{Current}}{\text{Area of cross section}} = \frac{5 \times 10^{-3}}{5 \times 10^{-6}} = 10^3 \, \text{A/m}^2$$

$$\text{Hall coefficient} = R_H = \frac{1}{ne} = \frac{E_y}{HJ} = \frac{20}{0.45 \times 10^3} = 0.044 \, \text{m}^3/\text{C}$$

$$\text{Electron mobility} = \mu_n = \frac{R_H}{\rho} = \frac{0.044}{0.135} = 0.33 \, \text{m}^2/\text{Vs}$$

Example 19.23 A copper strip (2 cm wide and 1 mm thick) is placed in a magnetic field with $B = 1.5 \, \text{Wb/m}^2$ perpendicular to the strip. Suppose a current of 200 A is set up in the strip; what Hall potential difference would appear across the strip? Given that electron concentration $N = 8.4 \times 10^{28}/\text{m}^3$.

Solution The following is the calculation:

$$\text{Hall potential} = V_H = \frac{I_x B_z}{e\,p\,d} = \frac{(200 \times 1.5)}{(8.4 \times 10^{28})(1.6 \times 10^{-19})(1 \times 10^{-3})} = 22 \, \text{mV}$$

Example 19.24 The Hall coefficient and conductivity of Cu at 300 K have been measured to be $0.55 \times 10^{-10} \, \text{m}^3/\text{As}$ and $5.9 \times 10^{15}/\Lambda\Omega\text{m}$, respectively. Calculate the drift mobility of the electrons in Cu.

Solution The following is the calculation:

$$\text{Drift mobility} = |R_H \sigma| = 0.55 \times 10^{-10} \times 5.9 \times 10^{15} = 3.2 \times 10^{-3} \, \text{m}^2/\text{Vs}$$

Example 19.25 Using the electron drift mobility from the Hall effect measurement ($\mu_d = 3.2 \times 10^{-3} \, \text{m}^2/\text{Vs}$), calculate the concentration of conduction electrons in Cu, and determine the average number of electrons contributed to the free electron gas per copper atom in a solid (given that $\sigma = 5.9 \times 10^{15}/\Omega\text{-m}$.)

Solution The conductivity of an n-type semiconductor is given as follows:

$$\sigma = ne\mu$$

$$n = \sigma/\mu e = \frac{5.9 \times 10^{15}}{3.2 \times 10^{-3} \times 1.6 \times 10^{-19}} = 1.15 \times 10^{29}/\text{m}^3$$

The concentration of free electrons in pure Cu:

$$= \frac{\text{Avogadro const.} \times \text{Density} \times \text{No. of free electrons per atom}}{\text{Atomic weight}}$$

$$= 8.44 \times 10^{28} \, \text{electrons/m}^3$$

Average number of electrons contributed per Cu atom:

$$= \frac{\text{Concentration of electrons in an n-type semiconductor}}{\text{Concentration of electrons in a pure Cu atom}}$$

$$= \frac{1.15 \times 10^{29}}{8.44 \times 10^{28}} = 1.36$$

Thus, the average number of electrons contributed per Cu atom is 1.

19.6 FORMATION OF P–N JUNCTION

We have discussed p- and n-type devices to understand the roles of the majority and minority carriers in the generation of a voltage gradient and hence the flow of current. A p–n junction is formed from a single-crystal intrinsic semiconductor. The intrinsic semiconductor can be doped with acceptor and a part with donor impurities. The rectifier is made on the basis of the p–n junction formation. In this section, we will not discuss the technique used for the formation of a p–n junction, but the interplay between acceptor and donor impurities that goes about when a p–n junction is formed.

A p–n junction is formed when a p-type doped semiconductor is made to join an n-type doped semiconductor. The variation in the impurity concentration occurs over a short distance, typically of the order of microns. As the p- and n-type materials are brought together, the acceptor concentration, N_a, becomes slightly greater than the donor concentration, N_d, such that a certain potential gradient is developed due to which there is a flow of carriers, even though no voltage is applied. In case of the majority carrier current, the majority holes diffuse out of the p-region, leaving behind negatively charged acceptor atoms bound to the lattice. A space charge is generated in a region that was neutral earlier. Similarly, the electrons diffusing from the n-region leave behind positively charged donor atoms bound to the lattice. Thus, a double space-charged layer is formed, which is of opposite sign to the majority carriers diffusing into them. This space-charged layer gradually starts to build up resistance for the flow of majority diffusing carriers across this boundary. The majority current gradually starts to decrease, as an electric field builds up due to the space charge and finally inhibits the flow of current. The space charge layer is referred to as the *depletion layer* or the *barrier layer*.

19.7 ENERGY DIAGRAM OF P–N JUNCTION

As the p and n junctions are brought together, equilibrium is established. This equilibrium indicates that the Fermi levels, E_F, of the two semiconductors (p- and n-type) have reached the same energy level. The Fermi level of the acceptor (p-type) E_{FA} lies just above the valence band and that of the donor (n-type) lies just below the conduction band. As the two p- and n-type junctions are brought together, the Fermi levels of each of the junctions gets readjusted. The holes diffusing from the p-type region leave a negative charge raising its energy level; similarly, the electrons diffuse from the n-type region, leaving behind a positive charge, thus lowering its level (refer to Fig.19.12). As the two Fermi levels reach the same level or equilibrium, a potential gradient (eV_0) is generated.

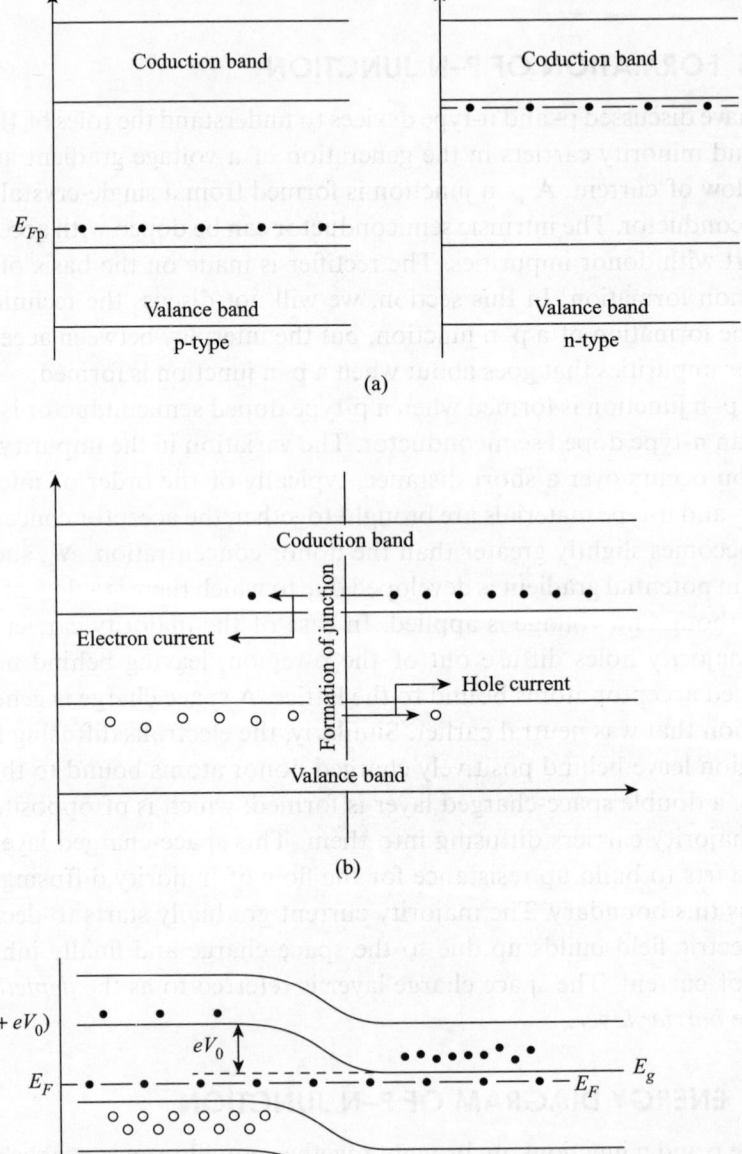

Fig. 19.12 Formation of p–n junction (a) Before contact (b) Immediately after contact (c) In equilibrium

The number of electrons in the conduction band of the n-region (N_n) can be calculated as follows:

$$N_n = N_c \exp\left[-\frac{E_g - E_F}{k_B T}\right] \qquad (19.81)$$

The number of electrons in the p-region N_p can be calculated as follows:

$$N_p = N_c \exp\left[-((E_g + eV_0) - E_F)/k_B T\right] \qquad (19.82)$$

Dividing the two equations, one gets the following relation:

$$\frac{N_n}{N_p} = \exp\left[\frac{eV_0}{k_B T}\right] \qquad (19.83)$$

Representing the mathematics in a reverse way, the barrier potential V_0 is represented as follows:

$$V_0 = \frac{k_B T}{e} \log_e\left[\frac{N_n}{N_p}\right] \qquad (19.84)$$

Equation (19.84) shows that barrier potential is related to the carrier concentrations N_n and N_p; the temperature T also plays a significant role.

19.8 DIODE EQUATION

In the preceding section, the concept of barrier potential has been established and its relation with the carrier concentrations highlighted. This barrier potential can be overcome and a current element can be created. Sections 19.8.1 and 19.8.2 discuss in detail the forward and reverse biases for the current element (p–n junction).

19.8.1 p–n Junction in Forward Bias

The potential that gradually builds across a p–n junction eventually stops the flow of majority carriers; hence, for the current to keep flowing one has to connect a battery across the junction, thus creating an external potential for the flow of carriers. The p–n junction is connected to an external battery of voltage V in a forward bias mode such that the positive terminal is connected to the p-region and the negative terminal to the n-region. This applied voltage acts directly across the depletion layer such that the effective voltage now drops to $V_0 - V$. Thus, the majority carriers are able to overcome this lowered potential and hence diffuse across the junction, causing a flow of large currents. Gradually as the position of the Fermi level is restored on both the sides of the junction, equilibrium or steady state is achieved. The majority carrier densities for the p and n junctions P_p and N_n are proportional to the majority carrier current due to diffusion, that is, J_{hp} and J_{hn}, respectively. When the forward bias voltage V is applied across the potential barrier, then using the Boltzmann statistics one can write the following equation:

$$e(V_0 - V) = -k_B T \log_e (P_{n0}/P_p)$$
(19.85)

where P_{n0} gives the minority carriers of the n-type at $V_0 = 0$. They indicate the hole carriers of the n-type.

Or
$$P_{n0} = P_p \exp\left[\frac{e(V - V_0)}{k_B T}\right]$$
(19.86)

This equation clearly indicates that the minority carriers increase exponentially along the depletion layer as they are injected from the p-side (majority carriers) to the n-side. When there is no initial voltage $V_0 = 0$; $P_{n0} = P_n$. Hence,

$$P_{n0} = P_n \exp\left[\frac{eV}{k_B T}\right]$$
(19.87)

Similarly, for the electrons injected into the p-region from the n-region, one gets the following equation:

$$N_{p0} = N_p \exp\left(\frac{eV}{k_B T}\right)$$
(19.88)

The energy band picture depicts that as the external voltage is applied, the bands tilt due to the lowering of the voltage across the depletion layer. This facilitates the flow of majority carriers across this lowered potential, and hence large currents flow. The currents, and the p and n junctions at equilibrium are given as follows:

$$J_p = J_{pdrift} + J_{pdiffusion} = J_{pe} + J_{ph} = 0$$
(19.89)

$$J_n = J_{ndrift} + J_{ndiffusion} = J_{nh} + J_{ne} = 0$$
(19.90)

The drift current for both p and n junctions, represented by J_{pe} and J_{nh}, respectively, is dependent on the respective mobilities μ_p and μ_n, and the electric field created due to the potential gradient at the junction. As the minority electrons and holes are injected into the p- and n-region, respectively, certain corresponding diffusion lengths L_p and L_n (usually of the order of 1 mm) are obtained such that the drift current can be expressed as follows:

$$J_{pe} = e\frac{D_h P_n}{L_h}; J_{nh} = e\frac{D_e N_p}{L_e}$$
(19.91)

where

$$D_e = (k_B T/e)\mu_e; D_h = (k_B T/e)\mu_h$$
(19.92)

The diffusion coefficients for electrons and holes indicate their ability to move through the crystal. For a typical crystal of germanium, $D_e = 0.0093 \text{ m}^2/\text{s}$; $D_h = 0.0044 \text{ m}^2/\text{s}$

One can easily infer from these equations that

$$\frac{D_e}{\mu_e} = \frac{D_h}{\mu_h} = \frac{k_B T}{e} \tag{19.93}$$

The diffusion current as the forward voltage is applied is given as follows:

$$J_{ph} = J_{pe} \exp(eV/k_B T) \tag{19.94}$$

Similarly,

$$J_{ne} = J_{nh} \exp(eV/k_B T) \tag{19.95}$$

Thus, the net hole current across the junction is given as follows:

$$J_h = J_{ph} - J_{pe} = J_{pe}(\exp(eV/k_B T) - 1) \tag{19.96}$$

The net electron current across the junction is given as follows:

$$J_e = J_{ne} - J_{nh} = J_{nh}(\exp(eV/k_B T) - 1) \tag{19.97}$$

The total junction current, J, is given as the sum of the hole and electron currents, that is,

$$J = J_h + J_e = (J_{pe} + J_{nh})\left(\exp\left(\frac{eV}{k_B T}\right) - 1\right) \tag{19.98}$$

$$J = J_0(\exp(eV/k_B T) - 1) \tag{19.99}$$

Equation (19.99) represents the diode equation or the rectifier equation J_0 represents the sum of the current densities due to the minority carriers across the p–n junction and is referred to as the saturation current density. Based on our discussion on the drift current, it can be concluded that the drift current is related to the following physical parameters:

$$J_0 = e\left(\frac{D_h P_n}{L_h} + \frac{D_e P_n}{L_e}\right) \tag{19.100}$$

For optimum working of a diode the exponential term far exceeds the drift current factor, and so one can safely conjecture the diode equation to take the following form:

$$J \cong J_0 \exp\left(\frac{eV}{k_B T}\right) \tag{19.101}$$

19.8.2 p–n Junction in Reverse Bias

In the previous section, the p–n junction has been connected so as to add to the diffusion current flow, physically giving an extra supply of the majority carriers. As has also been observed in the case of semiconductors, the drift current that arises due to the flow of minority carriers has a significant role.

This current arises due to thermal effect; due to the quantum tunnelling of particles, some carriers are drifted across the junction, giving a significant current density.

The battery that gives the external voltage (V) is now connected across the p–n junction in the reverse bias, that is, the negative terminal of the battery is connected across the p-end and the positive terminal across the n-end of the p–n junction. The effective voltage across the p–n junction is now greater than the equilibrium value V_0 such that it is now $V_0 + V$. Thus, it is now even more difficult for the majority carriers to cross the barrier, and hence one observes that the majority currents J_{hp} and J_{en} become zero. The flow of the minority carriers across the potential hill is due to thermal motion and is therefore unaffected by the increase in the barrier height. Hence, the flow of current in the reverse-biased p–n junction is owing to the minority carriers only, that is,

$$J \cong -(J_{pe} + J_{nh}) = -J_0 \qquad (19.102)$$

The reverse current is constituted by the saturation current density, physically indicated in Eq.(19.102). The negative sign indicates that the flow of the current is from the p to the n junction. The diode equation for reverse bias can be written as follows:

$$J = J_0 \left[\exp\left(-\frac{eV}{k_B T} \right) - 1 \right] = -J_0 \left[1 - \exp\left(-\frac{eV}{k_B T} \right) \right] \qquad (19.103)$$

Thus, as is evident, the sign of the external voltage is now negative ($-V$). As is evident from the equation, the exponential term is decreasing, and at room temperature it has a negligible contribution. Hence, the current contribution is from the saturation current only, as can be observed from Eq.(19.102).

19.9 CURRENT–VOLTAGE CURVE FOR P–N JUNCTION DIODE

The current–voltage characteristics of a p–n junction diode are shown in Fig. 19.13.

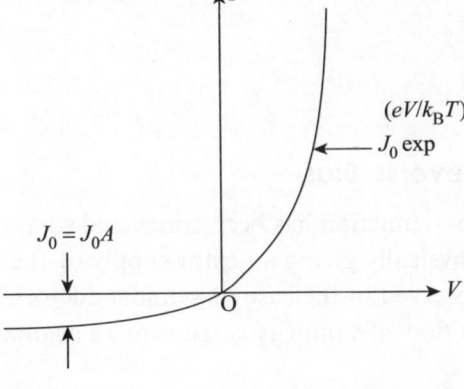

Fig. 19.13 Current–voltage characteristic of p–n junction diode

The diode equation is as follows:

$$J = J_0 \left[\exp\left(\frac{eV}{k_B T} \right) - 1 \right] \quad \text{(Forward bias)}$$

$$J = -J_0 \left[1 - \exp\left(-\frac{eV}{k_B T} \right) \right] \quad \text{(Reverse bias)}$$

This can be plotted for observing the *V–I* characteristics of a p–n junction. Depending on the material used for the formation of the p–n junction, the voltage drop across the depletion layer differs. If the conductivity of the material is high, then the diode equation gives almost the same result as seen in practical cases. For substances having low conductivity, the effective voltage across the depletion layer may differ from that in the diode equation and so must be accounted for.

19.10 LIGHT-EMITTING DIODES

A light-emitting diode (LED) is an integrated semiconductor device producing luminescence. It works on the regular p–n junction to emit light when recombination of an e–h pair takes place. An LED works on the principle of electroluminescence. The striking features of an LED, compared with other light emitting devices like a fluorescent bulb, are as follows:
1. Lower energy consumption
2. Longer lifetime
3. Better physical robustness
4. Smaller/handy size
5. High switching rate, which is used in remote control units and advanced communication technology

The disadvantages or challenges of LEDs that need to be overcome are as follows:
1. Requirement of precise current management and heat management
2. Relatively expensive

19.10.1 Working

An LED consists of a doped semiconducting material in the form of a p–n junction. Current flows, as the p–n junction is forward biased. As an electron recombines with a hole, it goes to a lower energy level, thus expending energy in the form of a light photon, as shown in Fig. 19.14. The colour or wavelength of the light photon emitted depends on the energy band gap. Figure 19.15 shows the current voltage graph for an LED.

It has been a challenge to get LEDs of different colours. Active research is going on in the field of coloured LEDs. In the year 2014, Isamu Akasaki, Hiroshi Amano and Shuji Nakamura were rewarded with the Nobel prize in

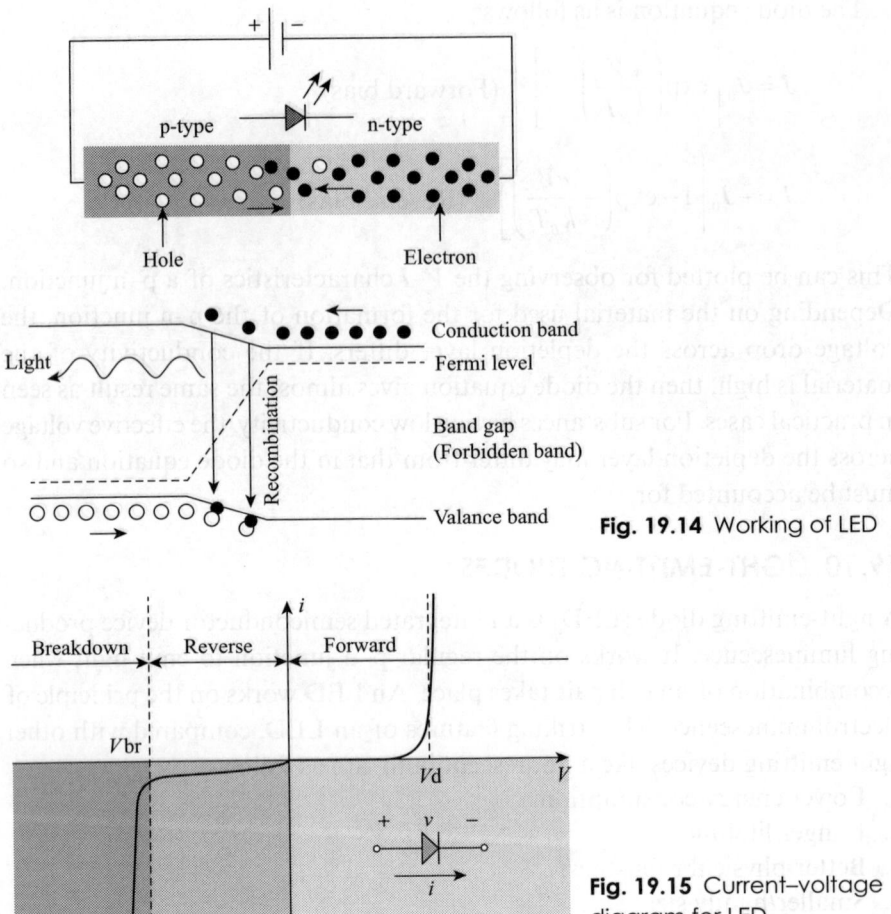

Fig. 19.14 Working of LED

Fig. 19.15 Current–voltage diagram for LED

Physics for inventing a new energy efficient and environment-friendly light source—the blue LED.

Gallium arsenide phosphide (GaAsP) is used to produce red LEDs. The first high-brightness blue LED was developed using indium gallium nitride (InGaN). The first white LED used a $Y_3Al_5O_{12}$:Ce phosphor coating to produce white light by mixing yellow light with blue light. The materials used for the LEDs have a direct band gap whose energies correspond to the near infrared, visible, or near ultraviolet light ranges.

The refractive index of the material used for making an LED is generally very high. This means that the light will be reflected back into the material at the material/air interface. For maximum light extraction from an LED, the semiconducting chip should ideally be a microsphere. In that case, all the emanating rays would be perpendicular to the surface and the electrodes can be kept in point contact with the chip.

An LED will begin to emit light when the on-voltage is exceeded. Typical on-voltages are in the range of 2–3 V.LED chips are placed in plastic sheets after fabrication, as they protect the semiconducting chip and also help in maximizing the light extraction.

The optimum power that an average light-emitting indicator gives is of the order of 30–60 mW at an optimum current of 350 mA. A conventional bulb of emits 60–100 W emits about 15 lumens/W in contrast to an LED that has a luminous efficacy of about 20 lumens/W.

Table 19.3 lists the Lumiled catalogue showing the best efficacy for each colour.

Table 19.3 Lumiled catalogue

Colour	Wavelength range (nm)	Typical efficacy (lm/W)
Red	$620 < \lambda < 645$	152
Red-orange	$610 < \lambda < 620$	98
Green	$520 < \lambda < 550$	93
Cyan	$490 < \lambda < 520$	155
Blue	$460 < \lambda < 490$	315

Light-emitting devices have a significantly long lifetime, ranging from 25,000 to 100,00 h, if operated under optimal (usually low) temperatures and current. Table 19.4 shows the available colours with wavelength range, voltage drop, and material.

Active and ongoing research in the upgradation of various challenges in this field has resulted in new advances, with the development of better white LEDs (WLEDs), organic LEDs (OLEDs), and polymer LEDs. They are used over a spectrum of devices due to their low driving voltage and better colour

Table 19.4 Colours with wavelength range, voltage drop, and semiconductor material

Colour	Wavelength (nm)	Voltage drop (ΔV)	Semiconductor material
Infrared	$\lambda > 1560$	$\Delta V < 1.63$	Gallium arsenide (GaAs), Aluminium gallium arsenide (AlGaAs)
Red	$610 < \lambda < 1560$	$1.63 < \Delta V < 2.03$	Aluminium gallium arsenide (AlGaAs), Gallium arsenide phosphide (GaAsP), Aluminium gallium indium phosphide (AlGaInP), Gallium(III) phosphide (GaP)
Orange	$590 < \lambda < 610$	$2.03 < \Delta V < 2.10$	Gallium arsenide phosphide (GaAsP), Aluminium gallium indium phosphide (AlGaInP), Gallium(III) phosphide (GaP)
Yellow	$5150 < \lambda < 590$	$2.10 < \Delta V < 2.18$	Gallium arsenide phosphide (GaAsP), Aluminium gallium indium phosphide (AlGaInP), Gallium(III) phosphide (GaP)

Table 19.4 (contd)

Colour	Wavelength (nm)	Voltage drop (ΔV)	Semiconductor material
Green	$500 < \lambda < 5150$	$1.9^{[59]} < \Delta V < 4.0$	Indium gallium nitride (InGaN), gallium(III) nitride (GaN), Gallium(III) phosphide (GaP), Aluminium gallium indium phosphide (AlGaInP), Aluminium gallium phosphide (AlGaP)
Blue	$450 < \lambda < 500$	$2.48 < \Delta V < 3.15$	Zinc selenide (ZnSe), Indium gallium nitride (InGaN), Silicon carbide (SiC) as substrateSilicon (Si) as substrate—under development
Violet	$400 < \lambda < 450$	$2.156 < \Delta V < 4.0$	Indium gallium nitride (InGaN)
Purple	Multiple types	$2.48 < \Delta V < 3.15$	Dual blue/red LEDs, blue with red phosphor, or white with purple plastic
Ultraviolet	$\lambda < 400$	$3.1 < \Delta V < 4.4$	Diamond (235 nm), Boron nitride (215 nm), Aluminium nitride (AlN) (210 nm), Aluminium gallium nitride (AlGaN), Aluminium gallium indium nitride (AlGaInN)—down to 210 nm
Pink	Multiple types	$\Delta V \sim 3.3^{[65]}$	Blue with one or two phosphor layers:yellow with red, orange or pink phosphor added afterwards,or white with pink pigment or dye
White	Broad spectrum	$\Delta V = 3.5$	Blue/UV diode with yellow phosphor

properties, such as in visual displays, computer monitors, car lights, mobile phones, digital cameras, television systems.

The recent advances in LEDs have resulted in the development of quantum dot LEDs. These diodes possess semiconductor nano-crystals that have unique optical properties. The colour spectrum for these LEDs ranges from the visible to the infrared region. In this case, quantum dots are placed in between electron and hole semiconducting materials. As an electric field is applied, electrons and holes recombine in the quantum dot layer to form an exciton, which in turn excites a quantum dot to produce emission.

19.11 LIQUID CRYSTAL DISPLAY

Liquid crystals are not capable of emitting light directly; instead, their light-modulating properties are used in a wide range of applications,such as in computer monitors, aircraft cockpit displays, televisions, digital clocks, telephones, calculators, and many others. They are electronically modulated optical devices that have a number of segments filled with liquid crystals. Liquid crystals are usually classified into three categories: nematics, smectics, and cholesterics. The most commonly used liquid crystal is nematics (discussed later). These segments are fixed in front of a light source or a reflector such that it produces coloured or black and white images. These liquid crystals were first discovered in 1888.

Advantages of LCDs are as follows:

1. LCDs are available in a wider range of screen sizes.
2. LCDs are used in flat panel displays, electronic visual displays, video displays, and plasma displays.
3. LCDs are more energy efficient, that is, they consume less electrical power and so can be used in battery-powered electronic equipment.

19.11.1 Working

The smallest illumination area (a pixel) of a liquid crystal display (LCD) consists of a layer of molecules aligned between two transparent electrodes and two polarizing filters. The axis of transmission of the electrodes and polarizing filters is perpendicular to each other. The liquid crystal is placed between the two crossed polarizing filters. The liquid crystal molecules are aligned to the surface of the electrodes (usually made of indium tin oxide) by rubbing a polymer in one direction. The liquid crystal molecules are aligned in the direction of rubbing. The orientation of the liquid crystal molecules is known by the alignment of the electrodes with the molecules; for example, in a nematic liquid crystal device, electrodes are perpendicular and so the liquid crystal molecules align themselves in a twisted or helical structure. When no voltage is applied, the incident light is polarized due to the twisted liquid crystals. Hence, the display on the device is grey in colour.

When a certain high voltage is applied, the liquid crystal molecules become completely untwisted. Thus, the incident light is no longer polarized and hence experiences no rotation as it passes through the liquid crystal. As the polarizing filters are crossed, this light is blocked as it is perpendicular to the filter. The pixel then appears dark and so is the display.

Thus, as the voltage is varied, liquid crystal molecules polarize the light. The device then displays various colours of the pixels.

An LCD is a simple light valve and is a passive device. LCD drivers are the circuits that have the data to be displayed on these devices.

The development of LCD can be attributed to George H. Heilmeier. In 1964, while working on the scattering effects in liquid crystals, he designed the first operational LCD based on the dynamic scattering mode (DSM). A DSM display converts the initially clear transparent liquid crystal layer into a milky turbid state. In the late 1960s, pioneering work on liquid crystals was undertaken by the Royal Radar Establishment at Malvem, England, UK. They discovered the cyanobiphenyl liquid crystals, which have appropriate stability and temperature properties for application in LCDs. On 4 December 1950, the twisted nematic effect in liquid crystals was filed for patent by Hoffmann-la Roche in Switzerland. In 1996, Samsung developed the optical patterning technique that enables the development of multi-domain LCDs, which remain the dominant LCD design through 2010.

The passive-matrix displays are usually not backlit, but the active-matrix displays are always backlit. The cold cathode fluorescent lamps (CCFL) that have been used for backlight illumination have now been replaced by the more energy-efficient LEDs. WLED is a full array of white LEDs. The LCD panel is lit by a row of WLEDs placed at one or more edges of the screen. A light diffuser is then used to spread the light evenly across the whole display. As of 2012, this is the most popular design used in desktop computer monitors.

19.12 REVERSE-BIAS DEVICES

The reverse bias for a p–n junction element can be used to design some special devices. These have been mentioned in the following sections.

19.12.1 Photodiodes

The term photodiode, as the name indicates, consists of the words photo + diode. Photo indicates an interface with light and diode, as we have discussed, is a semiconducting material p–n junction. When a p–n junction is exposed to light or photons, there is a flow of current. Thus, a photodiode acts as a photodetector.

In a photodiode, a PIN junction is preferred over a p–n junction for fast or sensitive action. In a PIN junction, the depletion region is nearly intrinsic, that is, it is very lightly doped and is quite wide, with the p- and n-region being very heavily doped for good contact. This helps increase the speed of response of photodiodes. Photodiodes are suitable for use in attenuators, fast switches, and high-voltage power electronics applications. Generally, photodiodes are used as photosensors, that is, for the measurement of light (as light meters), sensing the response to various light levels (as switches), and lighting street lights after dark.

19.12.1.1 Working

A photodiode operates in the reverse bias. As a packet of light (photons) strikes the PIN junction, an electron is excited. There is an electric field in the depletion region. As the free electron travels towards the depletion layer, its recombination with a hole takes place. The depletion region is designed to be of a significant width. Owing to this (wide depletion zone), recombination takes place either in the depletion layer or at one diffusion length from the depletion region. The electric field present in the depletion layer sweeps these carriers. The holes move towards the anode and the electrons towards the cathode, thus creating a current to flow, called a photocurrent. The total current through a photodiode is the sum of the dark current (current that flows with or without light) and the photocurrent. For a given spectral distribution, the photocurrent is linearly proportional to the illuminance and also to the irradiance. There are certain critical performance parameters of a photodiode:

Dark current   It is the current that flows in a photodiode due to the background radiation. In addition, a saturation current develops due to the semiconductor junction. These two currents together constitute the dark current. This is the superfluous current and restricts sensitivity of the photodiode towards the actual incident light. For the precise functioning of the photodiode, the dark current must be measured at the time of calibration. The dark current is accounted for in the photoconductive mode (or the reverse-bias mode) of the photodiode. It also accounts for the noise in the system.

Responsivity   The word responsivity generates from the word response. The generation of an excited electron when light is incident on a photodiode is akin to the photoelectric effect. The ratio of the number of photo generated carriers (or excited electrons) in the PIN junction to the incident photons is referred to as responsivity. In other words, one can say that it is the ratio of the generated photocurrent to the incident light power. If the first definition is considered, then the responsivity is a unitless quantity; otherwise, the unit is Ampere/Watt. Expressed as a new term, it can be called the 'quantum efficiency' of a photodiode.

Noise-equivalent power—NEP   It is akin to threshold frequency in photoelectric effect, that is, it is defined as the minimum optical power required to generate photocurrent in a photodiode. It can also be defined as the minimum detectable power. A term called 'detectivity' (D) is defined as the inverse of NEP, 1/NEP. The specific detectivity D^* is given as follows:

$$D^* = D\sqrt{A}$$

where A is the area of the photodetector, for a 1 Hz bandwidth. The specific detectivity helps us compare different systems independent of sensor area and system bandwidth. If the value of the detectivity is high, the system is supposedly said to have low noise. When a photodiode is used in an optical communication system, the above parameters contribute to the sensitivity of the optical receiver, which is the minimum input power required for the receiver to achieve a specified bit error rate.

Table 19.5 lists the materials commonly used to produce photodiodes.

Table 19.5 Materials used to produce photodiodes

Material	Electromagnetic spectrum wavelength range (nm)
Silicon	190–1100
Germanium	400–1500
Indium gallium arsenide	800–2600
Lead(II) sulphide	<1000–3500

Materials with a greater band gap are preferred, as they generate less noise. It is evident from this table that silicon- and germanium-based band gaps are preferred.

Photodiodes have a vast application in the medical industry. They are used as instruments to analyse samples, as immunoassays, as detectors for computed tomography, and as pulse oximeters.

The most recent advance in the use of photodiodes has been as a position sensor. A one-dimensional array of hundreds or thousands of photodiodes is designed, which helps in very fast parallel read-out. This type of fast parallel read-out is not possible with other electronic device sensors. A photodiode array is often referred to as a PDA.

19.12.2 Solar Cells

A solar cell is a large-area photodiode. A photodiode is made to work in the photovoltaic mode and in zero bias. Due to zero bias voltage, there is no flow of photocurrent out of the photodiode, and hence a voltage builds up inside the device. Essentially, sunlight is trapped inside the semiconducting device (photodiode) to produce an electric effect in the form of voltage; this is referred to as the 'photovoltaic effect' and happens to be the basis for solar cells.

19.12.2.1 Working

As a photodiode operates in the photovoltaic mode, the following features can be observed:
1. Absorption of light in the form of photons by a semiconducting element generates either hole–h pairs or excitons.
2. Opposite charge carriers (generally electrons) separate out, which causes an electric potential difference.
3. The separated charge carriers are made to flow out to an external circuit. Usually, an array of such cells (solar cells) converts solar energy into direct-current electricity.

Solar modules are made up of an array of solar cells that generate electrical power from sunlight. The cost associated with solar power has been a challenge to date. There has been a considerable improvement in the production cost, which has come down to under \$1 a watt. Large commercial arrays have been built at below \$3.40 a watt, fully commissioned with ongoing advance research to optimize solar power generation.

19.12.2.2 Efficiency of Solar Cell

The efficiency of a solar cell can be defined in terms of the following parameters:

Overall efficiency of a solar cell

= Reflectance efficiency × Thermodynamic efficiency

× Charge carrier separation efficiency × Conductive efficiency

In a solar cell, the efficiency parameters are also defined by voltage curves, temperature coefficients, and shadow angles. The other important parameters that can be measured are quantum efficiency (defined earlier), V_{oc} ratio, fill factor, and thermodynamic efficiency. For commercial solar cells, the value of the solar fill factor is calculated as follows:

$$\text{Fill factor} = \frac{\text{Actual maximum power obtained}}{(\text{Open circuit voltage}) \times (\text{Short circuit current})} > 0.150$$

Grade B cells have a fill factor that lies between 0.4 and 0.15. This factor indicates the performance of solar cells used for commercial purposes. Owing to advanced researches, solar cells have achieved very high power efficiency.

Materials used for solar cells have changed from regular mono-and polycrystalline silicon crystals to amorphous silicon, organic dyes, organic polymers, and nano-crystals. The recent advances in using nano-crystals as quantum dots is the pioneering area of research in solar cells, as it has led to the generation of multiple excitons in cells.

19.12.3 Zener Diodes

A Zener diode is designed to work in the reverse-bias mode. As discussed in Section 19.8.2, a saturation current is developed, which plays a significant role. A Zener diode is made to work in the breakdown region of the junction. In the reverse-bias mode, the diode can be made to cause three major breakdowns: (a) avalanche breakdown, (b) Zener breakdown, and (c) punch through. In the avalanche breakdown, moderately doped junctions are able to create scattering collisions, which give sufficient energy to the lattice to create an e–h pair, thus creating a chain reaction. This breakdown is dependent on temperature.

In a Zener breakdown, the junction is thin and abrupt. The doping level for such junctions is very high such that even a small reverse-bias voltage can create very large electric fields across the narrow depletion regions. The phenomenon of quantum tunnelling takes place across the thin junction, as it is designed to be abrupt as well. Electrons (or carriers) in the valence band do not possess sufficient energy to overcome the barrier; however, under quantum mechanical circumstances, due to a thin, abrupt depletion layer, the carriers cross this potential barrier to sneak into the conduction band, and under special circumstances, a sufficiently large number of carriers tunnel to the conduction band, resulting in the flow of large reverse currents. Zener breakdown is not dependent on temperature, but depends on the band gap energy, barrier thickness, and, in turn, doping of the junctions.

In a Zener diode, reverse breakdown of the junction is a reversible process.

19.12.4 Varactor Diodes

Varactor diodes are also instruments that are made to work in the reverse-bias region such that the voltage remains constant. They are also referred to as voltage-controlled capacitors. Varactor diodes are commonly used as tuners in electronic sets, in devices where voltage-controlled oscillations are required, as part of frequency synthesizers.

19.12.4.1 Working

As the applied bias voltage (V_B) governs the thickness of the depletion layer (T_D), there can be variation in the charge-storing capacity of a diode. In the reverse-bias state, the following relation holds: $T_D \propto \sqrt{V}$

In addition, we know that $C \propto \dfrac{1}{T_D}$

Co-relating the two relations: $C \propto \dfrac{1}{\sqrt{V}}$

Although this relation holds true for most of the diode operations, varactor diodes are so designed that there is an increase in the capacitance or storage of charge as the voltage is varied. The most extensive use of the varicap diodes is done in tuner circuits. Previously designed varicap diodes had a voltage range varying from 0 to 33 V, against which the capacitance range was very small, of the order of 1–10 pF. The varicap diodes now used in the electronic circuits have a higher variation in capacitance (100–500 pF) over a relatively small change in the reverse bias voltage (0–12 V).

19.13 PHOTOCONDUCTIVITY

As the term indicates, 'photo' means light and 'conductivity' deals with the movement of charges or carriers in a semiconducting material. Different materials create carriers depending on the temperature and their energy band gap. The rate of carrier production is given as follows:

$$\frac{dn}{dt} = P_a - r\,n\,p = P_a - rn_i^2$$

where P_a is the number of photons that are absorbed per second per unit volume, r is the recombination coefficient, n and p denote the carriers for the n- and p-type semiconductors, respectively, and n_i denotes the density of carriers for an intrinsic semiconductor. In the steady state of an intrinsic semiconductor, the rate of change of carriers will be equal to zero, that is,

$$\frac{dn}{dt} = 0 = P_a - r n_i^2 = 0$$

$$\Rightarrow P_a = r n_i^2 \Rightarrow n_i^2 = \frac{r}{P_a}$$

Hence, the conductivity for an intrinsic semiconductor is given as follows:

$$\sigma = n_i (\mu_e + n_h)e = \sqrt{\frac{r}{P_a}} (\mu_e + \mu_h)e$$

Thus, in this case, we have assumed that for an intrinsic semiconductor, each photon that is absorbed creates an e–h pair.

If we consider the holes in a semiconductor, then there are certain trapped holes that create a positive space charge $p_t e$ in the semiconductor. The conductivity for the holes is given as follows: $\sigma = en\mu_e = ep_t\mu_e$

The photoconductive gain is mathematically defined as follows:

$$G = \frac{I}{eP_a(Al)} = \frac{J}{eP_a l}$$

where J is the current density and I is the photocurrent carried in a length l of the semiconductor with a cross section A.

The example of a photoconductive material includes the conductive polymer polyvinylcarbazole used in photocopying. Some of the photoconductive materials can be used as photodetectors, particularly in the form of photoresistors, as in street lights, infrared detectors, and cameras.

Many of the devices discussed in the preceding section, such as photodiodes, solar cells, and others, are photodetectors and work based on the principle of photoconductivity.

IMPORTANT CONCEPTS

1. In a semiconductor, as the temperature is increased, some of the electrons reach the conduction band, thus leaving vacant places in the valence band; this leads to the process of conduction.
2. For an intrinsic semiconductor, $n = p = n_i$.
3. The product of electron and hole concentrations at equilibrium is a constant for a given material at a given temperature. This is also referred to as the law of mass action: $n_0 p_0 = n_i^2$

4. Distribution of the electrons over a range of allowed energy levels at thermal equilibrium is given by the *Fermi–Dirac distribution function* $f(E)$:

$$f(E) = \frac{1}{1 + \exp\left[\dfrac{E_F - E}{k_B T}\right]}$$

5. For an extrinsic semiconductor doped with a considerable amount of donor atoms, we have, at room temperature,

$$n_0 \gg (n_i, p_0)$$

Such a doped semiconductor is thus an n-type semiconductor.

6. Acceptor impurities can result in an equilibrium hole concentration p_0 that is much greater than the conduction band electron concentration n_0. Thus,

$$p_0 \gg (n_i, n_0)$$

Such an extrinsic semiconductor is called a p-type semiconductor.

7. $\dfrac{\partial}{\partial t}(\delta n) = -\dfrac{\delta n}{\tau_{Le}} + \mu_e E_x \dfrac{\partial}{\partial x}(\delta n) + D_e \dfrac{\partial^2}{\partial x^2}(\delta n)$

$\dfrac{\partial}{\partial t}(\delta p) = -\dfrac{\delta p}{\tau_{Lh}} + \mu_h E_x \dfrac{\partial}{\partial x}(\delta p) + D_h \dfrac{\partial^2}{\partial x^2}(\delta p)$

The continuity equation enables us to know about the excess density of electron or holes in time and space. The continuity equation and current density equation together explain the flow of carriers in semiconducting devices.

8. Hall effect is a process leading to the development of a voltage across one of the faces of a semiconductor slab when crossed electric and magnetic fields are applied across the other two pairs of faces of the slab. We can write that

$$V_H = \frac{I_x B_z}{epd} = \frac{I_x B_z}{d} R_H$$

9. A p–n junction is formed when a p-type doped semiconductor is made to join an n-type doped semiconductor. The variation in the impurity concentration occurs a short distance, typically of the order of microns.

10. The barrier potential is related to the carrier concentrations N_n and N_p; the temperature T also plays a significant role:

$$V_0 = \frac{k_B T}{e} \log_e \left[\frac{N_n}{N_p}\right]$$

11. The diffusion coefficient is defined as $\dfrac{D_e}{\mu_e} = \dfrac{D_h}{\mu_h} = \dfrac{k_B T}{e}$.

12. The diffusion current, as the forward voltage is applied, is given as follows:

$$J_{ph} = J_{pe} \exp(eV/k_B T)$$
$$J_{ne} = J_{nh} \exp(eV/k_B T)$$
$$J = J_0(\exp(eV/k_B T) - 1)$$

This represents the diode equation or the rectifier equation (forward bias).

13. $J = J_0 \left[\exp\left(-\dfrac{eV}{k_B T} \right) - 1 \right] = -J_0 \left[1 - \exp\left(-\dfrac{eV}{k_B T} \right) \right]$ (reverse bias).

14. An LED works on the principle of electroluminescence.

15. Liquid crystals are not capable of emitting light directly; instead, their light-modulating properties are used in a wide range of applications. Liquid crystals are usually classified into three categories: nematics, smectics, and cholesterics.
   Reverse Bias Devices

16. The term photodiode, as the name indicates, consists of two words photo + diode. Photo indicates an interface with light, and diode is a semiconducting material p–n junction.

17. Sunlight is trapped inside a semiconducting device (photodiode) to produce an electric effect in the form of voltage; this is referred to as the 'photovoltaic effect' and happens to be the basis for solar cells.

   Overall efficiency of a solar cell
   = Reflectance efficiency × Thermodynamic efficiency
   × Charge carrier separation efficiency × Conductive efficiency

   $$\text{Fill factor} = \frac{\text{Actual maximum power obtained}}{\text{(Open circuit voltage)} \times \text{(Short circuit current)}} > 0.150$$

18. Varactor diodes or voltage-controlled capacitors are instruments that are made to work in the reverse-bias region such that the voltage remains constant.

19. The term photoconductivity includes the words 'photo' meaning light and 'conductivity' that deals with the movement of charges or carriers in a semiconducting material. Different materials create carriers depending on the temperature and their energy band gap. The rate of carrier production is given as follows:

   $$\frac{dn}{dt} = P_a - rnp = P_a - rn_i^2$$

   where P_a is the number of photons that are absorbed per second per unit volume, r is the recombination coefficient, n and p denote the carriers for n- and p-type semiconductors, respectively, and n_i denotes the density of carriers for an intrinsic semiconductor.

20. The photoconductive gain is mathematically defined as follows:

   $$G = \frac{I}{eP_a(Al)} = \frac{J}{eP_a l}$$

   where J is the current density, I is the photocurrent carried in a length l of the semiconductor with a crosssection A.

APPLICATIONS

1. The p–n junction formulation has resulted in the formation of diodes that have revolutionized the electronic segment of research.

2. Some of the characteristic applications of diodes have been discussed in detail in this chapter. Some of them have been highlighted as follows:
   (a) LEDs
   (b) LCD
   (c) Photodiodes
   (d) Solar cells
   (e) Zener diodes
   (f) Varactor diodes

IMPORTANT FORMULAE

1. For an intrinsic semiconductor:
   $n = p = n_i$

2. The electron mobility:
   $\sigma = n_i e(\mu_n + \mu_p)$

3. The intrinsic concentration:
   $n_i = 2(2\pi k_B T/h^2)^{3/2}(m_n^* m_p^*)^{3/4}$
   $\exp(-E_g/2k_B T)$

4. For an intrinsic semiconductor:
   $n_i = N_c \exp[-(E_c - E_i)/k_B T]$

5. For a doped semiconductor:
   $$N_d = N_c \exp\left[-\frac{E_c - E_{Fd}}{k_B T}\right]$$

6. $\dfrac{D_e}{\mu_e} = \dfrac{D_h}{\mu_h} = \dfrac{k_B T}{e}$

7. The diode equation or the rectifier equation:
   $$J = J_0(\exp(eV/k_B T) - 1)$$

8. $D^* = D\sqrt{A}$

9. Overall efficiency of a solar cell
   = Reflectance efficiency ×
   Thermodynamic efficiency ×
   Charge carrier separation
   efficiency × Conductive efficiency

10. Photoconductive gain (G):
    $$G = \frac{I}{eP_a(Al)} = \frac{J}{eP_a l}$$

SELF-ASSESSMENT

Multiple-choice Questions

19.1 A semiconductor is different from a conductor because it possesses a forbidden energy band
   (a) that is of the order of 0.2–0.4 eV
   (b) that is overlapping the valence band
   (c) that does not exist
   (d) that is very far apart from valence band

19.2 For an intrinsic semiconductor,
   (a) the recombination rate is not the same as the generation rate
   (b) the recombination rate is the same as the generation rate
   (c) the recombination rate is not related to the generation rate
   (d) none of these

19.3 An intrinsic semiconductor is different from an extrinsic semiconductor as
   (a) there is addition of no external elements to the semiconducting element
   (b) there is addition of some external impurities to the pure semiconducting element
   (c) there is no change in the forbidden energy gap
   (d) all of these

19.4 In a p-type semiconductor,
- (a) electrons act as charge carriers
- (b) holes act as charge carriers
- (c) there are no charge carriers
- (d) there is an excess of hole carriers

19.5 In an n-type semiconductor,
- (a) electrons act as charge carriers
- (b) holes act as charge carriers
- (c) there are no charge carriers
- (d) there is an excess of electron carriers

19.6 The Fermi level in an intrinsic semiconductor
- (a) lies midway between the valence band and the conduction band
- (b) lies towards the conduction band
- (c) lies towards the valence band
- (d) does not exist

19.7 The Fermi level for an extrinsic n-semiconductor
- (a) lies midway between the valence band and the conduction band
- (b) lies towards the conduction band
- (c) lies towards the valence band
- (d) does not exist

19.8 The diode equation is valid
- (a) in forward bias only
- (b) in reverse bias only
- (c) in both forward and reverse biases
- (d) none of these

19.9 The diode current consists of
- (a) both the drift current and the diffusion current
- (b) the forward current
- (c) the reverse current
- (d) all of these

19.10 Zener diode is a
- (a) forward-current diode
- (b) reverse-current diode
- (c) voltage regulator
- (d) current regulator

19.11 A photodiode works on the principle of
- (a) a diode
- (b) a semiconductor
- (c) a conductor
- (d) an insulator

19.12 An LED uses
- (a) a conductor junction
- (b) an insulator junction
- (c) a semiconducting junction
- (d) a quasi-element junction

19.13 A varactor diode is so designed as the bias voltage
- (a) governs the thickness of the depletion layer
- (b) does not govern the depletion layer
- (c) has no role in the flow of carriers
- (d) is not related to the thickness of the depletion layer

Review Questions

19.1 How does the Fermi level plays a significant role in semiconductors? Explain with a minimum of three points of importance.

19.2 What is the charge carried by a hole?

19.3 Why are conduction electrons generally found at the bottom of the conduction band?

19.4 Enunciate the difference between ordinary voltage and Hall voltage.

19.5 Explain the significance of $(E_F - E_V)$.

19.6 Explain the significance of Fermi energy level. Mention its position in intrinsic and extrinsic semiconductors at 0 K.

19.7 Derive an expression for the density of electrons in the conduction band of an intrinsic semiconductor.

19.8 Derive an expression for the density of holes in the valence band of an intrinsic semiconductor.

19.9 Discuss the variation of the Fermi level with temperature in an intrinsic semiconductor.

19.10 Obtain an expression for the density of electrons in the conduction band of an n-type extrinsic semiconductor by assuming that they obey the Fermi–Dirac distribution function.

19.11 Obtain an expression for the density of holes in the valence band of a p-type extrinsic semiconductor.

19.12 Discuss the variation of carrier concentration with temperature in an n-type semiconductor.

19.13 Explain with a sketch the variation of the Fermi level and the carrier concentration with temperature in case of p- and n-type semiconductors for high and low doping levels.

19.14 What is the Hall effect? Derive an expression for the Hall coefficient. Describe an experimental set-up for the measurement of the Hall coefficient.

19.15 Explain the process of formation of holes in a semiconductor using a suitable schematic diagram.

19.16 What is the charge carried by a hole?

19.17 Why are conduction electrons generally found at the bottom of the conduction band?

19.18 What is an intrinsic semiconductor?

19.19 Explain the relation $r_i = g_i$.

19.20 What is the binding energy of an electron?

19.21 Define mobility of a charge carrier.

19.22 With a sketch, explain the Hall effect set-up.

19.23 What is the unit of Hall coefficient?

19.24 Explain the process of formation of e–h pairs.

19.25 How does a hole move in a semiconductor?

19.26 What is an extrinsic semiconductor?

19.27 State some column II impurities.

19.28 What are majority carriers?

19.29 Write an expression for the ground-state energy of an electron.

19.30 Write the units of mobility.

Numerical Problems

19.1 A sample of n-type silicon material has a donor concentration $N_d = 2.5 \times 10^{20}/m^3$. Calculate the temperature at which the Fermi level coincides with the edge of the conduction band. Assume the effective mass of electron to be equal to its rest mass. [*Hint:* If $N_d = N_c$, then $E_F = E_C$]

19.2 Calculate the resistivity of an n-type Ge sample at 300 K. The sample has a donor density $N_d = 10^{20}$ atoms/m³. Assume all donors to be ionized and $\mu_n = 0.38$ m²/Vs. [*Hint:* $\rho = 1/nq\mu_n$]

19.3 Mobilities of free electrons and holes in pure germanium are 0.38 and 0.18 m²/Vs, respectively. Assume that $n_i = 2.5 \times 19^{19}/m^3$ for Ge. Calculate the resistivity of pure Ge.

19.4 Find the intrinsic resistivity of Si from the following data: $n_i = 1.5 \times 10^{16}/m^3$, $\sigma_n = 0.13, \sigma_p = 0.05$ m²/Vs. $\left[\textit{Hint: } \rho_i = \left(\dfrac{1}{\sigma_i} \right) \right]$

19.5 A Si sample is doped with As to a concentration of 10^{16} atoms/cm³. The thickness of the sample is 500 μm. A current $I_x = 1$ mA is made to flow in the x-direction and $B_z = 5 \times 10^{-4}$ Wb/cm². Calculate the Hall voltage.

[*Hint: $V_H = (I_x B_z R_H)/d$*]

19.6 A Si wafer is doped with P with a donor concentration of $N_d = 10^{15}$ atoms/cm³. Calculate the equilibrium hole concentration p_0 at 400 K. How is E_F located relative to E_i? Sketch the resulting band diagram. Consider $n_i = 1.5 \times 10^9/cm^3$.

19.7 A particular sample of Ge has a donor density of $N_d = 10^{13}$ atoms/cm³. Assuming all donor atoms to be ionized, calculate the resistivity of the sample.

19.8 The intrinsic carrier density is $1.5 \times 10^{15}/m^3$. If the electron and hole mobilities are 0.13 and 0.05 m²/Vs, respectively, calculate its electrical conductivity.

19.9 Find the intrinsic resistivity of Ge at room temperature (300 K) if the carrier density is $2.15 \times 10^{13}/cm^3$. (The electron mobility $\mu_n = 3900$ cm²/Vs; hole mobility $\mu_p = 1900$ cm²/Vs.)

19.10 In a p-type Ge semiconductor, the intrinsic carrier concentration is $1.2 \times 10^{18}/m^3$ and the density of B is 5.0×10^{23} atoms/m³. The electron and hole mobilities are 0.3 and 0.5 m²/Vs, respectively. What is its electrical conductivity before and after the addition of B atoms?

19.11 An n-type Si wafer has been doped uniformly with Sb atoms, and the doped Si has a donor concentration of $10^{16}/cm^3$. Calculate the Fermi energy with respect to the Fermi energy in intrinsic Si. (For Si, $n_i = 1.45 \times 10^{10}/cm^3$.)

19.12 Mobilities of electrons and holes in a sample of intrinsic Ge at 300 K are 0.45 and 0.2 m²/Vs, respectively. If the resistivity of the specimen is 3.15Ω m, compute the forbidden energy gap for Ge.

19.13 The following data is given for an intrinsic Ge semiconductor at 300 K. Calculate the conductivity of the sample. Given that $n_i = 3.0 \times 10^{19}/m^3$, $\sigma_n = 0.40$ m²/Vs, $\sigma_p = 0.20$ m²/V.

19.14 In an n-type semiconductor the Fermi level lies 0.25 eV below the conduction band at 400 K. If the temperature is increased to 330 K, find the new position of the Fermi level.

19.15 The energy gap of Si is nearly 0.9 eV. The average electron effective mass is $0.25 m_e$, where m_e is the free electron mass. Calculate the electron concentration in the conduction band of Si at room temperature ($T = 400$ K). Assume that $E_F = E_g/2$.

19.16 Find the resistance of an intrinsic Ge rod (1.5 cm long, 0.5 mm wide, and 0.5 mm thick) at 400 K. For Ge, $n_i = 3.0 \times 10^{19}/m^3$; $\mu_n = 0.40$ m²/Vs; $\mu_p = 0.20$ m²/Vs.

19.17 For an Si semiconductor with a band gap 0.9 eV, determine the position of the Fermi level at 400 K if $m_n^* = 0.13 m_e$, $m_p^* = 0.30 m_e$.

19.18 The intrinsic carrier density is $2.0 \times 10^{15}/m^3$. If the electron and hole mobilities are 0.15 and 0.05 m²/Vs, respectively, calculate its electrical conductivity.

19.19   Find the intrinsic resistivity of Ge at room temperature (300 K), if the carrier density is $3.0 \times 10^{13}/cm^3$. Given that electron mobility $\mu_n = 3900\ cm^2/Vs$, hole mobility $= \mu_p = 1900\ cm^2/Vs$.

19.20   In a p-type Ge semiconductor, the intrinsic carrier concentration is $2.1 \times 10^{19}/m^3$ and the density of B is 4.5×10^{23} atoms/m^3. The electron and hole mobilities are 0.4 and 0.2 m^2/Vs, respectively. What is its electrical conductivity before and after the addition of B atoms?

19.21   An n-type semiconductor specimen has a Hall coefficient $R_H = 4.5 \times 10^{-11}\ m^3/As$. The conductivity of the specimen is found to be $112 \times 10^{15}/\Omega m$. Calculate the charge carrier density n and the electron mobility at room temperature.

19.22   A current of 50 A is established in a Cu slab (0.2 cm thick and 2 cm wide). The slab is placed in a magnetic field B of 1.5 T. The magnetic field is perpendicular to the plane of the slab and to the current. The free electron concentration in Cu is $8.4 \times 10^{28}/m^3$. What will be the magnitude of the Hall voltage across the width of the slab?

19.23   A sample of Si is doped with As to a level of 3×10^{16} atoms/cm^3. Calculate the resistivity of the sample. Determine the Hall coefficient and the Hall voltage if the thickness d of the sample is 3×10^{-2} cm, $I_x = 4\,mA$, $B_z = 3kG = \dfrac{3 \times 10^{-5}\ Wb}{cm^2}$. Assume $\mu_n = 700 cm^2/Vs \cdot s$.

19.24   Find the Hall coefficient and electron mobility of Ge for a given sample (length 1.5 cm, breadth 0.5 mm, and thickness 0.5 mm). A current of 3 mA flows from a 1.5 V supply and develops a Hall voltage of 25 mV across the specimen in a magnetic field of 0.55 Wb/m^2.

19.25   A copper strip (1 cm wide and 0.5 mm thick) is placed in a magnetic field with $B = 2.0\,Wb/m^2$ perpendicular to the strip. Suppose a current of 100 A is set up in the strip, what Hall potential difference would appear across the strip? Given that electron concentration $N = 7.0 \times 10^{28}/m^3$.

19.26   The Hall coefficient and conductivity of Cu at 400 K have been measured to be $0.45 \times 10^{-10}\ m^3/As$ and $6.5 \times 10^{15}/\Lambda\Omega m$, respectively. Calculate the drift mobility of the electrons in Cu.

Conducting Materials

Learning Objectives

After studying this chapter, students will be able to

- understand the Drude–Lorentz classical theory
- explain dc electrical conductivity
- elucidate thermal conductivity
- comprehend Lorentz's modification of the original classical theory of Drude
- demonstrate the drawbacks of the classical theory
- describe Sommerfeld's quantum theory
- explain Fermi energy, total energy, and density of states
- elucidate thermionic emission

List of Symbols

a = Acceleration

E = Electric field

e = Electronic charge

m_e = Mass of electron

v_d = Mean drift velocity

τ = Mean relaxation time

J = Net electric current density

n = Number of electrons per unit volume

σ = Conductivity

μ = Mobility

λ_m = Electron mean free path

T = Temperature

k = Boltzmann's constant

K = Thermal conductivity

L = Lorentz number

E_F = Fermi level

E_{F0} = Fermi level at absolute zero

V = Potential energy

ψ_n = Wave function

E_n = Kinetic energy of electron occupying the nth state

n_f = Principal quantum number

E_T = Total energy

$D(E)$ = Density of states

φ = Work function

p = Momentum

A = Emission coefficient

20.1 INTRODUCTION

Conducting materials constitute an important class of solids. Metals as conducting materials have applications in industry, medicine, defence, domestic use, etc. Properties such as mechanical strength, malleability, ductility, corrosion resistance, electrical conductivity, and thermal conductivity make metals an attractive choice for a whole range of applications. Why do metals have these properties and under what conditions do they possess these properties? This question begins our discussion of metals with the classical approach developed by Drude and Lorentz. Inadequacies of this approach would then be presented. A quantum model based on the Fermi–Dirac statistics can remove many drawbacks associated with the classical model.

20.2 CLASSICAL FREE ELECTRON THEORY OF METALS

In the year 1900, Drude presented a model for metals. In this model, valence electrons are assumed to behave like a gas, with the positive ions constituting the core of a metal. The electrons are confined to move within the metal due to the electrostatic attraction existing between them and the positive ion core. In 1909, Lorentz proposed that free electrons in a metal obey the Maxwell–Boltzmann statistics under equilibrium conditions. A combination of the salient points of both the Drude and the Lorentz theories constitutes the Drude–Lorentz classical theory.

The principal assumptions in Drude's approach are as follows:
1. Electron–Electron and electron–(positive ion) interactions are negligible in comparison with electron–(external field) interaction. Electrons, however, remain confined within the metal, due to the electron–(positive ion) attractive force.
2. Under the influence of an applied electric field, electrons experience a force in a direction opposite to the direction of the applied electric field. The moving electrons undergo collisions with immobile and impenetrable positive ion cores. Between successive collisions, the electrons move in straight lines and obey the classical laws of motion.
3. All free electrons move with the root mean square (RMS) speed of electrons, assuming the thermal velocity of electrons to have a Maxwell–Boltzmann distribution. Immediately after a collision, the average velocity of electrons becomes zero.

20.2.1 Electrical Conductivity—Drude's Theory

We will use Drude's model to derive an expression for the dc electrical conductivity. The acceleration a of an electron under the influence of an applied electric field e can be written in the following form:

$$a = \frac{-e\varepsilon}{m_e} \tag{20.1}$$

where m_e represents the mass of the electron and e the electronic charge. The mean drift velocity, assuming the average velocity to become zero immediately after a collision, is given by the following equation:

$$v_d = \left(\frac{-e\varepsilon}{m_e}\right)\tau \tag{20.2}$$

where τ is a parameter called the *mean relaxation time*. The negative sign indicates a motion accelerated in a direction opposite to the direction of the applied electric field. Let us assume that the number of electrons per unit volume is n, all of which are moving with a constant drift velocity v_d. The magnitude of net electric current density J can be expressed as follows:

$$J = nev_d \tag{20.3}$$

The use of Eq. (20.2) in Eq. (20.3) leads to the following expression:

$$J = \left(\frac{ne^2\tau}{m_e}\right)\varepsilon \tag{20.4}$$

Equation (20.4) can be rewritten in the following form:

$$J = \sigma\varepsilon \tag{20.5}$$

Equation (20.5) is a statement of Ohm's law, and the scalar quantity σ is defined to be the electrical conductivity of the metal. Electrical conductivity σ is mathematically expressed as follows:

$$\sigma = \frac{ne^2\tau}{m_e} \tag{20.6}$$

The reciprocal of σ is called the electrical resistivity and the symbol ρ is used for it. The electrical conductivity σ can be expressed in terms of the drift mobility using the following expression:

$$\sigma = ne\mu \tag{20.7}$$

where the mobility μ is defined as follows:

$$\mu = \frac{v_d}{\varepsilon} \tag{20.8}$$

Assuming the electrons in a metal to obey the kinetic theory, the relaxation time τ can be expressed in terms of the thermal velocity v_{th} through the following expression:

$$\tau = \frac{\lambda_m}{v_{th}} \tag{20.9}$$

where λ_m is the electron mean free path. Since electrons obey the kinetic theory, we can write the following equation:

$$\frac{3}{2}kT = \frac{1}{2}m_e v_{rms}^2 = \frac{1}{2}m_e v_{th}^2 \tag{20.10}$$

Substitution of τ from Eq. (20.9) into Eq. (20.6) yields the following equation:

$$\sigma = \frac{ne^2 \lambda_m}{m_e v_{th}} \qquad (20.11)$$

Substituting the value of v_{th} from Eq. (20.10) into Eq. (20.11) gives the following expression:

$$\sigma = \frac{ne^2 \lambda_m}{\sqrt{3km_e T}} \qquad (20.12)$$

One can conclude from Eq. (20.12) that the dc electrical conductivity of metals increases with a decrease in temperature.

20.2.2 Thermal Conductivity and Wiedemann–Franz Law—Lorentz Number

According to Drude's model, the major contribution to thermal conductivity of metals is made by conduction electrons. This is supported by the observed evidence that metals are better conductors of heat in comparison to insulators. Thus, Drude's model considers that the contribution of phonons to thermal conductivity is negligible.

The expression for thermal conductivity K of electrons has the following form:

$$K = \frac{3}{2}\left(\frac{nk^2 \tau}{m_e}\right)T \qquad (20.13)$$

Using Eqs (20.6) and (20.13), we get the following relation:

$$\frac{K}{\sigma} = \frac{3}{2}\left(\frac{nk^2 \tau}{m_e}\right)T \times \frac{m_e}{ne^2 \tau}$$

yielding $\dfrac{K}{\sigma} = \dfrac{3}{2}\left(\dfrac{k}{e}\right)^2 T$ $\qquad (20.14)$

or $\qquad \dfrac{K}{\sigma T} = \dfrac{3}{2}\left(\dfrac{k}{e}\right)^2 = L$ $\qquad (20.15)$

The parameter L is called the *Lorentz number*. Substituting the values of Boltzmann's constant and the electronic charge in Eq. (20.15), one gets the following results:

$$L = 1.11 \times 10^{-8} \text{ W}\Omega/\text{K}^2$$

$$L = 1.11 \times 10^{-8} \text{ (V/K)}^2 \qquad (20.16)$$

The experimental value of L is found to be around twice the value indicated in Eq. (20.16). Equation (20.14) is also called the Wiedemann–Franz law.

20.2.3 Lorentz's Modification

Lorentz made the following modifications in Drude's model for electrons:
1. All the electrons present in a metal do not possess the same thermal velocity.
2. Electric and thermal gradients perturb the classical Maxwell–Boltzmann law of distribution of velocity. This perturbation results in a displacement of the equilibrium velocity distribution; it also disturbs its symmetry.
3. The Boltzmann transport equation can be used to describe the transport of charge and kinetic energy of the mobile electrons constituting the electron gas.

Electrical conductivity of the Lorentz model, obtained after applying the aforementioned modifications to the Drude's model, can be expressed in the following form:

$$\sigma_L = \left(\frac{8}{3\pi}\right)^{\frac{1}{2}} \frac{ne^2\lambda_m}{\sqrt{3km_eT}} \tag{20.17}$$

Comparing Eqs (20.12) and (20.17), we can write the following relation:

$$\sigma_L = \left(\frac{8}{3\pi}\right)^{\frac{1}{2}} \sigma_D \tag{20.18}$$

where σ_D represents the conductivity according to Drude's model. Equation (20.18) can also be rewritten in the following form:

$$\sigma_D = 1.09\sigma_L \tag{20.19}$$

Equation (20.19) then relates the electrical conductivity as derived using the Drude's model with the electrical conductivity derived using the Lorentz's modification.

20.2.4 Drawbacks of Classical Theory

The Drude–Lorentz theory could correctly predict the room temperature resistivity of several metals. From Eqs (20.12) and (20.17), one can conclude that according to the Drude–Lorentz classical theory, resistivity of metals varies as $T^{\frac{1}{2}}$. Actual experimentation has, however, demonstrated the resistivity to vary linearly with temperature. This is one of the major drawbacks of the classical theory of metals. The classical theory assumes that the free electrons follow the Maxwell–Boltzmann statistics. Actually, the free electrons in a metal can interact with external sources of energy. These external sources can be thermal or magnetic in nature. Free electrons obeying the Maxwell–Boltzmann statistics can gain energy from these interactions. This exchange of energy can result in high values of heat capacity and paramagnetic susceptibility. Magnitudes of the heat capacity and paramagnetic susceptibility obtained using such a formulation are much higher than the experimentally observed values. This is another drawback of the classical theory of metals. Electrons in a metal have a long mean free path at low temperatures. The magnitude of the mean free path reaches values that are 10^8–10^9 times the interatomic spacing. Such high values of the mean free path cannot be explained by the classical theory of metals.

Example 20.1   Electrons in a metal have a Fermi velocity of 1×10^6 m/s. Calculate the Fermi energy for the electrons in the metal.

Solution   Assuming the electrons to have only kinetic energy, the Fermi energy can be expressed as follows:

$$E_F = \frac{1}{2}mv_F^2 \implies \frac{1}{2} \times \frac{9.1 \times 10^{-31} \times 10^{12}}{1.6 \times 10^{-19}} \text{eV}$$

or $E_F = 2.8$ eV

Example 20.2   The Fermi energy of copper at 0 K is 7.04 eV. Calculate the Fermi energy at 300 K.

Solution   The Fermi energy is calculated as follows:

$$E_F = E_{F0}\left[1 - \frac{\pi^2}{12}\left(\frac{kT}{E_{F0}}\right)^2\right] \tag{20.20}$$

Putting appropriate values in Eq. (20.20), one gets the following result:

$$E_F = 7.04\left[1 - \frac{\pi^2}{12}\left(\frac{1.38 \times 10^{-23} \times 300}{7.04 \times 1.6 \times 10^{-19}}\right)^2\right]$$

or $E_F = 7.0399$ eV

Example 20.3   Calculate the conductivity of Al at 25°C using the following data. Density of Al is 2.7 g/cm^3, atomic weight of Al is 27, and the relaxation time of electrons is 10^{-14} s.

Solution   Density of Al $d = 2.7 \times 10^3$ kg/m^3, atomic weight M_{at} of Al $= 27$, relaxation time $\tau = 10^{-14}$ s

$$n = \frac{dN_A}{M_{at}} \times \text{number of free electrons per atom}$$

where Avogadro constant is given as follows: $N_A = 15.022 \times 10^{23}$ and the number of free electrons/atom $= 3$. Hence,

$$n = \frac{(2.7 \times 10^3)(6.022 \times 10^{23}) \times 3 \times 10^3}{27}$$

$$= 18.066 \times 10^{28}/\text{m}^3$$

Conductivity of Al, $\sigma = \dfrac{ne^2\tau}{m_e} = \dfrac{18.066 \times 10^{28}(1.6 \times 10^{-19})^2(10^{-14})}{9.1 \times 10^{-31}}$

$$= 5.082 \times 10^7 \,\Omega\text{m}^{-1}$$

Conductivity of Al $= 5.082 \times 10^7 \,\Omega\text{m}^{-1}$

Example 20.4 Thermal and electrical conductivities of Cu at 20°C are 390 W/m K and $5.87 \times 10^7 (\Omega m)^{-1}$, respectively. Calculate the Lorentz number.

Solution We have the following values:

Electrical conductivity of Cu $\sigma = 5.87 \times 10^7 (\Omega m)^{-1}$

Thermal conductivity of Cu $K = 390$ W/mK

Temperature $T = 293$ K

The Lorentz number is given by the following relation: $L = \dfrac{K}{\sigma T}$

Substituting the given values of K, σ, and T gives the following result:

$$L = 2.267 \times 10^{-8} \text{ W } \Omega/K^2$$

Example 20.5 The density and atomic weight of Cu are 8900 kg/m³ and 63.5, respectively. The relaxation time of electrons in Cu at 300 K is 10^{-14} s. Calculate the electrical conductivity of copper.

Solution We have the following values:

Density of copper $d = 8900$ kg/m³

Atomic weight of Cu $= 63.5$

Relaxation time $\tau = 10^{-14}$ s

$$n = \frac{\text{avogadro's constant} \times \text{density} \times \text{number of free electrons per atom}}{\text{atomic weight}}$$

$$= \frac{6.022 \times 10^{23} \times 8900 \times 1 \times 10^3}{63.5} = 8.44 \times 10^{28} / m^3$$

Electrical conductivity $\sigma = \dfrac{ne^2 \tau}{m_e}$

$$= \frac{8.44 \times 10^{28} \times (1.6 \times 10^{-19})^2 \times 10^{-14}}{9.1 \times 10^{-31}} (\Omega m)^{-1} = 2.374 \times 10^7 \ (\Omega m)^{-1}$$

Example 20.6 A conduction wire has resistivity of 1.54×10^{-8} Ω m at room temperature. The Fermi energy for such a conductor is 5.5 eV. There are 5.8×10^{28} conduction electrons per m³. Calculate the following parameters:

(a) Relaxation time and the mobility of the electrons

(b) Average drift velocity of electrons when the electric field applied to the conductor is 1 V/cm

(c) Velocity of an electron with Fermi energy

(d) The mean free path of electrons

Solution We have the following values:

Resistivity $\rho = 1.54 \times 10^{-8} \Omega m$

Fermi energy $E_F = 5.5$ eV

Concentration of electrons $n = 5.8 \times 10^{28}/m^3$

Assuming the conductor to be isotropic, the conductivity is given by the following equation: $\sigma = \dfrac{ne^2 \tau}{m_e}$

(a) The relaxation time is given by the following equation: $\tau = \dfrac{\sigma m_e}{ne^2} = \dfrac{m_e}{\rho n e^2}$

   where $\sigma = 1/\rho$

$$\tau = \frac{9.1 \times 10^{-31}}{1.54 \times 10^{-8} \times 5.8 \times 10^{28} \times (1.6 \times 10^{-19})^2} = 3.98 \times 10^{-14}\, s$$

The mobility μ is given by [using Eqs (20.6) and (20.7)] the following relation:

$$\mu = \frac{e\tau}{m_e}$$

$$= \frac{1.6 \times 10^{-19} \times 3.98 \times 10^{-14}}{9.1 \times 10^{-31}} = 7 \times 10^{-3}\, m^2/V\, s$$

(b) Drift velocity is calculated as follows:

$$v_d = \frac{e\tau\varepsilon}{m_e}$$

$$v_d = \frac{1.6 \times 10^{-19} \times 3.98 \times 10^{-14}}{9.1 \times 10^{-31}} = 0.67\, m/s$$

(c) Fermi energy $E_F = 5.5\, eV = 5.5 \times 1.6 \times 10^{-19}\, J$. We know that $E_F = \frac{1}{2}mv_F^2$

or Fermi velocity

$$v_F = \sqrt{\frac{2E_F}{m}}$$

$$= \sqrt{\frac{2 \times 5.5 \times 1.6 \times 10^{-19}}{9.1 \times 10^{-31}}} = 1.39 \times 10^6\, m/s$$

(d) We know that the mean free path is expressed as follows:

$$\lambda_m = v_F\tau$$

$$= 1.39 \times 10^6 \times 3.98 \times 10^{-14} = 5.53 \times 10^{-8}\, m$$

Example 20.7   Electrical resistivity of copper at 27°C is $1.72 \times 10^{-8}\, \Omega m$. Compute its thermal conductivity if the Lorentz number is $2.26 \times 10^{-8}\, \Omega\, W/K^2$.

Solution   According to the Wiedemann–Franz law,

$$\frac{K}{\sigma} = LT \text{ or } K = \sigma LT \text{ or } K = \frac{LT}{\rho}$$

Substituting the given values, we have the following result:

$$K = \frac{2.26 \times 10^{-8} \times 300}{1.72 \times 10^{-8}}\, W/m\, K \text{ or } K = 394\, W/m\, K$$

Example 20.8   Thermal and electrical conductivities of copper at 20°C are 390 W/m K and $5.87 \times 10^7\, (\Omega m)^{-1}$, respectively. Calculate the Lorentz number.

Solution   We have the following values:
Electrical conductivity of copper $\sigma = 5.87 \times 10^7/(\Omega m)^{-1}$
Thermal conductivity of copper $K = 390\, W/m\, K$
Temperature $T = 20°C = 293\, K$
The Lorentz number is given by Eq. (20.15). Substituting the given values, we have
$L = 2.267 \times 10^{-8}\, W\, \Omega/K^2$.

Example 20.9 Calculate the electrical conductivity in copper if the mean free path of the electrons is 4×10^{-8} m, electron density is $8.4 \times 10^{28}/m^3$, and the average thermal velocity of electrons is 1.6×10^6 m/s.

Solution We have the following values:

Mean free path of electrons $\lambda = 4 \times 10^{-8}$ m, electron density $n = 8.4 \times 10^{26}/m^3$, average thermal velocity of the electrons $\overline{v}_{th} = 1.6 \times 10^6$ m/s, charge of electron $e = 1.6 \times 10^{-19}$ C, mass of electron $m_e = 9.11 \times 10^{-31}$ kg

We know from Eqs (20.6) and (20.9) that $\sigma = \dfrac{ne^2\lambda}{m\overline{v}_{th}}$

Substituting the given values, we get the following result:

$$\sigma = \frac{8.4 \times 10^{28} \times (1.6 \times 10^{-19})^2 \times 4 \times 10^{-8}}{9.11 \times 10^{-31} \times 1.6 \times 10^6}$$

$$= \frac{8.4 \times 2.56 \times 4 \times 10^{28} \times 10^{-38} \times 10^{-8}}{9.11 \times 1.6 \times 10^7} \; (\Omega m)^{-1} = 5.9 \times 10^7 (\Omega m)^{-1}$$

Example 20.10 Calculate the electrical and thermal conductivities for a metal with a relaxation time of 10^{-14} s at 300 K. In addition, calculate the Lorentz number using the aforementioned result (density of electrons = $6 \times 10^{28}/m^3$).

Solution We have the following values:

Relaxation time $\tau = 10^{-14}$ s, temperature $T = 300$ K, electron concentration $n = 6 \times 10^{28}/m^3$, mass of electron $m_e = 9.1 \times 10^{-31}$ kg, charge of electron $e = 1.6 \times 10^{-19}$ C, Boltzmann's constant $k = 1.38 \times 10^{-23}$ J/K,

We know from Eq. (20.6) that $\sigma = \dfrac{ne^2\tau}{m_e}$

Substituting the given values, we have the following result:

$$\sigma = \frac{6 \times 10^{28} \times (1.6 \times 10^{-19})^2 \times 10^{-14}}{9.1 \times 10^{-31}} = \frac{15.36 \times 10^{28} \times 10^{-52}}{9.1 \times 10^{-31}}$$

Thus, the electrical conductivity $\sigma = 1.69 \times 10^7 (\Omega m)^{-1}$.

From Eq. (20.13), the thermal conductivity is given as follows: $K = \dfrac{3}{2}\left(\dfrac{nk^2\tau}{m_e}\right)T$

Substituting the given values, we have the following result:

$$K = \frac{3}{2}\left(\frac{6 \times 10^{28} \times (1.38 \times 10^{-23})^2 \times 300 \times 10^{-14}}{9.1 \times 10^{-31}}\right)$$

or $\quad K = 58.2$ W/m K

From Eq. (20.15), the Lorentz number is expressed as follows: $L = \dfrac{K}{\sigma T}$

Substituting the given values, we have the following result:

$$L = \frac{58.2}{1.69 \times 10^7 \times 300} = 1.15 \times 10^{-8} \text{ W } \Omega/K^2$$

$$L = 1.15 \times 10^{-8} \text{ W } \Omega/K^2$$

Example 20.11   Find the relaxation time of conduction electrons in a metal of resistivity $1.54 \times 10^{-8}\,\Omega$ m if the metal has 5.8×10^{28} conduction electrons/m³.

Solution   Number of electrons/unit volume $n = 5.8 \times 10^{28}/\text{m}^3$
Resistivity of the metal $\rho = 1.54 \times 10^{-8}\,\Omega$ m
  We know from Eq. (20.6) that the conductivity of a metal is expressed as follows:

$$\sigma = \frac{ne^2\tau}{m_e}$$

We also know that $\rho = \dfrac{1}{\sigma}$

Therefore, $\tau = \dfrac{m}{ne^2\rho}$

Substituting the given values, we have the following result:

$$\tau = \frac{9.1 \times 10^{-31}}{5.8 \times 10^{28} \times (1.6 \times 10^{-19})^2 \times 154 \times 10^{-8}} \Rightarrow \tau = 3.297 \times 10^{-4}\,\text{s}$$

Example 20.12   A uniform silver wire has resistivity of $1.54 \times 10^{-8}\,\Omega$ m at room temperature. For an electric field of 1 V/cm, calculate the (a) drift velocity, (b) mobility, and (c) relaxation time of electrons, assuming that there are 5.8×10^{28} conduction electrons/m³ of the material.

Solution   We have the following values: $\rho = 1.34 \times 10^{-8}\,\Omega$m, $E = 1\,\text{V/cm} = 1 \times 10^2$ V/m, $n = 5.8 \times 10^{28}/\text{m}^3$
From Eq. (20.6), we know that the electrical conductivity is given by the following expression: $\sigma = \dfrac{ne^2\tau}{m_e}$

Therefore, the resistivity ρ is expressed as follows: $\dfrac{1}{\rho} = \dfrac{ne^2\tau}{m_e}$

Substituting the given values, we have the following result: $\tau = 3.97 \times 10^{-14}\,\text{s}$
The drift velocity is given by the following relation:

$$v_d = \frac{eE\tau}{m_e}$$

$$= \frac{1.6 \times 10^{-19} \times 1 \times 10^2 \times 3.97 \times 10^{-14}}{9.1 \times 10^{-31}} = 0.7\,\text{m/s}$$

The mobility $\mu = \dfrac{v_d}{E} = \dfrac{0.7}{10^2} = 0.7 \times 10^{-2}\,\text{m}^2/\text{V s}$

Example 20.13   Calculate the drift velocity of conduction electrons in copper at a temperature of 300 K, when a copper wire of length 2 m and resistance $0.02\,\Omega$ carries a current of 15 A. Given that $\mu = 4.3 \times 10^{-3}\,\text{m}^2/\text{V s}$.

Solution   We have the following values: $T = 300$ K, $l = 2$ m, $R = 0.02\,\Omega$, $\mu = 4.3 \times 10^{-3}$ m²/V s
The voltage drop V across the wire is given as follows: $V = IR = 15 \times 0.02 = 0.3$ V
The electric field E across the wire is, therefore, given by the following relation:

$$E = \frac{V}{l} = \frac{0.3}{2} = 0.15\,\text{V/m}$$

From Eq. (20.8), $\mu = \dfrac{v_d}{E} \Rightarrow v_d = \mu E$

$$= 4.3 \times 10^{-3} \times 0.15 = 0.65 \times 10^{-3} \, \text{m/s}$$

Example 20.14 Calculate the Fermi energy and Fermi temperature in a metal. The Fermi velocity of electrons in the metal is 0.86×10^6 m/s.

Solution We have the following values: Fermi velocity of electron $v_F = 0.86 \times 10^6$ m/s, mass of electron $m_e = 9.1 \times 10^{-31}$ kg, electronic charge $e = 1.6 \times 10^{-19}$ C, Boltzmann's constant $k = 1.38 \times 10^{-23}$ J/K

We know that the Fermi energy is expressed as follows: $E_F = \dfrac{1}{2}mv_F^2$

Substituting the given values, we have the following result:

$$E_F = \frac{1}{2}\frac{(9.1 \times 10^{-31})(0.86 \times 10^6)^2}{(1.6 \times 10^{-19})} \quad \text{or } E_F = 2.1 \, \text{eV}$$

We also know that the Fermi temperature is expressed as follows: $T_F = \dfrac{E_F}{k}$

Hence, $T_F = \dfrac{2.1 \times 1.6 \times 10^{-19}}{1.38 \times 10^{-23}} = 24{,}348 \, \text{K}$

Example 20.15 The Fermi temperature of a metal is 2460 K. Calculate the Fermi velocity.

Solution We have the following values: Fermi temperature of the metal $T_F = 2460$ K, mass of electron $m_e = 9.11 \times 10^{-31}$ kg

The relation between Fermi energy, Fermi velocity, and Fermi temperature is given by the following equation: $E_F = kT_F = \dfrac{1}{2}mv_F^2$

or $\qquad v_F^2 = \dfrac{2kT_F}{m}$

or $\qquad v_F = \sqrt{\dfrac{2 \times 1.38 \times 10^{-23} \times 2460}{9.11 \times 10^{-31}}} = 2.730 \times 10^5 \, \text{m/s}$

Example 20.16 Electrons in a metal have a Fermi velocity of 2×10^6 m/s. Calculate the Fermi energy of the electrons in the metal in joules.

Solution Fermi energy of the electrons is given by the following relation:

$$E_F = \frac{1}{2}mv_F^2 \tag{20.21}$$

The use of the given values in Eq. (20.21) leads to the following result:

$$E_F = \frac{1}{2} \times \frac{9.1 \times 10^{-31} \times 4 \times 10^{12}}{}$$

$$= 1.82 \times 10^{-18} \, \text{J}$$

Example 20.17 Conductivity of particular metal is $4 \times 10^7 \, (\Omega \text{m})^{-1}$. Carrier concentration in the metal is found to be $1 \times 10^{29}/\text{m}^3$. Calculate the relaxation time of electrons in the metal.

Solution   Conductivity σ, is given by the following expression:

$$\sigma = \frac{ne^2\tau}{m_e} \tag{20.22}$$

which can be rewritten as follows:

$$\tau = \frac{\sigma m_e}{ne^2} \tag{20.23}$$

Using these values in Eq. (20.23), we get the following relation:

$$\tau = \frac{4\times10^7 \times 9.1\times10^{-31}}{1\times10^{29} \times (1.6\times10^{-19})^2} \Rightarrow \tau = 1.42\times10^{-14}\,\text{s}$$

Example 20.18   Conductivity of a metal is 2×10^7 $(\Omega\text{m})^{-1}$. Relaxation time of electrons in the metal is 5×10^{-14} s. Calculate the electron concentration.

Solution   Conductivity σ, is given by the following relation:

$$\sigma = \frac{ne^2\tau}{m} \tag{20.24}$$

We can rewrite this expression as follows:

$$n = \frac{\sigma m}{e^2\tau} \tag{20.25}$$

Substitution of these values in Eq. (20.25) leads to the following relation:

$$n = \frac{2\times10^7 \times 9.1\times10^{-31}}{(1.6\times10^{-19})^2 \times 5\times10^{-14}} \Rightarrow n = 1.42\times10^{28}\,\text{m}^{-3}$$

20.3 QUANTUM THEORY

Sommerfeld assumed that the free electrons present in a metal obey the Fermi–Dirac statistics, and not the Maxwell–Boltzmann statistics. Sommerfeld also assumed the existence of a potential energy box bounded on either side by the boundaries of the crystal. Quantum statistics was then used to determine the possible electronic energy states and distribution of electrons within these energy states.

20.3.1 Fermi Distribution Function—Temperature Dependence

The Fermi–Dirac (FD) distribution function $f(E)$ is given as follows:

$$f(E) = \frac{1}{\exp\left[\dfrac{E - E_F}{kT}\right] + 1} \tag{20.26}$$

The highest energy level that is populated at absolute zero is called the Fermi level, E_F. The temperature dependence of $f(E)$ can be described as follows:

$$\text{At } T = 0, \quad f(E) = 0 \quad \text{for} \quad E > E_F \tag{20.27}$$

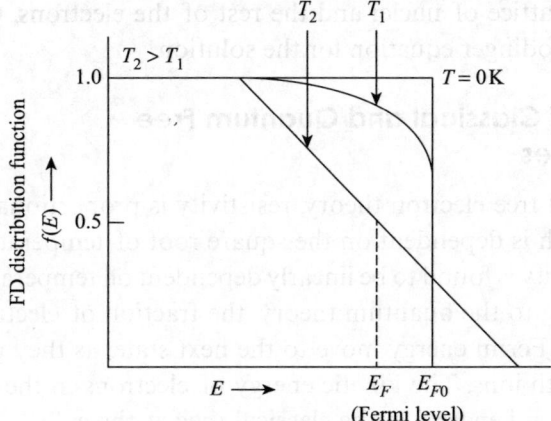

Fig. 20.1 Fermi–Dirac distribution function

$$\text{At } T > 0, \quad f(E) = \frac{1}{2} \quad \text{for} \quad E = E_F \tag{20.28}$$

Figure 20.1 shows a plot of $f(E)$ versus E for different temperatures. As evident from this figure, $f(E)$ is zero for $E > E_F$ at absolute zero. At other temperatures, $f(E)$ is equal to $1/2$ for $E = E_F$. The position of the Fermi level changes with temperature, and the dependence is given by the following approximate expression:

$$E_F = E_{F0}\left[1 - \frac{\pi^2}{12}\left(\frac{kT}{E_{F0}}\right)\right] \tag{20.29}$$

where E_{F0} is the Fermi level at absolute zero.

In the case of metals, the spacing between allowed energy levels is extremely small ($\sim 10^{-19}$ eV) and the highest filled energy level is taken to be the Fermi level.

20.3.2 Problems with Quantum Free Electron Theory

The quantum free electron theory assumes that valence electrons are completely detached from their ions. Another important assumption is the complete absence of electron–electron interactions. The quantum free electron theory does not take into account the crystal lattice. The theory uses Bloch's theorem, wherein, an unbound electron is assumed to move in a periodic potential as a free electron in vacuum. The electron mass, m, is replaced by an effective mass m^* that is different from m. The effective mass m^* is, in turn, dependent on the band structure. In reality, the electron moves through a crystal lattice that does not hinder its motion. The motion of electrons is, however, affected by scattering due to impurities and phonons. These two interactions determine both electrical and thermal conductivity. These issues can be resolved by assuming that each electron moves in a static, periodic potential created by the lattice of nuclei and the rest of the electrons. One then has to solve the Schrodinger equation for the solution.

20.3.3 Comparison of Classical and Quantum Free Electron Theories

According to the classical free electron theory, resistivity is proportional to the average velocity, which is dependent on the square root of temperature. In reality, however, resistivity is found to be linearly dependent on temperature for conductors. According to the quantum theory, the fraction of electrons that are within kT of the Fermi energy move to the next state, as they gain energy from collisions with ions. The kinetic energy of electrons in the last filled level is called the Fermi energy. In the classical theory, the collision of an electron with an ion is considered to be similar to that of a ball with a wall. In the quantum theory, an electron is viewed as a wave travelling through a medium. If the wavelength of the electron is larger than the crystal spacing, then the electron can propagate without collision within a conductor. Imperfections in a crystal lattice, however, result in scattering. The classical mean free path is replaced by the amplitude of vibration of ions. This results in a linear dependence of resistivity on temperature.

Example 20.19 Use the Fermi distribution function to obtain the value of $F(E)$ for $E - E_F = 0.01$ eV at 200 K.

Solution　We have the following values: $E - E_F = 0.01$ eV $= 0.01 \times 1.6 \times 10^{-19}$ J $= 1.6 \times 10^{-21}$ J,

Temperature $T = 200$ K, Boltzmann's constant $k = 1.38 \times 10^{-23}$ J/K

We know that $F(E) = \dfrac{1}{1 + e^{(E - E_F)/kT}}$

Substituting the given values, we have the following result:

$$F(E) = \frac{1}{1 + e^{(1.6 \times 10^{-21})/(1.38 \times 10^{-23} \times 200)}} = \frac{1}{1 + e^{0.5797}} = \frac{1}{1 + 1.7855} = \frac{1}{2.7855}$$

$$F(E) = 0.36$$

Example 20.20 Calculate the value of $(E - E_F)$, if the Fermi distribution function has a magnitude of 0.3 at 200 K.

Solution　Fermi distribution function $F(E)$ is given by the following relation:

$$F(E) = \frac{1}{1 + e^{(E - E_F)/RT}} \tag{20.30}$$

leading to $\left(\dfrac{E - E_F}{RT} \right) = \log_e \left(\dfrac{1}{F(E)} - 1 \right)$

resulting in

$$(E - E_F) = RT \log_e \left(\frac{1}{F(E)} - 1 \right) \tag{20.31}$$

Substituting the given values in Eq. (20.31), we get the following result:

$$(E - E_F) = \frac{(1.38 \times 10^{-23})(200)}{1.6 \times 10^{-19}} \log_e \left(\frac{1}{0.3} - 1 \right) = 0.01725 \log_e (2.33)$$

implying that $(E - E_F) = 0.015 \text{eV}$

20.4 FREE ELECTRON GAS

For the sake of mathematical simplicity, we shall discuss the one-dimensional model of free electron gas. Consider an electron of mass m to be bound to move in a one-dimensional crystal of length l. A large potential energy barrier exists $(V = \infty)$ at the bounded surfaces of the crystal, which prevents electrons from escaping from the crystal. Figure 20.2 shows a schematic diagram of the one-dimensional model.

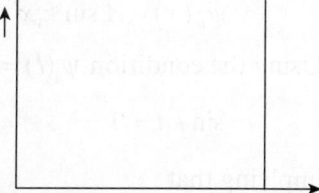

Fig. 20.2 One-dimensional model of free electron gas

The potential energy within the crystal is assumed to be zero. The boundary conditions for the wave function are as follows:

$$V(x) = 0, \quad \text{for } 0 \le x \le l \tag{20.32}$$

and $\quad V(x) = \infty, \quad \text{for } x \le 0 \text{ and } x \ge l \tag{20.33}$

The wave function ψ_n for the electron occupying the nth state can be obtained by Schrödinger's equation, as follows:

$$\frac{d^2 \psi_n}{dx^2} + \frac{2m}{\hbar^2}(E_n - V)\psi_n = 0 \tag{20.34}$$

where E_n represents the kinetic energy of the electron occupying the nth state and V gives the potential energy of the electron. Since $V = 0$ within the box, Eq. (20.34) can be rewritten in the following form:

$$\frac{d^2 \psi_n}{dx^2} + \frac{2m}{\hbar^2} E_n \psi_n = 0 \tag{20.35}$$

Equation (20.35) has solutions of the following form:

$$\psi_n(x) = A \sin k_1 x + B \cos k_1 x \tag{20.36}$$

where $\quad k_1 = \sqrt{\dfrac{2mE_n}{\hbar^2}} \tag{20.37}$

The constants A and B are determined using the following boundary conditions:

$$\psi_n(0) = 0 \quad \text{and} \quad \psi_n(l) = 0 \tag{20.38}$$

At $x = 0$ and $x = l$, $V \to \infty$. The product $V\psi_n$ in Eq. (20.34) also approaches ∞ at the boundaries. For the wave function $\psi_n(x)$ to be continuous, the kinetic energy E_n must also then approach infinity, which is not feasible practically. This implies that $\psi_n(x)$ must vanish at $x = 0$ and $x = l$, as indicated by the boundary conditions given by Eq. (20.38). Using the condition $\psi_n(0) = 0$ in Eq. (20.36), one gets the following relation:

$$\psi_n(0) = B \cos 0 = B \tag{20.39}$$

Substitution of Eq. (20.39) into Eq. (20.36) leads to the following expression:

$$\psi_n(x) = A \sin k_1 x \tag{20.40}$$

Using the condition $\psi_n(l) = 0$ in Eq. (20.40), we get the following relation:

$$\sin k_1 l = 0 \tag{20.41}$$

implying that

$$k_1 = \frac{n\pi}{l} \tag{20.42}$$

with $n = 1, 2, 3, \ldots$ for the allowed wave functions. Thus, Eq. (20.40) can be rewritten in the following form:

$$\psi_n(x) = A \sin\left(\frac{n\pi}{l} x\right) \tag{20.43}$$

From Eq. (20.37), we have

$$E_n = \frac{\hbar^2}{2m} k_1^2 \tag{20.44}$$

The use of Eq. (20.34) in Eq. (20.36) leads to the following relation:

$$E_n = \frac{\hbar^2}{2m}\left(\frac{n\pi}{l}\right)^2 = \frac{n^2 \hbar^2 \pi^2}{2ml^2} = \frac{n^2 h^2}{8ml^2} \tag{20.45}$$

A plot of E_n versus n is shown in Fig. 20.3. Although this plot has been drawn for a one-dimensional situation, it is a true representation of the real three-dimensional situation.

From Fig. 20.3 and Eq. (20.45), we can conclude that

$$E_n \propto n^2 \tag{20.46}$$

The quantity n is called the quantum number. The spacing between different energy levels is dependent on the values of n and l. The probability of finding an electron somewhere along the curve corresponding to a particular value of n is unity, implying that

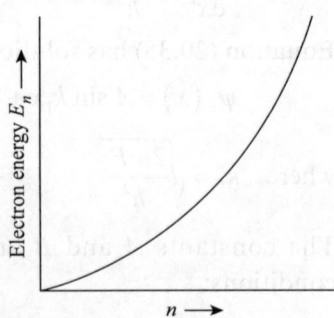

Fig. 20.3 Plot of E_n vs n

$$\int_0^l \psi_n^*(x)\psi_n(x)dx = 1 \qquad (20.47)$$

The use of Eq. (20.43) in Eq. (20.47) leads to the following relation:

$$A^2 \int_0^l \sin^2\left(\frac{n\pi}{l}x\right)dx = 1$$

which can be rewritten as follows: $\dfrac{A^2}{2}\int_0^l\left[1-\cos\left(\dfrac{2n\pi}{l}x\right)\right]dx = 1$

giving $\dfrac{A^2}{2}\int_0^l dx = 1\left[\because \int_0^l \cos\left(\dfrac{2n\pi}{l}x\right)dx = 1\right]$

This yields $\dfrac{A^2}{2}l = 1$

or $\qquad A = \sqrt{\dfrac{2}{l}} \qquad\qquad\qquad\qquad\qquad (20.48)$

Substituting the value of A from Eq. (20.48) into Eq. (20.43), one gets the following relation:

$$\psi_n(x) = \sqrt{\frac{2}{l}}\ \sin\left(\frac{n\pi}{l}x\right) \quad (20.49)$$

Equation (20.49) can be used to plot the wave functions corresponding to different values of quantum number n for different values of position x within the potential box of length l. Figure 20.4 shows the energy levels and wave functions corresponding to the first four states of a free electron.

The energy values indicated in Fig. 20.4 are also referred to as the *eigenvalues* and the corresponding wave functions as the *eigen functions*.

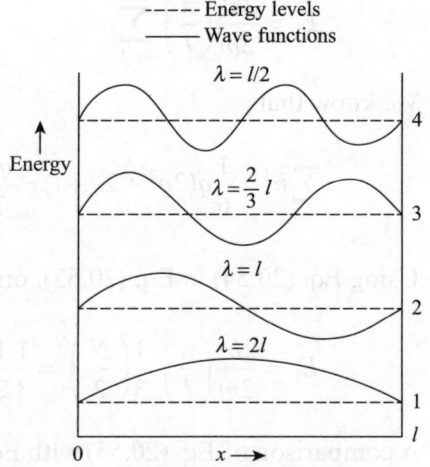

Fig. 20.4 Energy and wave functions for first four states of free electron

20.5 FERMI ENERGY AND CARRIER CONCENTRATION

According to Pauli's exclusion principle, no two electrons can have the same quantum state. Every energy level defined by a quantum number n can have two quantum states, having possible values of $+1/2$ and $-1/2$ (for spin up and spin down states of the electron, respectively). Thus, each energy level is doubly degenerate. The topmost filled energy level at $0\,K$ is called the *Fermi level* and the symbol E_F is used to represent it. The corresponding energy is called the *Fermi energy*. If n_F is the principal quantum number of the Fermi level, then the total number of electrons up to the Fermi level is given by the following relation:

$$N = 2n_F \tag{20.50}$$

The use of Eq. (20.50) in Eq. (20.45) results in the following relation:

$$E_F = \frac{\hbar^2}{2m}\left(\frac{n_F \pi}{l}\right)^2 = \frac{\hbar^2}{2m}\left(\frac{N\pi}{2l}\right)^2 \tag{20.51}$$

20.5.1 Total Energy

Suppose a system consists of N electrons in the ground state. The total energy E_T of all the electrons is given by the following expression:

$$E_T = 2\sum_{n=1}^{N/2} E_n \tag{20.52}$$

The factor 2 accounts for the fact that every energy level can be occupied by two electrons of either spin. Using Eq. (20.45), we get the following relation:

$$E_T = \frac{2\hbar^2}{2m}\left(\frac{\pi}{l}\right)^2 \sum_{n=1}^{N/2} n^2 \tag{20.53}$$

We know that

$$\sum_{n=1}^{q} n^2 = \frac{1}{6}q(2q^2 + 2q + 1) = \frac{q^3}{3} \quad (\text{for } q \gg 1) \tag{20.54}$$

Using Eq. (20.54) in Eq. (20.53), one gets the following relation:

$$E_T = \frac{2\hbar^2}{2m}\left(\frac{\pi}{l}\right)^2 \frac{1}{3}\left(\frac{N}{2}\right)^3 = \frac{1}{3}\frac{\hbar^2}{2m}\left(\frac{N\pi}{2l}\right)^2 N \tag{20.55}$$

A comparison of Eq. (20.55) with Eq. (20.51) yields the following expression:

$$E_T = \frac{1}{3}NE_F \tag{20.56}$$

Thus, the average kinetic energy for a one-dimensional crystal in the ground state is one-third of the Fermi energy.

20.5.2 Density of Energy States

The number of electronic states per unit energy is called the density of states, $D(E)$. Thus, we have the following relation:

$$D(E) = \frac{dn}{dE} \qquad (20.57)$$

Since each energy level can be occupied by two electrons, the density of states expression gets modified to the following form:

$$D(E) = 2\frac{dn}{dE} \qquad (20.58)$$

From Eq. (20.45), we have the following expression: $E = \dfrac{n^2 h^2}{8ml^2}$

or $\qquad n^2 = \dfrac{8ml^2}{h^2} E$

or $\qquad n = \left(\dfrac{8ml^2}{h^2}\right)^{\frac{1}{2}} E^{\frac{1}{2}} \qquad (20.59)$

Substituting for n in Eq. (20.58), we get the following relation:

$$\frac{dn}{dE} = \left(\frac{8ml^2}{h^2}\right)^{\frac{1}{2}} E^{\frac{1}{2}}$$

or $\qquad \dfrac{dn}{dE} = \dfrac{2l}{h}\left(\dfrac{m}{2E}\right)^{\frac{1}{2}} \qquad (20.60)$

The use of Eq. (20.60) in Eq. (20.58) results in the following expression:

$$D(E) = \frac{4l}{h}\left(\frac{m}{2E}\right)^{\frac{1}{2}} \qquad (20.61)$$

A plot of $D(E)$ versus E is shown in Fig. 20.5. All the energy levels below E_F are filled and those above E_F are empty. This situation exists in reality at $T = 0$ K, at which the Fermi level divides the filled and the unfilled levels.

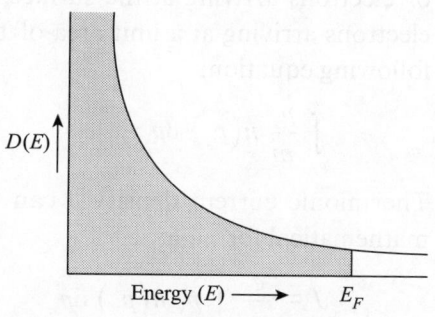

Fig. 20.5 Variation of $D(E)$ vs E

20.6 THERMIONIC EMISSION

Thermionic emission is the phenomenon of emission of electrons from the surface of a heated metal. According to the free electron model, at absolute zero, all energy levels up to the Fermi level E_F are filled. An energy barrier exists at the metal surface, preventing electrons with energy E_F from escaping. This energy barrier is called the work function φ. The situation is depicted schematically in Fig. 20.6.

As the temperature is raised, energy of the electrons increases and they start occupying levels above the Fermi level E_F. When the electrons attain energies above $(E_F + \phi)$, they escape from the surface of metals, resulting in thermionic emission.

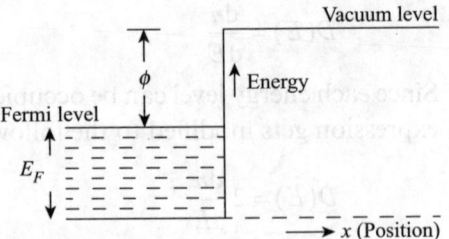

Fig. 20.6 Schematic of energy levels at metal surface

In Fig. 20.6, the x-axis is along a direction that is perpendicular to the surface of the metal. For electrons with just sufficient energy to escape from the metal surface, we can write the following relation:

$$E_F + \phi = \frac{P_{x0}^2}{2m} E_0 \qquad (20.62)$$

where p_{x0} denotes the momentum of the electrons satisfying Eq. (20.62) and E_0 is the surface energy.

Equation (20.62) leads to the following relation:

$$P_{x0}^2 = \sqrt{(E_F + \phi)2m} \qquad (20.63)$$

Thermionic emission current at a given temperature is the product of electronic charge and the number of electrons with a momentum in the x-direction striking a unit area of the surface in unit time. Suppose $n(p_x)$ represents the number of electrons per unit volume having momentum lying between p_x and $(p_x + dp_x)$ in the x-direction. The term (p_x/m) represents the velocity of electrons arriving at the surface, having momentum p_x. The number of electrons arriving at a unit area of the surface per unit time is given by the following equation:

$$\int_{p_{x0}}^{\infty} \frac{p_x}{m} n(p_x) \, dp_x \qquad (20.64)$$

Thermionic current density J can then be calculated using the following mathematical formula:

$$J = \frac{e}{m} \int_{p_{x0}}^{\infty} p_x \, n(p_x) \, dp_x \qquad (20.65)$$

The quantity $n(p_x)dp_x$ can be shown to be given by the following relation:

$$n(p_x)dp_x = \frac{4\pi \, mkT}{h^3} \exp\left(\frac{E_F}{kT}\right) \exp\left(\frac{-p_x^2}{2mkT}\right)dp_x \qquad (20.66)$$

Using Eq. (20.66) in Eq. (20.65), one obtains the following expression:

$$J = \frac{e}{m} \int_{p_{x0}}^{\infty} \frac{4\pi \, mkT}{h^3} \exp\left(\frac{E_F}{kT}\right) \exp\left(\frac{-p_x^2}{2mkT}\right) p_x \, dp_x \qquad (20.67)$$

Substituting $(p_x^2/2m) = E_F + \phi$ and $(p_x\, dp_x) = m\, d\phi$ in Eq. (20.67), we get the following relation:

$$J = \frac{e}{m} \int_\phi^\infty \frac{4\pi\, mkT}{h^3} \exp\left(\frac{E_F}{kT}\right) \exp\left(\frac{-E_F}{2mkT}\right) \exp\left(\frac{-\phi}{kT}\right) md\phi$$

leading to $J = \dfrac{e}{m}\ \dfrac{4\pi\, mkT}{h^3}\ m\int_\phi^\infty \exp\left(\dfrac{-\phi}{kT}\right) md\phi$

yielding $J = \dfrac{4\pi\, mek^2 T^2}{h^3} \exp\left(\dfrac{-\phi}{kT}\right)$ \hfill (20.68)

Equation (20.68) can be rewritten as follows:

$$J = AT^2 \exp\left(\frac{-\phi}{kT}\right) \tag{20.69}$$

Equation (20.69) is called the *Richardson–Dushman equation*. The quantity A is called the *emission coefficient* and is given by the following expression:

$$A = (4\pi\, mek^2)/h^3$$
$$A = 1.20 \times 10^6 \text{ A/m}^2 \text{ deg}^2 \text{ K}$$

20.7 CONDUCTING POLYMERS

Conductors, though useful, are sometimes very expensive for commercial applications. In addition to this, foldability and size can also present some limitations. In their quest for alternatives, scientists are increasingly focussing on a new class of materials called conducting polymers. Conducting polymers are organic polymers that have the ability to conduct electricity. They can have either metallic or semiconductor-like conductivity. What is more interesting is that their electrical properties can be modified and tuned using different methods of organic synthesis and advanced dispersion techniques. Some examples of conducting polymers include polyacetylene, polyphenylene vinylene, polythiophene, polyaniline, etc. Several processes have been developed for increasing the conductivity of polymers. Some of these are doping by oxidation, doping by reduction, and self-doping. At present, however, processability of conducting polymers is poor. Once this improves, the possible application areas would be organic solar cells, organic light-emitting diodes, supercapacitors, and biosensors.

20.7.1 Chargeable Polymers—Polymer Batteries

In polymer batteries, the liquid electrolyte is replaced with a polymer electrolyte. One example is the lithium polymer battery. The field is still in the research stage, with significant progress being reported. One approach is to use a polymer matrix (polyvinylidene fluoride or polyacrylonitrite), which is gelled with conventional salts and solvents. The ultimate aim is to develop solid polymer electrolytes for commercial applications.

Note: Conducting materials have a lot of practical applications. One interesting application is the *Faraday cage*. Electromagnetic interference (EMI) can be problematic for sensitive instruments. These EMIs can travel through different routes such as cables, power cords, or even electromagnetic radiation. A Faraday cage is basically a box made of a highly conducting material. The box acts as a hollow conductor. An electrically sensitive instrument or system is placed within this box. In the presence of any external electric field, electric charges within the conducting box rearrange themselves to result in a zero electric field inside the box. The Faraday cage, therefore, shields the system kept inside the box from the electromagnetic radiation in the surrounding area. One must keep the conducting cage well grounded to dissipate the charge accumulated on the external surface.

IMPORTANT CONCEPTS

1. In the Drude model, valence electrons are assumed to behave like a gas.
2. According to the Drude–Lorentz classical theory, all free electrons move with the RMS speed, with the thermal velocity of electrons assumed to have a Maxwell–Boltzmann distribution.
3. The dc electrical conductivity of metals increases with a decrease in temperature.
4. According to the Lorentz's modification of the Drude's model, the Maxwell–Boltzmann law of distribution of velocity is perturbed by electric and thermal gradients.
5. According to the quantum theory, the free electrons present in a metal obey the Fermi–Dirac statistics.
6. Thermionic emission is the phenomenon of emission of electrons from the surface of a heated metal.
7. The Richardson–Dushman equation is given as follows:

$$J = AT^2 \exp(-\phi/kT)$$

where $A = (4\pi mek^2)/h^3$

APPLICATIONS

1. Conducting materials are useful for the industries, where these are used to fabricate cables, wires, connectors, contact to semiconductor devices, etc.
2. Special conducting materials find applications in the field of defence, in the construction of aircraft, ship structures, artillery, armaments, etc.
3. Conducting materials are used in the construction of bridges, buildings, and structures, due to their high mechanical strength.
4. In electronic circuit boards, different components or sub-systems are sometimes joined together using a 'conducting ink'. Such inks usually contain powdered

or flaked conducting materials. Silver is the most commonly used conducting material. These conductive inks are far more economical to use than the conventional technique of using copper-plated substrates, where copper is etched to form conductive lines or patterns. Some common applications of a conductive ink include RFID tags, tickets, windshield defrosters, etc.

5. Electro-textiles are textiles that can conduct electricity. They are fabricated by blending or coating textiles with metallic fibres. These would form the basis of smart textiles that have embedded sensors in them. Some applications include wearable smart phones, electronics-enhanced garments, wearable computers, etc. Issues such as connectivity, choice of materials, bending, and stretching would, however, have to be fully sorted out before they become viable for a large scale commercial exploitation.

IMPORTANT FORMULAE

1. Drift velocity
$$v_d = (-eE/m_e)\tau$$

2. Current density
$$J = nev_d$$

3. Conductivity
$$\sigma = (ne^2\lambda_m)/\sqrt{(3k\ m_eT)}$$

4. Electrical conductivity
$$\sigma = (ne^2\lambda m)/\sqrt{(3km_eT)}$$

5. Thermal conductivity
$$K = 3/2(nk^2\tau/m_e)T$$

6. Lorentz number
$$K/\sigma T = L$$

7. Electrical conductivity (Lorentz's model)
$$\sigma_L = (8/3\pi)^{\frac{1}{2}}\sigma_D$$

8. Fermi–Dirac distribution function
$$f(E) = 1/\{\exp[(E - E_F)/kT] + 1\}$$

9. Fermi level
$$E_F = E_{F0}[1 - \pi^2/12(kT/E_{F0})]$$

10. Average kinetic energy
$$E_T = 1/3\ NE_F$$

11. Electron momentum
$$p_{x0} = \sqrt{(E_F + \phi)2m}$$

12. Thermionic current density
$$J = AT^2\exp(-\phi/kT),$$
with $A = 4\pi\ mek^2/h^3$

13. Richardson–Dushman equation
$$J = AT^2\exp(-\phi/kT)$$

SELF-ASSESSMENT

Multiple-choice Questions

20.1 In the Drude model, the core of a metal consists of
   (a) positive ions
   (b) negative ions
   (c) electrons
   (d) positrons

20.2 Electric current density is given by
   (a) $J = ne/v_d$
   (b) $J = nev_d$
   (c) $J = ev_d/n$
   (d) v_d/en

20.3 According to the classical theory, σ is proportional to

(a) $\sqrt{T}$ (b) T (c) $1/\sqrt{T}$ (d) $1/T$

20.4 Lorentz number L, is given by

(a) $\sigma K/T$ (b) σ/KT (c) $\sigma T/K$ (d) $K/\sigma T$

20.5 At $T = 0$ for $E > E_F$,

(a) $f(E) = 0$ (b) $f(E) > 0$ (c) $f(E) < 0$ (d) $f(E) < 1$

20.6 Select the correct form of Schrodinger's equation:

(a) $d^2\psi_n/dx^2 = 0$

(b) $d^2\psi_n/dx^2 + 2m/\hbar^2(E_n - V)\,\psi_n = 0$

(c) $d^2\psi_n/dx^2 + (E_n - V)\psi_n = 0$

(d) $d^2\psi_n/dx^2 + \hbar^2/2m(E_n - V)\psi_n = 0$

20.7 Fermi energy E_F is proportional to

(a) n_F (b) n_F^3 (c) n_F^2 (d) $1/n_F^2$

20.8 Total energy E_T is given by

(a) NE_F (b) $NE_F/2$ (c) $2NE_F$ (d) $NE_F/3$

20.9 Density of states $D(E)$ is proportional to

(a) $1/\sqrt{E}$ (b) $\sqrt{E}$ (c) $1/E$ (d) E

20.10 According to the Richardson–Dushman equation, the thermionic emission current density J is proportional to

(a) T (b) T^2 (c) $1/T$ (d) $1/T^2$

20.11 Select the correct expression:

(a) $J = \dfrac{\sigma}{E}$ (b) $J = \sigma E$ (c) $J = \dfrac{E}{\sigma}$ (d) $J = \sigma^2 E$

20.12 Unit of conductivity is

(a) $\Omega\,m$ (c) $\Omega^{-1}m^{-1}$

(b) $\Omega\,m^{-1}$ (d) $\Omega^{-1}m$

20.13 Conductivity σ is proportional to

(a) e (b) $\dfrac{1}{e}$ (c) $\dfrac{1}{e^2}$ (d) e^2

20.14 If electron relaxation time doubles, then conductivity

(a) doubles   (c) becomes four times

(b) becomes half   (d) does not change

20.15 The correct expression for mobility, μ, is

(a) $\mu = v_d E$ (c) $\mu = v_d^2 E$

(b) $\mu = \dfrac{v_d}{E}$ (d) $\mu = \dfrac{v_d}{E^2}$

20.16 Relaxation time is given by

(a) $\tau = \lambda_m v_{th}$

(c) $\tau = \dfrac{\lambda_m}{v_{th}}$

(b) $\tau = \lambda_m^2 v_{th}$

(d) $\tau = \dfrac{\lambda_m}{v_{th}^2}$

20.17 Lorentz number has units of

(a) $W \Omega K^2$ (b) $\dfrac{WK^2}{\Omega}$ (c) $\dfrac{WK}{\Omega}$ (d) $W\Omega/K^2$

20.18 σ_D and σ_L are related through the expression

(a) $\sigma_D = 1.09\sigma_L$

(c) $\sigma_L = \dfrac{2}{1.09}\sigma_D$

(b) $\sigma_L = 1.09\sigma_D$

(d) $\sigma_D = 2\sigma_L$

20.19 Kinetic energy E_n of an electron occupying the nth state is proportional to

(a) $\dfrac{1}{n}$ (b) n^2 (c) $\dfrac{1}{n^2}$ (d) n

20.20 The correct expression for density of states is

(a) $D(E) = \dfrac{dn}{dE}$

(c) $D(E) = \dfrac{2dn}{dE}$

(b) $D(E) = \dfrac{dE}{dn}$

(d) $D(E) = \dfrac{2}{dn/dE}$

Review Questions

20.1 Define mobility of electrons.

20.2 Define electrical conductivity.

20.3 What are the merits of the classical free electron theory?

20.4 Define the mean free path.

20.5 Define the relaxation time of an electron.

20.6 Define the drift velocity of an electron. How is it different from its thermal velocity?

20.7 State the Wiedemann–Franz law.

20.8 Define Lorentz number.

20.9 Write down the Fermi–Dirac distribution function.

20.10 Define Fermi level and state its significance. In addition, define Fermi energy.

20.11 How does Fermi function vary with temperature?

20.12 Define density of states. What is its significance?

20.13 What type of statistics do free electrons obey, according to the classical theory of metals?

20.14 What is the direction of motion of electrons in the presence of an electric field?

20.15 Define the RMS velocity.

20.16 Write an expression for current density in terms of conductivity.

20.17 How does electrical conductivity depend on temperature, according to the classical theory of metals?

20.18 Write the Fermi–Dirac distribution function.

20.19 Draw the potential versus position curve for a one-dimensional potential well.

20.20   Write down the Schrödinger's equation for the nth state of an electron.

20.21   Define Fermi energy.

20.22   On the basis of the free electron theory, derive an expression for electrical conductivity.

20.23   State and prove the Wiedemann–Franz law.

20.24   Deduce mathematical expressions for electrical conductivity and thermal conductivity of a conducting material, and from these derive the Wiedemann–Franz law.

20.25   Write the Fermi–Dirac distribution function. Explain how the Fermi function varies with temperature.

20.26   Derive an expression for the density of states, and based on that calculate the carrier concentration in metals.

20.27   With a neat diagram, derive an expression for the density of states.

20.28   Derive an expression for the total energy of a one-dimensional crystal.

20.29   Give the salient features of the Drude–Lorentz classical theory.

20.30   What is meant by relaxation time?

20.31   Define mobility of charge carriers.

20.32   Derive an expression for electrical conductivity in accordance with the classical theory.

20.33   Derive an expression for K/σ in accordance with the classical theory.

20.34   What is a potential energy box?

20.35   What is $f(E)$ for (a) $T = 0$, $E > E_F$ and (b) $T > 0$, $E = E_F$?

20.36   Write down the solution for the following equation:

$$\frac{d^2\psi_n}{dx^2} + \frac{2m}{\hbar^2} E_n \psi_n = 0$$

20.37   Derive an expression for Fermi energy.

20.38   Define density of states.

20.39   What is Fermi temperature?

Numerical Problems

20.1   Electrons in a metal have a Fermi velocity of 9×10^5 m/s. Determine the Fermi energy of the electrons in the metal. $\left[\text{Hint: } E_F = \frac{1}{2} m_e v_F^2 \right]$

20.2   The Fermi energy for electrons in a particular metal is 2.5 eV. Calculate the Fermi velocity of the electrons. $\left[\text{Hint: } E_F = \frac{1}{2} m_e v_F^2 \right]$

20.3   The Fermi energy of copper at 0 K is 7.04 eV. Determine the Fermi energy at 320 K. $\left[\text{Hint: } E_F = E_{F0} \left[1 - \frac{\pi^2}{12} \left(\frac{kT}{E_{F0}} \right)^2 \right] \right]$

20.4   The Fermi energy for a particular metal is 2 eV at 0 K. Calculate the Fermi energy at 300 K. $\left[\text{Hint: } E_F = E_{F0} \left[1 - \frac{\pi^2}{12} \left(\frac{kT}{E_{F0}} \right)^2 \right] \right]$

20.5 The Fermi energy of a metal at 0 K is 6 eV. Calculate the temperature at which

the Fermi energy is 5.999 eV. $\left[\text{Hint: } E_F = E_{F0}\left(1 - \frac{\pi^2}{E_{F0}}\left(\frac{kT}{E_{F0}}\right)^2\right)\right]$

20.6 Conductivity of a metal is $8 \times 10^7 \, (\Omega\text{m})^{-1}$. If the number density of free electron

is $20 \times 10^{28}/\text{m}^3$, calculate the relaxation time of electrons. $\left[\text{Hint: } \sigma = \frac{ne^2\tau}{m_e}\right]$

20.7 The number density of free electrons in a metal is $5 \times 10^{28}/\text{m}^3$. If the relaxation
time of free electrons is 2×10^{-14} s, calculate the conductivity of the metal.

$$\left[\text{Hint: } \sigma = \frac{ne^2\tau}{m_e}\right]$$

20.8 The electrical and thermal conductivities of a metal at $T = 300$ K are $\sigma = 7 \times 10^7$
$(\Omega\text{m})^{-1}$ and 430 W/m K, respectively. Determine the Lorentz number.
[Hint: $L = K/\sigma T$]

20.9 Lorentz number for a particular metal is $2 \times 10^{-8} \, \text{W}/\Omega \, \text{K}^2$. If the electrical conductivity is $4 \times 10^7 \, (\Omega\text{m})^{-1}$ at 300 K, find the thermal conductivity at the same

temperature. $\left[\text{Hint: } L = \frac{k}{\sigma T}\right]$

20.10 Density of a conductor is 7000 kg/m³ and its atomic weight is 72. If the number
of free electrons per atom is 1, calculate the number density of free electrons.

$$\left[\text{Hint: } n = \frac{(\text{Avogadro's constant}) \times (\text{Density}) \times (\text{Number of free electrons per atom})}{\text{Atomic weight}}\right]$$

20.11 Conductivity of a particular metal is $6 \times 10^7 (\Omega\text{m})^{-1}$ Carrier concentration in
the metal is found to be $8 \times 10^{28}/\text{m}^3$. Calculate the relaxation time of electrons

in the metal. $\left[\text{Hint: } \sigma = \frac{ne^2\tau}{m_e}\right]$

Superconducting Materials

21

Learning Objectives

After studying this chapter, students will be able to

- understand the Meissner effect
- define the transition temperature
- explain the isotope effect
- describe different types of superconductors
- elucidate the salient features of the BCS theory
- learn about high-T_c superconductors
- demonstrate important applications of superconductivity, such as SQUID, cryotron, and maglev

List of Symbols

H_C = Critical magnetic field

T_C = Critical temperature

λ_L = Penetration depth

k_B = Boltzmann constant

v_S = Velocity of superconducting electron

21.1 INTRODUCTION

The resistance offered by a material to the flow of electrons categorizes it as an insulator, a semiconductor, and a conductor. A new class of materials was discovered in the year 1911 by a Dutch physicist, which offered no resistance to the flow of electrons, that is, they possessed nearly infinite conductivity. This phenomenon, when it was discovered, occurred at very low temperatures. The specimen also showed some unique magnetic properties, for example, as the electrical resistance reaches zero, there is a complete expulsion of magnetic lines of force from the specimen. This unique phenomenon of zero electrical resistance and perfect diamagnetism of a certain class of materials (at specific temperatures) was termed as 'superconductivity'. This class of materials possesses certain unique properties, which will be discussed in this chapter.

Superconducting behaviour could not be explained on the basis of perfect conductors of classical physics, but was explained on the basis of quantum physics.

21.1.1 Basic Definitions

The superconducting behaviour of certain specific materials can be understood by understanding the following terminology. The basic definitions of these terms are as follows:

Transition temperature or critical temperature Generally, a low-temperature condition that leads to the formation of electron pairs such that a conductor offers zero resistance and magnetic field expulsion is referred to as the transition temperature (or critical temperature). The temperature at which electrical resistivity of a metal drops to zero is called the critical or transition temperature and is denoted by T_c. Figure 21.1 shows the various graphs explaining the concept of critical temperature and change in resistance values with temperature.

The critical temperature is inversely proportional to the square root of the atomic mass of the metal being considered.

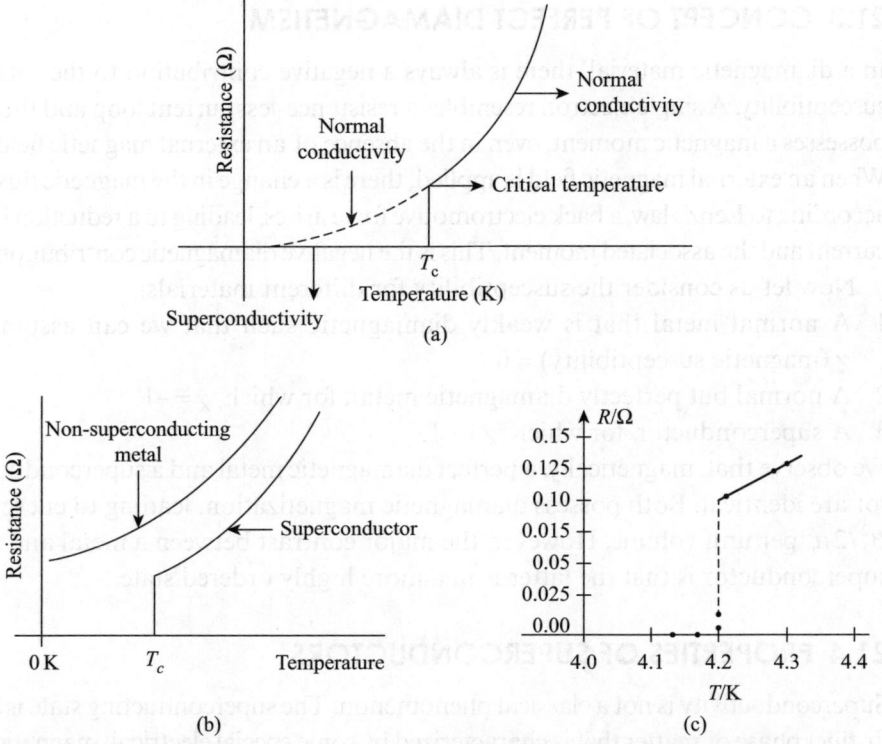

Fig. 21.1 Phenomenon of superconductivity (a) Graph of temperature vs resistivity showing critical temperature, normal conductivity, and superconductivity (b) Graph representing non-superconducting material and superconductor (c) Variation of resistance vs absolute temperature for mercury

Phonons Quantum lattice vibrations of the positive ions in a crystal are called phonons.

Cooper pair It refers to the electron–phonon–electron interaction at the critical or transition temperature. Such a pair of electrons is referred to as a Cooper pair.

21.2 PRINCIPLE OF SUPERCONDUCTIVITY

In the class of materials called 'superconductors', below the transition temperature T_c, there is no resistance to the flow of current-carrying electron pairs, called Cooper pairs. In a normal conductor, the flow of electrons through the crystal lattice is obstructed by the lattice symmetry or the phonons–lattice vibrations. In superconductors, Cooper pairs create an energy state that is different from the energy state of normal electrons. All Cooper pairs act as a single unit, as the space occupied by one pair contains many other pairs. As a result, the phonon–lattice vibrations or the lattice symmetry do not have the requisite amount of thermal energy to obstruct or slow down the motion of the Cooper pairs in the lattice. Thus, the flow of Cooper pairs in a lattice is without any resistance and hence persistent current is there in a superconducting material.

21.3 CONCEPT OF PERFECT DIAMAGNETISM

In a diamagnetic material, there is always a negative contribution to the total susceptibility. A single electron resembles a resistance-less current loop and thus possesses a magnetic moment, even in the absence of an external magnetic field. When an external magnetic field is applied, there is a change in the magnetic flux; according to Lenz's law, a back electromotive force arises, leading to a reduction in current and the associated moment. This is the negative diamagnetic contribution.

Now let us consider the susceptibility for different materials:

1. A normal metal that is weakly diamagnetic such that we can assume χ (magnetic susceptibility) $= 0$.
2. A normal but perfectly diamagnetic metal, for which $\chi = -1$.
3. A superconductor, for which $\chi = -1$.

We observe that, magnetically, a perfect diamagnetic metal and a superconductor are identical. Both possess diamagnetic magnetization, leading to energy $B_0^2/2\mu_0$ per unit volume. However, the major contrast between a metal and a superconductor is that the latter is in a more highly ordered state.

21.4 PROPERTIES OF SUPERCONDUCTORS

Superconductivity is not a classical phenomenon. The superconducting state is a distinct phase of matter that is characterized by some special electrical, magnetic, thermodynamic, isotopic, and energy state properties, which are as follows:

1. The electrical current in a superconducting sample persists for an infinitely long time.

2. As the current in the superconducting sample increases beyond a certain critical value I_c, the superconducting sample attains the properties of a normal conductor. This critical current produces a critical magnetic field H_c at the surface of the specimen. The superconducting state is broken not necessarily by the application of an external magnetic field, but by the generation of an internal magnetic field, due to the critical current at the surface of the superconducting wire. The critical current and the critical magnetic field are related through the following simple relation, which, after the name of Silsbee, is called the Silsbee's rule:

$$I_c(T) = 2\pi r H_c(T) \qquad (21.1)$$

3. The magnetic field gradually becomes zero inside the superconducting sample. This phenomenon of magnetic field, becoming zero inside the superconducting sample was observed by Meissner, and hence is called the Meissner effect. As the magnetic field attains a value greater than the critical magnetic field H_C, the superconducting properties of the specimen are lost and it becomes a normal conductor.

4. It is observed that the transition temperature T_c increases inversely with the atomic mass and high power of the atomic volume. This is called the isotope effect (discussed in Section 21.4.2).

5. The specific heat of the material shows an abrupt change at $T = T_c$, jumping to a large value for $T < T_c$. The heat capacity in the superconducting state varies with temperature in an exponential manner, that is, it is of the form $e^{-\Delta/k_B T}$. This indicates that there is thermal excitation across a gap in energy (Fig. 21.2).

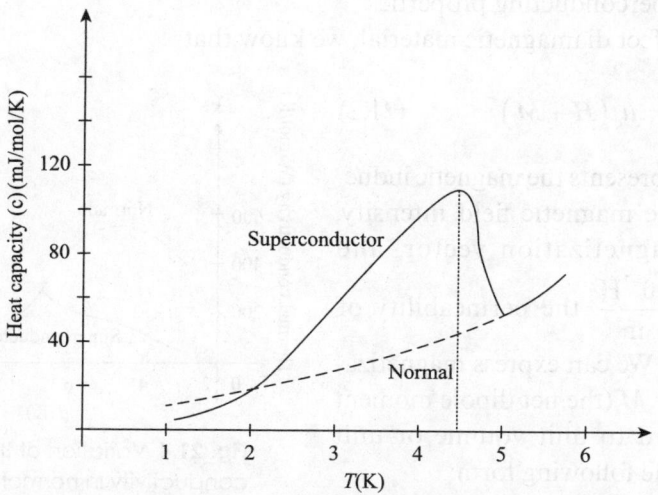

Fig. 21.2 Variation of heat capacity as normal metal goes through superconducting transition (Corak, et al. 1956)

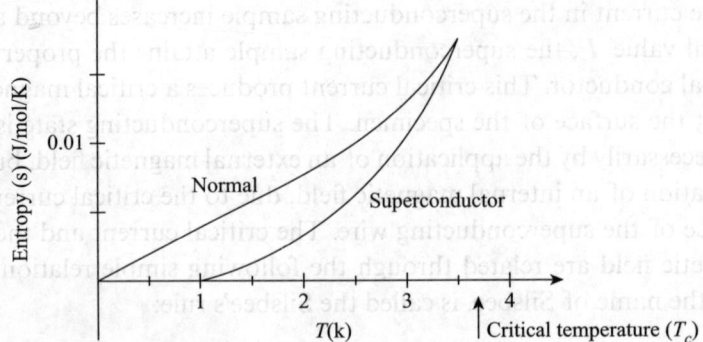

Fig. 21.3 Superconducting state has lower entropy than normal state (shown for Sn here) (Keesom and Van Laer 1938)

6. The superconducting state has a lower entropy than the normal state (refer to Fig 21.3).
7. The variation of thermal conductivity for a normal conductor is more than that of a superconductor (Fig. 21.4).

Some of the important properties of a superconductor are discussed in detail in the following sections.

21.4.1 Meissner Effect

It was observed by Meissner and Ochsenfield that in a superconducting sample, as the transition temperature T_c sets in and the surface currents appear, the magnetic field lines are completely expelled from the inside of the specimen as shown in Figs 21.5–21.7. The Meissner Effect is not observed in an ordinary conductor whose electrical resistance is nearly zero, but only in materials that possess superconducting properties.

In a perfect diamagnetic material, we know that

$$B = \mu_0 \left(H + M \right) \qquad (21.2)$$

where B represents the magnetic induction, H the magnetic field intensity, M the magnetization vector, and $\mu_0 = 4\pi \times \dfrac{10^{-7} \text{H}}{\text{m}}$ the permeability of free space. We can express magnetization vector M (the net dipole moment with regard to unit volume or unit mass) in the following form:

$$M = \chi H \qquad (21.3)$$

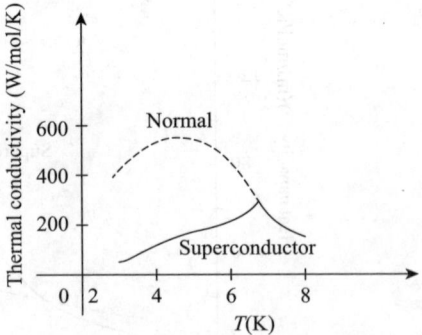

Fig. 21.4 Variation of thermal conductivity in normal and superconducting states for lead (Olsen 1952)

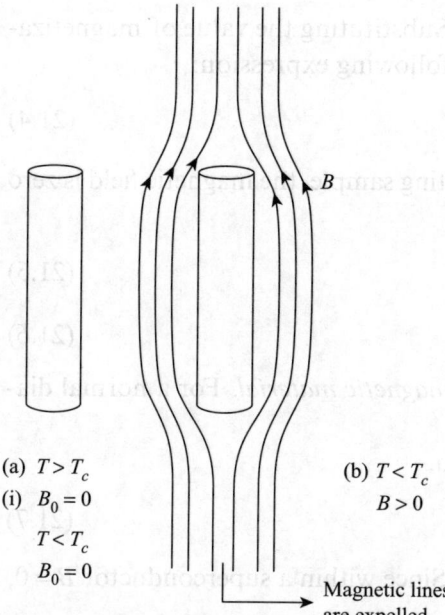

(a) $T > T_c$

(i) $B_0 = 0$

$T < T_c$

$B_0 = 0$

(b) $T < T_c$

$B > 0$

Magnetic lines are expelled

Fig. 21.5 For perfect conductor, magnetic field remains zero before and after application of magnetic field. In this case, magnetic field is applied after conductor has achieved its state of perfect conductivity

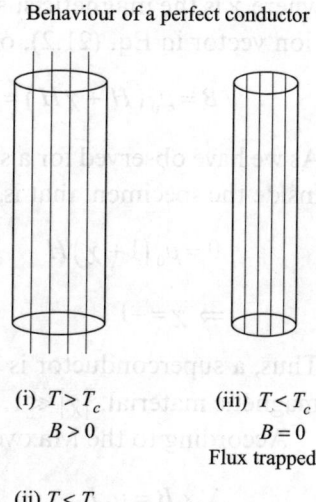

Behaviour of a perfect conductor

(i) $T > T_c$
$B > 0$

(ii) $T < T_c$
$B > 0$

(iii) $T < T_c$
$B = 0$
Flux trapped

Fig. 21.6 For perfect conductor, if magnetic field is applied before it achieves state of perfect conductivity, magnetic lines of force remain there even when source of magnetic field is removed

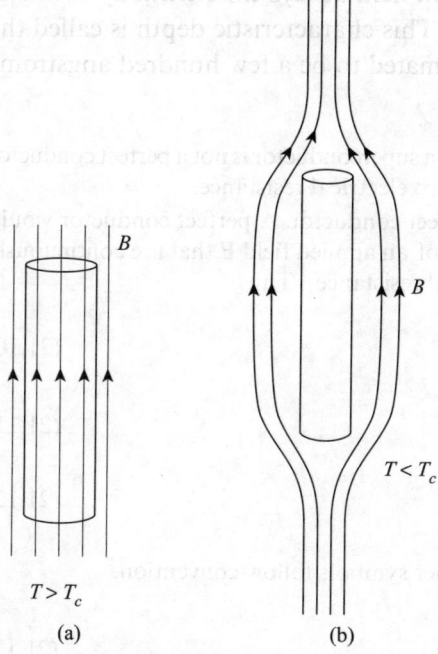

$T > T_c$

(a)

$T < T_c$

(b)

Fig. 21.7 Behaviour of magnetic field in a conductor and a super conductor (a) Applied magnetic field penetrates specimen when temperature of conductor is above critical temperature (b) Magnetic field lines due to application of magnetic field are expelled from conductor, as it acquires properties of superconductor

where χ is the magnetic susceptibility. Substituting the value of magnetization vector in Eq. (21.2), one gets the following expression:

$$B = \mu_0 \left(H + \chi H \right) = \mu_0 \left(1 + \chi \right) H \qquad (21.4)$$

As we have observed for a superconducting sample, the magnetic field is zero inside the specimen, that is,

$$0 = \mu_0 \left(1 + \chi \right) H \qquad (21.5)$$

$$\Rightarrow \chi = -1 \qquad (21.6)$$

Thus, a superconductor is an *ideal diamagnetic material*. For a normal diamagnetic material, $|\chi| \ll 1$.

According to the Maxwell's equation,

$$\nabla \times B = \mu_0 J \qquad (21.7)$$

where J represents the current density. Since within a superconductor $B = 0$,

$$0 = \mu_0 J \qquad (21.8)$$

$$\Rightarrow J = 0 \qquad (21.9)$$

Thus, the current density $J = 0$ in the interior of a superconductor. Surface currents, however, can exist in a superconductor.

In a superconducting pure metal, the magnetic flux is expelled from the sample, irrespective of its geometry and the sequence in which the magnetic field is applied. In reality, the induction field decays exponentially with distance from the surface of the sample. This characteristic depth is called the *penetration depth* and is generally estimated to be a few hundred angstroms $(10^{-7}\,\text{m})$.

Example 21.1 Show mathematically that a superconductor is not a perfect conductor but a perfect diamagnetic material with zero electrical resistance.

Solution I. Let us first look into the perfect conductor. A perfect conductor would have certain electrons under the influence of an applied field E that are continuously accelerated (under the absence of electrical resistance). Thus,

$$\frac{dv}{dt} = \frac{-eE}{m} \qquad (21.10)$$

$$J = -nev \qquad (21.11)$$

$$\Rightarrow E = \frac{m}{ne^2} \frac{dJ}{dt} \qquad (21.12)$$

where n is the density of electrons. All other symbols follow convention. From the Maxwell's equation,

$$\nabla \times B = \mu J \qquad (21.13)$$

$$E = \frac{m}{ne^2}\frac{1}{\mu}\frac{d}{dt}(\nabla \times B) = \frac{m}{ne^2}\frac{1}{\mu}\left(\nabla \times \frac{dB}{dt}\right) \tag{21.14}$$

$$\nabla \times E = \frac{m}{ne^2}\frac{1}{\mu}\nabla \times \left(\nabla \times \frac{dB}{dt}\right) \tag{21.15}$$

Thus, we obtain that

$$\nabla^2 \frac{dB}{dt} = \frac{\mu ne^2}{m}\frac{dB}{dt} = \frac{\omega_s^2}{c^2}\frac{dB}{dt} = \frac{1}{\lambda_l^2}\frac{dB}{dt} \tag{21.16}$$

Now, we observe that a particular solution of the above equation would be as follows:

$$\frac{dB}{dt} = 0 \tag{21.17}$$

As we were considering a perfect conductor, we observe that the state of the magnetic field is time dependent, that is, if the magnetic field is switched on after the material is in a perfectly conducting state, it cannot penetrate beyond the thickness of the order of λ_L; and if the magnetic field is applied before the material is in the perfectly conducting state, then the field remains there, even though the external field is removed.

II. In the case of a superconductor, as observed by Meissner and Ochsenfield, the magnetic flux is expelled out of the superconducting sample, irrespective of its geometry and the time at which the magnetic field is applied. This leads to the following mathematical equation:

$$B = 0 \tag{21.18}$$

$$\Rightarrow \chi = -1 \text{ (refer to the text for detailed mathematics)} \tag{21.19}$$

Hence, one can conclude that a superconductor is not a perfect conductor but a perfect diamagnetic material with zero electrical resistance.

Note: For a specimen to attain superconducting state, one has to establish independently, the two properties of zero electrical resistance and flux expulsion or Meissner effect.

21.4.2 Isotope Effect

In the year 1950, Reynolds discovered that the transition temperature is proportional to $M^{-1/2}$, where M is an isotopic mass of the ion of the superconducting metal. This is called the *isotope effect*. The experimental results fit the following relation:

$$M^\alpha T_c = \text{constant} \tag{21.20}$$

where α is about 0.5 for most metals and is also called the *isotope shift exponent*. If one isotope is substituted with another, this leads to a change in the mass of the nuclei, but does not produce any change in the electronic properties of the solid. This indicates that the transition to superconductivity, depends on the nuclear mass or the mass of the lattice ions. This helped Frohlich to propose that the electron–phonon interaction plays an important role in transition to the superconducting phase.

Example 21.2 The critical temperature T_c for Hg with an isotopic mass of 199.5 is 4.195 K. What will be its critical temperature, when its isotopic mass is increased to 203.4?

Solution We have that $T_{c1} = 4.195\,\text{K}; M_1 = 199.5; M_2 = 203.4; T_{c2} = ?$

$$T_{c1}M_1^{1/2} = T_{c2}M_2^{1/2}$$

$$4.195 \times (199.5)^{1/2} = T_{c2} \times (203.4)^{1/2} \Rightarrow 4.195 \times \frac{14.125}{14.26} = 4.146\,\text{K}$$

Thus, 4.146 K will be the increase in the temperature for the given situation.

21.4.3 Energy Gap in Superconductors

For a superconductor at 0 K, the width of the energy gap $E_g(0)$ is proportional to the critical temperature and the expression is as follows:

$$E_g(0) = 2\Delta = 3.53k_B T_c \tag{21.21}$$

where Δ is referred to as the energy gap parameter and is strongly dependent on the temperature. The mathematical relation for the energy gap parameter is as follows:

$$\Delta(T) = 1.74\Delta(0)\left(1 - \frac{T}{T_c}\right)^{1/2} \tag{21.22}$$

The energy gap parameter is usually of the order $10^{-3}\,\text{eV}$.

The electrons constituting the Cooper pair have a lower energy than two unpaired electrons. Figure 21.8 shows a schematic representation of the

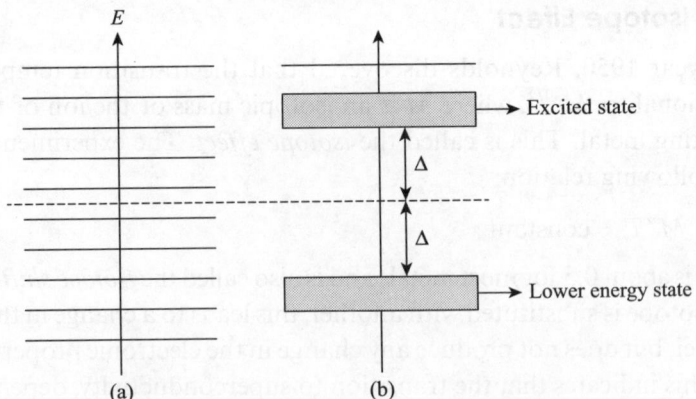

Fig. 21.8 Representation of energy spectrum (a) Energy bands in normal state of conducting material (b) Energy bands for superconducting state showing gap of 2Δ between lower energy state and excited state

energy spectrum of electrons in the normal [Fig 21.8(a)] and in the super-conducting state [Fig. 21.8(b)].

As there is a change in temperature, there is formation of a Cooper pair, due to which an energy gap is created. The Cooper pair occupies a lower energy state, and the existence of the band gap prevents the Cooper pair from breaking up. Energy 2Δ is required to dissociate a Cooper pair.

The existence of the energy gap has been verified experimentally. Suppose a superconductor maintained at $0\,\text{K}$ is exposed to a microwave radiation. Absorption of microwaves is observed only when the energy quantum of the incident radiation becomes equal to or greater than 2Δ.

Example 21.3 Evaluate the energy gap at absolute temperature for aluminium, if its critical temperature is $1.196\,\text{K}$.

Solution The calculations are as follows:

$$\text{Energy gap}\, E_g(0) = 3.53 k_B T_c = 3.53 \times 8.617 \times 10^{-5}\,\frac{\text{eV}}{\text{K}} \times 1.196\,\text{K} = 3.64 \times 10^{-4}\,\text{eV}$$

Example 21.4 The actual energy gap at $0\,\text{K}$ in lead is $2.73 \times 10^{-3}\,\text{eV}$.

(a) What is the prediction of the Bardeen–Cooper–Schrieffer (BCS) theory for this energy gap?

(b) Radiation of what minimum frequency can break apart Cooper pairs in lead at $0\,\text{K}$? In what part of the electromagnetic spectrum does such a radiation occur?

Solution The critical temperature for lead $T_c = 7.175\,\text{K}$

$$E_g(0) = \frac{3.53 \times 8.617 \times 10^{-5}\,\text{eV}}{\text{K}} \times 7.175\,\text{K} = 2.19 \times 10^{-3}\,\text{eV}$$

(a) According to the predictions of the BCS theory, at temperatures above $0\,\text{K}$, the Cooper pairs start breaking up. At the critical temperature T_C, there is no energy gap, that is, there are no Cooper pairs, and hence no superconductivity. Thus, as we move onwards from $0\,\text{K}$ towards the transition temperature, the Cooper pairs start interacting with the lone electrons that are present, and in turn reduce the energy gap.

(b) $h\vartheta \geq E_g$

$$\vartheta \approx \left(2.73 \times 10^{-3}\,\text{eV}\right)/(2.4 \times 10^{-16}\,\text{eV.s}) \approx 10^{13}$$

The frequency of the microwave is of the order of $10^{10} - 10^{13}\,\text{Hz}$. Hence, the microwave radiation should be capable of breaking the Cooper pair.

21.4.4 Coherence Length

Coherence length (ξ) is defined as the maximum distance up to which a Cooper pair exists to produce superconductivity. The magnitude of this $\left|\xi^2\right|$ represents the coherence volume. The coherence length is related to the energy gap as follows:

$$\xi = \frac{h v_F}{2\pi.2\Delta} \tag{21.23}$$

where $v_F = \left(\sqrt{2E_F/m} \right)$ is the Fermi velocity. The ratio of the London penetration depth to the coherence length is as follows:

$$K = \lambda_l / \xi \tag{21.24}$$

The dimensionless number K distinguishes between the Type-1 $\left(K < 1/\sqrt{2} \right)$ and the Type-2 $\left(K > 1/\sqrt{2} \right)$ superconductors.

21.4.5 Flux Quantization

Let us consider 'a ring' sample of the superconductor and observe its behaviour in the presence of a magnetic field. From our knowledge of the Maxwell's equation and also knowing that the vector A is a potential (see the derivation of London's equations—Section 21.9), one can define the flux through a hollow superconducting ring as follows:

$$\oint A.dl = \varnothing \tag{21.25}$$

Thus, the magnetic flux $\varnothing$ enclosed by a hollow superconducting ring is given as follows:

$$\varnothing = n\left(\frac{h}{q} \right) \qquad \text{where } n = 1, 2, 3, \ldots \tag{21.26}$$

The flux trapped inside the superconducting sample is quantized by a multiple of (h/q). In the case of a superconducting sample, the charge is contributed by a Cooper pair; thus, $q = 2e$. Hence, the flux trapped inside the superconductor is quantized as follows:

$$(h/2e) = 2 \times 10^{-15} \, \text{Wb} = 2 \times 10^{-7} \, \text{Maxwell} \tag{21.27}$$

The flux quantum is called a *fluxon*. The existence of a *fluxon* was predicted in the year 1957 by A.A. Abrikosov and verified experimentally by Deaver and Fairbank in 1961.

The derivation of the quantized flux has been verified by the London's equations and can be easily derived theoretically using the concepts of surface current and momentum.

21.4.6 Critical Magnetic Field

The magnetic field plays a significant role in the observation of the phenomenon of superconductivity. Superconductivity disappears at magnetic fields greater than a critical field H_c. The temperature dependence of the critical field is given by *Tuyn's law*, which is expressed in the following form:

$$H_c = H_0 \left[1 - \left(\frac{T}{T_c} \right)^2 \right] \tag{21.28}$$

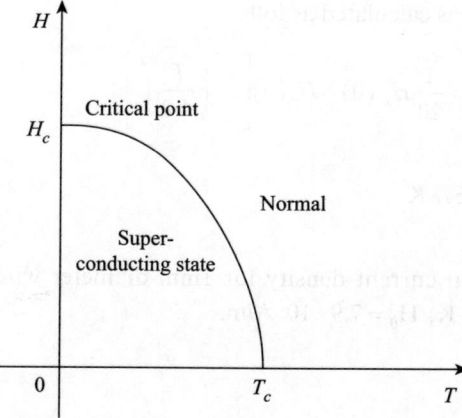

Fig. 21.9 Variation of magnetic field with temperature

where H_0 is the critical field at 0 K. (refer to Fig. 21.9). The following discussion on types of superconductors further highlight the role of magnetic field in the establishment of superconductivity.

Example 21.5 A superconducting sample has critical fields as 1.2×10^5 and 3.8×10^5 A/m for temperatures 2 and 12 K respectively. Calculate the transition temperature and critical fields at 0 and 3.2 K.

Solution From the formula for the critical field,

$$(H_c)_1 = H_0\left[1-\left(\frac{2}{T_c}\right)^2\right] = 1.2 \times 10^5 \, \text{A/m}$$

$$(H_c)_2 = H_0\left[1-\left(\frac{12}{T_c}\right)^2\right] = 3.8 \times 10^5 \, \text{A/m}$$

Dividing the two equations, $\dfrac{(H_c)_2}{(H_c)_1} = \dfrac{T_c^2 - (2)^2}{T_c^2 - (12)^2} = \dfrac{3.8}{1.2}$

Solving this equation, one gets the following result:

$$26 T_c^2 = 5424 \Rightarrow T_c = 14.44 \, \text{K}$$

Substituting the value of T_C in this equation, one gets the following result:

$$H_0\left[1-\left(\frac{2}{14.44}\right)^2\right] = 1.2 \times 10^5$$

$$H_0 = 1.224 \times 10^5 \, \text{A/m}$$

$$(H_C)_{3.2} = H_0\left[1-\left(\frac{3.2}{14.44}\right)^2\right] = 1.224 \times 10^5 \frac{\text{A}}{\text{m}} \times 0.95 = 1.163 \times 10^5 \, \text{A/m}$$

Example 21.6 Calculate the transition temperature for lead (Pb) if the critical magnetic field is $\dfrac{1}{20}$ th of that at 0 K if $T_c = 4.8$ K.

Solution   The transition temperature is calculated as follows:

$$H_C(T) = H_c(0)\left[1 - \left(\frac{T}{T_c}\right)^2\right] \Rightarrow \frac{1}{20}H_C(0) = H_c(0)\left[1 - \left(\frac{T}{4.8}\right)^2\right]$$

$$0.05 = \left[1 - \left(\frac{T}{4.8}\right)^2\right] \Rightarrow T = 4.678\,\text{K}$$

Example 21.7   Calculate the critical current density for 1mm diameter wire of aluminium at 1.0 K, where $T_c = 1.196$ K; $H_0 = 7.9 \times 10^3$ A/m.

Solution   We know that

$$H_c = H_0\left[1 - \left(\frac{T}{T_c}\right)^2\right]$$

We have that $H_0 = \dfrac{7.9 \times 10^3\,\text{A}}{\text{m}}$; $T_c = 1.196\,\text{K}$; $T = 1.0\,\text{K}$

$$H_c = \frac{7.9 \times 10^3\,\text{A/m}}{\left[1 - \left(\dfrac{1.0}{1.196}\right)^2\right]} = 2.38 \times 10^3\,\text{A/m}$$

$$I_c = 2\pi r H_C = 2 \times \frac{22}{7} \times 0.5 \times 10^{-3}\,\text{m} \times \frac{2.38 \times 10^3\,\text{A}}{\text{m}} = 7.48\,\text{A}$$

$$\text{Critical current density} = j_C = \frac{I_C}{A} = \frac{7 \times 7.48}{22 \times (0.5 \times 10^{-3})^2} = 9.52 \times 10^6\,\text{A/m}^2$$

Example 21.8   The penetration depth of mercury (Hg) at 3.5 K is about 750 Å. Estimate the values of penetration depth and number density as T tends to zero $(T_c = 4.12\,\text{K})$.

Solution   The calculations are as follows:

$$\lambda(T) = \frac{\lambda(0)}{\left[1 - \left(\dfrac{T}{T_C}\right)^4\right]^{1/2}}$$

Thus, we have $\lambda(3.5) = 750\,\text{Å}$; $T = 3.5\,\text{K}$; $T_c = 4.12\,\text{K}$

$$750\left(1 - \frac{3.5^4}{4.12^4}\right)^{1/2} = \lambda(0) = 519\,\text{Å}$$

$$\lambda(0) = \sqrt{\frac{mc^2}{4\pi n e^2}} \Rightarrow n = \frac{m}{\mu_0[\lambda(0)]^2(e)^2}$$

$$n = \frac{9.1 \times 10^{-31}}{4\pi \times 10^{-7} \times (519 \times 10^{-10})^2 \times (1.6 \times 10^{-19})^2} = 10^{28}\,\text{m}^3$$

Example 21.9 Calculate the transition temperature for lead (Pb) if the critical magnetic field is $\dfrac{1}{20}$ of that at $0\,\mathrm{K}$ if $T_c = 4.8\,\mathrm{K}$.

Solution The transition temperature for lead is calculated as follows:

$$H_C(T) = H_c(0)\left[1 - \left(\frac{T}{T_c}\right)^2\right] \Rightarrow \frac{1}{20} H_C(0) = H_c(0)\left[1 - \left(\frac{T}{4.8}\right)^2\right]$$

$$0.05 = \left[1 - \left(\frac{T}{4.8}\right)^2\right] \Rightarrow T = 4.678 \ \mathrm{K}$$

21.5 JOSEPHSON TUNNELLING

The word 'tunnelling' came into use when, with the advent of quantum mechanics, it was observed that electrons possess the unique property of sneaking through nano-sized barriers to contribute to the minority carriers in semiconductors. The probability of tunnelling is small; of the order 10^{-10}. The following question came to the mind of all scientists on observing and understanding superconductors: Do the Cooper pairs have this unique property of tunnelling? Josephson explored this probability of the newly-found cohesive pair of electrons, that is, the Cooper pair, to tunnel through, when two superconductors are joined together by a thin layer of oxide. This also led to many applications of superconductors.

Two superconductors are joined through a nano-dimensional insulating layer, forming coupled oscillators or coupled circuits. Josephson tunnelling can be explained in both DC and AC effects (refer to Fig. 21.10).

21.5.1 DC Effect

In the case of a conductor, the Fermi energy level marks the level for the Cooper pairs in a superconductor. As two superconductors are joined through

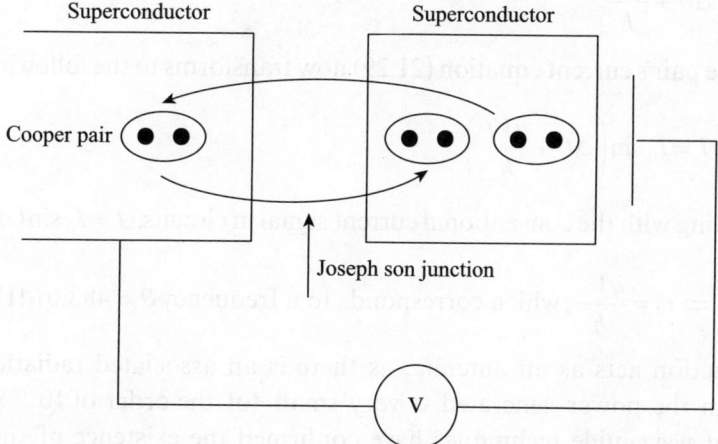

Fig. 21.10 Cooper pairs can tunnel through thin barriers (nano-size)

a nano-dimension, the Fermi level gets adjusted to have the same value for the entire system. This helps the flow of condensates from both the super-conductors to the thin insulating nano-sized oxide layer. Concentration of the condensates in the intermediate layer is much smaller than that on the superconductors at the sides. The continuous Fermi energy level helps in the free flow of condensates from one junction to another, without any transfer of energy. Thus, a current flows in the junction without any voltage difference. There is a distinct phase difference $\Delta\theta$ between the two superconductors, such that, the DC current flowing through this junction is given as follows:

$$I = I_0 \sin (\Delta\theta) \qquad (21.29)$$

The current I_0 is a characteristic of the given junction related to the tunnelling conductance between the junction metals in their normal states.

21.5.2 AC Effect

The Josephson AC effect comes into play, when a current through the super-conducting junction exceeds the critical value I_0. Owing to this, the super-conducting nature of the nano-sized oxide layer ceases to exist; as a result, normal electrons also start exhibiting themselves, in addition to the already existing condensates (or Cooper pairs). The presence of normal electrons causes a potential difference V to appear across the junction, generating an electric field that accelerates the motion of the Cooper pairs. The Cooper pairs acquire energy qV on crossing the junction, such that, their phase varies with time according to the following relation: $\hbar\dfrac{d\theta}{dt} = qV$

We know that Cooper pairs possess an initial phase difference of $\Delta\theta$, which changes to the following expression on the generation of a voltage V:

$$\Delta\theta + \frac{qVt}{\hbar}$$

Thus, the pair's current equation (21.29) now transforms to the following form:

$$I = I_0 \sin\left(\Delta\theta + \frac{qVt}{\hbar}\right)$$

Comparing with the conventional current signal in circuits, $I = I_0 \sin(\Delta\theta + \omega t)$

$$\Rightarrow \omega = \frac{qV}{\hbar}, \text{which corresponds to a frequency } \vartheta = 483.6\,\text{MHz}/\mu\text{V}^{-1}.$$

The junction acts as an antenna, as there is an associated radiation field, although the power generated is very small (of the order of $10^{-10}\,\text{W}$). The resonant waveguide techniques have confirmed the existence of these very small, yet significant radiation fields in coupled superconductors (Fig. 21.11).

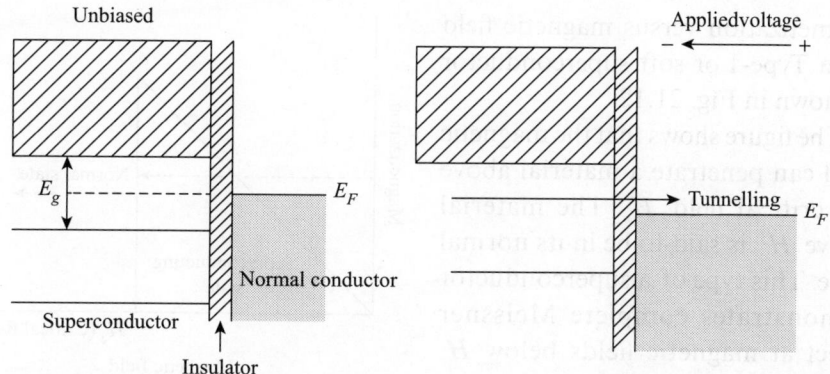

Fig. 21.11 Energy-band diagram using concept of Josephson junction to measure energy gap in superconductor

Example 21.10 If the critical magnetic field for a superconductor is $\dfrac{7.0 \times 10^3 \text{ A}}{\text{m}}$, estimate the critical current passing through a thin superconducting wire of radius 0.25 mm.

Solution The current is calculated as follows:

$$I_c = 2\pi r H_c = 2 \times \frac{22}{7} \times 0.25 \times 10^{-3} \text{ m} \times 7.0 \times 10^3 \text{ A/m} = 11 \text{ A}$$

Example 21.11 If the critical current passing through a 0.4 mm superconducting wire is 20 A, estimate the critical magnetic field for this superconductor.

Solution The critical magnetic field for the superconductor is

$$I_c = 2\pi r H_c \Rightarrow 20 \text{ A} = 2 \times \frac{22}{7} \times 0.4 \times 10^{-3} \times H_c$$

$$H_c = 20 A \times \frac{7}{22} \times \frac{1}{2 \times 0.4 \times 10^{-3} \text{ m}} = 7.95 \times 10^3 \text{ A / m}$$

21.6 TYPES OF SUPERCONDUCTORS

Superconductors can be classified into two categories depending on the properties of magnetization, under the influence of an external applied magnetic field. Superconductors can either be of Type 1 or of Type 2. Type-1 superconductors are called *soft superconductors* and Type-2 superconductors *hard superconductors*. We will be discussing these two types of superconductors in this section.

21.6.1 Soft Superconductors or Type-1 Superconductors

In soft superconductors, the magnetic field gets totally expelled from the interior of the material, below a certain critical magnetic field H_c. At the critical magnetic field, there is an abrupt loss of superconductivity. A plot of magnetization versus magnetic field for a Type-1 or soft superconductor is shown in Fig. 21.12.

The figure shows that the magnetic field can penetrate a material above the critical field H_c. The material above H_c is said to be in its normal state. This type of a superconductor demonstrates complete Meissner effect at magnetic fields below H_c where, it becomes an ideal diamagnetic material. Lead, tin, and mercury are common examples of soft superconductors. The transition to the superconducting state at H_c is reversible. Most superconducting metals exhibit Type-1 superconductivity at very low H_c values (around 10^{-1} T).

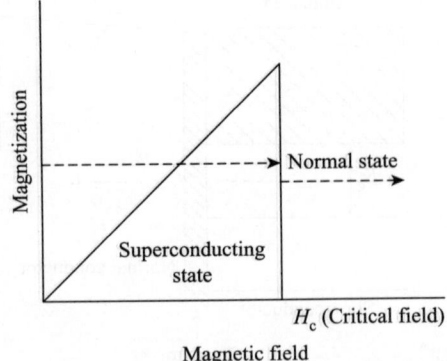

Fig. 21.12 Magnetization vs magnetic field for Type-1 superconductor

21.6.2 Hard Superconductors

Hard superconductors show two critical magnetic fields: H_{c1} and H_{c2}. Typical magnetization curves for hard superconductors are shown in Figs 21.13(a)–(c).

Below the first critical field H_{c1}, the material is superconducting, and hence exhibits ideal diamagnetic behaviour. Magnetic flux is expelled completely from within the material. Between the two critical fields H_{c1} and H_{c2} the material exists in a mixed state. The material still demonstrates superconducting property, but exclusion of flux from within the specimen is partial. The normal non-superconducting state is reached at the critical field H_{c2}.

Alloys or transition metals with high electrical resistivity in the normal state generally demonstrate Type-2 superconductivity. The values of H_{c2}

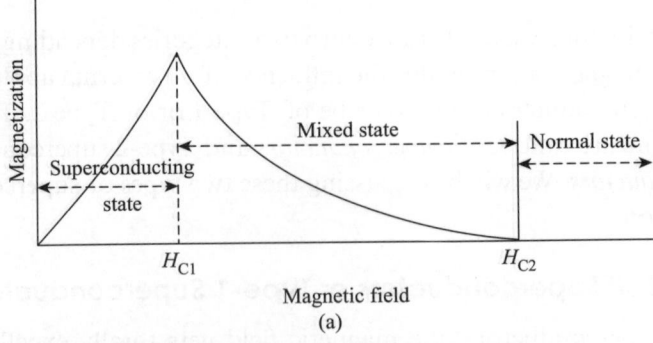

Fig. 21.13 Typical magnetization curves for a hard superconductor (a) Magnetization curve for hard superconductor (*Contd*)

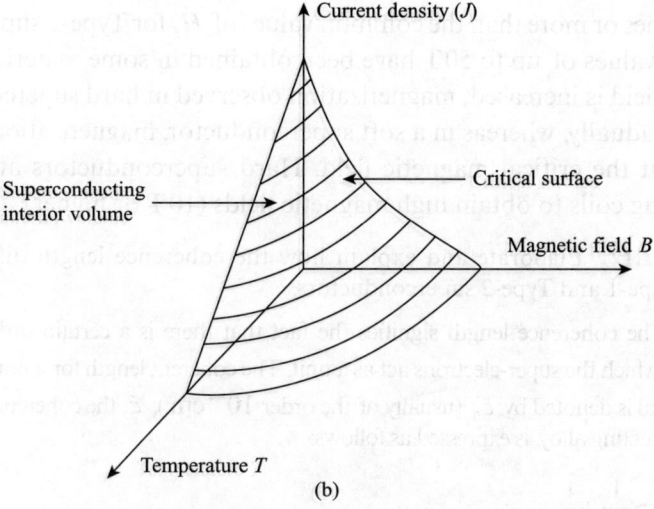

(b)

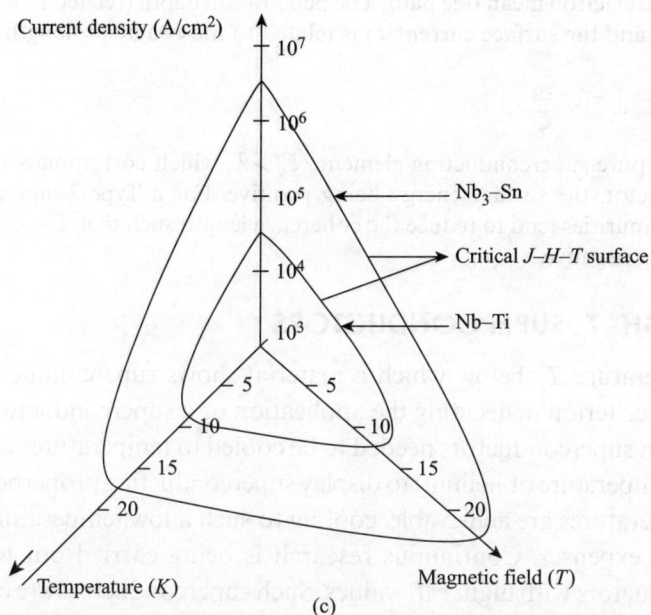

(c)

Fig. 21.13 (*Contd*) (b) Three-dimensional depiction of applied magnetic field (*B*), temperature (*T*), and current density (*J*), which should be maintained below critical surface to obtain superconducting sample (for Type-2 superconductor) (c) Quantitative diagram for Type-2 superconductor Nb–Ti and Nb_3–Sn: (i) critical magnetic fields: Nb–Ti—13 T; Nb_3–Sn—27 T; (ii) critical temperatures: Nb–Ti—10 K; Nb_3–Sn—18 K; (iii) current density: >10^5 A/cm^2 (for Copper wire: 10^3 A/cm^2) Picture Courtesy by Lyndon Evans, The Large Hadron Collider: Technology Marvel, CERN and EPFL Press, 2009

are 100 times or more than the common values of H_c for Type-1 superconductors. H_{c2} values of up to 50 T have been obtained in some materials. As the magnetic field is increased, magnetization observed in hard superconductors reduces gradually, whereas in a soft superconductor, magnetization vanishes abruptly at the critical magnetic field. Hard superconductors are used as magnetizing coils to obtain high magnetic fields (10 T or higher).

Example 21.12 Elaborate and explain how the coherence length differentiates between Type-1 and Type-2 superconductors.

Solution The coherence length signifies the fact that there is a certain order or coherence, ξ, for which the super-electrons act as a unit. The coherent length for a pure superconducting metal is denoted by ξ_0 (usually of the order 10^{-4} cm); ξ, the coherence length for a superconducting alloy, is expressed as follows:

$$\frac{1}{\xi} \approx \frac{1}{\xi_0} + \frac{1}{l}$$

where l is the electron mean free path. The penetration depth (related to the magnetic potential A and the surface current J) is related to the coherence length as follows:

$$\lambda_{\text{exptal}} = \lambda_L \frac{\xi_0}{\xi}$$

Thus, for a pure superconducting element, $\xi_0 > \lambda$, which corresponds to a Type-1 superconductor, the surface energy being positive. For a Type-2 superconducting alloy, the impurities tend to reduce the coherence length such that $\xi < \lambda$, the surface energy being negative.

21.7 HIGH- T_C SUPERCONDUCTORS

The temperature T_c below which a material shows superconductivity is an important criterion in deciding the application of a superconductor. The earliest known superconductors needed to be cooled to temperatures close to 4 K (boiling temperature of helium) to display superconducting properties. Though such temperatures are achievable, cooling to such a low temperature involves enormous expenses. Continuous research is being carried out to discover superconductors with higher T_c values. Such superconductors are collectively referred to as high- T_c superconductors (HTSCs). In 1986, K.A. Muller and J.G. Bednorz reported superconducting properties of lanthanum, barium, and copper oxides at temperatures higher than 30 K. This was followed by the synthesis of yttrium compounds by C.W. Chu in 1987. These yttrium compounds acquired the superconducting state with a T_c close to 90 K. Several HTSCs have since been reported with $T_c > 30$ K. The discovery of these high-temperature superconductors rekindled the interest in superconductivity from the application viewpoint. Several groups in the USA, Japan, USSR, India, and other countries are actively pursuing this field with renewed interest. Table 21.1 lists some superconductors in the family of layered cuprate HTSCs.

Table 21.1 Some layered cuprate HTSCs

Material	$T_c(K)$
$La_{2-x}M_xCuO_4$ (M = Ca, Sr, Ba)	20–40
$LnBa_2Cu_3O_7$	90
$LnBa_2Cu_4O_8$	90
$Bi_2(CaSr)_{n+1}Cu_nO_{2n+4}$ (n = 1–3)	90–110
$Tl_2Ca_{n-1}Cu_nO_{2n+4}$ (n = 1–4)	80–125
$TlCa_{n-1}Ba_2Cu_nO_{2n+3}$ (n = 2, 3)	90–117
$Tl_{1-x}Pb_x(Ca, Sr)_{n+1}Cu_nO_{6+x}$	70–90
$TlSr_{n+1-x}Ln_xCu_nO2_{n+2}$ (n = 1, 2)	40–90
$Pb_2Sr_2ACu_3O_8$	40–70
(A = Ln or Ln + Sr or Ca)	

The highest T_c recorded till now is 138 K, which has been obtained for a thallium-doped, mercuric-cuprate compound comprising elements such as mercury, thallium, barium, calcium, copper, and oxygen. The T_c of this ceramic superconductor was confirmed by Dr Ron Goldfarb at the National Institute of Standards and Technology—Colorado in February 1994. Under extreme pressure, its T_c can be coaxed up even higher—by approximately 25–30° more at 300,000 atm.

Note: Superconductors.ORG herein reports that the 30C superconductor discovered in December 2012 has been successfully reformulated to produce a high T_c of above 35°C (95°F, 308 K). This was accomplished by a simple substitution of tetravalent silicon in the magnesium atomic sites. The chemical formula thus becomes $Tl_5Pb_2Ba_2Si_{2.5}Cu_{8.5}O_{17+}$. This is the third material discovered with a critical transition temperature (T_c) above room temperature.

21.8 BCS THEORY

Bardeen, Cooper, and Schriffer in 1957 found that the electron–phonon interaction is a possible cause of formation of the superconducting state.

21.8.1 Electron–Phonon Interaction

The interaction of an electron with a lattice needs to be examined closely. Suppose an electron passes near an ion core; there is a mutual attraction between the electron and the ion core because of the Coulomb interaction, and as a result, the ion core is set into motion. Suppose that another electron passes through a nearby region; it is obvious that this electron will experience the effect of motion of the ion core. In effect, the ion motion provides the two electrons a means to interact with each other, despite their mutual Coulomb repulsion. Under very restricted circumstances (i.e., critical temperature), this interaction is attractive, and under even more restricted circumstances, this attraction exceeds their Coulomb repulsion. These circumstances are as follows:

1. Electrons entering into such an interaction have opposite momenta and opposite spin. Mathematically,

$$k_1 = -k_2 \qquad\qquad (21.30)$$

$$s_1 = -s_2 \qquad\qquad (21.31)$$

Such a pair of electrons is called the 'Cooper pair'. Such an attractive interaction between electrons is energetically favourable.

2. The paired stages are all situated within $k_B\theta_D$ of the Fermi energy, where θ_D is the Debye temperature and k_B is the Boltzmann constant.

3. The temperature is low enough, that is, $T \to 0$.

21.8.2 BCS Ground State

The two special features of the BCS ground state are as follows:

1. The total energy of the BCS state is lower than that of the Fermi state. The total energy of the BCS state comprises the kinetic energy and the attractive potential energy (whereas the energy of the Fermi state comprises only the kinetic energy.) The attractive kinetic energy acts to reduce the total energy of the BCS ground state. The consequence of this feature is that the BCS ground state is more stable than the Fermi state of the given material.

2. The one-particle state is occupied in pairs, that is, if the state with vector k has been occupied, then the state with vector $-k$ will also be occupied; similarly, if the state with vector $-k$ is vacant, then the state with vector k will also be vacant.

As understood by the BCS theory, superconductivity is due to the existence of an attractive force between the conduction electrons in a metal, which results in the formation of a bond between pairs of electrons. These pairs of bonded electrons are called Cooper pairs. Cooper pairs move through the lattice structure, without any scattering. Thus, these pairs do not encounter any resistance during their journey within the lattice. The negatively charged electrons, moving through a lattice of positive ions attract positively charged ions. This attraction distorts the lattice, giving rise to a wave of excess positive charge that attracts another electron of equal and opposite momentum and spin. These two electrons form a Cooper pair. Thus, Cooper pair formation is a result of electron–lattice–electron or electron–phonon–electron interaction, as shown in Fig. 21.14. During scattering, any change in the momentum of the electrons of a Cooper pair is accompanied

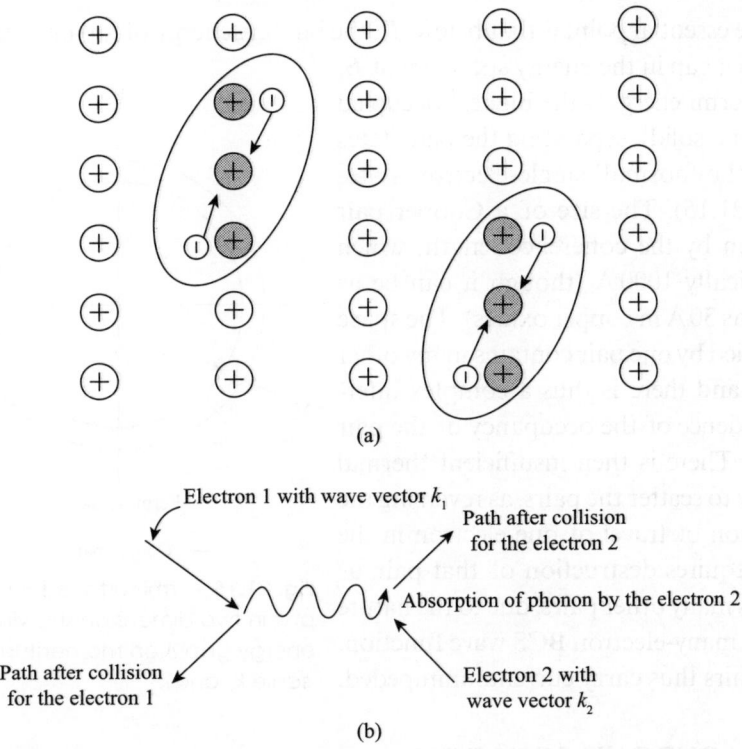

Fig. 21.14 Cooper pairs (a) Formation of Cooper pair in lattice
(b) Vector representation of formation of Cooper pair

by an equal and opposite change in momentum of the other electron of
the pair. A Cooper pair can, therefore, move through a material without
encountering any resistance. This makes it possible to achieve extremely
high conductivity in superconducting materials. The size of a Cooper pair
is given by the coherence length, which is typically $1000\,\text{Å}$ (though it can
be as small as $30\,\text{Å}$ in copper oxides).

Electrical resistance in metals arises because electrons propagating through
a solid are scattered due to deviation from perfect translational symmetry.
These are produced either by impurities (giving rise to a temperature independent contribution to the resistance) or by the phonon–lattice vibrations
in a solid.

In a superconductor below its transition temperature T_c, there is no resistance, because these scattering mechanisms are unable to impede the motion
of the current carriers. The current is carried in all known classes of superconductors by pairs of electrons known as Cooper pairs. The mechanism by
which two negatively charged electrons are bound together is still controversial
in 'modern' superconducting systems such as copper oxides or alkali metal
fullerides, but well understood in conventional superconductors such as aluminium, as explained by the mathematically complex BCS theory.

The essential point is that below T_c, the binding energy of a pair of electrons causes a gap in the energy spectrum at E_F (the Fermi energy—the highest occupied level in a solid), separating the pair states from the 'normal' single-electron states (Fig. 21.15). The size of a Cooper pair is given by the coherence length, which is typically $1000\,\text{Å}$ (though it can be as small as $30\,\text{Å}$ in copper oxides). The space occupied by one pair contains many other pairs, and there is thus a complex inter-dependence of the occupancy of the pair states. There is then insufficient thermal energy to scatter the pairs, as reversing the direction of travel of one electron in the pair requires destruction of that pair, as well as many other pairs, due to the nature of the many-electron BCS wave function. The pairs thus carry current, unimpeded.

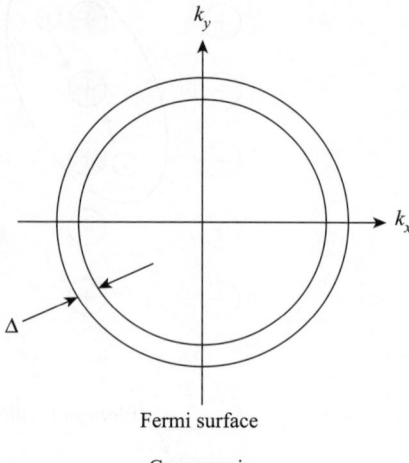

Fig. 21.15 Fermi surface for Cooper pair in two dimension showing energy gap Δ on momentum scale k_x and k_y

21.9 LONDON'S EQUATION

In laboratories, the Meissner effect was a special observation, in addition to zero electrical resistances in superconducting samples. The Meissner effect states that the entire magnetic flux is expelled out of the superconducting specimen at the transition temperature, but in fact, it is not so because some flux penetration has been observed in thin films of superconductors. This flux penetration means persistence of flux through a small infinite length along the surface of a superconductor. Thus, it is actually observed that the magnetic flux decays from a constant value at the surface to zero inside the specimen, somewhat gradually, but not abruptly. Hence, to account for this phenomenon of the magnetic flux weaning off gradually, a set of equations had to be derived. F. London and H. London successfully derived equations to effectively describe gradual decaying of magnetic flux in a given superconducting sample.

Let us assume that the density of electrons moving with velocity v_s in the presence of an electric field E and in the absence of any resistivity is n_S. The equation of motion is given as follows:

$$m\frac{dv_S}{dt} = F = -eE \tag{21.32}$$

The current density is defined as follows:

$$J_s = n_s e v_s \tag{21.33}$$

where n is the number of electrons per unit volume. Differentiating Eq. (21.33) with respect to time, we get the following relation:

$$\frac{dJ_s}{dt} = -n_s e \frac{dv_s}{dt} = -n_s e \left(\frac{eE}{m} \right) \tag{21.34}$$

$$\frac{dJ_s}{dt} = \frac{n_s e^2}{m} E \tag{21.35}$$

$$\nabla \times E = -\frac{dB}{dt} \tag{21.36}$$

$$\nabla \times E = -\mu_0 \frac{dH}{dt} \tag{21.37}$$

For a linear, isotropic, homogeneous medium, $B = \mu_0 H$. Thus, the curl of the electric field can be substituted in Eq. (21.38) from Eq. (21.37). Taking the curl of Eq. (21.35) one gets the following expression:

$$\nabla \times \frac{dJ_s}{dt} = \frac{n_s e^2}{m} (\nabla \times E) \tag{21.38}$$

$$\nabla \times \frac{dJ_s}{dt} = \frac{n_s e^2}{m} \left(-\mu_0 \frac{dH}{dt} \right) \tag{21.39}$$

$$\nabla \times \frac{dJ_s}{dt} = -\frac{\mu_0 n_s e^2}{m} \frac{dH}{dt} \tag{21.40}$$

Integrating this equation with time, one gets the following relation:

$$\text{curl}(J_s) = -\frac{\mu_0 n_s e^2}{m} (H \pm H_0) \tag{21.41}$$

where H_0 is a constant of integration. The magnetic field inside the superconducting sample is zero by the Meissner effect; therefore, H_0 is zero. Hence,

$$\text{curl}(J_s) = -\frac{\mu_0 n_s e^2}{m} H = -\frac{n_s e^2}{m} B \tag{21.42}$$

Let us assume that the magnetic field (H) can be expressed as follows:

$$H = \nabla \times A \tag{21.43}$$

where A is called the vector potential with a gauge, such that, divergence of the vector potential is zero, that is, $\nabla . A = 0$. Hence,

$$J_s = -\frac{n_s e^2 \mu_0}{m} A \tag{21.44}$$

Equations (21.38) and (21.40) together are known as the *London's equations*. Considering the Maxwell's equations, one gets the following relation:

$$\nabla \times H = \frac{4\pi}{c} J_s \qquad (21.45)$$

Taking curl on both the sides, one gets the following expression:

$$\nabla \times (\nabla \times H) = \frac{4\pi}{c} (\nabla \times J_s) \qquad (21.46)$$

Using the vector identity for curl of a quantity,

$$\nabla (\nabla . H) - \nabla^2 H = \frac{4\pi}{c} \times \left(\frac{\mu_0 n_s e^2}{m} \right) H \qquad (21.47)$$

$$\nabla^2 H = \frac{4\pi}{c} \times \left(\frac{\mu_0 n_s e^2}{m} \right) H = \frac{1}{\lambda_l^2} H \qquad (21.48)$$

This simple double-space differential equation can be solved using mathematical techniques to give the magnetic field, as follows:

$$H = H_0 e^{-x/\lambda_L} \qquad (21.49)$$

where x represents the distance from the surface to the specimen and H_0 is the field at the surface. The parameter λ_L has the dimensions of length and is called the *London's penetration depth*. Thus, physically one can see it as the depth to which the field penetrates before decaying to $1/e$ of its value at the surface.

Thus, as predicted by the Meissner effect, the magnetic field does not drop to zero abruptly at the surface, but travels to a certain distance below the surface (given by λ_L) and then gradually becomes zero. This is true for both the current and the magnetic field. The prediction of the London's equation that the current and the magnetic field penetrate to a small distance (or depth) inside the specimen and then disappear well inside the specimen gave new theoretical insights into the superconducting state of matter. This was later modified further by many other theories like the *Ginzburg–Landau* theory.

The London's penetration depth λ_L is dependent on the number of superconducting electrons n_s, and the transition temperature T_c. The corresponding mathematical relation is given as follows:

$$\lambda(T) = \frac{\lambda(0)}{\left[1 - \left(\dfrac{T}{T_c} \right)^4 \right]^{1/2}} \qquad (21.50)$$

where $\lambda(0)$ is the value of the penetration depth at absolute zero. We observe that when $T > T_c$, $n_s = 0 \Rightarrow \lambda \to \infty$, also at $T = 0$, $n_s = \max \Rightarrow \lambda \to$ minimum, and at $T = T_c$, λ is ∞. Thus, one can algebraically infer that

$$\frac{\lambda(T)}{\lambda(0)} = \left[1 - \left(\frac{T}{T_c} \right)^4 \right]^{-1/2} \tag{21.51}$$

If we develop Eq. (21.39) by taking curl on both sides, we get the following relation:

$$\nabla \times J_s = -\frac{n_s e^2 \mu_0}{m} \nabla \times A = -\frac{n_s e^2}{m} B \tag{21.52}$$

As we have developed for the magnetic field based on mathematics one can derive the following expression for the current density:

$$\nabla^2 J = \frac{J}{\lambda_L^2} \tag{21.53}$$

$$J = J(0) e^{-x/\lambda_L} \tag{21.54}$$

Equation (21.49) shows that the super-currents associated with a superconducting sample (in the presence of the magnetic field H_0) are also present on the surface of the specimen and exponentially decay, as we move inside the sample (this has been used in flux quantization). In addition, one can see that the surface current is related to the vector potential A.

Example 21.13 Determine the critical temperature of aluminium, if its penetration depth is 16 and 96 nm at 2.19 and 8.1 K, respectively.

Solution We know that

$$\lambda(T) = \lambda_0 \left[1 - \left(\frac{T}{T_c} \right)^4 \right]^{-1/2} \Rightarrow \lambda(8.1) = 96 \left[1 - \left(\frac{8.1}{T_c} \right)^4 \right]^{-1/2}$$

$$\lambda(2.19) = 16 \left[1 - \left(\frac{2.19}{T_c} \right)^4 \right]^{-1/2}$$

Thus, dividing the two equations, one gets the following relation:

$$6 = \sqrt{\frac{1 - \left(\frac{2.19}{T_c} \right)^4}{1 - \left(\frac{8.1}{T_c} \right)^4}} \Rightarrow 36 = \left[1 - \left(\frac{2.19}{T_c} \right)^4 \right] \left[1 - \left(\frac{8.1}{T_c} \right)^4 \right]^{-1}$$

$$T_c = 8.16 \text{ K}$$

Example 21.15   The density of lead is $\dfrac{11.4 \times 10^3 \text{ kg}}{\text{m}^3}$ with an atomic weight of 207.19 g. Estimate the London penetration depth at 0 K.

Solution   The London penetration depth at 0 K is calculated as follows:

$$\lambda(0) = \left(\frac{mc^2}{4\pi n_s e^2} \right)^{1/2}$$

Valency of lead = 2

Thus, the atomic weight 207.19 g contains 6.023×10^{23} atoms $\times 2 = 12.046 \times 10^{23}$ atoms

$$\text{Volume} = \frac{\text{Mass}}{\text{Density}} = \frac{207.19}{11.4} \text{cm}^3 = 18.174 \text{ cm}^3$$

18.174 cm^3 volume contains 12.046×10^{23} electron atoms

1cm^3 volume contains $\dfrac{12.046 \times 10^{23}}{18.174 \text{cm}^3}$ electron atoms

$$= 0.6628 \times 10^{23} \text{electron} \frac{\text{atoms}}{\text{cm}^3}$$

$$\lambda(0) = \sqrt{\frac{m}{\mu_0 n_s e^2}} = \sqrt{\frac{7 \times 9.1 \times 10^{-31} m^3}{4 \times 22 \times 10^{-7} \text{ H} \times 6.628 \times 10^{28} \text{atoms} \times \left(1.6 \times 10^{-19}\right)^2}}$$

$$= \sqrt{0.04266 \times 10^{-14}}$$

$$\lambda(0) = 0.206 \times 10^{-7} \text{ m} = 2.06 \times 10^{-8} \text{ m}$$

21.10 APPLICATIONS OF SUPERCONDUCTORS

The discovery of this new class of materials called superconductors led to many technological applications, some of which have been listed here.

21.10.1 Superconducting Quantum Interference Devices

As discussed in Section 21.3, the magnetic flux that can thread a superconducting loop is quantized, with one quantum of flux being 2.07×10^{-15} Wb. Superconducting quantum interference devices (SQUIDs) are suitably processed superconducting loops that can detect minute changes in the magnetic flux. These devices are, therefore, used as high-sensitivity magnetic flux detectors. Such detectors are used in precision instruments in several frontier areas such as advanced metrology.

21.10.2 Cryotron

A cryotron is a switch that utilizes superconductivity for its operation. We know that superconductors lose their superconductivity, if a magnetic field of sufficient strength is created around them. A simple cryotron consists of two superconducting wires (e.g., tantalum and niobium) having different critical temperatures T_c. The niobium wire (having a higher T_c) is wrapped around a straight wire of tantalum (lower-T_c component). An electrical insulation is provided between the two wires. The complete device is then immersed in a bath of liquid helium. Both the wires become superconducting and the tantalum wire is able to carry a large amount of current in the superconducting state, as long as no current is passing through the niobium wire. When a current of sufficient strength is passed through the niobium wire, the resultant magnetic field kills the superconductivity of the tantalum wire. This drastically reduces the current that can flow through the tantalum wire. Thus, the amount of current passing through the tantalum wire can be reduced from a high to a very low-value by passing a current through the coiled niobium wire. The tantalum wire thus acts as a *gate*, whereas the coiled niobium wire acts as a *control*.

21.10.3 Magnetic Levitation

If a superconducting material is placed in a magnetic field, electric currents are set up at its surface, screening out the external magnetic field. This results in a zero magnetic field inside the material. Thus, a superconducting material displays perfect diamagnetism. This property ensures that a magnet brought near a superconductor is repelled. This repulsion effect can be used to float a magnet above a superconductor. This phenomenon is called *magnetic levitation*.

Magnetic levitation (maglev) is used to operate levitated trains. These trains do not move on rails like the conventional trains, but float on an air cushion (3–4 inch thick) over strongly magnetized tracks. Superconducting coils in the trains result in magnetic repulsion that enables levitation. The absence of any mechanical friction allows the achievement of extremely high speeds. Speeds as high as 500 km/h have been reported. The following are some applications where this phenomenon is used:

1. A superconducting supercollider (SSC) other than maglev trains is a particle accelerator that uses strong magnetic fields generated by superconducting magnets. Energy in the range of tens of TeV can be generated and used for interactions.
2. Large superconducting magnets also find applications in magnetic resonance imaging systems used in medicine.

IMPORTANT CONCEPTS

1. In the class of materials called 'superconductors', below the transition temperature T_c, there is no resistance to the flow of current-carrying electron pairs called the Cooper pairs.

2. For a superconductor, $\chi = -1$. We observe that, magnetically, a perfect diamagnetic metal and a superconductor are identical. Both possess diamagnetic magnetization leading to energy $B_0^2 / 2\mu_0$ per unit volume. However, the major contrast between a metal and a superconductor is that the latter is in a more highly ordered state.

3. The superconducting state is a distinct phase of matter that is characterized by some special electrical, magnetic, thermodynamic, isotopic, and energy state properties.

4. The critical current and the critical magnetic field are related through the following simple relation, which, after the name of Silsbee, is called the Silsbee's rule:

$$I_C(T) = 2\pi r H_c(T)$$

5. The magnetic field is zero inside a superconducting sample, as observed by Meissner, and hence this phenomenon is called the Meissner Effect.

6. It is observed that the transition temperature T_c increases inversely with the atomic mass and with a high power of the atomic volume. This is called the isotope effect.

$$M^\alpha T_C = \text{constant}$$

7. Heat capacity in the superconducting state varies with temperature in an exponential manner, that is, it is of the form $e^{-\Delta / k_B T}$.

8. The energy gap for a superconductor is given as follows:

$$E_g(0) = 2\Delta = 3.53 k_B T_c$$

$$\Delta(T) = 1.74\Delta(0)\left(1 - \frac{T}{T_c}\right)^{1/2}$$

9. Coherence length (ξ) is defined as the maximum distance up to which a Cooper pair exists to produce superconductivity. Magnitude of $|\xi^2|$ represents the coherence volume. The coherence length is related to the energy gap as follows:

$$\xi = \frac{h v_F}{2\pi . 2\Delta}$$

where, $v_F = \left(\sqrt{2E_F / m}\right)$ is the Fermi velocity.

10. The flux trapped inside the superconducting sample is quantized by a multiple of (h / q). In the case of a superconducting sample, the charge is contributed by a Cooper pair; thus, $q = 2e$. Hence, the flux trapped inside the superconductor is quantized as follows:

$$(h / 2e) = 2 \times 10^{-15} \text{ Wb} = 2 \times 10^{-7} \text{ Maxwell}$$

11. The magnetic field plays a significant role in the observation of the phenomenon of superconductivity. Superconductivity disappears at fields greater than a critical field H_c. The temperature dependence of the critical field is given by Tuyn's law, which can be expressed in the following form:

$$H_c = H_0\left[1 - \left(\frac{T}{T_c}\right)^2\right]$$

where H_0 is the critical field at 0 K.

12. Bardeen, Cooper, and Schriffer in 1957 found that the electron–phonon interaction is a possible cause of formation of the superconducting state.

13. Under very restricted circumstances (i.e., critical temperature), the electron–phonon interaction is attractive, and under even more restricted circumstances, this attraction exceeds their Coulomb repulsion. Thus,

$$k_1 = -k_2$$
$$s_1 = -s_2$$

Such a pair of electrons is called the 'Cooper pair'. Such an attractive interaction between electrons is energetically favourable.

14. Two special features of the BCS ground state are as follows:
    (a) The total energy of the BCS state is lower than that of the Fermi state.
    (b) The one-particle state is occupied in pairs; if the state with vector k has been occupied, then the state with vector $-k$ will also be occupied; similarly, if the state with vector $-k$ is vacant, then the state with k will also be vacant.

15. The prediction of the London's equation that the current and the magnetic field penetrate to a small distance (or depth) inside the specimen and then disappear well inside the specimen gave new theoretical insights into the superconducting state of matter.

    The London's penetration depth λ_L is dependent on the number of superconducting electrons n_s, and the transition temperature T_c. The corresponding mathematical relation is as follows:

$$\lambda(T) = \frac{\lambda(0)}{\left[1 - \left(\dfrac{T}{T_c}\right)^4\right]^{1/2}}$$

where $\lambda(0)$ is the value of the penetration depth at absolute zero. We observe that when $T > T_c$, $n_s = 0 \Rightarrow \lambda \to \infty$, also at $T = 0$, $n_s = $ max $\Rightarrow \lambda \to$ minimum, and at $T = T_c$, λ is ∞. Thus, one can algebraically infer that

$$\frac{\lambda(T)}{\lambda(0)} = \left[1 - \left(\frac{T}{T_c}\right)^4\right]^{-1/2}$$

APPLICATIONS

1. Superconducting quantum interference devices are suitably processed superconducting loops that can detect minute changes in magnetic flux. These devices are, therefore, used as high-sensitivity magnetic flux detectors. Such detectors are used in precision instruments in several frontier areas like advanced metrology. An interferometer helps measure the magnetic flux to an accuracy of 10^{-11} G.

2. A cryotron is a switch that utilizes superconductivity for its operation.

3. Magnetic levitation (maglev) is used to operate levitated trains. These trains do not move on rails like the conventional trains, but float on an air cushion (3–4 inch thick) over strongly magnetized tracks. Superconducting coils in the

trains result in magnetic repulsion that enables levitation. The absence of any mechanical friction allows extremely high speeds to be achieved. Speeds as high as 500 km/h have been reported.

4. An SSC is a particle accelerator that uses strong magnetic fields generated by superconducting magnets. Energy in the range of tens of TeV can be generated and used for interactions.

5. Large superconducting magnets also find applications in magnetic resonance imaging systems used in medicine.

6. International laboratories use the AC Josephson effect for its accuracy in the measurement of the quantity (e/h) to within 1ppm. This has helped reassess various fundamental constants and voltage standards.

7. The Josephson junction can be used as a complex non-linear circuit element. The manifold applications of this junction include parametric amplifiers, mixers, or detectors of very-high-frequency radiation.

8. The radio-frequency power generated in the junction is between far infrared and the microwave regions. This results in the construction of efficient high-sensitivity low-noise amplifiers and detectors in the 50–100 GHz region particularly for use in radio-astronomy.

9. The most widespread use of the Josephson junction is in the basic unit of a computer. The bipolar element can be used as a switch, as it goes from the normal to the superconducting state, as a potential is developed across it. Two superconductors, let us say, lead (Pb) and tin (Sn) are used such that when one is actuated, it produces a magnetic field larger than the critical field of the other superconductor. This unit is called a cryotron. The Josephson junction is superior to an ordinary cryotron in its switching speed (of the order of 1ns).

IMPORTANT FORMULAE

1. Critical current:

$$I_C(T) = 2\pi r H_c(T)$$

2. Isotope effect:

$$M^\alpha T_C = \text{constant}$$

3. Energy gap:

$$E_g(0) = 2\Delta = 3.53 k_B T_c$$

4. Variation of energy gap with temperature:

$$\Delta(T) = 1.74\Delta(0)\left(1 - \frac{T}{T_c}\right)^{\frac{1}{2}}$$

5. Coherence length:

$$\xi = \frac{h v_F}{2\pi.2\Delta}, \text{ where } v_F = \left(\sqrt{2E_F/m}\right)$$

is the Fermi velocity

6. Flux quantization:

$$(h/2e) = 2 \times 10^{-15}$$

$$\text{Wb} = 2 \times 10^{-7} \text{ Maxwell}$$

7. Variation of magnetic field with temperature:

$$H_c = H_0\left(1 - \left(\frac{T}{T_c}\right)^2\right)$$

8. Variation of penetration depth with temperature:

$$\lambda(T) = \frac{\lambda(0)}{\left(1 - \left(\frac{T}{T_c}\right)^4\right)^{\frac{1}{2}}}$$

SELF-ASSESSMENT

Multiple-choice Questions

21.1 The first observation of the phenomenon of superconductivity was made in
   (a) liquid helium   (c) liquid nitrogen
   (b) liquid mercury   (d) liquid lead

21.2 The simultaneous occurrence of two phenomenon that make a superconductor different from a normal conductor are
   (a) zero resistance and Meissner effect
   (b) zero resistance and isotope effect
   (c) isotope effect and thermal conductivity
   (d) Cooper pair and zero resistance

21.3 The isotope effect in a superconducting element states that
   (a) $T_c M^{3/2} = \text{constt}$ (c) $T_c M^{5/2} = \text{constt}$
   (b) $T_c M^{1/2} = \text{constt}$ (d) $T_c M^{2/2} = \text{constt}$

21.4 The BCS theory is based on the
   (a) electron–electron interaction   (c) electron–phonon interaction
   (b) electron–spin interaction   (d) electron–lattice interaction

21.5 The electron molecule in the BCS theory refers to
   (a) an electron pair   (c) a Cooper pair
   (b) a lattice pair   (d) a Bardeen pair

21.6 The maximum distance of interaction between two electrons is said to be the
   (a) incoherence length   (c) experimental length
   (b) mean length   (d) coherence length

21.7 Transition temperature is controlled by the
   (a) binding energy 2Δ of the Cooper pair
   (b) electron–phonon interaction
   (c) zero resistance phenomenon
   (d) lattice interactions

21.8 A superconductor is a
   (a) purely paramagnetic substance
   (b) purely diamagnetic substance
   (c) purely ferromagnetic substance
   (d) none of these

21.9 The concept of flux quantization states that
   (a) there is no magnetic field inside a superconductor
   (b) there is no magnetic flux in a superconductor
   (c) the magnetic flux is a quantized integral multiple of (h/q)
   (d) none of these

21.10 The London's penetration depth is related to the extent to which
   (a) the magnetic field penetrates a superconducting element
   (b) the current penetrates a superconducting element
   (c) electrons remain coherently together
   (d) the magnetic field remains out of the superconducting element

21.11   The magnetic field for a superconducting element is given by the relation

(a) $H_c = H_0\left[1 - \left(\dfrac{T}{T_c}\right)^2\right]$

(c) $H_c = H_0\left[1 - \left(\dfrac{T}{T_c}\right)^{3/2}\right]$

(b) $H_c = H_0\left[1 - \left(\dfrac{T}{T_c}\right)^{1/2}\right]$

(d) $H_c = H_0\left[1 - \left(\dfrac{T}{T_c}\right)^4\right]$

21.12   The energy gap for a superconductor is given as

(a) $E_g(0) = 2\Delta = 3.53\,k_B T_c$

(c) $E_g(T) = 2\Delta = 3.53\,k_B T_c$

(b) $E_g(0.5) = 3\Delta = 3.53\,k_B T_c$

(d) $E_g(4) = 4\Delta = 3.53\,k_B T_c$

21.13   A superconducting sample is obtained when
  (a) the applied field, temperature, and current density are maintained below a critical surface
  (b) the applied field, temperature, and current density are maintained below a critical temperature
  (c) the entropy, temperature, and current density are maintained below a critical surface temperature
  (d) none of these

21.14   A superconductor has
  (a) lower entropy than the normal state
  (b) higher entropy than the normal state
  (c) same entropy as the normal state
  (d) all of these

21.15   The thermal conductivity for a superconductor:
  (a) is much higher than that for a normal conductor
  (b) is much lower than that for a normal conductor
  (c) is almost the same as that for a normal conductor
  (d) has no variation with respect to a normal conductor

Review Questions

21.1   Enunciate with salient features, how the superconducting phenomenon is a different state of matter, as compared with the normal conducting phenomenon.

21.2   Why are only surface currents present in a superconductor?

21.3   Cooper pairs are mutually interacting quantities forming a cooperatively stable system. Explain.

21.4   Why is superconductivity a low-temperature phenomenon?

21.5   Explain the term 'condensate' in relation to superconductivity.

21.6   What is the difference between a quasi-particle and a Cooper pair?

21.7   Discuss in detail the difference in the energy gap of a conductor, an insulator, a semiconductor, and a superconductor.

21.8   Explain graphically the difference between coherence length and penetration depth for a superconductor.

21.9   Describe the salient features of the phenomenon of superconductivity.

21.10 Write the name of four metals that demonstrate superconductivity.

21.11 Explain the concept of energy gap in superconductors.

21.12 Explain the term 'Cooper pair'.

21.13 State the difference between a hard superconductor and a soft superconductor.

21.14 Write the names of two HTSCs.

21.15 Explain the term 'fluxon'.

21.16 Explain why the magnetic susceptibility of a superconductor is negative.

21.17 Explain the concept of flux quantization.

21.18 Write in detail the salient features of the BCS theory.

21.19 Highlight the difference between an ordinary diamagnetic material and a superconductor.

21.20 Why are superconductors said to be ideal diamagnetic materials?

21.21 Plot a graph of magnetization versus magnetic field for a hard superconductor.

21.22 Why are high-temperature superconductors important for industrial applications?

21.23 Explain the principle of a cryotron.

21.24 Why do trains based on maglev have the potential to reach very high speeds?

21.25 Derive the London's equation. Explain how it has been able to analyse the superconducting phenomenon.

21.26 What do you understand by high-temperature superconductors?

Numerical Problems

21.1 Evaluate the energy gap at absolute temperature for lead, if its critical temperature is 7.19 K. $\left[Hint: E_g(0) = 3.53\, k_B T_c\right]$

21.2 The actual energy gap at 0 K in mercury is 1.08×10^{-3} eV.
   (a) What is the prediction of the BCS theory for this energy gap?
   (b) Radiation of what minimum frequency can break apart Cooper pairs in lead at 0 K? In what part of the electromagnetic spectrum does such a radiation occur? $\left[Hint: E_g(0) = 3.53\, k_B T_c\right]$

21.3 The critical temperature T_c for, Hg with isotopic mass 199.5 is 4.195 K. What will be its critical temperature, when its isotopic mass is increased to 203.4?
$$\left[Hint: T_{c1} M_1^{1/2} = T_{c2} M_2^{1/2}\right]$$

21.4 Find the critical current density for 1.5 mm diameter wire of lead at 4.2 K. For lead, the critical temperature is 7.19 K with a critical field 6.51×10^4 A/m.
$$\left[Hint: I_c = 2\pi r H_c\right]$$

21.5 The penetration depth of mercury (Hg) at 2.9 K is about 570 Å. Estimate the values of penetration depth and number density as T tends to zero $(T_c = 4.2 \text{ K})$.

$$\left[Hint: \lambda(T) = \frac{\lambda(0)}{\left(1 - \left(\dfrac{T}{T_C}\right)^4\right)^{1/2}}\right]$$

PART V
NUCLEAR PHYSICS AND NEW ENGINEERING MATERIALS

Nuclear Physics and Radioactivity

Learning Objectives

After studying this chapter, students will be able to

- understand that a nucleus is composed of nucleons
- comprehend that a nucleus with a higher nucleon number tends to be unstable and decay into more stable nuclei; this is the phenomenon of radioactivity
- elucidate the laws of radioactive disintegration; the most common form of nuclei decays into the alpha, beta, and gamma particles
- explain the laws of conservation of energy, leading to the concepts of binding energy and energy–mass equivalence
- demonstrate various models stated to explain the nuclear forces, in particular the liquid drop model
- explain that the theory of nuclear fission and fusion leads to the concept of a controlled chain reaction in a nuclear reactor
- discuss the concept of generation of high energy in the stellar thermonuclear reaction owing to fusion
- describe particle accelerators, for example, a linear accelerator such as a cyclotron that, under special conditions, accelerates elementary particles
- explain radiation detectors such as the ionization chamber, proportional counter, and Geiger–Muller counter that work on the principle of ionization of gas to detect the presence of radiation
- discuss the applications of nuclear radiation in detail, as well as the biomedical advantages and hazardous effects of radiation

List of Symbols

p = Proton number
N = Neutron
A = Atomic Number
R = Radii of nuclei
ρ = Nuclear density

S = Spin angular momentum
J = Orbital angular momentum
μ_B = Bohr magneton

E_B = Binding energy
$U_{\text{nucleus}}(r)$ = Potential energy

$P_{\Delta t}$ = Probability that a particle at time t will decay in the next short interval of time Δt

R = Rate of decay

$^{238}_{92}U$ = Unstable uranium isotope

$^{234}_{90}Th$ = Thorium

β^- = Beta-minus particle

ϑ_e = Electron neutrino

β^+ = Beta-plus particle

$^{294}_{96}Ba$ = Barium

22.1 INTRODUCTION

The atomic nature of matter has been established soundly. The atomic structure has been described in detail by Rutherford, Bohr, and Sommerfield, incorporating the classical, quantum, and relativistic concepts. At the core of an atom is the nucleus, consisting of protons and neutrons. The dimension of the nucleus is much smaller than the overall structure of the atom. The nucleus is extremely dense and possesses positive charge. There is a specific neutron–proton ratio for the nucleus to hold together, where the total number of protons is not too large. Protons and neutrons possess attractive nuclear forces, which need to be balanced by the electric repulsion of the protons for the stability of the nucleus. The nuclei where these forces are not balanced decay or transform themselves spontaneously into other stable nuclei (Figs 22.1 and 22.2). In this chapter, some of the characteristic features pertaining to nuclei are discussed.

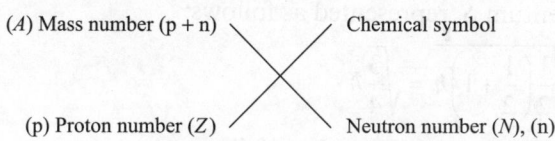

(A) Mass number (p + n) \\ Chemical symbol

(p) Proton number (Z) / Neutron number (N), (n)

Fig. 22.1 Nuclear notation

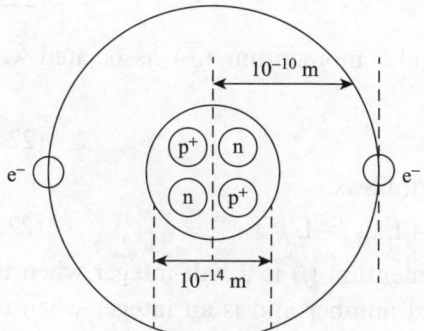

Fig. 22.2 Typical helium atom showing atomic dimensions

22.2 CHARACTERISTIC PROPERTIES OF NUCLEI

The scattering experiments that have been performed to understand the structure of nuclei show that the nucleus is a sphere with a radius R that depends on the total number of nucleons (neutrons and protons) present in the nucleus. The total number of neutrons and protons together are referred to as the nucleon number, A, also referred to as the mass number. The proton and neutron masses are both approximately referred to as

1u (where u is the unified atomic mass units),

$$1\,u = 1.6605402 \times 10^{-27}\ kg \tag{22.1}$$

The masses of the nuclei and the particles are the rest masses. Radii of the nuclei are represented as follows:

$$R = R_0 A^{\frac{1}{3}}, \text{ where } R_0 = 1.2 \times 10^{-15}\ m = 1.2\ fm \tag{22.2}$$

(where 'fm' represents Fermi)

The nuclear density is the mass per unit volume of a nucleus, that is,

$$\rho = \frac{m}{V} = \frac{m_N}{\left(\frac{4}{3}\pi R^3\right)} = \frac{m_N}{\left(\frac{4}{3}\pi R_0^3 A\right)} = \frac{A \times 1\,u}{\frac{4}{3}\pi R_0^3 A} = \frac{1\,u}{\frac{4}{3}\pi R_0^3} = 2.3 \times 10^{17}\ kg/m^3 \tag{22.3}$$

Thus, one observes that the nuclear density is constant for all nuclei. It is often observed that for a given atomic number (denoted as Z, the total number of electrons or protons in an atom), the number of neutrons (denoted as N, the total number of neutrons in the nucleus) in the nucleus may vary, that is, different masses due to variable neutrons; such nuclides of the same element are referred to as isotopes.

Nucleons, that is, protons and neutrons, are spin (1/2) particles with a spin angular momentum $\bar{S}$ represented as follows:

$$S = \sqrt{\frac{1}{2}\left(\frac{1}{2}+1\right)}\hbar = \sqrt{\frac{3}{4}}\hbar \tag{22.4}$$

The corresponding z-component is as follows:

$$S_z = \pm\frac{1}{2}\hbar \tag{22.5}$$

Nucleons also possess an orbital angular momentum ($\bar{J}$) associated with their motion within the nucleus:

$$J = \sqrt{j(j+1)}\hbar \tag{22.6}$$

The corresponding z-component is as follows:

$$J_z = m_j\hbar \quad \text{where } m_j = -j, -j+1, ..., j-1, j \tag{22.7}$$

The value of the orbital angular momentum (j) is a half-integer when the total number of nucleons A is an odd number and is an integer when the total number of nucleons is an even number. The orbital angular momentum is zero when both the atomic number (Z) and the number of neutrons (N) are even. Thus, pairing of particles with opposite spin components is an important consideration in determining the nuclear structure. Nuclear spin is the combination of the orbital and spin angular momenta of nucleons that form the nucleus.

Owing to the nuclear angular momentum, there exists a nuclear magnetic moment similar to the electron magnetic moment, that is, the Bohr magneton. The nuclear magnetic moment or the nuclear magneton is given as follows:

$$\mu_B = \frac{e\hbar}{2m_p} = \frac{5.05079 \times 10^{-27} \text{ J}}{\text{T}} = \frac{3.15245 \times 10^{-8} \text{ eV}}{\text{T}} \qquad (22.8)$$

where m_p is the proton mass. The nuclear magneton is 1836 times smaller than the Bohr magneton because of the significant difference between proton mass and the electron mass.

The positive charge on a proton is due to its spin magnetic moment $\bar{\mu}$ being parallel to its spin angular momentum $\bar{S}$, contrary to that for a neutron where the two are opposite. Figure 22.3 shows the nuclear magnetic resonance (NMR) in case of a hydrogen atom when they are in absolute alignment with the applied magnetic field.

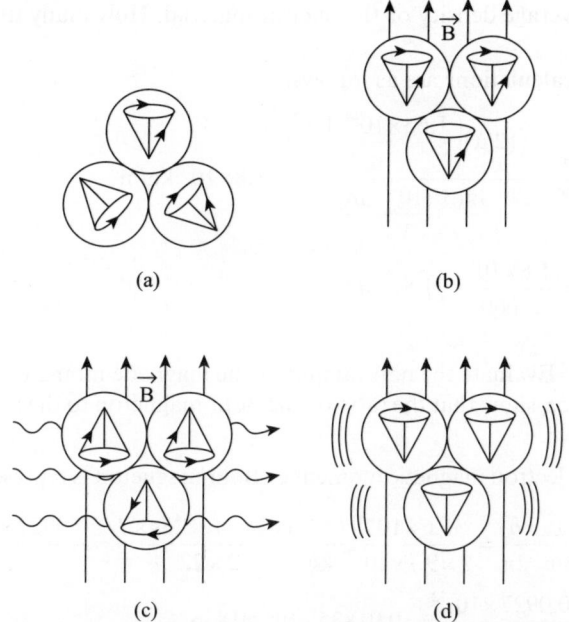

(a) (b)

(c) (d)

Fig. 22.3 Nuclear magnetic resonance (a) Random spin of hydrogen protons (b) Protons aligning themselves to uniform B̄-field as uniform magnetic field is applied (c) Resonant signal from electromagnetic wave causing protons to flip (d) Protons emitting signal as they re-align with field

Note: The NMR technique is used to measure precisely the nuclear magnetic moments owing to the spin flip properties of the elementary particles (protons, electrons, and others). Magnetic resonance imaging (MRI) is a non-invasive imaging technique that discriminates among the various body tissues on the basis of the differing environments of the protons in the tissues.

Example 22.1 The nucleon number for iron is $A = 56$. Find the radius, mass, and density of the nucleus.

Solution The radius of the nucleus can be calculated as follows:

$$R = R_0 A^{1/3} = (1.2 \times 10^{-15} \text{ m})(56)^{1/3} = 4.6 \times 10^{-15} \text{m} = 4.6 \text{ fm} \tag{22.9}$$

In addition,

$$m = (56)(1.66 \times 10^{-27} \text{ kg}) = 9.3 \times 10^{-26} \text{ kg} \tag{22.10}$$

The volume is calculated as follows:

$$V = \frac{4}{3}\pi R^3 = \frac{4}{3} \times \frac{22}{7} \times (4.6 \times 10^{-15} \text{ m})^3 = 4.1 \times 10^{-43} \text{ m}^3 \tag{22.11}$$

The density is calculated as follows:

$$\rho = \frac{m}{V} = \frac{9.3 \times 10^{-26} \text{kg}}{4.1 \times 10^{-43} \text{m}^3} = 2.268 \times 10^{17} \text{ kg/m}^3 \tag{22.12}$$

Example 22.2 The radius of a carbon nucleus is about 3×10^{-15} m and its mass is 12 u. Find the average density of the nuclear material. How many times denser than water is this?

Solution The calculations are as follows:

$$\rho_N = \frac{m}{V} = \frac{(12 \text{ u})\left(\dfrac{1.66 \times 10^{-27} \text{ kg}}{\text{u}}\right)}{\dfrac{4\pi(3 \times 10^{-15} \text{ m})^3}{3}} = 1.8 \times 10^{17} \text{kg/m}^3 \tag{22.13}$$

$$\frac{\rho_N}{\rho_{\text{water}}} = \frac{1.8 \times 10^{17}}{1000} = 1.8 \times 10^{14}$$

Example 22.3 Evaluate the natural unit of the magnetic moment, that is, the Bohr magneton. Hence, show that the ratio of the Bohr magneton to the nuclear magneton is 1836.

Solution The electron magnetic moment or Bohr magneton is represented as follows:

$$\mu_B = \frac{e}{2m_e} \frac{h}{2\pi} = \frac{1.6 \times 10^{-19} \text{ C}}{2 \times 9.1 \times 10^{-31} \text{ kg}} \frac{6.634 \times 10^{-34} \text{ Js} \times 7}{2 \times 22} = \frac{0.0927 \times 10^{-22} \text{ J}}{T}$$

$$\frac{\mu_B}{\mu_N} = \frac{0.0927 \times 10^{-22}}{5.050 \times 10^{-27}} = 0.01835 \times 10^5 \approx 1836$$

Note: The magnitude of the z-component of the spin magnetic moment of a proton is different from that of the nuclear magneton, that is, $|\mu_{SZ}|_{\text{proton}} = 2.7928\mu_e$, whereas that for the neutron, which has no charge, is $|\mu_{SZ}|_{\text{neutron}} = 1.913\mu_n$. This is owing to the fact that protons and neutrons are formed of still more elementary particles, called 'quarks'.

Example 22.4 Hydrogen atoms are placed in an external magnetic field. Protons can make transitions between states in which their spin component is parallel and anti-parallel to the field, by absorbing or emitting a photon. What is the magnitude of the magnetic field required for this transition to be induced by photons of frequency 36.9 MHz?

Solution The corresponding photon frequency is given as follows:

$$\vartheta = \frac{\Delta E}{h} = 36.9\,\text{MHz} = 36.9 \times 10^6\,\text{Hz}$$

$$\Rightarrow \Delta E = h\vartheta = 4.136 \times 10^{-15}\,\text{eV.s} \times 36.9 \times 10^6\,\text{Hz} = 152.6184 \times 10^{-21}\,\text{eV}$$

When the components are anti-parallel to the field, the energy difference between the two states is given as follows:

$$\Delta E = 2U \Rightarrow U = \frac{\Delta E}{2}$$

$$U = \frac{(152.6184 \times 10^{-21})\,\text{eV}}{2} = 76.3092 \times 10^{-9}\,\text{eV}$$

When the z-component of S (and μ) is parallel to the field, the interaction energy is calculated as follows:

$$U = -|u_z|B \Rightarrow B = \frac{76.3092 \times 10^{-9}\,\text{eV}}{\left(\dfrac{2.7928 \times 3.152 \times 10^{-8}\,\text{eV}}{\text{T}}\right)} = 0.8875\,\text{T}$$

Note: Spin flips in protons (or elementary particles) can be detected by the absorption of energy from the radiation. This is the principle of all the instruments that work based on the spin flip property of elementary particles.

22.3 NUCLEAR BINDING ENERGY

Binding energy E_B is the magnitude of the energy by which nucleons are bound together. The total rest energy E_0 of separated nucleons is greater than the rest energy $(E_0 - E_B)$ of the nucleus. The binding energy of a nucleus containing Z protons and N neutrons is defined as follows:

$$E_B = (ZM_H + Nm_N - {}_Z^A M)c^2 \tag{22.14}$$

where ZM_H is the mass of Z protons and Z electrons combined as Z neutral ${}_1^1\text{H}$ atoms to balance the Z electrons included in ${}_Z^A M$, the mass of neutral atom.

The mass defect for a nucleus is defined using the concept of energy–mass equivalence, as follows:

(Separated rest energy of the nucleons – Rest energy of the nucleus) $= M_D\,c^2$

$$E_0 - (E_0 - E_B) = M_D\,c^2 \tag{22.15}$$

$$M_D = \frac{E_B}{c^2} \tag{22.16}$$

Figure 22.4(a) shows the variation of binding energy per nucleon with the mass number. Alternatively, this can be understood using the concept of rest mass lost, that is,

Rest mass lost in the formation of a nucleus $= \Delta m = (m_p + m_n) - M_{\text{Nucleus}}$

Rest energy lost in the formation of nucleus $= \Delta E_0 = \Delta mc^2 \tag{22.17}$

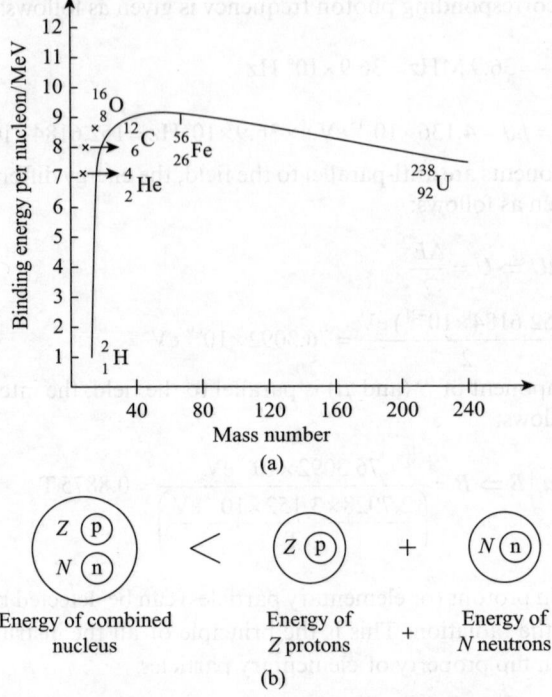

Fig. 22.4 Representation of binding energy for a nucleon (a) Variation of binding energy per nucleon with mass number (b) Concept of nuclear binding energy

Thus, the surroundings gain an amount of energy equivalent to ΔE_0 when a nucleus is formed. In addition, the same amount of energy, the binding energy, has to be supplied to the nucleus to tear such that the protons and neutrons are infinitely separated.

Figure 22.4(b) thus explains the concept of nuclear binding energy.

Example 22.5 The rest masses of proton, neutron, and deuteron are $m_p = 1.67261 \times 10^{-27}$ kg, $m_n = 1.67492 \times 10^{-27}$ kg, $m_d = 3.34357 \times 10^{-27}$ kg, respectively. How much energy should be liberated during the formation of a deuteron from a free proton and a free neutron, initially at rest 'at infinity'?

Solution Rest mass lost in the formation of a nucleus = $\Delta m = (m_p + m_n) - M_{\text{Nucleus}}$

$$\Delta m = (1.67261 \times 10^{-27} + 1.67492 \times 10^{-27}) \text{kg} - (3.34357 \times 10^{-27}) \text{kg} = 3.96 \times 10^{-30} \text{kg}$$

Rest energy lost in the formation of nucleus = $\Delta E_0 = \Delta mc^2$

$$\Delta E_0 = (3.96 \times 10^{-30}) \times (3 \times 10^8)^2 \text{kg} \frac{\text{m}^2}{\text{s}^2} = 3.564 \times 10^{-13} \text{J} = 2.22 \text{ MeV}$$

where 1 MeV = 1.6×10^{-13} J

Thus, the surroundings gain an amount of energy equivalent to $\Delta E_0 = 2.22$ MeV. In addition, the same amount of energy, referred to as the binding energy, has to be supplied to the nucleus to tear it apart into infinitely separated protons and neutrons.

Example 22.6 Show 1 u (atomic mass unit) is equivalent to 931.5 MeV of energy.

Solution The calculations are as follows:

$$E = mc^2 = (1.6605 \times 10^{-27})(3 \times 10^8)^2 = 14.9445 \times 10^{-11} \text{J}$$

$$1 \text{MeV} = 1.6 \times 10^{-13} \text{ J}$$

Thus, $$E = \frac{14.9445 \times 10^{-11} \text{J}}{1.6 \times 10^{-13} \text{J/MeV}} = 931.5 \text{ MeV}$$

Example 22.7 Find the binding energy of a deuteron. In addition, determine how tightly the nucleus is bound.

Solution The nucleus of a deuteron atom consists of a proton and a neutron bound together to form a particle called deuterium.

$$E_B = (ZM_H + Nm_N - {}_Z^A M)c^2; \quad E_B = (1 \times M_H + 1 \times m_N - {}_1^2 H)c^2$$

$$E_B = (1 \times 1.007825 \text{ u} + 1 \times 1.008665 \text{ u} - 2 \times 1.007051 \text{ u})c^2$$

$$E_B = (1.007825 \text{ u} + 1.008665 \text{ u} - 2.014102 \text{ u}) \times \left(\frac{931.5 \text{ MeV}}{\text{u}} \right)$$

$$E_B = (0.002388) \text{ u} \times 931.5 \ \frac{\text{MeV}}{\text{u}} = 2.224 \text{ MeV}$$

$$\text{Binding energy per nucleus} = \frac{E_B}{A} = \frac{2.24 \text{ MeV}}{2 \text{ nucleons}} = 1.112 \text{ MeV per nucleon}$$

Thus, ${}_1^2 H$ has the smallest binding energy per nucleon of all nuclides.

Example 22.8 Alpha particles are emitted from radium 224 (atomic mass of ${}_{88}^{224} Ra = 224.02022$ u) to produce radon 220 (atomic mass of ${}_{86}^{220} Rn = 220.01140$ u), according to the following equation:

$${}_{88}^{224} Ra \rightarrow {}_{86}^{220} Rn + {}_2^4 He$$

Calculate (a) the decrease in mass and (b) the energy released. (The atomic mass of ${}_2^4 He = 4.00260$ u).

Solution The following are the calculations:

(a) Decrease in mass is calculated as follows:

$$(\text{Atomic mass of } {}_{88}^{224} Ra) - (\text{Atomic mass of } {}_{86}^{220} Rn + \text{Atomic mass of } {}_2^4 He)$$

$$[224.02022 \text{ u} - (220.01140 \text{ u} + 4.00260 \text{ u})] = 0.00622 \text{ u}$$

(b) Energy released = (Decrease in mass) $\times c^2$

$$\text{Energy released} = (0.00622 \text{ u}) \times \left(\frac{931.5 \text{ MeV}}{\text{u}} \right) = 5.793 \text{ MeV}$$

Example 22.9 Calculate the energy released when 15 kg of ${}_{92}^{235} U$ undergoes fission according to the following equation:

$${}_{92}^{235} U + {}_0^1 n \rightarrow {}_{56}^{141} Ba + {}_{36}^{92} Kr + 3 {}_0^1 n$$

The atomic masses for ${}_{92}^{235} U = 235.04$ u; ${}_{56}^{141} Ba = 140.91$ u; ${}_{36}^{92} Kr = 91.91$ u; ${}_0^1 n = 1.01$ u.

Solution The calculations are as follows:

Mass defect $= (235.04\,u + 1.01\,u) - (140.91\,u + 91.91\,u + 3 \times 1.01\,u)$

Mass defect $= 0.20\,u$

Energy released $= 0.20\,u \times 931.5\,MeV = 186.3\,MeV$

235.04×10^{-3} kg of $^{235}_{92}U$ has 6.023×10^{23} atoms

15 kg of $^{235}_{92}U$ has $\dfrac{15 \times 6.023 \times 10^{23}}{235.04 \times 10^{-3}}$ atoms $= 0.384 \times 10^{26}$ atoms

Energy released by 15 kg of $^{235}_{92}U = 0.384 \times 10^{26} \times 186.3\,MeV$

$$= 71.61 \times 10^{26}\,MeV$$

22.4 NUCLEAR FORCES

The previous sections give us insight into the structure of a nucleus, which is primarily constituted of protons and neutrons, together referred to as nucleons. The forces that are present inside a nucleus have been intriguing scientists to date. As we move further into the chapter, the most popular model, referred to as the 'liquid drop model', is discussed. The following are certain basic characteristics possessed by the nuclear forces:

1. The nuclear force does not depend upon the charge, as neutrons and protons are bound together with the same binding force.
2. The force is short range, that is, 10^{-15} m or 1 fm. This is owing to the order of the nuclear dimension.
3. Nuclear force is a strong interaction. Electrostatic attraction is much weaker than the nuclear forces. This is why a nucleus is stable (see Table 22.1).
4. The density of nuclear matter (nuclear density) is nearly constant. The binding energy per nucleon for large nuclides is nearly constant. Thus, a particular nucleon does not interact simultaneously with all the other nucleons in a nucleus. Nucleons interact with their immediate few neighbours. Protons in the nucleus possess electrostatic forces. Thus, the nucleon force seems to be somewhat similar to the covalent bonding force in molecules and solids.
5. The nuclear force has a tendency to pair up protons or neutrons of opposite spins. This is exhibited in the stability of the alpha particle [refer to Fig. 22.4(b) for binding energy].
6. There is yet no single equation, similar to Coulomb's or Gravitational force equation, which defines the relation between the force and the separation

Table 22.1 Relative strengths of fundamental interactions

	Interaction	Relative strength	Range
1.	Strong	1	Short range about $10^{-16}\,m$
2.	Electromagnetic	10^{-22}	Infinite $(1/r^2)$
3.	Weak	10^{-13}	Extreme short range
4.	Gravitational	10^{-40}	Infinite $(1/r^2)$

r of the particles. Although by the law of attraction of two nucleons (known from the particle-scattering experiments), it is a short-range force (r_0) or distance.

Thus, the attractive nuclear interaction gives rise to the potential energy of the following form:

$$U_{\text{nucleus}}(r) = -C \frac{e^{-r/r_0}}{r} \tag{22.18}$$

where $r_0 \cong 10^{-15}\,\text{m}; C \cong 10^{-20}\,\text{J m}$.

Protons will experience a Columbic repulsion, but for $r \lesssim r_0$, the nuclear interaction is the dominant interaction. At a distance of nearly 10^{-12} m, which is 10^3 times r_0, the ratio of the nuclear potential energy to the electrostatic potential energy e^2/r is $(C/e^2)\exp(-10^3) \approx 10^{-400}$, which is a very small number. Thus, the exponential factor in the preceding equation controls the range of the nuclear interaction. The attractive nuclear force potential is of the following form:

$$U_{\text{nuclear}}(r) \propto \frac{e^{-\lambda r}}{r} \tag{22.19}$$

This falls rapidly to zero at great distances.

The inverse square forces (gravitational/electrostatic) are 'long range forces' unlike the attractive nuclear force potential $\sim -\dfrac{e^{-\lambda r}}{r}$ and falls rapidly to zero at great distances (Fig. 22.5).

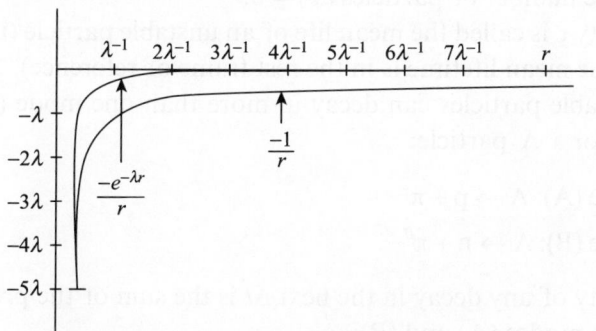

Fig. 22.5 Comparative curve of the potential (gravitational/electrostatic) with distance

22.5 LIFETIME OF UNSTABLE NUCLEI

The decay of an unstable particle is very different from the usual disintegration that one observes around us. Usually, disintegration is associated with those particles that have lived for a considerable duration of time. However, it has been observed experimentally that the decay probability of elementary particles, any kind of radioactive nuclei, excited atoms, or molecules is independent of the length of time the particle has lived. In other words, a free neutron is unstable, but a neutron that has already lived a long time is in no way different from the neutron that has been free for a short time. It is also impossible to predict when a given unstable particle will decay. Thus, we conventionally refer to the average lifetime.

The probability $P_{\Delta t}$ that a particle at time t will decay in the next short interval of time Δt (that is, $\Delta t \ll \tau$) is equal to the product of Δt and some constant $1/\tau$, which is characteristic of the particle but not of its history:

$$P_{\Delta t} = \frac{\Delta t}{\tau} \tag{22.20}$$

For a large number of particles N, the number of particles that decay in Δt is $NP_{\Delta t}$. The decay changes the number of particles by $\left[-\Delta t \left(\frac{dN}{dt} \right) \right]$. Thus, we have the following equation:

$$NP_{\Delta t} = \frac{N \Delta t}{\tau} = -\Delta t \frac{dN}{dt} \tag{22.21}$$

$$\frac{dN}{dt} = -\frac{N}{\tau} \tag{22.22}$$

The following is the solution of this one-dimensional differential equation:

$$N(t) = N_0 e^{-t/\tau} \tag{22.23}$$

where N_0 is the number of particles at $t = 0$.

The quantity τ is called the mean life of an unstable particle (in relativistic convention, the mean lifetime is in the rest frame of reference).

Many unstable particles can decay in more than one mode (Table 22.2), for example, for a Λ particle:

$$\text{Mode (A): } \Lambda \rightarrow p + \pi^- \tag{22.24}$$

$$\text{Mode (B): } \Lambda \rightarrow n + \pi^0 \tag{22.25}$$

The probability of any decay in the next Δt is the sum of the probability for decay through modes (A) and (B):

$$P = P_A + P_B \tag{22.26}$$

From Eq. (22.20) one infers $P_A = \frac{1}{\tau_A} \Delta t; P_B = \frac{1}{\tau_B} \Delta t$

Here, $-\Delta t \frac{dN}{dt} = PN \tag{22.27}$

$$\frac{dN}{dt} = -\frac{1}{\tau} N \tag{22.28}$$

Table 22.2 Radioactive nuclei

(a) Half-life of radioactive nuclides

S.No	Nuclide	Element Symbol	Decay mode	Half-life
1.	Beryllium–8	$^{8}_{4}\text{Be}$	α	1.0×10^{-6} s
2.	Polonium–213	$^{213}_{84}\text{Po}$	α	4×10^{-6} s
3.	Carbon–16	$^{16}_{6}\text{C}$	β	0.75 s
4.	Aluminium–28	$^{28}_{13}\text{Al}$	β	2.24 minutes
5.	Magnesium–28	$^{28}_{12}\text{Mg}$	β	21 hours
6.	Iodine–131	$^{131}_{53}\text{I}$	β	8 days
7.	Cobalt–60	$^{60}_{27}\text{Co}$	β	5.3 years
8.	Strontium–90	$^{90}_{38}\text{Sr}$	β	28 years
9.	Radium–226	$^{226}_{88}\text{Ra}$	α	1600 years
10.	Carbon–14	$^{14}_{6}\text{C}$	β	5730 years
11.	Uranium–238	$^{238}_{92}\text{U}$	β	4.5×10^{9} years
12.	Rubidium–87	$^{87}_{37}\text{Rb}$	β	4.7×10^{10} years

(b) Tabular display of first eight members of the Uranium series

S.No	Element	Nuclide	Half-life	Radiation	Energy of α or β in MeV
1.	Uranium	$^{238}_{92}\text{U}$	4.51×10^{9} years	α, γ	4.2
2.	Thorium	$^{234}_{90}\text{Th}$	24.1 days	β, γ	0.19
3.	Protactinium	$^{234}_{91}\text{Pa}$	6.75 hours	β, γ	2.3
4.	Uranium	$^{234}_{92}\text{U}$	2.47×10^{5} years	α, γ	4.77

S.No	Element	Nuclide	Half-life	Radiation	Energy of α or β in MeV
5.	Thorium	$_{90}^{230}\text{Th}$	8.0×10^4 years	α, γ	4.68
6.	Radium	$_{88}^{226}\text{Ra}$	1620 years	α, γ	4.78
7.	Radon	$_{86}^{222}\text{Rn}$	3.82 days	α	5.49
8.	Polonium	$_{84}^{218}\text{Po}$	3.05 minutes	α	6.0

(c) Common available radioactive sources

S.No	Element	Nuclide	Radiation	
1.	Americium	$_{95}^{241}\text{Am}$	α	It items γ rays. The γ-rays obtained are of low energy and no importance.
2.	Plutonium	$_{94}^{239}\text{Pu}$	α	It items γ rays. The γ-rays obtained are of low energy and no importance.
3.	Strontium	$_{38}^{90}\text{Sr}$	β	The decay product of strontium is Yitrium. Yitrium emits the β-particles
4.	Cobalt	$_{27}^{60}\text{Co}$	γ	It emits low-energy β-particles. These are absorbed by the foil surrounding the source.
5.	Radium	$_{88}^{226}\text{Ra}$	$\alpha\beta\gamma$	β-particles are produced by some of the decay products of the radium.

Equivalent dose (Sievert) = RBE × absorbed dose (Gy)
Equivalent dose (rem) = RBE × absorbed dose (rad)
1 rem = 0.01 Sv
Unit of RBE is 1 Sv/Gy or 1 rem/rad

Thus,

$$\frac{1}{\tau} = \frac{1}{\tau_A} + \frac{1}{\tau_B} \tag{22.29}$$

If one categorizes the particle decaying into the π^- and π^+ from the Λ particle to be mode A and mode B, respectively, then the rate R^- of π^- emission is expressed as follows:

$$R^- = \frac{N_0}{\tau_A} e^{-1/\tau} \tag{22.30}$$

The rate R^0 of π^0 emission is given as follows:

$$R^0 = \frac{N_0}{\tau_B} e^{-1/\tau} \tag{22.31}$$

The total decay rate R is given as follows:

$$R = R^- + R^0 = \frac{N_0}{\tau} e^{-t/\tau} \tag{22.32}$$

A particle has only a single mean life, although it can decay in a number of ways.

22.5.1 Half-life

Thus, we observe that the decay of unstable nuclei is a statistical process and has no effect on the physical external environment. The rate of a given reaction varies over a wide range for different nuclides. The decay of unstable nuclei is unpredictable.

As considered in the previous section, $N(t)$ represents a very large number of radioactive (unstable) nuclei at a given time t. If $dN(t)$ represents the negative change in the total number of particles in the short interval dt, then

$$\text{Number of decays during the interval } dt = -dN(t) \tag{22.33}$$

$$\text{Rate of change of } N(t) = -\frac{dN(t)}{dt} \tag{22.34}$$

This is also referred to as the decay rate or the activity of the specimen. The decay rate depends on the total number of unstable nuclei in a specimen, which can be mathematically expressed as follows:

$$-\frac{dN(t)}{dt} = \lambda N(t) \tag{22.35}$$

The constant λ is called the decay constant. The decay constant has different values for different nuclides. A larger value of λ corresponds to a very fast decay of the nuclide, whereas a smaller value corresponds to a slow decay. The preceding equation can be written in terms of the decay constant as follows:

$$\lambda = -\frac{1}{N(t)} \frac{dN(t)}{dt} \tag{22.36}$$

The decay constant is the ratio of the number of decays per unit time to the number of remaining radioactive nuclei; it is referred to as the probability per unit time that any individual nucleus will decay. The preceding equation is similar to the discharge of a capacitor, and thus reduces to the following form:

$$N(t) = N_0 e^{-\lambda t} \tag{22.37}$$

Thus, the plot of $N(t)$ as a function of time is an exponentially decaying graph (Fig. 22.6). In case of functions decaying exponentially with time, it is the convention to define half-life time periods. The half-life $T_{1/2}$ is the time required for a specific number (N_0) of radioactive nuclei to decrease to one-half of

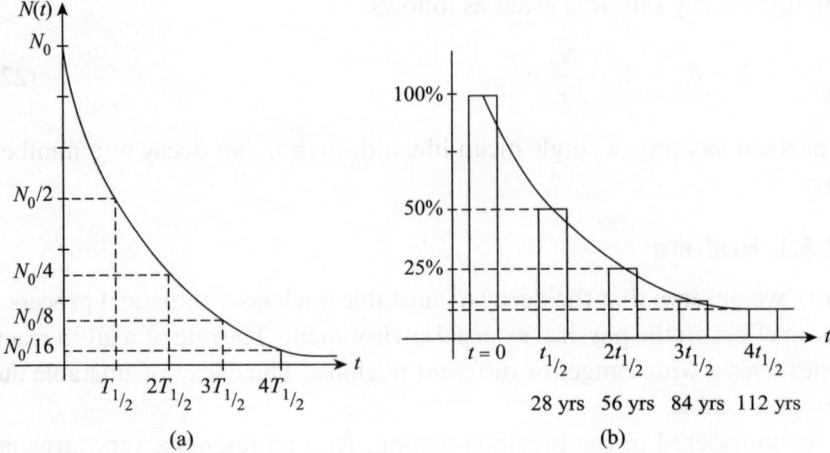

Fig. 22.6 Half-life of a radioactive element (a) Exponential decay of atoms in radioactive element with respect to time (b) Radioactive decay curve for ^{90}Sr. One half-life of ^{90}Sr (strontium-90), the time required for one-half of nuclei of given radioactive isotope to decay, is 28 years

the original number N_0. Then half of the remaining radioactive nuclei decay during a second interval $T_{1/2}$, and thus the process continues. From the preceding equation, it is evident that the number of decaying nuclei after each half-life can be evaluated to be $N_0/2, N_0/4, N_0/8, \ldots$ Thus, the time $T_{1/2}$ can be evaluated as follows:

$$\frac{N(t)}{N_0} = \frac{1}{2} = e^{-\lambda T_{1/2}} \tag{22.38}$$

Taking log on both the sides and solving for $T_{1/2}$:

$$\ln \frac{N(t)}{N_0} = \ln \frac{1}{2} = \ln(e^{-\lambda T_{1/2}}) \tag{22.39}$$

$$\ln 2 = \lambda T_{1/2} \tag{22.40}$$

$$T_{1/2} = \frac{\ln 2}{\lambda} = \frac{0.693}{\lambda} \tag{22.41}$$

The mean lifetime T_{mean}, can be defined in terms of the decay constant (λ) as follows:

$$T_{\text{mean}} = \frac{1}{\lambda} = \frac{T_{1/2}}{\ln 2} = \frac{T_{1/2}}{0.693} \tag{22.42}$$

Thus, the terms mean lifetime T_{mean}, half-life $T_{1/2}$, and decay constant λ define the time period of an unstable nucleus. The decay of a radioactive unstable nucleus follows an exponential graph, which is based on the following equation:

$$\frac{dN(t)}{dt} = -\lambda N(t) = -\lambda N_0 e^{-\lambda t} \tag{22.43}$$

The unit for the measurement of radioactivity has been named after the dedicated 'Curie' couple. The unit of radioactivity *Curie* is abbreviated as 'Ci', which is defined to be 3.7×10^{10} decays per second, which is approximately equal to the activity of 1 g of radium. The SI unit of activity is Becquerel, abbreviated as 'Bq', where 1 Bq is one decay per second, that is,

$$1 \, \text{Ci} = 3.70 \times 10^{10} \, \text{Bq} = 3.70 \times 10^{10} \, \text{decay/s} \tag{22.44}$$

Example 22.10   Plutonium decays as follows with a half-life of 24,000 years:

$$^{239}_{94}\text{Pu} \rightarrow \, ^{235}_{92}\text{U} + \, ^{4}_{2}\text{He}$$

If plutonium is stored for 48,000 years, what fraction of it is remaining?

Solution   The calculations are as follows:

$$\frac{48,000}{24,000} = 2 \text{ half-lives} \Rightarrow \left(\frac{1}{2}\right)^2 = \frac{1}{4}$$

Example 22.11   Derive the definition of Curie as a unit of activity. Assume that the half-life of radium is 1620 years having an atomic weight of $\dfrac{226 \, \text{kg}}{\text{kmol}}$.

Solution   The number of atoms contained in 1 g of radium is calculated as follows:

$$N = \left(\frac{0.001}{226} \text{kmol}\right) \times \left(\frac{6.023 \times 10^{(23+3)} \text{atoms}}{\text{kmol}}\right) = 0.0266 \times 10^{23} = 2.66 \times 10^{21} \text{ atoms}$$

$$\text{Decay constant} = \lambda = \frac{0.693}{T_{1/2}} = \frac{0.693}{(1620 \text{ years})\left(\dfrac{3.16 \times 10^7 \text{s}}{\text{year}}\right)} = 0.000135 \times 10^{-7}/\text{s}$$

$$= 1.35 \times 10^{-11}/\text{s}$$

$$\Delta N = \lambda N \Delta t = (1.35 \times 10^{-11}) \times (2.66 \times 10^{21})(1s) = 3.59 \times 10^{10} \cong 3.6 \times 10^{10}$$

Thus, ΔN represents the disintegrations per second in 1 g of radium. Curie is defined as the number of atoms that decay in 1 s in 1 g sample of radium which is calculated as $3.6 \times 10^{10} \dfrac{\text{disintegrations}}{\text{s}}$.

Example 22.12   The half-life of ^{60}Co is nearly 5.25 years. The activity for a radioactive element is proportional to the number of un-decayed atoms. Find the duration it will take for the activity of the sample to decrease to (a) (1/2) of its original value, (b) (1/4) of its original value, and (c) (1/3) of its original value.

Solution   The calculations are as follows:
(a) The duration for the sample of ^{60}Co to decay to one-half of its original value is 5.25 years.
(b) The duration for the sample of ^{60}Co to decay to one-fourth of its original value is evaluated as follows:

$$\left(\frac{1}{2} \times \frac{1}{2}\right) = \frac{1}{4}, \text{ that is, twice its half lives}$$

Hence, (2×5.25) years $= 10.5$ years.

(c) The graph is plotted for the half-life of ^{60}Co using the following equation:

$$N = N_0 e^{-\lambda t}$$

where the x-axis represents the t years and y-axis the number of atoms. Analysing the graph, one can find that for one-third of its original value, that is, 0.333 of its original value, the corresponding y-axis reading of the number of years is 8.3 years.

Example 22.13 The radioactive isotope ^{57}Co decays by the electron capture with a half-life of 272 days. (a) Find the decay constant and the lifetime. (b) How many radioactive nuclei does a ^{57}Co radiation source of 2.00 μCi activity of contain? (c) Give the radioactivity of this source after 1 year.

Solution The calculations are as follows:

(a) It is given that the half-life of the radioactive isotope ^{57}Co is

$$T_{1/2} = (272 \text{ days}) \times \left(\frac{86,400 \text{ s}}{\text{day}} \right) = 2.35 \times 10^7 \text{s}$$

$$T_{\text{mean}} = \left(\frac{T_{1/2}}{\ln 2} \right) = \frac{2.35 \times 10^7 \text{s}}{0.693} = 3.34 \times 10^7 \text{s}$$

The decay constant can be evaluated as follows:

$$\lambda = \frac{1}{T_{\text{mean}}} = \frac{1}{3.34 \times 10^7} = 2.99 \times 10^{-8} / \text{s}$$

(b) The decay of an unstable radioactive substance is given as follows:

$$-\frac{dN(t)}{dt} = 2.00 \, \mu\text{Ci} = (2.00 \times 10^{-6})(3.70 \times 10^{10} / \text{s}) = 7.40 \times 10^4 \text{ decays/s}$$

$$N(t) = -\frac{dN(t)/dt}{\lambda} = \frac{7.40 \times 10^4 / \text{s}}{2.99 \times 10^{-8} / \text{s}} = 2.51 \times 10^{12} \text{ nuclei} \cong 4.17 \times 10^{-12} \text{ mol}$$

$$\cong 2.38 \times 10^{-10} \text{g}$$

(c) $N(t)$ of nuclei remaining after 1 year is calculated as follows:

$$N(t) = N_0 e^{-\lambda t} = N_0 e^{(-2.95 \times 10^{-8} / \text{s}) \times (3.156 \times 10^7 \text{ s})} = 0.394 N_0 = (0.394) \times (2.00 \mu\text{Ci})$$

$$= 0.788 \, \mu\text{Ci}$$

Example 22.14 A sample of ^{215}At contains 5 mg of the element whose half-life is 100 μs; what is its activity (a) initially and (b) after 150 μs.

Solution The number of radioactive atoms present initially is estimated as follows:

$$N_0 = \frac{(5 \times 10^{-3} \text{g})}{\left(\frac{215 \text{ g}}{\text{mol}} \right)} (6.023 \times 10^{23} \text{ atoms/mol}) = 0.1400 \times 10^{20} \text{ atoms} = 1.40 \times 10^{19} \text{ atoms}$$

The decay constant of ^{215}At is $\lambda = \frac{\ln 2}{T_{1/2}} = \frac{0.693}{100 \times 10^{-6} \text{ s}} = 6930/\text{s}$

(a) Initial activity $= \lambda N_0 = (6930 \times 1.40 \times 10^{19})Bq= 9702 \times 10^{19} = 9.70 \times 10^{22}$Bq

(b) At $t = 1.5 T_{1/2}$, $N = N_0 e^{-\lambda t} = (9.70 \times 10^{22}) e^{-(6930 \times 1.5 \times 100 \times 10^{-6})} = 16.28 \times 10^{22}$

Example 22.15 Measurements indicate that 27.83% of all the rubidium atoms currently on the earth are radioactive ^{87}Rb isotope. The rest are the stable ^{85}Rb isotope. The half-life of ^{87}Rb is 4.89×10^{10} years. Assuming that no rubidium atoms have been formed since, what was the percentage of ^{87}Rb atoms when our solar system was formed 4.6×10^9 years ago?

Solution The calculations are as follows:

$$\lambda T = 2.303 \log \frac{N_0}{N_t} \tag{22.45}$$

$$\lambda T_{1/2} = 2.303 \log \frac{1}{2} = 2.303 \times 0.3010 \tag{22.46}$$

Dividing Eq. (22.45) by Eq. (22.46), one gets the following result:

$$\frac{T}{T_{1/2}} = \frac{2.303}{2.303 \times 0.3010} \log \frac{N_0}{N_t} \Rightarrow \frac{4.6 \times 10^9}{4.89 \times 10^{10}} = \frac{1}{0.3010} \log \frac{N_0}{N_t}$$

$$\frac{4.6 \times 0.3010}{4.89} = \log \frac{N_0}{N_t} \Rightarrow \text{antilog } (0.0283) = \frac{N_0}{N_t} = \frac{1}{30\%}$$

$$1.067 \times 30\% = 32.01\%$$

Example 22.16 The nucleus of $^{15}_{8}$O has a half-life of 122.2 s and $^{19}_{8}$O a half-life of 26.9 s. If at some time a sample contains equal amounts of $^{15}_{8}$O and $^{19}_{8}$O, what is the ratio of $^{15}_{8}$O to $^{19}_{8}$O after (a) 2.0 min and (b) 15.0 min?

Solution The calculations are as follows:

$$\lambda T = 2.303 \log \frac{N_0}{N_t} \tag{22.47}$$

$$\lambda T_{1/2} = 2.303 \log \frac{1}{2} = 2.303 \times 0.3010 \tag{22.48}$$

Dividing Eq. (22.47) by Eq. (22.48), one gets the following expression:

$$\frac{T}{T_{1/2}} = \frac{2.303}{2.303 \times 0.3010} \log \frac{N_0}{N_t}$$

(a)

$$\frac{2}{122.2} = \frac{1}{0.3010} \log \frac{N_0}{O_{15}} \tag{22.49}$$

$$\frac{2}{26.9} = \frac{1}{0.3010} \log \frac{N_0}{O_{19}} \tag{22.50}$$

Subtracting Eq. (22.50) from Eq. (22.49), one gets the following result:

$$2 \times \left(\frac{122.2 - 26.9}{122.2 \times 26.9} \right) = \frac{1}{0.3010} (\log N_0 - \log O_{15} - \log N_0 + \log O_{19})$$

$$0.0174 = \log \frac{O_{19}}{O_{15}} \Rightarrow \text{antilog } (0.0174) = \frac{O_{19}}{O_{15}} = 1.041$$

(b) Similarly, one gets the following result:

$$15 \times \left(\frac{122.2 - 26.9}{122.2 \times 26.9} \right) = \frac{1}{0.3010} (\log N_0 - \log O_{15} - \log N_0 + \log O_{19})$$

$$0.13089 = \log\frac{O_{19}}{O_{15}} \Rightarrow \text{antilog}(0.13089) = \frac{O_{19}}{O_{15}} = 1.352$$

Example 22.17   Estimate the age of an old bone if it has a carbon-14 activity of 4 beta emissions per minute per gram of carbon when it is excavated. Assume that the half-life of the bone element is 5730 years and there were 16 beta emissions per minute when the bone was a living organism.

Solution   Carbon-14 has a half-life of 5730 years

Carbon-14 has two half-lives, so 2(5730) years = 11,460 years

The Carbon-14 activity indicates there are four beta emissions per minute per gram of carbon. As there have been 16 beta emissions per minute, it implies that the bone has gone from $4 \rightarrow 8 \rightarrow 16$, which means two half-lives.

Hence, the estimated age of the bone is nearly 11,460 years.

22.6 RADIOACTIVITY

Radioactivity is the emission of electromagnetic radiation from unstable nuclides. The duration of this decay process can vary from a small fraction of a microsecond to billions of years. A graph, called the 'Segre chart', is plotted between the proton number (Z) and the neutron number (N) for each nuclide. It is observed that for nuclides with a low mass number, the number of protons is nearly equal to the number of neutrons, that is, $N = Z$. The ratio (N/Z) gradually increases with the atomic number (A) due to the increasing electrical repulsion between protons. Nuclides that have either a greater number of protons or a greater number of neutrons do not have nuclear stability. For nuclides having a greater number of protons compared to that of neutrons, the protons exert their electronic repulsion to break the nucleus apart. For nuclides with a greater number of neutrons as compared to protons, the neutrons generate an unbalanced force with respect to protons, leading to the decay of the nucleus. From the graph, it can be observed that nuclides with $A > 209$ or $Z > 83$ are not stable; generally, the nuclides that fall in the stability zone are light nuclides with a Z of up to 22.

The phenomenon of radioactivity can occur naturally for unstable nuclides. An unstable nuclide decay into different nuclides; they emit alpha (α) or beta (β) or gamma (γ) particles (Fig. 22.7). The other processes occurring owing to the nuclear instability are fission or fusion reactions. The number of unstable nuclei far exceeds that of the stable nuclei, in the ratio of 2500:300.

22.6.1 Natural Radioactivity

Henri Becquerel, near the year 1896, discovered radiations from uranium salts. Subsequently, Marie Curie and Pierre Curie, together with Rutherford, observed that the emissions from uranium consisted of positively and negatively charged particles as well as neutral rays. Later, these emissions were categorized as alpha, beta, and gamma rays, respectively, because of their different penetration characteristics.

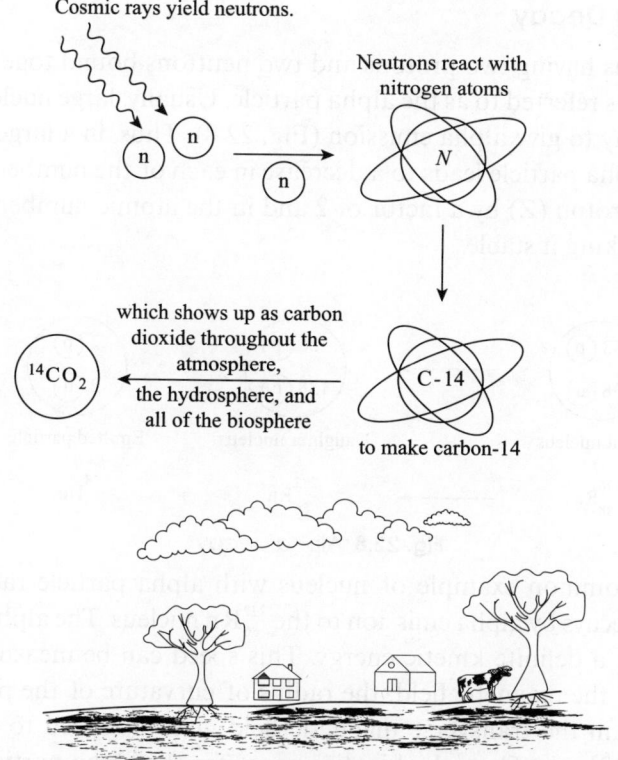

Cosmic rays yield neutrons.

Neutrons react with nitrogen atoms

which shows up as carbon dioxide throughout the atmosphere, the hydrosphere, and all of the biosphere

to make carbon-14

When any organism dies no fresh carbon-14 replaces the carbon-14 decaying in its tissues and the carbon-14 radioactivity decreases by half every 5730 years.

Fig. 22.7 Carbon-14 in environment

The decaying nucleus is called the 'parent nucleus', and the resulting nucleus is referred to as the 'daughter nuclei'. The daughter nuclei, through a series of successive decays, can convert into more stable nuclei.

The most abundant radioactive nucleus is the ^{238}U uranium isotope, which undergoes a series of 14 decays, including eight α emissions and six β^- emissions, that terminate to a stable isotope of lead ^{206}Pb.

The naturally occurring decays of the unstable uranium isotope can be represented by the following equations:

$$^{238}U \rightarrow {}^{234}Th + \alpha \tag{22.51}$$

$$^{234}U \rightarrow {}^{234}Pa + \beta^- \tag{22.52}$$

In Eq. (22.51), due to α particle emission, the atomic number decreases by a factor of 4. In the subsequent β^- emission, the neutron number (N) decreases by 1 and the atomic number increases by 1. The daughter nucleus of ^{234}Pa is in an excited state and decay by the emission of gamma ray photons is represented by the following expression:

$$^{234}Pa^* \rightarrow {}^{234}Pa + \gamma \tag{22.53}$$

22.6.2 Alpha Decay

The ^{4}He nucleus having two protons and two neutrons bound together with total spin zero is referred to as the alpha particle. Usually, large nuclei are not stable and decay to give alpha emission (Fig. 22.8). Thus, in a large nucleus, emission of alpha particle leads to a decrease in each of the numbers of neutron (*N*) and proton (*Z*) by a factor of 2 and in the atomic number (*A*) by a factor of 4, making it stable.

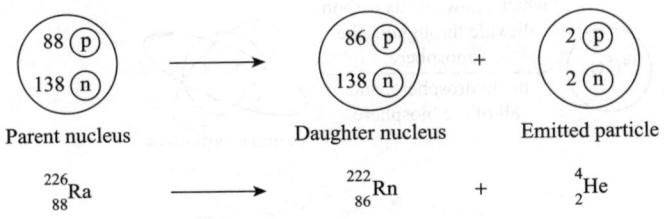

Fig. 22.8 Alpha decay

The most common example of nucleus with alpha particle radiation is $^{226}_{88}$Ra, which decays by alpha emission to the $^{222}_{86}$Rn nucleus. The alpha particle so emitted has a definite kinetic energy. This speed can be measured using the concept of the magnetic field: the radius of curvature of the path of an alpha particle in the transverse magnetic field gives its speed to be nearly about 1.52×10^7 m/s. Thus, the kinetic energy for the alpha particle can be evaluated using the non-relativistic mechanics, that is,

$$\frac{1}{2}mv^2 = \frac{1}{2} \times (1.66 \times 10^{-27}\,\text{kg}) \times \left(\frac{1.52 \times 10^7\,\text{m}}{\text{s}^2}\right)^2 \qquad (22.54)$$

$$= 7.7 \times 10^{-13}\,\text{J} = 4.8\,\text{MeV}$$

The alpha particles so emitted follow the principle of conservation of momentum and energy, as they possess a certain definite kinetic energy. Owing to their charge and mass, alpha particles can travel a certain distance, during which they undergo collisions and come to rest.

While quantum mechanics is discussed, tunnelling of alpha particles through a potential energy barrier can be observed and theorized. Alpha decay obeys the conservation of mass–energy. This decay is possible only for those nuclei for which the mass of the original neutral atom is greater than the sum of the masses of the final neutral atom and the neutral helium-4 atom.

22.6.3 Beta Decay

The beta decay process occurs for nuclides having a large or small neutron to proton ratio, that is, $\left(\dfrac{N}{Z}\right)$ is either too large or too small for the nucleus to be stable (Fig. 22.9). There are three different simple types of beta decay: beta minus, beta plus, and electron capture.

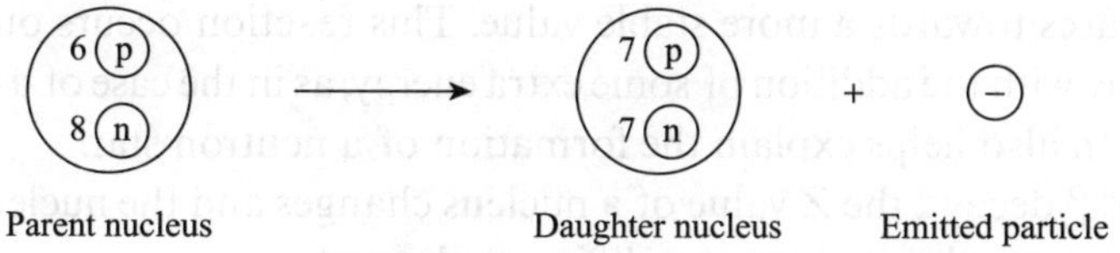

Parent nucleus   Daughter nucleus   Emitted particle

$$^{14}_{6}\text{C} \longrightarrow \, ^{14}_{7}\text{N} + \, ^{0}_{-1}\text{e}$$

Fig. 22.9 Beta decay

A beta-minus particle (β^-) is an electron. When there is a beta minus emission, a neutron and a proton transform to produce an electron; in addition, there is a third particle called an anti-neutrino (discussed in detail in Section 22.7). An anti-neutrino is a $-1/2$ spin neutral particle, taking into account the conservation of angular momentum and conservation of charge. A neutrino is an anti-particle for an anti-neutrino, and together the two possess zero charge and zero or very small mass. The equation for beta decay is as follows:

$$n \rightarrow p + \beta^- + \bar{\vartheta}_e \qquad (22.55)$$

This reaction occurs within the nucleus, although the decay of the neutron outside the nucleus proceeds as indicated in the preceding reaction. Beta particles are emitted with a continuous spectrum of energies. Velocities of the beta particles are near relativistic, that is, $0.9995c$. Beta-minus decay usually occurs when the original atom has a higher neutral atomic mass than the final atom.

A beta-plus particle (β^+) is a positron and ϑ_e is an electron neutrino. Beta-plus decay can occur whenever the neutral atomic mass of the original atom is at least two electron masses larger than that of the final atom. This can be explicitly seen using the conservation of mass–energy:

$$p \rightarrow n + \beta^+ + \vartheta_e \qquad (22.56)$$

This reaction is forbidden by the conservation of mass–energy for a proton outside the nucleus. Thus, this reaction takes place inside the nucleus. In this reaction, the value of N increases by 1 and that of Z decreases by 1 as the neutron–proton ratio increases towards a more stable value.

The third type of beta decay is electron capture. In some nuclides, β^+ emission is not energetically possible, but an orbital electron (usually in the K shell) can combine with a proton in the nucleus to form a neutron and a neutrino. The neutron remains in the nucleus and the neutrino is emitted. The corresponding equation is as follows:

$$p + \beta^- \rightarrow n + \vartheta_e \qquad (22.57)$$

Using the concept of mass–energy, electron capture occurs whenever the neutral atomic mass of the original atom is larger than that of the final atom. In addition, the value of N increases by 1 and Z decreases by 1 as the neutron–proton ratio increases towards a more stable value. This reaction occurs outside the nucleus only with the addition of some extra energy, as in the case of a collision. The reaction also helps explain the formation of a neutron star.

In α and β decays, the Z value of a nucleus changes and the nucleus of one element becomes the nucleus of a different element.

22.6.4 Recoilless Emission of Gamma Rays

A nucleus in an excited energy state may emit a photon (gamma ray) while making a transition to the ground or the unexcited state of the nucleus. The inverse process may also occur, that is, a nucleus in the ground state may absorb a photon, leaving the nucleus in an excited state.

To understand the phenomenon of emission of gamma rays, it is imperative to treat the nucleus in quantum mechanical physics, that is, the internal motion of the nucleus is said to be quantized leading to a set of discrete (allowed) energy bands.

Emission of gamma rays can be designed. A source containing excited nuclei is generated (prepared). Photons are made to strike an absorber that contains similar nuclei in their ground state. These nuclei will absorb the incident photons and will then remit photons. The phenomenon of absorption and re-emission is known as nuclear fluorescence. The photons emitted (by both the source and the absorber) will have a range of energy of approximate width Γ, as can be seen in Fig. 22.10.

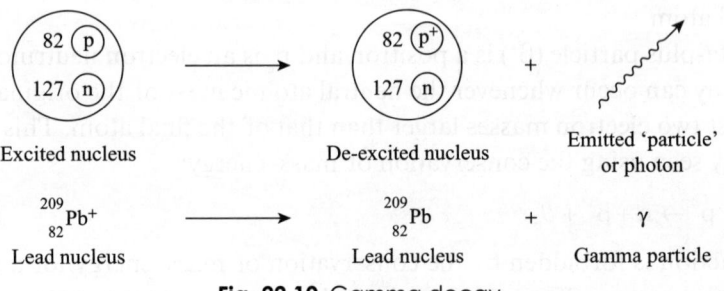

Fig. 22.10 Gamma decay

The energy involved in the decay or excitation of the nucleus is typically of the order of 10 keV to 5 MeV.

In a 'gamma decay' process, the element does not change; the nucleus merely goes from an excited state to a less excited state. Table 22.3 lists the detailed properties and contrasts the α, β, and γ particles.

22.7 CONSERVATION LAWS

Discovery of radioactivity and emission of the alpha, beta, and gamma particles in the then classical or Newtonian world (about the year 1930) led to many speculations on the credibility of the laws of conservation of energy and

Table 22.3 Detailed properties of α, β, and γ particles

Property	α particle	β particle	γ rays
Nature	Helium nucleus	Fast electron	Electromagnetic radiation
Charge	$+3.2 \times 10^{-19}$ C	-1.6×10^{-19} C	0
Rest mass	6.4×10^{-27} kg = 4.0015 u	9.1×10^{-31} kg = 0.00055 u	0
Velocity	~$0.06\,c$	Up to $0.98\,c$	c
Energy	~6 MeV	~1 MeV	hf ~0.01 MeV
Number of ion pairs per cm of air	~10^5	~10^3	~10
Penetration	~5 cm of air	~500 cm of air	~4 cm of lead reduces intensity to 10%
Path through matter	Straight	Tortuous	Straight
Ability to produce fluorescence	Yes	Yes	Yes (weak)
Ability to affect a photographic plate	Yes	Yes	Yes

momentum. Owing to the strong belief in these conservation laws, particle physicists started looking for particles other than the conventional electrons, protons, and neutrons.

Let us consider an unstable nucleus that comes apart into two fragments. If we assume that the total energy is conserved, the two fragments fly apart with velocities v_1 and v_2, following the laws of conservation of energy:

$$Q = \frac{1}{2}m_1v_1^2 + \frac{1}{2}m_2v_2^2 = K_1 + K_2 \tag{22.58}$$

The conservation law follows the fact that lighter particles get more kinetic energy than heavier ones. As the system is supposed to be isolated, the total momentum is zero both before and after the collision. Hence, the final momenta of the two particles must be equal and opposite [Figs 22.11(a) and (b)]:

$$m_1v_1 = -m_2v_2 \tag{22.59}$$

Squaring and dividing by 2, one gets the following expression:

$$\frac{1}{2}m_1^2v_1^2 = \frac{1}{2}m_2^2v_2^2 \tag{22.60}$$

$$m_1K_1 = m_2K_2 \Rightarrow \frac{m_1}{m_2} = \frac{K_2}{K_1} \tag{22.61}$$

$$Q = K_1 + K_2 = K_1 + \frac{m_1}{m_2}K_1 = \left(\frac{m_2 + m_1}{m_2}\right)K_1 \tag{22.62}$$

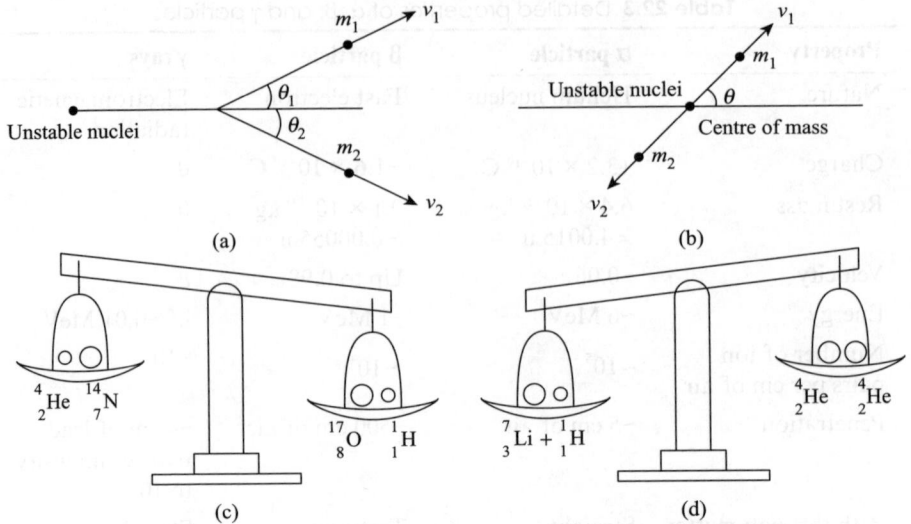

Fig. 22.11 Conservation of momentum for an unstable nuclei (a) After fragment formation in laboratory frame (b) After fragment formation in centre of mass frame (c) Endoergic reaction (mass of reactants < mass of products) (d) Exoergic reaction (mass of reactants > mass of products)

$$Q = K_1 + K_2 = \frac{m_2}{m_1} K_2 + K_2 = \left(\frac{m_2 + m_1}{m_1} \right) K_2 \qquad (22.63)$$

Thus, we can conclude that for a collision in which two bodies emerge after collision, the total available energy is divided in the aforementioned manner, depending on the masses of the emerging bodies. This can be experimentally seen in the case of the decay of the rest nucleus ^{22}Rn, resulting in the emergence of mono-energetic alpha particles (see the calculation of energy in Section 22.6.2).

Thus, the disintegration energy or the Q-value is the net energy, that is, the final energy minus the initial energy. A negative value of Q indicates an endoergic reaction and a positive value indicates an exoergic reaction.

In a collision process, the net number of emerging particles may be >2. It was observed by Henri Becquerel that in beta decay, the emerging particles are not mono-energetic. It became difficult to prove the conservation of energy and momentum unless new particles were introduced. Wolfgang Pauli suggested that an unseen electrically neutral third particle might be emitted in the beta decay process. The idea of a third particle led to several permutations and combinations for the energy and momentum distribution satisfying the conservation laws. Conservation of energy and momentum for three particles can be represented by the triangles of momentum vectors [Figs 22.11(a) and (b)]. Momentum and energy can be divided in such a way that the kinetic energies add up to Q. As seen in the beta decay procedure, a third particle is generated, which was initially named by Pauli as neutron. However, because Chadwick had by then named a massive neutral particle as neutron, it became known as a neutrino, with an anti-particle anti-neutrino. The presence of neutrino and anti-neutrino has helped us understand the celestial phenomenon of supernova. There is a shower of antineutrinos in a supernova explosion of the order of 10^{58}, with a total energy of 10^{46} J, which is over 100 times more than that the sun has emitted in the last 5 billion years.

22.8 DISINTEGRATION ENERGY

In a given nuclear reaction, whether spontaneous or through the particle–nuclear bombardment, the classical laws of conservation for charge, momentum, angular momentum, and energy are all obeyed. The conservation of charge requires that the sum of the initial atomic numbers must equal the sum of the final atomic numbers. In a given nuclear reaction, the nucleon number, that is, the number of protons and neutrons, is also conserved. Nuclear reactions are not elastic collisions; hence, the total initial mass is not equal to the total final mass. According to the principle of mass–energy equivalence from relativity, the difference in mass must be translated to energy. Hence, in a nuclear reaction, the difference between the masses before and after the reaction corresponds to the 'reaction energy' or the 'disintegration energy' (Q) [Figs 22.11(c) and (d)]. If the initial particles u and v interact to produce final particles X and Y, the reaction energy Q is mathematically represented as follows:

$$Q = (M_u + M_v - M_X - M_Y)c^2 \qquad \text{(reaction energy)} \qquad (22.64)$$

$$Q = +\text{ve}; \Rightarrow \Delta m = \text{decreases}; K = \text{increases} \quad \text{(exoergic reaction)} \quad (22.65)$$

$$Q = -\text{ve}; \Rightarrow \Delta m = \text{increases}; K = \text{decreases} \quad \text{(endoergic reaction)} \quad (22.66)$$

The experimental Q values of various reactions are used for the accurate and precise measurement of nuclear masses. The careful measurements of the Q value of reactions involving light nuclei have made it possible to assign mass values to all known light atoms without taking any recourse to the mass-spectroscopic data. This has been accomplished by the investigation of the various reaction chains that link light nuclei to the standard O^{16}. Thus, experimental Q values not only allow a valuable independent check on the mass-spectroscopic measurements, but also provide the only available method for measuring the masses of many of the unstable nuclides created during nuclear reactions.

Nuclear reactions for which masses of the reactants and end products are equal, that is, $Q = 0$, indicate cases of elastic collision. As mentioned earlier, usually nuclear reactions do not fall under the category of elastic collisions.

Various types of transmutation nuclear reactions can occur, depending upon the nature of the projectile particle and the outgoing particle, some of which are as follows:

1. Transmutation by the α particle: (α, p), (α, n)
2. Transmutation by protons: (p, α), (p, n), (p, d), (p, γ)
3. Transmutation by neutrons: (n, α), (n, p), (n, 2n), (n, γ), (n, n)

4. Transmutation by photons: (γ, n), (γ, p)
5. Transmutation by deuterons: (d, α), (d, p), (d, n), $(d, 2n)$

These symbols are as per the convention, where α, p, d, n, and γ represent the alpha particle, proton, deuteron, neutron, and gamma ray, respectively. The following are some popular examples for each one of these reactions:

1. (a) (α, p) reactions

$$_5B^{10} + _2He^4 \rightarrow _6C^{13} + _1H^1 \tag{22.67}$$

$$_{11}Na^{23} + _2He^4 \rightarrow _{12}Mg^{26} + _1H^1 \tag{22.68}$$

(b) (α, n) reactions

$$_3Li^7 + _2He^4 \rightarrow _5B^{10} + _0n^1 \tag{22.69}$$

$$_{11}Na^{23} + _2He^4 \rightarrow _{13}Al^{26} + _0n^1 \tag{22.70}$$

2. (a) (p, α) reactions

$$_3Li^7 + _1H^1 \rightarrow _2He^4 + _2He^4 \tag{22.71}$$

$$_9F^{19} + _1H^1 \rightarrow _8O^{16} + _2He^4 \tag{22.72}$$

(b) (p, n) reactions

$$_8O^{18} + _1H^1 \rightarrow _9F^{18} + _0n^1 \tag{22.73}$$

$$_{11}N^{23} + _1H^1 \rightarrow _{12}Mg^{23} + _0n^1 \tag{22.74}$$

(c) (p, d) reactions

$$_4Be^9 + _1H^1 \rightarrow _4Be^8 + _1H^2 \tag{22.75}$$

$$_3Li^7 + _1H^1 \rightarrow _3Li^6 + _1H^2 \tag{22.76}$$

(d) (p, γ) reactions

$$_5B^{11} + _1H^1 \rightarrow _6C^{12} + \gamma \tag{22.77}$$

$$_6C^{12} + _1H^1 \rightarrow _7Na^{13} + \gamma \tag{22.78}$$

3. (a) (n, α) reactions

$$_3Li^6 + _0n^1 \rightarrow _1H^3 + _2He^4 \tag{22.79}$$

$$_{11}Na^{23} + _0n^1 \rightarrow _9F^{20} + _2He^4 \tag{22.80}$$

(b) (n, p) reactions

$$_7Na^{14} + _0n^1 \rightarrow _6C^{14} + _1H^1 \tag{22.81}$$

$$_{30}Zn^{64} + _0n^1 \rightarrow _{29}Cu^{64} + _1H^1 \tag{22.82}$$

(c) $(n, 2n)$ reactions

$$_{29}Cu^{63} + _0n^1 \rightarrow _{29}Cu^{62} + _0n^1 + _0n^1 \tag{22.83}$$

$$_{92}U^{238} + _0n^1 \rightarrow _{92}U^{237} + _0n^1 + _0n^1 \tag{22.84}$$

(d) (n, γ) reactions

$$_{13}Al^{27} + _0n^1 \rightarrow _{13}Al^{28} + \gamma \tag{22.85}$$

$$_{92}U^{238} + _0n^1 \rightarrow _{92}U^{239} + \gamma \tag{22.86}$$

4. (a) (γ, n) reactions

$$_1H^2 + \gamma \to _1H^1 + _0n^1 \qquad (22.87)$$

$$_{15}P^{31} + \gamma \to _{15}H^{30} + _0n^1 \qquad (22.88)$$

(b) (γ, p) reactions

$$_{12}Mg^{25} + \gamma \to _{11}Na^{24} + _1H^1 \qquad (22.89)$$

$$_4Be^9 + \gamma \to _3Li^8 + _1H^1 \qquad (22.90)$$

5. (a) (d, α) reactions

$$_8O^{16} + _1H^2 \to _7N^{14} + _2He^4 \qquad (22.91)$$

$$_{20}Ca^{40} + _1H^2 \to _{19}K^{38} + _2He^4 \qquad (22.92)$$

(b) (d, p) reactions

$$_3Li^7 + _1H^2 \to _3Li^8 + _1H^1 \qquad (22.93)$$

$$_{15}P^{31} + _1H^2 \to _{15}P^{32} + _1H^1 \qquad (22.94)$$

(c) (d, n) reactions

$$_6C^{12} + _1H^2 \to _7N^{13} + _0n^1 \qquad (22.95)$$

$$_{83}Bi^{209} + _1H^2 \to _{84}Po^{210} + _0n^1 \qquad (22.96)$$

Example 22.18   Evaluate whether the following reaction is endoergic or exoergic:

$$_2He^4 + _7N^{14} \to _1H^1 + _8O^{17}$$

Solution   Neutral masses of the elements involved in the reaction are as follows:

$$M(_2He^4)(= 4.003873) + M(_7N^{14})(= 14.007515) \to$$

$$M(_1H^1)(= 1.008142) + M(_8O^{17})(= 17.004533)$$

$$Q = (4.003873 + 14.007515) - (1.008142 + 17.004533)$$

$$Q = (18.011388) - (18.012675) = -0.001287 \text{ MU} = -1.287 \text{ mMU}$$

$$Q = -\frac{1.287}{1.07395}\text{ MeV} = -1.198 \text{ MeV}$$

Thus, as the value of the threshold energy is negative, the nuclear reaction is an endoergic reaction.

Example 22.19   When an ^{7}Li nucleus is bombarded by a proton, two alpha particles (^{4}He) are produced. What is the reaction energy? State also the reaction name.

Solution   The following reaction takes place:

$$_1^1H + _3^7Li \to _2^4He + _2^4He$$

$$M(_1^1H)(= 1.007825) + M(_3^7Li)(= 7.016003) \to$$

$$M(_2^4He)(= 4.002603) + M(_2^4He)(= 4.002603)$$

$$Q = (1.007825 + 7.016003) - (4.002603 + 4.002603)$$

$$Q = (8.023828 - 8.005206) = 0.018622 \text{ MU} = 18.622 \text{ mMU}$$

$$Q = -\frac{18.622}{1.07395}\text{ MeV} = 17.35 \text{ MeV}$$

This is an exoergic reaction, as the final total kinetic energy of the two separating alpha particles is 17.35 MeV greater than the initial total kinetic energy of the proton and the lithium nucleus.

Note: It is customary to express mass decrements and packing fractions in milli mass units (mMU), whereas nuclear energies and binding energies are usually expressed in million electron volts (MeV).

$$1\,\text{mMU} = 0.93115\,\text{MeV}; 1\,\text{MeV} = 1.07395\,\text{mMU}$$

Example 22.20   Find the Q of the reaction $^{14}_{7}\text{N}(d, \alpha)\,^{12}_{6}\text{C}$.
Solution   The calculations are as follows:

Mass of $^{16}_{7}\text{N} = 14.00307$;   Mass of $^{2}_{1}\text{H} = 2.00410$

Total mass $= 16.00717\,\text{u}$

Mass of $^{12}_{6}\text{C} = 12.000$;   Mass of $^{4}_{2}\text{He} = 4.002603$

Total mass $= 16.002603\,\text{u}$

Net loss in mass $= (16.00717 - 16.002603)\,\text{u} = 0.01457\,\text{u}$

$$Q = (0.01457)\,\text{u}\,\frac{931.48\,\text{MeV}}{\text{u}} = 13.6\,\text{MeV}$$

Example 22.21   Carbon dating: A skull fragment was found in the cave that was inhabited by early humans. It contained 0.21 times as much ^{14}C as an equal amount of carbon in the atmosphere when the organism containing the skull died. Find the approximate age of the fragment.
Solution   The calculations are as follows:

$$\lambda T = 2.303 \log \frac{N_0}{N_t} = 2.0303 \log \frac{100}{21} \tag{22.97}$$

$$\lambda T_{1/2} = 2.303 \log \frac{1}{2} = 2.303 \times 0.3010 \tag{22.98}$$

Dividing Eq. (22.97) by Eq. (22.98), one gets the following result:

$$\frac{T}{T_{1/2}} = \frac{2.303}{2.303 \times 0.3010} \log \frac{100}{21} = \frac{T_{1/2}}{0.3010} \times 0.677780$$

$$T = \frac{5730}{0.3010} \times 0.67780 = 12,902\,\text{years} = 1.3 \times 10^4\,\text{years}$$

22.9 THRESHOLD ENERGY

In any given nuclear reaction, it is essential for the particles to possess a certain minimum kinetic energy, referred to as the 'threshold energy' disintegration or reaction energy. This energy is crucial for an endoergic reaction, as the initial kinetic energy of particles must be at least as great as the disintegration or reaction energy. Hence, a minimum 'threshold energy' or a minimum kinetic energy is required for an endoergic reaction to take place. Usually, such reactions are those where accelerated particles are made to bombard the nucleus.

The most common example is bombarding a stationary ^{14}N nucleus with alpha particles from an accelerator. The kinetic energy of the alpha particle must be greater than 1.91 MeV. This minimum kinetic energy is essential for the particle to overcome the potential energy barrier created by the repulsive electrostatic forces.

22.10 THEORY OF NUCLEAR FISSION

Various nuclear phenomena have been investigated using different models, in particular the liquid drop model, which is discussed in the following section.

22.10.1 Liquid Drop Model

The liquid drop model (Fig. 22.12) has been investigated by Weiszacker, Bethe, Bacher, Bohr, Wheeler, and Feenberg. This model has been exploited widely in nuclear physics as a basis for the prediction and interpretation of nuclear phenomena.

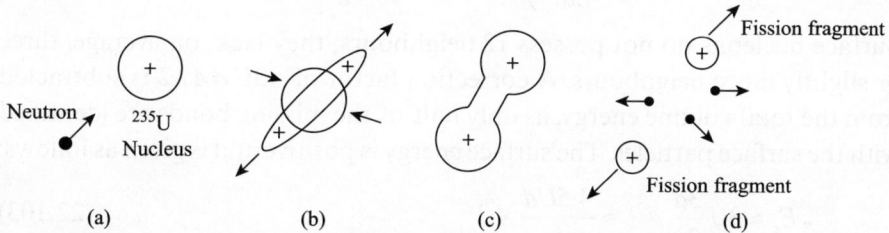

Fig. 22.12 Liquid drop model (a) Neutron is absorbed by nucleus (b) ^{235}U nucleus oscillates strongly and in highly excited state (c) Two fragments are separated due to electrical repulsion (d) Fission process is completed by emission of neutrons

Individuality of nucleons in complex nuclei is not taken into account owing to the number of nucleons. The simple classical considerations of the volume binding energy, surface energy, Coulomb energy, and symmetry energy together give a cohesive picture of the structure of a nucleus.

22.10.1.1 Volume Binding Energy

A complex nucleus is a collection of a large number of incompressible rigid spheres held together by charge-independent nuclear forces. The shape of the nucleus is spherical in nature due to its closest packing fraction. Nuclear forces are of short range such that every interior nucleon has 12 bonds, one with each neighbour in its contact. If the bonding energy of each bond is $-U$, then $-6U$ energy is associated with each interior nuclear particle, as two particles share $-U$ bond energy. Thus, the total volume binding energy is approximately given as follows:

$$E_V = -6UA \tag{22.99}$$

We will discuss the discrepancy in this equation due to the nucleons on the surface in the subsequent section of surface energy. The volume energy is negative, thus holding the nucleons together.

22.10.1.2 Surface Energy

The radius of the complex nuclei is given by the following equation:

$$R = bA^{\frac{1}{3}} \tag{22.100}$$

The number of particles per unit volume in a sphere of radius R containing A nucleons is given by the following equation:

$$\rho = \frac{A}{V} = \frac{3A}{4\pi R^3} = \frac{3}{4\pi b^3} \tag{22.101}$$

In a surface shell of thickness d, which is equal to the diameter of a nucleon, the number of particles is approximately given as follows:

$$\delta A = \rho d 4\pi R^3 = \left(\frac{3}{4\pi b^3}\right) d 4\pi (bA^{\frac{1}{3}})^2 = \frac{3d}{b} A^{\frac{2}{3}} \tag{22.102}$$

Surface nucleons do not possess 12 neighbours; they lack, on average, three or slightly more neighbours. A correction factor of $-3U\,\delta A/2$ is subtracted from the total volume energy, as only half of the missing bonds are identified with the surface particles. The surface energy is positive and is given as follows:

$$E_s \approx 3U \frac{3d}{2b} A^{\frac{2}{3}} \approx \frac{4.5Ud}{b} A^{\frac{2}{3}} \tag{22.103}$$

The concept of surface energy can also be understood using the concept of nuclear surface tension. If a nuclear surface tension of magnitude S_S exists, then the total surface energy associated is given as follows:

$$E_s = 4\pi R^2 S_s = 4\pi (bA^{\frac{1}{3}})^2 S_s = 4\pi b^2 A^{\frac{2}{3}} S_s \tag{22.104}$$

Thus, the concept of surface energy together with short-range forces maximizes the nuclear bonds, minimizing the surface area, leading to the spherical shape of complex nuclei.

22.10.1.3 Coulomb Energy

The atomic number Z of nucleons is the proton count of a complex nucleus, each of which carries a charge of $+e$. The nucleus is a sphere of radius $R = bA^{\frac{1}{3}}$, having a charge $q = Ze$ and a uniform charge density given by the following relation:

$$\rho_c = \frac{3q}{4\pi R^3} = \frac{3Ze}{4\pi b^3 A} \tag{22.105}$$

The electrostatic energy is the work done against the electrostatic forces in assembling a nuclear sphere of radius r. Let us suppose that we have assembled a sphere of radius r, and now add on a shell of thickness dr, which contains the charge given by the following equation:

$$dq = \rho_c 4\pi r^2 dr \qquad (22.106)$$

The entire charge of the sphere is concentrated at the centre; the work done in bringing dq from infinity to r is given as follows:

$$dE_c = \rho_c \frac{4\pi r^3}{3} (\rho_c 4\pi r^2 dr) \frac{1}{r} \qquad (22.107)$$

Integrating,

$$\int_0^R dE_c = \frac{16\pi^2 p_c^2}{3} \int_0^R r^4 dr \qquad (22.108)$$

$$E_c = \frac{\rho_c^2 16\pi^2 R^5}{15} = \left(\frac{3Ze}{4\pi b^3 A}\right)^2 \frac{16\pi^2}{15} R^5 = \frac{3}{5}\frac{Z^2 e^2}{R} = \frac{3}{5}\frac{Q^2}{bA^{1/3}} \qquad (22.109)$$

The coulomb energy is positive and increases with the atomic number. A complex nucleus minimizes the coulomb energy usually by converting into an all-neutron system, which is achieved by β^+ decay or K capture.

The stable light nuclei prefer the $\approx N$, which is indicative of the nuclear force that binds nucleons together.

22.10.1.4 Symmetry Energy

Nuclear forces are charge symmetric owing to the low-lying nuclear quantum states. Experimental evidence exists that had there been no columbic interaction between protons and neutrons and no neutron–proton mass difference, the energy states of protons and neutrons would have been identical. There is a tendency of the particles to acquire states with the minimum energy. This leads to the occurrence of nuclear energy degeneracy.

In the case of systems with $A = 4n$; the n, $(n + 1)$, $(n + 2)$, $(n + 3)$ states distribute themselves to minimize the coulomb energy and the effect of the proton–neutron mass difference. If the proton–neutron mass difference is ignored, then in an $A = 4n$ nucleus, the last four particles may go into the neutron state of the n, $(n + 1)$, $(n + 2)$, $(n + 3)$ states, producing a stable neutron-rich system. In the case of the $A = 4(n + 1)$ system, the last eight particles go into the up- and down-spin neutron states, giving a highly neutron-rich system. The last 12 particles in the $A = 4(n + 2)$ system or the last 16 particles in the $A = 4(n + 3)$ system would have to distribute themselves with eight in the neutron states and four in the proton states or eight in the neutron states and eight in the proton states, respectively. This argument can be modified somewhat when the neutron–proton mass difference is considered, but one expects the line of beta stability to have a very characteristic type of irregularity when we come across degenerate nuclear states. The irregularities observed in the line of beta stability do not conform to the simple model discussed earlier; actual evidence suggests that, in the hypothetical system, the tendency of nuclei to have equal N and Z persists despite degeneracy, as the tendency to fill the quantum states completely is much stronger than expected by this simple model.

Thus, a complex quantum-mechanical argument using exchange forces is capable of accounting for this strong tendency to equal N and Z. Qualitatively, an exchange force, which is attractive for all the four particles in the same quantum state but repulsive between identical particles in different states, even if these states are degenerate, has to be introduced. A stable nuclear system containing A nucleons would be filled in sets of four with $N = Z = A/2$, as such a system would have the greatest number of attractive bonds and the least number of repulsive bonds. Any other proportion of N and Z for A would involve fewer attractive bonds and more repulsive bonds, and would thus be less stable. The additional symmetry energy relative to an equally divided system is given by the following equation:

$$E_{\text{sym}} = 6kUA\left(\frac{D}{A}\right)^2 \tag{22.110}$$

where k is a constant relating the strength of the exchange force to the strength of the ordinary force.

22.10.1.5 Pairing Energy

Nuclear forces prefer paring of protons and neutrons. This energy term is positive (more binding) if both Z and N are even, negative (less binding) if both Z and N are odd, and zero if one of Z or N is even and the other odd. The pairing energy term that best fits the data is given as follows:

$$E_{\text{pairing}} = \pm A_5 A^{-4/3} \tag{22.111}$$

22.10.1.6 Semi-empirical Mass Surface

The five terms discussed earlier together give the total interaction energy of a complex nucleus, as follows:

$$E_B = -6UA + 4\pi b^2 A^{2/3} S_s + \frac{3}{5}\frac{Q^2}{bA^{1/3}} + 6kUA\left(\frac{D}{A}\right)^2 \pm A_5 A^{-4/3} \tag{22.112}$$

Values of the various constants in the equation should be ascertained, but due to a lack of understanding of the basic nature of the nuclear forces, there is no definite set of theoretical constants that are consistent with the nuclear energy data. The constants are adjusted to fit the experimental nuclear energy data. The nuclear energy equation can further be written as follows:

$$E_B = -A_1 A + A_2 A^{2/3} + A_3 \frac{Z^2}{A^{1/3}} + A_4 \frac{D^2}{A} \pm A_5 A^{-4/3} \tag{22.113}$$

This formula has been derived by Weiszacker, where the A and s are adjustable constants and the equation is semi-empirical. The values of the constants that are chosen to make this formula best fit the observed binding energies of nuclides are as follows:

$$A_1 = 15.75 \text{ MeV} \tag{22.114}$$

$$A_2 = 17.80 \text{ MeV} \tag{22.115}$$

$$A_3 = 0.7100 \text{ MeV} \tag{22.116}$$

$$A_4 = 23.69 \text{ MeV} \tag{22.117}$$

$$A_5 = 39 \text{ MeV} \tag{22.118}$$

The constant A_1 is the binding energy per nucleon due to the saturated nuclear force. The energy is nearly 16 MeV per nucleon, almost double the total binding energy per nucleon in most nuclides. The binding energy E_B can be estimated and subsequently used to ascertain the mass of any neutral atom, as follows:

$$_Z^A M = ZM_H + Nm_n - E_B / c^2 \tag{22.119}$$

This above equation is referred to as the semi-empirical mass formula. The word empirical is used because the constants have been determined experimentally. The formula has a sound theoretical basis.

22.10.1.7 Disadvantages of Liquid Drop Model

The liquid drop model, even with the best semi-empirical functions, does not conform to the detailed trend of the experimental data. The main assumptions that have led to the discrepancies are the statistical assumptions in connection to light nuclei, assumption of uniform charge density, incompressibility assumption, and assumption that shell effects are small and can be allowed by simply smoothing out the observed irregularities.

Thus, these assumptions do not correctly interpret the factual nuclear binding energy picture. The aspects of angular momentum and excited states are better understood by other models.

22.11 THEORY OF NUCLEAR FISSION

The process of fission is the breaking up of a heavy element into two equal fragments, resulting in the release of high reaction energy that is much greater than the reaction energy associated with any of the usual bombardment or decay reactions. This process was discovered by Hahn and Strassman in 1939.

The two most typical fission reactions are as follows:

$$_{92}^{235}\text{U} + {}_0^1\text{n} \rightarrow {}_{92}^{236}\text{U}^* \rightarrow {}_{56}^{144}\text{Ba} + {}_{36}^{89}\text{Kr} + 3{}_0^1\text{n} \tag{22.120}$$

$$_{92}^{235}\text{U} + {}_0^1\text{n} \rightarrow {}_{92}^{236}\text{U}^* \rightarrow {}_{54}^{140}\text{Xe} + {}_{38}^{94}\text{Sr} + 2{}_0^1\text{n} \tag{22.121}$$

The total kinetic energy of fission fragments is enormous. The semi-empirical mass formula can be used to estimate the energy released in the fission process. The binding energy curve [Fig. 22.3(a)] shows that the nuclides at the high end of the mass spectrum ($A > 240$) are less tightly bound than those near the middle ($A = 90–145$). The average binding energy per nucleon is about 7.6 MeV at $A = 240$ and about 8.5 MeV at $A = 120$.

$$\text{Increase in binding energy} = 8.5 \text{ MeV} - 7.6 \text{ MeV}$$
$$= 0.9 \text{ MeV per nucleon} \tag{22.122}$$

$$\text{Mass of uranium} = 235 \tag{22.123}$$

$$\text{Energy} = (235) \times (0.9 \text{ MeV}) = 200 \text{ MeV} \tag{22.124}$$

Thus, the nuclear fragments carry away an appreciable amount fraction of the fission energy as internal excitation energy; these fragments are actually energetically capable of emitting neutrons. These so-called 'prompt' neutrons are emitted almost instantaneously ($\sim 10^{-12}$ s) after the fission process. Generally, as indicated in the equation also, about two or three neutrons are instantaneously emitted per fission.

Fission fragments are usually neutron rich. As mentioned earlier, the neutron/proton N/Z ratio is usually 1 for light nuclides and 1.6 for the heaviest nuclides. For ^{235}U, the N/Z ratio is about 1.55, which is the same for fragments as well. Gradually, the fragments undergo further decay, wherein the N/Z ratio reduces and the nucleus becomes more stable by the process of β^- decays. This can be observed in the following reaction that shows decay of a xenon nucleus:

$$^{140}_{54}\text{Xe} \xrightarrow{\beta^-} {}^{140}_{55}\text{Cs} \xrightarrow{\beta^-} {}^{140}_{56}\text{Ba} \xrightarrow{\beta^-} {}^{140}_{57}\text{La} \xrightarrow{\beta^-} {}^{140}_{58}\text{Ce} \tag{22.125}$$

The nuclide ^{140}Ce is stable.

All nuclei having $A \geq 90$ are energetically unstable relative to decay by fission; nevertheless, the probability for spontaneous emission of naturally occurring heavy nuclei in their ground states is very small. The half-life of U^{235} relative to spontaneous fission is about 10^{17} years. There exists a highly effective barrier against fission, which holds any naturally occurring nucleus in a meta-stable state of equilibrium. A classical energy W^a, referred to as the activation energy, is required to bring a nucleus to the condition of unstable equilibrium, from which it will spontaneously dissociate into two parts. For heavy nuclei having $A \sim 230$, the activation energies are of the order of 4–6 MeV compared to medium-weight nuclei having activation energies of the order of 50 MeV. Looking at the activation energies, it is practically possible to observe the fission process for heavy nuclei.

The process of fission can be explained using the liquid drop model. A heavy nucleus is treated as a classical liquid droplet held together by nuclear surface tension. As the heavy nucleus acquires the excitation energy, it starts undergoing vibratory distortions. When the excitation energy exceeds W^a, the activation energy, the nucleus in its vibratory motion may reach a critical deformation state for fission before it loses part or all of its excitation energy by some radiation or neutron emission process. The nucleus then sponta- neously subdivides, releasing a very large amount of fission energy. The activation energies for heavy nuclei such as Th^{232}, U^{233}, U^{235}, U^{238}, and Pu^{239} range between 6.7 and 4.9 MeV.

The classical liquid drop model is capable of giving gross values for the acti- vation energy, but a more precise quantum model is needed for understanding the fine energy structure. The activation energy acts as a barrier against fission; in the quantum model, a heavy nucleus may undergo fission spontaneously by tunnelling through the barrier against fission, although the probability is small. If the fissioning nucleus is excited, the probability for fission increases rapidly with the energy of excitation, as the excitation energy approaches the activation energy. The capture of neutrons also leads to appreciable fission yields, provided that the excitation energy of the compound nucleus exceeds the activation energy required for the fission of the compound nucleus.

22.11.1 Sustained Fission Reactions

After Hahn and Strassmann correctly interpreted fission reaction, it was recog- nized by scientists that this reaction fulfilled the two basic requirements to serve as the basic process for a sustained nuclear-energy generator. The reaction is exoergic and automatically produces neutrons (initiators of the reaction). The basic nuclear reaction in a sustained fission process is the capture of a neutron by a very heavy nucleus, followed by splitting of the composite nucleus into two approximately equal parts, with the ejection, on an average, of about two and a half neutrons and liberation of energy. A nuclear reactor or pile is a device for harnessing nuclear power by providing the conditions for reproduction or regeneration of the fission reaction. In a reactor, the fuel or fissionable material is arranged in a manner to promote multiplication of neutrons, which are the initiators and products of the reaction. Some of the neutrons inevitably escape the surface of the pile; for a sustained process, the pile should be large enough.

For the reactor to be of a reasonable size, it is necessary that the fissionable fuel meet several mandatory requirements. U^{235} is the only naturally occurring substance that meets these requirements. Naturally occurring uranium samples contain 99.28% of U^{238} and 0.72% of U^{238}. It is possible to operate a controlled reactor using natural uranium based upon the fission of U^{235}. For such a pile, it is essential to use a moderator to slow down the fast neutrons ejected during fission and thereby improve their chances for fission capture with respect to radiative capture. At thermal energies, the radiative capture cross section of U^{238} is 2.8 barns and the fission cross section is vanishingly small.

$$\text{Neutron fission cross section} = 550 \times 0.72 = 3.9 \text{ barns} \tag{22.126}$$

$$(n, \gamma) \text{ cross section} = (2.8) \times (99.28) + (101 \times 0.72) = 3.5 \text{ barns} \tag{22.127}$$

A comparison of the neutron-fission cross section with the (n, γ) cross section makes it evident that at thermal energies more than half the neutrons moving through a mass of natural uranium will cause fission. On an average, 2.5 neutrons are emitted per thermal emission. For a sustained fission process, the loss of neutrons can be held to a minimum.

Other fissionable fuel requirements are fulfilled by breeding plutonium from a pile of naturally available uranium. Neutrons available from the neutron fission of U^{235} are captured by the compound nucleus U^{238}. The compound nucleus decays instantly by gamma emission, leaving the neutron-rich odd mass nuclide U^{239}. The reactions that take place are as follows:

$$_{92}U^{239} \rightarrow {}_{92}Np^{239} + \beta^- \quad (23 \text{ min}) \tag{22.128}$$

$$_{92}Np^{239} \rightarrow {}_{94}Pu^{239} + \beta^- \quad (23 \text{ days}) \tag{22.129}$$

Th^{232} can also be converted into fissionable U^{233}. The (n, γ) reaction leads to the neutron-rich Th^{233}, which decays as follows:

$$_{90}Th^{233} \rightarrow {}_{91}Pa^{233} + \beta^- \quad (23.5 \text{ min}) \tag{22.130}$$

$$_{91}Pa^{233} \rightarrow {}_{92}U^{233} + \beta^- \quad (27.4 \text{ days}) \tag{22.131}$$

Thus, thorium can be used to yield fissionable U^{233}. Several methods have been used for the separation of U^{235}, Pu^{239}, U^{233}. Such purified fissionable material in nuclear reactors increases the rate of the fission process.

22.11.2 Controlled Chain Reaction

A nucleus decays through the fission process. Fission is triggered by neutron bombardment. This neutron bombardment may trigger more fission reactions, indicating the chances of a sustained reaction or a chain reaction. A chain reaction can be made to proceed slowly and in a controlled manner in a nuclear reactor, or explosively in a bomb. The energy released in a nuclear chain reaction is enormous, much greater than that released in a chemical reaction. The chemical reaction for the formation of uranium oxide gives a heat of combustion of 4500 J/g in contrast to a fission reaction energy of about 200 MeV per atom.

22.11.3 Nuclear Reactor

A nuclear reactor is a system in which a controlled nuclear chain reaction is used to liberate energy. A fission ^{235}U nucleus produces about 2.5 free neutrons. A ^{235}U nucleus is likely to absorb a low-energy neutron (less than 1 eV) compared to one of the higher-energy neutrons (1 MeV) that are liberated during fission. As indicated earlier, for sustaining the chain reaction, about 40% neutrons are required. A moderator (usually cadmium) is used in a nuclear reactor to slow down the neutrons so that the fission process continues efficiently.

22.11.4 Fusion Reaction

The word 'fusion' means 'joining together'. Two nuclei join together to release a significant amount of energy. Light nuclei 'fuse' together, in an exoergic reaction, to form heavier nuclei with the release of energy; the most common example of fusion reaction is hydrogen fusion.

The sun is a stellar representative of a source of fusion energy. Four hydrogen nuclei are fused together to produce a helium nucleus having four protons. The process of fusion is not a one-step procedure, but takes various steps and requires a critical amount of energy, with the net result being represented by the following equation:

$$4{}^{1}_{1}\text{H} \rightarrow {}^{4}_{2}\text{He} + 2({}^{0}_{+1}\text{e}) \tag{22.132}$$

The new particle ${}^{0}_{+1}\text{e}$ is an anti-particle of electron and is referred to as a positron. The other simple reaction that represents energy-liberating fusion reactions are as follows:

$${}^{1}_{1}\text{H} + {}^{1}_{1}\text{H} \rightarrow {}^{2}_{1}\text{H} + (\text{or } {}^{0}_{+1}\text{e}) \, \beta^{+} + \vartheta_{e} \tag{22.133}$$

Here, two protons combine to form a deuteron (${}^{2}\text{H}$), with the emission of a positron (β^{+}) and an electron neutrino.

$${}^{2}_{1}\text{H} + {}^{1}_{1}\text{H} \rightarrow {}^{3}_{2}\text{He} + \gamma \tag{22.134}$$

Here, a deuteron and a proton combine to form a light isotope of helium with the emission of gamma ray.

$${}^{3}_{2}\text{He} + {}^{3}_{2}\text{He} \rightarrow {}^{4}_{2}\text{He} + {}^{1}_{1}\text{H} + {}^{1}_{1}\text{H} \tag{22.135}$$

Here, the two helium nuclei combine to form an alpha particle and two protons.

All these equations together make up the process called the proton–proton chain, where the four protons in the reactions, represented by the preceding equations, fuse to release one alpha particle, two positrons, two electron neutrinos, and gamma rays.

Using the mass–energy equivalence,

$$\text{Mass of four protons} = 4.029106 \text{ u} \tag{22.136}$$

$$\text{Mass of } {}^{4}\text{He} = 4.002603 \text{ u} \tag{22.137}$$

$$\text{Difference in mass} = 0.026503 \text{ u} \tag{22.138}$$

$$\text{Conversion of mass to energy} = 0.026503 \text{ u} \times \frac{931.5 \text{ MeV}}{\text{u}} \tag{22.139}$$

$$= 24.68 \text{ MeV}$$

Here the neutrinos and gamma rays have zero mass, but gamma rays have a certain energy that can be accounted for by the rest mass of the electrons. Thus, in the proton–proton chain reaction, two positrons collide with two electrons, resulting in the mutual annihilation of these four particles, which releases the following amount of energy:

$$4 \times 0.511 \text{ MeV} = 2.044 \text{ MeV} \tag{22.140}$$

$$\text{Total energy release} = (24.68 \text{ MeV} + 2.044 \text{ MeV}) \tag{22.141}$$

$$= 26.73 \text{ MeV}$$

Example 22.22 If the interior of the sun contains 4.5×10^{23} protons, calculate the energy released in KWh due to the fusion reaction taking place in the interior of the sun.

Solution Total energy released for four protons $= 26.73$ MeV

$$\text{Energy released for } 4.5 \times 10^{23} \text{ protons} = \frac{26.73 \times 10^6 \text{ eV} \times 4.5 \times 10^{23}}{4}$$

$$= \frac{30.07 \times 10^{29}}{3600 \times 1000} \times 1.6 \times 10^{-19} \text{ J} = \frac{48.114}{36} \times 10^5 \text{ kWh} = 1.336 \times 10^5 \text{ kWh}$$

Example 22.23 Consider the following fusion reaction:

$$^2_1\text{H} + ^2_1\text{H} \rightarrow ^3_2\text{He} + ^1_0\text{n}$$

Compute the energy liberated in this reaction in MeV and in joules. Give the name of the type of the fusion reaction that this equation represents.

Solution The energy liberated is calculated as follows:

Mass of $^2_1\text{H} = 2 * (2.014102) \text{ u} = 4.028202 \text{ u}$

Mass of $^3_2\text{He} = 3.0160267 \text{ u}$; Mass of $^1_0\text{n} = 1.008663 \text{ u}$

Energy liberated in this reaction $= (4.028202) \text{ u} - (3.0160267 \text{ u} + 1.008663 \text{ u})$

$$= (4.028202) \text{ u} - (4.0246897) \text{ u} = 0.0035123 \text{ u}$$

$$= 0.0035123 \text{ u} \times 931 \text{ MeV} = 3.27 \text{ MeV}$$

$$= 3.27 \times 10^6 \text{ eV} \times 1.6 \times 10^{-19} \text{ J}$$

$$= 5.23 \times 10^{-13} \text{ J}$$

This represents the D–D-type of fusion reaction. It can be represented by Fig. 22.13(a).

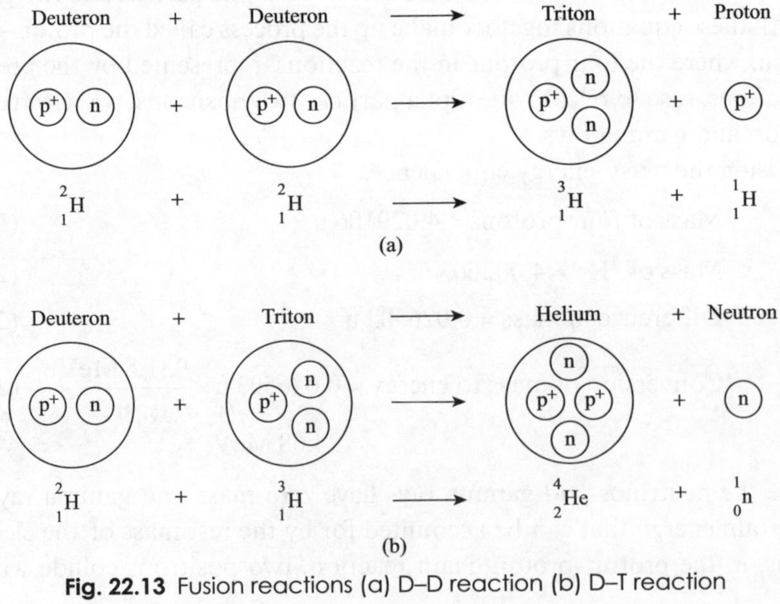

Fig. 22.13 Fusion reactions (a) D–D reaction (b) D–T reaction

Example 22.24 Consider the following fusion reaction:

$$^2_1H + {}^3_1H \rightarrow {}^4_2He + {}^1_0n$$

Compute the energy liberated in this reaction in MeV and in joules. Name the type of the fusion reaction that this equation represents.

Solution The energy liberated is as follows:

Mass of ^2_1H = (2.014102) u; Mass of ^3_1He = 3.0160267 u

Mass of ^4_2He = 4.003873 u; Mass of ^1_0n = 1.008663 u

Energy liberated in this reaction = $\{(2.014102)\,u + (3.0160267\,u)\} -$
$$\{(4.003873\,u + 1.008663\,u)\}$$
$$= \{(5.0301287)\,u - (5.012536)\,u\}$$
$$= 0.017592\,u$$
$$= 0.017592\,u \times 931\,\text{MeV} = 16.378\,\text{MeV}$$
$$= 16.378 \times 10^6\,\text{eV} \times 1.6 \times 10^{-19}\,\text{J}$$
$$= 26.205 \times 10^{-13}\,\text{J}$$

This represents the D–T type of fusion reaction. It can be represented by Fig. 22.13(b).

Note: In a fusion reactor, the preferred mode of reaction is the D–D fusion reaction.

Example 22.25 If a nuclear reactor consumes 5 kg of U^{235} per day, calculate the power output of the nuclear reactor if the average energy released per U^{235} fission is 200 MeV.

Solution Number of atoms in

235 kg of $^{235}U = 6.02 \times 10^{26}$, that is, the Avogadro's number $= \dfrac{6.02 \times 10^{26}}{235}$

Number of atoms in 5 kg $= \dfrac{6.02 \times 10^{26}}{235} \times 5 = 0.128 \times 10^{26} = 1.28 \times 10^{25}$

Energy released per fission = 200 MeV

Fission energy released by these atoms = $1.28 \times 10^{25} \times 200$ MeV
$$= 256 \times 10^{25} \times 1.6 \times 10^{-13}\,\text{J}$$
$$= 409.6 \times 10^{12}\,\text{J}$$

Time taken to consume 5 kg = 1 day = 24×3600 s

Power produced $= \dfrac{409.6 \times 10^{12}}{24 \times 3600} = 0.47 \times 10^{10}\,\text{W} = 4.7 \times 10^9\,\text{W}$

22.12 NUCLEAR FUSION REACTORS

In the fusion reactions discussed earlier, 'heavy hydrogen' or 'deuterium' plays an important role as a raw material. This is readily available on the earth in the form of molecules of heavy water. For every 6500 atoms of ordinary (light) hydrogen in water, there is one deuterium atom.

In a fusion mass reaction, the critical size or mass of the particle is imma-terial. The major hurdle is to obtain a self-sustaining reaction. The repulsive electric force between positively charged nuclei acts as a deterrent for the nuclei to fuse. For two nuclei to fuse, they must come together within the range of the nuclear force $\sim 2 \times 10^{-15}$ m. The potential energy corresponding to the electrical repulsion of the positive charges is $\sim 1.2 \times 10^{-13}$ J or 0.7 MeV, which is the threshold kinetic energy for the nuclei to fuse. This energy can be provided to a hydrogen atom using particle accelerators. The deuterium nuclei can be accelerated and collided with a solid deuterium target to produce D–D fusion reactions. The small fusion reaction that takes place produces energy much less than that required for the acceleration of the hydrogen particles. Another method of increasing the energy of the particle is to raise the temperature of the gas. For this the following equation is used:

$$E = \frac{3}{2}k_B T \Rightarrow T = \frac{2E}{3k_B} = \frac{2 \times (0.6 \times 10^{-13}\,\text{J})}{3 \times \left(\dfrac{1.38 \times 10^{-23}\,\text{J}}{\text{K}}\right)} = 3 \times 10^9 \text{ K} \qquad (22.142)$$

22.13 STELLAR THERMONUCLEAR REACTIONS

The sun happens to be the closest stellar star and is one of the major sources of energy. The source of energy is the nuclear burning of protons, forming helium atoms. The energy released per helium atom formed can be calculated from the net change in mass during the reaction.

The temperature at the core of the sun is 2×10^7 K. The hydrogen atoms present, in the presence of protons, converts into deuterium atoms, which further converts into helium atoms (Fig. 22.14). The corresponding reac-tions are as follows:

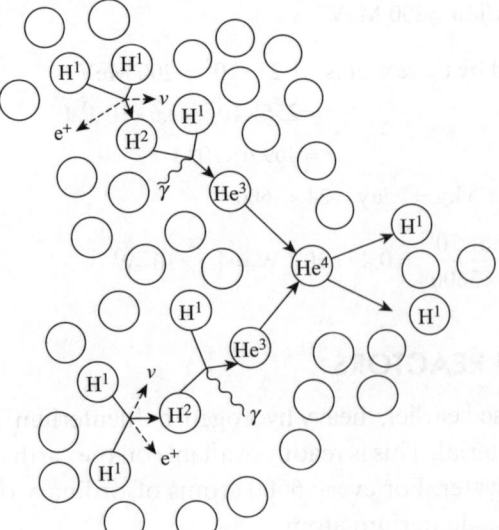

Fig. 22.14 Stellar energy reactions

$$H^1 + p \rightarrow H^2 + e^+ + \text{neutrino} \qquad (22.143)$$

$$H^2 + p \rightarrow He^3 + \gamma \qquad (22.144)$$

$$He^3 + He^3 \rightarrow He^4 + 2H^1 \qquad (22.145)$$

Using the concept of energy–mass equivalence, the energy released in this reaction can be calculated as follows:

$$4M_p + 2m_e - M(He^4) = 4 \times (1.6725 \times 10^{-24}) + 2 \times (0.911 \times 10^{-27}) \qquad (22.146)$$

$$- 6.647 \times 10^{-24}$$

$$\cong (6.69 \times 10^{-24}) + (1.822 \times 10^{-27}) - (6.647 \times 10^{-24})$$

$$\cong (0.043 \times 10^{-24})g \cong 50 \times m_e \cong 50 \times (0.511\ \text{MeV}) = 25\ \text{MeV}$$

Energy released = 25 MeV $\qquad (22.147)$

$$= (25 \times 10^6) \times (1.6 \times 10^{-19})\ J = 40 \times 10^{-13}\ J \qquad (22.148)$$

$$1\ J = 1\ Wh = \frac{1}{3600}\ Wh \qquad (22.149)$$

$$\text{Energy released} = \frac{40 \times 10^{-13}}{3600}\ \text{watt-hours}$$

$$= 0.011 \times 10^{-13}\ \text{watt-hours} \qquad (22.150)$$

Note: The solar constant is the flux of solar energy per square centimetre per second at the distance of the earth from the sun. Its value is measured to be 1.4×10^6 ergs/s-cm^2.

22.14 THEORIES FOR NUCLEAR FORCES

Nuclear forces seem to have a certain basic nature. The concepts of nuclear masses, binding energies, magnetic moments, energy levels, scattering cross sections, and transition probabilities exhibit important nuclear tendencies that help us understand the suitable theory of nuclear interactions.

A theory for the nuclear forces must account for the following important properties of a nucleus: (a) strong forces, (b) saturation, (c) incompressibility, (d) charge symmetry, (e) pairing, and (f) spin-orbit coupling.

Simple systems that help us understand the basic nature of nuclear forces, incorporating the salient characteristics, are the proton–proton, neutron–neutron, and neutron–proton systems. If one looks at a deuteron atom, it can be seen that it follows the neutron–proton system, which gives it a stable state of binding. Dwelling on the basics of the nucleon–nucleon interaction, it is spin dependent and is of the central-field type. To make it achieve the real dynamics, the existence of spin dependence and non-central forces is accounted for. The potential functions are fixed, which are usually the exponential well, Gaussian well, and Yukawa well. Once the potential is fixed (by ensuring a proper fit for the binding energy of the specific nucleus, e.g., deuteron), one proceeds to the analysis of the neutron–proton scattering, to account for the properties of the nuclei (deuteron) and the neutron–proton scattering cross sections. The results so obtained are in considerable agreement with the experimental results.

The neutron–proton interactions can be studied by the H(n, γ)D reactions, which are significant at very low energies. This process is similar to a transition from a continuous positive energy state of the neutron–proton system to the ^{3}S state of the deuteron. The D(n, γ)H reaction is the inverse of the radiative-capture reaction. The results of these findings indicate that the nuclear forces at low energies are essentially independent of the character of the charge.

The meson theory of nuclear forces is devised along the lines of the field theory of electromagnetic forces. A meson particle has a mass intermediate between that of an electron and a proton, and is an unstable particle. The meson theory followed the fundamental ideas of Yukawa and quantum electrodynamics. When a π^+ meson jumps from a neutron to a proton, it is converted to a neutron and vice versa.

$$p - \pi^+ \rightarrow n \text{ or } n + \pi^+ \rightarrow p \tag{22.151}$$

Similarly, when a π^- meson jumps from a neutron to a proton, it is converted to a proton and vice versa:

$$n - \pi^- \rightarrow p \text{ or } p + \pi^- \rightarrow n \tag{22.152}$$

A host of unanticipated mesons have been discovered in the cosmic rays. The multiple meson theories do not account for the internal structure of nucleons and other fundamental particles and are, thus, unable to unfold nuclear forces in the very high energy range. This area of nuclear physics still requires a considerable amount of work.

Conclusively, one can state that nuclear forces have the following characteristics: (a) these are short-range attractive forces ($\sim 10^{-15}$ m), (b) these forces are spin dependent, (c) these are the strongest forces in nature, (d) these forces are charge independent, and (e) each nucleon attracts the nucleons that are its immediate neighbours.

22.15 PARTICLE ACCELERATORS

Particle accelerators are designed for accelerating particles using precisely controlled beams of particles, ranging from electrons and positrons to heavy ions, with a wide range of energies. Particle accelerators produce beams that can be used to produce (a) new particles (an electron and a positron can collide to produce photons); (b) high-energy particles having short de-Broglie wavelengths, which can detect the interior structures of various particles, for example, an electron microscope; and (c) nuclear radiations for scientific or medical use.

22.15.1 Cyclotron

The cyclotron was invented by E.O. Lawrence. It is a device used for accelerating nuclear particles. Figure 22.15 shows a section of the cyclotron. The centre S is the ion source S. The hollow accelerating electrodes are marked as Dee_1 and Dee_2. There is a vertical (homogeneous) magnetic field ($\vec{B}$) (pointing downward). The entire apparatus is placed in this magnetic field. The particle orbits in a horizontal plane and is also the median plane of the dees. The accelerating radio frequency introduces an electric field inside the Dees.

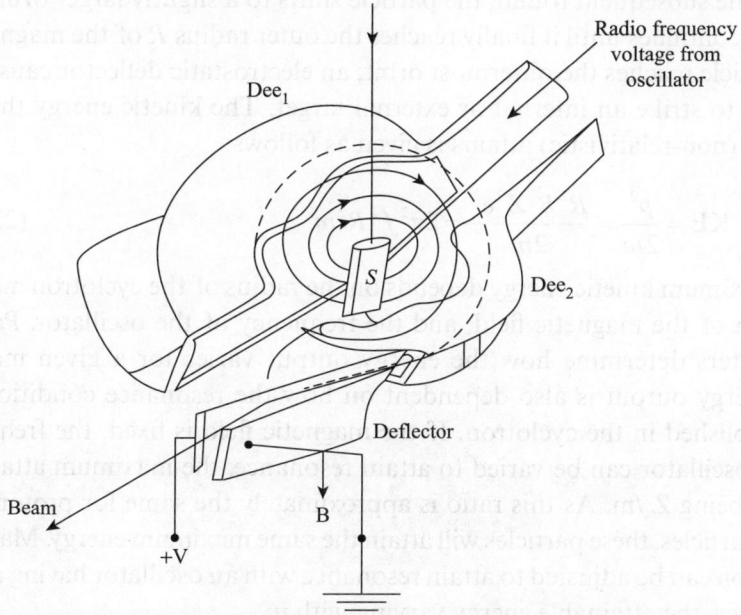

Fig. 22.15 Section of cyclotron

22.15.1.1 Principle

Particles are subjected to multiple voltage impulses. A high-frequency alternating voltage is applied between two hollow electrodes (dees), which are mounted between the poles of a huge magnet. An approximately uniform magnetic field bends the particle path into a circle having a radius expressed by the following equation:

$$r = \frac{p}{BZe} = \frac{mv\gamma}{BZe} \qquad (22.153)$$

$$\Rightarrow p = rBZe \qquad (22.154)$$

The momentum of the particle has been defined in terms of its radius and the magnetic field. The time interval is evaluated for the particle moving inside the dee in one complete rotation, that is, from the moment of entry of the particle into one dee to the moment of its emergence from it:

$$\tau_{1/2} = \frac{\pi r}{v} = \frac{\pi m \gamma}{BZe} \qquad (22.155)$$

The quantity defined as time interval $\tau_{1/2}$ is a constant for non-relativistic velocities ($\gamma \approx 1$). Thus, by adjusting the magnetic field or the radio frequency, resonance is established between the half-period ($1/2f$) of the oscillator and the natural half-period of the system (as given by the relation for $\tau_{1/2}$).

As the particle enters the dees of the cyclotron, the electric field is so aligned (in the proper direction) that it increases the kinetic energy of the particle. As the particle moves in the dees, it makes one complete rotation of a dee, and in the subsequent round, the particle shifts to a slightly larger orbit; this process continues until it finally reaches the outer radius R of the magnet. As the particle reaches the outermost orbit, an electrostatic deflector causes the particle to strike an internal or external target. The kinetic energy that the particle (non-relativistic) attains is given as follows:

$$\text{KE} = \frac{p^2}{2m} = \frac{R^2 B^2 Z^2 e^2}{2m} = 2\pi^2 f^2 R^2 m \tag{22.156}$$

The maximum kinetic energy depends on the radius of the cyclotron magnet, strength of the magnetic field, and the frequency of the oscillator. Particle parameters determine how the energy output varies for a given magnet. The energy output is also dependent on how the resonance condition can be established in the cyclotron. If the magnetic field is fixed, the frequency of the oscillator can be varied to attain resonance, the maximum attainable energy being Z^2/m. As this ratio is approximately the same for protons and alpha particles, these particles will attain the same maximum energy. Magnetic induction can be adjusted to attain resonance with an oscillator having a fixed frequency, the attainable energy varying with m.

The cyclotron can also be designed for relativistic energies, where the factor γ has a significant value. This creates considerable difficulty in the design aspect particularly as the particle reaches the periphery, that is, the outermost orbits; it arrives late at the gaps and gradually gets out of phase with the voltage pulses. For a linear accelerator, the lengths of the hollow drift tubes are adjusted to satisfy the relativistic equations. Alternatively, the shape of the pole pieces is modified such that the ratio B/γ remains approximately constant, that is, the field becomes relatively stronger at large radii. This curvature of the magnetic field tends to defocus the particle from the median plane of the dee chambers, resulting in the loss of the beam current. Practically, a cyclotron contains a magnetic field that decreases slightly with radius to ensure focusing. Thus, the maximum energies attainable with the cyclotron are around 25 MeV due to the focusing arrangement in addition to the relativistic effect.

A cyclotron produces an intense source of charged particles, which gives currents of the order of 1 mA in the internal beams and of the order of 50 μA in the external beams. These beams are used for the study of nuclear reactions, production of neutrons, and production of induced radioactivity.

Characteristics of a cyclotron at the Nobel Institute of Physics, Stockholm, Sweden, are as follows.

Weight of the magnet	: 400 tons
Weight of copper	: 27 tons
Pole-face diameter	: 88.5 in.
Power of 25-MeV deuteron	: 60 kW
Maximum power	: 240 kW
Radius of maximum orbit	: 90 cm
Field for 25-MeV deuteron	: 11,500 G
Maximum field	: 18,000 G
Oscillator frequency	: 8.7 Mc
Oscillator power	: 230 kW
Dee voltage	: 200 kV
Energy	: 25 MeV deuterons
Beam current (internal)	: 300 μA

Example 22.26 Find the frequency of the oscillating electric field if $B = 2.0$ T for accelerating deuterons of charge $+e$ and mass 3.3×10^{-27} kg.

Solution The oscillation period of the electric field must equal the orbital period, so the required oscillation frequency is calculated as follows:

$$\vartheta = \frac{1}{T} = \frac{qB}{2\pi m_0} = \frac{(1.6 \times 10^{-19}) \times (2.0) \times (7)}{(2 \times 22 \times 3.3 \times 10^{-27})} = 0.154 \times 10^8 = 15.4 \text{ MHz}$$

Example 22.27 How many times does a deuteron need to be accelerated to acquire 20 MeV of kinetic energy if the potential across the cyclotron dees are 50 kV?

Solution Thus, every time the deuteron completes one oscillation, it gains $50 \text{ keV} = 50 \times 10^3 \text{eV} = 5 \times 10^4 \text{ eV}$

Total gain in energy $= 20 \text{ MeV} = 20 \times 10^6 \text{ eV}$

$$\text{Deuteron undergoes cycles} = \frac{20 \times 10^6 \text{ eV}}{5 \times 10^4 \text{ eV}} = 400 \text{ cycles}$$

Example 22.28 Deuterons in a cyclotron travel in a circle of radius 32.0 cm just before emerging from the dees. The frequency of the applied alternating voltage is 11.0 MHz. Find (a) the magnetic field and (b) the energy and speed of the deuterons upon emergence.

Solution The calculations are as follows:
(a) Angular frequency of a deuteron in a cyclotron:

$$\vartheta = \frac{1}{T} = 11.0 \text{ MHz} = 11.0 \times 10^6 \text{ Hz} = \frac{qB}{2\pi m_0} = \frac{(1.6 \times 10^{-19}) \times (B) \times (7)}{(2 \times 22 \times 3.3 \times 10^{-27})}$$

$$\Rightarrow B = \frac{2 \times 22 \times 3.3 \times 10^{-27} \times 11 \times 10^6}{7 \times 1.6 \times 10^{-19}} = 1.42 \text{ wb/m}^2$$

(b) Speed of the deuterons $= v = \dfrac{Bq}{m} r = \dfrac{1.42 \times 1.6 \times 10^{-19}}{(3.3 \times 10^{-27})} \times (0.32 \text{ m}) = 2.2 \times 10^7 \text{ m/s}$

(c) Energy of the deuteron $=\frac{1}{2}mv^2 = \frac{1}{2}\times(3.3\times10^{-27})\times(2.2\times10^7)^2 = 7.986\times10^{-13}$ J

Example 22.29 A cyclotron is used to accelerate protons having a mass half that of the deuterons.
(a) If the magnetic field has an intensity of 2.0 T, what is the change in the frequency of the oscillating electric field?
(b) What is the maximum energy acquired by the protons if the potential applied across the dees of the cyclotron are 50 kV?

Solution The calculations are as follows:

$$m_p = \frac{1}{2}m_d$$

(a) $\vartheta_p = \frac{1}{T} = \frac{qB}{2\pi m_p} = \frac{2(qB)}{2\pi m_d} = 2\vartheta_d$

Thus, the frequency of the proton is double the frequency of the deuteron.

(b) Maximum energy acquired by the protons $=\frac{1}{2}mv^2_{max} = qV$

$$=1.6\times10^{-19}\times50\times10^3 = 80\times10^{-16}\,\text{J}$$

Example 22.30 If the magnetic field is directed upwards and the particles are moving counterclockwise in a cyclotron, what is the charge on the particles?

Solution The charge on the particles is negative. Thus, the force is directed towards the centre of the circle.

Example 22.31 A cyclotron of diameter 1 m is used to accelerate protons. An alternating voltage of 10 mc/s frequency and of a peak value of 5000 V is applied across it. Calculate the magnetic field needed and the energy of the emergent beam. The mass and the charge of the proton are 1.67×10^{-27} kg and 1.6×10^{-19} C, respectively.

Solution The calculations are as follows:

(a) $B = \frac{2\pi m f_0}{q} = \frac{2\times22\times1.67\times10^{-27}\,\text{kg}\times10\times10^6\,\text{c/s}}{7\times1.6\times10^{-19}\,\text{C}} = 0.66$ weber/m^2

(b) Energy of the deuteron $=\frac{1}{2}mv^2 = \frac{1}{2}mq^2B^2r^2 \cdot \frac{1}{m^2}$

Energy of the deuteron $=\frac{1}{2}\times(1.6\times10^{-19})^2\times(0.66)^2\times(1)^2\times\frac{1}{(1.67\times10^{-27})}$

$$=0.052\times10^{-11}\,\text{J}$$

Energy of the deuteron $=\frac{0.052\times10^{-11}}{1.6\times10^{-19}}\text{eV} = 5.2\times10^6\,\text{eV} = 5.2\,\text{MeV}$

Example 22.32 Calculate the energy of a proton when it emerges out of a cyclotron after 50 revolutions. The voltage applied across the dees is 20 kV.

Solution Thus, every time the deuteron completes one oscillation it gains 20 keV $= 20\times10^3$ eV $= 2\times10^4$ eV

$$\text{Deuteron undergoes cycles} = \frac{\text{Gain in energy of a proton}}{2 \times 10^4 \text{ eV}} = 50 \text{ cycles}$$

$$\text{Gain in energy} = 50 \text{ cycles} \times 2 \times 10^4 \text{ eV} = 1 \times 10^6 \text{ eV} = 1 \text{ MeV}$$

Example 22.33 The maximum path radius of a cyclotron is 0.30 m, and it has a magnetic field of magnitude 1.50 T. As the cyclotron is used to accelerate the protons, (a) calculate the frequency of the alternating voltage applied to the dees, and (b) find the maximum particle energy.

Solution The calculations are as follows:

(a) Frequency of alternating voltage $= f = \dfrac{qB}{2\pi m}$

$$= \frac{7 \times 1.6 \times 10^{-19} \text{C} \times 1.50 \text{ T}}{2 \times 22 \times 1.67 \times 10^{-27} \text{kg}} = 0.228 \times 10^8 \text{Hz}$$

$$= 23 \text{ MHz}$$

(b) Maximum kinetic energy $= \dfrac{q^2 B^2 R^2}{2m} = \dfrac{(1.6 \times 10^{-19}\text{ C})^2 \times (1.50 \text{ T})^2 \times (0.30\,m)^2}{(2 \times 1.67 \times 10^{-27}\text{kg})}$

$$= 0.155 \times 10^{-11} \text{ J}$$

$$K_{max} = \frac{0.155 \times 10^{-11}}{1.6 \times 10^{-19}} \text{eV} = 0.096 \times 10^8 \text{eV} = 9.6 \times 10^6 \text{eV} = 9.6 \text{ MeV}$$

Example 22.34 The magnetic field in a cyclotron that accelerates protons is 1.30 T.

(a) How many times per second should the potential across the dees reverse? (This is twice the frequency of the circulating protons.)
(b) The maximum radius of the cyclotron is 0.250 m. What is the maximum speed of the proton?
(c) Through what potential difference would the proton have to be accelerated to give it the same speed as that calculated in part (b)?

Solution The calculations are as follows:

(a) Frequency of alternating voltage $= f = \dfrac{qB}{2\pi m}$

$$= \frac{7 \times 1.6 \times 10^{-19} \text{ C} \times 1.30 \text{ T}}{2 \times 22 \times 1.67 \times 10^{-27}\text{kg}}$$

$$= 0.198 \times 10^8 \text{Hz} \cong 20 \text{ MHz}$$

The potential across the dees has to be reversed (2×20) MHz $= 40$ MHz times.

(b) Maximum kinetic energy $= \dfrac{q^2 B^2 R^2}{2m} = \dfrac{(1.6 \times 10^{-19}\text{ C})^2 \times (1.30 \text{ T})^2 \times (0.25 \text{ m})^2}{(2 \times 1.67 \times 10^{-27}\text{kg})}$

$$= 0.081 \times 10^{-11}\text{J} = \frac{1}{2}mv_{max}^2 = 0.081 \times 10^{-11}\text{ J}$$

$$v_{max}^2 = \frac{(2 \times 0.081 \times 10^{-11})}{(1.67 \times 10^{-27})} = 0.097 \times 10^{16}$$

$$v_{max} \cong \frac{0.311 \times 10^8 \text{m}}{\text{s}} \approx \frac{3.11 \times 10^7 \text{m}}{\text{s}}$$

(c) $eV = 0.081 \times 10^{-11}$ J

$$V = \frac{0.081 \times 10^{-11}}{1.6 \times 10^{-19}} = 0.051 \times 10^8 \text{ eV}$$

22.16 RADIATION DETECTORS

The radiation detector is usually a device that helps us detect, track, and identify the high-energy particles usually produced during nuclear decay, cosmic radiation, or reactions in a particle detector. In addition to detecting the nature of the high-energy particles, detectors are used to measure their energy, momentum, spin, and charge. Modern detectors or counters are huge in size and cost. The detection of the recent god particle or the Higgs boson has been done at CERN. Radiation detectors usually fall under the following categories:

Gaseous ionization detectors These detectors are usually categorized as gaseous ionization chambers, proportional counters, Geiger–Muller (GM) tubes, and spark chambers.

Solid-state detectors These are usually semiconductor detectors, solid-state track detectors, Cherenkov detectors, ring imaging Cherenkov detectors, scintillation detectors (photo-multipliers and photodiodes), Lucas cells, time-of-flight detectors, silicon vertex detectors, micro-channel plate detectors, and neutron detectors.

The most historic of these detectors are the bubble chamber, Wilson cloud chamber (diffusion chamber), and photographic plates.

The principle of operation of GM counters, ionization chambers, and proportional counters is that each of these detectors is works based on the ionization of a gas and separation and collection of the ions by means of an electrostatic field.

22.17 IONIZATION CHAMBER

The principle of operation of an ionization chamber (explained in Fig. 22.16) is to measure the charge of the ion pairs created within a gas when a radiation is incident within.

An ionization chamber consists of two electrodes, a cathode and an anode. The voltage potential applied across the two electrodes generates an electric field. As a radiation is incident inside the chamber, the gas is ionized and creates ion pairs. These ion pairs are constituted of dissociated electrons and positive ions, which move to the electrodes of opposite polarity under the influence of the electric field. This ionization current is measured by an electrometer circuit and is usually of the order of pico-amperes or femto-amperes. The magnitude of the current depends on the design of the chamber, radiation dose, and applied voltage. There is an accumulation of charge on the electrodes, which is proportional to the ion pairs created. There is a continuous flow of current due to the electric field, which continuously carries the electrons, preventing

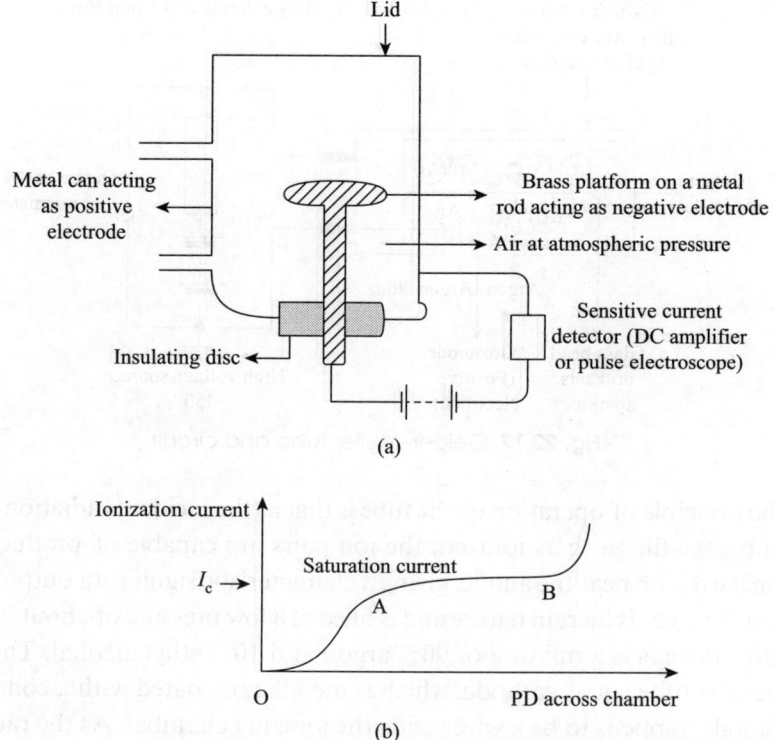

Fig. 22.16 Ionization chamber (a) Set-up (b) Variation of ionization current with potential difference across the chamber

saturation. The continuous flow of the current also ensures that the ions do not recombine, which can reduce the current. Thus, the ionization current is a direct measure of the amount of radiation dose entering the chamber.

However, the ionization chamber is not able to discriminate between the radiation type. Ionization chambers, such as free-air chambers, vented chambers, sealed low-pressure chambers, research and calibration chambers, and high-pressure chambers, are designed in different shapes, depending on how the gas or the ionizing medium is filled in the chamber.

The difference between a proportional counter or a GM counter and an ionization chamber is that the latter uses only discrete charges created by each interaction of the radiation particle with the gas. It does not use the gas multiplication mechanisms that are used by other radiation detectors. The output current is a continuous current, unlike in case of a Geiger–Muller counter, where we get a pulsed output.

22.18 GM COUNTERS

The Geiger–Muller counter/tube is named after two scientists Geiger and Walther Muller who invented the related principle and then developed the tube. This is shown in Fig. 22.17.

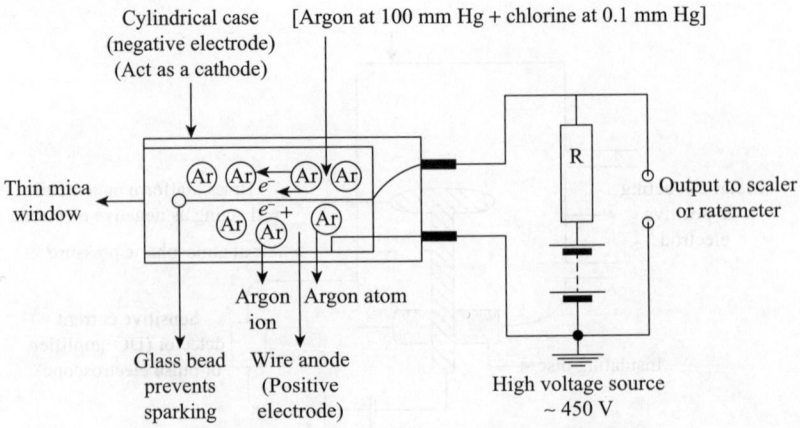

Fig. 22.17 Geiger–Muller tube and circuit

The principle of operation of the tube is that as the incident radiation strikes the tube, the fill gas gets ionized; the ion pairs are capable of producing an avalanche region near the anode, giving a characteristic significant output pulse. The ionizing gas is inert in nature and is filled at a low pressure of about 0.1 atm. Usually, the gas is a mixture of 90% argon and 10% ethyl alcohal. The walls of the GM tube act as cathode, which is metallic or coated with a conductor. The anode happens to be a wire inside the ionizing chamber. As the radiation strikes the ionizing chamber, some of the molecules of the fill gas are ionized; ionization can take place from the primary electrons (incident radiation) or the secondary electrons produced in the walls of the tube. The ion pairs so created consist of positively charged ions and electrons. The potential difference across the electrodes is of the order of several hundred volts, which creates an electric field, propelling the ion pairs to the opposite electrodes, that is, the positive ions towards the cathode and the electrons towards the anode. There is a gas multiplying effect in the GM tube, which takes place close to the anode and is referred to as the 'avalanche region'. The avalanche is created by the production of the ultraviolet photons in the initial avalanche. The ultraviolet photons are not affected by the electric field moving laterally to the anode, creating more collisions with the gas molecules. This triggers a chain reaction, generating a greater number of excited molecules (from 10^6–10^8 to 10^9–10^{10} ion pairs). The speed of propagation of the avalanches is typically 2–4 cm/μs. Complete ionization of the gas takes just a few microseconds, which is exhibited as a short intense pulse of current that is measured as a 'count event' in the form of a voltage pulse (Fig. 22.18). The voltage is of the order of volts and is measured across an external resistor. Usually, the GM counter is of two types:

Non-self-quenching counters   In general, these are filled with a simple gas; as the ions reach the cathode, they are neutralized by recombination with an electron from the metal. The difference between the ionization energy gained in the neutralization process and the energy spent to extract the electron

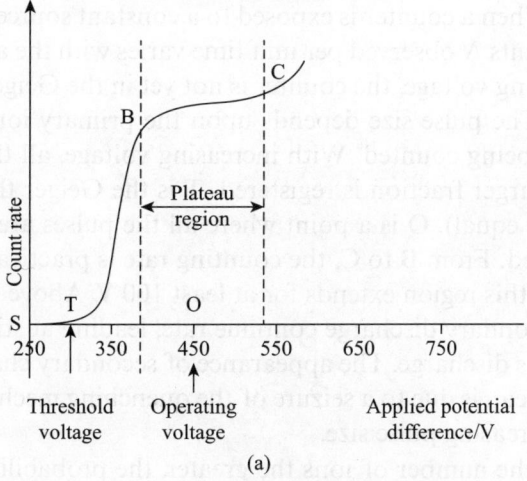

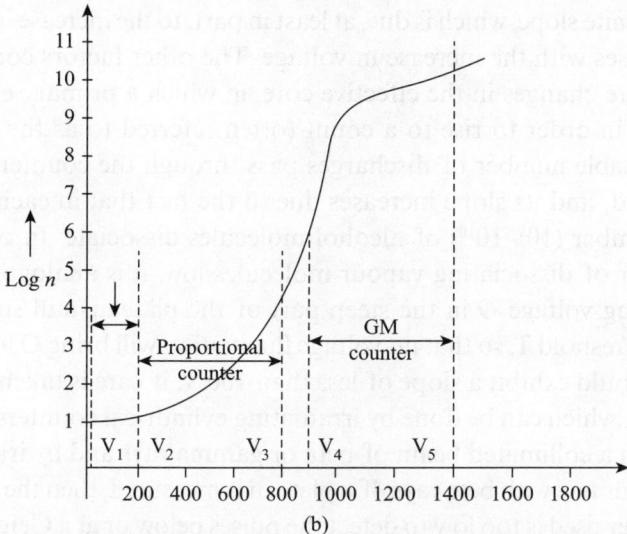

Fig. 22.18 Graphical representation of the variation of count rate with voltage (a) Plateau curve for a GM counter (b) Voltage axis differentiating the proportional counter to GM counter

from the cathode may be radiated as a photon, which then is able to release another electron from the metal by photo-effect. Alternatively, the energy may be used for a direct, radiation-less liberation of a second electron from the metal. In either case, the new electron will start a new discharge, unless the counter voltage is kept below the threshold value by external means until these electrons have reached the anode without multiplication. One of the common kinds of external quenchers is a resistor.

Self-quenching counters Such counters usually contain a polyatomic gas like methane, or a mixture of a simple gas and a polyatomic gas or vapour. In this case, a number of electrons are produced when the ion sheath reaches the cathode. When a counter is exposed to a constant source of radiation, the number of counts N observed per unit time varies with the applied voltage V. At S, the starting voltage, the counter is not yet in the Geiger region (refer to Fig. 22.18a). The pulse size depends upon the primary ionization, only the largest pulses being counted. With increasing voltage, all the pulses become bigger and a larger fraction is registered. T is the Geiger threshold (at O all pulses become equal). O is a point where all the pulses are of the same size and are counted. From B to C, the counting rate is practically constant. In a good counter, this region extends for at least 100 V. Above C, an appreciable number of secondary discharge continue rate, leading at still higher voltages to a continuous discharge. The appearance of secondary charges at the upper end of the plateau is due to a seizure of the quenching mechanism because of the rapidly increasing pulse size.

The larger the number of ions the greater, the probability of triggering a second discharge cycle. These considerations apply to the plateau region. It exhibits a finite slope, which is due, at least in part, to the increase in the number double pulses with the increase in voltage. The other factors contributing to the slope are changes in the effective core, in which a primary electron must be formed in order to rise to a count (often referred to as the end effects). An appreciable number of discharges pass through the counter, the plateau is shortened, and its slope increases due to the fact that in each discharge a certain number $(10^9–10^{10})$ of alcohol molecules dissociate. In order to keep the number of dissociating vapour molecules low, it is analogous to choose an operating voltage O in the steep part of the plateau, still sufficiently far from the threshold T, so that no voltage fluctuation will bring O lower. A good counter should exhibit a slope of less than 100 V, if care is taken to eliminate end effects, which can be done by irradiating cylindrical counters in the aerial region with a collimated beam of beta or gamma rays and by irradiating end window counters with beta rays. If a plateau is measured, then the sensitivity of the amplifier used is too low to detect the pulses below or at a Geiger threshold. The disadvantages are that the plateau is short and the pulses in the counter are large, such that its life becomes shortened. It is easy to check the fault in the apparatus using a cathode ray oscilloscope. The instrument should count at the Geiger threshold. If this is not the case, then the sensitivity setting of the pulse height selector may be wrong or the pulse coming in from the counter may be abnormally small. This often occurs when the counter is at a large distance from a specimen and connected with a higher-capacity shielded cable.

22.18.1 Loses in GM Counter

After the development of a discharge, the space charge of the sheath depresses the field at the wire, so as to make any further multiplication by collision impossible. A second particle entering the counter at this time will not produce a pulse. As the space charge moves towards the cathode, the field increases gain.

It becomes sufficiently large so that a second particle may produce a small pulse after a certain time, referred to as the dead time, following the passage of the first particle. It reaches the size of the original discharge at a time $(\tau_D + \tau_R), \tau_R$ being called the recovery time. The dead time and the recovery time can be measured by the time method. The pulses of the counter trigger the sweep of the cathode ray oscilloscope, which has a duration of 500–100 μs. The triggering pulses are seen to the left of the sweep. No pulses are seen within the dead time. During the recovery time, the pulse size increases to the original height.

Example 22.35 A GM Counter collects 10^6 electrons per discharge when the counting rate is 600 counts/min. What will be the average current in the circuit?

Solution The counting rate given = 600 counts/min

Total number of electrons collected/min = $600 \times 10^6 = 6 \times 10^8$

Charge/minute in the circuit = $6 \times 10^8 \times 1.6 \times 10^{-19}$ C/min

$$\text{Average current} = \frac{6 \times 10^8 \times 1.6 \times 10^{-19}}{60} = 1.6 \times 10^{-12} \text{ A}$$

Example 22.36 How many counts per minute are detected by a GM counter if the counter wire collects 10^6 electrons per discharge with an average current in the circuit of about 1.6×10^{-12} A.

Solution If R is the counting rate and about 10^6 electrons per discharge are produced, then

Total number of electrons collected in one minute = $n = R \times 10^6$

Charge per minute = $ne = (R \times 10^6) \times (1.6 \times 10^{-19})$ C/min

$$\text{Current} = I = \frac{q}{t}$$

$$1.6 \times 10^{-12} = \frac{(R \times 10^6) \times (1.6 \times 10^{-19})}{60} \Rightarrow R = 600 \text{ counts/min}$$

Example 22.37 A GM counter with a dead time of 200 μs records about 15,000 counts/min. What is the dead time loss in the counting rate?

Solution Observed counting rate = 15,000 counts/min

$$\text{True counting rate} = \frac{\text{Observed counting rate}}{1 - (\text{Observed counting rate} \times \text{Dead time})}$$

$$= \frac{15,000}{1 - \left(\frac{15,000 \times 200 \times 10^{-6}}{60}\right)} = 1.58 \times 10^4 \text{ counts/min}$$

Example 22.38 In an ionization chamber, a charged particle is completely stopped, producing an energy of 5 MeV per $\dfrac{2 \times 10^8 \text{ m}}{\text{s}}$. Estimate the pulse height if the total capacitance of the capacitor is 25 pF.

Solution The total charge produced = $Q = ne$

$$\text{Pulse height} = V = \frac{Q}{C} = \frac{ne}{C} = \frac{2 \times 10^8 \times 1.6 \times 10^{-19}}{30 \times 10^{-12}} = 1.067 \text{ V}$$

Example 22.39 In an ionization chamber, about 10×10^4 ion pairs are produced, resulting in the loss of about 40 eV of energy. Estimate the amount of charge collected by each plate. What is the kinetic energy of this alpha particle?

Solution Kinetic energy $= 10 \times 10^4 \times 40 \text{ eV} = 4 \times 10^6 \text{ eV} = 4 \text{ MeV}$

Amount of charge collected on each plate $= Q = ne = 10 \times 10^4 \times 1.6 \times 10^{-19}$

$$= 1.6 \times 10^{-14} \text{C}$$

IMPORTANT CONCEPTS

1. Protons and neutrons possess attractive nuclear forces, which need to be balanced by the electric repulsion of the protons for the stability of the nucleus.

2. The total number of neutrons and protons together are referred to as the nucleon number, A, also referred to as the mass number.

3. Radii of nuclei are represented as follows:

   $R = R_0 A^{1/3}$, where $R_0 = 1.2 \times 10^{-15}$ m $= 1.2$ fm (read fm as Fermi)

   The nuclear density is the mass per unit volume of a nucleus, that is,

   $$\rho = \frac{m}{V} = \frac{m_N}{\left(\frac{4}{3}\pi R^3\right)} = \frac{m_N}{\left(\frac{4}{3}\pi R_0^3 A\right)} = \frac{A \times 1\,\text{u}}{\frac{4}{3}\pi R_0^3 A} = \frac{1\,\text{u}}{\frac{4}{3}\pi R_0^3} = 2.3 \times 10^{17} \text{kg/m}^3$$

   Thus, one observes that nuclear density is constant for all nuclei.

4. The nuclear magneton is 1836 times smaller than the Bohr magneton because of the significant difference between the proton mass and the electron mass.

5. The surroundings gain an amount of energy equivalent to ΔE_0 when a nucleus is formed. In addition, the same amount of energy, the binding energy, has to be supplied to the nucleus to tear it apart into infinitely separated protons and neutrons.

6. The equation for the decay constant can be written as follows:

   $$\lambda = -\frac{1}{N(t)}\frac{dN(t)}{dt}$$

   The decay constant is the ratio of the number of decays per unit time to the number of remaining radioactive nuclei, that is, it is referred to as the probability per unit time that any individual nucleus will decay.

7. The mean lifetime T_{mean}, can be defined in terms of the decay constant (λ) as follows:

   $$T_{\text{mean}} = \frac{1}{\lambda} = \frac{T_{1/2}}{\ln 2} = \frac{T_{1/2}}{0.693}$$

8. Gamma rays are a form of electromagnetic radiation of a very short wavelength emitted by excited nuclei.

9. Alpha particles are helium nuclei.

10. Beta particles are electrons emitted by radioactive nuclei.

11. The common unit of radioactivity is Becquerel (Bq), which is equal to one nuclear disintegration per second.

12. The theory for the nuclear forces must account for the following important properties of a nucleus: (a) strong forces, (b) saturation, (c) incompressibility, (d) charge symmetry, (e) pairing, and (f) spin–orbit coupling.

13. The liquid drop model gives the following semi-empirical mass formula:

$$E_B = -6UA + 4\pi b^2 A^{\frac{2}{3}} S_s + \frac{3}{5}\frac{Q^2}{bA^{\frac{1}{3}}} + 6kUA\left(\frac{D}{A}\right)^2 \pm A_5 A^{-\frac{4}{3}}$$

APPLICATIONS

1. There are multiple applications of nuclear physics in numerous fields such as medicine, industry, and agriculture food processing.
2. Owing to nuclear physics, there has been path-breaking research in the field of medical diagnosis and treatment. A radiopharmaceutical is a drug containing a radioactive isotope, an unstable nucleus. This is administered to the human body; it emits a small amount of radiation, which can be sensed by a 'gamma camera'.
3. Radioimmunoassay is an *in vitro* procedure that combines radio-chemicals and antibodies to detect trace quantities of hormones, vitamins, or drugs in a patient's blood.
4. Positron emission tomography (PET) is a medical imaging technique, which exposes the dynamic effects of the human body like blood flow.
5. A new treatment, known as proton therapy, has been developed. It helps deposit the energy at a local point, although the extent of depth/penetration of the proton is low. Protons help carry the dose of medicine to the tumour, thus lessening it for the healthy tissues. The proton therapy is preferred around the eye and the spinal cord.
6. Accelerator mass spectrometer (AMS) is a technique that helps to know about the nucleus in concentrations below one part per trillion. This helps find the age of organic materials as old as 50,000 years. It also helps hydrologists determine the extent and source of the ground water.
7. Proton-induced X-ray emission (PIXE) can detect the haze in air down to one part per trillion so that the source of air pollution can be identified.

IMPORTANT FORMULAE

1. Radius of the nucleus:
$$R = R_0 A^{\frac{1}{3}}, \text{ where } R_0 = 1.2 \times 10^{-15}\,\text{m}$$
$$= 1.2\,\text{fm}$$

2. Binding energy:
$$E_B = (ZM_H + Nm_N - {}_Z^A M)c^2$$

3. Disintegration constant:
$$\lambda = -\frac{1}{N(t)}\frac{dN(t)}{dt}$$

4. Mean lifetime of a radioactive nucleus:
$$T_{\text{mean}} = \frac{1}{\lambda} = \frac{T_{1/2}}{\ln 2} = \frac{T_{1/2}}{0.693}$$

5. Reaction energy:
$$Q = (M_u + M_v - M_X - M_Y)c^2$$
(Reaction energy)
$$Q = +\text{ve}; \Rightarrow \Delta m = \text{decreases};$$
$$K = \text{increases (Exoergic reaction)}$$
$$Q = -\text{ve}; \Rightarrow \Delta m = \text{increases};$$
$$K = \text{decreases (Endoergic reaction)}$$

6. Semi-empirical mass formula from the liquid drop model:
$$E_B = -6UA + 4\pi b^2 A^{\frac{2}{3}} S_s + \frac{3}{5}\frac{Q^2}{bA^{\frac{1}{3}}}$$
$$+ 6kUA\left(\frac{D}{A}\right)^2 \pm A_5 A^{-\frac{4}{3}}$$

Multiple-choice Questions

22.1 The nucleus of an atom has the following properties:
   (a) A collection of the positive charge
   (b) A collection of neutrons to hold protons together
   (c) A nuclear force that is characteristic of the properties of nucleons
   (d) All of these

22.2 The radius of a nucleus is given as
   (a) $R = R_0 A^{1/3}$
   (c) $R = R_0 A^{5/3}$
   (b) $R = R_0 A^{2/3}$
   (d) none of these

22.3 In nuclear size, $r = r_0 A^{1/3}$; the size of the mass distribution is
   (a) 1.0 fm   (b) 2.0 fm   (c) 1.5 fm   (d) 1.8 fm

22.4 The binding energy of a nucleus is given as
   (a) $E_B = (ZM_H + Nm_N - {}^A_Z M)c^2$
   (c) $E_B = (ZM_H + Zm_N - {}^A_Z M)c^2$
   (b) $E_B = (NM_H + Nm_N - {}^A_Z M)c^2$
   (d) $E_B = (ZM_H + Nm_N - {}^A_Z M)$

22.5 A particle has only a single mean life, although it can decay in a number of ways:
   (a) $R = R^- + R^0 = \dfrac{N_0}{\tau} e^{-2t/\tau}$
   (c) $R = R^- + R^0 = \dfrac{N_0}{\tau} e^{-5t/\tau}$
   (b) $R = R^- + R^0 = \dfrac{N_0}{\tau} e^{-t/\tau}$
   (d) None of these

22.6 The mean lifetime of the particle is defined in terms of the decay constant (λ) as
   (a) $T_{mean} = \dfrac{1}{\lambda} = \dfrac{T_{1/2}}{\ln 2} = \dfrac{T_{1/2}}{0.693}$
   (c) $T_{mean} = \dfrac{1}{\lambda} = \dfrac{T_{1/2}}{\ln 2} = \dfrac{T_{1/2}}{3.303}$
   (b) $T_{mean} = \dfrac{1}{\lambda} = \dfrac{T_{1/2}}{\ln 2} = \dfrac{T_{1/2}}{2.303}$
   (d) $T_{mean} = \dfrac{1}{\lambda} = \dfrac{T_{1/2}}{\ln 4} = \dfrac{T_{1/2}}{4.693}$

22.7 An unstable nuclide decays into different nuclides; they
   (a) emit electromagnetic radiations
   (b) do not emit any radiations
   (c) emit alpha (α) or beta (β) or gamma (γ) particles
   (d) are discrete in the emission of radiations

22.8 The disintegration energy or the Q-value is the net energy, which represents
   (a) the final energy minus the initial energy
   (b) the initial energy minus the final energy
   (c) the total energy
   (d) none of these

22.9 A negative value of Q indicates
   (a) an exoergic reaction
   (c) an irreversible reaction
   (b) an endoergic reaction
   (d) a reversible reaction

22.10 A positive value of Q indicates
   (a) an endoergic reaction
   (b) a reversible reaction
   (c) an exoergic reaction
   (d) an irreversible reaction

22.11 A nuclear reactor is a system in which
    (a) an uncontrolled nuclear reaction does not take place
    (b) an uncontrolled nuclear reaction takes place to liberate energy
    (c) a controlled nuclear chain reaction is used to liberate energy
    (d) (a) and (c)

22.12 The major contributors to the nuclear force are
    (a) volume energy
    (b) surface energy
    (c) nuclear symmetry and pairing energy
    (d) all of these

22.13 The semi-empirical mass formula that is derived from the liquid drop model gives the total interaction energy of a complex nucleus as:

    (a) $E_B = -6UA + 4\pi b^2 A^{2/3} S_s + \dfrac{3}{5}\dfrac{Q^2}{bA^{1/3}} + 6kUA\left(\dfrac{D}{A}\right)^2 \pm A_s A^{-2/3}$

    (b) $E_B = -6UA + 4\pi b^2 A^{4/3} S_s + \dfrac{3}{5}\dfrac{Q^2}{bA^{1/3}} + 6kUA\left(\dfrac{D}{A}\right)^2 \pm A_s A^{-4/3}$

    (c) $E_B = -6UA + 4\pi b^2 A^{5/3} S_s + \dfrac{3}{5}\dfrac{Q^2}{bA^{1/3}} + 6kUA\left(\dfrac{D}{A}\right)^2 \pm A_s A^{-4/3}$

    (d) $E_B = -6UA + 4\pi b^2 A^{2/3} S_s + \dfrac{3}{5}\dfrac{Q^2}{bA^{1/3}} + 6kUA\left(\dfrac{D}{A}\right)^2 \pm A_s A^{-4/3}$

22.14 A stellar representative of a source of fusion energy is
    (a) the moon   (c) the sun
    (b) a star   (d) (b) and (c)

22.15 Which of the following forms are radiation detectors?
    (a) Proportional chamber   (c) GM counter
    (b) Ionization chamber   (d) All of these

22.16 Linear accelerators of particles are
    (a) cyclotrons   (c) (a) and (b)
    (b) synchrotrons   (d) none of these

22.17 Alpha rays emitted from a radioactive substance are
    (a) uncharged particles   (c) doubly ionized helium atoms
    (b) negatively charged particles   (d) ionized hydrogen nuclei

22.18 In a nuclear reaction, there is conservation of
    (a) energy   (c) momentum
    (b) mass   (d) all of these

22.19 Beta rays are electron particles that move with the velocity of
    (a) light   (c) air
    (b) sound   (d) none of these

22.20 There is a radioactive element whose original activity I_0 reduces to one-third over a period of 6 years. After a lapse of 6 years, the activity of the sample would be
    (a) I_0
    (b) $I_0/4$
    (c) $I_0/3$
    (d) $I_0/9$

Review Questions

22.1 The neutron–proton ratio plays a critical role in the stability of a nucleus. Enunciate the reasons that help in the stability of the nucleus.

22.2 On the basis of which parameter (i.e., the proton number, neutron number, or mass number of several nuclei) does one distinguish between the different nuclides?

22.3 Why are heavier nuclides not stable in terms of the forces acting in the nucleus?

22.4 Give the representation for a positron, a neutrino, and an anti-neutrino in a nuclear decay equation where an electron is represented as $_{-1}^{0}\beta$.

22.5 The spin flip properties of a nucleus can be used as a tool in imaging techniques. Explain in detail how this is done.

22.6 Beta particles have higher penetration power than alpha particles. Comment.

22.7 An unstable nuclide disintegrates into alpha, beta, or gamma particles. Explain why.

22.8 Is there any advantage in using protons or hydrogen nuclei instead of alpha particles in trying to initiate nuclear reactions?

22.9 Why does a nuclear reactor not blow up like an atom bomb.

22.10 Explain in detail the characteristic properties of a nucleus.

22.11 How can one define the dimensions of a nucleus. Describe in detail the constituents of a nucleus.

22.12 Explain in detail the functioning of the nuclear magnetic resonance (NMR) technique.

22.13 Derive the decay constant for the natural radioactive equations. Further, define the terms mean lifetime T_{mean} and $T_{1/2}$ (half-life) of a radioactive element.

22.14 Define one Becquerel. How many Becquerels are there in one Curie?

22.15 Explain in detail the processes of alpha decay, beta decay, and recoilless emission of gamma rays.

22.16 Explain the phenomenon of nuclear fluorescence.

22.17 Explain how the laws of conservation lead to the concept of endoergic or exoergic reactions.

22.18 Explain whether natural radioactivity is an endoergic reaction or an exoergic reaction.

22.19 Write the fission reaction for uranium-235 splitting into rubidium-94 and cesium-139.

22.20 Tritium can be obtained by neutron bombardment of lithium-6. Write the nuclear equation for this reaction.

22.21 The sun acts as a big fusion reactor. How do the fusion reactions remain confined within the sun in this case?

22.22 The energy release for each fusion reaction is less than 20 MeV, whereas about 200 MeV of energy is released from a fusion reaction. Explain. Besides this, discuss the advantages that fusion reactors would have over fission reactors.

22.23 Explain how the concept of liquid-drop model explains the binding together of nucleons in a nucleus. Hence, give the semi-empirical mass formula.

22.24 Particle accelerators are designed for accelerating particles. Explain in detail the principle and working of a cyclotron.

Numerical Problems

22.1 Plutonium decays as follows with a half-life of 24,000 years:

$$_{94}^{239}\text{Pu} \rightarrow \ _{92}^{235}\text{U} + \ _{2}^{4}\text{He}$$

If plutonium is stored for 72,000 years what fraction of it is remaining?

$$\left[Hint: \left(\frac{\text{Number of years stored}}{\text{Half–life of the nucleus}} \right) \right]$$

22.2 The nucleon number for iron is $A = 64$. Find the radius, mass, and density of the nucleus.

$$\left[Hint: = R_0 A^{1/3}; V = \frac{4}{3}\pi R^3; \rho = \frac{m}{V} \right]$$

22.3 Hydrogen atoms are placed in an external magnetic field. Protons can make transitions between states, in which their spin component is parallel and anti-parallel to the field, by absorbing or emitting a photon. What is the magnitude of the magnetic field that is required for this transition to be induced by photons of frequency 50 MHz?

$$\left[Hint: \vartheta = \frac{\Delta E}{h}; U = -|u_z| B \right]$$

22.4 A sample of ^{215}At contains 2.5 mg of the element whose half-life is 100 μs. What is its activity initially? $\qquad$ $\left[Hint: \text{Initial activity} = \lambda N_0 \right]$

22.5 The half-life of ^{60}Co is nearly 5.25 years. The activity of a radioactive element is proportional to the number of un-decayed atoms. Find the time it will take for the activity of the sample to decrease to (a) (1/8) of its original value and (b) (1/4) of its original value.

22.6 Find the frequency of the oscillating electric field if $B = 4.0$ T for accelerating deuterons of charge $+e$ and mass 3.3×10^{-27} kg.

$$\left[Hint: \vartheta = \frac{1}{T} = \frac{qB}{2\pi m_0} \right]$$

22.7 How many times does a deuteron need to be accelerated to acquire 15 MeV of kinetic energy if the potential across the cyclotron dees are 35 kV?

22.8 Deuterons in a cyclotron travel in a circle of radius 30.0 cm just before emerging from the dees. The frequency of the applied alternating voltage is 10.0 MHz. Find (a) the magnetic field, (b) the energy, and (c) speed of the deuterons upon emergence.

$$\left[Hint: = \frac{1}{T}; v = \frac{Bq}{m} r; E = \frac{1}{2} mv^2 \right]$$

22.9 A GM counter collects 10^7 electrons per discharge when the counting rate is 500 counts/min. What will be the average current in the circuit?

22.10 How many counts per minute are detected by a GM counter if the counter wire collects 10^5 electrons per discharge with an average current in the circuit of about 1.5×10^{-12} A.

$$\left[Hint: I = \frac{q}{t} \right]$$

22.11 A GM counter with a dead time of 210 μs records about 12,000 counts/min. What is the dead time loss in the counting rate?

$$\left[Hint: \text{True counting rate} = \frac{\text{observed counting rate}}{1 - (\text{observed counting rate } x \text{ dead time})} \right]$$

22.12 In an ionization chamber, a charged particle is completely stopped, producing an energy of 4 MeV per $\dfrac{3 \times 10^8 \, m}{s}$. Estimate the pulse height if the total capacitance of the capacitor is 35 pF.

$$\left[\text{Hint: Pulse height} = V = \frac{Q}{C} = \frac{ne}{C}\right]$$

22.13 In an ionization chamber, about 5×10^4 ion pairs are produced, resulting in the loss of about 20 eV of energy. Estimate the amount of charge collected by each of the plates. What is the kinetic energy of this alpha particle?

$$\left[\text{Hint: } Q = ne\right]$$

New Engineering Materials and Nanotechnology

Learning Objectives

After studying this chapter, students will be able to

- understand various new engineering materials such as metallic glasses, shape memory alloys, nanomaterials, and carbon nanotubes
- explain their physical and chemical properties
- describe the methods of their preparation
- demonstrate their applications in various fields

23.1 INTRODUCTION

Material science has progressed at a fast rate to keep pace with ever developing innovations and technologies. Gone are the days when tennis racquets and cricket bats were made of only wood. Polymers, carbon fibres, nanotechnology, and other such modern technologies now go into making modern sports tools. Advanced machines and systems such as aircraft, ships, and cars have also vastly benefitted from these innovations. Composite materials make up 50% of the primary structure of modern aircraft like the BOEING 787 Dreamliner, including the fuselage and wing. In addition, in the world's largest passenger aircraft, Airbus A380, composite materials account for 25% of the constituent structural weight. This allows modern-day advanced aircraft to be greener and more fuel efficient. We will deal with some new engineering materials in this chapter. The idea is to present some basic concepts and ideas on topics such as metallic glasses, shape memory alloys, nanotechnology, and similar areas, to give the students some initial information. This would enable readers to form ideas and take informed decisions.

23.2 METALLIC GLASSES

The word metal reminds us of materials such as Fe (iron), Cu (copper), and Al (aluminium). These materials have important properties such as toughness, malleability, ductility, and conductivity, which can be attributed to their uniform structural arrangement of atoms, namely, their crystalline nature.

Semi-transparent or opaque, brittle, bad conductor, ability to mould into various shapes—these are some of the features that come to our mind when we think of glasses. Glasses owe these properties to their amorphous nature.

Metallic glass, as the name suggests, is a combination of metal and glass. Metallic glasses are metallic alloys having an amorphous arrangement (resembling the amorphous structure of glass) of atoms. It is quite natural to wonder how a metallic alloy can be amorphous, a straight contradiction to the crystalline nature of a metallic substance. The answer to this question holds the key to the formation of metallic glasses, which are metallic substances in amorphous state.

23.2.1 Types of Metallic Glasses

Metallic glasses can be classified broadly into two types:

Metal–Metal metallic glasses   These are combinations of two or more metals. Examples include nickel–niobium alloy, magnesium–zinc alloy, and copper–zirconium alloy.

Metal–Metalloid glasses   These are combinations of metals and metalloids. Examples are iron, cobalt, and nickel metals in combination with boron, silicon, carbon, and phosphorous metalloids.

23.2.2 Preparation

Metallic glasses are made using the principle of ultrafast cooling, also called *rapid quenching*, a process that helps achieve the amorphous phase of a material. Molten alloys are fed onto high-thermal-conductivity metallic rollers moving at a high speed. This results in rapid cooling of the molten alloy and, subsequently, the formation of metallic glass sheets.

23.2.2.1 Melt-spinning Technique

Glass in the melt, on cooling, forms amorphous glass. When glass in the molten form is cooled, it results in the formation of amorphous glass. Metals, on the other hand, form crystalline substances. Hence, the cooling process needs to be controlled in such a way that crystallization does not occur, and formation of metallic glass can take place.

In this method, the molten alloy is cooled rapidly, the rate of cooling being about a million degrees per second. This ensures that there is not enough time available for crystallization. A metallic disc is maintained at very low temperatures and rotated at a high speed. The molten alloy is present in a container with a small nozzle at the bottom. Using pressure, the molten alloy is forced onto the disc, where it solidifies and a thin layer of metallic glass is obtained. This thin film is then extracted using appropriate arrangement. Figure 23.1 is a schematic representation of the melt-spinning technique.

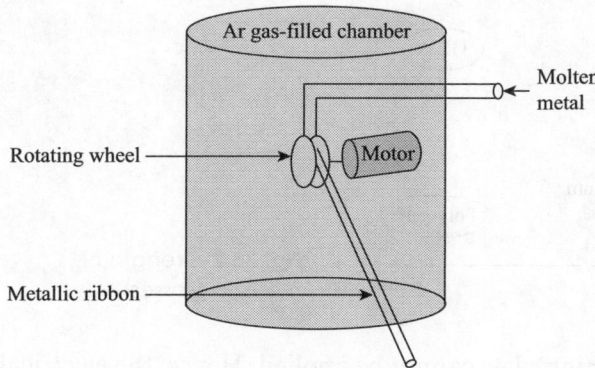

Fig. 23.1 Schematic diagram of melt-spinning technique

The following are other techniques of obtaining metallic glasses:

Radiation exposure A solid alloy is irradiated with electrons or ions. This leads to instability within the atoms, resulting in amorphization.

Annealing Annealing at moderate temperatures leads to inter-diffusion, yielding a metallic glass. This technique is suitable for two-phase crystalline materials only.

23.2.2.2 Deposition Technique

Molten alloy is converted to the vapour form; these vapours are then allowed to settle on the surface of a cold substrate atom by atom. This technique can be used to achieve very fast quenching.

23.2.3 Properties

Metallic glasses, especially transition metal–metalloid glasses, possess very high tensile strength. Their strength is comparable to the strongest steel, that is, in the range of 3000–4000 MPa. They have good ductility and high specific strength. They also show internal friction and sound attenuation. In case of zirconium (Zr)-based alloys, the strength can be as high as 2 GPa and the fracture toughness is 80 MPa. The absence of grain boundaries and dislocations makes them an ideal choice for micro-scale fabrication of mechanical parts.

The elastic modulus of metallic glasses is low. Poor plasticity of metallic glass composites can be attributed to the irregular interfacial morphology of the submicron-scale heterogeneous amorphous phases throughout the material. Figure 23.2 shows the strength of different metallic glasses.

23.2.3.1 Electrical Properties

In ideal metals, electrical resistivity becomes zero at $T = 0$ K (absolute zero temperature). In case of metallic glasses, because of the irregular

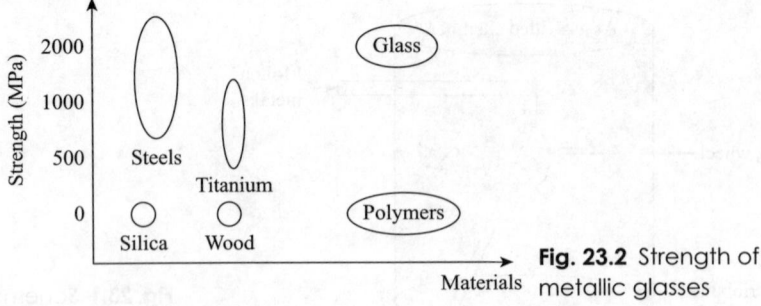

Fig. 23.2 Strength of metallic glasses

arrangement of atoms, Bragg's law cannot be applied. Hence, the electrical resistivity is high even at $T = 0$ K. The resistivity does not change much with temperature.

In the presence of transition metals of the 4D series of the periodic table, metallic glasses act as superconductors at low temperatures, with the critical magnetic fields being higher than those in the crystalline form.

According to some existing theories, metallic glasses display both positive and negative values of the resistivity coefficient. Thus, the minimum in the temperature dependence of electrical conductivity may not necessarily be related to the Kondo effect. The result is due to an interaction between the positive (metallic) and negative (semiconducting) contributions to the electrical conductivity, as observed in metallic glasses due to their inherent disorder.

23.2.3.2 Magnetic Properties

The presence of iron and cobalt gives rise to ferromagnetic properties, and hence anisotropy, in a material. Under suitable deposition conditions, the anisotropy can be strong enough to maintain the magnetization perpendicular to the film. Annealing can be used to reverse the process of magnetization. These materials exhibit the magnetostriction property. They also have high magnetic permeability.

23.2.3.3 Chemical Properties

A small percentage of chromium in a metallic glass results in very high corrosion resistance. This is due to the formation of uniform surface oxide layers. Metallic glasses also exhibit good catalytic property, since they are non-porous and reactions are generally confined to the surface.

The catalytic property is used in hydrogenation, hydrogenolysis, and dehydrogenation reactions. For example, $Ni_{28}Ti_{72}$ gives 100% selectivity for hydrogenation of acrolein to propanal. Activity of Raney nickel can be improved by enrichment of a glassy plate through appropriate treatments of precursors.

23.2.4 Applications of Metallic Glasses

The following are some application of metallic glasses:

1. The metallic glass $Ti_{40}Cu_{30}Pd_{14}Zr_{10}$ is non-carcinogenic and about three times stronger than titanium. Its elastic modulus nearly matches that of bones, and, therefore, it can be used for bone replacement. It also does not shrink on solidification.

2. Metallic glasses are used in fabrication of micro/nanoelectromechanical systems (MEMS/NEMS). As metallic glasses do not have the characteristics of crystals, their functional properties can be altered easily. This fabrication process is carried out in the super-cooled liquid region.

3. Metallic glasses are also used in aerospace industry due to their light weight, non-breakable nature, and corrosion resistance.

4. Another application of metallic glasses is in pressure sensors. Metallic glasses have high strength and a low elastic modulus. These lend high ductility with good spring property to metallic glasses, making them suitable for use in pressure sensors. These metallic glass-based pressure sensors have very high sensitivity. Due to the high strength of these materials, the developed sensors have a wide dynamic measurement range. Very large pressures can be measured with high accuracy. Metallic glass-based sensors are used in applications requiring high sensitivity and high resisting pressures.

23.3 SHAPE MEMORY ALLOYS

A shape memory alloy is an alloy that 'remembers' its shape and can be made to return to its initial shape after being deformed if subjected to an appropriate thermal procedure.

When the shape memory effect is correctly used, this material becomes a lightweight, solid-state alternative to conventional actuators such as hydraulic, pneumatic, and motor-based systems. Shape memory alloys have several applications in the medical and aerospace industries. Some examples of shape memory alloys are Ni–Ti, Cu–Al–Ni, and Cu–Zn–Al alloys.

23.3.1 Types of Shape Memory Alloy Effects

There are two types of shape memory alloy effects:

One-way shape memory alloy effect This shape memory alloy exhibits shape memory effect only on heating. On cooling, the shape memory alloy retains the shape that it had before heating.

Two-way shape memory alloy effect This shape memory alloy exhibits shape memory effect during heating and cooling. The shape change gets restored on cooling. The two-way shape memory alloy effect requires training of the material.

Some examples of shape memory alloys are Ni–Ti alloy (Nitinol), Cu–Al–Ni alloy, Cu–Zn–Al alloy, Au–Cd alloy, Ni–Mn–Ga alloy, and Fe-based alloys.

The following are some of the properties of Ni–Ti alloys:

1. Ni–Ti alloys are free from defects.
2. They have a negative thermal expansion coefficient at very low temperatures.
3. These materials have low eddy current losses.
4. They are highly stable.
5. Ni–Ti alloys show super-elastic behaviour just above the transformation temperature.
6. On loading, these alloys become deformed martensite, and on heating the deformed structure, they turn into twinned martensite.
7. While heating and cooling, these alloys deform and regain their original shape. This property is called *pseudoelasticity*.
8. These alloys show the hysteresis property during heating and cooling cycles.

23.3.2 Working of Shape Memory Alloys

Shape memory alloys owe their behaviour to a solid-state phase change that leads to a molecular rearrangement. Usually, a phase change is accompanied by a transformation of the material from solid to liquid, or liquid to gas, or vice versa. A solid-state phase change involves a molecular rearrangement, but the closely packed nature of the molecules is maintained and thus the material remains solid. A temperature of just about 10°C is sufficient to initiate the phase change in most shape memory alloys. These materials exist either in martensite or in austenite phases.

23.3.2.1 Martensite and Austenite Phases

Martensite is a relatively soft and easily deformed phase of shape memory alloys and exists at lower temperatures. The molecular structure in this phase is twinned. This configuration is shown in the middle portion of Fig. 23.3. Upon deformation, this phase takes on the second form shown in the right side of the figure.

Austenite, a stronger phase of shape memory alloys, occurs at higher temperatures. The shape of the austenite structure is cubic, as shown in the left side

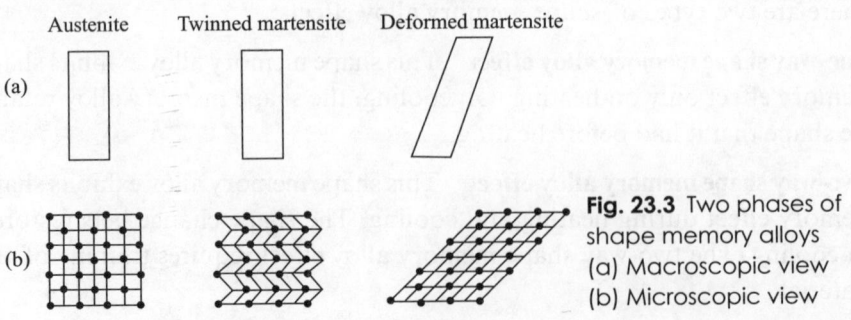

Austenite   Twinned martensite   Deformed martensite

(a)

(b)

Fig. 23.3 Two phases of shape memory alloys
(a) Macroscopic view
(b) Microscopic view

of the figure. The undeformed martensite phase is of the same size and shape as the cubic austenite phase on a macroscopic scale; therefore, no change in shape or size is visible in shape memory alloys until the martensite phase is formed.

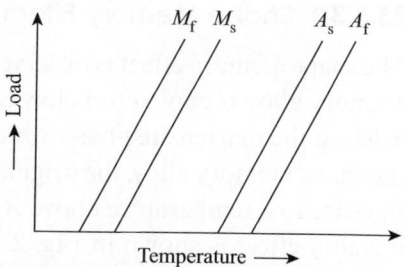

Fig. 23.4 Phase change temperature dependence on loading

The temperatures at which each of these phases begins to form and finishes formation are represented by the variables M_s, M_f, A_s, and A_f. The amount of loading placed on a piece of shape memory alloy increases the values of these four variables, as shown in Fig. 23.4.

A transition from the martensite phase to the austenite phase is dependent on temperature and stress. The transition is, however, not time dependent, unlike most phase changes. This is because diffusion does not play a role in this transition. The reversible, diffusion-less transition between the two phases results in imparting special properties to these materials. Transition reversibility is a crucial factor in the formation of shape memory alloys. For example, even in carbon steel, martensite can be formed from austenite by rapid quenching. However, this process is not reversible, and, therefore, steel does not have shape memory properties.

Figure 23.5 is a schematic representation of the thermal hysteresis curve for shape memory alloys. Here, $\xi(T)$ represents the fraction of the martensite phase. The difference between the heating and cooling transitions results in the hysteresis effect. The exact shape is dependent on the material properties of the shape memory alloy. These are, in turn, decided by process steps such as alloying and work hardening.

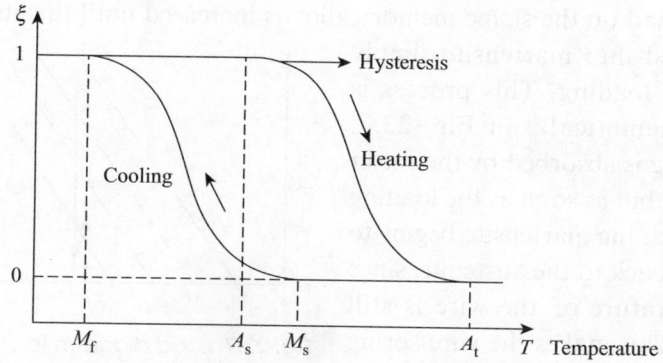

Fig. 23.5 Thermal hysteresis curve for shape memory alloys

23.3.3 Characteristic Properties of Shape Memory Alloys

The following subsections deal with the characteristic properties of shape memory alloys.

23.3.3.1 Shape Memory Effect

The shape memory effect is observed when the temperature of a piece of shape memory alloy is cooled to below the temperature M_f. At this stage, the alloy exists in the martensite phase, which can be deformed easily. After distorting the shape memory alloy, the original shape can be recovered by simply heating the wire to a temperature above A_f. A schematic representation of the shape memory effect is shown in Fig. 23.6. The heat transferred to the wire is the power driving the molecular rearrangement of the alloy. This is similar to heat melting of ice into water, although in this case the alloy remains solid. The deformed martensite is now transformed to the cubic austenite phase, which is configured in the original shape of the wire.

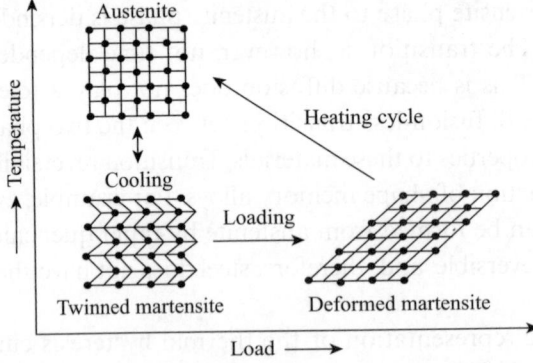

Fig. 23.6 Schematic representation of shape memory effect

23.3.3.2 Pseudoelasticity

Pseudoelasticity occurs in shape memory alloys when the alloy is completely composed of austenite phase at temperatures greater than A_f. Unlike the shape memory effect, pseudoelasticity occurs without a change in temperature. The load on the shape memory alloy is increased until the austenite is transformed into martensite simply due to the loading. This process is shown schematically in Fig. 23.7. The loading is absorbed by the softer martensite, but as soon as the loading is decreased, the martensite begins to transform back to the austenite, since the temperature of the wire is still above A_f. This makes the wire spring back to its original shape.

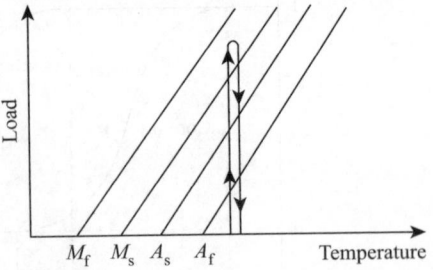

Fig. 23.7 Load diagram for pseudoelastic effect

23.3.3.3 Hysteresis

Figure 23.8 shows a typical hysteresis curve for shape memory alloys. Hysteresis represents the difference between the temperature at which the

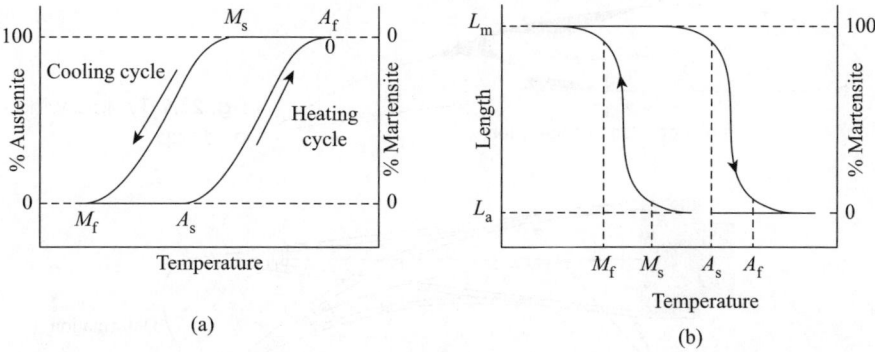

Fig. 23.8 Hysteresis curve for shape memory alloys (a) Composition dependence (b) Length dependence

material is 50% transformed to the austenite phase upon heating and that at which it is 50% transformed to martensite upon cooling. Generally, the difference in temperature is ~25°C.

23.3.4 Manufacturing Shape Memory Alloys

Shape memory alloys are typically manufactured by casting, using vacuum arc melting or induction melting. These specialized techniques are used to keep impurities at the minimum level in the manufactured alloy. These techniques also ensure a thorough mixing of the constituent metals. The ingot is then hot-rolled into longer sections and drawn into wires. The specific 'training' imparted to the alloy is dependent on the type of properties desired. The 'training' dictates the shape that the alloy will remember when it is heated. This is achieved by heating the alloy so as to allow the dislocations to reorder into stable positions, without allowing the temperature to be high enough for recrystallization to set in.

23.3.5 Applications of Shape Memory Alloys

Some applications of shape memory alloys are discussed in the following subsections.

23.3.5.1 Aeronautical Applications

Aircraft manoeuvrability depends heavily on the movement of flaps found at the rear or trailing edge of the wings. The efficiency and reliability of operation of these flaps are of critical importance. Most modern aircraft operate these flaps using extensive hydraulic systems or the fly-by-wire systems. This complex system of pumps and lines is often relatively difficult and costly to maintain. Many alternatives to the hydraulic systems are being explored by the aerospace industry. Some of the most promising alternatives are piezoelectric fibres, electrostrictive ceramics, and shape memory alloys.

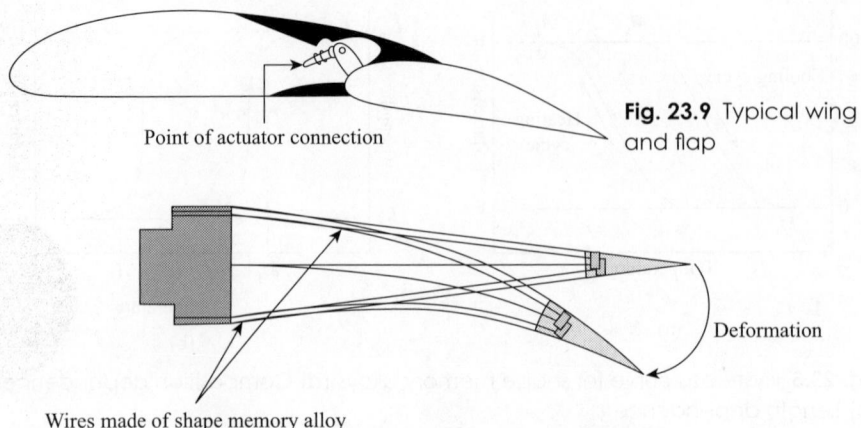

Fig. 23.9 Typical wing and flap

Point of actuator connection

Deformation

Wires made of shape memory alloy

Fig. 23.10 Hinge-less shape memory alloy flap

The flaps on a wing generally have the layout shown in the left portion of Fig. 23.9. A large hydraulic system is provided to it at the point of actuator connection, as shown in Fig. 23.9. 'Smart' wings based on shape memory alloys are typically like the wing shown in Fig. 23.10. This system is much more compact and efficient. The shape memory wires require small electrical currents to activate movement of the wings.

The shape memory wire is used to manipulate a flexible wing surface. The wire at the bottom of the wing gets shortened through the shape memory effect, whereas the top wire gets stretched, bending the edge downwards. The reverse movements take place when the wing needs to be bent upwards. The shape memory effect is induced in the wires simply by heating them with an electric current, which is easily supplied through electrical wiring. This eliminates the need for bulky hydraulic lines. By doing away with the hydraulic system, the aircraft weight, the maintenance costs, and the repair time can all be reduced significantly.

23.3.5.2 Dental Applications

Bone plates are surgical tools that are used to assist in the healing of broken and fractured bones (Fig. 23.11). The breaks are first set and then held in place using bone plates in circumstances where casts cannot be applied to the injured area. Bone plates are often applied to fractures occurring in areas such as the nose, jaws, or eye sockets. Such repairs fall under a discipline of medicine called osteosynthesis.

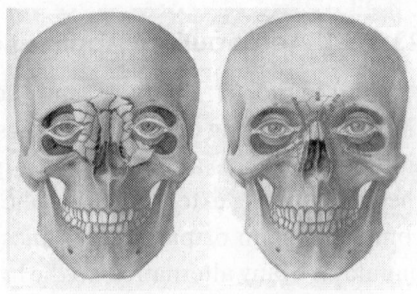

Fig. 23.11 Reconstruction of badly fractured face using bone plates

The present osteotomy procedure, shown schematically in Fig. 23.12, uses titanium and stainless steel. In this procedure, the broken bones are first surgically reset into their proper positions. Then a plate is screwed onto the broken bones to hold them in place, while the bones heal naturally. This method has been proved to be both successful and useful in treating all kinds of breaks. However, there are some drawbacks. After initially placing the plate on the break or fracture, the bones are compressed together and held under a small pressure that helps speed up the healing process of the bone. This tension is, however, only temporary and after a couple of days, the tension provided by the steel plate is lost and the break or fracture is no longer under compression. This greatly slows down the healing process.

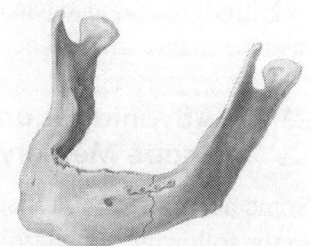

Fig. 23.12 Normal bone plate used to repair jaw fracture

This drawback can be overcome by using bone plates fabricated from shape memory alloys like nickel–titanium (transformation temperature A_f is much greater than 15°C). The surgical procedure followed for Ni–Ti plates is the same as that followed for the conventional bone plates. The Ni–Ti plates are first cooled to well below their transformation temperature; they are then placed on the set break just like titanium plates. However, at the normal body temperature, the Ni–Ti plates attempt to contract, applying sustained pressure on the break or fracture for significantly longer times than stainless steel or titanium. This steady pressure assists the healing process and leads to faster recovery.

23.3.5.3 Piping

The first commercial application of shape memory alloy material was in the field of coupling for piping, for example, oil line pipes for industrial applications, water pipes, and similar types of piping. The late 1980s saw the commercial introduction of Nitinol as an enabling technology in a number of minimally invasive endovascular medical applications. Though more expensive than stainless steel, the self-expanding properties of Nitinol alloys manufactured to body temperature response specifications have provided an attractive alternative to balloon expandable devices. About 50% of all peripheral vascular stents currently available in the markets worldwide are manufactured using Nitinol.

23.3.5.4 Robotics

Shape memory alloys have also been used in the field of robotics. They can be used to make very light robots. However, the present technology suffers from energy inefficiency, slow response times, and large hysteresis effects.

Nitinol wire is also used in robotics (e.g., the hobbyist robot Stiquito) and in some magic tricks. These involve heat and shape shifting processes.

23.3.6 Advantages and Disadvantages of Shape Memory Alloys

Some advantages and disadvantages of shape memory alloys are discussed in the following subsections:

Advantages   Biocompatibility, diverse fields of applicability, and good mechanical properties (strong, corrosion resistant) are the advantages of shape memory alloys.

Disadvantages   There are still some drawbacks of shape memory alloys that must be overcome before their full potential can be realized. These alloys are still relatively expensive to manufacture and machine compared to other conventional materials such as steel and aluminium. Most shape memory alloys have poor fatigue properties. Thus, under similar loading conditions (twisting, bending, and compression) a component made of steel can have at least two orders of magnitude higher useful cycles as compared to components manufactured using shape memory alloys.

23.4 NANOTECHNOLOGY

Nano stands for 10^{-9}. Nanomaterials form a branch of materials science that deals with the study of materials and their morphological characteristics on the nanoscale. Nanomaterials are composed of grains that may or may not be visible to the naked eye. Nanomaterials are mostly classified into the following categories:

1. Nanostructured materials like thin films
2. Nanoparticles
3. Nanotubes
4. Nanodots
5. Nanorods
6. Nanocomposites

All these materials, being nanomaterials, satisfy the criterion that they have structured components with dimensions in the range of 10–100 nm.

Nanoscience involves the manipulation of materials on atomic, molecular, or macromolecular scales. Nanotechnology is used to manufacture, design, and also alter the properties of these materials on the nanoscale.

23.4.1 Top-down and Bottom-up Approaches

Nanofabrication primarily relies on two different approaches, namely top-down and bottom-up approaches. We will discuss these approaches in brief in this section.

23.4.1.1 Top-down Approach

The top-down approach utilizes photolithography to pattern a layer of a material over a given substrate. This is then followed by patterning the bulk substrate by different etching techniques. Silicon is the most commonly used substrate. A typical top-down approach is shown in Fig. 23.13.

In the process, first a suitable substrate is realized. In the example illustrated, it consists of a silicon wafer with a top silicon dioxide layer [Fig. 23.13(a)]. A positive photoresist (a positive resist is a more popular choice than a negative resist) is then coated on the silicon dioxide layer and suitably treated. A mask with the requisite pattern is then used to expose the photoresist layer to ultra-violet rays in a selective manner, as shown in Fig. 23.13(b). The exposed portion of the resist is then developed using suitable solvents [Fig. 23.13(c)]. The windows opened in the resist are then used to etch the silicon dioxide to realize the desired pattern, as shown in Fig. 23.13(d).

In addition to the basic processes of photolithography and etching, the top-down approach also utilizes some additive processes such as deposition, growth, and ion implantation. The deposition process utilizes consumption of

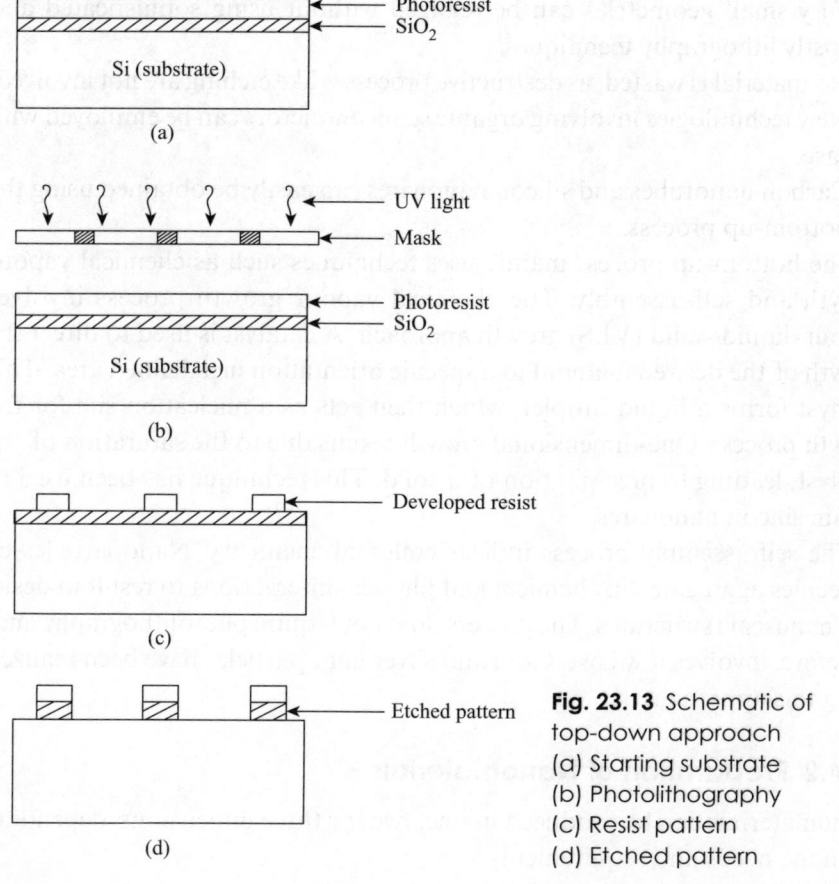

Fig. 23.13 Schematic of top-down approach
(a) Starting substrate
(b) Photolithography
(c) Resist pattern
(d) Etched pattern

energy to deposit a layer of the desired material. This can be achieved using physical vapour deposition techniques such as evaporation and sputtering, or chemical vapour deposition (CVD) techniques such as low-pressure CVD and plasma-enhanced CVD. Growth usually involves consumption of the substrate to create a new material. Very high temperatures are generally involved, and chemical reactions are important components. The ion-implantation process utilizes very-high-energy ions to modify the optical, mechanical, and electrical characteristics of a material. The process leads to stress creation and damage, and therefore necessarily involves a final annealing step.

23.4.1.2 Bottom-up Approach

The bottom-up approach uses a selective addition of desirable atoms to create nanostructures. It is pertinent to recollect that nature utilizes the bottom-up approach to build complex systems. Cells, crystals, animals, humans are all examples of the bottom-up approach. The growth process is controlled using chemistry and biology.

The bottom-up approach has the following distinct advantages over the top-down approach of realizing nanostructures:

1. Very small geometries can be realized without using sophisticated and costly lithography techniques.
2. No material is wasted, as destructive processes like etching are not involved.
3. New technologies involving organic semiconductors can be employed with ease.
4. Carbon nanotubes and silicon nanowires can easily be obtained using the bottom-up process.

The bottom-up process mainly uses techniques such as chemical vapour growth and self-assembly. The chemical vapour growth process involves vapour–liquid–solid (VLS) growth approach. A catalyst is used to direct the growth of the desired material to a specific orientation in a defined area. This catalyst forms a liquid droplet, which then acts as a nucleation site for the growth process. One-dimensional growth results due to the saturation of the catalyst, leading to precipitation of a solid. This technique has been used to obtain silicon nanowires.

The self-assembly process utilizes colloidal chemistry. Nanoparticles or molecules aggregate via chemical and physical interactions to result in desirable nanoscale structures. The process does not require photolithography and, therefore, involves low cost. Gold and silver nanoparticles have been realized using this approach.

23.4.2 Preparation of Nanomaterials

Nanomaterials can be produced in one, two, or three dimensions, depending upon the needs and requirements.

23.4.2.1 Plasma Arcing

A potential difference applied between two electrodes placed within a gaseous medium can, under suitable conditions, produce plasma. The material is first heated up to the evaporation point in induction plasma. The vapours are subsequently subjected to very rapid quenching in the quench/reaction zone. Quenching can be done using inert gases such as argon (Ar) and nitrogen (N_2), or reactive gases such as methane (CH_4) and ammonia (NH_3), depending upon the type of nanopowders to be synthesized. The nanometric powders produced are usually collected by porous filters. The typical size range of the nanoparticles produced is 20–100 nm, depending upon the exact quenching conditions.

23.4.2.2 Chemical Vapour Deposition and Chemical Vapour Condensation

Chemical vapour deposition is a process in which a solid is deposited on a heated surface using a chemical reaction from the vapour or gas phase. These reactions need activation energy to proceed. This energy can be provided in several ways. In thermal CVD, a high temperature of above 900°C activates the reaction. A typical apparatus comprises a gas supply system, a deposition chamber, and an exhaust system. In plasma CVD, the reaction is activated by plasma at temperatures between 300°C and 700°C. In laser CVD, pyrolysis occurs when laser thermal energy heats an absorbing substrate. In photolaser CVD, the chemical reaction is induced by an ultraviolet radiation, which has sufficient photon energy, to break the chemical bond in the reactant molecules. In this process, the reaction is photon-activated and deposition takes place at room temperature. Nanocomposite powders have been prepared by CVD. A composite powder of SiC/Si_3N has been prepared using SiH_4, CH_4, WF_6, or H_2 as a source of gas at a growth temperature of 1400°C.

23.4.2.3 Sol–Gel Techniques

In addition to the aforementioned techniques, the sol–gel processing techniques have also been used extensively for obtaining nanoparticles. Colloidal particles are much larger than normal molecules or nanoparticles. Upon mixing with a liquid, colloids appear bulky, whereas the nanosized molecules always look clear.

The sol–gel process uses colloidal suspension (sol) and gelatin to form a network in a continuous liquid phase (gel). The precursor for synthesizing these colloids consists of ions of metal alkoxides and aloxysilanes. The most commonly used materials are tetramethoxysilane (TMOS) and tetraethoxysilanes (TEOS), which form silica gels. Alkoxides are immiscible in water. They are organometallic precursors for silica, aluminium, titanium, zirconium, and many others. Mutual solvent alcohol is used. The sol–gel process initially uses a homogeneous solution of one or more selected alkoxides. These are organic precursors for silica, alumina, titania, and zirconia, among others. A catalyst is then used to start the reaction and control the pH. Formation of a sol–gel involves the following stages:

1. Hydrolysis condensation
2. Growth of particles
3. Agglomeration of particles

23.4.2.4 Electrodeposition

Nanostructured materials (thin films) can also be produced by electrodeposition. These films are mechanically strong and uniform. Substantial progress has been made in nanostructured coatings made using CVD. Many other non-conventional processes such as hypersonic plasma particle deposition (HPPD) have also been used to synthesize and deposit nanoparticles. It has been shown that certain properties of nanostructured deposits such as hardness, wear resistance, and electrical resistivity are strongly affected by their grain size. A combination of increased hardness and wear resistance results in an electrodeposit of superior quality.

23.4.2.5 Ball Milling

This technique uses the principle of mechanical crushing. Small balls inside a drum-like cavity are rotated at high speeds, and by gravity action, they settle on a solid layer where they are crushed into nanocrystals. This technique is mainly used to produce oxide-based nanoparticles (e.g., CeO_2).

The following are the various types of ball mills:

1. Attrition ball mill
2. Planetary ball mill
3. Vibrating ball mill
4. Low-energy tumbling mill
5. High-energy ball mill

23.4.3 Properties of Nanomaterials

The properties of nanomaterials are very different from those of the bulk materials. One important difference is the increased surface area to volume ratio of nanostructures. Nanostructures are also associated with quantum effects, which can be attributed to the size of the nanoparticles.

23.4.3.1 Physical Properties

In nanomaterials, the small size of the material results in an increase in its surface area to volume ratio, as compared to that of the normal bulk material. The interatomic spacing decreases with increase in size of the material. This happens due to the electrostatic forces over-balancing the repulsive forces of atomic centres. This in effect changes the surface free energy of the material and thereby its chemical potential. Thus, the thermodynamic properties undergo changes. We may note that the melting point also decreases with a decrease in size.

23.4.3.2 Chemical Properties

Variations in geometry, electronic arrangement, and surface area have a strong influence on the catalytic properties. The hydrogen-absorbing ability of a metal decreases with size. In nanomaterials, this ability is about 6–10 times more than that in the normal metal. This makes nanomaterials good candidates for hydrogen storage applications.

23.4.3.3 Electrical Properties

The size of nanomaterials leads to an increase in their ionization potential. Due to quantum confinement, electronic bands come together and become narrow. Energy states are transformed into localized molecular bands, which can be altered by the passage of a current or by application of a field. The change in electrical properties is material dependent. As an example, metals undergo an increase in conductivity, whereas in the case of non-metallic nanomaterials, a decrease in conductivity is observed.

23.4.3.4 Optical Properties

Depending upon their constituents, nanoparticles absorb a range of wavelengths and emit a characteristic wavelength. It is possible to alter the linear and nonlinear optical properties by altering the crystals. Nanomaterials are, therefore, used in electrochromic devices.

23.4.3.5 Magnetic Properties

Nanosized materials are more magnetic than their counterparts in the bulk. Nanoparticles of non-magnetic solids may also demonstrate magnetic properties.

23.4.3.6 Mechanical Properties

Nanomaterials contain a lot of grains and, therefore, have grain boundaries. These grains lead to drastic changes in the mechanical properties, which are as follows:

1. Nanomaterials have up to 50% lower elastic moduli than normal grain-sized materials.
2. They are up to seven times harder than materials in the micrometre range. It is also well established that Cr_2O_3 is one of the hardest oxides both on the mineralogical scale (8.5 Mohs) and on the microhardness scale (up to 29.5 GPa).
3. Superplastic nature in ceramics is observed on nanoscale.

23.5 CARBON NANOTUBES

Carbon nanotubes (CNTs) are molecular-scale tubes of graphitic carbon having outstanding properties. They are among the stiffest and strongest fibres known, and have remarkable electronic properties. They also have many other unique characteristics. They belong to the fullerene structural family.

CNTs are allotropes of carbon with a nanostructure that can have a length-to-diameter ratio of up to 28,000,000:1, which is significantly larger than any other material. These cylindrical carbon molecular tubes have novel properties that make them potentially useful in many applications in nanotechnology, electronics, optics, and other fields of materials science. These materials also have potential applications in architectural fields.

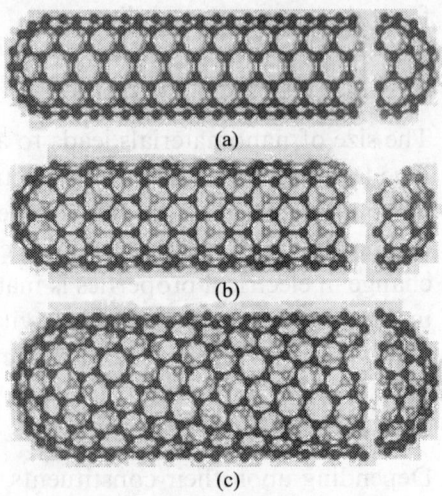

(a)

(b)

(c)

23.5.1 Structure

The bonding in CNTs is sp^2, with each atom joined to three neighbours, as in graphite. The tubes can, therefore, be considered as rolled-up graphene sheets (graphene is an individual graphite layer). There are three distinct ways in which a graphene sheet can be rolled into a tube. These are shown schematically in Fig. 23.14.

Fig. 23.14 Types of CNTs (a) Armchair (axis of CNT is parallel to C–C bonds of graphene) (b) Zigzag (axis of CNT is perpendicular to C–C bonds of graphene) (c) Chiral (axis of CNT is inclined to C–C bonds of graphene)

Electrical conductivity of a CNT depends on its specific structure. A single-walled nanotube can be obtained by wrapping one atomic layer of graphene into a seamless cylinder. A multiwalled CNT, on the other hand, has more than one surface within it.

23.5.2 Classification of CNTs

Based on the number of layers constituting them, CNTs can be classified as single walled or multiwalled. These two forms of CNTs are shown schematically in Fig. 23.15.

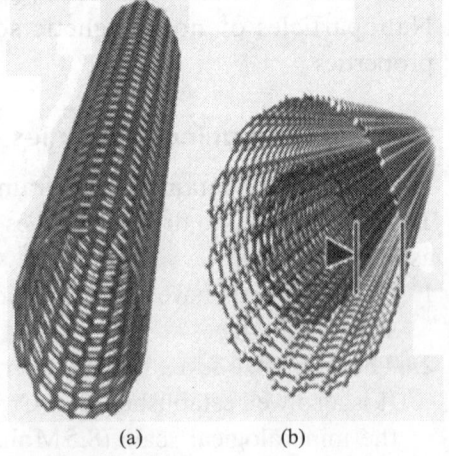

(a)   (b)

Fig. 23.15 CNTs (a) Single walled (b) Multiwalled

23.5.3 Fabrication

Techniques have been developed to produce nanotubes in sizeable quantities. Some common techniques include arc discharge, laser ablation, and CVD. Most of these processes take place in vacuum or with process gases. Chemical vapour deposition growth of CNTs can occur in vacuum or at atmospheric pressure. Large quantities of nanotubes can be synthesized by these methods. Advances in catalysis and continuous growth processes are making CNTs more commercially viable by the day.

23.5.3.1 Electric Arc Discharge Method

The carbon arc discharge method, initially used for producing C_{60} fullerenes, is the most common and easiest method used to produce CNTs. CNTs are produced from high-temperature plasma at around 3700°C. This method is also known as the DC arc discharge method.

Experimental set-up   The experimental set-up for the electric arc discharge method is shown in Fig. 23.16. Graphite rods with a diameter of 5–20 mm are fixed inside a discharge chamber at a separation of 1 mm. The discharge chamber is evacuated and filled with He gas at a pressure of 500 Torr. A direct current of 50–100 A, driven by a potential difference of approximately 20 V, leads to a high-temperature discharge between the two electrodes. Necessary cooling arrangements are also provided within the system.

Working   The discharge vaporizes the surface of one of the carbon electrodes and forms a small rod-shaped CNT deposit on the other electrode. The CNT yield is dependent on the uniformity of the plasma arc and the operating temperature for the deposits forming on the carbon electrode.

   Multiwalled CNTs can be produced using this method. The use of a catalyst such as Fe, Ni, or Co leads to the formation of single-walled CNTs. This

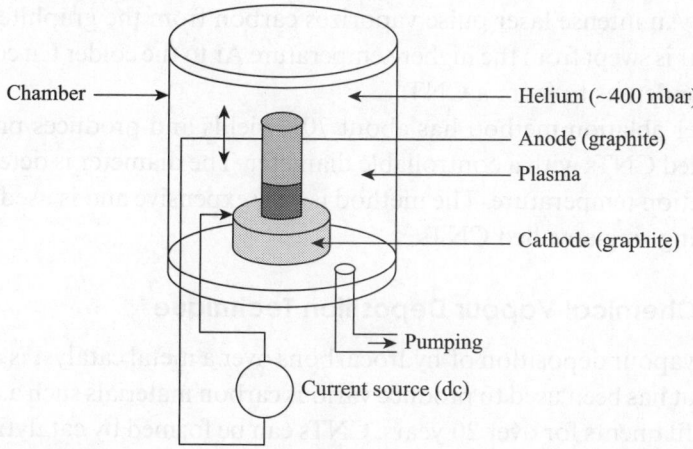

Fig. 23.16 Schematic diagram of electric arc discharge set-up

technique produces a complex mixture of components. Further purification is required to separate the CNTs from the soot and the residual catalytic metals present in the crude product. Nearly 75% of graphite is converted into CNTs.

23.5.3.2 Pulsed-laser Deposition Technique

Laser vaporization is an important method for obtaining small quantities of high-quality CNTs. A laser pulse of high intensity and energy is used to evaporate carbon from graphite. The evaporated carbon atoms are then condensed to form nanotubes.

Experimental set-up   The experimental set-up for the pulsed-laser deposition technique is shown schematically in Fig. 23.17.

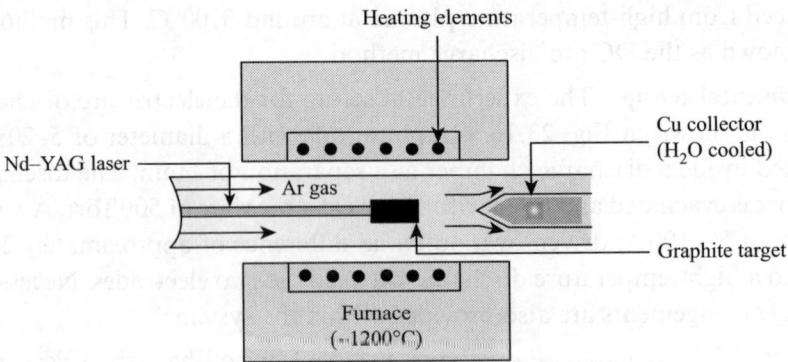

Fig. 23.17 Schematic diagram of pulsed-laser deposition system

A quartz tube, which contains the graphite target, is kept inside a muffle furnace. This tube is filled with Ar gas maintained at 1200°C. The graphite target contains a small amount of a catalyst mixture of Co and Ni (50:50). A water-cooled copper collector is fitted at one end.

Working   An intense laser pulse vaporizes carbon from the graphite target. This vapour is swept from the higher-temperature Ar to the colder Cu collector, where it condenses to form a CNT.

The laser ablation method has about 70% yields and produces primarily single-walled CNTs with a controllable diameter. The diameter is determined by the reaction temperature. The method is very expensive and is used mainly for obtaining single-walled CNTs.

23.5.3.3 Chemical Vapour Deposition Technique

Chemical vapour deposition of hydrocarbons over a metal catalyst is another method that has been used to produce various carbon materials such as carbon fibres and filaments for over 20 years. CNTs can be formed by catalytic CVD, with a very good overall yield.

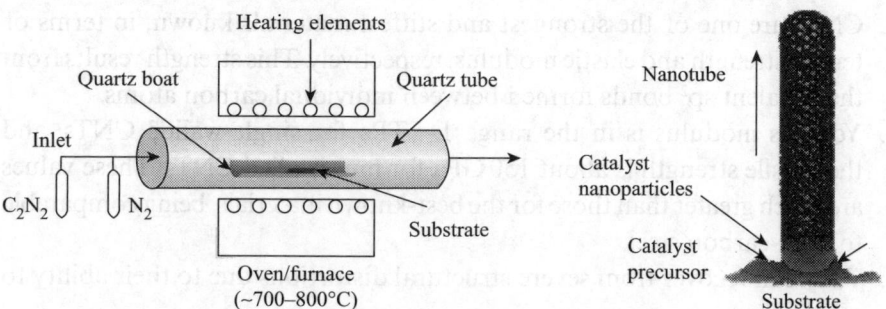

Fig. 23.18 Schematic diagram of CVD technique

Experimental set-up The experimental set-up is shown schematically in Fig. 23.18.

It consists of a quartz tube containing a graphite target placed on a quartz boat. The tube is heated to about 700–800°C using an oven. The tube is connected to sources of acetylene and nitrogen gas.

Working The hydrocarbon entering the quartz tube decomposes into carbon atoms (at about 720°C). These then condense on a substrate containing the catalyst, forming CNTs. The structure of multiwalled CNTs produced by CVD is shown in Fig. 23.19.

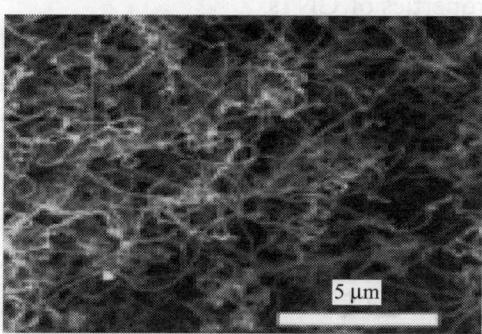

Fig. 23.19 Multiwalled CNTs produced by CVD

Of the various methods of nanotubes synthesis, CVD is most suited for industrial-scale deposition due to its price/unit ratio and its ability to grow nanotubes directly on a desired substrate. On the other hand, other growth techniques need an additional step involving collection of the produced nanotubes. Growth sites can be controlled by careful deposition of the catalyst.

23.5.4 Properties of CNTs

In the following sections, some of the properties of CNTs have been discussed.

23.5.4.1 Strength

Strength of CNTs can be expressed through the following features:

1. CNTs are one of the strongest and stiffest materials known, in terms of tensile strength and elastic modulus, respectively. This strength results from the covalent sp^2 bonds formed between individual carbon atoms.
2. Young's modulus is in the range 1–5 TPa for single-walled CNTs, and the tensile strength is about 150 GPa for multiwalled CNTs. These values are much greater than those for the best-known materials, being comparable to high-carbon steel.
3. They can recover from severe structural distortions due to their ability to rehybridize.

23.5.4.2 Electrical Properties

The following are the electrical properties of CNTs:
1. CNTs can be either metallic or semiconducting, depending upon the chirality. For example, the armchair form is metallic, whereas the zigzag form is found to be semiconducting.
2. The energy gap of semiconducting CNTs is inversely proportional to the diameter of the tube. The energy band gap can also be affected by localized defects.

23.5.4.3 Thermal Properties

The following are the thermal properties of CNTs:
1. CNTs are very good thermal conductors along the length of the tubes. The thermal conductivity of a CNT is found to increase with its length, as shown schematically in Fig. 23.20. CNTs exhibit ballistic conduction along their length. However, in the lateral direction with respect to the axis of the tube, they behave as good insulators.
2. It is predicted that CNTs will be able to achieve conductivities up to 20 times more than that of copper, a metal well known for its conductivity.
3. The temperature stability of CNTs is estimated to be up to 2800°C in vacuum and about 750°C in air.

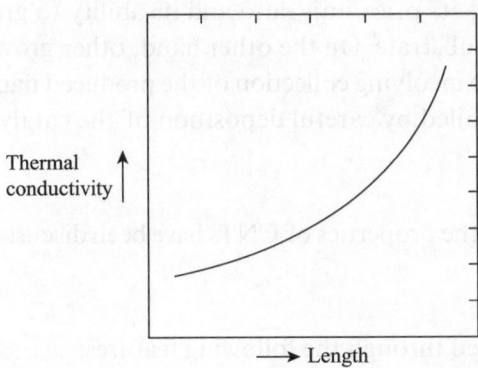

Thermal conductivity

→ Length

Fig. 23.20 Length-dependent thermal conductivity of CNTs

23.5.4.4 Chemical Properties

Chemical properties of CNTs are as follows:
1. CNTs are resistant to chemical reactions.
2. They can be used as catalysts in chemical reactions due to their high surface area.

23.5.5 Applications

CNTs have wide-ranging applications. The strength and flexibility of CNTs make them potentially useful in controlling other nanoscale structures. This suggests an important role for CNTs in nanotechnology engineering.

Structural Because of their superior mechanical properties, many CNT-based products have been proposed, ranging from everyday items such as clothes and sports gear to combat jackets and space elevators.

Electrical circuits Due to their unusual current conduction mechanism, CNTs make ideal components in electrical circuits. They are used in terahertz sources (switching instruments) and sensors. CNT-based transistors are capable of digital switching with a single electron.

Fibres and fabrics Fibres spun of pure CNTs, along with CNT composite fibres, have exceptional mechanical strength. Such super-strong fibres have many applications, such as in body and vehicle armour, transmission line cables, woven fabrics, and stain-resistant textiles.

CNT composite A plastic composite of CNT has been shown to act as a lightweight shielding material against electromagnetic radiations.

Drug delivery CNTs hold the potential to allow drug dosage to be lowered by localizing its distribution. This method is effective for controlled delivery and distribution of drugs inside the body.

Computer applications CNTs can be used to develop switching devices for use in computers.

Novel hard disk A novel data storage system capable of storing 10^{15} bytes/cm^2 is being explored using the unique properties of CNTs.

CNT gas sensor Based on their changing electrical properties, CNTs can detect very small concentrations (ppm) of O_2, NO_2, NH_3, etc. This opens up the possibilities of several applications in the field of extremely sensitive gas sensors.

23.6 FULLERENES

A *fullerene* is a molecule composed of carbon atoms arranged in the form of a hollow sphere, ellipsoid, or tube. The spherical form is called *bucky-ball* and the tubes are called *carbon nanotubes*. We have discussed CNTs

in Section 23.5. The first fullerene was prepared by
a team comprising Richard Smalley, Robert Curl,
James Heath, Sean O'Brien, and Harold Kroto at Rice
University. It was given the name *buckminsterfullerene*
in honour of Buckminster Fuller, who did pioneering
work on geodesic domes. The material discovered at
Rice University was called C_{60} and had a geodesic dome
structure. C_{60} is an allotrope of carbon with a structure
shown schematically in Fig. 23.21.

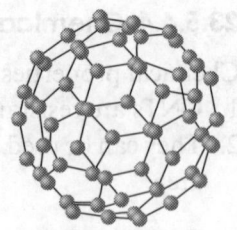

Fig. 23.21 Schematic of C_{60} molecule

Many structural variations of fullerenes have since been discovered. These
include *buckyball clusters*, *mega tubes*, *fullerene rings*, etc. Fullerenes have
found applications in diverse areas such as material science, electronics, nanotechnology, and tumour research.

Fullerenes were originally produced by laser vaporization of carbon in an
inert atmosphere. This process, however, results in a very low yield. Modern
methods use He gas in an evacuated chamber to produce a black soot-like
material on application of a suitable voltage between electrodes placed within
the chamber. The black soot is then scraped to obtain a high yield of C_{60}.
Several mechanisms have been proposed to explain the formation of C_{60}. One
of the most accepted proposals is the *pentagon road* mechanism. This mechanism explains the high yield as a consequence of clustering in a hot enough
region, followed by annealing to the minimum energy path.

23.7 GRAPHENE

A sheet of crystalline carbon that is one atom thick is called *graphene*. Such
a two-dimensional structure was long believed to be a physical impossibility
in terms of stability. This status was changed by a group led by Andre Geim
at University of Manchester in the year 2004, when they successfully realized
small fragments of graphene mono-layers. What is truly amazing is that the
group could achieve this seemingly impossible task with a childishly simple
technique. Called the 'Scotch tape method', the technique uses an adhesive
tape to rip off sheets of carbon from graphite. More sophisticated methods
like epitaxial growth have since followed, but the Scotch tape method still
remains immensely popular. Graphene has a single-layered honeycomb structure and can justifiably be called the mother of all carbon-based systems. This
is because all other forms such as graphite, CNTs, and buckminsterfullerene
can be derived from graphene, as seen in Fig. 23.22.

23.7.1 Importance of Graphene

What makes graphene such an important material that a large number of
groups are putting their best brains in it? The reasons are many; let us list
out a few significant ones:

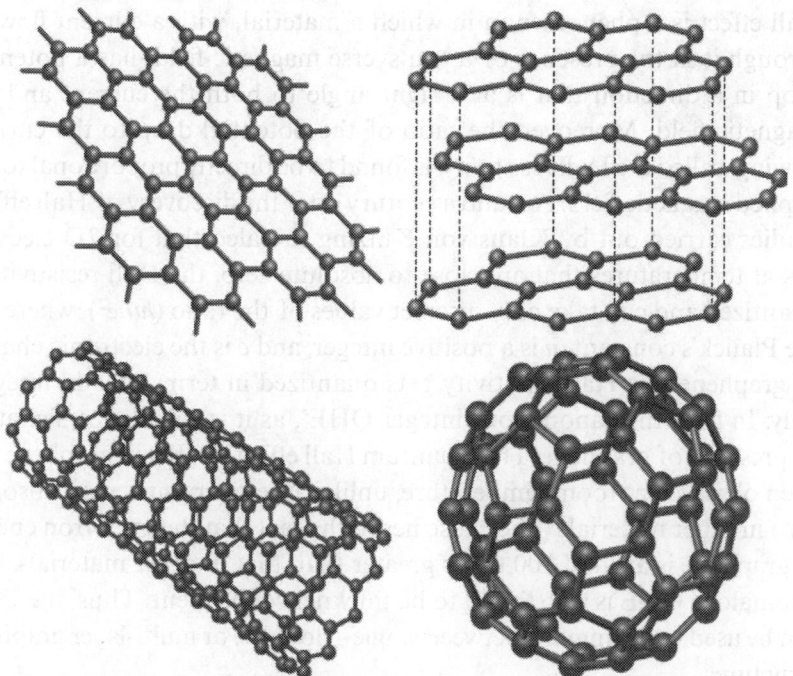

Fig. 23.22 Two-dimensional hexagonal lattice of carbon atoms and its transformation into graphite, CNT, and buckminsterfullerene

1. The honeycomb-lattice structure allows physicists to observe relativistic effects at speeds that are much below the speed of light. This behaviour is due to the strong interaction between the electrons in the graphene layer. These strong interactions ensure that the energy and momentum of electrons in graphene are related through the equation $E = vP$, where v is called the 'Fermi–Dirac velocity' and P is the momentum. In the special theory of relativity, energy, E and momentum P are related by the equation $E = \sqrt{(Mc^2 + P^2c^2)}$, which yields $E = cP$, as M turns to zero. Thus, electrons in graphene behave as if their mass is zero. Two things have to be remembered to appreciate this effect in graphene fully. Firstly, the Fermi–Dirac velocity in graphene is about 300 times less than the speed of light, and secondly, only electrons moving in a honeycomb lattice produce this energy–momentum relationship, whereas other lattices such as square or triangular ones always end up having electrons with a finite band mass. The massless electrons are, however, quite different from their particle-physics cousins, neutrinos. This is because unlike electrons, neutrinos carry no charge and, therefore, cannot interact strongly with any matter. The electrons in graphene can, however, interact with matter and can also be manipulated through externally applied electromagnetic fields. This manipulation makes graphene a very interesting candidate while trying to go beyond the technology limits set by silicon-based systems and devices.

2. Hall effect is a phenomenon in which a material, with a current flowing through it in the presence of a transverse magnetic field, has a potential drop in a direction that is at a right angle to both the current and the magnetic field. Moreover, the ratio of the potential drop to the current flowing, called the Hall resistivity, is found to be directly proportional to the applied magnetic field. Around a century after the discovery of Hall effect, studies carried out by Klaus von Klitzing revealed that for 2D electron gas at temperatures that are close to absolute zero, the Hall resistivity is quantized and can take only discreet values of the ratio (h/ne^2), where h is the Planck's constant, n is a positive integer, and e is the electronic charge. In graphene, the Hall resistivity gets quantized in terms of odd integers only. In fact, this 'anomalous integer QHE', as it is called, is a signature of presence of graphene. The Quantum Hall effect (QHE) in graphene has been observed at room temperature, unlike at temperatures near absolute zero in other materials. The cause lies in the fact that the cyclotron energy in graphene is around 100 times greater than that in other materials. The anomalous QHE is also found to be thickness dependent. Thus, the QHE can be used to distinguish between single-, double-, or multi-layer graphene structures.

3. In ordinary metals, electrons suffer scattering due to impurities in the crystal. This, in turn, leads to energy loss due to resulting electrical resistance. In graphene, the electrical resistance is not dependent upon the number of impurities, and the electrons can travel large distances without suffering collisions with impurities. This property of electrons in graphene makes it a promising material for developing a high-speed electronic switching device called a 'ballistic transistor'. Resistance offered by graphene is, in fact, lower than even silver, which is the least resistive metallic material. Graphene also conducts heat ballistically, since phonons also encounter very little resistance as they travel through graphene. This results in amazing thermal properties for graphene layers.

4. In the absence of an applied field, graphene has no band gap since the electrons in graphene are already in the conduction band. A bi-layer graphene, however, has a tunable band gap under the influence of an applied electric field. Studies have shown that the band gap of graphene can be modified or tuned in the range of 0–250 meV. This tunable band gap would enable device engineers to design extremely small chips with millions of graphene electronic devices and connections.

23.7.2 Applications of Graphene

The important properties and characteristics of graphene mentioned in the last section would make it a good choice for several applications. Some of these applications will be presented briefly in this section.

1. Graphene layers possess very high carrier mobility and low noise, and, therefore, can be used in the channel region of field effect transistors. Thus, graphene has the potential to replace silicon in future-generation integrated circuits. In fact, the smallest working transistor ever created uses graphene of one-atom thickness and 10-atom thick width. Due to the high area-to-mass ratio of graphene, it can be used for making the conducting plates in ultra-capacitors, which can have extremely high storage densities.

2. Graphene layers have very high electrical conductivity and optical transparency. This makes graphene a good candidate for applications requiring transparent electrodes. The devices that can benefit from its incorporation include touch-screen phones, handheld computers, organic photovoltaic cells, and organic light-emitting devices. Before these ideas become practically realizable, issues such as reproducible growth and stability would have to be suitably looked into.

3. Graphene would also find applications in the field of composite materials. Of course, questions regarding mass production and cost would have to be answered suitably. Applications may include gasoline tanks, food containers, sports equipment, aircraft and automobile parts, wind turbines, and medical implants, to name a few. Mechanical strength coupled with light weight would favour the use of graphene in many of these applications. Other important properties that would find use include high electrical conductivity and higher operating temperatures. The biggest competitor of graphene in this area would be CNTs. Graphene is, however, cheaper and does not suffer from toxicity issues that affect CNTs, which can penetrate lungs and cause complications including cancers.

4. Graphene has the ability to store a large amount of hydrogen and produce corresponding changes in the local electrical resistance. These properties make graphene a good choice for solid-state gas detection applications. Long-distance transportation of gases through pipes can gain immensely from graphene's properties. The inside surface of the pipes carrying gas can be coated with a graphene layer, and leaks, if any, can be detected by monitoring the electrical resistance across the entire length. Thus, the graphene layer would not only strengthen the transporting pipe, but also serve as a sensor element that is distributed throughout the length of the pipe.

5. Electron beam lithography techniques can be used to pattern graphene to realize electron waveguides. Properties of these waveguides can be controlled using externally applied voltages. Such waveguides can be put to novel applications in communication and other similar systems.

6. Graphene-based field-effect transistors fabricated on diamond-like carbon substrates have shown cut-off frequencies of approximately 155 GHz, with typical gate lengths of just 40 nm. The fabricated field-effect transistors have been shown to work down to temperatures as low as 4.3 K. Diamond-like carbon is a non-polar dielectric material, and, therefore, does not trap

charges or scatter charges the way the more conventionally used silicon dioxide does. The key to pushing the cut-off frequency up lies in improving the quality of the graphene layer that is obtained using the CVD process.

7. Graphene can be ripped to shreds to obtain graphene nanoribbons (GNRs). The structure at the edge of the ribbon can be either a zigzag or an armchair type. Interestingly, zigzag ribbons behave as conductors, whereas armchair ribbons behave as semiconductors. The band gap of armchair ribbons can be modulated by modulating the width of the ribbon, and experiments have shown that the band gap is inversely proportional to the width. Thus, thinner ribbons result in a higher band gap. Such graphene ribbons can be realized in practice by cutting open CNTs.

IMPORTANT CONCEPTS

1. Metallic glasses are metallic alloys having an amorphous arrangement of atoms.
2. Two types of metallic glasses are (a) metal–metal metallic glasses and (b) metal–metalloid metallic glasses.
3. Melt-spinning and deposition techniques are used in the fabrication of metallic glasses.
4. Metallic alloys that have the ability to return to their original structure or shape, when subjected to appropriate heat treatment are called shape memory alloys.
5. Pseudoelasticity of a material is a property that causes large strains to develop due to an applied stress.
6. Nanomaterials are materials that have structured components with dimensions in the range of 10–100 nm.
7. Methods such as plasma arcing, CVD, electrodeposition, sol–gel method, and ball milling are used to produce nanomaterials.
8. CNTs have various structures: armchair structure, zigzag structure, and chiral structure.
9. In single-walled CNTs, only one layer of CNT is present, whereas in multiwalled CNTs, more than one layer of CNTs is present.
10. Electric arc discharge method, pulsed-laser technique, and CVD are used to fabricate CNTs.
11. Important properties of CNTs include their high strength-to-weight ratio, high resistance to chemical reactions, and resistance to oxidations.

APPLICATIONS

1. New engineering materials, as already discussed, find applications in a variety of industries from medicine to aircraft, construction, and computing.
2. Nanotechnology can be used to fabricate nanotube transistors. These transistors are smaller and faster, and use less material as compared to conventional devices. Some designs use DNA strands to connect electrodes.
3. Nanotube-based circuits have been fabricated by IBM. NASA is developing nanomotors and gears for futuristic applications.

SELF-ASSESSMENT

Multiple-choice Questions

23.1 Metallic glasses can be classified under the categories of
(a) metal–metal and metal–semiconductor
(b) metal–metal and metal–metalloid
(c) metal–metal and metal–insulator
(d) metal–metal and metal–non-metal

23.2 Rapid quenching implies
(a) ultrafast cooling
(b) ultrafast heating
(c) first cooling and then heating
(d) first heating and then cooling

23.3 Tensile strength of metallic glasses is in the range of
(a) 300–400 MPa
(b) 300–500 MPa
(c) 3000–4000 MPa
(d) 30–100 MPa

23.4 Shape memory alloys owe their behaviour to
(a) liquid-state phase change
(b) gas-state phase change
(c) liquid–gas-state phase change
(d) solid-state phase change

23.5 Shape memory alloys demonstrate
(a) thermal hysteresis
(b) magnetic hysteresis
(c) electrical hysteresis
(d) no hysteresis

23.6 The top-down approach in nanotechnology involves
(a) drawing
(b) photolithography
(c) cutting
(d) slicing

23.7 Stress created during ion implantation is removed using
(a) etching
(b) polishing
(c) annealing
(d) scribing

23.8 Nature follows the
(a) top-down approach
(b) layered approach
(c) destruction approach
(d) bottom-up approach

23.9 One-dimensional growth is realized using the VLS approach due to
(a) saturation of catalyst
(b) photolithography
(c) anisotropic etching
(d) selective etching

23.10 In which of the following ways can a graphene sheet not be rolled to form a tube?
(a) Armchair
(b) Table
(c) Zigzag
(d) Chiral

23.11 In the absence of an applied field, band gap of graphene is
(a) 1 eV
(b) 0 eV
(c) 0.6 eV
(d) 0.1 eV

23.12 QHE is the short form for
(a) quick Hall effect
(b) quiescent Hall effect
(c) quantum Hall effect
(d) quiet Hall effect

23.13 According to special theory of relativity, the relationship between energy, E, and momentum, P, where mass is zero, is
(a) $E = cP$
(b) $E = c^2P$
(c) $E = cP^2$
(d) $E = c/P$

23.14 Modern methods of producing fullerenes use
(a) hydrogen
(b) nitrogen
(c) helium
(d) argon

23.15 Energy gap of semiconducting CNTs is
    (a) directly proportional to the diameter of tube
    (b) inversely proportional to the square of the diameter of tube
    (c) directly proportional to the square of the diameter of tube
    (d) inversely proportional to the diameter of tube

Review Questions

23.1   What are metallic glasses?
23.2   What are the applications of metallic glasses?
23.3   Mention the various properties of metallic glasses.
23.4   How many types of metallic glasses are there?
23.5   What are shape memory alloys?
23.6   Define the shape memory effect.
23.7   Mention the properties of shape memory alloys.
23.8   State the applications of shape memory alloys.
23.9   Mention the various structures of CNTs.
23.10 What are single-walled and multiwalled CNTs?
23.11 Mention the various methods of fabricating CNTs.
23.12 Mention a few physical properties of CNTs.
23.13 Mention any three applications of CNTs.
23.14 What are nanophase materials? Mention the various methods used to prepare them.
23.15 Discuss the mechanical and magnetic properties of nanomaterials.
23.16 Discuss briefly the methods of preparation of nanoparticles.
23.17 Using a diagram, describe the ball milling technique for making nanoparticles.
23.18 Explain how CNTs are fabricated using the laser ablation method.
23.19 What are metallic glasses? How are they prepared?

Units and Elements

SI UNITS AND PREFIXES

Quantity	Unit	Symbol
Mass	kilogram	kg
Length	metre	m
Time	second	s
Magnetic flux	weber (V s)	Wb
Magnetic flux density	tesla (Wb/m^2)	T
Force	newton (kg m/s^2)	N
Weight	newton	N
Pressure	pascal (N/m^2)	Pa
Energy	joule (N m)	J
Inductance	henry (Wb/A)	H
Power	watt (J/s)	W
Frequency	hertz (1/s)	Hz
PD, EMF (voltage)	volt (J/C)	V
Current	ampere	A
Resistance	Ohm (V/A)	Ω
Electric charge	coulomb (As)	C
Capacitance	farad (C/V)	F
Temperature	Kelvin	K
	degree Celsius	°C

Prefix	Meaning	
G (giga)	1 000 000 000	(10^9)
M (mega)	1 000 000	(10^6)
k (kilo)	1000	(10^3)
d (deci)	$\dfrac{1}{10}$	(10^{-1})
c (centi)	$\dfrac{1}{100}$	(10^{-2})
m (milli)	$\dfrac{1}{1000}$	(10^{-3})
μ (micro)	$\dfrac{1}{1000000}$	(10^{-6})
n (nano)	$\dfrac{1}{1000000000}$	(10^{-9})
p (pico)	$\dfrac{1}{1000000000000}$	(10^{-12})

Examples
1 μF (microfarad) = 10^{-6} F
1 ms (millisecond) = 10^{-3} s
1 km (kilometre)   = 10^3 m
1 MW (megawatt) = 10^6 W

Note: 'micro' means 'millionth'; 'milli' means 'thousandth'.

ELEMENTS

For simplicity, many of the rarer elements have been omitted from the table below.

Atomic number (Proton number)	Element	Chemical symbol
1	Hydrogen	H
2	Helium	He
3	Lithium	Li
4	Beryllium	Be
5	Boron	B
6	Carbon	C
7	Nitrogen	N
8	Oxygen	O
9	Fluorine	F

(Contd)

Table (Contd)

Atomic number (Proton number)	Element	Chemical symbol
10	Neon	Ne
11	Sodium	Na
12	Magnesium	Mg
13	Aluminium	Al
14	Silicon	Si
15	Phosphorus	P
16	Sulfur	S
17	Chlorine	Cl
18	Argon	Ar
19	Potassium	K
20	Calcium	Ca
22	Titanium	Ti
25	Manganese	Mn
26	Iron	Fe
27	Cobalt	Co
28	Nickel	Ni
29	Copper	Cu
30	Zinc	Zn
35	Bromine	Br
38	Strontium	Sr
47	Silver	Ag
48	Cadmium	Cd
50	Tin	Sn
53	Iodine	I
55	Caesium	Cs
74	Tungsten	W
78	Platinum	Pt
79	Gold	Au
80	Mercury	Hg
82	Lead	Pb
86	Radon	Rn
88	Radium	Ra
90	Thorium	Th
92	Uranium	U
94	Plutonium	Pu

Please refer to Online Resources (www.oupinheonline.com) for the complete periodic table.

Important Physical Constants

Quantity	Symbol	Value
Angstrom unit	Å	$1 \text{ Å} = 10^{-4} \text{ } \mu\text{m} = 10^{-10} \text{ m}$
Avogadro constant	N_{AVO}	$6.02204 \times 10^{23} \text{ mol}^{-1}$
Bohr radius	a_B	0.52917 Å
Boltzmann constant	k	$1.38066 \times 10^{-23} \text{ J/K } (R/N_{\text{AVO}})$
Elementary charge	q	$1.60218 \times 10^{-19} \text{ C}$
Electron rest mass	m_0	$0.91095 \times 10^{-30} \text{ kg}$
Electron volt	eV	$1 \text{ eV} = 1.60218 \times 10^{-19} \text{ J}$ $= 25.053 \text{ kcal/mol}$
Gas constant	R	$1.98719 \text{ cal mol}^{-1} \text{ K}^{-1}$ $(R = 8.315 \text{J/mol.K})$
Permeability in vacuum	μ_0	$4\pi \times 10^{-7} \text{ H-m}^{-1}$
Permittivity in vacuum	ε_0	$8.85418 \times 10^{-12} \text{ F/m}$
Planck's constant	h	$6.62617 \times 10^{-34} \text{ J s}$
Reduced Planck's constant	$\hbar$	$1.05458 \times 10^{-34} \text{ J s } (h/2\pi)$
Proton rest mass	M_p	$1.67264 \times 10^{-27} \text{ kg}$
Speed of light in vacuum	c	$2.99792 \times 10^{8} \text{ m/s}$
Standard atmosphere		$1.01325 \times 10^{5} \text{ N/m}^2$
Thermal voltage at 300 K	kT/e	0.0259 V
Wavelength of 1 eV quantum	λ	$1.2423 \text{ } \mu\text{m}$

Important Lattice Constants

Compound type	Element or compound	Name	Crystal structure	Lattice constant at 300 K (Å)
Element	C	Carbon (diamond)	Diamond	3.56683
	Ge	Germanium	Diamond	5.64613
	Si	Silicon	Diamond	5.43095
	Sn	Grey Tin	Diamond	6.48920
IV–IV	SiC	Silicon carbide	Wurtzite	$a = 3.086$, $c = 15.117$
III–V	AlAs	Aluminium arsenide	Zincblende	5.6605
	AlP	Aluminum phosphide	Zincblende	5.4510
	AlSb	Aluminum antimonide	Zincblende	6.1355
	BN	Boron nitride	Zincblende	3.6150
	BP	Boron phosphide	Zincblende	4.5380
	GaAs	Gallium arsenide	Zincblende	5.6533
	GaN	Gallium nitride	Wurtzite	$a = 3.189$, $c = 5.185$
	GaP	Gallium phosphide	Zincblende	5.4512
	GaSb	Gallium antimonide	Zincblende	6.0959

(Contd)

Table *(Contd)*

Compound type	Element or compound	Name	Crystal structure	Lattice constant at 300 K (Å)
	InAs	Indium arsenide	Zincblende	6.0584
	InP	Indium phosphide	Zincblende	5.8686
	InSb	Indium antimonide	Zincblende	6.4794
II–VI	CdS	Cadmium sulphide	Zincblende	5.8320
	CdS	Cadmium sulphide	Wurtzite	$a = 4.16,$ $c = 6.756$
	CdSe	Cadmium selenide	Zincblende	6.050
	CdTe	Cadmium telluride	Zincblende	6.482
	ZnO	Zinc oxide	Rock salt	4.580
	ZnS	Zinc sulphide	Zincblende	5.420
	ZnS	Zinc sulphide	Wurtzite	$a = 3.82,$ $c = 6.26$
IV–VI	PbS	Lead sulphide	Rock salt	5.9362
	PbTe	Lead telluride	Rock salt	6.4620

Properties of Some Common Semiconductors

Semiconductor		Band gap (eV)		Mobility at 300 K (cm²/V s)[1]		Band[2]	Effective Mass m^*/m_0		$\varepsilon_s/\varepsilon_0$
		300 K	0 K	Electrons	Holes		Elec.	Holes	
Element	C	5.47	5.48	1800	1200	i	0.2	0.25	5.7
	Ge	0.66	0.74	3900	1900	i	1.64[3]	0.64[5]	16.0
							0.082[4]	0.28[6]	
	Si	1.12	1.17	1500	450	i	0.98[3]	0.16[5]	11.9
							0.19[4]	0.49[6]	
	Sn		0.082	1400	1200	d			
IV–V	α-SiC	2.996	3.03	400	50	i	0.60	1.00	10.0
III–V	AISb	1.58	1.68	200	420	i	0.12	0.98	14.4
	BN	~7.5							
	BP	2.0							
	GaN	3.36	3.50	380			0.19	0.60	12.2
	GaSb	0.72	0.81	5000	850	d	0.042	0.40	15.7
	GaAs	1.42	1.52	8500	400	d	0.067	0.082	13.1
	GaP	2.26	2.34	110	75	i	0.82	0.60	11.1
	InSb	0.17	0.23	80000	1250	d	0.0145	0.40	17.7
	InAs	0.36	0.42	33000	460	d	0.023	0.40	14.6
	InP	1.35	1.42	4600	150	d	0.077	0.64	12.4
II–VI	CdS	2.42	2.56	340	50	d	0.21	0.80	5.4
	CdSe	1.70	1.85	800		d	0.13	0.45	10.0
	CdTe	1.56		1050	100	d			10.2
	ZnO	3.35	3.42	200	180	d	0.27		9.0
	ZnS	3.68	3.84	165	5	d	0.40		5.2

(Contd)

Table (Contd)

Semiconductor		Band gap (eV)		Mobility at 300 K (cm²/V s)[1]		Band[2]	Effective Mass $m*/m_0$		$\varepsilon_s/\varepsilon_0$
		300 K	0 K	Electrons	Holes		Elec.	Holes	
IV–VI	PbS	0.41	0.286	600	700	i	0.25	0.25	17.0
	PbTe	0.31	0.19	6000	4000	i	0.17	0.20	30.0

[1] The values are for drift mobilities obtained in the purest and most perfect materials available to date.
[2] i = indirect, d = direct.
[3] Longitudinal effective mass.
[4] Transverse effective mass.
[5] Light-hole effective mass.
[6] Heavy-hole effective mass.

Band Gaps of Some Semiconductors Relative to the Optical Spectrum

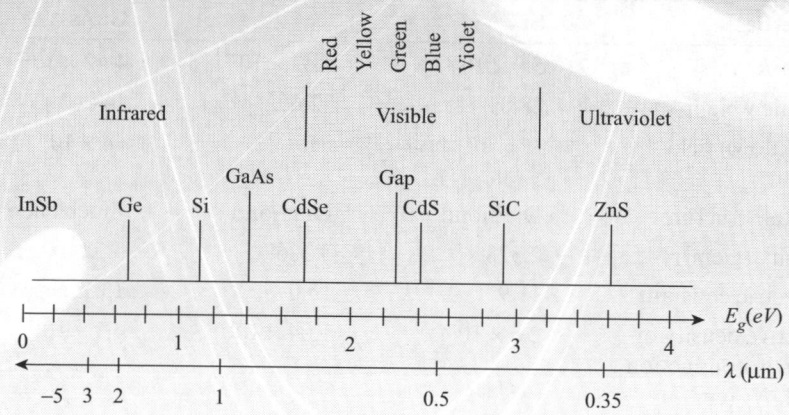

Properties of Silicon, Germanium, and Gallium Arsenide at 300 K

APPENDIX F

Properties	Si	Ge	GaAs
Atoms/cm³	5.0×10^{22}	4.42×10^{22}	4.42×10^{22}
Atomic weight	28.09	72.60	144.63
Breakdown field (V/cm)	~3×10^{5}	~10^{5}	~4×10^{5}
Crystal structure	Diamond	Diamond	Zincblende
Density (g/cm³)	2.328	5.3267	5.32
Dielectric constant	11.9	16.0	13.1
Effective density of states in conduction band, N_C (cm⁻³)	2.8×10^{19}	1.04×10^{19}	4.7×10^{17}
Effective density of states in valence band, N_V (cm⁻³)	1.04×10^{19}	6.0×10^{18}	7.0×10^{18}
Effective Mass, m^*/m_0			
Electrons	$m_l^* = 0.98$	$m_l^* = 1.64$	0.067
	$m_l^* = 0.19$	$m_l^* = 0.082$	
Holes	$m_{lh}^* = 0.16$	$m_{lh}^* = 0.044$	$m_{lh}^* = 0.082$
	$m_{hh}^* = 0.49$	$m_{hh}^* = 0.28$	$m_{lh}^* = 0.45$
Electron affinity, $\chi(V)$	4.05	4.0	4.07
Energy gap (eV) at 300 K	1.12	0.66	1.424
Intrinsic carrier concentration (cm⁻³)	1.45×10^{10}	2.4×10^{13}	1.79×10^{6}

(Contd)

Table (Contd)

Properties	Si	Ge	GaAs
Intrinsic Debye length (µm)	24	0.68	2250
Intrinsic resistivity (Ω cm)	2.3×10^5	47	10^8
Lattice constant (Å)	5.43095	5.64613	5.6533
Linear coefficient of thermal expansion, $\Delta L/L\Delta T$ (°C^{-1})	2.6×10^{-6}	5.8×10^{-6}	6.86×10^{-6}
Melting point (°C)	1415	937	1238
Minority carrier life-time (s)	2.5×10^{-3}	10^{-3}	$\sim 10^{-8}$
Mobility (drift) (cm^2/V s)	1500	3900	8500
	450	1900	400
Optical-phonon energy (eV)	0.063	0.037	0.035
Phonon mean free path λ_0 (Å)	76 (electron) 55 (hole)	105	58
Specific heat (J/g °C)	0.7	0.31	0.35
Thermal conductivity at 300 K (W/cm°C)	1.5	0.6	0.46
Thermal diffusivity (cm^2/s)	0.9	0.36	0.44
Vapour pressure (Pa)	1 at 1650°C 10^{-6} at 900°C	1 at 1330°C 10^{-6} at 760°C	100 at 1050°C 1 at 900°C

Answers to Multiple-choice Questions

Chapter 1

1.1 (a),	1.4 (a)	1.8 (b)	1.12 (c)	1.16 (c)	1.20 (c)	1.23 (c)
(c)	1.5 (c)	1.9 (c)	1.13 (b)	1.17 (d)	1.21 (c)	1.24 (b)
1.2 (c)	1.6 (b)	1.10 (d)	1.14 (a)	1.18 (a)	1.22 (b)	1.25 (a)
1.3 (d)	1.7 (c)	1.11 (d)	1.15 (b)	1.19 (b)		

Chapter 2

2.1 (c)	2.6 (d)	2.11 (a)	2.16 (c)	2.21 (c)	2.26 (b)	2.31 (c)
2.2 (d)	2.7 (d)	2.12 (b)	2.17 (a)	2.22 (b)	2.27 (c)	2.32 (b)
2.3 (a)	2.8 (c)	2.13 (c)	2.18 (c)	2.23 (a)	2.28 (b)	2.33 (a)
2.4 (c)	2.9 (b)	2.14 (b)	2.19 (d)	2.24 (c)	2.29 (a)	2.34 (b)
2.5 (a)	2.10 (c)	2.15 (b)	2.20 (c)	2.25 (c)	2.30 (b)	2.35 (c)

Chapter 3

3.1 (b)	3.5 (c)	3.9 (b)	3.13 (c)	3.17 (c)	3.20 (b)	3.23 (a)
3.2 (c)	3.6 (d)	3.10 (d)	3.14 (d)	3.18 (d)	3.21 (c)	3.24 (b)
3.3 (a)	3.7 (b)	3.11 (a)	3.15 (a)	3.19 (a)	3.22 (d)	3.25 (c)
3.4 (b)	3.8 (d)	3.12 (b)	3.16 (b)			

Chapter 4

4.1 (a)	4.4 (b)	4.6 (a)	4.8 (b)	4.10 (a)	4.12 (c)	4.14 (a)
4.2 (a)	4.5 (b)	4.7 (b)	4.9 (a)	4.11 (d)	4.13 (a)	4.15 (c)
4.3 (a)						

Chapter 5

5.1 (a)	5.4 (d)	5.7 (c)	5.10 (b)	5.12 (c)	5.14 (d)	5.16 (b)
5.2 (b)	5.5 (a)	5.8 (d)	5.11 (b)	5.13 (b)	5.15 (a)	5.17 (c)
5.3 (c)	5.6 (b)	5.9 (a)				

Chapter 6

6.1 (b)	6.5 (c)	6.8 (a)	6.11 (a)	6.14 (a)	6.17 (c)	6.20 (a)
6.2 (a)	6.6 (b)	6.9 (a)	6.12 (d)	6.15 (a)	6.18 (b)	6.21 (a)
6.3 (c)	6.7 (d)	6.10 (b)	6.13 (a)	6.16 (b)	6.19 (b)	6.22 (c)
6.4 (a)						

Chapter 7

7.1 (b)	7.3 (c)	7.5 (d)	7.7 (a),	7.8 (b)	7.10 (c)	7.12 (c)
7.2 (b)	7.4 (c)	7.6 (c)	(b)	7.9 (b)	7.11 (b)	7.13 (d)

7.14 (d)	7.18 (b)	7.21 (c)	7.24 (c)	7.27 (d)	7.30 (c)	7.33 (a)
7.15 (a)	7.19 (c)	7.22 (a)	7.25 (d)	7.28 (a)	7.31 (d)	7.34 (b)
7.16 (b)	7.20 (d)	7.23 (b)	7.26 (b)	7.29 (b)	7.32 (b)	7.35 (c)
7.17 (d)						

Chapter 8

8.1 (b)	8.3 (c)	8.5 (c)	8.7 (b)	8.8 (d)	8.9 (d)	8.10 (b)
8.2 (c)	8.4 (b)	8.6 (c)				

Chapter 9

9.1 (b)	9.5 (b)	9.10 (c)	9.15 (b)	9.19 (b)	9.23 (b)	9.27 (b)
9.2 (a)	9.6 (a)	9.11 (b)	9.16 (c)	9.20 (c)	9.24 (c)	9.28 (c)
9.3 (d)	9.7 (d)	9.12 (c)	9.17 (d)	9.21 (d)	9.25 (d)	9.29 (d)
9.4 (b),	9.8 (c)	9.13 (d)	9.18 (a)	9.22 (a)	9.26 (a)	9.30 (a)
(d)	9.9 (a)	9.14 (a)				

Chapter 10

10.1 (d)	10.4 (b),	10.5 (c)	10.8 (a)	10.10 (b)	10.12 (c)	10.14 (a)
10.2 (b)	(c),	10.6 (a)	10.9 (a)	10.11 (b)	10.13 (d)	10.15 (b)
10.3 (a)	(d)	10.7 (a)				

Chapter 11

11.1 (c)	11.5 (a)	11.9 (a)	11.13 (b)	11.16 (b)	11.19 (c)	11.22 (d)
11.2 (a)	11.6 (c)	11.10 (b)	11.14 (a)	11.17 (d)	11.20 (a)	11.23 (a)
11.3 (c)	11.7 (a)	11.11 (a)	11.15 (d)	11.18 (a)	11.21 (a)	11.24 (b)
11.4 (b)	11.8 (a)	11.12 (a)				

Chapter 12

12.1 (a)	12.4 (e)	12.7 (d)	12.10 (a)	12.13 (a)	12.16 (d)	12.19 (a)
12.2 (b)	12.5 (b)	12.8 (d)	12.11 (a)	12.14 (d)	12.17 (a)	12.20 (d)
12.3 (d)	12.6 (a)	12.9 (a)	12.12 (b)	12.15 (d)	12.18 (a)	12.21 (b)

Chapter 13

13.1 (a)	13.5 (d)	13.9 (a)	13.13 (c)	13.17 (b)	13.20 (a)	13.23 (a)
13.2 (a)	13.6 (c)	13.10 (c)	13.14 (a)	13.18 (c)	13.21 (d)	13.24 (a)
13.3 (a)	13.7 (a)	13.11 (b)	13.15 (c)	13.19 (a)	13.22 (d)	13.25 (a)
13.4 (b)	13.8 (c)	13.12 (a)	13.16 (a)			

Chapter 14

14.1 (b)	14.5 (c)	14.9 (b)	14.13 (c)	14.16 (a)	14.19 (a)	14.22 (d)
14.2 (a)	14.6 (a)	14.10 (a)	14.14 (a)	14.17 (c)	14.20 (b)	14.23 (b)
14.3 (a)	14.7 (b)	14.11 (d)	14.15 (b)	14.18 (a)	14.21 (d)	14.24 (a)
14.4 (a)	14.8 (c)	14.12 (a)				

Chapter 15

15.1 (a)	15.2 (b)	15.3 (b)	15.4 (d)	15.5 (c)	15.6 (a)	15.7 (b)

15.8 (a)   15.10 (c)   15.12 (b)   15.14 (b)   15.16 (b)   15.18 (c)   15.20 (a)
15.9 (a)   15.11 (a)   15.13 (a)   15.15 (c)   15.17 (c)   15.19 (c)   15.21 (a)

Chapter 16

16.1 (a)   16.4 (d)   16.7 (b)   16.10 (c)   16.13 (b)   16.15 (a)   16.17 (a)
16.2 (a)   16.5 (b)   16.8 (d)   16.11 (d)   16.14 (b)   16.16 (c)   16.18 (a)
16.3 (c)   16.6 (b)   16.9 (b)   16.12 (d)

Chapter 17

17.1 (a)   17.4 (d)   17.6 (b)   17.8 (d)   17.10 (b)   17.12 (c)   17.14 (a)
17.2 (b)   17.5 (a)   17.7 (c)   17.9 (a)   17.11 (b)   17.13 (d)   17.15 (b)
17.3 (c)

Chapter 18

18.1 (a)   18.4 (d)   18.6 (b)   18.8 (d)   18.10 (c)   18.12 (c)   18.14 (a)
18.2 (b)   18.5 (a)   18.7 (c)   18.9 (a)   18.11 (b)   18.13 (d)   18.15 (b)
18.3 (c)

Chapter 19

19.1 (a)   19.4 (d)   19.6 (a)   19.8 (c)   19.10 (b),   19.11 (a)   19.13 (a)
19.2 (b)   19.5 (d)   19.7 (b)   19.9 (a)   (c)   19.12 (c)
19.3 (b)

Chapter 20

20.1 (a)   20.4 (d)   20.7 (c)   20.10 (b)   20.13 (d)   20.16 (c)   20.19 (b)
20.2 (b)   20.5 (a)   20.8 (d)   20.11 (b)   20.14 (a)   20.17 (d)   20.20 (c)
20.3 (c)   20.6 (b)   20.9 (a)   20.12 (c)   20.15 (b)   20.18 (a)

Chapter 21

21.1 (a)   21.4 (c)   21.6 (d)   21.8 (b)   21.10 (a)   21.12 (a)   21.14 (a)
21.2 (a)   21.5 (c)   21.7 (a)   21.9 (c)   21.11 (a)   21.13 (a)   21.15 (b)
21.3 (b)

Chapter 22

22.1 (d)   22.4 (a)   22.7 (c)   22.10 (c)   22.13 (d)   22.16 (c)   22.19 (a)
22.2 (a)   22.5 (b)   22.8 (a)   22.11 (d)   22.14 (d)   22.17 (c)   22.20 (d)
22.3 (a)   22.6 (a)   22.9 (b)   22.12 (e)   22.15 (d)   22.18 (d)

Chapter 23

23.1 (b)   23.4 (d)   23.6 (b)   23.8 (d)   23.10 (b)   23.12 (c)   23.14 (c)
23.2 (a)   23.5 (a)   23.7 (c)   23.9 (a)   23.11 (b)   23.13 (a)   23.15 (d)
23.3 (c)

Answers to Numerical Problems

Chapter 1

1.1 16:9

1.2 $\dfrac{4}{3}$

1.3 4000 Å

1.4 5.33×10^{-5} cm

1.5 0.061 cm

1.6 67

1.7 6.3

1.8 8

1.9 5.89×10^{-9} m

1.10 5×10^{-5} cm

1.11 6.97×10^{-5} cm

1.12 1.33

1.13 4:1

1.14 0.068 cm

1.15 0.11 cm

Chapter 2

2.1 237 cm

2.2 78 cm

2.3 $u = -10.5$ cm

2.4 Does not change

2.5 0.1893 cm

2.6 0.0667 cm

2.7 5°44'

2.8 6.7', 10'

2.9 0.65 cm

2.10 6410×10^{-8} cm

2.11 9118

2.12 600

2.13 4800 Å

2.14 17%

2.15 1.25

2.16 23%

2.17 3.66×10^{-4} rad

Chapter 3

3.1 1.6753

3.2 56.3°

3.3 ±45°, ±135°

3.4 $\dfrac{3}{4} I_P$

3.5 37.5%

3.6 2.75×10^{-3} cm

3.7 1.375×10^{-3} cm

3.8 62.5°

3.9 1.356

3.10 2.3559

3.11 58.45°

3.12 61.1°, 118.9°

3.13 0.61

3.14 1.79

3.15 11%

3.16 14%

3.17 4.5%

3.18 43.6°

3.19 65.5

3.20 26.1°

3.21 1.5299

3.22 1.5317

3.23 2.21×10^{-5} cm

3.24 1.98×10^{-5} cm

3.25 1.5580

3.26 1.5563

3.27 14.5 cm

3.28 17.2°

Chapter 4

4.1 3.16×10^{-19}

4.2 1.2 eV

4.3 3.15×10^{-19}

4.4 1.16×10^{-20}

4.5 7.64×10^{17}

4.6 0.25 radians; 2×10^{-5} radians

4.7 $1.769 \times 10^{6}\,\text{W/m}^2$

4.8 $0.6 \times 10^{9}\,\text{Hz}$

4.9 $1.22 \times 10^{9}\,\text{V/m}$

4.10 0.69Å

4.11 $5 \times 10^{-13}; 10^{-4}$

4.12 $\dfrac{n_2}{n_1} = \exp(-2.89) = 0.0555$

4.13 5.33×10^{6}

4.14 1.79×10^{17} photons

4.15 0.378×10^{19} photons/s

4.16 688.88 nm

4.17 2.168×10^{24} eV

4.18 1.785 eV; 0.00432×10^{17} Hz; 28.658×10^{-20} J; 0.69 MW

4.19 0.11698 eV; 0.283×10^{14} Hz; 18.774×10^{-21} J; 45.23 kW

4.20 4.72×10^{-20} 4.22 133.75×10^{17}

4.21 3.169×10^{-18}

Chapter 5

5.1 0.25   5.3 6.76×10^{-11} s/m   5.5 76.83°

5.2 (a) 14.48°; (b) 9.59°   5.4 103%   5.6 1700 m^{-1}

Chapter 6

6.1 (a) $t = 2\tau$ s; (b) $t = \tau$ s; (c) $t = 4\tau$ s

6.2 $\dfrac{12.5}{Q^2}$ %

6.3 19.08 s

6.4 0.046

6.5 (a) 32.91 s; (b) −75.96 dynes; (c) 11.46 s; (d) 16.45 m

6.6 0.01 m; 3.185 Hz; 3.14 m; 10.0009 m s^{-1}

6.7 $y = 0.001 \sin(125.6t - 20.93x)$ m

6.8 (a) $A_0 = 0.663 A_0$ max; (b) 2°52′

6.9 0.085 s

Chapter 7

7.1 13 dB   7.6 4.1 s   7.11 32

7.2 4   7.7 6364 m³   7.12 6.5 dB

7.3 700% change   7.8 (a) 0.11; (b) 3.75 s   7.13 3.09 s

7.4 6 dB   7.9 0.48 bel

7.5 495 sabines   7.10 4.8 dB, 9 dB

Chapter 8

8.1 180 m   8.4 $\Delta v = -6.03$ kHz   8.7 2400 m/s

8.2 0.28 s   8.5 3.41×10^6 Hz

8.3 66.37 kHz   8.6 0.46 m

Chapter 9

9.1 $\sqrt[8]{3}$ N, $\tan \alpha = \dfrac{1}{\sqrt{3}} \Rightarrow \alpha = 30°$

9.2 15 min   9.5 $\theta = 120°$ 9.8 1.51×10^3 N

9.3 0.75 km   9.6 16 rad/s   9.9 41.3°

9.4 13.03   9.7 2.1×10^{-3} m/s   9.10 3.14 N

9.11 6.94 N	9.14 57.8 m/s	9.17 0.94 rad/s
9.12 20.3°	9.15 45.7 kg	9.18 0.36 rad/s
9.13 10.7 m/s	9.16 1.65 m	

Chapter 10

10.1 (a) 80 cm; (b) 69.28 cm

10.2 0.6×10^{10} cm/s

10.3 3.03×10^{-6} s

10.4 4.55×10^{-6} s

10.5 $0.995c$

10.6 $0.995c$

10.7 0.33×10^{10} cm/s

10.8 2.83×10^{10} cm/s

10.9 15.17×10^{-28} g

10.10 0.09

10.11 3.38×10^{32} eV

10.12 2.24×10^{10} cm/s

10.13 32.21×10^{-31} kg; 2.55×10^8 m/s

10.14 76 cm

10.15 41.4 cm

10.16 84.6 cm

10.17 13.4%

10.18 1.2×10^{10} cm/s

10.19 2.7×10^{10} cm/s

10.20 1.44×10^{10} cm/s

10.21 1.95×10^{10} cm/s

10.22 1.88×10^{-6} s

10.23 0.82×10^{-6} s

10.24 1.19×10^{-6} s

10.25 4.9×10^{-6} s

10.26 8.1×10^{-6} s

10.27 3.5×10^{-6} s

10.28 2.1×10^{10} cm/s

10.29 2.7×10^{10} cm/s

10.30 $0.99c$

10.31 5.5×10^{-8} s

10.32 3.1×10^{-8} s

10.33 $0.985c$

Chapter 11

11.1 2.067eV

11.2 $\vartheta_0 = 5.077 \times 10^{14}$ Hz; $\lambda = 5909$Å

11.3 2.15eV; $\vartheta_0 = 519T$ Hz

11.4 $27.3 \times 10^{19} \cong 28 \times 10^{19}$ photons/s

11.5 KE = 292.75keV; $v_p = 2.386c$; $v_g = 0.419c$

11.6 (a) $\lambda = 13.265 \times 10^{-29}$ m; (b) $\lambda = 0.729 \times 10^{-11}$ m

11.7 6.204×10^{-13} m

11.8 (a)(i) 4.39×10^6 m/s; (ii) 4.817×10^6 m/s; (b) $\lambda_{inside} = 0.166$ nm; $\lambda_{outside} = 0.138$ nm

11.9 $\geq 0.527 \times 10^{-29}$ J

11.10 5.45×10^{-4}

11.11 2.89×10^{-3} m

11.12 0.8333

11.13 (a) $0.333b^2$; (b) 0.75

11.14 (a) $b^2 0.19375$; (b) 0.828

11.15 Operator: $\dfrac{\partial^2}{\partial x^2}$; eigen function: e^{2x}; eigen value: 4

11.16 (a) 7.028; (b) 10^{-15}

11.17 10% of its length

11.18 $E_1 = 0.150 \times 10^{-17}$ J; $E_2 = 0.60 \times 10^{-17}$ J; $E_3 = 1.35 \times 10^{-17}$ J

11.19 For electron, $E_1 = 0.151 \times 10^{-17}$ J; For marble, $E_1 = 0.0137 \times 10^{-61}$ J

Chapter 12

12.1 (a) 6.97 m/s (b) 7.46 m/s (c) 10 m/s
12.2 $69 \times 10^{-29} \, \text{N/m}^2$
12.3 $110.4 \times 10^{-29} \, \text{N/m}^2$
12.4 142.33 m/s
12.5 0.388 s
12.6 4.912×10^{26}
12.7 (a) 472/199 K (b) 944 K/671 K

12.8 21.55 m/s
12.9 2979 K
12.10 3.72 W
12.11 0.20 W
12.12 $1.42 \, \text{MW/m}^2$
12.13 $T^4 = 1.058 \times 10^8 \, \text{K}$

Chapter 13

13.1 (a) 177pF; (b) $35.4 \times 10^{-9} \text{C}$; (c) 8.85pF; (d) 0.05; or 2
  (e) $17.7 \times 10^{-12} \text{C}^2 / \text{Nm}^2$; (f) $8.85 \times 10^{-12} \text{C}$; (g) $2.00 \times 10^5 \, \text{V/m}$; (h) $4.0 \times 10^5 \, \text{V/m}$

13.2 $8.85 \times 10^{-10} \, \text{C/m}$

13.3 3.9×10^{-5}

13.4 $1.954 \times 10^{-10} \, \text{C/m}^2$

13.5 $7.438 \times 10^{-36} \, \text{C} - \text{m}$

13.6 $8.85 \times 10^{-10} \, \text{C/m}^2$; $8.8415 \times 10^{-8} \, \text{C/m}^2$

13.7 $16.31 \times 10^{-12} \, \text{C/cm}^2$

13.8 (a) 118pF; (b) $26.55 \times 10^{-9} \text{C}$; (c) 8.85pF; (d) 0.075 or 2;
  (e) $17.7 \times 10^{-12} \, \text{C}^2/\text{Nm}^2$; (f) $8.85 \times 10^{-12} \text{C}$; (g) $2.00 \times 10^5 \, \text{V/m}$; (h) $2.0 \times 10^5 \, \text{V/m}$

13.9 $14.87 \times 10^{-36} \, \text{C} - \text{m}$
13.10 $3.516 \times 10^{-38} \, \text{C}^2 \text{N}^{-1} \text{m}^{-2}$
13.11 $1.646 \times 10^{-38} \, \text{C}^2 \text{N}^{-1} \text{m}^{-2}$
13.12 $4.326 \times 10^{-6} \, \text{C} - \text{m}$
13.13 $2.5 \times 10^4 \, \text{V}$
13.14 $6.6 \times 10^4 \, \text{V}$
13.15 33.3cm

13.16 167cm
13.17 $35.4 \times 10^{-10} \, \text{C/m}^2$; $17.7 \times 10^{-10} \, \text{C/m}^2$
13.18 $2.176 \times 10^{-40} \, \text{C}^2 \text{N}^{-1} \text{m}^{-2}$
13.19 1.000435
13.20 $5.63 \times 10^{-9} \, \text{C/cm}^2$
13.21 $6.78 \times 10^{-9} \, \text{C/cm}^2$

Chapter 14

14.1 $10^{-1} \, \text{Am}^2$
14.2 $4 \times 10^{-3} \, \text{Am}^2$
14.3 $8792 \times 10^{-40} \, \text{Am}^2$
14.4 $1.46 \, \mu_B$
14.5 $1.25 \times 10^{-3} \, \text{H/m}$
14.6 $1.0 \, \text{Am}^2$
14.7 $5.03 \times 10^{-39} \, \text{Am}^2$

14.8 $\mu_r = 1 + 900 = 901$
14.9 1999
14.10 (a) 1.6T (b) $20.11 \times 10^{-7} \text{T}$
14.11 $6.67 \times 10^{-4} \, \text{H/m}$
14.12 floppies, CDs, typical magnetic recording materials
14.13 hard magnets because of high coercivity and remanence

Chapter 15

15.1 (a) $4.2xyi + \left(4x^2 - 3y^2z^2\right)j - 2y^3zk$; (b) $-8i + j - 2k$;

15.2 $n \, (r^2)^{n/2-1} \boldsymbol{r} = n \, (r^2)^{n-2} \boldsymbol{r}$

15.3 $3r^{-5}\boldsymbol{r}.\boldsymbol{r} + 3r^{-3} = \boldsymbol{0}$

15.4 $a = -2$

15.5 $2(\omega_1\boldsymbol{i} + \omega_2\boldsymbol{j} + \omega_3\boldsymbol{k}) = 2\boldsymbol{\omega}$

15.6 $<\rho_e> = 3\times10^{11}; 10^3 = <\rho_m>$

15.7 (a) $E = 120\pi\,\text{V/m}$; (b) $240\pi\,\text{V/m}$

15.8 $0.098\,\text{m}^{-1}$

15.9 (a) $5.97\times10^2\,\text{Hz}$; (b) 6.67×10^{-2}

15.10 $3.3\times10^2\,\text{W/m}^2$

15.11 $E_0 = 0.163 \times 10^3$

15.12 4.88×10^4

15.13 $7.92\times10^{-6}\,\text{m}$

15.14 $\dfrac{E_0 k}{\omega}\sin(kx)\cos(\omega t)\boldsymbol{j}$

15.15 $2\times10^{-8}\,\text{J/m}^2$

15.17 $2.82\,\text{m}$

15.18 $3\,\text{m}$

15.19 $1.88\,\text{rad/m}$

15.20 8.892

15.21 $43.8\times10^{-6}\,\text{m}$

15.22 $\dfrac{120\pi}{\sqrt{\epsilon_r}} = 266.7\Omega; \dfrac{Z_1 - Z_2}{Z_1 + Z_2} = 0.172; \dfrac{2Z_1}{Z_1 + Z_2} = 1.171$

15.23 (a) $-\dfrac{10^3\sin(\omega t - \alpha z)}{120\pi}\,\text{A/m}$ (b) $i\,6.28\,\text{m}^{-1}$

15.24 2.73×10^5

15.25 1.13×10^5

Chapter 16

16.1 8

16.2 $\theta = 90°$

16.3 $9.192\,\text{Å}$

16.4 (a) $1.25 \times 10^{16}\,\text{electrons}$; (b) $0.71\times10^7\,\text{m/s}$

16.5 $0.413\,\text{Å}$

16.6 $3.60\,\text{Å}$

16.7 $0.479\,\text{MeV}$

16.8 (a) $50\,\text{mrem}$; (b) $50\,\text{mrad}, 0.50\,\text{mGy}$; (c) $6.25\times10^{10}\,\text{photons}$; (d) $1.5625\times10^{14}\,\text{eV}$

16.9 $280\,\text{rem}$

16.10 $0.2021\,\text{rad/s}$

16.11 (a) $8.0\times10^{-4}\,\text{J/kg}$; (b) $1.00\times10^9\,\text{protons}$; (c) $0.3125\times10^9\,\alpha - \text{particles}$

16.12 (a) $6.49\times10^{-6}\,\text{J}$; (b) $0.1298\,\text{rem}$

Chapter 17

17.1 74%   17.3 34%   17.5 (416)
17.2 52.4%   17.4 (236)

17.6

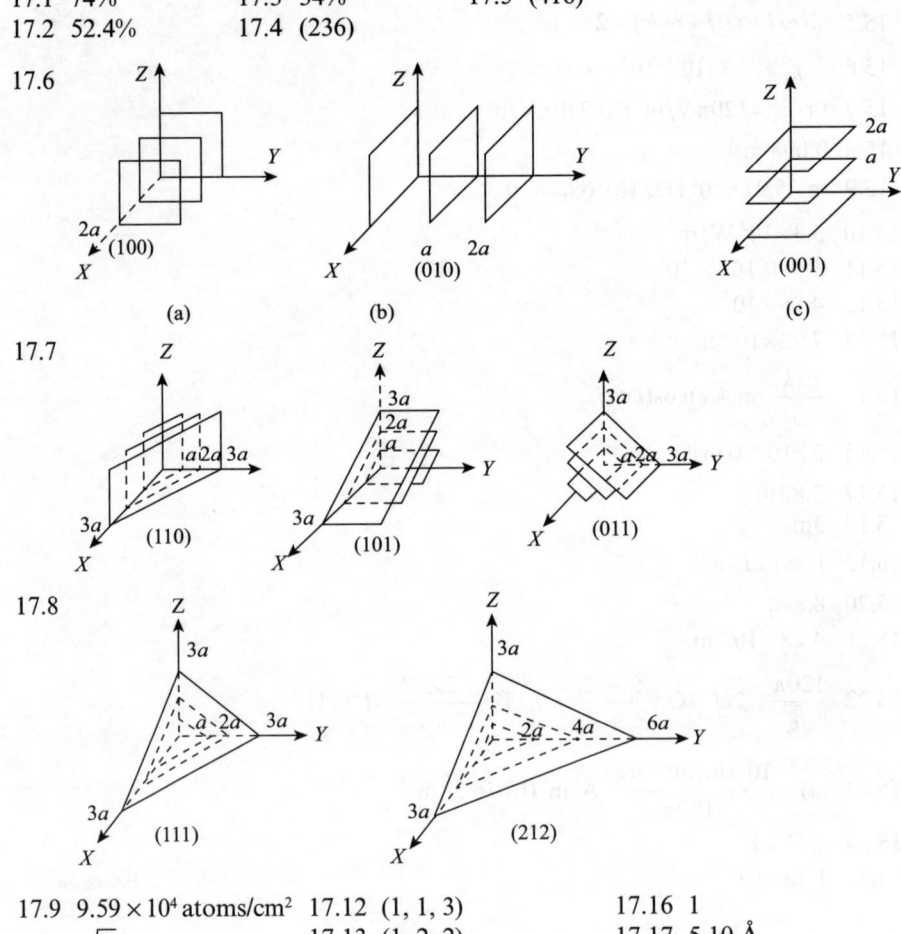

      (a)   (b)   (c)

17.7

17.8

17.9 9.59×10^4 atoms/cm^2 17.12 (1, 1, 3)   17.16 1

17.10 $\sqrt{\dfrac{3}{2}}$ 17.13 (1, 2, 2)   17.17 5.10 Å
                                        17.14 3.076 Å   17.18 16.9
17.11 (1, 1, 1)   17.15 3.903 Å   17.19 4.81 Å

Chapter 18

18.1 $\Delta E = 1.7 \times 10^{-7}$ eV
18.2 6.626×10^{-25} kg m/s, 1.51 eV
18.3 10^6 Å
18.4 (a) Fig. 18.20(a): electrons, Fig. 18.20(b): holes; (b) Fig. 18.20(a): effective mass of A < effective mass of B, Fig. 18.20(b): effective mass of A < effective mass of B
18.5 m_{eff} is positive at B, C and negative at A, D
18.6 0.28×10^{-35} kg   18.11 2.83×10^{-4} cm
18.7 0.12   18.12 0.53 nm
18.8 4×10^{10}/cm^3 18.13 $m_t = 0.23m_0$
18.9 18.2%   18.14 1.39 eV
18.10 $2.2kT$ above Fermi energy level   18.15 $1.92kT$

Chapter 19

19.1 $T = 0.14$ K

19.2 1.64×10^{-10} mVs

19.3 $0.446\,\Omega m$

19.4 $2.314\,\Omega m$

19.5 -6.25 mV

19.6 0.347 eV

19.7 $160.3\,\Omega\,cm$

19.8 $\sigma = 0.432 \times 10^{-4}/\Omega m$

19.9 $50\,\Omega\,cm$

19.10 $.01536/\Omega m$

19.11 0.348 eV

19.12 0.73 eV

19.13 $2.88/\Omega m$

19.14 0.25 eV

19.15 $1.92 \times 10^{25}/m^3$

19.16 0.459 eV

19.17 $0.64 \times 10^{-4}/\Omega m$

19.18 $35.97\,\Omega\,cm$

19.19 $2.016/\Omega m$

19.20 $50.72 \times 10^5\,m^2/As$

19.21 2.79×10^{-6} V

19.22 $0.29\,\Omega\,cm; -208\,cm^3/C;$
$\qquad -83.2 \times 10^{-5}$ V

19.23 $0.090\,m^3/C; 0.72\,m^2/Vs$

19.24 $71.43\,\mu V$

19.25 $2.92 \times 10^5\,m^2/Vs$

Chapter 20

20.1 2.3 eV

20.2 0.94×10^6 m/s

20.3 7.0399 eV

20.4 1.999 eV

20.5 1081 K

20.6 1.42×10^{-14} s

20.7 $2.81 \times 10^7\,\Omega$ m

20.8 2.05×10^{-8} W Ω/K^2

20.9 240 W/mK

20.10 $5.86 \times 10^{28}/m^3$

20.11 2.67×10^{-14} s

Chapter 21

21.1 2.18×10^{-3} eV

21.2 (b) 10^{13}

21.3 4.146 K

21.4 6.138 A

21.5 (a) 496.9Å (b) $1.64 \times 10^{27} m^3$

Chapter 22

22.1 $\dfrac{1}{9}$

22.2 $2.008 \times 10^{17}\,kg/m^3$

22.3 1.175 T

22.4 4.85×10^{14} Bq

22.5 (a) 15.75 years (b) 10.5 years

22.6 30.85 MHz

22.7 0.428×10^3 cycles

22.8 (a) $12.96 \times 10^2\,wb/m^2$; (b) (i) 1.88×10^{10} m/s; (ii) 5.83×10^{-7} J

22.9 1.34×10^{-11} A

22.10 5625 counts/min

22.11 1.25260×10^4 counts/min

22.12 1.37 V

22.13 1.6×10^{-13} C

Bibliography

A.J. Moulson and J.M. Herbert, *Electroceramics: Material, Properties, Application,* Chapman and Hall, 1990.

A.K. Ghatak and K. Thyagarajan, *Introduction to Fibre Optics,* Cambridge University Press, Cambridge, 1999.

A.Van der Ziel, *Solid State Physical Electronics,* Prentice Hall, New York, 1968.

Adrianus J. Dekker, *Solid State Physics,* Macmillan, London, 1969.

Andrew S. Grove, *Physics and Technology of Semiconductor Devices,* John Wiley, New York, 1967.

Arthur Beiser, *Concepts of Modern Physics,* McGraw Hill, 6th ed., 2003.

C. Kittel, *Solid State Physics,* John Wiley, 7th ed., 1995.

Charles P. Poole and Frank J. Owens, *Introduction to Nanotechnology,* Wiley India, New Delhi, 2006.

D.K. Bhattacharya and R. Sharma, *Solid State Electronic Devices,* Oxford University Press, 2007.

David Hailiday and Robert Resnick, *Physics Part II,* Wiley Eastern, New Delhi, 1972.

Donald A. Neamen, *Semiconductor Physics and Devices,* Tata McGraw Hill, New Delhi, 2005.

Donald E. Hall, *Musical Acoustics,* Brooks Cole, Thomson Learning, US, 2002.

Donald R. Askeland and Pradeep P. Phule, *The Science and Engineering of Materials,* Thomson, US, 2006.

E.H. Wichmann, *Quantum Physics,* Vol. 4, Berkeley Physics Course, McGraw Hill, 1971.

Edwin R. Jones and Richard L. Children, *Contemporary College Physics,* Addison–Wesley Publishing Company, New York.

F.K. Richtmyer, E.H. Kennard, John N Cooper, *Introduction to Modern Physics,* Tata McGraw-Hill, 2004.

Feynman, Leighton, Sands, Lectures on Physics, Narosa Publishing House, 1998.

F. Jorgensen, *Complete Handbook of Magnetic Recording,* 4th ed., McGraw-Hill, New York, 1996.

G.I. Epifanov, *Solid State Physics,* Mir Publishers, Moscow, 1979.

I. Kaplan, *Nuclear Physics,* Addison–Wesley, 1962.

J.L. Moll, *Physics of Semiconductors,* McGraw Hill, New York, 1964.

K. Boer (ed.), *Semiconductor Physics,* Vol. 1 and 2, Wiley, New York, 2001.

L.L. Hench and J.K. West, *Principles of Electronic Ceramics,* Wiley Intersects, 1990.

M.A. Duncan and D.H. Rouvray, *Microclusters,* Sci. Am. 110, Dec. 1989.

M.N. Avadhanulu and P.G. Kshirsagar, *A Textbook of Engineering Physics,* S. Chand & Company, New Delhi, 2005.

P. Bhattacharya, *Semiconductor Optoelectric Devices,* Prentice-Hall of India, New Delhi, 2002.

P.G. Collins and P. Avouris, *Carbon Nanotubes,* Sci. Am., 62, Dec. 2000.

P.S. Neelkanta, *Handbook of Electronic Materials,* CRC Press, Boca Raton, Florida, 1995.

Randall D. Knight, *Physics for Scientists and Engineers*, Pearson Addison Wesley, 2008.

R. Tewari and K. Thyagarajan, *J. Light Wave Technol*, Vol. 4, 1986.

R.E. Whan, *Materials Characterization*, Vol. 10, *Metals Handbook*, American Society of Metals Park, OH, 1986.

R.K. Gaur and S.L. Gupta, *Engineering Physics,* Dhanpat Rai Publications, New Delhi, 2006.

R.P. Khare, *Fibre Optics and Optoelectronics*, Oxford University Press, New Delhi, 2007.

S.C. Gupta, *Textbook on Optical Fibre Communication and its Applications*, Prentice Hall of India, New Delhi, 2004.

Subrayamanium, Brij Lal, and Avadhanulu, *A TextBook of Optics*, S. Chand, 2010.

S.L. Kakani and Shubhra Kakani, *Modern Physics,* Viva Books Private Limited, New Delhi, 2006.

S.L. Marshall, *Laser Technology and Applications*, McGraw-Hill International Edition, London, 1978.

S.M. Sze, *Physics of Semiconductor Devices*, John Wiley, New York, 2002.

S.O. Kasap, *Principles of Electrical Engineering Materials and Devices*, McGraw-Hill, New York, 1997.

S.P. Taneja, *Modern Physics for Engineers,* R. Chand & Co., New Delhi, 2007.

Steve Adams and Jonathan Allday, *Advanced Physics,* Oxford University Press, Great Clarendon Street, Oxford, 2000.

Vern O. Knudsen and Cyril M. Harris, *Acoustical Designing in Architecture,* John Wiley & Sons, New York, 1950.

W. Shockley, *Electrons and Holes in Semiconductors*, D. Van Nostrand, New York, 1950.

Y. Ando, *Concert Hall Acoustics,* Springer, New York, 1985.

Wilson, *Practical Physics*, CBS College Publishing, 1986.

Index

About the Authors

Dr D.K. Bhattacharya is Associate Director at Solid State Physics Laboratory Delhi, DRDO. A Ph D from the University of Delhi, he has over two decades of experience as a practising semiconductor scientist. He has had a long association with the MEMS Division at the Solid State Physics Laboratory, New Delhi, and currently heads the Hydrophone Group, Ion-implantation Group, Microwave Group, and Instrumentation Group there.

Dr Bhattacharya has more than 50 research papers to his credit. He has also co-authored *Solid State Electronic Devices* for OUP, India.

Dr Poonam Tandon is currently Associate Professor of Physics at Maharaja Agrasen Institute of Technology, New Delhi. She completed her bachelor's and master's degrees in Physics (Hons) from Hans Raj College, Delhi, and obtained her Ph D in Physics from the University of Delhi with CSIR (NET) scholarship. She has over fifteen years of teaching experience in Applied Physics and has been associated with the Delhi University and its affiliated colleges from the beginning of her career.

Her research interest lies in the field of condensed matter physics. Dr Poonam Tandon has published papers for international and national publications in this area.

Related Titles

Engineering Mechanics, 2/e
(9780198096320)

Basudeb Bhattacharyya, *Indian Institute of Engineering Science and Technology, Shibpur*

Engineering Mechanics is specially designed as a textbook for undergraduate students of engineering. It provides a detailed and holistic treatment of the basic theories and principles of both statics and dynamics.

Key Features
- Presents basic concepts of statics and dynamics using the vector approach
- Covers simple lifting machines, dynamics of particles, and dynamics of rigid bodies separately
- Provides a large number of solved examples and numerous illustrations to reinforce the understanding of theoretical concepts

Fundamentals of Computers
(9780199452729)

Reema Thareja, *Shyama Prasad Mukherjee College for Women, University of Delhi*

Fundamentals of Computers has been specifically designed for those who want to learn the basic concepts of computers.

Key Features
- Supports text with numerous well-labelled diagrams, screenshots, and notes to highlight important concepts to help in quick recapitulation
- Includes many solved examples and chapter-end exercises such as objective-type questions and review questions that enable students to check their understanding of concepts
- Includes a separate section on lab activities

Technical Communication: Principles and Practice, 2/e
(9780198065296)

Meenakshi Raman, *Department of Humanities and Social Sciences, BITS Pilani, Goa Campus and*

Sangeeta Sharma, *Department of Humanities and Social Sciences, BITS, Pilani*

Technical Communication, 2e, is an all-inclusive textbook aimed at undergraduate students of engineering. It conforms to the syllabi of courses such as communication skills, technical English, soft skills, and professional communication.

Key Features
- Provides comprehensive and up-to-date coverage of listening, speaking, reading, and writing techniques as well as contemporary communication media supported with plenty of practical tips and attractive technical document samples
- Contains ten new chapters on various topics such as phonetics, vocabulary, and English grammar
- Accompanying DVD contains interview and group discussion videos, listening and speaking practice, PowerPoint presentations, and text supplements

Engineering Drawing
(9780199455393)

N S Parthasarathy, *AU-FRG Institute of CAD/CAM, College of Engineering, Guindy, Anna University, Chennai and* **Vela Murali,** *College of Engineering, Guindy, Anna University, Chennai*

Engineering Drawing is a textbook designed for students of all engineering disciplines to develop a spatial bent of mind to observe, visualize, and understand the structure of objects from different perspectives.

Key Features
- Presents a detailed chapter on Visualization Concepts and Freehand Sketching with plenty of examples and exercises, and an exclusive chapter on AutoCAD 2013 tools
- Cites engineering applications and provides illustrative examples, multiple-choice questions, and work practice problems divided into multiple levels
- Accompanied by a CD containing animated drawings of 19 select problems from the book

Other Related Titles

9780198069324 **Kumar & Lata**: *Communication Skills*	9780198072089 **Rajagopalan**: *Environmental Studies, 2e*
9780198066217 **Mitra**: *Personality Development and Soft Skills*	9780198070047 **Thareja**: *Programming in C*
9780198068907 **Nagsarkar & Sukhija**: *Basic Electrical Engineering, 2e*	9780198070894 **Pal**: *Engineering Mathematics (forthcoming)*
9780198081807 **Nagsarkar & Sukhija**: *Basic Electrical and Electronics Engineering*	9780195677294 **Islam**: *Semiconductor Physics and Devices*
	9780195669305 **Khare**: *Fiber Optics and Optoelectronics*